Handbook of Research on Resource Management for Pollution and Waste Treatment

Augustine Chioma Affam
University College of Technology Sarawak, Malaysia

Ezerie Henry Ezechi
Universiti Teknologi Malaysia, Malaysia

A volume in the Advances in Environmental Engineering and Green Technologies (AEEGT) Book Series

IGI Global
DISSEMINATOR OF KNOWLEDGE

Published in the United States of America by
IGI Global
Engineering Science Reference (an imprint of IGI Global)
701 E. Chocolate Avenue
Hershey PA, USA 17033
Tel: 717-533-8845
Fax: 717-533-8661
E-mail: cust@igi-global.com
Web site: http://www.igi-global.com

Library of Congress Cataloging-in-Publication Data

Names: Affam, Augustine Chioma, 1979- editor. | Ezechi, Ezerie Henry, 1983- editor.
Title: Handbook of research on resource management for pollution and waste treatment / Augustine Chioma Affam and Ezerie Henry Ezechi, editors.

Description: Hershey PA : Engineering Science Reference, [2019] | Includes bibliographical references. | Summary: "This book examines resource management strategies for waste and pollution treatment methods"-- Provided by publisher.
Identifiers: LCCN 2019021271 | ISBN 9781799803690 (h/c) | ISBN 9781799803713 (eISBN)
Subjects: LCSH: Pollution--Control. | Environmental monitoring. | Refuse and refuse disposal. | Sewage--Purification.
Classification: LCC TD145 .R47 2019 | DDC 628.5--dc23
LC record available at https://lccn.loc.gov/2019021271

This book is published in the IGI Global book series Advances in Environmental Engineering and Green Technologies (AEEGT) (ISSN: 2326-9162; eISSN: 2326-9170)

British Cataloguing in Publication Data
A Cataloguing in Publication record for this book is available from the British Library.

For electronic access to this publication, please contact: eresources@igi-global.com.

Advances in Environmental Engineering and Green Technologies (AEEGT) Book Series

Sang-Bing Tsai
University of Electronic Science and Technology of China Zhongshan Institute, China
Ming-Lang Tseng
Lunghwa University of Science and Technology, Taiwan
Yuchi Wang
University of Electronic Science and Technology of China Zhongshan Institute, China

ISSN:2326-9162
EISSN:2326-9170

Mission

Growing awareness and an increased focus on environmental issues such as climate change, energy use, and loss of non-renewable resources have brought about a greater need for research that provides potential solutions to these problems. Research in environmental science and engineering continues to play a vital role in uncovering new opportunities for a "green" future.

The **Advances in Environmental Engineering and Green Technologies (AEEGT)** book series is a mouthpiece for research in all aspects of environmental science, earth science, and green initiatives. This series supports the ongoing research in this field through publishing books that discuss topics within environmental engineering or that deal with the interdisciplinary field of green technologies.

Coverage

- Green Transportation
- Radioactive Waste Treatment
- Industrial Waste Management and Minimization
- Electric Vehicles
- Green Technology
- Sustainable Communities
- Policies Involving Green Technologies and Environmental Engineering
- Renewable Energy
- Contaminated Site Remediation
- Pollution Management

IGI Global is currently accepting manuscripts for publication within this series. To submit a proposal for a volume in this series, please contact our Acquisition Editors at Acquisitions@igi-global.com or visit: http://www.igi-global.com/publish/.

The Advances in Environmental Engineering and Green Technologies (AEEGT) Book Series (ISSN 2326-9162) is published by IGI Global, 701 E. Chocolate Avenue, Hershey, PA 17033-1240, USA, www.igi-global.com. This series is composed of titles available for purchase individually; each title is edited to be contextually exclusive from any other title within the series. For pricing and ordering information please visit http://www.igi-global.com/book-series/advances-environmental-engineering-green-technologies/73679. Postmaster: Send all address changes to above address.

Titles in this Series

For a list of additional titles in this series, please visit:
https://www.igi-global.com/book-series/advances-environmental-engineering-green-technologies/73679

Implementation and Evaluation of Green Materials in Technology Development Emerging Research and Opportunities
M.T. Mastura (Universiti Teknikal Malaysia Melaka, Malaysia) and S. M. Sapuan (Universiti Putra Malaysia, Malaysia)
Engineering Science Reference • © 2020 • 215pp • H/C (ISBN: 9781799813743) • US $175.00 (our price)

Modern Techniques for Agricultural Disease Management and Crop Yield Prediction
N. Pradeep (Bapuji Institute of Engineering and Technology, India) Sandeep Kautish (Lord Buddha Education Foundation, Nepal & Asia Pacific University, Malaysia) C.R. Nirmala (Bapuji Institute of Engineering and Technology, India) Vishal Goyal (Punjabi University, Patiala, India) and Sonia Abdellatif (Seoul National University, South Korea)
Engineering Science Reference • © 2020 • 291pp • H/C (ISBN: 9781522596325) • US $195.00 (our price)

Novel Advancements in Electrical Power Planning and Performance
Smita Shandilya (Sagar Institute of Research, Technology and Science, India) Shishir Kumar Shandilya (Vellore Institute of Technology, India) Tripta Thakur (Maulana Azad National Institute of Technology, India) and Atulya K. Nagar (Liverpool Hope University, UK)
Engineering Science Reference • © 2020 • 388pp • H/C (ISBN: 9781522585510) • US $225.00 (our price)

Global Food Politics and Approaches to Sustainable Consumption Emerging Research and Opportunities
Luke Amadi (University of Port Harcourt, Nigeria) and Fidelis Allen (University of Port Harcourt, Nigeria)
Engineering Science Reference • © 2020 • 242pp • H/C (ISBN: 9781799801252) • US $185.00 (our price)

Fuzzy Expert Systems and Applications in Agricultural Diagnosis
A.V. Senthil Kumar (Hindusthan College of Arts and Science, India) and M. Kalpana (Tamil Nadu Agricultural University, India)
Engineering Science Reference • © 2020 • 335pp • H/C (ISBN: 9781522591757) • US $245.00 (our price)

Handbook of Research on Environmental and Human Health Impacts of Plastic Pollution
Khursheed Ahmad Wani (Government Degree College Bijbehara, India) Lutfah Ariana (Indonesian Institute of Sciences, Indonesia) and S.M. Zuber (Government Degree College Bijbehara, India)
Engineering Science Reference • © 2020 • 600pp • H/C (ISBN: 9781522594529) • US $285.00 (our price)

IGI Global
DISSEMINATOR OF KNOWLEDGE

701 East Chocolate Avenue, Hershey, PA 17033, USA
Tel: 717-533-8845 x100 • Fax: 717-533-8661
E-Mail: cust@igi-global.com • www.igi-global.com

List of Contributors

Table of Contents

Detailed Table of Contents

This chapter discusses the management of waste cooking oil (WCO) in a sustainable manner in order to protect the environmental pollution. Increasing consumption of edible oils worldwide leads to generation of substantial amount of waste cooking oil (WCO). While WCO is not considered toxic, large amount of WCO can contribute to environment pollution if not being handled properly. The huge generation of WCO in the world creates problem of collection, treatment and disposal. Due to its chemical features, the recycling of WCO not only provides a renewable feedstock for producing biofuels and bio-based products, but also alleviates environmental pollution arising from its improper handling. This chapter also provides an overview of some recent approaches in WCO recycling and applications.

Municipal solid waste (MSW) is an energy source that should not go untapped or unutilized. The waste must be properly utilized through combustion, anaerobic digestion, and landfill gas acquisition, as it represents material and energy content. This will reduce the effects of global warming, which is as a result of high concentration of carbon dioxide, methane, and other greenhouse gases (GHGs), in the atmosphere. This chapter focuses on the technologies for solid waste management and the thermodynamics involved in the process for sustainable and cleaner energy. The equations presented represent the thermal efficiency, conversion efficiencies, as well as possible work that can be derived from a power plant utilizing MSW as fuel. It is important that countries in Sub-Saharan Africa vigorously pursue sustainable waste management technologies, especially recycling and landfilling, while exploring and investing in waste-to-energy technologies that will perform optimally using the composition of the waste in Sub-Saharan Africa in the design of the waste-to-energy technology.

The chapter presented recent developments in fluidized bed gasification of oil palm empty fruit bunches (OPEFB). The bioenergy potential of OPEFB as an environmentally friendly biomass was presented. The chapter also highlighted the prospects of utilising biomass gasification technology as a practical method for valorising OPEFB. The fundamental theories, chemistry, operational parameters, classification of fluidized bed gasifiers were presented. The findings indicated that fluidized bed gasification is a practical approach for valorising OPEFB. The chapter concluded by highlighting the potential of OPEFB as a low cost, abundant, lignocellulosic feedstock for valorisation through fluidized bed gasification.

Microplastics are plastics with less than 5mm particle size and originate from sequential degradation of larger plastic molecules or manufactured which have found use in many realms of life. Their gradual degradability and ingestion by aquatic organisms have become a source of environmental concern. Microplastics are regarded as "tiny problem" requiring massive attention. Occurrences of microplastics have been detected in almost all environment matrices. This chapter views microplastics, occurrences, detection, and existing policies. The role of industry and individuals in preserving the ecosystems are deliberated. In summary, emphasis on the bottom-up strategy to curb the escalating amount of plastics waste in our environment is sought and adoption of the "avoid the avoidable" attitude for a more holistic approach in tackling the severity of the impending threat.

Biomass agricultural waste has a great potential for meeting part of the world energy need and it is environmentally friendly. One conversion method is thermochemical processes and specifically pyrolysis. Pyrolysis converts the lignocellulose waste to fuel and essential chemicals in the three-phase products: biogas, bio-oil and biochar. However, performance issues are limiting the potential of lignocellulose pyrolysis such as design and operation of pyrolysis reactor for effective heat transfer from the heat source to the biomass feedstock. Therefore, this study presents the necessary tool for pyrolysis scientist and engineer in determining the optimal operation and design of lignocellulose agricultural waste pyrolysis.

This chapter presents an overview of emerging contaminants in landfill leachate and their treatment methods. Emerging contaminants in the leachate such as poly-aromatic hydrocarbons (PAHs) and phthalate acid esters (PAEs) and their concentrations in various environmental matrices were discussed. Both PAHs and PAEs are highly carcinogenic, mutagenic and teratogenic substances widely known as endocrine disrupting compounds in animals and humans. The efficiency of physico-chemical, biological and advanced oxidation processes on leachate treatment were also discussed.

This chapter studies spatio temporal trends of air quality, its relation with urban forms, and Land Surface Temperature (LST) in Kolkata Metropolitan Area, the world's 55th most polluted city according to the World Health Organization. Air quality data were procured from 34 monitoring stations operated by West Bengal Pollution Control Board for 2005-18. Trend analysis showed declining NO2, SO2 values but a rise in PM10 concentration. Interpolation analysis showed high concentration of pollutants along the Hugli industrial belt, Dhulagarh, and Kolkata Municipal Corporation Area. Urban forms such as built-up density, distances from water bodies, parks, and bus stops demonstrated their reasonable influence on air pollution condition. LST generated from Landsat Thermal Infrared bands showed increase in temperature conditions from 2005-2018. Positive Correlation was identified between Land surface temperature and air pollution. Their relationship was assessed to have become stronger over the decades.

The paradigm shift towards waste reduction in the construction industry has produced revolutionary techniques in assessing and mitigating waste during the construction lifecycle. Unmanned Aerial Vehicle (UAV) technology presents potentials in real-time construction waste management for informed decision-making processes for site managers. Due to the dynamic and complex construction work environment, material and equipment, labour and management policies are subjected to latent and identified hazards. Therefore, this chapter focuses on the real time impact of construction waste. UAV technology was adopted. UAV flight mission and deployment time were designed to meet target requirements and snapshot checklist. Data images were captured from an 11-storey residential building and analysed for conformity to construction waste and safety requirements. The prevalence of exposed material and equipment, low labour awareness, and need for policy shift to circular economy was predominant. Further studies can explore multiple construction monitoring and other category of construction works.

Sustainable waste management of biodegradable waste is one of the serious concerns in developing countries. The energy production from organic waste material by anaerobic digestion (AD) is environmentally sound, energy-efficient, and economical technology that degrades substrate, with minimum production of harmful metabolites, and reuse of its byproducts as bio-fertilizers. The use of small-scale biogas digesters in developing countries and industrial scale or integrated biogas digesters in developed countries has been operationally upgraded. These advancements in the application of AD processes are due to multiple factors, including the economic and environmental value of AD technology, and the dynamics of a range of policies and measures related to agricultural sector, waste control, and green energy production. However, compared to other waste management and energy generation methods, AD requires fewer resources and cost investment, and is therefore sustainable for waste control and energy production.

Foreword

Waste management—collection, transport, treatment, disposal, or recycling and monitoring of waste materials—is an essential public service for every community and necessary for the protection of public health and the environment. It is no surprise that issues of waste management are often serious topics of concern for the government.

This book provides a valuable window on information regarding resource management approach for waste treatment and pollution control. It includes all aspects of the environment such as water, air, soil and the aquatic ecosystem. Waste management is a basic human need and can also be regarded as a basic human right. Ensuring proper sanitation and waste management ranks alongside the provision of potable water, shelter, food, energy, transport and communications; all are essential to the society and to the economy as a whole. Despite this, in most countries, attention given to waste management is often lower than that given to other infrastructural and utility services. The consequences of doing little or even nothing to address waste management and environmental degradation can be very costly to the society and to the economy. Uncontrolled waste is not managed and thus not measured, making it difficult to estimate the size of the problem and the scale of the associated costs. However, current evidence suggests that in a middle- or low-income city, the costs to society and the economy are about five to ten times of what sound solid waste management would cost per capita. It is dramatically cheaper to manage waste now in an environmentally sound manner by either reducing, reusing or recycling in order to clean up the past mismanagement practices which degraded the environment. The objective of this is not to eliminate waste totally, but to provide the best practices in terms of monitoring, treatment and handling or disposal, to keep the society safe from diseases as well as generate added value products from a stream of waste resources. This book gives a lead in the right direction for anyone who is ready to resolve the issues already highlighted.

Malay Chaudhuri
Independent Researcher, Malaysia

Malay Chaudhuri *graduated in Civil Engineering from the Jadavpur University in 1962. He obtained Master of Technology in Civil Municipal Engineering from the Indian Institute of Technology Kharagpur in 1963 and Doctor of Philosophy in Civil Environmental Engineering from the University of Illinois Urbana-Champaign in 1969. He served at Assistant Professor at the University of Illinois Urbana-Champaign, Assistant Professor and Professor at the Indian Institute of Technology Kanpur, Visiting Research Professor at the University of Ottawa, Visiting Professor at the Universiti Sains Malaysia and Professor at the Universiti Teknologi Petronas. His research areas: Physical-Chemical Processes of Water and Wastewater Treatment; Development of Low-Cost Adsorbents; Advanced Oxidation Processes; Low-Cost Water Treatment Alternatives; Removal of Arsenic from Water; and Health-Related Water Microbiology.*

Preface

Environmental waste treatment and resource management is one vital aspect most organisations and nations are hampering on improving. These include waste from soil, water/wastewater, and air pollution, its treatment strategies and/or minimisation. In addition, resources which constitute waste in the environment are considered for sustainable reuse, recycling and recovery as well as practical alternatives to traditional disposal techniques. Technology and management strategies associated with the chosen areas are dealt with in detailed review and research works and are presented herein. The book provides the research achievements and insights in the area of environmental waste treatment and sustainable monitoring and reuse of waste to add value to the society. It is an important reference for the academia, especially for the research field oriented and distributed amongst environmental and sanitary engineering experts and those in interdisciplinary areas of interest.

The field of environmental engineering is an obviously extended one with endless boundaries interwoven with other fields of engineering. It is therefore necessary to understand the extent of pollution in the environment and how harmful it could be for both humans and animals sanitary conditions in terms of the air, water and soil surrounding our environment. Poor waste treatment, domestic waste and construction wastes handling and pollution monitoring can lead to huge environmental issues, and cause a very significant impact on people, society, and the sustainability of waste valuable resources. The solution therefore is to imbibe preventive ideas, which leads to initiation of actions that prevent adverse effects. The reuse of waste from treatment processes or pollution of soil, air and water, production or manufacturing or construction wastes in other forms is important for sustainable waste management and this is presented in this book. This handbook is a collection of original works in these areas highlighted. This book provides sustainable insights and recent empirical research findings. It will be written for researchers who wish to improve their understanding of the waste and pollution treatment methods and how they can adopt appropriate resource management strategies for environmental related issues at local and internal levels. This would help organizations and individuals working in similar fields and those seeking solutions to their immediate environmental disasters.

To date, there are some literatures focused only on specialized characteristics of some types of waste monitoring and treatment including handling or disposal. However, this book attempted to be inclusive of a variety of waste which have not been well documented. There are a lot of developments in various fields of engineering requiring waste monitoring and treatment. However, there are still many gaps in knowledge in this area. This book will contribute in filling in some of the gaps. As far as we know, the book is advantage because they are original works from research and insights from specialists in their various fields. Hence, it is distinct and provides actual communications of work done. A few of them are collection strong innovative reviews, experimental works and proposals showing direction for new research studies that can be conducted.

The target audience of this book comprises of professionals and researchers whose work are related to environmental management, waste handling and treatment, natural resources and policy regulators in the broad field of engineering. It would also be a source of information and reference for libraries, educationists, technologists, postgraduate students, management of institutions and those seeking general knowledge to improve their society. Moreover, the book provides insights for experts in environmental engineering field whose knowledge are needed for various aspects of work categories in order to meet their challenges related to waste treatment and resource management.

This book presents an overview of the current state of knowledge and latest advances regarding waste and pollution in the environment and provides solutions to temporary and long term issues due to environmental waste generation and eventual degradation which reduces livability of communities. Future Resource Management for Pollution and Waste Treatment are presented in 26 chapters.

It is quite exciting to see interest from researchers globally making their contribution in various related backgrounds and sharing their experiences in this book. We remain indebted to all authors as well as reviewers who assisted during the making of this book in other to produce superb materials which can be referenced in areas of waste treatment and resource management practices.

ORGANIZATION OF THE BOOK

This handbook is organized into 26 chapters. A brief description of each chapter is presented as follows.

Chapter 1, "Waste Tyre Rubber Application in Semi-Rigid and Flexible Pavement," discusses the application and the benefits of waste tyre utilization in the production of rubberized concrete block pavement (RCBP) and rubberized asphalt concrete pavement (RACP). The characteristics of the pavements are discussed.

Chapter 2, "Utilization of Waste Cooking Oil (WCO) in Hot Mix Asphalt (HMA)," discusses the application of waste cooking oil (WCO) in hot mix asphalt (HMA). Physical and rheological tests conducted on the binder indicated that failure temperature of binder modified using the treated WCO increased to 70°C and the treated WCO mixture recorded superior performance by being less susceptible to permanent deformation as compared to the control mix.

Chapter 3, "Recent Advances in Waste Cooking Oil Management and Applications for Sustainable Environment," discusses the beneficial application and recycling of waste cooking oil (WCO). Recycling of WCO can reduce environmental pollution related to WCO generation. This chapter provides an overview of some recent approaches in WCO recycling and applications.

Chapter 4, "Sustainable Technologies for Solid Waste Monitoring and Treatment," focuses on the technologies for solid waste management and the thermodynamics involved in the process for sustainable and cleaner energy. The benefits, challenges and the current technologies in waste management were discussed. The major setbacks in waste to energy combustion as an attractive energy source were highlighted.

Chapter 5, "Recent Advances in Fluidized Bed Gasification of Oil Palm Empty Fruit Bunches," discusses recent developments in fluidized bed gasification of oil palm empty fruit bunches (OPEFB). The chapter highlighted the prospects of utilising biomass gasification technology as a practical method for valorising OPEFB. The fundamental theories, chemistry, operational parameters, classification of fluidized bed gasifiers were presented.

Chapter 6, "Microplastics: The Next Impending Threat to Mankind?" discusses the occurrences of microplastics in various environments and the potential danger it poses to these environments. It also discusses microplastic detection and existing policies. Anthropogenic role in protecting the ecosystems were also highlighted.

Chapter 7, "Fundamental Models and Theories for the Pyrolysis of Lignocellulose Biomass Waste," discusses the conversion methods of biomass agricultural waste and its potential as alternative energy source. Pyrolysis method was discussed using mathematical equations that govern the lignocellulose kinetics (model and model-free) and pyrolysis reactor macro and micro models.

Chapter 8, "Emerging Contaminants in Landfill Leachate and Their Treatment Methods," discusses emerging contaminants in landfill leachate and their physico-chemical, biological and advanced oxidation treatment methods. The deleterious effects of these emerging contaminants on animals and humans were emphasized.

Chapter 9, "Application of Graded Limestone as Roughing Filter Media for the Treatment of Leachate," discusses leachate treatment using limestone as a roughing media in a vertical upflow filtration column. It further highlights COD, BOD, NH3 and color removal efficiencies. The study indicates that limestone is an effective adsorbent that can be used for short term treatment of leachate

Chapter 10, "Review on Microbial Analysis Tools in POME Treatment" discusses the various palm oil mill treatment technologies such s anaerobic, aerobic, physico-chemical and membrane separation processes. It highlights the potential application of molecular biology technique in POME treatment.

Chapter 11, "Waste Activated Sludge Treatment Processes," discusses various waste activated sludge (WAS) treatment processes such as physico-chemical, biological and combined treatment processes with primary focus on anaerobic digestion and stabilization of WAS.

Chapter 12, "Biological Nutrient Removal by Suspended Growth Systems," discusses the sources, environmental, health and economic implications of waterbody pollution by nutrients. It also discusses various suspended biological nutrient removal processes and the factors that affect their performance.

Chapter 13, "Bioremediation: A Logical Approach for the Efficient Management o Textile Dye Effluent," discusses the potential application of bioremediation in the treatment of textile dye effluents. It emphasizes on the immobilization of fungi and bacteria to improve their stability during textile dye decoloration.

Chapter 14, "Industrial Wastewater Treatment: Recycled Paper Mill Effluent Treatment Using Modified Anaerobic Hybrid Baffled (Mahb) Bioreactors," discusses the application of a novel modified anaerobic hybridized baffled (Mahb) bioreactor modified with baffle for wastewater treatment. The bioreactor was effective for wastewater and comparable to the anaerobic baffled reactor. The compartmentalization of this bioreactor results in a buffering zone between the primary acidification zone and active methanogenesis zone, and provided the strong ability to resist shock loads which broaden the usage of multi-phase anaerobic technology for industrial wastewater treatment.

Chapter 15, "Reverse Osmosis Membrane Desalination Technology and Process: Case Study on Small Scale Brackish, City Polluted Water Treatment," discusses desalination technology with particular emphasis on reverse osmosis membrane processes. A case study involving a small-scale desalination plant that utilizes brackish city polluted water as source was illustrated.

Chapter 16, "Classical Fenton and Sequencing Batch Reactor Treatment of Pesticide Wastewater to Meet Discharge Effluent Requirement," discusses the application of Fenton and sequencing batch reactor

(SBR) for the treatment of chlorpyrifos, cypermethrin and chlorothalonil pesticide industrial wastewater. The Fenton-SBR process was suitable for the degradation of the pesticides.

Chapter 17, "Principles of Electrocoagulation and Application in Wastewater Treatment: Electrocoagulation in Wastewater Treatment," discusses electrocoagulation as an alternative and feasible method of wastewater treatment. Mechanism of electrocoagulation, components of electrocoagulation, types of electrodes, redox reaction during electrocoagulation, advantages and disadvantages of electrocoagulation were all discussed.

Chapter 18, "Coagulation-Flocculation Technology in Water and Wastewater Treatment," discusses coagulation and flocculation process of water treatment. principles of coagulation and flocculation such as charge neutralization and binding mechanisms such as interparticle bridging, sweeping coagulation and absorption were also discussed. Various coagulants and flocculants used in water treatment were also discussed.

Chapter 19, "Adsorption of Dyes in Aqueous Medium Using RHA and CFA: Effect of Preparation Methods and Process Optimization," discusses the potential application of an environmental friendly adsorbent derived from rice husk ash (RHA) and coal fly ash (CFA) for the adsorption of acid violet 7 and brilliant green dyes. The results indicated that the optimized values agree reasonably well with the validated experimental results

Chapter 20, "Application of Soil Washing Treatment Method for the Remediation of Petroleum Polluted Soil," discusses the application of soil washing treatment for the remediation of petroleum polluted soil using detergent. High percentage of aliphatic and polycyclic aromatic hydrocarbons (PAHs) removal was achieved.

Chapter 21, "Membrane Technology for Treatment of Pharmaceutical Wastewaters: A Novel Approach," discusses the application of membrane treatment technology for the treatment of pharmaceutical wastewaters. Reverse osmosis membrane technology was particularly emphasized.

Chapter 22, "Waste Management Practice in Malaysia and Future Challenges," discusses alternative municipal solid wastewater treatment solution that is environmentally friendly and sustainable.

Chapter 23, "Agricultural Runoff and Treatment Methods," discusses the state-of-the-art position on the treatment of agricultural runoff wastewater by various treatment methods.

Chapter 24, "Synergy Between Air Quality, Various Urban Forms and Land Surface Temperature: A Case Study of Kolkata Metropolitan Area," discusses air quality conditions in Kolkata Metropolitan Area in India. Air quality issues are a matter of global concern because it is the prime driving force for climate change. Air pollution not only contaminates the natural composition of the atmosphere but also hampers the health of the entire biota.

Chapter 25 discusses factors that influence anaerobic digestion (AD) process, biochemistry of AD and various reactor types applied in AD.

Chapter 26, "Real-Time Construction Waste Reduction Using Unmanned Aerial Vehicle," discusses the application of unmanned aerial vehicle (UAV) technology in real time construction waste management for informed decision-making processes for site managers.

Augustine Chioma Affam
University College of Technology Sarawak, Malaysia

Ezerie Ezechi Henry
Universiti Teknologi Malaysia, Malaysia

Chapter 1
Waste Tyre Rubber Application in Semi-Rigid and Flexible Pavement

Euniza Jusli
University College of Technology Sarawak, Malaysia

Hasanan Md. Nor
Universiti Teknologi Malaysia, Malaysia

Ramadhansyah Putra Jaya
https://orcid.org/0000-0002-5255-9856
Universiti Malaysia Pahang, Malaysia

Zaiton Haron
Universiti Teknologi Malaysia, Malaysia

Mastura Bujang
University College of Technology Sarawak, Malaysia

Wan Nur Aifa Wan Azahar
International Islamic University Malaysia, Malaysia

ABSTRACT

Increasing traffic requires durable and low-noise road surfaces. Urban residents complain about excessive traffic noise that leads to an unhealthy environment. Understanding techniques to produce durable, low-noise pavement has led to the development of rubberized concrete block pavement (RCBP) and rubberized asphalt concrete pavement (RACP). The chapter examines morphology and chemical properties of waste tyre rubber using FESEM, XRF, and TGA/DTA. Authors discuss characteristics of RCPB and RACP and conclude application of RCBP and RACP can lower traffic noise.

DOI: 10.4018/978-1-7998-0369-0.ch001

INTRODUCTION

In view of a conventional waste management system, waste tyres are basically either disposed in the landfill or burned. Tyre stockpiles caused severe health problem to human due to mosquito, vermin, and rats breeding. Fire hazard from the tyre burning activity also can cause uncontrollable burning and air pollution. However, tyre disposal and burning in landfills are banned in most countries (Martínez et al., 2013). According to Shah (2006), the current ''conservation of natural resource concept'', i.e. the reuse (retread) first, then reuse of rubber prior disposal, does not accommodate the ever-increased dumping of tyres. Due to the high cost of legal disposal for tyres, illegal dumping may increase. Disposal of tyres is becoming more expensive, while this trend is likely to continue as landfill spaces become insufficient. Tyres take up landfill space.

The challenge of scrap tyre management arises mainly from the technical and commercial issues relating to tyres both as a product and as a waste. What is tyre? Tyres are made of materials including synthetic and natural rubber, textiles, steel, carbon black, aromatic extender oils and various chemical additives, which are "vulcanised" at a high temperature during the manufacturing process (Chemsain, 2011). The main components of car and truck tyre as shown in Figure 1 is particularly a stable product that requires a great energy to properly break the material down to useful product. In this chapter, we'll look into the component of a tyre, the chemical and physical properties, waste tyre management, and waste tyre application in pavement industry.

WASTE TYRE MANAGEMENT APPROACH

Waste tyres are managed in various approaches to ensure the by-product can be used as a new energy sources or produced as new material. Szentannai *et al.* (2015) and Lopez *et al.* (2017) investigated ways to reclaim the virgin components and recover energy and new materials from waste tyres through the high technology oriented and excessive heat processes such as the de-vulcanisation, pyrolysis, gasification,

Figure 1. Components of a tyre
Source: https://kitchendecor.club/files/radial-tire-cross-section.html

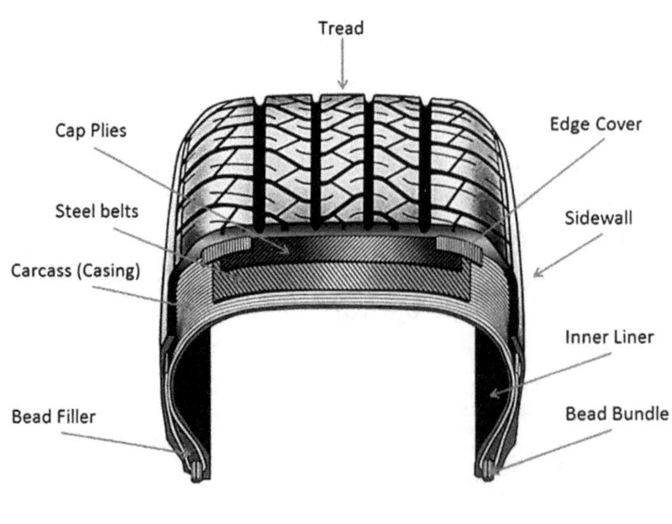

and hydrogenation processes. Besides that, Mavroulidou and Figueiredo (2010) also suggested several waste tyres recycling alternatives which include a) combustion/pyrolysis of waste tyres; b) shredding or grinding tyres; and c) re-treading and reuse of old tyres.

PYROLYSIS OF WASTE TYRES

Waste tyre pyrolysis is an endothermic process whereby the waste tyres were exposed and absorbed the surrounding heat provided in a reactor vessel containing an oxygen-free atmosphere. For the reactor, Lopez *et al.* (2017) suggest using conical spouted bed reactor (CSBR) that perform adequate pyrolysis for different waste materials. In addition, Kwon *et al.* (2015) recommend the utilization of CO_2 as the pyrolysis reaction medium. According to Miranda *et al.* (2013), WTR pyrolysis leads to the formation of non-condensable compounds, light liquids, and solid products. Similar observation found by Martínez *et al.* (2013) where 40% of the pyrolysis by-product is in the form of a solid fraction (char or carbon black and steel) and another 60% removed as a volatile fraction. These vapours can be burned directly to produce power or condensed into an oily type liquid, generally used as a fuel. Some molecules are too small to condense. They remain as a gas which can be burned as fuel. Figure 2 shows the primary and secondary products of WTR pyrolisis observed by Akhil et. al. (2018).

ACTIVATED CARBON

Activated carbon has an incredibly large surface area per unit volume and a network of submicroscopic pores where adsorption takes place. Activated carbon is a material that is produced from carbonaceous source materials, such as coal, coconuts, nutshells, peat, wood, lignite, and waste tyre. The carbon-based

Figure 2. Primary and secondary products of waste tyre rubber
Source: Akhil et. al. (2018)

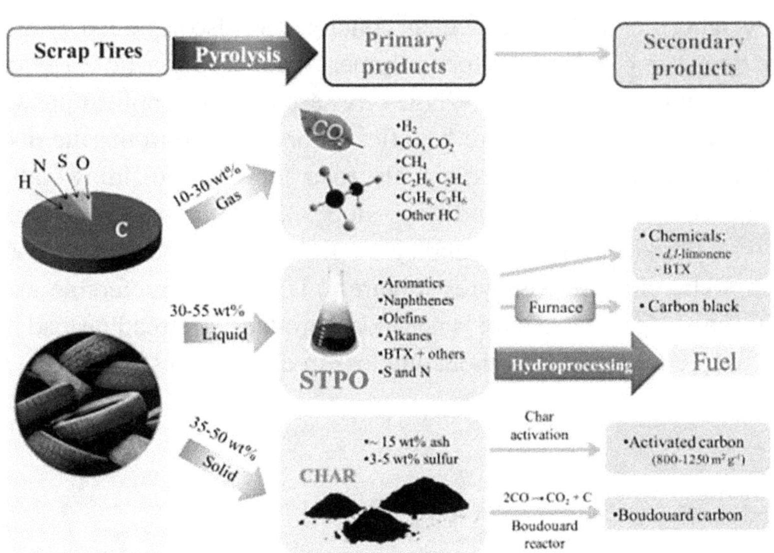

Figure 3. Activated carbon product derived from waste tyre rubber
Source: Shilpa (2017)

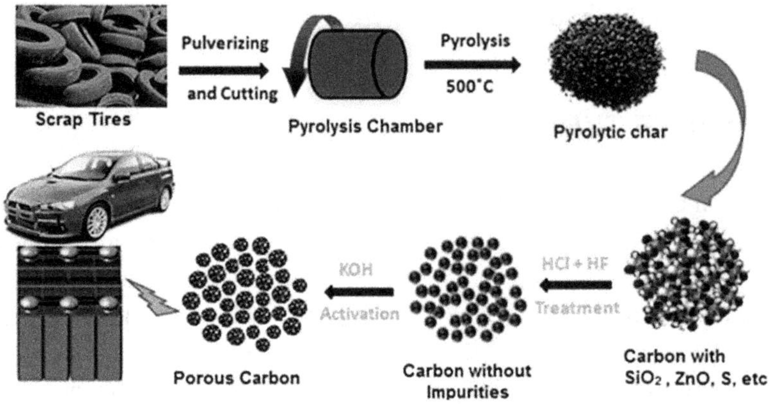

material is converted to activated carbon through physical modification and thermal decomposition in a furnace, under a controlled atmosphere and temperature (Figure 3). The finished product has a large surface area per unit volume and a network of submicroscopic pores where adsorption takes place. The potential of these products as possible adsorbents for various pollutants has been assessed and found to be very successful.

RECYCLING AND RETREATMENT OF WASTE TYRES

Waste tyre recycling may be used in many forms according to its suitable application. It is widely known that tyre recycling offers the most resource efficient strategy for used tyre recovery, saving both material and energy. Amari *et al.* (1999), Aniza *et al.* (2013), and Shu and Huang (2014) explained that the reuse of scrap tyres can be recycled whole or size-reduced for civil engineering applications and agricultural uses and for composite materials. The most common application in civil engineering practices are using scrap tyres as a lightweight filler in Portland cement concrete and also asphalt paving mixtures' modifier. Ground rubber from tyres can be added to other polymers (rubber or plastic) to extend or modify properties of thermoplastic polymeric materials. Whole tyres are used for applications where their physical form, resilience to impact, and longevity are beneficial, for example in marine docks, highway crash barriers, bumpers, landscape use, load-bearing walls, road sub-base, retaining wall, weights for silage cover sheets, erosion protection for walls and steep slopes, acoustic walls, and artificial reefs. Bound rubber product like a rubberized playground and sports field surfaces, roadways and drainage schemes also can be produced. Retreading of used tyres (Figure 4) is the most preferable way of making use of old tyres. On average, only 15% of crude oil is needed to produce a retread instead of a new tyre. Thus, the price of a tyre is reduced up to 45% without any loss in quality (Vest, 2000).

Figure 4. Waste tyre retreading
Source: http://www.bensontire.com/products.html

PROPERTIES OF WASTE TYRE

Physical Properties

Recent research has been devoted to the possibility of using rubber derived from scrap tyres to replace natural aggregates in concrete and asphalt. Research related to rubberized concrete and asphalt maybe varies in terms of rubber content, size, types, grinding method, manufacturing process and applications. For both rubberized concrete and rubberized asphalt, the effect of tyre rubber size and replacement ratio seems to be more dominant as compared to other factors. Since more aggregates were replaced with a softer material (rubber particles), it was predictable that the intensity of the concrete to resist compression forces will degrade. However, owing to the elastic nature of this material, it will help to increase the flexibility of the pavement.

In general, vehicles tyres can be divided into two types: passenger car and truck tyres. Waste tyres, especially for civil engineering application, were categorized based on the particles size and production or recycling process applied for that particular tyre. There are four broad categories of recycled or reprocessed waste tyre rubber (Figure 5) considered in most research civil engineering research.

Besides the shapes and sizes of tyre rubber particles, the density of the rubber may vary according to the source of the rubber as summarized in Table 1. Possible reasons for the variations in densities could be due to rubber quality, types of waste tyres (automobile or truck) and also types of shredding process.

The recycled waste tyres used in civil engineering application is mainly used as replacement of aggregate, filler or cement. Replacement of natural resources namely aggregate and cement with the WTR shows a very promising result. For every engineering structure, the strength cannot be compromised. Thus, it is also important to study the factors that may affect the strength of the concrete. First, we'll

Figure 5. Recycled waste tyre rubber

look into the surface texture or morphology of the WTR (Figure 6). The WTR surface is covered with layers and these layers contribute to hydrophobic nature of rubber which tends to trap air and bubbles. Obviously, air bubbles are easily trapped in between the pores and rough surface. This factor generally contributes to the increment of pore structures in hardened cement concrete, hence reducing the density of the concrete thus also affecting the strength of the cement concrete. For the case of asphaltic concrete, the roughness of WTR was found to increase the viscosity of asphalt that using WTR as filler.

CHEMICAL PROPERTIES

The chemical compositions of WTR (Figure 7) are distinguished using XRF analysis. From the results adopted from Euniza (2018), it is noticed that the WTR is mainly composed of 49% SBR synthetic rubber, produced from a copolymer of styrene and butadiene. Generally, SBR component is used in great quantities in automobile and truck tyres production as an abrasion-resistant replacement for natural rubber (produced from polyisoprene). Furthermore, the WTR is also made of 48% carbon black. Carbon black is used as a pigment and reinforcing phase in automobile tyres. In addition, carbon black also helps to conduct heat away from the tread and belt area of the tyre, reducing thermal damage and increasing tyre life.

As for extender oil, about 1.9% of the total tyre composition is also added to soften the rubber and improve workability. Extender oil is a mixture of an aromatic hydrocarbon. A small amount of sulfur (0.8%) was also detected. Sulfur is used in tyres production to cross-link the polymer chains within the rubber and also hardens and prevents excessive deformation at elevated temperature. The accelerator of 0.7% is typically an organosulfur compound which acts as a catalyst for the vulcanization process is also identified. Lastly, zinc oxide and stearic acid compounds help to control vulcanization and enhance

Table 1. Densities of waste tyre rubber

Categories	Density (g/cm³)	References
Rubber particles from recycled tyres with sizes ranging from 4.0 mm to 11.2 mm and obtained by mechanical grinding process.	1.15	Duarte *et al.* (2016)
Tyre rubber powder (600 μm) Tyre rubber (0.8 – 4 mm)	1.05 1.13	Thomas *et al.* (2016)
Granular samples of waste tyre rubber particles without any treatment or contaminants, sourced from a local recycling plant.	1.111 – 0.909	Su *et al.* (2015)
Crumb rubber from mechanical shredding of worn tyres.	1.15	Eiras *et al.* (2014)
Tyre-rubber particles obtained from the mechanical shredding of rubber automotive industry waste.	1.16	Bing and Ning (2014)
Rubber particles from shredding process.	1.089 ± 0.036	Aiello and Leuzzi (2010)
The crumb rubber produced by grinding recycled vehicle tyres into small particles.	0.62 – 0.96	Sukontasukkul (2009)
Coarse and fine Tyre-rubber particles produced by mechanical shredding.	1.16	Khaloo *et al.* (2008)
Tyre section without steel fibre by using cracker mill process.	1.05	Turgut and Yesilata (2008)
Rubber powder of ground tyre (2.36 mm)	0.56	Zheng *et al.* (2008)
Crushed rubber or used tyre chips (15 mm to 4 mm)	1.15	
Rubber particles of less than 1 mm in size and contains approximately 20% by volume of polypropylene fibers	0.43	Benazzouk *et al.* (2007)
Scrap and crumb tyres with size ranges 0.5 to 2 mm and 0.05 to 0.7 mm.	0.90	Bignozzi and Sandrolini (2006)
Crumb rubber of 3.35 mm and 0.85 mm.	0.88 – 0.97	Sukontasukkul and Chaikaew (2006)
Waste tyre chips or particles.	0.84	Li *et al.* (2004)
Crumb rubber produced from recapping truck tyres.	0.40	Pierce and Blackwell (2003)
Two types of rubber, obtained from the shredding processes of truck tyres.	0.46 – 0.525	Fattuhi and Clark (1996)
Shredded rubber tyres, produced by Envirocrete 2000®.	0.61	Toutanji (1996)
Rubber aggregate, obtained by mechanical grinding from the outer surface of scrap tyres.	0.656	Topcu (1995)
Two types of tyre chips: mechanical grinding (Edgar chips) and the other by cryogenic grinding (Preston rubber).	0.80 – 0.96	Eldin and Senouci (1993)

Source: Euniza (2018)

the physical properties of the rubber. Combinations of high abrasion-resistant and heat-resistant material component in tyre rubber obviously explain why waste tyre is extremely difficult to biodegrade compared to other types of waste.

The TGA application is for optimizing the acceleration of aging conditions and predictions of thermal aging of a material. Bystritskaya *et al.* (2013) emphasis by determining the lifetime of polymer and composite materials helps to provide the optimum combination of safety requirements and technical and economic considerations for manufacture. Thermal analysis of WTR alleviates determination of kinetic parameters related to the thermal aging of polymeric materials and to estimate the service life of these materials.

Figure 6. FESEM image of waste tyre rubber

Figure 7. Waste tyre rubber chemical compositions

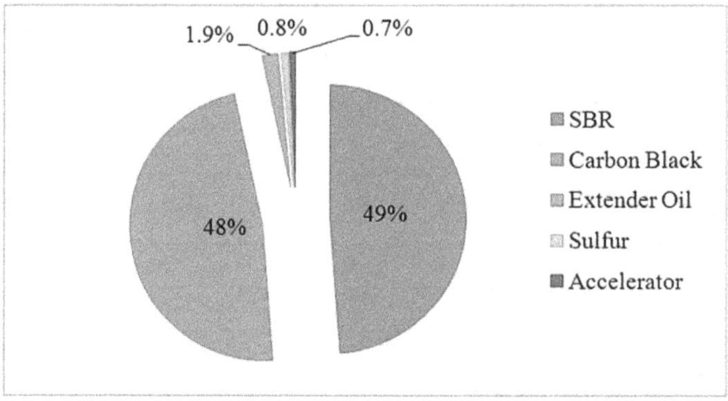

Figure 8. TGA/DTA curves of waste tyre rubber fired to various temperature regimes
Source: Euniza (2018)

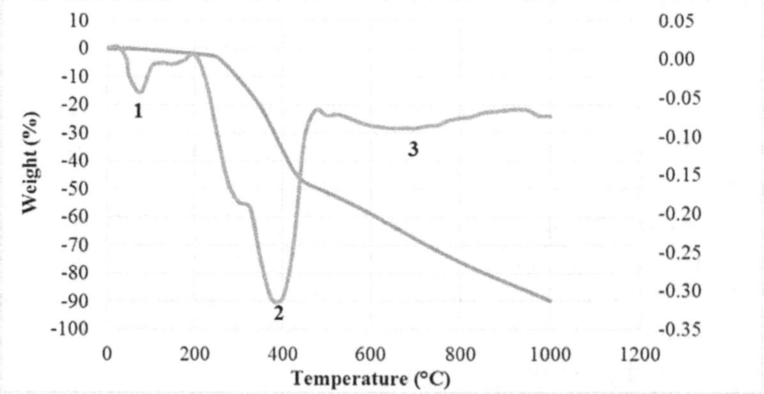

As reflected in Figure 8, there are three peaks detected from the TGA/DTA specimen. Decomposition of organic material is reflected in the first endothermic peak which located between 50 °C and 225 °C. As reported by Cui *et al.* (1999); and Knappe and Urso (1993), the process of decomposition begins with the emission of volatile plasticizers and residues of the vulcanization system and incorporated antioxidants. Fernández-Berridi *et al.* (2006) added that the first endothermic peak (at temperature 200 – 300 °C) is associated to volatilization of processing oil. At the temperature between 250 °C to 550 °C where the second peak takes place, it corresponds to the decomposition natural rubber (NR) and styrene rubber (BR or SBR). The result obtained is comparable to Belgacem *et al.* (2013) and Islam *et al.* (2009). Furthermore, Januszewicz *et al.* (2017) found at temperature 427 °C, the hydrocarbon oil is fully destroyed and pyrolitic oil transforms into gas. In addition, by referring to this finding, oil and gas derived from waste tyre can be extracted and used as alternative energy source. Lastly, the third endothermic peak spotted at the temperature of 600 °C to 900 °C, is attributed to the decay of carbon black and zinc oxide (Belgacem *et al.,* 2013). The detected weight loss and maximum peak were tabulated in Table 2.

From the TGA data in Table 2, the thermal decomposition starts at approximately 65 °C where 5% of weight loss is detected. It follows by two major losses of weight during the main devolatilization and pyrolysis which occurs at 385 °C and 675 °C, respectively. Most researchers also agree that major losses of 50 – 70% corresponds to decomposition NR, and BR or SBR (Januszewicz *et al.,* 2017; Belgacem *et al.,* 2013; Islam *et al.,* 2009; Fernández-Berridi *et al.,* 2006; Cui *et al.*, 1999). From the TGA-DTA result of WTR pyrolysis, it is found that the weight loss is relatively high due the presence of inorganic compounds.

PAVEMENT INDUSTRY

In general view of road pavement industry, the demand for producing low noise pavement has increased worldwide. Currently, optimum balance between the road primary function and higher noise reduction performance are essential, especially for the urban area. People living in the urban area begin to complain about the excessive noise disturbance which leads to an unhealthy environment (Tiesler *et al.,* 2013). Increasing road traffic requires road surfaces that are simultaneously durable and low noise. It is noted that tyre/road interaction noise is more dominant when a passenger car travel exceeding 55 km/h. The noise which derived from the engine and exhaust system is easier to handle since today's automobile technology has transformed engine and exhaust system to be more efficient and less noise. The remaining factor to be terminated is rolling noise. Heutschi *et al.* (2016) suggest rolling (tyre/road interaction) noise can be lowered by the installation of low noise road pavements to increase noise reduction at source.

Table 2. Weight Loss of tyre rubber compounds at different temperature

Compound	Temperature (°C)	Maximum Peak (°C)	Weight Loss (%)
Oil, plasticizer, organic additive	50 – 225	65	5
NR, BR, SBR	250 – 550	385	57
Fillers, Carbon Black	600 - 900	675	38

Source: Euniza (2018)

Traffic noise is a general term of unwanted sound generated by the traffic. It is all the sounds that are heard as a result of vehicles traveling down a road and includes the combination of all possible sources of noise on a vehicle. In particular, traffic noise has been spotted as an environmental stressor that causes sleep disturbance and annoyance to the society. According to Guski *et al.* (2016) questionnaire and quantitative meta-analysis results, it shows that people get annoyed about three times higher when the road traffic noise level increases by 10 dB. Furthermore, among the direct and most obvious effects of road traffic noise interferes negatively with the individual's speech communication, on their concentration ability, and consequently on the performance of tasks (Ouis, 2001).

TYRE/ROAD INTERACTION NOISE

Tyre/road interaction noise as shown in Table 3, arises from a combination of physical processes which consist of mechanical, aerodynamical, frictional and propagation. Mechanical effect involved the impacts and shocks resulting from the contact between the tyre tread and road surface that lead to vibrations. Moreover, the aerodynamical effect is related to the processes between, and within, the tyre tread and road surface patterns such as air turbulence, cavity resonance in tyre tube, air-pumping, and Helmholtz resonant radiation and pipe resonance. Frictional noise is due to the adhesion and micro-movement effects of the tread during its contact with the road, such as 'Stick-slip' and 'Stick-snap'. Lastly, propagation effect is an acoustical impedance related properties of the surface that influence transmission paths, such as horn effect, absorption, and directivity of sound frequency.

The tyre/road interaction noise is influenced by road surfacing characteristics as well tyre characteristics. According to Sandberg and Ejsmont (2002), road surfacing characteristics that influence tyre/road noise generation are megastructure, macrostructure, and porosity. Megastructure and macrostructure of road surface corresponding to the wavelength range of the texture between 50 mm and 0.5 mm, and 50 mm and 0.5 mm, respectively. Megastructure is also associated with the unevenness of road surface, while macrostructure is associated with the size of aggregates used in the pavement. Porosity is referring to the pore of the road surfacing. The percentage and connectivity of pore also influence the absorption of noise associated with the reflection and propagation of tyre/road interaction noise.

Propulsion noise which derived from the engine and exhaust system is easier to handle since today's automobile technology has transform engine and exhaust system to be more efficient and less noise. The remaining factor to be concerned is the rolling noise which mainly attributed to the tyre-road interaction noise. A noise barrier is also applicable to mitigate traffic noise along a busy road; however, that effectiveness of the noise barrier is somehow debatable. Height is an important variable both in regard to the barrier, the noise sources in need of attenuation, and the anticipated receiver point. Noise barrier must be taller than the tallest noise source in order to be effective. Hence, due to the limitation of noise barrier other noise reducing medium has to be used. Heutschi *et al.* (2016) suggested rolling noise can be lowered by the installation of low noise road pavements.

RESEARCH DEVELOPMENT

Praticò and Anfosso-lédée (2012) identify several types of low-noise pavement that currently available to mitigate traffic noise issue which consist of poro-elastic road surfacing (PERS) or rubberized

Table 3. Tyre/road noise generation

Interaction Effect	Mechanism
Mechanical	 *Running Deflection*  *Road texture impact* 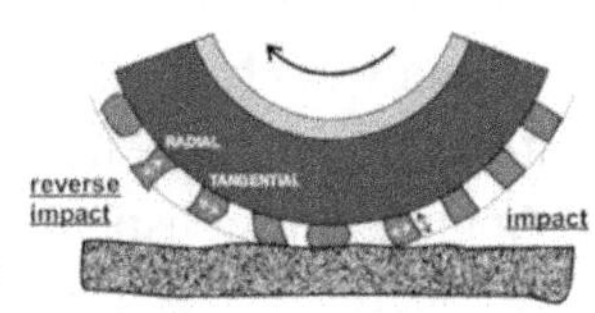 *Tread element impact*
Aerodynamical	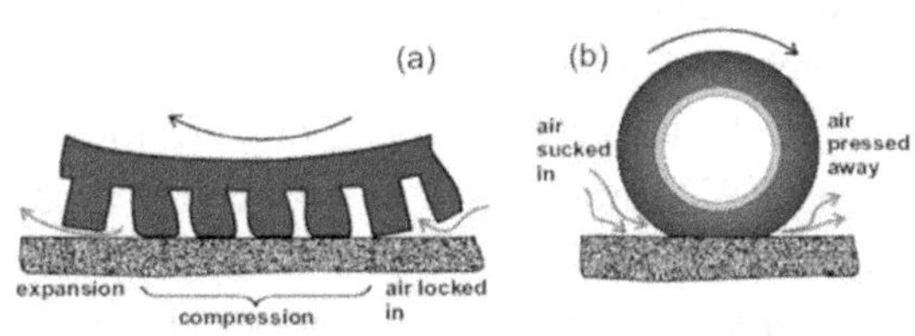 *Air displacement*  *Helmholtz resonator* 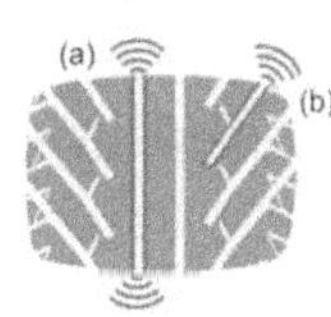 *Pipe resonator* 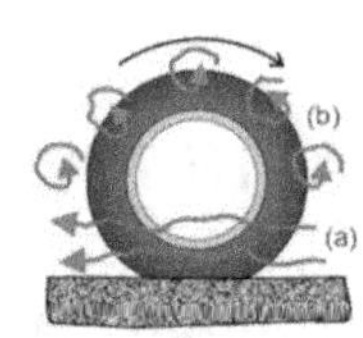 *Air turbulence*
Frictional	 *Stick – slip adhesion* 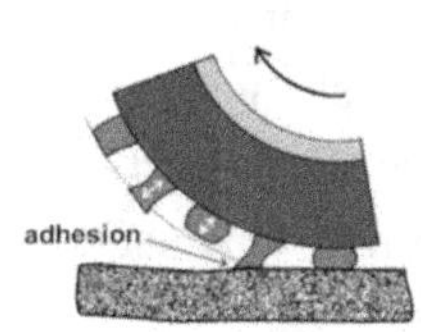 *Stick – snap adhesion*
Propagation	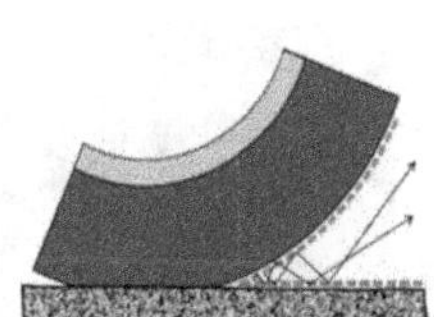 *Horn amplification*

Source: Sandberg and Ejsmont (2002); Kindt et al. (2016)

asphalt (RA); porous asphalt (PA); thin or very thin asphalt concrete (VTAC); and stone mastic asphalt (SMA). Most of the developed low-noise pavement is mainly of asphalt based. Goubert and Sandberg (2016) reported among all bituminous pavement, PERS or RA seems to generate impressive acoustic performance. According to Goubert (2014), the PERS is able to reduce noise up to 12 dB(A) while the others only reduce less than 8 dB(A). The PERS is designed with a minimum of 20% air voids and 20% rubber by weight of modified binder.

Besides asphalt pavement, cement concrete pavement has also commercially used as traffic noise reducer medium. Application of asphalt pavement always deals with durability issue. Durability could be best provided by concrete roads, but standard concrete road surfaces do not achieve noise reductions better than 2 dB, which is not sufficient to compensate the increase of noise levels due to increasing overall traffic volume. Kim and Lee (2010) stated that material such as concrete is classified as 'sound shielding materials' due to their acoustic reflection behaviour. It has been claimed that, at higher speed, where tyre noise is the dominant noise generating mechanism, segmental pavements may be associated with noise levels 5 to 8 dB(A), higher than asphalt surfaces (Ohio Department of Transportation, 2006).

Low-noise concrete pavement starts to evolve as a response to the limitation of asphaltic concrete especially in terms of durability. The WTR material was adopted as a pavement material in powder or granulated form. Research development in semi-rigid and flexible pavement has grown progressively and shows promising outcome. The implication of waste in producing new product will help in reducing waste tyre and saving the environment.

APPLICATION OF WASTE TYRE IN SEMI-RIGID PAVEMENT

Rubberized Concrete Block Pavement

The rubberized concrete paving block (RCPB) was developed by Ling (2011) to utilize waste material and reducing the use of natural material in concrete paving block (CPB). Developing an innovative RCPB product lead to better engineering properties and comparable service performance compared to the existing CPB. The feature of the blocks was shown in Figure 9 and Table 4. The highest percentage of replacement is up to 30% to tolerate the reduction of strength due to the substitution of WTR as aggregate. The WTR ranged from 1 – 5 mm was used to replace the fine aggregate in CPB.

Figure 9. Rubberized concrete paving blocks
Source: Ling (2011)

Table 4. Rubberized concrete paving blocks features

Mix. Notation	Rubber Content (%)	Depth (mm)	Top Surface (mm)	Weight (kg)	Strength (MPa)
CCPB	0	59.6	5.5	2.82	64
10-RCPB	10	59.8	5.0	2.82	63
20-RCPB	20	59.2	5.0	2.74	26
30-RCPB	30	59.8	3.3	2.68	23

Source: Ling (2011)

According to Shatanawi (2008), the peak of sound absorption coefficient for low and high-speed traffic occur at a frequency of approximately 600 Hz and 1000 Hz, respectively. In general, the RCPB containing crumb rubber was found to have slightly higher sound absorption coefficients than without crumb rubber over the entire frequency range (100–1600 Hz). This can be attributed to the fact that RCPB containing crumb rubber contributes a higher porous surface layer and less density, resulting in good characteristic in resonant absorbency.

DOUBLE-LAYER RUBBERIZED CONCRETE BLOCK PAVEMENT

The application of WTR in CPB was further improved by Euniza (2018) with the development of double-layer rubberized concrete paving block (DRCPB). The block was designed to provide an alternative type of block with better sound absorption properties. It consists of different facing layer (FL) thicknesses with different percentage of WTR replacement. The sound absorption level of the blocks was measured to simulate the level of tyre-pavement interaction noise that possibly be absorbed. The usage of waste material such as WTR was adopted as it produces positive effects on both environmental and economic prospect. The block and its cross-section are shown in Figures 10 and 11.

Figure 10. Double-Layer Rubberized Concrete Paving Block
Source: Euniza (2018)

Figure 11. Cross section of DRCPB
Source: Euniza (2018)

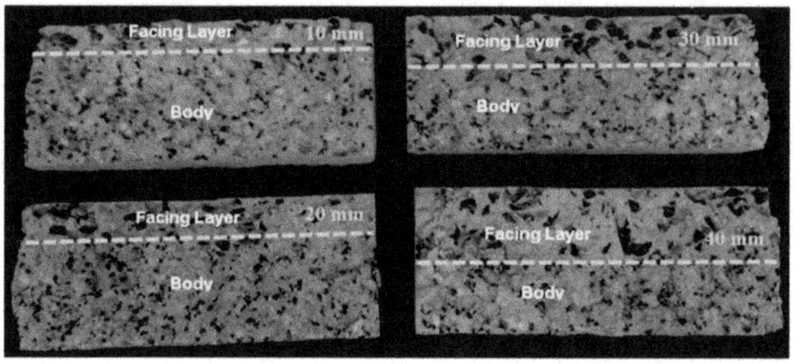

Acoustic parameters for DRCPB specimens showed that concrete pavement blocks have one peak of sound absorption located at a low frequency of 500 to 700 Hz. This indicates that it suitable for application of mitigation of low traffic speed condition. Two sound absorption prediction models are developed to predict absorption level at optimum and 600 Hz frequency.

Parameters of rubber content, FL thicknesses and porosity are statistically tested and proven to be significant ($p < 0.05$) for model 1. For model 2, only FL thicknesses and RG content give a significant effect on the prediction model. These two models are only applicable of predicting the optimum sound absorption and sound absorption at 600 Hz of DRCPB, if the FL thicknesses, the percentage of RG and the curing condition used are within the tested range. In addition, two equations as suggested by (Hamet, 2004) were used to indicate the prediction values of tyre-road noise on DRCPB. The predicted values using Equation 1 clearly showed that the only DRCPB specimens with TL thickness of 40 mm and RG of 40% able to reduce the tyre-block noise by 2 dB. This specimen block could be used as pathway application. However, if using Equation 2, the predicted noise reduction of DRCPB with TL thickness greater than 30 mm and percentage of RG equal or greater than 20% would be 3 dB which is the impact is noticeable.

$$\triangle L_{PB}\left(f,\theta\right) = 20\log\{\left|1 + Q\left(f,\theta\right)\right| / 2\}\ \text{dB} \tag{1}$$

$$\triangle L_{PB}\left(f_c, 81\right) = -11.9 * \left\langle \alpha_0\left(f_{sh}, 0\right)\right\rangle \text{dB} \tag{2}$$

Where, α_o is sound absorption coefficient at shifted frequency, $f_{sh} = f_c$ (1/3 octave band) /0.91.

COMPOSITE CONCRETE PAVING BLOCK

Composite CPB consist of PERS as the block facing and normal concrete as the block body. The idea of developing the PERS-CPB is as a medium to reduce tyre-road interaction noise for a semi-rigid pavement. It shows a very promising result, as the surfacing of the block was replaced with a more flexible

Table 5. Optimum sound absorption and sound absorption at 600 Hz and 1000 Hz.

Facing Layer Thickness (mm)	Rubber Granules (%)	Optimum Sound Absorption Coefficient (α_{opt})	Frequency at Optimum Sound (Hz)	Sound Absorption Coefficient at 600 Hz (α_{600})	Sound Absorption Coefficient at 1000 Hz (α_{1000})
10	10	0.16	600	0.16	0.06
10	20	0.19	559	0.19	0.06
10	30	0.22	543	0.20	0.07
10	40	0.25	540	0.23	0.06
20	10	0.18	634	0.19	0.07
20	20	0.20	550	0.19	0.07
20	30	0.23	531	0.21	0.06
20	40	0.29	559	0.26	0.06
30	10	0.19	500	0.15	0.09
30	20	0.22	625	0.21	0.08
30	30	0.26	643	0.24	0.08
30	40	0.30	618	0.29	0.07
40	10	0.21	500	0.16	0.10
40	20	0.25	613	0.25	0.06
40	30	0.28	634	0.26	0.09
40	40	0.34	640	0.31	0.07

Source: Euniza (2018)

material. The PERS is produced from the combination a minimum of 20% air voids and 20% rubber by weight modified binder. Figures 12 to 14 show the application of PERS on the concrete block in Japan and Slovenia, respectively.

APPLICATION OF WASTE TYRE IN FLEXIBLE PAVEMENT

Rubberized asphalt (RA) is a noise reducing pavement material that consists of regular asphalt concrete mixed with crumb rubber made from tyre rubber. Such pavement has been used in different countries such as the USA, Portugal, Spain, the Netherlands, and Japan. Praticò and Anfosso-lédée (2012) and Goubert (2014) discovered that the noise reduction result of RA ranged from 6.6 to 12 dB(A). According to Vázquez and Paje (2016), at the medium sound frequencies (around 800 Hz) it seems to be strongly correlated to the megatexture and the roughness of the surface studied. The researchers also indicate that high frequency sound could be attenuated inside the voids between particles whose upper surfaces form a nearly flat plane. A sound reduction is due to dispersion and multiple reflections of the tyre/pavement sound emitted.

Figure 12. Two types of concrete paving block with PERS tested in R46 Akita, Japan
Source: Sandberg et al. (2007)

Figure 13. Poro-elastic block pavement made of PERS glued on concrete blocks and test track in Nova Gorica, Slovenia
Source: Sandberg and Karman (2015)

PORO-ELASTIC ROAD SURFACING (PERS)

Incorporation of waste tyre rubber was implemented in SILVIA and PERSUADE projects, whereby road surface studied are poro-elastic road surface (PERS) and rubberized surfaces. The PERS is an improved rubberized asphalt incorporating voids and channels to reduce traffic noise (Sandberg *et al.*, 2014). The PERS is a surface with high elasticity where it is essentially made of aggregate and rubber granules or fibres bound with a polymer modified asphalt or polyurethane binder. In general, PERS is used as wearing course of a road and designed with high porosity. It was proposed by Sandberg and Ejsmont (2002), and Sandberg and Karman (2015) that the term PERS shall only be used for surfaces with a design air voids content of no less than 20% and a proportion of rubber of at least 20% by weight. Figures 15 and 16 show three type of prefabricated rubber panels produced by Tokai, Rosehill, and Spentab Rubber Company that was tested in SILVIA project. The rubber materials used for all cases are from waste tyres.

Figure 14. Solvenian full scale test track in Nova Gorica
Source: D. Kokot

In the laboratory test, it was found that the Rosehill PERS was about 9 dB(A) quieter than the other two PERS . However, a different result was observed for the field test. The Rosehill PERS was found to be less efficient in reducing noise as compared to the other two PERS.

The PERS as shown in Figure 17 provides an efficient combination of noise-reducing measures: (1) sound absorption and elimination of air pressure gradients by its high air voids content, (2) a smooth surface giving low vibration excitation to the tyre, (3) low contact angles in the leading and trailing edges of the tyre/road interface, and (4) low interaction impact between tyre tread and road surface elements. If the hysteresis losses in the deflection of the system can be made low (by material selection and design) the rolling resistance losses of such a system may be low or at least not higher than for a conventional tyre/road contact. Furthermore, according to Swedish-Japanese studies, PERS provides an effective reduction of tyre-pavement noise between 5 and 15 dB(A) compared with conventional dense asphalt surfaces (Praticò and Anfosso-lédée, 2012).

COST EVALUATION

A distinct advantage of using WTR as part of the materials used for pavement is its lower cost and its economic feasibility in comparison to the usage of natural aggregate. The WTR costs RM 40 – 42 per tonne while aggregate RM 70 per tonne (Malaysian supplier). From the material perspective, cost reduction up to 20 - 23% can be obtained.

Figure 15. Three types of PERS tested in SILVIA project
Source: Sandberg et al, (2007)

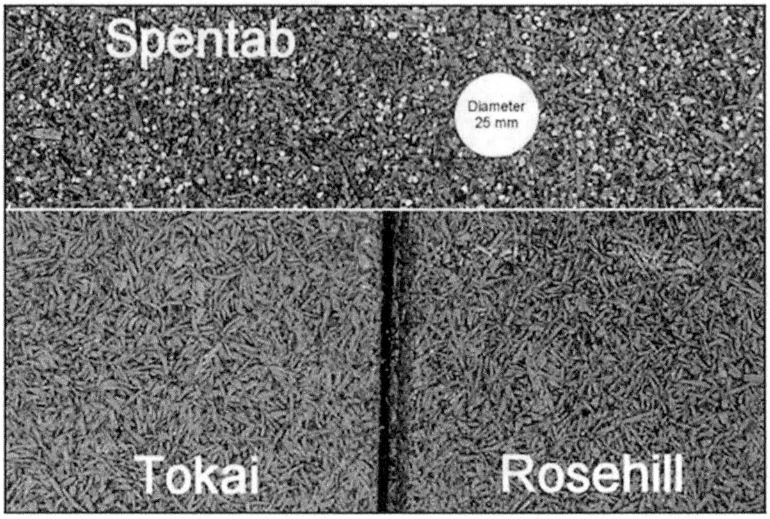

Figure 16. Test site in Stockholm with layout of different PERS surfaces
Source: Sandberg et al. (2010)

Figure 17. Close-up picture showing the PERS material texture
Source: Sandberg et al. (2013)

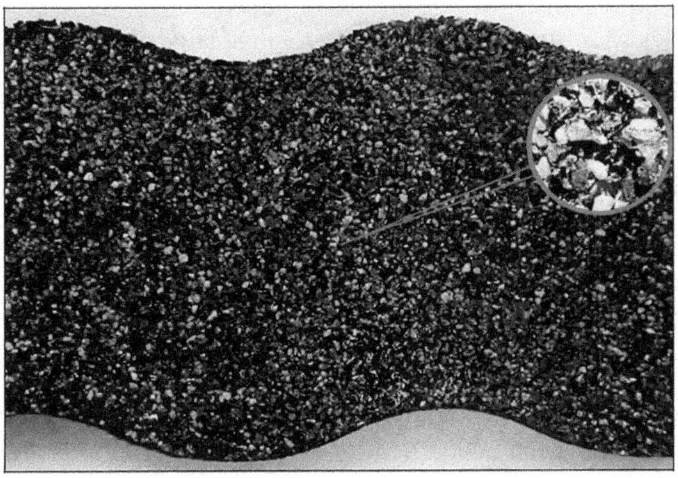

SUMMARY

A variety of research application lead to more innovative product derived from waste tyre. Application of waste tyre especially as part of pavement material gives very promising results. Incorporation of waste tyre rubber as one of the constituents used as pavement materials enhance the application of green and sustainable type of pavement. Improved performance of concrete, especially in terms of flexibility, can contribute to the utilization of RCPB and DRCPB as a substitution for the existing concrete block pavement. The rubberized concrete and rubberized asphalt show a positive impact in reducing tyre-road interaction noise thus, can be suggested to be used in the area where low noise is the main concern.

ACKNOWLEDGMENT

The authors would like to acknowledge the support of the Ministry of Higher Education (MOHE) and Universiti Teknologi Malaysia Research grant (RUG vote 03H47) for funding this research study.

REFERENCES

Akhil, M., Saikat, D., & Vasudeva, M. (2018, June 3-8). Studies on chemical composition and physico-chemical characterization of upgraded tire pyrolysis (UTPO) for the application in engine and furnaces. In *Proceedings of 9th International Freiberg Conference on IGCC & XtL Technologies closing the Carbon Cycle.* Berlin, Germany.

Amari, T., Themelis, N. J., & Wernick, I. K. (1999). Resource recovery from used rubber tires. *Resources Policy, 25*(3), 179–188. doi:10.1016/S0301-4207(99)00025-2

Aniza, A. A., Rao, S. P., & Elias, S. (2013). Waste tyres as heat sink to reduce the driveway surface temperatures in Malaysia. *Journal of Design and Built Environment*, *13*, 1–11.

Chemsain. (2011). *A study on scrap tyres management for peninsular Malaysia report*. September, 2011.

Euniza, J. (2018). Physical, mechanical, and acoustical performance of double-layer rubberized concrete paving blocks. *PhD Thesis*. Universiti Teknologi Malaysia.

Goubert, L. (2014, Nov. 16-19). Developing a durable and ultra low noise poroelastic pavement. In Proceedings of Internoise 2014. Melbourne, Australia. 1–10.

Goubert, L.,, & Sandberg, U. (2016, July 10-14). Development of the ultra low noise poroelastic road surface: the findings of the persuade project. In *Proceedings of 23rd International Congress on Sound and Vibration*. Athens, Greece. 1–8.

Guski, R., Schreckenberg, D., Schuemer, R., Gmbh, Z., & Gmbh, Z. (2016, Aug. 21-26). The WHO evidence review on noise annoyance 2000-2014. In *Proceedings of INTER-NOISE 2016 - 45th International Congress and Exposition on Noise Control Engineering: Towards a Quieter Future*. Hamburg, Germany. 2564–2570.

Hamet, J. F. (2004). Reduction of tire road noise by acoustic absorption : numerical evaluation of the pass-by noise level reduction using the normal incidence acoustic absorption coefficient reduction of tire road noise by acoustic absorption. *SILVIA* (silent roads). 1-25.

Heutschi, K., Bühlmann, E., & Oertli, J. (2016). Options for reducing noise from roads and railway lines. *Transportation Research Part A, Policy and Practice*, *94*, 308–322. doi:10.1016/j.tra.2016.09.019

Kindt, P., Vercammen, S., & Bianciardi, F. (2016, Aug. 21-24). Tire/road noise – characterization and potential further reductions of road traffic noise. In *Proceedings of the INTER-NOISE 2016 - 45th International Congress and Exposition on Noise Control Engineering: Towards a Quieter Future*. Hamburg, Germany. 2254–2264.

Kwon, E. E., Oh, J., & Kim, K. (2015). Polycyclic aromatic hydrocarbons (pahs) and volatile organic compounds (VOCS) mitigation in the pyrolysis process of waste tires using CO_2 as a reaction medium. *Journal of Environmental Management*, *160*, 306–311. doi:10.1016/j.jenvman.2015.06.033 PMID:26117814

Ling, T. C. (2011). Prediction of density and compressive strength for rubberized concrete blocks. *Construction & Building Materials*, *25*(11), 4303–4306. doi:10.1016/j.conbuildmat.2011.04.074

Lopez, G., Alvarez, J., Amutio, M., Mkhize, N. M., Danon, B., Van Der Gryp, P., & Olazar, M. (2017). Waste truck-tyre processing by flash pyrolysis in a conical spouted bed reactor. *Energy Conversion and Management*, *142*, 523–532. doi:10.1016/j.enconman.2017.03.051

Martínez, J. D., Puy, N., Murillo, R., García, T., Navarro, M. V., & Mastral, A. M. (2013). Waste tyre pyrolysis – a review. *Renewable & Sustainable Energy Reviews*, *23*, 179–213. doi:10.1016/j.rser.2013.02.038

Mavroulidou, M.,, & Figueiredo, J. (2010). Discarded tyre rubber as concrete aggregate: a possible outlet for used tyres. *Global NEST Journal*, *12*(4), 359–367.

Miranda, M., Pinto, F., Gulyurtlu, I., & Cabrita, I. (2013). Pyrolysis of rubber tyre wastes: a kinetic study. *Fuel*, *103*, 542–552. doi:10.1016/j.fuel.2012.06.114

Ouis, D. (2001). Annoyance from road traffic noise: a review. *Journal of Environmental Psychology*, *21*(1), 101–120. doi:10.1006/jevp.2000.0187

Praticò, F. G.,, & Anfosso-lédée, F. (2012). Trends and issues in mitigating traffic noise through quiet pavements. *Procedia: Social and Behavioral Sciences*, *53*, 203–212. doi:10.1016/j.sbspro.2012.09.873

Rodovalho, C.,, & De Tomi, G. (2017). Reducing environmental impacts via improved tyre wear management. *Journal of Cleaner Production*, *141*, 1419–1427. doi:10.1016/j.jclepro.2016.09.202

Sandberg, U., Goubert, L., Biligiri, K. P., & Kalman, B. (2010). State-of-the-art regarding poroelastic road surfaces. Deliverable D8.1 Project PERSUADE.

Sandberg, U., Bendtsen, H., Thomsen, S. N., Kragh, J., Kalman, B., & Kokot, D. (2007). *Possibilities to reduce tyre/road noise emission on paving stones and other block surfaces*. SILENCE Deliverable F.D3.

Sandberg, U.,, & Ejsmont, J. A. (2002). *Tyre/road noise reference book. INFORMEX*. Sweden.

Shah, J.,, & Rasul Jan, M. (2006). Conversion of waste tyres into carbon black and their utilization as adsorbent. *Journal of the Chinese Chemical Society (Taipei)*, *53*(5), 1085–1089. doi:10.1002/jccs.200600144

Shatanawi, K. (2008). *The effects of crumb rubber particles on highway noise reduction - a laboratory study*. Clemson, SC: Clemson University.

Shilpa, R. K.,, & Sharma, A. (2017). Morphologically tailored activated carbon derived from waste tires as high-performance anode for Li-ion battery. *Journal of Applied Electrochemistry*, 1–13.

Shu, X.,, & Huang, B. (2014). Recycling of waste tire rubber in asphalt and Portland cement concrete : an overview. *Construction & Building Materials*, *67*, 217–224. doi:10.1016/j.conbuildmat.2013.11.027

Szentannai, P., Bozi, J., Jakab, E., Osz, J., & Szucs, T. (2015). Towards the thermal utilisation of non-tyre rubbers – macroscopic and chemical changes while approaching the process temperature. *Fuel*, *156*, 148–157. doi:10.1016/j.fuel.2015.04.037

Tiesler, C. M. T., Birk, M., Thiering, E., Kohlböck, G., Koletzko, S., Bauer, C. P., ... Heinrich, J. (2013). Exposure to road traffic noise and children's behavioural problems and sleep disturbance: Results from the GINIplus and LISAplus studies. *Environmental Research*, *123*, 1–8. doi:10.1016/j.envres.2013.01.009 PMID:23498846

Tyre component: source. Retrieved from https://kitchendecor.club/files/radial-tire-cross-section.html

Vázquez, V. F.,, & Paje, S. E. (2016). Study of the road surface properties that control the acoustic performance of a rubberised asphalt mixture. *Applied Acoustics*, *102*, 33–39. doi:10.1016/j.apacoust.2015.09.008

Vest, H. (2000). *Recycling of used car tyres*. Retrieved from www.gtz.de/gate/gateid.afp

Chapter 2
Utilization of Waste Cooking Oil (WCO) in Hot Mix Asphalt (HMA)

Mastura Bujang
University College of Technology Sarawak, Malaysia

Wan Nur Aifa Wan Azahar
International Islamic University Malaysia, Malaysia

Euniza Jusli
University College of Technology Sarawak, Malaysia

ABSTRACT

The use of WCO in binder modification is widely explored in response to waste management issue. A chemical treatment was proposed to reduce the acidity of the WCO that causes its poor performance. Therefore, this chapter evaluates the performance of binders modified with untreated and treated WCO. The physical and rheological tests of binder were conducted to determine optimum percentages of untreated and treated WCO in the binder. The optimum WCOs were utilized for mechanical performance evaluation of Hot Mix Asphalt (HMA) mixture through resilient modulus and dynamic creep. The test showed the failure temperature of binder modified using the treated WCO has increased to 70°C and the treated WCO mixture recorded superior performance by being less susceptible to permanent deformation as compared to the control mix. In conclusion, the chemical treatment had improved the treated WCO performance in the modified binder as asphalt paving materials.

INTRODUCTION

Petroleum-based asphalt binder is derived from petroleum refinement process by-product (Wen *et al.*, 2013). In composition of asphalt mixture, the binder acts as an adhesive agent for coating process and binds the aggregate particles together. Generally, pure bitumen production through crude oil petroleum refining process is not desirable in road pavement application due to insufficient properties for pavement construction and need to be modified with various additives types such as carbonaceous materials, fine

DOI: 10.4018/978-1-7998-0369-0.ch002

minerals and polymers (Chebil *et al.*, 2000). An improvement of engineering properties for asphalt binder can be achieved with the application of a modifier by reducing temperature susceptibility and enhancing the rheological performance to withstand the environmental and traffic loading.

In recent years, a wide range of oil-based modifications have been introduced, especially involving waste cooking oil (WCO). The WCO is a waste grease oil which is categorized as by-product of fresh cooking oil produced during cooking and food processing. This oil source has recently gained widespread attention because of its satisfactory achievement as a potential waste material to enhance the physical and rheological performance of modified binder. The oil undergoes three types of common chemical reaction during frying, such as hydrolysis, oxidation and polymerisation. The chemical process in oil causes degradation in the physical and chemical properties (Cvengroš & Cvengrošová, 2004) which affects the WCO's physical properties, include alteration in foaming quantity, colour, viscosity, density and flavour. Meanwhile chemical properties are represented as total unsaturation compound, free fatty acid content, polar and polymeric material. The generation of huge quantities of WCO is attributed by the increasing numbers of human population and its usage in food preparation such as frying. It can be noticed that WCO production quantity is directly proportional to the frying rate.

According to Chhetri *et al.* (2008), enormous quantity of WCO that is generated worldwide is illegally dumped and released to the open environment. In Malaysia, the disposal of WCO is reported to be approximately around 50,000 tons, which was produced from the plant and animal-based fats source (Yaakob *et al.*, 2013). These wastes are disposed to the environment without undergoing any proper treatment (Kheang *et al.*, 2006). A survey conducted by Kabir *et al.* (2014) revealed that the majority of the residents (54.5%) discard WCO through the house sink. Meanwhile 22.2% stated that the WCO is dumped into outside drains. This coincides with the response on the level of awareness in WCO recycling. Unexpectedly, only 12% of the households recycled WCO, meanwhile about 88% of the residents did not implement a waste recycling approach. Such inappropriate action has consequently induced an undesirable impact to the entire environmental ecosystem, for instance distraction of aquatic life, contamination of water and soil, sewer system blockage and increased maintenance cost for water treatment and waste management (Chen *et al.*, 2009).

Abundant of WCO production can cause prominent adverse impact and threat to the environmental if no proper disposal management are taken. Therefore, recycling or reusing WCO in modified asphalt binder is considered as a sustainable technique, indirectly ensured an economic and environmental benefits to the nation (Patil *et al.*, 2012). It is noteworthy that, most researchers have focused on the superior performance of WCO as a rejuvenator for aged binder (Asli & Karim, 2011; Zargar *et al.,* 2012; Asli *et al.,* 2012; Zaumanis *et al.,* 2013; Chen *et al.,* 2014a; Binbin *et al.,* 2014; Chen *et al.,* 2014b), apart from replacing the WCO in modified binder in order to improve rheological performance. The WCO performance as a modifier at high and low temperatures was evaluated by Wen *et al.* (2013). The rheological findings indicate declination of the complex modulus (G*), which resulted in a low rutting resistance at high temperature. On the contrary, an increment in thermal cracking resistance performance at low-temperature was observed to occur linearly with the addition of WCO content.

This rheological result coincides with the study conducted by Maharaj *et al.* (2015) for un-aged sample, wherein an enhancement of fatigue cracking resistance was achieved at low temperature. Meanwhile, the high temperature performance showed an adverse effect as the rutting resistance decreased with the addition of WCO. This is also supported by Sun *et al.* (2016), which reveals the decrement of deformation resistance and improvement of thermal cracking resistance performance when applying the modified binder incorporated with bio-oil derived from WCO. According to Teymourpour *et al.* (2015), the application of WCO in conventional binder proved an enhancement of thermal cracking resistance but compromising the high temperature performance by the reduction of resistance to rutting. The superiority performance of fatigue cracking resistance was noticeable with the addition of WCO as reported in a previous research by improving the low temperature performance and deteriorating the rutting resistance performance at high temperature.

The WCO in asphalt binder softens the physical properties of modified asphalt binder, thereby unable to withstand the rutting exposure. According to Katamine (2000), the negative indicator in the quality of bitumen is expected with the addition of WCO in bitumen. This serves as the major reason why WCO is not suggested for optimal utilization, due to its poor performance in rutting resistance, which is not desirable especially for hot climatic region. Owing to the adverse and unsatisfactory performance in rheological properties, no further research work was conducted to evaluate the mechanical properties of asphalt mixture. This is due to the expectation of high tendency for rutting problem, thereby reducing the strength of asphalt mixture. Waste oil needs to be fully re-evaluated and assessed by undergoing further treatment before being recommended and applied in the asphalt binder to produce bituminous mixture (Borhan *et al.*, 2009).

However, despite the broadly practiced WCO application, the identification of fundamental parameters influencing WCO in binder modification and also the effectiveness of the process has not been investigated yet (Teymourpour *et al.*, 2015). In addition, the mechanism for modifying asphalt binder with WCO and the effect of WCO on the modified binder performance are not clearly documented. Obviously, the softer and less viscous modified binder with WCO depicts the weakness of internal chemical bonding of the material. Besides, there is the issue of chemical compatibility and interprets an assumption of incompatibility between these two materials due to chemical properties that should be further clarified and investigated. Thus, this chapter aims to evaluate Hot Mix Asphalt (HMA) incorporating untreated and treated WCO in modified binder.

The deficiency of the rutting resistance performance exhibited by modified binder and asphalt mixture when dealing with oil-based source can be solved by the improvement of untreated WCO properties through chemical modification. A comparison on the performance of binder and asphalt mix incorporating untreated and treated WCO, which has not been previously studied, is explored in this chapter in terms of binder test and mechanical asphalt mixture test.

ASPHALT BINDER

Physically, the asphalt binder, also known as bitumen, appears as a black and dark solid cementitious substance which consists of a large hydrocarbon chemical structure. Chemical elemental analysis has proven that bitumen contains approximately 82-88% carbon, 8-11% hydrogen, 0-6% sulphur, 0-1.5% oxygen and 0-1% nitrogen (Asphalt Institute, 2003). Bitumen can be categorized as a colloidal model in which asphaltenes and maltenes are the main constituents of bitumen. This colloidal structure depicts the dispersion of asphaltenes as solid, hard and brittle component with a radius of few nanometers in oily, liquid and resinous maltenes matrix appearance (Brownridge, 2010). Asphaltenes functional for binder viscosity is described as a polar compound which possesses a high molecular weight within 5% to 25% of bitumen component.

Meanwhile, maltenes provides stability to the asphaltenes component (Katamine, 2000). The three basic maltenes component are saturates, aromatic and resins. Saturates is predicted to form 5% to 20% of bitumen fraction. According to Read & Whiteoak (2003), the chemical structure of saturates consists of straight and branched-chain aliphatic hydrocarbons together with alkyl naphthalene and some alkyl-aromatic. The lowest molecular weight of aromatic and resin comprise 30-60% and 15-55% from bitumen mass, respectively. The binder viscosity characteristic is controlled by the aromatic polar nature, while resins provide the adhesive fraction component and ductility to bituminous materials (Robert, 2000).

A binder is a viscoelastic material which is partly an elastic solid and partly a viscous liquid. The binder viscoelastic property is affected by test temperature and loading. The binder viscoelastic property can be interchangeable depending on these temperatures and loading factors. When subjected to high temperature, binder acts as a low viscous liquid wherein the binder displays consistency almost entirely as fluid. In contrast, at low temperature exposure binder mostly behaves like an elastic solid with high viscous property. When load is applied on the binder, an elastic solid behaviour is exhibited, which indicates a recoverable deformation after subjected to the load. This implies the capability of binder to return to its original shape when exposed to unloading condition. Contrary to this, binder is a viscous liquid which depicts the non-recoverable deformation effects due to the applied load (Maharaj *et al.*, 2015). Hence, the viscoelastic property of binder is measured based on the temperature susceptibility performance.

WASTE COOKING OIL (WCO)

Waste cooking oil (WCO), is non-edible oil and can be defined as a by-product of fresh oil produced during food frying (Choe & Min, 2007), as shown in Figure 1. Fresh cooking oil is physically light yellow in colour with low viscosity, while used cooking oil has high viscosity and dark brown in colour. WCO source is collected from food industries, household disposables, restaurants and recycling centre by authorized companies (Singhabhandhu & Tezuka 2010, Zhang *et al.*, 2012). The annual WCO collection is approximately three billion gallons, as stated by the EPA (2016). Meanwhile, according to Math *et al.* (2010), about 15 million tonnes of WCO production per year was recorded based on the statistics. In Malaysia approximately 0.5 million tonnes of WCO is generated per year and derived from palm oil source (Gui *et al.*, 2008). An increasing demand for food consumption (Phan and Phan, 2008), besides rapid and tremendous growth of human population (Chen *et al.*, 2009) is identified as the main factor that attributed to the large quantity of WCO production.

The WCO recycling is at an empirical stage, in which only a small amount of used cooking oil is properly collected and recycled (Castellanelli, 2007) especially from residential area and food restaurants. In fact, enormous WCO quantity, without any appropriate treatment, is illegally dumped at the landfills and in rivers which has caused serious adverse towards environment, especially environmental pollution issues. The most common environmental issue that arise is the eutrophication process which is caused by the blockage of sunlight from passing through the river surface because of the presence of a thin layer of oil. Apart from sunlight, oxygen supply to aquatic life is also disrupted due to this process (Hamad *et al.*, 2003), thus leading to the excessive growth of micro-organisms, phytoplanktons and algae. River water quality is measured by the existence of these micro-organisms wherein their huge growth rate affects the intrinsic equilibrium of aquatic ecosystem.

Figure 1. Oil sample after frying process

At residential area, clogging problem has grabbed other human's attention because of blocked drains. This blockage happens when waste frying oil is poured through the kitchen sink and accumulates in the drains. This issue has affected wastewater treatment plants and indirectly maximised the processing costs since routine maintenance work must be done in order to mitigate and monitor this clogging problem (Sanli *et al.*, 2011). The Agriculture and Food Development Authority (2000) reported that, an estimation of about 40% of sewerage system blockages are contributed due to the waste cooking oil has been poured in the kitchen sinks.

Due to disposal problems and possible water and land resources pollution, supervision and management of waste cooking oil has created a significant challenge that need to be taken into consideration (Math *et al.*, 2010). Therefore, waste cooking oil recycling is becoming an effective option to mitigate pollution issues due to illegal dumping problem with the greatest concern in high construction cost and natural resource conservation (El-Fadel & Khoury, 2001). According to Alcantara *et al.* (2000), the mitigation of major environmental issues, such as waste oil disposal and wastewater treatment, can be encountered by the recyclability of WCO as a modifier in asphalt binder modification for pavement mixture.

CHARACTERIZATION OF WCO

Chemical Properties

The chemical compound possessed in WCO is determined by using Gas Chromatography Mass Spectrometry (GCMS). The majority of authors stated that the predominant compounds found in WCO are Oleic acid, Palmitic acid and Linoleic acid, as presented in Table 1.

Meanwhile, Udomsap *et al.* (2009) and Wen *et al.* (2010) reported that the major components of WCO is Linoleic acid, followed by Oleic acid and Palmitic acid. Ma & Hanna (1999) discovered that the dominant component found in WCO is Palmitic acid, Oleic acid and Linoleic acid. It can be observed that different fatty acids composition is recorded if different sources of waste cooking oil are used. These acids are described as Free Fatty Acids (FFA) that is naturally contained in fats, oils and greases, in which higher FFA component in WCO leads to higher acid value (Silitonga *et al.*, 2013). Theoretically,

Table 1. Free fatty acids in waste cooking oil

Fatty Acids	Chemical Formula	WCO (%)						
		Ma and Hanna (1999)	**Leung and Guo (2006)**	**Udomsap *et al.* (2009)**	**Wen *et al.* (2010)**	**Asli and Karim (2011)**	**Asli *et al.* (2012)**	**Zargar *et al.* (2012)**
Lauric acid	C12:0	0.1	0	0.1	0	0.34	0.34	0.34
Myristic acid	C14:0	1.0	0.9	0.1	0	1.03	1.03	1.03
Palmitic acid	C16:0	42.8	20.4	10.2	8.5	38.35	38.35	38.35
Stearic acid	C18:0	4.5	4.8	3.7	3.1	4.33	4.33	4.33
Oleic acid	C18:1n9c	40.5	52.9	22.8	21.2	43.47	43.67	43.67
Linoleic acid	C18:2n6c	10.1	13.5	53.7	55.2	11.39	11.39	11.39
γLinolenic acid	C18:3n6c	0	0	0	0	0.37	0.37	0.37
Linolenic acid	C18:3n3c	0.2	0.8	8.6	5.9	0.29	0.29	0.29
Arachidic acid	C20:0	0	0.12	0	0	0	0	0
Cis-11-Eicosenoic acid	C20:1n9c	0	0.84	0	0	0.16	0.16	0.16
Heneicosa-noic acid	C21:0	0	0	0	0	0.08	0.08	0.08
Behenic acid	C22:0	0	0.03	0	0	0	0	0
Erucic acid	C22:1n9c	0	0.07	0	0	0	0	0
Tetracosanic acid	C24:0	0	0.04	0	0	0	0	0

the fatty acids chain and number of double bonds affect oil physical and chemical properties (Zlatanic *et al.*, 2004) since the degree of unsaturated compound is more preferable for chemical reactions. The acid composition in WCO can be categorised as saturated, monounsaturated and polyunsaturated.

Basically, Lauric acid, Myristic acid, Palmitic acid and Stearic acid are categorized as the saturated fatty acids, while Linoleic acid, γLinolenic acid and Linolenic acid are identified as polyunsaturated fatty acids (Udomsap *et al.*, 2009). Oleic acid and Cis-11-Eicosenoic are monounsaturated fatty acids. The saturated component can be depicted as the non-existence double-bond in WCO fatty acids. This implies that maximum number of possible hydrogen atoms per atom carbon is contained in these chains (Lam *et al.*, 2010). Otherwise, the presence of double-bond in unsaturated component indicates insufficient maximum number of possible hydrogen atoms, thereby will induce the electronegativity to react during chemical reaction in which this unsaturated component tend to accept the hydrogen atoms. The comparison of FFA composition between fresh cooking oil and waste cooking oil reported by Maneerung *et al.* (2016) indicated that the reduction of unsaturated fatty acids consists of Oleic acid, Linoleic acid and Linolenic acid in WCO after frying process, as shown in Table 2. The decreasing trends of unsaturated fatty acids is attributed to the chemical degradation reaction during frying in which these acids tend for hydrogen acceptance thereby increase the saturated fatty acids.

Table 2. FFA composition in fresh and waste cooking oil

Free Fatty Acid Composition (%)	Fresh Cooking Oil	Waste Cooking Oil
Oleic ($C_{18}H_{34}O_2$)	43.9	43.2
Linoleic ($C_{18}H_{32}O_2$)	30.4	30.1
Palmitic ($C_{16}H_{32}O_2$)	20.3	19.4
Linolenic ($C_{18}H_{30}O_2$)	4.8	4.7
Stearic ($C_{18}H_{36}O_2$)	2.9	2.6

(Maneerung et al., 2016)

Table 3. Physiochemical properties of unused and used cooking oil

Properties	Unused Cooking Oil Values	Used Cooking Oil Values
Acid value (mg KOH/gm)	0.3	4.03
Calorific value (J/gm)	-----	39658
Peroxide value (meq/kg)	< 10	10
Density (gm/cm^3)	0.898	0.9013
Kinematic Viscosity (mm^2/s)	39.994	44.956
Dynamic Viscosity (mpa.s)	35.920	40.519
Flash point (°C)	161-164	222-224
Moisture content (wt %)	0.101	0.140

(Ullah et al., 2014)

Physiochemical Properties

Ullah *et al.* (2014) conducted test properties to compare the properties of unused and used cooking oil. The physiochemical properties result is presented in Table 3. The various properties test conducted were acid value, calorific value, peroxide value, density, kinematic viscosity, dynamic viscosity, flash point and moisture content. All these properties were measured to identify the WCO quality. The main parameter for physiochemical properties was evaluated based on the acid value test, density and viscosity. The acid value, density and viscosity (dynamic and kinematic) result obtained indicates an increment in value from unused to used cooking oil. Increasing acid value represents a higher value of free fatty acid content in WCO as these properties are related to each other. According to Chhetri *et al.* (2008), the free fatty acids contained in waste cooking oil is higher as compared to fresh cooking oil. These findings coincide with Choe & Min (2007), in which deep fat frying causes parameter alteration, where an increase in free fatty acids content, density and viscosity was observed in WCO.

In addition, the increasing value of density will cause the unsaturated fatty acids decreased due to the chemical reaction during frying oil that changed into saturated fatty acids. This can be observed by the physical WCO appearance wherein the unused cooking oil is light in condition while used cooking oil is thicker and more viscous. During frying process, the oil will undergo oil polymerisation, producing cyclic monomer and polymer. This process depends on the degree of unsaturation, in which the increase of cyclic compound formation is recorded as increase in the linolenic acid, since these types of acid is categorized as unsaturated fatty acid. The polymer compound produced from the polymerisation accelerates oil degradation, thereby will increase the WCO viscosity (Tseng *et al.*, 1996). Besides, the other factor that attributed to viscosity value is increase in polar formation and oxygen containing molecule. Therefore, it can be noticed that the used cooking oil viscosity is higher in comparison with unused cooking oil.

The peroxide value result also increased after frying due reaction in oxidation. Peroxide can be expressed in the unit of miliequivalents of active oxygen per kilogram of fat/oil (meq/kg). During frying, oil will undergo oxidation when exposed to air. Higher oxidation rate will induce peroxide value increment in WCO (Udomsap *et al.*, 2009). The increment of flash point in used cooking oil is affected by the high saturated compound possessed in the oil. The saturated compound contains a strong single bond in which higher temperature is gained to break the bond, therefore flash point value increased. The amount of moisture content in used cooking oil is higher than that of unused cooking oil. This implies that the presence of moisture in food during frying will increase the moisture content in used cooking oil.

Physical and Chemical Changes in WCO

The degradation of WCO quality parameter occurs when oil is repeatedly heated during frying, as shown in Figure 2. The reaction involved in frying at high temperature leads to WCO physical and chemical degradation. The alteration in physical properties can be identified as foaming, colour, viscosity, density and flavour, while the changes in major chemical properties include changes in free fatty acid content, total unsaturation compound, polar and polymeric materials. The obvious increment in properties is represented by the free fatty acids content which serve as the indicator for increased WCO acid value (Silitonga *et al.*, 2013; Andrade *et al.*, 2014), where the FFA value is higher in WCO when compared to fresh cooking oil (Chhetri *et al.*, 2008). Normally, WCO possesses FFA that ranges between 0.5% until 15% (Gerhard *et al.*, 2004). Frying time is identified as one of the vital factors that affect the acid value result. The longer the time WCO is used, the higher the resulting acid value will be (Mazza & Qi, 1992; Choe & Min, 2007). Besides, the increase in FFA content is attributed by the increased frying frequency (Mazza & Qi, 1992). The amount of FFA indicates the WCO quality parameter (Choe & Min, 2007) in which the increasing value of FFA or acidity represents the higher degradation rate experienced by WCO (Fernando *et al.*, 2007; Udomsap *et al.*, 2009). This coincides with the findings from Aladedunye & Przybylski (2009) and Knothe & Steidly (2009) who stated that the double bonds in WCO component are ruptured as WCO deteriorates through degradation and thereby causing changes in FFA composition which in turn increase the FFA level.

In addition, the decrement trend of unsaturation compounds is recorded with a higher frying rate. During the chemical reactions in frying, the higher tendency for unsaturated compounds is attacked by other molecules due to the increased amount of unsaturated composition and imbalanced electronegativity charges, thus leading to a high oxidation rate (Warner *et al.*, 1994). This implies that a high reactivity level in unsaturated composition is observed due to the presence of double bonds in WCO fatty acids. According to Leong *et al.* (2015), WCO quality is degraded during heating, which results in the formation of more saturated compounds, such as monomers, dimers, trimmers and high-molecular-weight compounds.

Figure 2. Degradation of WCO quality parameter
(Choe & Min, 2007)

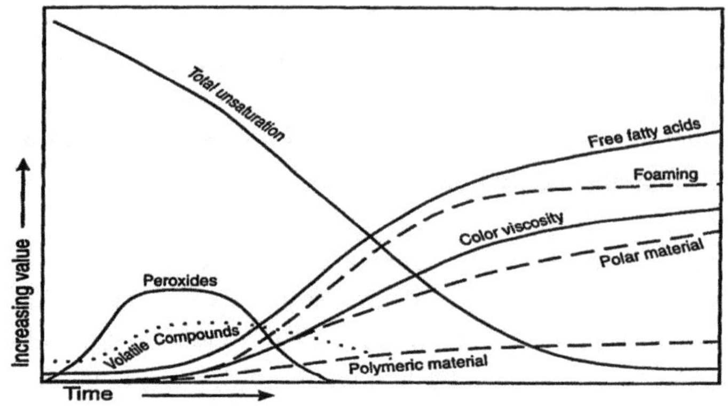

The amount of saturated compounds is inversely proportional to the unsaturated compounds, in which the increment of saturated composition indicates the decreased proportion of unsaturated compounds.

The other identified degraded parameter is antioxidant content, which is naturally possessed in WCO. The antioxidant is defined as a molecule capable of preventing and slowing down the rate of molecule oxidation. Apart from that, antioxidant can be recognised as a chemical agent that provides an inhibitive protection towards oxidation, where oxygen attack is retarded by the antioxidant content. This implies that the presence of antioxidant provides a natural resistance to oxidative deterioration in WCO. The antioxidant is operated by the inhibitive potential to donate hydrogen due to low activation energy in the antioxidant to interrupt the propagation of free radical chains. Basically, the depletion of this natural antioxidant contents is recorded in parallel with the repeated heating cycle process, thereby leading to the degradation in oil quality (Koh *et al.*, 2014; Leong *et al.*, 2015). This also indicates that the reduction in available antioxidant present in low quality oil and the antioxidant residue in oil after the depletion process cannot provide any protective effect against oxidative damage due to insufficient protection.

Minor changes are recorded by physical appearance, namely foam formation, WCO color becomes darken (brown), increased viscosity and density indicated by the thick condition from observation and an undesirable off-flavour produced. Alteration in physical properties is increased with repeated oil heating. Besides, the presence of uncomfortable odour and flavours occurs during linolenic acid oxidation (Choudhary & Grover, 2013). It can be summarised that the degradation reaction has reduced the WCO quality parameter and caused indirect negative impact to humans.

WCO as a Modifier for Binder Modification

A modifier can be defined as a potential material which enhances the service life performance and can withstand the loading factor exposure for improvement of asphalt mixture engineering properties (Teymourpour *et al.*, 2015). According to Arabani *et al.* (2010), the substitution of a modifier as part of the modification mechanism in asphaltic concrete mixture has reduced the deterioration rate due to presence of asphalt pavement distress. Based on the listed previous research, WCO was mostly applied as a rejuvenator since the fluidity factor in WCO has the capability for restoration aged binder by reducing its viscosity. Apart from these, the modification of binder by using WCO as a modifier was

Figure 3. G.sin δ versus WCO percentage*
(Maharaj et al., 2015)

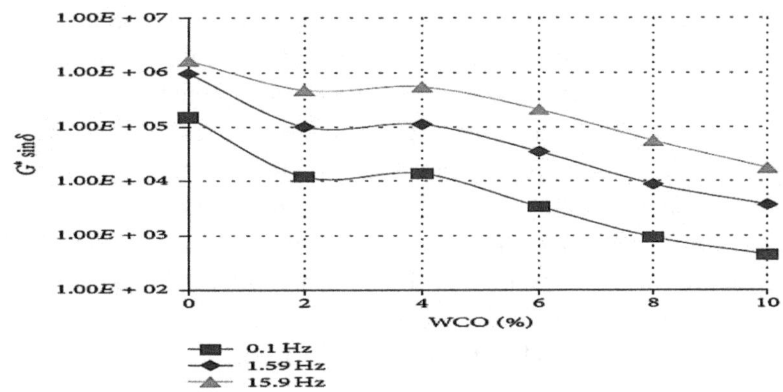

Figure 4. G/sin δ versus WCO percentage*
(Maharaj et al., 2015)

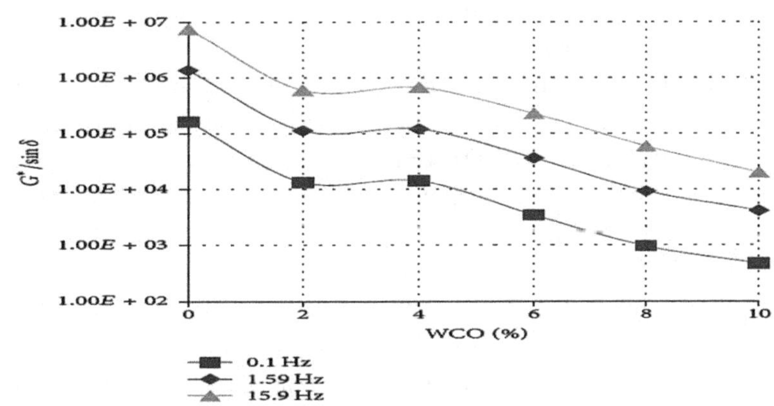

Figure 5. Rutting parameter (G/sin δ) versus temperature for unaged sample*
(Al-Omari et al., 2018)

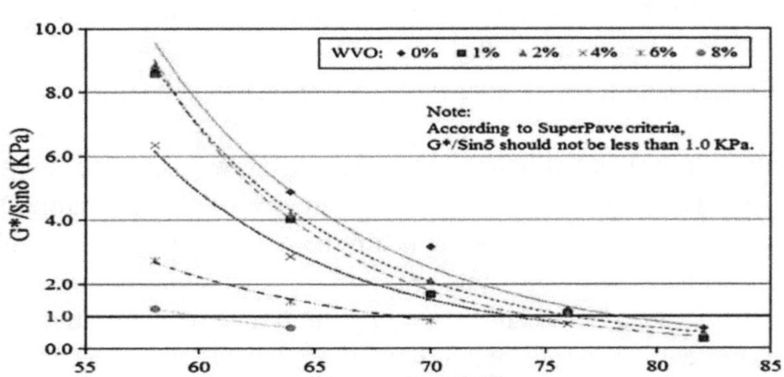

focused. Basically, the application of a rejuvenator is desirable to determine the cracking resistance at low temperature, while the function of a modifier is preferable for rutting resistance at high temperature.

Maharaj *et al.* (2015) performed an assessment for rutting and cracking resistance performance of modified asphalt binder by incorporating WCO as a modifier. The rheological performance demonstrated that the decreasing G*.sin δ was observed in Figure 3 which indicated an improvement in fatigue cracking resistance. The increasing trend of fatigue cracking resistance was recorded as the WCO concentration was increased in the modified binder. In contrast, with the increasing WCO concentration in control binder, the G*/sin δ was decreased which explained the decrement of rutting resistance performance, as shown in Figure 4. The rutting resistance displayed by the modified binder with WCO was considered poor performance since the rutting resistance of the control binder exhibited superior performance as compared to the modified binder. Based on the rheological findings, it can be summarised that the WCO application was noticeably significant in low temperature performance as compared to high temperature performance.

Wen *et al.* (2013) conducted a laboratory evaluation for binder containing bio-asphalt from the conversion of WCO through a thermochemical process. There were three different types of base binder used in this study which is PG 58-28, PG 76-22 and PG 82-16. About 30% and 60% of bio-asphalt was blended with PG 58-28 meanwhile the bio-asphalt was mixed at 10% and 30% (by the weight of total binder) for PG 76-22 and PG 82-16 binders. The rutting resistance performance achieved in this research coincided with the previous study, wherein increasing non-recoverable creep compliance (J_{nr}) value from multiple stress creep and recovery (MSCR) result was recorded with the addition of bio-asphalt. The increment of J_{nr} indicated a high susceptibility of modified binder with bio-asphalt to rutting exposure. Besides, the lower critical strain energy density (CSED) attained in this study resulted in the reduction of fatigue cracking resistance with the addition of bio-asphalt content in the conventional binder.

Al-Omari *et al.* (2018) evaluated the feasibility of 1%, 2%, 4%, 6% and 8% of waste vegetable oil (WVO) as a modifier for binder modification in terms of physical and rheological performance. The findings reported that increasing trend of penetration result was recorded with the addition of WVO content. In contrast, the decrement trend of softening point was observed when WVO was increasingly added into the base binder (PEN 60/70). A similar finding was demonstrated by Sun *et al.* (2016) while evaluating the effect of bio-oil derived from waste cooking oil on the physical and mechanical properties of modified binder. In addition, the viscosity result was consistently reduced with the increasing of WVO content as expected. This is due to the lubricating effect possessed in oil itself thereby causing the decrement of viscosity in modified binder. The rheological properties were assessed based on the low, intermediate and high temperature. For high temperature, the unaged and short term aged sample recorded the decreasing of G*/sin δ with the increasing amount of WVO added into the base binder and subjected to the higher temperature as illustrated in Figure 5. This implied that the modified binder incorporating WVO was susceptible to permanent deformation exposure thereby decreasing the rutting resistance performance.

On the contrary, an improvement of fatigue cracking resistance was recorded with the increasing of WVO dosage as evidenced by the reduction of G*.sin δ at intermediate temperature. Meanwhile, the low temperature performance obtained from bending beam rheometer (BBR) indicated that the decrement trend of creep stiffness with increasing m-value was noticeable with the addition of WVO. This finding is consistent with Sun *et al.* (2015) who claimed that the addition of biodiesel by-product produced from waste cooking oil refinement process decreased the stiffness result which depicts the improvement of thermal cracking resistance performance at low temperature as illustrated in Figure 6.

Figure 6. Effect of temperature at different percentage of biodiesel by-product from WCO on the stiffness (Sun et al., 2015).

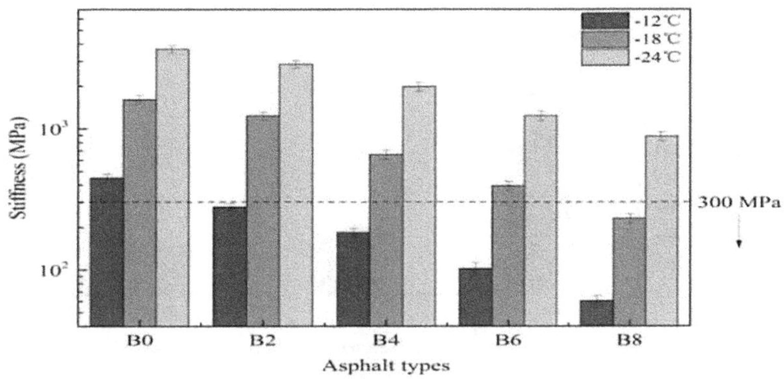

Hot Mix Asphalt (HMA)

In HMA, coarse and fine sizes of aggregate particles are bonded together by asphalt binder at certain temperature and is constructed by flexible pavement mode. The different sized aggregate particles exhibit as a skeleton that provides strength through the distribution of loading by the contacted aggregate particles. Meanwhile, bitumen acts as a binder to bind or hold the aggregate particles in position, which will enhance the asphalt mixture performance. HMA is constructed by heating the binder to reduce viscosity while the aggregate is dried to ensure effective mixing with the binder. Mixing process between aggregate and binder is performed when subjected to the 165°C temperature and compacted by using impact compactor at 75 blow/face compaction standards for the dense graded mixture.

Generally, a pavement structure is constructed in several layers which consist of subgrade, sub-base, road-base and surface course. Each layer functions for transmitting and distributing applied load from vehicles to the lower subgrade layer. Over the service life period of pavement structure, the transmittance of load increased the tendency of pavement in vulnerable to various types of pavement distresses. The most significant distress mode that existed in the pavement surface, which is known as rutting, fatigue cracking and thermal cracking which occur especially at high, intermediate and low temperature. Basically, the exposure of pavement to distress may affect the poor performance of road pavement, thereby reduces the pavement service life (Sengoz & Topal, 2005).

Modified Binder Incorporating with WCO in HMA

A laboratory evaluation conducted by Wen *et al.* (2013) reported that the addition of waste cooking oil – based bioasphalt to produce alternative binder for HMA reduced the dynamic modulus, wherein the stiffness was linearly decreased with the addition of bioasphalt concentration. The decrement of flow number with increasing amount of bioasphalt, as shown in Figure 7, also proved the reduction in rutting resistance performance, which is more prone to deflection than a control mixture. Another interpretation indicated that adding bio-asphalt reduces high-temperature performance as expected.

Figure 7. Flow number at different percentage of bio-asphalt from WCO
(Wen et al., 2013)

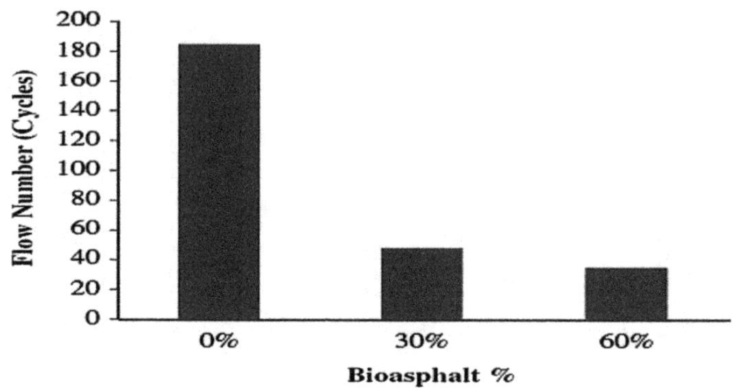

Table 4. CSED results at low temperature

PG	Bioasphalt (%)	Failure Strength (Pa)	CSED (Pa)
58-28	0	6,501,831	5,992
58-28	30	4,961,625	12,437
58-28	60	4,503,408	16,280
82-16	0	6,827,405	7,477
82-16	10	6,235,349	7,797
82-16	30	4,664,153	10,633
76-22	0	6,069,766	7,614
76-22	10	5,801,009	7,721
76-22	30	5,156,961	8,927

(Wen et al., 2013)

Figure 8. Reaction during transesterification process

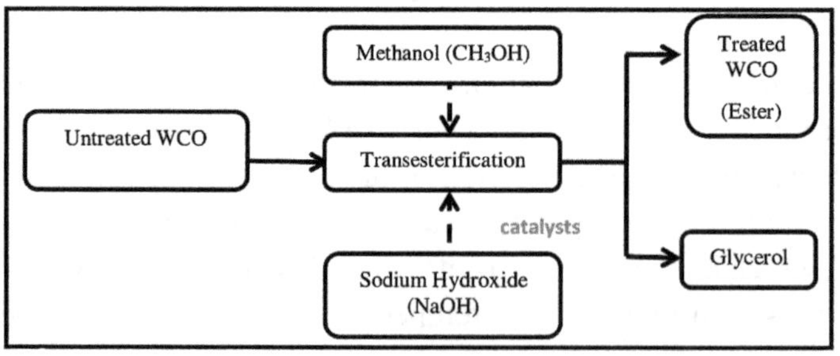

In contrast, the increment of critical strain energy density (CSED) has displayed a contrary result as presented in Table 4, wherein an increased in thermal cracking resistance at low temperature which is attributable by the improvement of ductility performance was indicated.

Proposed Potential Solution

The poor rheological performance exhibited by modified binder with WCO especially at high temperature has triggered a proposed an effective solution for WCO quality performance improvement. It is noteworthy that various attempts were recorded to modify the asphalt binder via physical blending and chemical modification (Tabatabaee *et al.* 2009). Mechanism for chemical modification is conducted by shifting the fatty acids component in oil. According to Hwang & Erhan (2001), the chemical modification is performed on epoxidised soybean oil, in which this oil is chemically reacted with alcohol in the presence of sulfuric acid as the catalyst. A better performance was recorded from this process and indicated that the chemical modification is identified as one promising solution to mitigate oxidative stability issue. Apart from preventing oxidative instability, the chemical modification through the trans-esterification process is proposed for minimising the high free fatty acids content in oil (Silitonga *et al.*, 2013). During transesterification chemical reaction, WCO is reacted with methanol, as an alcohol source, and catalyzed by NaOH as illustrated in Figure 8. The transesterification can be identified as alkali catalysts (homogeneous and heterogeneous), acid catalysts (homogeneous and heterogeneous), and enzymes which can be differentiated based on the types of catalyst used during the process. Basically, alkali-based catalyst is commercially used and is selected for transesterification (Rashid & Anwar, 2008).

Transesterification Process

Transesterification was performed in a 250 mL round–bottom flusk, equipped with a water-cooled reflux condenser, as shown in Figure 9. A total of 600 ml methanol and 1 ml NaOH were mixed to form methoxide. Thereafter, the methoxide solution was added to 100 ml WCO in a flask containing a magnetic stirring bar, to ensure there is a homogeneous solution during the reaction. The solution was heated on the hot plate at a temperature of 65 °C. Once the temperature was reached, the reaction was initiated for a duration time of 1 h. Temperature was maintained during the process, by a condenser for water flow and achieve equilibrium at 65 °C. Separation of the two major end products, namely treated WCO (esters) and glycerol, indicated that the reaction process has completed, which lasted after 24 hours at room temperature, as shown in Figure 10. The upper phase was recognised as treated WCO or ester, which was used for further modification of the asphalt binder, meanwhile the lower phase was

Figure 9. Transesterification process

Figure 10. Separation of the by-products

glycerol, which was denser than ester, thereby facilitating the separation process from ester (Leung *et al.*, 2010). The change from dark brown color for untreated WCO to the light yellow for treated WCO can be observed after the chemical process, as shown in Figure 11.

PERFORMANCE OF HMA

Binder Evaluation

Penetration test was performed on control and modified binder incorporated with untreated and treated WCO. These WCO were replaced in modified binder at 0%, 5%, 10%, 15% and 20% (by weight of binder) before being subjected to the penetration test. The control and modified binder were prepared under two types of conditions, which was fresh and short-term aging condition in which these samples were divided into un-aged and aged sample. The penetration results on un-aged and aged modified binder, containing untreated and treated WCO, are graphically presented in Figure 12. For un-aged condition, it can be observed that the increasing penetration trend was recorded as the higher amount of untreated WCO was replaced into the modified binder before it decreased at 20% of untreated WCO replacement. Meanwhile, a similar increasing trend was presented by treated WCO, wherein the penetration linearly increased as the percentage amount of treated WCO in modified binder was increased. On the other hand, the penetration was increased from control to untreated WCO before it decreased at treated WCO for all the percentage replacements.

For aged condition, a lower penetration was observed for untreated and treated WCO in comparison with un-aged condition. For untreated WCO, a similar trend with un-aged sample was recorded as the penetration was increased with the higher content replacement of untreated WCO before it decreased

Figure 11. Untreated WCO (left) and treated WCO (right)

Figure 12. Penetration for untreated and treated WCO

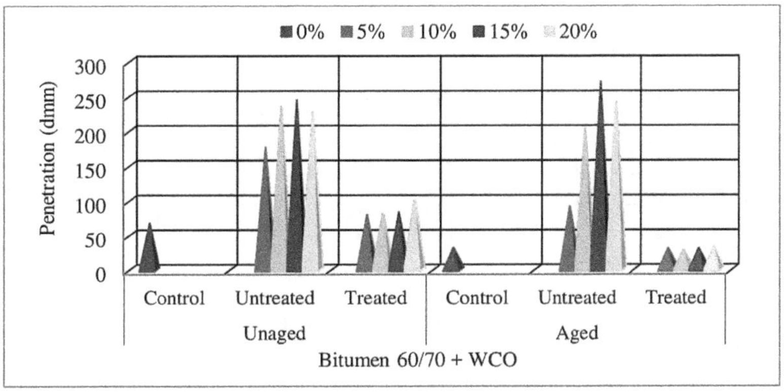

at 20% replacement percentage. Meanwhile, the inconsistent penetration data trends were observed for treated WCO in which the decreasing penetration was recorded from 5% to 10% before it increased further at 15% and 20% treated WCO replacement. The penetration was decreased from un-aged to aged condition for control and treated WCO samples due to the effect from aging process that caused physical hardening (Bahia & Anderson, 1993), which led to high stiffness and brittleness of modified asphalt binder. As a result, the hard binder from aged sample produced lower penetration as compared to un-aged sample for treated WCO.

The replacement of untreated WCO into the binder has softened the modified binder because of its polarity characteristic, which identified untreated WCO as a non-polar molecule and the binder as a polar molecule. The different polarity prevented the binder from properly mixing with WCO because the particles inside the material pushed one another and refused interaction to form a good bonding between materials, thereby producing the softer binder. The softer binder was produced due to the deterioration of low chemical bonding strength in modified binder with untreated WCO. After chemical treatment, this process changed the treated WCO polarity, wherein the treated WCO was recognised as a polar molecule. The similar polarity has induced the interaction between the particles in the binder and treated WCO for the formation of a strong bonding, thereby producing the hard binder.

Figure 13 presents the summary of softening point result subjected to the control and modified binder containing untreated and treated WCO at two different condition samples which were un-aging and aging conditions. An inversely proportional relationship with the penetration result was recorded by this test in which the softening point significantly decreased with an increase of WCO replaced in binder for un-aged condition. The similar decreasing trend of softening point was displayed by both modified binders with untreated and treated WCO. For untreated WCO, the softening point was linearly decreased with the higher replacement of untreated WCO content in modified binder. Meanwhile, the decreasing of softening point was recorded from 5% to 10% and showed a constant value at 15% before it decreased again at 20% treated WCO replacement. On the other hand, the softening point alteration trend was observed, in which the decreasing trends was recorded from control binder to untreated WCO and further indicated an improvement for treated WCO by increasing the softening point. The high softening point performance was desirable since this characteristic can withstand high temperature effect thereby minimizing the presence of permanent deformation deterioration in pavement.

Figure 13. Softening point for untreated and treated WCO

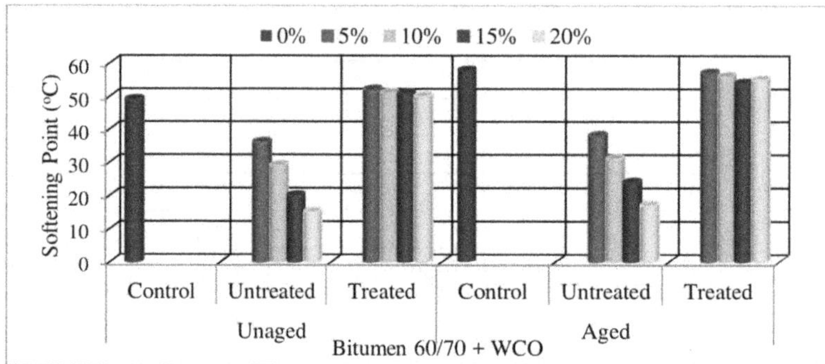

Besides, the higher softening point was achieved for untreated and treated WCO in modified binder at aged condition as compared to un-aged condition. This is due to the hard binder produced after aging, wherein during the aging this process has altered the chemical composition of binder which physically produced more brittle and stiffer material. Therefore, a higher softening point was achieved to soften the hard binder. A decreasing softening point trend was observed parallel with the higher amount of untreated WCO replaced in binder. Meanwhile, the decreasing softening point trend was recorded for treated WCO at 5% to 15% percentage replacement and increased again at 20% replacement with treated WCO. Overall, a similar trend of softening point was reported with un-aged condition, in which the resulting softening point was decreased from control binder to untreated WCO and showed a performance enhancement by increasing the softening point at all percentages with treated WCO replacement.

Figures 14 and 15 present the rutting resistance performance subjected to control and modified binder containing untreated and treated WCO, respectively. Based on the results as graphically illustrated in Figure 4.3, a lower rutting resistance was recorded with untreated WCO by reducing the failure temperature as compared to the control binder at 70 °C for un-aged condition. The rutting resistance of untreated WCO showed an inconsistency result in which the failure temperature was attained at 64 °C, 58 °C, 46 °C, 46 °C for 5%, 10%, 15% and 20% of untreated WCO replacement, respectively. In contrast, the superior rutting resistance performance was achieved with treated WCO by increasing the failure temperature, which was consistent at 70 °C for all the percentages of treated WCO replacement as presented in Figure 15. This implied that the modified binder with treated WCO exhibited the highest rutting

Figure 14. Rutting performance of WCO at untreated condition (unaged)

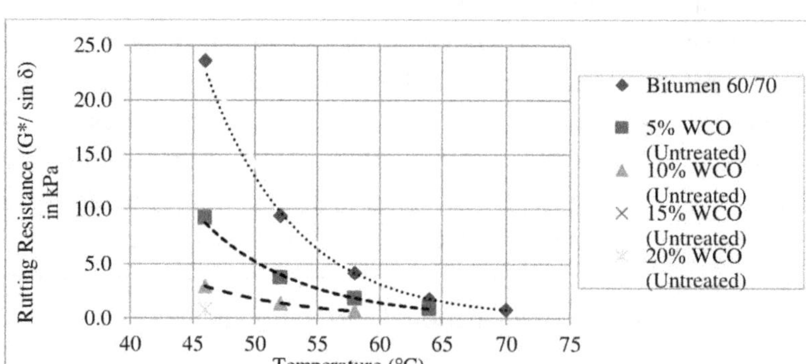

Figure 15. Rutting performance of WCO at treated condition (unaged)

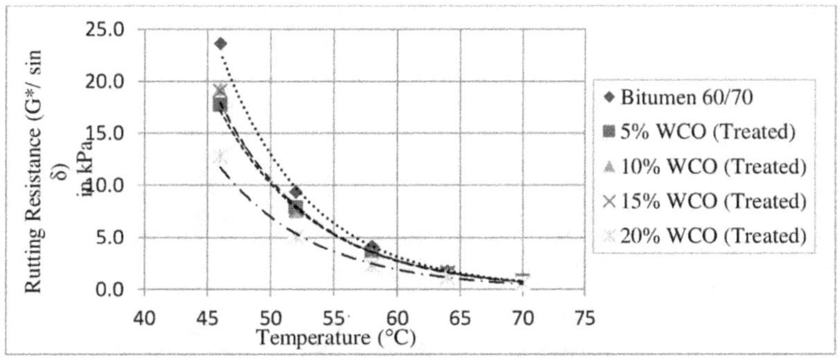

resistance performance based on the higher failure temperature attained as compared to the untreated WCO. In addition, according to Chen *et al.* (2014b) the increment of failure temperature indicated an improvement of rutting resistance for the binder. It can be noticed that the failure temperature was increased from untreated to treated WCO, thereby proved that the performance of rutting resistance was improved after the chemical modification. In addition, the failure temperature achieved by treated WCO was comparable with control binder which indicated the superiority performance and high potential as a modifier in binder modification.

Resilient Modulus

The summarized results for the resilient modulus of control, untreated WCO, and treated WCO mixtures which were tested at 25°C and 40°C are illustrated in Figure 16. At 25°C, it can be observed that the increasing trend of resilient modulus was recorded from 5% of untreated WCO, control and 5% of treated WCO in mixture. Meanwhile, at 40°C, a similar trend with test temperature of 25°C was recorded as the increasing of M_R was identified from untreated WCO, followed by control and treated WCO mixture. On the other hand, in comparison between both temperatures of 25°C and 40°C, a noticeable trend of resilient modulus result indicated that the superior performance was recorded at 25 °C as compared to that at 40°C. This implied that the resilient modulus was decreased from low temperature (25°C) to high temperature (40°C) which was presented by all the mixture. Based on the Figure 14, an increment of M_R was achieved from 2681 MPa for untreated WCO mixture to 5977 MPa for treated WCO mixture in which it indicated that the chemical modification enhanced the resilient modulus performance. The results showed that the mixes containing treated WCO exhibited superior results to the control mix sample by exceeding the resilient modulus for control mixture at 5006 MPa. This implied that treated WCO mixture exhibited the highest achievement of resilient modulus, while the lowest resilient modulus was recorded by untreated WCO mixture at both temperatures. Besides, the resilient modulus was reduced at 40°C temperature as compared to 25°C, which was recorded as 601 MPa for untreated WCO mixture, 915 MPa for control mixture and 1214 MPa for treated WCO mixture. This indicated that the treated WCO mixture achieved the highest resilient modulus when compared to other mixtures.

The high resilient modulus depicted the physical stiffening and brittleness of the mixture, which was attributable by the strong chemical bonding. The existence of similarity for internal polar – polar interaction between the particles of the modified binder containing treated WCO and aggregates in the HMA

Figure 16. Resilient modulus

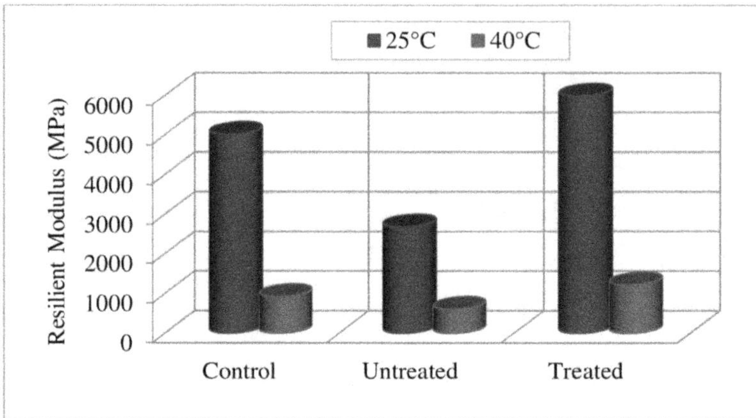

induced the high strength of interaction, thereby produced the good interlocking bonding. The superiority of chemical bonding results in hard and durable treated WCO mixture, which improved the tendency for high resistance impact to loading exposure from vehicles. In addition, high stiffness characteristic in this mixture enhanced the elastic property which had recoverable deformation potential when load was applied. In another interpretation indicated that, the capability of the mixture to return to its original condition after being subjected to the repetitive impact loading. In fact, the greater elasticity property in treated WCO mixture increased the recoverable deformation potential, thereby improved the rutting resistance performance when compared the other mixtures (Yao & You, 2016).

Dynamic Creep

The strength performance of bituminous mixture to resist permanent deformation was identified by the creep stiffness. The summarised results of the dynamic creep stiffness tested at both temperature of 40°C and 60°C are displayed in Figure 17. The result indicated that the creep stiffness at 40°C was increased from untreated WCO, followed by control and treated WCO mixtures. Meanwhile, a different result was reported at 60°C, in which the increasing trends of creep stiffness was observed from untreated WCO, treated WCO and control mixtures. This implied that the treated WCO mixture reduced the creep stiffness performance at a higher temperature. Overall, comparing between both temperatures at 40°C and 60°C, the creep stiffness was decreased from low temperature (40°C) to high temperature (60°C), which was displayed by all the mixtures. Additionally, the creep stiffness performance at 40°C depicted a superior performance as compared to the test temperature of 60°C. Based on the result, the increment of creep stiffness at 40°C was recorded from untreated WCO (196.15 MPa) mixture, control (264.29 MPa) and treated WCO mixtures (330.40 MPa). This implied that the treated WCO mixture exhibited the highest achievement of creep stiffness as compared to the control and untreated WCO mixtures. In addition, it proved that the rutting resistance was improved in treated WCO mixture as evidenced by the increased of creep stiffness strength. The improvement of creep stiffness in treated WCO mixture indicated a high rutting resistance, thereby reduced the tendency of permanent deformation. Besides, the decrement of creep stiffness was noticed at 60°C temperature as compared to 40°C. However, the result reported a conflicting finding with temperature 40°C, in which the control mixture achieved the highest creep stiff-

Figure 17. Dynamic creep stiffness

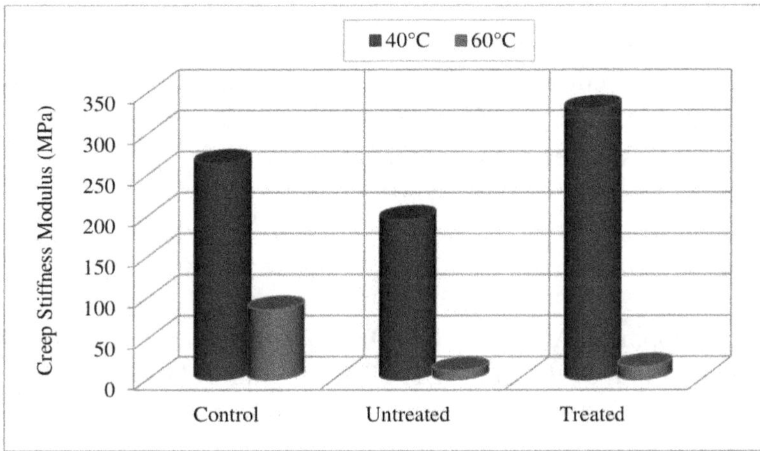

ness for 85.96 MPa at 60°C. In contrast, the creep stiffness result was recorded slightly lower for treated WCO (16.04 MPa) and untreated WCO mixtures (11.33 MPa) in comparison with the control mixture.

CONCLUSION

It is proven that an enhancement of physical and rheological performance was recorded in modified binder incorporating treated WCO as compared to the untreated WCO. The lower penetration with higher softening point was reported by the modified binder with treated WCO as compared to the untreated WCO. The rheological test indicated a noticeable improvement in which the higher failure temperature at 70°C was achieved by treated WCO for all percentages replacement relative with 5%, 10%, 15% and 20% of untreated WCO which was at 64°C, 58°C, 46°C and 46°C, accordingly. Besides, the superior failure temperature in modified binder with treated WCO was comparable with the control binder.

The mechanical performance for modulus test results showed that the mixtures containing 5% treated WCO exhibited higher M_R for both temperatures at 25°C and 40°C as compared to the control and 5% untreated WCO mixtures. This implied that a better resistance to rutting exposure was possessed in the treated WCO mixture. Meanwhile, the poor resilient modulus result indicated that the untreated WCO mixture has high susceptibility to rutting as evidenced by the lowest value of M_R. The creep resistance was improved in treated WCO mixture as evidenced by the highest creep stiffness with low permanent deformation as compared to the other mixtures. It proved that the greater achievement exhibited by treated WCO mixture with an increment of about 25% of creep stiffness relative to the control mixture. This implied that the mixture containing treated WCO was less susceptible to permanent deformation as compared to the control and untreated WCO mixtures.

The chemical modification improved treated WCO performance and offered a promising sustainable modifier in the modified binder as an effective paving material for road construction. Apart from promoting the sustainability approach, the usage of treated WCO in modified binder reduces dependencies against conventional binder due to the depletion of natural resources nowadays. Besides, the recyclability of WCO in modified binder reduces the inappropriateness of waste disposal thus also ensures environmental preservation.

Acid esterification and epoxidation treatment is recommended for the improvement of WCO properties and be compared with the current chemical treatment. Apart from that, further investigation on other types of asphalt mixture, such as porous asphalt mixture and stone mastic asphalt (SMA) mixture are recommended for the extensive research work in the future.

REFERENCES

Agriculture & Food Development Authority. (2000). *Waste Oils and Fats as Biodiesel Feedstocks. An Assessment of Their Potential in the EU*, ALTENER Program NTB-NETT Phase IV, Task 4, Final Report, March 2000.

Al-Omari, A. A., Khedaywi, T. S., & Khasawneh, M. A. (2018). Laboratory characterization of asphalt binder modified with waste vegetable oil using SuperPave specifications. *International Journal of Pavement Research and Technology*, *11*, 68–76.

Aladedunye, F. A.,, & Przybylski, R. (2009). Protecting oil during frying: a comparative study. *European Journal of Lipid Science and Technology*, *111*, 893–901.

Alcantara, R., Amores, J., Canoira, L., Fidalgo, E., Franco, M. J., & Navarro, A. (2000). Catalytic production of biodiesel from soybean oil: used frying oil and tallow. *Biomass and Bioenergy*, *18*, 515–527.

Andrade, I. C., Santiago, J. P., Sodre, J. R., Pathiyamattom, J. S., & Guerrero-Fajardo, C. A. (2014). Transesterification reaction of waste cooking oil and chicken fat by homogeneous catalysis. *Journal of Chemistry and Chemical Engineering*, *8*, 736–743.

Arabani, M., Mirabdolazimi, S. M., & Sasani, A. R. (2010). The effect of waste tire thread mesh on the dynamic behaviour of asphalt mixtures. *Construction & Building Materials*, *24*, 1060–1068.

Asli, H., Ahmadinia, E., Zargar, M., & Karim, M. R. (2012). Investigation on physical properties of waste cooking oil – rejuvenated bitumen binder. *Construction & Building Materials*, *37*, 398–405.

Asli, H.,, & Karim, M. R. (2011). Implementation of waste cooking oil as RAP rejuvenator. *Journal of the Eastern Asia Society for Transportation Studies*, *9*, 1336–1350.

Bahia, H. U.,, & Anderson, D. A. (1993). Glass transition behaviour and physical hardening of asphalt binders. *Journal Association Asphalt Paving Technology*, *62*, 93–129.

Binbin, L., Meizhu, C., & Shaopeng, W. (2014). Effect of waste edible vegetable oil on high temperature properties of different aged asphalts. *Key Engineering Materials*, *599*, 135–140.

Borhan, M. N., Suja, F., Ismail, A., & Rahmat, R. A. O. K. (2009). The effects of used cylinder oil on asphalt mixes. *European Journal of Scientific Research*, *28*(3), 398–411.

Brownridge, J. (2010). The role of an asphalt rejuvenator in pavement preservation: use and need for asphalt rejuvenation. *1st International Conference on Pavement Preservation*. April 13-15, 2010, Newport Beach, CA. pp. 351–364.

Castellanelli, C. A. (2007). Analyzes of the used oil under environmental perspective and its possibilities for production of biodiesel. Courtesy of UFSM – Federal University of Santa Maria.

Chebil, S., Chaala, A., & Roy, C. (2000). Use of softwood bark charcoal as a modifier for road bitumen. *Fuel*, *79*, 671–683.

Chen, M., Leng, B., Wu, S., & Sang, Y. (2014a). Physical, chemical and rheological properties of waste edible vegetable oil rejuvenated asphalt binders. *Construction & Building Materials*, *66*, 286–298.

Chen, M., Xiao, F., Putman, B., Leng, B., & Wu, S. (2014b). High temperature properties of rejuvenating recovered binder with rejuvenator, waste cooking and cotton seed oils. *Construction & Building Materials*, *59*, 10–16.

Chen, Y., Xiao, B., Chang, J., Fu, Y., Lv, P., & Wang, X. (2009). Synthesis of biodiesel from waste cooking oil using immobilized lipase in fixed bed reactor. *Energy Conversion and Management*, *50*(3), 668–673.

Chhetri, A. B., Watts, K. C., & Islam, M. R. (2008). Waste cooking oil as an alternate feedstock for biodiesel production. *Energies*, *1*(1), 3–18.

Choe, E.,, & Min, D. B. (2007). Chemistry of deep-fat frying oils. *Journal of Food Science*, *72*(5), 1–10. PMID:17995742

Choudhary, M.,, & Grover, K. (2013). Effect of deep-fat frying on physicochemical properties of rice bran oil blends. *IOSR Journal of Nursing and Health Science*, *1*, 1–10.

Cvengroš, J.,, & Cvengrošová, Z. (2004). Used frying oils and fats and their utilization in the production of methyl esters of higher fatty acids. *Biomass and Bioenergy*, *27*(2), 173–181.

El-Fadel, M.,, & Khoury, R. (2001). Strategies for vehicle waste-oil management: a case study. *Resources, Conservation and Recycling*, *33*, 75–91.

EPA. (2016). Learn about biodiesel. Retrieved from https://www3.epa.gov/region9/waste/biodiesel/questions.html

Fernando, S., Karra, P., Hernandez, R., & Jha, S. K. (2007). Effect of incompletely converted soybean oil on biodiesel quality. *Energy*, *32*(5), 844–851.

Gerhard, K., Jurgen, K., & Jon, V. G. (2004). *The biodiesel handbook* (2nd ed., p. 53). Urbana, IL: AOCS Press.

Gui, M. M., Lee, K. T., & Bhatia, S. (2008). Feasibility of edible oil versus non-edible oil versus waste edible oil as biodiesel feedstock. *Energy*, *33*, 1646–1653.

Hamad, B. S., Rteil, A. A., & El-fadel, M. (2003). Effect of used engine oil on properties of fresh and hardened concrete. *Construction & Building Materials*, *17*, 311–318.

Hwang, H.,, & Erhan, S. Z. (2001). Modification of epoxidized soybean oil for lubricant formulations with improved oxidative stability and low pour point. *Journal of the American Oil Chemists' Society*, *78*(12), 1179–1184.

Kabir, I., Yacob, M. R., & Radam, A. (2014). Households' awareness, attitudes and practices regarding waste cooking oil recycling in petaling, Malaysia. *IOSR-Journal of Environmental Science. Toxicology and Food Technology*, *8*(10), 45–51.

Katamine, N. M. (2000). Physical and mechanical properties of bituminous mixtures containing oil shales. *Journal of Transportation Engineering*, *126*(2), 178–184.

Kheang, L. S., May, C. Y., Foon, C. S., & Ngan, M. A. (2006). Recovery and conversion of palm olein-derived used frying oil to methyl esters for biodiesel. *Journal of Oil Palm Research*, *18*, 247–252.

Knothe, G.,, & Steidly, K. R. (2009). A comparison of used cooking oils: a very heterogeneous feedstock for biodiesel. *Bioresource Technology*, *100*(23), 5796–5801. PMID:19592239

Koh, W. F. E., Abdul Hamid, A., Pak Dek, M. S., Tan, C. P., & Mohd Zainudin, M. A. (2014). Changes of major antioxidant compounds and radical scavenging activity of palm oil and rice bran oil during deep-frying. *Antioxidants*, *3*, 502–515. PMID:26785067

Lam, M. K., Lee, K. T., & Mohamed, A. R. (2010). Homogeneous, heterogeneous and enzymatic catalysis for transesterification of high free fatty acid oil (waste cooking oil) to biodiesel: a review. *Biotechnology Advances*, *28*, 500–518. PMID:20362044

Leong, X. F., Ng, C. Y., Jaarin, K., & Mustafa, M. R. (2015). Effects of repeated heating of cooking oils on antioxidant content and endothelial function. *Austin Journal of Pharmacology and Therapeutics*, *3*(2), 1068.

Leung, D. Y. C., Wu, X., & Leung, M. K. H. (2010). A review on biodiesel production using catalyzed transesterification. *Applied Energy*, *87*(4), 1083–1095.

Ma, F.,, & Hanna, M. A. (1999). Biodiesel production: a review. *Bioresource Technology*, *70*, 1–15.

Maharaj, R., Harry, V. R., & Mohamed, N. (2015). Rutting and fatigue cracking resistance of waste cooking oil modified Trinidad asphaltic materials. *The Scientific World Journal*. doi:10.1155/2015/385013 PMID:26336652

Maneerung, T., Kawi, S., Dai, Y., & Wang, C. H. (2016). Sustainable biodiesel production via transesterification of waste cooking oil by using CaO catalysts prepared from chicken manure. *Energy Conversion and Management*, *123*, 487–497.

Math, M. C., Kumar, S. P., & Chetty, S. V. (2010). Technologies for biodiesel production from used cooking oil – a review. *Energy for Sustainable Development*, *14*(4), 339–345.

Mazza, G.,, & Qi, H. (1992). Effect of after-cooking darkening inhibitors on stability of frying oil and quality of french fries. *Journal of the American Oil Chemists' Society*, *69*, 847–853.

Patil, P. D., Gude, V. G., Reddy, H. K., Muppaneni, T., & Deng, S. (2012). Biodiesel production from waste cooking oil using sulfuric acid and microwave irradiation processes. *Journal of Environmental Protection*, *3*, 107–113.

Phan, A. N.,, & Phan, T. M. (2008). Biodiesel production from waste cooking oils. *Fuel*, *87*, 3490–3496.

Rashid, U.,, & Anwar, F. (2008). Production of biodiesel through optimized alkaline catalyzed transesterification of rapeseed oil. *Fuel*, *87*, 265–273.

Read, J. M., & Whiteoak, D. (2003). The shell bitumen handbook (5th ed.). Thomas Telford Publishing, Thomas Telford Ltd, 1 Heron Quay, London E144JD, 62-66, 136.

Robert, N. H. (2000). *Asphalt in roads construction. American Society of Civil Engineering* (p. 75). London, UK: Thomas Telford Publications.

Sanli, H., Canakci, M., & Alptekin, E. (2011). Characterization of waste frying oil obtained from different facilities. *World Renewable Energy Congress 2011*, May 8-13, 2011. Linkoping, Sweden, *Bioenergy Technology*, 479-485.

Sengoz, B.,, & Topal, A. (2005). Use of asphalt roofing shingle waste in HMA. *Construction & Building Materials*, *19*, 337–346.

Silitonga, A. S., Masjuki, H. H., Mahlia, T. M. I., Ong, H. C., Atabani, A. E., & Chong, W. T. (2013). A global comparative review of biodiesel production from Jatropha Curcas using different homogeneous acid and alkaline catalysts: study of physical and chemical properties. *Renewable & Sustainable Energy Reviews*, *24*, 514–533.

Singhabhandhu, A.,, & Tezuka, T. (2010). The waste-to-energy framework for integrated multi-waste utilization: waste cooking oil, waste lubricating oil, and waste plastics. *Energy*, *35*(6), 2544–2551.

Sun, Z., Yi, J., Huang, Y., Feng, D., & Guo, C. (2015). Investigation of the potential application of biodiesel by-product as asphalt modifier. *Road Materials and Pavement Design*, *17*(3), 737–752.

Sun, Z., Yi, J., Huang, Y., Feng, D., & Guo, C. (2016). Properties of asphalt binder modified by bio-oil derived from waste cooking oil. *Construction & Building Materials*, *102*, 496–504.

Tabatabaee, N., Tabatabaee, H. A., Sabouri, M. R., & Teymourpour, P. (2009). Evaluation of performance grading parameters for crumb rubber modified asphalt binders and mixtures. In *Proceedings of 7th International RILEM Symposium on Advanced Testing and Characterization of Bituminous Materials*. 1. May 2009. Rhodes, Greece, 1-11.

Teymourpour, P., Sillamäe, S., & Bahia, H. U. (2015). Impacts of lubricating oils on rheology and chemical compatibility of asphalt binders. *Road Materials and Pavement Design*, *16*, 50–74.

Tseng, Y. C., Moreira, R. G., & Sun, X. (1996). Total frying—use time effects on soybean oil deterioration and on tortilla chip quality. *International Journal of Food Science & Technology*, *31*, 287–294.

Udomsap, P., Chollacoop, N., Topaiboul, S., & Hirotsu, T. (2009). Effect of antioxidant on the oxidative stability of waste cooking oil based biodiesel under different storage conditions. *International Journal of Renewable Energy*, *4*(2), 47–59.

Ullah, Z., Bustam, M. A., & Man, Z. (2014). Characterization of waste palm cooking oil for biodiesel production. *International Journal of Chemical Engineering and Applications*, *5*(2), 134–137.

Warner, K., Orr, P., Parrott, L., & Glynn, M. (1994). Effects of frying oil composition on potato chip stability. *Journal of the American Oil Chemists' Society*, *71*, 1117–1121.

Wen, H., Bhusal, S., & Wen, B. (2013). Laboratory evaluation of waste cooking oil-based bioasphalt as an alternative binder for hot mix asphalt. *Journal of Materials in Civil Engineering*, *25*(10), 1432–1437.

Wen, Z., Yu, X., Tu, S. T., Yan, J., & Dahlquist, E. (2010). Biodiesel production from waste cooking oil catalyzed by TiO_2-MgO mixed oxides. *Bioresource Technology*, *101*, 9570–9576. PMID:20696572

Yaakob, Z., Mohammad, M., Alherbawi, M., Alam, Z., & Sopian, K. (2013). Overview of the production of biodiesel from waste cooking oil. *Renewable & Sustainable Energy Reviews*, *18*, 184–193.

Yao, H.,, & You, Z. (2016). Effectiveness of micro and nanomaterials in asphalt mixtures through dynamic modulus and rutting tests. *Journal of Nanomaterials*, 1–14. doi:10.1155/2016/2645250

Zargar, M., Ahmadinia, E., Asli, H., & Karim, M. R. (2012). Investigation of the possibility of using waste cooking oil as a rejuvenating agent for aged bitumen. *Journal of Hazardous Materials*, *233-234*, 254–258. PMID:22818590

Zaumanis, M., Mallick, R. B., & Frank, R. (2013). use of rejuvenators for production of sustainable high content RAP hot mix asphalt. *The XXVIII International Baltic Road Conference*. August, 2013. Vilnius, Lithuania, 1-10.

Zhang, H., Wang, Q., & Mortimer, S. R. (2012). Waste cooking oil as an energy resource: review of Chinese policies. *Renewable & Sustainable Energy Reviews*, *16*(7), 5225–5231.

Zlatanic, A., Lava, C., Zhang, W., & Petrovic, Z. S. (2004). Effect of structure on properties of polyols and polyurethanes based on different vegetable oils. *Journal of Polymer Science. Part B, Polymer Physics*, *42*(5), 809–819.

Chapter 3
Recent Advances in Waste Cooking Oil Management and Applications for Sustainable Environment

Ching Thian Tye
Universiti Sains Malaysia, Malaysia

ABSTRACT

This chapter discusses the management of waste cooking oil (WCO) in a sustainable manner in order to protect the environmental pollution. Increasing consumption of edible oils worldwide leads to generation of substantial amount of waste cooking oil (WCO). While WCO is not considered toxic, large amount of WCO can contribute to environment pollution if not being handled properly. The huge generation of WCO in the world creates problem of collection, treatment and disposal. Due to its chemical features, the recycling of WCO not only provides a renewable feedstock for producing biofuels and bio-based products, but also alleviates environmental pollution arising from its improper handling. This chapter also provides an overview of some recent approaches in WCO recycling and applications.

INTRODUCTION

The increase of population and living standards has led to higher demand of edible oils worldwide. This can be seen from the growth in global production of vegetable oils of 90.5 million metric tons in year 2000/2001 to 197.23 million metric tons in year 2017/2018 (Statista, 2019a). Edible oils are oils mainly extracted from plants or vegetables such as, oil palm fruit, soybean, rapeseed, sunflower seed, peanut, coconut and etc. Palm oil is the most common type of edible oil (Statista, 2019b). Edible oils consist mostly of triacylglycerides (96%) that composed of different fatty acids and some other compounds such as free fatty acids, phospholipids, phytosterols, tocopherols, other antioxidants or waxes (Matthäus, 2010).

All over the world, edible oil is essential in food preparation and substantial quantity of edible oils are used for food frying either in home, restaurants or in food industry. Waste cooking oils (WCO) are

DOI: 10.4018/978-1-7998-0369-0.ch003

bio-based oils that have been used for the same purpose. During the frying process, cooking oil undergoes various physical and chemical changes due to chemical reactions including hydrolysis, thermal degradation, oxidation, and polymerization (Panadare & Rathod, 2015). Repeated frying and usage of edible oil alters its physiochemical and nutrition properties, and leads to the formation of Total Polar Compounds (TPCs), which in high rates have been found to have a negative effect on a person's health. At certain point after reused multiple times, the oil becomes unfit for human consumption and may pose serious health risks such as hypertension, liver disease, increased cholesterol and atherosclerosis by consuming such oil (The news minute, 2019).

High amount of cooking oil consumption leads to significant wastage. WCO and fat generated from kitchens and food preparation have created serious problems for their disposal, due to its slow degradation. Proper waste management is thus necessary for WCO in order to preserve a sustainable environment.

WCO implies an economic loss in edible oils. Due to its chemical composition and physical nature, WCO can be considered as a potential waste which can be utilized as energy source and raw material for chemical or biological processes. It has been getting more attention as a low-cost feedstock for producing biodiesel or other biofuels and nonfuel products comparing to the use of edible oils as a food resource for human beings. The present article provides an overview regarding recent advancement in WCOs management and applications for recycled WCO.

DETRIMENTAL EFFECTS OF FOOD DERIVED USED OIL AND FAT

Fat and oil consumption per capita in developed countries was estimated at over 50 kg/annum compared to less than 20 kg/annum in less developed countries (Williams, Clarkson, Mant, Drinkwater, & May, 2012). It is estimated that 2.5 L of WCO are produced per person per month domestically (European Biomass Industry Association, 2015). In Japan, it is presumed that 100–140 ktonne WCO from household sector are discarded every year (Ministry of Environment, 2006). Oil cannot be removed from cooking operations as it is inveterate in many culinary customs. Used cooking oil is considered a waste upon being discharged into the sewer systems. The waste oil from multiple sites can accumulate in the sewer with other non-flushable waste to cause sewage blockage and overflow, leading to odor, nuisance and creating the corrosion of sewer lines under anaerobic conditions. It has been estimated that 50-75% of approximately 24,750 inline blockages per year in the UK (Arthur et al., 2008) are due to fat, oil and grease (FOG) deposits (Keener et al., 2008). In 2000, the Drainage Services Department of Hong Kong claimed that more than 60% of sewer blockages were due to excessive grease build-ups (Chan, 2010). These deposits can impact human health and the environment.

In addition, oil and fat that pass through the sewer system and enter wastewater treatment plants will increase the difficulties to treat. Additional techniques such as dissolved air flotation, centrifugation, filtration, biological removal and ultrafiltration are needed, in order to treat the oil contaminated water, which leading to increase operational and maintenance costs (Wallace et al., 2017).

Another concern is gutter oil, which is illicit cooking oil that has been recycled from waste oil collected from restaurant fryers, drains, grease traps, and slaughterhouse waste has emerged as a serious food-safety issue in China. Taking food prepared from oxidation and hydrogenation of this 'gutter oil' can cause health problems in humans (Lu & Wu, 2014). In September 2014, an incident of "food safety scandal" was reported in Taiwan (Wee, Budiman, Su, Chang, & Chen, 2016). It was related to the usage

Figure 1. Waste Hierarchy in relation to waste cooking oil management, adapted from European Parliament Council (2008)
Source: (Wallace et al., 2017)

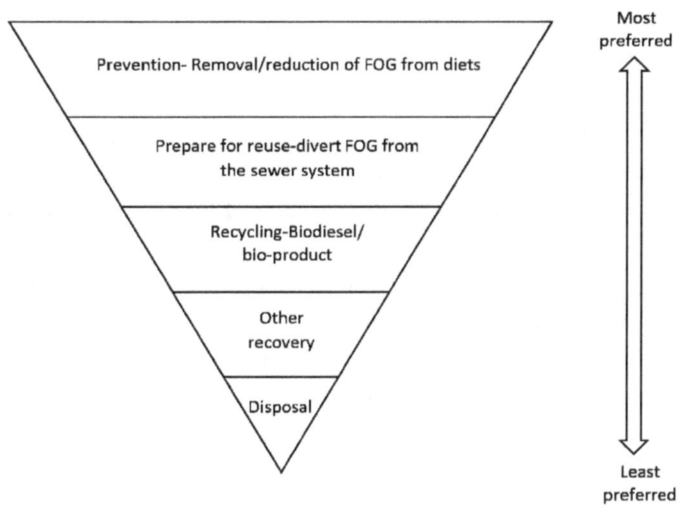

of gutter/tainted oil by several oil suppliers, in which the series of food safety incidents affected 1,256 businesses.

WASTE OIL MANAGEMENT

WCO does not contain toxic components or heavy metals commonly found in the used petroleum derived oil products. Nevertheless, used cooking oils should not be simply disposed and need to be handled properly. Appropriate waste oil management is important for sustainable development. European countries utilize the waste hierarchy, as depicted in Figure 1, to manage WCO (European Parliament Council, 2008). The waste hierarchy refers to prevention as the preferred method when tackling waste but when this is not feasible, the 3Rs (reduce, reuse or recycle) are the next option (Wallace, Gibbons, O'Dwyer, & Curran, 2017). Disposal of the used oil is to be totally avoided if possible.

The aim for any effective waste management is to minimize the production of waste and reuse waste produced. Education and awareness campaigns with the stakeholders are the foundation for all waste oil management initiatives. There are policies and regulations for how used cooking oils are to be collected, stored, transported and processed in different countries in order to ensure a sustainable living environment in long run. The following sections are some recent trends of international waste oil management approaches.

Public Awareness and Prevention

Public awareness is the key to successful waste oil management. With waste the result of human activities, a proper understanding of waste management issues is paramount, without which the success of even the best-conceived waste management plan becomes doubtful. A high level of public awareness regarding the issue can reduce the waste oil production at the top of the waste hierarchy. This can prevent and reduce the detrimental effects resulted from the waste oil generated. Public awareness can be achieved via public educational campaigns to mitigate domestic sewer deposits and inappropriate waste oil disposal. The public must be made aware of waste management issues to understand the consequences of improper management of waste and how it may ultimately pose a serious threat to their lives and well-being (Hasan, 2004).

In the meantime, attempts to increase public awareness and some regulatory preventive measurement have been carried out. All facilities with a trade effluent discharge are potential fat, oil and grease (FOG) producers. Many authorities around the world get to reduce FOG waste entering the sewerage system by requiring or enforcing the installation of grease trapping systems (GTSs) (He et al., 2013). Since the year 2000, the UK Building Regulations have required all new and converted premises to install grease management systems. Before the year 2000, food service outlets generally only had a GTSs if they had been identified as problematic. Even when GTSs were installed, maintenance was often poor, which resulted in FOG entering the sewer systems (Williams et al., 2012).

Waste Cooking Oil Recovery

When WCO generation is unavoidable, affords to recover the WCO are necessary in order to minimize the detrimental consequences. In terms of WCO recovery, it is found that an effective domestic WCO collection systems will reduce the WCO discharging down into the sewer system. Many countries' local authorities have started the system to collect the WCO from residential areas in addition to industry sector. For instance, in the United Kingdom, local authority has a legal duty under the Environmental Protection Act 1990 to collect household waste such as WCO (UK Environmental Law Association). The estimated WCO collected in Europe annually is 100,000-700,000 tonnes (Iglesias, Laca, Herrero, & Díaz, 2012).

In Asia, commencing from October 2014, the government of Republic of China (Taiwan) has made the recycling of WCO from residential and commercial sectors in Taiwan mandatory. The WCO produced by households and institutions can be collected by local environmental protection bureaus or sanitation teams, which are legally obliged to manage it. The WCOs that are produced by small-scale commercial stores (e.g., restaurant, snack bar, vendor, night market) can also be collected by municipal collection teams. In the meantime, in order to maintain effective WCO collection and to track its flow, applications for permits are henceforth to be reviewed and issued by local governments. The collected amounts of WCO was reported increased significantly from 1599 tonnes in 2015 to 12,591 tonnes in 2017, reflecting on the WCO recycling regulation effective since 2015 (Tsai, 2019).

A study using life cycle sustainability assessment (LCSA) methodology showed that domestic WCO collection systems through urban collection centers (UCC) has better environmental and economic performance compared to those collection systems through schools and door-to-door cities in Mediterranean countries (Vinyes, Oliver-Solà, Ugaya, Rieradevall, & Gasol, 2013). LCSA is the combination of life cycle assessment, life cycle costing, and social life cycle assessment.

European Biomass Industry Association carried out a multi country initiative RecOil project that involved Spain, Greece, Italy, Portugal, Belgium and Denmark. The RecOil project found that it was possible to collect 2.5 liters of WCO per household per month (European Biomass Industry Association, 2015). 60% of used cooking oil is found improperly disposed of. Information among 44 different WCO collection systems implemented was analyzed. 180 tonnes and 80 tonnes of WCO was collected from restaurants and private households, respectively, approximately 45% of the estimated potential of 400 tonnes per year from restaurants and 16% of the estimated potential of 500 tonnes from private households. Nearly €30,000 was saved from the cost of maintaining the wastewater treatment plants. The project highlighted the benefits of utilizing the waste stream and the potential saving available.

Apart from direct WCO collection, it is equally important to create an eco-system that encompasses the complete life cycle of the WCO recovery, in this case, from public education, to establishing effective oil collection to the producing of value-added end product, such as bio-diesel. With this in mind, on 1st of July 2018, the Food Safety and Standards Authority of India (FSSAI) announced the new regulations for monitoring WCO. The implementation of the regulations will require focus on consumer education, enforcement as well as creation of an eco-system for collection of WCO to produce biodiesel. FSSAI is in discussion with the Indian Biodiesel Association to establish a nation-wide eco-system for collection of used cooking oil and its conversion to bio-diesel (The Hindu Business line, 2018).

WASTE COOKING OIL RECYCLING AND APPLICATIONS

Concerns of world crude oil reserve depleting and impact of fossil fuels to the environment has led to searching for alternative renewable replacement. Use of WCO into the fuel mix has been an attractive one. Motivated by the aforementioned needs, the WCO recycling industry has grown over the past five years as both upstream and downstream markets have fared well (IBIS world, 2018). There are companies who offer services for WCO collection, grease trap maintenance, meat byproduct collection and deadstock pickup to the customers. This has directly promoting in WCO reused and recycling process. This is a good example of waste management not only solves the derived pollution problem, but also can become a resourceful feedstock for fuel production and a series of other applications. The following section states the various applications of WCO.

Conversion of Waste Cooking Oil Into Fuel

At least 8.9 million tons of WCO is produced worldwide annually. Many countries encourage the conversion of WCO to liquid fuel such as biodiesel, bio gasoline and green aviation fuel. China has been promoting this since 2006 and first implemented the Renewable Energy Law in 2006. Then, it is followed by the various medium- and long-term Renewable Energy Development Plans in later years. Liquid biofuels are important components in these legal initiatives to promote renewable energy (Liang, Liu, Xu, & Zhang, 2013).

WCO is one of the most prudent sources of energy because of its restrictions for food use, cheap, availability and appropriateness for fuel production. WCO can be converted into biofuel via different pathways or reactions. The most commons include transesterification, hydrocracking, catalytic cracking and pyrolysis (Table 1). With the different process, the properties and applications of the derived biofuel is also different. Some recent advances in WCO recycling via these processes are also discussed.

Table 1. Common processes used in WCO recycling and treating

Process	Operating Conditions	Major Product
Transesterification	React with alcohol at 60°C with the presence of base catalyst.	FAME (biodiesel), glycerol
Catalytic cracking	At 300-500 °C with selective catalyst.	Gasoline
Hydrodeoxygenation	With H_2 at 290-400 °C with metal catalyst.	C15-C18 hydrocarbons
Pyrolysis	At 550-800 °C in an inert atmosphere.	Bio-oil

Transesterification

Transesterification is the most widely used industrial process to produce biodiesel from vegetable or plant oil because the process is simple and cost effective. In this process, the fat or oil is first purified and then reacted with an alcohol, usually methanol (CH_3OH) or ethanol (CH_3CH_2OH), in the presence of a catalyst such as potassium hydroxide (KOH) or sodium hydroxide (NaOH) (Figure 2). When this process occurs, the triglycerides (oil or fat) is transformed to form esters and glycerol. The esters that remain are called biodiesel. In case of methanol used, a more common term used for biodiesel is fatty acid methyl ester (FAME). During the transesterification process, the reactant that reacts with methanol is triglyceride molecules (major component) in the plant or vegetable derived oil. Therefore, WCO with triglyceride as the major compound has been a very good candidate. In addition, there is another by–product, glycerol form in this process which can help in production economic. In the process, after the reaction, FAME is separated from glycerol.

Biodiesel have similar physical characteristics as those of fossil diesel fuels. The high viscosity of vegetable oils is reduced by the transesterification process. Biodiesel blend is the blend of fossil diesel and FAME (Khalid & Khalid, 2011). A different set of additives is required for biodiesel blend than fossil diesel which is for correcting low temperature behaviour and slowing down of the oxidation processes of biodiesel in storage (ETIP Bioenergy, 2019). Biodiesel is non-toxic and biodegradable. Using WCO as the feedstock in transesterification process is also found to have lower greenhouse gases compare to fresh cooking oil in life cycle assessment.

Over the year, effort is still going on to improve the process in order to optimize the yield from transesterification of WCO. This includes improvement from operating condition, catalysts used as well as using WCO from different sources as feedstock. Mohadesi et al. (2019) investigated the transesterification of WCO with methanol in the presence of potassium hydroxide as the catalyst using a semi-industrial

Figure 2. Transesterification of triglyceride (vegetable oil) into FAME and glycerol
R is the mixture of hydrocarbon parts of different fatty acids present in the triglyceride

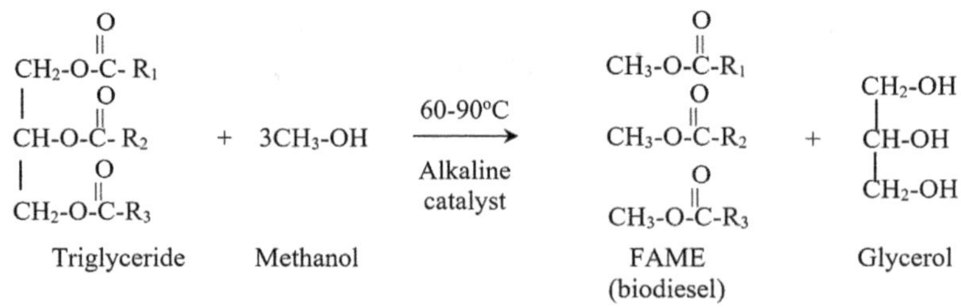

Figure 3. Esterification of free fatty acid into FAME (biodiesel) and water

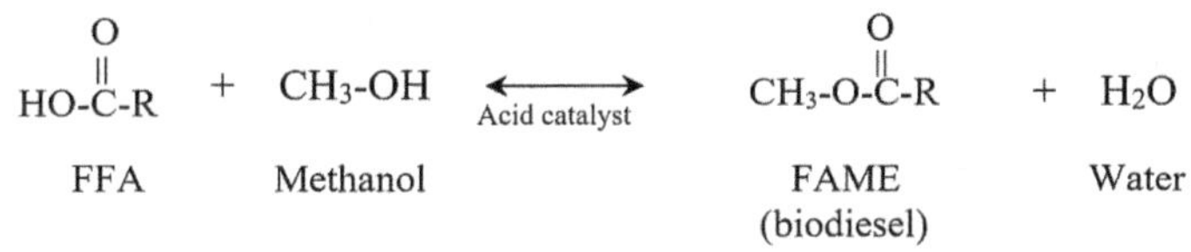

pilot of microreactor with 50 tubes to produce 5 L/h biodiesel. The authors claimed the highest level of biodiesel purity or FAMEs obtained at 98.26%.

In Indonesia, effects of alcohol type and base catalyst used during transesterification were investigated using chicken fat and WCO to produce methyl esters. The reaction process was run for 90 minutes at 60°C and 2000 rpm stirring rate. Maximum methyl esters yield of 91% was reported. The fuel properties of methyl esters produced with methanol and sugarcane spirits were close to each other. The production cost of biodiesels from chicken fat and WCO were reported cheaper than market biodiesel selling price (Soegiantoro, Chang, Rahmawati, Christiani, & Mufrodi, 2019).

Kataria et al. (2019) studied a heterogeneous base catalyzed transesterification under different reactant proportions: the molar ratio of alcohol to oil and mass ratio of catalyst to oil, for the optimum production of biodiesel. The optimum condition for base catalyzed transesterification of WCO was determined as 12:1 and 5 wt.% of zinc doped calcium oxide. The fuel properties of the produced biodiesel and their blend for different ratios were comparable with properties of fossil diesel and ASTM biodiesel standards. Tests have been conducted on a CI engine and the biodiesel produced using heterogeneous catalyst was found suitable to be used as diesel oil blends, having lesser emissions as compared to fossil diesel.

Esterification

Some WCOs have high free fatty acid (FFA) content, (>1%). In such cases, saponification hinders separation of the ester from glycerin and reduces the yield and formation rate of FAME during transesterification (Lucena, Silva, & Fernandes, 2008). Therefore, WCO with FFA contains higher than 1% needs to go through a pretreatment process before transesterification. Esterification is another process that converts component in WCO into biodiesel or FAME. The process is similar to transesterification as WCO is reacted with methanol to produce FAME. However, the component involve in the reaction is the free fatty acid in the WCO. FFA in WCO react with methanol to produce FAME and water. Figure 3 represents the reaction.

Pretreatment of WCO with esterification was found to be the most efficient using acidic catalyst. The FFA was reported to reduce up to 88.8% at 60 °C with 1:2.5 methanol to oil molar ratio (Sahar, Sadaf, Iqbal, Ullah, & Iqbal, 2018). In general, the process for biodiesel or FAME production is as shown in Figure 4.

Catalytic Cracking

Catalytic cracking has been a conventional process used to treat waste oil. It is considered to be inexpensive in terms of energy consumption to convert petroleum feedstocks to lighter fractions of gasoline boiling range. Catalytic cracking process is carried out with selective catalyst at temperature within 300-500°C. Conversion of WCO into green fuel via catalytic cracking has also been carried out. The

Figure 4. Transesterification process to produce FAME

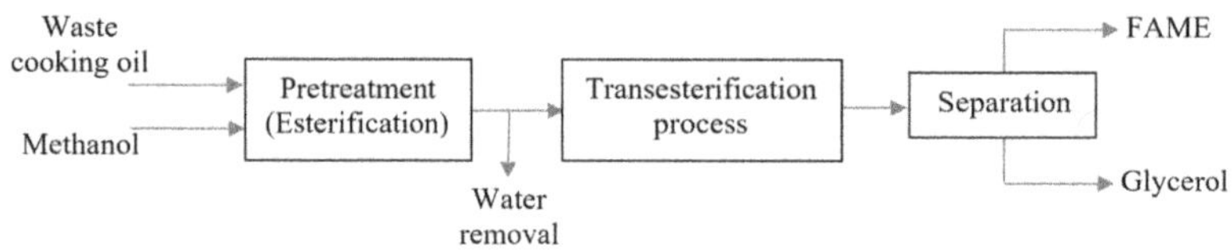

product obtained from catalytic cracking of vegetable oils includes organic liquid product, gas, coke and water. The distributions of products yield were influenced by various factors such as composition of feedstock, reaction temperature, residence/reaction time, nature and the type of catalyst used (Wakoc, Reshadb, Bhaleraoa, & Goud, 2018). During the conversion of vegetable oils under catalytic cracking conditions, the gasoline fraction can be produced in an amount of up to 40% (of feedstock mass), as well as 10–15 wt. % propane–propylene (PPF) and butane–butylene (BBF) fractions (Doronin, Potapenko, Lipin, Sorokina, & Buluchevskaya, 2012).

The major difference in catalytic cracking of WCO compare to the conventional petroleum catalytic cracking process is the feedstock. Therefore, studies of various potential catalysts such as composite zeolite, functionalized ZSM-5, metals etc. have been carried out (Chang & Tye, 2013; Chiam & Tye, 2013; Abdul Majed & Tye, 2018). The liquid fuel obtained from catalytic cracking process after water removal is mainly hydrocarbon, which has very similar properties as the fossil fuel with much lower oxygen content. This has made it a favorable process to convert WCO to fuel.

Hydrodeoxygenation

In order to improve the yield of liquid fuel in the cracking process, hydrogen is added during reaction. The process is called hydrodeoxygenation. There has been commercial production of renewable aviation fuel from plant/vegetable oil via hydrodeoxygenation. Successful green fuel production from vegetable has also leading to using WCO with similar properties as the feedstock. Li et al. (2018) claims a nearly complete conversion and 90% selectivity to C15-C18 hydrocarbons was achieved at 350 °C.

Pyrolysis

Pyrolysis is another process used to convert biomass into bio-oil or fuel. Pyrolysis is a thermal decomposition process that carried out at elevated temperatures (550-800 °C) in an inert atmosphere. Pyrolysis of WCO can produce bio-oil (main product) of high caloric value (high heating value ~ 8843 kg/Kcal) which can be further processed into liquid fuel.

The gas and solid byproducts of pyrolysis of waste oil are also have their respective applications as seen in Figure 5. The gases product (syngas) with heating value up to 8 MJ/kg is suitable as an energy source and the biochar which is rich in iron and organic carbon is suitable (Trabelsi, Zaafouri, Baghdadi, Naoui, & Ouerghi, 2018).

Figure 5. Pyrolysis of waste cooking oil

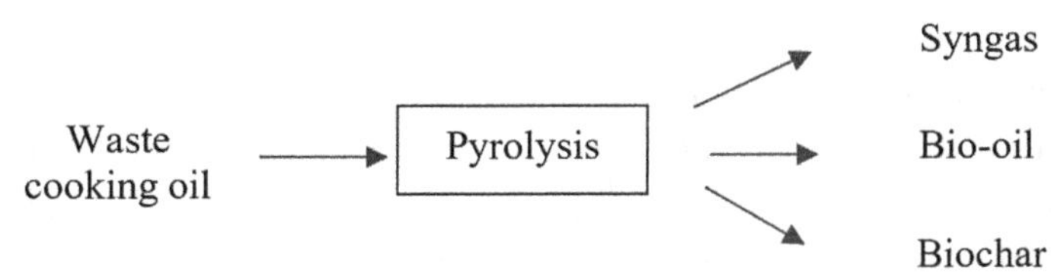

Waste Cooking Oil as Building Material

Other than fuel, WCO has also been used and tested for other applications such as construction and building material. It was added in the synthesis of lime mortars for restorative materials. Application of ordinary Portland cement (OPC) binders for repair and rehabilitation of old buildings is proven to be unsuitable, due to incompatibility problems frequently associated with the pathology origin in the restoration of historical buildings. The low affinity of OPC binders/mortars with historical substrates has been an issue (Grilo et al., 2014).

New lime mortars with addition of WCO was synthesized for restorative rendering applications. Air-, formulated- and hydraulic-lime mortars were synthesized with inclusion of used sunflower cooking oils and, brick waste powder as pozzolanic addition, when necessary. The results of WCOs addition in the mortars exhibited promising hydrophobic effects, such as sorptivity reduction up to 60 times and improvement of superficial durability. Addition of high-oleic acid cooking oil in 13 wt.% in the formulated and hydraulic lime mortars was reported to increase their hydrophobicity without worsening their mechanical strengths. All the investigated mortars exhibited appropriate indexes as restorative materials (Pahlavan, Manzi, Rodriguez-Estrada, Bignozzi, 2017). Later, another test: air and formulated lime mortars with pulverized brick waste were synthesized with addition of used cooking oil and albumen. Parallel additions of oil and albumen in lime mortars was found increased their hydrophobicity and superficial durability without retarding their setting time due to the role of fatty acids and unfolded proteins under alkaline conditions. Though there are still issues to which, (decreased in permeability values and setting kinetics), the parallel valorization showed high potentials in synthesis of restorative mortars design (Pahlavan, Manzi, Sansonetti, & Bignozzi, 2018).

On the other hand, the possibility of WCO as grinding aid in the production of Portland cement was also assessed. Cement clinker and gypsum were interground with various WCOs. In overall, WCO favorably improves cement grinding. Additionally, certain WCO increased the cement strength (Li, Zhao, Huang, Jiang, & Chen, 2016). All of these studies showed that WCO has its values as construction material.

Waste Cooking Oil in Pavement

Asphalt binder or bitumen plays an important role in the composition of bituminous pavements. Hot mix asphalt (HMA) is the most widely used binder for pavement. Application of recycled pavement materials to produce new pavement materials results in considerable savings of material, and energy. The main problem pertaining to the use of the recycled asphalt pavement (RAP) material in HMA is the ageing of bitumen, which limits the percentage of applied RAP in HMA.

In 2012, waste residential cooking oil was used to rejuvenate the used bituminous binder. The physical properties of the original bitumen, aged bitumen and rejuvenated bitumen were measured and compared by the conventional bitumen binder tests such as softening point, penetration and Brookfield viscosity. The results showed that the aged bitumen was rejuvenated by the WCO due to a change in its physical properties, which resemble the physical properties of original bitumen (80/100). The optimum percentage of WCO for the rejuvenated aged bitumen group of 50/60, 40/50, and 30/40 was reported by adding 1%, 3–4%, and 4–5% of WCO, respectively. The statistical analyses also confirms that there was no significant difference between the original asphalt binder and rejuvenated asphalt binder (Asli, Ahmadinia, Zargar, & Karim, 2012).

Later, similar tests were carried out, Sun et al. (2016) analyzed the properties of asphalt binder modified by bio-oil derived from WCO. The bio-oil, black oily liquid, used was the by-product of WCO refining for biodiesel. The bio-oil derived from WCO is found mainly composed of aromatics, resins and saturates. It was also reported that almost no chemical reactions between bio-oil and control asphalt was observed. The addition of bio-oil decreases the softening point and viscosity, increases the penetration and ductility of control asphalt binder. Furthermore, the addition of bio-oil was found could reduce the deformation resistance and elastic recovery performance of control asphalt and improve the stress relaxation property and thermal cracking resistance of control asphalt (Sun et al., 2016). Again, WCO residue was used in the preparation of HMA using recycled pavement material recently and encouraging results obtained (Ma et al., 2019). Therefore, WCO especially the residue actually is a cheap and widely available potential component, which can be applied in asphalt aggregate preparation for pavement.

Waste Cooking Oil as Cooling Medium

Vegetable oils, due to their attractive properties like renewability, low toxicity, biodegradability, high fire point and availability are considered as a reliable substitute for petroleum-based oils in transformer cooling system (Qiu et al., 2011). Hence, WCOs, with similar properties, instead of the fresh vegetable oil is considered a potential alternative cooling medium source for electrical transformer as well. The performance of WCO after transesterification process (mainly with fatty acid methyl ester) for application in electrical transformers as cooling medium was evaluated and compared to the mineral oil. Based on the study by Raeisian et al. (2019), the breakdown voltage and thermal conductivity (the two main indexes) of transformer that using WCO derived oil as cooling medium are, respectively, 48% and 33%, higher than those of the mineral oil. Furthermore, the transformer hotspot temperature with WCO derived oil is 3°C lower than the mineral oil and the transformer experiences lower temperature in the thermally critical region. Therefore, WCO derived oil was found to be an effective alternative cooling medium for liquid-filled transformers.

Waste Cooking Oil as Chemical Sources

Vegetable oils are made up predominantly of triesters of glycerol with fatty acids and commonly are called triglycerides. The many derivatives of fatty acids manufactures are used in surface coatings, plastics, detergents, lubricants, etc, where the long hydrocarbon chain confer needed plasticity, surface activity, or lubricity (Abdullah & Salimon, 2010). Therefore, large amount of WCOs that is generated daily widely can be considered a rich resource of fatty acid. This has leading to the eco-development of various WCO materials.

Figure 6. Epoxidation of palmitoleic acid

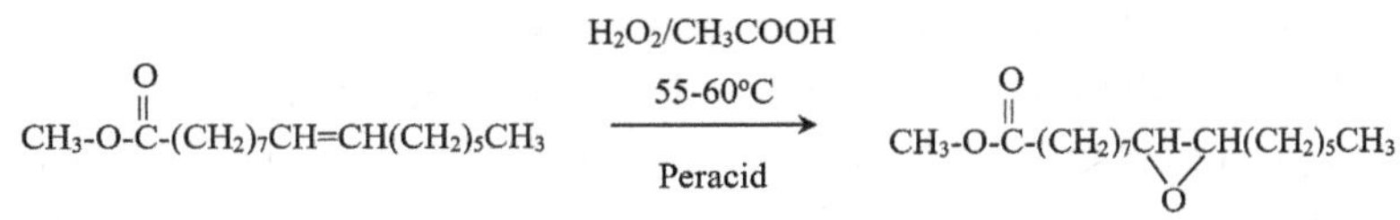

WCOs have the potential to be an economic alternative for synthesizing different biomaterials. WCO methyl esters were used as starting material and being structurally modified via epoxidation reaction to epoxide WCO esters. Epoxides are produced by reaction of double bonds with peracids. An example of epoxidation reaction is shown in Figure 6.

Zheng et al. (2018) tested the waste epoxide cooking oil 2-ethylhexyl esters as bio-plasticizer to substitute toxic plasticizer dioctyl phthalate used during poly(vinyl chloride) or PVC processing. Most of the plasticizers used during PVC processing are typically toxic petroleum-based phthalates. The epoxide WCO 2-ethylhexyl esters was found to have superior low temperature property, oxidation stability, and remarkably reduced the glass transition temperature of PVC, comparing to the conventional epoxy compound. It was proven to be an effective alternative to dioctyl phthalate, replacing about 40% of the total plasticizer (Zheng et al., 2018).

Another plasticizer (acetylated-fatty acid methyl ester-citric acid ester, ACFAME-CAE) for PVC was synthesized by using WCO and citric acid as raw materials. The mechanical properties of PVC films with AC-FAME-CAE plasticizer were found better than those of PVC films plasticized by ESO (epoxidized soybean oil) and as good as those of PVC films plasticized by dioctyl phthalate, the conventional plasticizer (Feng et al., 2018).

Waste cooking oil (WCO) has also been used as feedstock for different processes such as to produce aromatics from catalytic fast pyrolysis reaction (Wang, Zhong, Ding, Zhang, & Ruan, 2017). All these efforts have led to a more environmentally friendly process and product.

Waste Cooking Oil as Lubricant

WCO has been using as lubricant for some general work. It can also be used as a feedstock to produce specific lubricant formulation. An organic medium dispersible Cu nanoparticle was directly prepared in a WCO via a "wet chemistry" approach. Blending of nano-copper based additive and PAO6 base oil results in a stable formulation without segregation phenomena. The role of the copper nanoadditives as friction modifier was found to improve oxidative stability of a mixture (Sarno, Spina, & Senatore, 2019).

Waste Cooking Oil as Detergent

WCO can be used to make soap or detergent which can be further used as dish washing, laundry washing, house cleaning, animal or vehicle cleaning use. Low grade soaps can be directly obtained by having WCO reacts with alkali metal hydroxide (approximately 1-5%), via saponification method (Panadare, & Rathod, 2015). There is patented method to produce liquid soap or detergent from WCO in which amine derivatives is used to produce a mild liquid soaps which are safe to handle and without unpleasant odorous substances (Kazuo, 1989).

Recently, Junior et al. (2019) used WCO to synthesize an environmentally friendly detergent: anionic surfactant methyl ester sulfonate (MES). MES was combined with ZnO nanoparticles, producing nanofluidic detergent. WCO was filtered, neutralized and bleached before it underwent transesterification with methanol. The as-produced methyl ester was then undergone sulfonation process to become MES surfactant. The product was then further purified with methanol and neutralized by sodium hydroxide. Liquid detergent comprising of 15% MES concentration and 0.1% ZnO nanoparticles exhibits notable stability, while retaining 64.97% stain removal as well as 82.36% stain degradation.

Waste Cooking Oil as Solvents for Pollutants

WCO with its organic property is also explored for its tendency to solubilize small organic molecules. Tarnpradab et al. (2016) employed WCOs to treat the emission produced during the rice husk pyrolysis. In particular, WCO was found able to reduce the content of organic hydrocarbon contaminants with a molecular weight higher than that of benzene (which are generally referred to as tar). In the study, heavy tar was reported absorbed by WCO through a dissolution mechanism and several data about the saturation levels were reported (Mannu et al. 2019).

Waste Cooking Oil as a Component of Animal Feed

Oils and fats are essential components of animal diet as they serve high energy diets as well as some essential fatty acids are needed and are not synthesised by animals. Very economical WCOs that bind the other ingredients of animal feed together fulfilled this requirement (Panadare, & Rathod, 2015). Food provided to the animals like poultry farms and pigs, indirectly come to human beings via food chain. It should be free from any toxic and unsafe components. Due to food safety, UK government only allow WCO which is collected by a licensed waste carrier, treated well and certified to be used in animal feeds (Panadare, & Rathod, 2015).

Harmful chemicals, especially malondialdehyde and other peroxidation products, such as 2-thiobarbituric acid are generated due to thermal reaction during cooking or frying. This makes it unfit as a component for animal feed. Hence, necessary treating process is needed for WCO prior to be used in animal feed. However, conventional techniques like filtration, degumming, bleaching, and deodorizing processes failed in the removal of those harmful materials. Wei et al. (2011) developed a simple and reliable method to measure the content of malondialdehyde in WCO and studied the removal of malondialdehyde and other 2-thiobarbituric acid reactive substances via three methods: water extraction, physical adsorption, and chemical adsorption. Chemical adsorption using lysine or monosodium glutamate as chemical adsorbent was reported to be the most effective. The authors claimed that removal of 80% of those substances from WCO was achieved.

CONCLUSION

With proper management strategies, WCO can be converted into useful bio base material and helps in alleviating the environmental pollution. More noticeably, WCO represents a renewable resource to produce fuels or can be used as alternative feedstocks in replacements of petroleum-based chemicals. In some cases, it is utilized as a high energy feed additive in various livestock feeding products. WCO is normally a repeatedly used cooking oil. It is vital to understand that oil becomes too acidic when reused multiple times, and after a certain point, the oil becomes unsafe for consumption. This will cause health hazards when people consume it or its processing products. Noting that *repeated frying of oil leads to changes in physiochemical, nutritional, and sensory properties of edible oil,* the FSSAI has announced that restaurants will no longer be able to reuse cooking oil more than three times (The news minute, 2019). Therefore, the reuse of WCO as a feed additive without proper treatment should be banned to prevent it from re-entering the food chain. By the way, recycling of WCO is a sustainable way of using the waste resources that can decrease environmental pollution and promote social and economic benefits.

ACKNOWLEDGMENT

This work was conducted under FRGS 2019-1 by Ministry of Education Malaysia.

REFERENCES

Abdul Majed, M. A.,, & Tye, C. T. (2018). Catalytic cracking of used vegetable oil to green fuel with metal functionalized zsm-5 catalysts. *The Malaysian Journal of Analytical Sciences*, *22*(1), 8–16.

Abdullah, B. M., & Salimon, J. (2010). Epoxidation of vegetable oils and fatty acids: catalysts, methods and advantages. *Journal of Applied Sciences (Faisalabad)*, *10*(15), 1545–1553. doi:10.3923/jas.2010.1545.1553

Asli, H., Ahmadinia, E., Zargar, M., & Karim, M. R. (2012). Investigation on physical properties of waste cooking oil – Rejuvenated bitumen binder. *Construction & Building Materials*, *37*, 398–405. doi:10.1016/j.conbuildmat.2012.07.042

BioenergyE. T. I. P. (2019). *Transesterification to biodiesel.* Retrieved from http://www.etipbioenergy.eu/value-chains/conversion-technologies/conventional-technologies/transesterification-to-biodiesel

Chang, W. H.,, & Tye, C. T. (2013). Catalytic cracking of used palm oil using composite zeolite. *The Malaysian Journal of Analytical Sciences*, *17*(1), 176–184.

Chen, R.-X.,, & Wang, W.-C. (2019). The production of renewable aviation fuel from waste cooking oil. Part I: Bio-alkane conversion through hydroprocessing of oil. *Renewable Energy*, *135*, 819–835. doi:10.1016/j.renene.2018.12.048

Chiam, L. T.,, & Tye, C. T. (2013). Deoxygenation of plant fatty acid using NiSnK/SiO_2 as catalyst. *The Malaysian Journal of Analytical Sciences*, *17*(1), 129–138.

Doronin, V. P., Potapenko, O. V., Lipin, P. V., Sorokina, T. P., & Buluchevskaya, L. A. (2012). Catalytic cracking of vegetable oils for production of high octane gasoline and petrochemical feedstock. *Petroleum Chemistry*, *52*(6), 392–400. doi:10.1134/S0965544112060059

European Biomass Industry Association. (2015). Transformation of used cooking oil into biodiesel: from waste to resource. Promotion of used cooking oil recycling for sustainable biodiesel production (RecOil). Retrieved from http://www.eubren.com/UCO_to_Biodiesel_2030 _01.pdf

Feng, G., Hu, L., Ma, Y., Jia, P., Hu, Y., Zhang, M., ... Zhou, Y. (2018). An efficient bio-based plasticizer for poly (vinyl chloride) from waste cooking oil and citric acid: Synthesis and evaluation in PVC films. *Journal of Cleaner Production*, *189*, 334–343. doi:10.1016/j.jclepro.2018.04.085

Grilo, J., Faria, P., Veiga, R., Silva, A. S., Silva, V., & Velosa, A. (2014). New natural hydraulic lime mortars – physical and microstructural properties in different curing conditions. *Construction & Building Materials*, *54*, 378–384. doi:10.1016/j.conbuildmat.2013.12.078

Hasan, S. E. (2004). Public awareness is key to successful waste management. *Journal of Environmental Science and Health. Part A, Toxic/Hazardous Substances & Environmental Engineering*, *39*(2), 483–492. doi:10.1081/ESE-120027539 PMID:15027831

IBIS world. (2018). *Cooking oil recycling industry in the US*. Industry Market Research Report. September, 2018.

Iglesias, L., Laca, A., Herrero, M., & Díaz, M. (2012). A life cycle assessment comparison between centralized and decentralized biodiesel production from raw sunflower oil and waste cooking oils. *Journal of Cleaner Production*, *37*, 162–171. doi:10.1016/j.jclepro.2012.07.002

Junior, G. D., Ibadurrohman, M., & Slamet. (2019). Synthesis of eco-friendly liquid detergent from waste cooking oil and ZnO nanoparticles. In *AIP Conference Proceedings* (vol. 2085(1), 020075). doi:. doi:10.1063/1.5095053

Kataria, J., Mohapatra, S. K., & Kundu, K. (2019). Biodiesel production from waste cooking oil using heterogeneous catalysts and its operational characteristics on variable compression ratio CI engine. *Journal of the Energy Institute*, *92*(2), 275–287. doi:10.1016/j.joei.2018.01.008

Kazuo, S. US Pat. 4839089, Mimasu Oil Chemical Co., Ltd., (1989).

Khalid, K., & Khalid, K. (2011). Transesterification of palm oil for the production of biodiesel. *American Journal of Applied Sciences*, *8*(8), 804–809. doi:10.3844/ajassp.2011.804.809

Li, H., Zhao, J., Huang, Y., Jiang, Z., & Chen, Q. (2016). Investigation on the potential of waste cooking oil as a grinding aid in Portland cement. *Journal of Environmental Management*, *184*(3), 545–551. doi:10.1016/j.jenvman.2016.10.027 PMID:27793479

Li, Z., Huang, Z., Ding, S., Li, F., & Chen, C. (2018). Catalytic conversion of waste cooking oil to fuel oil: Catalyst design and effect of solvent. *Energy*, *157*, 270–277. doi:10.1016/j.energy.2018.05.156

Liang, S., Liu, Z., Xu, M., & Zhang, T. (2013). Waste oil derived biofuels in China bring brightness for global GHG mitigation. *Bioresource Technology*, *131*, 139–145. doi:10.1016/j.biortech.2012.12.008 PMID:23340111

Lucena, I. L., Silva, G. F., & Fernandes, F. A. N. (2008). Biodiesel production by esterification of oleic acid with methanol using a water adsorption apparatus. *Industrial & Engineering Chemistry Research*, *47*(18), 6885–6889. doi:10.1021/ie800547h

Ma, J., Sun, D., Pang, Q., Sun, G., Hu, M., & Lu, T. (2019). Potential of recycled concrete aggregate pretreated with waste cooking oil residue for hot mix asphalt. *Journal of Cleaner Production*, *221*, 469–479. doi:10.1016/j.jclepro.2019.02.256

Mannu, A., Ferro, M., Pietro, M. E. D., & Mele, A. (2019). Innovative applications of waste cooking oil as raw material. *Science Progress*, *102*(2), 153–160. doi:10.1177/0036850419854252

Matthäus, B. (2010). Oxidation in foods and beverages and antioxidant applications: management in different industry sectors. Chapter 6 - Oxidation of edible oils. Woodhead Publishing Series in Food Science, Technology and Nutrition, 183-238.

Ministry of Environment (2006). Report on promotion of eco-fuel for transportation.

Mohadesi, M., Aghel, B., Maleki, M., & Ansari, A. (2019). Production of biodiesel from waste cooking oil using a homogeneous catalyst: Study of semi-industrial pilot of microreactor. *Renewable Energy*, *136*, 677–682. doi:10.1016/j.renene.2019.01.039

Pahlavan, P., Manzi, S., Rodriguez-Estrada, M. T., & Bignozzi, M. C. (2017). Valorization of spent cooking oils in hydrophobic waste-based lime mortars for restorative rendering applications. *Construction & Building Materials*, *146*, 199–209. doi:10.1016/j.conbuildmat.2017.04.001

Pahlavan, P., Manzi, S., Sansonetti, A., & Bignozzi, M. C. (2018). Valorization of organic additions in restorative lime mortars: Spent cooking oil and albumen. *Construction & Building Materials*, *181*, 650–658. doi:10.1016/j.conbuildmat.2018.06.089

Panadare, D. C., & Rathod, V. K. (2015). Applications of waste cooking oil other than biodiesel: A review. *Iranian Journal of Chemical Engineering*, *12*, 55–76.

Qiu, F., Li, Y., Yang, D., Li, X., & Sun, P. (2011). Biodiesel production from mixed soybean oil and rapeseed oil. *Applied Energy*, *88*(6), 2050–2055. doi:10.1016/j.apenergy.2010.12.070

Raeisian, L., Niazmand, H., Ebrahimnia-Bajestan, E., & Werle, P. (2019). Feasibility study of waste vegetable oil as an alternative cooling medium in transformers. *Applied Thermal Engineering*, *151*, 308–317. doi:10.1016/j.applthermaleng.2019.02.010

Sahar, S., Sadaf, S., Iqbal, J., Ullah, I., Bhatti, H. N., Nouren, S., ... Iqbal, M. (2018). Biodiesel production from waste cooking oil: An efficient technique to convert waste into biodiesel. *Sustainable Cities and Society*, *41*, 220–226. doi:10.1016/j.scs.2018.05.037

Sarno, M., Spina, D., & Senatore, A. (2019). One-step nanohybrid synthesis in waste cooking oil, for direct lower environmental impact and stable lubricant formulation. *Tribology International*, *135*, 355–367. doi:10.1016/j.triboint.2019.03.025

Soegiantoro, G. H., Chang, J., Rahmawati, P., Christiani, M. F., & Mufrodi, Z. (2019). Home-made eco green biodiesel from chicken fat (CIAT) and waste cooking oil (PAIL). *Energy Procedia*, *158*, 1105–1109. doi:10.1016/j.egypro.2019.01.267

Solway Recycling Ltd. (2019). Some legal requirements for consideration when arranging a cooking oil collection. Retrieved from https://www.solwayrecycling.co.uk/recycling-services/oil-collection/legal-requirements-of-cooking-oil-collection

Statista. (2019a). *Global production of vegetable oils from 2000/01 to 2018/19 (in million metric tons)*. Retrieved from https://www.statista.com/statistics/263978/global-vegetable-oil-production-since-2000-2001/

Statista. (2019b). *Consumption of vegetable oils worldwide from 2013/14 to 2018/2019, by oil type (in million metric tons)*. Retrieved from https://www.statista.com/statistics/263937/vegetable-oils-global-consumption/

Sun, Z., Yi, J., Huang, Y., Feng, D., & Guo, C. (2016). Properties of asphalt binder modified by bio-oil derived from waste cooking oil. *Construction & Building Materials*, *102*, 496–504. doi:10.1016/j.conbuildmat.2015.10.173

Tarnpradab, T., Unyaphan, S., Takahashi, F., & Yoshikawa, K. (2016). Tar removal capacity of waste cooking oil absorption and waste char adsorption for rice husk gasification. *Biofuels*, *7*(4), 401–412. doi:10.1080/17597269.2016.1147919

The Hindu Business line. (2018). *New regulations for 'used cooking oil' come into effect*. Retrieved from https://www.thehindubusinessline.com/economy/policy/new-regulations-for-used-cooking-oil-come-into-effect/article24314377.ece

The news minute. (2019). *Restaurants can no longer reuse cooking oils as per new food safety regulations*. Retrieved from https://www.thenewsminute.com/article/restaurants-can-no-longer-reuse-cooking-oils-new-food-safety-regulations-96723

Trabelsi, A. B. H., Zaafouri, K., Baghdadi, W., Naoui, S., & Ouerghi, A. (2018). Second generation biofuels production from waste cooking oil via pyrolysis process. *Renewable Energy*, *126*, 888–896. doi:10.1016/j.renene.2018.04.002

Tsai, W.-T. (2019). Mandatory recycling of waste cooking oil from residential and commercial sectors in Taiwan. *Resources*, *8*(1), 38. doi:10.3390/resources8010038

UK Environmental Law Association. (2017). *The plain guide to environmental law*. Retrieved from http://www.environmentlaw.org.uk/rte.asp?id=85

Vinyes, E., Oliver-Solà, J., Ugaya, C., Rieradevall, J., & Gasol, C. M. (2013). Application of LCSA to used cooking oil waste management. *The International Journal of Life Cycle Assessment*, *18*(2), 445–455. doi:10.100711367-012-0482-z

Wakoc, F. M., Reshadb, A. S., Bhaleraoa, M. S., & Goud, V. V. (2018). Catalytic cracking of waste cooking oil for biofuel production using zirconium oxide catalyst. *Industrial Crops and Products*, *118*, 282–289. doi:10.1016/j.indcrop.2018.03.057

Wallace, T., Gibbons, D., O'Dwyer, M., & Curran, T. P. (2017). International evolution of fat, oil and grease (FOG) waste management - A review. *Journal of Environmental Management*, *187*, 424–435. doi:10.1016/j.jenvman.2016.11.003 PMID:27838205

Wang, J., Zhong, Z., Ding, K., Zhang, B., & Ruan, R. (2017). Successive desilication and dealumination of HZSM-5 in catalytic conversion of waste cooking oil to produce aromatics. *Energy Conversion and Management*, *147*, 100–107. doi:10.1016/j.enconman.2017.05.050

Wee, H. M., Budiman, S. D., Su, L. C., Chang, M., & Chen, R. (2016). Responsible supply chain management-An analysis of Taiwanese gutter oil scandal using the theory of constraint. *International Journal of Logistics Research and Applications*, *19*(5), 380–394. doi:10.1080/13675567.2015.1090964

Wei, Z., Li, X., Thushara, D., & Liu, Y. (2011). Determinations and removal of malodialdehyde and other 2-thiobarbituric acid reactive substances in waste cooking oil. *Journal of Food Engineering*, *107*(3-4), 379–384. doi:10.1016/j.jfoodeng.2011.06.032

Williams, J. B., Clarkson, C., Mant, C., Drinkwater, A., & May, E. (2012). Fat, oil and grease deposits in sewers: Characterisation of deposits and formation mechanisms. *Water Research*, *46*(19), 6319–6328. doi:10.1016/j.watres.2012.09.002 PMID:23039918

Zheng, T., Wu, Z., Xie, Q., Fang, J., Hu, Y., Lu, M., ... Ji, J. (2018). Structural modification of waste cooking oil methyl esters as cleaner plasticizer to substitute toxic dioctyl phthalate. *Journal of Cleaner Production*, *186*, 1021–1030. doi:10.1016/j.jclepro.2018.03.175

KEY TERMS AND DEFINITIONS

Biodiesel: Long chain alkyl esters that is made by catalytic reaction of triglycerides and alcohol.

Hydrodeoxygenation: A process for removing oxygen from oxygen containing compound by reacting with hydrogen.

Prevention: Action to stop something to happen.

Pyrolysis: A thermal decomposition process of materials at elevated temperatures in an inert atmosphere.

Recycling: A process to convert waste materials into new materials and objects.

Transesterification: A chemical reaction involves ester and alcohol.

Waste Cooking Oil: Used vegetable oils and animal fats derived from cooking.

Chapter 4
Sustainable Technologies for Solid Waste Monitoring and Treatment

Kafayat Olafunke Adeyemi
University of Abuja, Nigeria

Urbans Benywanira
Uganda Carbon Bureau, Uganda

ABSTRACT

Municipal solid waste (MSW) is an energy source that should not go untapped or unutilized. The waste must be properly utilized through combustion, anaerobic digestion, and landfill gas acquisition, as it represents material and energy content. This will reduce the effects of global warming, which is as a result of high concentration of carbon dioxide, methane, and other greenhouse gases (GHGs), in the atmosphere. This chapter focuses on the technologies for solid waste management and the thermodynamics involved in the process for sustainable and cleaner energy. The equations presented represent the thermal efficiency, conversion efficiencies, as well as possible work that can be derived from a power plant utilizing MSW as fuel. It is important that countries in Sub-Saharan Africa vigorously pursue sustainable waste management technologies, especially recycling and landfilling, while exploring and investing in waste-to-energy technologies that will perform optimally using the composition of the waste in Sub-Saharan Africa in the design of the waste-to-energy technology.

INTRODUCTION

Waste management (WM) is the collection, transport, processing, recycling or disposal, and monitoring of waste materials. The term "waste" usually relates to materials which are produced by human activity and is generally undertaken to reduce their effect on health, the environment, or aesthetics (Ioana, 2010). Through WM, it is also possible to recover fuel and energy. Today, WM remains one of the major problems in megacities and in most metropolitan areas. Globally, about 1.3 billion tons of garbage is

DOI: 10.4018/978-1-7998-0369-0.ch004

produced annually, but about 60% of it is not disposed of properly and ends up in landfills–yet, this is not a viable disposal method to the ever-increasing garbage disposal problems. As a result, solid wastes have become the biggest source of environmental pollution to soil, ground water, and air, due to increasing volumes of waste and absence of effective WM strategies. Whereas the technology of waste disposal has evolved during the past several decades, options are still limited, and many city administrators are indeed grappled with the challenge of raising the requisite financial resources towards WM services in a sustainable manner. The World Energy Council (2016) reported that, by 2025, global waste is estimated to reach 6 million tons per day.

Municipal solid waste (MSW) is composed of household, trade, and commercial waste. Wastes can also be categorized into biodegradable (food waste), recyclable (paper and glass bottles), inert (construction waste), electrical and electronic (computers screens), composite (toys and waste plastics), hazardous (paints and chemicals), toxic (pesticides and herbicides), and medical wastes (pharmaceutical drugs) (Dawei, 2015). It is called municipal waste because the municipality in which the waste is being generated is responsible for the disposal. Residential and institutional wastes are being disposed of by the local authorities in some capital cities in Sub-Saharan Africa (SSA). The disposal of industrial and commercial wastes which are generated in cities are not the responsibility of the local authorities, but of their generator. The composition and volume of MSW differs from location to location. In any waste-to-energy (WtE) conversion process, knowledge of energy content of wastes is important. Paper, plastics, and rubber have a high energy value and low moisture content, as compared to food waste with high moisture content and low energy value. Composition of paper in solid waste ranges from 1% to 5% in low income countries whereas, in high-income countries, it varies from 20% to 50%. Plastic composition in solid waste varies from 1% to 5% and 5% to 10% in low- and high-income countries, respectively (Sunil, 2016). Ash and fine material fractions in solid waste ranges from 15% to 60% and 3% to 10% in low- and high-income countries, respectively. In low-income countries, moisture content found in solid waste is around 30%–40% and, in high-income countries, it is 15%–30% (Sunil, 2016). The composition of MSW in SSA comprises over 50% in organic waste, and the trend will continue into 2025. Figure 1 shows the trends, constituents, and percentages of MSW in SSA. The generation of MSW is directly related to the level of socioeconomic development, as the development of a nation tends to increase waste generation. Other drivers of waste generation include population growth, urbanization and global trade.

In many SSA cities, waste collection is still overlooked with MSW collection rates ranging from 20% to 80% (Mohee & Bundhoo, 2015). There exist no proper waste treatment and disposal facilities and as a result, rudimentary and archaic waste disposal methods like illegal open dumping and uncontrolled burning practices are the day-to-day means employed to reduce the mountains of stinking refuse (Mohee & Bundhoo, 2015). Algeria, for example, dumps 80% of its MSW, while Egypt dumps 83.5% of generated MSW. Nigeria receives up to 2400 metric tonnes of waste each day at Olisosun landfill in an area of 42 hectares (Adegbola & Olawoyin, 2012).

Land filling is also commonly used, with Madagascar and Mauritius landfills accounting for 97% and 91% of their MSW respectively. Tunisia landfills 65% of its MSW (Hoornweg & Bhada-Tata, 2012) while Mauritania and Morroco landfill 37.3% and 28% of their MSW, respectively. This implies that a lot still needs to be done to encourage other options, such as composting and increased recycling, to minimize waste to the landfill sites (Sharholy, Ahmad, Mahmood & Trivedi, 2008, Chimuka & Ogola, 2015).

Apart from absence of proper waste treatment and disposal facilities, most cities in SSA countries lack appropriate policies and legislation that would support investments in waste recycling. In instances where these policies and legislative instruments exist, their application has proven to be inconsistent

Figure 1. Changing composition of wastes in African cities
(Hoqmweg & Bhada, 2012)

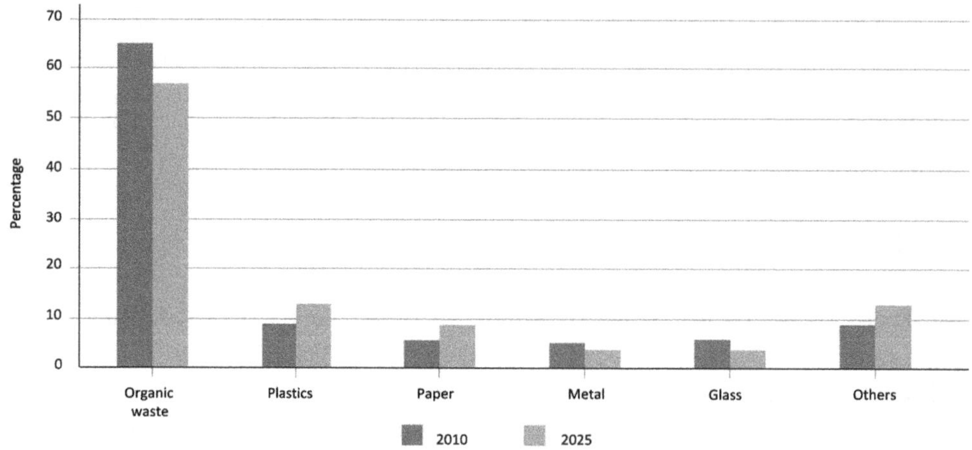

(Sentime, 2014). Equally, the lack of comprehensive policies which are well aligned has resulted in the fragmentation of strategies which, if properly developed would contribute towards reducing associated costs and bring out effective and sustainable resource management and socio-economic service delivery (Sentime, 2014). Another impediment is limited financial resources to undertake state-of-the art waste management facilities. In the eThekwini municipality in South Africa for example, the adopted technology for waste management is electricity intensive was costly to run. Chimuka & Ogolo (2015), observed that this situation created a scenario in which the local municipality increasingly turned to donors for financial help in order to continue operating and running the technology.

The tremendous rise in MSW in the fast-growing cities of developing and emerging countries has led to increasing public concerns with regards to the resultant health and environmental impacts. Today, the waste of about 3 billion people is still disposed of in an uncontrolled manner (Wilson et al., 2015). A project report on "sustainable waste management options for Africa" prepared for African Development Bank revealed that in most countries SWM is characterized by inefficient collection methods, insufficient coverage of the collection system, and improper disposal of MSW (Palcyzinsky & Scotia, 2002). The study also revealed that integrated WM is not implemented, and very little information is available on composting, controlled sanitary land fill, and recovery of the land fill gas. Poor WM practices and, in particular, the widespread dumping of wastes in water bodies and uncontrolled dump sites aggravate the problems of generally low sanitation levels across the African continent. Africa review report on WM prepared for United Nations Economic Commission for Africa explained that the gap between WM policy and legislation and actual WM practices is widening, due to ongoing capacity constraints or non-existence of WM facilities for the different waste streams (UNECA, 2009). However, it should be recognized that wastes represent nonnegligible renewable energy resources, contributing to a cleaner environment and development of business entrepreneurship, and offering security and jobs to local communities. For instance, there is a great potential for addressing methane emissions by reducing the amount of wastes that end up in a landfill. Worldwide, nearly 70% of the solid waste is land filled, a meager 19% is recovered through composting or recycling, whereas the remaining 11% is converted to energy through incineration or other WtE technologies (Derek, 2012).

Figure 2. Sustainable waste management
(Agbelie, Bawakyilenuo & lemarie 2015)

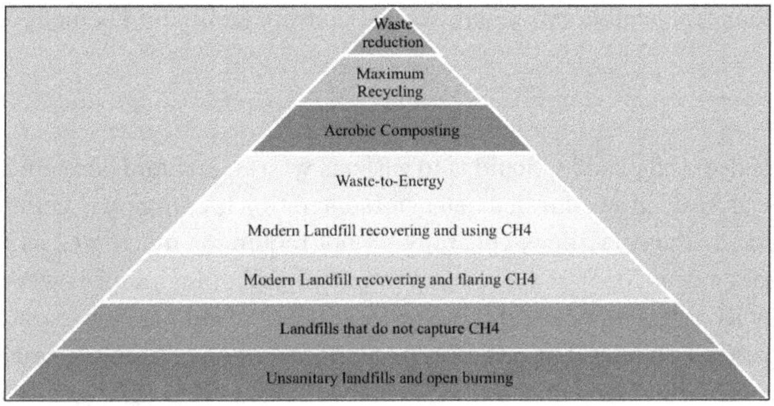

The purpose of sustainable WM is to reduce to a minimum the waste disposal and extract energy for economic activities in a sustainable and clean manner. In any WM policy, the priority is to reduce waste generation. The world today places much emphasis on "reduce" waste (zero waste generation) hence, there is a need to assess the effects of implementing such a system in SSA. Figure 2 shows the WM hierarchy. Between the two extreme options of waste reduction (most preferable) and unsanitary landfills and open burning (least desirable), there are options for effective WM.

The option of WtE is a process for removing, from MSW, energy that does not fall into the category of "reduce, reuse, and recycle". This energy can serve as replacement to fossil fuels. WtE technologies in use today are incineration, coprocessing, anaerobic digestion, land fill gas, and pyrolysis/gasification.

SOLID WASTE TREATMENT METHODS FOR ENERGY GENERATION

The solid waste treatment methods are classified and discussed below:

Mechanical and Thermal Treatment of Waste

The application of mechanical heat treatments (MHT) to treat MSW is an emerging technology with relatively few full-scale operational plants. A MHT is a process used to separate a mixed stream of waste into different components (classes) for further waste processing such as recycling and biological treatment. The thermal aspect of the Mechanical treatment of waste sanitizes the waste by destroying bacteria present, processes plastic waste further since the heat can cause the plastics to deform a bit which brings about easy removal of labels from plastic wastes. The thermal process may also reduce the moisture content in the waste and, thereby, increase the energy value at that stage. An example of a scenario in the application of MHT to waste treatment is in Europe. The European Union (EU) landfill directive prohibits the disposal of waste to a landfill without prior treatment. In line with this directive, countries within the EU have developed pretreatment options before disposal to a landfill. In the United Kingdom (UK), these options include recycling, anaerobic digestion (for further processing of food, agriculture, and animal wastes), and setting up of WtE plants to recover energy from wastes. In an article about the

UK's WM, 16.1% of the 31 million tons of solid waste that was generated in 2012 was processed in WtE plants. This accounted for about 5% of the UK's total renewable energy sources. In Scotland and Wales, however, only 0.02 tonnes per capita was sent to energy recovery facilities (Themelis & Bourtsalas, 2013)

Incineration

This is the process of disposal of solid, liquid, and gaseous wastes, including certain hazardous biowaste and biological medical waste in a controlled manner. Incineration is common in countries, such as Japan, where land is scarcer, as these facilities generally do not require as much area as landfills. In China, though power generation from MSW is a recent development, technologies for power generation from the incineration of MSW have developed rapidly as a result of the demand for a low-carbon economy and the encouragement of national policies. For instance, since the inception of the first incineration power plant in 1988, the number of solid waste incineration plants had increased to 166 by the end of 2013, and the incineration capacity of municipal waste to 158,488 tons per day (National Bureau of statistics, 2014).

One of the advantages of incineration is that it produces steam or hot flue gases that might be turned into heat and electrical energy when they are introduced into a thermodynamic cycle. However, the biggest disadvantage is that combustion in an incinerator produces micro pollutants in gaseous emissions from incinerator stacks, including some very persistent organics, such as dioxins. The implication of these releases are birth defects, cancer, and threat to air quality (Pachauri & Reisinger, 2007).

Hence, the flue gases which are generated must be treated by a flue gas scrubber which removes the acid materials. Then, bag filters are used to remove dust particles, so that the gas can meet the final emission standards. According to the National Research Council (2000), incineration facilities incorporate several general methods for ensuring proper combustion and reducing emissions. A steady situation, with no major fluctuations in the waste-feed supply rate, combustion-air flows, or other conditions, promotes efficient combustion. Nonetheless, inefficient combustion can result in higher levels of products of incomplete combustion. Likewise, the more often a facility is started up and shut down (for maintenance or because of inadequate or varying waste stream volume), the more uneven the combustion and the greater the potential for increased emissions.

Mechanical Biological Treatment

A Mechanical Biological Treatment (MBT) is a process of separating biodegradable wastes, such as paper and food waste, from the MSW being taken to a landfill. A MBT, then, involves conversion of the biodegradable waste into combustible biogas energy for energy recovery by subjecting the biodegradable waste to anaerobic digestion. MBT can also involve drying the biodegradable waste to produce a high calorific and organic-rich fuel, which is known as Refuse Derived Fuel. This RDF can also be produced through palletization of MSW. The palletization process involves drying, shredding, and densification (Jones, 2014). The production of RDF is not common, due to its disadvantage of having high ash content which can be corrosive to combustion plants. Additionally, MBT facilities in England processed 1.4% of the total MSW which was generated in the UK in 2012 to produce RDF (Department for Environment, Food and Rural Affairs, 2013).

Anaerobic Digestion

Anaerobic digestion is one of the oldest processes of extracting energy from waste. According to Yucheng & Pawloski (2012), anaerobic digestion is the most sludge stabilization technology currently on the market. The process transforms sludge organic solids into biogas, which is a mixture of methane, carbon dioxide, and traces of other gases, in an anaerobic environment (Ulas, Tandukar, & Pavlostathis, 2011). It involves the decomposition of organic waste (i.e., food and agricultural waste) in the absence of oxygen to produce biogas. Anaerobic digestion has many environmental benefits. For instance, the use of anaerobic digestion for the treatment of the organic fraction of MSWs would reduce the emission of carbon dioxide (Mata-Alvarez, Mace, & Llabres, 2000). Anaerobic digestion reduces the emission of landfill gas into the atmosphere, and is a net energy producing process, which produces renewable energy in form of biogas.

Biogas is rich in methane and is popularly used as a domestic fuel in East Africa and some Asian countries. Biogas is composed of methane, carbon dioxide, hydrogen, nitrogen, and traces of hydrogen sulphide. Based on research literature, experience and development, India appears prominent for anaerobic processing of organic waste (Vögeli & Zurbrügg, 2008).

The process of composting by anaerobic digestion reduces the amount of methane that would be emitted into the atmosphere from landfills. In East Africa and Asia, composting by anaerobic digestion is taken as a clean development mechanism project. Developed countries, which are commonly called Annex A countries that have exceeded their GHG emission limit target (Kyoto protocol) do set up clean development mechanism projects in least developed countries in order to offset their emissions or earn carbon credits. This can also be viewed from the angle of sustainable development as it gives access to clean and affordable energy to households in the rural communities of the developing countries. The most common digester type is the fixed dome type. This is attributed to ease of construction, ease of maintenance and lower construction costs when compared to other types of digesters. The construction basically consists of digging of pit and slurry discharge chamber based on size of digester to be constructed. The next stage is the form work for the concrete wall as shown in Figure 3 The dome vault is constructed by installation of wood guidepost and mesh network with inlet feed pipes and gas outlet pipes attached to the wood mesh network as shown in Figure 4. Finally, a concrete casting to cover the base of the dome vault as well as top of the dome. This is shown in Figure 5.

It can also be used to generate electricity and to heat homes or can be processed further into natural gas. Moreover, anaerobic digestion as a waste treatment option for urban settings, predominantly processing kitchen or market waste, still plays a big role, though little knowledge and information are available or accessible on technical and operational feasibility, challenges, and opportunities. However, Anaerobic digestion requires a high level of investment in large tanks and other process vessels especially when carried out at a commercial scale on farms and at wastewater treatment works.

Thermo Chemical Treatment Method

Pyrolysis

Pyrolysis is the thermal degradation of a substance or organic matter in the absence of oxygen. In order to reach decomposition, conditions in the heat reactor and sometimes steam need to be introduced in the reaction chamber. Figure 6 shows the process flow diagram. As compared to incineration, pyrolysis is

Figure 3. Digester wall formwork
(Oloche, Ozigis, Adeyemi & Ikpe 2017)

Figure 4. Wood mesh network for dome vault
(Oloche, Ozigis, Adeyemi & Ikpe 2017)

Figure 5. Completed dome vault on top of the digester
(Oloche, Ozigis, Adeyemi & Ikpe 2017)

Figure 6. Pyrolysis process flow diagram
(Gary, 2010)

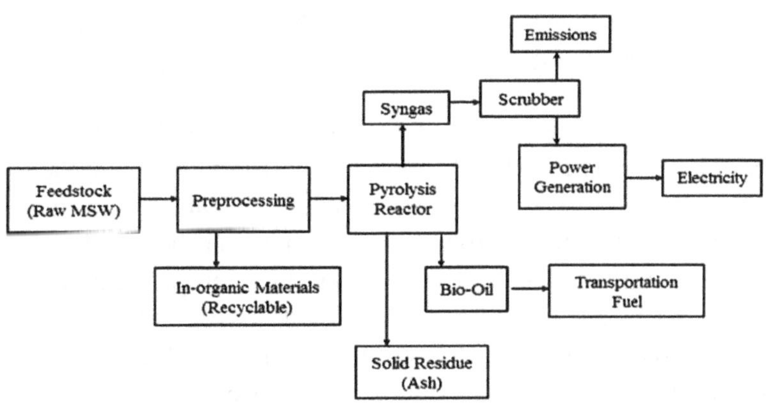

a conversion of solid waste to liquid or gaseous fuels, along with residue called char. Char is a mixture of noncombustible material and carbon (Kothari, Tyagi, & Pathak, 2010). The required temperature for the process ranges between 300 and 800°C (Kothari et al., 2010). The yield of char and chemical composition can be calculated as functions of pyrolysis temperature. Amount of char from pyrolysis of solid waste decreases with increasing pyrolysis temperature (Demirbas, 2007).

The product gas is known as syngas, which is the combination of volatile organic compounds, carbon monoxide (CO), hydrogen (H_2), and methane (CH_4). The volatilized gases and volatilized liquids are efficiently used to run a steam engine. The char can be used as fuel in power or cement plants. The pyrolysis gas has at least twice the energy content of gasification synthetic gas; both are produced from the same waste (fuel). Nevertheless, the overall energy efficiency is not always in favor of pyrolysis, as pyrolysis gas is partially utilized for heat and steam generation. Pyrolysis is an option for the treatment of specific solid wastes, such as used vehicle tires and clinical wastes. It is not ideal for mixed MSW. Pyrolysis and gasification are the most expensive technologies amongst WtE technologies. It can only be viable if there is a strong market for the process products (char and gas). Pyrolysis is advantageous in that it allows the recycling of waste mixed plastics that cannot be efficiently recycled by alternative means. The process also permits the recycling of unwashed and soiled plastics (e.g. agricultural plastics, mulch/greenhouse fills and dripper/irrigation tube. More so, pyrolysis enables recycling of plastic laminates, coextrusions and multilayer packaging films particularly those with aluminum foil layers that are difficult to recycle using traditional processing technologies. However, pyrolysis is dangerous because the gases, liquids and ashes released are potentially harmful to the environment. The produced ashes contain a high heavy metal content, depending on the concentrations in the to-be processed flow. Hence, an air purification installation is necessary in order to further treat flue gases from the pyrolysis Moreover, the pyrolysis process is very complex and requires high operational and investment costs.

Gasification

Gasification involves partial oxidation of a substance; it lies between combustion and pyrolysis. The required temperature is above 750°C.The products are almost the same as pyrolysis, that is syngas and low carbon ash. Syngas is composed of H_2, CO, and CH_4. These are the main components of the formed syngas. They have energy value and, together with CO_2, N_2, O_2, and H_2O, represent more than 93% of the components in syngas. The rest are higher order hydrocarbons (e.g., ethane, propane, butane, and benzene), some cyclical hydrocarbons (e.g., benzene and toluene) and other gases (e.g., HCl, HF, and SO_2). Higher-order hydrocarbons and cyclical hydrocarbons do have higher calorific value than H_2, CO, and CH_4, but are found in syngas in very low and negligible quantities (Kokalj &Samec, 2013)

Gasification is a reliable option, as it meets present emission standards and is helpful in maintaining the sustainability of landfill. The feedstock is fed into gasifiers, along with a limited amount of air. Many downstream gasification processes require syngas to be cleaned from trace level of impurities. The most common impurity is mercury. Carbon bed technology has been utilized for cleanup in recent years (Arena, Zaccariello, & Mastellone, 2010). The products of gasification are steam, chemicals, electricity, hydrogen, fertilizers, and natural gas (Arena et al., 2010). Different types of waste gasification methods are characterized on the basis of oxygen medium; two of them are steam gasification and plasma gasification. Recently, many new technologies have been developed, such as plasma melting gasification (Arena et al., 2010).

The utilization of synthetic gas in gas engine or turbine is only possible if high calorific synthetic gas is produced. Literature shows that the gases with calorific value between 4 and 6 MJ/Nm3 can be produced. The gas turbine can run on gases with calorific values as low as 2.5 MJ/m3 (Kokalj & Samec, 2013). Gasification is a useful process because it can produce different forms of energy and create useful byproducts that include sulfur and ammonia, both of which have important industrial applications, hence improving the supply of such commodities and helping to make gasification more economical for its operators. Importantly, the energy that is produced from the gasification process can be used immediately in form of electricity or stored as a fuel for future consumption.

Figure 7. Schematic of Rankine cycle [2015]
[Sharma & Dev, 2015] Used with permission

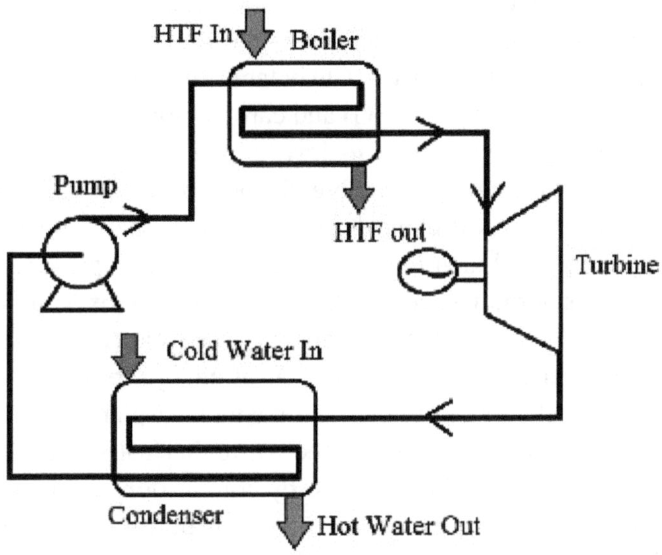

Figure 8. Schematic of reheat Rankine cycle
(Adapted from [Cengel & Boles, 2011])

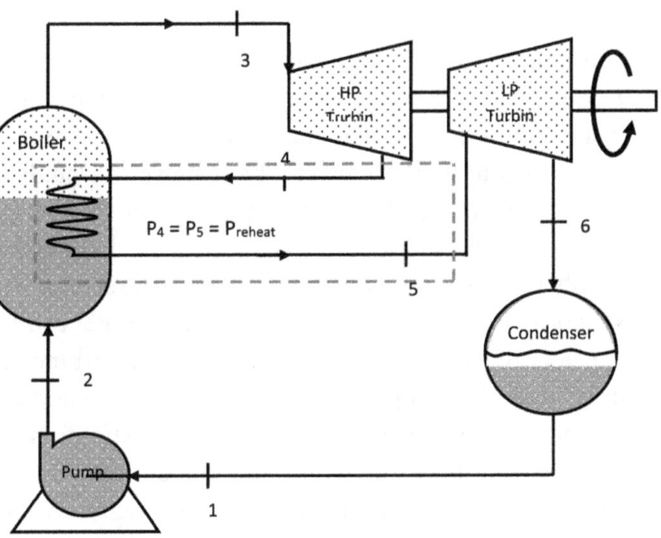

THERMODYNAMICS OF ENERGY GENERATION FROM SOLID WASTE

In an era of increasing energy demand and the need for sustainable energy generation, optimized use of energy and consumption management is pertinent. The application of WtE technology in heat and electricity generation reduces the amount of dumped waste in landfills, which invariably reduces the amount of methane (a very powerful greenhouse gas-GHG) and carbon dioxide emissions into the atmosphere. Most WtE plants are average to big size facilities, mainly for power generation, hot water, or steam with the energy of the waste. Different technologies enable distinct approaches for the utilization of the difference in inlet and oulet enthalpies in different thermal machines that are capable to transform these enthalpies into mechanical energy and then into power. These technologies are based on the principle of combustion, in which a fuel is burnt in excess or limited quantity of oxygen. Most WtE technologies use refuse derived fuel (RDF), which is produced from MSW. The main objective of a properly designed power plant is high thermal efficiency in the conversion of energy from waste to heat or electricity, with minimal emission of the pollutants to the environment. WtE plants combust the waste, thereby they decompose the hydrocarbons (fuel) present in the waste thermally and release heat energy from the waste for utilization. Thermal conversion of waste by combustion (incineration) can produce only hot water or steam as the products of complete combustion. The power production can only be achieved with the incorporation of a Rankine cycle (RC), with clear limitations of overall efficiency, as the inlet temperature of steam to the turbine must not exceed a certain value, to prevent the turbine blades from unsafe metallurgical temperature levels. The subsequent sections of the chapter will focus on the thermal efficiencies of incineration and RC. Even when combustion occurs in multiple stages (chambers), it does not improve the efficiency of power production. It only improves the environmental performance of the conversion process (Kokalj & Samec, 2013). The main components of the energy production system are the steam boiler, the steam turbine with the generator, air condenser and heat exchangers, and super heater, reheater, and economizer.

The power plant can be designed as a RC, a Reheat RC (RRC), or a Regenerative RC. The RC is the simplest theoretical vapor power cycle, and it is made up of two isentropic and two isothermal processes The schematic of the RC is shown in Figure 7. The main purpose of a RC is to convert part of the heat transferred to working fluid (steam) to work. This can be achieved through rotation of shaft located between generators with electromagnets thereby, converting mechanical energy to electrical energy.

One method of increasing the efficiency of the RC is to increase the boiler pressure. However, increasing the boiler pressure invariably increases the moisture content in the steam, which sometimes corrodes the turbine blades. The RRC corrects the increase in the moisture content of the steam and its adverse effects by incorporating a reheater in between a two-stage expansion turbine, as Figure 8 illustrates.

Heat addition to the liquid exiting the pump (in the RC) is at a low temperature. This has the implication of reduced thermal efficiency. A more technical solution is the extraction of steam from the turbine, at various points. The steam is used to heat the exiting liquid from the pump (feed water) in a device which is called the regenerator. This modified cycle is called the regenerative RC (Figure 9). Improving the efficiency of electrical power generation in a power plant involves the use of boilers that are highly corrosion resistant or of corrosion resistant plating on boiler tubes. The first and second law of thermodynamics are used to analyze the RC using MSW as fuel. The first law deals with energy conservation and management, while the second law is about irreversibility (entropy generation) and exergy.

Regeneration not only improves cycle efficiency, but also provides a convenient means of de-aerating the feedwater (i.e., removing the air that leaks in at the condenser) to prevent corrosion in the boiler. It

Figure 9. Schematic of regenerative Rankine cycle
(Adapted from [Cengel & Boles, 2011])

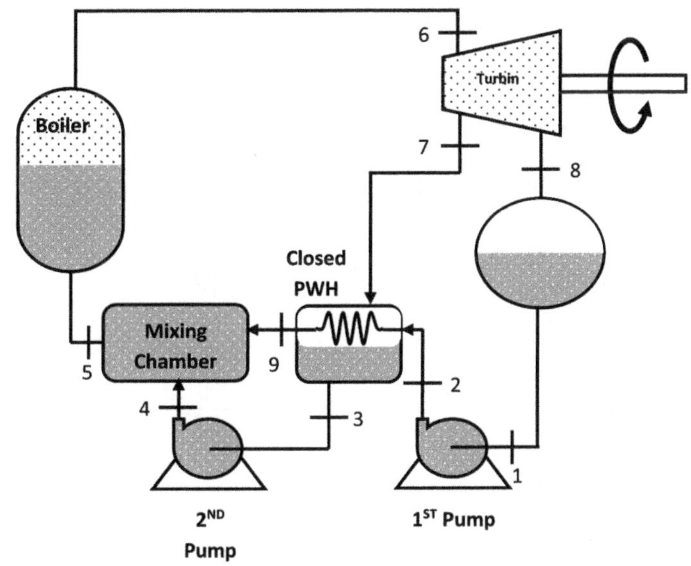

also helps control the large volume flow rate of the steam at the final stages of the turbine (due to the large specific volumes at low pressures). (Cengel & Boles, 2011).

Mass and Energy Balance for a Steam Power Cycle Used for Electricity Generation

In order to facilitate the application of the thermodynamic principle to power cycles, the authors assume steady flow process in which the properties at any point in the flow system are average values applicable to the entire cross section. A flow process constitutes an open system through which the working fluid enters and leaves the various components, such as the turbine and the boiler of a system. Energy interaction in the form of heat and work, internal energy, and gravitational and kinetic energy may also occur. The total energy of a system is the summation of the enthalpy, potential energy, and kinetic energy. Heat and work are defined as energy in transit.

A flow is said to be steady when the fluid parameters at any point in the control volume remain constant with respect to time. Thus, the volume V, mass m, and energy E remain constant. The authors adapted the equations from Cengel and Boles's (2011) work.

For mass balance (conservation of mass principle) in steady flow process:

$$m_1 = m_2 \tag{1}$$

The energy balance can be expressed as:

$$E_1 - E_2 = \frac{dE_{sys}}{dt} \tag{2}$$

$$\frac{dE_{sys}}{dt} = rate\,of\,change\,of\,internal, kinetic, and\,potential\,energies$$

The energy balance of the steady flow process involves heat transfer (Q), flow work (Pv), kinetic energy ($KE = \frac{v^2}{2}$), potential energy ($PE = gz$), internal energy (u), and shaft work (W_s). Assuming steady state for the process and excluding chemical changes that may occur in the working fluid, then:

$$m_1\left(u_1 + P_1v_1 + \frac{v_1^2}{2} + gz_1\right) + Q = m_2\left(u_2 + P_2v_2 + \frac{v_2^2}{2} + gz_2\right) + W_s \qquad (3)$$

For steady flows, $m_1 = m_2 = m$:

$$m\left(u_1 + P_1v_1 + \frac{v_1^2}{2} + gz_1\right) + Q = m\left(u_2 + P_2v_2 + \frac{v_2^2}{2} + gz_2\right) + W_s \qquad (4)$$

As enthalpy $h = u + pv$, by substitution:

$$m\left(h_1 + \frac{v_1^2}{2} + gz_1\right) + Q = m\left(h_2 + \frac{v_2^2}{2} + gz_2\right) + W_s \qquad (5)$$

On unit mass basis:

$$\left(h_1 + \frac{v_1^2}{2} + gz_1\right) + q = \left(h_2 + \frac{v_2^2}{2} + gz_2\right) + w_s \qquad (6)$$

$$\dot{m}\left(h_1 + \frac{v_1^2}{2} + gz_1\right) + \dot{Q} = \dot{m}\left(h_2 + \frac{v_2^2}{2} + gz_2\right) + \dot{W}_s \qquad (7)$$

$$\dot{m}\ rate\,of\,fliud\,flow\left(\frac{kg}{s}\right)$$

$$\dot{Q}\ rate\,of\,heat\,transfer\left(\frac{kJ}{s}\right)$$

$$\dot{W}_s \; rate\,of\;shaft\;work\left(\frac{kJ}{s}\right)$$

$$\dot{Q} - \dot{W}_s = \left(h_2 + \frac{v_2^2}{2} + gz_2\right) - \left(h_1 + \frac{v_1^2}{2} + gz_1\right) \tag{8}$$

The rate of shaft work (power) is also calculated from the difference between the enthalpies at inlet and outlet multiplied by the mass flow rate. Subscripts 1 and 2 represent inlet and outlet parameters in an isentropic process.

Exergy of a Power Plant

Exergy is a thermodynamic property and it is used to determine the quality of energy and the work potential from various sources of energy. Exergy of a thermodynamic process shows the efficiency or inefficiency of that process. Exergy provides a better understanding of processes for qualifying energy (Dhana, 2016). By analyzing the exergy destroyed for each component in a process, design engineers see areas on which efforts should be focused to improve the efficiency of the system. Exergy analysis always takes into consideration the reversible and the useful work which have been done. In ideal engineering processes, the reversible work done is equal to the useful work, while in actual processes the difference between the reversible work done and the actual work done is termed irreversibility. Irreversibility is termed as the wasted work potential or exergy destroyed. Exergy destruction occurs in the forms of friction, heat transfer through finite temperature difference, and processes that occur spontaneously such as chemical reactions and unrestrained expansion. An entropy generation process can be termed as an exergy destruction process. Exergy can be transferred during a system process by heat, work, and mass.

Exergy transfer by heat is expressed as:

$$X_{heat} = \eta_{rev} \times Q \tag{9}$$

η_{rev} represents the efficiency of a reversible engine, which is the maximum possible efficiency (Carnot engine). It is also the fraction of energy from a heat source at a temperature T that can be converted to work. Q represents the transferred heat. Therefore:

$$X_{heat} = \left(1 - \frac{T_0}{T}\right)Q \tag{10}$$

where

$$\left(1 - \frac{T_0}{T}\right) \tag{11}$$

represents the Carnot efficiency.

Exergy transfer by work W is expressed as:

$$X_{work} = W \tag{12}$$

For systems that perform no work, exergy transfer by work is equal to zero, that is:

$$X_{work} = 0 \tag{13}$$

Since the total energy of a control volume is the sum of the enthalpy, kinetic energy, and potential energy, then the exergy of the system is the sum of the work potentials of the enthalpy, potential, and kinetic energies. Therefore, the exergy transfer per unit mass is given as:

$$\psi = \left(h - h_0\right) - T_0\left(s - s_0\right) + \frac{V^2}{2} + gz \tag{14}$$

Exergy transfer by mass is:

$$X_{mass} = m\psi \tag{15}$$

The subscript 0 represents values of thermodynamic properties corresponding to the state of the environment. It is also referred to as dead state.

Exergy Balance of a Power Plant

The amount of exergy entering a steady flow system through heat, work, and mass transfer is equal to the amount of exergy leaving the system through heat, work, and mass transfer, in addition to the destroyed exergy. The exergy balance analyses for a power plant which consists of multiple components (e.g., turbines, boilers, condensers, heaters, and reheaters) gives the quantity of energy loss (i.e., the irreversibility or exergy destroyed) in any component of the plant. This gives plant engineers a better understanding of the location of irreversibilities, and how best to redesign the component or entire plant for increased thermal efficiency and make informed decision.

$$X_1 - X_2 - X_{destroyed} = \frac{dX_{system}}{dt} = 0 \tag{16}$$

This can be expressed more explicitly as:

$$X_{heat} - X_{work} + \left(X_{mass\ inlet} - X_{mass\ outlet}\right) - X_{destroyed} = \frac{dX_{system}}{dt} = 0 \tag{17}$$

Table 1. Parameters of a typical steam turbine

Parameters	Inlet of Turbine	Outlet of Turbine
Temperature (^{0}C)	538	94.1
Pressure (MPa)	12.5	0.0825
Enthalpy (kJ/kg)	3443.5	2393.1
Flow rate (kg/hr)	705000	495581

It is further stated in the form:

$$\left(1-\frac{T_0}{T}\right)Q - W + \left(m\psi_{inlet} - m\psi_{outlet}\right) - X_{destroyed} = 0 \tag{18}$$

From Equation 18, the destroyed exergy for a component of a power plant in which no work interaction occurs and the process is adiabatic is given by:

$$X_{destroyed} = m\left(\psi_{inlet} - \psi_{outlet}\right) \tag{19}$$

For other processes, destroyed exergy is:

$$X_{destroyed} = \left(1-\frac{T_0}{T}\right)Q - W + m\left(\psi_{inlet} - \psi_{outlet}\right) \tag{20}$$

From a given RC or organic RC processing solid waste, determination of the destroyed exergy from Equation (20) and making technical improvements will bring about higher thermal and conversion efficiency.

The possible electricity generation from MSW using the RC and technical specification of a turbine as illustrated in Table 1 is analyzed.

The total power the turbine generates is the difference in enthalpies multiplied by the flow rate. Therefore:

$$W_t = h_{inlet} - h_{outlet} = 3443.5 - 2393.1 = 1{,}050kJ\ /\ kg$$

where W_t is the power output of turbine.

The electrical power production from RC is presented as:

$$1-\frac{\frac{temperature\,of\,steam\,condensation\,(K)}{temperature\,of\,steam\,superheating\,(K)}}{cycle\,isentropic\,efficiency}$$

This implies:

$$1-\frac{\frac{367.1}{811}}{0.7}=1-\frac{0.45}{0.7}=1-0.64=0.36$$

The actual process of steam expansion in the turbine is not a reversible adiabatic (isentropic) process. Therefore, the entropy at inlet of the turbine is not equal to the entropy at the exit of the turbine. In other words:

$$S_{inlet\,turbine} \neq S_{outlet\,turbine}$$

Likewise, the process of increasing the pressure of saturated water by passing through a pump is not isentropic. Isentropic efficiencies are related to work producing and work consuming devices, and for a RC the components are the turbine and the pump. The cycle isentropic efficiency for Equation 27 is the product of the isentropic efficiencies of the turbine and the pump. The value of 0.7 is generally applied.

Energy Efficiency of Incineration

The possible maximum energy that can be recovered from MSW depends on the moisture content, sensible heat, and latent heat. This energy cannot be increased easily, unless by some cheap means of lowering the moisture content of the refuse without necessarily raising the amount of input energy. Furthermore, the actual energy A_e in MSW is equal to the total energy less the energy which is required to dry the waste:

$$A_e = T_e - D_e \tag{21}$$

The energy which is required to dry the waste D_e to a constant weight (i.e., remove all the moisture in the waste) is the sensible heat which is added to the waste S_H (to raise the temperature of water in the waste to boiling point) plus the latent heat L_H (to vaporize the water):

T_e = total energy

Therefore:

$$D_e = S_H + L_H \tag{22}$$

Also:

$$A_e = T_e - (S_H + L_H) = T_e - S_H - L_H \tag{23}$$

From physics, the quantity of heat which is required to raise the temperature of water at room temperature to boiling point (sensible heat) can be written as follows:

$$Q = m \times c \times \Delta t = S_H \tag{24}$$

where

m = mass of water

$c = specific\,heat\,capacity\left(\frac{kJ}{kgK}\right)$

Δt = change in temperature

Therefore:

$$A_e = T_e - mc\Delta t - L_H \tag{25}$$

The energy efficiency of the incinerator can then be computed from:

$$\frac{A_e}{T_e} \times 100 \tag{26}$$

Depending on the composition of the waste, Equation (26) can describe the percentage of the waste that is dumped in the incinerator that can be converted to energy. Incinerators have conversion efficiencies below 50%.

SUSTAINABILITY OF WASTE MONITORING AND TREATMENT

It is very important to understand that, aiming at sustainability, the most appropriate way to manage waste is reducing waste and recycling This can be achieved by adopting an integrated WM system, which is a holistic approach that considers how to reduce, reuse, recycle, and manage waste to protect human health and the natural environment. The planning of integrated WM is a dynamic tool including aspects that range from policymaking and institutional development to technical design of integrated solutions for the handling of waste.

Other approaches to ensure sustainability of waste monitoring and treatment include:

- Social support (in the form of stakeholder engagement in all stages of the WM process).
- As an initial step, the interests and opinions of various stakeholders be identified and a mapping exercise conducted, so as to create potential drivers of change and pave way for resource mobilization over the medium to long term.

The UN-Habitat (2010) stipulates that it is critical that relevant authorities demonstrate a range of good practices in issues such as consultation, communication and involvement of users, ensuring participatory and inclusive planning, and inclusivity in siting facilities, among others. Importantly, various stakeholder involvement at the planning stage will create an atmosphere of cooperation that will result in achieving the set objectives and support for the WM policy. Otherwise, developing a WM plan with limited or even no interaction with the involved stakeholders is a sure way of failure.

Policy approaches such as Extended Producer Responsibility (EPR) need to be enforced. EPR is an efficient resource management tool whereby producers take over the responsibility for the end of life management of their used products. The policy first appeared in the early 1990's in a few European member states, especially for packaging waste, and later expanded across the EU and beyond. With this approach, actors across the packaging value chain (i.e., manufacturers, importers, and retailers) assume a significant degree of responsibility for the environmental impact of their products throughout their lifecycle. This includes products' "upstream" impact linked to the selection of materials, product design, and production processes as such, including "downstream" impact relating to the use and disposal of the products. By doing so, producers accept their responsibility when designing their products so as to minimize their lifecycle environmental impact. As a result, they assume legal and economic liability for the environmental impact of their products, starting from the design phase. Thus, EPR is about extending producers' responsibility to the post-consumer stage of a product life cycle. This could provide incentives to prevent wastes at the source, promote product design for the environment, and support the achievement of public recycling and materials management. All procedures related to hazardous and nonhazardous wastes should have controls established, and these should be monitored to anticipate, prevent, identify, and verify unintentional releases. Minimum monitoring should include physical inspections of waste generation sources, storage areas, transport apparatuses, waste-related documentation, and proper labeling.

Finally, green technologies in WM need to be embraced. One of the most sustainable approaches to WM is the WtE endeavor. WtE technologies come in different forms, offer a variety of outputs, and are in various stages of development, but they have two common objectives: To both manage waste and generate energy. For instance, digestion-based WtE technology can be deployed to extract useful energy from biodegradable organic materials and from landfill gas that is captured to reduce pollutant and GHG emissions (Kaufman et al., 2010). Equally, incineration and advanced thermal conversion of the residual waste after recycling and composting represent environmentally sound SWM options. Conventional combustion-based processes transform solid wastes into heat for direct use or further conversion into steam and electricity, while advanced conversion processes convert solids into gaseous or liquids fuels, offering broader utility. In Thailand, for instance, the development of alternative energy sources is critical. In 2011, the government set the target for 8% of the nation's total energy reduction. This has given large rise to various large-scale biogas projects. For example, the Rayong municipality has constructed a wet fed-batch high solids plant for the treatment of the organic fraction of the MSW. The plant is comprised of two systems: A process that converts waste to biogas and fertilizer, and a biogas-fired cogeneration process. Beside the organic solid waste from the municipality, the Rayong plant uses food, vegetables, and fruit wastes and night soil as waste materials. The plant has a capacity to handle 60tons of waste per day, turning up into an output of 5800 tons organic fertilizer and electricity of about 5 Million kWh.

In India, Biotech-a nodal agency of the Ministry of Non-Conventional Energy Sources in Kerala, South India-has been developing and installing biogas plants that generate biogas from domestic biodegradable waste, such as cooked food waste, vegetable waste, and waste water from kitchen. It has also developed reactors for the decentralized anaerobic digestion of organic market waste or municipal solid or slaughterhouse waste. The electricity which is generated from the decentralized plants for market wastes is used for lighting and distributed to houses. In most sub-Saharan cities, local authorities have traditionally been at the centre of managing solid wastes in their jurisdictions, and these have repeatedly demonstrated a lack of capacity to effectively meet the demand for efficient and proper waste management practices (Okot-Okumu, 2012). In South Africa for example, informal waste recycling is mostly operated by waste pickers, and the informal sector is increasingly becoming part of waste management and recycling in the city of Johannesburg (Sentime, 2011, 2014). Sentime (2014) suggested that if waste pickers can be well integrated by the municipality and formalized by the private sector, informal waste recycling can immensely contribute to waste minimization and management. However, it is important to note that disposal and treatment technologies, such as engineered sanitary landfill, composting and incineration, are now emerging in developing countries and that the problem is decreasing with the introduction of proper regulations and infrastructure. In order to be successful, however, especially in Africa and in many developing countries, highly technical and skilled labor from the developed world would be required. However, it is the best avenue through which local people, particularly "waste scavengers" can be employed and their skills be utilized.

CONCLUSION

The benefits, challenges as well as the current technologies in WM has been discussed. The major setbacks in WtE combustion as an attractive energy source are the high cost of incineration facilities and the level of sophistication that is needed in their operation. SSA countries face the problem of setting up WtE facilities, because the waste constituent is mainly organic waste with high moisture content. This makes combustion of MSW for energy generation a great task. Most investors are not encouraged to set up WtE plant in SSA because of its lack of viability. The equations presented in the sections also enable design engineers to determine the conversion efficiencies of the WtE technology before investing in them. In general, effective WM can go a long way in mitigating the effects of climate change and also improve sanitation across the continent.

REFERENCES

Adegbola, A. A., & Oladeji, O. S. (2012). Investigating groundwater quality around Olusosun dump site, Lagos State Nigeria. *India Journal of Innovations and Development, 1*, 700–705.

Agbelie, I., Bawakyilenuo, S., & Lemarie, X. (2015). *Energy waste management, African cities Ghana.* Retrieved from https://www.urbanafrica.net/urban-voices/waste-to-energy-african-cities-can-transform-their-energy-landscapes/

Arena, U. L., Zaccariello, L., & Mastellone, M. L. (2010). Fluidized bed gasification of waste-derived fuels. *Waste Management (New York, N.Y.), 30*(7), 1212–1219. doi:10.1016/j.wasman.2010.01.038 PMID:20172708

Cengel, Y. A., & Boles, M. A. (Eds.). (2011). Thermodynamics an engineering approach. Singapore, McGraw-Hill.

Chimuka, L., & Ogola, J. (2015). Leading the way: directions of MSW management in South Africa. In T. Simelane & R. Mohee (Eds.), *Future directions of MSW management in Africa* (pp. 176–199). South Africa: Africa Institute of South Africa.

Dawei, H. (2015). Concise environmental engineering. Retrieved from bookboon.com

Department for Environment, Food and Rural Affairs. (2013). *Mechanical biological treatment of MSW.* Retrieved from https://assets.publishing.service.gov.uk/government/uploads/system/uploads/attachment_data/file/221039/pb13890-treatment-solid-waste.pdf

Derek, T. (2012, June 7). Where does the world trash go? *The Atlantic.* Retrieved from https://www.theatlantic.com

Dhana, V. R. (2016). Energy and exergy analysis of thermal power plants. *International Journal of Engineering Science and Computing, 6*(8), 2636-2644.

Gary, C. Y. (2010). *Municipal solid waste to energy conversion processes.* Canada: John Wiley & Sons.

Hoqmweg, D., & Bhada, T. (2012): World Bank Report. What a waste: A global review of solid waste management. Retrieved from http://documents.worldbank.org/curated/en/302341468126264791/pdf/68135-REVISED-What-a-Waste-2012-Final-updated.pdf

Ioana, I. (2010). *Clean technology from waste management.* Retrieved from https://www.researchgate.net/profile/Ioana_Ionel/publication/228905208/inline/jsViewer/00b7d5194fd1749cce000000

Jones, J. C. (Ed.). (2014). *Thermal processing of waste.* Retrieved from bookboon.com

Kokalj, F., & Samec, N. (2013). *Combustion of MSW for power generation.* doi:10.5772/55497

Kothari, R., Tyagi, V., & Pathak, A. (2010). Waste-to-energy: A way from renewable energy sources to sustainable development. *Renewable & Sustainable Energy Reviews, 14*(9), 3164–3170. doi:10.1016/j.rser.2010.05.005

Manish, S., & Dev, R. (2015). Working fluid selection for low temperature organic rankine cycle. In ISES Solar World Congress (vol. 205).

Mata-Alvarez, J., Mace, S., & Llabres, P. (2000). Anaerobic digestion of organic solid wastes. An overview of research achievements and perspectives. *Bioresource Technology*, *74*(1), 3–16. doi:10.1016/S0960-8524(00)00023-7

Mohee, R., & Bundhoo, M. A. (2015). A comparative analysis of solid waste management in developed and developing countries. In T. Simelane & R. Mohee (Eds.), *Future directions of MSW management in Africa* (pp. 6–29). South Africa: Africa Institute of South Africa. doi:10.2307/j.ctvh8r2sj.9

National Bureau of Statistics of China. (2014). *China statistical yearbook on the environment*. Beijing: Chinese Statistics Press.

National Research Council. (2000). Waste Incineration and Public Health. Washington D.C. The National Academies Press.

Okot-Okumu, J. (2015). Solid waste management in Uganda; challenges and options. In T. Simelane & R. Mohee (Eds.), *Future directions of MSW management in Africa* (pp. 107–135). South Africa: Africa Institute of South Africa.

Oloche, O. B., Ozigis, I. I, Adeyemi, K. O & Ikpe, O. E. (2017). Construction and leakage detection of a dome type biogas digester in a village at Abuja, Nigeria. *FUOYE Journal of Engineering & Technology, 2(1)*

Pachauri, R. K., & Reisinger, A. (Eds.). (2007). *Geneva. Climate change 2007: Synthesis Report*. Switzerland: IPCC.

Palcyzinsky, R., & Scotia, N. (2002). African Development Bank Report. *Study on solid waste management options for Africa*. Retrieved from https://sswm.info/sites/default/files/reference_attachments/AFRICAN%20DEVELOPMENT%20BANK%202002%20Study%20on%20Solid%20Waste%20Management%20Options%20for%20Africa.pdf

Sentime, K. (2011). Profiling solid waste pickers: A case study of Braamfontein-Greater Johannesburg. *Africanus*, *41*, 96–111.

Sentime, K. (2014). The impact of legislative framework governing waste management and collection in South Africa. *African Geographic Review*, *33*(1), 81–93. doi:10.1080/19376812.2013.847253

Sharholy, M., Ahmad, K., Mahmood, G., & Trivedi, R. C. (2008). MSW management in Indian cities: A review. *Waste Management (New York, N.Y.)*, *28*(2), 459–467. doi:10.1016/j.wasman.2007.02.008 PMID:17433664

Sunil, K. (2016). *Municipal solid waste management in developing countries*. Boca Raton, FL: CRC.

Themelis, N., & Bourtsalas, A. (2013). *UK waste management: Growing old or growing clean?* Retrieved from https://waste-management-world.com/a/uk-waste-management-growing-old-or-growing-clean

Ulas, T., Tandukar, M., & Pavlostathis, S. G. (2011). Anaerobic biotreatment of municipal sewage sludge. In M. M. Young (Ed.), *Comprehensive biotechnology* (pp. 447–461). UK: Pergamon Press.

UN-Habitat. (2010). *Solid waste management in the world's cities*. Retrieved from https://unhabitat.org/wp-content/uploads/2015/12/SolidWaste.pdf

United Nations Economic Commission for Africa. (2009). *Africa review report on waste management.* Retrieved from http://www1.uneca.org/Portals/3/documents/AfricanReviewReport-onWasteManagementMainReport.pdf

Vögeli, Y., & Zurbrügg, C. (2008, November). *Decentralized anaerobic digestion of kitchen and market waste in developing countries – state-of-the-art in South India.* Paper presented at the Second International Symposium on Energy from Biomass and Waste, Venice, Italy.

Wilson, D. C., Rodic, L., Modak, P., Soos, R., Carpintero, A., Velis, C., ... Simonett, O. (2015). Global waste management outlook. *Waste Management & Research, 33*(12), 1049–1051. doi:10.1177/0734242X15616055 PMID:26574579

World Energy Resource. (2016). 24th Edition report. Retrieved from World Energy Council database.

Yucheng, C., & Pawłowski, A. (2012). Sewage sludge-to-energy approaches based on anaerobic digestion and pyrolysis: Brief overview and energy efficiency assessment. *Renewable & Sustainable Energy Reviews, 16*(3), 1657–1665. doi:10.1016/j.rser.2011.12.014

ADDITIONAL READING

Anita, Z. (2003). A study of residential solid waste composition and management in a selected developing country – Guyana. *Journal of Solid Waste Technology Management, 29*(1), 1–7.

Anyaoku, C., & Baroutian, C. (2018). Decentralized anaerobic digestion systems for increased utilization of biogas from MSW. *Journal of Renewable and Sustainable Energy Reviews, 90*(C), 982–991. doi:10.1016/j.rser.2018.03.009

Bhatia, S. C. (2013). Environmental chemistry. India, CBS Publishers & Distributors Pvt Ltd.

Kumar, C. R., & Majid, M. A. (2019). Sustainable waste management through waste to energy technologies in India-Opportunities and environmental impacts. *International Journal of Renewable Energy Research, 9*(1), 309–342.

Litas, G. C., Lolos, T., & Tsompanidis, C. (2016, November). *Qualitative analysis of MSW in Armenia, Croatia, Cyprus, F.Y.R.O.M. and Ukraine – Methodology and results.* Paper presented at the International Solid Waste Association World Congress, Serbia.

Omer, A. M. (2018). Efficient use of biomass and biogas for sustainable energy generation: Recent development and perspectives. *Journal of Bioanalysis & Biomedicine*, 10.

Peavy, H. S., Rowe, D. R., & Tchobanoglous, G. (Eds.). (1985). Environmental engineering. London, UK: McGraw-Hill International Editions.

Tabasova, A., Kropac, J., Kermes, V., Nemet, A., & Stehlik, P. (2012). Waste-to-energy technologies: Impact on environment. *Energy, 44*(1), 146–155. doi:10.1016/j.energy.2012.01.014

KEY TERMS AND DEFINITIONS

Calorific Value: It is the energy that can be derived from a fuel when it is combusted.

Enthalpy: It is defined as the sum of the internal and flow energies of a process, and it is also referred to as the total energy per unit mass.

Isentropic Efficiency: This is the factor that takes into account the irreversibilities in work producing or consuming device.

Power Plant: An engine that operates based on the first and second laws of thermodynamics and producing energy per second.

Thermal Efficiency: This is the quantity of heat that an engine can to convert to useful work.

Turbine: A component of a power plant capable of generating power by extracting energy from steam to drive a rotor which is situated in between electromagnets and generator.

Waste to Energy: A process whereby energy is extracted from waste through combustion or other processes before dumping the ash in landfill.

Chapter 5
Recent Advances in Fluidized Bed Gasification of Oil Palm Empty Fruit Bunches

Bemgba Bevan Nyakuma
Universiti Teknologi Malaysia, Malaysia

Olagoke Oladokun
Universiti Teknologi Malaysia, Malaysia

ABSTRACT

The chapter presents recent developments in the gasification of oil palm empty fruit bunches (EFB) through fluidized bed gasifiers. The bioenergy potential of oil palm EFB as an environmentally friendly biomass is presented. Furthermore, the chapter highlights the prospects of utilising biomass gasification technology as a practical method for valorising EFB. The successful development and deployment of gasification for oil palm EFB depends on a critical understanding of the fundamental theories of the chemical reactions, classification, and operational parameters of biomass gasifiers. Hence, the potential use of fluidized bed gasifiers for oil palm empty fruit bunches (EFB) is highlighted in detail. Next, the analysis of the fundamental theories, assumptions, and equations of fluidization critical to fluidized bed gasification of EFB is presented. The chapter concludes by highlighting the potential of oil palm EFB as a low-cost, abundant, lignocellulosic feedstock for valorisation through fluidized bed gasification.

INTRODUCTION

The cultivation of oil palm (*E. guinnensis*) and the production of crude palm oil (CPO) generate large quantities of lignocellulosic wastes in Malaysia. Over the years, the inefficient disposal and management of oil palm wastes (OPW) have presented significant challenges to the oil palm industry. Besides, the rising costs of disposal, stricter environmental regulations, and low-efficiency conversion technologies have increased calls to address the problems associated with OPW in the country. Given this, the government of Malaysia (GoM) enacted the National Biomass Strategy (NBS-2020) to explore the potential

DOI: 10.4018/978-1-7998-0369-0.ch005

of valorising OPW into green chemicals, fuels, materials and energy in Malaysia. The policy also seeks to establish a bio-based energy economy (BEE), reduce greenhouse gas emissions (GHG) emissions, and create 66,000 jobs (AIM, 2013). However, this requires a comprehensive assessment of the biomass supply chain and bioenergy potentials of OPWs. In order to achieve the objectives of the NBS-2020, it is critical to identify and estimate the accessibility, acceptability, and abundance of OPW biomass. Secondly, it is crucial to analyse and select the most cost-effective, efficient, and environmentally friendly technologies for valorising OPW.

According to the Malaysian Palm Oil Board (MPOB), about 99.85 million tonnes of Oil Palm Fresh Fruit Bunches (OPFFB) are annually processed into CPO at over 400 oil palm mills (OPM) in Malaysia (Kamyab *et al.*, 2018; MPOB, 2016). The process results in a significant generation of lignocellulosic solid wastes. It is estimated that for energy 1 kg of CPO produced, approximately 4 kg of wastes are generated during the process. The composition of lignocellulosic wastes generated typically comprises; 23% oil palm empty fruit bunches (OPEFB, 15% oil palm fibres (OPF), and 7% palm kernel shells (PKS) (Basiron, 2007; Umikalsom *et al.*, 1997). Based on the findings, the most abundant OPWs generated in Malaysian OPM are oil palm empty fruit bunches (OPEFB). The OPEFB is the brown, spikey, and fibrous residue generated from stripping the oil palm fruits from OPFFB during CPO production. Due to its high moisture content, OPEFB is bulky and hence considered a low-value OPW that requires significant pre-treatment and processing before it can be utilised for further applications.

The current strategies for disposing and managing OPEFB involve open-air combustion, burying or landfilling, along with utilization as boiler fuel, organic manure, and mulching materials in plantations. However, large stockpiles of OPWs remain unutilized (De Souza *et al.*, 2010; Hamzah *et al.*, 2016). As a result, the wastes present significant waste disposal and management problems for the oil palm industry in Malaysia. Also, the long-term accumulation and poor management of OPW can potentially result in land degradation, GHG emissions, and loss of biodiversity (Mukherjee *et al.*, 2014). However, the potentially damaging effects of OPW wastes could be addressed through the design and development of novel, renewable, and sustainable technologies. It is envisaged that the deployment of these technologies could address the current challenges of OPW and effective valorise the wastes into useful products.

Consequently, numerous studies have examined the potential of valorising OPW through various biomass conversion technologies. Examples of conversion technologies typically employed include; gasification, pyrolysis, liquefaction, torrefaction and combustion (Hosseini *et al.*, 2014; Nizamuddin *et al.*, 2016; Samiran *et al.*, 2016; Sukiran *et al.*, 2017). Based on literature review, gasification is considered one of the most efficient, timely, and cost-effective technologies for the valorising oil palm EFB in literature. In addition, fluidized bed gasification of OPWs is regarded as one of the most promising technology for the valorisation of oil palm wastes, particularly OPEFB which is the most problematic yet abundant OPW from OPM in Malaysia (Lahijani *et al.*, 2011; Lahijani, P. *et al.*, 2014; Mohammed, M. A. A. *et al.*, 2011).

Therefore, this book chapter seeks to identify and highlight recent advances in fluidized bed gasification of oil palm empty fruit bunches (OPEFB). It will also present the fundamentals of gasification chemistry, operating parameters and types of biomass gasifiers. Lastly, an overview of the recent developments in biomass gasification and the fluidized bed gasification of OPEFB will be presented.

Table 1. Major biomass gasification reactions

Reaction Type	Reaction	Enthalpy	Equation
Boudouard	$C + CO_2 \leftrightarrow 2CO$	+ 172 kJ/mol	(1)
Water-Gas (steam)	$C + H_2O \leftrightarrow CO + H_2$	+ 131 kJ/mol	(2)
Methanation	$C + 2H_2 \leftrightarrow CH_4$	– 75 kJ/mol	(3)
Partial Combustion	$C + 0.5O_2 \rightarrow CO$	– 111 kJ/mol	(4)
Combustion	$C + O_2 \rightarrow CO_2$	– 394 kJ/mol	(5)
Steam Reforming	$CH_4 + H_2O \leftrightarrow CO + 3H_2$	+ 206 kJ/mol	(6)
Dry Reforming	$CH_4 + CO_2 \leftrightarrow 2CO + 2H_2$	+247 kJ/mol	(7)
Water Gas Shift	$CO + H_2O \leftrightarrow CO_2 + H_2$	– 41 kJ/mol	(8)

BIOMASS GASIFICATION

Biomass gasification is the partial oxidation of solid lignocellulosic feedstock into a gaseous fuel called synthesis gas (Syngas) (Basu, 2010; Higman *et al.*, 2008). The process involves the medium to high temperature (600 – 1600 °C) decomposition of biomass in specially designed equipment called gasifiers. Typically, biomass gasification occurs in the presence of a gasification (reaction) medium such as air, oxygen or steam, which determine the product gas yield, composition, and by-products. The process occurs in four (4) stages, namely; drying, devolatilization (pyrolysis), char gasification, and combustion (Basu, 2010).

The reaction pathway for biomass gasification begins with drying in the temperature range 50 – 200 °C. Next, the dried fuel undergoes devolatilization from 200 – 700 °C into char, condensable, and non-condensable gases. The pyrolysis products subsequently react with the gasification medium (GM) through char gasification and combustion reactions above 700 °C to produce syngas and by-products such as tar, fixed carbon and ash (Basu, 2006, 2010; Higman *et al.*, 2008). The significant advantage of gasification is its capacity to convert biomass into syngas, fuels, and electricity (Crocker, 2010; Sims, 2002).

Chemistry of Biomass Gasification

The chemistry of biomass gasification consists of a series of complex chemical reactions described in Table 1 (Basu, 2010; Higman *et al.*, 2008; Klass, 1998).

As observed in Table 1, biomass gasification is a largely endothermic process that requires energy input supplied by exothermic reactions influenced by gasification parameters (Basu, 2010). The four most crucial biomass gasification reactions are; Boudouard, water-gas, methanation and water gas shift (Basu, 2006).

Gasification Operating Parameters

The quality, quantity, and composition of the syngas and by-products of gasification are influenced by the; gasification medium (GM), gasification temperature (GT), gasifier operating pressure (GOP), and fuel composition (Basu, 2010; Speight, 2011). However, the most critical gasification operating

Table 2. Heating value and products of the gasification medium

Medium	Parameter	Bio-Syngas Heating Value (MJ/m^3)	Tar Yield (g/m^3)
Air	ER = Equivalence Ratio	4 - 7	2 - 20
Steam	SB = Steam to Biomass Ratio	10 - 18	30 - 80
Oxygen	GR = Gasification Ratio	12 - 28	4 - 30

(Basu, 2010)

parameters are; gasification temperature (GT) and gasification medium (GM), which are subsequently discussed in detail.

Gasification Medium

This parameter significantly influences the bio-syngas yield, composition, and by-products of gasification. Furthermore, it determines the gas-solid contact mode and mode in which thermal energy is delivered to the gasifier during the process of gasification. In practice, the GM reacts with char from the devolatilization (pyrolysis) stage along with heavier hydrocarbons to yield the gasification products. The properties of the most common GMs are presented in Table 2 (Basu, 2010).

As observed, the highest heating value (HHV) of bio-syngas is derived from oxygen gasification (oxy-gasification) compared to the air and steam processes (Basu, 2010). However, the high cost of producing oxygen along with safety issues make it an unattractive gasification medium (Sims, 2002). Similarly, steam gasification results in very high tar formation, despite the HHV of the syngas produced (Basu, 2010). Hence, air is considered the most effective and less problematic gasification medium for biomass gasification.

Equivalence Ratio (ER)

This is an important design parameter for biomass gasification using air as a gasification medium. It is defined as the ratio of actual to stoichiometric air-fuel ratio, which is generally expressed as;

$$ER = \frac{Actual\,Air}{Stoichiometric\,Air} \tag{1}$$

The typical *ER* values for biomass gasification range from 0.20 to 0.40. However, at values of $ER < 0.2$ operational problems such as high tar formation, incomplete char gasification, excessive char formation and low heating value product gas (bio-syngas) are severe. Similarly, the values of $ER > 0.4$ can result in combustion and the generation of flue gases (CO_2, H_2O). Therefore, bio-syngas production can be optimised at the conditions $0.2 < \lambda < 0.4$ although the theoretical optimum is $\lambda = 0.25$ (Basu, 2010).

Steam to Biomass (S/B)

This is an essential parameter during steam gasification. It is expressed mathematically as;

$$S / B = \frac{Feed\ rate\ of\ Steam}{Feed\ rate\ of\ Biomass} \tag{2}$$

In general, S/B ratios from 0.5 – 2.5 can increase the selectivity of hydrogen (Herguido *et al.*, 1992), gasifier bed temperature and reduce tar formation (García *et al.*, 1999). Therefore, the ER and SB are considered essential parameters in gasifier design, product yield, and selectivity (Basu, 2006, 2015; Gil *et al.*, 1999). Furthermore, the GMs are critical to the temperature distribution, thermal energy and efficiency of the gasification process. Therefore, the effect of gasification temperature (GT) is highlighted in the next section of this book chapter.

Gasification Temperature

This is the most critical design parameter for biomass gasification. The endothermic nature of biomass gasification requires thermal energy input to convert biomass into gasification products. As a result, gasification typically occurs above 600 °C, typically varying from 600 °C to 1600 °C (Basu, 2010; Higman *et al.*, 2008). In general, increasing the *GT* favours the endothermic reactions; Boudouard, water-gas, steam and dry reforming, as presented in Table 1. According to *Le Chatelier's* principle, higher GT favours syngas yield, selectivity, and the overall performance of the biomass gasification process.

Nevertheless, several studies have demonstrated that high temperature gasification can result in gasifier operational problems. For example, sintering, agglomeration, and fouling have been reported during the gasification of selected biomass species (Dahlin *et al.*, 2008; Fryda *et al.*, 2008; Lahijani, P *et al.*, 2014; Lahijani *et al.*, 2011; Mettanant *et al.*, 2009). The findings demonstrate that the high temperature gasification of biomass using silica sand generates eutectics due to ash melting. As a result, the bed materials undergo agglomeration during the high temperature gasification process. Bed agglomeration results in the generation of hot spots within the gasifier, which distort the dynamics of the fluidized bed. Consequently, this can cause defluidization, poor mixing, and low yield of syngas during biomass gasification (Basu, 2006; Kunii *et al.*, 2013; Rhodes *et al.*, 2008). To address these problems, several research groups have proposed the use of catalytic bed materials such as limestone, dolomite, and olivine as bed materials for biomass gasification (Corella *et al.*, 2004; De Andres *et al.*, 2011; Devi *et al.*, 2005). Other studies have shown that high temperatures enhance the reduction or elimination of tar and other high molecular weight wastes into syngas during gasification (Abu El-Rub *et al.*, 2004; Arena *et al.*, 2009; Devi *et al.*, 2005; Ma *et al.*, 2005; Shen *et al.*, 2013).

Classification of Biomass Gasifiers

Biomass gasifiers are chemical reactors designed for the valorisation of lignocellulosic feedstock into syngas and by-products based on the theory of gasification. In principle, gasifiers are classified according to four primary criteria (Basu, 2006);

Table 3. Classification of Gasifiers

Classification Criteria	Gasifier Type	Gasifier Sub-Class	Thermal Capacity	Industrial Examples
GSC Mode	Fixed or Moving Bed	Updraft (Counter Current)	1 – 10 MW_{th}	Lurgi Dry Bottom gasifier, BGL Slagging gasifier
		Downdraft (Co-Current)	0.01 – 10 MW_{th}	
	Fluidized Bed	Bubbling Fluidized Bed, Circulating Fluidized Bed	5 – 100 MW_{th}	Winkler, Transport, U-Gas, EBARA Gasifiers.
	Entrained Flow	Coaxial Downflow, Opposed Jet	> 50 MW_{th}	Koppers-Totzek, Seimens, E-gas, MHI Gasifier.
GM	Air-Blown, Oxygen-Blown, Steam-Blown	Air or Oxy-Steam Blown Gasifier	Depends on GSC Mode, Reaction conditions.	Shell, Siemens, Integrated Gasification Combined Cycle (IGCC) Gasifiers.
GOP	Pressurized Gasifier	Atmospheric Pressure	Depends on GSC Mode, Reaction conditions.	Sasol-Lurgi Dry Bottom Gasifiers.
		High-Pressure Gasifier		
MHG	Directly	Auto-thermal, Internal	Depends on GSC Mode, Reaction conditions.	Winkler,
	Indirectly	Allothermal, External		Transport, Twin Reactor (Double Fluidized Bed).

(Basu, 2010; Demirbas, 2009; Higman et al., 2008; Spliethoff, 2010).

- Gas-Solid Contact (GSC) mode,
- Gasification Medium (GM),
- Gasifier Operating Pressure (GOP),
- Mode of Heating Gasifier (MHG).

Table 3 presents an overview of the different classes, subclasses, and examples of gasifiers based on the various outlined classification criteria (Basu, 2010; Demirbas, 2009; Higman *et al.*, 2008; Spliethoff, 2010).

The gas-solid contacting (GSC) mode is the most commonly used criteria for classifying gasifiers (Basu, 2006; Higman *et al.*, 2008). The detailed comparison of the various gasifiers and their properties are presented in Table 4. The comparative assessment of the preceding indicates that the most suitable gasifier for biomass gasification is the fluidized bed gasifiers (FBG). This is mainly due to high fuel tolerance, sizeable thermal inertia, and vigorous mixing in FBGs (Basu, 2006). These features have enhanced the attractiveness of FBGs for the development of large-scale biomass gasification projects around the globe.

Types of Fluidized Beds

Biomass gasification in fluidized beds is based on the principle of fluidization. This is the gas-solid contact process that typically occurs in either Bubbling or Circulating mode. Therefore, there are two broad classifications; bubbling fluidised bed (BFB) and circulating fluidised bed (CFB) gasifiers.

Table 4. Comparison of Gasifiers

Parameters	Fixed Bed	Fluidized Bed	Entrained Bed
Feedstock Size	< 51 mm	< 6 mm	< 0.15
Tolerance for Fines	Limited	Good	Excellent
Tolerance for Coarse	Very Good	Good	Poor
Biomass Tolerance	Low	Excellent	Poor
Exit Gas Temperature	450 – 650 °C	800 – 1000 °C	1250 - 1600 °C
Oxidant Requirement	Low	Moderate	High
Steam Requirement	High	Moderate	Low
Reaction Zone Temperature	1090 °C	800 - 1000 °C	1990 °C
Nature of Ash	Dry	Dry	Slagging
Cold Gas Efficiency	80%	89%	80%
Bioenergy Application	Small	Medium	Large
Tar Production	Moderate-High	Moderate	Negligible
Operational Problems	Tar Production and fines.	Agglomeration, Sintering	Raw gas cooling

(Basu, 2010; Higman et al., 2008)

Bubbling Fluidized Bed (BFB) *Gasifiers*

These were first introduced in 1926 to convert various feedstocks in bubbling mode at low fluidizing velocities (0.5 – 1.5 m/s). The BFB gasifier consists of a riser and distributor plate (grate) designed to support the bed materials, solid fuel particles, and uniformly dispense the gasification medium during gasification. The bed materials and solid fuel particles are rapidly heated and transformed into a fluid-like state by the velocity of the upward moving gasification medium (or fluidizing gas) in the gasifier (Kunii *et al.*, 2013; Rhodes, 2008). Consequently, the gas-solid contact between the bed materials and fluidizing gas ensures the rapid drying and devolatilization of solid fuel particles into char and gases. The final stage involves the decomposition of pyrolysis products and char gasification into non-condensable gases comprising $CO + H_2$ (syngas), CH_4, CO_2 and C_1-C_3 hydrocarbons (Basu, 2010). In general, the advantages of BFB gasifier include a high degree of mixing and high thermal inertia (Basu, 2006), although it also suffers from operational challenges. These typically include back mixing of solids, slow fluidized gas diffusion, entrainment and bed agglomeration (Basu, 2010). Nevertheless, the simplicity and practicality of BFB gasifier design make it more attractive for small scale (< 25 MW_{th}) gasification (Demirbas, 2009; Sims, 2002).

Circulating Fluidized Bed (CFB) *Gasifiers*

These are relatively newer in existence compared to the BFB gasifiers. The CFB gasifier is designed to gasify feedstocks at higher fluidizing velocities (4 – 7 m/s). This condition ensures a higher intensity of mixing between solid fuel particles and bed materials (Basu, 2006; Higman *et al.*, 2008; Kunii *et al.*, 2013). However, it can result in the entrainment of bed materials (Basu, 2006). Consequently, the design of CFB gasifiers requires a "solids circulation loop" or "solids recycle loop" to redistribute entrained materials during gasification. This modification ensures longer residence time, higher carbon conversion, and "fast fluidization" during operation (Basu, 2010). Therefore, CFB gasifiers exhibit higher thermal efficiency and are easier to scale up for large gasification projects (~100 MW_{th}) (Basu, 2006; Kunii *et al.*, 2013).

In comparison, the CFB gasifiers are less prone to problems of elutriation, back mixing, and the slow diffusion rate of fluidizing gases. However, the CFB gasifiers are prone to operational problems, including erosion, low tolerance to particles above 400 μm, high operational costs and maintenance (Basu, 2006). Therefore, bubbling fluidized bed (BFB) gasifiers are considered more practical for small-medium scale biomass gasification, particularly for experimental laboratory set-ups.

OVERVIEW OF BIOMASS GASIFICATION

Based on the theory of gasification, numerous experimental studies have examined the gasification of biomass feedstocks using various reactors (Li *et al.*, 2004; Lv *et al.*, 2004; Lv *et al.*, 2007; Narvaez *et al.*, 1996; Rapagna *et al.*, 2000; Zainal *et al.*, 2002). Other researchers have explored the mathematical modelling approach to examine the gasification of various biomass in literature (Abdelouahed *et al.*, 2012; Baruah *et al.*, 2014; Beheshti *et al.*, 2015; Billaud *et al.*, 2016; Kaushal *et al.*, 2010; Mallick *et al.*, 2018; Nikoo *et al.*, 2008; Puig-Arnavat *et al.*, 2010). Similarly, the gasification of oil palm EFB has been examined in numerous studies (Erlich, 2009; Erlich *et al.*, 2011; Ismail *et al.*, 2014; Li *et al.*, 2018; Moon *et al.*, 2017; Ogi *et al.*, 2011; Ogi *et al.*, 2013). The effects of experimental and modelling parameters on biomass gasification, tar formation, and catalytic bed materials have been extensively reviewed in literature (Abu El-Rub *et al.*, 2004; Ahmad *et al.*, 2016; Ahmed *et al.*, 2012; Alauddin *et al.*, 2010; Anis *et al.*, 2011; Asadullah, 2014; Devi *et al.*, 2003; Kumar *et al.*, 2009; Puig-Arnavat *et al.*, 2010; Ruiz *et al.*, 2013; Sutton *et al.*, 2001; Taba *et al.*, 2012).

Typically, previous studies on biomass gasification have explored biomass gasification through various parameters. Consequently, the effects of temperature, pressure, equivalence ratio, steam-to-biomass ratio, catalytic bed materials and gasification agent have been examined on dry syngas yield (Nm^3/kg), volumetric composition (vol.%), and calorific value (LHV, MJ/Nm^3). Other parametric variables explored in these studies include gasification efficiency, cold gas efficiency (CGE), and carbon conversion efficiency (CCE). The findings have demonstrated that biomass gasification generates sufficient bio-syngas yields with syngas heating values (HHV) from 2.4–17 MJ/Nm^3, carbon conversion efficiencies (CCE) from 75–99%, and cold gas efficiencies (CCE) from 50% to 82%. Furthermore, the endothermic nature of biomass gasification requires high temperatures and low pressures to enhance syngas production. However, these conditions increase the setup and operating costs, materials requirement, and input energy for the process. The high-temperature gasification of high alkali content biomass can result in agglomeration, sintering, and corrosion particularly in fluidised beds (Bartels *et al.*, 2008; Lahijani, P *et al.*, 2014; Lahijani *et al.*, 2011; Mettanant *et al.*, 2009; Nijenhuis *et al.*, 2007). In addition, tar formation is a significant problem of biomass gasification in fix bed (updraft) gasifiers.

Addressing these challenges requires considerable investments in syngas cleaning and processing of tar (Devi *et al.*, 2003; Li *et al.*, 2009). Alternatively, the tar formation during biomass gasification can be addressed by utilising entrained flow reactors, which ensure rapid mixing at high operating conditions (1400 °C, 20 – 70 bar). However, the strict fuel particle size (< 75 μm), along with the high oxygen, material and energy conditions of EFR and other reactors indicate bioenergy utilisation is currently limited to large scale operations such as integrated gasification combined cycle (IGCC) plants (Basu, 2010). Lastly, biomass properties such as moisture content, particle size, and grindability are limitations in biomass gasification. Therefore, biomass requires pre-treatment or conditioning before gasification.

RECENT DEVELOPMENTS IN FLUIDIZED BED GASIFICATION OF OPEFB

Over the years, fluidized bed gasification has become an accessible technology for the valorisation of OPW such as OPEFB. The chronology of recent developments in the fluidized bed gasification of OPEFB along with the various products, gas yields and distribution are highlighted in this section of the book chapter.

The group of Salleh *et al.* (2010) examined the fluidized bed gasification of empty fruit bunch (EFB) biochar to examine the parametric effects of temperature on the carbon conversion, along with the syngas yield, quality, and composition. The results showed that the increase in gasification temperature favoured carbon conversion, which resulted in significant production of syngas during the process. Besides, the biochar possesses similar fuel characteristics to coal and can hence be utilized as a sustainable alternative to fossil fuel in future power plants (Salleh *et al.*, 2010).

Lahijani *et al.* (2011) investigated the gasification of oil palm EFB in a pilot-scale air blown bubbling fluidized bed gasifier. Similarly, the study examined the gasification performance and effects of the fuel properties on the agglomeration of the bed materials during the high-temperature process. The findings indicated that higher temperatures enhanced the higher heating values, dry gas yield, carbon conversion and cold gas efficiencies of syngas during EFB gasification. However, the authors observed that high-temperature gasification is prone to bed agglomeration resulting in de-fluidization. However, the study showed that the problem of agglomeration could be addressed by operating the gasification at lower temperatures and equivalent ratios (Lahijani *et al.*, 2011).

Mohammed, M., Salmiaton, A., Azlina, W. W., and Amran, M. M. (2011) also examined the gasification of EFB in a 300 g h^{-1} bench-scale fluidized bed gasifier. The objective of the study was to examine the parametric effects of temperature and particle size on hydrogen-rich gas production. The findings showed that high temperatures enhanced syngas yield, whereas small particle sizes enhanced the hydrogen (H_2) composition of the syngas and heating value of the syngas. In a more detailed study, Mohammed, M. A. A. *et al.* (2011) examined the fluidized bed gasification of oil palm EFB using air as a gasification medium for H_2-rich gas production. The study examined the effects of feedstock constituents and particle size along with the gasification parameters; temperature and equivalence ratio on the syngas yield during EFB gasification. The findings indicated that the distribution of syngas during EFB gasification consists of H_2, CO, CO_2 and CH_4. Furthermore, the results showed that higher gasification temperatures and smaller particle sizes significantly enhanced the gas yield, distribution, and heating value (Mohammed, M. A. A. *et al.*, 2011).

In a separate study, Lahijani, P. *et al.* (2014) examined the gasification of EFB in fluidized bed gasifier through the use of silica sand and calcined dolomite as bed materials. The study examined the effects of gasification temperature on the yield, composition, and heating value of syngas produced from the EFB gasification process. The findings showed that the heating value, carbon conversion, and cold gas efficiency were significantly enhanced by the addition of the catalyst bed materials during gasification. However, the authors noted that agglomeration remains an operational limitation of using sand as bed materials during high temperature gasification (Lahijani, P. *et al.*, 2014). The study of densified oil palm

EFB has also been examined in the literature (Nyakuma *et al.*, 2016). The authors examined the gasification of EFB in 4.5 kW allothermal fluidized bed gasifier to examine the effects of temperature and equivalence ratio on the syngas yield and composition. The findings showed that syngas composition was comprised mostly of H_2, CO, CO_2, and CH_4 along with solid biochar, which showed significantly high heating value (HHV). The results also showed that although the syngas yield was low, the problem of bed materials agglomeration reported by other studies in literature was non-existent during OPEFB pellet gasification.

Other studies have examined the gasification of oil palm EFB using temperature-programmed gasification (TPG) resulting in syngas yields and compositions of H_2, CO_2, CO, and CH_4 (Ismail *et al.*, 2011, 2012b). Ogi *et al.* (2013) investigated the gasification of OPEFB using an entrained-flow gasifier, whereas Ariffin *et al.* (2017) employed the use of a medium-scale downdraft gasifier. Mohammed *et al.* (2012) examined the effect of temperature, particle size and feedstock constituents on the kinetics and gasification of OPEFB through thermogravimetric analysis (TGA). The effect of catalyst materials and supercritical conditions on EFB gasification has also been examined in the literature (Sivasangar, Mastuli, *et al.*, 2015; Sivasangar, Zainal, *et al.*, 2015; Taufiq-Yap *et al.*, 2012). The summary of the reactors, operating parameters, empirical results and conclusions for OPEFB gasification in literature are presented in Table 5.

Based on the review of literature, it can be observed that a gasification is a practical approach for the valorisation of OPEFB. The review of the literature indicates that fluidized bed gasifiers are the commonly utilized reactors for the gasification of EFB.

Recently, the use of oxygen, steam and carbon dioxide (CO_2) has also been employed to enhance the EFB gasification process along with the metal oxide catalysts. Other researchers have also demonstrated that fixed bed, downdraft, and entrained flow gasifiers are also potentially practical reactors for the gasification of OPEFB. The outlined reactors can be operated in either continuous or batch mode under the conditions that favour high syngas, low tar, and residual products. Furthermore, the studies demonstrate that other thermochemical conversion techniques, such as torrefaction and pyrolysis, can be integrated into the energy recovery process. Likewise, the gasification of EFB is a potential route for the sustainable production of green chemicals, low carbon fuels, and renewable energy generation.

CONCLUSION

The chapter presented the fundamental theories and recent advances in fluidized bed gasification of oil palm empty fruit bunches (EFB). The detailed summary of the oil palm biomass and its bioenergy potential was also presented. Furthermore, the chapter highlighted the fundamental theories on biomass gasification, chemistry, gasification operating parameters, and classification of biomass gasifiers. The potential use of fluidized bed gasifiers for oil palm empty fruit bunches (EFB) was also presented in detail. The review of the literature demonstrated that OPEFB is a potentially low cost, abundant, lignocellulosic feedstock for valorisation into clean energy and sustainable fuels. Besides, the fluidized bed gasification is a practical route for the valorisation of EFB for future energy generation and other industrial applications.

Table 5. Summary of EFB Gasification reactors, parameters and results

Reference	Feedstock	Reactor	Reactions	Results	Findings/Conclusions
Salleh *et al.* (2010)	OPEFB Biochar	Fluidized bed gasifier	Gasification Temperature (GT) = 500-850 °C, Pressure (P) = 1 atm	Enhanced H_2 production from 5.5 to 28% (v/v), LHV Gas = 30 kJ/g, Carbon Conversion 76-84%	High temperature gasification of OPEFB favours syngas production. It is a good feedstock for renewable energy generation.
Lahijani *et al.* (2011)	OPEFB	Fluidized bed gasifier (pilot scale)	GT=650-1050 °C; P=1 atm; ER = 0.17, 0.21, 0.25 and 0.32	High calorific values 5.37, dry gas yield = 2.04, carbon conversion = 93%, cold gas efficiency = 72% obtained at 1050 °C and ER = 0.25	OPEFB gasification at high temperatures favoured syngas production, but process is prone to bed materials agglomeration. Gasification at 770 °C and ER = 0.21 can prevent bed agglomeration during gasification.
Mohammed, M., Salmiaton, A., Azlina, W. W., Amran, M. M., *et al.* (2011)	OPEFB	Fluidized Bed gasifier (bench scale)	GT=700-1000 °C; Particle size = 0.3-1.0; ER= 0.15-0.35	Gas yield = 92 wt%, H_2 = 38.02 vol%, CO = 36.36 vol%; LHV=15.25 MJ/ m^3; H_2 costs from EFB gasification is RM6.70/kg.	OPEFB gasification in bench-scale gasifier feasible, despite low yields. Cost of producing syngas by EFG gasification is practical but can be improved.
Ismail *et al.* (2011)	OPEFB /Premixed with Calcium oxide (CaO)	Temperature programmed gasification (TPG) instrument	GT=50 - 550 °C and 50-700 °C in 5% oxygen in He; Time = 1 hour.	Syngas produced along with CO_2, and CH_4. The nanocatalyst was more influential in OPEFB gasification.	The size of the catalyst significantly influenced the syngas yield. CaO resulted in low CO_2 during gasification.
Ismail *et al.* (2012a)	OPEFB /Premixed with CaO/MgO	TPG instrument	GT=50 - 700 °C in 5% oxygen (O_2) in He; Time = 1 hour	Change in a mass ratio of metal oxide to EFB significantly influenced the syngas yield.	Syngas yield was influenced by mass ratio and particle nanosize of metal oxide. Nanosized MgO significantly increased H_2 production.
Ogi *et al.* (2013)	OPEFB	Entrained flow gasifier in steam and Oxygen (O_2)	GT = 900 °C	The conversion rate in H_2O 95% and 99% in O_2.	High syngas produced from steam and oxy-fuel gasification; syngas (H_2/CO) ratios of 1.8-3.9. Low residue formation and low yield of tar comprising naphthalene and five other compounds.
Lahijani, P *et al.* (2014)	OPEFB (using sand and calcined dolomite catalyst)	Fluidized bed gasifier (pilot scale)	GT=600-850 °C	Syngas heating value = 4.48-5.33 MJ/m^3, gas yield = 1.62 - 1.68 m^3 gas/kg, carbon conversion = 82.5-83%, cold gas efficiency = 50.1-58.9% for sand and dolomite at 850 °C, respectively.	Syngas quality was enhanced at high temperatures; however, the process is prone to the agglomeration of bed materials above 800 °C using silica sand.
Nyakuma *et al.* (2016)	OPEFB pellets	Fluidized bed gasifier	GT=600-800 °C, ER = 0.25	Pelletization improved the moisture content and HHV properties of EFB. H_2 yield was higher at 600 °C with LHV of syngas.	Gasification of pelletized EFB is a practical route for the production of bioenergy, bio-syngas, and biochar.
Rahmat *et al.* (2017)	OPEFB	Entrained flow Gasifier using CO_2	GT = 700-900 °C; O_2 to Fuel ratio = 0.6-0.8	CO_2 increased the gasification reactivity, syngas ratio (H_2/CO) from 30-80%.	The CO_2 gasification process was highly efficient in terms of carbon conversion, syngas yield, and potential for liquid yield production.
Ariffin *et al.* (2017)	OPEFB	Downdraft gasifier (medium scale gasifier)	Feed rate = 126 kg/h, GT = 850-950 °C	OPEFB gasification was favoured by high temperatures resulting in high calorific syngas and cold gas efficiency.	The process was thermally efficient with low tar yield, high syngas at the selected feed rate, and temperature.
Li *et al.* (2018)	OPEFB /torrefied OPEFB	Fixed bed (Updraft), Batch Gasifier	Gasification Temp = 800 °C in Nitrogen gas	The temperature of gasification played a significant role in the process, particularly syngas yield, and the H_2/ CO was 50% higher at 680-780 ° C.	OPEFB gasification in a fixed bed reactor operated in batch mode is feasible. The process can be further enhanced by the addition of steam, which increases the H_2/ CO ratio by around 3.4%

REFERENCES

Abdelouahed, L., Authier, O., Mauviel, G., Corriou, J.-P., Verdier, G., & Dufour, A. (2012). Detailed modeling of biomass gasification in dual fluidized bed reactors under Aspen Plus. *Energy & Fuels*, *26*(6), 3840–3855. doi:10.1021/ef300411k

Abu El-Rub, Z., Bramer, E. A., & Brem, G. (2004). Review of catalysts for tar elimination in biomass gasification processes. *Industrial & Engineering Chemistry Research*, *43*(22), 6911–6919. doi:10.1021/ie0498403

Ahmad, A. A., Zawawi, N. A., Kasim, F. H., Inayat, A., & Khasri, A. (2016). Assessing the gasification performance of biomass: A review on biomass gasification process conditions, optimization and economic evaluation. *Renewable & Sustainable Energy Reviews*, *53*, 1333–1347. doi:10.1016/j.rser.2015.09.030

Ahmed, T. Y., Ahmad, M. M., Yusup, S., Inayat, A., & Khan, Z. (2012). Mathematical and computational approaches for design of biomass gasification for hydrogen production: A review. *Renewable & Sustainable Energy Reviews*, *16*(4), 2304–2315. doi:10.1016/j.rser.2012.01.035

AIM. (2013). *National Biomass Strategy 2020*. Retrieved from Kuala Lumpur, Malaysia: http://docs.wixstatic.com/ugd/43b4fe_00564483596c43d5a1619e82131f1832.pdf

Alauddin, Z. A. B. Z., Lahijani, P., Mohammadi, M., & Mohamed, A. R. (2010). Gasification of lignocellulosic biomass in fluidized beds for renewable energy development: A review. *Renewable & Sustainable Energy Reviews*, *14*(9), 2852–2862. doi:10.1016/j.rser.2010.07.026

Anis, S., & Zainal, Z. (2011). Tar reduction in biomass producer gas via mechanical, catalytic and thermal methods: A review. *Renewable & Sustainable Energy Reviews*, *15*(5), 2355–2377. doi:10.1016/j.rser.2011.02.018

Arena, U., Zaccariello, L., & Mastellone, M. L. (2009). Tar removal during the fluidized bed gasification of plastic waste. *Waste Management (New York, N.Y.)*, *29*(2), 783–791. doi:10.1016/j.wasman.2008.05.010 PMID:18693006

Ariffin, M. A., Mahmood, W. M. F. W., Harun, Z., & Mohamed, R. (2017). Medium-scale gasification of oil palm empty fruit bunch for power generation. *Journal of Material Cycles and Waste Management*, *19*(3), 1244–1252. doi:10.100710163-016-0518-8

Asadullah, M. (2014). Barriers of commercial power generation using biomass gasification gas: A review. *Renewable & Sustainable Energy Reviews*, *29*, 201–215. doi:10.1016/j.rser.2013.08.074

Bartels, M., Lin, W., Nijenhuis, J., Kapteijn, F., & Van Ommen, J. R. (2008). Agglomeration in fluidized beds at high temperatures: Mechanisms, detection and prevention. *Progress in Energy and Combustion Science*, *34*(5), 633–666. doi:10.1016/j.pecs.2008.04.002

Baruah, D., & Baruah, D. (2014). Modeling of biomass gasification: A review. *Renewable & Sustainable Energy Reviews*, *39*, 806–815. doi:10.1016/j.rser.2014.07.129

Basiron, Y. (2007). Palm oil production through sustainable plantations. *European Journal of Lipid Science and Technology*, *109*(4), 289–295. doi:10.1002/ejlt.200600223

Basu, P. (2006). *Combustion and gasification in fluidized beds*. Burlington, MA: CRC Press. doi:10.1201/9781420005158

Basu, P. (2010). Biomass gasification and pyrolysis: practical design and theory. Burlington, MA: Academic Press (Elsevier).

Basu, P. (2015). *Circulating fluidized bed boilers: design, operation and maintenance* (1st ed.). Switzerland: Springer International.

Beheshti, S., Ghassemi, H., & Shahsavan-Markadeh, R. (2015). Process simulation of biomass gasification in a bubbling fluidized bed reactor. *Energy Conversion and Management*, *94*, 345–352. doi:10.1016/j.enconman.2015.01.060

Billaud, J., Valin, S., Peyrot, M., & Salvador, S. (2016). Influence of H2O, CO2 and O2 addition on biomass gasification in entrained flow reactor conditions: Experiments and modelling. *Fuel*, *166*, 166–178. doi:10.1016/j.fuel.2015.10.046

Corella, J., Toledo, J. M., & Padilla, R. (2004). Olivine or dolomite as in-bed additive in biomass gasification with air in a fluidized bed: Which is better? *Energy & Fuels*, *18*(3), 713–720. doi:10.1021/ef0340918

Crocker, M. (2010). Thermochemical conversion of biomass to liquid fuels and chemicals. London, UK: Royal Society of Chemistry (RSC). doi:10.1039/9781849732260

Dahlin, R. S., Dorminey, J. R., Peng, W., Leonard, R. F., & Vimalchand, P. (2008). Preventing ash agglomeration during gasification of high-sodium lignite. *Energy & Fuels*, *23*(2), 785–793. doi:10.1021/ef800568z

De Andres, J. M., Narros, A., & Rodríguez, M. E. (2011). Behaviour of dolomite, olivine and alumina as primary catalysts in air–steam gasification of sewage sludge. *Fuel*, *90*(2), 521–527. doi:10.1016/j.fuel.2010.09.043

De Souza, S. P., Pacca, S., De Avila, M. T., & Borges, J. L. B. (2010). Greenhouse gas emissions and energy balance of palm oil biofuel. *Renewable Energy*, *35*(11), 2552–2561. doi:10.1016/j.renene.2010.03.028

Demirbas, A. (2009). Biofuels securing the planet's future energy needs. *Energy Conversion and Management*, *50*(9), 2239–2249. doi:10.1016/j.enconman.2009.05.010

Devi, L., Ptasinski, K. J., & Janssen, F. J. (2003). A review of the primary measures for tar elimination in biomass gasification processes. *Biomass and Bioenergy*, *24*(2), 125–140. doi:10.1016/S0961-9534(02)00102-2

Devi, L., Ptasinski, K. J., & Janssen, F. J. (2005). Pretreated olivine as tar removal catalyst for biomass gasifiers: Investigation using naphthalene as model biomass tar. *Fuel Processing Technology*, *86*(6), 707–730. doi:10.1016/j.fuproc.2004.07.001

Erlich, C. (2009). *Comparative study of residue pellets from cane sugar and palm-oil industries with commercial wood pellets, applied in downdraft gasification.* (Doctoral Thesis), KTH Royal Institute of Technology, Sweden. Retrieved from http://bit.ly/1YwDTh4

Erlich, C., & Fransson, T. H. (2011). Downdraft gasification of pellets made of wood, palm-oil residues respective bagasse: Experimental study. *Applied Energy*, *88*(3), 899–908. doi:10.1016/j.apenergy.2010.08.028

Fryda, L., Panopoulos, K., & Kakaras, E. (2008). Agglomeration in fluidised bed gasification of biomass. *Powder Technology*, *181*(3), 307–320. doi:10.1016/j.powtec.2007.05.022

García, X. A., Alarcón, N. A., & Gordon, A. L. (1999). Steam gasification of tars using a CaO catalyst source. *Fuel Processing Technology*, *58*(2), 83–102. doi:10.1016/S0378-3820(98)00087-3

Gil, J., Corella, J., Aznar, M. P., & Caballero, M. A. (1999). Biomass gasification in atmospheric and bubbling fluidized bed: Effect of the type of gasifying agent on the product distribution. *Biomass and Bioenergy*, *17*(5), 389–403. doi:10.1016/S0961-9534(99)00055-0

Hamzah, A., Salleh, S. N. M., Wong, K. K., & Sarmani, S. (2016). Nutrient amendments of inorganic fertiliser and oil palm empty fruit bunch and their influence on bacterial species dominance and degradation of the associated crude oil constituents. *Soil and Sediment Contamination: An International Journal*(just-accepted).

Herguido, J., Corella, J., & Gonzalez-Saiz, J. (1992). Steam gasification of lignocellulosic residues in a flidized bed at a small pilot scale: Effect of the type of feedstock. *Industrial & Engineering Chemistry Research*, *31*(5), 1274–1282. doi:10.1021/ie00005a006

Higman, C., & van der Burgt, M. (2008). Gasification (2nd ed.). Gulf Professional Publishing/Elsevier.

Hosseini, S. E., & Wahid, M. A. (2014). Utilization of palm solid residue as a source of renewable and sustainable energy in Malaysia. *Renewable & Sustainable Energy Reviews*, *40*, 621–632. doi:10.1016/j.rser.2014.07.214

Ismail, K., Yarmo, M. A., Taufiq-Yap, Y., & Ahmad, A. (2012a). The effect of particle size of CaO and MgO as catalysts for gasification of oil palm empty fruit bunch to produce hydrogen. *International Journal of Hydrogen Energy*, *37*(4), 3639–3644. doi:10.1016/j.ijhydene.2011.05.100

Ismail, K., Yarmo, M. A., Taufiq-Yap, Y. H., & Ahmad, A. (2011). Calcium oxide as potential catalyst for gasification of palm oil empty fruit bunch to produce syngas. In Z. A. Ahmad, M. A. Yarmo, F. H. A. Aziz, M. Y. M. Sulaiman, B. Ahmad, K. N. Ismail, & N. A. Rejab (Eds.), *X-Ray and Related Techniques* (Vol. 173, p. 178).

Ismail, K., Yarmo, M. A., Taufiq-Yap, Y. H., & Ahmad, A. (2012b). The effect of particle size of CaO and MgO as catalysts for gasification of oil palm empty fruit bunch to produce hydrogen. *International Journal of Hydrogen Energy*, *37*(4), 3639–3644. doi:10.1016/j.ijhydene.2011.05.100

Ismail, N., & Ani, F. (2014). Syngas production from microwave gasification of oil palm biochars. *Carbon*, *64*(1), 74–90.

Kamyab, H., Chelliapan, S., Din, M. F. M., Rezania, S., Khademi, T., & Kumar, A. (2018). Palm oil mill effluent as an environmental pollutant. *Palm Oil*, 13.

Kaushal, P., Abedi, J., & Mahinpey, N. (2010). A comprehensive mathematical model for biomass gasification in a bubbling fluidized bed reactor. *Fuel*, *89*(12), 3650–3661. doi:10.1016/j.fuel.2010.07.036

Klass, D. L. (1998). *Biomass for renewable energy, fuels, and chemicals*. Academic press.

Kumar, A., Jones, D. D., & Hanna, M. A. (2009). Thermochemical biomass gasification: A review of the current status of the technology. *Energies*, *2*(3), 556–581. doi:10.3390/en20300556

Kunii, D., & Levenspiel, O. (2013). Fluidization Engineering (2nd ed.). Stoneham MA: Reed-Elsevier.

Lahijani, P., & Zainal, Z. (2014). Fluidized bed gasification of palm empty fruit bunch using various bed materials. *Energy Sources. Part A, Recovery, Utilization, and Environmental Effects*, *36*(22), 2502–2510. doi:10.1080/15567036.2011.572129

Lahijani, P., & Zainal, Z. A. (2011). Gasification of palm empty fruit bunch in a bubbling fluidized bed: A performance and agglomeration study. *Bioresource Technology*, *102*(2), 2068–2076. doi:10.1016/j.biortech.2010.09.101 PMID:20980143

Lahijani, P., & Zainal, Z. A. (2014). Fluidized bed gasification of palm empty fruit bunch using various bed materials. *Energy Sources Part a-Recovery Utilization and Environmental Effects, 36*(22), 2502-2510. doi:10.1080/15567036.2011.572129

Li, C., & Suzuki, K. (2009). Tar property, analysis, reforming mechanism and model for biomass gasification—An overview. *Renewable & Sustainable Energy Reviews*, *13*(3), 594–604. doi:10.1016/j.rser.2008.01.009

Li, X., Grace, J., Lim, C., Watkinson, A., Chen, H., & Kim, J. (2004). Biomass gasification in a circulating fluidized bed. *Biomass and Bioenergy*, *26*(2), 171–193. doi:10.1016/S0961-9534(03)00084-9

Li, Y.-H., & Chen, H.-H. (2018). Analysis of syngas production rate in empty fruit bunch steam gasification with varying control factors. *International Journal of Hydrogen Energy*, *43*(2), 667–675. doi:10.1016/j.ijhydene.2017.11.117

Lv, P., Xiong, Z., Chang, J., Wu, C., Chen, Y., & Zhu, J. (2004). An experimental study on biomass air–steam gasification in a fluidized bed. *Bioresource Technology*, *95*(1), 95–101. doi:10.1016/j.biortech.2004.02.003 PMID:15207301

Lv, P., Yuan, Z., Ma, L., Wu, C., Chen, Y., & Zhu, J. (2007). Hydrogen-rich gas production from biomass air and oxygen/steam gasification in a downdraft gasifier. *Renewable Energy*, *32*(13), 2173–2185. doi:10.1016/j.renene.2006.11.010

Ma, L., Verelst, H., & Baron, G. (2005). Integrated high temperature gas cleaning: Tar removal in biomass gasification with a catalytic filter. *Catalysis Today*, *105*(3-4), 729–734. doi:10.1016/j.cattod.2005.06.022

Mallick, D., Buragohain, B., Mahanta, P., & Moholkar, V. S. (2018). Gasification of mixed biomass: analysis using equilibrium, semi-equilibrium, and kinetic models. In *Coal and Biomass Gasification* (pp. 223–241). Springer. doi:10.1007/978-981-10-7335-9_9

Mettanant, V., Basu, P., & Butler, J. (2009). Agglomeration of biomass fired fluidized bed gasifier and combustor. *Canadian Journal of Chemical Engineering*, *87*(5), 656–684. doi:10.1002/cjce.20211

Mohammed, M., Salmiaton, A., Azlina, W. W., & Amran, M. M. (2011). Gasification of empty fruit bunch for hydrogen rich fuel gas production. *Journal of Applied Sciences (Faisalabad)*, *11*(13), 2416–2420. doi:10.3923/jas.2011.2416.2420

Mohammed, M., Salmiaton, A., Azlina, W. W., Amran, M. M., Fakhru'l-Razi, A., & Taufiq-Yap, Y. (2011). Hydrogen rich gas from oil palm biomass as a potential source of renewable energy in Malaysia. *Renewable & Sustainable Energy Reviews*, *15*(2), 1258–1270. doi:10.1016/j.rser.2010.10.003

Mohammed, M. A. A., Salmiaton, A., Azlina, W. A. K. G. W., & Amran, M. S. M. (2012). Gasification of oil palm empty fruit bunches: A characterization and kinetic study. *Bioresource Technology*, *110*, 628–636. doi:10.1016/j.biortech.2012.01.056 PMID:22326334

Mohammed, M. A. A., Salmiaton, A., Azlina, W. A. K. G. W., Amran, M. S. M., & Fakhru'l-Razi, A. (2011). Air gasification of empty fruit bunch for hydrogen-rich gas production in a fluidized-bed reactor. *Energy Conversion and Management*, *52*(2), 1555–1561. doi:10.1016/j.enconman.2010.10.023

Moon, J., Jo, W., Jeong, S., Bang, B., Choi, Y., Hwang, J., & Lee, U. (2017). Gas cleaning with molten tin for hydrogen sulfide and tar in producer gas generated from biomass gasification. *Energy*, *130*, 318–326. doi:10.1016/j.energy.2017.04.121

MPOB. (2016). *Malaysian Oil Palm Statistics*. Malaysia: Retrieved from Selangor.

Mukherjee, I., & Sovacool, B. K. (2014). Palm oil-based biofuels and sustainability in southeast Asia: A review of Indonesia, Malaysia, and Thailand. *Renewable & Sustainable Energy Reviews*, *37*, 1–12.

Narvaez, I., Orio, A., Aznar, M. P., & Corella, J. (1996). Biomass gasification with air in an atmospheric bubbling fluidized bed. Effect of six operational variables on the quality of the produced raw gas. *Industrial & Engineering Chemistry Research*, *35*(7), 2110–2120. doi:10.1021/ie9507540

Nijenhuis, J., Korbee, R., Lensselink, J., Kiel, J., & van Ommen, J. R. (2007). A method for agglomeration detection and control in full-scale biomass fired fluidized beds. *Chemical Engineering Science*, *62*(1-2), 644–654. doi:10.1016/j.ces.2006.09.041

Nikoo, M. B., & Mahinpey, N. (2008). Simulation of biomass gasification in fluidized bed reactor using ASPEN PLUS. *Biomass and Bioenergy*, *32*(12), 1245–1254. doi:10.1016/j.biombioe.2008.02.020

Nizamuddin, S., Shrestha, S., Athar, S., Ali, B. S., & Siddiqui, M. A. (2016). A critical analysis on palm kernel shell from oil palm industry as a feedstock for solid char production. *Reviews in Chemical Engineering*, *32*(5), 489–505. doi:10.1515/revce-2015-0062

Nyakuma, B. B., Ahmad, A., Johari, A., Abdullah, T. A. T., Oladokun, O., & Alkali, H. (2016). Gasification of oil palm empty fruit bunches (OPEFB) briquettes for bio-syngas production. *Jurnal Teknologi*, *78*(8-3), 83-88.

Ogi, T., Nakanishi, M., & Fukuda, Y. (2011). Gasification of empty fruit bunch and bagasse using an entrained-flow mode reactor. *Journal of the Japan Institute of Energy*, *90*(9), 886–894. doi:10.3775/jie.90.886

Ogi, T., Nakanishi, M., Fukuda, Y., & Matsumoto, K. (2013). Gasification of oil palm residues (empty fruit bunch) in an entrained-flow gasifier. *Fuel*, *104*, 28–35. doi:10.1016/j.fuel.2010.08.028

Puig-Arnavat, M., Bruno, J. C., & Coronas, A. (2010). Review and analysis of biomass gasification models. *Renewable & Sustainable Energy Reviews*, *14*(9), 2841–2851. doi:10.1016/j.rser.2010.07.030

Rahmat, N. F. H., & Rasid, R. A. (2017). Gasification of empty fruit bunch with carbon dioxide in an entrained flow gasifier for syngas production. In A. Saptoro, W. S. Khur, L. S. Wei, W. P. Q. Ng, M. Anwar, C. Yeo, & K. E. Huey (Eds.), *29th Symposium of Malaysian Chemical Engineers* (Vol. 206). 10.1088/1757-899X/206/1/012013

Rapagna, S., Jand, N., Kiennemann, A., & Foscolo, P. (2000). Steam-gasification of biomass in a fluidised-bed of olivine particles. *Biomass and Bioenergy*, *19*(3), 187–197. doi:10.1016/S0961-9534(00)00031-3

Rhodes, M. J. (2008). Introduction to particle technology (2nd ed.). West Sussex, UK: John Wiley & Sons. doi:10.1002/9780470727102

Ruiz, J., Juárez, M., Morales, M., Muñoz, P., & Mendívil, M. (2013). Biomass gasification for electricity generation: Review of current technology barriers. *Renewable & Sustainable Energy Reviews*, *18*, 174–183. doi:10.1016/j.rser.2012.10.021

Salleh, M. A. M., Kisiki, N. H., Yusuf, H. M., & Ghani, W. A. W. A. K. (2010). Gasification of biochar from empty fruit bunch in a fluidized bed reactor. *Energies*, *3*(7), 1344–1352. doi:10.3390/en3071344

Samiran, N. A., Jaafar, M. N. M., Ng, J.-H., Lam, S. S., & Chong, C. T. (2016). Progress in biomass gasification technique–With focus on Malaysian palm biomass for syngas production. *Renewable & Sustainable Energy Reviews*, *62*, 1047–1062. doi:10.1016/j.rser.2016.04.049

Shen, Y., & Yoshikawa, K. (2013). Recent progresses in catalytic tar elimination during biomass gasification or pyrolysis—A review. *Renewable & Sustainable Energy Reviews*, *21*, 371–392. doi:10.1016/j.rser.2012.12.062

Sims, R. E. (2002). The brilliance of bioenergy. In *Business and in Practice*. Earthscan.

Sivasangar, S., Mastuli, M. S., Islam, A., & Taufiq-Yap, Y. H. (2015). Screening of modified CaO-based catalysts with a series of dopants for the supercritical water gasification of empty palm fruit bunches to produce hydrogen. *RSC Advances*, *5*(46), 36798–36808. doi:10.1039/C5RA03430B

Sivasangar, S., Zainal, Z., Salmiaton, A., & Taufiq-Yap, Y. H. (2015). Supercritical water gasification of empty fruit bunches from oil palm for hydrogen production. *Fuel*, *143*, 563–569. doi:10.1016/j.fuel.2014.11.073

Speight, J. G. (2011). *The biofuels handbook*. Cambridge, UK: Royal Society of Chemistry. doi:10.1039/9781849731027

Spliethoff, H. (2010). Power generation from solid fuels. Munchen, Germany: Springer Science+Business Media. doi:10.1007/978-3-642-02856-4

Sukiran, M. A., Abnisa, F., Daud, W. M. A. W., Bakar, N. A., & Loh, S. K. (2017). A review of torrefaction of oil palm solid wastes for biofuel production. *Energy Conversion and Management*, *149*, 101–120. doi:10.1016/j.enconman.2017.07.011

Sutton, D., Kelleher, B., & Ross, J. R. (2001). Review of literature on catalysts for biomass gasification. *Fuel Processing Technology*, *73*(3), 155–173. doi:10.1016/S0378-3820(01)00208-9

Taba, L. E., Irfan, M. F., Daud, W. A. M. W., & Chakrabarti, M. H. (2012). The effect of temperature on various parameters in coal, biomass and CO-gasification: A review. *Renewable & Sustainable Energy Reviews*, *16*(8), 5584–5596. doi:10.1016/j.rser.2012.06.015

Taufiq-Yap, Y. H., Sivasangar, S., & Salmiaton, A. (2012). Enhancement of hydrogen production by secondary metal oxide dopants on NiO/CaO material for catalytic gasification of empty palm fruit bunches. *Energy*, *47*(1), 158–165. doi:10.1016/j.energy.2012.09.026

Umikalsom, M. S., Ariff, A. B., Zulkifli, H. S., Tong, C. C., Hassan, M. A., & Karim, M. I. A. (1997). The treatment of oil palm empty fruit bunch fibre for subsequent use as substrate for cellulase production by Chaetomium globosum Kunze. *Bioresource Technology*, *62*(1–2), 1–9. doi:10.1016/S0960-8524(97)00132-6

Zainal, Z., Rifau, A., Quadir, G., & Seetharamu, K. (2002). Experimental investigation of a downdraft biomass gasifier. *Biomass and Bioenergy*, *23*(4), 283–289. doi:10.1016/S0961-9534(02)00059-4

Chapter 6
"Microplastics": The Next Threat to Mankind?

Asha Embrandiri
University Malaysia Kelantan, Malaysia

Shlrene Quaik
Universiti Sains Malaysia, Malaysia

Madu Ijanu Emmanuel
https://orcid.org/0000-0002-2976-8334
Universiti Sains Malaysia, Malaysia

Mariyam Rahma
Kannur University, India

Parveen Fatemeh Rupani
Jiangsu University, PR China

Mohd Hafiz Jamaludin
Universiti Malaysia Kelantan, Malaysia

Mohd Azrul Naim
International Islamic University Malaysia, Malaysia

ABSTRACT

Microplastics are plastics with smaller than 5mm particle size and they originate from sequential degradation of larger plastic molecules or are manufactured, and they have found use in many realms of life. Their gradual degradability and ingestion by aquatic organisms have become an environmental concern. Microplastics are regarded as a "tiny problem" requiring massive attention. Occurrences of microplastics have been detected in almost all environment matrices. Although several committees have taken steps towards handling the menace, most of the regulations' guidelines refer to "all wastes" in general, leaving many loopholes. This chapter views microplastics, occurrences, detection, and existing policies. The roles of industry and individuals in preserving the ecosystems are deliberated. In summary, emphasis on the bottom-up strategy to curb the escalating amount of plastics waste in our environment is sought and adoption of the "avoid the avoidable" attitude for a more holistic approach in tackling the severity of the impending threat.

DOI: 10.4018/978-1-7998-0369-0.ch006

INTRODUCTION

Occurrence of Microplastics

Plastics are undeniably a key component of our contemporary world, confronted every day at home, offices, grocery shops to the hospitals, schools and almost everywhere. However Fendell and Sewell, (2009) observed that unfortunately, over 50% large plastic items culminate in the oceans and over the course of time form macroplastic debris.

Global plastic production in 2017 according to Statistica (2017), was estimated at a whopping 348 million metric tonnes. This trend is still likely to rise as depicted in Figure 1.

Sources of Microplastics

Microplastics are defined as plastic fragments with the size of less than 5 mm. Browne et al., (2015) had however proposed that the definition should take into account smaller fragments as well (<1 mm). Their gradual degradability, ingestion by aquatic organisms and carriers of persistent organic pollutants from environment to aquatic organisms has become a major source of growing environmental concern.

Figure 2 illustrates the classification of microplastics based on shape, size and polymer type. Generally, microplastics in the marine environment are typically found as pellets, fragments, or fibers and are composed of diverse polymers (Smith et al., 2018, Galgani et al., 2013). Denser particles such as PVC, polyester, polyamide tend to settle at the bottom of the ocean bed while the lighter (polyethylene, polypropylene, and polystyrene) are found floating on the surface of the oceans.

Mega debris fall in the 100 mm range while micro-debris is below 5 mm (Ryan et al., 2009; Thompson et al., 2009). Also, they are categorized based on their production. Primary microplastics are produced for direct consumption which is found in our everyday lives whereas the secondary microplastics occur as a result of the degradation of larger plastic molecules. Table 1 shows a compilation of the worldwide production of plastic polymers in million tonnes, estimated decomposition rate and recyclability. It reveals that production of LPDE (used in production of containers for milk, cleaning agents, shampoo, boxes) is 57 million tonnes yearly, closely followed by PP (55 million tonnes). However, a good reprieve is that they are recyclable and can be used for other things long after their primary uses are over. It is alarming to know that polyamide polymers take an estimated time of 600 years to degrade and is not recyclable.

Microplastics in Fresh Water Ecosystems

Thompson et al. (2009) observed that there is limited documentation on the accumulation of plastics in freshwater organisms unlike in marine systems. Eleven streams in France were reported to have microplastics in the digestive tracts of gudgeon fish (Sanchez et al., 2014). A study conducted in the Mongolian lakes, revealed that the lakes were heavily polluted with microplastics (Christopher et al., 2014) with fragments and films as the most abundant microplastic type. Moore et al. (2005) discovered that in three Californian rivers, an average of 30-109 items m^{-3} microplastics particles were found. Even though Africa is a continent with largest and deepest freshwater lakes, till 2017 there has been no reports on studies with regards to microplastics (Khan et al., 2017). Racchid et al.'s study (2017) in the Metropolis of Paris, revealed the heavy presence of fibers which include plastic fibers as well as also other

Figure 1. Global trends of plastic production (metric tonnes).
Source: Plastics Europe 2015 (http://grid-arendal.herokuapp.com/resources/6923)
Maphoto/Riccardo Pravettoni

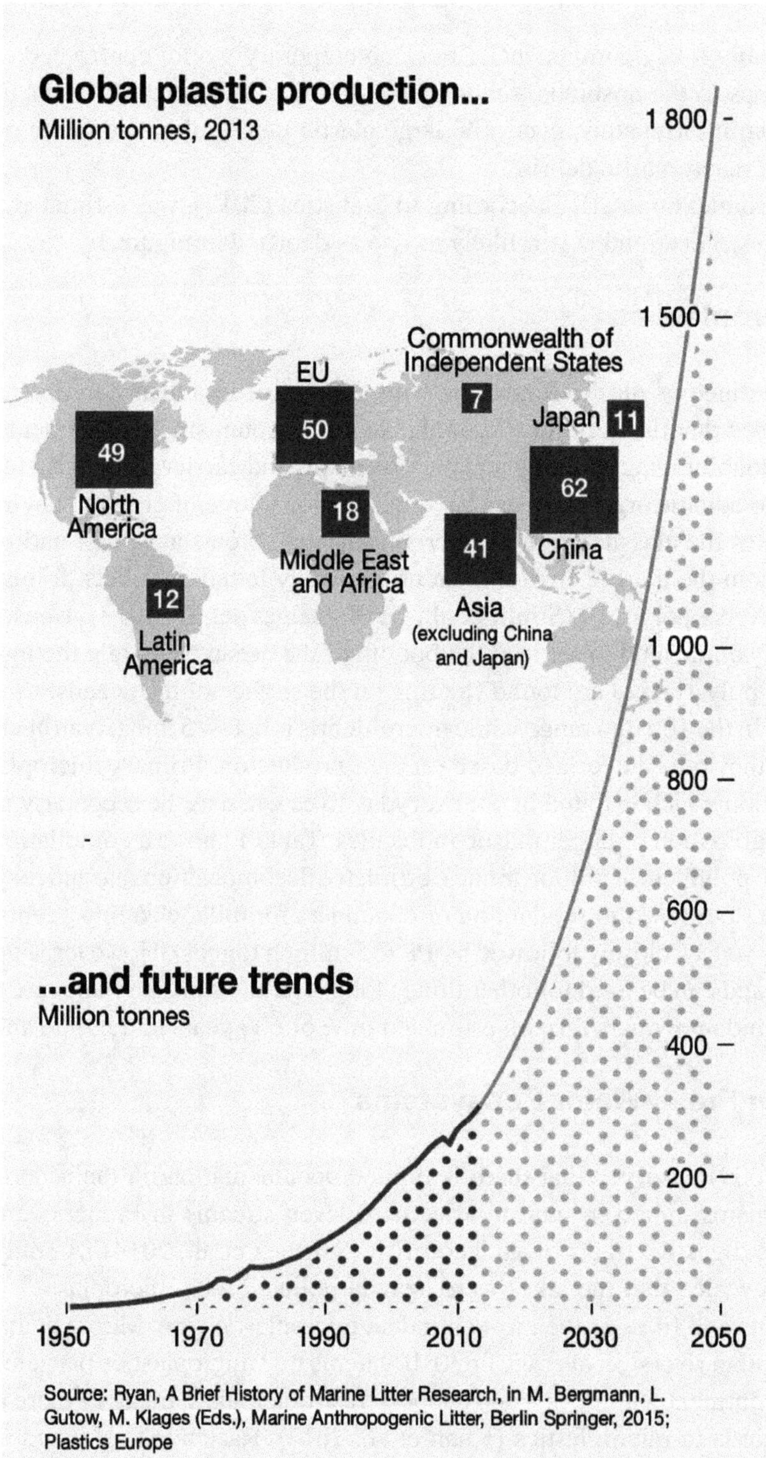

Figure 2. Classification of microplastics based on shape, size and polymer type.

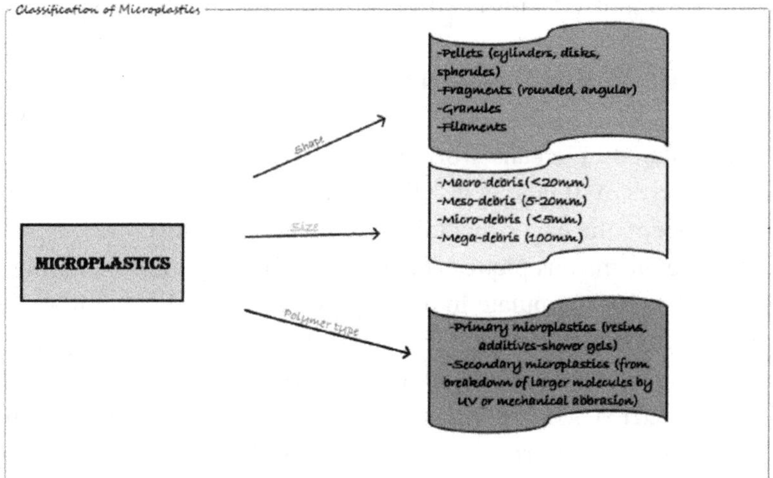

Table 1. Worldwide production of polymers (million tonnes), estimated decomposition (years) time and recyclability

Type of Microplastics Present	Production in Million Tonnes	Commercial Application	Estimated Time for Decomposition (yrs)	Recyclability
Polyamide (PA)/Nylon	42	Fishing nets and ropes, carpets, sportswear, textile	600	No
Polyvinyl chloride (PVC)	15	Film, Pipe, Insulation, roofing materials,	>450	No
Polypropylene (PP), Polyterephthalate (PET), Polyester(PES)	55 32	Rope, bottle caps, gear. Strapping, bottles, boats, textiles	450	Yes
Polystyrene (PS)	17	Cool boxes, floats, cups, utensils, take away packs.	50	Rarely
Polyethylene PE HPDE LPDE	40 57	Plastic bags, storage, Containers for milk, cleaning agents, shampoo, boxes, straws,	20	Yes
Cellulose acetate(CA)	11	Cigarette butts	10-12	No

Modified from: Enders et al.,2015, Geyer et al., 2017

synthetic fibers such as rayon. Wastewater from washing machines was postulated as a major source of these fibers which could be eliminated with an efficient wastewater treatment system (Racchid et., 2017).

Microplastics in Marine Water Systems

Microplastics have accumulated in oceans worldwide in recent years, with maximum concentrations reaching over 100,000 particles m^3 (Gallo et al., 2018). Although the intake of microplastics by marine organisms is a familiar concept, the experimental demonstration of the consequences of these ingested microplastics on marine organisms has piqued the interest of the scientific community in recent years. Low density microplastic debris accumulate in ocean gyres and pelagic invertebrates dwelling in these regions while the benthic community is a probable sink for high density microplastics (Duis and Coors 2016). Benthic suspension and deposit feeders are therefore likely to ingest sinking and sedimentary microplastics (Wright et al., 2013). Microplastics were discovered in the guts of seabirds in the 60's and have been on the increase since then (Barnes et al., 2009). A study by Amanda et al. (2014), divulged that the microplastics colonized by bacterial biofilms possess distinct taxonomic composition as compared to the bacteria present in the water surrounding them. The circulation of microplastics in the marine ecosystem is correlated to their polymer density (Graca et al., 2017). This would imply that low density plastic debris most likely recirculates between beach sediments and seawater to a greater extent unlike high density debris. In the 1990s, it was recognized that a minor source of microplastics pollution was derived from liquid hand-cleansers which was still new to the markets. However over a decade after, the average household now uses microplastic-containing products on a daily basis, as the majority of facial cleansers and personal care products are found to contain polyethylene (93%) microplastics (Lisa et al., 2009, Gouin et al., 2012). These are not often caught by sewage treatment plants and thus end up in the oceans.

Microplastics in Beach Sediments

Global occurrence of microplastics in marine and freshwater habitats has been so far documented in mostly in sandy beaches and coastal/transitional sediments (Browne et al., 2010). Zbyszewski and Corcoran. (2011) conducted one the first thorough investigation on the types, distribution, types and degradation processes of plastic particles in Lake Huron, Canada. Results indicated that polypropylene degraded much faster whereas polyethylene was more resistant to degradation under natural conditions. Likewise, Rowshyra et al. (2014) revealed that sediment samples obtained from St Lawrence river, also in Canada contained polyethylene microbeads of 0.40-2.16 mm diameter. An Italian study in 2015 reported that the beach sediments of subalpine lake was littered with plastic particles and the northern beach was more contaminated with microplastics than the southern beach due to wind actions (Imhof et al., 2015). A recent investigation of the river sediments in Shanghai, China indicated approximately 800 particles/kg dry weight of sediment (Guyu et al., 2018).

These and many more documented studies have identified benthic sediments as the major sink for microplastics and the interactions between this plastic debris and marine organisms play a key role in the circulation of plastics across the oceans (Clark et al., 2016). Nonetheless, the effects of microplastics on soil aggregation are still uncertain with regards to clogging of macropores and movement (Devriese et al., 2015).

Microplastics in Biota

Monitoring the intake of microplastics in the wild is methodologically a herculean task; nonetheless there have been an increasing number of researches reporting microplastics ingestion across the trophic levels (Browne et al., 2008). The existence of microplastics has been reported in a large number of fish species captured in the oceans, seas and freshwater (Alomar et al., 2017; Bessa et al., 2018; Jabeen et al., 2017; Morgana et al., 2018). Research also suggests that the effects of the ingestion of microplastics on smaller animals are quite similar to that of microplastics in larger animals as plastic fragments block the feeding appendages resulting in reduced food intake (Barnes et al.,2009; Fendall and Sewell, 2009). Nonetheless, a number of marine organisms (polycheate worms) possess the ability to excrete unwanted materials such as microplastics, sediments and particulates from their body without it being detrimental to them (Thomson et al., 2004). Translocation of polystyrene microspheres was first reported in rodents, humans and also mussels (Browne et al., 2008). The biological effects of some of these plasticizers in invertebrates, fish and amphibians was investigated by Oehlmann et al. (2009) revealing the impairment of development in the species. They also observed that mollusks, crustaceans and amphibians appear to be especially sensitive to these compounds. In the African Great Lakes, polyethylene, polypropylene and silicone rubber were recovered from the gastrointestinal tracts of Nile perch (*Lates niloticus*) and Nile tilapia (*Oreochromis niloticus*) fished from Lake Victoria (Farhan, 2017). The likely sources of these plastics were postulated to be from human activities such as fishing, tourism and urban waste.

Several duck species in South Africa were reported to have been exposed to microplastics detected in their faeces and feather brushings (Reynolds and Ryan, 2016). On a sadder note, Lima et al. (2015) disclosed that at certain time intervals in the north of Brazil, microplastics were of higher density than the ichthyoplankton in the Goiana Estuary. In another context, the presence of earthworms in contaminated soils increased the presence of micro plastic particles (PE) at 3.5 cm depth (Rillig et al., 2017). This is because of the feeding mechanism of earthworms which aid in breaking down of the microparticles.

During the production of plastics, certain additives referred to as "*plasticizers*" are used to alter their physical properties. These plasticizers are harmful to the living organisms and the environment. The presence of phthalate group (high percentage) and others such as epoxies, aliphatic polybrominated diphenyl ethers (PBDEs), bisphenol A etc. can result in the imbalance of hormones in aquatic life as well as the effects on mortality, abnormal reproductive and neurological development depending on the dosage of intake (Barnes et al., 2009; Lithner et al., 2009, Tullo 2015). These additives have become an environmental concern since they both extend the degradation times of plastic and in addition, leach out, introducing potentially hazardous chemicals to biota and eventually humans who are the major consumers of fish and other marine organisms.

Wastewater Treatment Plants (WWTP) as Hotspots of Microplastics

Another major source of microplastics is wastewater treatment plants (WWTP) giving rise to environmental and health concerns. Studies have reported that plastics harbor distinct microbial communities and the risk of introduction of unknown and potentially pathogenic taxa from wastewaters to freshwaters and eventually marine habitats is very high (Harrison, 2014, Mirka et al., 2018). In this context, microplastics may act as vectors for harmful additives and contaminants (Teuten et al., 2009) and they have the capability of being transferred within the food web (Setala et al., 2014). This has a direct impact on the fate of numerous toxic substances in the ecosystem. In the last decade, global surveys with regards to the occurrence and fate of microplastics (MPs) in aquatic environments have amplified. This is as a result of the increase in plastic production (Plastics Europe, 2016) and subsequently the threat of microplastics contamination in the environment. As discussed in previous sections, the sources of microplastics are numerous and most wastewater treatment plants do not have any specific system in place to check the contamination or removal of microplastics. Microplastics have been detected in both the primary and secondary treatment stages of the plants even though study showed that most microplastics are actually removed during the primary treatment stage during the skimming and sludge settling stages (Car et al., 2016). Proper functioning of the treatment plants would result in lesser contamination of water ways with microplastics. Nonetheless, in several countries, the sludge from the WWTPs is used as soil fertilizer which exposes microplastics present to different environmental conditions such as extreme heat and cold or other biological factors resulting in fragmentation of these polymers. Subsequently, the resulting microplastics most frequently wind up in drains and ultimately into water bodies. As time goes by the fate of these microplastics are becoming known to man. In addition to that an important reason for the urgent need to promote research studies on microplastics is their potential to cause eco-toxicological issues in the environment (Hidalgo-Ruz et al., 2012). Marcus et al., (2017) concluded that of all analyzed polymers in a WWTP, Polyethylene terephthalate (PET) was revealed to be most abundant, constituting 79% of the entire microplastics load and being present in all studied stages of the wastewater treatment plants.

Methods of Detection of Microplastics

Small particles of plastic were first reported in the marine environment in the early 1970s (Hidalgo-Ruz et al., 2012). The abundance of microplastics may vary with sampling sites, which is not necessarily reflective of the magnitudes of microplastics pollution, but may also be related to the pore sizes of sampling tools. Identification of micro plastics is based on the physical and chemical characterization of isolated particles in mixtures of inorganic and organic remnant particles after the extraction and clean-up steps. It is difficult to identify micro plastics of various sizes, shapes, and polymer types fully and reliably from complex environmental matrices using a single analytical method (Shim et al. 2017). Therefore, the combination of more than two analytical techniques has been widely recommended.

Visual Sorting

After separation and purification, target MPs need to be sorted from the remaining surrounding substance. Large plastics can be sorted out directly, while smaller-sized ones need further observation under a microscope (usually a stereomicroscope) after the samples are dried under cover in an oven, generally at 60 °C (Mai et al 2018). Potential target MPs are thus magnified, sorted, and counted. This method is suitable for the sorting of large MPs, but it is not always applicable and is necessary to use other instrumental techniques (e.g. Fourier Transform-Infrared Spectroscopy (FT-IR) and Raman spectroscopy) for further identification. Many researchers reported that only 68% of visually sorted particles were MPs after identification by Raman spectroscopy, and the success rates of different particles or fibers were dependent on the particle color. Fibers are easier to be identified than particles.

Thermal Analysis

The thermo-analytical technique, which measures changes in the physical and chemical properties of polymers depending on their thermal stability, has been recently tested for micro plastic identification. Differential scanning calorimetry (DSC) is a useful method for studying the thermal properties of polymeric materials (Shim et al., 2017)

Pyrolysis-Gas Chromatography/Mass Spectrophotometry (P-GC/MS) and Thermal Desorption SamplingGas Chromatography/Mass Spectrophotometry (TDS-GC/MS)

PGC/MS can simultaneously analyze both the polymer type and organic additives of MPs. Upon being extracted from environmental matrices, plastic particles are thermally deconstructed before the polymer composition of each particle is determined with GC/MS (Mai et al 2018). This technique allows only one particle at a time to go through the pyrolysis tube, which is both time consuming and limited by the aperture size of the tube.

FT-IR Spectroscopy

FTIR spectroscopy provides information on the specific chemical bonds of particles. Carbon-based polymers are easily identified with this method, and different bond compositions produce unique spectra that discriminate plastics from other organic and inorganic particles (Shim et al 2017). FT-IR and its optimized technologies, such as micro FT-IR, attenuated total reflectance (ATR) FTIR, and focal plane array detector-based micro FT-IR imaging, are also used in MP studies.

Raman Spectroscopy

Raman spectroscopy is another promising analytical technique frequently used for MP detection. The main benefits of Raman spectroscopy are that small particles down to 1 µm can be examined and that it has better responses to non-polar plastic functional groups than other analytical methods. To minimize false signals with Raman micro-spectrometry, rigorous sample purification is strongly recommended (Mai et al 2018).

Due to the non-uniformity of protocols in the detection of microplastics currently in practice, the reliability of the data is often a cause for debate (Hidalgo-Ruz et al. 2012). Therefore in order to rectify this, the standardization of methodologies for identification, quantification and ensuring the formulation of SOPs of microplastics in the marine environment must be put in place. The whole cycle of the assessment of microplastics from sampling procedures to purification and identification of microplastic particles must be considered (Loder and Gerdts 2015). This would guarantee for a reliable appraisal and the quantification of potential impacts and risks of microplastics in the oceans globally.

Degradation of Microplastics by Bacteria

Biodegradation of synthetic polymers has come into the limelight in recent years as a result of the ever increasing pressure on scientists to come up with solutions to this plastic '*menace*'. Several bacteria species have been reported to degrade plastic polymers on the assumption that bacteria are very opportunistic and have the potential to invade/adapt in harsh conditions. (Sangeetha Devi, 2019). Singh et al., (2016) reported the degradation of polyethylene by 3 strains of bacteria; *Staphylococcus sp., Pseudomonas sp.*, and *Bacillus sp.*, isolated from soil while Sivan (2011) testified that *Brevibacillus spp.* and *Bacillus spp.* were capable of degrading polyethylene as well. In other studies; Mor and Sivan (2008) and Caruso, (2015) noted that polystyrene and polyvinyl chloride (PVC) were broken down by *Rhodococcus ruber* and *Pseudomonas putida* respectively. *Arthrobacter sp. and Pseudomonas sp.* were identified as having the potential to degrade HDPE by Balasubramanian et al. (2010). In 2016, *Ideonella sakaiensis*201-F6 was discovered to have used PET as a carbon source for feeding Yoshida et al. (2016). Most recently, Alka Kumari et al., (2019) observed the destabilization of LDPE, HPDE and PVC) by AIIW2 strain.

It is therefore evident that the process of biodegradation does show potential in the reduction of plastic waste or harmful effects of the waste. This however should be given more consideration and support in order to seek for a long-lasting solution to our impending threat.

Revisiting the Existing Policies on Microplastics

Potential consequences of the increasing release of microplastics into the environment are not covered by current environmental risk assessment acts which were typically developed for chemical substances. Steensgaard et al., 2017 advocated in their analysis of macro to micro plastics that polymers should not be exempted from the Registration, Evaluation Authorisation & Restriction of Chemicals (REACH) (EC regulation, 1907/2006). If considered in the same category as hazardous waste there is bound to be stricter regulations as to their processing and lifecycle. However, the fate and effects in water and soil, and size distributions in all environmental compartments need to be analysed and quantified before adopting of the risk procedure statutes.

Thorough assessments of the potential environmental risks caused by microplastics remain to be vetted despite many damning reports and visualisation from news and social media. The UNEP came forward with several actions towards ridding our oceans of the humungous amounts of plastic debris found in the marine ecosystem. This further led to the creation of awareness and promotion of recycling plastic, proper disposal measures. In the same vein there have been several attempts by various governmental and non-governmental organizations (the United Nations Environment Program/ Mediterranean Action Plan (UNEP–MAP), the Oslo/Paris convention (for the protection of the marine environment of the North-East Atlantic (OSPAR), and the Baltic Marine Environment Protection Commission-Helsinki Commission (HELCOM) in setting guidelines for the assessment of marine litter taking account of microplastics (Lassen et al., 2015).

Cosmetics including toothpaste as well as the UK ban on microbeads in cosmetics by 2017 (UK Department for Environment, Food & Rural Affairs 2016) was another initiative towards the goal. The UNEP's report on "marine plastic debris and microplastics," included the issue in the European Marine Strategy Framework Directive, as "marine litter" (Brennholt et al., 2018)

In 2015, the California Microbead Ban bill was approved followed by the House Representatives Legislation for prohibiting the manufacture, sale, or distribution of microbeads. In France, a decree was created aimed at stopping the use of microplastic particles in cosmetics effective from 1st January 2018 and from January 2020, the use of plastic ear buds will be included to the decree. In addition, Ministry of Environment and Food of Denmark in conjunction with a chemicals Agency (KEMI) proposed a microbeads ban as well. The Joint Group of Experts on the Scientific Aspects of Marine Environmental Protection (GESAMP) encourages all nations to adopt the *Reduce– Reuse–Recycle* policy as cost effec-

Figure 3. Hierarchy of waste pyramid
Modified from US EPA (2016) and Quaik et al., (2017)

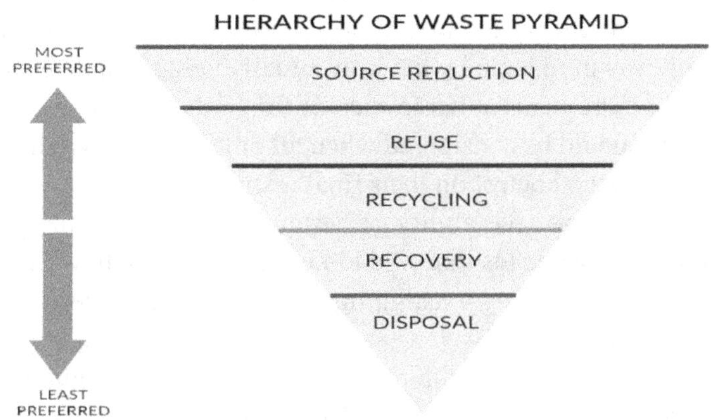

tive way of lessening the quantity of plastic objects and microplastics particles find way into the ocean (GESAMP, 2015)

In spite of all the efforts made by different countries, microplastic litter in freshwater systems has not yet been explicitly addressed in the respective regulations. Brennholt et al., (2018) observed that most of the decrees, regulations, agreements, strategies, programs, and guidelines refer to "all wastes" in general which still points to the deficit in the policy.

THE WAY FORWARD

The invasion of plastics and its derivatives in our lives has reached unprecedented levels. It has reached a point that almost everything around us is made of plastic. News and social media have also played active roles in enlightening people across the globe. From whales found with kilograms of plastic in their guts (Thailand), to the deplorable condition of Midway Atoll, North Pacific Ocean a.k.a 'Plastic island' where birds where found choked to death by plastics, "developed" countries sending off their plastic wastes in tonnes to poorer countries and so many more. Experts have located bits of degraded plastic along with synthetic fibres and microbeads lurking around our waterways, soil and even in the air. Despite all this, the question that lingers is what has been done? Has it reduced? What can be done either as an individual or collectively? These questions lead us back to two basic questions? What is being produced? How is it disposed?

There is no single or definite solution. Policy makers all over the world have instigated that the key to plastic waste problems i.e. microplastics inclusive is to go back to fundamental rethinking (3R strategy). This is supported by UNEP's 4- tiered waste management hierarchy as a guide for decision makers in waste management as illustrated in Figure 3. The efforts to combat this menace begin from the source and are the most cost effective and preferred method.

As immediate measures, the curtailing of avoidable single-use plastic items such as shopping bags, straws, food packaging, plates, spoons and water bottles would be a start in the right direction. Cutting down of single-use plastics should cut across various industries other than the food packaging industry (highest user of plastic polymers) such as construction, electronics and fabrics. Reusable items can be retrieved by fortifying the waste disposal systems and more emphasis be placed on the segregation of plastic wastes. As mentioned in the earlier section, recyclable plastics if separated at the source would help reduce the problem by great margins. In addition to this, some of these measures if practiced at home would also go a long way in reducing what is eventually being dumped. There are so many ways we can repurpose our used plastic items at home such as for gardening, storage, etc. The bottom of the pyramid lays disposal which should be considered when all other alternatives have been exhausted. It is least preferred because we have no control on their final resting place.

Although studies have mentioned the ability of certain bacteria and enzymes that are capable of breaking down certain types of plastic (as discussed in earlier section), how much of this can work for the humungous amounts of waste already existing in our oceans and if there are any side effects such as production of greenhouse gases.

In recent years, with all the hype and attention given to microplastics, multinationals have also made some efforts in this regard. IKEA has incorporated EPR (Extended Producer Responsibility) policies into their business model by endorsing material reuse and recycling throughout its production process and consumer experience. Adidas likewise partnered with Parley for the Oceans (a committee that addresses major threats towards our oceans) in 2015, to manufacture sneakers and clothing from plastic pollution in the Maldives.

Another approach to mitigation is beach cleanup programs. The International Coastal Cleanup (ICC) coordinated by the Ocean Conservancy, a US NGO, started in 1986 educates and empowers citizens all over the world to take action on behalf of the ocean. In 2018, under the umbrella of Ocean Conservancy a Beach Cleanup at Port Dickson, Malaysia resulted in 1,939.50 kilograms of trash. Reef Check Malaysia (RCM), as part of the International year of the Reef 2018 organised a cleanup campaign that recorded 14,000 plastic bottles, 6,200 plastic bags and 1,700 straws. In India, the Dadar Beach Clean Up Campaign in 2018 spearheaded by Aarna Foundation took it upon themselves to clean up the beaches of Mumbai after the Ganesh Pooja and cyclone Okhi. 100 tonnes of garbage was collected after 19 weeks of the exercise.

As much as these efforts are definitely going to reduce the overall burden of microplastics on our ocean, the solution still lies in our hands. The most effective ways will be avoidance at all costs.

Stealth Microplastics... 'Discovering the Unknown'

Many are oblivious of so called "stealth microplastics" which are hidden in plain sight but are ever present in our daily life. Right from tea bags, wet wipes, glitter, laundry detergents, to our disposable paper cups are lined with microplastics. The misconception that "disposable coffee cups" are considered recyclable is common. We automatically dispose them in the paper section of recycling bins without realizing that it is not made of 100% paper. Currently, only special recycling facility is able to process such waste which renders it unproductive for the copious amounts discarded all over the world. As expected, they end up in the landfills, which take hundreds of years to degrade. Figure 4 provides an illustration of the sources of these hidden microplastics which are still very much present in our lives despite the efforts to ban the visible ones.

Figure 4. Stealth microplastics.

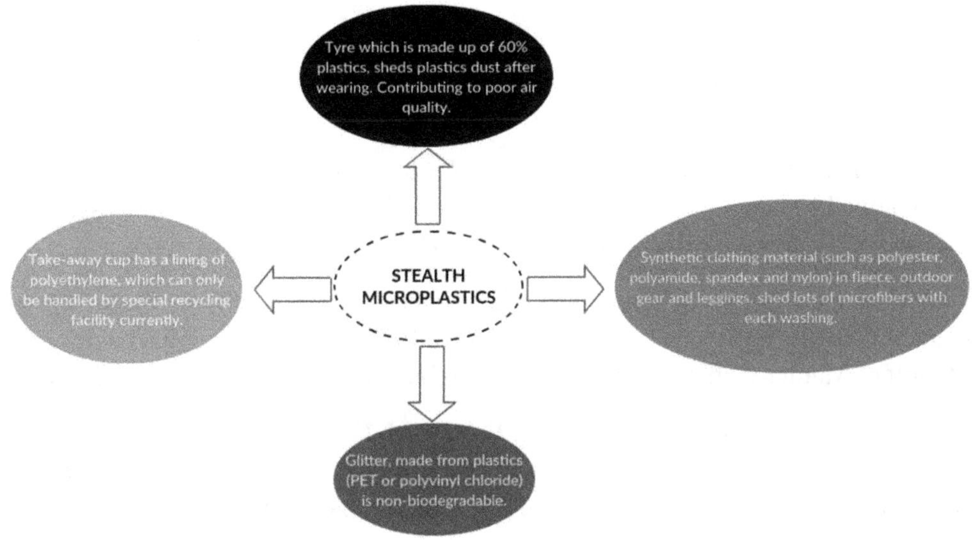

CONCLUSION

With the advancement of technology and population rise, increase in waste is eminent. The vast application of plastics in our lives has led to accumulation of huge amounts of waste which most often ends up in our oceans. The degradation of these plastics over the years has led to micro plastic pollution which has become an impending threat to mankind. A menace of huge proportions which if left to unattended would lead to severe problems all over the world. It is often referred to as tiny problems requiring massive solution.

Microplastics have invaded every nooks and crannies across the world as a result of human activities such as tourism, marine activities and normal everyday life. From densely populated regions such as China and India, to the sparsely populated areas such as the Polar Regions the presence of microplastics has been known. The distribution of microplastics is determined by environmental factors such as winds and current directions, and natural calamities. The penetration into food chains and ecosystem has resulted in the disruption in the biological systems of fauna leading to poor growth, reproductive and developmental calamities. However, knowledge on the toxicity of microplastics to humans has not been documented and there is little or no evidence to ascertain the direct effects on man other than the implied ones.

Members of the scientific community and other stakeholders have suggested the inclusion of microplastics to the current waste management handling rules and implored people to treat plastics or microplastics with great concern.

Above all, despite all the cautions and rules, the most important measure would be to follow the "avoid the avoidable" principle. Bring our own bottles, use steel cups/glass, biodegradable utensils, and alternate food packing methods. In addition, there is need for a more holistic approach in understanding the severity of the impending threat and plan for the future.

REFERENCES

Shahul Hamid, F., Bhatti, M. S., Anuar, N., Anuar, N., Mohan, P., & Periathamby, A. (2018). Worldwide distribution and abundance of microplastic: How dire is the situation? *Waste Management & Research, 36*(10), 873–897. PMID:30103651

Alomar, C., Sureda, A., Capó, X., Guijarro, B., Tejada, S., & Deudero, S. (2017). Microplastic ingestion by Mullus surmuletus Linnaeus, 1758 fish and its potential for causing oxidative stress. *Environmental Research, 159*, 135–142. doi:10.1016/j.envres.2017.07.043 PMID:28800471

Balasubramanian, V., Natarajan, K., Hemambika, B., Ramesh, N., Sumathi, C. S., Kottaimuthu, R., & Rajesh Kannan, V. (2010). High-density polyethylene (HDPE) degrading potential bacteria from marine ecosystem of Gulf of Mannar, India. *Letters in Applied Microbiology, 51*(2), 205–211. PMID:20586938

Barnes, D. K. A., Galgani, F., Thompson, R. C., & Barlaz, M. (2009). Accumulation and fragmentation of plastic debris in global environments. *Philosophical transactions of the Royal Society of London. Series B, Biological Sciences, 364*(1526), 1985–1998. doi:10.1098/rstb.2008.0205 PMID:19528051

Bessa, F., Barría, P., Neto, J. M., Frias, J. P. G. L., Otero, V., Sobral, P., & Marques, J. C. (2018). Occurrence of microplastics in commercial fish from a natural estuarine environment. *Marine Pollution Bulletin, 128*, 575–584. doi:10.1016/j.marpolbul.2018.01.044 PMID:29571409

Biginagwa, F. J., Mayoma, B. S., Shashoua, Y., Syberg, K., & Khan, F. R. (2015). First evidence of microplastics in the African Great Lakes: Recovery from Lake Victoria Nile perch and Nile tilapia. *Journal of Great Lakes Research.* doi:10.1016/j.jglr.2015.10.012

Brennholt, N., Heß, M., & Reifferscheid, G. (2018). Freshwater microplastics: challenges for regulation and management. In M. Wagner & S. Lambert (Eds.), *Freshwater microplastics. The handbook of environmental chemistry* (Vol. 58). Cham, Switzerland: Springer. doi:10.1007/978-3-319-61615-5_12

Browne, M. A., Underwood, A. J., Chapman, M. G., Williams, R., Thompson, R. C., & van Franeker, J. A. (2015). Linking effects of anthropogenic debris to ecological impacts. *Proceedings of the Royal Society B: Biological Sciences, 282*(1807), 20142929. doi:10.1098/rspb.2014.2929 PMID:25904661

Carr, S. A., Liu, J., & Tesoro, A. G. (2016). Transport and fate of microplastic particles in wastewater treatment plants. *Water Research, 91*, 174–182. doi:10.1016/j.watres.2016.01.002 PMID:26795302

Clark, J., Cole, M., Lindeque, P. K., Fileman, E., Blackford, J., Lewis, C., ... Galloway, T. S. (2016). Marine microplastic debris: A targeted plan for understanding and quantifying interactions with marine life. *Frontiers in Ecology and the Environment, 14*(6), 317–324. doi:10.1002/fee.1297

Devi, R. S., Ramya, R., Kannan, K., Antony, A. R., & Kannan, V. R. (2019). Investigation of biodegradation potentials of high density polyethylene degrading marine bacteria isolated from the coastal regions of Tamil Nadu, India. [PubMed]. *Marine Pollution Bulletin, 138*, 549–560.

Devriese, L. I., Van der Meulen, M. D., Maes, T., Bekaert, K., Paul-Pont, I., Frère, L., ... Vethaak, A. D. (2015). Microplastic contamination in brown shrimp (Crangon crangon, Linnaeus 1758) from coastal waters of the Southern North Sea and Channel area. *Marine Pollution Bulletin, 98*(1-2), 179–187. doi:10.1016/j.marpolbul.2015.06.051 PMID:26456303

Duis, K., & Coors, A. (2016). Microplastics in the aquatic and terrestrial environment: Sources (with a specific focus on personal care products), fate, and effects. *Environmental Sciences Europe*, *28*(1), 2. doi:10.118612302-015-0069-y PMID:27752437

Enders, K., Lenz, R., Stedmon, C. A., & Nielsen, T. G. (2015). (2015) Abundance, size and polymer composition of marine microplastics ≥10μm in the Atlantic Ocean and their modelled vertical distribution. *Marine Pollution Bulletin*, *100*(1), 70–81. doi:10.1016/j.marpolbul.2015.09.027 PMID:26454631

Eriksen, M., Thiel, M., Prindiville, M. M., & Kiessling, T. (2017) Microplastic: What are the Solutions. The Handbook of Environmental Chemistry. Volume 58, Freshwater Microplastics 58: 273-298.

Europe, P. (2016). *Plastics - the Facts 2016. An analysis of European plastics production, demand and waste data.* Available at http://www.plasticseurope.org/Document/plastics–the-facts-01615787.aspx?FolID1/42

European Parliament and the Council, 2006. Regulation (EC) No. 1907/2006 of the European Parliament and of the Council of 18 December 2006 Concerning the Registration, Evaluation, Authorisation and Restriction of Chemicals (REACH), Establishing a European Chemicals Agency, Amending Directive 1999/45/EC and Repealing Council Regulation (EEC) No 793/93 and Commission Regulation (EC) No 1488/94 as Well as Council Directive 76/769/EEC and Commission Directives 91/155/EEC, 93/67/EEC, 93/105/EC and 2000/21/EC. OJ L, vol. 396, pp. 1e849.

Fendall, L. S., Mary, A., & Sewell, M. A. (2009). Contributing to marine pollution by washing your face: Microplastics in facial cleansers. *Marine Pollution Bulletin*, *58*(8), 1225–1228. doi:10.1016/j.marpolbul.2009.04.025 PMID:19481226

Galgani, F., Hanke, G., Werner, S., & Oosterbaa, L. (2013). *Guidance on monitoring of marine litter in European seas. Scientific and Technical Research series*. Luxembourg: Publications Office of the European Union.

Gallo, F., Fossi, C., Weber, R., Santillo, D., Sousa, J., Ingram, I., ... Romano, D. (2018). Marine litter plastics and microplastics and their toxic chemicals components: The need for urgent preventive measures. *Environmental Sciences Europe*, *30*(1), 13. doi:10.118612302-018-0139-z PMID:29721401

Geyer, R., Jambeck, J. R., & Law, K. L. (2017). Production, use, and fate of all plastics ever made. *Science Advances*, *3*(7), e1700782. doi:10.1126ciadv.1700782 PMID:28776036

Gouin, T., Avalos, J., Brunning, I., Brzuska, K., de Graaf, J., Kaumanns, J., … Wolf, T. (2015). Use of micro-plastic beads in cosmetic products in Europe and their estimated emissions to the North Sea environment. *SOFW-J* 141, 40–46.

Guyu, P., Xu, P., Zhu, B., & Li, D. (2017). Microplastics in freshwater river sediments in Shanghai,China: A case study of risk assessment in megacities. *Environmental Pollution*, *234*, 448–456. PMID:29207296

Harrison, J. P., Schratzberger, M., Sapp, M., & Osborn, A. M. (2014). Rapid bacterial colonization of low-density polyethylene microplastics in coastal sediment microcosms. *BMC Microbiology*, *14*(1), 232. doi:10.118612866-014-0232-4 PMID:25245856

Hidalgo-Ruz, V., Gutow, L., Thompson, R. C., &Thiel, M. (2012). Microplastics in the marine environment: a review of the methods used for identification and quantification. *Environ. Sci. Technol. 46*(6), 3060-3075.

Imhof, H. K., Ivleva, N. P., Schmid, J., Niessner, R., & Laforsch, C. (2015). Contamination of beach sediments of a subalpine lake with microplastic particles. *Current Biology, 23*(19), R867–R868. doi:10.1016/j.cub.2013.09.001 PMID:24112978

Jabeen, K., Su, L., Li, J., Yang, D., Tong, C., Mu, J., & Shi, H. (2017). Microplastics and mesoplastics in fish from coastal and fresh waters of China. *Environmental Pollution, 221*, 141–149. doi:10.1016/j.envpol.2016.11.055 PMID:27939629

Jiang, J. (2018) Occurrence of microplastics and its pollution in the environment: A review. *Sustainable Production and Consumption* (13), 16–23.

Kershaw, P. J. (Ed.). (2015). Sources, fate, and effects of microplastics in the marine environment: a global assessment (IMO/FAO/UNESCO-IOC/UNIDO/WMO/IAEA/UN/UNEP/UNDP Joint Group of Experts on the Scientific Aspects of Marine Environmental Protection). Rep. Stud. GESAMP No. 90.

Khan, F. R., Mayoma, B. S., Biginagwa, F. J., & Syberg, K. (2017). Microplastics in inland African waters: presence, sources, and fate. *Freshwater microplastics,* 101-124.

Kumari, A., Chaudhary, D. R., & Jha, B. (2019). Destabilization of polyethylene and polyvinylchloride structure by marine bacterial strain. *Environmental Science and Pollution Research International*, 26(2), 1507-1516.

Lithner, D., Larsson, Å., & Dave, G. (2011). Environmental and health hazard ranking and assessment of plastic polymers based on chemical composition. *The Science of the Total Environment, 409*(18), 3309–3324. doi:10.1016/j.scitotenv.2011.04.038 PMID:21663944

Löder, M. G. J., & Gunnar Gerdts, G. (2015) Methodology used for the detection and identification of microplastics—a critical appraisal. In M. Bergmann, et al. (Eds.), Marine Anthropogenic Litter, 8, 201-227.

Mai, L., Bao, L. J., Shi, L., Wong, C. S., & Zeng, E. Y. (2018). A review of methods for measuring micro plastics in aquatic environments. *Environmental Science and Pollution Research International, 25*(12), 11319–11332. doi:10.100711356-018-1692-0 PMID:29536421

Mirka, L., Mohamed, C., Sillanpää, M., & Sillanpää, M. (2018). Occurrence, identification and removal of microplastic particles and fibers in conventional activated sludge process and advanced MBR technology. *Water Research, 133*, 236–246. doi:10.1016/j.watres.2018.01.049 PMID:29407704

Mor, R., & Sivan, A. (2008). Biofilm formation and partial biodegradation of polystyrene by the actinomycete *Rhodococcus ruber. Biodegradation, 19*(6), 851–858. doi:10.100710532-008-9188-0 PMID:18401686

Morgana, S., Ghigliotti, L., Estévez-Calvar, N., Stifanese, R., Wieckzorek, A., Doyle, T., ... Garaventa, F. (2018). *Microplastics in the Arctic: A case study with sub-surface water and fish samples off Northeast Greenland.* Environ Pollut. 242 (Pt B)

Oehlmann, J., Schulte-Oehlmann, U., Kloas, W., Jagnytsch, O., Lutz, I., Kusk, K. O., ... Tyler, C. R. (2009). A critical analysis of the biological impacts of plasticizers on wildlife. *Philos Trans R Soc B, 364*(1526), 2047–2062. doi:10.1098/rstb.2008.0242 PMID:19528055

Quaik, S., Embrandiri, A., Hossain, K., & Ismail, N. (2017). Giving waste a second chance. Public awareness and youth empowerment. In Environmental Issues Surrounding Human Overpopulation (pp. 1–18). Hershey, PA: IGI Global. doi:10.4018/978-1-5225-1683-5.ch009

Rillig, M. C., Ziersch, L., & Hempel, S. (2017). Microplastic transport in soil by earthworms. *Scientific Reports, 7*(1), 1362. doi:10.103841598-017-01594-7 PMID:28465618

Rowshyra, A., & Castaneda, S. A. (2014). Microplastic pollution in St Lawrence river sediments, NRC press. *Canadian Journal of Fisheries and Aquatic Sciences, 71.*

Ryan, P. G., Moore, C. J., Van Franeker, J. A., & Moloney, C. L. (2009). Monitoring the abundance of plastic debris in the marine environment. *Philos Trans R Soc B, 364*(1526), 1999–2012. doi:10.1098/rstb.2008.0207 PMID:19528052

Sanchez, W., Bender, C., & Porcher, J. M. (2014). Wild gudgeons(Gobio gobio)from French rivers are contaminated by microplastics: Preliminary study and first evidence. *Environmental Research, 128*, 98–100. doi:10.1016/j.envres.2013.11.004 PMID:24295902

Setala, O., Fleming-Lehtinen, V., & Lehtiniemi, M. (2014). Ingestion and transfer of microplastics in the planktonic food web. *Environ. Pollut, 185*, 77–83. doi:10.1016/j.envpol.2013.10.013 PMID:24220023

Shim, W. J., Hong, S. H., & Eo, S. E. (2017). Identification methods in micro plastic analysis: A review. *Analytical Methods, 9*(9), 1384–1391. doi:10.1039/C6AY02558G

Singh, G., Singh, A. K., & Bhatt, K. (2016). Biodegradation of polythenes by bacteria isolated from soil. *International Journal of Research and Development in Pharmacy and Life Sciences, 5*(2), 2056–2062.

Sivan, A. (2011). New perspectives in plastic biodegradation. *Current Opinion in Biotechnology, 22*(3), 422–426. doi:10.1016/j.copbio.2011.01.013 PMID:21356588

Smith, M., Love, D. C., Rochman, C. M., & Neff, R. A. (2018). Microplastics in seafood and the implications for human health. *Current Environmental Health Reports, 5*(3), 375–386. doi:10.100740572-018-0206-z PMID:30116998

Statistica. (2017). Retrieved from https://www.statista.com/statistics/282732/global-production-of-plastics-since-1950/

Steensgaard, I. M., Syberg, K., Rist, S., Hartmann, N. B., Boldrin, A., & Hansen, S. F. (2017). From macro- to microplastics - Analysis of EU regulation along the life cycle of plastic bags. *Environmental Pollution, 224*, 289–299. doi:10.1016/j.envpol.2017.02.007 PMID:28222979

Teuten, E. L., Saquing, J. M., Knappe, D. R. U., Barlaz, M. A., Jonsson, S., Bjorn, A., ... Takada, H. (2009). Transport and release of chemicals from plastics to the environment and to wildlife. *Philos. Transac. Roy. Soc. B Biol. Sci., 364*(1526), 2027–2045. PMID:19528054

Thompson, R. C., Swan, S. H., Moore, C. J., & vom Saal, F. S. (2009b). Our plastic age. *Philos Trans R Soc B. Biol. Sci.*, *364*(1526), 1973–1976. doi:10.1098/rstb.2009.0054

Tullo, A. H. (2015). Plasticizer makers want a piece of the Phthalates pie. *Chemical and Engineering News*, *93*(25), 16–18.

UK Department for Environment, Food, & Rural Affairs (2016) Microbead ban announced to protect sealife. Retrieved from https://www.gov.uk/government/news/microbead-ban-announced-to-protect-sealife

US Environmental Protection Agency (US EPA). (2016). Sustainable materials management: nonhazardous materials and waste management hierarchy. Retrieved from https://www.epa.gov/smm/sustainable-materials-management-non-hazardous-materials-and-waste-management-hierarchy

US Congress. (2016). Public Law 114–114, 114th Congress an act, H.R.1321 – Microbead free waters act 2015. *Public Law*, 3129–3130.

Yoshida, S., Hiraga, K., Takehana, T., Taniguchi, I., Yamaji, H., Maeda, Y., ... Oda, K. (2016). A bacterium that degrades and assimilates poly (ethylene terephthalate). *Science*, *351*(6278), 1196–1199. doi:10.1126cience.aad6359 PMID:26965627

Zbyszewski, M., & Corcoran, P. L. (2011). Distribution and degradation of fresh water plastic particles along the beaches of Lake Huron, Canada. *Water, Air, & Soil Pollution, 220*(1-4), pp. 365–372.

Chapter 7
Fundamental Theories and Kinetic Models for the Pyrolysis of Lignocellulosic Biomass Wastes

Olagoke Oladokun
University of Maiduguri, Nigeria

Bemgba B. Nyakuma
Universiti Teknologi Malaysia, Malaysia

Arshad Ahmad
Universiti Teknologi Malaysia, Malaysia

ABSTRACT

Biomass agricultural waste has a great potential for meeting part of the world energy need and is completely environmentally friendly. One conversion method is thermochemical processes and specifically, pyrolysis. Pyrolysis converts the lignocellulose waste to fuel and essential chemicals into three products: biogas, bio-oil, and biochar. However, performance issues limit the potential of lignocellulose pyrolysis such as design and operation of pyrolysis reactor for effective heat transfer from the heat source to the biomass feedstock. Therefore, this study presents the necessary tools for pyrolysis scientists and engineers in determining the optimal operation and design of lignocellulose agricultural waste pyrolysis. The tools consist of mathematical equations that govern the lignocellulose kinetics (model and model-free) and pyrolysis reactor macro and micro models. A practical model for hydrogen production from pyrolysis bio oil solidifies the viability of biomass as an energy source.

DOI: 10.4018/978-1-7998-0369-0.ch007

INTRODUCTION

Biomass is an abundant, renewable and sustainable source of energy for the future. Biomass typically includes agricultural waste residues, forest waste products, municipal solid waste and energy crops such as perennial grasses (Hayes, 2013; Liew, Hassim, & Ng, 2014; Mohammadi et al., 2011). Biomass has been burnt directly to generate heat for household, gasified to power engines or converted to biofuels and syngas by pyrolysis and other thermochemical processes (Ahmed et al., 2015; Basu, 2013; A. V. Bridgwater, 2012; Miller & Bellan, 1997). A vast amount of biomass is produced every year and of that only 5% (13.5 billion metric tonnes) are effectively converted to energy, which is equivalent to about 26% of the world's energy consumption (Basu, 2013). Figure 1 gives the biomass resources and their energy contribution (Balat, 2008; Demirbas, 2007; L. Zhang, Xu, & Champagne, 2010)).

However, the conversion of agroforest and agricultural product into bioenergy and biofuels presents significant challenges for the environment and human food supply (Oladokun et al., 2017; Sims, Mabee, Saddler, & Taylor, 2010). Consequently, the search for alternative biomass feedstocks such as perennial energy grasses must continue. Numerous studies have investigated the carbonization, torrefaction, gasification and pyrolysis potential of energy grasses such Miscanthus, Switchgrass, and Canary Reed (Floudas, Elia, & Baliban, 2012; Heaton et al., 2010; Lemus et al., 2002; Moutsoglou, 2012; Woli et al., 2011; Yang, Yan, Chen, Lee, & Zheng, 2007). Other potential energy grasses such Elephant grass, and *Imperata cylindrica* found around the world can also be utilised for bioenergy applications.

The conversion of biomass resources into solid, liquid and gaseous products such as biohydrogen through biochemical or thermochemical routes is widely reported in the literature (Monlau et al., 2013; Nyakuma et al., 2014; X. Zhang, Wang, Ma, Zhang, & Jiang, 2013). The biochemical route involves the breakdown of biomass molecules by enzymes or acid (enzymatic or acid hydrolysis) into sugar and subsequent fermentation to bioethanol (Balan et al., 2012; Basu, 2013). The process is well established and cost effective due to advances in research and optimisation of fermentation. However, biochemical processing of biomass into fuels cannot be easily integrated into existing commercial refining operation compared to thermochemical routes (Anthony V. Bridgwater, 2012). Combustion, pyrolysis and gasifi-

Figure 1. Biomass resources and energy contribution

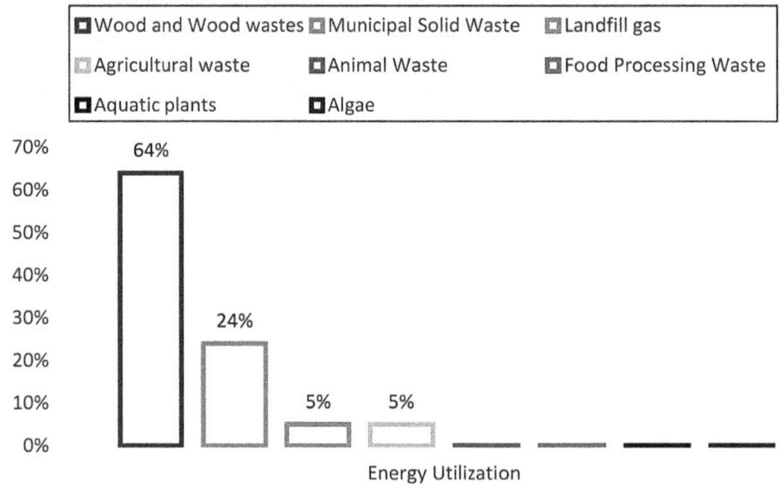

cation are thermochemical processes where lignocellulosic components of biomass are breakdown at high temperatures into gaseous (biohydrogen, biogas), liquid (biooil) and solid (char) products (Goyal, Seal, & Saxena, 2008).

Pyrolysis is an important process in biomass decomposition since it is the first step in combustion and gasification (Damartzis, Vamvuka, Sfakiotakis, & Zabaniotou, 2011). Biomass pyrolysis into solid, liquid, and gas essentially occurs at a temperature between 400 and 550 °C with the product selectivity dependent on the factors such feedstock type, temperature, pressure and heating rate (Anca-Couce & Zobel, 2012; Basu, 2013). The volatile products of pyrolysis can be either condensable or non-condensable. The non-condensable yields small proportions of hydrogen, methane, carbon monoxide, carbon dioxide. The condensable (biooil) contains large hydrogen derivable species, precursor chemicals of some important products (Basu, 2013). Biooil is a major product of fast pyrolysis and could be upgraded to a conventional transport fuel such as gasoline, diesel, kerosene, methane, LPG, alcohol and hydrogen. Furthermore, biooil contains water soluble components suitable as feedstock for steam reforming into biohydrogen (H. Xie, Yu, Wang, Shi, & Li, 2014).

Biomass Characterization

Biomass conversion to biofuel could be achieved by thermochemical conversion in combustion, gasification, and pyrolysis (Basu, 2013). Pyrolysis is the precursor of all thermochemical processes. It utilises heat energy to convert biomass into liquid (tar), gas and solid (char) products. Biomass devolatilization involves the liquefying of biomass, an important process that is vital to pyrolysis science and optimising the production of tar and char. The physical and chemical properties of a biomass, such as proximate analysis and ultimate analysis, could reveal the fuel properties (Vassilev, Baxter, Andersen, & Vassileva, 2010). Furthermore, the atomic ratio of hydrogen-carbon (H / C) and oxygen-carbon (O / C) ratio could give an indication (Basu, 2013).

Biomass Pyrolysis Kinetic Parameters

Many researchers have studied and determined the kinetic parameters of different biomass pyrolysis reactions using the kinetic models as shown in Figures 2, 3 and 4. The competing and parallel reaction model is the first of many semi-global models where there are primary and secondary reactions with lumped products such as tar, gas and char. The semi-global model for wood suggested by Shafizadeh and Chin (1977) is given in Figure 2.

Thurner and Mann (1981) simplified the mechanism by proposing that there are no secondary reactions over a narrow temperature range of 300 – 400 °C as shown in Figure 3.

Both mechanisms mentioned above assume that the overall rate is controlled by the pyrolysis reaction alone. However, this assumption limits the application of the model to low temperatures. At temperature, higher than 400 °C, the rates are high for the three primary reactions.

The single particle global reaction mechanism proposed by Broido-Shafizadeh kinetic mechanism for pyrolysis lignocellulose was modified for the char and gas ratio as suggested by Xue, Heindel, and Fox (2011) and given in Figure 4.

The model involves a series and parallel reactions with the initial conversion of cellulose, hemicellulose, and lignin from inactive to an active form at lower pyrolysis temperature by a high activation

Figure 2. Mechanism of wood by Shafizadeh and Chin (1977)

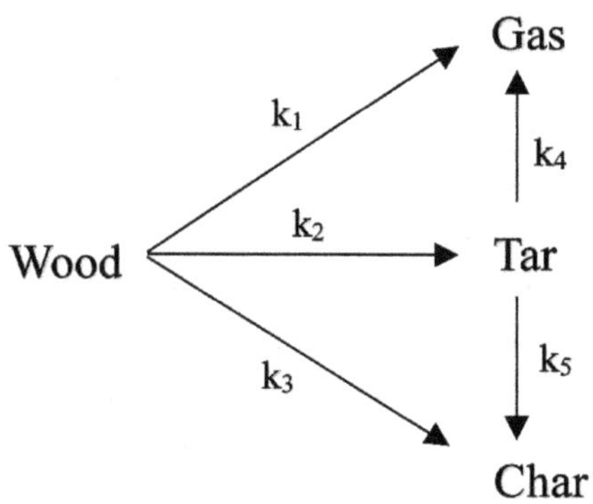

Figure 3. Mechanism of wood by Janse, de Jong, Prins, and van Swaaij (1999)

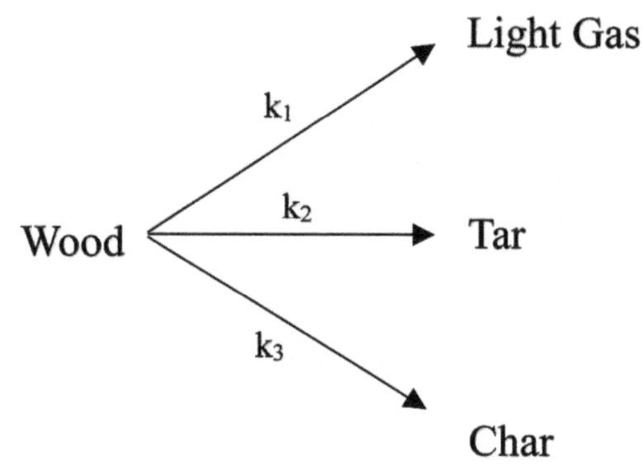

Figure 4. Biomass pyrolysis kinetic mechanism with char ratio

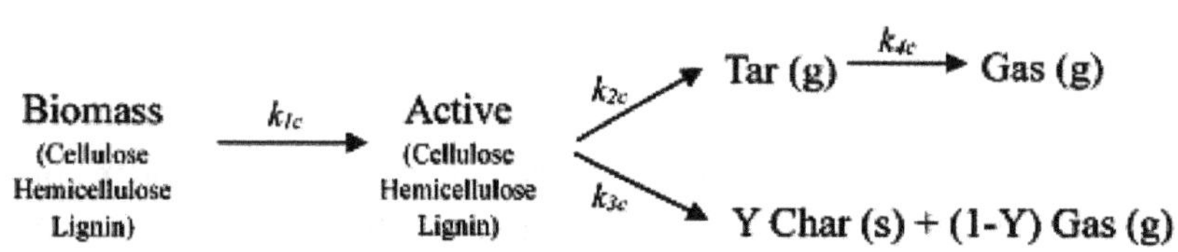

energy process. This step is immediately followed by two parallel and competitive reactions. One reaction produces tar and the tar convert to gases and the second reaction formed char and gases.

However, only one component mechanism, Figure 3, is considered in this work, and the kinetic constants for some biomass feedstock are listed in Table 1. Table 1, present feedstocks, the particle sizes used in the experiment to determine the overall devolatilization constant k and each product kinetic constants ki.

Biomass Thermal Degradation

The thermal degradation behaviour of biomass is examined through non-isothermal thermogravimetric (TG) analysis under pyrolysis conditions. Thermal analysis was carried out using the high precision NETZSCH 209 F3 thermo-microbalance. For each TGA run approximately 8-10 mg of powdered *I. cylindrica* was heated in an alumina (Al_2O_3) crucible in the temperature range 30 – 1000 °C. The heating rates was 5, 10, 15 and 20 °C/min and Nitrogen gas was used as the purge gas flowing at 20 ml/min. The resulting thermogravimetric data was analysed using the proprietary NETZSCH Proteus 6.1 thermal analysis software to acquire the derivative thermogravimetric (DTG) data as a function of temperature (T) and time (t).

Literature studies on the devolatilization process kinetics and mechanisms have generally applied "model free" or "model fitting" methods, which use non-isothermal thermogravimetric analysis (TGA) data to predict the devolatilization kinetic parameters. The model-free methods include Friedman, Kissinger, Kissinger-Akahira-Sunose (KAS) and Flynn-Wall-Ozawa (FWO), which are also known as isoconversional methods. Friedman is differential isoconversion and isothermal model (Friedman, 1964). The accuracy is limited by not using a mathematical approximation in determining equivalent conversion temperature for each heating rates (Starink, 2003). The Kissinger model uses a mathematical approxi-

Table 1. Kinetic parameters for one component mechanism

Feedstock	Experimental System	Tr (K)	Kinetic Constants: $\left[k = k_o \exp\left(-E_a / RT\right)\right]$ E_a (kJ/mol), k_0 (s-1)	Author (Ref)
Oak, Diameter: 650mm	Isothermal tube furnace	573-673	$k = 2.47 \times 10^6 \exp\left(-106.5 / RT\right)$ $k_1 = 4.12 \times 10^6 \exp\left(-112.7 / RT\right)$ $k_2 = 1.43 \times 10^4 \exp\left(-88.6 / RT\right)$ $k_3 = 7.4 \times 10^5 \exp\left(-106.5 / RT\right)$	(Thurner & Mann, 1981)
Harwood, Diameter: 300-350 μm	Isothermal entrained-flow reactor	677-822	$k = 1.483 \times 10^6 \exp\left(-89.52 / RT\right)$	(Knight, Gorton, & Kovac, 1984)
Beech, Diameter: < 80 μm	Tube Furnace	573-708	$k = 2.4 \times 10^5 \exp\left(-95.4 / RT\right)$ $k = 4.4 \times 10^9 \exp\left(-141 / RT\right)$ $k_1 = 1.1 \times 10^{10} \exp\left(-148 / RT\right)$ $k_2 = 4.4 \times 10^9 \exp\left(-153 / RT\right)$ $k_3 = 3.3 \times 10^6 \exp\left(-112 / RT\right)$	(Di Blasi & Branca, 2001)
Almond shells, Diameter: 300-500 μm	Pyroprobe 100	733-878	$k = 1.885 \times 10^6 \exp\left(-121 / RT\right)$ $k_1 = 5.85 \times 10^6 \exp\left(-119 / RT\right)$ $k_2 = 1.52 \times 10^7 \exp\left(-139 / RT\right)$ $k_3 = 2.98 \times 10^3 \exp\left(-73 / RT\right)$	(Font, Marcilla, Verdu, & Devesa, 1990)
Pine, Diameter: 100-125 μm	TGA Drop tube	553-673 773-873	$k = 1.4 \times 10^{10} \exp\left(-150 / RT\right)$ $k_1 = 9.28 \times 10^9 \exp\left(-149 / RT\right)$ $k_2 = 1.11 \times 10^{11} \exp\left(-177 / RT\right)$ $k_3 = 3.05 \times 10^7 \exp\left(-125 / RT\right)$	(Wagenaar, Prins, & van Swaaij, 1993)

mation to identify peak temperature for conversion. It assumes a constant conversion for each heating rate making it popular and easy to apply. Kissinger suffers by being non-isoconversion making it over simplified (Blaine & Kissinger, 2012). Kissinger-Akahira-Sunose (KAS) improves Kissinger method, using isoconversion with several kinetic parameters determined at different mathematically approximated conversion temperature (Blaine & Kissinger, 2012). FWO is similar to KAS with computation in reduce time and accuracy (Ozawa, 1992). The popular model fitting methods are Freeman and Carroll (1958) and Coats and Redfern (1965). The two model fitting methods require the model equation and reaction order be fixed, the activation energies and other kinetic parameters are estimated from regression by adjusting the differential rate equation to get a straight-line profile.

The model fitting method has been widely used to analyse biomass decomposition and thermochemical conversion processes and has recently found new use in coal and coal-biomass mixture kinetic studies (Várhegyi, 2007). Aboyade et al. (2011) found a hybrid application by combining model fitting and model free (Friedman method) in predicting the decomposition kinetics of corn cobs and sugar. This hybrid method deconvoluted the TGA data using 3 pseudocomponents which normally doesn't give a good quality of fit (Aboyade, Carrier, Meyer, Knoetze, & Görgens, 2012). In a study of several coal samples, Hattingh et al. (2014) applied model fitting to 3-8 pseudocomponents and found the best fit to be for 8 pseudocomponents for all the coal samples considered. Furthermore, Oladokun et al. (2015) applied the multicomponent modelling to biomass and obtained a good fit with 8 pseudocomponents.

Biomass Kinetic Model Theory

Typically, biomass kinetic analysis is carried out using the model-free isoconversion kinetic models of Flynn-Wall-Ozawa (FWO) and Kissinger-Akahira-Sunose (KAS). The models are applied to obtain the kinetic parameters; activation energy (*Ea*) and pre-exponential factor (k_o) for the degrees of conversion α = 0.10 to 0.90 for the pyrolytic decomposition of *I. cylindrica* during TG-DTG analysis. The governing equations for the FWO and KAS kinetic models are presented in Eq 1-10.

Isoconversion Model Theory

The pyrolysis decomposition can be investigated by applying the kinetic models of *Flynn-Wall-Ozawa* (FWO) and *Kissinger-Akahira-Sunose* (KAS) (Slopiecka, Bartocci, & Fantozzi, 2012). The pyrolysis of biomass species is typically described by the generic Eq (1) (Maia & de Morais, 2016);

$$Biomass \xrightarrow{k(T)} Char + Gas + Tar \tag{1}$$

Where the term *k*(T) represents the rate constant dependent on temperature described by the Arrhenius equation, Equation (2);

$$k\left(T\right) = k_o \exp\left(-\frac{E_a}{RT}\right) \tag{2}$$

where Ea, activation energy in kJ mol-1; k_o, represents the pre-exponential factor in min-1; R - universal gas constant in J. K-1 mol-1 and absolute temperature in degree K. Equation (2) typically describes the temperature dependence of biomass pyrolysis on the temperature at different conversions, α, during TG analysis. Consequently, the reaction model f(α) for the pyrolytic decomposition of biomass during TGA can be represented as;

$$\frac{d\alpha}{dt} = k(T)f(\alpha) \quad (3)$$

The degree of conversion, α is obtained by normalising TGA sample weight loss and expressed as the quotient of the difference in initial weight (m_i) and actual weight (m_t) divided by difference in initial weight (m_i) and final weight (m_f). Reaction model $f(\alpha)$ is a function representing the degree of conversion as observed in TGA weight loss. The expression of $f(\alpha)$ is not explicitly stated in Equation (4) - (8) given rise to the concept of model free in Equation (8). By separating variables and rearranging like terms Equation (3) can be re-written as;

$$\frac{d\alpha}{f(\alpha)} = k(T)dt \quad (4)$$

Combining Equation (2) – (4) the temperature dependence of rate constant on the reaction model for pyrolytic biomass decomposition is Equation (5):

$$\frac{d\alpha}{f(\alpha)} = k_o \exp\left(-\frac{E_a}{RT}\right)dt \quad (5)$$

However, to describe the non-isothermal pyrolysis of biomass at different degrees of conversion (α), the term β that describes the isoconversion decomposition at different heating rates is introduced. The term β is defined as;

$$\beta = \frac{dT}{dt} \quad (6)$$

Next, substituting Eq (6) in Eq (5) yields an expression for the pyrolysis of biomass at different heating rates is Eq (7);

$$\frac{d\alpha}{f(\alpha)} = \frac{k_o}{\beta} \exp\left(-\frac{E_a}{RT}\right)dT \quad (7)$$

By integrating Eq (7) the conversion function $g(\alpha)$ which describes the kinetic decomposition of biomass pyrolysis at different heating rates can be achieved;

$$g(\alpha) = \int_0^\alpha \frac{d\alpha}{f(\alpha)} = \frac{k_o}{\beta}\int_0^T \exp\left(-\frac{E_a}{RT}\right)dT \quad (8)$$

The expression for $g(\alpha)$ represents the fundamental model equation for the *isoconversional* kinetic decomposition of biomass species and deducing the parameters Ea, and k_o. The solution to Eq (8) can be obtained from Doyle's approximation, which forms the theoretical basis for the *isoconversional* kinetic models of *Flynn-Wall-Ozawa* (FWO) and *Kissinger-Akahira-Sunose* (KAS) (Doyle, 1965).

By implementing Doyle's approximation, the final expression for the FWO model is described in Eq (9). Consequently, the kinetic parameters *Ea*, and k_o can be deduced from the plot of *In* (β) against (1/T) at different heating rates during TG analysis (Ma, Gao, Zhong, & Xing, 2016).

$$\ln\left(\beta\right) = \ln\left(\frac{k_o E_a}{g\left(\alpha\right) R}\right) - 5.331 - 1.052\left(\frac{E_a}{RT}\right) \tag{9}$$

The value of E_a is calculated from the slope *–1.052 Ea/R*, where the value of *R* is 8.314 $J.K^{-1}.mol^{-1}$ while the pre-exponential factor, k_o, is obtained from the relation *ln* [*AR/Ea*)] as described by Damartzis et al. (2011).

Similarly, using Doyle's approximation the KAS model for determining the pyrolysis decomposition of biomass is Equation (10) (Xu, Ma, Yu, & Cai, 2014)

$$\ln\left(\frac{\beta}{T^2}\right) = \ln\left(\frac{k_o R}{g\left(\alpha\right) E_a}\right) - \left(\frac{E_a}{RT}\right) \tag{10}$$

Subsequently, the kinetic parameters *Ea*, and k_o can be deduced from the plot of *ln* (β/T^2) against (*1/T*) for a given degree of conversion (α) at different heating rates, where the slope is *–Ea/R*.

Multi-Pseudo Components Model Theory

Devolatilization kinetic model applies the Arrhenius rate of decomposition for the solid-state equation. The equation relates fractional conversion, α, pre-exponential constant, k_o (s^{-1}), activation energy, *Ea* ($J\ mol^{-1}$), temperature, T (*K*), universal gas constant, R (8.314 $J\ mol^{-1}\ K^{-1}$) and time, t (s) Equation (11) (Hattingh et al., 2014):

$$\frac{d\alpha}{dt} = k_0 \exp\left(-E_a \big/ RT\right)\left(1-\alpha\right) \tag{11}$$

where

$$\alpha = \frac{m_o - m_t}{m_0 - m_f} \tag{12}$$

The mass parameters in Eq (12) represent sample masses (in mg) and can be defined as $m_{o,}$ initial mass; $m_{t,}$ mass at a specific time; and m_f, final mass after devolatilization.

The thermal decomposition of any biomass species can generally be attributed to its major components hemicellulose, cellulose and lignin degradation. As a result, biomass degradation is the sum total of all the multiple pseudo-components degrading during thermal analysis. Therefore, the kinetics of multicomponent devolatilization of biomass can be modelled by assuming the pseudocomponents decomposes as independent and parallel first order reactions in Eq (13).

$$\frac{d\alpha_i}{dt} = \xi_i k_{0,i} \exp\left(\frac{-E_a}{RT}\right) \cdot \left(1 - \alpha_i\right) \tag{13}$$

$$\frac{d\alpha_n}{dt} = \sum_{i=1}^{n} \frac{d\alpha_i}{dt}, \Rightarrow \alpha_n = \sum_{i=1}^{n} \alpha_i \tag{14}$$

Since TGA is a function of heating rate, (°C min^{-1}), rearranging Equation (13) to reflect heating rate by substituting Eq (15) and introducing the integral limit in Eq (16):

$$\frac{d\alpha}{dt} = \frac{d\alpha}{dT} \cdot \beta \tag{15}$$

$$\int_0^{\alpha_i} \frac{d\alpha_i}{\left(1 - \alpha_i\right)} = \xi_i \cdot \frac{k_{0,i}}{\beta} \cdot \int_{T_0}^{T} \exp\left(\frac{-E_a}{RT}\right) \cdot dT \tag{16}$$

Consequently, the analytical solution of Equation (16) by integration by part and substitution gives Equation (17) for n, numbers of pseudo-reactions.

$$\alpha_n = \sum_{i=1}^{n} \xi_i \cdot \left(1 - \exp\left\{\frac{k_{0,i}}{\beta} \cdot \frac{E_{a,i}}{R} \cdot \left[\frac{\exp\left(-y_i\right)}{y_i} - \int_y^{\infty} \frac{\exp\left(-y_i\right)}{y_i} dy\right]\right\}\right) \tag{17}$$

where $y_i = \frac{E_{a,i}}{RT}$.

The kinetic parameters ξ_i, $k_{o,i}$ and $Ea_{,i}$ for each pseudocomponent of biomass devolatilization can then be obtained by minimising the objective function (OBF) described in Eq (18) using multidimensional nonlinear regression (MNR) using the *lsqlcurvefit* function in MATLAB R2013a.

$$OBF = \sum_{k=1}^{N_k} \sum_{m=1}^{N_m} \left[\left(\frac{d\alpha_n}{dt}\right)_{\exp,k,m} - \left(\frac{d\alpha_n}{dt}\right)_{cal,k,m}\right] \tag{18}$$

where N_k and N_m are the number of heating rates and numbers of TGA measured respectively.

Conversely, minimising Eq (18) requires a mathematical evaluation of Equation (17) using the *expint* function and numerical differentiation of Eq (15) in MATLAB R2013a.

Since the simulation considers a good number of pseudocomponents The model quality of fit (QOF) is determined by Equation (19) for each overall simulation process and their heating rates (Aboyade et al., 2011).

$$QOF = 100 \times \sum_{m=1}^{N_m} \frac{\sqrt{\left[\left(\frac{d\alpha_n}{dt}\right)_{\exp,m} - \left(\frac{d\alpha_n}{dt}\right)_{cal,m}\right]^2 \Big/ N_m}}{\max\left|\frac{d\alpha}{dt}\right|_{\exp,m}} \tag{19}$$

Biomass and Biohydrogen

Biomass is an abundant, renewable and sustainable source of energy for the future. However, biomass conversion technologies are perceived as inefficient compared to fossil fuels. Biomass typically includes agricultural waste residues, forest waste products, municipal solid waste and energy crops such as perennial grasses (Ceylan & Topcu, 2014; Kafle et al., 2016). The conversion of biomass resources into solid, liquid and gaseous products such as biohydrogen through biochemical or thermochemical routes is widely reported in the literature (Monlau et al., 2013; Nyakuma et al., 2014; X. Zhang et al., 2013). The biochemical route involves the breakdown of biomass molecules by enzymes or acid (enzymatic or acid hydrolysis) into sugar and subsequent fermentation to bioethanol (Balan et al., 2012; Basu, 2013). The process is well-established and cost-effective due to advances in research and optimisation of fermentation. However, biochemical processing of biomass into fuels cannot be easily integrated into existing commercial refining operation compared to thermochemical routes (Anthony V. Bridgwater, 2012).

Pyrolysis, gasification and combustion involve the breakdown of the lignocellulosic components of biomass at high temperatures into gaseous (biohydrogen, biosyngas), liquid (biooil) and solid (char) products (Goyal et al., 2008). Pyrolysis is an important process in biomass decomposition since is the first step in combustion and gasification (Damartzis et al., 2011). Biomass pyrolysis into solid, liquid, and gas essentially occurs at a temperature between 400 – 550 °C with the product selectivity dependent on the feedstock type, temperature, pressure and heating rate (Anca-Couce & Zobel, 2012; Basu, 2013). The volatile products of pyrolysis can be either condensable or non-condensable. The non-condensable yields small proportions of hydrogen, however, the condensable (biooil) contains large hydrogen derivable species (Basu, 2013). This biooil is the major product of fast (flash) pyrolysis of biomass and contains water soluble components suitable as feedstock for steam reforming into biohydrogen (Vagia & Lemonidou, 2007).

Biohydrogen From Steam Reforming Thermodynamics

Experimental and thermodynamics studies on biohydrogen production from steam reforming of biomass derived biooil have been investigated by several research groups (Goicoechea, Ehrich, Arias, & Kockmann, 2015; Mei, Wu, & Liu, 2016; Resende, Ávila-Neto, Rabelo-Neto, Noronha, & Hori, 2015b). Vagia and Lemonidou (2007) investigated the thermodynamic yield of hydrogen from three functional group components of biooil (acetic acid, acetone and ethylene glycol) at various operating conditions and obtained a good yield of 85%. Similarly, H. Xie et al. (2014) considered phenol in addition to acetic acid, acetone and ethylene glycol components as well as the addition of CaO as CO_2 sorbent with an average yield of 95% (H. Xie et al., 2015). Resende, Ávila-Neto, Rabelo-Neto, Noronha, and Hori (2015a) used acetic acid in modelling and experimental study with $LaNiO_3$ catalyst and found a good agreement between experiment and thermodynamic routes for converting acetic acid via acetone and acetate species. Remón, Broust, Volle, García, and Arauzo (2015) discovered that the chemical components of pyrolysis products are unique to specific biomass sources and pyrolysis unit. Pine and poplar wood were used as feedstock in fluidised and spouted bed pyrolyzer and the amount of propanoic acid from pine in fluidised and spouted bed were 0.5 and 0.9 mg/g respectively and from poplar in fluidised and spouted bed were 1.2 and 1.3 mg/g respectively.

However, the conversion of wood and agroforestry products into bioenergy and biofuels presents significant challenges for the environment (Sims et al., 2010). The high rate of deforestation, loss of biodiversity, and land degradation resulting from wood fuels and agroforestry products necessitates the search for alternative biomass feedstocks such as perennial energy grasses (Chen, Liu, & Scott, 2016; Felix & Tilley, 2009; Hodgson et al., 2011; Vyn, Virani, & Deen, 2012). Numerous studies have investigated the carbonisation, torrefaction, pyrolysis potential of energy grasses such Miscanthus, Switchgrass, and Canary Reed (Moutsoglou, 2012). Other potential energy grasses such Elephant grass, and *Imperata cylindrica* found around the world can also be utilised for bioenergy applications.

Thermodynamic and Stoichiometry Theory

The simulation flowsheet adopted the RGibbs reactor from the ASPEN Plus library and calculations were carried out by minimising the Gibbs free energy to obtain steam reforming equilibrium compositions. The governing equations for Gibbs free energy, G_{total} for N species is given by (Basu, 2013):

$$G_{total} = \sum_{i=1}^{N} n_i \Delta G_{f.i}^{0} + \sum_{i=1}^{N} n_i RT \ln\left[\frac{n_i}{\sum n_i}\right] \tag{20}$$

where $i = 1 \ldots N$; n- moles of species, i; $\Delta G_{f.i}^{0}$ is the Gibbs free energy of formation of species i at standard pressure of 1 bar; R- universal gas constant and T- Temperature. The objective function can be minimized by solving for n_i in Equation 2.20. Ultimate analysis carbon equals the gas mixture total carbon. Therefore, the expression for *j*th element can be written as;

$$\sum_{i=1}^{N} a_{i,j} n_i = A_j \tag{21}$$

Here $a_{i,j}$ is the number of atoms of the *j*th element in the *i*th species and A_j is the total number of atoms of element *j* entering the reactor. Using Lagrange multiplier, λ methods:

$$L = G_{total} - \sum_{i=1}^{N} \lambda_j \left(\sum_{i=1}^{N} a_{i,j} n_i - A_j \right) \tag{22}$$

Consequently, the extreme points can be obtained by substituting Equation 2.22 into Equation 2.20;

$$\frac{\partial L}{\partial n_i} = \frac{\Delta G_{f.i}^0}{RT} + \sum_{i=1}^{N} \ln\left(\frac{n_i}{n_{total}} \right) + \frac{1}{RT} \sum_{j=1}^{K} \lambda_j \left(\sum_{i=1}^{N} a_{i,j} n_i \right) \tag{23}$$

The major products of steam reforming considered were H_2, CO, CO_2, CH_4, C (graphite) and excess steam. In addition, potential secondary components such as ethane, ethylene, acetylene and ethanol reported to be negligible by Vagia and Lemonidou (2007) were considered. However, the high carbon content in the reacting components was considered in this study from biooil obtained by Kamaroddin (2014). Furthermore, the presence of these secondary products even in low amount in practice is found to be coke precursors.

The three most common reaction in the steam reform of biooil are steam reforming, water shift and methanation reactions as presented in the generic reactions, respectively (Liu, Chen, Hu, Wang, & Kong, 2013):

$$C_aH_bO_c + (a - c)H_2O \leftrightarrow aCO + (a + b/2 - c)H_2 \tag{24}$$

$$CO + H_2O \rightarrow CO_2 + H_2 \tag{25}$$

$$CO + 3H_2 \rightarrow CH_4 + H_2O \tag{26}$$

Combining Equation 2.24 and Equation 2.25 gives the overall steam reforming reaction

$$C_aH_bO_c + (2a - c)H_2O \leftrightarrow aCO_2 + (2a + b/2 - c)H_2 \tag{27}$$

Pyrolysis Reactors

There are many pyrolysis reactors and most are named by the mechanism through, which heat is transferred to the biomass or the transport of the biomass and heat source this includes (Basu, 2013; A. V. Bridgwater, 2012; Bridgwater, Meier, & Radlein, 1999; Sharma, Pareek, & Zhang, 2015):

1. Fixed or moving bed
2. Bubbling fluidized bed
3. Circulating fluidized bed
4. Ultra-rapid, Rotating cone
5. Ablative
6. Vacuum

Fluidized Bed

Small size particles of biomass (< 2-3 mm) are fed into a reactor of high-temperature sand bed and both are fluidized by inert gas that is mainly the non-condensable pyrolysis gas. The process heat was delivered by heating coils and was obtained from the combustion of the gas and char by-products of the pyrolysis process (see Figure 5). It is a well-studied and understood process and can easily be scaled up and has been known to achieve about 70-75% biooil production from dry wood (Basu, 2013; Venderbosch & Prins, 2010).

Circulating Fluid Beds and Transported Beds

This pyrolyzer is similar to the fluidized bed pyrolyzer system. However, unlike the fluidized bed, the char residence time is almost the same as that of gas and biooil vapours. In addition, in the circulating fluid bed (CFB) system, the heat is directly supplied from the recirculating sand to the biomass, the sand is reheated by char combustion (see Figure 6).

Rotating Cone

The rotating cone is the most recent pyrolyzer. In this system, biomass particles and hot sands are fed into the cone reactor and are effectively mixed and transported by centrifugal force due to the rotating of the cone (see Figure 7). The heating rate is very high due to excellent mixing with the hot sand (heat carrier) and biomass. The residence time for solid and gas phase are 0.5 seconds and 0.3 seconds respectively. Due to this short residence time the reactor is optimised for biooil production of about 60-70% yield (Basu, 2013; Wagenaar, Kuipers, Prins, & Swaaij, 1993).

Reactors Types and Heat Transfer

Every reactor has its advantages and disadvantages, but in pyrolysis, the lookout for advantage is high heat transfer rate. Among the reactors, fluidized bed, circulating fluidized bed and transported bed are characterised by high heat transfer rates. In addition, fluidized beds have very good solids mixing and

Figure 5. Fluid bed pyrolyzer

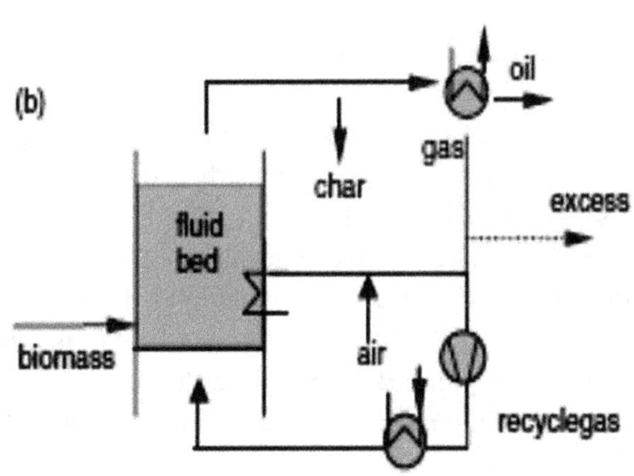

Figure 6. Circulating fluid beds and transported beds
(Venderbosch & Prins, 2010)

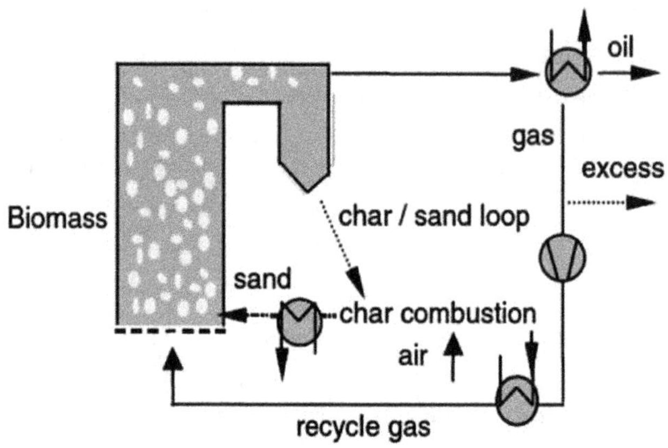

Figure 7. Rotating cone pyrolyzer
(Venderbosch & Prins, 2010)

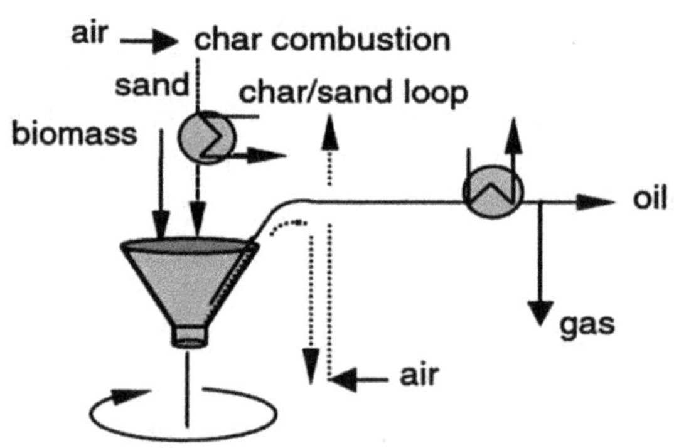

simple reactor configuration, but are ineffective for particle sizes greater than 2 mm. However, the circulating fluid bed and transported bed can support up to 6 mm particle size. Table 2 shows the summary of reactors with the mode of heat transfer (Bridgwater, 1999).

Pyrolysis Reactor Performance Review

Biooil is the major product of fast pyrolysis. The process needs rapid heating rates and volatile residence time should be short (Xue et al., 2011). In optimising biomass heat distribution and volatile residence time for fast pyrolysis many reactors have been designed. Studies show that reactors of fluidized bed, vacuum, ablative, circulating fluidized bed and auger technologies are of commercial application (Mohan, Pittman, & Steele, 2006). Bubbling fluidized bed, circulating fluidized bed and transported bed (CFB-TB) and ablative reactor gave a biooil yield of 75% compared to rotating cone reactor 65%, vacuum reactor 35-50% and auger reactor 30-50% biooil yield. Both bubbling fluidized bed and circulating fluidized bed and transported bed (CFB-TB) reactors are more matured technologies with better particle heat transfer. However, the process hydrodynamic and reactors operation are complex.

Table 2. Reactor types and heat transfer

Reactor	Heat Transfer Mechanism	Remarks	References
Ablative	Conduction 95% Convection 4% Radiation 1%	Advantages: 1. Large size feedstock acceptable 2. Compact design Disadvantage 1. Heat supply problematic. 2. Very high mechanical char abrasion from biomass. Features 1. Heat transfer gas not required 2. Particulate transport gas not always required.	(Bridgwater, 1999)
Circulating fluid bed and transported bed	Conduction 80% Convection 19% Radiation 1%	Advantages: 1. High heat transfer rates 2. Compact design 3. Accept up to 6mm particle size. Disadvantage 1. High char abrasion from biomass and char erosion leading to high char in product 2. Char/Solid heat carrier separation required 3. Greater reactor wears possible. Features 1. Solid recycle required. 2. Possible liquid cracking by hot solids. 3. Possible catalytic activity from hot char.	(Bridgwater, 1999)
Entrained flow	Conduction 4% Convection 95% Radiation 1%	Advantages: Disadvantage 1. Low heat transfer rates 2. Particle size limit < 2mm. Features 1. Limited gas/solid mixing.	(Bridgwater, 1999)
Fluid bed	Conduction 90% Convection 9% Radiation 1%	Advantages: 1. High heat transfer rates 2. Heat supply to fluidizing gas or to bed directly 3. Limited char abrasion 4. Very good solids mixing Disadvantage 1. Particle size limit < 2mm. Features 1. Simple reactor configuration.	(Bridgwater, 1999)

The other high yield reactor is the ablative reactor, which operates at low temperatures (< 600 °C). However, the reaction rates are limited by heat transfer to the reactor and not the biomass, the reactor is costly and not easily scaled. Due to the disadvantages associated with the mentioned fast pyrolysis reactors, Scott, Majerski, Piskorz, and Radlein (1999) designed a improved bubbling fluidized bed system developed by RTI laboratory Limited. The RTI process gave equal or higher yield of biooil for hardwood sawdust, wheat straw and sugar cane bagasse feed at lower temperatures, compared to regular bubbling fluidized bed reactor. The energy efficiency and cost of production of RTI were greatly improved compared to regular reactors. However, the reactor complex hydrodynamics and operating condition, as well as high cost of running fluidized bed, are still a factor to be dealt with.

Therefore, a simple and cost-effective reactor (transported bed) was developed in Universiti Teknologi Malaysia (UTM). The transported bed reactor was successfully employed in the pyrolysis of *Imperata*

Table 3. Selected CFD analysis application to Biomass Pyrolysis

Authors	Reactor / Model	Performance Equations	Remarks
Anca-Couce, Zobel, and Jakobsen (2013)	**Type**: Fixed-bed **Approach**: Euler (RPM) Dimension: 1D Solver: SIMPLE **Scheme:** Finite Volume **Simulation:** Splitting Algorithms	Mass balance $\frac{\partial \rho_g}{\partial t} = -\frac{\partial(w_z \rho_g)}{\partial z} + \frac{(1-\epsilon_b)}{\epsilon_b}(\rho_g v)'$ $\frac{\partial(\rho_g Y_i)}{\partial t} = -\frac{\partial(w_z \rho_g Y_i)}{\partial z} + \frac{\partial}{\partial z}\left(\rho_g D_{z,i} \frac{\partial Y_i}{\partial t}\right) + s$ $+\frac{(1-\epsilon_b)}{\epsilon_b}\left\{(\rho_g Y_i v)' + \zeta_i \beta_i' \left[(\rho_g Y_i)' - (\rho_g Y_i)\right]\right\}$ Momentum balance $\frac{\partial(\rho_g w_z)}{\partial t} = -\frac{\partial(\rho_g w_z w_z)}{\partial z} - \frac{\partial p}{\partial z}$ $-f_1 w_z - \epsilon_b f_2 w_z \lvert\vec{w}\rvert + 2\times\frac{\partial}{\partial z}\left(\mu_{eff}\frac{\partial w_z}{\partial z}\right)$ Energy balance $\rho_g c_{p,g}\frac{\partial T_g}{\partial t} = -\rho_g c_{p,g} w_z \frac{\partial T_g}{\partial z} + \frac{\partial}{\partial z}\left(\Lambda_z \frac{\partial T_g}{\partial t}\right) - s$ $+\frac{(1-\epsilon_b)}{\epsilon_b}\left[\left((\rho_g c_{p,g} v)' + \zeta_e \alpha_i'\right)(T_g - T_p')\right]$	Agreement with Experiment: Good Limitation: Fixed bed and two Eulerian phases considered.
	where	z = axial axis w_z = reactor axial velocity v = superficial velocity ρ = density Y = mass fraction ϵ = porosity D = mass dispersion-diffusivity ζ = strefan correction β = mass transfer f_1 = first friction factor f_2 = second friction factor μ_{eff} = viscosity effective c_p = specific heat capacity T = temperature t = time Λ = thermal dispersion conductivity α = heat transfer coefficient s = ratio surface: volume of the particle g = gas b = biomass, particles	

continued on following page

Table 3. Continued

Authors	Reactor / Model	Performance Equations	Remarks
Papadikis, Bridgwater, and Gu (2008); (Papadikis, Gu, & Bridgwater, 2009a)	**Type:** Fluidized-bed **Approach:** Euler-Euler Dimension: 2D and 3D **Solver:** ANSYS Fluent 6.2 **Scheme:** Finite Volume **Grid:** 1mm x 1mm Time Step: 10^{-5} s **Simulation:** Splitting Algorithms	Mass conservation Gas phase: $\frac{\partial\left(\varepsilon_g\rho_g\right)}{\partial t}+\nabla\cdot\left(\varepsilon_g\rho_g v_g\right)=0$ Solid phase: $\frac{\partial\left(\varepsilon_s\rho_s\right)}{\partial t}+\nabla\cdot\left(\varepsilon_s\rho_s v_s\right)=0$ Momentum conservation Gas phase: $\frac{\partial\left(\varepsilon_g\rho_g v_g\right)}{\partial t}+\nabla\cdot\left(\varepsilon_g\rho_g v_g\otimes v_g\right)$ $=-\varepsilon_g\cdot\nabla p-\nabla\cdot\bar{\bar{\tau}}_g+\varepsilon_g\rho_g g+K_{gs}\left(u_g-u_s\right)$ Solid phase: $\frac{\partial\left(\varepsilon_s\rho_s v_s\right)}{\partial t}+\nabla\cdot\left(\varepsilon_s\rho_s v_s\cdot v_s\right)$ $=-\varepsilon_s\cdot\nabla p-\nabla p_s-\nabla\cdot\bar{\bar{\tau}}_s+\varepsilon_s\rho_s g+K_{gs}\left(u_g-u_s\right)$ Fluctuation energy conservation of solid particles $\frac{3}{2}\left[\frac{\partial}{\partial t}\left(\varepsilon_s\rho_s\Theta_s\right)+\nabla\cdot\left(\varepsilon_s\rho_s u_s\Theta_s\right)\right]$ $=\left(-p_s\bar{\bar{I}}_s+\bar{\bar{\tau}}_s\right):\nabla\cdot u_s+\nabla\cdot\left(k_{\Theta s}\cdot\nabla\cdot\Theta_s\right)-\gamma_{\Theta s}$ Force on discrete particles $\frac{du_d}{dt}=\frac{f}{\tau_u}\left(u_c-u_d\right)+g\left(1-\frac{\rho_c}{\rho_d}\right)+F_{vm}$ Heat transfer $\frac{\partial}{\partial t}\left(\rho C_{peff}T\right)=\frac{1}{r^2}\frac{\partial}{\partial r}\left(k_{eff}r^2\frac{\partial T}{\partial r}\right)+\left(-\Delta H\right)\left(\frac{\partial\rho}{\partial t}\right)$	Agreement with Experiment: Satisfactory Limitation: Fluidized bed and two Eulerian phases considered.
	where	g = gravitational acceleration p = pressure K_{gs} = gas-solid momentum exchange coefficient, dimensionless u, v = velocity $\bar{\bar{\tau}}$ = stress tensor Θ = granular temperature $\bar{\bar{I}}$ = stress tensor, dimensionless $k_{\Theta s}$ = diffusion coefficient for granular energy $\gamma_{\Theta s}$ = collision dissipation of energy f = drag factor, dimensionless F = force k = thermal conductivity ΔH = heat of reaction	

continued on following page

Table 3. Continued

Authors	Reactor / Model	Performance Equations	Remarks
(Papadikis, Gu, & Bridgwater, 2009b)	**Type:** Fluidized-bed **Approach:** Euler-Euler Dimension: 3D **Solver:** ANSYS Fluent 6.2 **Scheme:** Finite Volume **Grid:** 1mm x 1mm Time Step: 10^{-5} s **Simulation:** Splitting Algorithms	Shrinkage parameters $\frac{V_s}{V_{w0}} = \frac{m_w}{m_{w0}} + \frac{\propto m_c}{m_{w0}}$ $V_g = V_{g0} + \beta\left(V_{w0} - V_s\right)$ $V = V_s + V_g = \eta V_{gi} + \left(1-\eta\right)\gamma V_{gi} + \beta\left(V_{w0} - V_s\right) + V_s$	Agreement with Experiment: Satisfactory Limitation: Fluidized bed and two Eulerian phases considered.
	where	V = volume m = mass $\propto, \beta, \gamma$ = shrinkage parameter w = wood	
(Papadikis, Gu, Bridgwater, et al., 2009)	**Type:** Entrained Flow Reactor **Approach:** Lagrange-Euler Dimension: 2D **Solver:** ANSYS Fluent 6.2 **Scheme:** Finite Volume **Grid:** 1mm x 1mm Time Step: 10^{-5} s **Simulation:** Splitting Algorithms	Mass balance in multiphase (i) $m_i = \iiint V\left(t\right)\varepsilon_i\rho_i dV$ $\frac{dm_i}{dt^i} = \frac{d}{dt^i}\iiint \varepsilon_i\rho_i dV = \iiint m_i' dV$ Momentum balance $\frac{\partial\left(\varepsilon_i\rho_i v_i\right)}{\partial t} + \nabla\cdot\left(\varepsilon_i\rho_i v_i \cdot v_i\right) = \nabla\cdot T_i + \varepsilon_i\rho_i f_i + p_{ij} + m_i' v_i$ Energy balance $\frac{du_p}{dt}_i = F_D\left(u - u_p\right) + \frac{g_x\left(\rho_p - \rho\right)}{\rho_p}$ Reaction kinetics $K_i = A_i \exp\left(-\frac{E_i}{RT}\right)$	Agreement with Experiment: Satisfactory Limitation: Entrained Flow Reactor and two Eulerian phases considered.

continued on following page

cylindrica with yield range: char (20-30%), gas (50-70%) and biooil (7-21%) at a temperature (450-650 °C) (Kamaroddin, 2014). The results show that the low yield of biooil could be due to any of the factors mentioned by Mohan et al. (2006) especially inefficient particle mixing and heat transfer due to reactor design.

Optimal Performance Using CFD Model

Computational Fluid Dynamics (CFD) modelling is increasingly and steadily becoming an important tool in chemical plant design and process development. It is also an increasingly useful tool in research work (Mellin, Zhang, Kantarelis, & Yang, 2013). The combination of CFD model with kinetic model would be applied to biomass pyrolysis to investigate parameters such as heat distribution, product distribution, gas and solid phase temperature as well as vapor and particle residence time (Mellin, Kantarelis, & Yang, 2014; Papadikis, Gu, Bridgwater, & Gerhauser, 2009; Xue, Dalluge, Heindel, Fox, & Brown, 2012).

Table 3. Continued

Authors	Reactor / Model	Performance Equations	Remarks
	where	F_D = drag force per unit particle mass	
(Papadikis, Gu, Fivga, & Bridgwater, 2010)	**Type:** Fluidized-bed Approach: Euler-Euler Dimension: 2D **Solver:** ANSYS Fluent 6.3 **Scheme:** Finite Volume **Grid:** 1mm x 1mm Time Step: 10^{-5} s **Simulation:** Splitting Algorithms	The drag model considered. Gidaspow: $K_{gs} = \frac{3}{4} C_d \frac{\varepsilon_s \varepsilon_g \rho_g \lvert u_s - u_g \rvert}{d_s} \varepsilon_g^{-2.65} \; for\, \varepsilon_g > 0.8$ $K_{gs} = 150 \frac{\varepsilon_s^2 \mu_g}{\varepsilon_g d_s^2} + 1.75 \frac{\varepsilon_s \rho_g \lvert u_s - u_g \rvert}{d_s} \; for\, \varepsilon_g \leq 0.8$ $C_d = \frac{24}{\varepsilon_g \mathrm{Re}_s} \left[1 + 0.15 \left(\varepsilon_g \mathrm{Re}_s \right)^{0.687} \right]$ Syamlal-O'Brien: $K_{gs} = \frac{3 \varepsilon_s \varepsilon_g \rho_g}{4 u_{r,s}^2 d_s} C_d \left(\frac{\mathrm{Re}_s}{u_{r,s}} \right) \lvert u_s - u_g \rvert$ $C_d = \left(0.63 + \frac{4.8}{\sqrt{\mathrm{Re}_s / u_{r,s}}} \right)$ Wen-Yu: $K_{gs} = \frac{3}{4} C_d \frac{\varepsilon_s \varepsilon_g \rho_g \lvert u_s - u_g \rvert}{d_s} \varepsilon_g^{-2.65}$ $C_d = \frac{24}{\varepsilon_g \mathrm{Re}_s} \left[1 + 0.15 \left(\varepsilon_g \mathrm{Re}_s \right)^{0.687} \right]$	Agreement with Experiment: Good Limitation: Fluidized bed and two Eulerian phases considered.

continued on following page

The application of CFD analysis on many reactors involved in thermochemical processes is summarised in Table 3. The nomenclature differences were added progressively after each model. Many of the developed and highlighted model focuses on fluidized bed but this research is to model a transported bed reactor. Furthermore, many models are based on two Eulerian model (Solid and Gas). However detailed CFD analysis should involve three Eulerian model: gas or primary phase and the solid (disperse) phase should be considered as two separate phases of biomass and sand phase. (Syamlal, Rogers, & O'Brien, 1993; Xue et al., 2012)

CONCLUSION

The kinetic reaction mechanism of biomass pyrolysis is widely recognized. This is because of the nature of its complexity and difficulty, which is due to the large numbers of intermediates and end products formed Therefore, over time the entire process product components have been lumped into biooil, biogas and biochar as shown in many schemes. The schemes and the methods of determining the kinetic

Table 3. Continued

Authors	Reactor / Model	Performance Equations	Remarks
	where	C_d = drag coefficient, dimensionless μ = viscosity Re = Reynolds number, dimensionless	
Bruchmüller, van Wachem, Gu, Luo, and Brown (2012)	**Type:** Fluidized-bed **Approach:** Euler / Lagrangian (DPM) Dimension: 3D Solver: Multiflow **Scheme:** Discrete element **Simulation:** Variable time-stepping routine	Gas-Phase Conservation Equations and Interphase Exchanged Equations Used $\frac{\partial(\varepsilon\rho_f)}{\partial t}+\nabla\cdot(\varepsilon\rho_f\vec{u}_f)=S$ $\frac{\partial(\varepsilon\rho_f\vec{u}_f)}{\partial t}+\nabla\cdot(\varepsilon\rho_f\vec{u}_f\vec{u}_f)=-\varepsilon\nabla P+\nabla\cdot(\varepsilon\overline{\overline{\tau}})$ $+\rho_f\varepsilon\vec{g}-\sum_{i=1}^{n}\frac{V_{p,i}\beta}{(1-\varepsilon)}(\vec{u}_f-\widetilde{\vec{u}_{f,i}})\delta(x-x_{P,i})+S\vec{u}_f$ $\frac{\partial(\varepsilon\rho_f c_{P,f}T_f)}{\partial t}+\nabla\cdot(\varepsilon\rho_f\vec{u}_f c_{P,f}T_f)=\nabla\cdot(\varepsilon\lambda_f\nabla T_f)+Q_f$ $\frac{\partial(\varepsilon\rho_f Y_i)}{\partial t}+\nabla\cdot(\varepsilon\rho_f\vec{u}_f Y_i)=\nabla\cdot(\varepsilon\rho_f\Gamma_f\nabla Y_i)+S_i$ $\beta=\frac{3}{4}C_D\lvert\widetilde{\vec{v}_f}-\vec{v}_P\rvert\frac{\rho_f(1-\varepsilon)(\varepsilon)^{-1.65}}{d_P}$ $C_D=\begin{cases}\frac{24}{\varepsilon\mathrm{Re}_P}\left(1+0.15(\varepsilon\mathrm{Re}_P)^{0.687}\right) & if\ \varepsilon\mathrm{Re}_P<1000\\0.44 & if\ \varepsilon\mathrm{Re}_P\geq 1000\end{cases}$ Particle-Phase modelling $m_P\frac{d\vec{u}_P}{dt}=\vec{F}_g-V_P\nabla P+\frac{V_P\beta}{(1-\varepsilon)}(\widetilde{\vec{u}_f}-\vec{u}_P)+\sum_{col}(\vec{F})$ $I_P\frac{d\vec{\omega}_P}{dt}=\sum_{col}(\vec{T})$ $m_Pc_P\frac{dT_P}{dt}=-hA(T_P-\tilde{T}_f)-\omega\sigma A(T_P^4-T_{bed}^4)$ $-H_{evap}\cdot S_{H_2O}-H_{k_1/k_2/k_3}\cdot S_{1/2/3}$ $\frac{\partial\Phi}{\partial t}=S_i$	Agreement with Experiment: Good Limitation: Fluidized bed Future work: Optimisation of the complex processes

continued on following page

Table 3. Continued

Authors	Reactor / Model	Performance Equations	Remarks
	Where	S = species source term P = local normal pressure δ = interphase momentum transfer coefficient λ = thermal heat conductivity Q = energy source term Y = mass fraction Γ = diffusion coefficient ω = emissivity A = surface area	
(Mellin et al., 2013)	**Type:** Fluidized-bed Approach: Euler-Euler Dimension: 3D **Solver:** ANSYS Fluent 14.5 **Scheme:** Finite Volume **Grids:** 4.11mm x 9.5mm x 4.11mm Time Step: 2 x 10^{-4} **Simulation:** Splitting Algorithms	Mass conservation $\frac{\delta}{\delta t}\left(\alpha_f \rho_f\right) + \nabla \cdot \left(\alpha_f \rho_f \vec{v}_f\right) = \dot{m}_f + S_f$ $\frac{\delta}{\delta t}\left(\alpha_f \rho_f Y_i\right) + \nabla \cdot \left(\alpha_f \rho_f Y_i \vec{v}_f\right) = \dot{m}_f + S_f$ Momentum conservation $\frac{\delta}{\delta t}\left(\alpha_f \rho_f \vec{v}_f\right) + \nabla \cdot \left(\alpha_f \rho_f Y_f \vec{v}_f \vec{v}_f\right)$ $= \alpha_f \nabla p + \nabla \overline{\overline{\tau}}_f + \alpha_f \rho_f \vec{g} + K_{fs}\left(\vec{v}_f - \vec{v}_s\right) + \dot{m}_{fs} \vec{v}_f$ $\frac{\delta}{\delta t}\left(\alpha_s \rho_s \vec{v}_s\right) + \nabla \cdot \left(\alpha_s \rho_s Y_s \vec{v}_s \vec{v}_s\right) = -\alpha_s \nabla p - \nabla p_s$ $+ \nabla \overline{\tau}_s + \alpha_s \rho_s \vec{g} + K_{fs}\left(\vec{v}_s - \vec{v}_f\right) + \dot{m}_{fs} \vec{v}_s$ Energy conservation $\frac{\delta}{\delta t}\left(\alpha_f \rho_f h_f\right) + \nabla \cdot \left(\alpha_f \rho_f Y_f \vec{v}_f h_f\right)$ $= -\alpha_f \frac{\partial p}{\partial t} + \overline{\overline{\tau}}_f : \nabla \vec{v}_f - \nabla \vec{q}_f + S + Q + \dot{m}_{fs} h_{fs}$	Agreement with Experiment: Acceptable Limitation: Considered only Fluidized bed and two Eulerian phases. It also focuses mainly on gas product yields.

continued on following page

Table 3. Continued

Authors	Reactor / Model	Performance Equations	Remarks
		α = volume fraction $\dot{m}$ = mass exchange between phases S = source term	
Xue et al. (2012); Xue et al. (2011); Syamlal et al. (1993)	**Type:** Fluidized-bed **Approach:** Euler-Euler Three phase Dimension: 2D and 3D Solver: MFIX **Scheme:** Finite Volume **Grid:** 2mm x 2mm; 4mm x 4mm; 8mm x 8mm **Simulation:** Splitting time approach. First Hydrodynamics follow by Reaction. (Pope, 1997; N. Xie, Battaglia, & Fox, 2004)	Volume fraction $\epsilon_g + \sum_{a=1} \epsilon_{b_a} = 1$ Gas Phase Continuity $\frac{\partial(\epsilon_g \rho_g)}{\partial t} + \nabla \cdot (\epsilon_g \rho_g \mathcal{V}_g) = R_g$ Species $\frac{\partial}{\partial t}(\epsilon_g \rho_g X_{gi}) + \nabla \cdot (\epsilon_g \rho_g X_{gi} \mathcal{V}_g) = R_{gi} + \mathfrak{D}_{gi} \nabla^2 X_{gi}$ Momentum $\frac{\partial(\epsilon_g \rho_g \mathcal{V}_g)}{\partial t} + \mathcal{V}_g \nabla \cdot (\epsilon_g \rho_g \mathcal{V}_g) = -\nabla p_g + \epsilon_g \rho_g f_g + \sum_{a=1} I_{gb_a}$ Energy $\epsilon_g \rho_g C_g \left(\frac{\partial T_g}{\partial t} + \mathcal{V}_g \cdot \nabla T_g \right)$ $= -\nabla \cdot q_g + Q_{rg} + h_{wall}(T_{wall} - T_g) - \sum_{a=1} H_{gb_a}$ Solid Phase Continuity $\frac{\partial(\epsilon_{b_a} \rho_{b_a})}{\partial t} + \nabla \cdot (\epsilon_{b_a} \rho_{b_a} \mathcal{V}_{b_a}) = R_{b_a}$ Species $\frac{\partial}{\partial t}(\epsilon_{b_a} \rho_{b_a} X_{b_a i}) + \nabla \cdot (\epsilon_{b_a} \rho_{b_a} X_{b_a i} \mathcal{V}_{b_a}) = R_{b_a i}$ Momentum $\frac{\partial(\epsilon_{b_a} \rho_{b_a} \mathcal{V}_{b_a})}{\partial t} + \mathcal{V}_{b_a} \nabla \cdot (\epsilon_{b_a} \rho_{b_a} \mathcal{V}_{b_a})$ $= -\nabla p_{b_a} + \epsilon_{b_a} \rho_{b_a} f_{b_a} + I_{gb_a} - \sum_{i=1, l \neq a} I_{b_a b_l}$ Energy $\epsilon_{b_a} \rho_{b_a} C_{b_a} \left(\frac{\partial T_{b_a}}{\partial t} + \mathcal{V}_{b_a} \cdot \nabla T_{b_a} \right)$ $= -\nabla \cdot q_{b_a} + Q_{rb_a} + \sum_{a=1} H_{gb_a}$	Agreement with Experiment: Acceptable Limitation: Fluidized bed Future work: Optimisation of the reactor.

continued on following page

Table 3. Continued

Authors	Reactor / Model	Performance Equations	Remarks
	where	R = rate of reaction X = mass fraction $\mathfrak{D}$ = diffusivity I = momentum exchange q = heat of conduction Q_r = heat of reaction H = fluid-solid interphase heat transfer	

parameters such as empirical method, model and model-free are developed and presented Furthermore, the reactor hydrodynamics models were developed for micromodel and micromodel for the optimization of the design of the reactors. The micromodel make use of computational fluid dynamics (CFD) in other to determine the optimal heat transfer scenario to the biomass particle in the reactor.

REFERENCES

Aboyade, A. O., Carrier, M., Meyer, E. L., Knoetze, J. H., & Görgens, J. F. (2012). Model fitting kinetic analysis and characterisation of the devolatilization of coal blends with corn and sugarcane residues. *Thermochimica Acta, 530*, 95–106. doi:10.1016/j.tca.2011.12.007

Aboyade, A. O., Hugo, T. J., Carrier, M., Meyer, E. L., Stahl, R., Knoetze, J. H., & Görgens, J. F. (2011). Non-isothermal kinetic analysis of the devolatilization of corn cobs and sugar cane bagasse in an inert atmosphere. *Thermochimica Acta, 517*(1-2), 81–89. doi:10.1016/j.tca.2011.01.035

Ahmed, S. I., Johari, A., Hashim, H., Mat, R., Lim, J. S., Ngadi, N., & Ali, A. (2015). Optimal landfill gas utilization for renewable energy production. *Environmental Progress & Sustainable Energy, 34*(1), 289–296. doi:10.1002/ep.11964

Anca-Couce, A., & Zobel, N. (2012). Numerical analysis of a biomass pyrolysis particle model: Solution method optimized for the coupling to reactor models. *Fuel, 97*, 80–88. doi:10.1016/j.fuel.2012.02.033

Anca-Couce, A., Zobel, N., & Jakobsen, H. A. (2013). Multi-scale modeling of fixed-bed thermo-chemical processes of biomass with the representative particle model: Application to pyrolysis. *Fuel, 103*, 773–782. doi:10.1016/j.fuel.2012.05.063

Balan, V., Kumar, S., Bals, B., Chundawat, S., Jin, M., & Dale, B. (2012). Biochemical and thermochemical conversion of switchgrass to biofuels. In A. Monti (Ed.), *Switchgrass, green energy and technology* (pp. 153–185). London, UK: Springer. doi:10.1007/978-1-4471-2903-5_7

Balat, M. (2008). Mechanisms of thermochemical biomass conversion processes. Part 1: Reactions of pyrolysis. *Energy Sources Part a-Recovery Utilization and Environmental Effects, 30*(7), 620-635. doi:10.1080/15567030600817258

Basu, P. (2013). *Biomass gasification, pyrolysis and torrefaction: practical design and theory*. Academic press.

Blaine, R. L., & Kissinger, H. E. (2012). Homer Kissinger and the Kissinger equation. *Thermochimica Acta, 540*, 1–6. doi:10.1016/j.tca.2012.04.008

Bridgwater, A. V. (1999). Principles and practice of biomass fast pyrolysis processes for liquids. *Journal of Analytical and Applied Pyrolysis, 51*(1-2), 3–22. doi:10.1016/S0165-2370(99)00005-4

Bridgwater, A. V. (2012). Review of fast pyrolysis of biomass and product upgrading. *Biomass and Bioenergy, 38*, 68–94. doi:10.1016/j.biombioe.2011.01.048

Bridgwater, A. V. (2012). Upgrading biomass fast pyrolysis liquids. *Environmental Progress & Sustainable Energy, 31*(2), 261–268. doi:10.1002/ep.11635

Bridgwater, A. V., Meier, D., & Radlein, D. (1999). An overview of fast pyrolysis of biomass. *Organic Geochemistry, 30*(12), 1479–1493. doi:10.1016/S0146-6380(99)00120-5

Bruchmüller, J., van Wachem, B. G. M., Gu, S., Luo, K. H., & Brown, R. C. (2012). Modeling the thermochemical degradation of biomass inside a fast pyrolysis fluidized bed reactor. *AIChE Journal. American Institute of Chemical Engineers, 58*(10), 3030–3042. doi:10.1002/aic.13705

Ceylan, S., & Topcu, Y. (2014). Pyrolysis kinetics of hazelnut husk using thermogravimetric analysis. *Bioresource Technology, 156*, 182–188. doi:10.1016/j.biortech.2014.01.040 PMID:24508656

Chen, T., Liu, R., & Scott, N. R. (2016). Characterization of energy carriers obtained from the pyrolysis of white ash, switchgrass and corn stover—Biochar, syngas and bio-oil. *Fuel Processing Technology, 142*, 124–134. doi:10.1016/j.fuproc.2015.09.034

Coats, A., & Redfern, J. (1965). Kinetic parameters from thermogravimetric data. II. *Journal of Polymer Science. Part B: Polymer Letters, 3*(11), 917–920. doi:10.1002/pol.1965.110031106

Damartzis, T., Vamvuka, D., Sfakiotakis, S., & Zabaniotou, A. (2011). Thermal degradation studies and kinetic modeling of cardoon (Cynara cardunculus) pyrolysis using thermogravimetric analysis (TGA). *Bioresource Technology, 102*(10), 6230–6238. doi:10.1016/j.biortech.2011.02.060 PMID:21398116

Demirbas, A. (2007). Combustion systems for biomass fuel. *Energy Sources. Part A, Recovery, Utilization, and Environmental Effects, 29*(4), 303–312. doi:10.1080/009083190948667

Di Blasi, C., & Branca, C. (2001). Kinetics of primary product formation from wood pyrolysis. *Industrial & Engineering Chemistry Research, 40*(23), 5547–5556. doi:10.1021/ie000997e

Doyle, C. D. (1965). Series approximations to the equation of thermogravimetric data. *Nature, 207*(4994), 290–291. doi:10.1038/207290a0

Felix, E., & Tilley, D. R. (2009). Integrated energy, environmental and financial analysis of ethanol production from cellulosic switchgrass. *Energy, 34*(4), 410–436. doi:10.1016/j.energy.2008.10.013

Floudas, C. A., Elia, J. A., & Baliban, R. C. (2012). Hybrid and single feedstock energy processes for liquid transportation fuels: A critical review. *Computers & Chemical Engineering, 41*, 24–51. doi:10.1016/j.compchemeng.2012.02.008

Font, R., Marcilla, A., Verdu, E., & Devesa, J. (1990). Kinetics of the pyrolysis of almond shells and almond shells impregnated with cobalt dichloride in a fluidized bed reactor and in a pyroprobe 100. *Industrial & Engineering Chemistry Research*, *29*(9), 1846–1855. doi:10.1021/ie00105a016

Freeman, E. S., & Carroll, B. (1958). The application of thermoanalytical techniques to reaction kinetics: The thermogravimetric evaluation of the kinetics of the decomposition of calcium oxalate monohydrate. *Journal of Physical Chemistry*, *62*(4), 394–397. doi:10.1021/j150562a003

Friedman, H. L. (1964). Kinetics of thermal degradation of char-forming plastics from thermogravimetry. application to a phenolic plastic. *Journal of Polymer Science Part C: Polymer Symposia*, *6*(1), 183–195. doi:10.1002/polc.5070060121

Goicoechea, S., Ehrich, H., Arias, P. L., & Kockmann, N. (2015). Thermodynamic analysis of acetic acid steam reforming for hydrogen production. *Journal of Power Sources*, *279*, 312–322. doi:10.1016/j.jpowsour.2015.01.012

Goyal, H. B., Seal, D., & Saxena, R. C. (2008). Bio-fuels from thermochemical conversion of renewable resources: A review. *Renewable & Sustainable Energy Reviews*, *12*(2), 504–517. doi:10.1016/j.rser.2006.07.014

Hattingh, B. B., Everson, R. C., Neomagus, H. W. J. P., Bunt, J. R., van Niekerk, D., & Ashton, B. P. (2014). Modeling the nonisothermal devolatilization kinetics of typical South African coals. *Energy & Fuels*, *28*(2), 920–933. doi:10.1021/ef402124f

Hayes, D. J. M. (2013). Second-generation biofuels: Why they are taking so long. *Wiley Interdisciplinary Reviews. Energy and Environment*, *2*(3), 304–334. doi:10.1002/wene.59

Heaton, E. A., Dohleman, F. G., Migucz, A. F., Juvik, J. A., Lozovaya, V., Widholm, J., . . . Long, S. P. (2010). Chapter 3 - Miscanthus: A Promising Biomass Crop. In K. Jean-Claude & D. Michel (Eds.), Advances in botanical research (Vol. 56, pp. 75-137): Academic Press.

Hodgson, E. M., Nowakowski, D. J., Shield, I., Riche, A., Bridgwater, A. V., Clifton-Brown, J. C., & Donnison, I. S. (2011). Variation in miscanthus chemical composition and implications for conversion by pyrolysis and thermo-chemical bio-refining for fuels and chemicals. *Bioresource Technology*, *102*(3), 3411–3418. doi:10.1016/j.biortech.2010.10.017 PMID:21035327

Janse, A., de Jong, X. A., Prins, W., & van Swaaij, W. P. (1999). Improvement and validation of the fluoroptic temperature-measurement method. *AIChE Journal. American Institute of Chemical Engineers*, *45*(10), 2087–2096. doi:10.1002/aic.690451007

Kafle, S., Parajuli, R., Euh, S. H., Oh, K. C., Choi, Y. S., Adhikari, K., ... Kim, D. H. (2016). Potential biomass supply for agro-pellet production from agricultural crop residue in Nepal. *Energy Sources. Part A, Recovery, Utilization, and Environmental Effects*, *38*(1), 149–153. doi:10.1080/15567036.2015.1043474

Kamaroddin, M. F. A. (2014). *Bio-oil production by pyrolysis of lalang grass. (Master of Chemical Engineering)*. Johor Bahru: Universiti Teknologi Malaysia.

Knight, J. A., Gorton, C. W., & Kovac, R. J. (1984). Oil production by entrained flow pyrolysis of biomass. *Biomass*, *6*(1), 69–79. doi:10.1016/0144-4565(84)90009-X

Lemus, R., Brummer, E. C., Moore, K. J., Molstad, N. E., Burras, C. L., & Barker, M. F. (2002). Biomass yield and quality of 20 switchgrass populations in southern Iowa. *Biomass and Bioenergy*, *23*(6), 433–442. doi:10.1016/S0961-9534(02)00073-9

Liew, W. H., Hassim, M. H., & Ng, D. K. S. (2014). Review of evolution, technology and sustainability assessments of biofuel production. *Journal of Cleaner Production*, *71*, 11–29. doi:10.1016/j.jclepro.2014.01.006

Liu, S., Chen, M., Hu, Q., Wang, J., & Kong, L. (2013). The kinetics model and pyrolysis behavior of the aqueous fraction of bio-oil. *Bioresource Technology*, *129*, 381–386. doi:10.1016/j.biortech.2012.11.006 PMID:23262015

Ma, C., Gao, J., Zhong, L., & Xing, S. (2016). Experimental investigation of the oxidation behaviour and thermal kinetics of diesel particulate matter with non-thermal plasma. *Applied Thermal Engineering*, *99*, 1110–1118. doi:10.1016/j.applthermaleng.2016.01.165

Maia, A. A., & de Morais, L. C. (2016). Kinetic parameters of red pepper waste as biomass to solid biofuel. *Bioresource Technology*, *204*, 157–163. doi:10.1016/j.biortech.2015.12.055 PMID:26773950

Mei, Y., Wu, C., & Liu, R. (2016). Hydrogen production from steam reforming of bio-oil model compound and byproducts elimination. *International Journal of Hydrogen Energy*, *41*(21), 9145–9152. doi:10.1016/j.ijhydene.2015.12.133

Mellin, P., Kantarelis, E., & Yang, W. (2014). Computational fluid dynamics modeling of biomass fast pyrolysis in a fluidized bed reactor, using a comprehensive chemistry scheme. *Fuel*, *117*, 704–715. doi:10.1016/j.fuel.2013.09.009

Mellin, P., Zhang, Q., Kantarelis, E., & Yang, W. (2013). An Euler–Euler approach to modeling biomass fast pyrolysis in fluidized-bed reactors – Focusing on the gas phase. *Applied Thermal Engineering*, *58*(1-2), 344–353. doi:10.1016/j.applthermaleng.2013.04.054

Miller, R. S., & Bellan, J. (1997). A generalized biomass pyrolysis model based on superimposed cellulose, hemicellulose and liqnin kinetics. *Combustion Science and Technology*, *126*(1-6), 97–137. doi:10.1080/00102209708935670

Oladokun, O., Ahmad, A., Abdullah, T. T., Nyakuma, B. B., Al-Shatri, A. H., & Bello, A. A. (2015). Modelling multicomponent devolatilization kinetics of imperata cylindrica. *Chemical Engineering Transactions*, *45*, 919–924. doi:10.3303/CET1545154

Mohammadi, M., Najafpour, G. D., Younesi, H., Lahijani, P., Uzir, M. H., & Mohamed, A. R. (2011). Bioconversion of synthesis gas to second generation biofuels: A review. *Renewable & Sustainable Energy Reviews*, *15*(9), 4255–4273. doi:10.1016/j.rser.2011.07.124

Mohan, D., Pittman, C. U. Jr, & Steele, P. H. (2006). Pyrolysis of wood/biomass for bio-oil: A critical review. *Energy & Fuels*, *20*(3), 848–889. doi:10.1021/ef0502397

Monlau, F., Barakat, A., Trably, E., Dumas, C., Steyer, J.-P., & Carrère, H. (2013). Lignocellulosic materials into biohydrogen and biomethane: impact of structural features and pretreatment. *Critical Reviews in Environmental Science and Technology*, *43*(3), 260–322. doi:10.1080/10643389.2011.604258

Moutsoglou, A. (2012). A comparison of prairie cordgrass and switchgrass as a biomass for syngas production. *Fuel, 95*, 573–577. doi:10.1016/j.fuel.2011.12.016

Nyakuma, B. B., Mazangi, M., Tuan Abdullah, T. A., Johari, A., Ahmad, A., & Oladokun, O. (2014). Gasification of empty fruit bunch briquettes in a fixed bed tubular reactor for hydrogen production. *Applied Mechanics and Materials, 699*, 534-539. doi:10.4028/www.scientific.net/AMM.699.534

Oladokun, O., Ahmad, A., Abdullah, T. A. T., Nyakuma, B. B., Kamaroddin, M. F. A., & Nor, S. H. M. (2017). Biohydrogen production from Imperata cylindrica bio-oil using non-stoichiometric and thermodynamic model. *International Journal of Hydrogen Energy, 42*(14), 9011–9023. doi:10.1016/j.ijhydene.2016.05.200

Ozawa, T. (1992). Estimation of activation-energy by isoconversion methods. *Thermochimica Acta, 203*, 159–165. doi:10.1016/0040-6031(92)85192-X

Papadikis, K., Bridgwater, A. V., & Gu, S. (2008). CFD modelling of the fast pyrolysis of biomass in fluidised bed reactors, Part A: Eulerian computation of momentum transport in bubbling fluidised beds. *Chemical Engineering Science, 63*(16), 4218–4227. doi:10.1016/j.ces.2008.05.045

Papadikis, K., Gu, S., & Bridgwater, A. V. (2009a). CFD modelling of the fast pyrolysis of biomass in fluidised bed reactors. Part B Heat, momentum and mass transport in bubbling fluidised beds. *Chemical Engineering Science, 64*(5), 1036–1045. doi:10.1016/j.ces.2008.11.007

Papadikis, K., Gu, S., & Bridgwater, A. V. (2009b). CFD modelling of the fast pyrolysis of biomass in fluidised bed reactors: Modelling the impact of biomass shrinkage. *Chemical Engineering Journal, 149*(1-3), 417–427. doi:10.1016/j.cej.2009.01.036

Papadikis, K., Gu, S., Bridgwater, A. V., & Gerhauser, H. (2009). Application of CFD to model fast pyrolysis of biomass. *Fuel Processing Technology, 90*(4), 504–512. doi:10.1016/j.fuproc.2009.01.010

Papadikis, K., Gu, S., Fivga, A., & Bridgwater, A. V. (2010). Numerical comparison of the drag models of granular flows applied to the fast pyrolysis of biomass. *Energy & Fuels, 24*(3), 2133–2145. doi:10.1021/ef901497b

Pope, S. B. (1997). Computationally efficient implementation of combustion chemistry using in situadaptive tabulation. *Combustion Theory and Modelling, 1*(1), 41–63. doi:10.1080/713665229

Remón, J., Broust, F., Volle, G., García, L., & Arauzo, J. (2015). Hydrogen production from pine and poplar bio-oils by catalytic steam reforming. Influence of the bio-oil composition on the process. *International Journal of Hydrogen Energy, 40*(16), 5593–5608. doi:10.1016/j.ijhydene.2015.02.117

Resende, K. A., Ávila-Neto, C. N., Rabelo-Neto, R. C., Noronha, F. B., & Hori, C. E. (2015a). Hydrogen production by reforming of acetic acid using La–Ni type perovskites partially substituted with Sm and Pr. *Catalysis Today, 242*, 71–79. doi:10.1016/j.cattod.2014.07.013

Resende, K. A., Ávila-Neto, C. N., Rabelo-Neto, R. C., Noronha, F. B., & Hori, C. E. (2015b). Thermodynamic analysis and reaction routes of steam reforming of bio-oil aqueous fraction. *Renewable Energy, 80*, 166–176. doi:10.1016/j.renene.2015.01.057

Scott, D. S., Majerski, P., Piskorz, J., & Radlein, D. (1999). A second look at fast pyrolysis of biomass - the RTI process. *Journal of Analytical and Applied Pyrolysis*, *51*(1-2), 23–37. doi:10.1016/S0165-2370(99)00006-6

Shafizadeh, F., & Chin, P. P. (1977, June 1). Thermal deterioration of wood. Paper presented at the *ACS Symposium Series American Chemical Society*. 10.1021/bk-1977-0043.ch005

Sharma, A., Pareek, V., & Zhang, D. (2015). Biomass pyrolysis—A review of modelling, process parameters and catalytic studies. *Renewable & Sustainable Energy Reviews*, *50*, 1081–1096. doi:10.1016/j.rser.2015.04.193

Sims, R. E., Mabee, W., Saddler, J. N., & Taylor, M. (2010). An overview of second generation biofuel technologies. *Bioresource Technology*, *101*(6), 1570–1580. doi:10.1016/j.biortech.2009.11.046 PMID:19963372

Slopiecka, K., Bartocci, P., & Fantozzi, F. (2012). Thermogravimetric analysis and kinetic study of poplar wood pyrolysis. *Applied Energy*, *97*, 491–497. doi:10.1016/j.apenergy.2011.12.056

Starink, M. J. (2003). The determination of activation energy from linear heating rate experiments: A comparison of the accuracy of isoconversion methods. *Thermochimica Acta*, *404*(1-2), 163–176. doi:10.1016/S0040-6031(03)00144-8

Syamlal, M., Rogers, W., & O'Brien, T. J. (1993). MFIX documentation: Theory guide. *National Energy Technology Laboratory, Department of Energy, Technical Note DOE/METC-95/1013 and NTIS/DE95000031.*

Thurner, F., & Mann, U. (1981). Kinetic investigation of wood pyrolysis. *Industrial & Engineering Chemistry Process Design and Development*, *20*(3), 482–488. doi:10.1021/i200014a015

Vagia, E., & Lemonidou, A. (2007). Thermodynamic analysis of hydrogen production via steam reforming of selected components of aqueous bio-oil fraction. *International Journal of Hydrogen Energy*, *32*(2), 212–223. doi:10.1016/j.ijhydene.2006.08.021

Várhegyi, G. (2007). Aims and methods in non-isothermal reaction kinetics. *Journal of Analytical and Applied Pyrolysis*, *79*(1-2), 278–288. doi:10.1016/j.jaap.2007.01.007

Vassilev, S. V., Baxter, D., Andersen, L. K., & Vassileva, C. G. (2010). An overview of the chemical composition of biomass. *Fuel*, *89*(5), 913–933. doi:10.1016/j.fuel.2009.10.022

Venderbosch, R. H., & Prins, W. (2010). Fast pyrolysis technology development. *Biofuels, Bioproducts & Biorefining*, *4*(2), 178–208. doi:10.1002/bbb.205

Vyn, R. J., Virani, T., & Deen, B. (2012). Examining the economic feasibility of miscanthus in Ontario: An application to the greenhouse industry. *Energy Policy*, *50*, 669–676. doi:10.1016/j.enpol.2012.08.009

Wagenaar, B. M., Kuipers, J. A. M., Prins, W., & Swaaij, W. P. M. (1993). The rotating cone flash pyrolysis reactor. In A. V. Bridgwater (Ed.), *Advances in Thermochemical Biomass Conversion* (pp. 1122–1133). Dordrecht, The Netherlands: Springer Netherlands. doi:10.1007/978-94-011-1336-6_87

Wagenaar, B. M., Prins, W., & van Swaaij, W. P. M. (1993). Flash pyrolysis kinetics of pine wood. *Fuel Processing Technology*, *36*(1-3), 291–298. doi:10.1016/0378-3820(93)90039-7

Woli, K. P., David, M. B., Tsai, J., Voigt, T. B., Darmody, R. G., & Mitchell, C. A. (2011). Evaluating silicon concentrations in biofuel feedstock crops Miscanthus and switchgrass. *Biomass and Bioenergy*, *35*(7), 2807–2813. doi:10.1016/j.biombioe.2011.03.007

Xie, H., Yu, Q., Wang, K., Shi, X., & Li, X. (2014). Thermodynamic analysis of hydrogen production from model compounds of bio-oil through steam reforming. *Environmental Progress & Sustainable Energy*, *33*(3), 1008–1016. doi:10.1002/ep.11846

Xie, H., Yu, Q., Yao, X., Duan, W., Zuo, Z., & Qin, Q. (2015). Hydrogen production via steam reforming of bio-oil model compounds over supported nickel catalysts. *Journal of Energy Chemistry*, *24*(3), 299–308. doi:10.1016/S2095-4956(15)60315-1

Xie, N., Battaglia, F., & Fox, R. O. (2004). Simulations of multiphase reactive flows in fluidized beds using in situ adaptive tabulation. *Combustion Theory and Modelling*, *8*(2), 195–209. doi:10.1088/1364-7830/8/2/001

Xu, Q., Ma, X., Yu, Z., & Cai, Z. (2014). A kinetic study on the effects of alkaline earth and alkali metal compounds for catalytic pyrolysis of microalgae using thermogravimetry. *Applied Thermal Engineering*, *73*(1), 357–361. doi:10.1016/j.applthermaleng.2014.07.068

Xue, Q., Dalluge, D., Heindel, T. J., Fox, R. O., & Brown, R. C. (2012). Experimental validation and CFD modeling study of biomass fast pyrolysis in fluidized-bed reactors. *Fuel*, *97*, 757–769. doi:10.1016/j.fuel.2012.02.065

Xue, Q., Heindel, T. J., & Fox, R. O. (2011). A CFD model for biomass fast pyrolysis in fluidized-bed reactors. *Chemical Engineering Science*, *66*(11), 2440–2452. doi:10.1016/j.ces.2011.03.010

Yang, II., Yan, R., Chcn, H., Lcc, D. H., & Zheng, C. (2007). Characteristics of hemicellulose, cellulose and lignin pyrolysis. *Fuel*, *86*(12-13), 1781–1788. doi:10.1016/j.fuel.2006.12.013

Zhang, L., Xu, C., & Champagne, P. (2010). Overview of recent advances in thermo-chemical conversion of biomass. *Energy Conversion and Management*, *51*(5), 969–982. doi:10.1016/j.enconman.2009.11.038

Zhang, X., Wang, T., Ma, L., Zhang, Q., & Jiang, T. (2013). Hydrotreatment of bio-oil over Ni-based catalyst. *Bioresource Technology*, *127*, 306–311. doi:10.1016/j.biortech.2012.07.119 PMID:23138057

Chapter 8
Emerging Contaminants in Landfill Leachate and Their Treatment Methods

Pradeep Kumar Singa
Universiti Teknologi PETRONAS, Malaysia

Jun-Wei Lim
Universiti Teknologi PETRONAS, Malaysia

Mohamed Hasnain Isa
Universiti Teknologi Brunei, Brunei

Yeek-Chia Ho
https://orcid.org/0000-0002-2820-9696
Universiti Teknologi PETRONAS, Malaysia

ABSTRACT

This chapter presents an overview of emerging contaminants in landfill leachate and their treatment methods. In addition to introducing the conventional contaminants present in the leachate, the chapter also details emerging contaminants such as poly-aromatic hydrocarbons (PAHs) and phthalate acid esters (PAEs) and their concentrations in various environmental matrices. PAHs and PAEs are highly carcinogenic, mutagenic, and teratogenic substances which is why they have attracted a lot of concern in the studies of water, air, and soil pollution. They affect the endocrinal activities in animals and humans, therefore they are known as endocrine disrupting compounds. Drawing on the treatment methods of leachate, the chapter explains physico-chemical, biological, and advanced oxidation processes. The chapter also advances the discussion on their importance and efficiency in the leachate treatment.

DOI: 10.4018/978-1-7998-0369-0.ch008

INTRODUCTION

This chapter discusses emerging micro-pollutants detected in the leachate mainly poly-aromatic hydrocarbons (PAHs) and phthalate acid esters (PAEs) and their treatment methods. At the beginning of the chapter, landfill composition, significance of landfill leachate, landfill age and various micro-pollutants were discussed. Various leachate treatment methods viz. physico-chemical, biological and advanced oxidation processes (AOPs) are discussed. The reaction mechanism of various AOPs such as Fenton, Photo-Fenton and E-Fenton are discussed in detail along with studies conducted on leachate treatment using these AOPs.

Landfill Leachate and Micro-Pollutants

Uncontrolled generation of leachate is the inevitable problem in landfilling. Leachate is the contaminated wastewater generated in the landfills due to percolation of rainwater through the layers of landfill. The layers of landfill consist of various types of hazardous waste including medical waste, sludge from treatment plants, lab chemicals, waste ash, refractory materials and halogenated organics. The hazardous landfill is closely monitored to avoid challenges like emission of landfill gases into atmosphere and damage of membrane which leads to contamination of ground water and other environmental media. The landfill gas contains many obnoxious gases and chemicals detrimental to environment promoting global warming. On the other hand, leachate contains high concentration of ammoniacal nitrogen, heavy metals and posses low BOD_5/COD ratio (Biglarijoo, Mirbagheri, Bagheri, & Ehteshami, 2017). Recently with technological improvements in the analytical methods, various micro-pollutants/emerging pollutants are being detected in the landfill leachate. The compounds which are found in the environment in low concentrations ranging from parts per billion (ppb) to parts per trillion (ppt) are said to be micro-pollutants or emerging pollutants (Clara et al., 2005). Many compounds are expected to be present in leachate in concentrations far below the quantification limits of standard analytical methods. On the other hand these compounds are hazardous and they do not overrule the environmental threats even at low concentrations (Öman & Junestedt, 2008). These pollutants include cosmetics, industrial compounds, pharmaceuticals, perfumes, waterproofing agents, insulating foams, plasticizers and drugs. The frequently detected micro-pollutants are pharmaceuticals and endocrine disrupting compounds. The age of the landfill also has significant impact on pollutants. In one of the studies conducted by Kulikowska and Klimiuk (2008) reported that the organic content expressed in terms of COD decreased from 1800 mg/L to 610 mg/L over a period of 4 years. Also few other studies reported that young landfills have higher COD concentrations as compared to older landfills. Similarly, the concentration of aromatic hydrocarbons decreases over a period of time depending on degradation with the landfill gas of each compound (Christensen et al., 2001). Previous studies also have reported the presence of micro-pollutants in surface water, groundwater, drinking water, industrial and domestic wastewater, hospital effluents, and runoff from agriculture (Kim, Cho, Kim, Vanderford, & Snyder, 2007; Ribeiro, Nunes, Pereira, & Silva, 2015). The detailed study of the micro-pollutants (PAHs and PAEs) is as described below.

POLYCYCLIC-AROMATIC HYDROCARBONS (PAHs)

Characteristics and Sources of PAHs

Polycyclic aromatic hydrocarbons (PAHs) are organic compounds that are generally colorless, white, or pale-yellow solids. They are a set of above one hundred organic compounds which are toxic and persistent in the environment (Abdel-Shafy & Mansour, 2016). United States Environmental Protection Agency (USEPA) has listed 16 PAHs as priority pollutants, owing to their carcinogenic, mutagenic and toxic properties (Rubio-Clemente, Torres-Palma, & Peñuela, 2014). They are composed of carbon and hydrogen assembled in two or more stable benzene rings. They are grouped into two categories namely low and high molecular weight. Low molecular weight has two or more benzene rings and high molecular weight have four or more condensed benzene rings. Also based on their structure they are categorized as alternant and non-alternant, alternant is entirely made of benzene rings and non-alternant has four to six non-aromatic rings (Wick, Haus, Sukkariyah, Haering, & Daniels, 2011). PAHs are found in both biotic and abiotic components of environment due to various natural and anthropogenic activities. Natural activities include forest fires, volcanic eruptions, natural oil seeps and anthropogenic include combustion process, industrial discharge, biomass burning, fossil fuel burning and waste incineration (Ravindra, Sokhi, & Van Grieken, 2008). Table 1 shows the characteristics of USEPA classified 16 priority PAHs.

Table 1. Characteristics of 16 priority PAHs

PAHs	Abbreviation	Chemical Formula	Benzene Rings	Molecular Weight	Solubility (mg/L)	Retention Time
Naphthalene	Nap	$C_{10}H_8$	2	128.17	32.0	7.57
Acenaphthylene	Acy	$C_{12}H_8$	3	152.2	3.93	11.19
Acenapthene	Ace	$C_{12}H_{10}$	3	154.2	3.40	11.61
Fluorene	Flu	$C_{13}H_{10}$	3	166.22	1.90	12.84
Phenanthrene	Phe	$C_{14}H_{10}$	3	178.23	1.30	15.12
Anthracene	Ant	$C_{14}H_{10}$	3	178.23	0.07	15.13
Fluoranthene	Fla	$C_{16}H_{10}$	4	202.26	0.265	17.96
Pyrene	Pyr	$C_{16}H_{10}$	4	202.26	1.132	18.48
Benz(a) anthracene	BaA	$C_{18}H_{12}$	4	228.28	0.014	21.38
Chrysene	Chr	$C_{18}H_{12}$	5	228.28	0.002	21.46
Benzo(b) fluoranthene	BbF	$C_{20}H_{12}$	5	252.31	0.0012	23.77
Benzo(K) fluoranthene	BkF	$C_{20}H_{12}$	5	252.31	0.0005	23.83
Benzo(a) pyrene	BaP	$C_{20}H_{12}$	5	252.31	0.110	24.45
Indeno (1,2,3,-CD) pyrene	Ind	$C_{22}H_{12}$	6	276.33	0.062	27.28
Dibenz (a,h) anthracene	DbA	$C_{22}H_{14}$	6	278.34	0.005	27.4
Benzo (g,h,i) perylene	BPer	$C_{22}H_{12}$	6	276.33	0.0026	28.05

(Pakpahan, Isa, Kutty, Chantara, & Wiriya, 2013)

Toxicity of PAHs

PAHs are carcinogenic, mutagenic and teratogenic which can cause various types of cancer such as lung, bladder, larynx, scrotum, breast, esophageal, prostate, kidney, skin, and pancreas. They are also known to be endocrine disrupters and suppress the immune system (Manoli & Samara, 2008). PAHs can have harmful effects on plants and animals of affected habitats, resulting in the uptake and accumulation of toxic chemicals in food chains (Liu, Han, Pan, & Riley, 2001). Naphthalene, is a commonly found micro-pollutant in drinking water which is known to be cataractogenic. Naphthalene binds covalently to molecules in liver, kidney and lung tissues, thereby enhancing its toxicity, it is also an inhibitor of mitochondrial respiration. Acute naphthalene poisoning in humans can lead to hemolytic anemia and nephrotoxicity. In addition, dermal and ophthalmological changes have been observed in workers occupationally exposed to naphthalene (Falahatpisheh, Donnelly, & Ramos, 2001; Goldman et al., 2001). Phenanthrene is known to be a photosensitizer of human skin, a mild allergen and mutagenic to bacterial systems under specific conditions. Not much information is available for other PAHs such as acenaphthene, fluranthene and flourene with respect to their toxicity in mammals. However, toxicity of benzo(a) pyrene, benzo(a)anthracene, benzo(b)fluoranthene, benzo(k)fluranthene, dibenz(a,h)anthracene, and indeno(1,2,3-c,d)pyrene has been experimented to show that they are carcinogenic (Šrám et al., 1999).

PHTHALIC ACID ESTERS (PAES)

The term "phthalate esters" is used to describe the mono-and di-esters of the isomeric benzene dicarboxylic acid, which is called phthalic acid. The most widely used of these, and the most studied toxicologically, are the esters of ortho-phthalic (o-phthalic) acid. Diesters may be symmetrical (e.g. diethyl phthalate) or asymmetrical (e.g. butyl benzyl phthalate). There are other types of phthalate esters such as the iso-phthalic acid and terephthalic acid but these are not produced or used in the same large volumes as those of o-phthalic acid (Watts, 1998). Phthalates are colorless or yellowish fluids of low viscosity and exist as liquids at room temperature. However, their viscosities increase with increasing length of the side chain. The majority of phthalates have low melting points (-25 to - 47°C) with the exception of DMP which has a melting point o f + 5.5°C. These phthalates have relatively high boiling points (280 to 370°C). The low melting point and high boiling point of these phthalate esters contribute to their usefulness as plasticizers, heat transfer fluids, and carriers. Specific gravities of these compounds range from 0.98 to 1.192. The diesters have low vapor pressures generally below 0 .01 mmHg (1.3 Pa) at room temperature (Ostrowski et al., 1999).

PAEs are produced both by natural and anthropogenic activities, however anthropogenic activities produce quite high amounts of PAEs as compared to natural origins. Phthalates or Phthalic acid esters have wide industrial applications since 1920s in the production of plastic goods as plasticizers (Serôdio & Nogueira, 2006). The total quantity of plastic production and consumption of PAEs throughout the world is 150 million tons and 8 million tons per annum respectively. Every year the consumption of PAEs is increasingly rapidly and so the production. In 1975 the production was 1.8 million tons which drastically increased to 8 million tons in the year 2011. PAEs have wide range of applications and their content could be up to 10-60% by weight (Sopheak Net, Sempéré, Delmont, Paluselli, & Ouddane, 2015). The various products which contains PAEs are building materials, PVC products, personal care products, detergents and surfactants, medical devices, children's toys, food products, pharmaceuticals, textiles, household appliances and cleaning materials.

PAEs have been found in air, municipal solid waste compost, sludge and lentic and lotic water bodies and drinking water throughout the world (Gao, Li, Wen, & Ren, 2014; S Net, Dumoulin, El-Osmani, Rabodonirina, & Ouddane, 2014). The concentrations of PAEs range from trace levels to high concentrations up to 500 mg/L in surface water, 10 mg/kg dw in sediments, 100 mg/kg dw in sludge, 1000 ng/m^3 in air with respect to the location and the activities nearby (Sopheak Net et al., 2015). PAEs have become a matter of concern due to their environmental risks and potential health issues throughout the world. US-EPA, Chinese water list and the European Union (EU) have included six PAEs (DMP, DEP, DnBP, BBzP, DEHP and DnOP) as priority pollutants. As per WHO the concentration of DEHP in drinking water should not exceed 8 µg/L (WHO, 2004). The maximum concentration of DEHP allowed in drinking water in the countries like Australia, New Zealand, Japan and USA is 9, 10, 100 and 6 µg/L respectively (Maycock & Inspectorate, 2008). The concentration of DEHP in marine and fresh waters should not exceed 1.3 µg/L as per EU guidelines to maintain the environmental safety and quality.

Studies Conducted on Phthalic Acid Esters

Tan (1995) investigated the concentration levels of PAEs in water and sediment samples from Klang River basin Malaysia. It was found that the concentrations of PAEs agree well with those reported from North America and Western Europe. DHEP concentration in freshwater sediment samples ranged from 180 mg/L to 30 mg/L. However, the levels of DHEP in sediment samples ranged from 493 ppb to 15.014 ppm. PAEs concentrations were high at sites where direct discharge points were located.

Bauer (1998) determined the concentrations of PAEs in leachate in a number of municipal landfills. It was found that the maximum levels of DEHP and DBP were 235.9 ppb and 62.7 ppb respectively. The maximum concentrations of DHEP and DBP in suspended particles were found to be 167.6 µg/g and 126.0 µg/g respectively. It was also reported that 85% of total PAEs and 87% of the hydrophobic DEHP was found in the dissolved phase. The concentrations of phthalates in solution and dissolved organic carbon (DOC) were correlated. Moreover, there was a negative correlation of the contents of suspended solids with PAEs in dissolved form. The study concluded that in municipal landfill leachates the DOC plays a vital role in transporting the hydrophobic PAEs as compared to the suspended particles.

Bauer et al. (1998) published the results of their work on degradation of phthalates in municipal landfill leachates. They observed that large amounts of phthalates were leached from plastic dumped at municipal landfills. The leachate transports phthalates either adsorbed on particulate matter or in the dissolved phase. Dissolved organic materials such as humic substances enhance the solubility of PAEs. The results of their work also showed that in the biochemical environment of municipal landfills, short chain PAEs can be degraded by microorganisms, however, there was no cleavage of the aromatic ring. Long chain PAEs are neither degraded abiotically nor by microorganisms and hence they can be leached and washed out of the landfills into the groundwater and continue to be a threat to the environment.

Stringer et al. (2000) conducted a survey on the presence of phthalate esters and other components in children's toys. The results of their survey indicated that the majority of soft PVC toys contain substantial proportions of phthalate esters that have the potential to leach out during use and therefore present an exposure risk to infants. Numerous studies have been carried out to investigate the concentration levels of PAEs in baby food and infant formula in plastic packaged food such as milk, cream, butter and cheese (Renner, 2000). Studies have also been conducted on migration of PAEs from PVC films during domestic use and cooking in microwave oven (Burke, 1999; Murphy & Wadey, 1988; Stringer et al., 2000).

LEACHATE TREATMENT METHODS

The suitable treatment method of leachate depends on the composition and characteristics, organic matter present as well as the age and structure of the landfill. Various treatment techniques have been developed in recent years, not only to degrade the contaminants present in the leachate but also to meet the discharge limits in various countries. Following are some treatment techniques commonly used in leachate treatment:

Biological Treatment

Biological treatment (suspended/attached growth) is reliable, simple and cost effective. It is commonly used for degrading the leachate containing high concentrations of BOD (Uygur & Kargı, 2004). When employed for young leachate, this technique produces good treatment efficiency with respect to COD, ammonia nitrogen and heavy metals. However, when employed to treat non-biodegradable leachate, it may not be able to degrade the COD levels to permitted standards for direct discharge in to receiving water bodies. Due to which the search for other effective techniques for the treatment of leachate has grown in recent decades (Kurniawan, Lo, & Chan, 2006). There are various biological treatment methods available and well known for treatment of various kinds of wastewaters. Among them are rotating biological treatment and activated sludge treatment.

Rotating Biological Contactors (RBCs)

Rotating biological contactors is the aerobic biological treatment method involving the attached growth systems. The system involves the circular discs mounted centrally over the horizontal shaft. During operation of RBCs, over 40% of the discs are submerged in to the wastewater which are mechanically driven for slow rotation. Over a period of time, microorganisms develop on the surface of disc and form slime layer of 1-2 mm thickness. When the discs are submerged in wastewater the microorganisms come in contact with organics and cause the oxidation. When the discs move out of wastewater they get oxygenated from atmosphere and thereby creating aerobic conditions. When the thickness of the biofilm becomes critical, portions of biofilm slough off the discs. The optimum rotation speed is about 2 rpm for 3 m diameter discs (Wiszniowski, Robert, Surmacz-Gorska, Miksch, & Weber, 2006).

Aerobic Activated Sludge

In case of activated sludge treatment, the reactor is continuously supplied with microorganisms along with oxygen and organic matter. These micro-organisms' intake the organic matter supplied and convert it by aerobic process. One portion of the transformation forms new microbial biomass and the other into carbon-dioxide, minerals and water. The water flow washes the microorganisms from reactor to the settler. Here, the growth of microorganisms is enhanced having sufficient density to decant, are retained and then removed with the underflow. Portion of this sludge is recycled to provide biomass for the treatment of new influent. The excess portion is discharged. The advantages of this method include easy implementation, adaption to larger large community and higher efficiency in the removal of all pollution parameters such as suspended solids, BOD and COD (Tchobanoglous, Burton, & Stensel, 2003).

Physico-Chemical Treatment

Physico-chemical treatment methods have been found to be appropriate not only for the degradation of refractory substances from stabilized leachate, also as an additional refining treatment for biologically treated leachate before discharging into the environment (Kurniawan et al., 2006).

Coagulation -Flocculation

This treatment is usually adopted for the removal of non-biodegradable organic compounds and heavy metals (Urase et al., 1997). Coagulation process involves addition of coagulant which neutralizes the charged colloidal particles. It is followed by flocculation of particles into heavy particles which makes the particles settle easily at the floor. This technique simplifies the removal of suspended solids and colloidal particles from the aqueous solution (G. Ayoub, Semerjian, Acra, Fadel, & Koopman, 2001).

Chemical Precipitation

This is a simple process that requires inexpensive equipment. This method is found be highly efficient in the removal of non-biodegradable organic compounds such as ammonia nitrogen and heavy metals from the leachate. During this process, dissolved ions get converted to insoluble solid phase. Usually the metal precipitate is in the hydroxide form. Lime is usually used as the precipitant but may also depend on the target of removal (Calli, Mertoglu, & Inanc, 2005).

Ammonia Stripping

It is quite effective and widely used treatment method for the removal of ammonia nitrogen from leachate. Leachate containing ammonia nitrogen and air phase are allowed to interact in a countercurrent flow in a stripping tower: ammonia nitrogen is transferred from the waste stream into the air and is then absorbed from the air into a strong acid or directly flux into the ambient air (Bonmatı & Flotats, 2003; Marttinen, Kettunen, Sormunen, Soimasuo, & Rintala, 2002).

Reverse Osmosis

Reverse osmosis (RO) is a technology wherein a solution is forced from high concentration region to a low concentration region passing through semipermeable membrane by applying external pressure which is higher than osmotic pressure. The significant application of this technology is in the purification of sea water. The salt laden sea water is pressurized against the semipermeable membrane and potable water free from salts is emerged on the other side of the membrane. This technique can also be applied for removal of suspended/colloidal materials and heavy metals from various kinds of wastewater (Jenkins, Mannapperuma, & Bakker, 2003).

Activated Carbon Adsorption

This is one of the most widely adopted techniques for the removal of recalcitrant organic compounds present in landfill leachate. It is a mass transfer phenomenon in which a substance is shifted from the liquid phase to the surface of solid and gets attached by physical or chemical interactions. Owing to the huge surface area, micro-porous structure and relatively high adsorption capacity adsorption using granular activated carbon (GAC) or powder activated carbon (PAC) has been used for the removal of organic and inorganic compounds from wastewater (Aziz, Adlan, Zahari, & Alias, 2004; Babel & Kurniawan, 2004).

Ion Exchange

Ion exchange is a technique where reversible interchange of ions takes place between solid and liquid phases with no permanent change in the structure of the solid. This method is effective in removing metal impurities. Usually ion exchange is preceded with biological treatment of leachate. Nevertheless, ion exchange is not usually used for the leachate treatment but it has received significant interest in countries like Germany for the removal of non-biodegradable compounds that contains humic substances (Fettig, 1999).

ADVANCED OXIDATION PROCESSES (AOPS)

The reason why AOPs are frequently being employed by many researchers is that they offer various possible processes for hydroxyl radical (OH•) generation, thus obeying with the precise treatment requirements. The following methods are most frequently used in AOPs 1. Fenton oxidation 2. Photo-Fenton 3. E-Fenton and 4. Ozonation.

Researchers started showing interest in AOPs only around 1995. A huge number of treatments devoted to its application to treat water, wastewater and soil is still rising extensively.

AOPs are considered to be "environmental friendly" as they do not indulge in transfer of pollutants from one phase to another (as in chemical precipitation and adsorption) and produce very negligible quantity of hazardous sludge (K. Ayoub, van Hullebusch, Cassir, & Bermond, 2010). AOPs are efficient in degrading almost all types of organic compounds into harmless products and rely on the generation of reactive hydroxyl radicals with a redox potential of 2.8 V. OH• is considered to be a highly reactive species after fluorine atom. It degrades most of organic pollutants with rate constants in the order of 10^6 -10^9 $M^{-1}s^{-1}$ (Esplugas, Yue, & Pervez, 1994). In these processes, OH initiates a chain of oxidation reactions yielding carbon-dioxide and water as final mineralization products. The production of OH• can be accomplished by adopting different AOPs based on different fundamentals which allows one to adopt suitable AOP in accordance with the specific characteristics of water/wastewater and treatment requirements (Y. Zhang, Han, Zhang, Dionysiou, & Nadagouda, 2015). Owing to these peculiar properties, several works have been carried out to investigate the application of AOPs in the degradation of various types of pollutants in aqueous solutions and soils (Cheng et al., 2016).

Post treatment may not be required in case of contaminants and their intermediate products are completely degraded by the mineralization and secondary wastes are not being generated (Andreozzi, Caprio, Insola, & Marotta, 1999). However, if total mineralization is not attained, a post treatment may be required before final disposal into the environment. The reaction byproducts in AOPs are usually highly biodegradable and less toxic but in few cases it may be vice-versa. They can be employed as post or pre-treatment for biological processes. The combination of various types of AOPs in an order of complementary processes is also a usual method to attain a biodegradable effluent that can be further treated by economical and conventional biological processes, dropping the residence time and reagent consumption in comparison with AOP alone. However, it is significant to attain complete removal of oxidizing agent before any biological treatment, as they can hinder the growth of micro-organisms (Oller, Malato, & Sánchez-Pérez, 2011).

Conventional Fenton Oxidation Process

The classical Fenton process was named after Henry Fenton following his studies in 1894 (Fenton, 1894). This method uses environmentally friendly reagents (H_2O_2 and Fe^{2+}) at room temperature and pressure and doesn't need any special equipment. The efficiency of Fenton process is dependent on the pH of the solution due to hydrogen peroxide and iron speciation factors. Several researchers have worked upon the optimum pH range for Fenton reaction and it was found to be 3, irrespective of the target substrate (Babuponnusami & Muthukumar, 2014).

The combination of the two reagents hydrogen peroxide and iron generates highly oxidative hydroxyl radicals which degrade the organic compounds present in wastewater. The generated hydroxyl radicals completely destruct the organic compounds to form simpler compounds of CO_2, inorganic salts and water. Fenton process involves the sequence of following reactions.

$$Fe^{2+} + H_2O_2 \rightarrow Fe^{3+} + OH^- + OH\bullet \quad (1)$$

Fe^{3+} is again converted to Fe^{2+}

$$H_2O_2 + Fe^{3+} \rightarrow Fe^{2+} + HO + H^+ \quad (2)$$

Additionally, hydroxyl radicals rapidly destroy ferrous ions.

$$Fe^{2+} + OH\bullet \rightarrow Fe^{3+} + OH^- \quad (3)$$

Thereafter the organic matter would be degraded by hydroxyl radicals generated in the landfill leachate:

$$RH + OH\bullet \rightarrow H_2O + R\bullet \quad (4)$$

$$R\bullet + Fe^{3+} \rightarrow R^+ + Fe^{2+} \quad (5)$$

The advantages of this process are easy implementation and operation, low cost reagents, high efficiency and doesn't require energy for H_2O_2 activation. However, this process also has a few limitations which include fast consumption of Fe^{2+} relative to its rate of regeneration, generation of sludge that requires post treatment, narrow pH range to operate, the complexity of some iron species and the possible waste of oxidants. The efficiency of Fenton process can be enhanced by optimizing the operational conditions such as pH, dosage of Fenton reagents, reaction time, feeding mode, agitation rate and temperature. The optimum conditions differ based on the characteristics of water or wastewater.

Studies on Leachate Treatment by Fenton Process

Numerous studies have been conducted using Fenton oxidation process whereas as the study conducted by Mohajeri et al. (2010) was typical which included optimization of all the operational conditions of Fenton oxidation process for the treatment of municipal landfill leachate. The highest COD and color removal attained under optimum conditions were 58.1% and 78.3% respectively. The best operating conditions in their study observed were pH=3, molar ratio (H_2O_2/Fe^{2+}) = 3, agitation rate = 400 rpm, reaction time =120 min, dosage of Fe= 560 mg/L, dosage of H_2O_2 = 1020 mg/L and temperature = 30oC. It was concluded that step wise addition of Fenton reagents was more effective compared to addition of all the volume in one step. Likewise, Amiri et al. (2014) studied the Fenton oxidation treatment of leachate using response surface methodology. The conclusion drawn were molar ratio (H_2O_2/Fe^{2+}) = 17.72, dosage of Fe^{2+} = 195 mM were optimum which led to highest COD removal of 69%. Chen et al. (2014) compared their study on treatment of leachate by Fenton process with other researchers adopting various advanced oxidation processes. The study reports that Fenton oxidation is a better method in comparison to other advanced oxidation process. The importance of various operational conditions and their effect on Fenton process is as discussed below.

Factors Affecting Fenton Process

pH

pH has a major role in the Fenton process since it has influence on the production of hydroxyl radicals and the concentration of ferrous ions in the solution (H. Zhang, Choi, & Huang, 2005). When the pH increases beyond 5, iron ions particularly the Fe^{3+} precipitates, which hinders the regeneration of ferrous ions. Consequently, the quantity of the catalyst in the Fenton's process reduces leading to reduction in the process efficiency. In addition to that reduction in the removal efficiency at pH more than 5 can be due to production of lesser amounts of hydroxyl radicals as compared to the production rate at pH 3. Higher pH values enhance the production of CO_3^{2-} and HCO_3^- which scavenge the hydroxyl radicals necessary for the oxidation of organics and also reduces the oxidation potential of hydroxyl radicals which ultimately leads to reduction in the efficiency. It should be noted that hydrogen peroxide is not stable in basic solution and it may auto-decompose to oxygen and water at pH more than 5 (Deng & Englehardt, 2006). It has been reported that stable hydroxyl radicals are generated at pH values between 2-5 and high efficiency can be obtained within this pH range (Sedlak & Andren, 1991). On the other hand, if the pH is less than 2, hydrogen peroxide cannot be decomposed to form potential hydroxyl radicals by Fe^{2+}. In this particular case, hydrogen peroxide converts to form H_3O^{2+} by attaining additional proton. H_3O^{2+} is electrophilic which inhibits the reaction between hydrogen peroxide and ferrous ions. Ultimately leading to reduction in the efficiency.

Dosage of Ferrous Ion

Optimum dosage of Ferrous ion helps to attain maximum removal efficiency. Higher dosage of ferrous ion scavenges the highly reactive hydroxyl radicals required for the degradation of organic contaminants. On the other hand, lower dosages are not sufficient to undergo the degradation process efficiently

(Kang & Hwang, 2000; S. H. Lin & Leu, 1999). Additionally, excessive amounts of ferrous ions added increases the total dissolved solids concentration which is quite undesirable. Optimum dosage varies with the matrix of wastewater and initial concentration of pollutant. Hence conducting laboratory scale experiments is necessary to fix optimum dosages of Fenton reagents.

Dosage of Hydrogen Ion

The dosage of hydrogen peroxide is as important as ferrous ion, even in this case the degradation efficiency increases with increase in hydrogen peroxide dosage. Whereas excessive addition of hydrogen peroxide contributes to the COD. To be more specific each mg/L of residual H_2O_2 equals to 0.27 mg/L of COD (Saber, Hasheminejad, Taebi, & Ghaffari, 2014). The unused hydrogen peroxide scavenges the hydroxyl radicals reducing the efficiency (S. H. Lin & Lo, 1997). In addition, hydrogen peroxide is toxic to many organisms (Ito et al., 1998). Hence laboratory scale experiments are required to determine the optimum dosage required for the particular target pollutants and wastewater to be studied.

Reagents Feeding Mode

Many researchers have reported the Fenton reagents feeding mode (Hermosilla, Cortijo, & Huang, 2009; Mohajeri, Aziz, Isa, Bashir, et al., 2010). All of them concluded that step-wise addition of reagents increased the process efficiency by around 10% as compared to addition of total volume in a single step (Umar, Aziz, & Yusoff, 2010). It was also investigated by Yoo et al. (2001) that continuous addition of reagents led to 25% less consumption of reagents and increase in the COD removal efficiency by around 5%. Similarly another study conducted by Primo et al. (Primo, Rivero, & Ortiz, 2008) observed that COD removal increased by around 8% by adding the reagents in multiple doses.

Temperature

Studies on the effect of temperature on Fenton process are quite limited. The effect of temperature on process efficiency is not significant as compared to other operating conditions. Temperature less than 8.3 °C. and higher than 50 °C. reduces the efficiency (Umar et al., 2010). However, many researches have reported that good efficiency can be obtained with ambient temperatures. whereas Lin and Lo (1997) found that the optimum temperature for Fenton process is 30 °C and Rivas et al. (Rivas, Beltran, Frades, & Buxeda, 2001) reported that the efficiency of degradation remains same up to the temperature of 40 °C. Nevertheless, cooling is required if the temperature is more than 40 °C because hydrogen peroxide cannot be utilized completely due to breakdown of hydrogen peroxide into water and oxygen (Babuponnusami & Muthukumar, 2014).

Photo-Fenton Process

Photo-Fenton process is a combination of Fenton reagents (H_2O_2 and Fe^{2+}) and UV–vis radiation ($\lambda <$ 600 nm) that gives rise to extra OH• radicals by two additional reactions: (i) photo-reduction of Fe^{3+} to Fe^{2+} ions as shown in Eq. (6) (Faust & Hoigné, 1990) and (ii) peroxide photolysis via shorter wavelengths as shown in Eq. (7):

$$Fe(OH)^{2+} + h\nu \rightarrow Fe^{2+} + OH\bullet, \lambda < 580 \text{ nm} \quad (6)$$

$$H_2O_2 + h\nu \rightarrow 2\,OH\bullet, \lambda < 310 \text{ nm} \quad (7)$$

The ferrous ions produced by UV light enter the Fenton reaction to produce extra hydroxyl radicals. Furthermore, the photo-Fenton process has higher oxidation rate as compared to conventional Fenton process (Ortega-Liébana et al., 2012). Additionally, there is enormous reduction in the consumption of iron and generation of sludge. In addition to that it has been reported that photo-Fenton process using UV light or solar light has given results in the killing of the microorganisms in polluted water bodies for potable or irrigation purpose (Gárcia-Fernández, Polo-López, Oller, & Fernández-Ibáñez, 2012). The main advantage of decreasing the quantity of catalyst required leads to a compensation in the higher costs of UV as well as to reduction of the volume of sludge. There are several factors which effect the Photo-Fenton process as described below.

Factors Effecting Photo-Fenton Process

The significant factor which effects the Photo-Fenton process is UV intensity and other factors remains same as conventional Fenton process.

UV Radiation

Literature reports studies carried out on photolysis of PAHs under artificial conditions (Jacobs, Weavers, & Chin, 2008; Shemer & Linden, 2007). Under these conditions, the emission spectrum of the light source may overlap the absorption spectrum of the substance of interest thereby causing its degradation (Shemer & Linden, 2007). The wavelength that can be absorbed by a compound depends on its structure (Sanches et al., 2011). The absorption spectra of PAHs range from 210 nm to 386 nm (Dabestani & Ivanov, 1999). The most widely used artificial light for studies are LP-UV and MP-UV. Apart from the absorption spectrum of the compound, photolysis efficiency also depends on the intensity of radiation, pH, temperature and constituents of water or wastewater. It is well known that an increase in intensity of radiation increases the photon absorption and the subsequent PAH photolysis (Bertilsson & Widenfalk, 2002; Fasnacht & Blough, 2003; Jacobs et al., 2008; Shemer & Linden, 2007).

Studies on Leachate Treatment by Photo-Fenton Process

Many authors have reported studies conducted on treatment of leachate using Photo-Fenton process. Primo et al. (2008) evaluated the efficiency of various AOPs. In this study it was observed that addition of hydrogen peroxide alone did not contributed much to the COD removal efficiency. While addition of UV radiation to the system drastically increased the efficiency up to 36%. And finally attained COD removal efficiency of around 80% within reaction time of one hour. Morais et al. (2005) compared the UV/H_2O_2 and photo-Fenton process for the treatment of mature leachate. The performance of Photo-Fenton process was slightly higher with a COD removal of 57.5% as compared to the UV/H_2O_2 process with 55.5% for a reaction time of one hour. Sanitary leachate treatment was conducted by Vilar et al. (2011) using solar photo-Fenton process. In this study leachate was mineralized by 57% with a reduction of aromatic content by 86%. Hermosilla et al. (2009) compared the Fenton and Photo-Fenton process and observed no difference in the efficiency of both processes. However, it was noticed that Photo-Fenton process consumed lesser amounts of ferrous ions in comparison to Fenton process to yield same results. Additionally, the sludge generation in the Photo-Fenton process was drastically reduced to 1% from 25%. Heterogeneous (TiO_2/UV, $TiO_2H_2O_2$/UV) and homogeneous (H_2O_2/UV, Fe^{2+}/H_2O_2/UV) photo-catalytic process for treatment of mature leachate was studied by (Rocha, Vilar, Fonseca, Saraiva, & Boaventura, 2011). It was found that degradation efficiency of Photo-Fenton process was 20 times more than heterogeneous photo-catalytic process. However, acidification of heterogeneous photo-catalytic process increased the degradation efficiency by 10 folds. Heng et al. (2012) adopted Photo-Fenton as a pretreatment method for mature landfill leachate, where in it was observed that UV lamp of 6W emitting wavelength of 365nm was efficient enough to achieve COD and color removal of 70% and 80% respectively.

Electro-Fenton (E-Fenton) Process

Electro-Fenton process is fusion of electrochemical oxidation and Fenton process. As two processes are combined it yields higher efficiency as compared to conventional Fenton process. The only drawback of this technic is production of iron-sludge which can be minimized by application of optimum amounts of Fenton reagents. As this method is combination of two processes, it requires lesser dosages of Fenton reagents and still produces higher efficiency. The efficiency of this process can be further enhanced by passing UV light through the wastewater which is known as photo-electro-Fenton (Nidheesh & Gandhimathi, 2012; Ribeiro et al., 2015). The factors pH, Fenton reagent dosages, temperature and agitation rate remain same as Fenton process. Whereas the factors current intensity, electrode distance and concentration of electrolyte are few more additional factors which effect the Electro-Fenton process.

Factors Effecting E-Fenton Process

Current Density

In this method current density plays a significant role in enhancing the process efficiency. Increase in current density increases the production of hydroxyl radicals up to a certain point and beyond which further increase in the current density decreases the efficiency of the system (Eq.8) (Sankara Narayanan, Magesh, & Rajendran, 2003). Oxygen get discharged at anode at higher current densities (Eq.9) and hydrogen evolves at cathode (Eq. 10). These reactions affect the process leading to reduction in the efficiency of the E-Fenton process (H. Zhang, Fei, Zhang, & Tang, 2007). Literature reported that degradation rate of organic pollutants reaches equilibrium beyond 300 mA, this may be due to formation of H_2O as in Eq.11 (Özcan, Şahin, Koparal, & Oturan, 2008). Previous studies have also discussed that maximum current density should be 10 A/m^2 in the E-Fenton process (Fang, Guangming, Xiuhua, Huikang, & Juwen, 2004).

$$Fe^{3+} + e^- \rightarrow Fe^{2+} \tag{8}$$

$$2H_2O \rightarrow 4H^+ + O_2 + 4e^- \tag{9}$$

$$2H^+ + 2e^- \rightarrow H_2 \tag{10}$$

$$4H^+ + O_2 + 4e^- \rightarrow 2H_2O \tag{11}$$

Distance Between Electrodes

Distance between electrodes is a factor effecting the efficiency of E-Fenton process. Too short distance or long distances between the electrodes leads to lesser efficiency. Hence optimum electrodes distance has to be maintained to prevent Fe^{2+} oxidation to Fe^{3+} at the anode and limiting mass transfer of Fe^{3+} to the cathode surface where regeneration of Fe^{2+} takes place (H. Zhang, Zhang, & Zhou, 2006). Zhang et al. (2006) in their study on COD removal from landfill leachate reported that there was no change in the efficiency when electrode distance was changed from 1.3 cm to 2.1 cm. The results of this study agreed with the study conducted by (Atmaca, 2009) who observed that change in electrode distance has insignificant effect on the efficiency of the process, and also reported that longer distances in the electrode leads to higher energy consumption in the E-Fenton process.

Supporting Electrolyte

The electrolyte used in E-Fenton process enhances the conductivity in the solution and also prompts electron transfer thus providing positive effect on the reaction (Jiang & Zhang, 2007). In case of E-Fenton process, sodium sulphate is widely used as electrolyte. (Zhou, Yu, Lei, & Barton, 2007) observed that increase in the sodium sulphate dosage increased the current density which ultimately increased the efficiency. According to Ghoneim et al. (2011) the optimum dosage of sodium sulphate for sunset yellow FCF is 0.05 M and sodium sulphate has the highest rate of de-colorization of sunset yellow FCF compared to chloride electrolytes of the same concentration . Previous studies have also reported that efficiency of the process dropped when the concentration of sodium sulphate used was 0.2 M. This may be due to scavenging of hydroxyl radicals by high concentration of SO_4 as explained in Eq.12 (Zhou et al., 2007).

$$HO^{\bullet} + SO_4^{2-} \xrightarrow{\text{yields}} HO^- + SO_4^{\bullet -} \tag{12}$$

Electrode Materials

Various types of electrode materials have been used by researchers in the electrochemical oxidation process. It is important to choose appropriate electrode material in this process because unstable anode can lead to deterioration of electrode (H. Zhang et al., 2007). Pt has been used as electrode material by many of the researches due to its good conductivity and chemical stability even at high potentials and in corrosive media (Panizza & Cerisola, 2009). Whereas Pt electrode is expensive as compared to other electrodes which makes the treatment process uneconomical. Boron-doped diamond (BDD) electrode has been used as anode for the treatment of various kinds of wastewaters (Da Pozzo et al., 2005; Sires et al., 2007). Titanium is one of the widely used electrode materials in the E-Fenton process. Titanium coated with IrO_2/RuO_2 was used by (Huang, Huang, Chang, & Chen, 2008) and (Prabhakaran, Kannadasan, & Basha, 2009) in the E-Fenton process. Cathode is known to be working electrode in E-Fenton process because efficiency of the process depends on the type of cathode used (Nidheesh & Gandhimathi,

2012). Various materials used as cathode in the treatment of wastewater are commercial graphite felt (Khataee, Vatanpour, & Ghadim, 2009), carbon felt (Guivarch, Trevin, Lahitte, & Oturan, 2003), BDD plate (Cruz-González et al., 2010), graphite (Rosales, Pazos, Longo, & Sanromán, 2009; C.-T. Wang, Hu, Chou, & Kuo, 2008), reticulated vitreous carbon (RVC) (Ghoneim et al., 2011), activated carbon fiber (ACF) (A. Wang, Qu, Ru, Liu, & Ge, 2005), titanium (Yaqub, Isa, Kutty, & Ajab, 2014), carbon sponge (Özcan, Şahin, Koparal, & Oturan, 2009), Pt flakes (Yuan, Fan, Zhang, Tong, & Liao, 2011), and stainless steel (Prabhakaran et al., 2009).

Studies on Leachate Treatment by E-Fenton Process

Studies on leachate treatment by E-Fenton process are very few compared to other advanced oxidation processes. A study conducted by (Altin, 2008) achieved more than 90% color removal. Similarly, Atmaca (2009) reported 26% and 87% removal of NH_4-N and PO_4-P respectively under optimum conditions. Additionally, 72% COD and 90% color removal was reported. These studies agree well with studies conducted by Mohajeri et al. (2010), achieving 94% COD removal and 95.8% color removal by E-Fenton process. Zhang et al. (2006) investigated treatment of high strength leachate using RuO_2 and IrO_2 as anode electrodes. It was observed that the process was rapid initially up to 30 min and then slowed down to complete the process within 75 min.

Studies on Application of Advanced Oxidation Processes for Removal of PAHs

Advanced oxidation (Fenton, Photo-Fenton and Electrochemical oxidation) processes have been extensively used for degradation of organic pollutants viz. pharmaceuticals, dyes, poly-aromatics, pesticides, phenols and chloro-phenols (Rubio-Clemente et al., 2014). Rocha et al. (2013) reported that about 92% of PAHs were removed from petrochemical wastewater using solar photo-Fenton. Tran et al. (2009) and Duarte et al. (2011) conducted electrochemical oxidation of synthetic water containing 16 PAHs achieving 80% and 85% removal efficiency respectively. Engwall et al. (1999) studied the photo-Fenton treatment of synthetic water containing mixture of 85% PAHs, 10% phenolic compounds and 5% heterocyclic compounds and reported that 2 and 3 ring PAHs were effectively degraded whereas 4 and 5 ring PAHs were not completely degraded. Yaqub et al. (2014) studied electrochemical degradation of synthetic and produced water containing 16 PAHs. They have achieved maximum removal efficiency of 82% and 93% without and with electrolyte respectively. Lin et al. (2016) in their study on degradation of 16 PAHs in textile dying sludge using Fenton process observed removal efficiency of 83.5%.

CONCLUSION

In this chapter, various micro-pollutants were discussed. Conventional leachate treatment methods such as physico-chemical and biological treatment methods were discussed along with advanced oxidation processes. The mechanism of various advanced oxidation processes such as Fenton, Photo-Fenton, E-Fenton and Ozonation were discussed in detail along with application of these methods by various researchers. In summary, review conducted indicates that micro-pollutants (PAHs and PAEs) are among the contaminants which are quite complex and are not amenable to treat by the conventional methods. Physico-chemical characteristics such as high molecular weight and hydrophobicity cause complexities in treatment of these chemicals. The conventional methods (air stripping and adsorption) do not completely destroy the contaminants, but only transfer it from one phase to another whereas biological methods are prone to poisoning of microorganisms by the heavy metals and the phthalates themselves. Hence advanced oxidation processes are the only best options to degrade these micro-pollutants in to simpler compounds.

ACKNOWLEDGMENT

The authors would like to express their sincere gratitude to UTP for funding this project under YUTP grant (0153AA-E15).

REFERENCES

Abdel-Shafy, H. I., & Mansour, M. S. (2016). A review on polycyclic aromatic hydrocarbons: Source, environmental impact, effect on human health and remediation. *Egyptian Journal of Petroleum*, *25*(1), 107–123. doi:10.1016/j.ejpe.2015.03.011

Altin, A. (2008). An alternative type of photoelectro-Fenton process for the treatment of landfill leachate. *Separation and Purification Technology*, *61*(3), 391–397. doi:10.1016/j.seppur.2007.12.004

Amiri, A., & Sabour, M. R. (2014). Multi-response optimization of Fenton process for applicability assessment in landfill leachate treatment. *Waste Management (New York, N.Y.)*, *34*(12), 2528–2536. doi:10.1016/j.wasman.2014.08.010 PMID:25212172

Andreozzi, R., Caprio, V., Insola, A., & Marotta, R. (1999). Advanced oxidation processes (AOP) for water purification and recovery. *Catalysis Today*, *53*(1), 51–59. doi:10.1016/S0920-5861(99)00102-9

Atmaca, E. (2009). Treatment of landfill leachate by using electro-Fenton method. *Journal of Hazardous Materials*, *163*(1), 109–114. doi:10.1016/j.jhazmat.2008.06.067 PMID:18657358

Ayoub, G., Semerjian, L., Acra, A., Fadel, M. E., & Koopman, B. (2001). Heavy metal removal by coagulation with seawater liquid bittern. *Journal of Environmental Engineering*, *127*(3), 196–207. doi:10.1061/(ASCE)0733-9372(2001)127:3(196)

Ayoub, K., van Hullebusch, E. D., Cassir, M., & Bermond, A. (2010). Application of advanced oxidation processes for TNT removal: A review. *Journal of Hazardous Materials*, *178*(1-3), 10–28. doi:10.1016/j.jhazmat.2010.02.042 PMID:20347218

Aziz, H. A., Adlan, M. N., Zahari, M. S. M., & Alias, S. (2004). Removal of ammoniacal nitrogen (N-NH3) from municipal solid waste leachate by using activated carbon and limestone. *Waste Management & Research*, *22*(5), 371–375. doi:10.1177/0734242X04047661 PMID:15560441

Babel, S., & Kurniawan, T. A. (2004). Cr (VI) removal from synthetic wastewater using coconut shell charcoal and commercial activated carbon modified with oxidizing agents and/or chitosan. *Chemosphere*, *54*(7), 951–967. doi:10.1016/j.chemosphere.2003.10.001 PMID:14637353

Babuponnusami, A., & Muthukumar, K. (2014). A review on Fenton and improvements to the Fenton process for wastewater treatment. *Journal of Environmental Chemical Engineering*, *2*(1), 557–572. doi:10.1016/j.jece.2013.10.011

Bauer, M., & Herrmann, R. (1998). Dissolved organic carbon as the main carrier of phthalic acid esters in municipal landfill leachates. *Waste Management & Research*, *16*(5), 446–454. doi:10.1177/0734242X9801600507

Bauer, M., Herrmann, R., Martin, A., & Zellmann, H. (1998). Chemodynamics, transport behaviour and treatment of phthalic acid esters in municipal landfill leachates. *Water Science and Technology*, *38*(2), 185–192. doi:10.2166/wst.1998.0135

Bertilsson, S., & Widenfalk, A. (2002). Photochemical degradation of PAHs in freshwaters and their impact on bacterial growth–influence of water chemistry. *Hydrobiologia*, *469*(1-3), 23–32. doi:10.1023/A:1015579628189

Biglarijoo, N., Mirbagheri, S. A., Bagheri, M., & Ehteshami, M. (2017). Assessment of effective parameters in landfill leachate treatment and optimization of the process using neural network, genetic algorithm and response surface methodology. *Process Safety and Environmental Protection*, *106*, 89–103. doi:10.1016/j.psep.2016.12.006

Bonmatı, A., & Flotats, X. (2003). Air stripping of ammonia from pig slurry: Characterisation and feasibility as a pre-or post-treatment to mesophilic anaerobic digestion. *Waste Management (New York, N.Y.)*, *23*(3), 261–272. doi:10.1016/S0956-053X(02)00144-7 PMID:12737968

Burke, M. (1999). Phthalate standard endorsed by industry. *Environmental Science & Technology*, *33*(19), 398A–399A. doi:10.1021/es993032z PMID:21657395

Calli, B., Mertoglu, B., & Inanc, B. (2005). Landfill leachate management in Istanbul: Applications and alternatives. *Chemosphere*, *59*(6), 819–829. doi:10.1016/j.chemosphere.2004.10.064 PMID:15811410

Chen, Y., Liu, C., Nie, J., Wu, S., & Wang, D. (2014). Removal of COD and decolorizing from landfill leachate by Fenton's reagent advanced oxidation. *Clean Technologies and Environmental Policy*, *16*(1), 189–193. doi:10.100710098-013-0627-1

Cheng, M., Zeng, G., Huang, D., Lai, C., Xu, P., Zhang, C., & Liu, Y. (2016). Hydroxyl radicals based advanced oxidation processes (AOPs) for remediation of soils contaminated with organic compounds: A review. *Chemical Engineering Journal*, *284*, 582–598. doi:10.1016/j.cej.2015.09.001

Christensen, T. H., Kjeldsen, P., Bjerg, P. L., Jensen, D. L., Christensen, J. B., Baun, A., . . . Heron, G. J. A. G. (2001). Biogeochemistry of landfill leachate plumes. *16*(7-8), 659-718.

Clara, M., Strenn, B., Gans, O., Martinez, E., Kreuzinger, N., & Kroiss, H. (2005). Removal of selected pharmaceuticals, fragrances and endocrine disrupting compounds in a membrane bioreactor and conventional wastewater treatment plants. *Water Research*, *39*(19), 4797–4807. doi:10.1016/j.watres.2005.09.015 PMID:16242170

Cruz-González, K., Torres-López, O., García-León, A., Guzmán-Mar, J., Reyes, L., Hernández-Ramírez, A., & Peralta-Hernández, J. (2010). Determination of optimum operating parameters for Acid Yellow 36 decolorization by electro-Fenton process using BDD cathode. *Chemical Engineering Journal*, *160*(1), 199–206. doi:10.1016/j.cej.2010.03.043

Da Pozzo, A., Merli, C., Sirés, I., Garrido, J. A., Rodríguez, R. M., & Brillas, E. (2005). Removal of the herbicide amitrole from water by anodic oxidation and electro-Fenton. *Environmental Chemistry Letters*, *3*(1), 7–11. doi:10.100710311-005-0104-0

da Rocha, O. R. S., Dantas, R. F., Bezerra Duarte, M. M., Lima Duarte, M. M., & da Silva, V. L. (2013). Solar photo-Fenton treatment of petroleum extraction wastewater. *Desalination and Water Treatment*, *51*(28-30), 5785–5791. doi:10.1080/19443994.2013.792136

Dabestani, R., & Ivanov, I. N. (1999). A compilation of physical, spectroscopic and photophysical properties of polycyclic aromatic hydrocarbons. *Photochemistry and Photobiology*, *70*(1), 10–34.

De Morais, J. L., & Zamora, P. P. (2005). Use of advanced oxidation processes to improve the biodegradability of mature landfill leachates. *Journal of Hazardous Materials*, *123*(1-3), 181–186. doi:10.1016/j.jhazmat.2005.03.041 PMID:15878233

Deng, Y., & Englehardt, J. D. (2006). Treatment of landfill leachate by the Fenton process. *Water Research*, *40*(20), 3683–3694. doi:10.1016/j.watres.2006.08.009 PMID:17045628

Duarte, J. P. S., Martínez-Huitle, C. A., & da Silva, D. R. (2011). Electrochemical treatment for removing petroleum polycyclic aromatic hydrocarbons(PAHs) from synthetic produced water using a DSA-type anode: Preliminary results. *Sustainable Environment Research*, *21*(5), 329–335.

Engwall, M. A., Pignatello, J. J., & Grasso, D. (1999). Degradation and detoxification of the wood preservatives creosote and pentachlorophenol in water by the photo-Fenton reaction. *Water Research*, *33*(5), 1151–1158. doi:10.1016/S0043-1354(98)00323-6

Esplugas, S., Yue, P., & Pervez, M. I. (1994). Degradation of 4-chlorophenol by photolytic oxidation. *Water Research*, *28*(6), 1323–1328. doi:10.1016/0043-1354(94)90297-6

Falahatpisheh, M., Donnelly, K., & Ramos, K. (2001). Antagonistic interactions among nephrotoxic polycyclic aromatic hydrocarbons. *Journal of Toxicology and Environmental Health. Part A.*, *62*(7), 543–560. doi:10.1080/152873901300007833 PMID:11289703

Fang, Z., Guangming, L., Xiuhua, Z., Huikang, H., & Juwen, H. (2004). Study status and progress in wastewater treatment by electro-Fenton method [J]. *Industrial Water Treatment, 12*, 003.

Fasnacht, M. P., & Blough, N. V. (2003). Kinetic analysis of the photodegradation of polycyclic aromatic hydrocarbons in aqueous solution. *Aquatic Sciences*, *65*(4), 352–358. doi:10.100700027-003-0680-7

Faust, B. C., & Hoigné, J. (1990). Photolysis of Fe (III)-hydroxy complexes as sources of OH radicals in clouds, fog and rain. *Atmospheric Environment. Part A, General Topics*, *24*(1), 79–89. doi:10.1016/0960-1686(90)90443-Q

Fenton, H. J. H. (1894). LXXIII.—Oxidation of tartaric acid in presence of iron. *Journal of the Chemical Society Transactions*, *65*(0), 899–910. doi:10.1039/CT8946500899

Fettig, J. (1999). Removal of humic substances by adsorption/ion exchange. *Water Science and Technology*, *40*(9), 173–182. doi:10.2166/wst.1999.0470

Gao, D., Li, Z., Wen, Z., & Ren, N. (2014). Occurrence and fate of phthalate esters in full-scale domestic wastewater treatment plants and their impact on receiving waters along the Songhua River in China. *Chemosphere*, *95*, 24–32. doi:10.1016/j.chemosphere.2013.08.009 PMID:24001662

Gárcia-Fernández, I., Polo-López, M. I., Oller, I., & Fernández-Ibáñez, P. (2012). Bacteria and fungi inactivation using Fe3+/sunlight, H2O2/sunlight and near neutral photo-Fenton: A comparative study. *Applied Catalysis B: Environmental*, *121*, 20–29. doi:10.1016/j.apcatb.2012.03.012

Ghoneim, M. M., El-Desoky, H. S., & Zidan, N. M. (2011). Electro-Fenton oxidation of Sunset Yellow FCF azo-dye in aqueous solutions. *Desalination*, *274*(1-3), 22–30. doi:10.1016/j.desal.2011.01.062

Goldman, R., Enewold, L., Pellizzari, E., Beach, J. B., Bowman, E. D., Krishnan, S. S., & Shields, P. G. (2001). Smoking increases carcinogenic polycyclic aromatic hydrocarbons in human lung tissue. *Cancer Research*, *61*(17), 6367–6371. PMID:11522627

Guivarch, E., Trevin, S., Lahitte, C., & Oturan, M. A. (2003). Degradation of azo dyes in water by electro-Fenton process. *Environmental Chemistry Letters*, *1*(1), 38–44. doi:10.100710311-002-0017-0

Heng, G. C., Elmolla, E. S., & Chaudhuri, M. (2012). Optimization of photo-Fenton treatment of mature landfill leachate. *Nature Environment and Pollution Technology*, *11*(1), 65–72.

Hermosilla, D., Cortijo, M., & Huang, C. P. (2009). Optimizing the treatment of landfill leachate by conventional Fenton and photo-Fenton processes. *The Science of the Total Environment*, *407*(11), 3473–3481. doi:10.1016/j.scitotenv.2009.02.009 PMID:19278717

Huang, Y.-H., Huang, Y.-F., Chang, P.-S., & Chen, C.-Y. (2008). Comparative study of oxidation of dye-Reactive Black B by different advanced oxidation processes: Fenton, electro-Fenton and photo-Fenton. *Journal of Hazardous Materials*, *154*(1-3), 655–662. doi:10.1016/j.jhazmat.2007.10.077 PMID:18063472

Ito, K., Jian, W., Nishijima, W., Baes, A. U., Shoto, E., & Okada, M. (1998). Comparison of ozonation and AOPs combined with biodegradation for removal of THM precursors in treated sewage effluents. *Water Science and Technology*, *38*(7), 179–186. doi:10.2166/wst.1998.0291

Jacobs, L. E., Weavers, L. K., & Chin, Y. P. (2008). Direct and indirect photolysis of polycyclic aromatic hydrocarbons in nitrate-rich surface waters. *Environmental Toxicology and Chemistry*, *27*(8), 1643–1648. doi:10.1897/07-478.1 PMID:18275255

Jenkins, B., Mannapperuma, J., & Bakker, R. (2003). Biomass leachate treatment by reverse osmosis. *Fuel Processing Technology*, *81*(3), 223–246. doi:10.1016/S0378-3820(03)00010-9

Jiang, C., & Zhang, J. (2007). Progress and prospect in electro-Fenton process for wastewater treatment. *Journal of Zhejiang University. Science A*, *8*(7), 1118–1125. doi:10.1631/jzus.2007.A1118

Kang, Y. W., & Hwang, K.-Y. (2000). Effects of reaction conditions on the oxidation efficiency in the Fenton process. *Water Research*, *34*(10), 2786–2790. doi:10.1016/S0043-1354(99)00388-7

Khataee, A., Vatanpour, V., & Ghadim, A. A. (2009). Decolorization of CI Acid Blue 9 solution by UV/Nano-TiO2, Fenton, Fenton-like, electro-Fenton and electrocoagulation processes: A comparative study. *Journal of Hazardous Materials*, *161*(2-3), 1225–1233. doi:10.1016/j.jhazmat.2008.04.075 PMID:18524478

Kim, S. D., Cho, J., Kim, I. S., Vanderford, B. J., & Snyder, S. A. (2007). Occurrence and removal of pharmaceuticals and endocrine disruptors in South Korean surface, drinking, and waste waters. *Water Research*, *41*(5), 1013–1021. doi:10.1016/j.watres.2006.06.034 PMID:16934312

Kulikowska, D., & Klimiuk, E. J. B. T. (2008). The effect of landfill age on municipal leachate composition. *99*(13), 5981-5985.

Kurniawan, T. A., Lo, W., & Chan, G. Y. (2006). Physico-chemical treatments for removal of recalcitrant contaminants from landfill leachate. *Journal of Hazardous Materials*, *129*(1-3), 80–100. doi:10.1016/j.jhazmat.2005.08.010 PMID:16314043

Lin, M., Ning, X., An, T., Zhang, J., Chen, C., Ke, Y., & Liu, J. (2016). Degradation of polycyclic aromatic hydrocarbons (PAHs) in textile dyeing sludge with ultrasound and Fenton processes: Effect of system parameters and synergistic effect study. *Journal of Hazardous Materials*, *307*, 7–16. doi:10.1016/j.jhazmat.2015.12.047 PMID:26795704

Lin, S. H., & Leu, H. G. (1999). Operating characteristics and kinetic studies of surfactant wastewater treatment by Fenton oxidation. *Water Research*, *33*(7), 1735–1741. doi:10.1016/S0043-1354(98)00403-5

Lin, S. H., & Lo, C. C. (1997). Fenton process for treatment of desizing wastewater. *Water Research*, *31*(8), 2050–2056. doi:10.1016/S0043-1354(97)00024-9

Liu, K., Han, W., Pan, W.-P., & Riley, J. T. (2001). Polycyclic aromatic hydrocarbon (PAH) emissions from a coal-fired pilot FBC system. *Journal of Hazardous Materials*, *84*(2-3), 175–188. doi:10.1016/S0304-3894(01)00196-0 PMID:11406305

Manoli, E., & Samara, C. (2008). The removal of polycyclic aromatic hydrocarbons in the wastewater treatment process: Experimental calculations and model predictions. *Environmental Pollution*, *151*(3), 477–485. doi:10.1016/j.envpol.2007.04.009 PMID:17548140

Marttinen, S., Kettunen, R., Sormunen, K., Soimasuo, R., & Rintala, J. (2002). Screening of physical–chemical methods for removal of organic material, nitrogen and toxicity from low strength landfill leachates. *Chemosphere*, *46*(6), 851–858. doi:10.1016/S0045-6535(01)00150-3 PMID:11922065

Maycock, D., & Inspectorate, G. B. D. W. (2008). Review of England and Wales monitoring data for which a national standard has been set: final report to DWI.

Mohajeri, S., Aziz, H. A., Isa, M. H., Bashir, M. J., Mohajeri, L., & Adlan, M. N. (2010). Influence of Fenton reagent oxidation on mineralization and decolorization of municipal landfill leachate. *Journal of Environmental Science and Health Part A*, *45*(6), 692–698. doi:10.1080/10934521003648883 PMID:20390917

Mohajeri, S., Aziz, H. A., Isa, M. H., Zahed, M. A., & Adlan, M. N. (2010). Statistical optimization of process parameters for landfill leachate treatment using electro-Fenton technique. *Journal of Hazardous Materials*, *176*(1-3), 749–758. doi:10.1016/j.jhazmat.2009.11.099 PMID:20022166

Murphy, S. R., & Wadey, B. L. (1988). DEHP and toxicity. *Journal of Vinyl and Additive Technology*, *10*(3), 121–124. doi:10.1002/vnl.730100306

Net, S., Dumoulin, D., El-Osmani, R., Rabodonirina, S., & Ouddane, B. (2014). Case study of PAHs, Me-PAHs, PCBs, phthalates and pesticides contamination in the Somme River water, France. *International Journal of Environmental of Research*, *8*(4), 1159–1170.

Net, S., Sempéré, R., Delmont, A., Paluselli, A., & Ouddane, B. (2015). Occurrence, fate, behavior and ecotoxicological state of phthalates in different environmental matrices. *Environmental Science & Technology*, *49*(7), 4019–4035. doi:10.1021/es505233b PMID:25730609

Nidheesh, P., & Gandhimathi, R. (2012). Trends in electro-Fenton process for water and wastewater treatment: An overview. *Desalination*, *299*, 1–15. doi:10.1016/j.desal.2012.05.011

Oller, I., Malato, S., & Sánchez-Pérez, J. (2011). Combination of advanced oxidation processes and biological treatments for wastewater decontamination—A review. *The Science of the Total Environment*, *409*(20), 4141–4166. doi:10.1016/j.scitotenv.2010.08.061 PMID:20956012

Öman, C. B., & Junestedt, C. (2008). Chemical characterization of landfill leachates–400 parameters and compounds. *Waste Management (New York, N.Y.)*, *28*(10), 1876–1891. doi:10.1016/j.wasman.2007.06.018 PMID:17869498

Ortega-Liébana, M., Sánchez-López, E., Hidalgo-Carrillo, J., Marinas, A., Marinas, J., & Urbano, F. (2012). A comparative study of photocatalytic degradation of 3-chloropyridine under UV and solar light by homogeneous (photo-Fenton) and heterogeneous (TiO2) photocatalysis. *Applied Catalysis B: Environmental*, *127*, 316–322. doi:10.1016/j.apcatb.2012.08.036

Ostrowski, S. R., Wilbur, S., Chou, C.-H. S. J., Pohl, H. R., Stevens, Y.-W., Allred, P. M., ... Tylenda, C. A. (1999). Agency for toxic substances and disease registry's 1997 priority list of hazardous substances. Latent effects—carcinogenesis, neurotoxicology, and developmental deficits in humans and animals. *Toxicology and Industrial Health*, *15*(7), 602–644. doi:10.1177/074823379901500702 PMID:10677885

Özcan, A., Şahin, Y., Koparal, A. S., & Oturan, M. A. (2008). Degradation of picloram by the electro-Fenton process. *Journal of Hazardous Materials*, *153*(1-2), 718–727. doi:10.1016/j.jhazmat.2007.09.015 PMID:17935883

Özcan, A., Şahin, Y., Koparal, A. S., & Oturan, M. A. (2009). A comparative study on the efficiency of electro-Fenton process in the removal of propham from water. *Applied Catalysis B: Environmental*, *89*(3-4), 620–626. doi:10.1016/j.apcatb.2009.01.022

Pakpahan, E. N., Isa, M. H., Kutty, S. R. M., Chantara, S., & Wiriya, W. (2013). Polycyclic aromatic hydrocarbon removal from petroleum sludge cake using thermal treatment with additives. *Environmental Technology*, *34*(3), 407–416. doi:10.1080/09593330.2012.698648 PMID:23530354

Panizza, M., & Cerisola, G. (2009). Direct and mediated anodic oxidation of organic pollutants. *Chemical Reviews*, *109*(12), 6541–6569. doi:10.1021/cr9001319 PMID:19658401

Prabhakaran, D., Kannadasan, T., & Basha, C. A. (2009). Treatability of resin effluents by electrochemical oxidation using batch recirculation reactor. *International Journal of Environmental Science and Technology*, *6*(3), 491–498. doi:10.1007/BF03326088

Primo, O., Rivero, M. J., & Ortiz, I. (2008). Photo-Fenton process as an efficient alternative to the treatment of landfill leachates. *Journal of Hazardous Materials*, *153*(1-2), 834–842. doi:10.1016/j.jhazmat.2007.09.053 PMID:17961917

Ravindra, K., Sokhi, R., & Van Grieken, R. (2008). Atmospheric polycyclic aromatic hydrocarbons: Source attribution, emission factors and regulation. *Atmospheric Environment*, *42*(13), 2895–2921. doi:10.1016/j.atmosenv.2007.12.010

Renner, R. (2000). Human phthalate study changes exposure picture. *Environmental Science & Technology*, *34*(21), 451A–452A. doi:10.1021/es003481v PMID:21662257

Ribeiro, A. R., Nunes, O. C., Pereira, M. F., & Silva, A. M. (2015). An overview on the advanced oxidation processes applied for the treatment of water pollutants defined in the recently launched Directive 2013/39/EU. *Environment International*, *75*, 33–51. doi:10.1016/j.envint.2014.10.027 PMID:25461413

Rivas, F. J., Beltran, F. J., Frades, J., & Buxeda, P. (2001). Oxidation of p-hydroxybenzoic acid by Fenton's reagent. *Water Research*, *35*(2), 387–396. doi:10.1016/S0043-1354(00)00285-2 PMID:11228990

Rocha, E. M., Vilar, V. J., Fonseca, A., Saraiva, I., & Boaventura, R. A. (2011). Landfill leachate treatment by solar-driven AOPs. *Solar Energy*, *85*(1), 46–56. doi:10.1016/j.solener.2010.11.001

Rosales, E., Pazos, M., Longo, M., & Sanromán, M. (2009). Electro-Fenton decoloration of dyes in a continuous reactor: A promising technology in colored wastewater treatment. *Chemical Engineering Journal*, *155*(1-2), 62–67. doi:10.1016/j.cej.2009.06.028

Rubio-Clemente, A., Torres-Palma, R. A., & Peñuela, G. A. (2014). Removal of polycyclic aromatic hydrocarbons in aqueous environment by chemical treatments: A review. *The Science of the Total Environment*, *478*, 201–225. doi:10.1016/j.scitotenv.2013.12.126 PMID:24552655

Saber, A., Hasheminejad, H., Taebi, A., & Ghaffari, G. (2014). Optimization of Fenton-based treatment of petroleum refinery wastewater with scrap iron using response surface methodology. *Applied Water Science*, *4*(3), 283–290. doi:10.100713201-013-0144-8

Sanches, S., Leitão, C., Penetra, A., Cardoso, V., Ferreira, E., Benoliel, M., ... Pereira, V. (2011). Direct photolysis of polycyclic aromatic hydrocarbons in drinking water sources. *Journal of Hazardous Materials*, *192*(3), 1458–1465. doi:10.1016/j.jhazmat.2011.06.065 PMID:21784577

Sankara Narayanan, T., Magesh, G., & Rajendran, N. (2003). Degradation of O-chlorophenol from aqueous solution by electro-Fenton process. *Fresenius Environmental Bulletin*, *12*(7), 776–780.

Sedlak, D. L., & Andren, A. W. (1991). Oxidation of chlorobenzene with Fenton's reagent. *Environmental Science & Technology*, *25*(4), 777–782. doi:10.1021/es00016a024

Serôdio, P., & Nogueira, J. (2006). Considerations on ultra-trace analysis of phthalates in drinking water. *Water Research*, *40*(13), 2572–2582. doi:10.1016/j.watres.2006.05.002 PMID:16797669

Shemer, H., & Linden, K. G. (2007). Aqueous photodegradation and toxicity of the polycyclic aromatic hydrocarbons fluorene, dibenzofuran, and dibenzothiophene. *Water Research*, *41*(4), 853–861. doi:10.1016/j.watres.2006.11.022 PMID:17217979

Sires, I., Garrido, J. A., Rodriguez, R. M., Brillas, E., Oturan, N., & Oturan, M. A. (2007). Catalytic behavior of the Fe3+/Fe2+ system in the electro-Fenton degradation of the antimicrobial chlorophene. *Applied Catalysis B: Environmental*, *72*(3-4), 382–394. doi:10.1016/j.apcatb.2006.11.016

Šrám, R. J., Binková, B., Rössner, P., Rubeš, J., Topinka, J., & Dejmek, J. (1999). Adverse reproductive outcomes from exposure to environmental mutagens. *Mutation Research. Fundamental and Molecular Mechanisms of Mutagenesis*, *428*(1), 203–215. doi:10.1016/S1383-5742(99)00048-4 PMID:10517994

Stringer, R., Labunska, I., Santillo, D., Johnston, P., Siddorn, J., & Stephenson, A. (2000). Concentrations of phthalate esters and identification of other additives in PVC children's toys. *Environmental Science and Pollution Research International*, *7*(1), 27–36. doi:10.1065/espr199910.007 PMID:19153836

Tan, G. (1995). Residue levels of phthalate esters in water and sediment samples from the Klang River basin. *Bulletin of Environmental Contamination and Toxicology*, *54*(2), 171–176. doi:10.1007/BF00197427 PMID:7742623

Tchobanoglous, G., Burton, F. L., & Stensel, H. D. J. I. E. M. (2003). Metcalf & Eddy wastewater engineering: treatment and reuse. *International Edition. McGrawHill, 4*, 361-411.

Tran, L.-H., Drogui, P., Mercier, G., & Blais, J.-F. (2009). Electrochemical degradation of polycyclic aromatic hydrocarbons in creosote solution using ruthenium oxide on titanium expanded mesh anode. *Journal of Hazardous Materials*, *164*(2-3), 1118–1129. doi:10.1016/j.jhazmat.2008.09.012 PMID:18926633

Umar, M., Aziz, H. A., & Yusoff, M. S. (2010). Trends in the use of Fenton, electro-Fenton and photo-Fenton for the treatment of landfill leachate. *Waste Management (New York, N.Y.)*, *30*(11), 2113–2121. doi:10.1016/j.wasman.2010.07.003 PMID:20675113

Urase, T., Salequzzaman, M., Kobayashi, S., Matsuo, T., Yamamoto, K., & Suzuki, N. (1997). Effect of high concentration of organic and inorganic matters in landfill leachate on the treatment of heavy metals in very low concentration level. *Water Science and Technology*, *36*(12), 349–356. doi:10.2166/wst.1997.0464

Uygur, A., & Kargı, F. (2004). Biological nutrient removal from pre-treated landfill leachate in a sequencing batch reactor. *Journal of Environmental Management*, *71*(1), 9–14. doi:10.1016/j.jenvman.2004.01.002 PMID:15084355

Vilar, V. J., Rocha, E. M., Mota, F. S., Fonseca, A., Saraiva, I., & Boaventura, R. A. (2011). Treatment of a sanitary landfill leachate using combined solar photo-Fenton and biological immobilized biomass reactor at a pilot scale. *Water Research*, *45*(8), 2647–2658. doi:10.1016/j.watres.2011.02.019 PMID:21411117

Wang, A., Qu, J., Ru, J., Liu, H., & Ge, J. (2005). Mineralization of an azo dye Acid Red 14 by electro-Fenton's reagent using an activated carbon fiber cathode. *Dyes and Pigments*, *65*(3), 227–233. doi:10.1016/j.dyepig.2004.07.019

Wang, C.-T., Hu, J.-L., Chou, W.-L., & Kuo, Y.-M. (2008). Removal of color from real dyeing wastewater by Electro-Fenton technology using a three-dimensional graphite cathode. *Journal of Hazardous Materials*, *152*(2), 601–606. doi:10.1016/j.jhazmat.2007.07.023 PMID:17707581

Watts, R. J. (1998). Hazardous wastes: sources, pathways, receptors.

Wick, A. F., Haus, N. W., Sukkariyah, B. F., Haering, K. C., & Daniels, W. L. (2011). Remediation of PAH-contaminated soils and sediments: a literature review. *CSES Department, internal research document, 102*.

Wiszniowski, J., Robert, D., Surmacz-Gorska, J., Miksch, K., & Weber, J. J. E. L. (2006). Landfill leachate treatment methods. *RE:view*, *4*(1), 51–61.

Yaqub, A., Isa, M. H., & Ajab, H. (2014). Electrochemical degradation of polycyclic aromatic hydrocarbons in synthetic solution and produced water using a Ti/SnO 2-Sb 2 O 5-RuO 2 anode. *Journal of Environmental Engineering*, *141*(4), 04014074. doi:10.1061/(ASCE)EE.1943-7870.0000900

Yaqub, A., Isa, M. H., Kutty, S. R. M., & Ajab, H. (2014). Electrochemical degradation of PAHs in produced water using Ti/Sb2O5-SnO2-IrO2 anode. *Electrochemistry*, *82*(11), 979–984. doi:10.5796/electrochemistry.82.979

Yoo, H.-C., Cho, S.-H., & Ko, S.-O. (2001). Modification of coagulation and Fenton oxidation processes for cost-effective leachate treatment. *Journal of Environmental Science and Health. Part A*, *36*(1), 39–48. PMID:11381784

Yuan, S., Fan, Y., Zhang, Y., Tong, M., & Liao, P. (2011). Pd-catalytic in situ generation of H2O2 from H2 and O2 produced by water electrolysis for the efficient electro-Fenton degradation of rhodamine B. *Environmental Science & Technology*, *45*(19), 8514–8520. doi:10.1021/es2022939 PMID:21866953

Zhang, H., Choi, H. J., & Huang, C.-P. (2005). Optimization of Fenton process for the treatment of landfill leachate. *Journal of Hazardous Materials*, *125*(1-3), 166–174. doi:10.1016/j.jhazmat.2005.05.025 PMID:16011874

Zhang, H., Fei, C., Zhang, D., & Tang, F. (2007). Degradation of 4-nitrophenol in aqueous medium by electro-Fenton method. *Journal of Hazardous Materials*, *145*(1-2), 227–232. doi:10.1016/j.jhazmat.2006.11.016 PMID:17161909

Zhang, H., Zhang, D., & Zhou, J. (2006). Removal of COD from landfill leachate by electro-Fenton method. *Journal of Hazardous Materials*, *135*(1-3), 106–111. doi:10.1016/j.jhazmat.2005.11.025 PMID:16359785

Zhang, Y., Han, C., Zhang, G., Dionysiou, D. D., & Nadagouda, M. N. (2015). PEG-assisted synthesis of crystal TiO2 nanowires with high specific surface area for enhanced photocatalytic degradation of atrazine. *Chemical Engineering Journal*, *268*, 170–179. doi:10.1016/j.cej.2015.01.006

Zhou, M., Yu, Q., Lei, L., & Barton, G. (2007). Electro-Fenton method for the removal of methyl red in an efficient electrochemical system. *Separation and Purification Technology*, *57*(2), 380–387. doi:10.1016/j.seppur.2007.04.021

Chapter 9
Application of Graded Limestone as Roughing Filter Media for the Treatment of Leachate

Augustine Chioma Affam
University College of Technology Sarawak, Malaysia

Ezerie Henry Ezechi
https://orcid.org/0000-0003-4968-5541
Universiti Teknologi Malaysia, Johor Bahru, Malaysia

ABSTRACT

This chapter examined limestone performance as a roughing media for pollutant removal from leachate using a filtration column. Limestone with a density of 2554 kg/m3 was crushed and graded between 4-6, 8-12 and 12-18 mm as filter media. The length of run for each experiment of the various individual media sizes and combination sizes was 10 days. Test analysis was performed at intervals of 24 hours to ascertain the percentage removal efficiency for the parameters desired. The observed best removal occurred when a combination of 4 – 8mm, 8 – 12mm and 12 – 18mm, media top to bottom was used. This obtained a BOD removal in the range of 22 to 81%, COD was 22 to 75%, and turbidity was 32 to 86% while the colour was 36 to 62%, respectively. A general decline in removal efficiency was observed after the sixth day showing maximum adsorption and breakthrough had been achieved. The study indicates that limestone is an effective adsorbent that can be used for short-term treatment of leachate.

BACKGROUND OF STUDY

Land disposal of solid wastes has been practiced for centuries, dating back to prehistoric times. Municipal, industrial, agricultural, and urban activities produce huge amounts of wastes which require permanent disposal. Returning some of the solid wastes to the land is a practical approach for waste disposal. Human

DOI: 10.4018/978-1-7998-0369-0.ch009

population rate increases every year and as such the solid waste generated increases. As the amount of waste produced rapidly increases, space for permanent disposal becomes crucial. Since the production of solid waste is increasing much more rapidly than it degrades, land space for disposal has become more difficult and expensive to attain. There are several waste management options that can be used to reduce the amounts of waste requiring land disposal. Incineration of solid waste can be used but this is expensive and the emissions are of health concern. This is why landfills remain the major solid waste disposal option for most countries. Solid waste in a landfill is degraded through aerobic and anaerobic processes. Stabilization of the wastes is a very complex and variable event due to the site-specific characteristics of each landfill. The degradation products generated from the stabilization process include leachate and gas. Landfill gas is generated due to the anaerobic biological degradation of organic material (Wei et al., 2013; Tchobanoglous and Kreith, 2003). Leachate is formed from the contact of water with refuse or solid waste. The water mainly from precipitation, dissolves soluble organics and inorganics including some toxic compounds if present in the landfill material. Solid waste should be properly disposed in order to protect environmental quality and human health, as well as to preserve natural resources. With the rapid socio-economic development, the contradiction between increasing waste generation rates and decreasing waste-disposal capacities is becoming more and more acute. In response to this concern, effective solid waste management models are desired to be developed. This will help to provide sound management strategies with satisfactory economic and environmental efficiencies. Solid waste management is a complex, multidisciplinary problem involving economic and technical aspects, normative constraints about the minimum requirements for recycling and sustainable development issues. Most industrialized countries have adopted the philosophy of the 'Waste Management Hierarchy' (prevention/ minimization, materials recovery, incineration and landfill) as a guide for developing municipal or industrial solid waste management strategies (Van Ewijk & Stegemann, 2016; Sakai et al., 2000).

A leachate stream can be compared to a complex wastewater stream with varying characteristics. Leachate characteristics not only vary because of the different kinds of waste present, but vary according to the landfill age. Usually leachate from old landfills is rich in ammonia nitrogen due to the hydrolysis and fermentation of the nitrogenous fractions of the biodegradable wastes (Siti and Aziz, 2018; Kamaruddin et al., 2013; Onay and Pohland, 1998). Leachate from young landfills contains high dissolved solids as well as high concentrations of organic matter compared to domestic wastewater (Abd El-Salam and Abu-Zuid, 2015; Reinhart and Townsend, 1998). The leakage of leachate which contains high organic, inorganic suspended solids, heavy metals and other pollutants will contaminate the ground water and surface water sources (Naveen et al., 2018; Qasim and Chiang, 1994). Contamination of water resources by landfill leachate is a growing problem. Waste management companies often struggle with the challenge of containing and controlling leachate migration. Leachate containment structures built in recent years have improved water quality conditions however problem areas within the site may continue to exist. Analysis have found that high water table levels on site are causing the leachate management structures to under-perform. Leachate from the landfill is migrating through and bypassing containment walls and re-circulation systems resulting in local and downstream surface and groundwater contamination. Landfill leachate migration is a long-standing problem that is expected to increase with the growth of urban populations and resulting landfills. (Izabela and Biedka, 2016).

Although attention has been drawn to increase recycling, waste reduction and incineration, the sanitary landfill will remain dominant in solid waste disposal for many more decades due to rapid population growth (Oasim and Chiang, 1994). The high level of chemical oxygen demand (COD), biochemical oxygen demand (BOD), turbidity and other impurity components of landfill leachate make it very important to devise treatment methods. There are many different landfill leachate treatment options. The options include complex and expensive ones from physical-chemical and biological processes for the treatment of high- strength organics and inorganics. These separate treatment processes can result in large costs. Studies have shown that roughing filter technology can be used for treatment of wastewater to remove BOD, COD, turbidity and suspended solids. This time for landfill leachate it should be considered a prime objective especially since it will be a way to reduce the higher or additional cost required to treat leachate from landfill. This study focuses on the removal of COD, BOD, turbidity and colour from landfill leachate obtained from Penang State, Malaysia. The exact location was Pulau Burung landfill site using limestone media in a vertical roughing filter column. At test condition of the room temperature of about 28°C, standard laboratory tests were carried out to determine the percentage removal efficiency of limestone media in the roughing filter. Hence, its effectiveness for the treatment of leachate was determined. Leachate is heavily polluted and likely hazardous liquids that percolates through solid wastes generated industrially and domestically. Leachate is formed due to biodegradation of organic components of such wastes and thus produces liquid. The process can be accelerated by increased precipitation or rainfall, nearby drainage and presence of high water table or ground water. Therefore, leachate will require the best treatment before it is discharged or allowed to be released to surface water or rivers.

LANDFILL

Landfilling is the term used to describe the process by which solid waste and solid waste residuals are placed in a landfill. In the past, the term sanitary landfill was used to denote a landfill in which the waste placed in the landfill was covered at the end of each day's operation. Today, sanitary landfill refers to an engineered facility for the disposal of municipal solid waste (MSW) designed and operated to minimize public health and environmental impacts. Landfills for individual waste constituents such as combustion ash, asbestos, and other similar wastes are known as monofills. Landfills for the disposal of hazardous wastes are called secure landfills. Those places where waste is dumped on or into the ground in no organized manner are called uncontrolled land disposal sites or waste dumps (Chen et al., 2019; Tchobanoglous and Kreith, 2003).

Landfill liners are materials (both natural and man-made) that are used to line the bottom area and below-grade sides of a landfill. Liners usually consist of successive layers of compacted clay and/or geosynthetic material designed to prevent migration of landfill leachate and landfill gas. The final landfill cover layer is applied over the entire landfill surface after all landfilling operations are complete. Landfill covers consist of successive layers of compacted clay and/or geosynthetic material designed to prevent the migration of landfill gas and to limit the entry of surface water into the landfill. In developing countries, the implementation of improved land disposal practices is progressing at varying rates dependent upon the available resources and national regulatory standards. The need to improve land disposal practices is being forced along by consolidating populations, where rural residents are moving to cities resulting in rapid urban population growth. This has created an ever increasing need for better solid waste disposal practices. In a number of instances, less than adequate disposal practices have resulted in accidents that have led to loss of life. Large, uncontrolled dumps in urban areas are also a significant source of air pollution and water contamination and, as such, over time these facilities will need to be closed and replaced with landfills that meet conventional standards (Florin-Constantin and Mohammad, 2017). Tchobanoglous and Kreith, 2003).

Table 1. Categories of landfill

Class	Type of Waste	Category
I	Hazardous waste	Secure landfill
II	Designated waste	Monofills
III	Municipal solid waste (MSW)	Sanitary landfill

CLASSIFICATION OF LANDFILLS

The classification system adopted by the state of California in 1984 is perhaps the most widely accepted classification system for landfills. In the California system, the following three classifications are used. Table 1 shows the categories of landfill. Hazardous waste is waste that is dangerous or potentially harmful to our health or the environment. Hazardous wastes can be liquids, solids, gases, or sludge. They can be discarded commercial products, like cleaning fluids or pesticides, or the by-products of manufacturing processes. Designated wastes are nonhazardous wastes that may release constituents in concentrations that are in excess of applicable water quality objectives established by various state and federal agencies. Combustion ash, asbestos, and other similar wastes often identified as designated wastes are typically placed in lined monofills to isolate them from materials placed in municipal landfills. Municipal Solid waste (MSW), also called urban solid waste, is a waste type that includes predominantly household waste (domestic waste) with sometimes the addition of commercial wastes collected by a municipality within a given area. They are in either solid or semisolid form (Tchobanoglous and Kreith, 2003).

In Malaysia, a large number of landfills are available for solid waste reception and handling. As at year 2008, the per capita generation of solid waste in Malaysia varied from 0.45 to 1.44 kg/day. In general, the per capita generation rate was about 1kg/day. Malaysian solid wastes contain very high organic waste and consequently high moisture content and bulk density of above 200 kg/m^3. A study conducted in Kuala Lumpur revealed that the amount of organic wastes for residential area range from 62 to 72% of the total waste. Disposal of solid waste is done almost solely through landfill method. There were about 177 disposal sites in Peninsular Malaysia at that time and in most cases, open dumping was practiced and this took place at about 50% of the total landfills (DOE, 2008). Currently, over 23,000 tonnes of waste is produced each day in Malaysia. However, this amount is expected to rise to 30,000 tonnes by the year 2020. The amount of waste generated continues to increase due to the increasing population and development, and only less than 5% of the waste is being recycled (Global Environment Centre report, 2019).

TYPES OF LANDFILL

There are three types of landfills namely aerobic landfill, semi aerobic and anaerobic landfill (Hecham, 2016; Matsufuji, 1990).

Aerobic Landfill System

The aerobic landfill system operation in excess of oxygen to the waste and recirculation of leachate is usually done to maintain nutrient and moisture content for the microbes. The outcome of this process will convert most biodegradable waste into bacteria mass, CO_2 and water (Hecham, 2016; Matsufuji, 1990).

Anaerobic Landfill System

Anaerobic landfill is the common landfill type, which mainly depends on natural decomposition of solid waste. This design will usually generate high volume of methane gas, which poses the danger of fire with bad smell due to H_2S production (Hecham, 2016; Matsufuji, 1990).

Semi-Aerobic Landfill System

The semi-aerobic landfill is designed with proper filling method and leachate collection pipe system. These collection pipes are capable of drawing in large amount of air to provide oxygen to the decomposing micro-organism. The fast removal of leachate has the effect of enhancing the aerobic condition and improves decomposition of solid waste and leachate quality (Hecham, 2016; Matsufuji, 1990).

GENERATION OF LANDFILL GASES

The generation of the principal gases, the variation in their rate of generation with respect to time and the sources of traces gases in landfills can be said to occur in five sequential phases. Each of these phases is described below with the illustration in Figure 1 (Hecham, 2016; Tchobanoglous and kreith, 2003). The general anaerobic transformation of the organic portion of the solid waste placed in a landfill can be described by Equation (1).

$$\text{Organic matter} + H_2O + \text{nutrients} \rightarrow \text{new cells} + \text{resistant organic matter}$$

$$+ CO_2 + CH_4 + NH_3 + H_2S + \text{heat} \quad (1)$$

Figure 1, describes the activities in a typical landfill where leachate is generated. The phase I is characterized by initial waste placement from residence, preliminary moisture accumulation, initial subsidence of the solid waste usually accompanied by presence of oxygen. As the phase II begins, the field capacity is exceeded, leachate is formed, electron acceptor shifts from oxygen to nitrates and then to sulphates and thereafter volatile acids appear. In the phase III, the volatile fatty acids, predominantly becomes leachate as pH declines, enhanced by substrate conversion which occurs as well. Phase IV, is followed as methane and carbon dioxide production is obvious with pH at minimum and high nutrient consumption occasioned by precipitation of metals. Leachate BOD/COD ratio declines. Phase V is the biological dormancy whereby nutrients are reduced or limited at the same time. Gas production ceases as well as oxygen slowly reappears and more humic substances are produced. Table 2. Compares different landfill scenarios.

Figure 1. Generalized phases in the generation of landfill gases (I—Initial Adjustment, II—Transition Phase, III—Acid Phase, IV—Methane Fermentation, and V—Maturation Phase).

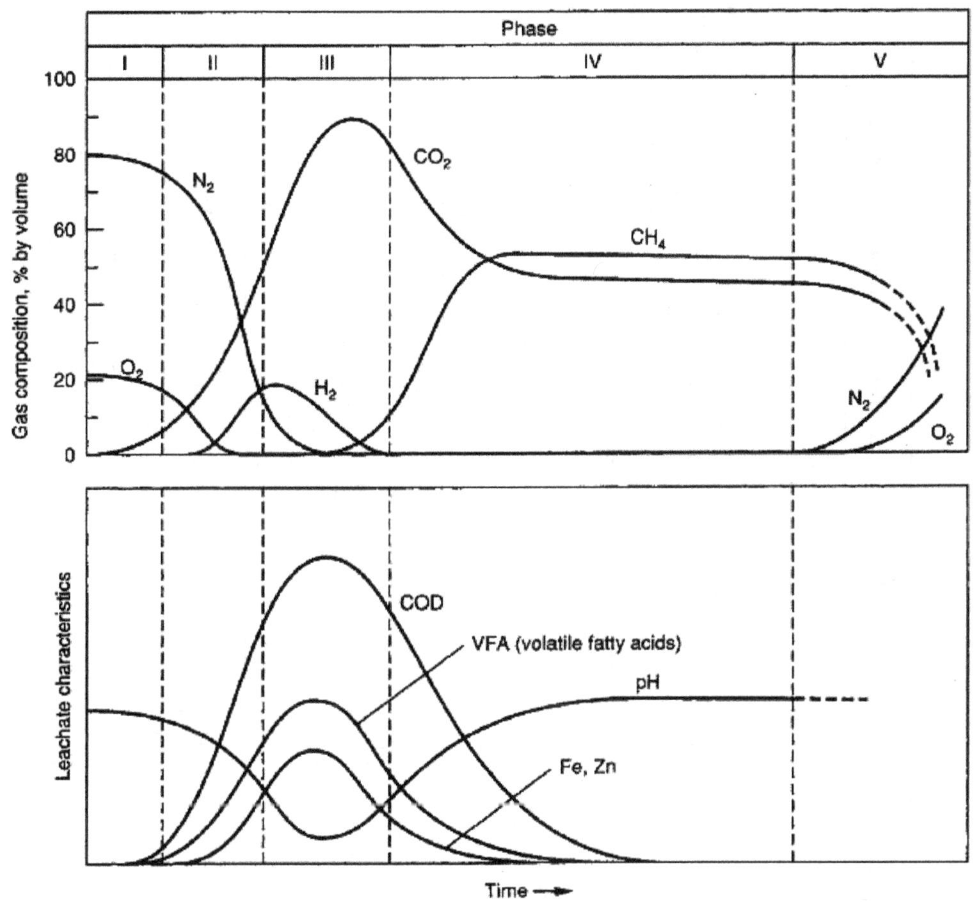

In Table 2 above, Nicole et al., (2009), proposes an economic model which evaluates the impact of various operational (anaerobic, aerobic or hybrid) and construction (retrofit and as-built) bioreactor landfill strategies on project economies as against the traditional landfill. While, the retrofit bioreactor is most expensive, followed by the as-built and aerobic, the traditional is least expensive. But the long term gain such as extension of landfill life and reduced leachate treatment cost, recovery of gas for generation of electricity is sufficient for the bioreactor type to be recommended for any new landfill being considered in future.

COMPARISON OF TRADITIONAL LANDFILL, LEACHATE RECIRCULATION LANDFILL AND BIOREACTOR LANDFILL

The modern municipal solid waste (MSW) landfill has evolved into a sophisticated treatment and storage facility. Landfill bioreactors have emerged as one of the new generation methods of managing solid wastes and they are used to minimize environmental impacts while optimizing the degradation and stabilization

Table 2. Comparisons of landfill scenarios

Landfill Type	Landfill Definition	Potential Monetary Benefits	Operation and Maintenance	Leachate Volume Reduction	Closure and post-Closure care
Traditional	Operated with intent of entombing waste	Utilization of methane for electricity generation	Typical costs are low	None	Typical costs are low
Anaerobic retrofit bioreactor	Bioreactor operation commences following active filling(6th year)	Leachate treatment savings; utilization of methane for electricity; airspace recovery.	Typical costs are low	Yes, but begins in year 6. Only leachate not recirculated requires treatment. The landfill reaches field capacity in year 9.	A temporary final cap is installed at the end of the active filling period. The final cap is constructed at the end of 15th year. Bioreactor operation and maintenance costs are incurred during the period.
Anaerobic as-built bioreactor	Bioreactor operation commences following active filling (6th year). Air injection occurs for periods ranging between 0.25 and 5th year	Leachate treatment savings; utilization of methane for electricity; airspace recovery.	In addition to typical costs, leachate recirculation costs are incurred	Yes, begins immediately	A temporary final cap is installed at the end of the active filling period. The final cap is constructed at the end of 10th year. Bioreactor operation and maintenance costs are incurred during the period
Retrofit hybrid bioreactor	Bioreactor operation commences during active filling. Air injection occurs for periods ranging between 0.25 and 5th year	Leachate treatment savings; utilization of methane for electricity; airspace recovery.	In addition to typical costs, leachate recirculation costs are incurred	Yes, but begins in year 6. Only leachate not recirculated requires treatment. The landfill reaches field capacity in 9th year. During air injection, 50% of the leachate produced is evaporated.	A temporary final cap is installed at the end of the active filling period. The final cap is constructed at the end of 15th year. Bioreactor operation and maintenance costs are incurred during the period
Anaerobic bioreactor	Bioreactor operation commences during active filling. Air	Leachate treatment savings; utilization of methane for electricity; airspace recovery.	In addition to typical costs, leachate recirculation costs are incurred	Yes, but begins in 6th year. Only leachate not recirculated requires treatment. The landfill reaches field capacity in year 3. During air injection, 50% of the leachate produced is evapourated.	A temporary final cap is installed at the end of the active filling period. The final cap is constructed at the end of 15th year. Bioreactor operation and maintenance costs are incurred during the period

(Nicole et al., 2009)

processes. Many of the old sanitary landfills have been converted into bioreactor type landfills because of the many advantages that the bioreactors offer compared to the old systems. Recirculation landfills are often confused to be the same as bioreactor landfills, but this is not true. The major difference is that bioreactor landfills must operate at approximately 45% moisture content. Moisture from other sources, as well as leachate recirculation, drives the moisture content to high levels. Recirculation landfills do not operate at such high moisture content levels (Pohland, 1995).

The new systems are operated and controlled to rapidly accelerate the biological stabilization of the stored waste. Leachate generation, collection, and in situ recirculation are what drive the bioreactor processes. The recirculation is what separates the two different types of landfills. As moisture accumulates and becomes more uniformly distributed with leachate recirculation, waste stabilization progresses sequentially through initial, transition, acid formation, methane fermentation, and final maturation phases. Compared to traditional sanitary landfills, landfill bioreactors initiate and provide more rapid, complete, and predictable conversion of readily degradable solid waste products, therefore enhancing the potential for gas recovery and utilization, decreasing management time and process uncertainty, and reducing the potential for negative health and environmental impacts and attendant liabilities. The bioreactors also increase resource recovery and site reuse opportunities (Pohland, 1995).

LEACHATE PRODUCTION AND CHARACTERISTICS

Rainfall is the main contributor to the generation of leachate. The precipitation percolates through the waste and gains dissolved and suspended components from the biodegrading waste through several physical and chemical reactions. Other contributors to leachate generation include groundwater inflow, surface water runoff, and biological decomposition (Reinhart and Townsend, 1998). Liquid fractions in the waste will also add to the leachate as well as moisture in the cover material. Moisture can be removed from the landfill by water consumed in the formation of landfill gas, water vapour removed in the landfill gas, and leachate leaking through the liner (Tchobanoglous et al., 1993). Moisture content appears to be the variable that is associated with the greatest stabilization enhancement potential because it provides an aqueous environment that facilitates the transport of nutrients and microbes within the landfill. This transport mechanism dilutes the concentration of inhibitors and enhances the microorganisms' access to their substrates and hence improves gas generation and stabilization. However, in some instances, increased moisture content could over-stimulate the rate of hydrolysis, and consequently fermentative organisms would grow at a much faster rate than the methanogens, promoting acid accumulation and decrease in pH of the waste material (Barlaz et al., 1990). A balance of activity amongst the populations of hydrolytic, acetogenic, and methanogenic bacteria is needed to ensure adequate stabilization of the organic material. The addition of a buffer, such as calcium carbonate, has been found to adjust the pH of the environment and enhance the methanogenic activity, therefore improving the chances of attaining the required balance between acidogenic and methanogenic reactions. In previous studies in which such a balance was reached, common leachate quality variations typically observed include a relatively sharp increase in pH, associated with a decrease in the organic content of the leachate (Barlaz et al., 1990; Reinhart and Townsend, 1998). Since for short term leachate quantity depends heavily on precipitation, it is sometimes tough to predict. Long term leachate quantity is not as difficult to predict. Leachate quality is difficult to predict because each landfill is unique and the wastes vary widely. The major factors that affect leachate quantity and quality are; the type of disposed waste, hydrogeolic and climactic conditions, the age of the landfill, the phase of waste decomposition occurring, and the chemical and physical properties of the precipitation. Leachate quantity and quality is site specific. In arid regions, leachate quantity can be zero, while in areas of wet climate, 100% of precipitation can become leachate. Once the adsorptive capacity of the trash field capacity has been achieved, continuous leachate flow will occur (Bagchi, 1990).

The characteristics of leachate from landfills vary according to the operational stage of the landfill and the climatic features of the location of the landfill. Landfill leachates from old sites are usually highly contaminated with ammonia resulting from the hydrolysis and fermentation of nitrogen containing fractions of biodegradable refuse substrates (Carley and Mavinic, 1991). As stabilization of the waste proceeds, the accumulating concentration of ammonia is also influenced by washout as leachate is collected and removed for offsite treatment. However, in bioreactor landfills with leachate containment, collection, and in situ recirculation to accelerate decomposition of readily available organic fractions of the wastes, leachate ammonia nitrogen concentrations may accumulate reaching a much higher level when compared to traditional landfills. Recirculation of leachate will produce stabilized leachates containing relatively low concentrations of degradable carbon compounds but high concentrations of ammonia (Knox, 1985); therefore, COD and BOD will be removed, but ammonia concentrations will increase.

LANDFILL LEACHATE COMPOSITION

According to Qasim and Chiang (1994), prior to 1965 very few people were aware of the fact that water passing through solid waste in a sanitary landfill will become highly contaminated. This water termed 'leachate' was generally not a matter of concern because few cases of water pollution were noted where leachate had caused harm. It is now known that leachates from sanitary landfill maybe an important source of ground water contamination. It also reported that, many contaminants released from a sanitary landfill, if allowed to migrate, may pose a severe threat to surface and ground water. Federal as well as State and local governmental regulations in various countries, now call for effective control of leachates from sanitary landfills. Effective pollution control requires an understanding of the quantity and quality of landfill leachates.

The composition of landfill leachate and pollutants in municipal solid waste (MSW) landfill leachate can be divided into four groups namely:

- **Dissolved Organic Matter**: these are quantified as chemical oxygen demand (COD) or total organic carbon (TOC), volatile fatty acids which accumulates during the acid phase of the waste stabilization and more refractory compounds such as fulvic-like and humic-like compounds.
- **Inorganic Macrocomponents**: These includes calcium (Ca^{2+}), magnesium (Mg^{2+}), sodium (Na^{+}), potassium (K^{+}), ammonium (NH_4^{+}), iron (Fe_2^{+}), manganese (Mn^{2+}), chloride (Cl^{-}), sulphate (SO_4^{2-}) and hydrogen carbonate (HCO_3^{-}).
- **Heavy Metals**: Includes cadmium (Cd^{2+}), chromium (Cr^{3+}), copper (Cu^{2+}), lead (Pb^{2+}), nickel (Ni^{2+}) and zinc (Zn^{2+}).

- **Xenobiotic Organic Compounds (XOCS)**: These originates from household or industrial chemicals and are present in relatively low concentrations (usually less than 1 mg/L of individual compounds). These compounds include among others a variety of aromatic hydrocarbons, phenols, chlorinated aliphatic, pesticides and plastizers. Other compounds may be found in leachate from landfills include borate, sulphide, arsenate, selenate, barium, lithium, mercury, and cobalt. However, these compounds are found in very low concentrations and are only of secondary importance. Leachate composition may also be characterized by different toxicological tests, which provide indirect information on the content of pollutants that may be harmful to a class of organisms (Qasim and Chiang, 1994). Table 3, gives a range of the composition of leachate. Solid wastes placed in a sanitary landfill may undergo a number of biological, chemical, and physical changes. Aerobic and anaerobic decomposition of the organic matter results in both gaseous and liquid end products. Volatile organic chemicals (VOC) often contribute to the contamination of groundwater because they dissolve in the leachate as it passes through the landfill. Generally, as the years increase constituents of the leachate composition decreases (Kjeldsen etal, 2002; Davis and Cornwells, 1988).

Table 3. Typical leachate composition from new and mature landfills

Constituent	Value, mg/L		
	New Landfill (Less Than 2 Years)		Mature Landfill (Greater than 10 Years)
	Range of Values	Average Value	Range of Values
BOD5	2000-30,000	10,000	100-200
TOC	1,500-20,000	6,000	80-160
COD	3000-60,000	18,000	100-500
TSS	200-2,000	500	100-400
Organic Nitrogen	10-800	200	80-120
Ammonia Nitrogen	10-800	200	20-40
Nitrate	5-40	25	5-1
Total phosphate	5-100	30	4-8
Ortho phosphate	4-80	20	2001,000
Alkalinity as $CaCO_3$	1,000-10,000	3,000	6.6-7.5
pH (no unit)	4.5-7.5	6	6.67.5
Total Hardness as $CaCO_3$	300-10,000	3,500	200-500
Calcium	200-3,000	1,000	100-400
Magnesium	50-1,500	250	50-200
Potassium	200-1,000	300	50-400
Sodium	200-2,500	500	100-200
Chloride	200-3,000	500	100-400
Sulphate	50-1,000	300	20-50
Total iron	50-1,200	60	20-00

(Davis and Cornwells, 1988)

FACTORS AFFECTING LEACHATE QUALITY

Detailed review for landfill composition from various studies indicated that, leachate quality was highly variable. Leachate contains larger pollutant loads than raw sewage or many industrial wastes. The variation in leachate quality are generally attributed to a myriad of interacting factors such as type and depth of solid waste, age of fill, rate of water application, landfill design and operations, and the interaction of air which provides oxygen for aerobic activities by micro organisms. The chemical composition of leachate generally dominates the quality of the leachate. However, this composition depends on several parameters. It also includes those concerning waste mass and site location and those derived from design and management of landfill (Qasim and Chiang, 1994). The major factors affecting leachate quality are:

1. **Waste Composition**: The nature of the waste organic fraction influences considerably the degradation of waste in the landfill and thus also the quality of the leachate produced. In particular, the presence of substances which are toxic to bacteria flora may slow down or inhibit biological degradation processes with consequences for the leachate. The inorganic content of the leachate depends on the contacts between waste and leaching water as well as on pH and the chemical balance at the solid–liquid interface. In particular, the majority of metals are released from the waste mass under acid conditions.
2. **PH**: pH influences chemical processes which are the basis of mass transfer in the waste leachate system, including precipitation, dissolution, redox and sorption reactions. It will also affect the speciation of most of the constituents in the system.
3. **Redox Potential**: Reducing conditions, corresponding to the second and third phases of the anaerobic degradation, will influence solubility of nutrients and metals in leachate.
4. **Landfill Age**: Variations in leachate composition and in quantity of pollutants removed from waste are often attributed to landfill age, defined as time measured from the disposition of waste or time measured from the appearance of leachate. Landfill age obviously plays an important role in the determination of leachate characteristics governed by the type of waste stabilization processes. It should be underlined that variations in composition of leachate do not depend exclusively on the landfill age but on the degree of waste stabilization, and volume of water which infiltrates into the landfill. The pollutant load in leachate generally reaches maximum value during the early years of operation of the landfill (2-3 years) and then gradually decreases over the following years. This trend is generally applicable to organic components, main indicators of organic pollution (COD, BOD, TOC), microbiological population and to main inorganic ions (heavy metals, Cl, SO_4, etc) (Qasim and Chiang, 1994).

LEACHATE MANAGEMENT STRATEGIES

There are two leachate management strategies used by modern landfills. These two processes are single pass leaching and leachate recirculation. Most traditional landfills use the single pass leaching strategy where the generated leachate is collected and treated to remove all the contaminants before it is discharged. There are several physical, chemical, and biological processes that can be used for treatment. The recirculation management strategy includes leachate containment, collection, and recirculation. Using this strategy, the leachate that is produced is collected, and then redistributed back over the landfill.

Recirculation turns the traditional landfill into an anaerobic bioreactor. There have been numerous studies which have proven the effectiveness of bioreactors. The fundamental process used for waste treatment in a bioreactor landfill is leachate recirculation. Recycling or recirculation of the leachate back to the landfill creates the perfect environment for rapid microbial decomposition of the biodegradable waste products. Not only does the system remain a storage facility for the solid waste, it also becomes a treatment system. The accelerated breakdown and stabilization of the waste can make the landfill a reusable system, and increase the operating life dramatically compared to the traditional landfill. This space that results from rapid stabilization can be used to store more solid waste instead of having to purchase more land. Laboratory and pilot scale studies have clearly demonstrated that operation of a landfill as a bioreactor accelerates waste degradation, provides in situ treatment of leachate, enhances gas production rates, and promotes rapid settling (Peng, 2017; Reinhart and Townsend, 1998).

One of the most important factors that control solid waste biodegradation is moisture content. This parameter can be controlled by leachate recirculation. Leachate recirculation optimizes environmental conditions within the landfill to initiate stabilization of the contents as well as treatment of the moisture flowing through the landfill. The numerous advantages of leachate recirculation include distribution of nutrients and enzymes, pH buffering, dilution and precipitation of inhibitory compounds, recycling and distribution of methanogens, liquid storage, and evaporation opportunities at low additional construction and operating cost (Peng, 2017; Reinhart and Townsend, 1998). Not only does recirculation of the leachate accelerate rapid degradation, it also treats the leachate at no extra capital costs. It has been suggested that leachate recirculation can reduce the time required for landfill stabilization from several decades to two or three years (Pohland, 1995).

Other methods include, physical reduction of organic or biodegradable waste going to the landfills, keep the wastes dry by controlling the water entering the landfilled wastes, though it is a fact that all materials such as the mineral clay liners, synthetic lining material such as the high density polyethylene (HDPE), breakdown over a geological time scale, therefore waste cannot be isolated for an infinite period and it is important to deal with any hazards within the landfills before problems arise. Landfill operation and after closure management should be therefore designed so that degradation processes are completed before the material lifetime ends (Ministry of Housing and Local Government, Malaysia 2004; Cossu, 2003).

Leachate that is collected and removed from a landfill must be managed with care. Some types of treatment, either at the landfill site or at a treatment plant offsite, must be performed. Treated leachate must meet the required regulatory limits for discharge to the environment as treated wastewater. There are many different landfill leachate treatment options. The options include complex and expensive routines of in-situ physical-chemical and biological processes for the treatment of organic and inorganic constituents. A simple approach to managing leachate would be to discharge the leachate to a nearby sewage treatment plant. If the landfill had a sewer connection, the leachate could be directly discharged from the storage containers. Since most landfills are located in sparsely populated areas, sewer connections are not usually available. Therefore, leachate usually is hauled by tanker trucks to treatment facilities. Also, sewage treatment plants often refuse to treat landfill leachate because the leachate may contain high concentrations of inhibitory chemicals that might interfere with the facilities treatment process (Mulamoottil et al., 1999).

If a landfill does not transport and treat the leachate offsite, a treatment facility can be constructed on site. There are a number of studies reporting various treatment alternatives for leachate from landfills. Processes that have been evaluated include biological treatment (aerated lagoons, activated sludge, anaerobic filters, and stabilization ponds), and physical-chemical processes such as adsorption, chemical oxidation, coagulation/precipitation, and reverse osmosis (Pohland, 1995). Other treatment options researched include trickling filters (Knox, 1985), and suspended-carrier biofilm processes (Welander et al., 1997). The types of constituents in leachate are different from typical domestic wastewater. Not only does leachate contain organic compounds that require biological treatment, it also contains inorganic dissolved solids (sodium, chloride, etc.) which cannot be removed by biological treatment (Reinhart and Townsend, 1998). The leachate characteristic will be the determining factor in choosing the eventual treatment method. In practice, more than one method is used to enhance the leachate quality before discharge or recirculation into the landfill.

LEACHATE TREATMENT PROCESSES

A wide range of processes are in place for the treatment of leachate both in-situ and offsite location. Types of leachate treatment process according to Tchobanoglous et al, (2003) is shown in Table 4.

ROUGHING FILTER

The roughing filter is quite an age long technology in water and wastewater treatment process. Roughing filters are used in separating solid matter from water. It can also reduce the number of pathogens in the water, as well as the amount of iron and manganese. They significantly improve the efficiency of sedimentation tanks. Most times the fine solid particle in suspension in the water is so light that, most of them will not settle in a normal sedimentation tank. The major difference between the roughing filter and other filter is that it does a physical filtration where no biological or chemical reaction takes place in the process. The fine solids crossing a rectangular sedimentation tank have to overcome a vertical settling distance before coming into contact with the bottom of the tank. Due to small settling velocities, a large portion of the fine solids might not reach the tank bottom and hence, will not be separated. The same sedimentation tank can be filled with rough filter material. The fine solid particles flowing through the filter touches the gravel or filter material surface already after a few millimetres of settling distance. Since the settling distance is drastically reduced by the filter material, roughing filtration is thus a more effective process for solids removal than plain sedimentation. Simple, low cost, low-maintenance, multi-stage roughing filters for household and community use have been described and characterized (Wegelin, 1996; Schertenlieb et al., 1987). Typically, these filters are rectangular, multi-compartment basins constructed of concrete or other materials. They require modest skills for operation and maintenance, and therefore, are best suited for use by communities or at least multiple households. However, it is possible for these multi-compartment tanks to be centrally fabricated and distributed at low cost for placement and final installation at their locations of use. Many of these filters are designed to use two different sizes of low cost, coarse granular media in two or three compartments or stages, and such media are generally locally available. For highly turbid water containing settleable solids, a horizontal or vertical sedimentation basin to remove this coarse material prior to filtration precedes the filter. The filter has provision for backwashing the medium from a valve inlet (at the bottom of the filter medium chamber in the horizontal and down flow filter designs). Roughing filters usually consist of differently sized filter material decreasing successively in size in the direction of flow. Most of the solids are separated by the

Table 4. Typical leachate composition from new and mature landfills

Treatment Process	Application	Comments
	Biological Processes	
Activated Sludge	Removal of Organics from Leachate	Defoaming activities may be necessary; separate clarifier needed.
Sequencing Batch Reactors	Removal of Organics	Similar to activated sludge, but no separate clarifier needed; applicable only to relatively low flow rates
Aerated stabilization basins	Removal of Organics	Requires large land area
Fixed film processes (trickling filters, rotating similar to leachates, biological contactors)	Removal of Organics	Commonly used on industrial effluents but untested on actual landfill leachates
Anaerobic lagoons and contactors	Removal of Organics	Lower power requirements and sludge
Nitrification/denitrification	Removal of nitrogen	Denitrification accomplished simultaneously with the removal of organics
	Physical/Chemical	
Sedimentation/flotation	Removal of suspended matter	Of limited applicability alone used in conjunction with other treatment processes
Filtration	Removal of suspended matter	Useful only as a polishing step
Air stripping	Removal of ammonia or volatile organics	May require air pollution control equipment
Steam stripping	Removal of volatile organics	High energy costs; condensate steam organics requires further treatment
Adsorption	Removal of organics	Proven technology; variable costs depending on leachate
Ion exchange	Removal of dissolved inorganics	Useful only as a polishing step
Ultrafiltration	Removal of bacteria and high-molecular-weight organics	Subject to fouling; of limited applicability to leachate
Reverse osmosis	Dilute solutions of inorganics	Costly; extensive pretreatment necessary
Neutralization	pH control	Of limited applicability to most leachates
Precipitation	Removal of metals and some anions	Produces a sludge, possibly requiring disposal as a hazardous waste
Oxidation	Removal of organics; detoxification of inorganic species	Works best on dilute waste streams; use of chlorine can result in formation of chlorinated hydrocarbons
Evaporation	Where leachate discharge is not permissible	Resulting sludge may be hazardous; can be costly except in arid regions
Wet air oxidation	Removal of organics	Costly; works well on refractory organics

coarse filter medium near the filter inlet, with additional removal by the subsequent medium and fine granular media in subsequent compartments. Roughing filters are operated at relatively low hydraulic loads or flow rates. Regular backwashing is required to maintain flow rates and achieve efficient particulate removals, and therefore, some skill and knowledge is required to properly operate and maintain a roughing filter. Removal of indicator bacteria by roughing filters has been reported to be 90-99%. It is expected that compared to bacteria removals, virus removals would be lower and parasite removals would be similar to or higher (WHO, 2005). A simple comparison of various types of filter in terms of their advantages and consequent disadvantages can be found in Table 5 (WHO, 2005).

Table 5. Advantages and disadvantages of different granular medium filters for household use

Filter Design or Type	Advantages	Disadvantages
Bucket filter	Useable on a small scale at household level; simple; can use local, low cost media and buckets; simple to operate manually.	May require fabrication by user; initial education and training in fabrication and use needed; requires user maintenance.
Barrel or drum filter	Useable on a small scale at household or community level; relatively simple; can use local, low cost media and barrels or drums.	Requires some technical know-how for fabrication and use; initial education and training needed; requires user maintenance.
Roughing filter	Useable on a small scale at community level; relatively simple; can use local, low cost construction material and media.	Less amenable to individual household use because of scale; requires some technical know-how for construction and use.

(World Health Organization, 2005).

Although viruses, bacteria and the smaller protozoans are too small to settle, these waterborne pathogens are often associated with larger particles or they are aggregated (clumped). Aggregated or particle-associated microbes are easier to remove by physical processes than the free or dispersed microbes. Consequently, observed reductions of waterborne microbes by physical removal processes are sometimes greater than expected or anticipated based strictly on their individual sizes. In some situations, efforts are made to promote the association of pathogens with larger particles, such as by coagulation-flocculation, to promote their physical removal (WHO, 2005).

RATIONALE FOR ROUGHING FILTER APPLICATION

Slow sand filtration applied as surface water treatment is particularly effective in improving the microbiological water quality. However, efficient application of this treatment process requires raw water of low turbidity. Pre-treatment of surface water with a high load of solid matter is usually necessary. Chemical flocculation, combined with sedimentation for solid matter reduction, is mostly inapplicable in rural water supplies of developing countries as these schemes generally face serious chemical water treatment problems (EAWAG, 1996).

Pre-filtration is a simple and efficient alternative treatment process used mainly for solid matter separation. It does not require the use of chemicals and also improves the microbiological water quality. Since pre-filters generally contain different fractions of rough filter material, they are known as roughing filters. Similar to slow sand filters, they make ample use of local resources and hardly require mechanical equipment. Consequently, roughing filters are often an appropriate pre-treatment technology for rural water supply schemes. Different filter types have been developed to meet various raw water qualities. Intake and dynamic filters are often applied as first pre-treatment stage followed by roughing filters operated either as vertical or horizontal-flow filters. These filters are usually cleaned hydraulically by fast filter drainage. A sequence of different pre-filtration stages is frequently the most cost-effective option, applying the multi-barrier concept and, hence, providing an efficient way of improving the microbiological water quality. Pre-filters and roughing filters are now extensively used both in water supply schemes of many developing countries and in artificial groundwater recharge plants of industrialised countries. According to practical experience, intake filters are capable of reducing the solid matter content by 50-70%, and roughing filters are able to separate particulate matter by 90% or more. Pre-filters and roughing filters also improve the bacteriological water quality; a 1-2 log reduction of faecal coliforms is often recorded. The filters also contribute to reducing the colour of dissolved organics and other substances in surface water. However, since stable suspensions rich in colloidal matter are difficult to treat by roughing filters, the addition of coagulants is usually required. In combination with slow sand filters, pre-filters and roughing filters present a reliable and sustainable treatment process particularly appropriate for developing countries. However, implementation of the technology alone may possibly fail. Hardware always has to be complemented by software. Close involvement of the future users in the planning phase, adequate treatment of plant operators and post project support to enhance a sustainable use of any type of water treatment process is, therefore, of key importance (EAWAG, 1996).

TYPES OF ROUGHING FILTERS

There are many types of roughing filters, but depends on the flow directions. Such as down flow, up flow and horizontal flow filters. Upflow roughing filters are relatively cheap and easier to clean than downflow or horizontal flow filters. An upflow filter box can be made of bricks, concrete, etc. It can have a round or rectangular shape, with vertical or partially inclined walls, and it is usually about 1.5 m deep. Water flows in through an underdrain system from the bottom, usually a perforated polyvinyl chloride (PVC) pipe, which also permits rapid abstraction during cleaning when the flow direction is reversed i.e. backwashing (Wegelin, 1996).

VERTICAL-FLOW ROUGHING FILTERS

Vertical-flow roughing filters operate either as downflow or upflow filters. They are hence either supplied by inflowing water at the filter top or at the filter bottom. The vertical flow roughing filters incorporates a simple self-cleaning mechanism and occupies minimal floor space when compared to horizontal flow roughing filters. The filter material of vertical-flow roughing filters is completely submerged. A water volume of about 10 cm depth usually covers the gravel and other local available materials like coconut fiber and broken burnt bricks. The top should be covered by a layer of coarse stones to shade the water and thus prevent algal growth often experienced in pretreated water exposed to the sun. Drainage facilities, consisting of perforated pipes or a false filter bottom system, are installed on the floor of the filter boxes (Nkwonta and Ochieng, 2009a). Vertical flow roughing filters usually consist of three units arranged in series as shown in Figure 2. The water to be treated flows in sequence through the three filter compartments filled with coarse, medium and fine filter material. The size of the three distinct filter material fractions is generally between 20 and 4mm, and graded, for example, into fractions of 12-18mm, 8-12mm and 4-8mm.

HORIZONTAL FLOW ROUGHING FILTERS

Horizontal roughing filters have a large silt storage capacity. Solids settle on top of the filter medium surface and grow to small heaps of loose aggregates with progressive filtration time. Part of the small heaps will drift towards the filter bottom as soon as they become unstable. As shown in Figure 3, un-

Figure 2. Layout and design of vertical flow roughing filter

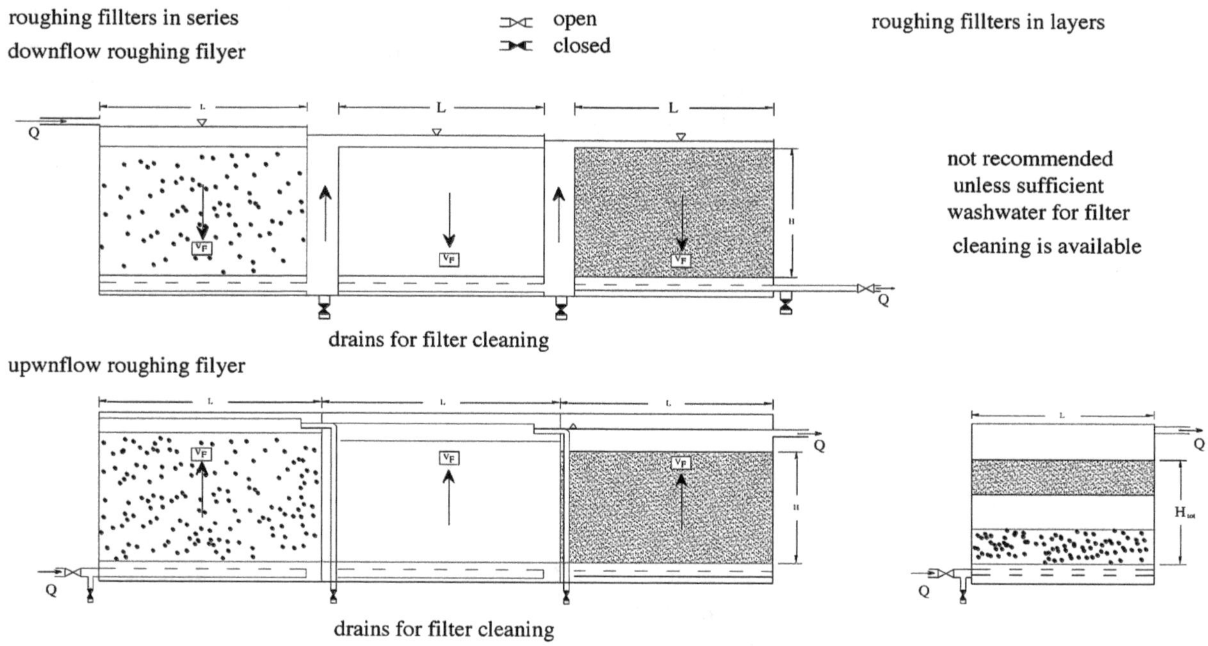

limited filter length and simple layout are the main advantages of horizontal roughing filters. This drift regenerates filter efficiency at the top and slowly silts the filter from bottom to top. Horizontal-flow roughing filters also react less sensitively to filtration rate changes, as clusters of suspended solids will drift towards the filter bottom or be retained by the subsequent filter layers. Horizontal-flow roughing filters are thus less susceptible than vertical-flow filters to solid breakthroughs caused by flow rate changes. However, they may react more sensitively to short circuits induced by a variable raw water temperature (Wegelin, 1996).

CLASSIFICATION OF ROUGHING FILTERS

The two criteria for filter classification are size of filter media and rate of filtration. Rapid sand filter and slow sand filter are different from intake filter and roughing filter according to their filter media size. The coarse filter media and the low flow rates are applied to roughing filtration. Table 6 explains the differences between the material sizes and flow rates of each filter. The department of water and sanitation in developing countries (SANDEC) classifies the roughing filters according to the following (Wegelin, 1996).

- Location within the water supply scheme
- Main application purpose
- Flow direction
- Filter design
- Filter cleaning technique.

Figure 3. Layout and design of horizontal, upflow and down flow roughing filter

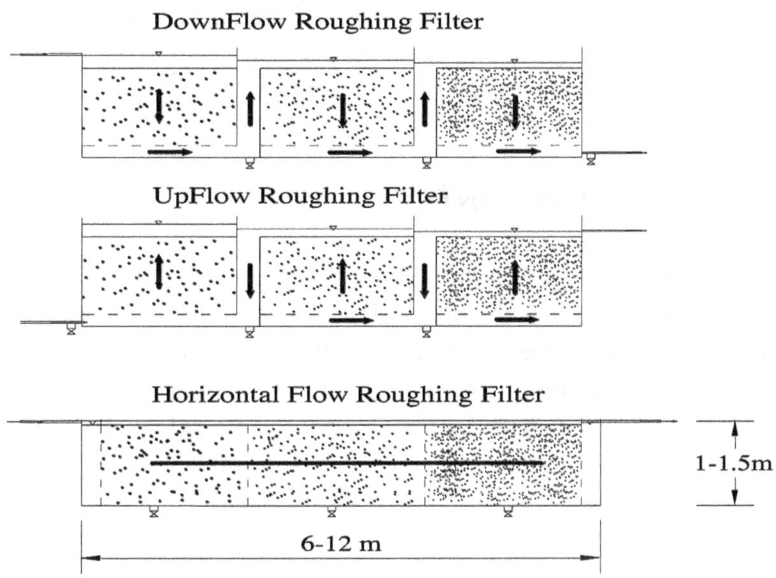

Table 6. Classification of filters

Filter Type	Size of Filter (mm)	Filtration Rate (m/h)
Rock filter	> 50mm	1 – 5m/h
Roughing filter	20 – 4 mm	0.3 -1.5 m/h
Rapid sand filter	4 – 1 mm	5 – 15 m/h
Slow sand filter	0.35 – 0.15 mm	0.1 – 0.2 m/h

CLEANING OF ROUGHING FILTER

Roughing filters need to be cleaned for the purpose of removing all of the accumulated particulate matter in the filter. Roughing filters are cleaned by opening a drain valve at the bottom of the filter and allowing the draining under gravity. It is suggested that the height of roughing filters should be limited to 1.5 m to facilitate easy cleaning. Infact, Galvis et al. (1998) found that cleaning roughing filter with a final head loss of 30 cm returned the head loss to 5 - 7 cm. The frequency of cleaning is actually very much dependent on the loading of particulate matters and biological activity in the filter and there are no general recommendations available as water quality conditions are site specific (Wegelin, 1996).

ROUGHING FILTER THEORY

Filtration consists of the transport mechanism and attachment of suspended particles in water or wastewater to the media. The flow through the media obeys Darcy's law which states that in all cases of filtration the flow is in the laminar range and that the pressure drop is governed by the flow rate (Belfort et al., 1995). Necessary variable parameters that should be considered in practice include media porosity, depth of bed, filter size in terms of area or surface area and filtration rate will be sufficient to express the filter efficiency (Wegelin, 1996). A number of processes such as physical, chemical and or biological processes may also be involved to ensure effective performance.

FLOW RATE AND HEAD LOSS CONTROL

Flow rate is an important factor affecting removal in roughing filter. Wegelin (1996) had also earlier observed and noted that, sedimentation and biological mechanisms depend on the filtration rates. Lower filtration rates will amount to less turbulent conditions in the filter media and will promote gravitational sedimentation and increase the hydraulic retention time in biologically active regions of the filter. The main objective of roughing filter is to reduce the amount of solid matter from the raw water or wastewater. Filters are usually operated at the filtration rate of up to 1.5 $m^3/m^2/h$ and size of filter material ranged between 4 mm and 20 mm. Head loss in a roughing filter is usually small. Head loss can be generally recorded as, water level difference between the inlet and outlet water level (Wegelin, 1996). The head loss in a filter can be calculated as in Equation (1);

$$h = 1.07L\frac{C_D V^2}{\psi g d f^4} \qquad (1)$$

Where,

h = head loss (m)
L = filter depth (m)
$C_D = (24/Re) + (3/\sqrt{Re}) + 0.34$ (the drag coefficient)
Re = Vd/μ (dimensionless Reynolds Number)
μ = kinematics viscosity of fluid (m^2/s)
d = particle diameter (m)
Ψ = dimensionless shape or sphericity factor (<1)
g = gravitational constant (m^2/s)
V = flow rate ($m^3/m^2/h$)
f = porosity the ratio of (volume voids)/(total bed volume)

The development of head loss in the filter is small during the initial period of operation, however, it later increases with most of the head loss occurring on the top dirty surface of the filter media (Wegelin, 1996). The hydraulic conditions in roughing filters are determined by the filtration rate V_F (m/h), calculated as flow rate Q (m^3/h) divided by the active cross-sectional filter area A(m^2) given in Equation (2).

$$V_F \text{ (m/h)} = Q \text{ (m}^3\text{/h)} / A \text{ (m}^2\text{)} \qquad (2)$$

SETTLING VELOCITY ESTIMATION IN ROUGHING FILTER

Sedimentation of particles in the voids of filter media is part of the treatment process for the removal of suspended particles in roughing filter. The process is known as settling or clarification. The efficiency of this process is measured by turbidity removal. It depends on size of particle and settling rate. Sedimentation is a treatment process in which the velocity of the water is lowered below the suspension velocity and the suspended particles settle out of the water due to gravity. The settling velocity is influenced by mass density, size and shape of particle, as well as by viscosity and hydraulic conditions of water. There is a basic formula to calculate settlement velocity for a spherical particle by using Stoke's law and it is given in Equation (3).

$$V = \frac{gd^2(P_p - P_w)}{18\mu} \tag{3}$$

Where:

v = settling velocity in (cm/s)
g = acceleration due to gravity in (cm/s^2)
d = diameter of particle in (cm)
P_ρ = particle density in (g/cm^3)
P_w = fluid density in (g/cm^3) and
μ = fluid viscosity in (g/cm/s)

Stoke's Law gives the relationship between the settling rate, particle size and density. From the Equation (3), for all other parameters being constant, dense particles settle faster, larger particles settle faster, and more viscous water causes particles to settle slower. Interception is the process which enhances particle removal in the filter. Interception occurs when deposited particles accumulate on the filter media that gradually reduce the pore size. Suspended particles travelling in roughing filter are obstructed to continue forward due to reduction of pore size. Therefore, particles will collide with each other and are packed closely and lose energy to go forward. Finally, they will deposit on filter media (Wegelin, 1996).

FACTORS AFFECTING REMOVAL IN ROUGHING FILTERS

The basic parameters that usually affect suspended solids removal by roughing filters include; filter media size, filtration rate and bed depth. Generally, removal efficiency increases with decreasing filter media size, decreasing filtration rate and increasing filter bed depth (Collins et al., 1994). The filter media size is an important variable. An increased efficiency in the treatment has been observed with decreasing filter media size, which indicates the importance of straining. Higher removals can be obtained due to smaller interstices between smaller media, as well as the larger surface area available, which allow more adsorption. A smaller size of filter media will have a larger total surface area available for biofilms to grow on, and therefore more biofilm can be exposed to raw water. Therefore, removal efficiency increases. The empty space or pore size within a filter medium is important for determining the right filter size and efficiency. Pore size is a measure of how much of the medium consists of empty space. The filter efficiency depends on the ratio of filter media surface area to its volume, which means total specific surface area per cubic metre. Despite this enormous specific surface area, sand would make a poor filter medium because the small particle size would soon lead to blockages. Because of the dense packing, any flow through the sand would be very slow. Therefore, despite its massive surface area, the volume of water that could be treated per hour would actually be quite small. For a medium such as gravel, it is larger in size and less in specific surface area that would make it less prone to blocking. Special media such as filter matting, plastic or sintered glass, have both a large specific surface area (SSA) and a generous void space. In fact, many of them are more than 90% void or empty space. This makes blockage almost impossible (Wegelin, 1996).

Filtration rate also has a significant influence on the particle removal. Various reports described that good removals in the roughing filters were achieved at low filtration rates. It is attributable that low filtration rates give support to retain particles that are gravitationally deposited to the upper side of filter media. It is important to have laminar flow conditions. Wegelin (1996) affirmed that, filtration rate greatly influences the filter efficiency. Flow conditions are described by Reynolds number. At Reynolds number less than 10, laminar flow can be expected. Removal efficiency increases with decreasing Reynolds Number (Re). Accordingly, turbidity removal was 40% at a Re of 8, whereas removal was greater than 80% at Re of less than 3. Wegelin (1996) described that at increased filtration rates, coarse particles penetrated deeper into the bed, clogged the finer gravel media, and re-suspended pre-deposited particles resulting in decrease in filter efficiency.

Filter bed depth also affects efficiency of roughing filters. While particles deposits on the filter bed, pore spaces becomes smaller. As suspended particles, accumulate on a filter bed, the pressure drop through the filter will be increased. Operating with high-pressure drop may increase the chance of detachment and penetration of detached solids will move deeper into the filter bed. Therefore, increasing filter bed's depth will improve overall performance and coliform removal (Wegelin, 1996). Pressure drop can be reduced with the use of multiple media sizes. The larger solids particles are removed by coarse media, medium size filter media can polish and the finest media could remove the remaining traces of solid matters. Removal of particulate matters in roughing filter is also dependent on raw water characteristics. Thus, it is important to study the characteristics of the given source of water in designing a roughing filter. Wegelin (1996) showed that roughing filters were good for removal of major solid particles and for highly turbid waters. It also described that, filtration performance depends on the source of water quality (i.e. types and concentration of natural organic matter and suspended particles) would affect filter's performance. In a nutshell, performance of roughing filter depends on influent solids concentration, particle size, filter media size, bed depth and filtration rate. Roughing filter design becomes more critical when trying to determine the optimal combination of media size and bed depth for particular source of water or wastewater as the case maybe (Wegelin, 1996).

FILTER MEDIA SIZE

Media types commonly used in roughing filtration are quartz sands and gravels but can be replaced by any clean, insoluble and mechanically resistant material. Previous work by Wegelin (1996) showed that the effect of surface porosity and roughness of filter media on particle removal efficiency in roughing filtration was insignificant compared to the size and shape of macropores in the filter. Improved removal efficiencies are generally correlated to smaller media sizes (Collins et al., 1994). The use of multiple grades of filter media in a roughing filter promotes the penetration of particles throughout the filter bed and takes advantage of the large storage capacities offered by larger media and high removal efficiencies offered by small media. The size of filter media decreases successively in the direction of water flow and ideally the uniformity of filter media fractions is maximized to increase filter pore space (storage capacity) and aid in filter cleaning (Boller, 1993).

FILTER MEDIA

The filter material should have a large specific surface to enhance the sedimentation process taking place in the roughing filter and high porosity to allow the accumulation of the separated solids. Generally speaking, any inert, clean and insoluble material meeting the above two criteria can be used as filter medium. Filtration tests revealed that neither the roughness nor the shape or structure of the filter material have a great influence on filter efficiency.

The following material could therefore be used as filter media:

1. Gravel from a riverbed or from the ground.
2. Broken stones or rocks from a quarry.
3. Broken burnt clay bricks.
4. Plastic material either as chips or modules (e.g. used for trickling filters) may be used if the material is locally available.
5. Burnt charcoal, although there is a risk of disintegration when cleaning the filter material, it should only be considered in special cases (e.g. for removal of dissolved organic matter)
6. Coconut fibre, however, due to the risk of flavouring the water during long filter operation, it should be used with care (Nkwonta and Ochieng, 2009b).

In addition, Nkwonta and Ochieng (2009b) noted that broken burnt bricks and improved agricultural waste (charcoal maize cobs), can also be effectively used as pretreatment media and therefore could serve as alternative where natural gravel is not readily available. It was observed that in general both broken burnt bricks and charcoal performed better than gravel. This observation could have resulted from the reason that both charcoal maize cobs and broken burnt brick have a slightly higher specific surface area and porosity respectively to enhance the sedimentation and other filtration processes compared to gravel.

FILTRATION RATE

Filtration rate also has a significant influence on the treatment removal. According to Nkwonta and Ochieng, (2009a) good removal in roughing filters are best achieved with low filtration rate, because low filtration rates are critical to retain particles that are gravitationally deposited to the surface of the media. Pretreatments used for removal of iron and manganese were able to operate at filtration rates of 1.5 - 3 m/h. In addition, horizontal flow roughing filter was capable of removing metals such as iron, manganese, turbidity and colour however at a filtration rate of 1.8 m/h. Wegelin (1996) found that at increased filtration rates (2 m/h), coarse particles penetrate deeper into the bed and these will cause decrease in filter efficiency. Whereas at 1 m/h there was good distribution of solids loading throughout the bed. Nkwonta and Ochieng, (2009a) also suggested that normal filtration rate of horizontal roughing filters is between 0.3 and 1.5 m/h.

FILTER LENGTH

Improved cumulative removal efficiencies are typically correlated to longer filter lengths (Collins et al., 1994). However, incremental removal efficiencies tend to decrease with increasing filter length due to the preferential removal of larger particles early in the filter (Wegelin, 1996). The rate of decline is dependent on filter design variables and the size and nature of particles in suspension. The use of different media sizes may allow for treatment targets to be met by a shorter filter with multiple media sizes compared with long filter packed with one media size.

ROUGHING FILTER OPERATION AND MAINTENANCE

The treatment performance of a roughing filter over time can be divided into two phases. The first phase represents a period when the particle removal efficiency remains relatively constant (steady-state) with increasing solids deposition, whereas the second phase represents a period of decreasing removal efficiency due to increasing particle deposition and penetration through the filter(Collins et al., 1994). In another study (Wegelin, 1996), noted that particle removal efficiency and particle penetration play a key role in determining the length of filter run. Particles in a horizontal roughing filter (HRF) during a filter run, drift deeper in the direction of flow and also downward by gravity. Unlike in a HRF, particle drift in vertical flow roughing filter, occurs only in the direction of water flow allowing for deeper penetration of particles in the filter and generally shorter filter run lengths (Collins et al., 1994).

The end of a filter run is typically determined when the quality of filter effluent deteriorates due to increasing solids deposit until minimum water treatment targets are exceeded. Drainage facilities located at the base of roughing filter compartments allow for rapid downflow drainage, a common maintenance procedure used to remove solids accumulated in the filter at the end of a filter run cycle (Galvis et al., 1998). Because of the reliance on gravity drainage for filter cleaning, a dual roughing filter (in layers) configuration, wherein large media is placed over smaller media in the same filter compartment, is not recommended. As a general guide, filters should be drained when deposited solids reach ~10 g of solids per length of filter (Collins et al., 1994). Increase in filter resistance or head loss are typically small (<5 cm) in roughing filtration due to the relatively coarse filter media used and low hydraulic loading rates applied. A condition, whereby actual headloss exceeds the available headloss prior to decreases in removal efficiency (i.e. the filter storage capacity is not exhausted before filter cleaning is required) signals the need to re-evaluate the filter design (Wegelin, 1996).

ADVANTAGES OF ROUGHING FILTERS OVER CONVENTIONAL METHODS

Conventional system is quite demanding in chemical use, energy input and mechanical parts as well as skilled manpower that are often unavailable, especially in rural areas of developing countries like Tanzania, Kenya and Sri Lanka. But roughing filters does not require chemical use, energy input and mechanical parts. Additionally, conventional methods demand high operating costs (Nkwonta and Ochieng, 2009a).

DISADVANTAGES OF ROUGHING FILTERS OVER CONVENTIONAL METHODS

Colour removal is fair to poor and in some cases it requires a large area of land for effective treatment. It can handle only relatively low strength wastes compared to conventional methods. It also can handle only very low organic loads compared to conventional treatment methods such as activated sludge process (Nkwonta and Ochieng, 2009a).

ADSORPTION

Adsorption is the accumulation of atoms or molecules on the surface of a material. This process creates a film of the adsorbate (the molecules or atoms being accumulated) on the adsorbent's surface. It is different from absorption, in which a substance diffuses into a liquid or solid to form a solution. The term sorption encompasses both processes, while desorption is the reverse process of "adsorption". In simple terms, adsorption is "the collection of a substance onto the surface of adsorbent solids." It is a removal process where certain particles are bound to an adsorbent particle surface by either chemical or physical attraction (Reynolds and Richards, 1996).

Adsorption can be classified into two major types based on the process. There is the physical adsorption and chemical adsorption. Physical adsorption is due to Van der Waals force and it is a reversible process. The molecular force of attraction between the solute and absorbent are greater than the force of attraction between the solvent hence the solute will be adsorbed onto the surface. In the case of the chemical adsorption, a chemical reaction occurs between the solid and the adsorbed solute, and the reaction is usually irreversible. However, such process is rarely used in environmental engineering. Adsorption is present in many natural physical, biological, and chemical systems, and it is widely used in industrial applications such as activated charcoal, capturing and using waste heat to provide cold water for air conditioning and other process requirements adsorption chillers, synthetic resins, increase storage capacity of carbide-derived carbons for tunable nonporous carbon, and water purification. Adsorption, ion exchange, and chromatography are sorption processes in which certain adsorbates are selectively transferred from the fluid phase to the surface of insoluble, rigid particles suspended in a vessel or packed in a column. Similar to surface tension, adsorption is a consequence of surface energy. In a bulk material, all the bonding requirements (be they ionic, covalent, or metallic) of the constituent atoms of the material are filled by other atoms in the material. However, atoms on the surface of the adsorbent are not wholly surrounded by other adsorbent atoms and therefore can attract adsorbates (Reynolds and Richards, 1996).

FACTORS ON WHICH ADSORPTION DEPENDS

A number of conditions must be satisfied for adsorption to occur. Amongst these include the following which shall either be in favour or against the process depending on whether it is chemical or physical adsorption process.

1. **Temperature**: Adsorption increases at low temperature conditions. Adsorption process is exothermic in nature. According to Le Chatleir principle, low temperature conditions would favour the forward direction.
2. **Pressure**: As depicted by adsorption isotherm, with the increases in pressure, adsorption increases up to a certain extent till saturation level is achieved. After saturation level is achieved, no more adsorption takes place no matter how high the pressure is applied.
3. **Surface Area**: Adsorption is a surface phenomenon therefore; it increases with increase in surface area. Surface area offered by adsorbent becomes an important factor for consideration.
4. **Activation of Adsorbent:** Activation of adsorbent surface is done so as to provide more vacant sites on surface of adsorbent. This can be done by breaking solid crystal in small pieces, heating charcoal at high temperature, breaking lump of solid into powder or other methods suitable for particular adsorbent.
5. **Particle Size of ADSORBENT:** SMALLER particle sizes reduce internal diffusional and mass transfer limitation to the penetration of the adsorbate inside the adsorbent (i.e., equilibrium is more easily achieved and nearly full adsorption capability can be attained).
6. **Contact Time or Residence Time:** The longer the time the more complete the adsorption will be. However, the equipment will be larger.
7. **Solubility of Solute (adsorbate) in Liquid:** Substances slightly soluble in water will be more easily removed from water (i.e., adsorbed) than substances with high solubility. Also, non-polar substances will be more easily removed than polar substances since the latter have a greater affinity for water.
8. **Affinity of the Solute for the Adsorbent (Carbon):** The surface of activated carbon is only slightly polar. Hence non-polar substances will be more easily picked up by the carbon than polar ones.
9. **Number of Carbon Atoms**: For substances in the same homologous series a larger number of carbon atoms is generally associated with a lower polarity and hence a greater potential for being adsorbed (e.g., the degree of adsorption increases in the sequence formic-aceticpropionic- butyric acid).
10. **Size of the Molecule with Respect to Size of the Pores:** Large molecules may be too large to enter small pores. This may reduce adsorption independently of other causes

Degree of Ionization of the Adsorbate Molecule: More highly ionized molecules are adsorbed to a smaller degree than neutral molecules.

11. **pH**: The degree of ionization of a species is affected by the pH e.g., a weak acid or a weak basis. Adsorption exhibits an inverse relationship with pH. Thus, as pH is increased adsorption is reduced and vice versa. It is noteworthy to mention that acidic compounds are better removed at lower pH. Hence, it could be better a lower pH is maintained for increased adsorption to occur (Srdjan 2007; Tchobanoglous et al., 2003; Cooney, 1999; Davis and Cornwell, 1998). In contrast

to adsorption towers where the pollutant is continuously removed by flowing liquid, the collected pollutant remains in the adsorption bed. Thus, while the bed has sufficient capacity, no pollutants are emitted. At some point in time, the bed will become saturated with pollutants. This is called breakthrough. When the bed is exhausted, the influent and effluent concentration will be equal (Davis and Cornwell, 1998).

LIMESTONE

Limestone is a sedimentary rock composed mostly of calcite and aragonite, both comprised of calcium carbonate ($CaCO_3$), but assembled in a trigonal and orthorhombic system, respectively. Limestone can also contain quartz, sand, chert and clay. The presence of the mineral dolomite in higher proportions is responsible for the designation of a particular type of limestone: dolomitic limestone or dolomite. Some of the features that turn carbonate rocks (limestones in particular) into good targets for biodeterioration are the same features that make them an historic favourite for the construction and ornamentation of our cultural heritage – a soft, bright and easy to carve rock (Pinheiro et al., 2018).

Harrington et al., (1968) attempted to explain the wide range of reactivity's exhibited by the different test stones in view of their varying chemical composition. Single component and multiple-component correlation studies between the reactivity of a material and its calcium, magnesium, silicon, and aluminum content failed to establish any composition dependency. The only correlation obtained was a very weak relationship between increasing reactivity and increasing content of iron oxide. Relatively high reactivity's of stones low in iron content indicated that other characteristics of the stones are equally important. Potter (1969) observed that amongst all the chemical components, only iron is significantly related to the $S0_2$ capacity of limestone at or above the 90% confidence level. Whether the contribution of the iron is physical or chemical has not been established. Falkenberry and Slack (1969) noted that high-calcium limestones are more reactive than dolomitic limestones. Silica was found to be the most undesirable impurity, as it appeared to fuse and glass-over the calcine and thus reduced the effectiveness of the limestone particles in reacting with Sulphur dioxide ($S0_2$). This was reaffirmed by Moropoulou et al., (2001) in their study.

There is a grading of limestone according to their density by the United States geological classification (Marble Institute of America, 2019). Limestone is divided into three sub-classifications that describe their densities in approximate ranges as follows:

- **Low Density:** Limestone having a density ranging from 110 through 135/lb/ft^3 (1760 through 2160 kg/m^3).
- **Medium Density:** Limestone having a density greater that 135 and not greater than 160 lb/ft^3 (2160 through 2560 kg/m^3).
- **High Density:** Limestone having a density greater than 160 lb/ft^3 (2560 kg/m^3).

As a result of the effectiveness of limestone in various treatment processes, it has been used for removal of heavy metals in leachate. A study indicated that limestone is capable of removing heavy metals including Cu, Zn, Cd, Pb, Ni, Cr, Fe and Mn through a batch process or by filtration technique. The removal capability is reported to reach up to 90% (Aziz et al, 2004).

Smith et al., (1994) used limestone filters to treat contaminated groundwater containing iron concentration of 5 mg/ L and reported final concentration of iron of 0.2 mg/ L. Xu et al, (1997) conducted batch experiments using calcite and quartz grains as filter media and reported an iron removal efficiency of 99.8%. Sun (2004) conducted batch experiments in which limestone was used as a filter medium to treat an iron acid solution (27.9 mg iron/L) and reported an iron removal of 100% after 150 min. Aziz et al. (2004) reported 90% removal of iron by limestone filter from landfill leachate containing 19.5 mg/L iron. The highest average concentration of iron remaining in solution in this study was 0.1 mg/L which is lower than the standard guidelines of 0.3 mg/L for the protection of aquatic life (Ghaly et al,. 2007).

The roughing filter principles has been used for the pre-treatment of water before supply to communities. A wide number of experiences are available in almost all continents. A summary of the efficiency of various parameter removal is given in Table 7.

The performance of horizontal flow roughing filter using limestone media for the removal of turbidity, suspended solids, biochemical oxygen demand and coliform organisms from wastewater using different sizes of limestone roughing filter has been reported. Results indicated that removal efficiencies depended on the size of the filter medium and applied flow rates. Turbidity, suspended solids, BOD and coliform organisms' removals were between 75 and 92%, 79 and 88%, 51 and 67% and 67 and 96%, respectively using a combination of three filter media with particle sizes between 1.91 mm and 16.28 mm. Removal efficiency was found to increase with lower flow rates (Aziz et al., 2006).

LEACHATE COLLECTION

Leachate used in this research was generated and subsequently collected from Pulau Burung Landfill Site (PBLS); which is situated in Penang, Malaysia. PBLS has a semi-aerobic system and it is one of the only three sites of its kind found in Malaysia. A semi-aerobic system can be achieved through a convection process. The latter involves the decomposition of organic matter inside the landfill and will cause an increase in temperature. The difference in temperature between inside and outside of the landfill will generate a heat convection current into the landfill through the leachate pipe. PBSL has been developed semi-aerobically into a sanitary landfill Level II by establishing a controlled tipping technique in 1991. It was further upgraded to a sanitary landfill Level III employing controlled tipping with leachate recir-

Table 7. Performance of roughing filter

Reference	Filtration Rates (m/h)	Parameter	Mean Percentage Removed (%)
Pacini (2005)	1.20	Iron & Manganese	85 and 95
Dome (2000)	0.3	Algae & turbidity	95 and 90
Mahvi (2004)	1.5	Turbidity	90
Ochieng and Otieno (2004)	0.75	Turbidity & algae	90 and 95
Dastania (2007)	1.8	Turbidity, TSS & Coliforms	63.4, 89 and 94
Jayalath (1994)	1.5	Colour & Turbidity	50 and 80
Rabindra (2008)	1.0	TSS & Turbidity	95 and 95
Mukhopadhay (2008)	0.75	Turbidity	75

Table 8. Composition of leachate from Pulau Burung landfill site

Parameters	Range of Values, mg/l
BOD_5, mg/l	48 – 1120
COD, mg/l	1533-3600
Suspended Solid, mg/l	159-1220
pH value (no unit)	7.8-9.4
Zinc, mg/l	0.1 – 1.8
Manganese, mg/l	0.6 – 1.1
Iron, mg/l	0.32 – 7.5
Copper, mg/l	0.1 – 0.4
Cadmium, mg/l	<0.04
Colour	2430 – 8180

All units in mg/L except for pH (no unit) and colour (Platinum unit)

culation in 2001. It was found that the leachate from a semi-aerobic system has slightly lower organic contaminants compared with an anaerobic landfill in terms of BOD and COD (Aziz et al., 2001). This site receives 1500 tons of solid waste daily. The composition of the leachate is provided in Table 8.

EXPERIMENTAL APPROACH

Limestone samples used in this study were obtained from the quarry industry located at Ipoh, Malaysia. The limestone chips composed of 95.5% $CaCO_3$, 3.0% $MgCO_3$ and 1.5% impurities (Aziz et al., 2006). The limestone samples were crushed and graded into the desired size ranges (4-20mm) and was not treated with any chemical.

A pressure pump in the category of the peristaltic pump was used to generate the pressure for flow of the leachate from the residual tank through the vertical roughing filter in an upflow manner from the bottom. The capacity of the peristaltic pump was 0 to 2000 ml/min. The grain density of a filter bed is the key parameter in deep-bed filtration. Therefore, the influence of media size on treatment performance may be assessed only if the uniformity of different sized media is consistent and preferably high. A roughing filter packed with a media of low uniformity is likely to perform better than a filter packed with media of similar average size of high uniformity (Nkwonta and Ochieng, 2009a). This phenomenon occurs because the pore spaces in the filter packed with media of low uniformity are smaller due to filling of pore spaces by smaller grains. Although removal efficiencies may be improved by using filter media of low uniformity, such design may promote undesired cake and straining filtration leading to shorter filter run and increased filter cleaning difficulties (Collins et al., 1994). For this study the filter column was packed with limestone which were initially crushed from bolder sizes to reduce them to the range of 4 to about 20 mm sizes. Thereafter, the filter media was packed into the column with three layers, decreasing sizes from bottom to top. The three media size ranges (4-8mm, 8-12mm, and 12-18 mm) were used to assess the influence of flow rate, pore size and media density on BOD, COD, colour and turbidity removal efficiency. After packing the column, the 900 mm filter media bed was washed with 20 liters of distilled water for some time and then was ready for the experiment to be performed.

Table 9. Experiment phases and media packing in column

Flow Rate mL/min	Media Size (mm)	Dilution	Duration (days)	Sampling & Analysis (hours)
Phase 1- Initial Trials				
100	12-18,8-12 & 4-8	No	1	24
80	12-18,8-12 & 4-8	No	1	24
60	12-18,8-12 & 4-8	No	1	24
40	12-18,8-12 & 4-8	No	1	24
20	12-18,8-12 & 4-8	No	1	24
Phase 2 - Dilution of Leachate				
20	12-18,8-12 & 4-8	50% dilution of leachate	10	24
Phase 3 – Main Experiment (Various Sizes of Media)				
20	12-18,8-12 & 4-8	No	10	24
	12-18	No	10	24
	12 -18 & 4-8	No	10	24
	8-12	No	10	24
	4-8	No	10	24
Phase 4 – Re-trial of Higher Flow Rate				
50	12-18,8-12 & 4-8	No	10	24

NB: Media size arrangements were in the decreasing order from bottom to top.

Leachate was then run through the upflow vertical roughing filter. Five hydraulic loading rates (100mL/min, 80mL/min, 60mL/min, 40mL/min and 20mL/min) were initially assessed in this study to determine the influence of interstitial fluid velocity on removal efficiency of the various parameters. The capacity of the peristaltic pump was 0 to 2000mL/min. Three collection and monitoring ports at (300 mm, 600 mm and 900 mm) were provided on the filter length. Sampling was done after every 24 h and analysis was performed after the treated sample was filtered through a 0.45μm membrane filter. There were four phases in the experiment to ensure that data obtained were consistent as shown in Table 9.

In the initial stage, various flow rates and filtration rates were used over a constant bed depth of 900mm in the filter column, there was no dilution. The leachate was passed through the roughing filter column in the upflow direction and in each case a minimum of 24 hour was allowed to ensure proper action between the filter media and the leachate. Afterwards, the most suitable flow rate was used all through the remaining experiments for consistency of results. The effluents from pre-treated and filtered leachate were collected through sampling points each day to study the filtration efficiency by carrying out laboratory tests for COD, BOD, colour and turbidity. Sampling was done from bottom towards the top. Three readings from port 1, 2 and 3 in the column (Figure 4). This shows the sectional diagram together with the dimensions of the upflow vertical column used in the experiment.

Figure 4. Sectional diagram of vertical upflow roughing filter column used
(Geometric Similarity; 1:3) (Source: Wegelin, 1996)

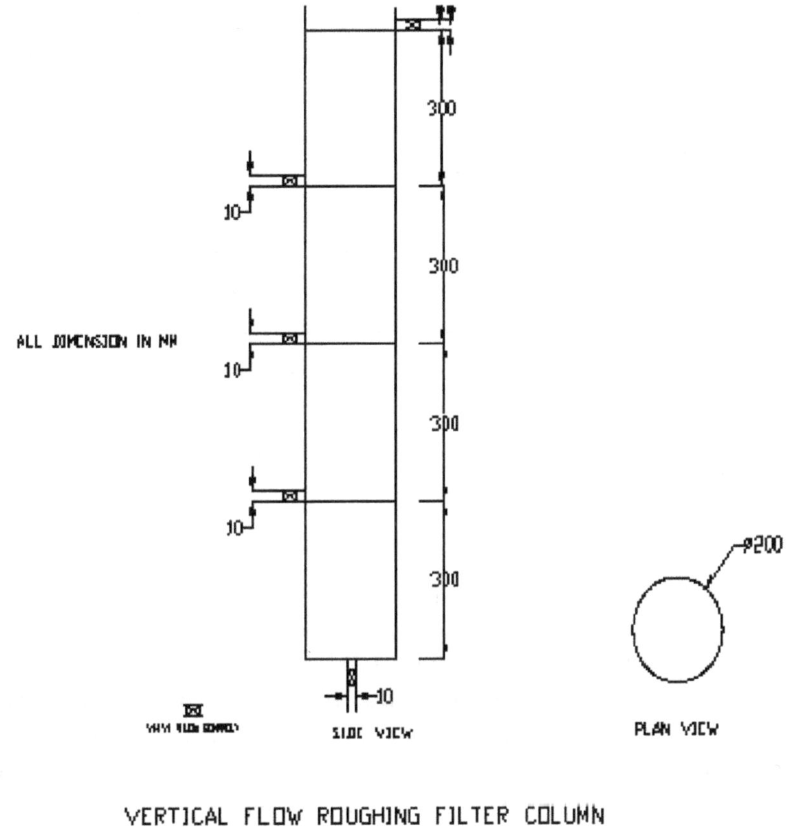

ANALYTICAL METHODS

Leachate was collected from Pulau Burung Landfill Site (PBLS) was transported and stored at 4°C in a cold room since weekly sampling was adopted. Limestone samples used in this study were obtained from the quarry industry located at Ipoh, Malaysia. COD and colour were analyzed using DR/2800 Hach Spectrophotometer, method 8000 and 8025 respectively. BOD was analyzed using standard methods (APHA, 1998) Method 5210B for wastewater analysis. The pH was measured by pH meter (CyberScan 20) while turbidity was measured using 2020 Turbidimeter (LaMotte). Analysis of variance (ANOVA) and Boxplot was performed using MINITAB version 14.0.

CHARACTERISTICS AND OPERATIONAL PARAMETERS OF THE ROUGHING FILTER COLUMN

The characteristics of the experimental column and set up is shown in Table 10. This was used throughout the study.

Table 10. Characteristics of the materials used for experimental phase

Parameter	Data	Unit
Flow Rate	20	mL/min
Column Height	1500	mm
Internal Diameter	200	mm
Surface Area of Column	314	cm^2
Column Material	Perspex Plastic	
Total Bed Volume *net	28260	cm^3
Particle Size Range	12-18,8-12 & 4-8	mm
Limestone Density	2554	Kg/m^3
Retention Time	31.4	hours
Contact Time	24	hours
Filtration Rate	0.2	m/hr
Mode of Flow	Upflow	

Raw Leachate Quality

Initial analysis of the raw leachate sample collected from the landfill is shown in Table 11. The characteristics are representation of the methane fermentation phase. This phase is usually characterized by microorganisms, which converts the acetic acid and hydrogen gas formed by the acid formers in the acid phase to methane (CH_4) and CO_2, which become more predominant. Both methane and acid fermentation proceed simultaneously, but the rate of acid fermentation is considerably reduced compared to the former. Because the acids and the hydrogen gas produced by the acid formers have been converted to CH_4 and CO_2, the pH within the landfill will rise to more neutral values in the range of 6.8 to 8. In turn, the pH of the leachate, if formed, will rise, and the concentration of BOD_5 and COD and the conductivity value of the leachate will be reduced. With higher pH values, fewer inorganic constituents are solubilized; as a result, the concentration of heavy metals present in the leachate will also be reduced (Tchobanoglous and Kreith, 2003).

Table 11. Raw leachate quality

Parameter	Readings
COD (mg/L)	2100 - 2530
BOD (mg/L)	271 - 370
pH	7.94 - 8.12
Turbidity(NTU)	226 - 274
Colour (Pt-Co)	3310 - 3920

Performance of Various Media Sizes

The removal efficiency of the limestone media (12 –18mm, 8-12mm & 4-8mm) during the trial runs is shown in Figure 5. Each flow rate was experimented for 24 h using different batch of limestone each day. The highest removal was observed at 20 mL/min. It showed that BOD, COD, colour and turbidity removed were 23, 22, 38 and 30%, respectively. Hence, it was used for the subsequent experiments performed in phase three.

The removal efficiencies when 50 mL/min flow rate was applied to media 12 – 18mm, 8 – 12mm and 4 - 8mm size is shown in Figure 6. It indicates removal in the range of 7 to 18%, 32 to 40%, 20 to 63% and 28 to 40% for BOD, COD, turbidity and colour, respectively. The limestone media experienced a breakthrough except for turbidity.

The BOD, COD, turbidity and colour removal by the 12 – 18 mm 8 -12 mm and 4 – 8 mm media combination is illustrated in Figure 7. It indicates BOD removal of 22 to 81%, COD removal of 22 to 75%, turbidity removal of was 32 to 86% and colour removal of 36 to 62%. Except for turbidity, a decline in removal of the other parameters was observed after the sixth day owing to the optimum adsorption reached.

The limestone media 12 – 18mm, 8-12mm & 4 – 8mm sizes was experimented using a 50:50 (leachate: water) dilution. It indicated removal of BOD, COD, turbidity and colour of 46 to 68%, 27 to 46%, 58 to 82% and 60 to 70%, respectively. Physical adsorption is due to van der Waals' forces, which are weak bonds compared to chemical adsorption. However, because of the weak nature of these bonds, adsorbed molecules can be easily removed with a change in the solute concentration. The low removal of COD was probably due to a lower degree of attraction of the adsorbate molecules onto the media (i.e. the molecules being accumulated on the surface of the adsorbent) caused by the weaker vander waal's forces between the leachate and limestone media. Figure 8 illustrates limestone media 12 – 18mm size removal efficiency. BOD removal was from 22 to 41%, COD removal was from 9 to 27%, turbidity removal was from 62 to 66% and colour removal was from 33% to 56%. Low removal in BOD and COD can be attributed to large media size which resulted in reduced specific surface area of the media adsorption.

Figure 9 shows the pattern of removal that occured in the 12 – 18mm & 4 – 8mm, media. In the experiment, BOD removal was 34% to 60%, COD removal was 19% to 54%, turbidity removal was 32%

Figure 5. Trial runs in terms of removal efficiency of various parameters and flow rates.

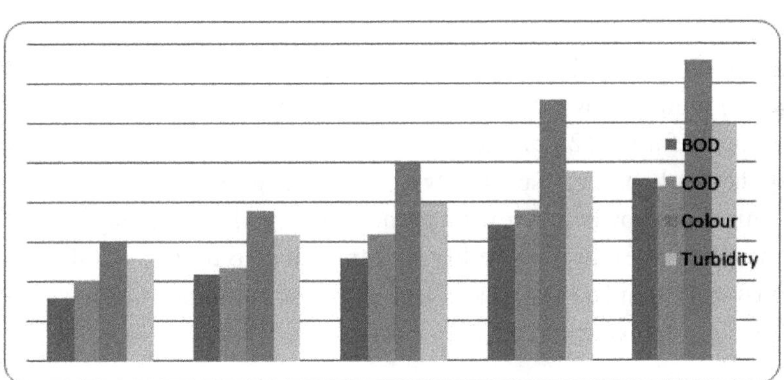

Figure 6. 12-18mm, 8-12 & 4-8mm removal efficiency (%) of various parameters versus days at 50 mL/min

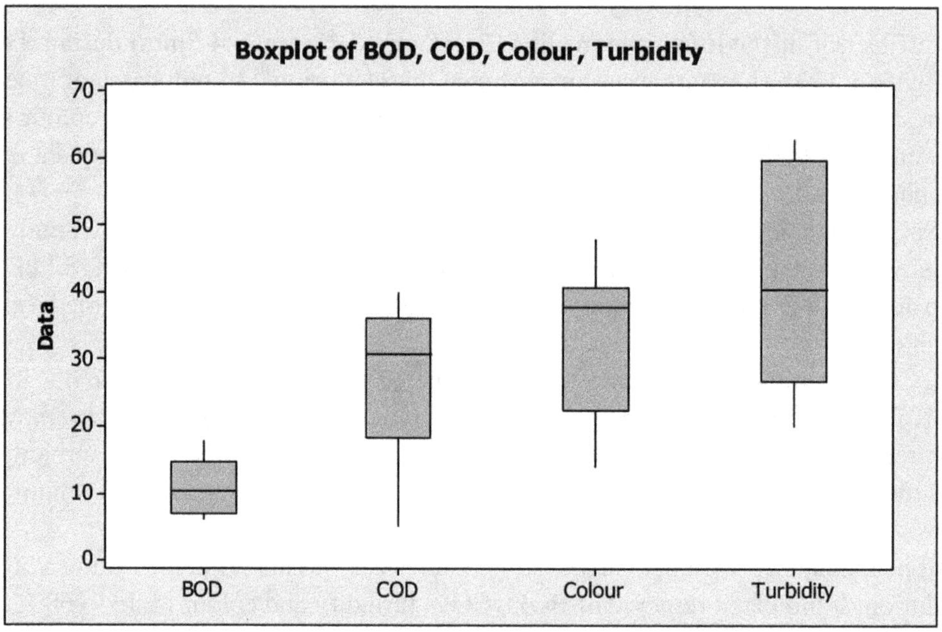

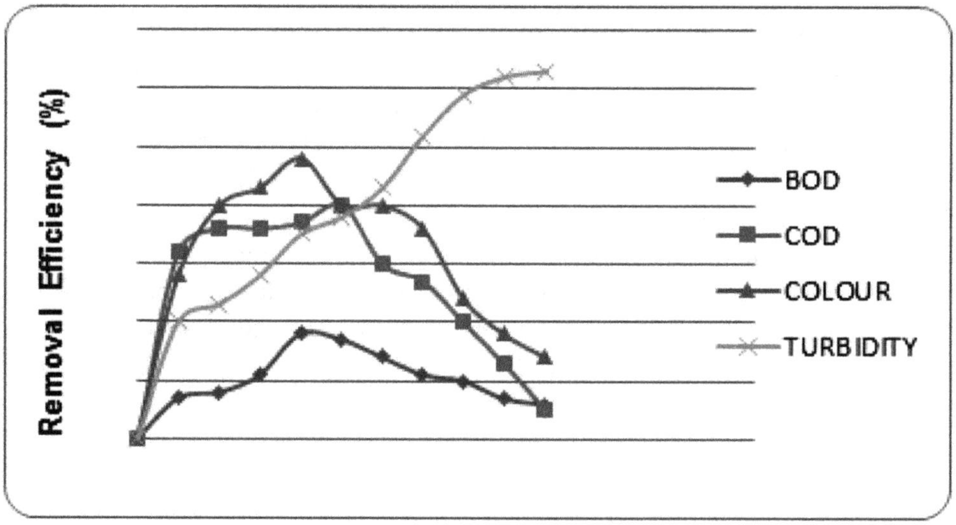

to 86% and colour removal was 35% to 52%. After the seventh day, the media was fully adsorbed and hence showed reduction in its removal capacity except for turbidity.

The percentage removal for 8 – 12mm size media was 14 to 61%, 32 to 62%, 17 to 83% and 34 to 61% for BOD, COD, turbidity and colour, respectively . Breakthrough was experienced in all the parameters except in turbidity which was however not very pronounced on the tenth day of the experiment.

The 8 – 12mm media size (Figure 10) had BOD removal of 6 to 23%, COD removal of 24 to 42%, turbidity removal of 18 to 37% and colour was 39 to 45%. Flow continued until media pores were blocked at end of the fifth day causing experiments to terminate.

Figure 7. 12 – 18mm, 8 -12mm & 4 – 8mm removal efficiency (%) of various parameters versus days at 20mL/min

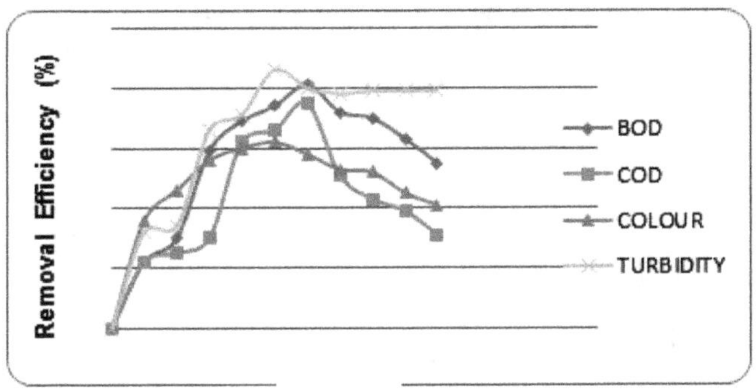

Figure 8. 12-18mm) removal efficiency (%) of various parameters versus days at 20 mL/min

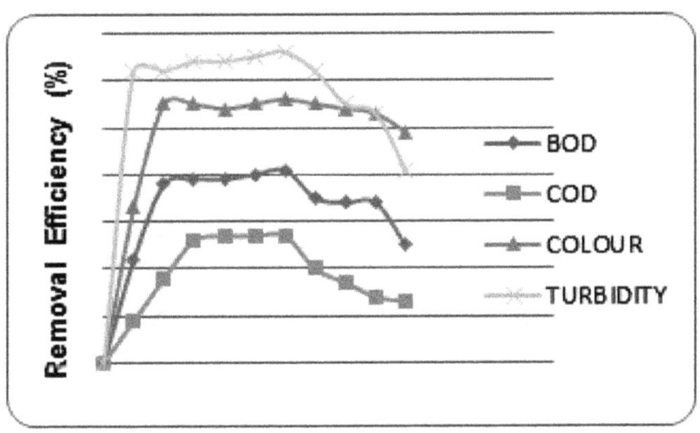

One-way ANOVA analysis performed on the experimental data is shown in Table 12. The *p* column indicates the significance level: a p-value less than 0.05 indicates that the variable (factor) is significant to a level of 95.0%. The *p*- value of the media size was 0.317 > 0.05. This affirms that media 8-12mm size was not significant for the pre-treatment of the leachate. In several other figures not included in this work, box plots were also performed to indicate which observations, might be considered outliers in the distribution efficiency.

TURBIDITY REMOVAL

Effect of Filter Media Size on Turbidity Removal

The most important factor enhancing turbidity removal with respect to filter media size is the reduced pore spaces between the grains of the media in the filter bed. Smaller grain sizes have larger adsorption area and perform better in treatment processes (Tamar, 2008). In this experiment, the specific surface area and particle size arrangement for limestone media 12 – 18 mm, 8 -12 mm & 4 – 8 mm were sufficient to reduce turbidity of the leachate compared to others.

Figure 9. 12 – 18mm & 4 – 8mm) removal efficiency (%) of various parameters versus days at 20 mL/min.

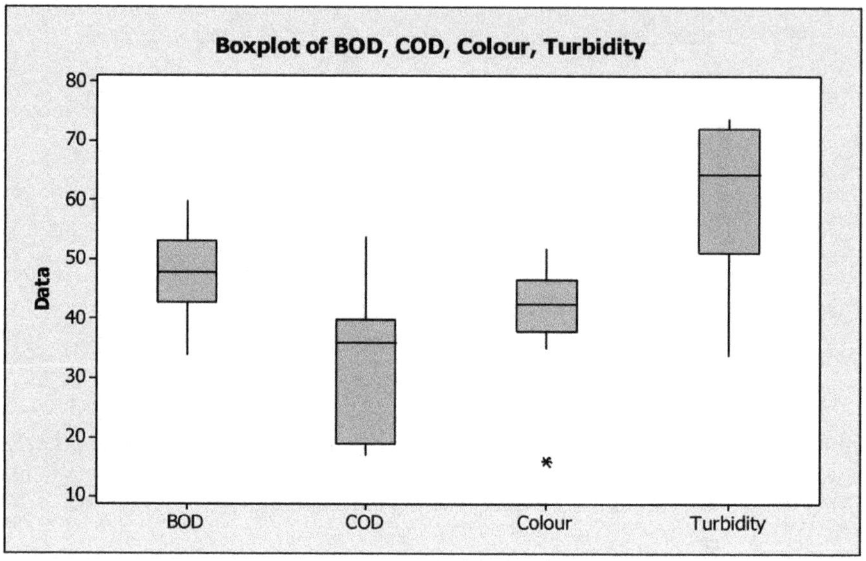

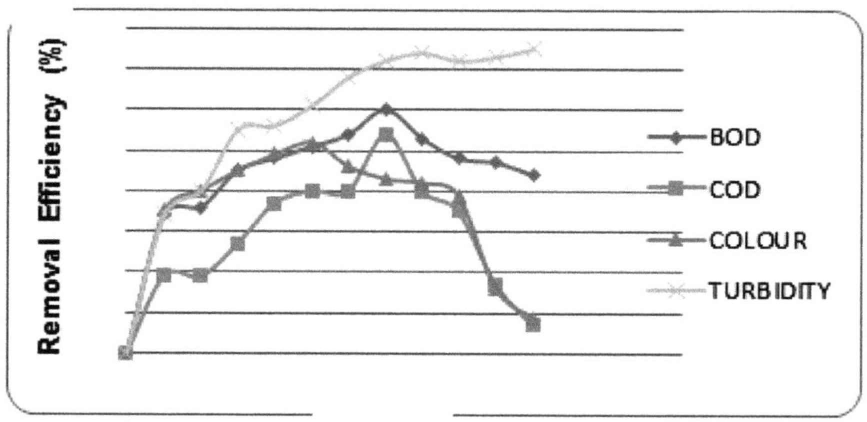

Effect of Flow Rates on Turbidity Removal

The higher pilot flow rates strongly affected the resultant effluent turbidity values because the main mechanism for removing turbidity is filtration. The higher the flow rate, the less time a particle has to travel the settling distance and stick onto the media's surface and layers or be adsorbed. Higher flow rates would have been desirable if they produced greater quantities of treated leachate with better removal efficiency. Results obtained by Nkwonta and Ochieng, (2009b) and Muhammad et al., (1996) showed that flow rates when lower will remove turbidity more effectively. However, Wegelin, (1996) upholds that pressure drop will be a setback for such filtration unit. With this in mind, operational flow rate for a desired turbidity to be achieved was considered and implemented.

Figure 10. 8-12mm Removal efficiency (%) of various parameters versus days @ 0.2m/hr Loading.

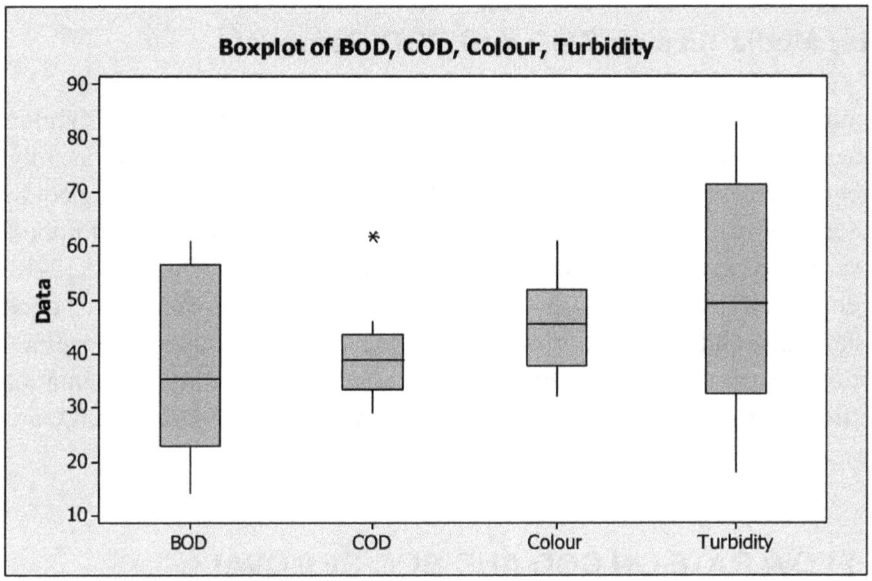

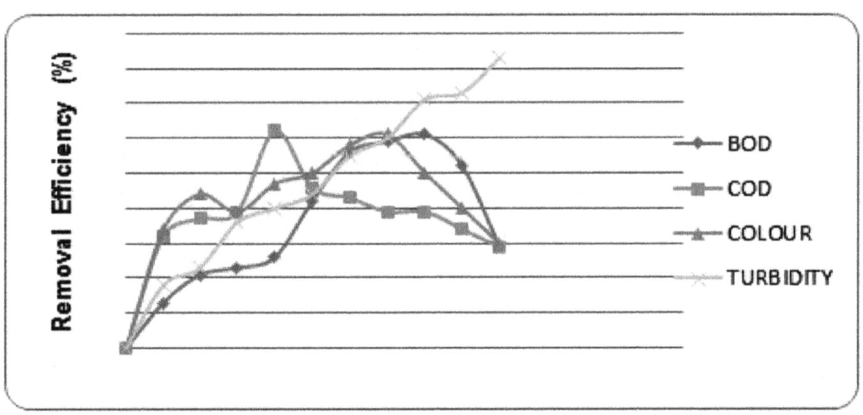

Table 12. One- Way ANOVA summary for phase 2, 3 & 4 experiments

Media Size (mm)		Flow Rate (mL/min)	*P* Value	Phase
12-18,8-12&4-8		50	0.016	4
12-18,8-12 & 4-8		20	0.011	3
12-18,8-12 & 4-8	1:1 ratio (water:leachate)	20	0.090	2
12 – 18		20	0.043	3
12-18 & 4-8		20	0.082	3
8-12		20	0.317	3
4- 8		20	0.094	3

COD AND BOD REMOVAL

Effect of Filter Media Size on COD and BOD Removal

Biological "ripening" of filter media may improve particle removal efficiency in roughing filters due to increased stickiness of filter media (Collins et al, 1994). Exposure of the media to sunlight was avoided to minimize potential errors which can be introduced by biological activity into the column. To evaluate the potential in removal of BOD and COD, the experiment was non-recycle and upon final use of the media it was observed to be very sticky. In addition, media 4-8 mm, 8-12 mm and 12-18 mm size had a higher removal compared to 8-12 mm, 4-8 mm alone and other combinations. This could be as a result of increased surface area and hence adsorption sites. This could be similar to Wegelin (1996) study in slow sand filtration experiments where removal of pathogens was by the sedimentation or deposition and effective settling of particles on the grains of the media. Thus, COD and BOD_5 are possible to be reduced from leachate by limestone media.

EFFECT OF FLOW RATE ON COD AND BOD REMOVAL

The flow rate which removed optimum COD and BOD was observed at a flow rate of 20 ml/L. This was obvious when higher flow rates indicated lower removal of COD and BOD. However, a more critical factor that can be attributed to the low removal was the media saturation point that must have been reached. Lower flow rate allows an increased pathogen removal, which is especially important in colder climates where biological activity is more time dependent (Huisman and Wood, 1974; Wegelin, 1996; Nkwonta and Ochieng, 2009). Research carried out by Jenkins *et al.*, (2009) on intermittently-operated sand filters has highlighted the importance of sand size and hydraulic loading both of which directly affect microbiological removal. However, it is possible to increase the filtration rate if effective pretreatment has been given and if an effective disinfection stage follows after filtration (Ellis, 1987).

COLOUR REMOVAL

Effect of Flow Rate on Colour Removal

A study observed that colour removal was poor when higher filtration rates were applied in continually-operated sand filters, although the filtrate quality remained reasonably good (Muhammad et al, 1996). In our study, it was justified and reasonable to apply lower flow rates for optimum removal to be obtained. The limestone media was not capable of removing the colour from the leachate. It would therefore require a chemical or biological treatment for effective removal of colour.

EFFECT OF FILTER MEDIA SIZE ON COLOUR REMOVAL

The removal of dissolved colour or true colour by varying filter media size in roughing filtration has not been properly documented. Since colour is related to humic substances, it is expected that true colour exists in relatively stable suspension and it is more difficult to remove. Collins et al., (1994) reported that removal of true colour in roughing filters compares favourably to that achieved by slow sand filtration. Wegelin (1996) observed true colour removal in the range of 20 to 50%. There has been some reports of removal of apparent colour, which is the colour attributed to undissolved particulate matter. Wolters et al., (1989) and Barret (1989) both found removal of apparent colour of 45 to 80%. In this study, the optimum true colour removal was 62% using the media 12–18 mm, 8-12 mm & 4-8mm size combination.

LENGTH OF RUN TIME ON REMOVAL EFFICIENCIES

The length of run for each set of media combination study was ten days. The limestone media experienced an average breakthrough in a period of six days as found from the laboratory experiments. The length of run was basically dependent on the performance of the media in terms of its removal and adsorption capacity. In each study, when breakthrough was experienced the experiment was terminated.

CHANGE OF pH VALUE DURING THE ROUGHING FILTRATION

During the roughing filtration treatment the pH showed a little variation. The raw leachate sample was between pH 7.94 and pH 8.12 but increased up to 8.42 after filtration. This is due to the presence of CO_3 in limestone (Aziz, *et al.*, 2004) an unstable oxide of carbon and declined as removal efficiency reduced. The dissolution of calcium carbonate can increase the concentration of alkalinity.

HEADLOSS DURING LENGTH OF RUN TIME

The headloss was also considered in this study. Although, the main essence of conducting the research was to ascertain the effectiveness of the limestone media in pre-treatment of leachate, a documentary of the headloss during each run duration for every media batch was randomly done. To affirm the postulation by Boller and Kavanaugh (1995) which demonstrated that the rate of headloss build-up in a granular media filter, for a constant mass of solids being removed, is strongly dependent on the size of the particulates in suspension and the size of the granular media. In addition, the principal cause of the rapid increase in headloss observed for smaller particles in many filtration studies is due to the nature of the aggregation or deposition process inside the pore spaces of the porous media. Observations during this experiment indicated that maximum headloss was 0.5 cm at 20 mL/min flow rate. According to Amin and Aziz (2002), negative heads do not occur in the upflow vertical roughing filter bed, however a crack occurs at the bottom layers of filter bed when headloss reaches maximum. It is noteworthy that during the filtration treatment process for media 4 – 8 mm size, there was an entire blockage and flow stopped abruptly.

CONCLUSION AND RECOMMENDATIONS

The optimum removal efficiency of BOD, COD, colour and turbidity using the limestone media in the vertical upflow roughing filter stacked with 12-18 mm, 8-12 mm and 4-8 mm size from bottom towards the top of the column at 20 mL/min and with no dilution of leachate was 81, 75, 62 and 86%, respectively. During the 1:1 water-leachate dilutions, colour removal was 70%. Increasing filtration rate and reducing contact time simultaneously reduced adsorption and hence removal efficiency. The vertical upflow roughing filter can be used for pre-treatment or polishing of leachate before further biological treatment. The possibility of carrying out tests with limestone media from various sources to ascertain that improved chemical composition and quality of media can increase the breakthrough period for the filtration process is recommended. Again, the theoretical adsorption capacity of the limestone media for leachate treatment should be determined by developing its adsorption isotherm and any other suitable process is recommended to have a wider view of the extent of limestone performance in the removal of COD, BOD, turbidity and colour from landfill leachate. In addition, it is recommended that the leachate should go through other treatment processes such as coagulation and flocculation before it comes to the filtration stage. In this research, it is noteworthy that colour removal is not very satisfactory as evident in the appearance of the leachate after the optimised treatment was done and this indicated an apparent colour which was observed.

REFERENCES

Abd El-Salam, M. M., & Abu-Zuid, G. I. (2015). Impact of landfill leachate on the groundwater quality: A case study in Egypt. *Journal of Advanced Research*, *6*, 579–586. doi:10.1016/j.jare.2014.02.003 PMID:26199748

American Public Health Association. (1998). *APHA* (20th ed.). Washington, DC: Standard Methods for Examination of Water and Wastewater.

Amin, K. N., & Aziz, S. Q. (2002). Pressure distribution in filter media in conventional filters. *Journal of Dohuk University*, *5*, 56–59.

Aziz, H. A., Othman, N., Yusuff, M. S., Basri, D. R. H., Asaari, F. A. H., Adlan, M. N., ... Perwira, M. (2001). Removal of copper from water using limestone filtration technique-determination of mechanism of removal. *Environment International*, *26*(5-6), 395–399. doi:10.1016/S0160-4120(01)00018-6 PMID:11392757

Aziz, H. A., & Salina, A. (2006). Colour removal from landfill leachate by coagulation and flocculation processes. *Bioresource Technology*, *98*(1), 218–220. doi:10.1016/j.biortech.2005.11.013 PMID:16386895

Aziz, H. A., Yusoff, M. S., Mohd, N. A., Adnan, N. H., & Salina, A. (2004). Physico-chemical removal of iron from semi-aerobic landfill leachate by limestone filter. *Waste Management (New York, N.Y.)*, *24*(4), 353–358. doi:10.1016/j.wasman.2003.10.006 PMID:15081062

Bagchi, A. (1990). *Design, construction, and monitoring of sanitary landfill.* New York, NY: John Wiley & Sons.

Barlaz, M. A., Schaefer, D. M., & Ham, R. K. (1990). Bacterial population development and chemical characteristics of refuse decomposition in a simulated sanitary landfill. *Applied and Environmental Microbiology*, *55*, 55–65. PMID:16347835

Barrett, J. M. (1989). Improvement of slow sand filtration of warm water by using coarse sand. (Doctoral Thesis), University of Colorado. Colorado.

Belfort, G., Dolecek, P., & Mikulasek, P. (1995). The performance of a rotating filter 1. Theoretical analysis of the flow in an annulus with a rotating inner porous wall. *Journal of Membrane Science*, *99*(3), 241–248. doi:10.1016/0376-7388(94)00221-J

Boller, M. (1993). Filter mechanisms in roughing filters. *Journal of Water Supply: Research & Technology - Aqua*, *42*, 174–185.

Boller, M. A., & Kavanaugh, M. C. (1995). Particle characteristics and headloss increase in granular media filtration. *Water Research*, *29*(4), 1139–1149. doi:10.1016/0043-1354(94)00256-7

Carley, B. N., & Mavinic, D. S. (1991). The effects of external carbon loading on nitrification and denitrification of a high ammonia landfill leachate. *Journal - Water Pollution Control Federation*, *63*, 51–59.

Chen, W., Zhang, A., Jiang, G., & Li, Q. (2019). Transformation and degradation mechanism of landfill leachates in a combined process of SAARB and ozonation. *Waste Management (New York, N.Y.)*, *85*, 283–294. doi:10.1016/j.wasman.2018.12.038 PMID:30803582

Collins, M. R., Westersund, C., Cole, J., & Roccaro, J. (1994). *Evaluation of roughing filtration design variables. American Water Works Association Research Foundation (AWWARF) and America Water Works Association* (pp. 77–88). Denver, CO: AWWA.

Cooney, D. O. (1999). *Adsorption design for wastewater treatment*. Boca Raton, FL: CRC Press.

Cossu, R., Raga, R., & Rossetti, D. (2003). The PAF model: An integrated approach for landfill sustainability. *Waste Management (New York, N.Y.)*, *23*(1), 37–44. doi:10.1016/S0956-053X(02)00147-2 PMID:12623100

Davis, M. L., & Cornwell, D. A. (1998). *Environmental engineering* (3rd ed.). New York, NY: McGraw Hill.

Department of Environment (DOE). (2008). Malaysia Country Waste Management Report for the year 2008. Available at http:// www. doe. gov. my

EAWAG. (1996). Swiss Federal Institute for Environmental Science and Technology.

Enden, V. J. C. & Calvert, K. C. (2002) *Limit environmental damage by basic knowledge of coffee wastewaters*. Report on GTZ-PPP Project-Improvement of coffee quality and sustainability of coffee production in Vietnam.

Falkenberry, H. L., & Slack, A. V. (1969). SO_2 removal by limestone injection. *Chemical Engineering Programme*, *65*, 61–66.

Florin-Constantin, M., & Mohammad, J. T. (2017). *Introductory chapter: rural waste management issues at global level* (pp. 1–10). Solid Waste Management in Rural Areas, Florin-Constantin Mihai, IntechOpen. doi:10.5772/intechopen.70268

Galvis, G., Visscher, J. T., & Latorre, J. (1998). Multi-stage filtration and innovation water treatment technology. Report from International reference centre for community water supply and sanitation. The Hague, The Netherlands and Universidad del valle instituto Cinara, Cali, Colombia.

Ghaly, A. E., Kamal, M. A., Mahmoud, N. S., & Cote, R. (2007). Treatment of landfill leachate using limestone/sandstone filters under aerobic batch conditions. *American Journal of Environmental Sciences*, *3*(2), 43–53. doi:10.3844/ajessp.2007.43.53

Global Environment Centre report. (2019). Available at http:// www. gecnet. info/ index. cfm?&menuid=83

Harrington, R. E., Borgwardt, R. H., & Potter, A. E. (1968). Reactivity of selected limestones and dolomites with sulfur dioxide. *American Industrial Hygiene Association Journal*, *29*(2), 152–158. doi:10.1080/00028896809343289 PMID:5656035

Hecham, M. O. (2016). Treatment of landfill waste, leachate and landfill gas: modelling/simulation and experimental studies. (Master's Thesis), The University of Western Ontario, Canada.

Huisman, L., & Wood, W. E. (1974). *Slow sand filtration* (p. 44). Geneva, Switzerland: WHO.

Izabela, A. T., & Biedka, P. (2016). Use of the landfill water pollution index (LWPI) for groundwater quality assessment near the landfill sites. *Environmental Science and Pollution Research International*, *24*, 24601–24613. PMID:27640059

Jenkins, M. W., Tiwari, S. K., Darby, J., Nyakash, D., Saenyi, W., & Langenbach, K. (2009). *The bio-sand filter for improved drinking water quality in high risk communities in the Njoro watershed, Kenya. Research Brief 09-06-SUMAWA, Global Livestock Collaborative Research Support Program.* Davis, CA: University of California.

Kamaruddin, M. A., Yusoff, M. S., Aziz, H. A., & Basri, N. K. (2013). Removal of COD, ammoniacal nitrogen, and colour from stabilized landfill leachate by anaerobic organism. *Applied Water Science*, *3*(2), 359–366. doi:10.100713201-013-0086-1

Kjeldsen, P., Barlaz, P. M., Rooker, A. P., Baun, A., Ledin, A., & Christensen, T. H. (2002). Present and long-term composition of MSW landfill leachate: a review. *Critical Reviews in Environmental Science and Technology*, *32*(4), 297–336. doi:10.1080/10643380290813462

Knox, K. (1985). Leachate treatment with nitrification of ammonia. *Water Research*, *19*(7), 895–904. doi:10.1016/0043-1354(85)90148-4

Marble Institute of America. (2019). Available at http://www.atlastileandstone.com/ Limestone %20 Geological%20Classification.html

Matsufuji, Y. (1990). *What is Fukuoka method? Report on road to sanitary landfill.* Japan: Faculty of Engineering Fukuoka University.

Ministry of Housing and Local Government, Malaysia. (2004). The final report of "The study of the safe closure and rehabilitation of landfill sites in Malaysia": *Guideline for safe closure and rehabilitation of MSW landfill sites.* Vol. 3, pp. 1 – 126.

Moropoulou, A., Bakolas, A., & Aggelakopoulou, E. (2001). The effects of limestone characteristics and calcination temperature on the reactivity of quicklime. *Cement and Concrete Research, 31*(4), 633–639. doi:10.1016/S0008-8846(00)00490-7

Muhammad, N., Ellis, K., Parr, J., & Smith, M. D. (1996) Optimization of slow sand filtration. In *22nd WEDC Conference Reaching the unreached: challenges for the 21st century. New Delhi*, India.

Mulamoottil, G., McBean, E., & Rovers, F. (1999). *Constructed wetlands for the treatment of landfill leachates.* Boca Raton, FL: CRC Press.

Naveen, B. P., Sumalatha, J., & Malik, R. K. (2018). A study on contamination of ground and surface water bodies by leachate leakage from a landfill in Bangalore, India. *International Journal of Geo-Engineering,* (9), 27. doi:10.118640703-018-0095-x

Nicole, D. B., Reinhart, D. R., & Batarseh, E. S. (2009). An assessment of bioreactor landfill costs and benefits. *Waste Management (New York, N.Y.), 29*(5), 1558–1567. doi:10.1016/j.wasman.2008.12.010 PMID:19167875

Nkwonta, O. I., & Ochieng, G. M. (2009a). Passive treatment of mine water using roughing filters as a pre-treatment option. In *Abstracts of the International Mine Water Conference*, Pretoria, South Africa.

Nkwonta, O. I., & Ochieng, G. M. (2009b). Roughing filter for water pre-treatment technology in developing countries: A review. *International Journal of Physical Sciences, 4*, 455–463.

Onay, T. T., & Pohland, F. G. (1998). In situ nitrogen management in controlled bioreactor landfills. *Water Research, 4*(5), 1383–1392. doi:10.1016/S0043-1354(97)00392-8

Patil, V. B., Kulkarni, G. S., & Kore, V. S. (2012). Performance of horizontal roughing filters for wastewater: A review. *International Research Journal of Environment Sciences, 1*, 53–55.

Peng, Y. (2017). Perspectives on technology for landfill leachate treatment. *Arabian Journal of Chemistry, 10*, 2567–2574. doi:10.1016/j.arabjc.2013.09.031

Pinheiro, A. C., Mesquita, N., Trovao, J., Soares, F., Tiago, I., Coelho, C., ... Portugal, A. (2018). Limestone biodeterioration: A review on the Portuguese cultural heritage scenario. *Journal of Cultural Heritage (In-Press).*

Pohland, F. G. (1995) Landfill bioreactors: historical perspective, fundamental principles, and new horizons in design and operations. In *Landfill Bioreactor Design and Operation Sem. Proc.*, EPA/600/R-95/146, pp. 9-24.

Potter, A. E. (1969). Sulfur oxide capacity of limestones. *American Ceramic Society Bulletin, 48*, 855–858.

Qasim, S. R., & Chiang, W. (1994). *Sanitary landfill leachate-generation, control and treatment.* Texas: Technomic Publishing.

Reinhart, D. R., & Townsend, T. G. (1998). *Landfill bioreactor design & operation*. Boca Raton, FL: CRC Press.

Reynolds, T. D., & Richards, P. A. (1996). *Unit operations and processes in environmental engineering (2nd Ed.)*. Boston, MA: PWS.

Sakai, S., Katsuura, H., Inoue, T., & Hiraoka, M. (2000). State-of-the-art treatment processes for municipal solid waste incineration residues. *Waste Management & Research*, *18*(1), 41–51. doi:10.1177/0734242X0001800106

Schertenleib, R., Wegelin, M., & Boller, M. (1987). Particle removal by horizontal-flow roughing filtration. *Journal of Water Supply: Research & Technology - Aqua*, *36*, 80–90.

Siti, N. F. Z., & Aziz, H. A. (2018) IOP Conference Series: Earth and Environmental Science, (140), 012013.

Smith, K. S., Plumlee, G. S., & Ficklin, W. H. (1994) Predicting water contamination from metal mines and mining wastes. International Land Reclamation and Mine Drainage Conference and Third International Conference on the Abatement of Acidic Drainage. U.S. Geological Survey Open-File Report No. 94-264.

Srdjan, M. B. (2007). *Handbook of flotation reagents: chemistry, theory and practice; flotation of sulphide ores*. Oxford, UK: Elsevier.

Sun, Q. (2004) Iron and acid removal from acid mine drainage in open limestone systems. (Doctoral dissertation), West Virginia University.

Tamar, R. L. (2008) Pilot study of horizontal roughing filtration in Northern Ghana as pretreatment for highly turbid dugout water. (MSc. Thesis), Massachusetts Institute of Technology.

Tchobanoglous, G., Franklin, B. H., & Stensel, D. (2003) Wastewater engineering: treatment and reuse (4th ed.). Metcalf & Eddy, Inc., Santa Maria, CA.

Tchobanoglous, G., & Kreith, F. (2003) Handbook of solid waste management (2nd edition). Metcalf & Eddy, Inc., Santa Maria, CA.

Tchobanoglous, G., Theisen, H., & Vigil, S. (1993). *Integrated solid waste management: engineering principles and management issues*. New York, NY: McGraw-Hill.

Van Ewijk, S., & Stegemann, J. A. (2016). Limitations of the waste hierarchy for achieving absolute reductions in material throughput. *Journal of Cleaner Production*, *132*, 122–128. doi:10.1016/j.jclepro.2014.11.051

Wegelin, M. (1996) Surface water treatment by roughing filters. *A Design, Construction and Operation Manual*. SANDEC Report No. 2/96.

Wei, Z., Khamphe, P., Fan, L. M. S., & Pin, J. H. (2013). Evaluation of a classification method for biodegradable solid wastes using anaerobic degradation parameters. *Waste Management (New York, N.Y.)*, *33*(12), 2632–2640. doi:10.1016/j.wasman.2013.08.015 PMID:24045172

Welander, U., Henrysson, T., & Welander, T. (1997). Nitrification of landfill leachate using suspended-carrier biofilm technology. *Water Research*, *31*(9), 2351–2355. doi:10.1016/S0043-1354(97)00080-8

Wolters, H., Smet, J. E. M., & Galvis, J. (1989). *Pre-treatment methods for community water supply.* Ch. 8, International Water Sanitation Center (IRC).

World Health Organization. (2005). Available at http://www.who.int/water_ sanitation_health/ wsh0207/ en/index5.html

Xu, C. Y., Schwartz, F. W., & Traina, S. J. (1997). Treatment of acid-mine water with calcite and quartz sand. *Environmental Engineering Science*, *14*(3), 141–152. doi:10.1089/ees.1997.14.141

Chapter 10
Review on Microbial Analysis Tools in POME treatment

Mohamad Mokhtar Ibrahim
Faculty of Chemical and Natural Resources Engineering, University Malaysia Pahang, Malaysia

Zulkifly Jemaat
https://orcid.org/0000-0001-6602-5292
Faculty of Chemical and Natural Resources Engineering, University Malaysia Pahang, Malaysia

Abdurahman Hamid Nour
Faculty of Chemical and Natural Resources Engineering, University Malaysia Pahang, Malaysia

ABSTRACT

Palm oil mill effluent (POME) is one of the major sources of water pollution in Malaysia. POME is produced in large volumes by many palm oil mills and has acidic pH and high concentrations of COD, BOD, and suspended solids, which have adverse effect to the environment. Currently, the technology to treat POME is either physical, chemical, or biological. About 80% of palm oil mills treat their POME by using biological method. Recent studies have indicated that understanding the microbial community structure is of great importance to improve and control the biological treatment performance. Currently, the most popular molecular biology tools for microorganism community analysis are fluorescence in situ hybridisation (FISH), cloning of 16S rDNA, and denaturing gradient gel electrophoresis (DGGE). This chapter aims to review the current and ongoing treatments of POME (mainly anaerobic, aerobic, physicochemical, and membrane separation) and discuss the potential of using the molecular biology techniques in POME treatment. The importance and effectiveness of the microbiology tools are also discussed. The ability to monitor microorganisms and understand their ecology is essential to effectively control the startup and operation of biological treatment system in treating POME and eventually producing effluent of acceptable quality.

DOI: 10.4018/978-1-7998-0369-0.ch010

INTRODUCTION

Palm oil is an important commodity especially in Malaysia and Indonesia. The global production of palm oil in 2014 was 62.34 million tonnes and 85% of the productions came from these two countries. In 2016, the palm oil industry in Malaysia generated approximately 58 million tonnes of palm oil mill effluent (POME) which arose from the processing of oil palm fresh fruits bunch (OPFFB) for palm oil extraction (Zainal et al., 2017). Currently, Malaysia has produced 21,000 metric tonnes of palm oil (Mundi, 2017). Table 1 shows the world palm oil production in 2017 (Mundi, 2017). The demand for palm oil will increase steadily with the increasing world population that is predicted to grow from 7 billion tonnes in 2011 to 9 billion tonnes by 2043. Therefore, the world demand for fats and oils will reach 360 million tonnes by 2043 (Basiron, 2013). Palm oil have been recognised as the most utilised vegetable oil globally with the total production reaching up to 40% compared to other vegetable oils (Hansen et al., 2015; Oosterveer, 2015).

The different steps in palm oil extraction process generate various types of waste by-products such as oil palm shell (OPS), fibres, empty fruit bunches (EFB), palm oil clinker (POC), palm oil fuel ash (POFA), and palm oil mill effluent (POME). Figure 1 shows a simplified flow diagram of palm oil extraction processes. The approximate generation of POME is also depicted in Figure 1. POME is a thick brownish liquid discharged at temperature between 80 and 90 °C, with pH ranging from 4.0 to 5.0, biological oxygen demand (BOD) in the range of 10.25–43.75 g/L, chemical oxygen demand (COD) in the range of 15–100 g/L, high salt content, and high suspended solids (SS) around 5–54 g/L (Ahmad et al., 2003; Baranitharan et al., 2013). If not managed properly, POME could have a great impact on the environment (Kanadasan & Razak, 2015). About 1.5 m^3 of water is used for processing of 1 tonne of fresh fruit bunches (FFB). A study has shown that half of the processing water used ends up as POME (Chang, 2014). In Malaysia, palm oil mills have been recognised as the second largest source for the greenhouse gas (38%) next to waste landfills (53%) (Chan & Chong, 2019). The treatment of POME is crucial to conserve the environment due to the emissions of biogas and the concern on the quality of the final discharge to the watercourse, which might cause environmental deterioration (Basri et al., 2010; Harsono et al., 2014).

The information of microbial profiling of biological treatment could improve treatment performance and control strategy of wastewater treatment plant. Modern molecular biology techniques have replaced traditional microbiological methods. Traditional methods possess several disadvantages such as the impossibility to isolate pure cultures and cultivate certain kinds of bacteria in artificial conditions, e.g., activated sludge bacteria. Bacterial strains grown using the traditional methods do not possess particular phenotypes, grow slowly, and are affected by environmental factors. In addition, studies in natural biodiversity and engineered ecosystem are not adequate because the procedures in conventional methods need more laborious characterisation experiments (Sanz & Kochling, 2007). On the other hand, modern molecular biology techniques are easier and faster compared to conventional methods. In modern techniques, growth medium is not required, because the probes used in the analysis are taken directly from the environment. These tools enable researchers to study all types of bacteria, including nonculturable ones, under laboratory conditions. Moreover, samples for analysis are small and results are repeatable. Thus, molecular methods have become more popular in technological and microbiological laboratories, and can be used to monitor the biological community in activated sludge and other biological research systems.

Figure 1. Simplified flow diagram of the production of crude palm oil
Adapted from Lam and Lee (2011)

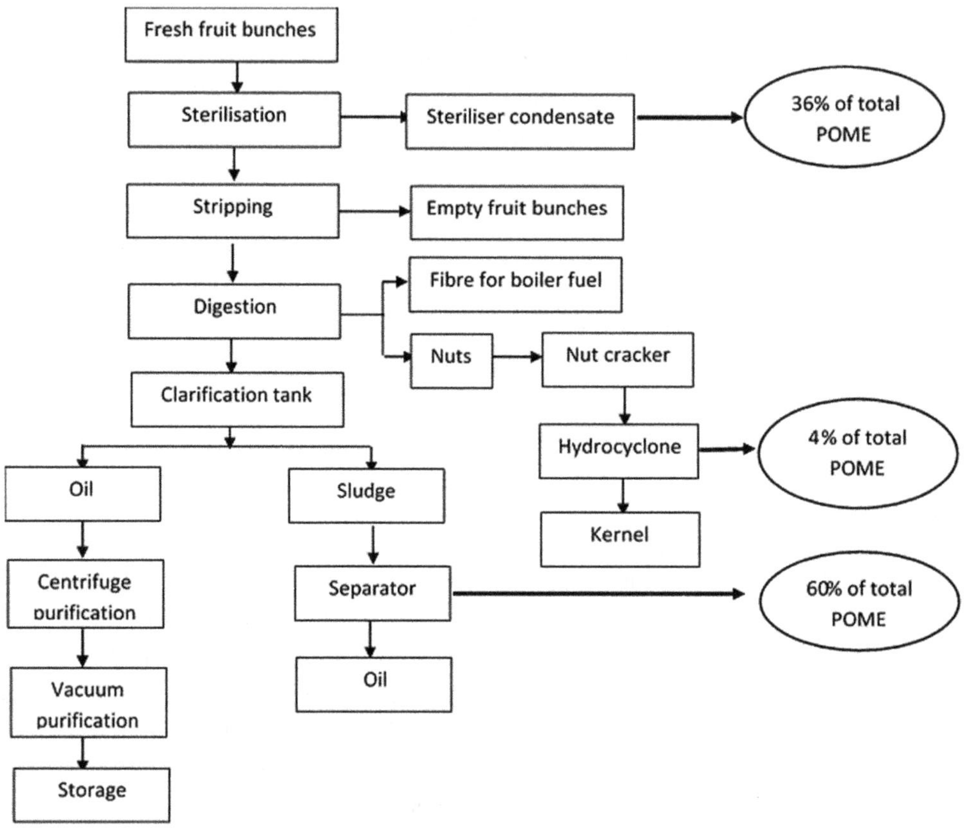

Table 1. World palm oil production in 2017

Country	Production (1000 MT)	Share (%)
Indonesia	36000	54
Malaysia	21000	32
Thailand	2200	3
Colombia	1320	2
Nigeria	970	1
Others	5365	8

Adapted from Mundi (2017)

POME TREATMENT METHODS

Several methods or technologies are utilised in treating POME. Table 2 shows the comparison between various methods of POME treatment such as aerobic, anaerobic, physicochemical, membrane, and integrated treatments. Some of these methods are discussed in the following section.

Table 2. Comparison between various methods of POME treatment

Methods	Advantages	Disadvantages	References
Membrane treatment system	Excellent removal efficiency	Costly treatment system High maintenance High pressure requirement	(Ahmad et al., 2003; Subramaniam et al., 2018)
Biological membrane integrated system	Excellent removal performances	Costly treatment method	(Zhang et al., 2008)
Microbial fuel cell	Generation of electricity High removal performance	Difficult to scale up Costly treatment method	(Cheng et al. 2010)
Upflow anaerobic sludge fixed film (UASFF)	Small footprint Produces and captures methane gas Able to retain biomass in the reactor for higher loading treatment	Poor separation between treated effluent and biomass	(Zainal et al., 2019; Najafpour et al., 2006)
Ponding system	Low cost Less/no maintenance needed	Large land area required Long HRT (>50 days) Direct release of methane and hydrogen sulphide to the environment	(Chin et al., 1996)
Upflow anaerobic sludge blanket (UASB)	Small footprint Produces and captures methane gas	Unable to retain biomass for higher organic loading treatment Poor separation between treated effluent and biomass	(Maneerat et al., 2012; Uddin et al., 2019)
Coagulation–flocculation	Low cost	Removes only suspended solids and residual oil Release of residual elements from coagulants pollutes water	(Hossain et al., 2019; Uzir et al., 2019)
Sequencing batch reactor-aerobic granules	No formation of pollution-causing gases Excellent separation between treated effluent and biomass Small footprint Able to be operated at higher loading rate	High energy for aeration is required to break down the organic content of POME No production of methane	(Angenent et al., 2004; Najib et al., 2017)
Ozone-anaerobic digestion integrated system	Efficient in converting recalcitrant pollutants into biodegradable organics	Costly treatment method	(Chan et al., 2012)
Aerobic–anaerobic integrated system	Small footprint Can treat high organic loading of POME	Poor separation between treated effluent and biomass	(Chan et al., 2012; Farid et al., 2019)
Anaerobic filtration integrated system	Can treat high organic loading of POME	Requires frequent maintenance to sustain desirable product	(Choi et al., 2013)
Aerobic digestion	No formation of pollution-causing gas	High energy for aeration required to break down organic content of POME Poor separation between treated effluent and biomass	(Vijayaraghavan et al., 2007)

Anaerobic Treatment

Anaerobic treatment has grown rapidly due to increasing awareness on global warming and the need for sustainable wastewater treatment. Anaerobic treatment is regarded as an effective method for treating wastewater containing a huge amount of organic substances such as POME (Borja et al., 1996). Anaerobic digestion has been employed by most palm oil mills as their primary treatment of POME (Poh et al., 2010) with the main method of treatment systems being in ponds and/or open facultative tank systems (Yacob et al., 2005). The ponding system for POME treatment has several disadvantages such as requiring long retention time and large treatment areas, producing bad odour, and causing other environmental issues due to gas emissions and contamination of groundwater system (Tong & Jaafar, 2004; Yacob et al., 2005).

Anaerobic digestion does not require oxygen in the degradation process of organic and inorganic matters. This digestion process includes a diverse community or species of anaerobic microbes, which are responsible for the degradation of the organic matter. The performance of anaerobic system mainly depends on a few factors such as the structure of microbial community and suitable environmental factors such as pH and temperature (Weiland, 2010). Generally, the anaerobic process involves four degradation stages of organic matters, namely, hydrolysis, acidogenesis, acetogenesis, and methanogenesis, to obtain methane (CH_4) and carbon dioxide (CO_2) by the action of a group of microorganisms (Guerrero et al., 1999; Karakashev et al., 2005). Figure 2 shows the four degradation stages involved in anaerobic process. The compounds possibly produced at each stage from various types of substrate are also illustrated in Figure 2. In each degradation stage, specific bacteria are responsible to consume each substrate and they have different physiological and nutritional requirements to be effective in the treatment (Olvera & Lopez, 2012).

Various studies have been performed on anaerobic treatment of POME by utilising different reactor configurations. Among recommended reactor setups include anaerobic filters and anaerobic fluidised bed reactors, upflow anaerobic sludge blanket reactor (UASB), expanded granular sludge blanket (EGSB), anaerobic baffled reactor (ABR), anaerobic sequencing batch reactor (ASBR), continuous stirred tank reactor (CSTR), and upflow anaerobic sludge-fix film reactor (UASFF). A few researchers reported the application of an upflow anaerobic sludge blanket reactor, upflow anaerobic filtration, and fluidised bed reactor (Borja & Banks, 1994; Borja et al., 1996), while Najafpour et al. (2006) reported the application of upflow anaerobic sludge fix film reactor. Anaerobic contact digester (Ibrahim et al., 1984) and continuous stirred tank reactor, CSTR (Chin, 1981) have also been studied for treatment of POME.

Figure 2. Four stages of anaerobic digestion/biogas processes
Adapted from Cheng & Wang (2013)

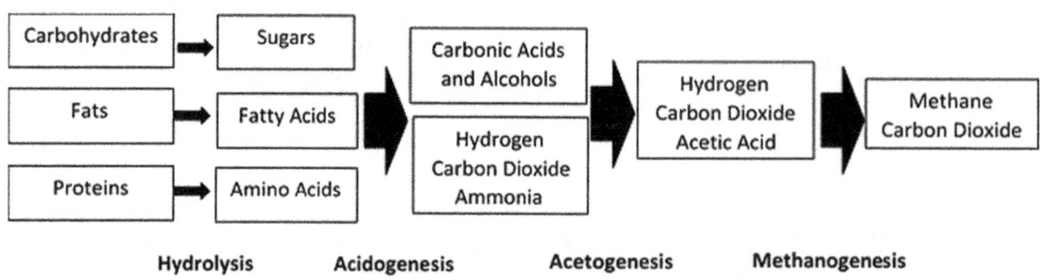

Others methods include sequencing batch reactors (Najafpour et al., 2006) and anaerobic membrane bioreactors (Ahmad & Chan, 2009; Edgar et al., 2006). Researchers found that the UASB process is the most promising and practical technology in the anaerobic treatment of wastewater (Pol et al., 2004; Zhang et al., 2011).

Aerobic Treatment

Aerobic treatment is a process where biodegradable organic matter is hydrolysed and converted into CO_2, water, and active biomass by the action of heterotrophic bacteria in the presence of oxygen (Leslie et al., 2011). Among the aerobic treatment systems are rotating biological contactor (RBC), trickling filter, sequencing batch reactor (SBR), and oxidation pond. In one study, by using RBC to treat POME, a removal efficiency of total nitrogen (TN), SS, COD, and BOD_5 up to 80%, 89%, 88%, and 91%, respectively, were obtained (Najafpour et al., 2005). An 83% removal efficiency was obtained when a trickling filter was used as a model of aerobic attached-growth system for the treatment of POME supernatants (Norulaini et al., 2004). Ho & Tan (1988) used pressurised activated sludge process to treat the secondary anaerobically digested POME. Meanwhile, Vijayaraghavan et al. (2007) used aerobic oxidation to investigate the treatment efficiency of anaerobically digested and diluted raw POME. Other researchers investigated the aerobic treatment of anaerobically digested POME by using SBR reactor (Chan et al., 2010). They examined the practicability of aerobic biological treatment under thermophilic conditions for anaerobically treated POME. Nowadays, due to its high removal activity of BOD and SS, SBR is widely used as cost-effective treatment system.

Physicochemical Treatment

Although most palm oil mills employ numerous anaerobic and aerobic treatment systems, they still fail to achieve the effluent discharge standard. Researchers are still in searching for alternatives methods to treat POME and thus, meeting the discharge standard limits prescribed by regulatory bodies. Physicochemical treatment is one of the options in which wastewater treatment is predominantly focused on the separation of colloidal particles. The treatment utilises various coagulants and flocculants that change the physical state of the colloids, allowing them to remain in an indeterminately stable form and therefore, form particles or flocs. The recommended physicochemical treatment processes for POME are coagulation and flocculation, electrocoagulation, sedimentation and centrifugation, flotation, and adsorption, among others.

Many researchers have studied the potential of using physicochemical treatment on POME. For example, Khadidi et al. (2013) studied the effectiveness of coagulation and flocculation process in POME treatment. Meanwhile, Sontaya et al. (2013) investigated the efficiency of electrocoagulation process to remove colour and COD in POME. Ahmad et al. (2002) studied the sedimentation and centrifugation in POME, Ng et al. (1988) used dissolved air flotation (DAF) for the separation of liquid–solid from POME, while Ho & Chan (1986) used electrofloatation process to treat COD and suspended solid of POME. Chitosan was used by Ahmad et al.(2006) as an adsorbent in their various POME experiments to remove oil. Nowadays, many researchers use an environmentally friendly coagulant for the treatment of POME. Bhatia et al. (2007a, 2007b) developed a natural coagulant from *Moringa oleifera* for POME treatment. Photocatalytic degradation technology has also been applied by some researchers to improve POME treatment (Charles & Chin, 2019; Ng & Cheng, 2015).

Numerous physicochemical treatment processes have been investigated at bench or pilot scale to treat POME. However, none of the processes could be upscaled to a commercial level because huge volumes of adsorbents and coagulants are required. Consequently, the entire treatment is impractical and uneconomic.

Membrane Separation Processes (MSPs)

Membrane separation has gained great attention to eliminate the huge volume of chemicals and microorganisms from wastewater (Judd, 2011). To date, different membrane separation systems have been effectively used for wastewater treatment such as anaerobic membrane system, ultrafiltration (UF), microfiltration (MF), nanofiltration (NF), and reverse osmosis (RO) (Ellouze et al., 2012). Several studies have been conducted to evaluate the efficiency of POME treatment by using membrane systems. For instance, Ahamed (2015) studied an integrated system, a combination between an anaerobic UASFF reactor and membrane system, and a higher quality effluent was achieved. In his study, in the first biological stage (UASFF), organic wastes were converted into biogas and produced low strength effluent through the different ranges of organic loading rate (OLR) at short hydraulic retention time (HRT). The produced effluent improved the membrane fouling and extended the membrane lifetime. In the subsequent stage, i.e., membrane treatment unit, the ultrafiltration (UF) membrane reduced the amount of most of the suspended solid materials from the first biological stage by removing dissolved solids and inorganic salts. Finally, by using reverse osmosis (RO) membrane, a crystal-clear effluent was obtained. The final effluent could be recycled as boiler feed water and the biogas obtained from the first stage could be recovered for the generation of heat or electricity in the mill. These sustainable steps could help reduce operating costs and global warming.

A recent research work has shown the potential of a novel hybrid polyvinylidene fluoride/titanate nanotube (PVDF–TNT) photocatalytic membranes for the colour removal of aerobically treated palm oil mill effluent (AT-POME) in a submerged membrane photo reactor (SMPR) (Subramaniam et al., 2018). The performance analysis indicates that up to 67.3% of colour removal was achieved by this method. In addition, the hybrid technology between membrane filtration and photocatalysis remarkably improved the colour removal efficiency compared to filtration alone, with 34.2% colour removal. However, utilisation of membrane separation systems alone without integration with other methods will result in high cost and impracticality.

MOLECULAR BIOLOGY TECHNIQUES USED IN WASTEWATER TREATMENT

The conventional methods in identification of microorganisms require the isolation of pure cultures followed by laborious characterisation experiments. These procedures are inadequate for studying the biodiversity of a natural or engineered ecosystem. Traditional microscopy observations (Curds, 1975; Eikelboom, 1975) or culture-dependent techniques (Ueda & Earle, 1972) were used in initial investigations of wastewater microbial communities. These techniques have led to the identification of numerous eukaryotic organisms (Esteban et al., 1991; Madony et al., 1993) and the association of filamentous organisms with sludge bulking (Eikelboom, 1975; Seviour et al., 1997). The majority of the bacterial population cannot be distinguished from one another under the microscope. A new set of molecular techniques developed during the 1990s has revolutionised microbial ecology research. The possibility of identifying specific populations of microorganisms in their native habitat or environment without the need to isolate them is revolutionising microbial ecology. This molecular technique has given rise to various new applications in numerous research fields. Molecular methods can be generally classified into three groups (Tabatabaei et al., 2010):

1. Analysis of small subunit ribosomal RNA (SSU rRNA) clone libraries;
2. Community fingerprinting techniques using SSU rRNA gene such as denaturing gradient gel electrophoresis (DGGE), terminal restriction fragment length polymorphism (T-RFLP), ribosomal intergenic spacer analysis (RISA), single-strand conformation polymorphism (SSCP), and length heterogeneity PCR (LH-PCR); and
3. Dot blotting, fluorescence in situ hybridisation (FISH), and stable isotope probing.

Most of the molecular approaches used so far are based on the analysis of SSU rRNA. Recent studies also use quantification of functional gene expressions. Several quantitative methods have been developed such as quantitative real-time PCR assay and quantitative FISH. Among these techniques, cloning and the creation of a gene library, DGGE, and FISH with DNA probes are the most popular.

Cloning of 16S rDNA

The cloning and sequencing of the genes that code for 16S rRNA is still the most widely used in the field of microbial ecology. This methodology involved the extraction of nucleic acids, amplification, cloning of the 16S rRNA genes followed by sequencing, and finally identification and affiliation of the isolated clone with the aid of phylogenetic software (Sanz & Kochling, 2007). Generally, cloning and construction of rRNA gene library has been applied in combination with other techniques in wastewater treatment. Cloning of the whole gene library yields a more exact phylogenetic information in microorganisms than other molecular techniques such as FISH and DGGE in various bioreactors in wastewater treatment.

Denaturing Gradient Gel Electrophoresis (DGGE)

DGGE is based on the differing mobility of denatured DNA fragments of the same size but with different nucleic acid sequences on a gel. This generates band patterns that directly reflect the genetic biodiversity of the sample. The number of bands corresponds to the number of dominant species. Coupled with sequencing and phylogenetic analysis of the bands, this method gives a good overview of the composition of a microbial community. DGGE has been employed in the characterisation of a wide array of habitats, such as soil, bacterioplankton, hot springs, and continental waters (Sanz & Kochling, 2007). The technique is less used in anaerobic wastewater treatment. However, in recent years, DGGE seems to be increasing in popularity. The most important application of DGGE is monitoring dynamic changes in microbial communities, especially when many samples have to be processed. There are multiple applications of DGGE related to anaerobic digestion processes. The DGGE method is repeatable, fast, and relatively cheap. These features contribute to its popularity in the molecular analysis of environmental samples (Muyzer & Smalla, 1998). DGGE has been used to study bacterial groups like ammonia-oxidising bacteria (AOB), nitrite-oxidising bacteria (NOB), and polyphosphate-accumulating organisms (PAO). The tool allows for the separation of the most common groups in activated sludge bacteria and their categorisation into relevant subclasses according to genetic relatedness (Snaidr et al., 1997). DGGE has also been used to find the dominant species in the microbial community and the biological structure of activated sludge (Chonticha et al., 2019; Maaroff et al., 2019).

Fluorescent In Situ Hybridization (FISH)

One of the ways to overcome some of the problems of studying microbial populations of a microcosm without resorting to traditional methodology is to use fluorescent probes. These are short sequences of DNA (16–20 nucleotides) labelled with a fluorescent dye. These sequences recognise 16S rRNA sequences in fixed cells and hybridise with them in situ (DNA–RNA matching). Figure 3 depicts the typical steps used in FISH protocols. Through this method, microorganisms can be identified, localised, and quantified in almost every ecosystem (Amann et al., 1995). The specificity of the probe enables bacterial detection or identification on any desired taxonomic level, from the domain level down to species level. Previous knowledge of the expected microorganisms in the sample is often required to apply this method successfully. To target a particular species, a specific probe must be ready or its 16S rRNA sequence must be available (Sanz & Kochling, 2007). The use of oligonucleotide probes targeting 16S rRNA presents a revolution in microbial ecology, both for basic research and practical applications. Within the area of wastewater treatment, hybridisation techniques are by far the most extensively used ones.

Alternative and New Methods

Several new and alternative methods or technologies have arisen from the existing molecular methods. Some of these methods are discussed in the following section.

Amplified Ribosomal DNA Restriction Analysis (ARDRA)

This technique involves the amplification of the 16S ribosomal gene followed by enzymatic restriction and gel electrophoresis. The DNA fingerprints for the whole community can be produced using PCR with universal primers, whereas fingerprints for subgroups in the community can be generated with species-specific primers.

The microbial diversity of activated sludge from various wastewater plants has been described using ARDRA by first generating 16S rRNA gene clone libraries (Blackall et al., 1998). The plasmid inserts from the clones are then grouped by ARDRA (also referred to as 16S-RFLP or REA) and subjected to partial sequencing.

Figure 3. Typical steps in fluorescence in situ hybridisation (FISH) protocol
Adapted from Jemaat (2013)

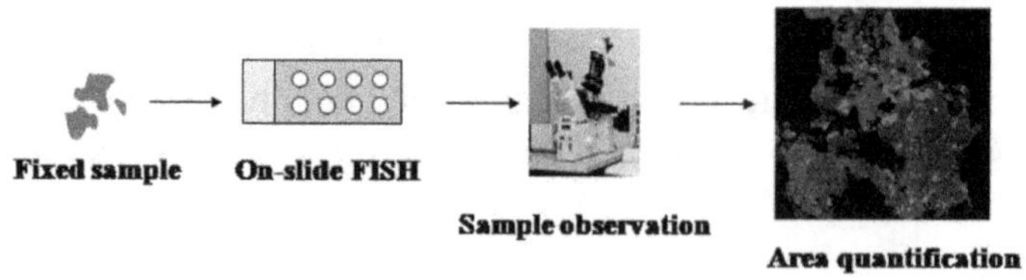

Ribosomal Intergenic Spacer Analysis (RISA)

Ribosomal intergenic spacer analysis (RISA) takes advantage of the DNA length variability as well as sequence diversity of the region between 16S and 23S rRNA genes. It has been successfully used to describe different environmental communities.

RISA is a rapid and effective freshwater bacterial community analysis technique. It has been successfully used to detect community shifts due to perturbations in the bacterial community (Fisher & Triplett, 1999). Furthermore, RISA, along with 16S rRNA sequencing, has been used to investigate the bacterial diversity and community structure in plug-flow lagoons treating pulp and paper wastewater. The study found that temporal differences in community structure were greater than spatial differences and that the dominant phylotypes belong to the genera *Methylobacillus* and *Azospirillum* (Yu & Mohn, 2001).

Terminal-Restriction Fragment Length Polymorphism (t-RFLP)

Terminal-restriction fragment length polymorphism (t-RFLP) is also a ribosomal RNA gene-targeted microbial community analysis tool. In this case, one or both primers is/are fluorescently labelled to tag the PCR product. The amplicons are then digested with a restriction endonuclease and the fragments are separated by capillary electrophoresis. The labelled fragments are detected and an electrophoretic profile is produced as a unique signature for each microbial community.

Multiplex PCR

Multiplex PCR is a process where multiple primer pairs are used to simultaneously amplify several sequences in a single reaction. Time and effort can be saved when analysing different target regions by using this method. However, it can be restrictive since all the combinations of primer pairs must be able to function in a single set of reaction conditions. Additionally, primer dimer formation between primer pairs can occur that leads to poor sensitivity and preferential amplification of certain targets. However, multiplex PCR can be a rapid and convenient screening assay for the detection of viral, bacterial, and/or other microorganisms in samples.

Nucleic Acid Microarrays

Microarray technology is based on the ability of complementary sequences of nucleic acids to hybridise or bind to one another (Lucchini et al., 2001). Microarrays are one such tool that can be used for genome composition analysis or genotyping. Categorisation of bacteria on the basis of their genotypic characteristics instead of phenotypic markers provides deeper insights into the relationships between species (Murray et al., 2001; van Leeuwen et al., 2003). Therefore, it can be used to characterise previously uncultured microorganisms (Sebat et al., 2003). Microarray technology has become a standard laboratory tool for both pathogen detection and enumeration and genome wide analysis due to the increasing availability of genomic sequence (Ye et al., 2001).

Generally, microarrays can be classified into two main types: i). microarrays for genotyping and detecting single nucleotide polymorphisms, and ii). microarrays for analysis of genome diversity and gene expression profiling. The former type of microarray has been used extensively for the identification and characterisation of microbial pathogens and the detection of genetic variation in human genetic disorders. Meanwhile, the latter form of microarray has been developed to better understand the diversity and functioning of populations within complex environmental samples.

EFFECTIVENESS OF MODERN MOLECULAR BIOLOGY TECHNIQUES FOR POME TREATMENT

Various recent research in POME treatment have used molecular biology techniques to find the correlation between microbial communities and the performance of POME treatment. However, only a few studies have been conducted on the microbial aspects of POME, especially POME anaerobic treatment.

For instance, Tabatabaei et al. (2010) conducted a comprehensive study on the methanogenic diversity during the anaerobic treatment of POME in a closed digester tank (CDT). The majority (>99%) of the methanogens counted by using FISH in their study belonged to the genus *Methanosaeta*. However, 16S rRNA cloning along with DGGE analysis showed that *M. concilii* was the only member of the genus present. *Methanosarcina* accounted for <1% of the whole methanogenic population. The high number of *M. concillii* was attributed to the highly favourable concentration of acetate in POME.

Siriporn et al. (2012) also used PCR–DGGE to study the effect of initial pH, nutrients, and temperature on hydrogen production from POME using a thermotolerant consortia. Microbial community studies reveal that *Clostridium* spp. was the dominant species and *C. paraputrificum* was present at all temperatures tested (25–55 °C), indicating its thermotolerant properties. The study reveals that both initial pH and nutrient optimisation factors improved hydrogen production from POME under mesophilic to thermophilic conditions.

Furthermore, Badiei et al. (2012) studied the microbial community analysis of mixed anaerobic microflora in suspended sludge of ASBR producing hydrogen from POME. The microbial community was determined using PCR by isolating cultivable bacteria and at each operating HRT. The PCR products were sequenced and sequence identification was performed using the BLAST algorithm and GenBank database. The findings reveal that about 50% of the isolates present were members of the genus *Streptococcus*, about 30% were from *Lactobacillus* species, and around 20% were identified as species from the genus *Clostridium*. Scanning electron microscopy (SEM) images were used to distinguish between communities found in the bioreactor and identify the dominant microbial morphology of each community to further confirm the results of PCR and correlate them to the hydrogen productivity of the bioreactor.

Moreover, Soleimaninanadegani & Manshad (2014) investigated the microorganisms associated with POME. The subcultured POME microorganisms were characterised by using biochemical tests. Identification of fungi was based on microscopic morphology and cultural characteristics and the bacterial isolates were characterised based on biochemical test (Harrigan, 1998; Ayyachamy et al., 2012). The obtained characteristics were compared with those known using Manual of Determinative Bacteriology (Holt et al., 2000) and isolated bacteria were compared with the culture characteristics.

In another study, the microbial species isolated from dairy origin were similar to those found in local POME. However, *Bacillus* sp., *Pseudomonas* sp., and *Staphylococcus* sp. are lipase-producing organisms associated with pathogenicity (Griffths et al., 1981). Meanwhile, spores make POME microbial species inactive and highly resistant to the lethal effect of boiling, dry heating, and ultraviolet radiation (Okechalu et al., 2011). Other researchers discovered that POME is a potential habitat for thermophilic lipolytic bacteria (Ohimain et al., 2012a; Ohimain et al., 2012b). Other studies identified *Aspergillus* sp. as the fungal species associated with lipase production that accelerates the hydrolysis of lipid, causing its consequent breakdown into fatty acid and alcohol (Ghosh et al., 1996; Guehi et al., 2007). Meanwhile, *Penicillium* sp., *Fusarium* sp., and *Mucor* sp. in the POME were observed to survive in hostile environment (Nwuche & Oghonna, 2011; Ohimain et al., 2012). According to previous reports, these microorganisms can degrade oily wastewater effectively.

Bala et al. (2014) studied the biodegradation potential of bacteria isolated from POME to find the most suitable strain(s) for use in biological treatment of POME. The isolates were identified by sequence analysis of 16S rRNA genes as *Micrococcus luteus* 101 PB, *Stenotrophomonas maltophilia* 102 PB, *Bacillus cereus* 103 PB, *Providencia vermicola* 104 PB, *Klebsiella pneumonia* 105 PB, and *Bacillus subtilis* 106 PB. In their study, *Bacillus cereus* 103 PB produced the highest activity in reducing TSS (71.63%) and oil and grease (O&G) (85.14%). These results suggest that *Bacillus cereus* 103 PB might be applicable for use in wastewater treatment for the removal of TSS and O&G. The strain also produces extracellular lipase and cellulase which stimulate better waste treatment. The study demonstrated that POME-degrading microorganisms can be isolated from POME-polluted area and the degrading ability of these microorganisms is a clear indicator that these bacteria can be applied in the bioremediation techniques for biodegradation of POME to enhance treatment.

DGGE was used by Husna et al. (2015) to monitor the progression of the microbial community and revealed the microbial community composition inside palm oil mill effluent (POME) wastewater. Sequences related to *Bacillus subtilis* Rhizo-03 dominated in the starting material of the treatment. PCR–DGGE was used to study the isolated microorganisms that play an important role in the biodegradation of the pollutants in the anaerobic POME treatment. In this treatment, six sequences of *Firmicutes* (the most predominant phylum), five sequences of *Proteobacterium* (the second largest phylum), and two sequences of *Bacteroidetes* (the third most predominant phylum) were found through DGGE method.

Maneerat et al. (2012) studied the performance of UASB and microbial adaptation during the transition from mesophilic to thermophilic treatment of POME. The PCR–DGGE analysis (Figure 4) shows that the bacterial population profiles significantly changed with the temperature transition from mesophilic to thermophilic conditions. The results suggest that even though the thermophilic temperature of 57 °C is suitable for a number of hydrolytic, acidogenic, and acetogenic bacteria, it may not be suitable for some *Methanosaeta* spp. which have acclimatised at 37 °C.

Overall, all studies show the importance of molecular biology techniques to monitor the microbial community and link that information to the performance of POME treatment. The ability to monitor microorganisms and understand their ecology is essential to effectively control aerobic or anaerobic treatment system in treating POME during the startup, operation, and maintenance periods.

CONCLUSION

This review highlights available treatment methods for POME. Anaerobic process is the most popular approach being opted by palm oil mills. This process has low cost and produces CH_4 that could be utilised as boiler fuel or electricity generation, yet requires large land area and has a slow bioreaction rate. Molecular biology techniques have a big potential to provide detailed profile of the microbial community and establish the phylogenetics of microorganisms in bioreactors used for POME treatment. Recent advances have demonstrated that combining data obtained from microbial analysis with the performance of POME treatment could give an indicator on how the treatment plant should be operated. Microbial data also allows us to evaluate microorganism–microorganism and microorganism–environment interac-

Figure 4. DGGE profiles of the bacterial community in UASB
Adapted from Maneerat et al. (2012)

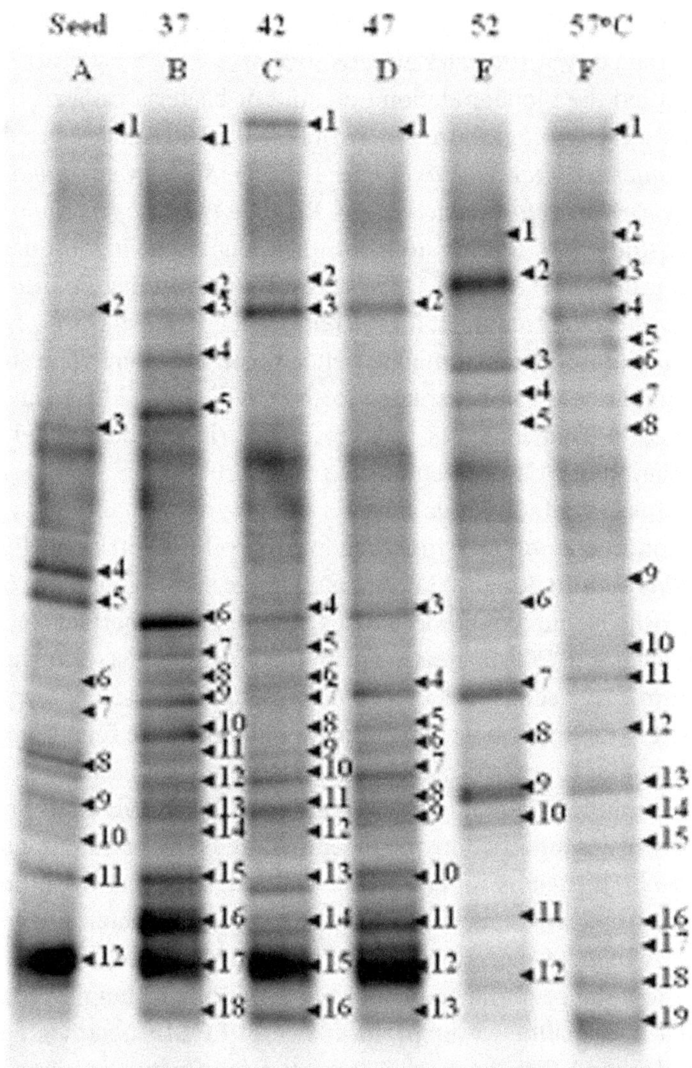

tions. Several molecular biology techniques highlighted in this review could be employed to meet the abovementioned purposes. The quick information gathered from the molecular methods could be used for better monitoring of bioreactor performance and eventually producing acceptable effluent quality.

ACKNOWLEDGMENT

This research was supported by the Ministry of Science and Innovation (MOSTI), Malaysia under RAGS grant and Universiti Malaysia Pahang [grant number RDU131419].

REFERENCES

Ahamed, A., Chen, C.-L., Rajagopal, R., Wu, D., Mao, Y., Ho, I. J. R., ... Wang, J.-Y. (2015). Multi-phased anaerobic baffled reactor treating food waste. *Bioresource Technology, 182*, 239–244. doi:10.1016/j.biortech.2015.01.117 PMID:25704096

Ahmad, A., Ibrahim, N., Ismail, I., & Bhatia, S. (2002). Coagulation-sedimentation-extraction pretreatment methods for the removal of suspended solids and residual oil from palm oil mill effluent (POME). *IIUM Eng J, 3*, 25–33.

Ahmad, A., Ismail, S., & Bhatia, S. (2003). Water recycling from palm oil mill effluent (POME) using membrane technology. *Desalination, 157*(1-3), 87–95. doi:10.1016/S0011-9164(03)00387-4

Ahmad, A., Sumathi, S., & Hameed, B. H. (2006). Coagulation of residue oil and suspended solid in palm oil mill effluent by chitosan, alum and PAC. *Chemical Engineering Journal, 118*(1-2), 99–105. doi:10.1016/j.cej.2006.02.001

Ahmad, A. L., & Chan, C. Y. (2009). Sustainability of palm oil industries: An innovative treatment via membrane technology. *Applied Science, 9*(17), 3074–3079. doi:10.3923/jas.2009.3074.3079

Ahmad, J. L., Ismail, S., & Bhatia, S. (2003). Water recycling from palm oil mill effluent (POME) using membrane technology. *Desalination, 157*(1-3), 87–95. doi:10.1016/S0011-9164(03)00387-4

Amann, R., Ludwig, W., & Schleifer, K. (1995). Phylogenetic identification and *in situ* detection of individual microbial cells without cultivation. *Microbiological Reviews, 59*, 143–169. PMID:7535888

Angenent, L. T., Karim, K., Al-Dahlan, M. H., Wrenn, B. A., & Domiguez-Espinosa, R. (2004). Production of bioenergy and biochemicals from industrial and agricultural wastewater. *Trends in Biotechnology, 22*(9), 477–485. doi:10.1016/j.tibtech.2004.07.001 PMID:15331229

Ayyachamy, M., Turner, K., O'Donovan, A., Gupta, V., & Tuohy, M. (2012). *Laboratory protocols in fungal biology: current methods in fungal biology*. New York, NY: Springer.

Badiei, M., Jahim, J., Anuar, N., Sheik Abdullah, S., Su, L., & Kamaruzzaman, M. (2012). Microbial community analysis of mixed anaerobic microflora in suspended sludge of ASBR producing hydrogen from palm oil mil effluent. *International Journal of Hydrogen Energy, 37*(4), 3169–3176. doi:10.1016/j.ijhydene.2011.11.063

Bala, J., Lalung, J., & Ismail, N. (2014). Biodegradation of palm oil mill effluent (POME) by bacterial. *Int J Scientific and Researchs Publications, 4*(3).

Baranitharan, E., Khan Maksudur, R., & Prasad, D. M. R. (2013). Treatment of palm oil mill effluent in microbial fuel cell using polyacrylonitrile carbon felt as electrode. *Journal of Medical Bioengineering, 2*(4), 252–256. doi:10.12720/jomb.2.4.252-256

Basiron, Y. (2013). Malaysia supply and demand of palm oil: challenges and opportunities until 2020. In *Proceedings of the Oil World Outlook Conference 2013*, Hamburg, Germany.

Basri, M. F., Yacob, S., Hassan, M. A., Shirai, Y., Wakisaka, M., Zakaria, M. R., & Phang, L. Y. (2010). Improved biogas from palm oil mill effluent by a scaled-down anaerobic treatment process. *World Journal of Microbiology & Biotechnology, 26*(3), 505–514. doi:10.100711274-009-0197-x

Bhatia, S., Othman, Z., & Ahmad, A. (2007a). Coagulation-flocculation process for POME treatment using *Moringa oleifera* seeds extract; optimization studies. *Chemical Engineering Journal, 133*(1-3), 205–212. doi:10.1016/j.cej.2007.01.034

Bhatia, S., Othman, Z., & Ahmad, A. (2007b). Pretreatment of palm oil mill effluent (POME) using *Moringa oleifera* seeds as natural coagulant. *Journal of Hazardous Materials, 145*(1-2), 120–126. doi:10.1016/j.jhazmat.2006.11.003 PMID:17141409

Blackall, L., Burrell, P., Gwilliam, H., Bradford, D., Bond, P., & Hugenholtz, P. (1998). The use of 16S rRNA gene clone libraries to describe the microbial diversity of activated sludge communities. *Water Science and Technology, 37*(4-5), 451–454. doi:10.2166/wst.1998.0691

Borja, R., & Banks, C. J. (1994). Anaerobic digestion of Palm Oil Mill Effluent using an up-flow anaerobic sludge blanket (UASB) reactor. *Biomass and Bioenergy, 6*(5), 381–389. doi:10.1016/0961-9534(94)E0028-Q

Borja, R., Banks, C. J., & Sanchez, E. (1996). Anaerobic treatment of palm oil mill effluent in a two-stage up-flow anaerobic sludge blanket (UASB) reactor. *Journal of Biotechnology, 45*(2), 125–135. doi:10.1016/0168-1656(95)00154-9

Chan, Y., Chong, M., & Law, C. (2010). Biological treatment of anaerobically digested palm oil mill effluent (POME) using a lab-scale sequencing batch reactor (SBR). *Journal of Environmental Management, 91*(8), 1738–1746. doi:10.1016/j.jenvman.2010.03.021 PMID:20430515

Chan, Y. J., & Chong, M. F. (2019). Palm oil mill effluent (POME) treatment-Current technologies, biogas capture and challenges. In D. C. Y. Foo & M. K. Tun Abdul Aziz (Eds.), *Green technologies for the palm oil industry* (pp. 71–92). Springer Singapore, Singapore. doi:10.1007/978-981-13-2236-5_4

Chan, Y. J., Chong, M. F., & Law, C. L. (2012). An integrated anaerobic–aerobic bioreactor (IAAB) for the treatment of palm oil mill effluent (POME): Start-up and steady state performance. *Process Biochemistry, 47*(3), 485–495. doi:10.1016/j.procbio.2011.12.005

Chang, S. H. (2014). An overview of empty fruit bunch from oil palm as feedstock for bio-oil production. *Biomass and Bioenergy, 62*, 174–181. doi:10.1016/j.biombioe.2014.01.002

Charles, A., & Chin, K. K. (2019). Photocatalytic treatment of palm oil mill effluent by visible light-active calcium ferrite: Effect of catalyst preparation technique. *Journal of Environmental Management, 234*, 404–411. doi:10.1016/j.jenvman.2019.01.024 PMID:30640165

Cheng, H., & Wang, L. (2013). Lignocelluloses feedstock biorefinery as petrorefinery substitutes. In M. D. Matovic (Ed.), *Biomass now-sustainable growth and use.* Retrieved from http://www.intechopen.com/books/biomass- now-sustainable-growth-and use/lignocelluloses-feedstock-biorefinery-as-petrorefinery-substitutes)

Cheng, J., Zhu, X., Ni, J., & Borthwick, A. (2010). Palm oil mill effluent treatment using a two-stage microbial fuel cells system integrated with immobilized biological aerated filters. *Bioresource Technology, 101*(8), 2729–2734. doi:10.1016/j.biortech.2009.12.017 PMID:20042327

Chin, K. K. (1981). Anaerobic treatment kinetics of palm oil sludge. *Water Research, 15*(2), 199–202. doi:10.1016/0043-1354(81)90112-3

Chin, K. K., Lee, S. W., & Mohammed, H. H. (1996). A study of palm oil mill effluent treatment using ponding system. *Water Science and Technology, 34*(11), 119–123. doi:10.2166/wst.1996.0270

Choi, W. H., Shin, C. H., Son, S. M., Ghorpade, P. A., Kim, J. J., & Park, J. Y. (2013). Anaerobic treatment of palm oil mill effluent using combined high-rate anaerobic reactors. *Bioresource Technology, 141*, 138–144. doi:10.1016/j.biortech.2013.02.055 PMID:23489567

Curds, C. (1975). Protozoa. In C. R. Curds & H. A. Hawkes (Eds.), Ecological aspects of used-water treatment. Vol. 1. The organisms and their ecology. London, UK: Academic Press. 203-268.

Edgar, F. C. M., Diego, E. C., & Victor, A. A. (2006). Study of the operational conditions for anaerobic digestion of urban solid wastes. *Waste management (New York, N.Y.), 26*(5), 546–556. doi:10.1016/j.wasman.2005.06.003 PMID:16321515

Eikelboom, D. (1975). Filamentous organisms observed in activated sludge. *Water Research, 9*(4), 365–388. doi:10.1016/0043-1354(75)90182-7

Ellouze, E., Tahri, N., & Amar, R. (2012). Enhancement of textile wastewater treatment process using nanofiltration. *Desalination, 286*, 16–23. doi:10.1016/j.desal.2011.09.025

Esteban, G., Tellez, C., & Bautista, L. (1991). Dynamics of ciliated protozoa communities in activated sludge process. *Water Research, 25*(8), 967–972. doi:10.1016/0043-1354(91)90145-G

Farid, M. A. A., Zakaria, M. R., Hassan, M. A., Ali, A. A. M., Othman, M. R., Ibrahim, I., & Shirai, Y. (2019). A holistic treatment system for palm oil mill effluent by incorporating the anaerobic-aerobic-wetland sequential system and a convective sludge dryer. *Chemical Engineering Journal, 369*, 195–204. doi:10.1016/j.cej.2019.03.033

Fisher, M., & Triplett, E. (1999). Automated approach for ribosomal intergenic spacer analysis of microbial diversity and its application to freshwater bacterial communities. *Applied and Environmental Microbiology, 65*, 4630–4636. PMID:10508099

Ghosh, P., Saxena, R., Gupta, R., & Yadav, R., & S, D. (1996). Microbial lipases: Production and applications. *Science Progress, 79*, 119–158. PMID:8828407

Griffths, M., Phillps, J., & Muir, D. (1981). Thermostability of proteases and lipases from a number of species of psychrotropic bacteria of dairy origin. *The Journal of Applied Bacteriology, 50*(2), 289–303. doi:10.1111/j.1365-2672.1981.tb00894.x PMID:7019181

Guehi, T., Dingkunh, M., & Cros, E. (2007). Identification and lipase-producing abilities of moulds isolated from Ivorian raw cocoa beans. *Research Journal of Agriculture and Biological Sciences, 3*, 838–843.

Guerrero, J., Omil, F., Mendez, R., & Lema, J. (1999). Anaerobic hydrolysis and acidogenesis of wastewaters from food industries with high content of organic solids and protein. *Water Research*, *33*(15), 3281–3290. doi:10.1016/S0043-1354(99)00041-X

Hansen, S. B., Padfield, R., Syayuti, K., Evers, S., Zakaria, Z., & Mastura, S. (2015). Trends in global palm oil sustainability research. *Journal of Cleaner Production*, *100*, 140–149. doi:10.1016/j.jclepro.2015.03.051

Harrigan, W. (1998). *Laboratory methods in food microbiology*. Gulf Professional Publishing.

Harsono, S. S., Grundmann, P., & Soebronto, S. (2014). Anaerobic treatment of palm oil mill effluents: Potential contribution to net energy yield and reduction of greenhouse gas emissions from biodiesel production. *Journal of Cleaner Production*, *64*, 619–627. doi:10.1016/j.jclepro.2013.07.056

Ho, C., & Chan, C. (1986). The application of lead dioxide-coated titanium anode in the electrofloation of palm oil mill effluent. *Water Research*, *20*(12), 1523–1527. doi:10.1016/0043-1354(86)90116-8

Ho, C., & Tan, Y. (1988). The treatment of anaerobically digested palm oil mill effluent by pressurized activated sludge. *Journal of Chemical Technology and Biotechnology (Oxford, Oxfordshire)*, *41*(1), 75–84. doi:10.1002/jctb.280410110

Holt, J. G., Krieg, N. R., Sneath, P. H. A., Staley, J. T., & Williams, S. T. (2000). *Bergey's manual of determinative bacteriology* (9th ed.). Philadelphia, PA: Lippincott Williams & Wilkins.

Hossain, M. S., Omar, F., Asis, A. J., Bachman, R. T., Sarker, M. Z. I., & Kadir, M. O. A. (2019). Effective treatment of palm oil mill effluent using FeSO4.7H2O waste from titanium oxide industry: Coagulation adsorption isotherm and kinetic studies. *Journal of Cleaner Production*, *219*, 86–98. doi:10.1016/j.jclepro.2019.02.069

Husna, A., Luqman, C., & Thomas, S. (2015). Microbial community analysis in anaerobic palm oil mill effluent (POME) wastewater by denaturing gradient gel electrophoresis (DGGE). *International Journal of Research in Engineering and Technology*, *4*(8), 1–9. doi:10.15623/ijret.2015.0408001

Ibrahim, A., Yeoh, B. G., Cheah, S. C., Ma, A. N., Ahmad, S., Chew, T. Y., & Wahid, M. J. A. (1984). Thermophilic anaerobic contact digestion of palm oil mill effluent. *Water Science and Technology*, *17*(2-3), 155–165. doi:10.2166/wst.1985.0127

Jemaat, Z. (2013). *Treatment of complex industrial wastewaters containing ammonium and phenolic compounds using granular sludge in continuous airlift reactors.* (Doctoral thesis). Universitat Autonoma de Barcelona, Barcelona, Spain.

Judd, S. (2011). *The MBR book: principles and applications of membrane bioreactors for water and wastewater treatment.* UK: Elsevier Science.

Kanadasan, J., & Razak, H. A. (2015). Engineering and sustainability performance of selfcompacting palm oil mill incinerated waste concrete. *Journal of Cleaner Production*, *89*, 78–86. doi:10.1016/j.jclepro.2014.11.002

Karakashev, D., Batstone, D., & Angelidaki, I. (2005). Influence of environmental conditions on methanogenic compositions in anaerobic biogas reactors. *Applied and Environmental Microbiology*, *71*(1), 331–338. doi:10.1128/AEM.71.1.331-338.2005 PMID:15640206

Khadidi, M. H. J., Al-Shorgani, N. K., Ali, E., Hamid, A. A., & Kalil, M. (2013). A new flocculant-coagulant with potential use for industrial wastewater treatment. In *Proceedings of the 2nd international Conference on Environment, Energy and Biotechnology (ICEEB), Singapore; IACSIT Press*, 139-142.

Lam, M. K., & Lee, K. T. (2011). Renewable and sustainable bioenergies production from palm oil mill effluent (POME): Win–win strategies toward better environmental protection. *Biotechnology Advances*, *29*(1), 124–141. doi:10.1016/j.biotechadv.2010.10.001 PMID:20940036

Leslie, C., Grady, J., Daigger, G., & Lim, C. (2011). *Biological wastewater treatment* (3rd ed.). Boca Raton, FL: CRC Press.

Lucchini, S., Thompson, A., & Hilton, J. (2001). Microarrays for microbiologists. *Microbiology*, *147*(6), 1403–1414. doi:10.1099/00221287-147-6-1403 PMID:11390672

Maaroff, R. M., Jahim, J. M., Azhar, A. M., Abdul, P. M., Masdar, M. S., Nordin, D., & Nasir, M. A. A. (2019). Biohydrogen production from palm oil mill effluent (POME) by two stage anaerobic sequencing batch reactor (ASBR) system for better utilization of carbon sources in POME. *International Journal of Hydrogen Energy*, *44*(6), 3395–3406. doi:10.1016/j.ijhydene.2018.06.013

Madony, P., Davoli, D., & Chirerici, E. (1993). Comparative analysis of the activated sludge microfauna in several sewage treatment works. *Water Research*, *27*(9), 1485–1491. doi:10.1016/0043-1354(93)90029-H

Mamimin, C., Kongjan, P., O-Thong, S., & Prasertsan, P. (2019). Enhancement of biohythane production from solid waste by co-digestion with palm oil mill effluent in two-stage thermophilic fermentation. *International Journal of Hydrogen Energy*, *44*(32), 17224–17237. doi:10.1016/j.ijhydene.2019.03.275

Maneerat, K., Boonyarit, N., Somkiet, T., & Chantaraporn, P. (2012). UASB performance and microbial adaptation during a transition from mesophilic to thermophilic of palm oil mill effluent. *Journal of Environmental Management*, *103*, 74–82. doi:10.1016/j.jenvman.2012.03.004 PMID:22466006

Mundi, I. (2017). *Malaysia palm oil production by year.* Retrieved from http://www.indexmundi.com/agriculture/?country=my&commodity=palm&oilgraph=production

Murray, A., Lies, D., Li, G., Nealson, K., Zhou, J., & Tiedje, J. (2001). DNA/DNA hybridization to microarrays reveals gene specific differences between closely related microbial genomes. *Proceedings of the National Academy of Sciences of the United States of America*, *98*(17), 9853–9858. doi:10.1073/pnas.171178898 PMID:11493693

Muyzer, G., & Smalla, K. (1998). Application of denaturing gradient gel electrophoresis and temperature gradient gel electrophoresis (TGGE) in microbial ecology. Mini review. *Anton Leeuw Int J G*, *73*(1), 127–141. doi:10.1023/A:1000669317571 PMID:9602286

Najafpour, G., Yieng, H. A., Younesi, H., & Zinatizadeh, A. (2005). Effect of organic loading on performance of rotating biological contactors using palm oil mill effluents. *Process Biochemistry*, *40*(8), 2879–2884. doi:10.1016/j.procbio.2005.01.002

Najafpour, G., Zinatizadeh, A., Mohamed, A. R., Isa, M. H., & Nasrollahzadeh, H. (2006). High-rate anaerobic digestion of palm oil mill effluent in an up-flow sludge fixed film bioreactor. *Process Biochemistry*, *41*(2), 370–379. doi:10.1016/j.procbio.2005.06.031

Najib, M. Z. M., Ujang, Z., Salim, M. R., & Ibrahim, Z. (2017). Developed microbial granules containing photosynthetic pigments for carbon dioxide reduction in palm oil mill effluent. *International Biodeterioration & Biodegradation*, *116*, 163–170. doi:10.1016/j.ibiod.2016.10.031

Ng, K. H., & Cheng, C. K. (2015). A novel photomineralization of POME over UV-responsive TiO_2 photocatalyst: Kinetics of POME degradation and gaseous product formation. *RSC Advances*, *5*(65), 53100–53110. doi:10.1039/C5RA06922J

Ng, W., Goh, A., & Tay, J. (1988). Palm oil mill effluent treatment-liquid-solid separation with dissolved air floatation. *Biological Wastes*, *25*(4), 257–268. doi:10.1016/0269-7483(88)90087-0

Norulaini, N., Zuhairi, A., Hakimi, I., & Omar, A. (2004). Treatment of palm oil mill effluent (POME) supernatants using aerobic attached-growth system: trickling filter as a case study. *J Teknol, 40*(F), 77-90.

Nwuche, C., & Oghonna, J. (2011). Isolation of lipase producing fungi from palm oil mill effluent (POME) dump sites at Nsukka. *Brazilian Archives of Biology and Technology*, *54*(1), 113–116. doi:10.1590/S1516-89132011000100015

Ohimain, E., Daokoru-Olukole, C., Izah, S., Eke, R., & Okonkwo, C. (2012a). Microbiology of palm oil mill effluents. *J Microbiology and Biotechnology Research*, *2*(6), 852.

Ohimain, E., Seiyaboh, E., Izah, S., Oghenegueke, V., & Perewarebo, T. (2012b). Some selected physico-chemical and heavy metal properties of palm oil mill effluents. *Greener Journal of Physical Sciences, 2*, 131-137.

Okechalu, J., Dashen, M., Lar, P., Okechalu, B., & Gushop, T. (2011). Microbiological quality and chemical characteristics of palm oil sold within Jos Metropolis, Plateau State, Nigeria. *J Microbiology and Biotechnology Research*, *1*, 107–112.

Olvera, J., & Lopez, A. (2012). *Biogas production from anaerobic treatment of agro-industrial wastewater*. Rijeka, Croatia: Intech.

Oosterveer, P. (2015). Promoting sustainable palm oil: Viewed from a global networks and flows perspective. *Journal of Cleaner Production*, *107*, 146–153. doi:10.1016/j.jclepro.2014.01.019

Poh, P. E., Yong, W. J., & Chong, M. F. (2010). Palm oil mill effluent (POME) characteristic in high crop season and the applicability of high-rate anaerobic bioreactors for the treatment of POME. *Industrial & Engineering Chemistry Research*, *49*(22), 11732–11740. doi:10.1021/ie101486w

Pol, L. W. H., Lopes, S. I. D., Lettinga, G., & Lens, P. N. L. (2004). Anaerobic sludge granulation. *Water Research*, *38*(6), 1376–1389. doi:10.1016/j.watres.2003.12.002 PMID:15016515

Sanz, J. L., & Kochling, T. (2007). Molecular biology techniques used in wastewater treatment: An overview. *Process Biochemistry*, *42*(2), 119–133. doi:10.1016/j.procbio.2006.10.003

Sebat, J., Colwell, F., & Crawford, R. (2003). Metagenomic profiling: Microarray analysis of an environmental genomic library. *Applied and Environmental Microbiology*, *69*(8), 4927–4934. doi:10.1128/AEM.69.8.4927-4934.2003 PMID:12902288

Seviour, E., Blackall, L., Christensson, C., Hugenholtz, P., Cunningham, M., Bradford, D., & Seviour, R. (1997). The filamentous morphotype Eikelboom Type 1863 is not a single genetic entity. *Journal of Applied Microbiology*, *82*(4), 411–421. doi:10.1046/j.1365-2672.1997.00111.x PMID:9134715

Siriporn, Y., Sompong, O., & Poonsuk, P. (2012). Effect of initial pH, nutrients and temperature on hydrogen production from palm oil mill effluent using thermotolerant consortia and corresponding microbial communities. *International Journal of Hydrogen Energy*, *37*(18), 13806–13814. doi:10.1016/j.ijhydene.2012.03.151

Snaidr, J., Amann, R., Huber, I., Ludwig, W., & Schleifer, K. (1997). Phylogenetic analysis and in situ identification of bacteria in activated sludge. *Applied and Environmental Microbiology*, *63*, 2884–2896. PMID:9212435

Soleimaninanadegani, M., & Manshad, S. (2014). Enhancement of biodegradation of palm oil mill effluents by local isolated microorganisms. *Int Scholarly Research Notices*, *2014*, 8. doi:10.1155/2014/727049 PMID:27433516

Sontaya, K., Pitiyont, B., & Punsuvon, V. (2013). Decolorization and COD removal of palm oil mill wastewater by electrocoagulation. *Int J Environ Sci Eng*, *7*, 370–374.

Subramaniam, M. N., Goh, P. S., Lau, W. J., Ng, B. C., & Ismail, A. F. (2018). AT-POME colour removal through photocatalytic submerged filtration using antifouling PVDF-TNT nanocomposite membrane. *Separation and Purification Technology*, *191*, 266–275. doi:10.1016/j.seppur.2017.09.042

Tabatabaei, M., Rahim, R., Abdullah, N., Wright, A., Shirai, Y., Sakai, K., & Hassan, M. A. (2010). Review: Importance of the methanogenic archaea populations in anaerobic wastewater treatments. *Process Biochemistry*, *45*(8), 1214–1225. doi:10.1016/j.procbio.2010.05.017

Tong, S., & Jaafar, A. B. (2004). Waste to energy: Methane recovery from anaerobic digestion of palm oil mill effluent. *Energy Smart*, *14*, 11456.

Uddin, M., Rahman, M., Taweekun, J., Techato, K., Mofijur, M., & Rasul, M. (2019). Enhancement of biogas generation in up-flow sludge blanket (UASB) bioreactor from palm oil mill effluent (POME). *Energy Procedia*, *160*, 670–676. doi:10.1016/j.egypro.2019.02.220

Ueda, S., & Earle, R. (1972). Microflora of activated sludge. *The Journal of General and Applied Microbiology*, *18*(3), 239–248. doi:10.2323/jgam.18.239

Uzir, N. M., Aziz, M. M. A., Ismail, S. B., Mahmood, N. A. N., Umor, N. A., & Muhammad, S. A. F. A. S. (2019). Optimization of coagulation-flocculation process for the palm oil mill effluent treatment by using rice husk ash. *Industrial Crops and Products*, *139*, 111482. doi:10.1016/j.indcrop.2019.111482

van Leeuwen, C., Sniders, S., Durin, N., Lacroix, B., Verbrugh, H., Enright, M., & van Belkum, A. (2003). Multilocus sequence typing of *Staphylococcus aureus* with DNA array technology. *Journal of Clinical Microbiology*, *41*(7), 3323–3326. doi:10.1128/JCM.41.7.3323-3326.2003 PMID:12843083

Vijayaraghavan, K., Ahmad, D., & Aziz, M. (2007). Aerobic treatment of palm oil mill effluent. *Journal of Environmental Management*, *82*(1), 24–31. doi:10.1016/j.jenvman.2005.11.016 PMID:16584834

Weiland, P. (2010). Biogas production: Current state and perspectives. *Applied Microbiology and Biotechnology*, *85*(4), 849–860. doi:10.100700253-009-2246-7 PMID:19777226

Yacob, S., Hassan, M. A., Shirai, Y., Wakisaka, M., & Subash, S. (2005). Baseline study of methane emission from open digesting tanks of palm oil mill effluent treatment. *Chemosphere*, *59*(11), 1575–1581. doi:10.1016/j.chemosphere.2004.11.040 PMID:15894045

Ye, R., Wang, T., Bedzyk, L., & Croker, K. (2001). Applications of DNA microarrays in microbial systems. *Journal of Microbiological Methods*, *47*(3), 257–272. doi:10.1016/S0167-7012(01)00308-6 PMID:11714516

Yu, Z., & Mohn, W. (2001). Bacterial diversity and community structure in an aerated lagoon revealed by ribosomal intergenic spacer analyses and 16S ribosomal DNA sequencing. *Applied and Environmental Microbiology*, *67*(4), 1565–1574. doi:10.1128/AEM.67.4.1565-1574.2001 PMID:11282606

Zainal, B. S., Akhbari, A., Zinatizadeh, A. A., Mohammadi, P., Danaee, M., Mohd, N. S., & Ibrahim, S. (2019). UASFF start-up for biohydrogen and biomethane production from treatment of palm oil mill effluent. *International Journal of Hydrogen Energy*, *44*(37), 20725–20737. doi:10.1016/j.ijhydene.2018.07.037

Zainal, N. H., Jalani, N. F., Mamat, R., & Astimar, A. A. (2017). A review on the development of palm oil mill effluent (POME) final discharge polishing treatments. *Journal of Oil Palm Research*, *29*(4), 528–540. doi:10.21894/jopr.2017.00012

Zhang, Y., An, X., & Quan, X. (2011). Enhancement of sludge granulation in a zero valence iron packed anaerobic reactor with a hydraulic circulation. *Process Biochemistry*, *46*(2), 471–476. doi:10.1016/j.procbio.2010.09.021

Zhang, Y., Yan, L., Chi, L., Long, X., Mei, Z., & Zhang, Z. (2008). Startup and operation of anaerobic EGSB reactor treating palm oil mill effluent. *Journal of Environmental Sciences (China)*, *20*(6), 658–663. doi:10.1016/S1001-0742(08)62109-9 PMID:18763558

Chapter 11
Waste-Activated Sludge Treatment Processes

Chin Heng Gan
https://orcid.org/0000-0003-1007-3353
Universiti Tunku Abdul Rahman, Malaysia

Siew Hoong Shuit
https://orcid.org/0000-0003-3379-6597
Universiti Tunku Abdul Rahman, Malaysia

ABSTRACT

Biological processes are the most common methods to treat wastewater but they produce huge amounts of waste-activated sludge (WAS) as a by-product. WAS generated during wastewater treatment must be stabilized sufficiently prior to ultimate disposal. The most common methods of sludge stabilization are biological processes (aerobic or anaerobic digestion) due to simplicity of the operation, and lower operating and maintenance costs for similar wastes. Anaerobic digestion remains the principal process for the stabilization of sludge solids as organic waste is converted into methane-rich biogas and fertilizer in an oxygen-free environment. However, WAS treatment processes are often a challenging problem. Therefore, the characteristics of WAS are discussed to have a better understanding so as to treat it efficiently. Various kinds of treatment processes such as physicochemical treatment, biological treatment, and combined treatment are the focus of this chapter; however, many concepts developed in this chapter apply to anaerobic digestion to stabilize the WAS.

INTRODUCTION

Sludge production in the industrialized world has been steadily increasing during the last decades, with biological wastewater treatment processes being one of the major contributors (Feng et al., 2009; Neyens, Baeyens, Dewil, & De heyder, 2004). Sludges of biological origin such as waste activated sludge (WAS) are known to exhibit high organic and solids content as well as poor dewaterability. The particulate organic matters of WAS are difficult to convert into soluble substances, therefore its efficiency of

DOI: 10.4018/978-1-7998-0369-0.ch011

treatment is generally low (Xu et al., 2017). Together with more stringent disposal regulations, this has caused a demand for more efficient sludge conditioning and dewatering techniques to reduce the volume of sludge, resulting in lower costs of further processing as well as transport and disposal of sludge. The management of sludge, or biosolids, requires major investment and operating costs. Typically, about 50–60% of the operating cost of a wastewater treatment plant caters for sludge management (Mininni, Laera, Bertanza, Canato, & Sbrilli, 2015; Tomei et al., 2016). Thus, sludge dewatering is of paramount importance in wastewater treatment systems. Organic polymers are the most widely employed conditioners in the wastewater treatment industry. The polymers mainly consist of acrylamide and acrylate, which are potential toxins to aquatic animals and human beings at certain concentrations, even though they are sometimes biodegradable (Bolto & Gregory, 2007). The residual polymers in dewatered sludge cakes may also pose a long-term risk to surrounding environment when the cakes are subject to landfill as the final disposal and it is usually very expensive. Despite conditioning, a considerable amount of water is retained within the sludge, leading to extreme compressibility and moderate dewaterability.

Many wastewater treatment facilities employ biological processes such as activated sludge process as they are an effective way of treating wastewater. The activated sludge (biomass) serves as a catalyst for the conversion of pollutants originating from wastewater to harmless gaseous products or biomass, which can be removed by sedimentation. The chemical conversions in a biological wastewater treatment system are mainly carried out by microorganisms. Under substrate limiting conditions, part of the microorganisms live on available substrate to survive and grow, whereas a huge fraction of them die and contribute to the WAS which also include the living excess biomass. A typical lifecycle for microorganisms consists of the hydrolysis of slowly degradable matter (e.g. in particulate form) into easily degradable matter, which may be soluble. This is further hydrolyzed into very easily degradable matter, i.e. substrate (acetic acid, methanol, propionic acid, glucose, ammonium, nitrite etc.) that are used for the growth of the biomass. Therefore, biological processes in WWTPs have the serious drawback of continually producing huge amounts of WAS.

Municipal and industrial wastewater treatment plants (WWTPs) generate large amounts of sludge as a by-product of the physical, chemical and biological processes used during treatment. Biological processes, such as activated sludge process are the most common methods to treat wastewater but at the same time, they produce huge amounts of waste activated sludge (WAS). Hence, the WAS is an inevitable drawback inherent to the activated sludge process and must undergo some pretreatment or post-treatment in order to reduce its associated volumes, improve its characteristics for reducing the associated health problems as well as handling and disposal. However, sludge treatment is one of the most difficult and expensive problems in WWTPs. The treatment and disposal cost of WAS may be as high as 50% of the operating cost of the WWTPs (Mininni et al., 2015; Tomei et al., 2016). Generally, pretreatments are applied to (i) firstly reduce the water content of the WAS (dewatering), (ii) transform refractory organic material into biodegradable residue, and (iii) finally transfer the residue to the subsequent treatments.

OVERVIEW OF WASTEWATER TREATMENT PROCESSES

Activated sludge, aerated lagoons, rotating biological contactors and trickling filters are the most common wastewater treatment processes being used in Malaysia ("Sewage," 2016). In an activated sludge system, the wastewater first passes through the preliminary process (screening, grease and grit removal),

Figure 1. Complete overall process flow scheme of a conventional large-scale activated sludge system

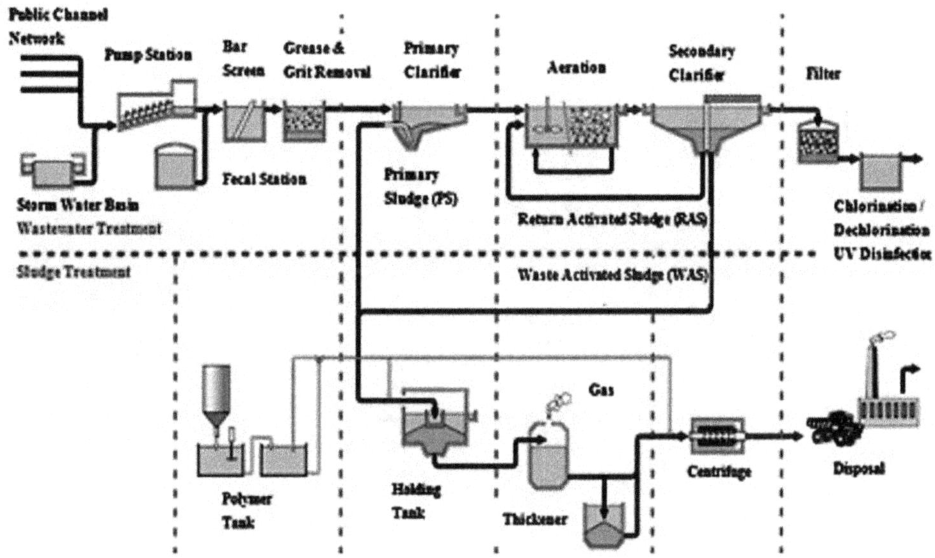

followed by primary clarifier, aeration, secondary clarifier, filtration and eventually disinfection process before being discharged as show in Figure 1.

A secondary clarifier produces the dischargeable effluent as overflow and return activated sludge (containing 98–99% water), that is partly recycled to the aeration tank to maintain the concentration of the biomass to food supplied at the required level, to keep the system in balance. Excess sludge, also known as WAS is evacuated to the sludge treatment units of the WWTP (Appels, Baeyens, Degreve, & Dewil, 2008).

The aim of sludge treatment is to remove the organic matter, excess water, degradable material, odors and pathogens prior to ultimate disposal (USEPA, 1993). The basic options for sludge treatment include thickening, dewatering, drying, stabilization and composting. The abovementioned methods normally depend on the initial characteristics of the WAS and its final use. Main methods for sludge disposal are landfilling, land application and incineration. However, these methods have some limitations.

CHARACTERISTICS OF WASTE ACTIVATED SLUDGE

The characteristics of WAS are greatly dependent on the nature of the wastewater treated in the WWTP. For instance, WAS originating from municipal and industrial wastewater generally contains pathogenic organisms, toxic organic substances and heavy metals, and inorganic nutrients such as phosphate and ammonium, which causes harmful effect on the environment and public health (Campbell, 2000). Meanwhile, the WAS originating from domestic wastewater is generally free from toxic substances and heavy metals. However, a common characteristic from different types of WAS is having very high water content and affinity of the sludge solids. Besides that, WAS has low biodegradability and hydrolysis limited, leading to slower process in anaerobic digestion (requires longer retention time) (Zhou, Wang, Jiang, Liu, & Yuan, 2015).

Figure 2. Schematic diagram of EPS
(Wang et al., 2019)

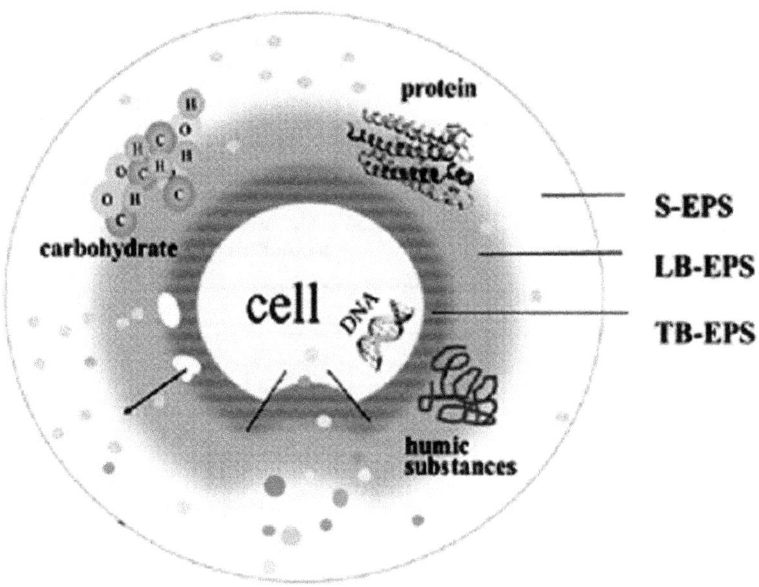

The microbial activity (secretion and lysis of cells) during aeration processes in wastewater treatment plants results in a microbial matrix (flocs) composed of microbial and exopolymeric substances. These microbial-originated EPS are a complex mixture of biopolymers comprising different classes of organic macromolecules such as polysaccharides, proteins (Wilén, Jin, & Lant, 2003; Comte, Guibaud, & Baudu, 2007), nucleic acids, uronic acids, humic substances (Frohlund, Griebe, & Nielsen, 1995), and lipids, amongst others, which have been found to occur in the intracellular space of microbial aggregates, more specifically at or outside the cell surface (Flemming & Wingender, 2001) as shown in Figure 2.

The production of EPS is a general property of microorganisms in natural environments and it forms a three-dimensional, gel-like, highly hydrated and often charged biofilm matrix, in which the microorganisms are embedded and more or less immobilized (Wingender, Jäger, & Flemming, 1999). They form a protective layer for the cells against the harsh external environment, and also serve as carbon and energy reserves during starvation. The total mass of EPS (EPS and water held within the EPS-structure) has been found to represent up to 80% of the mass of activated sludge (Frølund, Palmgren, Keiding, & Nielsen, 1996). Polysaccharides have often been assumed to be the most abundant components of EPS in early biofilm research (approximately 60% of the dry solids content). However, proteins, nucleic acids and amphiphilic compounds including (phospho)-lipids have also been shown to appear in significant amounts (up to 40% of the DS content) (Wingender et al., 1999).

EPS are thought to be of considerable importance in the removal of pollutants from wastewater, in bioflocculation and settling (Lü, Wang, Shao, & He, 2017), and in the dewatering of WAS (Urbain, Block, & Manem, 1993). The fact that EPS are highly hydrated and are able to bind a large volume of water was discovered by (Forster & Lewin, 1972; Li et al., 2016). EPS therefore play an important ecological role in protecting biofilm organisms from desiccation (Flemming & Wingender, 2001). Increasing relatively low levels of EPS can aid sludge dewaterability by improving the level of sludge flocculation (Sanin & Vesilind, 1994). It is envisaged that this is due to the highly hydrated nature of the EPS retaining water within the sludge matrix, which counteracts the benefit of flocculation. According to Flemming (Flemming, 1996), two types of binding mechanisms between water molecules and the EPS-structure must be considered, i.e. electrostatic interactions and hydrogen bonds. Considering the high water binding properties of EPS (Flemming & Wingender, 2001), together with the finding that the mass of total EPS represents up to 80% of the mass of activated sludge (Frølund et al., 1996), it is concluded that the dewatering efficiency of activated (biological) sludge can be increased by degrading the EPS.

It is generally accepted that the strong water retention in sludge is related to the presence of EPSs, which form a negatively charged polymer network. To date, many investigations have been carried out on the role of EPS in bioflocculation, sludge settleability and dewaterability. Apart from that, various studies have reported EPS as 30–50% biodegradable (Wang, Liu, & Tay, 2007) and specific compounds in EPS are known to be recalcitrant to both aerobic and anaerobic activity (Park, Helm, & Novak, 2008).

WASTE ACTIVATED SLUDGE TREATMENT

Current legal constraints, the rising costs and public sensitivity to municipal sludge disposal have provided considerable impetus to explore and develop strategies and technologies for minimization of sludge production. In short, an ideal way to solve the sludge associated problems is to reduce the sludge production in the wastewater treatment rather than the post-treatment of the sludge produced.

The selection for sludge treatment methods mainly based on relatively low energy consumption for example dewatering, mechanical processes such as filter presses or centrifuges are usually used to carry out the process. But, WAS dewatering is a difficult process, mainly because of the colloidal nature of particles and the gel-like structure of the flocculated system in EPS (Dursun & Dentel, 2009). Therefore, an effective dewatering technique requires the attack of the microorganisms which live in aggregates such as films, flocs and sludges.

Stabilization is a favored method to (1) reduce pathogens, (2) eliminate offensive odors, and (3) inhibit, reduce, or eliminate the potential for putrefaction. It is mainly related to the biological reduction of the volatile content and the addition of chemicals to the sludge to render them unsuitable for the survival of microorganisms. Apart from that, stabilization can be used for volume reduction, production of usable gas (methane) and improving the dewaterability of sludge (Eddy, 2003). The principle methods used for stabilization of sludge are alkaline stabilization (usually with lime), aerobic/anaerobic digestion and composting. Anaerobic digestion is a favored stabilization method compared to aerobic digestion, due to its lower cost, lower energy footprint, and moderate performance, especially for stabilization (Appels et al., 2008).

There are different sludge conditioning techniques which have been tested such as salt content modification (B. Jin, Wilen, & Lant, 2004), acid/alkaline treatment (Apul, Atalar, Zorba, & Sanin, 2010; Heng, Chen, & Isa, 2016), chemical oxidation (Heng & Isa, 2016; Heng et al., 2016), sonication (Apul et al., 2010), thermal treatment (Apul et al., 2010), electrochemical treatment (Heng & Isa, 2014; Yuan, Zhu, & Song, 2011) and freeze-thaw treatment (Omerci & Vesilind, 2001). However, over dosage in chemical oxidation, high energy usage in electrochemical treatment or thermal treatment and prolonged reaction time might not only disrupt floc structures and cell membranes but also reduce floc size and deteriorate sludge dewaterabilty. Therefore, an optimum condition could be achieved by considering the abovementioned factors, to destroy the sludge surface and degrade some organic substance by releasing the interstitial water of sludge into solution, improving the sludge disintegration, dewaterability as well as digestibility.

Sludge pretreatment processes that were studied by other researchers are discussed in detail in the following section.

Physicochemical Treatment

Different types of sludge disintegration methods have been tested such as thermal treatment, electrochemical treatment, acid/alkaline treatment, chemical oxidation treatment and sonication treatment.

Thermal Treatment

Thermal treatment was first applied to improve the sludge dewaterability (Haug, Stuckey, Gossett, & Carty, 1978). Extreme temperature can cause the sludge to lose its natural shape (denaturation) and hence allows degradation of the sludge gel structure and release of linked water. Results of Neyens & Baeyens (2003b) revealed that thermal treatment alters the structure of the sludge, reduces the specific resistance and transforms some of the suspended organic solids into soluble compounds. (Everett, 1974) confirmed that solubilization of sludge occurred when the COD concentration increased in the water phase, producing organic macromolecules (amino acids, volatile acids and simple sugars) as a result of thermal hydrolysis.

Semi-pilot-scale experiments done by the authors (Neyens, Baeyens, Weemaes, & Heyder, 2002) demonstrated that thermal hydrolysis most effectively reduces the residual sludge amounts and enhances the dewaterability when performed at temperatures of 120 °C for 60 min reaction time. The dewaterability of the sludge, measured as CST values, decreased compared to the untreated sludge sample. The dried solids (DS) of the dewatered sludge cake increased from 30% (untreated sludge) to approximately 43% after thermal treatment.

Thermal hydrolysis leads to partial solubilization of sludge, which enhances anaerobic digestion, as can be seen in numerous studies on thermal hydrolysis for pretreatment of anaerobic digestion (Bougrier, Delgenès, & Carrère, 2008). Thermal disintegration at high temperatures (>100 °C) was found to be superior in terms of volatile solids (VS) reduction, biogas production and pathogens destruction. However, thermal treatment requires a huge demand of energy and longer contact time to achieve high temperature (Carrere et al., 2010).

However, treatment at excessively high temperatures (higher than 170–190 °C) lead to decreased sludge biodegradability in spite of achieving high solubilization efficiencies. This is usually ascribed to the so called Maillard reactions (Dwyer et al., 2008), involving carbohydrates and amino acids in the formation of melanoidins, which are difficult or impossible to biodegrade (Bougrier et al., 2008). Melanoidins can also increase color in the final effluent (Dwyer et al., 2008).

Electrochemical Treatment

The disintegration of sludge flocs and disruption of microbial cell walls of sludge which lead to the lysis of organic macromolecules associated with microbial cells could also be achieved by electrochemical treatment. The cells of microorganisms in sludge burst and the bound water could be released during the electrolysis process (Song, Zhu, Yuan, Hong, & Ding, 2010). Electrochemical treatment shows great versatility, high removal efficiency, lower temperature requirement and environmental compatibility.

The behavior of anode materials is a major experimental concern during electrochemical process. Various anode materials have been investigated for electrochemical process, including ternary Sn-Pd-Ru oxide coated titanium (SPR), binary oxide-coated titanium Ru-Ti oxide (DSA), PbO_2-coated titanium (PbO_2/Ti), graphite, SnO_2-coated titanium (SnO_2/Ti), iron (Fe) and aluminum (Al). Chiang, Chang, & Wen (1995) found that COD removal efficiencies when using the anode materials followed the order of SPR > DSA > PbO_2/Ti > graphite.

Electrochemical treatment has been applied on industrial effluents which contain bio-refractory organic pollutants such as landfill leachate (Deng & Englehardt, 2006), phenol (Yavuz & Koparal, 2006), and wastewater (Vlyssides, Papaioannou, Loizidoy, Karlis, & Zorpas, 2000; Körbahti, Aktaş, & Tanyolaç, 2007). Although the literature on electrochemical treatment of activated sludge is scarce, the capability of electrochemical treatment to decompose organic macromolecules to smaller molecules observed from above mentioned studies may justify its application in WAS treatment. (Yuan et al., 2011) studied the effect of different types of anode on waste activated sludge dewaterability and sludge disintegration degree by using Ru/Ir or Ti/RuO_2 mesh as anode (10 cm × 10 cm) and Ti/RuO_2 mesh as cathode. They found that Ti/RuO_2 anode performs better in sludge dewaterability at 50 V and 5 min, reducing CST by 35.4%. Also, the EPS concentration had more positive relation with the normalized CST values. Subsequently, the authors concluded that low electrolysis voltages (< 20 V) and short electrolysis time (< 20 min) enhanced sludge dewaterability. However, sCOD increased gradually with the increase in electrolysis voltage and time. In another study which was done by (Song et al., 2010), the sCOD concentration, removals of VS and VSS after electrochemical treatment with a pair of Ti/RuO2 mesh plate electrodes, increased with the increase of electrolysis time and electric power, but decreased when sludge concentration was increased. The optimum conditions for VS and VSS removal to achieve 2.75% and 7.87%, respectively, were pH 10, 30 min electrolysis time, electric power of 5 W and initial sludge concentration of 12.9 g/L.

Although temperature changes resulting from the heat generated during electrolysis may warrant attention during experimentation. According to (Yuan et al., 2011), the observed rise in sludge temperature during electrolysis was insignificant. Therefore, no effort was made to control the sludge temperature.

Alkali Treatment

Alkali treatment is relatively effective in sludge solubilisation, with the order of efficacy being NaOH >KOH>$Mg(OH)_2$ and $Ca(OH)_2$ (Kim et al., 2003). However, too high concentrations of Na^+ or K^+ may cause subsequent inhibition of anaerobic digestion (Mouneimne, Carrère, Bernet, & Delgenès, 2003). Sludge solubilization increase with alkali dose and temperature, with an upper limit (Valo, Carrère, & Delgenès, 2004; Kim et al., 2003; Kim et al., 2009) studied on alkaline pretreatment to enhance the efficiency of hydrogen peroxide oxidation of sludge by evaluations of total solids concentration and particle size distribution. The authors found that total solids removal was 49% and the median diameter of sludge particle decreased from 34.5 µm to 10.8 µm under the conditions of pH 11 and 1800 g H_2O_2/kg TS dose.

Another study showed that alkali treatment by sodium hydroxide aided with hydrogen peroxide achieved 20% of VSS and CST removals, together with a significant increment of sCOD (Heng, Isa, & Han, 2018). The sCOD was increased from 1100 mg/L to about 6000 mg/L.

Chemical Oxidation Treatment

Chemical oxidation typically involves the use of oxidizing agents such as ozone (O_3), hydrogen peroxide (H_2O_2), permanganate ($MnO4^-$), chloride dioxide (ClO_2), chlorine (Cl_2) or hypochlorous acid (HOCl) and oxygen (O_2), to bring about change in the chemical composition of a compound or a group of compounds (Eddy, 2003). In advanced oxidation processes (AOPs), free hydroxyl radical (•OH) is used as a strong oxidant to destroy specific organic constituents and compounds that cannot be oxidized by conventional oxidants such as ozone and chlorine. Ozonation treatment is not favored because the cost of oxidant is high, needs ozone off-gas treatment for ozone destruction and bromate formation. More importantly, ozonation treatment has been proved to produce fewer amounts of hydroxyl radicals compared to AOPs. Thus, it will lower the efficiency of the treatment process.

(Pere, Alen, Viikari, & Eriksson, 1993) indicated that peroxidation enhances the dewaterability of sludge. The effects of temperature, hydrogen peroxide concentration, pH and reaction time on the dewaterability of the sludge were tested by(Neyens et al., 2002; Neyens et al., 2004) and (Neyens & Baeyens, 2003a). The results showed that Fenton's peroxidation can be considered as a useful sludge treatment, yielding a considerable reduction of dried solids (DS) and organic dried solids (ODS) in the filter cake of approximately 20%, an improved dewaterability with a 30% reduction of the sludge volume, and a 30% increase of the cake DS-content when compared with the untreated sludge sample. (Neyens et al., 2002) concluded that peroxidation is able to degrade EPS (polysaccharides and proteins), reducing the water retention properties and promotes flocculation which subsequently reduces the amount of fine flocs, and hence enhances sludge dewaterability.

Tokumura, Sekine, Yoshinari, Znad, & Kawase (2007) studied sludge disintegration using photo-Fenton treatment; varying the initial solids concentration of sludge (2000–4000 mg/L), Fe(II) dosage (1–40 mg/L), H_2O_2 dosage (0–4 g/L), initial pH of sludge (pH 3 and 6.8), and irradiation time (0–45 h). The authors extended the reaction time to investigate the changes in sCOD and H_2O_2 concentration in each experiment and found that the optimum H_2O_2/Fe^{2+} molar ratio was around 100 and pH 3 works much better than unadjusted pH at 6.8 in terms of mineralization of sCOD. (Kim et al., 2009) studied alkaline pretreatment to enhance the efficiency of hydrogen peroxide oxidation of sludge by evaluation of total solids concentration and particle size distribution. The authors found that total solids removal was 49% and the median diameter of sludge particle decreased from 34.5 μm to 10.8 μm at pH 11 and 1800 g H_2O_2/kg TS dose.

Biological Treatment

Aerobic and anaerobic digestion processes are the two main biological treatment methods for the stabilization and reduction of WAS (Liu et al., 2010). Aerobic digestion is applied in small- and medium-sized WWTPs for economic considerations (Andreadakis, 1993). On the other hand, anaerobic digestion is commonly used in large-sized WWTPs to convert WAS to biogas (e.g., methane or hydrogen) or value-added products (e.g., organic acids) (Wang, Deng, Ma, & Ge, 2013). Mesophilic anaerobic digestion is more commonly used than thermophilic digestion because of its higher stability. Thermophilic digestion, because of the process instability, is not listed as one of the processes to further reduce pathogens (USEPA, 1993).

Aerobic Digestion

Aerobic digestion involves the oxidation of biodegradable and microbial cellular matter by aerobic microorganisms resulting in the overall reduction in the mass of sludge and generation of finite amount of stabilized cell mass (McFarland, 2001). In aerobic digestion process, long-term aeration is used to hydrolyze the biodegradable particulate organic matter and converting it into biodegradable soluble organic matter, releasing ammonia-N and phosphate. The biodegradable soluble organic matter thus produced is then converted into water, carbon dioxide and active biomass through the action of heterotrophic bacteria.

Aerobic digestion has many advantages, such as, low capital cost, stabilized sludge with no odor and easy operation. Also, aerobic digestion gives a higher reduction in pathogenic indicator organisms and pathogens than storage of the sludge (Rao, Shankha, & Chen, 1993). Some materials can be degraded under aerobic but not under anaerobic conditions (Subramanian, Kumar, Murthy, & Novak, 2007). However, this process has no energy recovery and is costly regarding the energy costs associated with continued aeration (Bernard & Gray, 2000).

Previous studies reported that aerobic digestion could significantly affect WAS properties or features that influence settleability, flocculability, dewaterability and the downstream treatment efficiency. Murthy & Novak (1999) found that aerobic digestion caused poor dewatering properties and increased biopolymer content. (Murthy & Novak, 1998) also reported that aerobic digestion reduced sludge dewaterability due to the increase in dissolved extracellular polymeric substances (EPS) and caused by the increase of proteins; similar to anaerobic digestion. Moreover, the increase of dissolved EPS could also be attributed to carbohydrates (L.H. Mikkelsen & Keiding, 2002).

Anaerobic Digestion

Anaerobic digestion is recognized as a cost-effective sludge stabilization technique that reduces the amount of sludge solids before final disposal (typically by 30–50% for domestic wastewater sludge) and allows energy to be recovered in the form of methane gas (CH_4) (Eddy, 2003; de la Rubia, Perez, Romero, & Sales, 2006). Although anaerobic digestion is a mature and widely applied technology, the design of digester systems is still generally performed by rule-of-thumb since no tools are currently available for an accurate evaluation of performance. This is mainly due to the fact that the process is not yet fully understood and an optimization of the current technology is still needed. It is clear that a further development and optimization of anaerobic digestion requires additional fundamental knowledge on the mechanisms and information on the microbial community composition (microscale), which in turn can be linked to the macroscale system performance and behavior.

During single stage anaerobic digestion, the hydrolysis of large organic molecules associated with microbial cells is known to be the rate limiting step. Hydrolysis controls the degree of COD conversion to CH_4 of organic solids, as CH_4 cannot be formed faster than the rate at which complex organic matter is converted to acetate and hydrogen gas (Valentini, Garuti, Rozzi, & Tilche, 1997; Li & Noike, 1992). Methanogenesis can be the rate-limiting step if methanogens are washed out when the solids retention time (SRT) is too low. Methanogens are slow growers that are also inhibited by low pH which may develop if organic acids accumulate in the digester. Most organic matter in municipal wastewater sludge is solid and not readily bio-available to fermentative and methanogenic bacteria (Rittmann, Krajmalnik-Brown, & Halden, 2008).

Most anaerobic digesters in WWTP are operated with a solids retention time (SRT) of 15–30 days under mesophilic temperature conditions. Even though shorter SRT can make the volumetric production rate of methane higher; the fractional conversion of COD to CH_4 declines as the extent of hydrolysis declines (Miron, Zeeman, van Lier, & Lettinga, 2000). Furthermore, shortening the SRT too much can lead to washout of methanogens, and the subsequent build-up of organic acids causes the failure of digestion (Stamatelatou, Vavilin, & Lyberatos, 2003). Also, mesophilic anaerobic digestion is more commonly used than thermophilic digestion because of its higher stability and lower cost.

In a conventional mesophilic anaerobic digestion process with an HRT of 20 days, conversion of organics to biogas is typically 25–60% (Bhattacharya, Madura, Walling, & Farrell, 1996). Usually retention time would be maintained at 30–40 days for a VSS degradation of 40–50% (Appels et al., 2008).

Combined Physicochemical and Biological Treatment

Electrochemical and Aerobic Digestion

Electrochemical pretreatment with a pair of Ti/RuO_2 mesh plate electrode was applied to waste activated sludge (WAS) prior to aerobic digestion in a study (Song et al., 2010). In the subsequent aerobic digestion, VS and VSS reductions at solids retention time (SRT) of 17.5 d were 34.3% and 39.6%, respectively. However, an SRT of 23.5 d was required to achieve equivalent reductions for the control run. The application of electrochemical pretreatment on WAS was shown to enhance the subsequent aerobic digestion.

Thermal and Anaerobic Digestion

Thermal hydrolysis for pretreatment of sludge for anaerobic digestion has been widely reported in literature. Most studies report an optimal temperature in the range of 160–180 °C and treatment times from 30 to 60 min. The increase of methane gas (CH_4) production depends on the initial biodegradability of the sludge, with higher impacts on hardly biodegradable sludge (Carrere, Bougrier, Castets, & Delgenes, 2008) and better results on waste activated sludge than on primary sludge (Eskicioglu, Kennedy, & Droste, 2008). Haug et al. (1978) compared the methane gas production of three types of sludge, i.e. activated sludge, primary sludge and mixed sludge, under the same thermal pretreatment conditions (175 °C, 30 min) which were then treated by anaerobic digestion in continuous stirred-tank reactors at 35 °C and HRT 15 d. The authors found that activated sludge showed the highest CH_4 production of 62% after pretreatment but the pretreated primary sludge had no improvement on CH_4 production. Whereas the mixed sludge showed 14% increment on CH_4 production compared to the untreated sludge. (Kim et al., 2003) studied the effect of thermal pretreatment at 121 °C for 30 min on activated sludge, followed by anaerobic digestion at 37 °C for 7 d. The biogas production was increased by 32% from 3657 L/m^3 to 4843 L/m^3 compared to the untreated sludge.

Thermochemical and Anaerobic Digestion

Kim et al. (2003) studied various pretreatments (thermal, chemical, ultrasonic and thermochemical) for enhancing the anaerobic digestion of waste activated sludge in terms of COD solubilization, particle size reduction, soluble protein and biogas production. Under the optimum operating conditions for thermal (121 °C for 30 min), chemical (7 g/L NaOH addition), ultrasonic (42 kHz for 120 min) and thermochemical (121 °C for 30 min and 7 g/L NaOH addition), pretreated WAS were anaerobically digested for 7 d under mesophilic condition at 37 °C to investigate the sCOD reduction, VS reduction and gas production. Thermochemical pretreated WAS showed the greatest effect on sCOD removal and VS reduction, 67.8% and 46.1%, respectively. The production of methane also increased by more than 34.3% compared to the untreated WAS.

Another study showed that a maximum COD solubilization of about 26% was obtained at pH 10 by 5 N of NaOH and incubation at 105 °C for 2 h (Feki, Khoufi, Loukil, & Sayadi, 2015). The pretreated WAS was subsequently treated by anaerobic digestion under mesophilic conditions (37 °C ± 1°C) for 35 d. A significant increase in biogas yield of up to 40% was observed in thermo-alkaline pretreated WAS compared to raw sludge.

(Dogan & Sanin, 2009) found improvement in the dewaterability (measured by CST) by about 22% after anaerobic digestion of pretreated sludge (pH of 12, 160 °C microwave) compared to anaerobic digestion of untreated waste activated sludge.

Electrochemical and Anaerobic Digestion

(Feki et al., 2015) investigated the effect of electrochemical pretreatment on WAS by using a pair of iron plate electrodes with effective area of 16 cm^2. The authors found that the maximum COD solubilization (17.5%) was achieved with the operating conditions of pH 3.0 and current density 25 mA/cm^2. After anaerobic digestion, COD was further removed up to 21.5%, compared to 17.3% with raw sludge. The biogas yield was also improved from 68.2 to 82.6 mL/g VS.

Another study showed that with electrochemical pretreatment by Ti/RuO_2 mesh plate electrodes at 20 V and electrolysis time of 40 min, the VS removal efficiency of 38% was obtained after 28 d of anaerobic digestion while the same VS removal efficiency was achieved after 40 d for the control (Yu et al., 2014). The cumulative biogas production from electrochemical pretreated WAS of 647 L/kgVS was obtained compared to only 396 L/kgVS for control.

(Yuan et al., 2016) studied the effects of electrochemical (Ti/RuO_2 mesh plate electrodes) and sodium hypochlorite (NaClO) pretreatment on the performance of waste activated sludge pilot-scale anaerobic digestion. The VS of pretreated WAS was reduced by 49% and got stabilized 9 d earlier than the control. The biogas yield increased by 1.83 times compared to the control.

Combined Biological Treatments

In order to have anaerobically digested effluent meet the discharge standards, some researchers have focused on combinations of anaerobic and aerobic processes. The aerobic post-treatment of sludge after anaerobic digestion has advantages such as simple design technology and minimization of sludge production (Jenicek, Dohanyos, & Zabranska, 1999). Apart from that, the use of combined anaerobic and aerobic digestion of sludge can eliminate the need for a separate sludge stabilization units (Motta La, 2007).

In combined anaerobic/aerobic systems, the influent of aerobic reactor is pretreated in the anaerobic digester. Thus a large fraction of organic matter is already biodegraded and eliminated. So the residual organic matter from the anaerobic digester requires a lower oxidation capacity in the aerobic digester for nitrification and further degradation of the residual organics. (Castillo, 1997) proposed that combined anaerobic/aerobic digestion of sludge has lower energy consumption and less excess sludge production than a single conventional anaerobic digestion. In a later study by (Tapana, 2000), it was reported that pre-aerobic and post-aerobic treatment of anaerobic sludge reduced VS up to approximately 65%, while a single mesophilic anaerobic digestion had only about 51% VS reduction, at SRT of 15 d. Similar results were obtained by (Pagilla, 2000). In a batch experiment on digestibility of waste activated sludges, (Park, 2006) found that combined anaerobic/aerobic or aerobic/anaerobic digestion of sludge yielded more VS destruction (63%) than single anaerobic digestion process. This implies that with single digestion (be it anaerobic or aerobic), some degradable organic matter remains in the floc. This degradable organic fractions are further degraded when an additional digestion system is applied, which accounts for the additional VS destruction in a combined digestion. Organic solids in mesophilic anaerobically digested sludge, with 30 d SRT, were reduced by 20% through post-aerobic digestion (5 d SRT, 30 °C) (Parravicini, Svardal, & Kroiss, 2004). Later in 2006, in another study, Parravicini, Smidt, Svardal, & Kroiss (2006) investigated further stabilization of digested sludge under both anaerobic and aerobic conditions. They found that aerobically stabilizing/digesting the digested sludge could degrade the residual VSS significantly. Parravicini, Svardal, & Kroiss (2008) reported that post-aerobic digestion (6 d SRT, 36 °C) of mesophilic anaerobically digested sludge (30 d SRT) reduced additional 16% organic solids. (Tapana, 2000), in their research, also demonstrated that the pre-aerobic and post-aerobic treatments reduced COD by 56% while only 44% reduction was obtained in single mesophilic anaerobic digestion. In sequential anaerobic/aerobic digestion of sludge, at 3 day SRT of aerobic digestion, up to 60% to 70% COD removal can be achieved (Kumar, 2006).

COMPARISON BASED ON REMOVAL EFFICIENCIES AND BENEFITS

WAS treatment can be classified into four major groups – physicochemical treatment, biological treatment, combined physicochemical and biological treatment and combined biological treatments.

Hydrolysis has been identified as the rate-limiting step in anaerobic digestion. Hence, various pretreatments can be applied to accelerate the digestion and enhance the production of biogas at the same time. Efficient sludge pretreatment can alter sludge structure and physical states of water in sludge, and change the bound water in sludge into free water, and eventually improve sludge dewatering characteristics to achieve a high solids content. Studies reporting the effects of treatment methods on WAS are summarized in Table 1.

The literature showed parameters such as solids removal, degree of COD solubilization, CST, EPS and biogas (methane) production were measured to determine the efficiency of the particular treatment. Diameter of sludge particle was also measured in certain cases to examine the degree of sludge disintegration but this measurement is not popular and will not reflect the sludge digestibility improvement. Most studies reported a positive impact of treatment methods on WAS i.e. lowered VS concentration, increased biogas production, and enhanced COD reduction. CST results, however, do not always show a clear trend. The reading of CST might show a negative effect (higher than the raw sludge sample) due to a number of factors, for instance, reaction time and insufficient dosage of chemicals used in the treatment. In general, single stage treatment is less efficient as compared to combined treatment, resulting in poor solids degradation.

The major drawbacks of sludge stabilization by biological processes (aerobic and anaerobic digestion) are long retention times (typical 15–30 days) and low digestion efficiency for organic matter, which are partly caused by the inability of the microorganisms to degrade certain biomass components efficiently in each sludge stabilization (Jin, Li, Mahar, Wang, & Nie, 2009). The organics reduction through diges-

Table 1. Overview of WAS treatment methods

Treatment Method	Treatment Condition	Comments	Reference
Thermal	121 °C, 30 min	30% increase of VS reduction	Kim et al., 2003
	170 °C, 15 min	59% increase of TS reduction 92% higher gas production	Valo et al., 2004
	70–134 °C, 90 min–9 h	50% increase of biogas production at 70 °C and 9 h	Climent et al., 2007
	135–190 °C	25% increase of methane production at 190 °C	Bougrier, Delgenès, & Carrère, 2007
	121 °C, 15 min	EPS and sCOD solubilization occurred at higher temperature	Sun et al., 2016
Electrochemical	Ti/RuO_2, pH 10, 30 min electrolysis time, 5 W, 12.9 g/L sludge concentration	2.75% and 7.87% increase of VS and VSS removal, respectively	Song et al., 2010
	Ti/RuO_2 anode, 50V, 5 min electrolysis time	35.4% increase of CST reduction	Yuan et al., 2011
	Ti/RuO_2 electrodes, hypochlorite	38.1% increase of organic matter solubilization	Yu et al., 2014
Alkali (thermal, H_2O_2)	121 °C, 7 g/L NaOH, 30 min	85.4% COD solubilization 34.3% more methane production over the control	Kim et al., 2003
	pH 11, 1800 g H_2O_2/kg TS	49% increase of TS removal Median diameter of sludge particle decreased from 34.5 μm to 10.8 μm	Kim et al., 2009
Chemical Oxidation	Ozonation, 0.1 g O_3/g TS	20–25% increase of COD solubilisation Negative effect on CST	Bougrier, Albasi, Delgenès, & Carrère, 2006
	Fenton, pH 3, 50 g H_2O_2/kg DS, 0.07 g Fe^{2+}/g H_2O_2	26.6% increase of VS reduction	Dewil, Baeyens, & Neyens, 2005
	Fenton, pH 3, 60 g H_2O_2/kg DS, 0.067 g Fe^{2+}/g H_2O_2	25.4% increase of VS reduction	Kaynak & Filibeli, 2008
	Solar photo-Fenton, pH 3, 40 mg/L of ferrous ions and 4000 mg/L of H_2O_2 dosage	40% MLSS removal	Tokumura, Katoh, Katoh, Znad, & Kawase, 2009
Electrochemical + Aerobic Digestion	Ti/RuO_2, pH 10, 30 min electrolysis time, 5 W, 12.9 g/L sludge concentration, 17.5 d SRT	34.3% and 39.6% increase of VS and VSS removal, respectively	Song et al., 2010
Electrochemical + Anaerobic Digestion	Ti/RuO_2 electrodes, hypochlorite	63.4% higher cumulative biogas production than in control (647 L/ kg•VS)	Yu et al., 2014
	Ti/RuO_2 electrodes, 0.6% (v/v) of hypochlorite, pH 8, 20 V, 40 min	61.1% increase of methane production	Ye, Yuan, Lou, & Zhu, 2016
Fenton pretreated + Anaerobic Digestion	pH 3, 60 g H_2O_2/kg TS, 0.07 g Fe^{2+}/g H_2O_2	Methane production by 15% (average production rate of 0.496 m^3 CH_4/ kg·VS_{fed})	Pilli, More, Yan, Tyagi, & Surampalli, 2016

tion of WAS from biological wastewater treatment plants is limited to about 50%, even after a residence time of 20 days. Therefore, pretreatment (sludge disintegration and dewaterability) studies are becoming popular in order to enhance the sludge digestibility. These methods disrupt cell walls which results in lysis or disintegration of sludge cells. Slowly degradable, particulate organic material is converted to low molecular weight and readily biodegradable compounds, thus bypassing the rate-limiting hydrolysis stage. The bottleneck of the sludge handling system is the dewatering operation. Advanced sludge treatment processes have been developed and investigated in order to improve sludge dewatering and facilitate the ultimate disposal. Possible pretreatments include mechanical, thermal, chemical and biological processes. Although these methods may have high dewatering potentials, their application has been limited by factors including sludge volume increase and high energy requirement for operation, as well as complexity of implementation.

Combinations of two or more treatments have been proved to be more efficient and effective than individual treatment in improving the effluent quality. This could be due to the fact that a two or more-step treatment has the ability to synergize the advantages of individual treatments, while overcoming their respective limitations. The pretreatment for WAS is commonly followed by biological process. Thus, there exists the possibility of performance improvement by adding an appropriate thermal, mechanical or chemical pretreatment leading to the breakage of flocs and particulate solubilization, which accelerates the hydrolysis step (Gianico, Braguglia, Cesarini, & Mininni, 2013; Braguglia, Gianico, Gallipoli, & Mininni, 2015), and improve the EPS degradation (dewaterability). Subsequently, the microbes are able to remove the remaining (or nearly total) COD, suspended solids, nitrogen, phosphorus and other constituents more easily. In addition, VS removal efficiency of 38% is recommended for the assessment of sewage sludge stabilization according to USEPA Standards for the use or disposal of sewage sludge. When VS removal efficiency meets the USEPA standard, it might save almost 30% capital cost by the reduction of reactor.

ECONOMICAL FEASIBILITY OF WAS TREATMENT PROCESSES

As shown in Table 2, thermal hydrolytic pretreatment followed by anaerobic digestion appeared to be the most economical feasible method to treat WAS, requiring only 0.04 USD in treating 1 m^3 of WAS. It is interesting to note that the overall treatment cost of thermal hydrolytic pretreatment followed by anaerobic digestion was lower than the cost required for thermal hydrolytic of WAS. This was due to the production of methane gas during the anaerobic digestion of WAS which can be used to generate heat and electricity. On the other hand, the totally opposite condition was observed for aerobic digestion by electrochemical pretreatment. Owing to the inability of aerobic process to generate methane gas and at the same time energy was required for the aerobic digestion, the overall cost of electrolysis pretreatment followed by aerobic digestion was higher than other treatment processes. However, the efficiency of each treatment process in terms of sludge reduction was incomparable, therefore the net cost which includes sludge handling and disposal after the treatment could not be justified.

Table 2. Cost for various WAS treatment process

Treatment Process	Energy Required, kWh/m³	*Treatment Cost, USD/ m³	References
Dark Fenton	0.02	**0.50	(Banu et al., 2019)
Combined dark Fenton and disperser	1.63	**0.13	(Banu et al., 2019)
Thermal hydrolytic	0.76	0.17	(Vicentin, Fdz-Polanco, & Fdz-Polanco, 2019)
Thermal hydrolytic pretreatment followed by anaerobic digestion	0.16	0.04	(Vicentin et al., 2019)
Electrolysis	5	1.15	(Song et al., 2010)
Electrolysis pretreatment followed by aerobic digestion	14	3.22	(Song et al., 2010)

*Treatment costs were calculated based on electricity price at 0.23USD/kWh (J. R. Banu et al., 2019)

**Treatment costs were included energy and chemical cost.

CONCLUSION

Sludge disintegration can be investigated through the changes of solids and sCOD concentrations, whereas sludge dewaterability can be investigated through the changes of CST. Since most waste activated sludge microorganisms would be killed and oxidized to organic substances, the effect of sludge reduction were determined in term of changes in sCOD. Therefore, the main goal of applying pretreatment is to remove sludge solids (VSS), improve the concentration of sCOD (maximize biogas production in anaerobic digestion) and reduce CST (improve sludge dewaterability). At the same time, EPS concentration was measured as monitoring purpose and kept as low as possible to prevent the adverse effect on the sludge dewaterability.

Presumably, an effective pretreatment system for WAS is able to improve the sludge disintegration and digestibility as well as enhancing biogas production.

ACKNOWLEDGMENT

The authors are thankful to the Universiti Teknologi PETRONAS, Perak, Malaysia and Universiti Tunku Abdul Rahman, Selangor, Malaysia for providing facilities and support for this research.

REFERENCES

Andreadakis, A. D. (1993). Physical and chemical properties of activated sludge floc. *Water Research*, *27*(12), 1707–1714. doi:10.1016/0043-1354(93)90107-S

Appels, L., Baeyens, J., Degreve, J., & Dewil, R. (2008). Principles and potential of the anaerobic digestion of waste-activated sludge. *Progress in Energy and Combustion Science*, *34*(6), 755–781. doi:10.1016/j.pecs.2008.06.002

Apul, O. G., Atalar, I., Zorba, G. T., & Sanin, F. D. (2010). The dewaterability of disintegrated sludge samples before and after anaerobic digestion. *Drying Technology*, *28*(7), 901–909. doi:10.1080/07373937.2010.490764

Banu, J. R., Sharmila, V. G., Devi, M. G., Kumar, S. A., Kumar, G., Nguyen, D. D., & Saratale, G. D. (2019). Cost effective sludge reduction using synergetic effect of dark fenton and disperser treatment. *Journal of Cleaner Production*, *207*, 261–270. doi:10.1016/j.jclepro.2018.09.252

Bernard, S., & Gray, N. F. (2000). Aerobic digestion of pharmaceutical and domestic wastewater sludges at ambient temperature. *Water Research*, *34*(3), 725–734. doi:10.1016/S0043-1354(99)00234-1

Bhattacharya, S. K., Madura, R. L., Walling, D. A., & Farrell, J. B. (1996). Volatile solids reduction in two-phase and conventional anaerobic sludge digestion. *Water Research*, *30*(5), 1041–1048. doi:10.1016/0043-1354(95)00252-9

Bolto, B., & Gregory, J. (2007). Organic polyelectrolytes in water treatment. *Water Research*, *41*(11), 2301–2324. doi:10.1016/j.watres.2007.03.012 PMID:17462699

Bougrier, C., Albasi, C., Delgenès, J. P., & Carrère, H. (2006). Effect of ultrasonic, thermal and ozone pre-treatments on waste activated sludge solubilisation and anaerobic biodegradability. *Chemical Engineering and Processing*, *45*(8), 711–718. doi:10.1016/j.cep.2006.02.005

Bougrier, C., Delgenès, J. P., & Carrère, H. (2007). Impacts of thermal pre-treatments on the semi-continuous anaerobic digestion of waste activated sludge. *Biochemical Engineering Journal*, *34*(1), 20–27. doi:10.1016/j.bej.2006.11.013

Bougrier, C., Delgenès, J. P., & Carrère, H. (2008). Effects of thermal treatments on five different waste activated sludge samples solubilisation, physical properties and anaerobic digestion. *Chemical Engineering Journal*, *139*(2), 236–244. doi:10.1016/j.cej.2007.07.099

Braguglia, C. M., Gianico, A., Gallipoli, A., & Mininni, G. (2015). The impact of sludge pre-treatments on mesophilic and thermophilic anaerobic digestion efficiency: Role of the organic load. *Chemical Engineering Journal*, *270*, 362–371. doi:10.1016/j.cej.2015.02.037

Campbell, H. W. (2000). Sludge management-future issues and trends. *Water Science and Technology*, *41*(8), 1–8. doi:10.2166/wst.2000.0135

Carrere, H., Bougrier, C., Castets, D., & Delgenes, J. P. (2008). Impact of initial biodegradability on sludge anaerobic digestion enhancement by thermal pretreatment. *Journal of Environmental Science and Health. Part A, Toxic/Hazardous Substances & Environmental Engineering*, *43*(13), 1551–1555. doi:10.1080/10934520802293735 PMID:18821243

Carrere, H., Dumas, C., Battimelli, A., Batstone, D. J., Delgenes, J. P., Steyer, J. P., & Ferrer, I. (2010). Pretreatment methods to improve sludge anaerobic degradability: A review. *Journal of Hazardous Materials*, *183*(1-3), 1–15. doi:10.1016/j.jhazmat.2010.06.129 PMID:20708333

Castillo, A., Cecchf, F., & Alvarez, J. M. (1997). A combined anaerobic aerobic system to treat domestic sewage in coastal areas. *Water Research*, *31*(12), 3057–3063. doi:10.1016/S0043-1354(96)00064-4

Chiang, L.-C., Chang, J.-E., & Wen, T.-C. (1995). Indirect oxidation effect in electrochemical oxidation treatment of landfill leachate. *Water Research*, *29*(2), 671–678. doi:10.1016/0043-1354(94)00146-X

Climent, M., Ferrer, I., Baeza, M. M., Artola, A., Vazquez, F., & Font, X. (2007). Effects of thermal and mechanical pretreatments of secondary sludge on biogas production under thermophilic conditions. *Chemical Engineering Journal*, *133*(1-3), 335–342. doi:10.1016/j.cej.2007.02.020

Comte, S., Guibaud, G., & Baudu, M. (2007). Effect of extraction method on EPS from activated sludge: An HPSEC investigation. *Journal of Hazardous Materials*, *140*(1-2), 129–137. doi:10.1016/j.jhazmat.2006.06.058 PMID:16879910

de la Rubia, M. A., Perez, M., Romero, L. I., & Sales, D. (2006). Effect of solids retention time (SRT) on pilot scale anaerobic thermophilic sludge digestion. *Process Biochemistry*, *41*(1), 79–86. doi:10.1016/j.procbio.2005.03.073

Deng, Y., & Englehardt, J. D. (2006). Treatment of landfill leachate by the Fenton process. *Water Research*, *40*(20), 3683–3694. doi:10.1016/j.watres.2006.08.009 PMID:17045628

Dewil, R., Baeyens, J., & Neyens, E. (2005). Fenton peroxidation improves the drying performance of waste activated sludge. *Journal of Hazardous Materials*, *117*(2–3), 161–170. doi:10.1016/j.jhazmat.2004.09.025 PMID:15629575

Dogan, I., & Sanin, F. D. (2009). Alkaline solubilization and microwave irradiation as a combined sludge disintegration and minimization method. *Water Research*, *43*(8), 2139–2148. doi:10.1016/j.watres.2009.02.023 PMID:19285332

Dursun, D., & Dentel, S. K. (2009). Toward the conceptual and quantitative understanding of biosolids conditioning: The gel approach. *Water Science and Technology*, *59*(9), 1679–1685. doi:10.2166/wst.2009.202 PMID:19448301

Dwyer, J., Starrenburg, D., Tait, S., Barr, K., Batstone, D. J., & Lant, P. (2008). Decreasing activated sludge thermal hydrolysis temperature reduces product colour, without decreasing degradability. *Water Research*, *42*(18), 4699–4709. doi:10.1016/j.watres.2008.08.019 PMID:18834611

Eddy, M. (2003). Wastewater engineering: treatment and reuse (4th ed.). New York, NY: McGraw-Hill.

Eskicioglu, C., Kennedy, K. J., & Droste, R. L. (2008). Initial examination of microwave pretreatment on primary, secondary and mixed sludges before and after anaerobic digestion. *Water Science and Technology*, *57*(3), 311–317. doi:10.2166/wst.2008.010 PMID:18309206

Everett, J. G. (1974). The effect of pH on the heat treatment of sewage sludges. *Water Research*, *8*(11), 899–906. doi:10.1016/0043-1354(74)90104-3

Feki, E., Khoufi, S., Loukil, S., & Sayadi, S. (2015). Improvement of anaerobic digestion of waste-activated sludge by using H_2O_2 oxidation, electrolysis, electro-oxidation and thermo-alkaline pretreatments. *Environmental Science and Pollution Research International*, *22*(19), 14717–14726. doi:10.100711356-015-4677-2 PMID:25982985

Feng, X., Deng, J., Lei, H., Bai, T., Fan, Q., & Li, Z. (2009). Dewaterability of waste activated sludge with ultrasound conditioning. *Bioresource Technology*, *100*(3), 1074–1081. doi:10.1016/j.biortech.2008.07.055 PMID:18783942

Flemming, H. C. (1996). The forces that keep biofilms together. In W. Sand (Ed.), Biodeterioration and Biodegradation (Vol. 133, pp. 311–316). Weinheim, Germany: Dechema Monographs, VCH Verlagsgesellschaft.

Flemming, H. C., & Wingender, J. (2001). Relevance of microbial extracellular polymeric substances (EPSs)-Part I: Structural and ecological aspects. *Water Science and Technology*, *43*(6), 1–8. doi:10.2166/wst.2001.0326 PMID:11381954

Forster, C. F., & Lewin, D. C. (1972). Polymer interactions at activated sludge surfaces. *Effluent and Water Treatment Journal*, *12*, 520–525.

Frohlund, B., Griebe, T., & Nielsen, P. H. (1995). Enzyme activity in activated-sludge floc matrix. *Applied Microbiology and Biotechnology*, *43*, 755–761. doi:10.1007/BF00164784 PMID:7546613

Frølund, B., Palmgren, R., Keiding, K., & Nielsen, P. H. (1996). Extraction of extracellular polymers from activated sludge using a cation exchange resin. *Water Research*, *30*(8), 1749–1758. doi:10.1016/0043-1354(95)00323-1

Gianico, A., Braguglia, C. M., Cesarini, R., & Mininni, G. (2013). Reduced temperature hydrolysis at 134 °C before thermophilic anaerobic digestion of waste activated sludge at increasing organic load. *Bioresource Technology*, *143*, 96–103. doi:10.1016/j.biortech.2013.05.069 PMID:23792658

Haug, R. T., Stuckey, D. C., Gossett, J. M., & Carty, P. L. M. (1978). Effect of thermal pretreatment on digestibility and dewaterability of organic sludges. *Journal - Water Pollution Control Federation*, *50*, 73–85.

Heng, G. C., Chen, K. W., & Isa, M. H. (2016, Aug. 15-17). *Enhancement of waste activated sludge disintegration and dewaterability by H_2O_2 oxidation.* Paper presented at the 3rd International Conference on Civil, Offshore and Environmental Engineering, ICCOEE 2016, KLCC.

Heng, G. C., & Isa, M. H. (2014). Electrochemical disintegration of activated sludge using Ti/RuO_2 anode. *Applied Mechanics and Materials*, *567*, 44–49. doi:10.4028/www.scientific.net/AMM.567.44

Heng, G. C., & Isa, M. H. (2016). Enhancement of activated sludge disintegration and dewaterability by Fenton process. *IOP Conference Series: Earth and Environmental Science*, *36*, 012018. 10.1088/1755-1315/36/1/012018

Heng, G. C., Isa, M. H., & Han, L. M. (2018). Effect of H2O2 oxidation/alkaline hydrolysis on waste activated sludge disintegration and dewaterability. In *Proceedings International Conference on Civil and Environmental Engineering*. 10.1051/e3sconf/20186505021

Jenicek, P., Dohanyos, M., & Zabranska, J. (1999). Combined anaerobic treatment of wastewaters and sludges. *Water Science and Technology*, *40*(1), 85–91. doi:10.2166/wst.1999.0019

Jin, B., Wilen, B. M., & Lant, P. (2004). Impacts of morphological, physical and chemical properties of sludge flocs on dewaterability of activated sludge. *Chemical Engineering Journal*, *98*(1-2), 115–126. doi:10.1016/j.cej.2003.05.002

Jin, Y., Li, H., Mahar, R. B., Wang, Z., & Nie, Y. (2009). Combined alkaline and ultrasonic pretreatment of sludge before aerobic digestion. *Journal of Environmental Sciences (China)*, *21*(3), 279–284. doi:10.1016/S1001-0742(08)62264-0 PMID:19634436

Kaynak, G. E., & Filibeli, A. (2008). Assessment of Fenton process as a minimization technique for biological sludge: Effects on anaerobic sludge bioprocessing. *Journal of Residuals Science and Technology*, *5*, 151–160.

Kim, J., Park, C., Kim, T. H., Lee, M., Kim, S., Kim, S. W., & Lee, J. (2003). Effects of various pretreatments for enhanced anaerobic digestion with waste activated sludge. *Journal of Bioscience and Bioengineering*, *95*(3), 271–275. doi:10.1016/S1389-1723(03)80028-2 PMID:16233404

Kim, T.-H., Lee, S.-R., Nam, Y.-K., Yang, J., Park, C., & Lee, M. (2009). Disintegration of excess activated sludge by hydrogen peroxide oxidation. *Desalination*, *246*(1-3), 275–284. doi:10.1016/j.desal.2008.06.023

Körbahti, B. K., Aktaş, N., & Tanyolaç, A. (2007). Optimization of electrochemical treatment of industrial paint wastewater with response surface methodology. *Journal of Hazardous Materials*, *148*(1-2), 83–90. doi:10.1016/j.jhazmat.2007.02.005 PMID:17374443

Kumar, N. (2006). *Sequential anaerobic-aerobic digestion: A new process technology for biosolids product quality improvement.* (MS thesis), Virginia Tech.

Li, J., Liu, L., Liu, J., Ma, T., Yan, A., & Ni, Y. (2016). Effect of adding alum sludge from water treatment plant on sewage sludge dewatering. *Journal of Environmental Chemical Engineering*, *4*(1), 746–752. doi:10.1016/j.jece.2015.07.021

Li, Y.-Y., & Noike, T. (1992). Upgrading of anaerobic digestion of waste activated sludge by thermal pretreatment. *Water Science and Technology*, *26*(3-4), 857–866. doi:10.2166/wst.1992.0466

Liu, X. M., Sheng, G. P., Luo, H. W., Zhang, F., Yuan, S. J., Xu, J., ... Yu, H. Q. (2010). Contribution of extracellular polymeric substances (EPS) to the sludge aggregation. *Environmental Science & Technology*, *44*(11), 4355–4360. doi:10.1021/es9016766 PMID:20446688

Lü, F., Wang, F., Shao, L., & He, P. (2017). Deciphering pretreatment-induced repartition among stratified extracellular biopolymers and its effect on anaerobic biodegradability and dewaterability of waste activated sludge. *Journal of Environmental Chemical Engineering*, *5*(3), 3014–3023. doi:10.1016/j.jece.2017.05.047

McFarland, M. J. (2001). *Biosolids Engineering*. New York, NY: McGraw-Hill.

Mikkelsen, L. H., & Keiding, K. (2002). Physico-chemical characteristics of full-scale sewage sludges with implications to dewatering. *Water Research*, *36*(10), 2451–2462. doi:10.1016/S0043-1354(01)00477-8 PMID:12153011

Mininni, G., Laera, G., Bertanza, G., Canato, M., & Sbrilli, A. (2015). Mass and energy balances of sludge processing in reference and upgraded wastewater treatment plants. *Environmental Science and Pollution Research International*, *22*(10), 7203–7215. doi:10.100711356-014-4013-2 PMID:25598155

Miron, Y., Zeeman, G., van Lier, J. B., & Lettinga, G. (2000). The role of sludge retention time in the hydrolysis and acidification of lipids, carbohydrates and proteins during digestion of primary sludge in CSTR systems. *Water Research*, *34*(5), 1705–1713. doi:10.1016/S0043-1354(99)00280-8

Motta La, E. J., Silva, E., Bustillos, A., Padron, H., & Luque, J. (2007). Combined anaerobic/aerobic secondary municipal wastewater treatment: pilot-plant demonstration of the UASB/aerobic solids contact system. *Journal of Environmental Engineering*, *133*(4), 397–403. doi:10.1061/(ASCE)0733-9372(2007)133:4(397)

Mouneimne, A. H., Carrère, H., Bernet, N., & Delgenès, J. P. (2003). Effect of saponification on the anaerobic digestion of solid fatty residues. *Bioresource Technology*, *90*(1), 89–94. doi:10.1016/S0960-8524(03)00091-9 PMID:12835063

Murthy, S. N., & Novak, J. T. (1998). Effects of potassium ion on sludge settling, dewatering and effluent properties. *Water Science and Technology*, *37*(4-5), 317–324. doi:10.2166/wst.1998.0655

Murthy, S. N., & Novak, J. T. (1999). Factors affecting floc properties during aerobic digestion: Implications for dewatering. *Water Environment Research*, *71*(2), 197–202. doi:10.2175/106143098X121879

Neyens, E., & Baeyens, J. (2003a). A review of classic Fenton's peroxidation as an advanced oxidation technique. *Journal of Hazardous Materials*, *B98*(1-3), 33–50. doi:10.1016/S0304-3894(02)00282-0 PMID:12628776

Neyens, E., & Baeyens, J. (2003b). A review of thermal sludge pre-treatment processes to improve dewaterability. *Journal of Hazardous Materials*, *98*(1-3), 51–67. doi:10.1016/S0304-3894(02)00320-5 PMID:12628777

Neyens, E., Baeyens, J., & Dewil, R. (2004). Advanced sludge treatment affects extracellular polymeric substances to improve activated sludge dewatering. *Journal of Hazardous Materials*, *106*(2–3), 83–92. doi:10.1016/j.jhazmat.2003.11.014 PMID:15177096

Neyens, E., Baeyens, J., Weemaes, M., & Heyder, B. D. (2002, Aug. 25–29). *Treatment of sewage sludge: The economy of using advanced sludge treatment methods for sewage sludge.* Paper presented at the CHISA, Prague.

Omerci, B., & Vesilind, P. A. (2001). Effect of dissolved organic material and cations of freezeethaw conditioning of activated and alum sludges. *Water Research*, *25*, 4299–4306.

Pagilla, K. R., Kim, H., & Tapana, C. (2000). Aerobic thermophilic and anaerobic mesophilic treatment of swine waste. *Water Research*, *34*(10), 2747–2753. doi:10.1016/S0043-1354(00)00012-9

Park, C., Abu-Orf, M. M., & Novak, J. T. (2006). The digestibility of waste activated sludges. *Water Environment Research*, *78*(1), 59–68. doi:10.2175/106143005X84521 PMID:16553167

Park, C., Helm, R. F., & Novak, J. T. (2008). Investigating the fate of activated sludge extracellular proteins in sludge digestion using sodium dodecyl sulfate polyacrylamide gel electrophoresis. *Water Environment Research*, *80*(12), 2219–2227. doi:10.2175/106143008X325791 PMID:19146099

Parravicini, V., Smidt, E., Svardal, K., & Kroiss, H. (2006). Evaluating the stabilization degree of digested sewage sludge: Investigations at four municipal wastewater treatment plants. *Water Science and Technology*, *53*(8), 81–90. doi:10.2166/wst.2006.238 PMID:16784192

Parravicini, V., Svardal, K., & Kroiss, H. (2004). *Further stabilisation of digested sewage sludge under anaerobic or aerobic conditions*. Paper presented at the 4th IWA World Water Congress and Exhibition, Marrakech, Morocco.

Parravicini, V., Svardal, K., & Kroiss, H. (2008). Post-aeration of anaerobically digested sewage sludge for advanced COD and nitrogen removal: Results and cost-benefit analysis at large-scale. *Water Science and Technology*, *57*(7), 1087–1094. doi:10.2166/wst.2008.211 PMID:18441437

Pere, J., Alen, R., Viikari, L., & Eriksson, L. (1993). Characterization and dewatering of activated sludge from the pulp and paper industry. *Water Science and Technology*, *28*(1), 193–201. doi:10.2166/wst.1993.0046

Pilli, S., More, T. T., Yan, S., Tyagi, R. D., & Surampalli, R. Y. (2016). Fenton pre-treatment of secondary sludge to enhance anaerobic digestion: Energy balance and greenhouse gas emissions. *Chemical Engineering Journal*, *283*, 285–292. doi:10.1016/j.cej.2015.07.056

Rao, Y. S., Shankha, K. B., & Chen, J. C. (1993). Microbiological stabilization of sludge by aerobic digestion and storage. *Journal of Environmental Engineering*, 119.

Rittmann, B. E., Krajmalnik-Brown, R., & Halden, R. U. (2008). Pre-genomic, genomic and post-genomic study of microbial communities involved in bioenergy. *Nature Reviews. Microbiology*, *6*(8), 604–612. doi:10.1038/nrmicro1939 PMID:18604223

Sanin, F. D., & Vesilind, P. A. (1994). Effect of centrifugation on the removal of extracellular polymers and physical properties of activated sludge. *Water Science and Technology*, *30*(8), 117–127. doi:10.2166/wst.1994.0394

Sewage. (2016). Retrieved from https://www.iwk.com.my/do-you-know/sewage-treatment-system

Song, L. J., Zhu, N. W., Yuan, H. P., Hong, Y., & Ding, J. (2010). Enhancement of waste activated sludge aerobic digestion by electrochemical pre-treatment. *Water Research*, *44*(15), 4371–4378. doi:10.1016/j.watres.2010.05.052 PMID:20580054

Stamatelatou, K., Vavilin, V., & Lyberatos, G. (2003). Performance of a glucose fed periodic anaerobic baffled reactor under increasing organic loading conditions. 1: Experimental results. *Bioresource Technology*, *88*(2), 131–136. doi:10.1016/S0960-8524(02)00276-6 PMID:12576006

Subramanian, S., Kumar, N., Murthy, S., & Novak, J. T. (2007). Effect of anaerobic digestion and anaerobic/aerobic digestion processes on sludge dewatering. *Journal of Residuals Science and Technology*, *4*, 17–23.

Sun, J., Guo, L., Li, Q., Zhao, Y., Gao, M., She, Z., & Wang, G. (2016). Structural and functional properties of organic matters in extracellular polymeric substances (EPS) and dissolved organic matters (DOM) after heat pretreatment with waste sludge. *Bioresource Technology*, *219*, 614–623. doi:10.1016/j.biortech.2016.08.042 PMID:27544911

Tapana, C., & Pagilla, K. R. (2000). Aerobic thermophilic and anaerobic mesophilic treatment of sludge. *Journal of Environmental Engineering*, 790–795.

Tokumura, M., Katoh, H., Katoh, T., Znad, H. T., & Kawase, Y. (2009). Solubilization of excess sludge in activated sludge process using the solar photo-Fenton reaction. *Journal of Hazardous Materials*, *162*(2-3), 1390–1396. doi:10.1016/j.jhazmat.2008.06.026 PMID:18639985

Tokumura, M., Sekine, M., Yoshinari, M., Znad, H. T., & Kawase, Y. (2007). Photo-Fenton process for excess sludge disintegration. *Process Biochemistry*, *42*(4), 627–633. doi:10.1016/j.procbio.2006.11.010

Tomei, M. C., Bertanza, G., Canato, M., Heimersson, S., Laera, G., & Svanström, M. (2016). Techno-economic and environmental assessment of upgrading alternatives for sludge stabilization in municipal wastewater treatment plants. *Journal of Cleaner Production*, *112*, 3106–3115. doi:10.1016/j.jclepro.2015.10.017

Urbain, V., Block, J. C., & Manem, J. (1993). Bioflocculation in activated sludge: An analytic approach. *Water Research*, *27*(5), 829–838. doi:10.1016/0043-1354(93)90147-A

USEPA. (1993). *Standards for the Use or Disposal of Sewage Sludge 40CFR PART 503*. Washington, DC: United States Environmental Protection Agency.

Valentini, A., Garuti, G., Rozzi, A., & Tilche, A. (1997). Anaerobic degradation kinetics of particulate organic matter: A new approach. *Water Science and Technology*, *36*(6-7), 239–246. doi:10.2166/wst.1997.0596

Valo, A., Carrère, H., & Delgenès, J. P. (2004). Thermal, chemical and thermo-chemical pretreatment of waste activated sludge for anaerobic digestion. *Journal of Chemical Technology and Biotechnology (Oxford, Oxfordshire)*, *79*(11), 1197–1203. doi:10.1002/jctb.1106

Vicentin, R., Fdz-Polanco, F., & Fdz-Polanco, M. (2019). Energy integration in wastewater treatment plants by anaerobic digestion of urban waste: a process design and simulation study. *International Journal of Chemical Engineering, 2019, 2621048*. doi:10.1155/2019/2621048

Vlyssides, A. G., Papaioannou, D., Loizidoy, M., Karlis, P. K., & Zorpas, A. A. (2000). Testing an electrochemical method for treatment of textile dye wastewater. *Waste Management (New York, N.Y.)*, *20*(7), 569–574. doi:10.1016/S0956-053X(00)00028-3

Wang, H., Deng, H., Ma, L., & Ge, L. (2013). Influence of operating conditions on extracellular polymeric substances and surface properties of sludge flocs. *Carbohydrate Polymers*, *92*(1), 510–515. doi:10.1016/j.carbpol.2012.09.055 PMID:23218328

Wang, X., Zhang, L., Peng, Y., Zhang, Q., Li, J., & Yang, S. (2019). Enhancing the digestion of waste activated sludge through nitrite addition: Insight on mechanism through profiles of extracellular polymeric substances (EPS) and microbial communities. *Journal of Hazardous Materials*, *369*, 164–170. doi:10.1016/j.jhazmat.2019.02.023 PMID:30776599

Wang, Z.-W., Liu, Y., & Tay, J.-H. (2007). Biodegradability of extracellular polymeric substances produced by aerobic granules. *Applied Microbiology and Biotechnology*, *74*(2), 462–466. doi:10.100700253-006-0686-x PMID:17103158

Wilén, B. M., Jin, B., & Lant, P. (2003). The influence of key chemical constituents in activated sludge on surface and flocculating properties. *Water Research*, *37*(9), 2127–2139. doi:10.1016/S0043-1354(02)00629-2 PMID:12691899

Wingender, J., Jäger, K. E., & Flemming, H.-C. (1999). Interaction between extracellular polysaccharides and enzymes. In J. Wingender & T. R. Neu (Eds.), *Microbial extracellular polymeric substances* (pp. 1–19). Heidelberg, Germany: Springer.

Xu, Q., Li, X., Ding, R., Wang, D., Liu, Y., Wang, Q., ... Li, H. (2017). Understanding and mitigating the toxicity of cadmium to the anaerobic fermentation of waste activated sludge. *Water Research*, *124*, 269–279. doi:10.1016/j.watres.2017.07.067 PMID:28772139

Yavuz, Y., & Koparal, A. S. (2006). Electrochemical oxidation of phenol in a parallel plate reactor using ruthenium mixed metal oxide electrodes. *Journal of Hazardous Materials*, *136*(2), 296–302. doi:10.1016/j.jhazmat.2005.12.018 PMID:16427192

Ye, C., Yuan, H., Lou, Z., & Zhu, N. (2016). Combined electrochemical and hypochlorite pretreatment for improving solubilization and anaerobic digestion of waste activated sludge: Effect of hypochlorite dosage. *Energy & Fuels*, *30*(4), 2990–2996. doi:10.1021/acs.energyfuels.5b02884

Yu, B., Xu, J., Yuan, H., Lou, Z., Lin, J., & Zhu, N. (2014). Enhancement of anaerobic digestion of waste activated sludge by electrochemical pretreatment. *Fuel*, *130*, 279–285. doi:10.1016/j.fuel.2014.04.031

Yuan, H., Yu, B., Cheng, P., Zhu, N., Yin, C., & Ying, L. (2016). Pilot-scale study of enhanced anaerobic digestion of waste activated sludge by electrochemical and sodium hypochlorite combination pretreatment. *International Biodeterioration & Biodegradation*, *110*, 227–234. doi:10.1016/j.ibiod.2016.04.001

Yuan, H., Zhu, N., & Song, F. (2011). Dewaterability characteristics of sludge conditioned with surfactants pretreatment by electrolysis. *Bioresource Technology*, *102*(3), 2308–2315. doi:10.1016/j.biortech.2010.10.065 PMID:21041078

Zhou, X., Wang, Q., Jiang, G., Liu, P., & Yuan, Z. (2015). A novel conditioning process for enhancing dewaterability of waste activated sludge by combination of zero-valent iron and persulfate. *Bioresource Technology*, *185*, 416–420. doi:10.1016/j.biortech.2015.02.088 PMID:25804531

Chapter 12
Biological Nutrient Removal by Suspended Growth Systems

Ezerie Henry Ezechi
https://orcid.org/0000-0003-4968-5541
Universiti Teknologi Malaysia, Johor Bahru, Malaysia

Augustine Chioma Chioma Affam
University College of Technology Sarawak, Malaysia

Khalida Muda
Universiti Teknologi Malaysia, Johor Bahru, Malaysia

ABSTRACT

Nutrients contribute to the wellbeing of water bodies. However, excessive enrichment of water bodies by nutrients could harm aquatic organisms. Some of the severe environmental problems caused by nutrients on water bodies include eutrophication and algal bloom. In children, consumption of nitrate-polluted water causes methaemoglobinemia. Consumption of food irrigated with nutrient-polluted wastewater has been associated with several health implications. These nutrients primarily originate from point and non-point sources. Several biological-suspended treatment systems have been developed to reduce nutrients to acceptable limits prior to discharge into water bodies. These treatment systems are associated with several merits and demerits. This chapter explores the implications of water body enrichment by nutrients and their impact on health, economy, and environment. Furthermore, some suspended growth treatment systems applied for the treatment of nutrient polluted wastewater were explored.

INTRODUCTION

Water is an essential commodity commonly used in almost all activities. Fresh water resources are the main source of water supply in many Countries. However, recent projections have indicated that by 2025, about 60% of the global population could suffer water scarcity (Seckler, 1998). Higher living standards, urbanization and population growth will increase the demand for good quality water. At the

DOI: 10.4018/978-1-7998-0369-0.ch012

same time, sewage flows will increase and more irrigation water will be required in agriculture to meet the increasing demands for the growing populations. Environmental concerns such as wildlife refuges, recreation, scenic values and riparian habitats will also require more water (Bouwer, 2000; Qadir et al., 2007). As a result, a significant freshwater shortage could affect economic development, human health and well-being of many countries.

Wastewater treatment has been used to supplement limited freshwater resources in many countries. Various industries generate different contaminated wastewater which is characterized in terms of their physical, biological and chemical compositions. Three significant methods can be used to treat these wastewaters and they include primary, secondary and tertiary/advanced treatment methods. Primary methods are used to separate suspended solids, floating and settleable materials from water; secondary methods are used to remove biodegradable waste and some suspended solids while tertiary methods are used for the removal residual suspended solids after secondary treatment. Non-biodegradable contaminants present severe health and environmental challenges. Tertiary wastewater treatment processes can be used to treat these contaminants. However, most of these conventional methods are associated with some level of demerits and setbacks which affect their efficiency (Sagiv & Semiat, 2004).

Biological Treatment processes are aged long, well established techniques for wastewater treatment (EH Ezechi et al., 2015; Mann & Stephenson, 1997). They are primarily used for the removal of biodegradable compounds, organic matter and nutrients from wastewater. Its principle depends on the interaction of microorganisms with wastewater contaminants. It is reliable particularly in nutrient removal from wastewater (Obaja et al., 2005) and biogas production (Poh & Chong, 2009). It is a dynamic process because it could be configured in various designs according to the objective of the operator. It has numerous advantages such as low capital and operating cost, high organic matter removal efficiency and biogas production. The generation of highly polluted wastewater from various industries and the advent of emerging contaminants have continuously led to the modification of biological systems to meet the increasingly stringent discharge standards. Although a significant number of successes have been achieved with various system modifications, on-going efforts have focused on optimizing these systems to achieve greater benefits. The performance of these suspended growth systems has significant impact on the receiving water bodies. Thus, it is imperative that these systems achieve high nutrient removal from wastewater before discharge to avoid water body pollution.

From the availability standpoints, a higher percentage (97%) of the planet's water occurs as saltwater in the oceans, whereas two-third of the remaining 3% occurs as snow and ice in polar and mountainous regions. Liquid fresh water represents only about 1% of the global water. Consequently, about 98% of the liquid fresh water are present as groundwater while less than 2% occurs in the form of streams and lakes (Bouwer, 1978). These limited available water resources (lakes and rivers) are constantly exposed to the dangers of pollution from natural disaster such as flood and anthropogenic sources such as poorly treated wastewater effluents. These challenges require immediate operational re-adjustment to protect the limited water bodies from dangers of nutrient pollution. The objective of this chapter is to identify the primary sources of nutrients in waterbodies, highlight the environmental impacts of waterbody pollution and discuss the primary method of nutrient removal from wastewater. The discussion on nutrient removal from wastewater is limited to aerobic and anaerobic suspended growth processes earlier developed and the recent modified configurations.

SOURCES OF NUTRIENTS IN WATERBODIES

Water bodies require some nutrients to be healthy. However, excessive enrichment of nutrient could be harmful (EH Ezechi et al., 2016). The primary sources of nutrients could be classified into point and non-point sources. The primary nutrients causing waterbody deterioration and impairment are nitrogen and phosphorus compounds arising from point and non-point sources(Ezechi et al., 2014). Nutrients essentially contribute to the survival and existence of aquatic organisms. However, excessive enrichment of nutrient constitutes the most significant water quality issues. The impacts of nutrient related water body deterioration include high gelatinous zooplankton bloom, high toxin level from algal species, lower available carbon to food webs, reduced fisheries, oxygen depletion, low habitat biodiversity and coral reef reduction (Badruzzaman et al., 2012; Lapointe et al., 2017). Furthermore, wastewater generated from various sectors including municipal, industrial and agricultural activities consists of high nutrient (nitrogen and phosphorus) concentrations (Cai et al., 2013). These sources constitute a significant water quality issue if not adequately addressed.

Point Sources

Anthropogenic activities are the major point source of water body pollution (E Ezechi et al., 2015). The release of nutrients from most of these sources is inevitable. For instance, a study in Thailand observed that the daily average human nitrogen and phosphorus excreta were in the range of 7.9 and 1.7 g/cap/day, respectively (Schouw et al., 2002). These macro and micro-nutrients require adequate treatment before their effluents are discharged. In 2008, the Department of Environment (DoE) Malaysia found that sewage treatment plants and manufacturing industries were responsible for about 54.05% and 38.73% of the total water body pollution in Malaysia (Figure 1). Several other anthropogenic sources of nutrients such as industrial run-offs, agricultural run-offs, storm water run-off, animal farms, and agro-based industries cause water body pollution (Camargo et al., 2005; Schoen et al., 2011). These anthropogenic activities do not only pollute water bodies but increases the mobility of various forms of nitrogen through air and water. The movement of these compounds has tremendously altered the nitrogen cycle. The annual release of fixed nitrogen from anthropogenic activities such as fertilizer manufacture and application, leguminous crops, fossil fuel burning, wood fuels and forest burning, wetland draining and land clearing is in the range of 80, 40, 20, 40, 10 and 20 teragrams, respectively (Vitousek et al., 1997). Approximately, 60% of the annual global nitrogen in the environment is from anthropogenic sources and have wider health and environmental implications.

Non-Point Sources

Non-point sources of water body pollution are mainly from earthworks, storm-runoff after a downpour and other sources that are unidentifiable or difficult to control. They include natural disasters, soil bacteria and lighting. Natural sources contribute about 40% of the total annual global nitrogen in the environment (Vitousek et al., 1997).

Figure 1. Composition of water pollution sources by sector in Malaysia
(Doe) 2008

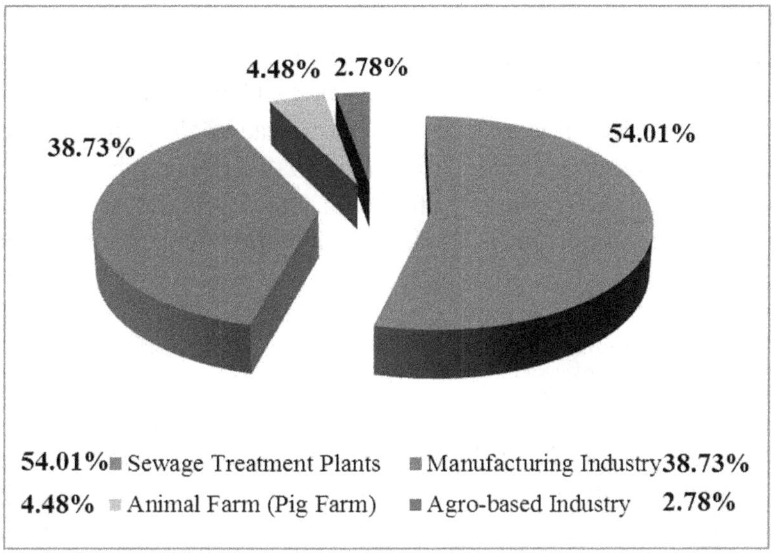

Impacts of Nutrients on Water Quality

The presence of excessive nutrients in water bodies have negative impact on the entire biota and can be examined from the environmental, health and economic point of view.

Environmental Perspective

The entrance of nitrogenous compounds (ammonia and nitrate) into waterways has a devastating effect on the receiving biota and exposes the environment to dangers of various kinds of pollution including air and water. About 25% of all water body impairments are attributed to nutrient related pollutions (Usepa, 2007). The major environmental problem associated with nutrient pollution of water bodies is eutrophication.

1. **Eutrophication**: Eutrophication is caused by the excessive enrichment of water bodies by nutrients. It is a potential danger for aquatic organisms because it deteriorates water quality (Ghafari et al., 2008). On a global scale, the discharge of inadequately treated wastewater into water bodies poses a serious threat to the limited freshwater resources (De-Bashan & Bashan, 2004). Algae bloom is the most prevalent effect of eutrophication. As the water body is clogged with excessive nutrients, algae and submerged plants die and decompose due to the obstruction of sunlight. This phenomenon consumes the oxygen required for the well-being of aquatic organisms and result to algae bloom ((Mpca), 2008). Algae bloom impairs the vision of aquatic organisms and reduce their ability to find food due to water cloudiness. Some algae also produce toxins which could be harmful to pets, marine mammals, fish and shellfish ((Epa), 2012).

Health Perspective

The health implication of the consumption of nutrient polluted water is multifarious. Most of these health issues such as cholera, hepatitis A, cyclosporiasis and ascaris were caused by foods irrigated with polluted and inadequately treated water (Kirby et al., 2003). Consumption of nutrient polluted water has pronounced effect in infants as it is associated with methaemoglobinemia. Methaemoglobinemia, otherwise also known as blue baby syndrome, is a condition that occurs when the oxygen carrying pigments (hemoglobin, hemocyanin) is converted to forms that are unable to carry oxygen (methemoglobin) in infants by nitrate as shown in Equation 1 (Camargo et al., 2005; Ghafari et al., 2008).

$$\underset{(can\,carry\,oxygen)}{Haemoglobin}\left(Fe^{2+}\right) \xrightarrow{NO_2^-} \underset{(cannot\,carry\,oxygen)}{Methaemoglobin} \quad (1)$$

The anaerobic conditions that develop in the gut converts ingested nitrate to nitrite which subsequently blocks the oxygen carrying capacity of hemoglobin (Camargo et al., 2005). This toxic condition increases with increasing nitrate concentrations and exposure time and could result in a bluish color of infant's skin (Camargo et al., 2005; Ghafari et al., 2008). In addition, nitrate plays a potential role in the cancers of the digestive tract through its contribution to the formation of nitrosamines, a prominent carcinogen in mammals (Camargo et al., 2005).

Economic Perspective

The economic loss associated with nutrient pollution of water bodies is enormous. However, such economic losses are not well documented in literature. Reports from few literatures suggest that such economic loss could be viewed from three major sectors, namely tourism and recreation, commercial fishing and property devaluation.

1. **Tourism and Recreation**: In many countries, the tourism and recreation industry significantly contribute to the annual gross domestic product (GDP). Nonetheless, nutrient pollution of water bodies can have a devastating effect on the annual revenue of this industry. In the United State, it is reported that the tourism industry loses close to one billion dollars annually from fishing and recreational activities due to nutrient polluted water bodies ((Epa), 2012). As a result of the impact of algae blooms in the States of Ohio, Florida, Texas and Washington, restaurant sales, lakeside business, and tourism associated spending in local areas have declined (Anderson, 1989).
2. **Commercial Fishing**: Commercial fishing in the coastal areas of the United States is negatively impacted by algae blooms. This have led to reduced fish harvest, increased processing cost and fishery closure. It is estimated that economic loss in the fishing industry due to nutrient pollution could reach up to $2.5 million in soft shell clam harvests and $460,000 in mussel harvests ((Epa), 2015).
3. **Property Depreciation**: Generally, any type of pollution (air, soil or water) affects the normal existence of people and could cause property depreciation within the environment. For instance, studies have shown that in New England, a 1-meter difference in water clarity equates up to $61,000 in property depreciation and $85,000 in Minnesota ((Epa), 2015).

BIOLOGICAL TREATMENT CONCEPTS

In this section, nitrification, denitrification and anaerobic concepts of biological wastewater treatment as well as factors that influence these processes are discussed. These three concepts are discussed.

Anaerobic Processes

Several anaerobic processes such as upflow anaerobic sludge blanket reactor (UASB) (Najafpour et al., 2006), plug flow reactor (PFR) (Lansing et al., 2010), continuously stirred tank reactor (CSTR) (Luo et al., 2010), fixed bed reactors (De Smet et al., 2001) and fluidized bed reactors (Tomishige, 2004) have been used for wastewater treatment. These bioreactors have similar degradation mechanisms involving hydrolysis, acidogenesis, acetogenesis and methanogenesis. Despite its great advantages, anaerobic bioreactors do not readily produce effluent concentrations below stringent limits in terms of nitrogenous compound removal. In a study of domestic wastewater treatment using an upflow anaerobic sludge blanket reactor (UASB), Banu et al. (Banu et al., 2007) observed that ammonia removal was negative and in the range of 42-48.1 mg/L from an initial concentration of 23-28.5 mg/L. In anaerobic systems without an aerobic post-treatment, high effluent nitrogenous compounds are usually common and have been attributed to the mineralization of organic compounds and ammonification (Banu et al., 2007). Ammonia could remain in an anaerobic compartment un-metabolized (Chernicharo, 2006). Additionally, the decomposition of organic matter containing nitrogen such as protein and amino acids within the anaerobic compartment contribute significantly to high ammonia concentration (Sponza & Ağdağ, 2004). In order to overcome this limitation, post treatment of anaerobic treated effluents have been conducted with aerobic compartment (Bernet et al., 2000; Del Pozo & Diez, 2005).

Nitrification

Biological nitrification and denitrification processes are commonly used for biological nutrient removal (BNR). Nitrification is the biological oxidation of ammonia to nitrite, then to nitrate and the rate of such oxidation depend on the activities of microorganisms within the bioreactor. Torresi et al. (2019) noted that a moving bed biofilm reactor was suitable for the removal of nine micro-pollutants from wastewater. Jin et al. (2019) achieved total nitrogen removal efficiency of 95.45% from domestic wastewater using a simultaneous heterotrophic nitrifying aerobic denitrifying bacterium in a purification tank bioreactor (PTBR) at DO > 4 mg/L. The authors noted that nitrate and nitrite did not accumulate in the bioreactor. Cai et al. (2019) achieved COD and TN removal of 91.6% and 88.07% from ship sewage using a novel aerobic-anaerobic micro-sludge membrane bioreactor (O-AMSMBR). The authors noted that the modified bioreactor achieved 10% higher TN removal rate compared to the conventional membrane bioreactor (MBR). Gu et al. (2018) reported a nitrogen removal of 54% from synthetic dye wastewater using an integrated hydrolysis/acidification and multiple anoxic/aerobic (AO) process. Iorhemen et al. (2019) evaluated the performance of a submerged aerobic granular sludge membrane bioreactor (AGMBR) for the removal of organics and nutrients at HRTs of 6, 8 and 10 hours and found that COD, NH_3-N, TN and PO_4-P removal exceeded 96%, 97%, 50% and 35%, respectively. Henriet et al. (2016) evaluated phosphorus removal in aerobic granular sludge sequential batch reactors (AGS-SBR) in two approaches and found that the overall performance of the system exceeded 90% phosphorus removal. Modification of existing systems to address existing challenges has been largely employed in aerobic processes. This has significantly contributed to the success of most aerobic processes. Nitrification usually takes place in the presence of oxygen as electron acceptor and could be represented by the energy yielding two step oxidation reaction expressed below:

Nitrosomonas:

$$2NH_4^+ + 3O_2 \rightarrow 2NO_2^- + 4H^+ + 2H_2O \tag{2}$$

Nitrobacter

$$2NO_2^- + O_2 \rightarrow 2NO_3^- \tag{3}$$

Total oxidation reaction

$$NH_4^+ + 2O_2 \rightarrow NO_3^- + 2H^+ + H_2O \tag{4}$$

Nitrification is carried out by ammonium oxidizing bacteria (AOB) and nitrate oxidizing bacteria (NOB), a group of microorganisms including Nitrosomonas and Nitrobacter.

Several factors can affect nitrification. Unionized ammonia or free ammonia (FA) can inhibit nitrification at high concentration. Such inhibitory effects depend on the total nitrogen species concentration, temperature and pH. FA can exert an inhibitory effect on Nitrosomonas at a concentration of 10-150 mg/L and 0.1-1.0 mg/L for Nitrobacter (Ford et al., 1980). Thus, knowledge of the concentration of FA present in wastewater is necessary to prevent inhibition. A theoretical calculation of the free ammonia concentration could be made using the following expression (Kim et al., 2008):

$$FA\left(\frac{mg}{L}\right) = \frac{\left[Total\,NH_3\right] \times 10^{pH}}{Exp\left[\frac{6334}{273+T}\right] + 10^{pH}} \tag{5}$$

Where FA is free ammonia (mg/L), T is temperature (°C).

Several other factors that affect nitrification include pH, dissolve oxygen (DO), temperature, alkalinity, solid retention time (SRT), heavy metal toxicity and organic loading rate (OLR). These factors were selected based on their influence on bioreactor performances reported in literatures. In order words, they are primary factors that influence bioreactor performance during nitrification process.

pH

Nitrifying bacteria are very sensitive to pH. pH plays a crucial role during nitrification. However, the optimum pH for nitrification has been a subject of discussion. For instance, several optimum pH values such as 6.8-8.0 (Eddy, 2004) and 6.45-8.95 (Ruiz et al., 2006) have been reported in literature. At optimum pH, nitrifying bacteria are activated, alkalinity is effectively consumed and free ammonia and nitrous acid inhibition diminish. Ruiz et al. (Ruiz et al., 2006) investigated nitrification with high nitrite accumulation for the treatment of wastewater with high ammonia concentration using a 2.5 L activated sludge bioreactor and found out that nitrification took place within pH of 6.45 to 7.85 but was inhibited at pH of 9.05 and below 6.45. Jianlong and Ning (Jianlong & Ning, 2004) observed that ammonia oxidation was high between pH of 6.5 to 7.5, slightly decreased at pH 8.5 and severely inhibited at pH 9.5 in their study of partial nitrification under limited dissolve oxygen conditions. In the treatment of high strength ammonium industrial wastewater with two sludge systems, Carrera et al. (Carrera et al., 2003) achieved ammonia removal of about 90-100% at fixed pH of 7.5. Thus, pH should be carefully monitored during any bioreactor operation.

Dissolve Oxygen

Dissolve oxygen (DO) is an important parameter and the electron acceptor during nitrification. When oxygen is limiting, metabolic activities of nitrifying organisms drastically reduce and could lead to partial denitrification, a phenomenon that rarely occur during nitrification. The influence of DO on nitrification has been extensively investigated. Ruiz et al. (Ruiz et al., 2006) obtained high nitrification rates at DO range of 1.7-5.1 mg/L. Nitrite accumulation was observed at a DO range of 0.7-1.4 mg/L. At DO concentration of 0.5 mg/L, complete loss of nitrification was observed. Jianlong and Ning (Jianlong & Ning, 2004) also reported that ammonia oxidation increased when DO was raised from 0.5 to 1.5 mg/L at fixed pH of 7.5.

Temperature

Temperature is a significant nitrification parameter. Biochemical and metabolic processes accelerates with increasing temperature until their threshold limit is attained (Chen et al., 2006). Jianlong and Ning (Jianlong & Ning, 2004) observed an increase in ammonia oxidation rate (4.5 times) with increasing temperature from 12 to 30 °C at fixed pH of 7.5 and DO concentration of 1.5 mg/L. On the other hand, anaerobic digestion is sensitive to temperature and can operate in the psychrophilic (0-20 °C), mesophilic (20-42 °C) and thermophilic (42-75 °C) states, respectively. A fall in temperature of a given state will cause a decrease in the digestion rate and an increase of the digestion time.

Alkalinity

Alkalinity is necessary during nitrification process due to its buffering capacity to prevent changes in pH. The oxidation of ammonia to nitrate consumes alkalinity. Theoretically, for each gram of ammonia nitrogen oxidized, 4.25 g of O_2 are utilized, 0.16 g of new cells are formed, 7.07 g of alkalinity as $CaCO_3$ are consumed and 0.08 g of inorganic carbon are utilized in the formation of new cells (Eddy, 2004).

Solid Retention Time (SRT)

Solid retention time (SRT) is an important parameter for bioreactor design. It represents the average period of time the active biomass is required to stay within the bioreactor and could be represented by the expression below:

$$SRT = \frac{VX}{\left(Q - Q_w\right)X_e + Q_w X_R} \tag{6}$$

SRT for different bioreactors differ and it affects process performance, sludge production and oxygen requirement. For instance, the typical design SRT for complete mix activated sludge bioreactor (CMAS) is 3-15 days, 20-40 days for extended aeration processes, 10-30 days for sequencing batch reactors (SBR) (Eddy, 2004). SRT is usually controlled or maintained through periodic wasting of excess sludge. Prolonged SRT could have negative effect on biological nutrient removal (BNR) due to changes in properties of the mixed liquor such as viscosity, concentration, composition of microbial products and cell surface properties (Shin & Kang, 2003). Han et al. (Han et al., 2005) observed a nitrogen removal in the range of 89, 96 and 87% with decreasing SRT in the range of 100, 70 and 50 days in a membrane bioreactor coupled to SBR, treating synthetic wastewater. A similar observation was made for denitrification. The mixed liquor suspended solids (MLSS) in the reactor increased to about 18,000 mg/L. Thus, the decrease of nitrogen removal at too high SRT was attributed to the impediment in oxygen and substrate utilization by the sludge viscosity and inert matter accumulation. Ouyang and Junxin (Ouyang & Junxin, 2009) observed a nitrogen removal in the range of 89.6 and 91.7% whereas the COD removal was in the range of 91.2 and 88.2% at SRT of 10 and 40 days, respectively in a three-membrane bioreactor (MBR). The sludge concentration increased and macromolecules (protein and polysaccharides) and extracellular polymeric substances in the mixed liquor decreased at longer SRT. Huang et al. (Huang et al., 2001) observed that nitrogen removal reached 90% with increasing SRT in the range of 5, 10, 20 and 40 days at a fixed hydraulic retention time (HRT) of 5 hours in their investigation of the effect of sludge retention time on microbial behavior in submerged MBR. The influence of SRT was not significant on nitrogen removal. A similar observation was made for COD. From every indication, SRT is an important parameter that requires effective monitoring during any bioreactor operation.

Toxicity

Nitrifying organisms are sensitive to a wide range of organic and inorganic toxic compounds at concentrations above their minimal inhibitory threshold. Several compounds such as tannins, heavy metals, phenolic compounds and many other chemicals are known to exhibit toxic effects on nitrifying organisms. The concentration threshold of inhibitory effect of heavy metals on nitrifying organisms is presented in Table 1.

Liu et al. (Liu et al., 2005) investigated the influence of various phenol concentrations on nitrification using microbial granules in a sequencing batch reactor (SBR). The authors observed higher nitrifying activities at phenol concentrations lower than 10 mg/L. Although phenol was degraded by the aerobic granules, oxidation of ammonia and production of nitrate decreased significantly in comparison with the control reactor. The nitrifying activity of aerobic granules completely recovered after exposure to phenol concentration of 10 mg/L but was partially irreversible when exposed to higher phenol concentrations (15-20 mg/L). Kim et al. (Kim et al., 2008) examined the inhibitory effects of toxic compounds such as free ammonia (FA), thiocyanate, free cyanide and phenolic compounds in the nitrification process for the treatment of Coke wastewater. Free cyanide was found to exert a significant inhibitory effect on the nitrification process at concentrations of 0.1-1.0 mg/L. Nitrification inhibition by free ammonia was not observed below 350 mg/L. Thiocyanate (200 mg/L) was completely removed within 18 hours

Table 1. Inhibitory threshold of heavy metals on nitrifying organisms

Metals	Concentration Threshold of Inhibitory Effect on Heterotrophic Organisms (mg/L)
Arsenic (As)	0.05
Cadmium (Cd)	1.0
Copper (Cu)	1.0
Lead (Pb)	0.1
Mercury (Hg)	0.1
Nickel (Ni)	1.0
Zinc (Zn)	1.0

(Eddy, 2004)

and no inhibitory effect was noticed. Phenolic compounds exerted inhibitory effects only at concentrations above 200 mg/L and loss of nitrification was observed at concentrations of 500 mg/L. Dyreborg and Arvin (Dyreborg & Arvin, 1995) examined the inhibition of nitrification by creosote contaminated wastewater in leaching experiment conducted in a saturated sand column. The authors observed an inhibitory threshold for the compounds at 10.7 mg/L (benzene), 8.4 mg/L (o-xylene), 3.7 mg/L (phenol) and 1.3 mg/L (o-cresol).

In order to overcome the inhibition effect of toxic compounds, physical processes could be added to support nitrifying organisms. Introduction of physical processes into biological systems provides microorganisms an environment suitable to resist toxicity. Such technique has been examined and proved effective. Kutty et al. (2014) added microwave incinerated rice husk ash (MIRHA) into an activated sludge process for copper degradation. The authors observed copper toxicity in the control and MIRHA bioreactors at concentrations of 2 mg/L and 5 mg/L, respectively. The authors concluded that MIRHA improved bacteria resistance to toxicity up to a copper concentration of 5 mg/L. MIRHA provided large surface area and acted as a support for the bacteria to resist copper toxicity at higher concentration.

Denitrification

Denitrification involves the sequential biological reduction of nitrate to nitrite, nitric oxide, nitrous oxide and nitrogen gas, respectively and could be expressed as:

$$NO_3^- \rightarrow NO_2^- \rightarrow NO \rightarrow N_2O \rightarrow N_2 \tag{7}$$

Biological denitrification requires an electron donor in the form of carbon source either from wastewater or methanol, glucose and acetate as represented below (Eddy, 2004).

Wastewater:

$$C_{10}H_{19}O_3N + 10NO_3^- \rightarrow 5N_2 + 10CO_2 + 3H_2O + NH_3 + 10OH^- \tag{8}$$

Methanol:

$$CH_3OH + 6NO_3^- \rightarrow 3N_2 + 5CO_2 + 7H_2O + 6OH^- \quad (9)$$

Acetate:

$$5CH_3COOH + 8NO_3^- \rightarrow 4N_2 + 10CO_2 + 6H_2O + 8OH^- \quad (10)$$

Several factors such as carbon source, substrate availability, presence of active denitrifiers, DO, carbon source and temperature play significant role in denitrifier growth, metabolism, gene expression and denitrification rate (Saleh-Lakha et al., 2009).

Dissolve Oxygen

The presence of dissolve oxygen could inhibit the activities of denitrifiers. Thus, in most treatment process, mechanical mixing is used to achieve denitrification. In a pre-denitrifying system, high residual oxygen may be found in the anoxic compartment at high recirculation. High presence of residual oxygen could inhibit denitrifying activities. This scenario has been reported by Ramos et al. (Ramos et al., 2007). A similar observation was made by Galvez et al. (Gálvez et al., 2003) who noted that the presence of dissolve oxygen at the bottom of the anoxic column in their study of nitrogen removal with submerged fixed film reactor caused a decrease of denitrification activities and an increase in methanol consumption. In contrast, Wang et al. (Wang et al., 2005) noted that denitrifying activities were not interrupted at low DO concentrations of 0.3-0.5 mg/L in their study of nitrogen and carbon removal from wastewater by anoxic/aerobic MBR. Rather, the authors concluded that such DO concentration can support nitrogen removal in the anoxic compartment. Therefore, extensive care is required in the management of DO in the anoxic unit.

Carbon Source

The addition of carbon source such as acetate, glucose or methanol in denitrification process is a common age long practice for biological nutrient removal (BNR) system. The carbon source could be externally or internally added to the system, depending on the existing process technology. Recent studies have preferred the internal use of carbon source for reasons of better economy, less sludge production and enhanced use of the wastewater organic compounds. The nature of carbon source is considered vital because higher denitrification rates are achieved with the most easily degraded forms of carbon. Additionally, most water pollutants can have a preference for carbon source with respect to their chemistry. For instance, it is reported that phosphate removal is preferentially achieved with acetate whereas glucose or other carbohydrates are needed for energy generation used in poly-hydroxybutyrate (PHB) and polyphosphate synthesis (Kargi & Uygur, 2003). In their investigation of nutrient removal from synthetic wastewater by SBR using different carbons (acetate, glucose and a 50/50 combination of both), Kargi and Uygur (Kargi & Uygur, 2003) noted that nutrient removal was more prominent when both carbon sources were combined. Ammonia removal efficiency was in the range of 83, 89 and 86%, whereas nitrate removal was in the range of 63, 58 and 81% for glucose, acetate and mixed carbon sources, respectively.

Similar observations were made for COD removal. The authors recommended acetate for high COD and ammonium-N removal. Peng et al. (Peng et al., 2007) observed a denitrification rate of 3.2, 9.6 and 12 mgN/(g VSS.h) for methanol, ethanol and acetate, respectively. They concluded that ethanol was the best external carbon source due to quick adaptation time, cost and availability.

Efforts to operate bioreactors based on the utilization of the internal carbon source or low cost materials have intensified and are aimed at cost reduction. The sludge addition system (SAS) (Ra et al., 2000) and non-digested pig manure (Obaja et al., 2005) have been reportedly effective as carbon source for denitrification.

Biological nitrification/denitrification processes have been widely used in biological wastewater treatment specifically for nutrient removal. These biological wastewater treatment systems have been operated in various configurations. More researches into the modification of the process to meet the objectives of the operators are still on-going. The advantages and disadvantages of the anaerobic and nitrification/denitrification processes are summarized in Table 2.

Summary of Section

Anaerobic, nitrification and denitrification processes are suitable techniques for nutrient elimination from wastewater. While nitrification techniques have higher nitrogen removal capacity than the anaerobic systems, both systems are largely affected by several operating factors. These influencing factors can be reduced by closely monitoring the system performance and bacteria activity. Biological processes consist primarily of microorganisms which are required in a good state to adequately conduct bio-degradative activities. Thus, it is important to closely monitor bacteria activities in these systems.

SUSPENDED GROWTH PROCESSES

Biological wastewater treatment processes are basically secondary treatment technologies used for the transformation of biodegradable constituents in wastewater into acceptable end products. Microorganisms are used in all biological wastewater treatment methods. As a result, biological wastewater treatment processes are classified based on their metabolic functions. For instance, during nitrification, nitrifiers (Nitrosomonas and Nitrobacter) oxidize ammonia to nitrite and nitrate, respectively in the presence of oxygen as an electron acceptor, whereas denitrifying bacteria (Pseudomonas, Micrococcus, Archromobacter and Bacillus) reduce nitrate to a gaseous form in the presence of a carbon source (wastewater, methanol, ethanol). These biological treatment operations can be achieved through a series of processes such as suspended growth, attached growth, combined and lagoon processes. In this section, a detailed evaluation of suspended and attached growth processes is presented.

Suspended growth biological wastewater treatment processes have been widely used for the treatment of wastewater. The distinctive feature of a suspended growth process is that microorganisms are maintained in liquid suspension within the bioreactor through appropriate mixing strategy. Thus, microorganisms in suspension interact with the influent wastewater within the bioreactor and convert the biodegradable constituents into acceptable end products. Suspended growth processes can be operated in the aerobic, anoxic and anaerobic concepts. The biodegradation of organic matter causes an increase of the microbial population within the bioreactor. Different forms of suspended growth biological wastewater treatment processes have been developed and are discussed below.

Table 2. Advantages and disadvantages

	Anaerobic	Nitrification	Denitrification	References
Advantages	No aeration	Shorter retention time	No pH control	(Eddy, 2004)
	Biogas production	Effective against nutrients	Produce alkalinity	(Eddy, 2004; Poh & Chong, 2009; Seghezzo et al., 1998; Van Rijn, 1996)
	Flexibility	Stable performance	No aeration	(Eddy, 2004; Poh & Chong, 2009; Seghezzo et al., 1998; Van Rijn, 1996)
	Low sludge production	Indoor and outdoor use	-	(Eddy, 2004; Poh & Chong, 2009; Seghezzo et al., 1998; Van Rijn, 1996)
Disadvantages	Slow start-up	pH sensitive	Carbon source needed	(Eddy, 2004; Poh & Chong, 2009; Seghezzo et al., 1998; Van Rijn, 1996)
	Long retention time	Sludge production	Ammonia accumulation	(Eddy, 2004; Poh & Chong, 2009; Seghezzo et al., 1998; Van Rijn, 1996)
	Large space area required for conventional digesters	Nitrate accumulation	Requires agitation	(Eddy, 2004)
	Low nutrient removal	Require alkalinity		(Eddy, 2004; Poh & Chong, 2009; Seghezzo et al., 1998; Van Rijn, 1996)
	Require post treatment	Sensitive to toxic compounds		(Eddy, 2004; Poh & Chong, 2009; Seghezzo et al., 1998; Van Rijn, 1996)

Completely Mixed Activated Sludge Process (CMAS)

An ideal completely mixed activated sludge (CMAS) or continuously stirred tank reactor (CSTR) process is primarily characterized with homogenous and instantaneous mixing which ensures substrate uniformity in all areas of the bioreactor. The oxygen demand, organic loading and substrate concentration are uniform throughout the entire bioreactor due to proper mixing. The aeration tank is equipped with air diffusers that transfer the required oxygen into the biomass and enable homogenous mixing (Mittal, 2011). CMAS set-up usually consists of an aeration tank and in most cases, equipped with baffles for thorough mixing whereas the effluent can settle in the clarifier as shown in Figure 2. In the design of a CMAS, care is needed in the selection of the SRT, kinetic and stoichiometric coefficients and mass balance application (Grady Jr et al., 2011). Padoley et al. (Padoley et al., 2006) evaluated the performance of a CMAS for the biodegradation of pyridine using a bacteria (*Pseudomonas Pseudoalcaligenes-KPN*) isolated from garden soil and found that pyridine degradation increased in the range of 45 to 99.6% with increasing HRT from 4 to 24 hours. However, pyridine residual concentration increased from 0.2 to 7.2 mg/L when loading rate was raised from 0.042 to 0.251 kg pyridine kg $MLSS^{-1}d^{-1}$ at fixed HRT of 24 hours. Additionally, the authors observed that CMAS has good mixing characteristics. Pala and Tokat (Pala & Tokat, 2002) evaluated the performance of a CMAS aided with different adsorbents such as specific organic flocculant (Marwichem DEC), powdered activated carbon (PAC), bentonite, activated

Figure 2. Completely mixed activated sludge process
(Padoley et al., 2006)

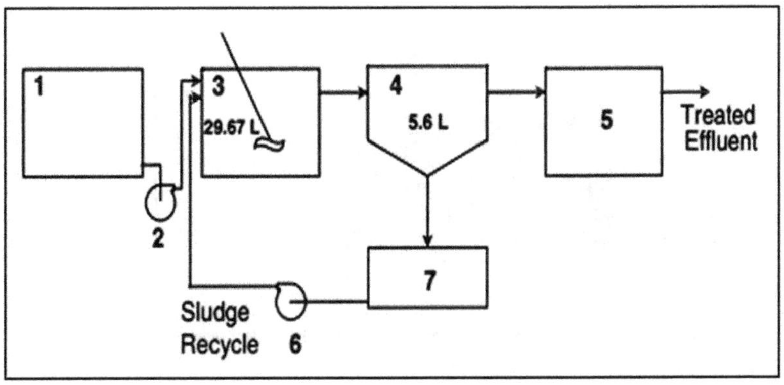

clay and synthetic inorganic clay (Macrosorb) for color removal from cotton textile wastewater and found that COD and color removal was in the range of 94 and 36% at SRT and HRT of 30 and 1.6 days, respectively. Specific organic flocculant (Marwichem DEC) proved more effective and produced low sludge compared to other adsorbents. Agdag and Sponza (Ağdağ & Sponza, 2005) also evaluated the performance of two UASB coupled to an aerobic CSTR in series for the treatment of landfill leachate and observed that COD removal increased with increasing influent concentration in the first reactor. The COD removal was in the range of 79, 98 and 98% for the first UASB, second UASB and CSTR reactors, respectively at highest organic loading rate. However, ammonia removal was in the range of 2, 7 and 99.6% for the first, second and third reactors, with an increase of nitrate concentration in the effluent of each compartment. The CMAS system is summarized in Table 3.

Sequencing Batch Reactors (SBR)

Sequencing batch reactor (SBR) is a set of fill and draw activated sludge technique for wastewater treatment. It has been successfully used in the treatment of various wastewater such as aquaculture (Fontenot et al., 2007), tannery soak (Lefebvre et al., 2005), saline (Woolard & Irvine, 1995), piggery (Obaja et al., 2005), domestic (Blackburne et al., 2008) and landfill leachate (Spagni & Marsili-Libelli, 2009). Each

Table 3. Summary of completely mixed activated sludge process

	CMAS	References
Advantages	• Good nitrification. • Withstands peak-load by diluting toxic substances. • Could be applied in smaller plants.	(Eddy, 2004)
Limitations	• Higher aeration cost at larger volume	(Eddy, 2004)
	• Low flexibility, nitrate production	(Thompson et al., 2001; Von Sperling, 1996)
	• Sludge bulking due to biomass instability	(Thompson et al., 2001; Von Sperling, 1996)
	• Shock loading	(Padoley et al., 2006)

Figure 3. Sequencing batch reactor
(Fontenot et al., 2007)

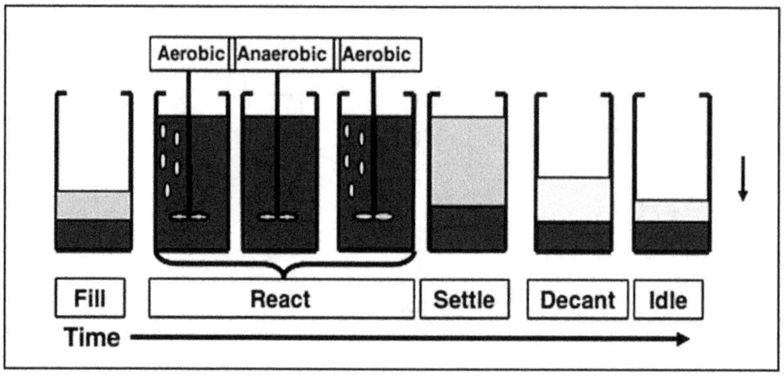

compartment of the SBR is filled within a certain period of time and operated as a batch system after which the mixed liquor settles and the supernatant is drawn from the tank. The automation of SBR enable it to alternate from one treatment mechanism (aerobic, anoxic and anaerobic) to another (Fontenot et al., 2007). The major advantage of the SBR system over conventional wastewater treatment process is that all reactions occur in one single batch reactor in a timed sequence rather than in space sequence. SBR systems are operated automatically in five distinct steps of fill, react, settle, draw and idle mechanisms, each of which are sequentially timed and accomplished as shown in Figure 3 (Epa, 1999). SBR eliminates the need for separate primary and secondary clarifiers through alternate mixing and aeration technique.

The success of SBR in wastewater treatment has been widely reported in literature.

Obaja et al. () examined the performance of a SBR for nitrification, denitrification and biological phosphorus removal from piggery wastewater and found that nitrogen and phosphorus removal reached 99.7 and 97.3% within 3 operational cycles. The flexibility of SBR has also been demonstrated by some studies. Uygur et al. () evaluated the performance of a three cycle SBR for the treatment of landfill leachate, pre-treated with coagulation-flocculation and air stripping using lime at pH 12. Each of the three SBR cycle consisted of 3, 4 and 5 steps in the following order: (An)/anoxic, (Ax/oxic, (Ox); (An/Ox/Ax/Ox) and (An/Ax/Ox/Ax/Ox) for the first, second and third cycles and was operated at HRT and SRT of 7 hours and 10 days, respectively. The authors found that the effluent concentration was in the range of 2110, 2275 and 1400 mg/L from an initial concentration of 5750 mg/L for COD, 107, 116 and 107 mg/L for ammonia and 70, 76 and 65 mg/L for phosphorus, respectively for the first, second and third cycles. This study clearly demonstrated the flexibility of the SBR system to alternate from one treatment mode to another. In terms of COD and phosphorus removal, the five-step process was effective, but the three processes showed an effective ammonia removal capacity. (Blackburne et al.) utilized the aerobic duration control strategy (ADCS) which automatically terminates aeration as soon as ammonium oxidation is completed while nitrite concentration accumulates in a SBR. Nitrification and denitrification mechanisms were operated through the alternation of oxygen. The authors achieved 80% nitratation during the study and attributed the success to purposeful aerobic control. In the phase without ADCS, the SBR achieved over 95% ammonium oxidation. Morling () also reported high nitrification capacity of almost 100% for two landfill leachate treatment plants using the SBR technology, whereas the total nitrogen removal was about 90-95%. The SBR system is summarized in Table 4.

Table 4. Summary of SBR

	SBR	References
Advantages	• Robust flexibility and effluent concentration control.	(Lefebvre et al., 2005)
	• Smaller footprint, cost savings	(Epa, 1999)
	• Single reactor operation	(Epa, 1999; Lefebvre et al., 2005; Wang et al., 2010)
Limitations	• High sophisticated control systems	(Lefebvre et al., 2005)
	• Higher maintenance, potential plugging of aeration device	(Epa, 1999)
	• Complex settling systems	(Epa, 1999; Lefebvre et al., 2005; Wang et al., 2010)

Modified Ludzack Ettinger (MLE)

The activated sludge processes discussed above have one thing in common, which is the absence of a well-defined anoxic process. Those systems (CMAS and SBR) are primarily aerobic processes but could be modified to other system modes through the alternation of dissolve oxygen and mixing. The carbon source for denitrification (in most cases, partial denitrification) of these processes is from the hydrolysis of particulate substrates entrapped in the mixed liquor suspended solids, biomass death and cell lysis. However, the modified Ludzack-Ettinger Process (MLE) presents a different scenario in terms of configuration and system operation as shown in Figure 4.

The MLE is a post-aerobic two stage activated sludge system which consists of anoxic, aerobic and a settling tank. The return activated sludge (RAS) is conducted from the settling tank into the anoxic tank whereas the mixed liquor is recirculated from the aerobic tank into the anoxic tank (Eddy, 2004). The fundamental key to the success of the MLE is the effective mixed liquor recirculation (MLR) into the anoxic tank. Park et al. (Park et al., 2012) evaluated the performance of a combined MLE and MBR for the treatment of food waste disposer wastewater at SRT of 45 days and overall HRT of 18 hours. The authors observed a high substrate removal in the range of 72-88% for COD, 86% for TN whereas 50-71% total phosphorus (TP) was achieved at higher SRT. The system performance increased with increasing recirculation rate from 1-4. Effluent nitrate concentration below 5 mg/L was achieved at recirculation rate of 3 and 4. The simplicity and flexibility of MLE can be incorporated into various treatment ideas.

Figure 4. Modified ludzack ettinger process
(Eddy, 2004).

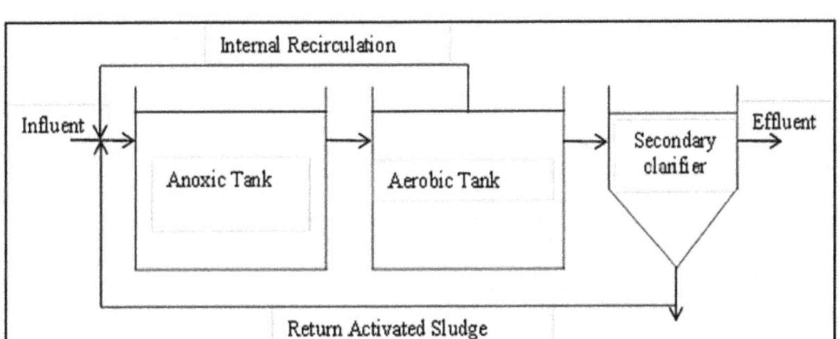

Gajaraj and Hu (Gajaraj & Hu, 2014) incorporated a microbial fuel cell (MFC) into MLE and MBR processes respectively in a bench scale study. The two bioreactors have a total working volume of 7.2 L and were divided with baffles to yield a three-step treatment mode of anaerobic/anoxic (2.1 L), aerobic (3.3 L) and internal settling chamber (1.8 L). The experiments were conducted in two phases using synthetic wastewater. The MFC module consisted of a graphite anode and cathode of commercial quality. They were inserted into the MFC reactor systems. The authors observed no significant difference between the MLE-MFC and MBR-MFC systems with respect to NH_4^+-N removal. Both reactors achieved more than 99.9% of NH_4^+-N removal. For COD removal, the MLE-MFC reactor achieved higher removal (89%) compared to the MBR-MFC reactor (86%). However, both reactors (MLE-MFC and MBR-MFC) achieved a poor nitrate removal of 31 and 20% respectively. Both the MLE-MFC and MBR-MFC modules were able to generate an average voltage of about 0.13 (±0.03) V during the study. Hafez et al. (Hafez et al., 2010) evaluated the performance of a pilot scale MLE and a full-scale CAS system, treating low strength municipal wastewater at temperature of 12 °C. The HRT of the aerobic, anoxic and settling tanks of the MLE were 4.6, 1.5 and 2.6 hours whereas it was 5.5 hours for CAS. The full-scale plant has a primary clarification prior to CAS. The HRT of the aerobic tank was 6.5 hour, operated at an average daily flow (ADF) of 80,600 m^3.d, SRT of 3.6 d and settling time of 3.3 hours. The authors observed that both treatment systems achieved full nitrification with higher instability in CAS. Nitrogen removal was about 75-80% in MLE, about 56% higher nitrification rate compared to CAS. The higher nitrification rate of the pilot MLE over the full-scale CAS was attributed to the absence of the anoxic compartment in the CAS which resulted in the inhibition of nitrification in the aerobic compartment. In the pilot MLE, both the anoxic and aerobic compartments effectively utilized the readily biodegradable organic matter. The pilot MLE was also observed to reduce the net observed biomass yield by 28% relative to the fullscale plant. Thus, at an optimum combination of process parameters, MLE could be used in various forms for wastewater treatment. The summary of the MLE system is presented in Table 5.

Bardenpho Process (BP)

The Bardenpho process (BP) operates a post-aerobic technique similar to the MLE. The BP is an extension of MLE modified with the inclusion of extra aerobic and anoxic compartments in order to prevent high effluent nitrate concentration. Denitrification occurs in the second anoxic compartment through biomass decay and the utilization of slowly biodegradable substrate. However, at unstable and low mixed liquor recirculation (MLR), BP can produce high effluent nitrate concentration (C.P. Leslie Grady, 1999). The BP has been modified into 4 and 5 unit configurations. In the 4 unit configuration (Figure 5), the system consist of anoxic-aerobic-anoxic-aerobic sequence whereas it consist of anaerobic-anoxic-aerobic-anoxic-aerobic sequence in the 5 unit configuration (Federation, 2007). The anaerobic zone of the 5-stage BP was incorporated for phosphorus removal. Remarkably, BP configurations can be easily alternated

Table 5. Summary of MLE

	MLE	References
Advantages	• Easily adaptable to existing systems • Good nitrification	(Eddy, 2004)
Limitations	• Potential effluent nitrate concentration • Strict internal recycle	(Eddy, 2004; Jeyanayagam, 2005)

Figure 5. 4-Stage Bardenpho Process
(Eddy, 2004)

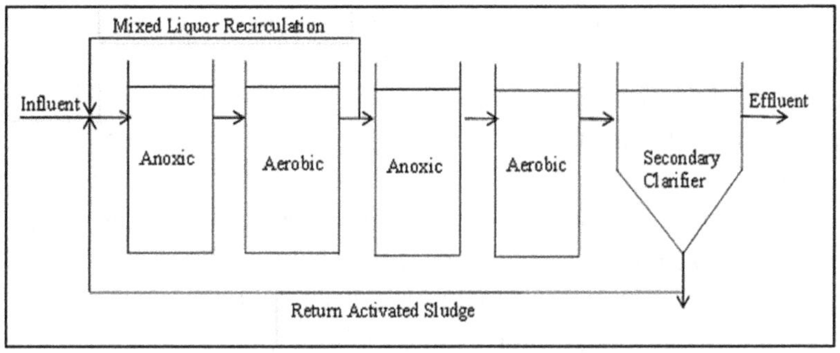

based on the operational objective. In both the 4 and 5 staged configurations, mixed liquor is recycled from the first aerobic compartments into the first anoxic compartments whereas RAS is recycled from the clarifier into the first anoxic and anaerobic compartments, respectively.

Mostafa M. Emaraa (2014) evaluated the performance of a laboratory scale 4 and 5 stage BP for nitrogen and phosphorus removal and found that substrate removal was in the range of 98 and 99% for COD, 97 and 99% for total nitrogen (TN), 69.5 and 88% for phosphorus removal in the 4 and 5 stage BP, respectively. P Ilies (2001) evaluated the performance of a 4 stage BP for nitrification and denitrification treating high ammonia landfill leachate with two parallel pre-denitrifying (anoxic-aerobic-anoxic-aerobic) and post-denitrifying (aerobic-anoxic-aerobic-anoxic) reactor setups at SRT, MLR and temperature in the range of 20 days, 3:1 and 10-20 °C. The ammonia concentration was varied in the range of 1200-2200 mg N/L by the addition of ammonium chloride. The authors observed that the ammonia removal efficiency drastically decreased with reducing temperature from 20-10 °C. However, both systems achieved ammonia removal of about 50% at all experimental conditions. Oldham and Stevens (1984) evaluated the performance of a 5 stage (anaerobic-anoxic-aerobic-anoxic-aerobic) BP, treating domestic wastewater at Kelowna WWTP, British Columbia (a 260 L/s facility) and found that the effluent substrate concentration was in the range of 5, 0.3, 10, and 4 mg/L for nitrogen, phosphorus, BOD_5 and TSS, from an influent nitrogen and phosphorus concentration of 30 and 6 mg/L, respectively. The summary of the BP system is presented in Table 6.

Table 6. Summary of BP

	BP	References
Advantages	• High nitrification • High organic matter removal	(Eddy, 2004)
Limitations	• Large space requirement. • Inefficiency of the second anoxic unit. • Denitrification inhibition in the first anoxic unit due to oxygen transfer.	(C.P. Leslie Grady, 1999; Eddy, 2004).

Oxidation Ditch

Oxidation Ditch (OD) system is a modified activated sludge process with a central reactor and a settling tank. The central reactor is used for mixing and aeration. OD could be referred to as a complete mix closed loop reactor with surface aerators providing oxygen and mixing. The aeration tank usually has an oval shape and equipped with a horizontal brush or rotor. The OD effluent settles in the secondary clarifier whereas the return activated sludge (RAS) is recycled back into the ditch through the pumping system. Oxidation ditch was developed to reduce the cost of conventional wastewater treatment. Several types of OD have been developed over the years such as Pasveer OD, Orbal OD, Carrousel OD, Total Barrier OD and Jet Aeration Channel OD (Fan et al., 2010). OD systems have been globally applied in small-medium scale municipal wastewater treatment plants. One significant characteristics of OD systems is their ability to achieve nitrification and denitrification in one reactor through periodic alternation of the aerobic and anoxic zones (Zhou et al., 2013). OD has shown great flexibility in simultaneous nitrification and denitrification (SND) for the treatment of domestic wastewater. The system eliminates the anoxic compartment for denitrification but regulates the dissolve oxygen to create a spatial anoxic condition. The influence of operational parameters such as DO, temperature and hydraulic retention time was evaluated in a two 120 L Carrousel OD systems used for both laboratory and pilot scale studies (Liu et al., 2010). The authors found that operating an SND in OD can achieve high COD removal of about 86-94%. In terms of nitrification, an almost 100% removal was achieved at DO concentration greater than 1.2 mg/L. Dissolve oxygen have less effect on the operation of the SND in OD and a concentration of 1.2 mg/L was sufficient for effective organic matter removal. This could be as a result of the low biomass concentration in the reactor (1620 mg/L). A low DO concentration of about 1.2 mg/L can adequately diffuse through the low biomass concentration in the reactor. Temperature was found to influence the performance of SND in OD and a value above 11 °C is recommended. Denitrification efficiency improved at lower ammonia loading and shorter HRT (4 hours). The authors concluded that shorter HRT provided sufficient and rapid biodegradable substrates to the biomass (Liu et al., 2010). A bench-scale integrated oxidation ditch process with vertical circle was investigated for carbonaceous substrate and ammonia nitrogen removal from domestic wastewater (Xia & Liu, 2004). The modified OD was made of plexiglass and has a single channel, a Kesner brush and an integral clarifier. Three different zones were established with volumes of 10% (anaerobic zone), 50-70% (anoxic zone) and 20-35% (aerobic zone). The integral clarifier is located at the opposite end of the brush in the OD and sludge returns to the down-ditch automatically without any pumping system. HRT and temperature were varied. The new system achieved an average COD removal of about 95% and maximum of 98.5% in the aerobic zone and a 98% BOD_5 removal was achieved in the anoxic zone. Effluent Ammonium concentration was low (0.5 -1.5 mg/L) but fluctuated (5.6 mg/L) at low DO concentrations (0.9-1.5 g/L). Total nitrogen (TN) removal was unstable in the range of 40-60%. COD removal increased with increasing HRT from 6 to 16.5 hours but decreased with increasing organic loading. However, no change was observed for ammonia removal when HRT or organic loading was raised. Thus, for effective biodegradation of organic matter (COD), an increase of HRT was recommended. The integration of the OD process was reported to have economic benefits. Space requirement (land area) and energy consumption reduced by almost 50 and 40% respectively in comparison with the conventional process (Xia & Liu, 2004). Yongzhen et al. (2008) monitored the performance of an OD (simultaneous nitrification, denitrification and phosphorus removal) treating municipal wastewater in three step sequence of anaerobic (7 L), anoxic (21 L), secondary clarifier and OD system (280 L) respectively at a flowrate of 14 L/h under varying HRT of 0.5, 1.5 and 18 h. The OD was operated in four different SND phases of stabilization, damaging, recovering and recovered phases respectively. The return activated sludge was pumped into the anaerobic zone (30%) and anoxic zone (70%) respectively. At steady state, high nitrogen and phosphorus removal were achieved with TN and TP removal rates of about 80% and 85% respectively. The system stability was damaged at excessive aeration and high TP removal was achieved at nitrate concentrations below 6 mg/L. In phase 1 at DO concentration of 0.5 m^3/h, effluent NH_4^+-N, NO_3-N, COD and TN concentrations were about 3.63 mg/L, 4.31 mg/L, 46 mg/L and 9.36 mg/L from initial concentration of 60.21 mg/L (TN) and 390 mg/L (COD) respectively. The effluent NH_4^+-N, NO_3-N, COD and TN concentrations in phase II at elevated DO concentration of 0.7 m^3/h were 1.46 mg/L, 11.18 mg/L, 14.29 mg/L and 43 mg/L from initial influent concentrations of 42.34 g/L (TN) and 265 mg/L (COD). In phase III and IV, the DO concentrations were 0.5 m^3/h and 0.52 m^3/h and the effluent NH_4^+-N, NO_3-N, COD and TN concentrations were in the range 4-6.38 mg/L, 4-11 mg/L, 45-50 mg/L and 11-17 mg/L respectively. The reactor successfully recovered from the damage after phase II through oxygen control (Yongzhen et al., 2008). A modified pilot scale Orbal OD system was investigated for the treatment of domestic wastewater in a simultaneous nitrification/denitrification process varying DO concentrations and internal recirculation under two experimental phases at different run times (Zhou et al., 2013). The OD working volume was 11.43 m^3. The authors reported that DO concentrations of 0.15-0.25 mg/L were sufficient to achieve high nitrification of about 97% NH_4^+-N removal in phases 1 and 2 of the experiment. For TN removal, a fluctuation was observed in the range of 59, 64 and 42% for runs 1, 4 and 3 respectively in phase 1, indicating the reliance of TN removal on the denitrification efficiency of nitrates. In phases 2, TN removal improved with increasing recycling ratio. Oxidation Ditch systems have shown significant capacity for wastewater treatment. The advantages and disadvantages of OD are summarized in Table 7.

Integrated Biological Systems

The integrated biological systems are techniques developed by modifying existing conventional biological treatment systems. Integrated systems primarily focus on improving the limitations or drawbacks of existing biological systems. These modifications could involve the inclusion of baffles, creation of smaller and compact treatment chambers, application of programmable timers and efficient control system. The objective of integrated systems is primarily to achieve high bioreactor performance at reduced cost. Several modifications of existing systems have shown that among other things, integrated systems can reduce large space requirement for installation of wastewater treatment plant (Ezechi et al., 2019).

Table 7. Advantages and disadvantages of OD

	OD	References
Advantages	• Resistance to load variations due to large reactor volume and long HRT • Simple construction • Good nitrogen removal capacity • Potential application in smaller plants • Less sludge production	(Lei & Ni, 2014)
Limitations	• Skill requirement • Knowledge of control methods • Relatively high effluent TSS • Large space requirement limits the installation of OD in urban, sub-urban and other areas where land is a scare commodity	(Jeyanayagam, 2005; Us Environmental Protection Agency, 2000; Xia & Liu, 2004)

Furthermore, cost reduction and reduction of sludge production are synonymous in integrated systems (Azhdarpoor et al., 2018). E Ezechi et al. (2016) evaluated the performance of an integrated aerobic/anoxic bioreactor (IAAB) for the treatment of simulated synthetic domestic wastewater at overall HRT of 12 and 7.2 days and found that the ammonia and nitrate removal exceeded 90% and 80%, respectively. The authors noted that IAAB was stable during bioreactor operation and the bacteria population increased simultaneously with removal efficiency. The results obtained in this study demonstrate that IAAB could be a suitable integrated system for nutrient removal from wastewater. However, this study was conducted with synthetic wastewater which clearly has differing characteristics from real domestic wastewater. A better and reliable assessment of the system should be with real domestic wastewater. Ezechi et al. (2019) further evaluated an integrated suspended bioreactor (ISB) coupling aerobic, anoxic and settling units for the treatment of industrial wastewater at overall HRT of 36, 12, 7.2, 5.1 and 4 days, respectively and found that ISB achieved an effluent ammonia-nitrogen concentration of 0.3-4.5 mg/L with decreasing HRT. The authors also noted absence of nitrate and nitrite accumulation in the bioreactor, indicating that ISB achieved complete nitrification and denitrification. The stability of ISB at short HRT treating industrial wastewater from beverage industry shows that the bioreactor is suitable for nutrient removal. Additionally, ISB occupied smaller space compared to conventional systems. Coupling of the aerobic, anoxic and settling units in Fig. 6 was responsible for the small space occupied by ISB. Abyar et al. (2018) evaluated the performance of a high rate up-flow anaerobic, anoxic and oxic (A_2O) for the treatment of meat-processing wastewater. Three variables including HRT, COD/TN ratio and aerated volume fraction were evaluated for CNP removal. The authors found that under optimum conditions, COD, TKN, nitrate-nitrogen (NO_3-N), total nitrogen (TN) and phosphorus (PO_4^{3-}-P) reached 98.33, 92.06, 91.97, 90.48 and 83.48%, respectively. This bioreactor presents significant advantage for nutrient removal due to the integration of important biological treatment components. Large molecular compounds can be decomposed by the anaerobic component while less complex biodegradable compounds are decomposed by the aerobic component. The oxidized nitrogen compounds are reduced in the anoxic component. Ye et al. (2018) evaluated the performance of a novel divisional influent dual-anaerobic-anoxic/oxic (D-$A_2$0) system for municipal wastewater treatment and found that COD, TN, NH_3-N and TP removal reached 95.23%, 80.64%, 90.42% and 90.03%, respectively. The D-A_2O has a reasonable stability in terms of sludge yield and energy cost. Furthermore, carbon from anaerobic decomposition was sufficient to operate the system, excluding the addition of external carbon. This seems like one of the unique dissimilarities between D-A_2O and A_2O. High sludge production is a major drawback of conventional biological processes due to sludge management challenges. Systems with moderate sludge generation are required to mitigate sludge related issues. The features of integrated systems are summarized in Table 8.

Summary of Section

The suspended growth processes discussed in this chapter are detailed evaluation of performance from the earlier developed systems such as the modified Ludzack-Ettinger Process (MLE) and Bardenpho process (BP) to the recently modified systems such as IAAB and D-A_2O. There is a significant progression of success in modifying existing conventional systems. The drawbacks of conventional systems such as MLE and BP are mitigated through continued modification and re-configuration. The sequencing batch reactor (SBR) completely demonstrates the significant improvement in the modification of existing systems. Integrated systems discussed in this chapter demonstrated remarkable achievement in terms of sludge reduction, cost reduction, space reduction and efficiency improvement. These modified systems

Table 8. Features of integrated bioreactors

Advantages	Disadvantages
Smaller footprint	High recirculation
Small space requirement	Addition of carbon source
Easy operation	Anaerobic unit is sensitive to temperature
Less sludge waste	Long start of the anaerobic unit
No sludge bulking or scum formation	Potential odor generation in the anaerobic unit
Installable in various environments	-
High shock load tolerance	-

(Chan et al., 2012a, 2012b)

Figure 6. Integrated suspended bioreactor
(Ezechi et al., 2019)

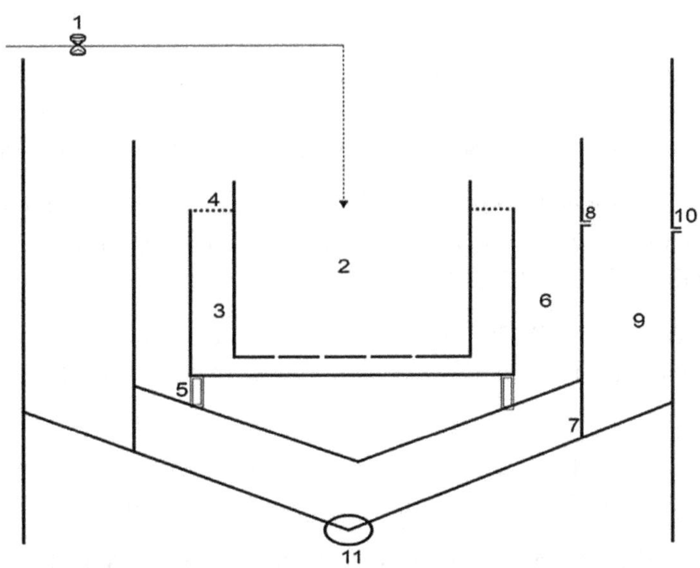

[1] Influent pump [2] Aerobic unit [3] Baffle unit [4] Baffle filter [5] Aerobic support
[6] Anoxic unit [7] Anoxic support [8] Anoxic outlet [9] Settling unit [10] Effluent outlet
[11] WAS, IER

are capable of achieving lower effluent nutrient (nitrogen and phosphorus) concentrations acceptable for disposal by many regulating agencies in-spite of their stringent limits. Although these modified systems have shown tremendous success, efforts to develop and improve on the success achieved so far is still on-going.

CONCLUSION

Discharge of inadequately treated wastewater into water bodies causes severe dangers of pollution. Eutrophication and algal bloom are caused when water bodies are enriched with nutrients. These point and non-point water body pollution are a significant threat to the existing water resources. Adequate care must be taken in ensuring that wastewater treatment meets the stringent discharge regulation to avoid water body pollution. Suspended treatment processes have been used in nutrient removal from wastewater. These systems have their merits and demerits. However, it is imperative to improve these systems in order to achieve complete nutrient removal before wastewater discharge into water body. Furthermore, monitoring the wastewater treatment facilities by the regulatory authorities to ensure compliance is important. The presence of point source pollution in water bodies can reduce if wastewater facilities are improved to achieve low effluent nutrient concentrations.

ACKNOWLEDGMENT

The authors are grateful to Universiti Teknologi Malaysia (UTM) Johor Bahru for support this study.

REFERENCES

Ağdağ, O. N., & Sponza, D. T. (2005). Anaerobic/aerobic treatment of municipal landfill leachate in sequential two-stage up-flow anaerobic sludge blanket reactor (UASB)/completely stirred tank reactor (CSTR) systems. *Process Biochemistry*, *40*(2), 895–902. doi:10.1016/j.procbio.2004.02.021

Anderson, E. E. (1989). Economic benefits of habitat restoration: Seagrass and the Virginia hard-shell blue crab fishery. *North American Journal of Fisheries Management*, *9*(2), 140–149. doi:10.1577/1548-8675(1989)009<0140:EBOHRS>2.3.CO;2

Banu, J., Kaliappan, S., & Yeom, I. T. (2007). Treatment of domestic wastewater using upflow anaerobic sludge blanket reactor. *International Journal of Environmental Science and Technology*, *4*(3), 363–370. doi:10.1007/BF03326295

Bernet, N., & Delgenes, N. (2000). Combined anaerobic–aerobic SBR for the treatment of piggery wastewater. *Water Research*, *34*(2), 611–619. doi:10.1016/S0043-1354(99)00170-0

Emaraa, M. M., Ahmeda, F. A., El-Azizc, F. M. A., & El-Razek, A. M. A. (2014). Biological aspects of the wastewater treatment plant "Mahala Marhoom" in Egypt and modified with Bardenpho processes. *Nature and Science*, *12*(10), 41–51.

Blackburne, R., Yuan, Z., & Keller, J. (2008). Demonstration of nitrogen removal via nitrite in a sequencing batch reactor treating domestic wastewater. *Water Research*, *42*(8), 2166–2176. doi:10.1016/j.watres.2007.11.029 PMID:18155266

Bouwer, H. (1978). Groundwater hydrology (Vol. 480). New York, NY: McGraw-Hill.

Bouwer, H. (2000). Integrated water management: Emerging issues and challenges. *Agricultural Water Management*, *45*(3), 217–228. doi:10.1016/S0378-3774(00)00092-5

Bricker, S. B., Longstaff, B., Dennison, W., Jones, A., Boicourt, K., Wicks, C., & Woerner, J. (2008). Effects of nutrient enrichment in the nation's estuaries: A decade of change. *Harmful Algae*, *8*(1), 21–32. doi:10.1016/j.hal.2008.08.028

Camargo, J. A., Alonso, A., & Salamanca, A. (2005). Nitrate toxicity to aquatic animals: A review with new data for freshwater invertebrates. *Chemosphere*, *58*(9), 1255–1267. doi:10.1016/j.chemosphere.2004.10.044 PMID:15667845

Carrera, J., Baeza, J. A., Vicent, T., & Lafuente, J. (2003). Biological nitrogen removal of high-strength ammonium industrial wastewater with two-sludge system. *Water Research*, *37*(17), 4211–4221. doi:10.1016/S0043-1354(03)00338-5 PMID:12946903

Chen, S., Ling, J., & Blancheton, J.-P. (2006). Nitrification kinetics of biofilm as affected by water quality factors. *Aquacultural Engineering*, *34*(3), 179–197. doi:10.1016/j.aquaeng.2005.09.004

De-Bashan, L. E., & Bashan, Y. (2004). Recent advances in removing phosphorus from wastewater and its future use as fertilizer (1997–2003). *Water Research*, *38*(19), 4222–4246. doi:10.1016/j.watres.2004.07.014 PMID:15491670

de Chernicharo, C. A. L. (2006). Post-treatment options for the anaerobic treatment of domestic wastewater. *Reviews in Environmental Science and Biotechnology*, *5*(1), 73–92. doi:10.100711157-005-5683-5

De Smet, C. R. H., De Croon, M. H. J. M., Berger, R. J., Marin, G. B., & Schouten, J. C. (2001). Design of adiabatic fixed-bed reactors for the partial oxidation of methane to synthesis gas. Application to production of methanol and hydrogen-for-fuel-cells. *Chemical Engineering Science*, *56*(16), 4849–4861. doi:10.1016/S0009-2509(01)00130-0

Del Pozo, R., & Diez, V. (2005). Integrated anaerobic–aerobic fixed-film reactor for slaughterhouse wastewater treatment. *Water Research*, *39*(6), 1114–1122. doi:10.1016/j.watres.2005.01.013 PMID:15766966

DOE, Malaysia. Environmental Quality (Sewage) Regulations 2009, Environmental Quality Act 1974, Environmental Quality (Control of Pollution from Solid Waste Transfer Station and Landfill) Regulations., 1-82.

Dyreborg, S., & Arvin, E. (1995). Inhibition of nitrification by creosote-contaminated water. *Water Research*, *29*(6), 1603–1606. doi:10.1016/0043-1354(94)00247-5

Ezechi, E., Kutty, M., Isa, M., Malakahmad, A., & Aminu, N. (2015). An integrated attached growth bioreactor for the treatment of wastewater. *Research Journal of Applied Sciences, Engineering and Technology*, *11*(10), 1066–1070. doi:10.19026/rjaset.11.2120

Ezechi, E., Kutty, S., Isa, M., Malakahmad, A., Udeh, C., & Menyechi, E. (2016). Treatment of synthetic domestic wastewater by integrated aerobic/anoxic bioreactor (IAAB). *Nature Environment and Pollution Technology*, *15*(3), 943.

Ezechi, E., Kutty, S., Muda, K., & Yaqub, A. (2019). A comparative evaluation of an integrated hybrid bioreactor treating industrial wastewater. *Journal of Water Process Engineering*, *31*, 100805. doi:10.1016/j.jwpe.2019.100805

Ezechi, E. H., Kutty, S., Isa, M. H., Malakahmad, A., Ude, C. M., Menyechi, E. J., & Olisa, E. (2016). Nutrient removal from wastewater by integrated attached growth bioreactor. *Research Journal of Environmental Toxicology*, *10*(1), 28–38. doi:10.3923/rjet.2016.28.38

Ezechi, E. H., Kutty, S. R. M., Isa, M. H., Malakahmad, A., & Ibrahim, S. U. (2015). Chemical oxygen demand removal from wastewater by integrated bioreactor. *Journal of Environmental Science and Technology*, *8*(5), 238–243. doi:10.3923/jest.2015.238.243

Ezechi, E. H., Kutty, S. R. M., Isa, M. H., & Rahim, A. F. A. (2014). Treatment of wastewater using an integrated submerged attached growth system. Paper presented at *Applied Mechanics and Materials.* 10.4028/www.scientific.net/AMM.567.167

Fan, L., Xu, N., Wang, Z., & Shi, H. (2010). PDA experiments and CFD simulation of a lab-scale oxidation ditch with surface aerators. *Chemical Engineering Research & Design*, *88*(1), 23–33. doi:10.1016/j.cherd.2009.07.013

Federation, Water Environment. (2007). *Biological nutrient removal processes* (pp. 22:23-65).

Fontenot, Q., Bonvillain, C., Kilgen, M., & Boopathy, R. (2007). Effects of temperature, salinity, and carbon: Nitrogen ratio on sequencing batch reactor treating shrimp aquaculture wastewater. *Bioresource Technology*, *98*(9), 1700–1703. doi:10.1016/j.biortech.2006.07.031 PMID:16935499

Ford, D. L., Churchwell, R. L., & Kachtick, J. W. (1980). Comprehensive analysis of nitrification of chemical processing wastewaters. *Journal - Water Pollution Control Federation*, 2726–2746.

Gajaraj, S., & Hu, Z. (2014). Integration of microbial fuel cell techniques into activated sludge wastewater treatment processes to improve nitrogen removal and reduce sludge production. *Chemosphere*, *117*, 151–157. doi:10.1016/j.chemosphere.2014.06.013 PMID:25014565

Gálvez, J. M., Gómez, M. A., Hontoria, E., & González-López, J. (2003). Influence of hydraulic loading and air flowrate on urban wastewater nitrogen removal with a submerged fixed-film reactor. *Journal of Hazardous Materials*, *101*(2), 219–229. doi:10.1016/S0304-3894(03)00173-0 PMID:12927737

Ghafari, S., Hasan, M., & Aroua, M. K. (2008). Bio-electrochemical removal of nitrate from water and wastewater—A review. *Bioresource Technology*, *99*(10), 3965–3974. doi:10.1016/j.biortech.2007.05.026 PMID:17600700

Grady, C. P. L., Jr., Glent, T. D., & Lim, H. C. (1999). Biological wastewater treatment (2nd Ed., Revised and Expanded). Marcel Dekker, Inc.

Grady Jr, C. L., Daigger, G. T., Love, N. G., & Filipe, C. D. (2011). Biological wastewater treatment. Boca Raton, FL: CRC Press.

Hafez, H., Elbeshbishy, E., Chowdhury, N., Nakhla, G., Fitzgerald, J., Van Rossum, A., & Gauld, G. (2010). Pushing the hydraulic retention time envelope in Modified Ludzack Ettinger systems. *Chemical Engineering Journal*, *163*(3), 202–211. doi:10.1016/j.cej.2010.07.033

Han, S.-S., Bae, T.-H., Jang, G.-G., & Tak, T.-M. (2005). Influence of sludge retention time on membrane fouling and bioactivities in membrane bioreactor system. *Process Biochemistry*, *40*(7), 2393–2400. doi:10.1016/j.procbio.2004.09.017

Huang, X., Gui, P., & Qian, Y. (2001). Effect of sludge retention time on microbial behaviour in a submerged membrane bioreactor. *Process Biochemistry*, *36*(10), 1001–1006. doi:10.1016/S0032-9592(01)00135-2

Ilies, P., & Mavinic, D. S. (2001). The effect of decreased ambient temperature on the biological nitrification and denitrification of a high ammonia landfill leachate. *Water Research*, *35*(8), 2065–2072. doi:10.1016/S0043-1354(00)00477-2 PMID:11337855

Jeyanayagam, S. (2005). True confessions of the biological nutrient removal process. *Florida Water Resources Journal*, *1*, 37–46.

Jianlong, W., & Ning, Y. (2004). Partial nitrification under limited dissolved oxygen conditions. *Process Biochemistry*, *39*(10), 1223–1229. doi:10.1016/S0032-9592(03)00249-8

Kargi, F., & Uygur, A. (2003). Effect of carbon source on biological nutrient removal in a sequencing batch reactor. *Bioresource Technology*, *89*(1), 89–93. doi:10.1016/S0960-8524(03)00031-2 PMID:12676505

Kim, Y. M., Park, D., Lee, D. S., & Park, J. M. (2008). Inhibitory effects of toxic compounds on nitrification process for cokes wastewater treatment. *Journal of Hazardous Materials*, *152*(3), 915–921. doi:10.1016/j.jhazmat.2007.07.065 PMID:17804160

Kirby, R. M., Bartram, J., & Carr, R. (2003). Water in food production and processing: Quantity and quality concerns. *Food Control*, *14*(5), 283–299. doi:10.1016/S0956-7135(02)00090-7

Kutty, S. R. M., Ezechi, E. H., Khaw, S. G., Lai, C. L., & Isa, M. H. (2014). Evaluation of copper removal using MIRHA as an adsorbent in a continuous flow activated sludge system. *Water Pollution XII*, *182*, 233–244. doi:10.2495/WP140211

Lansing, S., Martin, J. F., Botero, R. B., Da Silva, T. N., & Da Silva, E. D. (2010). Methane production in low-cost, unheated, plug-flow digesters treating swine manure and used cooking grease. *Bioresource Technology*, *101*(12), 4362–4370. doi:10.1016/j.biortech.2010.01.100 PMID:20153173

Lefebvre, O., Vasudevan, N., Torrijos, M., Thanasekaran, K., & Moletta, R. (2005). Halophilic biological treatment of tannery soak liquor in a sequencing batch reactor. *Water Research*, *39*(8), 1471–1480. doi:10.1016/j.watres.2004.12.038 PMID:15878018

Lei, L., & Ni, J. (2014). Three-dimensional three-phase model for simulation of hydrodynamics, oxygen mass transfer, carbon oxidation, nitrification and denitrification in an oxidation ditch. *Water Research*, *53*, 200–214. doi:10.1016/j.watres.2014.01.021 PMID:24525069

Liu, Y., Shi, H., Xia, L., Shi, H., Shen, T., Wang, Z., ... Wang, Y. (2010). Study of operational conditions of simultaneous nitrification and denitrification in a Carrousel oxidation ditch for domestic wastewater treatment. *Bioresource Technology*, *101*(3), 901–906. doi:10.1016/j.biortech.2009.09.015 PMID:19818603

Liu, Y.-Q., Tay, J.-H., Ivanov, V., Moy, B. Y.-P., Yu, L., & Tay, S. T.-L. (2005). Influence of phenol on nitrification by microbial granules. *Process Biochemistry*, *40*(10), 3285–3289. doi:10.1016/j.procbio.2005.03.018

Luo, G., Xie, L., Zou, Z., Wang, W., Zhou, Q., & Shim, H. (2010). Anaerobic treatment of cassava stillage for hydrogen and methane production in continuously stirred tank reactor (CSTR) under high organic loading rate (OLR). *International Journal of Hydrogen Energy*, *35*(21), 11733–11737. doi:10.1016/j.ijhydene.2010.08.033

Mann, A. T., & Stephenson, T. (1997). Modelling biological aerated filters for wastewater treatment. *Water Research, 31*(10), 2443–2448. doi:10.1016/S0043-1354(97)00095-X

Metcalf & Eddy. (2004). *Wastewater engineering, treatment and reuse* (4th Ed.).

Minnesota Pollution Control Agency (MPCA). (2008). Nutrients: Phosphorus, Nitrogen Sources, Impact on Water Quality - A General Overview. 1-2. Retrieved from www.pca.state.mn.us/water/standards/index.html

Mittal, A. (2011). Biological wastewater treatment. *Water Today, 1*, 32–44.

Morling, S. (2010). Nitrogen removal and heavy metals in leachate treatment using SBR technology. *Journal of Hazardous Materials, 174*(1), 679–686. doi:10.1016/j.jhazmat.2009.09.104 PMID:19836884

Najafpour, G. D., Zinatizadeh, A. A. L., Mohamed, A. R., & Isa, M. (2006). High-rate anaerobic digestion of palm oil mill effluent in an upflow anaerobic sludge-fixed film bioreactor. *Process Biochemistry, 41*(2), 370–379. doi:10.1016/j.procbio.2005.06.031

Obaja, D., Mace, S., Costa, J., Sans, C., & Mata-Alvarez, J. (2003). Nitrification, denitrification and biological phosphorus removal in piggery wastewater using a sequencing batch reactor. *Bioresource Technology, 87*(1), 103–111. doi:10.1016/S0960-8524(02)00229-8 PMID:12733583

Obaja, D., Mace, S., & Mata-Alvarez, J. (2005). Biological nutrient removal by a sequencing batch reactor (SBR) using an internal organic carbon source in digested piggery wastewater. *Bioresource Technology, 96*(1), 7–14. doi:10.1016/j.biortech.2004.03.002 PMID:15364074

Oldham, W. K., & Stevens, G. M. (n.d.). S Initial operating experiences of a nutrient removal process (Modified Bardenpho) at Kelowna, British Columbia. *Canadian Journal of Civil Engineering*, 474-479.

Ouyang, K., & Liu, J. (2009). Effect of sludge retention time on sludge characteristics and membrane fouling of membrane bioreactor. *Journal of Environmental Sciences (China), 21*(10), 1329–1335. doi:10.1016/S1001-0742(08)62422-5 PMID:19999985

Padoley, K. V., Rajvaidya, A. S., Subbarao, T. V., & Pandey, R. A. (2006). Biodegradation of pyridine in a completely mixed activated sludge process. *Bioresource Technology, 97*(10), 1225–1236. doi:10.1016/j.biortech.2005.05.020 PMID:16084081

Pala, A., & Tokat, E. (2002). Color removal from cotton textile industry wastewater in an activated sludge system with various additives. *Water Research, 36*(11), 2920–2925. doi:10.1016/S0043-1354(01)00529-2 PMID:12146882

Park, C., Lee, J. W., Moon, S., Park, K. Y., & Jutidamrongphan, W. (2012). Advanced treatment of wastewater from food waste disposer in modified Ludzack-Ettinger type membrane bioreactor. *Environmental Engineering Research, 17*(2), 59–63. doi:10.4491/eer.2012.17.2.059

Peng, Y.-Z., Yong, M.A., & Wang, S.-Y. (2007). Denitrification potential enhancement by addition of external carbon sources in a pre-denitrification process. *Journal of Environmental Sciences, 19*(3), 284-289.

Poh, P. E., & Chong, M. F. (2009). Development of anaerobic digestion methods for palm oil mill effluent (POME) treatment. *Bioresource Technology*, *100*(1), 1–9. doi:10.1016/j.biortech.2008.06.022 PMID:18657414

Qadir, M., Sharma, B. R., Bruggeman, A., Choukr-Allah, R., & Karajeh, F. (2007). Non-conventional water resources and opportunities for water augmentation to achieve food security in water scarce countries. *Agricultural Water Management*, *87*(1), 2–22. doi:10.1016/j.agwat.2006.03.018

Ra, C. S., Lo, K. V., Shin, J. S., Oh, J. S., & Hong, B. J. (2000). Biological nutrient removal with an internal organic carbon source in piggery wastewater treatment. *Water Research*, *34*(3), 965–973. doi:10.1016/S0043-1354(99)00189-X

Ramos, A. F., Gomez, M. A., Hontoria, E., & Gonzalez-Lopez, J. (2007). Biological nitrogen and phenol removal from saline industrial wastewater by submerged fixed-film reactor. *Journal of Hazardous Materials*, *142*(1), 175–183. doi:10.1016/j.jhazmat.2006.08.079 PMID:17070991

Ruiz, G., Jeison, D., Rubilar, O., Ciudad, G., & Chamy, R. (2006). Nitrification–denitrification via nitrite accumulation for nitrogen removal from wastewaters. *Bioresource Technology*, *97*(2), 330–335. doi:10.1016/j.biortech.2005.02.018 PMID:16171689

Sagiv, A., & Semiat, R. (2004). Analysis of parameters affecting boron permeation through reverse osmosis membranes. *Journal of Membrane Science*, *243*(1), 79–87. doi:10.1016/j.memsci.2004.05.029

Saleh-Lakha, S., Shannon, K. E., Henderson, S. L., Goyer, C., Trevors, J. T., Zebarth, B. J., & Burton, D. L. (2009). Effect of pH and temperature on denitrification gene expression and activity in Pseudomonas mandelii. *Applied and Environmental Microbiology*, *75*(12), 3903–3911. doi:10.1128/AEM.00080-09 PMID:19376915

Schoen, M. E., Soller, J. A., & Ashbolt, N. J. (2011). Evaluating the importance of faecal sources in human-impacted waters. *Water Research*, *45*(8), 2670–2680. doi:10.1016/j.watres.2011.02.025 PMID:21429551

Schouw, N. L., Danteravanich, S., Mosbaek, H., & Tjell, J. C. (2002). Composition of human excreta—A case study from southern Thailand. *The Science of the Total Environment*, *286*(1), 155–166. doi:10.1016/S0048-9697(01)00973-1 PMID:11886091

Seckler, D. W. (1998). *World water demand and supply, 1990 to 2025: Scenarios and issues* (Vol. 19). Iwmi.

Seghezzo, L., Zeeman, G., van Lier, J. B., Hamelers, H. V. M., & Lettinga, G. (1998). A review: The anaerobic treatment of sewage in UASB and EGSB reactors. *Bioresource Technology*, *65*(3), 175–190. doi:10.1016/S0960-8524(98)00046-7

Shin, H.-S., & Kang, S.-T. (2003). Characteristics and fates of soluble microbial products in ceramic membrane bioreactor at various sludge retention times. *Water Research*, *37*(1), 121–127. doi:10.1016/S0043-1354(02)00249-X PMID:12465793

Spagni, A., & Marsili-Libelli, S. (2009). Nitrogen removal via nitrite in a sequencing batch reactor treating sanitary landfill leachate. *Bioresource Technology*, *100*(2), 609–614. doi:10.1016/j.biortech.2008.06.064 PMID:18707876

Sponza, D. T., & Ağdağ, O. N. (2004). Impact of leachate recirculation and recirculation volume on stabilization of municipal solid wastes in simulated anaerobic bioreactors. *Process Biochemistry*, *39*(12), 2157–2165. doi:10.1016/j.procbio.2003.11.012

Thompson, G., Swain, J., Kay, M., & Forster, C. F. (2001). The treatment of pulp and paper mill effluent: A review. *Bioresource Technology*, *77*(3), 275–286. doi:10.1016/S0960-8524(00)00060-2 PMID:11272013

Tomishige, K. (2004). Syngas production from methane reforming with CO 2/H 2 O and O 2 over NiO–MgO solid solution catalyst in fluidized bed reactors. *Catalysis Today*, *89*(4), 405–418. doi:10.1016/j.cattod.2004.01.003

United States Environmental Protection Agency. (2012). The facts about nutrient pollution. 1-2. Retrieved from www.epa.gov/nutrientpollution

US EPA. (1999). Wastewater technology fact sheet—sequencing batch reactors: EPA 832-F-99-073.

US EPA. Office of Water, Washington. (2000). Wastewater technology fact sheet: Oxidation ditches. EPA 832-F-00-013, 1-6.

US EPA. (2007). National section 303(d) list fact sheet. Retrieved from Information Online at www.iaspub.epa.gov/waters/national_rept.control

US EPA. (2015). A compilation of cost data associated with the impacts and control of nutrient pollution. 1-110. Retrieved from http://www2.epa.gov/nutrient-policy-data/compilation-cost-data-associated-impacts-and-control-nutrient-pollution

Uygur, Ahmet, & Kargı, F. (2004). Biological nutrient removal from pre-treated landfill leachate in a sequencing batch reactor. *Journal of Environmental Management*, *71*(1), 9–14.

Van Rijn, J. (1996). The potential for integrated biological treatment systems in recirculating fish culture—A review. *Aquaculture (Amsterdam, The Netherlands)*, *139*(3), 181–201. doi:10.1016/0044-8486(95)01151-X

Vitousek, P. M., Aber, J. D., Howarth, R. W., Likens, G. E., Matson, P. A., Schindler, D. W., & Tilman, D. G. (1997). Human alteration of the global nitrogen cycle: Sources and consequences. *Ecological Applications*, *7*(3), 737–750.

von Sperling, M. (1996). Comparison among the most frequently used systems for wastewater treatment in developing countries. *Water Science and Technology*, *33*(3), 59–72. doi:10.2166/wst.1996.0059

Wang, L. K., Shammas, N. K., & Hung, Y.-T. (2010). *Advanced biological treatment processes* (Vol. 9). Springer Science & Business Media.

Wang, Y., Huang, X., & Yuan, Q. (2005). Nitrogen and carbon removals from food processing wastewater by an anoxic/aerobic membrane bioreactor. *Process Biochemistry*, *40*(5), 1733–1739. doi:10.1016/j.procbio.2004.06.039

Woolard, C. R., & Irvine, R. L. (1995). Treatment of hypersaline wastewater in the sequencing batch reactor. *Water Research*, *29*(4), 1159–1168. doi:10.1016/0043-1354(94)00239-4

Xia, S., & Liu, J. (2004). An innovative integrated oxidation ditch with vertical circle for domestic wastewater treatment. *Process Biochemistry*, *39*(9), 1111–1117. doi:10.1016/S0032-9592(03)00216-4

Yongzhen, P., Hongxun, H., Shuying, W., Youwei, C., & Zhiguo, Y. (2008). Nitrogen and phosphorus removal in pilot-scale anaerobic-anoxic oxidation ditch system. *Journal of Environmental Sciences (China)*, *20*(4), 398–403. doi:10.1016/S1001-0742(08)62070-7 PMID:18575122

Zhou, X., Han, Y., & Guo, X. (2013). Enhanced total nitrogen removal performance in a modified Orbal oxidation ditch system with internal nitrate recycle. *Chemical Engineering Journal*, *228*, 124–131. doi:10.1016/j.cej.2013.04.096

Chapter 13
Bioremediation:
A Logical Approach for the Efficient Management of Textile Dye Effluents

Sneha Unnikrishnan
B. S. Abdur Rahman Cresecent Institute of Science and Technology, India

Nagamani Bora
University of Nottingham, UK

Karthikeyan Ramalingam
https://orcid.org/0000-0002-9334-427X
B. S. Abdur Rahman Cresecent Institute of Science and Technology, India

ABSTRACT

Synthetic dyes are extensively used in several industries and the dyes are great concern for the ecosystem. During the dyeing process, a certain percentage of the used dye is released into the wastewater, causing severe environmental and health hazards. The physiochemical methods are not sustainable due to high-cost, high-energy requirements and hazardous by-products. Among all the alternative technologies to conventional wastewater treatment, bioremediation has emerged as the most desirable approach to clean up the environment and to restore its original status. The ability of microbes in decolorizing the textile effluents is significant. The decolorization can be further enhanced using immobilization techniques. Immobilization increases the stability and reusability of the microorganisms. The microorganisms can be entrapped in calcium alginate beads, that can be used to construct a packed bed reactor in which dye decolorization can be carried out on a large scale. Thus, bioremediation serves as an effective, eco-friendly solution for the pollution caused by textile dye effluents.

DOI: 10.4018/978-1-7998-0369-0.ch013

INTRODUCTION

The wastewater effluent from the textile industry is considered as one of the most polluting agents among all other industries. The environmental and health effects of textile industry wastewater have been a subject to scientific scrutiny for a long time. The presence of even very low concentrations of dyes in the effluent is highly visible and undesirable (Nigam *et al.,* 2000). There are more than 100,000 commercially available dyes with over 700,000 ton of dyestuff produced annually (Al Fawaaz and Abdullah., 2016). Synthetic dyes are chemically diverse and can be divided into azo, triphenylmethane or heterocyclic/polymeric structures (Gregory, 1993; Liu *et al.*, 2004). They are used extensively in the biochemical, foodstuff, plastic and textile industries; where it is estimated that 10-14% of the dye is lost in the effluents during the dyeing process. The textile industry plays an important role in the world economy, but at the same time, it consumes a large quantity of water and generates a huge amount of wastewater (Wouter *et al.*, 1998; Hai *et al.*, 2007). Synthetic dyes are recalcitrant and are persistent when discharged into the environment and many of them are also toxic. Many dyes are difficult to decolorize due to their chemical structure and synthetic origin. The classification of industrial dyes is given in figure 1.

TREATMENT PROCESSES

The treatment of textile effluents is of paramount importance due to toxic and aesthetic impacts on receiving waters. The various methods of dye decolorization are listed in figure 2.

Various physical and chemical methods for the elimination of dyestuff havebeen developed during the past decades (Kandelbauer *et al.*, 2004). The removal of color from textile industry effluents are mainly based on physical and chemical methods (Banat *et al.*, 1996). The color removal techniques usually followed in these methods are coagulation or adsorption of dyes, ultra-filtration, ion-exchange, chemical oxidation, electrolysis, etc. However, these methods are not very much applied because of their high cost, high energy requirements and hazardous by-products. Also, these techniques generate a huge volume of sludge and cause secondary pollution due to the formation of sludge and hazardous by-products (Maier *et al.*, 2004; Ramya *et al.*, 2007).

The various physical and chemical methods of dye decolorization along with their advantages and disadvantages have been listed in table 1.

In the past few decades, there is an immense effort to develop a cost-effective and eco-friendly alternative to conventional waste treatment methods. Among all the technologies decolorization by bioremediation has emerged as the most desirable approach to clean up the environment and to restore its original status.

BIOREMEDIATION

Bioremediation is important, as it is cost effective and eco-friendly, and produces less sludge (Robinson *et al.*, 2001; Chen *et al.*, 2003; Binkley and Kandelbauer, 2003).

Figure 1. Classification of industrial dyes
Symbols are given in the figure:
1st alphabet- D: direct, V: vat, S: sulfur, D': disperse, A: azo, R: reactive, B: basic, A': acidic
2nd alphabet- B: blue, B': black, B": Brown, R: red, Y: yellow, O: orange, G: green, V: violet
ADC 35: blue, ADC 2: orange, ADC 1: violet

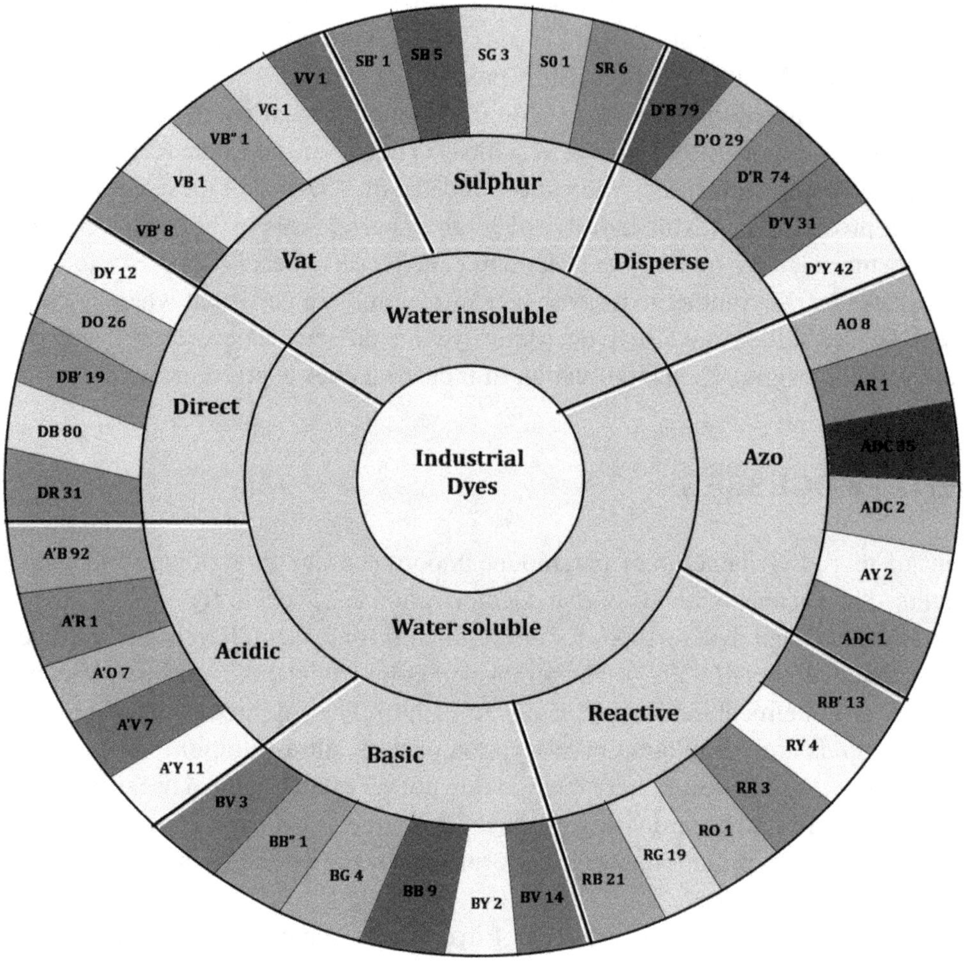

Factors Influencing Bioremediation

The effectiveness of microbial decolorization depends on the activity and the adaptability of the microorganisms. The nature of the environmental contaminant is another influencing factor. Physiochemical parameters like nutrients, presence of oxygen, moisture, pH, temperature and media are important in deciding the rate of dye degradation (Kensa, 2011). Most microbes degrade optimally in the pH range of 6-8 and temperature range of 25-40°C. Reductive cleavage of the azo bond by a wide variety of microorganisms, sediments, and sludges occurs readily under anaerobic conditions (Field *et al.*,1993; Stolz 2001).

Figure 2. The various treatment processes used for dye degradation

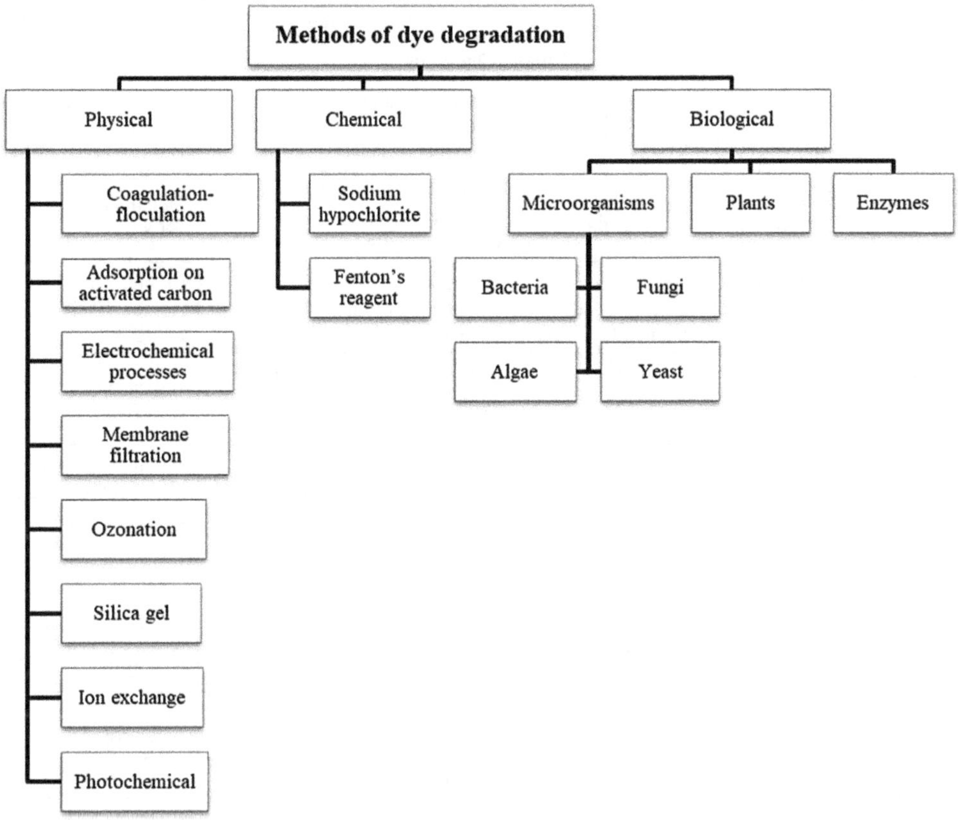

Microorganisms in Bioremediation

Many microorganisms are capable of decolorizing the azo dyes, including various Gram-positive and Gram-negative bacteria (Sani and Banerjee, 1999; Kodam *et al.*, 2005; Moosvi *et al.*, 2005), fungi (Balan and Monteneiro, 2001; Verma and Madamwar, 2005), actinomycetes and algae (Banat *et al.*, 1996). Fungi and algae have been successfully employed in dye decolorization. White rot edible mushroom can grow well on sludge and produce a large amount of laccase that decreases the toxicity of the sludge. However, moderate decolorization rate, the complexity of textile effluents and long growth cycle limits the performance of fungal systems. On the contrary, bacteria can degrade and mineralize many reactive azo dyes faster (Banat *et al.*, 1996). Table 2 lists the bacteria, fungi, yeast, and algae which have been reported to degrade various dyes.

Bacteria

Bacteria are capable of degrading and completely mineralizing various reactive dyes under certain specific conditions (Moosvi *et al.*, 2005; Wang *et al.*, 2009). Entire biodegradation of a compound is referred to as mineralization. There have been reports on the mineralization of dyes with bacteria (Haug *et al.*, 1991; Ramalingam and Shobana 2011; Singh *et al.*, 2011). Extensive research has been performed

Table 1. Various methods used for dye degradation along with their advantages and disadvantages

Processes	Advantages	Disadvantages	References
Physical Methods			
Coagulation-flocculation	Elimination of insoluble dyes Good removal efficiency	Sludge formation that blocks filter Cost of coagulants	Aguilar *et al.*, 2005; Golob and Ojstrsek., 2005 Dubrow *et al.*, 1996
Adsorption on activated carbon	Good removal efficiency Removes suspended solids	Large amounts of solid waste generated Cost of activated carbon	Slokar and Majcen Le Marechal., 1998
Electrochemical processes	Capacity to adapt to different volumes and pollution load End products are non-hazardous	High cost Formation of iron hydroxide sludges	Carneiro *et al.*, 2005 Bizuneh Adinew., 2012; Lin *et al.*, 1994; Laing, 1991
Membrane filtration	Removes all types of dyes Recovery and reuse of chemicals	Concentrated sludge production High running cost Dissolved solids are not removed	Anjaneyulu *et al.*, 2005; Chakraborty *et al.*, 2003
Ozonation	No alteration of volume	The short half-life of ozone Releases aromatic amines	Muthukumar *et al.*, 2005; Aplin& Wait., 2000; Langlais *et al.*, 1991; Adams *et al.*, 1995; Scott *et al.*, 1995; Park *et al.*, 1999
Silica gel	Effective for basic dye removal	Side reactions prevent commercial application	Robinson *et al.*, 2001
Ion exchange	No adsorbent loss due to regeneration	Not suitable for all dyes	Robinson *et al.*, 2001
Photochemical	No sludge production	Byproduct formation	Gogate& Pandit, 2004; Forgacs *et al.*, 2004
Chemical Methods			
Sodium hypochlorite	Initiates and accelerates azo bond cleavage	Release of aromatic amines	Robinson *et al.*, 2001
Fenton's reagent	Effective in decolorizing both soluble and insoluble dyes	Sludge disposal problems	Hao *et al.*, 2000; Nesheiwat& Swanson, 2000; Sun *et al.*, 2009; Ay *et al.*, 2009; Alshamsi *et al.*, 2007
Biological Methods			
Aerobic bacterial biodegradation	Decolorizes both azo and anthraquinone dyes	Low decolorization rates Requires specific oxygen catalyzed enzymes Requires additional carbon and energy sources	Stolz, 2001 Husain, 2006
Anaerobic bacterial biodegradation	Suitable for large scale application Allows obligate and facultative bacteria to reduce azo dyes	Generation of toxic substance Requires post treatment	Blumel *et al.*, 2002; Pinheiro *et al.*, 2004
Fungal biodegradation	Decolorizes anthraquinone and indigo-based dyes at higher rates	Decolorization rate is very low for azo dyes Needs acidic pH Inhibition by mixture of dyes and chemical in textile effluents	Libra *et al.*, 2004

Table 2. List of dyes decolorized by certain microorganisms

	Name of the Microorganism	Dye	Reference
Bacteria	*Klebsiella pneumoniae*	Methyl Red	Wong and Yuen 1998
	Bacillus gordonae	Tectilon Blue 4R-01	Walker and Weatherley 2000
	Klebsiella sp.	Orange 3R	Ponraj *et al.* 2011
	Clostridium perfringens	Bromophenol Blue	Kim *et al.* 2002
	Bacillus thuringiensis	Acid Red-119	Dave and Dave 2009
	Kocuria roseaMTCC 1532	Malachite Green	Parshetti *et al.* 2006
	Bacillus cereus strain DCII	Anthraquinone dye	Deng *et al.* 2008
	Pseudomonas sp. SUK-1	Reactive Red- 22	Kalyani *et al.* 2009
	Escherichia coli	Reactive Red- 22	Chang and Lin 2001
	Pseudomonas desmolyticum	Direct Blue-6	Kalme *et al.* 2007
Bacterial consortia	*Alcaligenes faecalis, Sphingomonas sp. EBD, Bacillus subtilis, Bacillus thuringienesis, Enterobacter cancerogenus*	Direct blue- 15	Kumar *et al.*, 2007
	Chryseo bacterium sp, Flavobacterium sp	Procion blue HERD	Khadijah *et al.*, 2009
	Providencia sp SDS, Pseudomonas aeruginosa BCH	Red HE3B	Phugare *et al.*, 2010
	Providencia rettgeri HSL1, Pseudomonas sp SUK1	Reactive black 5, Reactive Orange 16, Disperse Red 78, Direct Red 81	Harshad *et al.*, 2015
	Paenibcillus polymyxa, Micrococcus luteus, Micrococcus sp	Reactive violet 5R	Moosvi *et al.*, 2007
	Proteus vulgaris, Micrococcus glutamicus	Scarlet R	Saratale *et al.*, 2009
	Proteus mirabilis, Aeromonas caviae, Rhodococcus globerulus	Acid orange 7	Joshi *et al.*, 2008
Fungi	*Mucor mucedo*	Congo red	Gupta and Kumari, 2012
	Trichoderma harzianum	Direct green, Acid red, Bromophenol blue, Basic blue	Singh and Singh, 2010
	Phanerocheate chrysosporium	Orange II, Tropaeolin	Cripps *et al.*, 1990, Singh and Singh, 2010
	Aspergillus flavus	Bromophenol blue, Congo red	Singh and Singh, 2010
	Fusarium solani	Methylene blue	Al-Jawhari, 2015
	Gliocladium virens	Congo red	Singh, 2008
	Penicillium funigulosum	Methylene blue	Al-Jawhari, 2015
Yeast	*Candida krusei*	Basic Violet 3	Charumathi and Das 2011.
	Saccharomyces cerevisiae	triphenylmethane dyes	Jadhav and Govindwar, 2008
	Galactomyces geotrichum	methyl red	Jadhav *et al.*, 2008
	Candida tropicalis	reactive textile dyes	Donmez, 2002
	Candida tropicalis	Acid Blue 93, Direct red 28 and Basic Violet 3	Charumathi and Das, 2010
	Galactomyces geotrichum	Remazol Red	Waghmode *et al.*, 2012
	Candida zeylanoides	azo dyes	Martins *et al.* 1999, Ramalho *et al.* 2002
Algae	*Chlorella tropicalis*	basic violet 3	Charumathi and Das, 2010
	Cosmarium sp	Malachite Green	Daneshvar *et al.*, 2007
	Chlorella vulgaris	Tectilon Yellow 2G (TY2G), and di-azo dye, Erionyl Navy (ENR)	Acuner and Dilek 2002
	Chlorella pyrenoidosa, Chlorella vulgaris and *Oscillateria tenius*	Azo dyes	Yan and Pan 2004

on biodegradation of dyes by bacteria. Diverse studies have proved the ability of bacteria to degrade dyes both aerobically and anaerobically (Banat *et al.* 1996). Dye decolorization by bacteria has been examined under different conditions. A high alkali thermostable *Bacillus* sp. strain SF isolated from the wastewater drain of a textile finishing company was capable of degrading an azo dye (Maier *et al.*, 2004). In reactive azo dyes like Remazol Black B, Remazol Brilliant Orange 3R, and Remazol Brilliant Violet 5R, total degradation was attained using an anaerobic-aerobic treatment (Supaka *et al.*, 2003). Degradation of azo dyes occurs under combined anaerobic/aerobic conditions if co-substrate and oxygen are balanced (Tan, 2001). Certain specific strains of aerobic bacteria have developed the capability to utilize azo dyes as the only source of carbon and nitrogen (Barragan *et al.*, 2006).

Halophiles have been reported to take part in dye decolorization (Salah Uddin *et al.*, 2007). Decolorization may occur as a result of the high metabolic diversity in the halophiles owing to their extremophilic nature (Oren *et al.*, 1992; Ventosa *et al.*, 1998).

In phthalocyanine dyes, there is reversible reduction and decolorization under anaerobic conditions (Van der Zee, 2002). Research is focused on the bacteria which can achieve a high rate of discoloration under anoxic conditions and detoxify aromatic products under aerobic situations (Stolz, 2001; Upadhyay, 2002; Chen *et al.*, 2007). Several strains of *Pseudomonas, Alcaligenes, Acinetobacter,* and *Bacillus* are some of the vital bacteria which are useful in bioremediation of halogenated organic compounds (Pandey *et al.*, 2011). *Arthobacter sp.* isolated from textile effluents is capable of almost completely degrading pararosaniline chloride dye (Dwivedi *et al.*, 2012). Actinomycetes, especially *Streptomyces* species can produce extracellular peroxidases that take part in the degradation of lignin and dyes (Bhaskar *et al.*, 2003). Under aerobic conditions, azo dyes cannot be metabolized readily; even though it was found that the bacteria with specific reducing enzymes were able to aerobically degrade some azo dyes (Stolz, 2001). Besides, numerous bacteria can degrade anthraquinone dyes (Fontenot *et al.*, 2001). Moreover, studies on bacterial species were also carried on the degradation of reactive dyes. *Rhodopseudomonas palustris*51ATA strain is an excellent bacterium for the removal of reactive red 195 dye (Ozturk *et al.*, 2012).

Fungi

The most extensively studied group of microorganisms with respect to dye decolorization and degradation are the fungi (Agarry and Ajani, 2011). Among these the most widely researched fungi are the ligninolytic fungi (Zille, 2005). Various classes of fungi produce ligninolytic enzymes, but white-rot fungi are capable of extensively producing these enzymes and have the unique ability to transform lignin to CO_2 (Lopez *et al.*, 2012). This extent of degradation is attributed to the strong oxidative activity and low substrate specificity of their ligninolytic system, which primarily contains laccase (Lac), Lignin peroxidase (LiP), and manganese peroxidase (MnP) (Martinez *et al.*, 2009). The other commonly studied white rot fungi are *Phanerochaete chrysosporium* (Chagas and Durrant 2001; Martins *et al.*, 2002), *Trametes versicolor* (Kapdan *et al.*, 2000; Kapdan and Kargi, 2002) and *Bjerkandera adusta* (Heinfling-Weidtmann *et al.*, 2001). Various other non-white rot fungi like *Aspergillus niger* (Abd El-Rahim and Moawad, 2003), *Geotrichum candidum* (Kim *et al.*, 1995; Shoda and Kim, 1999), *Pleurotus oestreatus* (Martins *et al* 2003; Palmieri *et al.*, 2005) and *Cunninghamella elegans* (Cha *et al.*, 2001; Ambrosio and Campos –Takaki 2004), can also effectively decolorize dyes. The use of different *Aspergillus* species (*Aspergillus niger, Aspergillus foetidus, Aspergillus fumigatus* and *Aspergillus terreus*) for decolorization of various classes of dyes have been reported (Sumathi and Manju, 2000; Ali *et al.*, 2007; Jin *et al.*, 2007; Andleeb et al., 2010). Various fungi such as *Aspergillus niger* (Fu and Viraraghavan, 1999, 2000),

Rhizopus arrhizus (Zhou and Banks, 1991, 1993), *Rhizopus oryzae* (Gallagher *et al.*, 1997) which can also decolorize and/or biosorb diverse dyes.

Fungican degrade although not completely, the structures that are difficult for the bacteria to handle, as a result of their extracellular enzymes excretion (Forss and Welander, 2009).

The primary mechanism of decolorization is the adsorption of dyes to the microbial cell surface (Knapp *et al.*, 1995). There is not much data available on the mineralization of dyes by fungi. The fungal action rarely results in mineralization of dyes and it depends on the chemical structure (Pavko, 2011).

The fungal mycelia have an additive advantage over single cell organisms by solubilizing the insoluble substrates by producing extracellular enzymes, due to an increased cell-to-surface ratio; fungi have a greater physical and enzymatic contact with the environment. The extracellular nature of the fungal enzymes is also advantageous in tolerating high concentrations of the toxicants (Kaushik and Malik, 2009).

Yeast

The yeasts have many advantages compared to bacteria and filamentous fungi. Yeasts are a better raw biosorbent material for the removal of reactive dye due to their unicellular nature and high growth rate. Yeast cells can be easily cultivated into inexpensive growth media, and are a readily available source of biomass that has the potential for bioremediation of wastes at lower pH values. Yeasts are the inexpensive and readily available source. Yeasts can adapt and grow in various extreme conditions of pH, temperature, and availability of nutrients as well as high concentrations of pollutant. The cell walls of Baker's yeast are composed of glucan and mannan with traces of chitin (Zhou *et al*, 2008; Abd El-Rahim *et al.*, 2009; Ertugrul *et al.*, 2009; El-Hennawi *et al.*, 2013; Kelewou *et al.*, 2014; Sadeghi *et al.*, 2014)

The ability of yeasts to degrade dyes is described only in a few reports. According to a study, the ascomycete yeast *Candida zeylanoides* isolated from contaminated soil can reduce model azo dyes (Ramalho *et al.*, 2002). The enzymatic activity characterization is described in studies using the yeast *Issatchenkia occidentalis* (Ramalho *et al.*, 2004), and the enzymatic system involved is explained in a study using *Saccharomyces cerevisiae* (Ramalho *et al.*, 2005).

Dye decolorization by yeasts usually occurs by three mechanisms, viz., biosorption, bioaccumulation and biodegradation (Das and Charumathi, 2012). *Galactomyces geotrichum, Saccharomyces cerevisiae and Trichosporon beigelii,* etc. (Jadhav *etal.,* 2008) can decolorize several dyes.

Algae

Algae are one of the most adaptive and diverse groups of organisms on earth. Algae develop themselves to adapt in a polluted environment and not only that they eventually flourish well. Some azo dyes present in water bodies exhibit high toxicity to aquatic life but do not significantly reduce algal growth (Acuner, 2004). The one more advantage of using algae for bioremediation is that they do not require any carbon sources for their growth, derives energy from sunlight and carbon from the air, and scavenges atmospheric nitrogen (Saha *et al.*, 2010). The phenomena of degradation by algae are likely to associate with enzymatic activity either, absorption or both (Solis *et al.*, 2012). Algae used in both living and non-living form for dye degradation. The mechanism for degradation by living algae may be due to the enzymatic while non algae remove dyes by bio-absorption.

Studies have indicated that in addition to providing oxygen for aerobic bacterial biodegraders, microalgae can also biodegrade organic pollutants directly (Mallick 2002; Tam *et al.*, 2002). The use of algae in dye degradation is mentioned in only a few reports and is achieved by *Chlorella* (Jinqi and Houtian, 1992; Acuner and Dilek, 2004), *Oscillateria* (Jinqi and Houtain, 1992) and *Spirogyra* (Mohan *et al.*, 2002) species. All these reports used azo dyes (Jinqi and Houtain, 1992) state that some of the tested azo compounds could be used as sole sources of carbon and nitrogen by the algae. This could mean that algae can play an important role in the removal of azo dyes and aromatic amines in stabilization ponds (Banat *et al.*, 1996). Usage of algae for dye degradation is rare. Both viable and non-viable algae have been used in textile wastewater treatment. This may be achieved via bioconversion and biosorption (Maghraby, 2013). It is achieved by *Chlorella* (Jinqi and Houtian, 1992; Acuner and Dilek, 2004), *Oscillateria* (Jinqi and Houtain, 1992) and *Spirogyra* (Mohan *et al.*, 2002) species. Algae have been studied for the decolorization of industrial effluents (Jianwei *et al.,* 1997; Semple *et al.*, 1999; Daneshvar *et al.*, 2007; Vijayaraghavan and Yun, 2008).

DYE DECOLORIZATION USING GENETICALLY MODIFIED ORGANISMS

Genetically engineered bacteria can be used efficiently to degrade dyes. This can be done by incorporating the dye-degrading gene from bacteria known to degrade dyes into rapidly growing microbial cells which lack the capacity to degrade dyes. The gene is usually a dye degrading enzyme like azoreductase. Some of the genetically engineered bacteria used for dye degradation are given in table 3.

DYE DECOLORIZATION USING IMMOBILIZED MICROBES

The use of freely suspended microbial cells for dye removal is limited owing to their inherent disadvantages such as small particle size, possible clogging and low mechanical strength of the biomass. Immobilized cells offer advantages over-dispersed cells such as high cell density, strong endurance of toxicity, lower operating costs, simple maintenance management, and lower residual sludge. Numerous studies have proved that the mechanisms for the decolorization of dyes using immobilized microorganisms may include both biodegradation and adsorption (Chen *et al.*, 2003; Pazarlioglu *et al.*, 2005). Table 4 lists the immobilized bacteria and immobilized fungi which have been reported to degrade various dyes.

Table 3. List of Genetically engineered bacteria for dye degradation

Sl. No.	Donor Microbe	Acceptor Microbe	Function	Reference
1	*Pseudomonas luteola*	*Escherichia coli*	Azoreductase	Chang *et al.*, 2000
2	*Bacillus sp*	*Escherichia coli*	Azoreductase	Suzuki *et al.*, 2001
3	*Rhodobacter sphaeroides*	*Escherichia coli*	Azoreductase	Jin *et al.*, 2009
4	*Clostridium perfringens*	*Escherichia coli*	Azoreductase	Rafii and Coleman., 1999
5	*Rhodococcus sp*	*Escherichia coli*	Azoreductase	Chang and Lin., 2001
6	*Caulobacter subvibrioides*	*Escherichia coli*	Azoreductase	Govind *et al.*, 1993

Table 4. List of immobilized microbes found to degrade dyes

Sl. No.	Microbe	Dye	Reference
Immobilized Bacteria			
1	*Pseudomonas putida* and *Bacillus licheniformis* immobilized using sodium alginate and polyacrylamide gel beads	Reactive Red 195, Reactive Orange 72, Reactive Yellow 17, Reactive Blue 36	Suganya and Revathi, 2016
2	*Desmodesmus* species	methylene blue and malachite green	Al-Fawwaz and Abdullah, 2016
3	*Bacillus firmus* in tubular polymeric gel	C.I. Direct red 80,	Ogugbue *et al.*, 2011
Immobilized Fungi			
1	White-rot basidiomycetous fungi	Various Dyes	Abarnadevi *et al.*, 2013
2	White-rot fungus *Phanerochaete chrysosporium*	Various Dyes	Susana 2009

Figure 3. Different physical, chemical, advanced oxidation processes and biological methods of dye degradation which could be used in various combinations for efficient dye degradation

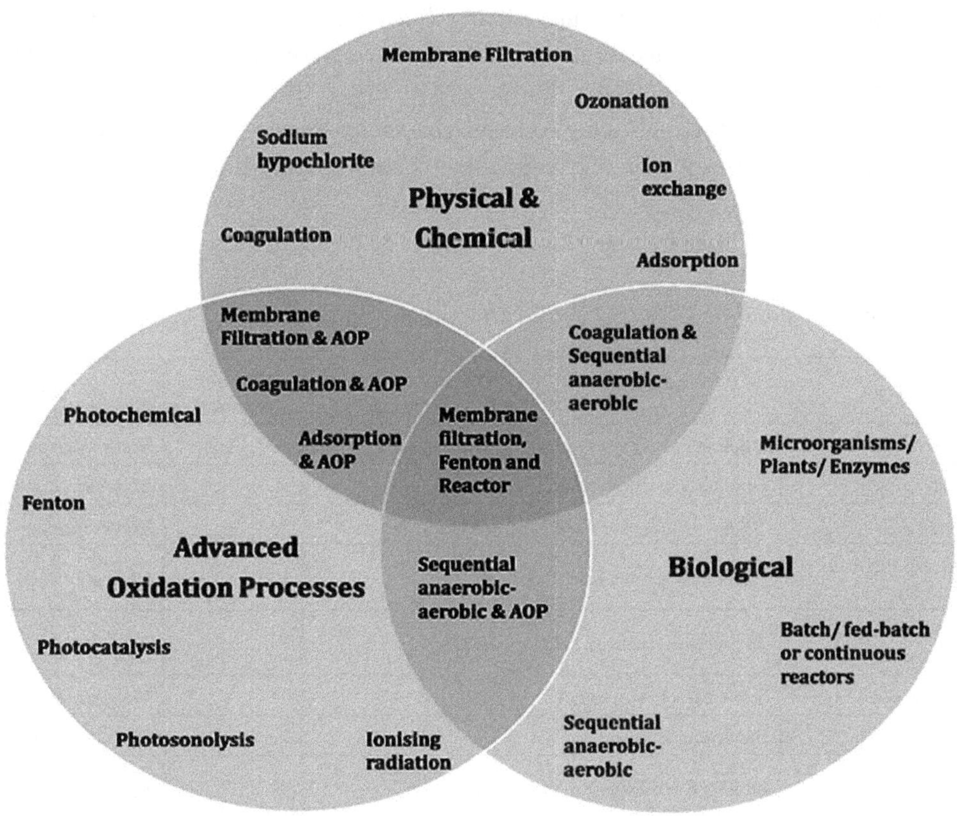

REACTORS USED FOR LARGE SCALE DYE DEGRADATION

Fixed film bioreactors, packed bed bioreactors, anaerobic/ aerobic rotating biological contactors, fluidized bed reactors, etc. have been used in dye decolorization (Kapdan and Kargi., 2002). Different reactor configurations using mixed microbial consortia like upflow anaerobic sludge blanket (UASB), anaerobic/ aerobic sequential batch reactor (SBR), expanded granular sludge bed (EGSB), activated sludge process (ASP) have been reported for the dye decolorization (Dos Karatas *et al.*, 2010; Da Silva *et al.*, 2012; Yasar *et al.*, 2012). The effects of various operating conditions like bed expansion, cell-bead number, density, initial dye concentration, hydraulic retention time and diameters of immobilized cell beads on azo dye decolorization using fluidized bed reactor have been reported (Wu *et al.*, 2005).

Figure 3 represents how all the physical, chemical and biological methods can be used in combination for efficient dye degradation.

LIMITATIONS OF BIOREMEDIATION

The major drawback of bioremediation is that only materials which are biodegradable can be degraded by this method. Treatment time is usually longer than other methods. If the conditions are not carefully monitored the breakdown might be partial resulting in more toxic by-products. Hence the physiochemical parameters need to be maintained precisely (Sharma and Reddy, 2004). Extensive research is required to extrapolate bioremediation from bench scale to industrial scale. Research should focus on site specific requirements for degradation of specific contaminants, its nature, conditions needed for its degradation etc. (Kensa, 2011).

Table 5. List of dyes degraded in various bioreactors

Sl. No.	Dye	Reactor	Dye Concentration (mg/L)	% Color Removal	Reference
1	Reactive red 120	Sulphidogenic bioreactor	1000	50	Mutambanengwe *et al.*, 2007
2	Reactive black 5	Sulphidogenic bioreactor	1000	50	Mutambanengwe *et al.*, 2007
3	Orange 2	Sulphidogenic bioreactor	1000	53	Mutambanengwe *et al.*, 2007
4	Remazol Brilliant Violet 5R	Anaerobic-aerobic SBR	30	100	Cirik *et al.*, 2012
5	Simulated mixed azo dye waste	Two-phase anaerobic PBR	1200	73	Talarposhti *et al.*, 2001
6	Disperse blue 79	SBR anaerobic- aerobic	533	94	Melgoza and Buitron., 2004
7	Acid orange 7	Batch anaerobic-aerobic	100	99	Manu and Chaudhari., 2002
8	Remazol Brilliant Violet 5R	Fluidized bed reactor	150	99	Sen and Demierer 2003
9	Direct black 38	UASB- CSTR	3200	1000	Isik and Sponza., 2004
10	Acid red 88	FB-SBBR	100	72	Koupaie *et al.*, 2013

CONCLUSION

Thus, bioremediation is a very good alternative to the physiochemical methods of dye decolorization and can serve towards reducing the industrial dye effluents. Immobilization of the microorganisms increases their reusability in decolorization and the beads can be used for several rounds of decolorization. This provides scope for their use in reactors. Thus, the immobilized microorganisms can be used in packed bed reactors for dye decolorization of textile dye effluents. A major drawback of bioremediation is that the growth of the microorganism might be hindered by the dye product or the activity of other unwanted microorganisms. A detailed study on the mechanism by which the microorganism degrades the dye would throw light on how this drawback could be overcome.

ACKNOWLEDGMENT

This work was supported by the DST Science and Engineering Research Board (SERB, Grant No. SERB/LS-267/2014) and the Extra Mural Research Funding of Ayurveda, Yoga and Naturopathy, Unani, Siddha and Homoeopathy (AYUSH, Grant No. Z. 28015/209/2015-HPC).

REFERENCES

Abarnadevi, J., Anu, M., Bharani, M., & Prabha, L. P. (2013). Treatment of dye industry effluent by free and immobilized fungi. *International Journal of Pharmacy & Life Sciences*, *4*, 2340–2346.

Abd El-Rahim W. M., El-Ardy, O. A. M., Mohammad, F. H. A. (2009). The effect of pH on bioremediation potential for the removal of direct violet textile dye by *Aspergillus niger Desalination*, 249, 1206–1211.

Abd El-Rahim, W. M., & Moawad, H. (2003). Enhancing bioremoval of textile dyes by eight fungal strains from media supplemented with gelatin wastes and sucrose. *Journal of Basic Microbiology*, *43*(5), 367–375. doi:10.1002/jobm.200310267 PMID:12964179

Abdullah, A. T., & Abdullah, M. (2016). Decolorization of methylene blue and malachite green by immobilized *desmodesmus* sp. Isolated from North Jordan. *International Journal of Environmental Sciences and Development*, *7*(2), 95–99. doi:10.7763/IJESD.2016.V7.748

Acuner, E., & Dilek, F. B. (2004). Treatment of tectilon yellow 2G by *Chlorella vulgaris*. *Process Biochemistry (Barking, London, England)*, *39*(5), 623–631. doi:10.1016/S0032-9592(03)00138-9

Adams, C. D., Fusco, W., & Kanzelmeyer, T. (1995). Ozone hydrogen peroxide/ ozone and UV/ozone treatment of chromium and copper complex dyes: Decolorization and metal release. *Ozone Science and Engineering*, *17*(2), 149–161. doi:10.1080/01919519508547543

Adinew, B. (2012). Textile effluent treatment and decolorization techniques – a review. *Chemistry. Bulgarian Journal of Science Education*, *21*(3).

Agarry, S. E., & Ajani, A. O. (2011). Evaluation of microbial systems for biotreatment of textile waste effluents in nigeria: biodecolorisation and biodegradation of textile dye. *Journal of Applied Science & Environmental Management*, *15*(1), 79–86. doi:10.4314/jasem.v15i1.65680

Aguilar, M. I., Sáez, J., Lloréns, M., Soler, A., Ortuno, J. F., Meseguer, V., & Fuentes, A. (2005). Improvement of coagulation-flocculation process using anionic polyacrylamide as coagulant aid. *Chemosphere*, *58*(1), 47–56. doi:10.1016/j.chemosphere.2004.09.008 PMID:15522332

Al-Jawhari, I. F. H. (2015). Decolorization of methylene blue and crystal violet by some filamentous fungi. *International Journal of Environmental Bioremediation & Biodegradation*, *3*, 62–65.

Ali, N., Hameed, A., Ahmed, S., & Khan, A. G. (2007). Decolorisation of structurally different textile dyes by *Aspergillus niger* SA1. *World Journal of Microbiology & Biotechnology*, *24*(7), 1067–1072. doi:10.100711274-007-9577-2

Alshamsi, F., Albadwawi, A., Alnuaimi, M., Rauf, M., & Ashraf, S. (2007). Comparative efficiencies of the degradation of crystal violet using UV/hydrogen peroxide and Fenton's reagent. *Dyes and Pigments*, *74*(2), 283–287. doi:10.1016/j.dyepig.2006.02.016

Ambrosio, S. T. (2004). Decolorisation of reactive azo dyes by Cunninghamella elegans UCP 542 under co-metabolic conditions. *Bioresource Technology*, *91*(1), 69–75. doi:10.1016/S0960-8524(03)00153-6 PMID:14585623

Andleeb, S., Atiq, N., Ali, M. I., Razi-Ul-Hussain, R., Shafique, M., Ahmed, B., ... Ahmed, S. (2010). Biological treatment of textile effluent in stirred tank bioreactor. *International Journal of Agriculture and Biology*, *12*, 256–260.

Anjaneyulu, Y., Chary, N. S., & Raj, D. S. S. (2005). Decolourization of industrial effluents - available methods and emerging technologies - a review. *Reviews in Environmental Science and Biotechnology*, *4*(4), 245–273. doi:10.100711157-005-1246-z

Aplin, R., & Waite, T. D. (2000). Comparison of three advanced oxidation processes for the degradation of textile dyes. *Water Science and Technology*, *42*(5-6), 345–354. doi:10.2166/wst.2000.0534

Ay, F., Catalkaya, E. C., & Kargi, F. (2009). A statistical experiment design approach for advanced oxidation of direct red azo-dye by photo-Fenton treatment. *Journal of Hazardous Materials*, *162*(1), 230–236. doi:10.1016/j.jhazmat.2008.05.027 PMID:18555599

Balan, D. S. L., & Monteneiro, R. T. R. (2001). Decolorization of textile indigo dye by ligninolytic fungi. *Journal of Biotechnology*, *89*(2-3), 141–145. doi:10.1016/S0168-1656(01)00304-2 PMID:11500207

Banat, I. M., Nigam, P., Singh, D., & Marchant, R. (1996). Microbial decolorization of textile-dye containing effluents: A review. *Bioresource Technology*, *58*(3), 217–227. doi:10.1016/S0960-8524(96)00113-7

Barragan, B. E., Costa, C., & Marquez, M. C. (2007). Biodegradation of azo dyes by bacteria inoculated on solid media. *Dyes and Pigments*, *75*(1), 73–81. doi:10.1016/j.dyepig.2006.05.014

Bhaskar, M., Gnanamani, A., Ganeshjeevan, R. J., Chandrasekar, R., Sadulla, S., & Radhakrishnan, G. (2003). Analyses of carcinogenic aromatic amines released from harmful azo colorants by Streptomyces. *Journal of Chromatography. A*, *1018*(1), 117–1123. doi:10.1016/j.chroma.2003.08.024 PMID:14582632

Binkley, J., & Kandelbauer, A. (2003). Effluent treatment/enzymes in activated sludge, In A. Cavaco-Paulo & G. M. Gubitz, (Eds.), Textile processing with enzymes (Woodhead Publishing, Cambridge), 199-221.

Blümel, S., Knackmuss, H. J., & Stolz, A. (2002). Molecular cloning and characterization of the gene coding for the aerobic azoreductase from *Xenophilusazovorans*KF46F. *Applied and Environmental Microbiology*, *68*(8), 3948–3955. doi:10.1128/AEM.68.8.3948-3955.2002 PMID:12147495

Carneiro, P. A., Osugi, M. E., Fugivara, C. S., Boralle, N., Furlan, M.,, & Zanoni, M. V. B. (2005). 'Evaluation of different electrochemical methods on the oxidation and degradation of Reactive Blue 4 in aqueous solution'. *Chemosphere*, *59*(3), 431–439. doi:10.1016/j.chemosphere.2004.10.043 PMID:15763096

Cha, C. J., Doerge, D. R., & Cerniglia, C. E. (2001). Biotransformation of malachite green by the fungus *Cunninghamella elegans*. *Applied and Environmental Microbiology*, *67*(9), 4358–4360. doi:10.1128/AEM.67.9.4358-4360.2001 PMID:11526047

Chagas, E. P., & Durrant, L. R. (2001). Decolorization of azo dyes by Phanerochaetechrysosporium and Pleurotussajorcaju. *Enzyme and Microbial Technology*, *29*(8-9), 473–477. doi:10.1016/S0141-0229(01)00405-7

Chakraborty, S., Purkait, M. K., DasGupta, S., De, S., & Basu, J. K. (2003). Nanofiltration of textile plant effluent for color removal and reduction in COD. *Separation and Purification Technology*, *31*(2), 141–151. doi:10.1016/S1383-5866(02)00177-6

Chang, J. S., & Lin, C. Y. (2000). Fed-batch bioreactor strategies for microbial decolorization of azo dye using a Pseudomonas luteola strain. *Biotechnology Progress*, *16*(6), 979–985. doi:10.1021/bp000116z PMID:11101324

Chang, J. S., & Lin, C. Y. (2001). Decolorization Kinetics of a recombinant *Escherichia coli* strain harboring azo-dye-decolorizing determinants from Rhodococcus sp. *Biotechnology Letters*, *23*(8), 631–636. doi:10.1023/A:1010306114286

Charumathi, D., & Das, N. (2010). Removal of synthetic dye basic violet 3 by immobilized *Candida tropicalis* grown on sugarcane bagasse extract medium. *International Journal of Engineering Science and Technology*, *2*, 4325–4335.

Chen, C. C., Liao, H. J., Cheng, C. Y., Yen, C. Y., & Chung, Y. C. (2007). Biodegradation of crystal violet by Pseudomonas putida. *Biotechnology Letters*, *29*(3), 391–396. doi:10.100710529-006-9265-6 PMID:17206376

Chen, K. C., Wu, J. Y., Liou, D. J., & Hwang, S. C. J. (2003). Decolorization of the textile azo dyes by newly isolated bacterial strains. *Journal of Biotechnology*, *101*(1), 57–68. doi:10.1016/S0168-1656(02)00303-6 PMID:12523970

Cirik, K., Kitis, M., & Cinar, O. (2012). The effect of biological sulfate reduction on anaerobic color removal in anaerobic-aerobic sequencing batch reactors. *Bioprocess and Biosystems Engineering*. doi:10.1007/ s00449-012-0813-2 PMID:23277271

Couto, S. R. (2009). Dye removal by immobilized fungi. *Biotechnology Advances*, *27*(3), 227–235. doi:10.1016/j.biotechadv.2008.12.001 PMID:19211032

Cripps, C., Bumpus, J. A., & Aust, S. D. (1990). Biodegradation of azo and heterocyclic dyes by Phanerochaete chrysosporium. *Applied and Environmental Microbiology*, *56*, 1114–1118. PMID:2339873

Da Silva, M., Firmino, P., De Sousa, M., & Dos Santos, A. (2012). Sequential anaerobic/aerobic treatment of dye—containing wastewaters: Color and COD removals and Ecotoxicity tests. *Applied Biochemistry and Biotechnology*, *166*(4), 1057–1069. doi:10.100712010-011-9493-7 PMID:22238010

Daneshvar, N., Khataee, A. R., Rasoulifard, M. H., & Pourhassan, M. (2007). Biodegradation of dye solution containing malachite green: Optimization of effective parameters using the Taguchi method. *Journal of Hazardous Materials*, *143*(1-2), 214–219. doi:10.1016/j.jhazmat.2006.09.016 PMID:17052836

Das, N., & Charumathi, D. (2012). Remediation of synthetic dyes from wastewater using yeast – An overview. *Indian Journal of Biotechnology*, *11*, 369–380.

Dave, S. R., & Dave, R. H. (2009). Isolation and characterization of *Bacillus thuringiensis* for Acid Red-119 dye decolorization. *Bioresource Technology*, *100*(1), 249–253. doi:10.1016/j.biortech.2008.05.019 PMID:18590958

Deng, D., Guo, J., Zeng, G., & Sun, G. (2008). Decolorization of anthraquinone, triphenylmethane and azo dyes by a new isolated Bacillus cereus strain DC11. *Int. Biodetor. Biodegrad*, *62*(3), 263–269. doi:10.1016/j.ibiod.2008.01.017

Dönmez, G. (2002). Bioaccumulation of the reactive textile dyes by Candida tropicalis growing in molasses medium. *Enzyme and Microbial Technology*, *30*(3), 363–366. doi:10.1016/S0141-0229(01)00511-7

Dos Santos, A., Cervantes, F., & Vanlier, J. (2007). Review paper on current technologies for decolorization of textile wastewaters: Perspectives for anaerobic biotechnology. *Bioresource Technology*, *98*(12), 2369–2385. doi:10.1016/j.biortech.2006.11.013 PMID:17204423

Dubrow, S., Boardman, G., & Michelsen, D. (1996). Chemical pretreatment and aerobic-anaerobic degradation of textile dye wastewater. John Wiley & Sons, 75–102.

Dwivedi, P., Kaur, J., Dhanjal, N. I. K., & Cameotra, S. S. (2012). Biodecolorization of pararosaniline chloride dye using a bacterial isolate, Arthrobacter species. *J. Adv. Dev. Res*, *3*, 77–81.

El-Hennawi, H. M., Shahin, A. A., & Abd El-Thalouth, I. (2013). Evaluation of dried biomass of baker's yeast as reactive dye adsorbent using ultrasonic technique. *Journal of Applied Sciences Research*, *9*, 1401–1408.

Ertugrul, S., San, N. O., & Dönmez, G. (2009). Treatment of dye (Remazol Blue) and heavy metals using yeast cells with the purpose of managing polluted textile wastewaters. *Ecological Engineering*, *35*(1), 128–134. doi:10.1016/j.ecoleng.2008.09.015

Field, J., & Brady, J. (2003). Riboflavin as a redox mediator accelerating the reduction of the azo dye Mordant Yellow 10 by anaerobic granular sludge. *Water Science and Technology*, *48*(6), 187–193. doi:10.2166/wst.2003.0393 PMID:14640217

Field, J. A., DeJong, E., Feijoo, C. G., & DeBont, J. A. M. (1993). Screening for ligninolytic fungi application to the biodegradation of xenobiotics. *Trends in Biotechnology*, *11*(2), 44–49. doi:10.1016/0167-7799(93)90121-O

Fontenot, E. J., Lee, Y. H., Matthews, R. D., Zhu, G., & Pavlostathis, S. G. (2003). Reductive decolorization of a textile reactive dye bath under methanogenic conditions. *Applied Biochemistry and Biotechnology, 109*(1-3), 207–225. doi:10.1385/ABAB:109:1-3:207 PMID:12794295

Forgacs, E., Cserhati, T., & Oros, G. (2004). Removal of synthetic dyes from wastewaters: A review. *Environment International, 30*(7), 953–971. doi:10.1016/j.envint.2004.02.001 PMID:15196844

Forss, J., & Welander, U. (2009). Decolorization of reactive azo dyes with microorganisms growing on soft wood chips. *Int BiodeterBiodegr, 63*, 752–758.

Fu, Y., & Viraraghavan, T. (1999). Removal of acid blue 29 from aqueous solution by fungus Aspergillus niger. In *Proceedings of the 31st Mid-Atlantic Industrial and Hazardous Waste Conference.* Storrs, Connecticut. Technomic Publishing Company, Lancaster, PA. pp. 510-519.

Fu, Y., & Viraraghavan, T. (2000). Removal of a dye from an aqueous solution by fungus Aspergillus niger. *Water Quality Research Journal of Canada, 35*(1), 95–111. doi:10.2166/wqrj.2000.006

Gallagher, K. A., Healy, M. G., & Allen, S. J. (1997). Biosorption of synthetic dye and metal ions from aqueous effluents using fungal biomass. In D. L. Wise (Ed.), *Global Environmental Biotechnology* (pp. 27–50). UK: Elsevier.

Gogate, P. R., & Pandit, A. B. (2004). A review of imperative technologies for wastewater treatment II: Hybrid methods. *Advances in Environmental Research, 8*(3-4), 553–597. doi:10.1016/S1093-0191(03)00031-5

Golob, V., & Ojstršek, A. (2005). Removal of vat and disperse dyes from residual pad liquors. *Dyes and Pigments, 64*(1), 57–61. doi:10.1016/j.dyepig.2004.04.006

Govind, A. M., Schmidt, T. M., White, D. C., & Loper, J. C. (1993). Phylogenetic analysis of bacterial aerobic degrader of azo dyes. *Journal ofBacteriology, 175*(18), 6062–6066. doi:10.1128/jb.175.18.6062-6066.1993 PMID:8376354

Gregory, P. (1993). Dyes and dyes intermediates. In J. I. Kroschwitz (Ed.), *Encyclopedia of chemical technology* (Vol. 8, pp. 544–545). New York, NY: John Wiley & Sons.

Hai, F. I., Yamamoto, K., & Fukushi, K. (2007). Hybrid treatment system for dye wastewater. *Critical Reviews in Environmental Science and Technology, 37*(4), 315–377. doi:10.1080/10643380601174723

Hao, O., Kim, H., & Chiang, P. C. (2000). Decolorization of wastewater. *Critical Reviews in Environmental Science and Technology, 30*(4), 449–505. doi:10.1080/10643380091184237

Haug, W., Schmidt, A., Nortemann, B., Hempel, D. C., Stolz, A., & Knackmuss, H. J. (1991). Mineralization of the sulfonated azo dye Mordant Yellow 3 by a 6-aminonaphthalene-2-sulfonate degrading bacterial consortium. *Applied and Environmental Microbiology, 57*, 3144–3149. PMID:1781678

Heinfling-Weidtmann, A., Reemtsma, T., Storm, T., & Szewzyk, U. (2001). Sulfophthalimide as major metabolite formed from sulfonated phthalocyanine dyes by the white-rot fungus Bjerkanderaadusta. *FEMS Microbiology Letters, 203*(2), 179–183. doi:10.1111/j.1574-6968.2001.tb10838.x PMID:11583845

Husain, Q. (2006). Potential applications of the oxidoreductive enzymes in the decolorization and detoxification of textile and other synthetic dyes from polluted water: A review. *Critical Reviews in Biotechnology*, *26*(4), 201–221. doi:10.1080/07388550600969936 PMID:17095432

Isik, M., & Sponza, D. (2004). Monitoring of toxicity and intermediated of C. I. Direct Black 38 azo dye through decolorization in an anaerobic/aerobic sequential reactor system. *Journal of Hazardous Materials*, *114*(1-3), 29–39. doi:10.1016/j.jhazmat.2004.06.011 PMID:15511571

Jadhav, S. U., Jadhav, M. U., Kagalkar, A. N., & Govindwar, S. P. (2008). Decolorization of brilliant blue G Dye mediated by the degradation of the microbial consortium of Galactomyces geotrichum and Bacillus sp. *J Chin Inst Chem Engrs*, *39*(6), 563–570. doi:10.1016/j.jcice.2008.06.003

Jianwei, M., Xingdong, Y., & Gulan, Z. (1997). Studies on the decolorization of azo dyes in the aqueous system using metal ion together with algae (*Saragassum). Huanj. Huax. Chin*, *16*, 238–240.

Jin, R., Yang, H., Zhang, A., Wang, J., & Liu, G. (2009). Bioaugmentation on decolorization of C.I.Direct Blue 71 by using genetically engineered strain *Escherichia coli* JM109 (PGEX-AZR). *Journal of Hazardous Materials*, *163*(2-3), 1123–1128. doi:10.1016/j.jhazmat.2008.07.067 PMID:18755538

Jin, X., Liu, G., Xu, Z., & Yao, W. (2007). Decolorization of a dye industry effluent by Aspergillus fumigatus XC6. *Applied Microbiology and Biotechnology*, *74*(1), 239–243. doi:10.100700253-006-0658-1 PMID:17086413

Jinqi, L., & Houtain, L. (1992). Degradation of azo dyes by algae. *Environ. Pollut*, *75*(3), 273–278. doi:10.1016/0269-7491(92)90127-V PMID:15092015

Joshi, T., Iyengar, L., Singh, K., & Garg, S. (2008). Isolation, Identification, and application of novel bacterial consortium TJ-1 for the decolorization of structurally different azo dyes. *Bioresource Technology*, *99*(15), 7115–7121. doi:10.1016/j.biortech.2007.12.074 PMID:18289845

Kalme, S. D., Parshetti, G. K., Jadhav, S. U., & Govindwar, S. P. (2007). Biodegradation of Benzidine based dye Direct Blue-6 by Pseudomonas desmolyticum NCIM 2112. *Bioresource Technology*, *98*(7), 1405–1410. doi:10.1016/j.biortech.2006.05.023 PMID:16822666

Kalyani, D. C., Telke, A. A., Dhanve, R. S., & Jadhav, J. P. (2009). Eco-friendly biodegradation and detoxification of Reactive Red-2 textile dye by newly isolated Pseudomonas sp. SUK1. *Journal of Hazardous Materials*, *163*(2-3), 735–742. doi:10.1016/j.jhazmat.2008.07.020 PMID:18718713

Kandelbauer, A., Erlacher, A., Cavaco-Paulo, A., & Guebitz, G. M. (2004). Laccase-catalyzed decolorization of the synthetic Azo-dye Diamond Black PV 200 and of some structurally related derivatives. *Biocatalysis and Biotransformation*, *22*(5-6), 331–339. doi:10.1080/10242420400024573

Kapdan, I., Kargi, F., Mcmullan, G., & Marchant, R. (2000). Comparison of white rot fungi cultures for decolorization of textile dyestuffs. *Bioprocess Engineering*, *22*(4), 347–351. doi:10.1007004490050742

Kapdan, I. K., & Kargi, F. (2002). Biological decolorization of textile dyestuff containing wastewater by Coriolus versicolor in a rotating biological contactor. *Enzyme and Microbial Technology*, *30*(2), 195–199. doi:10.1016/S0141-0229(01)00468-9

Kaushik, P., & Malik, A. (2009). Fungal dye decolorization: Recent advances and future potential. *Environment International*, *35*(1), 127–141. doi:10.1016/j.envint.2008.05.010 PMID:18617266

Kelewou, H., Merzouki, M., & Lhassani, A. (2014). Biosorption of textile dyes Basic Yellow 2 (BY2) and Basic Green 4 (BG4) by the live yeast S.cerevisiae. *J. Mater. Environ. Sci*, *5*, 633–640.

Kensa, V. M. (2011). Bioremediation-an overview. *Jr. of Industrial Pollution Control*, *27*(2), 161–168.

Khadijah, O., Lee, K. K., & Faiz, F. M., & Abdullah, M. F. F. (2009). Isolation, screening, and development of local bacterial consortia with azo dyes decolorizing capability. *Malaysian Journal of Microbiology*, *5*(1), 25–32.

Kim, D. J., An, H. Y., Yoon, J. H., Park, Y. H., Kawai, F., & Jung, C. M. (2002). Identification of Clostridium perfringens AB and J and its uptake of Bromophenol Blue. *Journal of Microbiology and Biotechnology*, *12*, 544–552.

Kim, S. J., Ishikawa, K., Hirai, M., & Shoda, M. (1995). Characteristics of a newly isolated fungus, Geotrichumcandidum Dec 1, which decolorizes various dyes. *Journal of Fermentation and Bioengineering*, *79*(6), 601–607. doi:10.1016/0922-338X(95)94755-G

Knapp, J. S., Newby, P. S., & Reece, L. P. (1995). Decolorization of dyes by wood-rotting basidiomycete fungi. *Enzyme and Microbial Technology*, *17*(7), 664–668. doi:10.1016/0141-0229(94)00112-5

Kodam, K. M., Soojhawon, I., Lohande, P. D., & Gawai, K. R. (2005). Microbial decolorization of reactive azo dyes under aerobic conditions. *World Journal of Microbiology & Biotechnology*, *21*(3), 367–370. doi:10.100711274-004-5957-z

Koupaie, E., Moghaddam, A. M., & Hashemi, S. (2013). Evaluation of integrated anaerobic/aerobic fixed-bed sequencing batch biofilm reactor for decolonization and biodegradation of azo dye Acid Red 18: Comparison of using two types of packing media. *Bioresource Technology*, *127*, 415–421. doi:10.1016/j.biortech.2012.10.003 PMID:23138064

Kumar, K., Devi, S. S., Krishnamurthi, K., Dutta, D., & Chakrabarti, T. (2007). Decolorization and detoxification of Direct Blue-15 by a microbial consortium. *Bioresource Technology*, *98*(16), 3168–3171. doi:10.1016/j.biortech.2006.10.037 PMID:17321737

Lade, H., Kadam, A., Paul, D., & Govindwar, S. (2015). Biodegradation and detoxification of textile azo dyes by a bacterial consortium under sequential microaerophilic/aerobic processes. *EXCLI Journal*, *14*, 158–174. PMID:26417357

Laing, I. G. (1991). The impact of effluent regulations on the dyeing industry. *Rev. Prog. Coloration*, *21*(1), 56–71. doi:10.1111/j.1478-4408.1991.tb00081.x

Langlais, B., Reckhow, D. A., & Brink, D. R. (1991). *Ozone in water treatment, application and engineering* (1st ed.). Lewis Publishers.

Libra, A. J., Borchert, M., Vigelahn, L., & Storm, T. (2004). Two-stage biological treatment of a diazo reactive textile dye and the fate of the dye metabolites. *Chemosphere*, *56*(2), 167–180. doi:10.1016/j.chemosphere.2004.02.012 PMID:15120563

Lin, S. H., & Peng, C. F. (1994). Treatment of textile wastewater by electrochemical method. *Water Research*, *28*(2), 277–282. doi:10.1016/0043-1354(94)90264-X

Liu, W., Chao, Y., Yang, X., Bao, H., & Qian, S. (2004). Biodecolourization of azo, anthraquinonic and triphenylmethane dyes by white-rot fungi and a laccase secreting engineered strain. *Journal of Industrial Microbiology & Biotechnology*, *31*(3), 127–132. doi:10.100710295-004-0123-z PMID:15069603

Lopez, J. M., Guisado, G., Garcia-Vargas, C. M., & Estrella-Suarez, F. (2012). Pseudallescheriaangusta, a ligninolytic microorganism or wood fibres biomodification. *BioResources*, *7*, 464–474.

Maghraby, D. M. E. (2013). Evaluation of non-viable biomass of Laurencia papillosa for decolorization of dye wastewater. *African Journal of Biotechnology*, *12*(17), 2215–2223. doi:10.5897/AJB2013.11975

Maier, J., Kandelbauer, A., Erlacher, A., Cavaco-Paulo, A., & Gübitz, G. M. (2004). A new alkali-thermostable azoreductase from *bacillus* sp. Strain SF. *Applied and Environmental Microbiology*, *70*(2), 837–844. doi:10.1128/AEM.70.2.837-844.2004 PMID:14766562

Mallick, N. (2002). Biotechnological potential of immobilized algae for wastewater N, P, and metal removal: A review. *Biometals*, *15*(4), 377–390. doi:10.1023/A:1020238520948 PMID:12405533

Manu, B., & Chaudhari, S. (2002). Anaerobic decolorization of simulated textile wastewater containing azo dyes. *Bioresource Technology*, *82*(3), 225–231. doi:10.1016/S0960-8524(01)00190-0 PMID:11991070

Martinez, A. T., Ruiz-Duenas, F. J., Martinz, M. J., Del Rio, J. C., & Gutierrez, A. (2009). Enzymatic delignification of plant cell wall: From nature to mill. *Current Opinion in Biotechnology*, *20*(3), 348–357. doi:10.1016/j.copbio.2009.05.002 PMID:19502047

Martins, M. A., Lima, N., Silvestre, A. J., & Queiroz, M. J. (2003). Comparative studies of fungal degradation of single or mixed bioaccessible reactive azo dyes. *Chemosphere*, *52*(6), 967–973. doi:10.1016/S0045-6535(03)00286-8 PMID:12781230

Martins, M. A., Queiroz, M. J., Silvestre, A. J., & Lima, N. (2002). Relationship of chemical structures of textile dyes on the pre-adaptation medium and the potentialities of their biodegradation by Phanerochaetechrysosporium. *Research in Microbiology*, *153*(6), 361–368. doi:10.1016/S0923-2508(02)01332-3 PMID:12234010

Martins, M. A. M., Cardoso, M. H., Queiroz, M. J., Ramalho, M. T., & Oliveira-Campos, A. M. (1999). Biodegradation of azo dyes by the yeast Candida zeylanoides in batch aerated cultures. *Chemosphere*, *38*(11), 2455–2460. doi:10.1016/S0045-6535(98)00448-2 PMID:10204232

Meera, G., & Manisha, K. (2012). Biodegradation of textile dye congo red by fungus mucor mucedo. *Indian Journal of Fundamental and Applied Life Sciences*, *2*, 251–255.

Melgoza, R., & Buitron, C. (2004). Anaerobic/aerobic treatment of colorants present in textile effluents. *Water Science and Technology*, *50*(2), 149–155. doi:10.2166/wst.2004.0111 PMID:15344785

Mohan, S. V., Rao, N. C., Srinivas, S., Prasad, K. K., & Karthikeyan, J. (2002). Treatment of stimulated Reactive Yellow 22 (azo) dye effluents using Spirogyra species. *Waste Management (New York, N.Y.)*, *22*(6), 575–582. doi:10.1016/S0956-053X(02)00030-2 PMID:12214968

Moosvi, S., Kehaira, H., & Madamwar, D. (2005). Decolorization of textile dye Reactive Violet 5 by a newly isolated bacterial consortium RVM 11.1. *World Journal of Microbiology & Biotechnology*, *21*(5), 667–672. doi:10.100711274-004-3612-3

Moosvi, S., Kher, X., & Madamwar, D. (2007). Isolation, characterization, and decolorization of textile dyes by a mixed bacterial consortium JW-2. *Dyes and Pigments*, *74*(3), 723–729. doi:10.1016/j.dyepig.2006.05.005

Mutambanengwe, C., Togo, C., & Whiteley, C. (2007). Decolourization and degradation of textile dyes with bio sulfidogenic hydrogenases. *Biotechnology Progress*, *23*, 1095–1100. PMID:17880103

Muthukumar, M., Sargunamani, D., Senthilkumar, M., & Selvakumai, N. (2005). Studies on decoloration, toxicity and the possibility for recycling of acid dye effluents using ozone treatment. *Dyes and Pigments*, *64*(1), 39–44. doi:10.1016/j.dyepig.2004.03.012

Nesheiwat, F. K., & Swanson, A. G. (2000). Clean contaminated sites using Fenton's reagent. *Chemical Engineering Progress*, *96*(40), 61–66.

Nigam, P., Armour, G., Banat, I. M., Singh, D., & Marchant, R. (2000). Physical removal of textile dyes from effluents and solid-state fermentation by dye-adsorbed agricultural residues. *Bioresource Technology*, *72*(3), 219–226. doi:10.1016/S0960-8524(99)00123-6

Ogugbue, C. J., Morad, N., Sawidis, T., & Oranusi, N. A. (2011). Decolorization and partial mineralization of a polyazo dye by Bacillus firmus immobilized within tubular polymeric gel. *3 Biotech*, 2, 67–78.

Oren, A., Gurevich, P., Azachi, M., & Henis, Y. (1992). Microbial degradation of pollutants at high salt concentrations. *Biodegradation*, *3*(2-3), 387–398. doi:10.1007/BF00129095

Ozturk, A., Celik, L., & Abdullah, I. M. (2012). Biodegradation of reactive red 195 azo dye by the bacterium Rhodopseudomonas Palustris 51ATA. *African Journal of Microbiological Research*, *6*, 120–126.

Palmieri, G., Cennamo, G., & Sannia, G. (2005). Remazol Brilliant Blue R decolorisation by the fungus Pleurotusostreatus and its oxidative enzymatic system. *Enzyme and Microbial Technology*, *36*(1), 17–24. doi:10.1016/j.enzmictec.2004.03.026

Pandey, R. R., Pandey, B. V., & Upadhyay, R. S. (2011). Characterisation and identification of Pseudomonas fluorescens NCIM 2100 degraded metabolic products of crystal violet. *African Journal of Biotechnology*, *10*, 624–631.

Park, T. J., Lee, K. H., Jung, E. J., & Kim, C. W. (1999). Removal of refractory organics and color in pigment wastewater with Fenton oxidation. *Water Science and Technology*, *39*(10-11), 189–192. doi:10.2166/wst.1999.0653

Parshetti, G., Kalme, S., Saratale, G., & Govindwar, S. P. (2006). Biodegradation of malachite green by Kocuriarosea MTCC 1532. *Acta Chimica Slovenica*, *53*, 492–498.

Pavko, A. (2011). Fungal decolorization and degradation of synthetic dyes. Some Chemical Engineering Aspects. In Waste Water-Treatment and Reutilization, Intech, Croatia.

Pazarlioglu, N. K., Urek, R. O., & Ergun, F. (2005). Biodecolourization of Direct Blue 15 by immobilized *Phanerochaete Chrysosporium. Process Biochemistry (Barking, London, UK)*, *40*(5), 1923–1929. doi:10.1016/j.procbio.2004.07.005

Phugare, S. S., Kalyani, D. C., Patil, A. V., & Jadhav, J. P. (2011). Textile dye degradation by bacterial consortium and subsequent toxicological analysis of dye and dye metabolites using cytotoxicity, genotoxicity and oxidative stress studies. *Journal of Hazardous Materials*, *186*(1), 713–723. doi:10.1016/j.jhazmat.2010.11.049 PMID:21144656

Pinheiro, H. M., Tourand, E., & Thomas, O. (2004). Aromatic amines from azo dye reduction: Status review with emphasis on direct UV spectrophotometric detection in textile industry wastewaters. *Dyes and Pigments*, *61*(2), 121–139. doi:10.1016/j.dyepig.2003.10.009

Ponraj, M., Gokila, K., & Zambare, V. (2011). Bacterial decolorization of textile dye-Orange 3R. *Int. J. Adv. Biotechnol. Res.*, *2*, 168–177.

Rafii, F., & Coleman, T. (1999). Cloning and expression in E.coli of an azoreductase gene from Clostridium perfringens and comparison with azoreductase genes from other bacteria. *Journal of Basic Microbiology*, *39*, 29–35. doi:10.1002/(SICI)1521-4028(199903)39:1<29::AID-JOBM29>3.0.CO;2-W PMID:10071864

Ramalho, P. A., Cardoso, M. H., Cavaco-Paulo, A., & Ramahlo, M. T. (2004). Characterization of azo reduction activity in a novel ascomycete yeast strain. *Applied and Environmental Microbiology*, *70*(4), 2279–2288. doi:10.1128/AEM.70.4.2279-2288.2004 PMID:15066823

Ramalho, P. A., Paiva, S., Cavaco-Paulo, A., Casal, M., Cardoso, M. H., & Ramalho, M. T. (2005). Azoreductase activity of intact saccharomyces cerevisiae cells is dependent on the Frelp component of the plasma membrane ferric reductase. *Appl. Environ. Technol*, *71*(7), 3882–3888. doi:10.1128/AEM.71.7.3882-3888.2005 PMID:16000801

Ramalho, P. A., Scholze, H., Cardoso, M. H., Ramalho, M. T., & Oliveira-Campos, A. M. (2002). Improved conditions for the aerobic reductive decolorisation, of azo dyes by Candida zeylanoides. *Enzyme and Microbial Technology*, *31*(6), 848–854. doi:10.1016/S0141-0229(02)00189-8

Ramya, M., Anusha, B., Kalavathy, S., & Devilaksmi, S. (2007). Biodecolorization and biodegradation of Reactive Blue by Aspergillus sp. *African Journal of Biotechnology*, *6*, 1441–1445.

Robinson, T., McMullan, G., Marchant, R., & Nigam, P. (2001). Remediation of dyes in textile effluent: A critical review on current treatment technologies with a proposed alternative. *Bioresource Technology*, *77*(3), 247–255. doi:10.1016/S0960-8524(00)00080-8 PMID:11272011

Sadeghi, A., Bazardehi, M. A. K., Raffe, S., & Zarif, B. (2014). Biotransformation of Carmoisine and Reactive Black 5 Dyes using S. cerevisiae. *Health*, *6*(10), 859–864. doi:10.4236/health.2014.610108

Saha, S. K., Swaminathan, P., Raghavan, C., Uma, L., & Subramanian, G. (2010). Ligninolytic and antioxidative enzymes of a marine cyanobacterium Oscillatoria willei. BDU 130511 during Poly R-478 decolourization. *Bioresource Technology*, *101*(9), 3076–3084. doi:10.1016/j.biortech.2009.12.075 PMID:20061142

Sani, R. K., & Banerjee, U. C. (1999). Decolorization of triphenylmethane dyes and textile and dyestuff effluent by Kurthia sp. *Enzyme and Microbial Technology*, *24*(7), 433–437. doi:10.1016/S0141-0229(98)00159-8

Saratale, R. G., Saratale, G. D., Kalyani, D. C., Chang, J. S., & Govindwar, S. P. (2009). Enhanced decolorization and biodegradation of textile azo dye Scarlet R by using developed microbial consortium-GR. *Bioresource Technology*, *100*(9), 2493–2500. doi:10.1016/j.biortech.2008.12.013 PMID:19157864

Scott, J. P., & Ollis, D. (1995). Integrated chemical and biological oxidation processes for water treatment, review and recommendations. *Environment and Progress*, *14*(2), 88–103. doi:10.1002/ep.670140212

Semple, K. T., Cain, R. B., & Schmidt, S. (1999). Biodegradation of aromatic compounds by microalgae. *FEMS Microbiology Letters*, *170*(2), 291–300. doi:10.1111/j.1574-6968.1999.tb13386.x

Sen, S., & Demierer, G. (2003). Anaerobic treatment of synthetic textile wastewater containing a reactive azo dye. *Journal of Environmental Engineering*, *129*(7), 595–601. doi:10.1061/(ASCE)0733-9372(2003)129:7(595)

Sharma, H. D., & Reddy, K. R. (2004). *Geoenvironmental Engineering*. Hoboken, NJ: John Wiley & Sons.

Shoda, M., & Kim, J. S. (1999). Purification and characterization of a novel peroxidase from Geotrichumcandidum Dec. 1 involved in decolorisation of dyes. *Applied and Environmental Microbiology*, *65*, 1029–1035. PMID:10049859

Singh, L., & Singh, V. P. (2010). Microbial degradation and decolourization of dyes in semi-solid medium by the fungus – Trichoderma harzianum. *Environ. We Int. J. Sci. Tech*, *5*, 147–153.

Singh, S., Awasthi, S. K., Iyenger, L., Ashfaq, M., & Singh, P. (2011). Mineralisation of aminobenzene sulfonates by a newly isolated bacterial co-culture (AS1 and AS2). *Biol. Med*, *3*, 53–59.

Singh, S., Kang, S. H., Mulchandani, A., & Chen, W. (2008). Bioremediation: Environmental clean-up through pathway engineering. *Current Opinion in Biotechnology*, *19*(5), 437–444. doi:10.1016/j.copbio.2008.07.012 PMID:18760355

Slokar, Y. M., & Majcen Le Marechal, A. (1998). Methods of decoloration of textile wastewater. *Dyes and Pigments*, *37*(4), 335–356. doi:10.1016/S0143-7208(97)00075-2

Solis, M., Solis, A., Perez, H. I., Norberto, M., & Flores, M. (2012). Microbial decolouration of azo dyes: A review. *Process Biochemistry*, *47*(12), 1723–1748. doi:10.1016/j.procbio.2012.08.014

Stolz, A. (2001). Basic and applied aspects in the microbial degradation of azo dyes. *Applied Microbiology and Biotechnology*, *56*(1-2), 69–80. doi:10.1007002530100686 PMID:11499949

Suganya, K., & Revathi, K. (2016). Decolorization of reactive dyes by immobilized bacterial cells from textile effluents. *International Journal of Current Microbiology and Applied Sciences*, *5*(1), 528–532. doi:10.20546/ijcmas.2016.501.053

Sumathi, S., & Manju, B. S. (2000). Uptake of reactive textile dyes by Aspergillus foetidus. *Enzyme and Microbial Technology*, *27*(6), 347–355. doi:10.1016/S0141-0229(00)00234-9 PMID:10938413

Sun, S., Li, C., Sun, J., Shi, S., Fan, M., & Zhou, Q. (2009). Decolorization of an azo dye Orange G in aqueous solution by Fenton oxidation process: Effect of system parameters and kinetic study. *Journal of Hazardous Materials*, *161*(2-3), 1052–1057. doi:10.1016/j.jhazmat.2008.04.080 PMID:18538927

Suzuki, Y., Yoda, T., Rahul, A., & Sagiura, W. (2001). Molecular cloning and characterization of the gene encoding azoreductase from Bacillus sp. OY 1-2 isolated from soil. *The Journal of Biological Chemistry*, *246*(12), 9059–9065. doi:10.1074/jbc.M008083200 PMID:11134015

Talarposhti, A., Donnelly, T., & Andersonm, G. (2001). Colour removal from a simulated dye wastewater using a two-phase anaerobic packed bed reactor. *Water Research*, *35*(2), 425–432. doi:10.1016/S0043-1354(00)00280-3 PMID:11228995

Tam, N. F., Chong, A. M., & Wong, Y. S. (2002). Removal of tributyltin (TBT) by live and dead microalgal cells. *Marine Pollution Bulletin*, *45*(1-12), 362–371. doi:10.1016/S0025-326X(02)00184-4 PMID:12398407

Uddin, M. S., Zhou, J., Qu, Y., Guo, J., Wang, P., & Zhao, L. (2007). Biodecolourization of Azo-dye acid red B under high salinity condition. *Bulletin of Environmental Contamination and Toxicology*, *79*(4), 440–444. doi:10.100700128-007-9260-1 PMID:17721733

Upadhyay, R. S. (2002). Microbial bioremediation of textile effluents. In V. P. Singh & R. D. Stapleton Jr., (Eds.), *Biotransformation: bioremediation technology for health and environmental protection* (pp. 331–348). Amsterdam, The Netherlands: Stapleton, Elsevier Science BV. doi:10.1016/S0079-6352(02)80018-2

Ventosa, A., Nieto, J., & Oren, A. (1998). Biology of moderately halophilic aerobic bacteria. *Microbiology and Molecular Biology Reviews*, *62*, 504–544. PMID:9618450

Verma, P., & Madamwar, D. (2005). Decolorization of Azo dyes using Basidiomycete strain PV 002. *World Journal of Microbiology & Biotechnology*, *21*(4), 481–485. doi:10.100711274-004-2047-1

Vijayaraghavan, K., & Yun, Y. (2008). Biosorption of C. I. Reactive Black 5 from aqueous solution using acid treated biomass of brown seaweed Laminaria sp. *Dyes Pig*, *76*(3), 726–732. doi:10.1016/j.dyepig.2007.01.013

Waghmode, T. R., Kurade, M. B., Kabra, A. N., & Govindwar, S. P. (2012). Degradation of Remazol Red dye by *Galactomycesgeotrichum* MTCC 1360 leading to increased iron uptake in sorghum vulgare and phaseolus mungo from soil. *Biotechnology and Bioprocess Engineering; BBE*, *17*(1), 117–126. doi:10.100712257-011-0307-0

Walker, G. M., & Weatherley, L. R. (2000). Biodegradation and biosorption of acid anthraquinone dye. *Environ. Pollut.*, *108*(2), 219–223. doi:10.1016/S0269-7491(99)00187-6 PMID:15092952

Wang, H., Su, J. Q., Zheng, X. W., Tian, Y., Xiong, X. J., & Zheng, T. L. (2009). Bacterial decolorisation and degradation of the reactive dye Reactive Red 180 by Citrobacter sp. CK3. *International Biodeterioration & Biodegradation*, *63*(4), 395–399. doi:10.1016/j.ibiod.2008.11.006

Wong, P. K., & Yuen, P. Y. (1998). Decolorization and biodegradation of N, N, - dimethyl-P- Phenylenediamine by Klebsiella pneumoniae RS-13 and Acetobacter liquefaciens S-1. *Journal of Applied Microbiology*, *85*(1), 79–87. doi:10.1046/j.1365-2672.1998.00479.x PMID:9721658

Wouter, D., Cliona, O., Freda, R. H., & Helena, M. P. (1998). Anaerobic treatment of textile effluents: A review. *Journal of Chemical Technology and Biotechnology (Oxford, Oxfordshire)*, *73*(4), 323–335. doi:10.1002/(SICI)1097-4660(199812)73:4<323::AID-JCTB976>3.0.CO;2-S

Wu, J. Y., Hwang, S. J., Chen, C., & Chen, K. (2005). Decolorization of azo dye in a FBR reactor using immobilized bacteria. *Enzyme and Microbial Technology*, *37*(1), 102–112. doi:10.1016/j.enzmictec.2005.02.012

Yan, H., & Pan, G. (2004). Increase in biodegradation of dimethyl phthalate by Closterium lunula using inorganic carbon. *Chemosphere*, *55*(9), 1281–1285. doi:10.1016/j.chemosphere.2003.12.019 PMID:15081769

Yasar, S., Cirik, K., & Cinar, O. (2012). The effect of cyclic anaerobic– aerobic conditions on biodegradation of azo dyes. *Bioproc Biosyst Eng*, *35*, 449–457. PMID:21858702

Zhou, J. L., & Banks, C. J. (1991). Removal of humic acid fraction by Rhizopus arrhizus: Uptake and kinetic studies. *Environmental Technology*, *12*(10), 859–869. doi:10.1080/09593339109385080

Zhou, J. L., & Banks, C. J. (1993). Mechanism of humic acid color removal from natural waters by fungal biomass biosorption. *Chemosphere*, *27*(4), 607–620. doi:10.1016/0045-6535(93)90096-N

Zhou, W., & Zimmermann, W. (1993). Decolorization of industrial effluents containing reactive dyes by actinomycetes. *FEMS Microbiology Letters*, *107*(2-3), 157–162. doi:10.1111/j.1574-6968.1993.tb06023.x PMID:8472899

Zhou, Y., Liang, Z., & Wang, Y. (2008). Decolorization and COD removal of secondary yeast wastewater effluents by coagulation using aluminum sulfate. *Desalination*, *225*(1-3), 301–311. doi:10.1016/j.desal.2007.07.010

Zille, A. (2005). *Laccase reactions for textile applications*. (Thesis submitted to Universidade do Minho). pp. 168.

Chapter 14
Industrial Wastewater Treatment:
Recycled Paper Mill Effluent Treatment Using Modified Anaerobic Hybrid Baffled (MAHB) Bioreactor

Siti Roshayu Binti Hassan
Universiti Malaysia Kelantan, Malaysia

Mohamad Johari Abu
Universiti Malaysia Kelantan, Malaysia

Irvan Dahlan
Universiti Sains Malaysia, Malaysia

ABSTRACT

The development of effective and simple methods for treating wastewater is a challenging task for environmental engineers. In this chapter, a novel modified anaerobic hybridized baffled (MAHB) bioreactor, which is a combination of regular suspended-growth and fixed biofilm systems together with the modification of baffled-reactor configurations, was proven to be a modest bioreactor for wastewater treatment rather than the commercial anaerobic baffled reactor (ABR). The significant advantages of this bioreactor were its ability to nearly realize the multi-stages anaerobic theory, allowing different bacterial groups to develop under more favourable conditions, reduced sludge bed expansion, no special gas or sludge separation required, and high stability to organic and toxic shock loads. The compartmentalization of this bioreactor results in a buffering zone between the primary acidification zone and active methanogenesis zone, and provided the strong ability to resist shock loads which broaden the usage of multi-phase anaerobic technology for industrial wastewater treatment.

DOI: 10.4018/978-1-7998-0369-0.ch014

INTRODUCTION

With the rapid change of world economic conditions and human populations, water resources are becoming progressively deficient and the quality of environment in the world is constantly becoming worse in most regions. Disposal of wastewater directly into the lands and rivers is presently one of the serious environmental problem contributors. Therefore, there is a dire need to develop reliable technologies for wastewater treatment. In general, anaerobic process has been extensively applied to treat wastewater due to its design simplicity, high treatment efficiency, low excess sludge production and low operating cost (Moawad et al., 2009; Vicentin et al., 2019). In addition, it involves biological conversions in a stepwise fashion of organic materials to various end-products including methane (CH_4) and carbon dioxide (CO_2). A well-managed anaerobic digestion system has the ability to produce maximum methane production without discharging any gases into the atmosphere (Paolini et al., 2018). There are various types of anaerobic reactors, such as up-flow anaerobic sludge blanket (UASB), expanded granular sludge bed (EGSB), strengthened circulation anaerobic reactor (SCAR), anaerobic membrane bioreactors (An MBRs), anaerobic sequential batch reactor (ASBR), anaerobic side-stream reactor (ASSR) and anaerobic baffled reactor (ABR) have been employed to process dyeing wastewater. Their results demonstrated that the treatment on the dyeing wastewater by pilot-scale could be achieved chemical oxygen demand (COD) removal rate around 10 – 45.8% with an organic loading rate (OLR) at 0.53–7.4 kg COD/(m^3.h) (Li et al., 2017; Qi et al., 2019; Yang et al., 2018). Among these technologies, ABR takes advantage of the baffle structure to form several separate compartments up to eight inside the reactor, where the flow regime in which is mainly plug flow. It also forms multiphase anaerobic environment, and has the fine buffer capability resistant to shock loads and adaptability to influent toxicants. It uses two-way direction (up and down) of water stream to stir anaerobic sludge to mix with wastewater make a certain microorganism community cultivated in each of its compartments. Thus, better biological population distribution will be formed to improve treatment efficiency and operation stability due to the unique structure and flow regime of the reactor. However, most ABR reactors displays large solid loss and slow start-up period (usually more than a month) due to slow-degree adaptation of anaerobic microorganisms (Hassan et al., 2013b). Successful start-up period in the shortest time is an important indicator of a successful start-up process of an anaerobic reactor. Moreover, various trophic groups may not be in balance during the start-up process and the reactor may fail. Hence, the development of the optimum start-up condition is a very crucial. During start-up, external fluctuations such as pH, temperature and hydraulic retention time (HRT) must be avoided whilst organic loading rates (OLR) must be fixed. In the initial phase of start-up, the loading rate must be in the low region of 0.1 g CODgVSS^{-1}d^{-1}, which corresponds to 1.2 g CODL^{-1}d^{-1} for a system with biomass of 12 g VSS L^{-1} (Barber & Stuckey, 1998). Some modification on the ABR is required to treat wastewater in a very short of time. Besides that, there is a combination of both suspended and attached-growth processes in a single reactor such as hybrid anaerobic filters (HAF) like plastic media, are designed to take advantage of huge solid loss (Hassan et al., 2013a). The improvement of the structure led to the adherence of the microorganisms on the packing to form biofilms, and further increased the microbial richness and diversity (Donlan, 2002). Therefore, in this chapter, a modify ABR reactor with a combination of regular suspended growth and fixed biofilm systems, together with the modification of baffled-reactor configurations was developed to treat effluent from local recycled paper mill industry.

Figure 1. Schematic diagram of 58 L capacity bench-scale continuous feed MAHB reactor

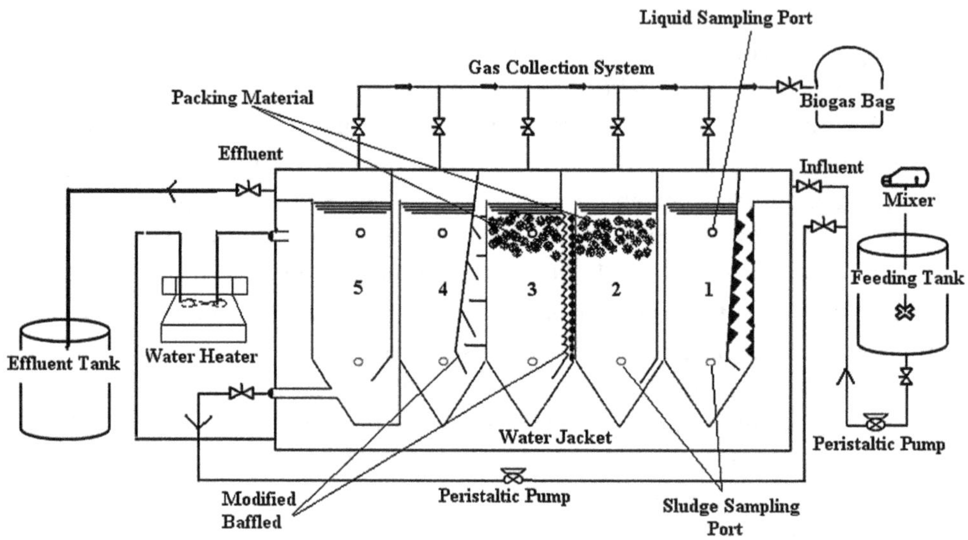

MODIFIED ANAEROBIC HYBRID BAFFLED (MAHB) REACTOR

Figure 1 shows the basic design of the MAHB reactor was rectangular in shape which contains five compartments and each compartment was separated by a modified vertical baffle. The reactor has a total active volume of 58 L. Polypropylene ring-packing was used as media for supporting the biofilm formation and these packing materials were located at the under surface of compartments 2 and 3. Sampling ports were present on the top of each compartment and the reactors were maintained at constant temperature of 35 °C (±0.2 °C). Samples were collected from each compartment for analysis, together with the effluent.

Start-Up

MAHB reactor was seeded with anaerobic granular sludge (10% v/v) and mixed with 750 mg L^{-1} COD of RPME (10% v/v) as a substrate, respectively. The anaerobic digesters were filled up with 1 L of effluent wastewater, during the feeding phase, daily into each compartment until the reactor became full. The RPME was then added continuously into the same reactor during the stabilization phase at a flow rate of 0.1757 mLs^{-1}. The OLR was maintained at 0.1963 $gCODL^{-1}day^{-1}$at HRT of 4 days. The pH of the digesters was kept without any adjustment during this period. The acclimatization of sludge with wastewater during the start-up period was monitored by the daily measurement of biogas production and effluent COD until steady state is achieved.

Bioreactor Performance During Start-Up

As the start-up period reached its steady state, all samples from compartments 1 to 5, influent and effluent were analysed for different analyses to screen out the behaviour of wastewater achieved. The results were tabulated in Table 1. It shows that pH, COD removal, ammonia (NH_3), BOD and alkalinity gave increasing patterns as it goes from compartments 1 to 5. Similar trends were reported by Chang et al.,[9] where the COD removal at last compartment was higher (96%) than first compartment (61%). In this study, the pH increased from 5.92 to 6.61, COD removal percentage increased from 62.37% to 69.92%, ammonia from 0 to 0.5 mg L^{-1} and alkalinity slightly increased from 150 to 200 mg L^{-1}. Floc size demonstrates an interesting profile through the reactor. The floc size increased from compartments 1 to 4 (from 115.93 to 307.10μm) and reduced in compartment 5 (189.26μm). The increase in floc size might be due to flocculation occurred inside each compartment. In compartment 5, low substrate levels resulting in low bacterial growth, which contributes to smaller flocs size [9]. As for BOD, it is decreased from 324 to 255 mgL^{-1}, as it goes from compartments 1 to 5. This indicates that this MABR successfully removed the COD and BOD inside the RPME. Fatty acid content also decreased due to increasing pH. Wastewater effluents released from Muda Paper Mills Sdn Bhd have high COD (4328 mg L^{-1}) and BOD (669 mg L^{-1}) contents. This highlights the characteristics of wastewater and suggests that it is not safe to dispose effluent water into streams or nearby fields. Result also demonstrates the performance of start-up period of semi-continuous MAHB reactor under anaerobic mesophilic conditions (35 ± 3 °C) with semi-continuous feeding. The maximum methane content (33.138%) of biogas collected was achieved at the end of feeding phase with low OLR of 0.1963 $gCODL^{-1}$ day^{-1}. This shows that the intermittent feeding (1000 mL) everyday with COD concentration of 750 mg L^{-1} are efficient during feeding phase because it gives the microorganisms enough time to acclimate with the new environment. High COD removal (85%) indicates that MABR was successfully used to treat RPME at a shorter time. This study showed that steady state was achieved within a period of less than 30 days. However, further study should be done in order to reduce other pollutants in RPME.

Bioreactor Performance at Different Start-Up

At the end of start-up 1, COD removal efficiency was in a range of 61 – 69% for almost one week. This indicated that the reactor could be operated continuously until it reached steady state condition. The RPME wastewater was pumped into the MAHB reactor with an increased in seeding arrangement. For every change in seeding arrangement, various parameters were monitored until reached steady state (defined by constant COD values in the effluent for 4 days) condition. COD removal efficiency, pH, methane production rate, alkalinity and floc size at steady state condition were shown in Table 2. From the table, it could be seen that the COD removal efficiencies increased from 61 – 69%, 68 – 78% and

Table 1. Steady state performance of MAHB bioreactor

Details	pH	COD removal (%)	NH_3 (mgL^{-1})	BOD_5 (mgL^{-1})	VFA (mgL^{-1})	Alkalinity (mg $CaCO_3L^{-1}$)	DO (mgL^{-1})	Floc size (μm)	Biogas (L)
Influent	7.18	-	0.40	747	-	165	5.68	-	2.35
Comp. 1	5.92	62.37	0.00	324	191	150	5.66	115.93	
Comp. 2	5.90	63.57	0.10	183	192	163	5.76	126.20	
Comp. 3	5.95	63.62	0.30	216	165	172	5.69	132.64	
Comp. 4	5.95	61.22	0.70	360	153	200	5.83	307.10	
Comp. 5	6.61	69.92	0.50	255	124	200	5.51	189.26	
Effluent	6.62	68.12	0.35	201	-	150	7.14	-	

69 – 87% for start-up 1, 2 and 3, respectively after each increased in seeding arrangement at steady state condition. During the first few days of start-up, the effluent obtained was quite turbid because some light suspended solids in the seed sludge had been washout. For start-up 1, COD concentration of the effluent quite high which resulted in low COD removal efficiencies and only 33.90% (0.0566 L CH4/day) of methane was produced. This indicates that the volume of seed sludge was not enough to digest the substrate inside the RPME wastewater. However, with the increasing ratio of seed sludge in subsequent start-up (start-up 2), a significant increase in COD removal efficiencies (as high as 78%) and methane content (> 0.2 L/day) were achieved. During start-up 3, the MAHB reactor was able to remove as high as 87% of influent COD and produce biogas containing 70% (0.7967 LCH4/day) methane. Figure 2 shows the top view of opened MAHB reactor at the end of start-up 3. Observation shows that a thick film was develop on the surface of polypropylene ring on second and third compartments which indicates that inoculated digested sludge was fully acclimatized and the reactor could be operated for further study. The biogas bubbles were also clearly observed between the polypropylene pall ring in compartment 2 and 3. During the experimental phase, the influent pH varied from 6.8 to 7.1, alkalinity from 90 to 200 mg/L, while effluent pH and alkalinity were 7.1-7.8 and 155 to 240 mg/L, respectively as shown in Table 2. As the seeded ratio increased, the system pH in all five chambers remained at 5.5 – 7.0, which indicates that buffering capacity of RPME wastewater is satisfactory enough to neutralise the produced VFA during the anaerobic digestion. This is then capable to maintain the digestion liquor pH at the optimum value for methane-producing bacteria, thus guaranteeing the stable operation of the reactor. L/day) were achieved. During the anaerobic digestion. This is then capable to maintain the digestion liquor pH at the optimum value for methane-producing bacteria, thus guaranteeing the stable operation of the reactor. Initial sludge inoculum had an average floc size of 92.805 μm. Increasing size may due to flocculation process where several small particles were combines to form larger floc that leads to granules formation. This clarification supported by report stating that proteinaceous feed may add to the polymers present in the MAHB reactor and encourages polymer bridging between particles. Results clearly demonstrate a considerable increase in floc size (Table 2) over time with an interesting profile through the MAHB reactor. However, the floc size was slightly decreased in compartments 5 due to possible washout and less nutrients at the rear end of the MAHB reactor.

Compartment 1 -5 had floc size of average diameter of 105 -160 μm, 110 – 190 μm and 150 – 210 μm for start-up 1, 2, and 3 respectively. This result compared with previous result obtained by Barber and Stuckey (Barber & Stuckey, 1998). This indicates that start-up strategy does affect the floc size profile in the MAHB reactor. The results presented shows that the larger floc size was found in compartment 4 for each start-up. The reduced floc size in front compartment may be attributed to high COD levels which result in high gas production and thus disrupt the flocs and reducing their size while the reduced size towards the rear ends of the MAHB reactor were due to low level of substrates, which results in low bacterial growth (Jirka & Carter, 1975). Subsequently, the floc size of the middle compartment (3 and 4) were attributed to sufficient substrate for bacterial growth and slightly lower gas production reduced effects of gas breakage. From the result obtained, it is concluded that start-up 3 provide better performance in terms of COD removal efficiencies, pH, alkalinity, floc size and methane content since it has the highest ratio of seeds during the acclimatization process.

Table 2. Steady state analysis for different start-up of MAHB reactor

Start-up	Description	pH	COD Removal (%)	Methane Production (L/day)	Alkalinity	Floc Size (μm)
1	Influent	6.82	-	0.0566	165.00	-
	Comp. 1	5.92	63.37		150.00	105.9
	Comp. 2	5.90	63.57		163.00	126.2
	Comp. 3	5.95	63.62		172.00	132.6
	Comp. 4	5.95	61.22		200.00	175.4
	Comp. 5	6.61	69.92		200.00	157.6
	Effluent	7.18	68.12		150.00	-
2	Influent	6.23	-	0.2042	203.00	-
	Comp. 1	5.56	69.24		185.00	114.1
	Comp. 2	5.57	69.33		170.00	136.7
	Comp. 3	5.55	68.89		165.00	156.9
	Comp. 4	5.58	69.46		160.00	207.1
	Comp. 5	6.04	77.90		175.00	189.2
	Effluent	7.40	78.12		155.00	-
3	Influent	7.18	-	0.7967	94.17	-
	Comp. 1	6.37	69.03		158.33	156.9
	Comp. 2	6.40	80.62		185.00	173.3
	Comp. 3	6.46	85.30		208.75	208.6
	Comp. 4	6.50	85.75		221.25	218.3
	Comp. 5	6.70	87.10		223.75	206.6
	Effluent	7.78	86.41		236.25	-

COMPARTMENT-WISE BEHAVIOUR OF MAHB REACTOR

Generally, treatments of low strength wastewater were done using aerobic processes. However, these processes have several drawbacks which are high operational and maintenance costs, high energy consumption, high sludge production and considerable investment. Highlighted disadvantages limit the application of aerobic processes in the developing country. Due to this, an alternative process which is anaerobic treatment seems to be an ideal and reasonable solution for environmental protection (Arshad & Hashim, 2012; Thompson et al., 2001), due to not only consume lower quantity of energy, but also produce less sludge which indicates less treatment cost required. Bachmann and McCarty firstly develop an anaerobic baffled reactor (ABR) which was described as a series of Upflow Anaerobic Sludge Blanket (UASBs) (Shaw et al., 2002). It consists of a series of vertical baffles which forces the wastewater to flow upwards through the series of compartments that contains mixed anaerobes as the wastewater passed from the inlet to the outlet of the reactor. This system enables high degree of sludge retention under high hydraulic load, simple construction, low operation and maintenance cost (Eaton et al., 1998; Hassan & Dahlan, 2013). Although ABRs were broadly used to treat low strength wastewater, a thorough literature

Figure 2. Thick film developed on the surface of polypropylene ring

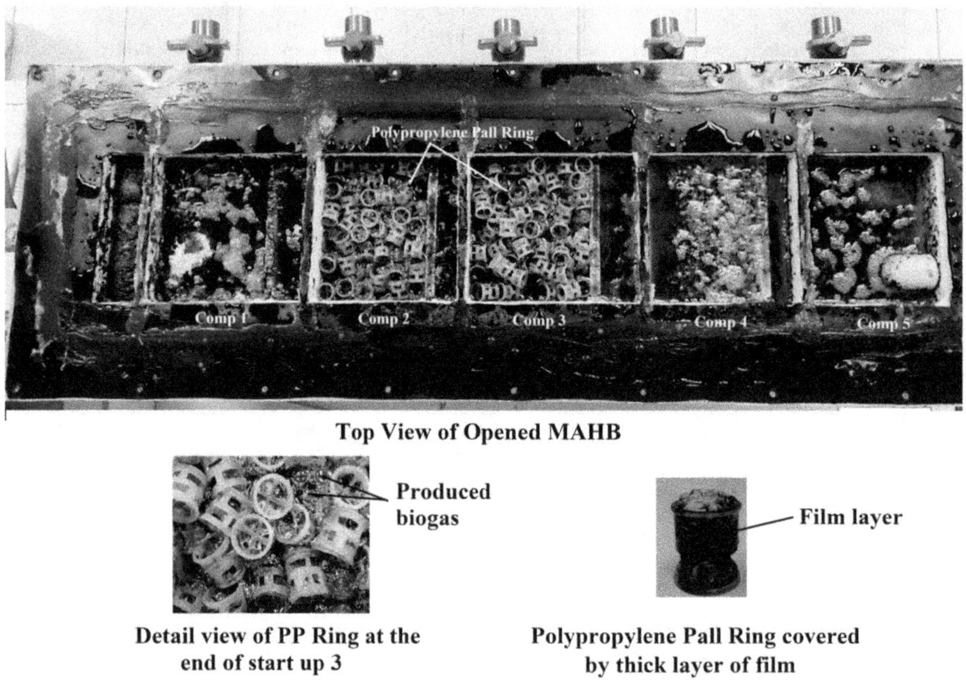

survey (Hassan et al., 2013b; Hutnan et al., 2005) revealed that there is no application of novel modified anaerobic hybrid baffled (MAHB) reactor for the treatment of low strength wastewater discharged from recycled paper mill industry. This novel MAHB reactor is a combination of regular suspended and fixed biofilm systems in a single reactor to take advantages of both biomass types together with the modification of baffled-reactor configurations seems to be a good candidate to be used in this study to treat recycled paper mill effluent (RPME). The most significant benefits of this design was its capability to nearly perfectly realize the staged multi-phase anaerobic theory, allowing different bacterial groups to develop under more favourable conditions, low costs and without the associated control problems (Hassan et al., 2013a). The purpose of this present study is to examine the performance and characteristics of MAHB reactor treating low strength RPME wastewater in terms of chemical oxygen demand (COD) removal, volatile fatty acid (VFA) concentration, pH value, methane content and biogas production.

Chemical Oxygen Demand (COD) Removal

MAHB reactors were subjected to four different changes in the feed concentration which corresponds on the changes of feeding COD concentration. COD variations within the MAHB reactor were shown in Figure 3. For each increment in OLR, COD removal efficiency showed increment gradually with the passage of time. Throughout the first 28 days of start-up, the highest COD removal rate was 86%. From then on, COD removal efficiencies kept above 94% at each steady state condition. Result shows that a change in feeding COD concentrations did not have any noticeable effect on the stability and performance of MAHB reactor. This indicates that MAHB reactor had high ability to resist shocking loading rate. Results showed that 96% COD removal were achieved within 31 days at OLR of 0.29 g/L/dy (Stage

Figure 3. COD variation and COD removal efficiency of MAHB reactor against various loading rates

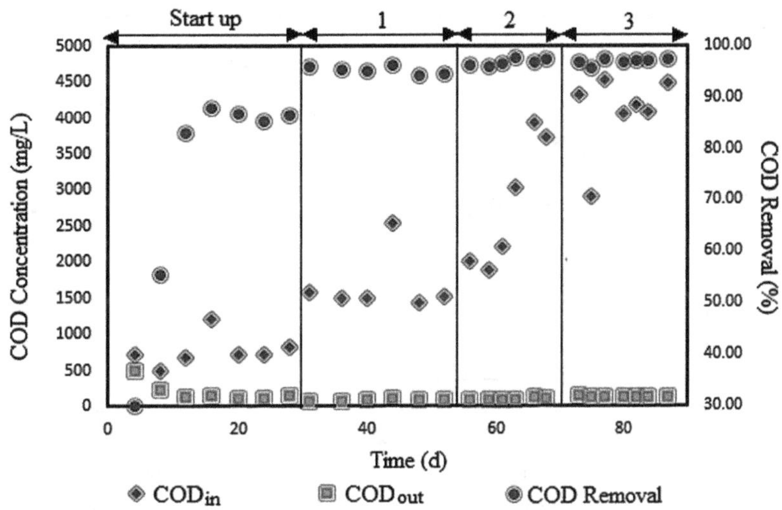

1). The removal of COD from RPME wastewater was generally good at 0.43 and 0.57 g/L/dy (Stage 2 and 3) with the highest removal efficiency achieved was 97.49% at day 63. Not just provides high COD removal efficiency, the effluent COD concentration were less than 150 mg/L. During start-up stage, under a constant and long retention time coupled with a steady increase in substrate concentration, a positive effect on development of appropriate microbial culture contributes to the better performance of MAHB reactor (Gopala Krishna et al., 2008). Ever since the loading rates were only increased when steady state conditions were achieved for the present loading condition, the microorganisms in individual compartments have acclimatized to increasingly OLR, hence great reactor stability and superior performance was provided.

pH Profile

Values of pH for each different compartment for all different OLRs conditions tested were illustrated in Figure 4. Levels of pH of each individual compartment show a narrow change regardless of the OLRs. At OLR of 0.14 g/L/dy, the pH of all different compartment shows declining behaviour.

When the OLR was further increased to 0.57 g/L/dy, a downward tendency of pH was observed. This is mainly due to increment of VFA concentration as the influent COD concentration increased. The change characteristics of pH along the MAHB reactor can indicate whether the anaerobic system is working normally after each change of OLR, also as a control parameter of hydrolysis acidogenesis phases in MAHB reactor. pH profile shows an increment along the reactor from Compartment 1 to 5. This shows that Compartment 5 could maintain the acidity in terms of pH and VFA concentration at pH level in a range of 6.3 to 7.3 which provide for effective anaerobic digestion to occur. The pH values in this compartment were maintained at nearly neutral and constant level due to VFA consumption. The considerably increased in pH level in the rear end compartment of the reactor provides a better condition to generate a methane rich biogas as the methanogens attained maximum activity in terms of acetate conversion under pH level of 6.8 and 7.0 (Chen et al., 2019). Similar pattern was also recorded by Dama et al (Akunna & Clark, 2000).

Figure 4. pH variation against various loading rates

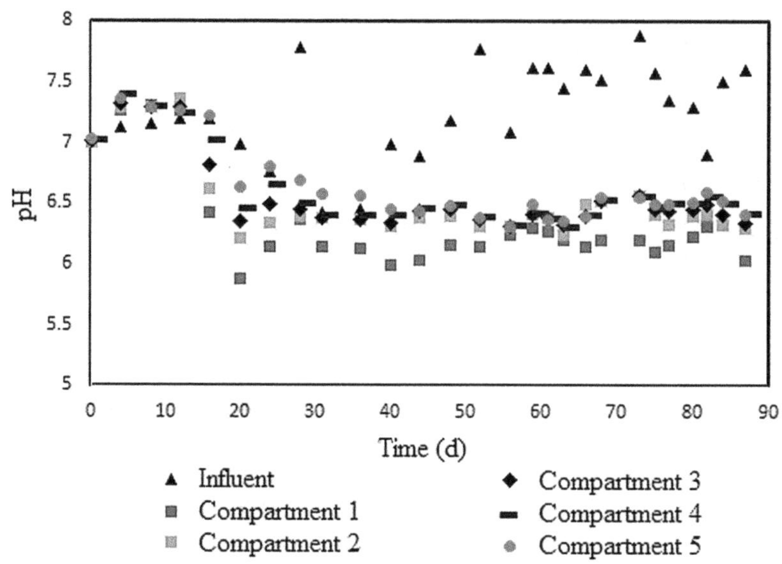

Biogas Production and Composition

Biogas production rate shows increment with the increasing load, increased from 5.31 to 7.21 L/dy for loading rate of 0.14 to 0.57 g/L/dy (see Figure 5). This indicates that moderate substrate feeding is the most effective approach to improve gas production rates. As the OLR rising, the anaerobic bacteria in the MAHB reactor will rapidly reproduce which resulted in apparent increased in biogas yield (Baloch & Akunna, 2003). Once the OLR increased, the substrate was excessive compared to the biomass in the first compartment, so more substrate flowed into the subsequent compartments.

Consequently, all compartment yields more biogas that contributes to the increasing of total biogas produced. Methane content in biogas shows gradual increment as the OLR increased. This indicates the growth characteristics of the microbial populations inside the MAHB reactor. High OLR gaves more biomass for the microbes to convert them as a substrate to methane. Acidogenic bacteria hydrolyze complex polymer substrate to organic acids, alcohol, hydrogen, and carbon dioxide, followed by the conversion of the above intermediates to acetate by hydrolysis and acidogenesis. In final stage, the methanogens converts simple compounds (i.e. acetate, methanol, and combining carbon dioxide with hydrogen) into methane. Actually, this pattern follows the accepted understanding of multi-staged anaerobic digestion, i.e. hydrolysis, acidogenesis, acetogenesis and methanogenesis (Wang et al., 2004). Based on the result obtained, it is concluded that MAHB reactor is suitable for anaerobic process treating low strength recycled paper wastewater with a performance;

- The COD removal efficiency achieved more than 90% after a successful and short start-up.
- As the organic loading rate increased to 0.57 g/L/dy, excellent stability performance were achieved as high as 97% COD removal, methane content of 85.95% and biogas produce of 7.21 L/dy.
- Rapid acidification of recycled paper mill effluent (RPME) in the front compartment contributes to the decrease of pH values.

Figure 5. Variation of biogas against various loading rates

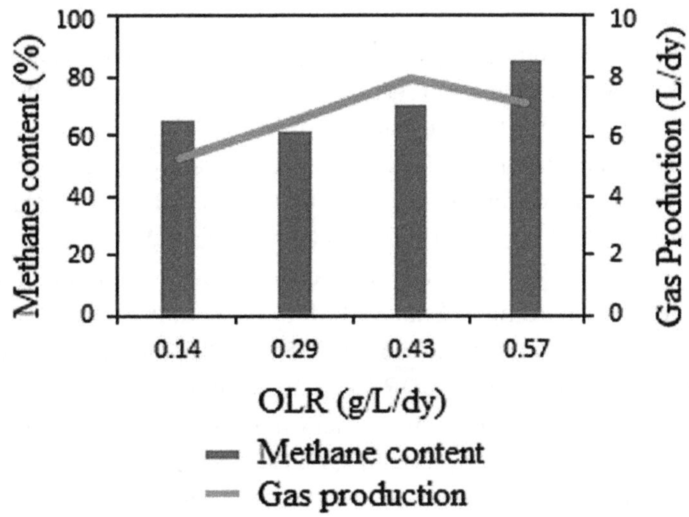

EFFECT OF ORGANIC LOADING RATE

The most significant advantages of Anaerobic Baffled Reactor (ABR) is its ability to separate acidogenesis and methanogenesis longitudinally down the reactor, in which allowing the different bacterial groups to develop under most favorable conditions (Bassuney et al., 2013). It is recommended in many literature sources that the ABR should be equipped with at least 3 chambers (Bassuney et al., 2013; Malakahmad et al., 2011). The requirement of shallow reactor to maintain acceptable liquid and gas velocities is one of the disadvantages of traditional ABR (Barber & Stuckey, 1999). This also can lead to bacterial washout and delay of startup stage. In addition, potential problems can arise during start-up as a result of plug-flow characteristics, the accumulation of Volatile Fatty Acid (VFA) and low pH, exposition sensitive bacteria in front compartments to toxic levels of inorganic and organic compounds in the high strength feeding wastewater (Liu et al., 2010). To overcome these difficulties, some approaches such as low Organic Loading Rate (OLR), feed dilution, periodic feeding and effluent recycling have been recommended (Skiadas & Lyberatos, 1998) but they may not be the best solution to the prompt start-up of ABR because compartmentalized structure of ABR bringing with another problems such as nutrient limits in the final compartment and elimination of phase separation. In order to stabilize ABR during start-up by establishment of the most suitable microbial population and eliminate disadvantages of traditional ABR, a modification was achieved by using Modified Anaerobic Hybridized Baffled (MAHB) reactor with five compartments to provide adequate Hydraulic Retention Time (HRT) and therefore Organic Loading Rate (OLR). OLR is an important parameter because it indicates the amount of volatile solids to be fed into the reactor each day (Mattocks, 1984). In a case of long hydraulic retention time, overloading can results in biomass washout for nonattached biomass which will lead to process failure. Volatile solids represent portion of the organic-material solids that can be digested, while the remainder of the solids is fixed. The 'fixed' solids and a portion of the volatile solids are non-biodegradable. The actual loading rate depends on the types of wastes fed into the reactor (Mattocks, 1984) because the types of waste determine the level of biochemical activity that will occur in the reactor. The aims of this study were to investigate the influence of OLR on performance and treatment efficiency of RPME digestion.

Bioreactor Operation

The inoculum for seeding was anaerobic granular sludge (10% v/v) taken from anaerobic pond of Malpom Industries Berhad that were mixed with 750 mg/lit COD of RPME taken from Muda Paper Mills Sdn Bhd. The characteristics of RPME used and the startup of MAHB reactor have been reported previously (Hassan et al., 2013b). Different dilutions of RPME were prepared using tap water. The steady state performance was evaluated under different influent COD concentrations (1000-4000 mg/L) and feed flow rates (HRT of 1-7 days and OLR of 0.14-4.00 gm COD/L/day). Variation of ± 5% in effluent COD concentration at each condition was considered as the criterion for steady state conditions. The following parameters were analyzed according to Standard Methods (Eaton et al., 1998) alkalinity, pH, Total Suspended Solids (TSS), Volatile Suspended Solids (VSS) and Biological Oxygen Demand (BOD_5). COD was measured using Spectrophotometer DR-2800 according to the reactor digestion method (Jirka & Carter, 1975). Volatile Fatty Acids (VFAs) were measured using esterification methods while biogas composition was determined using Shimadzu GC-FID with propack N column. Carrier gas was helium set at a flow rate of 50 ml/min, column temperature of 28°C, detector temperature of 38°C and injector temperature of 128°C. Triplicate samples were collected for each parameter reading to increase the precision of the results, and only the average value was reported throughout this study. The repeatability of the analysis data was found to be sufficiently high with relative error between repeated runs less than 5%.

Steady State Performance for Different Compartment

Table 3 illustrated the performance of MAHB reactor treating the RPME at different OLRs. As can be seen, a steady increase in specific methane yield was achieved, to a maximum value of 8.47 L CH_4/day at organic loading of 1.33 gm COD/L/day. From there on, the incremental increase in specific methane yield declined and eventually decreased to 6.20 L CH_4/day at an OLR value of 4.00 gm COD/L/day. The results are in similarities trend with the work reported by [20]. This apparent instability was mainly due to attachment of gas bubbles at high OLR values which leads to biomass suspension and cell washout as methane production rate increases. Although the system shows lower efficiency in terms of specific methane yield at high OLR values, however there is no drastic VSS degradation were reported (Torkian et al., 2003). The maximum VSS degradation of 16.90% was observed at OLR of 2.00 gm COD/ L/day, at around 146 mg/L of the effluent VSS (Babaei & Shayegan, 2011). Reactor showed stable performance at steady state in terms of methane content at different OLRs. Highest methane content of 91.63% with a volume of 6.43 L biogas/day were recorded at loading rate of 0.4 gm COD/L/day while the lowest methane content of 51.58% with total volume collected of 12.51 L biogas/day were observed at OLR of 2.00 gm COD/L/day. The measurement of the composition and quantity of methane and other gas content is of fundamental importance to determine the performance of the whole system. The decrease of methane content in biogas means that the acidifying microorganisms are prevailing on the methanogens which may lead to volatile fatty acids accumulation (Babaei & Shayegan, 2011). Daily biogas production at steady state condition shows a fluctuating result. VSS degradation was fluctuating throughout the experiments in the range of 2.26% to 16.90%. VSS increment was relatively slow when compared to the higher OLRs. The MAHB bioreactor in terms of HRTs, OLRs and substrate pH as operated in this study, was more efficient when compared to results reported by others, (Austermann-Haun et al., 1997) where a UASB was used to treat a fruit-juice effluent.

Table 3. Performance of MAHB at different OLRs at steady state condition

OLR (gm COD/L/day)	Methane Content (%)	Highest Specific Methane Yield (L CH_4/day)	Highest vol. Prod. (L biogas/day)	pH Value	VSS Degradation (%)
0.14	65.97	3.50	5.31	7.2	8.34
0.20	64.98	4.07	6.27	6.6	14.35
0.29	62.08	4.12	6.63	6.8	12.13
0.33	57.78	4.35	7.54	7.4	11.17
0.40	91.63	5.89	6.43	7.2	9.49
0.43	71.55	5.72	7.99	7.1	4.25
0.57	85.95	6.20	7.21	6.7	6.28
0.60	88.97	6.65	7.47	7.1	1.14
0.67	54.13	4.67	8.62	7.5	13.15
0.80	78.58	6.23	7.93	7.4	8.41
1.00	60.03	7.92	11.90	7.2	9.06
1.33	66.21	8.47	12.80	6.7	6.13
1.50	61.47	7.97	12.96	6.1	8.56
2.00	51.58	6.45	12.51	6.9	16.90
3.00	66.29	6.76	10.20	6.5	2.26
4.00	63.17	6.20	9.81	6.3	3.10

Process Performance and Stability

Figure 6 illustrates the variation of VFA profiles, pH and alkalinity of the effluent at different organic loading rates. The characteristics of the effluent represent the stability of the reactor performance. In anaerobic processes, methanogenic bacteria are more sensitive to environmental conditions than hydrolytic and acidogenic bacteria. Among the environmental condition, pH is the most sensitive parameters. For example, the pH of MAHB reactor effluent indicates the stability of the system and its variation also depends on the buffering capacity of the system (Mata Alvarez et al., 2000). The pH of effluent from the MAHB reactor remained stable to the range of 6.32-7.51 during the experiments. This shows that the system was well buffered. When the loading rate was increased from 0.14 to 4.00 gm COD/L/day, the pH value was slightly fluctuated but still in the methanogenic range which results on methane content in the biogas above 50% throughout the whole experiments. Since the pH is controlled by the volatile organic acid's concentration, the alkalinity showed similar trends. VFA concentration in the effluent was increased with increasing loading rate, achieving a maximum value of 35 mg/L for an OLR value of 2.00 gm COD/L/day. In addition, a reduction of fatty acid removal was observed at OLR of 2.00 gm COD/L/day. This also could be due to the insufficient amount of acetoclastic methanogens or due to their inhibition (Wijekoon et al., 2010). The decrease of specific methane yield at the highest OLR value might be attributed to inhibition of methanogens, which caused an increase in effluent VFA before its reach its limitations process at OLR of OLR of 2.00 gm COD/L/day. As can be seen, further increment in OLR result a fast decreased in VFA value to a minimum value of 3.25 mg/L. This result indicates the increase in nonacetic volatile acids concentration in the effluent when the OLR was increased, which

Figure 6. Variation of pH, Volatile Fatty Acids (VFA) and Alkalinity with OLR

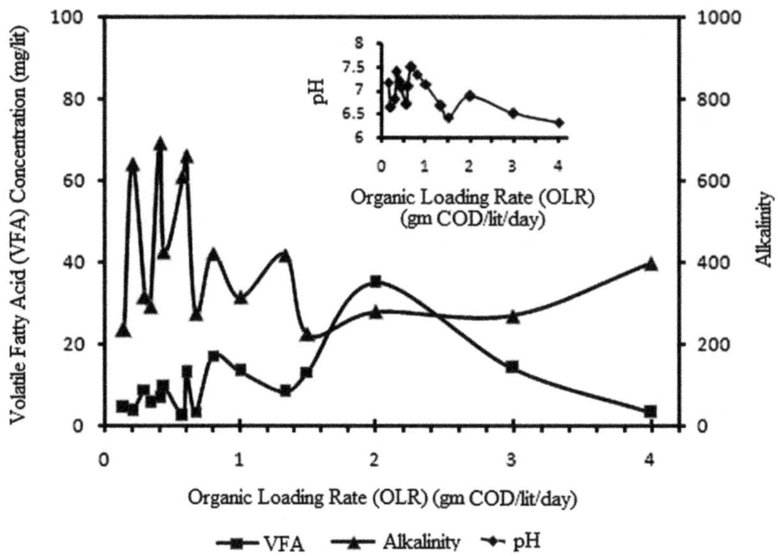

consequently inhibit the process at OLR values higher than 2.00 gm COD/L/ day. This result was in similarities to the previous finding reported by (Rincón et al., 2007). Although VFA concentration increased with increasing OLR, the concentration of other higher molecule weight volatile acids was increased faster, resulting the relative decreased in the VFA concentration. This phenomenon clearly indicates the process inhibition and instability at high OLR (Raposo et al., 2003). MAHB reactor showed a stable performance with high specific methane yield, COD removal and alkalinity (8.47 L CH4/day, 97.69%, and 419.2) at OLR of 1.33 gm COD/L/day. Further increase in the OLR to 4.00 gm COD/L/day results in further drop in the COD removal efficiency to 78.37%. This deleterious response might be because high organic loadings greater than 1.33 gm COD/L/day brought about a decrease in the biomass concentration and accumulation of inorganics in the reactor causing destabilization of the reactor and process failure, and thereby significantly affects the reactor performance in terms of organic removal. From the results, it may be concluded that the biological treatment with MAHB reactor at an OLR of 1.33 gm COD/L/day is a potential alternative for the treatment of RPME. Effective implementation of anaerobic digestion as the method of wastewater treatment leads to the utilization of renewable energy resources.

EFFECT OF SEED LOADS

Successful start-up of anaerobic reactors is an important indication for a stable operation and high removal efficiencies within a shorter possible time (Hassan et al., 2013b; Lee et al., 2009). Variety of bacteria groups during start-up process may cause unbalance and hence contributes to reactor failure. Few factors were extremely important in getting successful start-up such as biomass activity, growth rate, volume of inoculum, size and properties of granules, loading rate, hydraulic retention time (HRT) and mixing characteristics, environmental parameters (pH and temperature) and finally wastewater, composition, strength and fluctuation in these parameters. Therefore, to achieve an optimum start-up strategy can be a time consuming and complex process (Bakonyi et al., 2014; Rajakumar & Meenambal, 2008). The fluctuations in external parameters such as temperature, HRT and pH should be avoided while organic loading rates (OLR) should be consistent throughout the process. Adequate seeding is crucial since anaerobic bacteria may undergo slow growth process which leads to a long start-up process. Considerable literatures are available on the start-up due to the need to develop sludge with certain properties, especially good settle ability. However, extensive variation is seen in the time taken for the sludge agglomerates to appear. Anaerobic responses to seed volume depend on many factors such as presence of toxic materials, activity of the biomass, characteristics of the seed and duration of overloading (Stamatelatou et al., 2003; Whittaker et al., 2016). The imbalance or upset of the reactor usually occurred due to the complexity of microbial interactions that involved. A build up on the residual chemical oxygen demand (COD) concentration owing to volatile acid (VFA) accumulation is one of the most common symptoms that caused by the overloading of organic. This in which can leads to inhibition of methane production (acidosis) (Sterling et al., 2001). An active methanogens biomass with high digesting capacity were develop in order to determine the importance of seeding concentration on reactor start-up and to monitor certain key parameters such as solids loss and floc size that had not been monitored yet. Apart from that, reactor configuration also plays a significant role during anaerobic start-up. Hence, a novel MAHB reactor which is a combination of regular suspended growth systems and fixed biofilm together with the modification of baffled-reactor configurations has been developed. Each different modified baffles have its own characteristics (form/shape) to assist a better contact and superior mixing of wastewater and anaerobic microorganisms. Furthermore, literature survey designates that treatment of effluents that released from recycled paper mills have not been reported using MAHB reactor. The objectives of this present work are to investigate whether different seed loading could enhance the MAHB reactor performance and to explore the ability of the reactor to handle different seed ratio. For that reason, the main operating parameters studied includes COD removal efficiencies, system pH, total volatile fatty acids (VFA) concentration, solids analysis, and reactor performance at each steady state. The bacterial morphology also was observed at each steady state condition.

COD Removal Efficiency

MAHB reactor shows stable performance for each start-up period. Overall process took about 75 days to complete all 3-different start-up where the startup process for all three-different seed loading achieved stable performance regarding the COD removal and methane production rate in less than 30 days. However, extra attention should be focused on their changes in removal efficiencies (Figure 1) and performance behaviors. The fluctuating removal efficiencies were recorded as continuous feeding started for each start-up. During each start-up period, an increase in COD removal efficiency was observed where highest removal efficiencies were recorded in startup 3 of value of 87%. The results show similarities with the research work reported by a researcher (Barber & Stuckey, 1998). They found out that more than 85% COD removal were achieved during start-up period. High COD removal indicates that the organic matter in the supernatant was hydrolyzed by microorganisms available that grow well from the provided seed until the number of microbes become limited to digest further the organic matter. Start-up 3 was accomplished after 28 days of continuous operation (day 73) with a subsequent stable performance (Hassan et al., 2013b). The early variations in COD removal efficiency in all three start-up were due to overloading slower methanogens and environmental conditions favoring acidogenic bacteria. The fluctuating efficiencies are clear in first two start-up before it reached stabilization in startup 3. This indicates that the MAHB reactor has a good potential for treating RPME wastewater at ambient temperature.

The design of MAHB reactor itself contributes to the high removal of COD. The hanging and standing baffles that were bent 3 cm above the reactor's base at a 45° angle route the flow to the center of the up comer, thus allowing for better contact and biosolid mixing at the base of each compartment (Vossoughi et al., 2003). The advantage of this design is that the liquid flow makes repeated passes through the sludge, ensuring good contact between liquid and solid and reducing bypassing of the sludge (Singh et al., 2006). Hence, the microorganisms able to decompose complex organic substances into simple, chemically stabilized compounds mainly methane and CO2 effectively. Furthermore, this hybrid system (attached and suspended growth system) was better compared to conventional system. The difference is due to the existence of the biofilm in the hybrid reactors and these results confirm the competence of the hypotheses and agree well with data from the literature (Li & Li, 2010).

System pH and Total VFAs Present in Reactor for Different Start-Up Strategies

Figure 7 illustrated concentration of volatile fatty acids (VFA) and system pH found in each compartments of different start-up (start-up 1, 2 and 3). High level of VFA concentration (191 – 66 mg/L) and slightly lower pH values in between 5.9 to 6.37 in front compartment were noted for each start-up 1, 2 and 3, respectively. This might be due to high activity of hydrolytic and acidogenic bacteria compared to methanogenic bacteria. The later compartments show lower VFA concentrations (decreased from 192 to 124 mg/L) and increasing pH up to pH of 6.7 in compartment 5. This is due to high activity of methanogenic bacteria producing methane. This indicates the different microenvironments select the dominant bacteria population. The similar behaviour was noted for all start-up 1, 2 and 3. It was observed that for each start-up strategies, the VFA in first three compartments dropped rapidly, while other compartments (compartment 4 and 5) were found to be the least affected.

This phenomenon indicates accumulation of fatty acids in first three compartments while methanogenesis occurred in later compartments. This was due to the fact that the hydrolysis, acidogenesis and acetogenesis dominantly occurred in the initial chambers (Ge et al., 2015). Result shows that pH profiles were uniformly below 7 for each start-up in the whole system. It was observed that pH value was slightly constant in front compartment which designates the effective consumptions of volatile acids by methanogens. The earlier compartments had lower pH as acidogenesis and acetogenesis occurred in these compartments (Dama et al., 2002). Thus, due to the high concentration of volatile fatty acids (VFAs), the sudden drop in the pH (pH of 5.92 to 6.37), especially in the first compartment is quite noticeable. Fig. 3 shows that the pH in the first compartment was the lowest. The pH values increased along the reactor to pH of 6.61, 6.04 and 6.70 for each start-up 1, 2 and 3, respectively due to the conversion of intermediate products, i.e. the VFAs in the latter compartments which also contributes to a high methanogenic activity (Gopala Krishna et al., 2007; Zwain et al., 2018). This increase happened because COD removal increased up to 85% which means no sufficient food was left for microorganisms to survive. Similar trend results were reported by some researcher which are supported by an increased pattern of MLVSS loss (Alrawi et al., 2010; Barber & Stuckey, 1998). Furthermore, the environmental change for these bacteria from the original environment to the new environment of the anaerobic digester can explain the steep decrease of the microorganisms in general and these methane-forming bacteria in specific, during the start-up period (Zwain et al., 2017). These methane-forming bacteria are oxygen-sensitive, fastidious anaerobes and are free-living terrestrial and aquatic organisms. Although these bacteria are oxygen sensitive, this is not a significant disadvantage where it is usually found in habitats that are rich in degradable organic compounds. Through microbial activity, a rapid removed of

Figure 7. Volatile fatty acids (VFA) concentration and system pH for different start-up

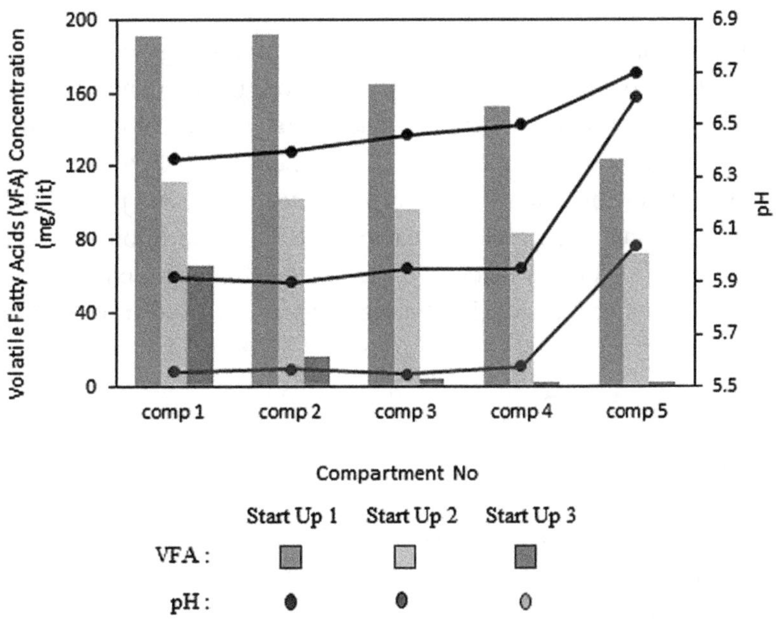

oxygen occurred in this habitat (Gerardi, 2003). From the data obtained, it shows that start-up 3 have more stable VFA and pH variation which indicates better performance of the MAHB reactor in terms of methanogenic activity. This is proving by high pH value and quite low VFA concentration (< 10 mg/L) at the rear end of the MAHB reactor that provides superior environmental condition for methanogenic bacteria to become more active in converting acetate and hydrogen to methane. This data is supported from previous study (Hassan et al., 2013b).

FUTURE RESEARCH DIRECTIONS

Modified anaerobic hybrid bioreactor (MAHB) shows that it is capable to treat recycled paper milled effluent of varying strength. The physical structure of the reactors itself allows various modifications to be made that providing the capability to treat different types of wastewaters that currently require at least two separate units, therefore substantially reducing capital costs. The systems can be further qualified based on process rate, which were low-rate systems and high rate systems. Both, these approaches are efficient at retaining flocculent sludge owing to lower organic and hydraulic loading rates than the high-rate systems. Particularly they are very well suited for industrial applications that do not granulate well or have higher than desirable amounts of troublesome constituents. It is highly recommended for further knowledge transfer from lab scale into industrial scale MAHB bioreactor that not only treating wastewater but can be a source in generating biogas as a new biofuel source.

CONCLUSION

An increasing level of attention has focused on the emerging technology of modified anaerobic hybrid baffled bioreactors (MAHB), owing to its merits such as low maintenance cost, easy control of gas separation, low energy consumption and sludge production, as well as biogas production. The process performance of the MAHB for recycled paper effluent treatment is then reviewed with regard to start up procedure, COD removal, sludge production, biogas production, methane production rate and potential advantages over the conventional anaerobic bioreactor (ABR). In addition, the important parameters affecting process performance are briefly discussed. Lastly, the challenges encountered and perspectives regarding the future development of the MAHB bioreactor process to promote its practical applications are presented.

ACKNOWLEDGMENT

This research was supported by the Research University Grant of Universiti Sains Malaysia [grant number: RU-I A/C.1001/PJKIMIA/814148]; the short-term research grant scheme of Universiti Malaysia Kelantan [grant number: R/SGJP/A13.00/01555A/001/2018/000533] and the Fundamental Research Grant Scheme (FRGS) of Malaysia [grant number: R/FRGS/A1300/01367A/006/2018/00566].

REFERENCES

Akunna, J. C., & Clark, M. (2000). Performance of a granular-bed anaerobic baffled reactor (GRABBR) treating whisky distillery wastewater. *Bioresource Technology*, *74*(3), 257–261. doi:10.1016/S0960-8524(00)00017-1

Alrawi, R., Ahmad, A., Ismail, N., & Kadir, M. (2010). Methane production during start-up phase of mesophilic semi-continues suspended growth anaerobic digester. *International Journal of Chemical Reactor Engineering*, *8*(1), 1–13. doi:10.2202/1542-6580.2034

Arshad, A., & Hashim, N. H. (2012). Anaerobic digestion of NSSC pulping effluent. *International Journal of Environmental of Research*, *6*(3), 761–768.

Austermann-Haun, U., Seyfried, C. F., & Rosenwinkel, K.-H. (1997). UASB-reactor in the fruit juice industry. *Water Science and Technology*, *36*(6), 407–414. doi:10.2166/wst.1997.0617

Babaei, A., & Shayegan, J. (2011). Effect of organic loading rates (OLR) on production of methane from anaerobic digestion of vegetables waste. In *Proceedings of World Renewable Energy Congress* (pp. 411-417). Sweden: Linköping University Electronic Press. 10.3384/ecp11057411

Bakonyi, P., Nemestóthy, N., Simon, V., & Bélafi-Bakó, K. (2014). Review on the start-up experiences of continuous fermentative hydrogen producing bioreactors. *Renewable & Sustainable Energy Reviews*, *40*, 806–813. doi:10.1016/j.rser.2014.08.014

Baloch, M. I., & Akunna, J. C. (2003). Granular bed baffled reactor (Grabbr): Solution to a two-phase anaerobic digestion system. *Journal of Environmental Engineering*, *129*(11), 1015–1021. doi:10.1061/(ASCE)0733-9372(2003)129:11(1015)

Barber, W. P., & Stuckey, D. C. (1998). The influence of start-up strategies on the performance of an anaerobic baffled reactor. *Environmental Technology*, *19*(5), 489–501. doi:10.1080/09593331908616705

Barber, W. P., & Stuckey, D. C. (1999). The use of the anaerobic baffled reactor (ABR) for wastewater treatment: A review. *Water Research*, *33*(7), 1559–1578. doi:10.1016/S0043-1354(98)00371-6

Bassuney, D., Ibrahim, W., & Moustafa, M. (2013). Performance of an anaerobic baffled reactor (ABR) treating high-strength food industrial wastewater with fluctuating pH, *Proc. of the Intl. Conf. on Advances in Civil, Structural and Environmental Engineering - ACSEE 2013*, 98-102.

Chen, X., Ottosen, L. D. M., & Kofoed, M. V. W. (2019). How low can you go: methane production of methanobacterium congolense at low CO2 concentrations. *Frontiers in Bioengineering and Biotechnology*, *7*(34). PMID:30899758

Dama, P., Bell, J., Foxon, K. M., Brouckaert, C. J., Huang, T., Buckley, C. A., ... Stuckey, D. (2002). Pilot-scale study of an anaerobic baffled reactor for the treatment of domestic wastewater. *Water Science and Technology*, *46*(9), 263–270. doi:10.2166/wst.2002.0255 PMID:12448477

Donlan, R. M. (2002). Biofilms: Microbial life on surfaces. *Emerging Infectious Diseases*, *8*(9), 881–890. doi:10.3201/eid0809.020063 PMID:12194761

Eaton, A. D., Clesceri, L. S., Greenberg, A. E., & Franson, M. A. H. (1998). *Standard methods for the examination of water and wastewater*. Washington, DC: American Public Health Association.

Ge, S., Usack, J., Spirito, C., & Angenent, L. (2015). Long-term n-caproic acid production from yeast-fermentation beer in an anaerobic bioreactor with continuous product extraction. *Environmental Science & Technology*, 49. PMID:25941741

Gerardi, M. H. (2003). *The microbiology of anaerobic digesters*. Hoboken, NJ: John Wiley & Sons. doi:10.1002/0471468967

Gopala Krishna, G. V. T., Kumar, P., & Kumar, P. (2007). Complex wastewater treatment using an anaerobic baffled reactor. *Environment and Progress*, *26*(4), 391–398. doi:10.1002/ep.10239

Gopala Krishna, G. V. T., Kumar, P., & Kumar, P. (2008). Treatment of low-strength soluble wastewater using an anaerobic baffled reactor (ABR). *Journal of Environmental Management*, *90*(1), 166–176. doi:10.1016/j.jenvman.2007.08.017 PMID:18096298

Hassan, S. R., & Dahlan, I. (2013). Anaerobic wastewater treatment using anaerobic baffled bioreactor: A review. *Central European Journal of Engineering*, *3*(3), 389–399.

Hassan, S. R., Zwain, H. M., & Dahlan, I. (2013a). Development of anaerobic reactor for industrial wastewater treatment: an overview, present stage and future prospects. *Journal of Advanced Scientific Research*, *4*(1), 7–12.

Hassan, S. R., Zwain, H. M., Qamaruz Zaman, N., & Dahlan, I. (2013b). Recycled paper mill effluent treatment in a modified anaerobic baffled reactor: Start-up and steady-state performance. *Environmental Technology*, 1–6.

Hutnan, M., Drtil, M., Derco, J., & Mrafkova, L. (2005). Biodegradation of hexamethylenetetramine in anaerobic baffled reactor. *Polish Journal of Environmental Studies*, *14*(5), 585–591.

Jirka, A. M., & Carter, M. J. (1975). Micro semiautomated analysis of surface and waste waters for chemical oxygen demand. *Analytical Chemistry*, *47*(8), 1397–1402. doi:10.1021/ac60358a004 PMID:1147260

Lee, C., Kim, J., Hwang, K., O'Flaherty, V., & Hwang, S. (2009). Quantitative analysis of methanogenic community dynamics in three anaerobic batch digesters treating different wastewaters. *Water Research*, *43*(1), 157–165. doi:10.1016/j.watres.2008.09.032 PMID:18945471

Li, H.-T., & Li, Y.-F. (2010). Performance of a hybrid anaerobic baffled reactor (HABR) treating brewery wastewater. In *Proceedings of 2010 International Conference on Mechanic Automation and Control Engineering* (pp. 2032-2037). Wuhan, China: IEEE

Li, Y., Zhang, J., Zhang, Y., & Quan, X. (2017). Scaling-up of a zero valent iron packed anaerobic reactor for textile dye wastewater treatment: A potential technology for on-site upgrading and rebuilding of traditional anaerobic wastewater treatment plant. *Water Science and Technology*, *76*(3-4), 823–831. doi:10.2166/wst.2017.270 PMID:28799929

Liu, R. R., Tian, Q., Yang, B., & Chen, J. H. (2010). Hybrid anaerobic baffled reactor for treatment of desizing wastewater. *International Journal of Environmental Science and Technology*, *7*(1), 111–118. doi:10.1007/BF03326122

Malakahmad, A., & Noor Ezlin, A. B., & Shahrom, M. Z. (2011). Study on performance of a modified anaerobic baffled reactor to treat high strength wastewater. *Journal of Applied Sciences (Faisalabad)*, *11*(8), 1449–1452.

Mata Alvarez, J., Macé, S., & Llabrés, P. (2000). Anaerobic digestion of organic solid wastes. An overview of research achievements and perspectives. *Bioresource Technology*, *74*(1), 3–16. doi:10.1016/S0960-8524(00)00023-7

Mattocks, R. (1984). *Understanding biogas generation (Technical paper No. 4. volunteers in technical assistance)*. Virginia: I. K. Intern Publishing House.

Moawad, A., Mahmoud, U. F., El-Khateeb, M. A., & El-Molla, E. (2009). Coupling of sequencing batch reactor and UASB reactor for domestic wastewater treatment. *Desalination*, *242*(1), 325–335. doi:10.1016/j.desal.2008.05.011

Paolini, V., Petracchini, F., Segreto, M., Tomassetti, L., Naja, N., & Cecinato, A. (2018). Environmental impact of biogas: A short review of current knowledge, *Journal of Environmental Science and Health. Part A*, *53*(10), 899–906.

Qi, Z., Xiang, G., & Xiong, D. (2019). Performance evaluation of pilot-scale hybrid anaerobic baffled reactor (HABR) to process dyeing wastewater based on grey relational analysis. *Applied Sciences*, *9*(10), 1974. doi:10.3390/app9101974

Rajakumar, R., & Meenambal, T. (2008). Comparative study on start - up performance of HUASB and AF reactors treating poultry slaughterhouse wastewater. *International Journal of Environmental of Research*, *2*(4), 401–410.

Raposo, F., Borja, R., Sanchez Hernandez, E., Martín, M. A., & Martín, A. (2003). Inhibition kinetics of overall substrate and phenolics removal during the anaerobic digestion of two-phase olive mill effluents (TPOME) in suspended and immobilized cell reactors. *Process Biochemistry*, *39*(4), 425–435. doi:10.1016/S0032-9592(03)00095-5

Rincón, B., Travieso, L., Sánchez, E., de los Ángeles Martín, M., Martín, A., Raposo, F., & Borja, R. (2007). The effect of organic loading rate on the anaerobic digestion of two-phase olive mill solid residue derived from fruits with low ripening index. *Journal of Chemical Technology and Biotechnology (Oxford, Oxfordshire)*, *82*(3), 259–266. doi:10.1002/jctb.1663

Shaw, C. B., Carliell, C. M., & Wheatley, A. D. (2002). Anaerobic/aerobic treatment of coloured textile effluents using sequencing batch reactors. *Water Research*, *36*(8), 1993–2001. doi:10.1016/S0043-1354(01)00392-X PMID:12092574

Singh, K., Viraraghavan, T., & Bhattacharyya, D. (2006). Sludge blanket height and flow pattern in UASB reactors: temperature effects. *Journal of Environmental Engineering*, *132*(8), 895–900. doi:10.1061/(ASCE)0733-9372(2006)132:8(895)

Skiadas, I. V., & Lyberatos, G. (1998). The periodic anaerobic baffled reactor. *Water Science and Technology*, *38*(8), 401–408. doi:10.2166/wst.1998.0831

Stamatelatou, K., Vavilin, V., & Lyberatos, G. (2003). Performance of a glucose fed periodic anaerobic baffled reactor under increasing organic loading conditions: 1. Experimental results. *Bioresource Technology*, *88*(2), 131–136. doi:10.1016/S0960-8524(02)00276-6 PMID:12576006

Sterling, M. C. Jr, Lacey, R. E., Engler, C. R., & Ricke, S. C. (2001). Effects of ammonia nitrogen on H2 and CH4 production during anaerobic digestion of dairy cattle manure. *Bioresource Technology*, *77*(1), 9–18. doi:10.1016/S0960-8524(00)00138-3 PMID:11211082

Thompson, G., Swain, J., Kay, M., & Forster, C. F. (2001). The treatment of pulp and paper mill effluent: a review. *Bioresource Technology*, *77*(3), 275–286. doi:10.1016/S0960-8524(00)00060-2 PMID:11272013

Torkian, A., Eqbali, A., & Hashemian, S. J. (2003). The effect of organic loading rate on the performance of UASB reactor treating slaughterhouse effluent. *Resources, Conservation and Recycling*, *40*(1), 1–11. doi:10.1016/S0921-3449(03)00021-1

Vicentin, R., Fdz-Polanco, F., & Fdz-Polanco, M. (2019). Energy integration in wastewater treatment plants by anaerobic digestion of urban waste: a process design and simulation study. *International Journal of Chemical Engineering*, *2019*, 11. doi:10.1155/2019/2621048

Vossoughi, M., Shakeri, M., & Alemzadeh, I. (2003). Performance of anaerobic baffled reactor treating synthetic wastewater influenced by decreasing COD/SO4 ratios. *Chemical Engineering and Processing: Process Intensification*, *42*(10), 811–816. doi:10.1016/S0255-2701(02)00107-1

Wang, J., Huang, Y., & Zhao, X. (2004). Performance and characteristics of an anaerobic baffled reactor. *Bioresource Technology*, *93*(2), 205–208. doi:10.1016/j.biortech.2003.06.004 PMID:15051083

Whittaker, C., Hunt, J., Misselbrook, T., & Shield, I. (2016). How well does Miscanthus ensile for use in an anaerobic digestion plant? *Biomass and Bioenergy*, *88*, 24–34. doi:10.1016/j.biombioe.2016.03.018

Wijekoon, K., Visvanathan, C., & Abeynayaka, A. (2010). Effect of organic loading rate on VFA production, organic matter removal and microbial activity of a two-stage thermophilic anaerobic membrane bioreactor. *Bioresource Technology*, *102*(9), 5353–5360. doi:10.1016/j.biortech.2010.12.081 PMID:21256737

Yang, B., Xu, H., Yang, S., Bi, S., Li, F., Shen, C., ... Liu, Y. (2018). Treatment of industrial dyeing wastewater with a pilot-scale strengthened circulation anaerobic reactor. *Bioresource Technology*, *264*, 154–162. doi:10.1016/j.biortech.2018.05.063 PMID:29803085

Zwain, H., Aziz, H. A., Ng, J., & Dahlan, I. (2017). Performance and microbial community analysis in a modified anaerobic inclining-baffled reactor treating recycled paper mill effluent. *Environmental Science and Pollution Research International*, *24*(14), 13012–13024. doi:10.100711356-017-8804-0 PMID:28378314

Zwain, H. M., Abdul Aziz, H., & Dahlan, I. (2018). Performance of modified anaerobic inclining-baffled reactor treating recycled paper mill effluent: Effects of influent chemical oxygen demand concentration and hydraulic retention time. *Environmental Technology*, *39*(12), 1557–1565. doi:10.1080/09593330.2017.1332692 PMID:28514902

KEY TERMS AND DEFINITIONS

Acetogenesis: A process through which acetate is produced from CO_2 and an electron source (e.g., H_2, CO, formate, etc.) by anaerobic bacteria via the reductive acetyl-CoA or Wood-Ljungdahl pathway.

Acclimate: Respond physiologically or behaviourally to a change in an environmental factor under controlled conditions.

Attached Growth: Wastewater treatment processes in which the microorganisms and bacteria treating the wastes are attached to the media in the reactor.

Methanogenesis: The production of methane by bacteria or other living organisms.

Seeding: To spread a defined amount (volume or cell number) of a cell suspension into a new flask or onto a plate, etc.

Suspended: Wastewater treatment processes in which the microorganisms and bacteria treating the wastes are suspended in the wastewater being treated.

Washout: The sudden erosion of soft soil or other support surfaces by a gush of water, usually occurring during a heavy downpour of rain (a flash flood) or other stream flooding.

Chapter 15
Reverse Osmosis Membrane Desalination Technology and Process:
Case Study on Small-Scale Brackish City Polluted Water Treatment

Man Djun Lee
https://orcid.org/0000-0002-1250-9459
University College of Technology Sarawak, Malaysia

Pui San Lee
https://orcid.org/0000-0001-9631-0983
University College of Technology Sarawak, Malaysia

ABSTRACT

This chapter gives an overview about reverse osmosis membrane desalination technology and process. Desalination process can be considered as one of the crucial processes in obtaining fresh water to meet the increasing fresh water demand throughout the world. Desalination process begins with the intake of seawater or brackish water. The intake system usually comprises a pump and piping system. Then, the seawater goes through pre-treatment process. From there, the treated seawater will go through desalination process. The most widely used desalination is membrane desalination utilizing reverse osmosis membrane. After desalination process, the fresh water will go through more filtration and a series of post-treatment. Post-treatment consists of conditioning and stabilizing the water for distribution. This chapter concludes with a case study to illustrate the operation and sustainability of a small-scale desalination plant that utilizes brackish city polluted water as source.

DOI: 10.4018/978-1-7998-0369-0.ch015

INTRODUCTION

This chapter gives an overview about reverse osmosis membrane desalination technology and process. Water is one of the important necessities for living organisms on this planet Earth. The whole population on Earth keep increasing from time to time. As a result, the demand for freshwater supply will increase and the current freshwater resources and reservoirs will not be able to keep up with the high demand soon. Alternatively, seawater can be used as new water supply. However, seawater consists of high content of salts and other impurities. To overcome this, desalination technology has been developed to separate the pure, fresh water from the high content of salts and impurities. Desalination refers to the removal process of salt and other minerals from the seawater to produce freshwater supply for human consumptions and industrial use. Many countries are now considering desalination as an important source of water supply as desalination process able to produce large quantities of water in excellent quality. To meet the demands for freshwater, worldwide capacity in desalination has increased significantly overtime. In 2000, the capacity has rose up to 130 million cubic meters per day. Desalination process can be considered as one of the crucial processes in obtaining fresh water to meet the increasing fresh water demand throughout the world. Without it, areas with fresh water scarcity will have to transport water from other places or find new freshwater alternatives. Desalination process begins with the intake of seawater or brackish water. The intake system is usually comprised of pump and piping system. This intake is crucial in water desalination plant whereby it can be a method of controlling the quality of the seawater that actually goes into the plant itself. This can be as the first step of pre-treatment which is known as filtering or screening. The seawater will go through pre-treatment process. In pre-treatment processes, the seawater is then again treated and filtered to remove any debris, sedimentations and microscopic organisms. From there, the treated seawater will go through desalination process. Common types of desalination processes are distillation (thermal-based) or membranes desalination process. The most widely used desalination is membrane desalination utilizing reverse osmosis membrane. Membrane desalination is popular due to many reasons, the main one is being more energy efficient than thermal desalination which utilize fossil fuel as energy source. Membrane desalination can be paired with renewable energy source such as solar, wind or wave energy for more environmentally sustainable. After desalination process, the freshwater will go through more filtration and a series of post-treatment. Post-treatment consists of conditioning and stabilizing the water for distribution. The post-treatment processes may contain of pH adjustment and disinfection of product water. If the product water is blending with other sources, it is essential to ensure similar water quality in both water sources. This is to ensure that the blended water meets health, aesthetic and anti-corrosion standards. This chapter concludes with a case study to illustrate the operation and sustainability of a small-scale desalination plant that utilizes brackish city polluted water as source.

CURRENT AND FUTURE OF WATER AVAILABILITY

Global warming is a global issue and may be humanity's greatest challenge (Karl & Trenberth, 2003). Indeed, global warming has already negatively impacted the environment, economy, and human health (Bosello et al., 2006; Luber & Prudent, 2009; O'Brien & Leichenko, 2000; Reuveny, 2007), and particularly on fresh water availability (Arnell, 1999; Intergovernmental Panel on Climate Change, 2008; Martínez-Alvarez et al., 2016; Payne et al., 2004). The mechanics of global warming and climate changes is well-defined in relevant published journals and shown in Figure 1.1. Global warming leads to an increase in extreme climate events such as rising sea level, droughts and floods which would jeopardize our fresh water supply (Jeppesen et al., 2009). On the other hand, increasing population, industrial and agricultural demand also means there is an increase in water demand. Most of the water are not returned back to the mainstream due to water holding by plants, human and animal bodies (consumptive water). Ultimately, all these activities contribute to water supply constraint. On top of that, the main runoff fresh water source is not properly distributed as per density of the world population (Gizelis and Wooden, 2010). For example, Asia has 69% of world population but it has only 36% of global runoff; yet about 5% of world population in South America is living with 25% runoff water. To address this issue, a sustainable water source is required. The most abundant source of water comes from seawater and brackish water which makes up about 96.4% of total water in the world (Khublaryan, 2009). Due to the salinity of sea and brackish water [1,000 mg/L <Total Dissolved Solid (TDS) < 60,000 mg/L], desalination technology is required to process this water to drinkable water standard (<250mg/L) (Voutchkov, 2017).

Water Source Distribution and Conflict

Water is an essential resource needs for all activities of human beings. It is an essential element for daily life and economy. For examples, water needs for production of food, agriculture, electricity and industrial product processing like medicine, textiles, paper etc. Most water supply comes from surface runoff and ground water (Adzlan et al., 2013; Gopi et al., 2019). A part of used water returns back to the mainstream for recycling. A significant part of water which does not come back after use is known to be consumptive water. The water on Earth surface is not evenly distributed and the distribution is not based on population size and land surface. About 50% of all available water is transboundary which means water located in the rivers, lakes or groundwater systems is sharing by two or more countries. The current water sharing scenario becomes the source of water conflict; it is important to devise strategy for reducing water sharing conflict among water sharing nations.

The current water consumption pattern has a substantial effect on water management. Based on water consumption pattern, it could be divided into two parts. Firstly, recyclable water which is water that coming back to mainstream after consumption. Secondly, non-recyclable water that part of water is not coming back to mainstream after consumption. Fact is that the water on earth is not fully recyclable, because a major percentage of water is consumptive in nature (Postel et al., 1996). Mainly, consumptive water belongs to agriculture, human and animal bodies that contribute to reduce the availability of the fresh water. Consumptive water is the main reason of water shortage in our planet. However, our current knowledge and skills are only allowing us to manage recyclable water.

The water on Earth surface is not distributed evenly. The main fresh water source is runoff that is not properly distributed as per density of the world population (Gizelis & Wooden, 2010). For example, Asia has 69% of world population but it has only 36% of global runoff. On the other hand, about 5% of world population in South America is living with 25% runoff. The Amazon River contains about 15% of runoff and only 0.4% of world's population are able to accesses to this water source. However, Giordano, et al. (2002) reports that about 27% of global runoff is river based renewable water source. All this information indicate that water is poorly distributed to the world population. In future, fresh water will be a critically limited resource for many regions of the world. About one-third of the world's population live in the countries with shortage of water resource (Yoffe et al., 2004). For example, Africa region is a historically water shortage area. In Asia, 69% of world population are living there with 36%

Table 1. Transboundary Water Conflict

Time	Events
1948-1960	Disputes over Indus basin water sharing lead to Indus Water Treaty between India and Pakistan.
1951-1953	Sporadic fire exchange over Huleh basin, demilitarized zone between Syria and Israel.
1958	Tensions between Sudan and Egypt over Nile waters lead to Nile Water Treaty.
1963-1964	Border skirmishes between Somalia and Ethiopia over critical water and oil resources in Ogaden desert. Hundreds were killed before cease-fire negotiations.
1965-1966	Israel and Syria exchange fire over an out of basin diversion plan from Sea of Galilee. Construction plan of Syria is halted in July 1966.
1975	Tension risen between Iraq and Syria over low flow year along the Euphrates. Military troops were reportedly transfer between their mutual borders.
1989-1991	Sporadic violence broke out as the result of grazing rights along Senegal River, boundary between Mauritania and Senegal.

of water source. In the Middle East especially Arabian countries belong to the lowest amount of fresh water. Basically, this poor water distribution becomes the source of global water conflict and appears as a barrier to economic growth.

Transboundary water sharing is one of the identified causes of political conflict and war. It is evident that natural surface water is mostly transboundary which means available surface water is being shared by two or more countries. For examples; Indo-Bangladesh, Indo-Pakistan water basins. A few transboundary water conflicts are listed in Table 1 (Lee et al., 2015; Wolf, 1998). It is also reported that around two thirds of the world's transboundary rivers are lack of proper agreements for water sharing among the countries. This situation might be the cause of regional political conflicts and wars (Reuveny, 2007). On the other hand, water sharing could be a good mean of regional co-operation (Bernauer & Kalbhenn, 2010). For example, India and Pakistan have worked together in managing the Indo River water. Although it was observed that in the last fifty years, three wars have been occurred between two countries. Another example, Israel and Palestine also cooperate each other in water sharing in spite of their political conflict. Thus, management approach and co-operation attitude towards transboundary water conflict could be effective tools for cooperation in regional economic, environmental and social development (Sarsons, 2015).

Human rights, economic growth and environmental sustainability would be the good means of water sharing. Effort towards strengthening the human rights in the aspect of water sharing could also create regional cooperation and peace (Wolf et al., 2005). Global water sources should be distributed fairly among the sharing countries based on criteria such as population density, climatic conditions, degree of pollution and environment sustainability instead of military and political strength. In this regard, the Helsinki Rules could be used as a guideline for proper utilization and administration of international rivers (El-Fadel et al., 2003). Diplomatic process could also useful to build infrastructure projects for water distribution. Relevant stakeholder's cooperation for upholding human rights, economy growth and to maintain environmental sustainability would be the strong ways to reduce water conflict.

In water distribution human rights, economy and environment have not get full priority. As a result, social and political conflicts appear, sometimes the conflicts turn into military intervention and war. The lessons learned from this are:

1. Human rights must get the priority during water distribution.
2. Water must be distributed based on the population density, economy, and environment sustainability.
3. Social and political conflicts, and war could be reduced to zero level.
4. Holistic approach is essential including redefining the water use, water productivity performance, capacity utilization of water production resources, and sustainability in economy and environment.
5. As seawater is the most abundant water sources, the future research should be focused on water desalination technology aiming to increase portable water supply.

Impact of Global Warming on Fresh Water Availability

Global warming's effects on fresh water availability are expected to be significant by the 2050s. Climate change will affect all components of the freshwater system, and the estimated water stress area will be doubled in next three decades compare to the present status (Arora & Boer, 2001; Bosello et al., 2006). Global warming's effects on the water stress model are shown in Figure 1. The most observable effects would be thermal expansion of sea water and melting of ice and glaciers, which leads to increase in sea level. This scenario has direct impact on water supply at coastal areas and islands (Deke et al., 2001) as follows:

- Vulnerable to flood and damage of water treatment plant thus reducing quality of potable water
- Intrusion of saltwater (>35,000 ppm) to land fresh water supply such as ground water and river water.
- Surface fresh water and underground water will be replaced by saltwater and this will create water scarcity.

On the other hand, global warming also increases evaporation rate and this accelerates land-surface drying in arid areas and more atmospheric water vapour in wet areas. Accelerated drying increases the frequency and severity of droughts and wildfire. Likewise, intensified water vapour increases the risk of floods (Karl & Trenberth, 2003). Drought reduce runoff water supply; floods reduce water quality and destroy livelihood. Another scenario is that global warming leads to reduction of precipitation that falls as snow and duration of snow cover particularly in northern America and Eastern Europe. This scenario has significant effect on timing of stream flow in those region, shifting from spring snow melt to winter runoff (Arnell, 1999). In this aspect, it is evident that global climate change will contribute to increase of water stress problems.

Damages done to the environment, whether due to housing development, mining and manufacturing operations, are often resulted from engineering activities. At the same time, efforts to remediate the damage, to a large extent, also require the efforts of engineers. The roles of engineers to meet global warming challenges are shown in the following:

- Solving Water Problem. Global climate change has huge impact on fresh water supply; and in the near future, water stress will increase. To address this issue world needs economic and environmentally sustainable solutions. To find out ways, engineering community needs to pay attention on R&D to optimize water use in economic activities. Speed up the research programs to device an economic way of industrial and domestic wastewater recycling. Innovation on developing water and energy efficient kitchen and toilet equipment. Climate change will likely to increase rainfall and engineer role is to device system to use rainwater in industrial and agriculture applications.

- Engineering in Devising Efficient Desalination Process. The most abundant water source is seawater, but consists high salts content and other impurities. To process this water to produce potable and drinking water, energy efficient desalination and demineralization water technology is essential. The current available desalination technology is not energy efficient and environmentally friendly. The engineering role is to develop affordable water processing technology to provide solution to current and future water problem.
- Green House Gases (GHGs) Reduction. Manufacturing industries are the major contributors of GHGs emission. Lower energy efficiency of fossil fuel combustion and industrial waste discharging to environment are responsible for GHGs emission. By devising energy efficient manufacturing processes machinery, and waste recycling techniques it is possible to reduce GHGs emission from industry.
- Engineering Role in Remodeling Lifestyle. The present lifestyle demands a significant amount of water and energy per capita; and current architecture of development and township are responsible for higher utility consumptions. Engineering community may conduct R&D towards developing water-and energy-efficient development and township.
- Engineering Role in Global Actions in Managing Climate Change Effects. The Paris summit (December 2015) agreed to limit global warming within 2°C of pre-industrial levels (Green, 2014).

To achieve this goal, effective public and private investment for R&D is essential to remodel current architecture of infrastructure, lifestyle, production process, recycling utilities, and energy-efficient economic operations. In this aspect, the role of engineering is to be dedicated in R&D activities for developing environmentally sustainable technology to reduce GHGs emission in order to minimize global warming effects.

Figure 1. Water problem caused by global warming
(Oki & Kanae, 2006)

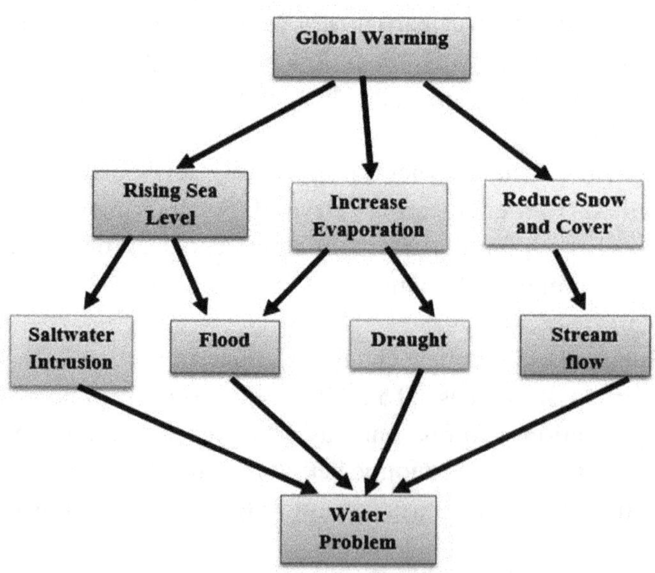

Water Demand Scenario

The increasing water demand contributes to increasing water stress. According to the Millennium Development Goals (MDGs) Report 2015, 783 million people or 11% of the global population, remain without access to safe water supply source. Such sources include household connections, public standpipes, boreholes, protected dug wells, protected springs and rainwater collections. The world has yet to meet the drinking water target. Also, there are regions such as Sub-Saharan Africa where over 40% of all people are living without improved drinking water. The challenges of safe water supply are affordable cost (water costs ≤3% of household income) and physically accessible to source (≤ 1,000 meters of the home and collection time ≤30 minutes) (Gopi et al., 2019; United Nations, 2015). To meet the safe water supply conditions mentioned above, MDG objectives could be achieved. However, demand is continuing to increase with population growth and other factors, hence we have to plan accordingly. The increasing global population growth and other factors such as agricultural and industrial usage have significantly impacted water demand, and this scenario is illustrated in Table 2.

Urbanization, economy migration from agriculture to industrialization and changing lifestyle all contribute to increasing water demand. Furthermore, not all water is returned back to the mainstream because of water holding by plant, human and animal bodies (consumptive water). This contributes to reduce the availability of fresh water too. Ultimately, all these issues have created water supply constraint (Gizelis and Wooden, 2010).

Literature suggests that about 70.8% (1.39 billion km3) of the Earth's surface is covered by water; Ocean and seas make up about 96.4%, snow and ice or Artic trapped in Antarctic and Greenland is about 2% and the other sources are about 1.6% (Khublaryan, 2009). In future, fresh water will be a critically limited resource for many regions of the world. About one-third of the world's population live in the countries with shortage of water resource. To address this issue, abundant seawater source is a good option. The most abundant water source is seawater, brackish water and underground water. However, seawater has high salt content and other impurities. Brackish water contains sands, industrial and domestic pollutants. Underground water contains high level of arsenic, iron and other heavy metals (Jiang et al., 2015). To pro-

Table 2. Drivers of water demand scenario

Drivers of Water Demand	Scenario
Population Pressure	World population is growing by 80 million people per year (United States Census Bureau, 2015)
	2.4 billion people living in Sub-Saharan Africa, the region with unevenly distributed water resources (United Nations, 2012)
Industrial Usage	Water for industry is directly related to economy but this water is not readily available on the surface and underground. Water for industry basically is the water for machine and water for product processing. However, higher degree of technological input is essential to produce water for machine and product processing.
	Water demand for manufacturing industry is expected to increase by 400% from year 2000 to 2050 (United Nations World Water Assessment Programme, 2015)
Agriculture	Water for agriculture is readily available on the surface and underground, if it is free of arsenic and other toxic heavy metal.
	Largest user of water resources, accounting for roughly 70% of all freshwater withdrawals globally, and over 90% in most of the world's least-developed countries (United Nations World Water Assessment Programme, 2014)

cess this water to produce potable and drinking water, desalination and demineralization water technology has been developed (Shahidul and Sadrul, 2002). By applying affordable water processing technology, available water could be processed to provide solution to current and future water problem. As seawater is the most abundant source available, research effort should be put in sustainable desalination process.

FUNDAMENTAL OF MEMBRANE DESALINATION

Principle and Practice of Reverse Osmosis Process

Reverse osmosis (RO) is currently the most important desalination technology and it is the most widely used desalination technology globally. The RO membrane technique is considered the most efficient way for brackish and seawater desalination (Qasim et al., 2019). Generally, osmosis is the tendency of a fluid, usually water, to pass through a semi-permeable membrane into a solution of higher solvent concentration such as salt solution. In reverse osmosis (RO), it uses dynamic pressure to overcome the osmotic pressure of the salt solution. This causes the water-selective permeation from the saline side of a membrane to the freshwater side. Then, salts will be rejected from the membrane and this indicates that separation process is accomplished (Figure 2). The RO membranes used for desalination processes are usually semi-permeable polymeric thin layers, adhering to a thick support layer. These membranes are often made of cellulose acetates, polyamides, polyimides and polysulfones. However, these RO membranes are sensitive to changes in pH, small concentration of oxidized substances like chlorine and chlorine oxides, organic materials and well as the presence of algae and bacteria (Belila et al., 2016). Therefore, careful pre-treatment is needed in order to prevent membrane contamination and fouling which will then affect the membrane performance.

The solvent (permeate) and solute transport inside desalination membrane is present in Figure 3.

The solvent mass flux J_w, which is generally water, can be expressed by Fick's law. It depends on ΔP, the pressure difference across the membrane and the osmotic pressure difference of the solution across the membrane (Abbes, 2005; Al-Bastaki & Abbes, 1999):

$$J_w = A_w(\Delta P - \Delta\pi)) \quad (1)$$

Figure 2. Working principle of osmosis, equilibrium and reverse osmosis

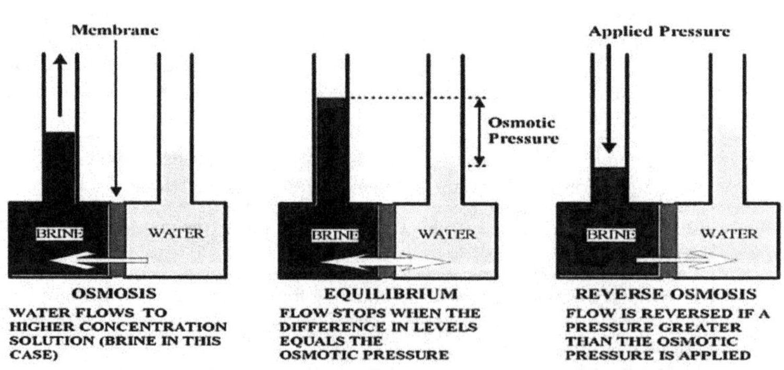

Figure 3. Solvent and solute transport inside desalination membrane
(Abbes, 2005)

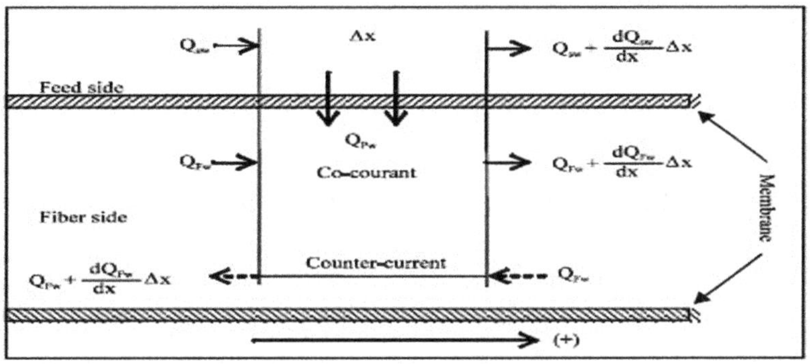

where, A_w is the water permeability constant.

The difference in osmotic pressure $\Delta\pi$ on both sides of the membrane is expressed as:

$$\Delta\pi = \pi_F - \pi_P \qquad (2)$$

where, the subscript F refers to the feed side and the subscript P to the permeate side.

For not too high a solute concentration, the osmotic pressure is approximately a linear function of solute concentrations (Senthilmurugan et al., 2005):

$$\pi = Ac \qquad (3)$$

where, α is a proportionality coefficient (Abbes, 2005). By substituting Equation 2.45 into Equation 2.43, the new equation becomes:

$$J_w = A_w\left(\Delta P - \alpha\Delta C\right) \qquad (4)$$

where, ΔC is the difference of the solute concentration across the membrane, expressed as:

$$\Delta C = C_F - C_P \qquad (5)$$

where, C_F and C_P are the concentrations in the feed and permeate side, respectively.

The volumetric flow rate can be expressed as:

$$Q_w = \frac{J_w Sa}{\rho_w} \qquad (6)$$

where, Sa is the membrane surface area and ρ_w is the water density.

For the solute flux, it is assumed that the chemical potential difference due to pressure is negligible and so the driving force is almost entirely due to concentration differences. From Fick's law, the solute mass flux is:

$$J_s = B_s\left(\Delta C\right) \tag{7}$$

where, B_s the solute permeability coefficient which is a function of the solute composition and membrane structure.

The solute mass flow rate is expressed as:

$$Q_s\left(t\right) = J_s Sa = B_s Sa\left(C_F - C_P\right) \tag{8}$$

The membrane rejection is defined as the fraction of solute present in the solution which is stopped by the membrane:

$$T_R = \frac{C_F - C_P}{C_F} = 1 - \frac{C_P}{C_F} \tag{9}$$

These models are useful for evaluating the product water flow rate and salt rejection rate to understand the behavior and characteristics of these parameters.

RO Membrane Desalination

RO membrane desalination technology is proved to be the most reliable, cost effective and energy efficient in producing fresh water compared to other available desalination methods (Fane, 2018). Membrane desalination is a pressure intensive process and, the seawater can be pressurized to pressures as 40–80 bar pressure is required to produce fresh water. It is the fastest-growing research area in the word. Scientist have put huge R&D efforts to make membrane technology efficient, resulting in stable, long-lived membrane elements (Anis et al., 2019). Component parts membrane desalination machinery has been improved, as well, reducing maintenance and down time. Additional advancements in pretreatment have been made in recent years, further extending membrane life and improving performance. Membrane desalination delivers product water having essentially the same temperature as the raw water source. This is more desirable than the hot water produced by evaporation technologies. Membrane desalination systems can be designed to deliver virtually any required product water quality. For these and other reasons, RO membrane desalination is usually the preferred method of water desalination (Qasim et al., 2019). A typical membrane desalination process is pictured is Figure 4.

A disadvantage of MDT is the need for significant pre-conditioning of the feed water to protect the membranes. The extent of pre-treatment requirements depends on a variety of factors, such as seawater composition and temperature, seawater intake, membrane materials, and recovery ratio, non-organic matter and biomass, concentration of salt, humic acid and pH value. Typical electricity consumption of MDT plants is in the range of 4 to 7 kWh/m^3, depending on a few factors such as water salinity, recovery ratio, water quality, plant configuration, and energy recovery system in the brine water circuit.

Figure 4. Example of membrane desalination

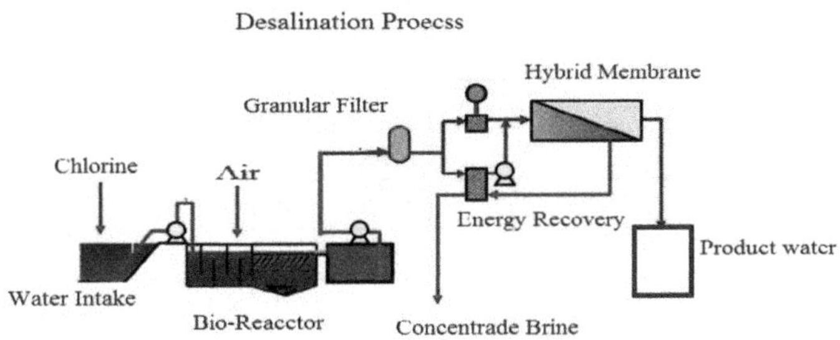

In RO membrane desalination under constant flow rate or permeate flux is described by generalized form of Darcy equation (Karabelas & Sioutopoulos, 2014):

$$J = \frac{\Delta P}{\mu\left(R_m + R_c + R_f\right)} \tag{10}$$

where J is permeate flux; ΔP is transmembrane pressure; μ is viscosity of the permeate; R_m is intrinsic membrane resistance; R_c is external fouling resistance formed by cake layer, and R_f is internal resistance due to materials absorbed into the pores. Power consumption by the system which includes energy for feed water pumping, high pressure pumping, booster, and chemical treatment could be calculated using model developed by Darwish et al. (2002):

$$Pw_n = \frac{Q_n \times \mathrm{Pr}_n}{E_n} \tag{11}$$

Where, Pw_n(kW) = Power consumed by feed, low and high pressure, booster and chemical water treatment pumps, Q_n (m^3/s) = Rates of feed water, Pr_n (kPa) = Feed pressure, boosted pressure, rejection pressure, and E_n (Net efficiency of feed pump) = Ep (pump efficiency) x En (motor efficiency) for high pressure pump (booster) and energy recovery turbine. According to Gilau and Small (2006), energy recovery system should be used if brine exits the system at 300 psig or more and if system recovery is less than 80%. The model to estimate energy recovery is shown below:

$$Pw_n = Q_n \times \mathrm{Pr}_n \times E_t \tag{12}$$

Where, Et is turbine efficiency, Prn(kPa) is the feed pressure, and Qn(m^3/s) is rates of the feed brine. The model is important to determine cost-effective and feasible energy option to produce targeted output in a desalination plant.

Table 3. Classification of water based on dissolved solids

Table Classification	Dissolved Solids (mg/l)
Drinking Water	< 1,000
Mildly Brackish	1,000 – 5,000
Moderately Brackish	5,000 – 15,000
Heavily Brackish	15,000- 35,000
Seawater	35,000

(National Research Council, 2004)

CHARACTERISTICS OF SALINE WATER

The most abundant source of water comes from seawater and brackish water which makes up about 96.4% of total water in the world. Due to the salinity of sea and brackish water [1,000 mg/L <Total Dissolved Solid (TDS) < 60,000 mg/L], desalination technology is required to process this water to drinkable water standard (<250mg/L) (Suwaileh et al., 2019). The dissolved substances contribute to the flavours of water. The dissolved substances are measured commonly by dissolved solids or mg/l. The classification of water based on dissolved solids are shown in Table 3.

The primary solutes in seawater are cations and anions. Majority of them comprised of chloride (Cl^-) and sodium (Na^+) with more than 85.65 percent. When these two ions bond together, they form halite and give seawater its saltiness. The rest of the dissolved substances are shown in Table 4. A simple way to get the salt from sea water is simply evaporate water from a container of seawater. However, that method is not acceptable due to its precision and accuracy. The standard definition of "salinity" is the total mass expressed in grams of all the substances dissolved in 1 kilogram of seawater, when all the carbonate has been converted to oxide, all bromine and iodine have been replaced by chlorine and all organic compounds have been oxidized at a temperature of 480o, i.e. total grams of dissolved salts in 1 kilogram of seawater expressed as %o (parts per thousand).

Source Water Characterization for Desalination

A comprehensive raw water characterization at the planned intake site must include an analysis of data covering physical, microbial and chemical characteristics, meteorological and oceanographic and aquatic biology (Khan et al., 2018). A suitable intake design must consider the potential effects of fouling, continuous or intermittent pollution, and navigation, and necessary steps must be taken to minimize source water contamination, environmental and operational risks. Seasonal variations ought to be characterized, ideally before the design phase of the desalination plant. Typically, desalination plants operate most efficiently and predictably with relative constant feedwater characteristics and are fluctuating too rapidly or dramatically. In this aspect, the water quality review should take into account both seasonal and diurnal fluctuations. The review also should include constituents that may possessed negative impact to plant operation and process performance which includes: water temperature, total dissolved solids (TDS), total suspended solids (TSS), membrane scaling factors (calcium, silica, magnesium, barium, etc.) and total organic carbon (TOC) (Belila et al., 2016). Desalination intake water requirements and quality vary depending on the desalination process.

Table 4. Constitutes of Salt Ions

Salt Ion	Concentration (%o parts per thousand, g/kg)	Weight Percent (%)	Percent Total Dissolved
Chloride (Cl-)	18.980	55.04	55.04
Sodium (Na+)	10.556	30.61	85.65
Sulfate (SO42+)	2.649	7.68	93.33
Magnesium (Mg2+)	1.272	3.69	97.02
Calcium (Ca2+)	0.400	1.16	98.18
Potassium (K+)	0.380	1.10	99.28
Bicarbonate (HCO3-)	0.140	0.41	99.69
Bromide (Br-)	0.065	0.19	99.88
Boric Acid (H3BO3)	0.026	0.07	99.95
Strontium (Sr2+)	0.013	0.04	99.99
Fluoride (F-)	0.001	0.00	99.99
Total	34.482	99.99	99.99

Feedwater volume requirements typically ranges from twenty-five percent more than the production capacity of typical brine reverse osmosis (BWRO) plants to two times the plant production capacity for seawater reverse osmosis (SWRO) systems. The feedwater should be available whenever the plant is operating to meet targeted productivity (Ruiz-Garcia & Nuez, 2016). On top of that, raw water quality determines the location and feasibility of the intake for desalination plant. Plant failure might be due to huge fluctuations in intake water salinity, and variations or large presence of water quality contaminants such as silica, manganese, iron, radionuclides, or scaling compounds that might cause membrane fouling or operational problems which consequently increase treatment costs or reduce the availability of the plant (Younos, 2005).

PRE-TREATMENT AND MAINTENANCE FOR MEMBRANE DESALINATION

Feedwater pre-treatment has long been an important production performance issue in the membrane desalination process, especially for open intake systems such as river or sea water. This is because the ineffective pre-treatment of feedwater causes higher operating pressure and increases the membrane cleaning frequency. Inefficient pre-treatment reduces the membrane operating life and capacity utilization (CU) of plant machinery (Lee et al., 2018). Ineffective pre-treatment is associated with five specific types of membrane fouling: particulate fouling, scaling, bio-fouling, organic and colloidal fouling. Fundamentally, feedwater is the source of membrane fouling. Therefore, the efficiency of the pre-treatment process is vital for membrane desalination production performance (Ang et al., 2016). Pre-treatment for desalination is a challenging issue and will likely continue to be the dominant challenge in designing water desalination plants to a targeted performance. The literature has reported that brackish saline water quality is not stable and its salinity fluctuates greatly. For this reason, water industries are struggling to establish a standard pre-treatment model. Nowadays, pre-treatment plants are, indeed, site-specific and

Figure 5. Typical processes in desalination plant

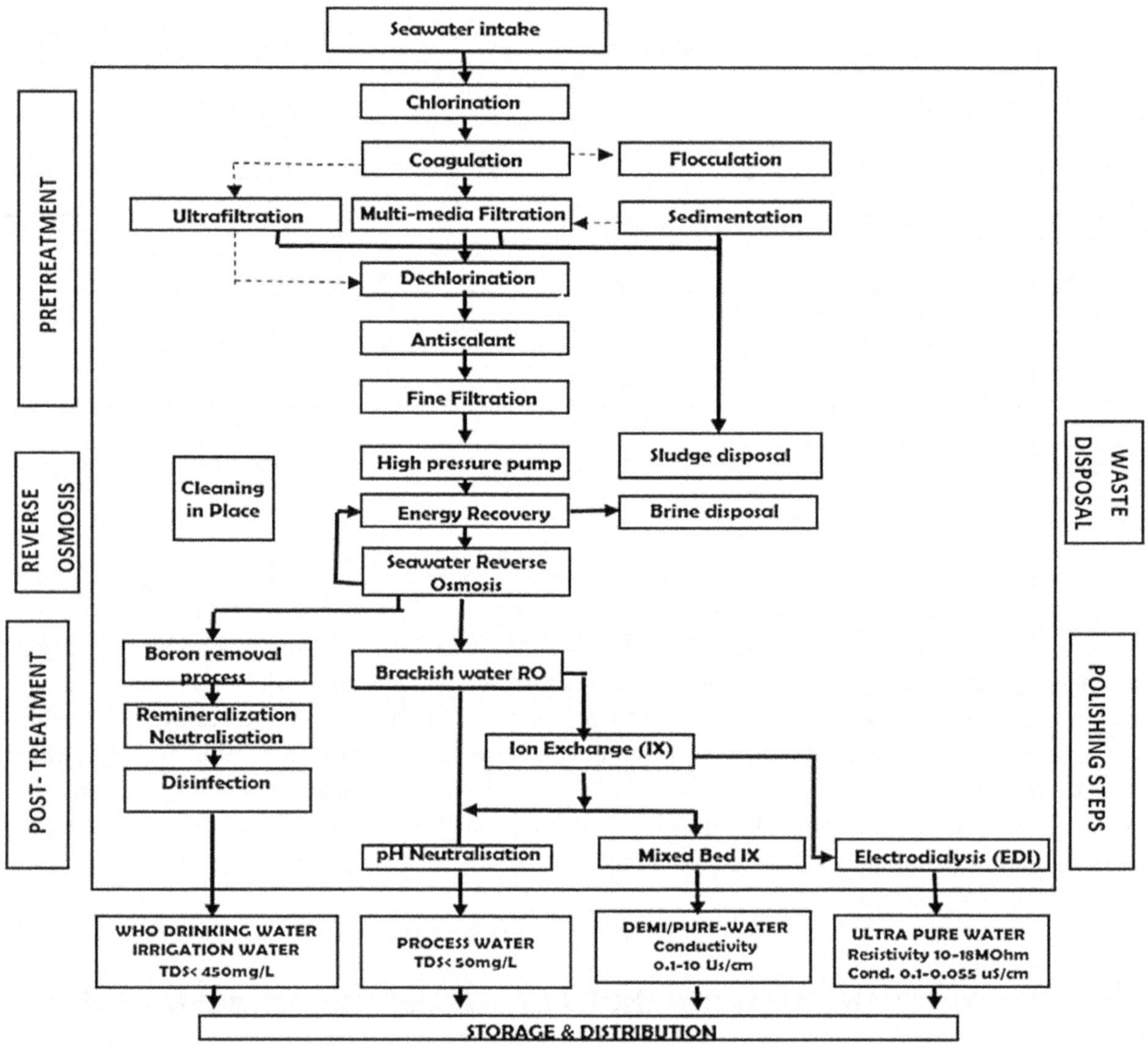

depend on feedwater constituents. It appears that accessing historical intake quality data and conducting pilot testing are key to designing an efficient pre-treatment plant (Williams et al., 2011). Example of pre-treatment process and other processes in desalination plant is shown in Figure 5.

Sources Contamination

Source selection and source protection are the best ways to avoid contamination of finished water by certain organics, surface runoff, ship discharges, and chemical and sanitary waste outfalls near the intake to the desalination plant. When contamination occurs, pretreatments may be necessary and these can involve enhanced disinfection, an adsorption process using granular activated carbon or more frequently powdered activated carbon for intermittent contamination (Williams et al., 2011). Of course, contaminants in blending waters will be transported to the finished water thus appropriate pretreatment of blending water may also be required.

Petroleum and Petroleum Products

The molecular weight cut off for RO membrane performance is typically in the range of 100-300 daltons for organic chemicals. RO membranes can remove larger molecules, however significant fouling would impede operations (Lu, Hu, Xu, & Wu, 2006); small molecules pass through the membrane. Distillation processes can theoretically separate any substance by fractionation based upon boiling point differences, however distillation for desalination is not designed to be a fractionating system, thus substances with boiling points lower than waters would be carried over in the vapours and should be vented out.

Disinfection and Microbial Control in Drinking Water

Similar to fresh waters, sea and brine waters can contain pathogenic microorganisms including bacteria, protozoa and viruses. Disinfection can be applied at several points during the treatment process. The question is what is the adequate level of disinfection to protect public health from exposure to pathogenic microbes and are there any unique risks that may be associated with desalination practices. During pretreatment a disinfectant, often chlorine, will be added to reduce biofouling and protect the membrane from degradation. However, chlorination also oxidizes membrane materials. The maximum allowable dosage is therefore 1000 ppmh (Belila et al., 2016). The typical dosage of chlorine is about 3 mg/L of active chlorine. A common way of de-chlorination is by injecting sodium bisulfite or meta bisulfate which is known as chlorine scavenger. Membranes also have the capacity to remove microorganisms by preventing their passage to the finished water. So long as the membranes are intact virtually complete removals of microorganisms can occur, however some bacteria can grow through the membrane (Belila et al., 2016).

Ultrafiltration Membranes

Ultrafiltration membranes which have pores (~0.001 to 0.1 µm) have been demonstrated to achieve significant reductions of virus and protozoa (Williams et al., 2011). Better performance would be expected from RO membranes. Several challenge tests employing giardia lamblia and cryptosporidium cysts and MS2 bacteriophage with an ultra- filtration membrane of nominal pore size of 0.035 µm, and 0.1 µm absolute have demonstrated very effective removals (Pangarkar et al., 2011). Guardia cysts can vary from 4 to 14 µm in length and 5 to 10 µm in width; cryptosporidium cysts range from about 4 to 6 µm. These intact ultrafiltration membranes (0.1-micron nominal) should completely remove the cysts. MS2 bacteriophage size is approximately 0.027 µm, which is smaller than the pore size of the UF membrane. However, substantial removal can be achieved probably due to adsorption of the virus on suspended particles, adsorption on the membrane or from the secondary filtration due to fouling of the membrane surface.

Coagulation-Flocculation

The coagulation-flocculation processes facilitate the removal of suspended solid and colloidal particles. It's used in the first stage of solids-liquids separation: settling (sedimentation), flotation or filtration (Camacho, Sousa, Bergamasco, & Ribau Teixeira, 2017). Coagulation is the destabilization of colloidal particles brought about by the addition of a chemical reagent called as coagulant. Flocculation is the agglomeration of destabilized particles into micro floc and after into bulky floccules which can be settled called floc. The addition of another reagent called flocculant or a flocculant aid may promote the formation of the floc. The factors, which can promote the coagulation-flocculation, are the velocity gradient, the time, and the pH. The time and the velocity gradient are important to increase the probability of the particles to come together. Moreover, the pH is a prominent factor in the removal of colloids. The pH necessary for coagulation may be adjusted by addition of an acid or a base. Typical example of pH for aluminium based are between pH 6 to 7.4 and pH more than 5 for ferric based coagulants or flocculants (Obiora-Okafo & Onukwuli, 2015).

Membrane Desalination Plant Maintenance

Production performance of plant machinery is associated positively with performance of maintenance activities (Ting et al., 2013). Literature suggests that maintenance activities are responsible to restore plant machineries back to or close to original health condition (Liyanage, 2007). Maintenance activities cannot be separated from production machineries as it is needed a manufacturing plant for utilizing its capacity to optimize production performance. Good implementation of maintenance strategy not only improves machinery efficiency and effectiveness but also brings significant improvements in plant capacity utilization. Consequently, the production system being benefited by becoming more productive. The typical maintenance activities of desalination plant include (Avlonitis et al., 2003; Mon, 2013):

- Membrane Fouling and scaling prevention
- Calibration of instruments such as gauges and meters
- Function of alarms and safety equipment
- Corrosion prevention
- Leak prevention
- Back wash
- Flushing of the system before shutting down and operating
- Replacement of parts due to mechanical failures
- Routine preventive maintenance
- Keep track of data for all the maintenance activities

The production efficiency outputs at the beginning is almost same as designed value. Over time, the outputs start to reduce due to various fouling effects. Specially Bio film, colloidal materials, organic materials make a semi solid layer on membrane surface that contributed to reduce outputs of desalination plant. The recovery percentage is a measure of water production efficiency. The overall recovery increases linearly with time up to 60 h (Figure 6). Beyond this time, the increase is slightly constant due to a decrease in the solvent flux caused by an increase in the osmotic pressure. The overall recovery could reach a value of over 90% at the end of a production cycle.

Figure 6. Water production efficiency

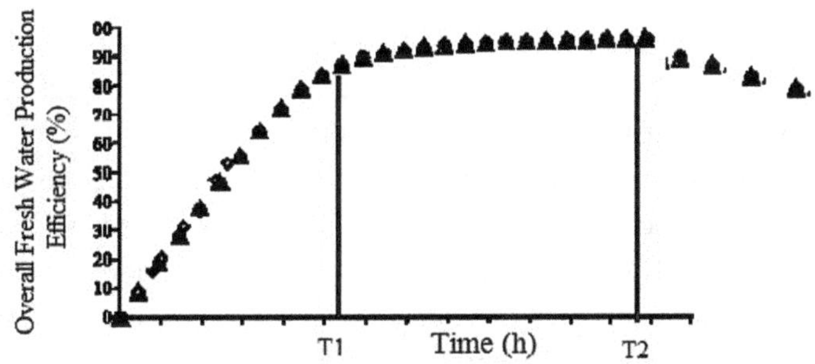

In membrane desalination, maintenance activities are essential to maintain product water flow rate and quality by eliminating membrane fouling for achieving sustainable water production (Jiang et al., 2014; Warsinger et al., 2014). Because Murphy's Law always prevails and no amount of pre-treatment will completely eliminate fouling, cleaning ultimately will become necessary. The typical membrane fouling factors and solutions are listed in Table 5. Details will be explained in the next subsections.

Membrane Bio-Fouling and Solutions

Surface water contains various microorganisms such as bacteria and protozoa that contribute to biofilm formation on the membrane surface and grow if carbon source and nutrients are available (Monnot et al., 2016). Killing bacteria and microorganisms before the water enters the membrane is essential to prevent bio-fouling. A very cost-effective way to avoid biological fouling is chlorination but chlorine oxidizes the membrane material and the majority of available membranes can only tolerate 1000 ppm chlorine (Belila et al., 2016). The typical chlorine dose is about 3 mg/L of active chlorine. A common de-chlorination process is the injection of sodium bisulfite or meta-bisulfite, classified as chlorine scavengers, but this process contributes to environmental problems (Dawoud & Mulla, 2012). At most large-scale desalination plants, electro-chlorination is being used to produce sodium hypochlorite (NaOCl) from the excess of sodium chloride salt (NaCl) present in seawater. For the efficient use of active chlorine to produce NaOCl, the pH should be between 5.5 and 6.5 (Avlonitis et al. 2003).

Table 5. Membranes fouling and solutions

Membrane Fouling	Causes	Solutions
Biological fouling	Bacteria, microorganism, viruses and protozoan	Chlorination
Particle fouling	Sand, clay (turbidity, suspended solids)	Filtration
Colloidal fouling	Organic and inorganic complex, colloidal particles and microalgae	Coagulation, filtration, flocculation/Sedimentation
Organic fouling	Natural organic matter, humic acid and fulvic acid and biopolymer	Coagulation, ultra-filtration, activated carbon
Mineral fouling	Calcium, magnesium, Barium sulfates and carbonate	Anti-sealant dosing and acid dosing
Oxidant fouling	Chlorine, ozone and KMnO4	Sodium bi-sulfate dosing, Activated carbon

Membrane Particulate, Natural Fouling and Solution

Natural organic materials (NOM) are present in all sources of water including seawater. One type of particulate fouling refers to the deposition of various materials such as organics, suspended matter and colloids on the high-pressure side of the membrane, building up a thin film on the membrane surface. Particulate fouling is a persistent problem in most membranes and a secondary problem arising from this is a reduction in water production, which translates into increased operating costs (Kang et al., 2016). Reliable methods of preventing particulate fouling employ efficient pre-treatment processes prior to membrane filtration. The characteristics of each raw water source such as salinity, total suspended solids, total dissolved solids and other pollutants are different and require a custom approach (Zaiedy et al., 2016). This particulate material can be removed through gravity filter media with a coagulation process. To measure the degree of this fouling problem, silt density index (SDI) is used as described in American Society for Testing and Materials (ASTM) 4189-95. To achieve high economic operating efficiency, the SDI limit should be maintained at less than 3 (SDI < 3) (Monnot et al. 2016). Membrane cleaning is essential in achieving better production performance but will contribute to increased machinery downtime, energy and chemical use. Sources of NOM include pollutants with carboxyl and hydroxyl functional groups (Woo et al., 2015). NOM is considered a major contributor to membrane fouling in a desalination process and can significantly reduce production performance and membrane life and increase operating costs (Ju et al. 2015). Understanding fouling mechanisms and developing ways to control them are critical to the economy of desalination technology. Ferric chloride ($FeCl_3$) dosing in the pre-treatment phase is an effective way of removing organic compounds from feedwater. The coagulation by $FeCl_3$ at optimum dosage removed about 95% of hydrophobic compounds. The application of a UFM in pre-treatment has also shown an extremely good result (Löwenberg et al., 2015).

Mineral Fouling and Solution

Mineral fouling of membranes is predominantly composed of divalent metal ions. Monovalent metals such as sodium and potassium are nearly completely soluble, whereas in the presence of sulfate, phosphate or carbonate, divalent ions such as calcium, iron, magnesium, barium, strontium, radium, beryllium, lead and silicon are nearly insoluble. These molecules precipitate on the membrane surface and form crystals. Common forms of mineral scale are calcium carbonate, calcium sulfate, calcium, phosphate, barium sulfate, strontium sulfate, iron hydroxide and silicon dioxide (silica) (Sahachaiyunta et al. 2002). To prevent mineral fouling, a small but significant dosage of anti-scalants are being used to reduce the mineral fouling threat (Zhu et al. 2009).

Oxidant Fouling and Solution

Chlorine and its associated products have been used for many years to destroy various of microorganisms. Usually, feedwater passes through a series of dedicated water pre-treatment equipment before being fed into the membrane. To reduce the membrane bio-fouling threat, free residual chlorine concentrations of 0.50 mg/L to 1.0 mg/L have been maintained through the entire pre-treatment line (Belila et al., 2016). When the chlorinated water of such concentrations feed into the filtration membrane, the membrane's structural materials will degrade. Common chemicals that could cause significant damage to membrane materials include chlorine dioxide, hydrogen peroxide, ozone and permanganate. The desalination process should be carried out under acidic conditions ($5.5 < pH < 6.5$) to remove free chlorine (Belila et al., 2016).

POST-TREATMENT PROCESS FOR MEMBRANE DESALINATION

Desalination plants use salt water containing other contaminants as their sources. It appears that performance, operating and product quality specifications have evolved over time depending on the source and usage. Most drinking water applications comply with World Health Organization drinking water standards as finished water quality specifications. WHO Guidelines for Drinking Water Quality (GDWQ) cover a broad range of contaminants ranging from inorganic and synthetic organic chemicals, disinfection byproducts, microbial indicators and radionuclides, and are aimed at typical drinking water sources and technologies. Due to the nature of desalination which does not uses typical water source as feed water, WHO Guidelines may not be fully applicable to some factors that can be encountered during intake, production and distribution of desalinated water. On the other hand, the quality and safety of the product water, numerous other health and environmental protection issues are also evident when considering the impacts of desalination processes. This section addresses post treatment of drinking water quality to assist both proposed and existing desalination facilities to achieve sustainability.

Disinfection of Output Water

Chlorine in numerous forms (sodium hypochlorite, chlorine gas) is usually used for disinfection due to its recognized potency as a disinfectant and since a reduced level of disinfection by-product precursors is generated in permeate. However, alternative options such as chlorine dioxide or even chloramines as a secondary disinfectant, and ozone or UV could be used in combination with chlorination to control microbial regrowth depending on specific conditions.

Chlorination

Chlorination is the most widely accepted method for water disinfection. Chlorine gas and sodium hypochlorite are the two most well-known chemicals used for chlorination. The desalination process has high possibility to eliminate pathogens and other undesirable microorganisms. The typical target chlorine dose that provides sufficient disinfection depends on two main factors – permeate temperature and contact time. Typical chlorine dosage used for disinfection is between 1.5 to 2.5 mg/litre (Belila et al., 2016). The use of chlorine gas is linked with other potential hazardous and safety considerations might linked to accidental gas releases. Hence, chlorine gas disinfection facilities must be equipped with gas detection, containment and treatment facilities that provide sufficient protection to the public health. A five to fifteen percent solution of sodium hypochlorite is safer to use, handle and store than chlorine gas. Sodium hypochlorite could be transfer to the desalination plant site as a commercial product or produced on site using low bromide sodium chloride (salt). However, electrolysis of seawater to produce hypochlorite is not recommended due to its high bromide content which in turn produces large amounts of bromate.

Chloramination

Chloramination is another alternative of disinfection method. This disinfection method contains sequential addition of chlorine and ammonia to the output water to form chloramines. Chloramines have a considerably slower rate of decay than free chlorine and thus are usually favoured (Watson et al., 2003), particularly for product water delivered to massive distribution systems with high temperatures and long retention times and high potential for chlorine residual loss. Chloramination generally results in a creation of lower levels and different kind of disinfection by-products than free chlorine disinfection. It might contribute to the production of nitrite under various conditions. Since desalted seawater has a low organic content, the use of chloramines for seawater desalination is not as recommended as it may be for disinfection of drinking water that uses brackish or fresh water sources and hence, is not widely practiced. However, chloramination of desalinated water may be needed if the output water is to be blended with other water sources disinfected with chloramines. If chlorinated desalinated water is blended with chloraminated drinking water produced from fresh surface water source, blending of the water may accelerates the decay of chlorine residual from the two waters. To avoid such decay the desalinated water is usually recommended to be chloraminated at higher dose than the fresh-source potable water/s with which is it are blended.

Ozonation

The usage of ozone gas in product water disinfection is a widely accepted practice (Aparicio et al., 2007). However, ozonation of desalted water is related to potential formation of huge amounts of assimilable organic carbon (AOC), and bromate because of the relatively high content of bromide in the feed water as compared to that in potable water from different surface water sources.

Ultraviolet Light Disinfection

Usage of ultraviolet (UV) irradiation in desalination plant permeate is a viable disinfection alternative. The disinfection of desalinated water can generally need require lower UV dosages than those used for UV disinfection of different surface water sources because of the lower turbidity and lower content of pathogens in the desalted product water. Another advantage of UV in product water disinfection is that it does not require the addition of any chemicals to the product water and thus will not create any chemical disinfection byproducts (Neyens & Baeyens, 2003). However, it does not leave any disinfectant residual in the product water to combat regrowth, which can be added later in the form of chlorine or chloramine.

CASE STUDY

Description of Small-Scale Desalination Plant

The small-scale water desalination plant located in Kampung Pendam, Sadong Jaya, Sarawak, Malaysia was built to solve local water shortage problem. Sadong Jaya water desalination plant has put into operation on January 31, 2015. Sadong Jaya is small down under Samarahan division of Sarawak Malaysia. Local community had suffered water shortage supply. The only means of getting water supply is from

Figure 7. Desalination Plant Location

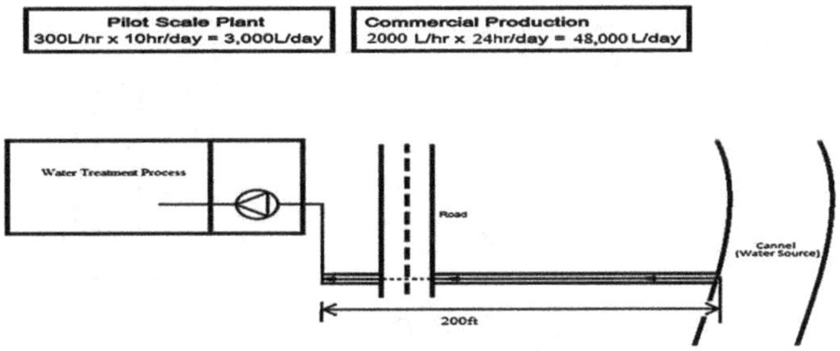

river of Batang Sadong. The small-scale desalination plant aims to contribute to solve local water shortage problem. The village has been suffering from water shortage problem and the nearest river nearby is water with high turbidity (3100NTU), salt content of 35,000 mg/litre (ppm). These characteristics of feed water is classified as brackish and city polluted water (BWCP) with reference from Shahidul et al. (2014). To support the local water demand, the plant aimed to supply water about 3000 L/day as pilot study. The location of the plant from the source water are shown in Figure 7.

Components of the Desalination Plant

The basic components of the plant consist of water intake, bio-reactor, dual media filter, ultrafiltration membrane and production membrane pack. There are two (2) main stages of water treatment process: pre-treatment and filtration process. Figure 8 presents the two stages of water treatment process, namely pre-treatment process for removing settle-able contaminants and desalination process for separating dissolved salts and other minerals from feed water.

Stage 1: Pre-Treatment Process

Brackish and city polluted water (BWCP) from the Sadong Jaya River is used as water source for fresh water production (Figure 9). Source water is pumped through non-return valve by high pressure pump into bio-reactor tank. HDPE pipes are used as water delivering channels. After tightening the pipe connections, the pump is turned on. This operation is constantly monitored to ensure no leakage or damage in the pipeline. At full operating capacity, the pump is able to deliver water at the rate of 600 litres per minute (LPM). The pump consumed about 1 kWh when turned on based on the manufacturer's manual. The source of electrical power for the whole plant operations is from certified local energy provider (SESCO). The pump is turned on when it is needed only.

Ferric chloride ($FeCl_3$) are used as coagulant to bind smaller suspended solids. As the smaller suspended solids are beginning to coagulate and gaining size and weight, they settle to the bottom of tank. This scenario is known as sedimentation process. The sedimentation process is shown in Figure 10.

Calcium hypochlorite [$Ca(ClO)_2$] is commonly used as disinfectant. BWCP contains microorganisms such as bacteria, algae, fungi and viruses that not only harmful to human but also can cause serious biofouling (Prihasto et al., 2009) to the filtration membranes. The mixing chamber of chemical and water is designed using simple and cost-saving 'waterfall' effect. The unique design of the mixing chamber is enhancing the chemical mixing effectiveness. Feed water is allowed to flow up and down from one chamber

Figure 8. Schematic diagram for small scale desalination plant

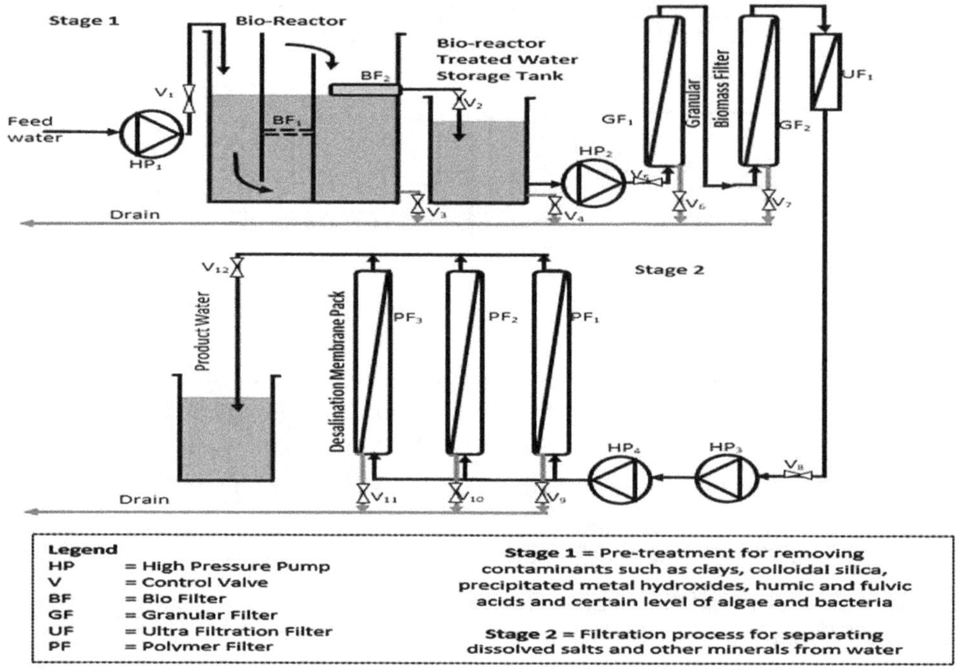

Figure 9. Water source and intake system of the desalination plant

Figure 10. Sedimentation process for pre-treatment

to another and finally create waterfall-like water passage to maximize mixing process. Full bio-reactor tank of water (approx. 15 m^3) is left for few hours for coagulations and sedimentation processes to occur.

Macro-cylindrical filter is used to eliminate floating contaminant such as tree branches and leaves. Heavy particles such as mud and sand are settled at the bottom of bio-reactor tank during pre-treatment and then discharged through drainage. Most common practice throughout the world utilized activated carbon in this stage to remove unpleasant odours. However, previous research done by Shahidul et al. (2014) showed that the use of activated carbon in pre-treatment process is not feasible and its contribution is not significant. Therefore, activated carbon is not being used in this research to eliminate non-value-added inputs. The quality of pre-treated feed water is a key factor for successful operations. High quality feed water leads to longer membrane lifespan and longer operating time; contributes to less maintenance works. In this stage, the maximum allowable contaminants are: turbidity lower than 1.0 NTU; SDI <3.0; oil and grease <1.0 mg/L (Alawadhi, 1997).

Stage 2: Filtration Process

In stage 2 of desalination process, pre-treated water that stored temporarily in pre-treated water tank are pumped through coarse and fine granular biomass filter by high pressure pump (Figure 11). In this part, pre-treated water is pumped through granular biomass filters to remove organic particles in water. Initially, the pump is turned on to channel the pre-treated water to wash and rinse the filters. Rinsing and washing of the filters are carried out by changing the direction of water flow in the filtration using the valves located on the top of the filters. Rinse and wash procedures of the filters are carried out for a few minutes until the discharged water appears to be clean.

Pre-treated water usually are not allowed to be stored for more than 2 hours to prevent significant occurrence of bacterial growth in water phase (Flemming, 1997). As water passes through granular media filter, the suspended solids contact and stick onto the surface of grains inside the filter. This is a common practice in many desalination plant to prolong membrane lifespan by reducing fouling agents (Woo et al., 2015). The dual media filtered water are then pass through ultra-filter prior entering polymer membrane. Two pumps are used to create high pressure feed water to enter parallel polymer membranes for filtration process

Figure 11. Granular media filter

Figure 12. RO filtration membrane

Water undergone granular biomass filtration process will be pumped through a series of membrane to remove the salt content in water (Figure 12). At the beginning of this process, water is pumped to rinse and wash the concentrated side of the membranes to remove residue adhered. By controlling the rate of concentrated water, water will be pushed through the membrane pores and filters out salt to produce potable water at the permeate side of the membrane.

Higher feed water pressure tends to yield higher permeate flow rate and lower permeate salinity (Djebedjian et al., 2007). Throughout the process, brines are rejected through drainage. The feed pressure recorded throughout the study was between 30 psi to 60 psi. The product water is then being tested for quality control before public distribution. Since the output water is distributed directly to the local village for immediate consumption, no post treatment is needed.

Preventive Maintenance for Membrane Cleaning

Membrane maintenance activities requires two parts of cleaning: chemical and physical cleaning. In physical cleaning, it generally involves one or more of the following cleaning forces which are hydraulic, mechanical and electrical. In hydraulic and mechanical cleaning, both of these cleaning alternate the shear forces on the membrane surface, allowing the deposits to loosen up and dislodge itself from it. As for the electrical cleaning, a voltage is applied across the membrane which in return pushing all the charged deposits away. Only when the physical cleaning gives no effect to the dropping output flow of the water production, chemical cleaning is the only option to restore back the flux to an acceptable level. For membrane cleaning by using chemical, the amount of sodium hypochlorite needed throughout the entire membrane cleaning was calculated first. In order to complete the membrane cleaning procedure, about 400 ppm of chlorine is needed.

Maintenance Procedures

1. Calculate the amount of cleaning agent (sodium hypochlorite) needed for the entire membrane cleaning process.
2. Using a measuring jar and weighing scales, measure and weigh the needed amount precisely. Then, pour the sodium hypochlorite solution into the 800 litres tank and mix it with water.

Figure 13. First stage membrane cleaning

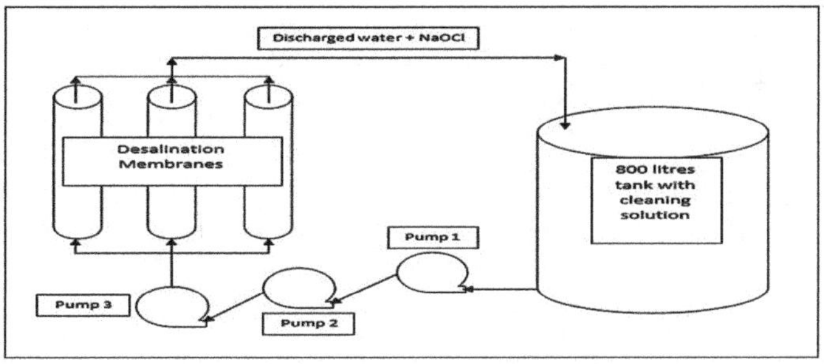

3. Only desalination membrane is involved during this cleaning operation. Water containing the sodium hypochlorite is recycled during this operation.
4. This cleaning operation involved three stages. The first and second stage was conducted for an hour whereas for the third stage, it was conducted until all the water was drained out completely from the 800 litres tank.
5. Each stage will be described in the Figure 13, Figure 14 and Figure 15 respectively.

First Stage Membrane Cleaning

1. Switch on pump 1. Then, switch on pump 2 followed by pump 3. Do not switch on all of these three pumps simultaneously. All of these three pumps are crucial in maintaining the steady flow of the water containing the sodium hypochlorite (NaOCl) into the desalination membranes.
2. Water containing sodium hypochlorite goes through the outer layer of the desalination membrane. At the same time, the mixed water will clean and removes and deposits on the outer layer of the membrane.
3. The discharged water will be recycled back and goes back into the 800 litres tank. The whole process is conducted for an hour.
4. Switch off pump 3, followed by pump 2 and pump 1. Prepare the membranes for the second stage of membrane cleaning.

Second Stage Membrane Cleaning

1. Switch on pump 1. Then, switch on pump 2 followed by pump 3. Do not switch on all of these three pumps simultaneously. All of these three pumps are crucial in maintaining the steady flow of the water containing the sodium hypochlorite (NaOCl) into the desalination membrane.
2. Water containing sodium hypochlorite goes through the inner layer of the desalination membrane. At the same time, the mixed water will clean and removes and deposits on the inner layer of the membrane.
3. The discharged water will be recycled back and goes back into the 800 litres tank. The whole process is conducted for an hour.
4. Switch off pump 3, followed by pump 2 and pump 1. Prepare the membranes for the third stage of membrane cleaning.

Figure 14. Second stage membrane cleaning

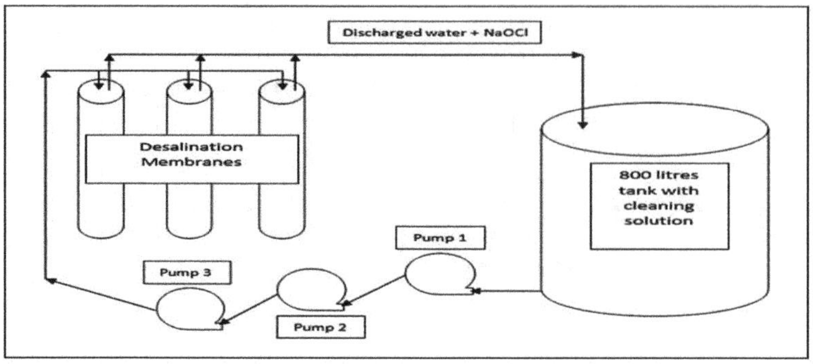

Figure 15. Third stage membrane cleaning

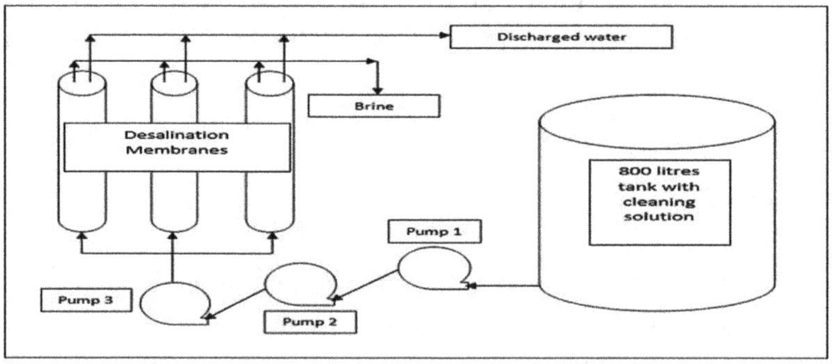

Third Stage Membrane Cleaning

1. Switch on pump 1. Then, switch on pump 2 followed by pump 3. Do not switch on all of these three pumps simultaneously. All of these three pumps are crucial in maintaining the steady flow of the water containing the sodium hypochlorite (NaOCl) into the desalination membrane.
2. Water containing sodium hypochlorite goes through the membrane pores. At the same time, the mixed water will clean and removes and deposits on the membrane.
3. The discharged water will be disposed. This process is conducted until all of the mixed water is completely drained out as brine from the 800-litres tank.
4. After the mixed water is completely drained out, the membranes are rinsed using only water to get rid of the excess sodium hypochlorite which is still inside the system. Then, switch off pump 3, followed by pump 2 and pump 1. This completes the membrane cleaning procedure.

Production Performance of the Plant

Scaling and fouling on high pressure sides of membrane create resistance to fresh water production. The thickness of the scale and degree of fouling in membrane are determinants of feed water quality and capacity utilization of desalination machinery. The declining water production rate is an indicator of scale formation on the membrane surface and fouling of membrane materials. However, the water production performance models are listed below:

Plant Capacity Utilization [CU]

$$CU = \frac{Actual\,Capacity\,Q_A}{Potential\,Capacity\,Q_p} \tag{13}$$

Benefit-Cost Ratio [$R_{b/c}$]

$$R_{b/c} = \frac{Value\,of\,Water}{Production\,Cost\,of\,Water} \tag{14}$$

Plant Operating Reliability [R(t)]

$$R(t) = \int_0^t f(t)\,dt = 1 - e^{\propto t} \tag{15}$$

α = failure rate of machine;
t = failure time

Plant Economic Efficiency $\left[\eta_{EC}\right]$

$$\eta_{EC} = \frac{Value\,of\,Water}{Production\,Cost\,of\,Water} \tag{16}$$

Slit Density Index [SDI]

$$SDI = \left[\frac{1 - \frac{Ti}{Tf}}{T}\right] \times 100\% \tag{17}$$

T = elapsed time;
T_i = initial time
T_f = final time

Plant Availability [At]

$$At = \frac{T_{op} - T_d}{T_{Op}} \tag{18}$$

T_{op} = design operating time

T_d = downtime for maintenance

Relevant input-output data of production have collected directly from the operating log of the plant machinery operation and production performance models have estimated by using plant operating data. The performance parameters of this plant are shown in Table 6.

The estimated values of parameters are demonstrating that performance indicators are within the sustainable range. These findings indicate that Sadong Jaya water desalination plant is economically and environmentally sustainable. The estimated CU of the plant machinery is found 69%, it indicates that 69% potential capacity of the plant machinery is being used to produce water; and it is within the acceptable limit. The estimated economic efficiency of the plant is found 1.25. As per the basic economic theory, it indicates that the value of outputs is more than the expenditure; the plant will sustain. The estimated benefit cost ratio is found 1.5; meaning is benefit of the plant is 1.5 times more than cost. It indicates plant is feasible to operate and to be sustainable. The water production rate is found 360 litter per square meter of membrane which is within the standard value. The estimated plant down time is found 12%, plant availability for production is 69% and operating reliability is found 85%. These findings also indicate that the plant operations are feasible because all values are within the acceptable limit. The chemical consumption is found less than 3 mg/L of water which is the average value being used in other plants. The lessons learned from this case study are described below:

1. Production performance of Sadong Jaya water desalination plant is economically and environmentally sustainable.

Table 6. Plant performance parameters and sustainability index

Performance Parameters	Performance	Sustainability Index*	Comments
CU	69%	>50%	Achievement is significant
η_{EC}	1.25	>1	Achievement is sustainable
$R_{b/c}$	1.5	>1	Economically feasible
Q/m^2-hr	360 liter	>350 liter	Achievement is significant
SDI_{15}	2.81	3<	Achievement is significant
T_d	12%	15%<	Achievement is significant
A(t)	69%	>60%	Achievement is significant
R(t)	0.85	>0.75	Achievement is significant
$FeCl_3$	2.25 mg/l	5%<	Achievement is significant

A(t)- Plant Availability, R(t)-Reliability, Td-down time, * is the reference from published literature

2. The general operating characters of the plant are:
    a. Efficient pre-treatment plant contributed to achieve a sustainable operating condition of Sadong Jaya water desalination plant.
    b. Feed water to membrane contained about 0.5 mg/L of chlorine that contributed to prevent the formation of biofilm in membrane.
    c. Bio-reactor that used in Sadong Jaya to separate colloid materials and pollutant contributed to achieve sustainable production performance.
    d. Preventive maintenance strategy that adopted at Sadong Jaya water desalination plant also contributed to achieve sustainable production performance.

REFERENCES

Abbes, A. (2005). Simulation and anlysis of an industrial water desalination plant. *Chemical Engineering and Processing*, *44*(9), 999–1004. doi:10.1016/j.cep.2004.12.001

Adzlan, A., Khairi, F., Shahidul, M. I., Tarmizi, S., Shazali, S., & Yassin, A., … Ting, C. H. (2013). Investigation on water consumption pattern in traditional Malaysian Kampung : An insight study on water and environmental issues. In *6th Engineering Conference*. doi:10.3850/978-981-07-6059-5_048

Al-Bastaki, N. M., & Abbes, A. (1999). Modelling an industrial reverse osmosis unit. *Desalination*, *126*(1-3), 33–39. doi:10.1016/S0011-9164(99)00152-6

Alawadhi, A. A. (1997). Pretreatment plant design - key to a successful reverse osmosis desalination plant. *Desalination*, *1*(1-2), 1–10. doi:10.1016/S0011-9164(97)00079-9

Ang, W. L., Nordin, D., Mohammad, A. W., Benamor, A., & Hilal, N. (2016). Effect of membrane performance including fouling on cost optimization in brackish water desalination process. *Chemical Engineering Research & Design*. doi:10.1016/j.cherd.2016.10.041

Anis, S. F., Hashaikeh, R., & Hilal, N. (2019). Reverse osmosis pretreatment technologies and future trends: A comprehensive review. *Desalination*, *452*(October 2018), 159–195. doi:10.1016/j.desal.2018.11.006

Aparicio, M. A., Eiroa, M., Kennes, C., & Veiga, M. C. (2007). Combined post-ozonation and biological treatment of recalcitrant wastewater from a resin-producing factory, *143*, 285–290. doi:10.1016/j.jhazmat.2006.09.025

Arnell, N. W. (1999). Climate change and global water resources. *Global Environmental Change*, *9*, S3–S19. doi:10.1016/S0959-3780(99)00017-5

Arora, V. K., & Boer, G. J. (2001). Effects of simulated climate change on the hydrology of major river basins. *Journal of Geophysical Research*, *106*(D4), 3335–3348. doi:10.1029/2000JD900620

Avlonitis, S., Kouroumbas, K., & Vlachakis, N. (2003). Energy consumption and membrane replacement cost for seawater RO desalination plants. *Desalination*, *157*(1-3), 151–158. doi:10.1016/S0011-9164(03)00395-3

Belila, A., El-Chakhtoura, J., Otaibi, N., Muyzer, G., Gonzalez-Gil, G., Saikaly, P. E., … Vrouwenvelder, J. S. (2016). Bacterial community structure and variation in a full-scale seawater desalination plant for drinking water production. *Water Research, 94*, 62–72. doi:10.1016/j.watres.2016.02.039

Bernauer, T., & Kalbhenn, A. (2010). The politics of international freshwater resources. *International Studies Encyclopedia, IX*, 5800–5821.

Bosello, F., Roson, R., & Tol, R. S. J. (2006). Economy-wide estimates of the implications of climate change: Human health. *Ecological Economics, 58*(3), 579–591. doi:10.1016/j.ecolecon.2005.07.032

Camacho, F. P., Sousa, V. S., Bergamasco, R., & Ribau Teixeira, M. (2017). The use of Moringa oleifera as a natural coagulant in surface water treatment. *Chemical Engineering Journal, 313*, 226–237. doi:10.1016/j.cej.2016.12.031

Darwish, M., Asfour, N., & Najem, A. (2002). Energy consumption in equivalent work by different desalination methods: Case study for Kuwait. *Desalination, 152*(1-3), 83–92. doi:10.1016/S0011-9164(02)01051-2

Dawoud, M. (2012). Environmental impacts of seawater desalination : Arabian Gulf case study. *International Journal of Environment and Sustainability, 1*(3), 22–37. doi:10.24102/ijes.v1i3.96

Deke, O., Hooss, K. G., Kasten, C., Klepper, G., & Springer, K. (2001). Economic impact of climate change: simulations with a regionalized climate-economy model (No. 1065). Kiel, Germany.

Djebedjian, B., Gad, H., Khaled, I., & Rayan, M. A. (2007). Reverse osmosis desalination plant in Nuweiba city. In *Eleventh International Water Technology Conference*. Egypt: IWTC11.

El-Fadel, M., El-Sayegh, Y., El-Fadl, K., & Khorbotly, D. (2003). The Nile River Basin: A case study in surface water conflict resolution. *Journal of Natural Resources and Life Sciences Education, 32*, 107–117.

Fane, A. G. (2017, September). A grand challenge for membrane desalination: More water, less carbon. *Desalination, 426*, 155–163. doi:10.1016/j.desal.2017.11.002

Flemming, H.-C. (1997). Reverse osmosis membrane biofouling. *Experimental Thermal and Fluid Science, 14*(4), 382–391. doi:10.1016/S0894-1777(96)00140-9

Gilau, A. M., & Small, M. J. (2006). Designing cost-effective sea water reverse osmosis system under optimal energy options for developing countries. In *International Conference on Renewable Energy for Developing Countries* (pp. 1–18).

Gizelis, T.-I., & Wooden, A. E. (2010). Water resources, institutions, & intrastate conflict. *Political Geography, 29*(8), 444–453. doi:10.1016/j.polgeo.2010.10.005

Gopi, G., & Arthanareeswaran, G. (2019). Perspective of renewable desalination by using membrane distillation. *Chemical Engineering Research & Design, 144*, 520–537. doi:10.1016/j.cherd.2019.02.036

Green, F. (2014). "This time is different": The prospects for an effective climate agreement in Paris 2015. London, UK.

Intergovernmental Panel on Climate Change. (2008). Climate change and water. (B. Bates, Z. W. Kundzewicz, J. Palutikof, & S. Wu, Eds.), IPCC Technical Paper VI (Vol. 403). Geneva.

Jeppesen, E., Kronvang, B., Meerhoff, M., Søndergaard, M., Hansen, K. M., & Andersen, H. E., … Olesen, J. E. (2009). Climate change effects on runoff, catchment phosphorus loading and lake ecological state, and potential adaptations. *Journal of Environmental Quality*, *38*(5), 1930. doi:10.2134/jeq2008.0113

Jiang, W., Wei, Y., Gao, X., Gao, C., & Wang, Y. (2015). An innovative backwash cleaning technique for NF membrane in groundwater desalination: Fouling reversibility and cleaning without chemical detergent. *Desalination*, *359*, 26–36. doi:10.1016/j.desal.2014.12.025

Ju, Y., Hong, I., & Hong, S. (2015). Multiple MFI measurements for the evaluation of organic fouling in SWRO desalination. *Desalination*, *365*, 136–143. doi:10.1016/j.desal.2015.02.035

Kang, J.-S., Sung, S. C., Lee, J. J., & Kim, H.-S. (2016). Application of ceramic membrane for seawater desalination pretreatment. *Desalination and water treatment*, *3994*(November), 1–6. doi:10.1080/19443994.2016.1189702

Karabelas, A. J., & Sioutopoulos, D. C. (2014). Toward improvement of methods for predicting fouling of desalination membranes— The effect of permeate flux on specific fouling resistance. *Desalination*, *343*, 97–105. doi:10.1016/j.desal.2013.10.008

Karl, T. R., & Trenberth, K. E. (2003). Modern global climate change. *Science*, *302*(5651), 1719–1723. doi:10.1126cience.1090228 PMID:14657489

Khan, M. A. M., Rehman, S., & Al-Sulaiman, F. A. (2018). A hybrid renewable energy system as a potential energy source for water desalination using reverse osmosis: A review. *Renewable & Sustainable Energy Reviews*, *97*(January), 456–477. doi:10.1016/j.rser.2018.08.049

Khublaryan, M. G. (2009). Types and properties of water. Encyclopedia of Life Support Systems (EOLSS), 1.

Lee, M. D., Dayou, S., & Karunakaran, P. (2018). Determinants of a pre-treatment model in achieving economic and environmental sustainability in membrane desalination. [JKUKM]. *Journal of Engineering*, *30*(2). doi:10.17576/jkukm-2018-30(2)

Lee, M. D., Islam, S., Ting, C. H., & Begum, S. (2015). Current and future water scenario. Retrieved from http://www.leemandjun.blogspot.my/2015/08/current-and-future-water-scenario.html

Liyanage, J. P. (2007). Operations and maintenance performance in production and manufacturing assets: the sustainability perspective. *Journal of Manufacturing Technology Management*, *18*(3), 304–314. doi:10.1108/17410380710730639

Löwenberg, J., Baum, J. A., Zimmermann, Y. S., Groot, C., van den Broek, W., & Wintgens, T. (2015). Comparison of pre-treatment technologies towards improving reverse osmosis desalination of cooling tower blow down. *Desalination*, *357*, 140–149. doi:10.1016/j.desal.2014.11.018

Lu, Y. Y., Hu, Y. D., Xu, D. M., & Wu, L. Y. (2006). Optimum design of reverse osmosis seawater desalination system considering membrane cleaning and replacing. *Journal of Membrane Science*, *282*(1–2), 7–13. doi:10.1016/j.memsci.2006.04.019

Luber, G., & Prudent, N. (2009). Climate change and human health. *Transactions of the American Clinical and Climatological Association*, *120*, 113–117. doi:10.5124/jkma.2008.51.8.764 PMID:19768168

Martínez-Alvarez, V., Martin-Gorriz, B., & Soto-García, M. (2016). Seawater desalination for crop irrigation— A review of current experiences and revealed key issues. *Desalination*, *381*, 58–70. doi:10.1016/j.desal.2015.11.032

Mon, R. (2013). Seawater desalination operation maintenance and troubleshooting. Retrieved from http://www.slideshare.net/RajeshMon/seawater-desalination-operation-maintainence-and-trouble-shooting

Monnot, M., Laborie, S., & Cabassud, C. (2016). Granular activated carbon filtration plus ultrafiltration as a pretreatment to seawater desalination lines: Impact on water quality and UF fouling. *Desalination*, *383*(April), 1–11. doi:10.1016/j.desal.2015.12.010

National Research Council. (2004). *Review of the desalination and water purification roadmap*. National Academy of Sciences.

Neyens, E., & Baeyens, J. (2003). A review of classic Fenton's peroxidation as an advanced oxidation technique. *Journal of Hazardous Materials*, *98*(1–3), 33–50. doi:10.1016/S0304-3894(02)00282-0 PMID:12628776

O'Brien, K. L., & Leichenko, R. M. (2000). Double exposure: Assessing the impacts of climate change within the context of economic globalization. *Global Environmental Change*, *10*(3), 221–232. doi:10.1016/S0959-3780(00)00021-2

Obiora-Okafo, I. A., & Onukwuli, O. D. (2015). Optimization of coagulation-flocculation process for colour removal from synthetic dye wastewater using natural organic polymers: Response surface methodology applied. *International Journal of Scientific & Engineering Research*, *6*(12), 693–704.

Oki, T., & Kanae, S. (2006). Global hydrological cycles and world water resources. *Science*, *313*(5790), 1068–1071. doi:10.1126cience.1128845 PMID:16931749

Pangarkar, B. L., Sane, M. G., & Guddad, M. (2011). Reverse osmosis and membrane distillation for desalination of groundwater: a review. *ISRN Materials Science*, *2011*, 1–9. doi:10.5402/2011/523124

Payne, J. T., Wood, A. W., Hamlet, A. F., Palmer, R. N., & Lettenmaier, D. P. (2004). Mitigating the effects of climate change on the water resources of the Columbia River Basin. *Climatic Change*, *62*(1–3), 233–256. doi:10.1023/B:CLIM.0000013694.18154.d6

Postel, S. L., Daily, G. C., & Ehrlich, P. R. (1996). Human appropriation of renewable fresh water. *Science*, *271*(5250), 785–788. doi:10.1126cience.271.5250.785

Prihasto, N., Liu, Q.-F., & Kim, S.-H. (2009). Pre-treatment strategies for seawater desalination by reverse osmosis system. *Desalination*, *249*(1), 308–316. doi:10.1016/j.desal.2008.09.010

Qasim, M., Badrelzaman, M., Darwish, N. N., Darwish, N. A., & Hilal, N. (2019). Reverse osmosis desalination: A state-of-the-art review. *Desalination*, *459*(February), 59–104. doi:10.1016/j.desal.2019.02.008

Reuveny, R. (2007). Climate change-induced migration and violent conflict. *Political Geography*, *26*(6), 656–673. doi:10.1016/j.polgeo.2007.05.001

Ruiz-Garcia, A., & Nuez, I. (2016). Long-term performance decline in a brackish water reverse osmosis desalination plant. Predictive model for the water permeability coefficient. *Desalination*, *397*, 101–107. doi:10.1016/j.desal.2016.06.027

Sahachaiyunta, P., Koo, T., & Sheikholeslami, R. (2002). Effect of several inorganic species on silica fouling in RO membranes. *Desalination*, *144*(1–3), 373–378. doi:10.1016/S0011-9164(02)00346-6

Sarsons, H. (2015). Rainfall and conflict : a cautionary tale. *Journal of Development Economics*, *115*, 62–72. doi:10.1016/j.jdeveco.2014.12.007

Senthilmurugan, S., Ahluwalia, A., & Gupta, S. K. (2005). Modeling of spiral-wound module and equation of model parameters using numerical techniques. *Desalination*, *173*(3), 296–286. doi:10.1016/j.desal.2004.08.034

Shahidul, M. I., & Sadrul, A. K. M. (2002). Industrial water treatment technology: an overview. *The Journal of Environment*, *2*, 598–603.

Shahidul, M. I., Shazali, S. T. S., Abdullah, Y., Ting, C. T., Lee, M. D., Hishamuddin, A. H. … Adzlan, A. F. K. (2014). Determinants of desalination machinery capacity utilization: evaluate in the aspect of energy, environment and economic sustainability. In Advances in Process Engineering & Green Energy. Kuching, Malaysia: EnCon.

Suwaileh, W., Johnson, D., & Hilal, N. (2019). Brackish water desalination for agriculture: Assessing the performance of inorganic fertilizer draw solutions. *Desalination*, *456*(January), 53–63. doi:10.1016/j.desal.2019.01.014

Ting, C. H., Shahidul, M. I., Shazali, S. T. S., & Abdullah, Y. (2013). Measuring contextual impact of maintenance on machinery capacity utilization and environment. In *6th Engineering Conference*. Kuching, Malaysia. doi:10.3850/978-981-07-6059-5_034

United Nations. (2012). *The Millennium Development Goals Report 2012*. New York. doi:10.1177/1757975909358250

United Nations. (2015). *The Millennium Development Goals Report 2015. United Nations*. New York.

United Nations World Water Assessment Programme. (2014). *Water and Energy. The United Nations World Water Development Report 2014* (Vol. 1). United Nations Educational, Scientific and Cultural Organization.

United Nations World Water Assessment Programme. (2015). *Water for a sustainable world*. Paris, France: United Nations Educational, Scientific and Cultural Organization.

United States Census Bureau. (2015). International Programs. Retrieved from https://www.census.gov/population/international/data/idb/worldpopgraph.php

Voutchkov, N. (2017, October). Energy use for membrane seawater desalination - current status and trends. *Desalination*, *431*, 2–14. doi:10.1016/j.desal.2017.10.033

Warsinger, D. M., Swaminathan, J., Guillen-Burrieza, E., Arafat, H. A., & Lienhard, V. J. H. (2014). Scaling and fouling in membrane distillation for desalination applications: A review. *Desalination*, *356*, 294–313. doi:10.1016/j.desal.2014.06.031

Watson, I. C., Morin, O. J., & Henthorne, L. (2003). *Desalting handbook for planners. Desalination and Water Purification Research and Development Program Report No. 72* (3rd Edit.). Virginia. doi:10.3109/15360288.2011.620691

Williams, M. E., Wagner, J., Eng, B. S. C., Stoughton, M., Duan, X., Wendel, E., ... Cotruvo, B. J. A. (2011). Reverse osmosis technology for water treatment: State of the art review. *Desalination*, *267*(August), 1–8. doi:10.1016/j.desal.2010.09.001

Wolf, A. (1998). Conflict and cooperation along international waterway. *Water Policy*, *1*(2), 251–265. doi:10.1016/S1366-7017(98)00019-1

Wolf, A. T., Kramer, A., Carius, A., & Dabelko, G. D. (2005). Managing water conflict and cooperation. In *State of the World 2005* (pp. 80–99). Redefining Global Security. Retrieved from http://tbw.geo.orst.edu/publications/abst_docs/wolf_sow_2005.pdf

Woo, Y. C., Lee, J. J., Tijing, L. D., Shon, H. K., Yao, M., & Kim, H.-S. S. (2015). Characteristics of membrane fouling by consecutive chemical cleaning in pressurized ultrafiltration as pre-treatment of seawater desalination. *Desalination*, *369*, 51–61. doi:10.1016/j.desal.2015.04.030

Yoffe, S., Fiske, G., Giordano, M., Giordano, M., Larson, K., Stahl, K., & Wolf, A. T. (2004). Geography of international water conflict and cooperation: data sets and applications. *Water Resources Research*, *40*(5), 1–12. doi:10.1029/2003WR002530

Younos, T. (2005). Environmental issues of desalination. *Journal of Contemporary Water Research & Education*, *132*(1), 11–18. doi:10.1111/j.1936-704X.2005.mp132001003.x

Zaiedy, N. I., Karim, O. A., & Mutalib, N. A. A. (2016). Water quality of surface runoff in loop two catchment area in UKM. *Jurnal Kejuruteraan*, *28*, 65–72. doi:10.17576/jkukm-2016-28-07

Zhu, A., Christofides, P. D., & Cohen, Y. (2009). Effect of thermodynamic restriction on energy cost optimization of RO membrane water desalination. *Industrial & Engineering Chemistry Research*, *48*(13), 6010–6021. doi:10.1021/ie800735q

Chapter 16
Classical Fenton and Sequencing Batch Reactor Treatment of Pesticide Wastewater

Augustine Chioma Affam
University College of Technology Sarawak, Malaysia

Ezerie Henry Ezechi
https://orcid.org/0000-0003-4968-5541
Universiti Teknologi Malaysia, Johor Bahru, Malaysia

Malay Chaudhuri
Independent Researcher, Malaysia

ABSTRACT

This chapter examined Fenton and sequencing batch reactor (SBR) treatment of pesticide industrial wastewater. The optimum operating conditions for Fenton pretreatment of the pesticide wastewater were H2O2/COD molar ratio 3.0, H2O2/Fe2+ molar ratio 10, pH 3 and reaction time 60 min. The COD and TOC removal were 58.51 and 39.76%, respectively and biodegradability (BOD5/COD ratio) increased from 0.02 to ~ 0.30 after 60 min reaction time. The reaction followed pseudo-first order kinetics with a rate constant (k) of 0.0083 min−1. In the post-treatment by aerobic SBR, five different Fenton operating conditions were investigated and H2O2/COD molar ratio 3.0, H2O2/Fe2+ molar ratio 25, pH 3 and reaction time 60 min appeared to be the most significant (p < 0.05) operating conditions. The Fenton–SBR treatment at 12 hr HRT achieved COD, TOC and BOD5 removal efficiency of 96.7, 97.7 and 93.3%, respectively. The Fenton-SBR process was effective for the treatment of pesticide wastewater.

DOI: 10.4018/978-1-7998-0369-0.ch016

INTRODUCTION

Extensive use of pesticides and their contamination potential have become a major concern. Although the benefits of pesticide application are evident in terms of increased agricultural productivity, the presence of pesticide residues in the soil, water and air has created potential risks to the natural environment (Barron et al., 2003). Major sources of pesticide in the water stream include wastewater from agricultural industries and pesticide formulating and manufacturing plants, rinsing and washing of spray equipment and run off during application in crop fields. Wastewater from these sources may have pesticide contamination levels as high as 500-1000 mg/L (Malato et al., 2000). Pesticide active ingredients have been reportedly found in inland waterways and rivers in Malaysia. Organophosphate pesticides (chlorpyrifos and diazinon) and organochlorine pesticides (lindane, heptachlor, endosulfan, dieldrin, endosulfan sulfate, DDT and DDE) have been detected in the Selangor river (Leong et al., 2007). The presence of pesticide active ingredients in surface water is presumably due to the inability of conventional wastewater treatment plant to completely remove them. Organochlorine pesticides such as heptachlor, dieldrin and pp-DDT (52–159 mg/kg) have been detected in sewage sludge in three locations in South Johore, Johor State (Ahmad et al., 2004). Pesticides in the sewage sludge may be incorporated into agricultural lands during application (Tadeo et al., 2010).

Advanced oxidation process (AOP) is a promising technology whereby chemical reactions produce hydroxyl radical ($OH^{\bullet}$) which oxidizes the organic contaminants in wastewater under mild operating conditions (Zhou et al., 2001). Oxidation with Fenton's reagent is based on ferrous ion, hydrogen peroxide (H_2O_2) and $OH^{\bullet}$ produced by the catalytic decomposition of hydrogen peroxide in acidic solution (Chamarro et al., 2001). Sequencing batch reactor (SBR) is based on the principles of the activated sludge process. The wastewater is treated in batches, with aeration and settlement both occurring in the same tank. The SBR process consists of several phases: filling, aeration-reaction, settling, decant and idle phase. With respect to application, SBR has been successfully used in both municipal and industrial wastewater treatment (Mohan et al., 2005). SBR has several advantages such as combining the reactor and the setting tank in the same vessel and this easily controls performance with respect to reaction time and sludge solids maintenance. Aeration of the mixed liquor occurs in the second stage, sludge is formed and ammonia is oxidized to nitrites and nitrates. In the settling stage, settlement of the sludge takes place and denitrification process may occur due to continuous consumption of nitrogen (Neczaj et al., 2008). The SBR process is easy to operate and the process saves more than 60% of the expenses when compared with the conventional activated sludge process (Chang et al., 2000). Recalcitrant organics such as pesticides may be toxic to the microorganisms of the activated sludge process. Hence, the use of Fenton pretreatment to produce non-toxic and biodegradable intermediates that could be treated in a biologically has been recommended. Complete mineralization of the contaminants may be achieved with partial use of Fenton pretreatment and complementary SBR treatment (Farré et al., 2007). Chlorpyrifos and chlorothalonil were removed from aqueous mixture in constructed wetlands (Sherrard et al., 2004), and cypermethrin and chlorfenvinphos residual water by microwave-assisted photo-Fenton process (Gromboni et al., 2007).

For a complete treatment of industrial pesticide wastewater that would meet effluent discharge standard as required by Malaysia (EQA, 1974), there is a need to combine Fenton and SBR treatment of a chlorpyrifos, cypermethrin and chlorothalonil pesticide wastewater. The purpose of this study was to examine Fenton–sequencing batch reactor (SBR) treatment of a chlorpyrifos, cypermethrin and chlorothalonil pesticide wastewater. The first phase of this study examined the effect of operating conditions

(H_2O_2/COD molar ratio and H_2O_2/Fe^{2+}molar ratio) of Fenton pretreatment of the pesticide wastewater. The second phase of the study examined SBR treatment of the Fenton pretreated pesticide wastewater. Effect of Fenton pretreatment operating conditions and hydraulic retention time (HRT) on SBR performance, dissolved oxygen (DO) and pH profile in SBR, and Fenton-SBR process efficiency were evaluated.

ADVANCE OXIDATION PROCESS

The advance oxidation process (AOP) is widely known for generation of hydroxyl radicals. The hydroxyl radicals are highly reactive and are used to degrade toxic organic compounds in wastewater and soils. As reported by Crittenden et al. (1997), the main advantages of AOP is that they aid destruction of pollutants. Unlike the conventional methods such as flotation, filtration and adsorption, using activated coal are non-destructive physical separation process and they only remove the pollutants by transferring them to the next phase (Bill Grote, 2012). Advanced oxidation processes (AOPs) are known for their promising options for degrading and mineralizing organic pollutants. The AOP makes use of the high strength oxidizing power which is the hydroxyl radicals in reducing the organic compounds to non-hazardous end product such as water and carbon dioxide. For some water treatment plants, the level of treatment is as of now not regarded to be important to comply with the standard requirements. However, this situation may change due to the concerns regarding the water quality arising from the retention and accumulation of organic compounds present in water systems (Bill Grote, 2012).

The oxidation-reduction (Redox) is the measure of the ability for a chemical species to gain electrons and become reduced. The redox potential is measured in volts, E. The process is measured at standard state conditions, such that, temperature is 25°C and 1 atmospheric pressure, and one unit activity for all species. Redox potential consists of two parts which are the oxidation potential and the reduction potential (Banchuen, 2003). Table 1 shows the standard reduction potentials of half-cells.

Iron metals are commonly used in activating oxidants such as hydrogen peroxide, (H_2O_2) and persulfate, (S_2O_8). The reactive species produced from the reactions are hydroxyl radicals and sulfate radicals. Although the hydroxyl radicals and sulfate radicals are highly reactive species, both radicals have different reaction patterns.

Table 1. Standard reduction potentials of half-cells reducing agents

Oxidizing Agents	Reducing Agents	E° (Volts)
$Cl_{2(g)} + 2e^-$	$2Cl^-_{(aq)}$	+1.36
$O_{2(g)} + 4H^+_{(aq)} + 4e^-$	$2H_2O_{(l)}$	+1.23
$Br_{2(aq)} + 2e^-$	$2Br_{(aq)}$	+1.07
$Fe^{3+}_{(aq)} + e^-$	$Fe^{2+}_{(aq)}$	+0.77
$2H^+_{(aq)} + 2e^-$	$H_{2(g)}$	0.00
$Fe^{2+}_{(aq)} + 2e^-$	$Fe_{(s)}$	-0.45
$Al^{3+}_{(aq)} + 3e^-$	$Al_{(s)}$	-1.66
$Mg^{2+}_{(aq)} + 2e^-$	$Mg_{(s)}$	-2.37
$Na^+_{(aq)} + e^-$	$Na_{(s)}$	-2.71

CLASSICAL ADVANCED OXIDATION PROCESSES AND MODIFIED FENTON REACTIONS

Classical Fenton

Classical Fenton is known as the activation of hydrogen peroxide (H_2O_2) by ferrous ion (Fe^{2+}) to generate hydroxyl radicals as shown in Equations (1-3) below;

$$Fe^{2+} + H_2O_2 \rightarrow Fe^{3+} + HO^{\cdot} + OH^{-} \quad (1)$$

$$Fe^{2+} + HO^{\cdot} \rightarrow Fe^{3+} + OH^{-} \quad (2)$$

$$Fe^{3+} + H_2O_2 \rightarrow Fe^{2+} + HO_2^{\cdot} + H^{+} \quad (3)$$

Ferric ions, Fe^{3+} produced from Equation (1) to (3) (Fenton, 1894) in neutral and near-neutral pH also reacts with hydroxide ion to form insoluble ferric hydroxide precipitate (Neyens and Baeyens, 2003). This will reduce the overall oxidation efficiency and additional ferrous ion salts maybe needed.

PHOTO-FENTON PROCESS

The photo-Fenton process is a process where the irradiation of Fe^{2+} and H_2O_2 solution using the ultra-violet (UV) or solar light takes place (Zepp *et al.*, 1992). The process then enhances the overall $HO^{\cdot}$ production by photo-reduction of Fe^{3+} to Fe^{2+} with combination of H_2O_2 photolysis as in Equations (4-5) below (Malato *et al.*, 2009);

$$Fe\,(OH)^{2+} + hv \rightarrow Fe^{2+} + HO^{\cdot} + H^{+} \quad (4)$$

$$H_2O_2 + hv \rightarrow 2HO^{\cdot} \quad (5)$$

ELECTRO-FENTON PROCESS

The electro-Fenton process is where the efficiency of Fenton process increases by electrochemical reduction of O_2. The process is conducted by continuous generation of hydrogen peroxide (H_2O_2) at the cathode (Brillas *et al.*, 2009).

SONO-FENTON PROCESS

The sono-Fenton process is where the ultrasonic dissociation of water and molecular oxygen aids in generation of HO as in Equations (6-7) and H_2O_2 as seen in Equation (8) (Adewuyi, 2005).

$$H_2O +))) \rightarrow H^{\cdot} + HO^{\cdot} \quad (6)$$

$$O_2 +))) + H_2O \rightarrow 2HO^{\cdot} \quad (7)$$

$$2H^{\cdot} + O_2 \rightarrow H_2O_2 \quad (8)$$

Advantages of Fenton-based advanced oxidation process is summarized in Table 2.

Table 3 shows a summarized lists of some redox potentials of reactive radical species.

Persulfate may be activated by adding ferrous iron (Fe^{2+}) as it is non-toxic nature, gives high reactivity and very cost effective. Ferrous ion produces a powerful oxidant known as the sulfate free radical (SO_4^{2-}) with a redox potential of 2.6 V, which can potentially destroy organic contaminants (Bazrafshan *et al.*, 2012).When adding ferrous iron to persulfate, the Fe^{2+} will donate electron and start the generation of SO_4^{2-}. The overall reaction between ferrous ion and persulfate is expressed in the Equations (8-10) (Kolthoff *et al.*, 1951).

$$2Fe^{2+} + S_2O_8^{2-} \rightarrow 2Fe^{3+} + 2SO_4^{2-} \quad (8)$$

The Equation (9) above is derived from the Equations (10 -11).

$$Fe^{2+} + S_2O_8^{2-} \rightarrow Fe^{3+} + SO_4^{\cdot -} + SO_4^{2-} \quad (9)$$

$$SO_4^{\cdot -} + Fe^{2+} \rightarrow Fe^{3+} + SO_4^{2-} \quad (10)$$

Table 2. Advantages of fenton-based AOPs

Process	Advantages
Classical Fenton	• Requires only Fe^{2+} salts and H_2O_2 • Works in ambient conditions • No mass transfer limitations
Photo-Fenton (UV/Solar light)	• Reduces large amount of iron sludge waste • Helps in decomposition of H_2O_2
Photo-Fenton (Visible light)	• Regenerates Fe^{2+} easily • Requires low energy visible light
Electro-Fenton	• In situ formation of H_2O_2 and Fe^{2+} reduces the external dosage • Fe^{3+} salts or sacrificial iron electrode can be used as iron source
Sono-Fenton	• Use of water and oxygen as additional precursors for $HO^{\cdot}$ Generation • Fe^{2+} catalyst is easily regenerated
Sono-Photo-Fenton	• Decreases mass transfer limitation in case of heterogeneous catalytic system • Reduces the initial Fe^{2+} dosage required
Sono-Electro-Fenton	• High rate of mixing and enhanced mass transfer of oxygen to the electrode • Continuous cleaning of the electrode surface by dissolution of inhibiting layers
Photo-Electro-Fenton (UV/Solar light)	• Continuous generation of Fe^{2+} • Higher yield of HO. • Complete photolysis of residual ferric carboxylate complexes

Table 3. Redox potentials of reactive radical species

Reactive Species		Redox Potential (V)
Sulfate Radical	$S_2O_8^{2-}$	+ 2.5 to 3.1
Persulfate	$SO_4^{\cdot-}$	+ 2.1
Hydroxyl radical	H_2O_2	+ 1.9 to 2.7
Hydrogen Peroxide	$HO^{\cdot}$	+ 1.8
Monopersulfate	HSO_5^-	+ 1.4

(Negash, 2014)

The persulfate-ferrous ion reaction will produce high amount of sulfate radicals in a short time interval (Liang *et al.*, 2004). The main advantage of the pursulfate activation by ferrous ions process is the degradation of organic and inorganic pollutions leading to high mineralization levels (Krzemińska *et al.*, 2015).

The Equations (11-17) shows the classic Fenton's reactions between hydrogen peroxide and Fe^{2+} mixed in an acidic aqueous medium.

$$Fe^{2+} + H_2O_2 \rightarrow Fe^{3+} + {}^{\cdot}OH + OH^- \quad (11)$$

$K_1 = 76\ M^{-1}\ s^{-1}$

$$Fe^{3+} + H_2O_2 \rightarrow Fe^{2+} + {}^{\cdot}OOH + H^+ \quad (12)$$

$K_2 = 0.01\text{-}0.02\ M^{-1}\ s^{-1}$

$$Fe^{3+} + {}^{\cdot}OOH \rightarrow Fe^{2+} + H^+ + O_2 \quad (13)$$

$K_3 = 3.1 \times 10^5\ M^{-1}\ s^{-1}$

$$^{\cdot}OH + H_2O_2 \rightarrow {}^{\cdot}OOH + H_2O \quad (14)$$

$K_4 = (1.2\text{-}4.5) \times 10^7\ M^{-1}\ s^{-1}$

$$^{\cdot}OH + Fe^{2+} \rightarrow Fe^{3+} + OH^- \quad (15)$$

$K_5 = 4.3 \times 10^8\ M^{-1}\ s^{-1}$

$$^{\cdot}OH + {}^{\cdot}OH \rightarrow H_2O_2 \quad (16)$$

$K_6 = 5.3 \times 10^9\ M^{-1}\ s^{-1}$

$$^{\cdot}OH + {}^{\cdot}OOH \rightarrow O_2 + H_2O \quad (17)$$

$K_7 = 1 \times 10^{10}\ M^{-1}\ s^{-1}$

The degradation of an organic compound by Fenton's reagent is said to be complex. However, Sun *et al.*, (2007) reported the formation of the hydroxyl radicals as an active oxidant in the Fenton system follows several techniques. One of the techniques mentioned was the electron paramagnetic resonance (EPR) spin-tapping technique. The hydroxyl radicals can start up the oxidation process of organic pollutant molecule (R) following the mechanisms shown in Equations (18-20).

$$^{\cdot}OH + R\text{-}H \rightarrow H_2O + R^{\cdot} \quad (18)$$

$K_8 = 10^7$ to 10^{10} M^{-1} s^{-1}

$$R^{\cdot} + Fe^{3+} \rightarrow Fe^{2+} + R^{+} \quad (19)$$

$$R^{\cdot} + H_2O_2 \rightarrow ROH + {}^{\cdot}OH \quad (20)$$

The principle of pseudo steady state assumption was mainly utilized when concerning the concentrations of reactive radicals in kinetic studies. Based on this principle, at certain time, the concentration of reactive hydroxyl radicals will not differ with response to time. Sun *et al.*, (2007) also reported that in the Fenton process, pollutants can be degraded effectively by providing sufficient dosage of hydrogen peroxide in order to produce more hydroxyl radicals.

BIOLOGICAL WASTEWATER TREATMENT

Biological treatment is the most common method adopted for degradation of contaminants in wastewater. Biological degradation refers to the removal of contaminants by the metabolic activity of living organisms. Usually microorganisms and in particular bacteria and fungi that are found in natural water and soil are have such capacity (Oller et al., 2011). The major categories of biological wastewater treatment are aerobic and anaerobic treatments. In the aerobic treatment, oxygen is present in the reactor whereas, in the anaerobic treatment there is complete absence of oxygen supply. The aerobic wastewater treatment is advantageous in so many ways. These include, faster degradation rates, rapid inactivation of pathogenic microorganisms, and low sludge yields (Stephenson and Blackburn, 1998). Also, elevated biodegradation rates reduce the necessary detention time for treatment, and subsequent capital and running cost of facility. Furthermore, the aerobic treatment is well suited for high strength/low flow wastewater with toxicity concerns stemming from high salinity levels and/or the presence of hazardous compounds (Rozich and Colvin, 1997).

AEROBIC BIOLOGICAL WASTEWATER TREATMENT

Aerobic biological wastewater treatment is basically based on the bacteria's capability of assimilating biodegradable organic matter present in a polluted effluent. During aerobic degradation, organic compounds are oxidized to carbon dioxide and oxygen is reduced to water in the presence of bacteria which enhances the wastewater degradation (Zitomer and Speece, 1993). Bacteria can be classified by the way they obtain carbon for their development. Autotrophic organisms obtain their carbon from inorganic sources, whereas heterotrophic organisms require organic matter as food source. Most heterotrophic organisms obtain energy from stored energy in chemical bonds, while autotrophic bacteria obtain theirs from the light through photosynthesis (Tchobanoglous et al., 2002). Among the various aerobic wastewater treatments, activated sludge process is the most popular biological treatment (Gonzalez et al., 2006). The basic process begins with aeration in the reactor where microorganisms grow by assimilating the contaminants in the wastewater. The solids are settled out in a separate clarifier or in the same reactor, depending on the configuration and are re-used to degrade more contaminants. The fundamental classification of biological system is available in different configurations and it is fully described in some other literature (Tchobanoglous et al., 2002). The complex microbial community used in this process is known as activated sludge and the aerobic heterotrophic bacteria process is described as in Equations (21-22).

$$\text{Organic matter} + O_2 + \text{bacteria} \rightarrow CO_2 + H_2O \text{ products} + \text{energy} \tag{21}$$

$$\text{Organic matter} + \text{bacteria} + \text{energy} \rightarrow \text{new bacteria cells} \tag{22}$$

Other nutrients and trace elements availability in the wastewater increase the treatability of the wastewater such as potassium, sodium, calcium, magnesium etc. The ratio of nutrients supplied with COD: N: P ratios of 100:20:1 or 250:7:1 and 100:10:1 and trace sulphur are all required (Tchobanoglous et al., 2002). Apart from nutrient requirements, other physical factors such as pH and temperature remain the most influential environmental factors that affect biodegradability. Moreover, metabolic reactions occur faster at optimum pH, usually between 6 and 5.5 and temperature between 25 and 33°C (Stephenson and Blackburn, 1998). The final product of this process is the sludge. This is the main drawback of the biological treatment since post management prior to eventual disposal is required. The posterior management includes stabilizing the sludge, reducing odour, removing water, to decompose the organic matter and kill disease causing organisms. A typical sludge treatment process involves thickening, dewatering and stabilization process followed by a main disposal method such as incineration, agricultural land application or landfill disposal (Suh and Resseaux, 2002).

SEQUENCING BATCH REACTOR

Sequential batch reactor (SBR) is a type of activated sludge process and the oldest type of batch biological treatment (Stephenson and Blackburn, 1998). The design was started in 1914 (Arden and Lockett, 1914) however actual research began in the 1970's (Mace and Mata-Alvarez, 2002). The SBR works under non-steady state conditions with unit operations that involves aeration and decantation phases based on a fill-and-draw cycle in the same reactor (Widerer et al., 1999). The SBR technology has five phases as described herein.

SEQUENCING BATCH REACTOR OPERATION

The SBR operation is executed in the following sequential stages;

- **First Stage:** Fill - Raw wastewater is filled and mixed with the biomass in the reactor.
- **Second Stage:** React/Aeration – The metabolic and biological reactions whereby microorganisms in the presence of oxygen utilize the organics in the reactor as food to degrade contaminants under controlled conditions e.g. temperature, pH and dissolved oxygen monitoring.
- **Third Stage:** Settle – The aeration and mixing is stopped and the sludge is allowed to separate from the supernatant which remains at the upper portion of the reactor. Duration is subject to the number of cycles per day.
- **Fourth Stage:** Decant – The treated effluent or supernatant is removed from the reactor.
- **Fifth Stage:** Idle: This is the time for the reactor to be idle and excess sludge can be removed. In some case, the idle stage may not be necessary and it could be omitted.

ADVANTAGES OF SEQUENCING BATCH REACTOR

The most important advantages of the sequencing batch reactor technology when compared with activated sludge continuous flow process includes its simple operation, manual control and operation flexibility, ease of managing unexpected surges or effects like poor settling i.e. sludge bulking and lower economy in terms of operating cost. In addition, the fill-and-draw model prevents the washed out biomass, hence reducing the need for a clarifier as in the case of traditional activated sludge process (Mace and Mata-Alvarez, 2002).

Figure 1. Principles of sequencing batch reactor operation

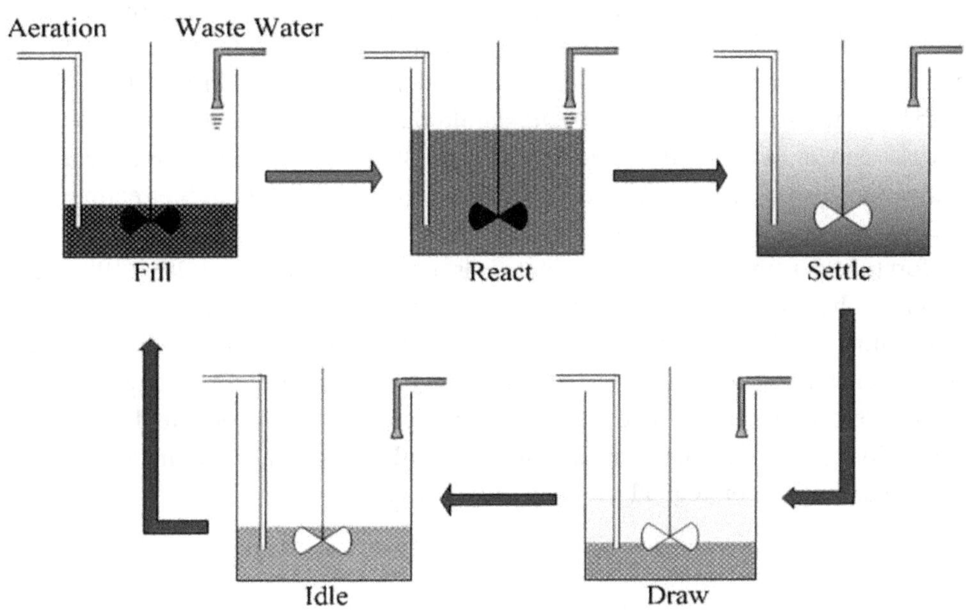

ADVANCED OXIDATION PROCESS AND BIOLOGICAL TREATMENT

Biological processes are the commonly used methods for the treatment of wastewater containing biodegradable compounds. The presence of inhibitory or recalcitrant organic compounds having high toxicity impede the application of the biological treatment process as the only method in the removal of such compounds from high concentration wastewater. Due to reported improvement in biodegradability and complete removal of recalcitrant organics from a variety of wastewater, AOPs have been considered to be viable pretreatment before application of biological treatment (Moussavi et al., 2009). The AOP increases the intermediates and the wastewater becomes less toxic and biodegradable, thus making combined AOP and biological treatment suitable for intercellular enzymes which cannot metabolize the parent contaminants (Ballesteros- martin et al., 2009) and/or if they have high-molecular-weight polymer in the wastewater they could be reduced (Ollis, 2001).

The combination technique of AOP-SBR has been reported and proven to achieve degradation of targeted pollutants in an economical and environmental-friendly way. The main objective being to modify the structure of organic pollutants by making them less toxic and biodegradable by means of AOP and then subsequently complete mineralization in a shorter time duration by the SBR. Sarria et al. (2002) and Oller et al. (2011) developed and proposed general strategies for planning a combined AOP and biological process for wastewater treatment. Another development of control strategies proposed by the EU consist of an independent simulation environment defining a plant layout, a simulation model, influent loads, test procedures and evaluation criteria have obtained results using several simulation platforms such as GPS-X™, Simulink™, Simba™, WestÔ, FORTRAN code and have been developed as bench mark (Alex, et al., 1999).

To carry out a combined chemical and biological process, a few steps are desired to be followed. On one hand, chemical oxidants and bio-culture cannot mix because the chemical oxidants used can cause damaging effects to the microorganisms. Again, detrimental effects can be produced by an excessively low or acid pH i.e. < 7. As a result, pH adjustment to about neutral pH 7 is necessary because chemically treated effluents are acidic in most AOPs. On another hand, large oxidation times are not necessary when combining AOP-biological treatments because excessive chemical oxidation could imply high electrical cost and increased chemical reactants consumption and this may result in highly oxidized products possessing little metabolic value for the microorganism (Sarria et al., 2002; Scott and Ollis, 1995). Nevertheless, the solution from the AOP should be less toxic and biodegradable. To meet this condition, the assessment of the biodegradability such as BOD_5/COD ratio and maybe toxicity together with the chemical process efficiency is necessary to determine the optimum pretreatment reaction time that could lead to a successful AOP-biological combined system.

The first effort to review the combined treatment was reported by Scott and Ollis (1995). Other organic compounds such as pharmaceutical, polymers, oils and specific industrial wastewater were investigated. Since 1990 until now much effort has been put into development of this strategy. Different chemical (AOPs) amongst them are Fenton, UV photo-Fenton, solar photo-Fenton and heterogeneous photocatalysis amongst others have been reported to be effective in combining with aerobic and anaerobic biological treatments. In addition, microorganism or pure microbial culture, whether acclimated or non-acclimated can be used in enhancing the biological process (Sarria et al., 2002).

PESTICIDE WASTEWATER TREATMENT BY ADVANCED OXIDATION PROCESS

Several studies have reported removal and/or degradation of some pesticides and other pollutants by advanced oxidation processes (AOPs). It is noteworthy that the studies were carried out using either aqueous solutions or contaminated soil. However, there is no reported study on the degradation of the combined chlorpyrifos, cypermethrin and chlorothalonil pesticide wastewater by Fenton process. In addition, there is no report on degradation of this mixture of pesticides in wastewater by combined AOP and biological treatment such as SBR. Further, the efficiency of these methodologies has not been assessed under real conditions, i.e., in the presence of a mixture of several pesticides and their formulating agents at concentrations over 100 mg/L (Zapata et al., 2009). Therefore, studies with higher total organic content (TOC) (200 to 500 mg/L) in wastewater polluted by commercial pesticides are required to be carried out integrating sequencing batch bioreactors or immobilized biomass reactors. The need to apply AOPs in combination with biodegradation is considered important because a treatment is usually not considered complete until mineralization is nearly complete. Thus, combining both methods will completely eliminate the organic content of the pesticide wastewater. Also, the assessment of the combined technology should take into account higher pesticide concentrations and how it affects the AOP and subsequent biological oxidation efficiencies (Oller et al., 2011).

MATERIALS AND ANALYTICAL METHODS

Chemicals

Hydrogen peroxide (30%, w/w) and ferrous sulphate heptahydrate ($FeSO_4{\cdot}7H_2O$) were purchased from R&M Marketing, Essex, U.K. Analytical grade of chlorpyrifos was obtained from Dr. Ehrenstorfer, Germany, and cypermethrin and chlorothalonil from Sigma-Aldrich, Germany. They were used for analytical determination of pesticide concentration by gas chromatography–mass spectroscopy (GCMS) examined Fenton–sequencing batch reactor (SBR) treated chlorpyrifos, cypermethrin and chlorothalonil pesticide wastewater.

Analytical Methods

Chemical oxygen demand (COD) was determined according to the Standard Methods (APHA, 2005). When the sample contained hydrogen peroxide (H_2O_2), to reduce interference in COD determination pH was increased to above 10 so as to decompose hydrogen peroxide to oxygen and water (Talinli and Anderson, 1999; Kang et al., 1999). TOC analyzer (Model 1010; O & I Analytical) was used for determining total organic carbon (TOC). The pH measurement was made using a pH meter (HACH sension 4) and a pH probe (HACH platinum series pH electrode model 51910, HACH Company, USA). Biodegradability was measured by 5-day biochemical oxygen demand (BOD_5) test according to the Standard Methods (APHA, 2005). The seed for BOD_5 test was obtained from a municipal wastewater treatment plant. Dissolved oxygen (DO) was measured using a YSI 5000 dissolved oxygen meter. Other parameters such as mixed liquor suspended solids (MLSS) and mixed liquor volatile suspended solids (MLVSS) were measured according to the Standard Methods (APHA, 2005). Sludge volume index (SVI) was analyzed before end of reaction phase by measuring the volume in ml occupied by 1 g of the MLSS after 30 min of settling. Pesticide concentration in untreated and treated samples were determined by GCMS. The analyses were carried out on a GCMS-QP5050A Gas Chromatograph–Mass Spectrometer system (Europe) equipped with a column (30 m × 0.25 mm, 0.25 μm). The mass spectrometer employed was an ion trap (20 μA) with 0.82 s of scan time and mass spectra between 35 and 500 nm. The injector was maintained at 300°C and operated in the splitless mode with the split closed for 0.75 min. Helium (>99.999% pure) was used as the carrier gas at a flow rate of 1.0 ml/min. The column oven was initially set at 50°C for 1 min, programmed to 190°C at 20°C/min rate for 5 min, and finally to 280°C at 3.5°C/min rate. The column flow was 2.4 ml/min and injected volume of 1 μl. The interface temperature was set at 300°C and the detector voltage at 4 V. A 35-min time was allowed for all analyses. The base peak ion and two other significant ions of each analyte were chosen as the quantifying ions. Prior to quantification, the identification of all target compounds was based on their mass spectra and GC retention times.

PESTICIDE WASTEWATER

Pesticide wastewater was collected from a formulating and production company in Malaysia, transported to the laboratory and stored in a cold room at 4°C until used. The wastewater contained the pesticides chlorpyrifos, cypermethrin and chlorothalonil, with COD of 3350 mg/L and TOC of 2960 mg/L. The characteristics of the wastewater used in this study is presented in Table 4.

Figure 2 shows chlorpyrifos, cypermethrin, and chlorothalonil pesticides structure and GCMS chromatograph of the wastewater. Possible contaminant peak from formulation substances are also seen in the chromatograph. These unknown peaks may be solvents which were used as emulsifier, stabilizer, adjuvant and solvent in the production (Senthilnathan and Philip, 2012).

Table 4. Characteristics of pesticide wastewater

Parameter	Unit	Range
Chlorpyrifos		805.56 ± 10.0
Cypermethrin	mg/L	105.75 ± 10.0
Chlorothalonil	mg/L	692.08 ± 10.0
Chemical oxygen demand (COD)	mg/L	3350.0 ± 100
Total organic carbon (TOC)	mg/L	2960.0 ± 100
Biochemical oxygen demand (BOD_5)	mg/L	63.0 ± 5.0
BOD_5/COD ratio	mg/L	0.02 ± 0.01
Ammonia-nitrogen (NH_3-N)	mg/L	57.4 ± 2.0
Nitrate (NO_3^--N)	mg/L	13.1 ± 0.5
Total phosphate (TP)	mg/L	79.0 ± 0.5
Chloride (Cl^-)	mg/L	1280 ± 0.5
Fluoride (F^-)	mg/L	5.0 ± 0.5
Sulfate (SO^{2-}_4)	mg/L	500.0 ± 10
Conductivity	(μS/cm)	680.0 ± 10
Total suspended solids (TSS)	mg/L	130.0 ± 5.0
Total volatile suspended solids (TVSS)	mg/L	105.0 ± 5.0
Total nitrogen	mg/L	114.8 ± 2.0
pH		5.08 ± 0.5

Figure 2. Chemical structure and GCMS chromatograph of (a) Chlorpyrifos, (b) Chlorothalonil and Cypermethrin of pesticide wastewater

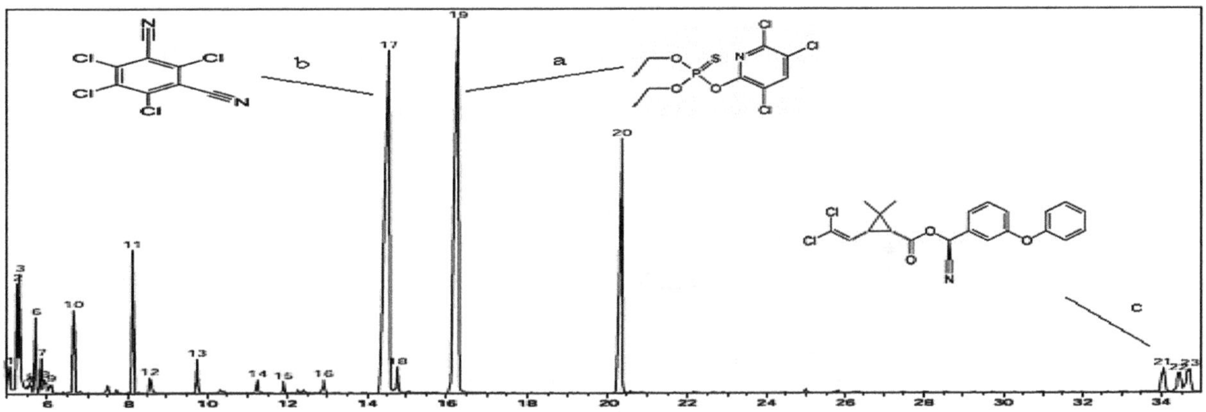

EXPERIMENTAL SET-UP

The experiment was conducted and data were analyzed as described in the sub-sections 3.4.1 – 3.4.3.

Fenton Treatment

Fenton treatment was carried out in a 6000 ml glass reactor with 5000 ml of the pesticide wastewater at room temperature (23±2°C). The required amount of iron ($FeSO_4{\cdot}7H_2O$) was added to the wastewater and the pH was adjusted to the required value using 1 M sulfuric acid (H_2SO_4) or 1 M sodium hydroxide (NaoH) and mixed by a magnetic stirrer to ensure complete homogeneity. Thereafter, the necessary amount of H_2O_2 was added to the mixture. After 60 min reaction time, the reaction was stopped and the treated wastewater was filtered through 0.45 μm membrane filter for measurement of COD, TOC and BOD_5, and filtered through 0.20 μm membrane filter for determination of pesticide concentration by GCMS.

Design of the Sequencing Batch Reactor Treatment

Sequencing batch reactor treatment was carried out in a Plexiglass reactor (15 cm ID, 25 cm height, total volume 2 L and working volume 1.5 L) at room temperature (23±2°C). The working volume was divided into 750 mL decanting volume and 750 mL sludge volume. Air was introduced into the reactor through a submerged diffuser at the bottom of the reactor and dissolved oxygen was maintained at higher than 3 mg/L by mixing with stirrers. Feeding and decanting were carried out by calibrated variable flow peristaltic pumps. An initial 10 days were used for acclimation of the seeded sludge with the Fenton pretreated pesticide wastewater. The Fenton treated pesticide wastewater and municipal wastewater was mixed at a ratio of 1:1 during the acclimation period. Intelligent system control was used for timely operation of the fill, react, settle and draw stages of the reactor. Three hydraulic retention times (HRT) were investigated (6, 12 and 24 hr). The 6 hr HRT was divided into 30 min filling, 4 hr aeration, 1 hr settling, 15 min decanting and 15 min idle period. The 12 hr HRT was divided divided into 30 min filling, 10 hr aeration, 1 hr settling, 15 min decanting and 15 min idle period. The 24 hr HRT was divided

Figure 3. Schematic of Fenton-SBR process

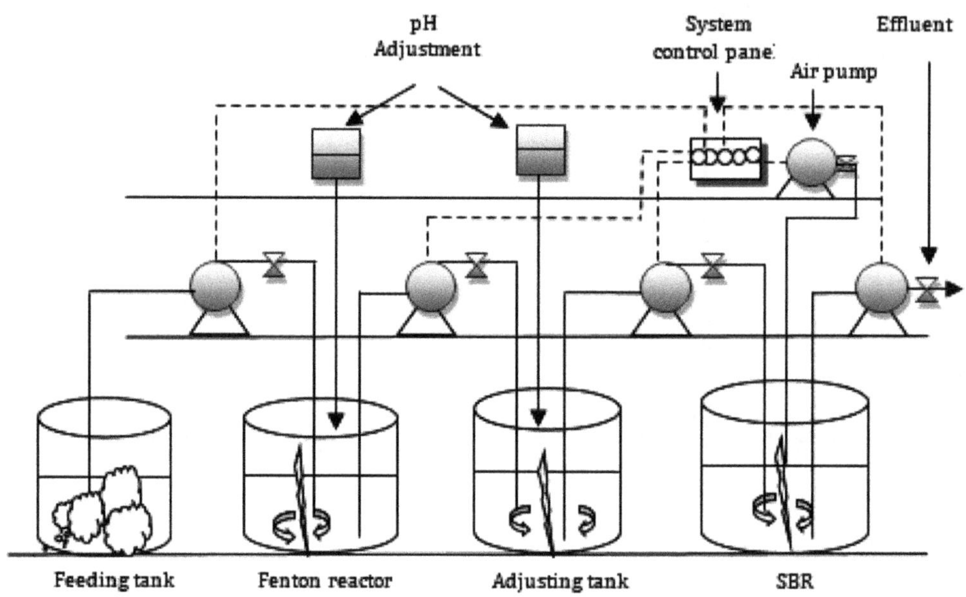

divided into 30 min filling, 22 hr aeration, 1 hr settling, 15 min decanting and 15 min idle period. The supernatant collected after settling was filtered through 0.45 μm membrane filter for measurement of COD, TOC, BOD_5, MLSS and MLVSS. Figure 3 shows a schematic of the Fenton-SBR process.

Data Analysis

Statistical analysis (ANOVA) was conducted using MINITAB software version 13.0 and Microsoft excel. One-way ANOVA was applied to the results to determine the significance of each treatment process and to compare between various batch treatments data obtained from the laboratory experiments.

FENTON TREATMENT

Effect of H_2o_2/Cod Molar Ratio

To study the effect of H_2O_2/COD molar ratio on degradation and biodegradability improvement, the H_2O_2 concentration was varied between 157.03 and 366.45 mM. The other operating conditions were pH 3, initial COD 3350 mg/L (104.69 mM), TOC 2960 mg/L, reaction time 60 min and H_2O_2/Fe^{2+} molar ratio 10. An increase in H_2O_2/COD molar ratio from 1.5 produced sufficient $OH^{\bullet}$ radical for degradation of the pesticides, but above H_2O_2/COD molar ratio 3.0 there was decrease in degradation. The COD removal was 35.9, 41.1, 49.9, 56.1 and 53.4% for H_2O_2/COD molar ratio 1.5, 2.0, 2.5, 3.0 and 3.5, respectively. The TOC removal was 18.5, 22.3, 28.3, 36.2 and 31.7% for H_2O_2/COD molar ratio 1.5, 2.0, 2.5, 3.0 and 3.5, respectively. The BOD_5/COD ratio was 0.15, 0.17, 0.20, 0.28 and 0.23 for H_2O_2/COD molar ratio 1.5, 2.0, 2.5, 3.0 and 3.5, respectively (Figure 4). An addition of excess H_2O_2 decreased the degradation

Figure 4. Effect of H_2O_2/COD molar ratio on pesticide degradation in Fenton process in terms of COD and TOC removal and BOD_5/COD ratio

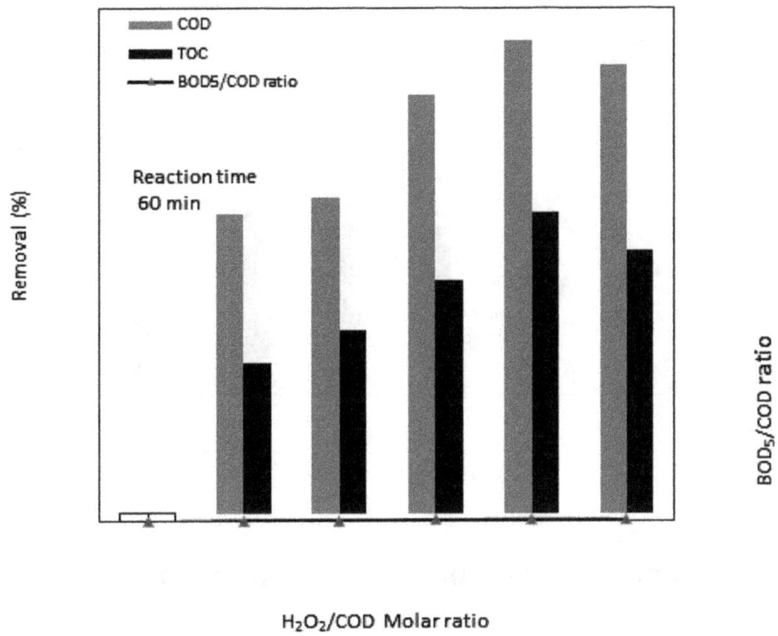

and mineralization of the pesticides. This may be due to scavaging of $OH^\bullet$ radicals by H_2O_2 as in reaction Equation (23) (Kevitha and Palanivelu, 2005). It was observed that H_2O_2/COD molar ratio 3.0 was optimum for degradation and biodegradability improvement.

$$OH^\bullet + H_2O_2 \rightarrow H_2O + HO_2^\bullet \tag{23}$$

Effect of H_2O_2/FE^{2+} Molar Ratio

To study the effect of H_2O_2/Fe^{2+} molar ratio, Fe^{2+} concentration was varied between 6.28 and 157 mM. The other operating conditions were pH 3, initial COD 3350 mg/L (104.69 mM), TOC 2960 mg/L, BOD_5/COD ratio 0.02, reaction time 60 min, H_2O_2/COD molar ratio 3.0. Degradation increased for H_2O_2/Fe^{2+} molar ratio 5 to 10 but decreased at H_2O_2/Fe^{2+} molar ratio 25. The COD removal was 45.2, 51.9, 58.5, 54.9 and 41.3% for H_2O_2/Fe^{2+} molar ratio 2.0, 5.0, 10.0 and 25.0 and 50.0, respectively. The TOC removal was 23.9, 30.1, 39.8, 33.6 and 20.7% for H_2O_2/Fe^{2+} molar ratio 2.0, 5.0, 10.0, 25.0, 50.0, respectively. The BOD_5/COD ratio was 0.17, 0.22, 0.30, 0.26 and 0.14 for H_2O_2/Fe^{2+} molar ratio 2.0, 5.0, 10.0, 25.0 and 50.0, respectively (Figure 5). An increase in H_2O_2/Fe^2 molar ratio above 10 decreased degradation. This was presumably due to direct reaction of $OH^\bullet$ radical with Fe^{2+} ion at high concentration as in reaction Equation (24) (Joseph et al., 2000). Lower Fe^{2+} concentration yielded less $OH^\cdot$ required for degradation of the pesticides. It was observed that H_2O_2/Fe^{2+} molar ratio 10.0 was optimum for degradation and biodegradability improvement.

$$Fe^{2+} + OH^\bullet \rightarrow Fe^{3+} + OH^- \tag{24}$$

Figure 5. Effect of H_2O_2/Fe^{2+} molar ratio on pesticide degradation in Fenton process in terms of COD and TOC removal and BOD_5/COD ratio

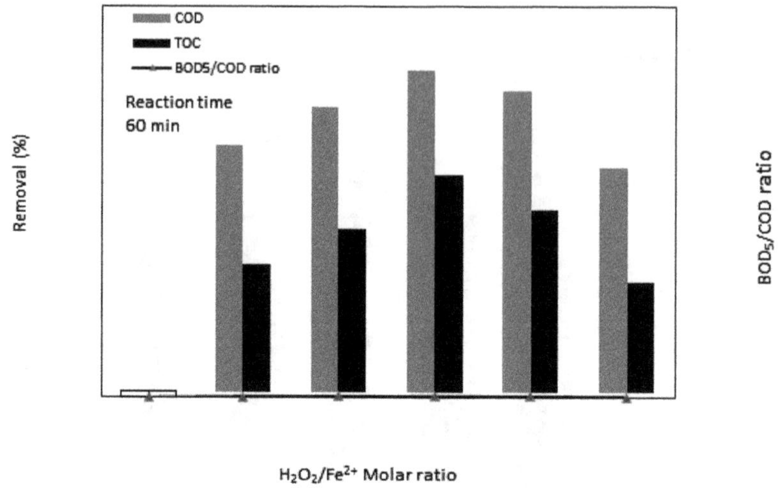

Degradation of Pesticide Under Optimum Operating Conditions

In order to obtain the optimum time required for the Fenton process to completely degrade the pesticides in the wastewater, experiments were conducted under optimum operating conditions (H_2O_2/COD molar ratio 3.0, H_2O_2/Fe^{2+} molar ratio 10.0, pH 3). As shown in Figure 6, chlorpyrifos, cypermethrin and chlorothalonil were completely degraded in 1 min. In another study, 100% of methomyl was degraded in 30 min (Tamimi et al., 2008).

KINETICS OF PESTICIDES DEGRADATION

The kinetics of degradation of substrates by the Fenton treatment was studied. The kinetics of substrates is an important parameter for prediction of the fate and in design of wastewater treatment plant for optimization and low-cost efficiencies (Scott and Ollis, 1995). To study the kinetics of degradation of the

Figure 6. Degradation of pesticides in wastewater under optimum Fenton operating conditions.

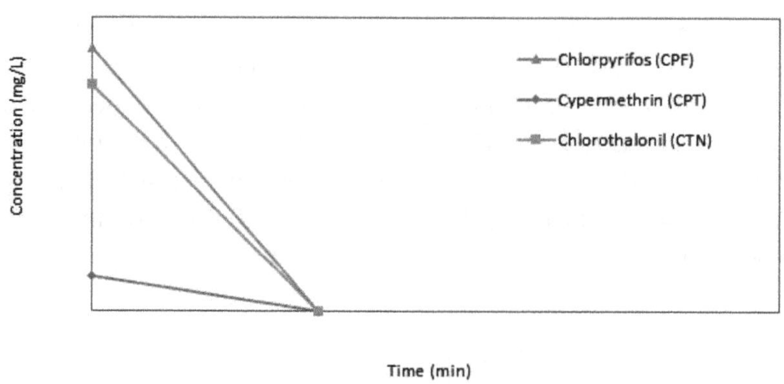

pesticide wastewater, experiments were conducted under the optimum operating conditions. The approach to the study was in accordance with similar studies by Zhang et al. (2010) and can be represented by the expression as in Equation (25).

$$In\left(\frac{C_t}{C_o}\right) = -k * t \tag{25}$$

where, k is a pseudo-first-order rate constant, t is the irradiation time in min; C_0 the initial concentration of TOC in aqueous solution and C_t is the residual concentration of TOC at time t. By plotting Eq. (3), the rate constant (k) was determined from the slope of the straight line. The mineralization of organic carbon in the pesticide wastewater as a function of reaction time is shown in Figure 7. The reactions followed pseudo-first order kinetics with rate constant (k) 0.0083 min^{-1} and R^2 was 0.99.

FENTON-SBR PERFORMANCE

The Fenton pretreated pesticide wastewater was fed into the SBR to study the effect of H_2O_2/COD molar ratio and H_2O_2/Fe^{2+} molar ratio on SBR performance. In addition, the effect of SBR hydraulic retention time (HRT) was also studied. Following the completion of the Fenton preatreatment of the wastewater, five varying H_2O_2/COD and H_2O_2/Fe^{2+} molar ratios as in Table 5 were selected for the Fenton-SBR (F-SBR) process.

Figure 7. Kinetics of pesticide degradation by Fenton process

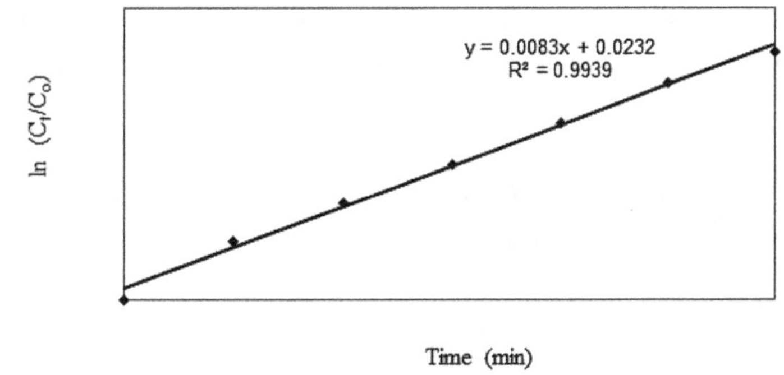

Table 5. Fenton operating conditions for F-SBR

	F-SBR 1	F-SBR 2	F-SBR 3	F-SBR 4	F-SBR 5
H_2O_2/COD Molar Ratio H_2O_2/Fe^{2+} Molar Ratio	3 25	3 10	3 5	2.5 10	3.5 10

EFFECT OF FENTON OPERATING CONDITIONS

The experiments performed considered the effect of Fenton operating conditions such as H_2O_2/COD molar ratio and H_2O_2/Fe^{2+} molar ratio on the performance of the F-SBR 1 to F-SBR 5. To assess the degradation and biodegradability in the SBR, target parameters monitored were chemical oxygen demand (COD), total organic carbon (TOC) and BOD_5/COD ratio. The SBR was operated considering the effect of increasing and decreasing H_2O_2 and Fe^{2+} concentration from the obtained optimum operating conditions in the pretreatment. An adapted population can be indicated by the effluent COD concentration reaching constant values, and the reactor MLVSS concentration gradually increasing (Morgan-sagastume and Allen, 2003). The increase in MLSS and MLVSS concentrations indicate the active growth of bacteria and thus a successful startup. This was because the bacteria consumed organic matter and multiplied to form new cells. In the 6 hr HRT, after acclimatization, the SBR operation started up with MLSS and MLVSS increasing from 2670-2700 mg/L and 1380-1420 mg/L to 2860-2990 mg/L and 1500-1560 mg/L, respectively in the 7th cycle. Thereafter, decreased to 2200-2000 mg/L and 1200-1230 mg/L, respectively. Effluent COD and TOC decreased to 39-90 mg/L and 26-79 mg/L and then increased to 104-139 mg/L and 107-145 mg/L, respectively in the 13th cycle. The COD removal efficiency of the F-SBR 1, F-SBR 2, F-SBR 3, F-SBR 4 and F-SBR 5 processes were 98.9, 98.2, 95.2, 97.4 and 96.6%, respectively. The TOC removal efficiency were 98.7, 98.5, 94.5, 95.6 and 95.8%, and the BOD_5/COD ratio were 0.49, 0.45, 0.40, 0.40 and 0.41 for F-SBR 1, F-SBR 2, F-SBR 3, F-SBR 4 and F-SBR 5, respectively (Figure 8). The COD and TOC removal efficiency and BOD_5/COD ratio in the 7th cycle is shown in Figure 9. From the outcome, the F-SBR 1 performed most effectively.

Similar experiments were also performed at 12 hr HRT for F-SBR 1 to F-SBR 5 and were operated for 15 cycles. After acclimatization, the SBR operation began with MLSS 2680-2730 mg/L and MLVSS 1400-S1430 mg/L and reached 2880-2950 mg/L and 1530-1570 mg/L, respectively in the 8th cycle. Thereafter, a decrease in MLSS to 2260-2310 mg/L and MLVSS to 1200-1150 mg/L was observed. Effluent COD and TOC decreased to 38-65 mg/L and 32-81 mg/L and then increased to 129-145 mg/L and 116-133 mg/L, respectively in the 15th cycle (Figure 10). However, The reduction in MLSS concen-

Figure 8. Performance of F-SBR 1 to F-SBR 5 in terms of COD and TOC concentration, and BOD_5/ COD ratio as a function of Fenton operating conditions at 6 hr HRT

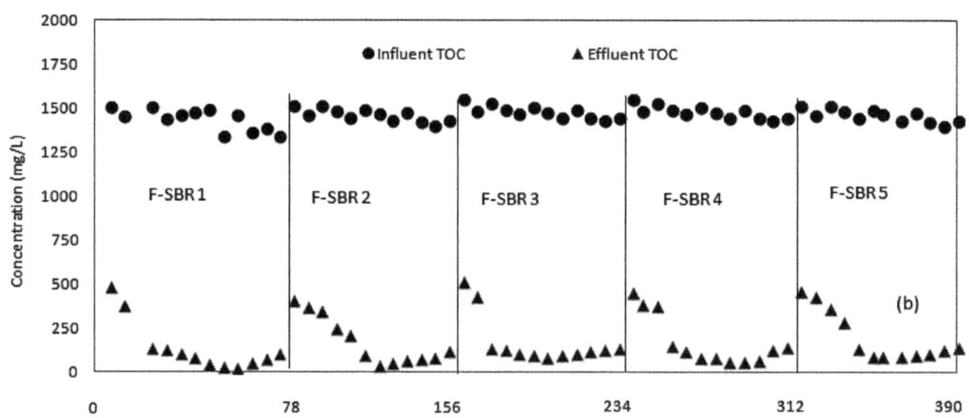

Figure 9. COD and TOC removal efficiency and BOD_5/COD ratio of F-SBR 1 to F-SBR 5 at 6 hr HRT

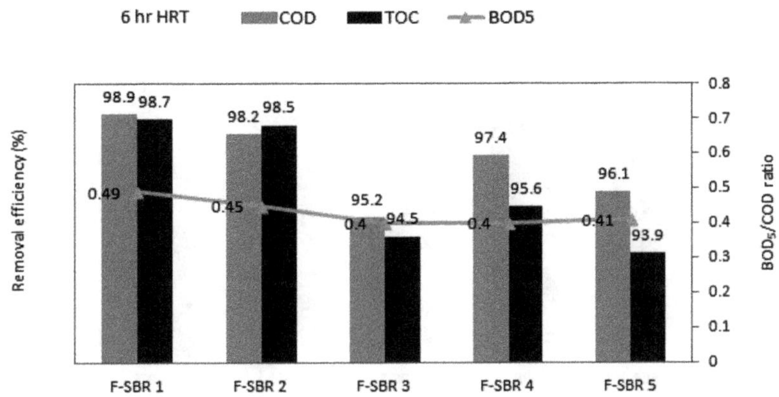

tration may be due to inhibition of aerobic oxidation by the pesticide intermediates. The COD removal efficiency of the combined F-SBR 1, F-SBR 2, F-SBR 3, F-SBR 4 and F-SBR 5 processes were 97.0, 96.9, 95.7, 96.4 and 94.2%, respectively. The TOC removal were 97.7, 98.2, 98.1, 97.7 and 97.0% and BOD_5/COD ratio were 0.53, 0.50, 0.41, 0.45 and 0.42 for F-SBR 1, F-SBR 2, F-SBR 3, F-SBR 4 and F-SBR 5, respectively. Wastewater is considered biodegradable where BOD_5/COD ratio is 0.40 (Al-Momani et al., 2003). The COD and TOC removal efficiency and BOD_5/COD ratio in the 8th cycle is shown in Figure 11. The F-SBR 1 and F-SBR 2 performed effectively.

The HRT was further increased to 24 hr in an effort to obtain the best HRT for effective performance of the F-SBR process. After acclimatization, the SBR operation started with MLSS increasing from 2700 - 2750 mg/L and MLVSS 1420-1460 mg/L and reached 2900-2980 mg/L and 1550-1580 mg/L, respectively in the 8th cycle. Thereafter, MLSS decreased to 2340-2290 mg/L and MLVSS to 1230-1180 mg/L. Effluent COD and TOC decreased from 26-73 mg/L and 30-73 mg/L in the 8th cycle and then increased to 127 and 134 mg/L and 86 and 140 mg/L, respectively in the 15th cycle (Figure 12).

Figure 10. Performance of F-SBR 1 to F-SBR 5 in terms of COD and TOC concentration, and BOD_5/COD ratio as a function of Fenton operating conditions at 12 hr HRT

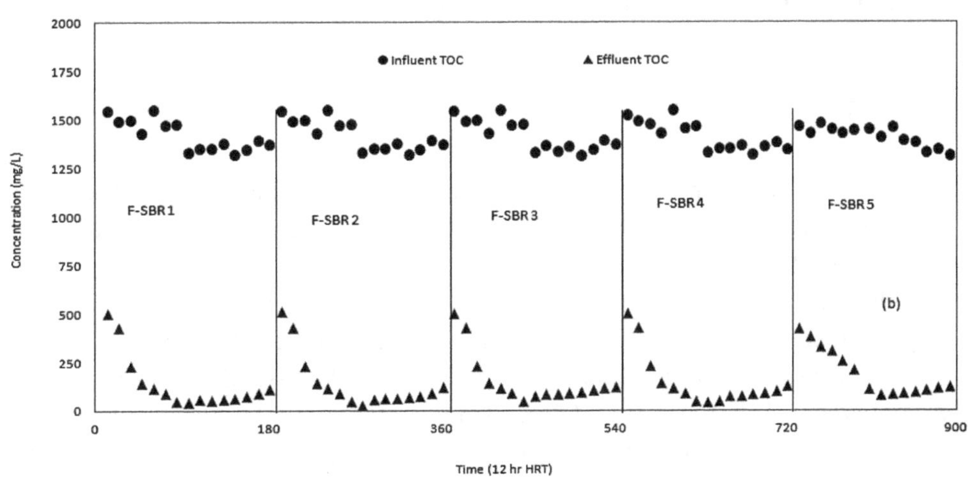

Figure 11. COD and TOC removal efficiency and BOD_5/COD ratio of F-SBR 1 to F-SBR 5 at 12 hr HRT

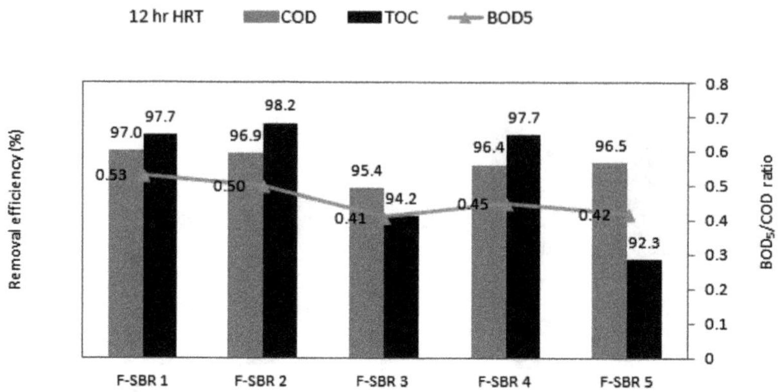

The COD removal efficiency of the combined F-SBR 1, F-SBR 2, F-SBR 3, F-SBR 4 and F-SBR 5 were 97.8, 97.6, 95.9, 97.7 and 96.1%, respectively. The TOC removal efficiency were 98.3, 98.5 and 94.9, 96.9 and 95.0% and BOD_5/COD ratio were 0.58, 0.53, 0.43, 0.47 and 0.45 for F-SBR 1, F-SBR 2, F-SBR 3, F-SBR 4 and F-SBR 5, respectively. The COD and TOC removal efficiency and BOD_5/COD ratio in the 8th cycle is shown in Figure 13.The

F-SBR 1 and F-SBR 2 performed effectively. It was found in all three HRTs that an increase above H_2O_2/COD molar ratio 3 and an increase above H_2O_2/Fe^{2+} molar ratio 50 decreased the overall performance in F-SBR process. This may be due to the fact that, excess H_2O_2 reacts with OH• radicals and contributes to the OH• radicals and hole scavenging as in Equation (23) and the direct reaction of OH• radical with Fe^{2+} ion at high concentration as in reaction Equation (24). The results are comparable with that reported in the literature. Hadjiniculaou (1998) reported COD removal of 97% from slaughterhouse effluent wastewater. Ganjidoust and Ayati (2004) reported SBR efficiency of 92, 84 and 52% for COD, turbidity and TSS removal, respectively in the treatment of wood industry effluent.

Since the target was to minimize the use of chemicals (Fe^{2+} or H_2O_2) in the pretreatment, it was important to know if increasing or decreasing Fe^{2+} or H_2O_2 in the Fenton reaction significantly affected the overall SBR performance especially with respect to HRT. However, From the available results it is obvious that a decrease in H_2O_2/Fe^{2+} molar ratio from 3.0 to 2.5 and H_2O_2/COD molar ratio 10 was sufficient to yield the desired results in the F-SBR process.

EFFECT OF HRT

The effect of HRT on F-SBR performance was studied at 6, 12 and 24 hr HRTs. The F-SBR removal efficiencies were almost at par in terms of COD and TOC removal except with some variations in BOD_5/COD ratio as seen in Table 6.

It is apparent that most of the substrate degradation occurred during the first 6 hr or between 6 and 12 hr and a very small portion during the 24 hr HRT. To ascertain the true position, statistical analysis (one-way ANOVA analysis at 95% confidence interval (5% level of significance) was performed. The statistical analysis indicated that increase of HRT from 6 (p value > 0.05) to 12 hr (p value > 0.05) significantly improved the degradation and mineralization of the pesticides, but decreased at 24 hr (p in

Figure 12. Performance of F-SBR 1 to F-SBR 5 in terms of COD and TOC concentration, and BOD_5/COD ratio as a function of Fenton operating conditions at 24 hr HRT

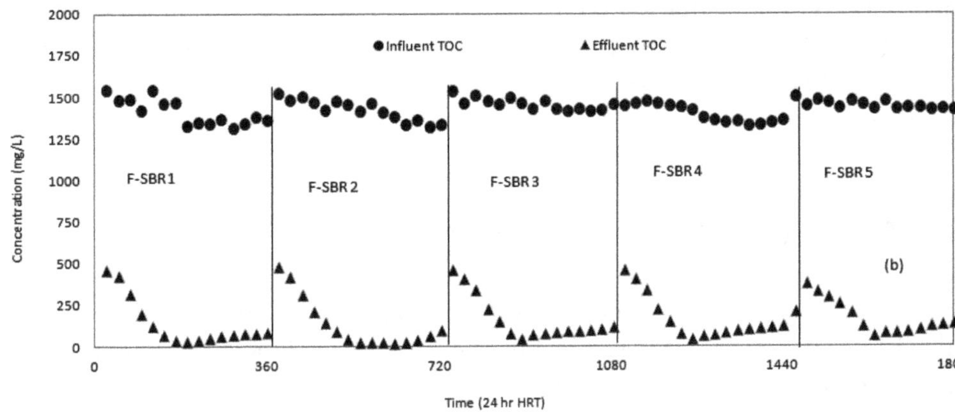

Figure 13. COD and TOC removal efficiency and BOD_5/COD ratio of F-SBR 1 to F-SBR 5 at 24 hr HRT

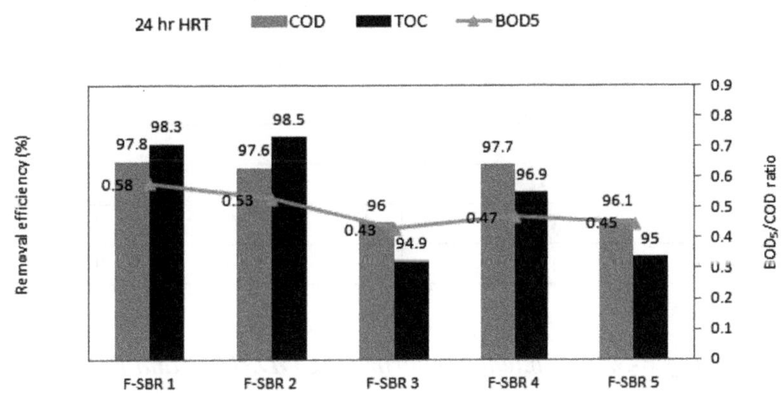

Table 6. Effect of cycle period on F-SBR performance in terms of HRT and biodegradability (BOD_5/COD ratio)

F-SBR	Biodegradability (BOD_5/COD Ratio)		
	6 hr	**12 hr**	**24 hr**
1	0.49	0.53	0.58
2	0.45	0.50	0.53
3	0.40	0.41	0.43
4	0.40	0.45	0.47
5	0.41	0.42	0.45

terms of COD, TOC and BOD_5 removal for F-SBR 1 to F-SBR 5 processes. The 12 hr HRT was found to be the most significant ($p < 0.05$) amongst the three cycle periods (6, 12 and 24 hr) as presented in Tables 7, 8 and 9. In conventional wastewater treatment, 5-14 hr HRT is recommended (Grady et al., 2004). The F-SBR 1 at 12 hr HRT performed most effectively. Based on the results, the F-SBR 1 (H_2O_2/COD molar ratio 3 and H_2O_2/Fe^{2+} molar ratio is 25) combination is the most significant according to the statistical analysis.

SUSTENANCE OF FENTON-SBR PROCESS

Performance of the Fenton-SBR deteriorated, and MLSS and MLVSS in the SBR decreased after the 8th cycle in the 12 hr HRT. This was presumably due to nutrient deficiency and/or accumulation of non-biodegradable pesticide degradation products. Ballesteros et al. (2009) reported that pretreatment time needed for pesticide degradation increased with organic load up to 120 to 257 min. Specifically, at

Table 7. One-Way ANOVA of F-SBR efficiency in terms of COD, TOC and BOD_5 removal at 6 hr HRT

	F	p-Value	F-Crit.
6 hr HRT, F-SBR 1	3.157	0.0545	3.259
6 hr HRT, F-SBR 2	3.003	0.0622	3.259
6 hr HRT, F-SBR 3	2.969	0.0640	3.259
6 hr HRT, F-SBR 4	2.935	0.0659	3.259
6 hr HRT, F-SBR 5	2.552	0.0909	3.259

Table 8. One-Way ANOVA of F-SBR efficiency in terms of COD, TOC and BOD_5 removal at 12 hr HRT

	F	p-Value	F-Crit.
12 hr HRT, F-SBR 1	9.881	0.00055	3.219
12 hr HRT, F-SBR 2	5.249	0.0092	3.219
12 hr HRT, F-SBR 3	4.592	0.0157	3.219
12 hr HRT, F-SBR 4	4.788	0.0133	3.219
12 hr HRT, F-SBR 5	4.370	0.0188	3.219

Table 9. One-Way ANOVA of F-SBR efficiency in terms of COD, TOC and BOD_5 removal at 24 hr HRT

	F	p-Value	F-Crit.
24 hr HRT, F-SBR 1	2.998	0.0606	3.219
24 hr HRT, F-SBR 2	2.987	0.0612	3.219
24 hr HRT, F-SBR 3	2.489	0.0951	3.219
24 hr HRT, F-SBR 4	2.818	0.0711	3.219
24 hr HRT, F-SBR 5	2.054	0.1109	3.219

200 and 500 mg/L, the best point to stop the AOP pretreatment and complete degradation by biological treatment was 66 and 57%, respectively. On the other hand, nutrients, rather than carbon or energy sources, may at times be the limiting material for microbial cell synthesis and growth. Required organic nutrients, known as growth factors, are sometimes needed by organisms as percursors for organic cell material, which cannot be synthesized from carbon sources. For municipal wastewater, sufficient nutrients are generally present, but for industrial wastewater, nutrients may need to be added for the biological treatment (Metcalf and Eddy, 2002). Therefore, pretreatment of the pesticide wastewater was conducted with higher Fenton reaction time (120 min) and the Fenton pretreated wastewater was combined with a municipal wastewater in the SBR.

In the F-SBR (C), Fenton pretreatment operating conditions were H_2O_2/COD molar ratio 3, H_2O_2/Fe^{2+} molar ratio 25, pH 3 and reaction time 120 min. Fenton pretreated pesticide wastewater was combined with municipal wastewater in ratio 3:1. In the F-SBR (C), after acclimation and start up, MLSS and MLVSS increased from 2580 and 1700 mg/L to 3250 and 2210 mg/L, respectively and the SBR was operated for 40 days at 12 hr HRT. The removal efficiency was 96.7, 97.7 and 93.3% for COD, TOC and BOD_5 removal, respectively. Effluent COD, BOD_5 and TOC reached a final value of 26, 19 and 20 mg/L, respectively and BOD_5/COD ratio 0.73 (Figure 14 and 15).

Sludge volume index (SVI) is one of the important parameters to assess the stability of sludge in any aerobic suspended growth system such as SBR. An SVI of 150 mg/L is often considered to be the dividing line between good (SVI < 150) and poor (SVI greater than 150) settling sludge (Grady et al., 2004). SVI was monitored during the end of each cycle before withdrawal. The F-SBR (C) process, the

Figure 14. COD concentration and BOD_5/COD ratio of F-SBR (C) at 12 hr HRT

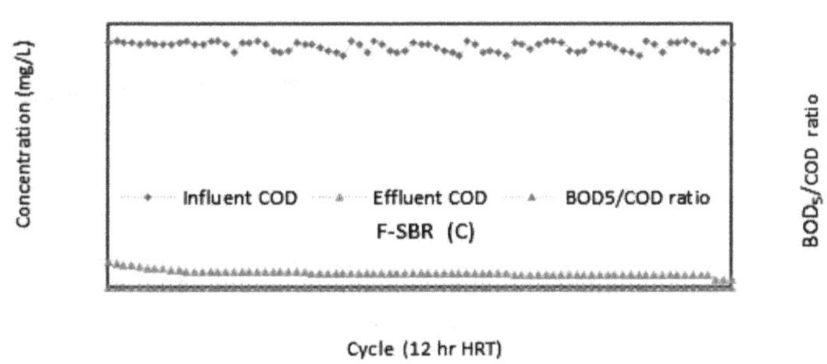

Figure 15. TOC concentration of F-SBR (C) at 12 hr HRT

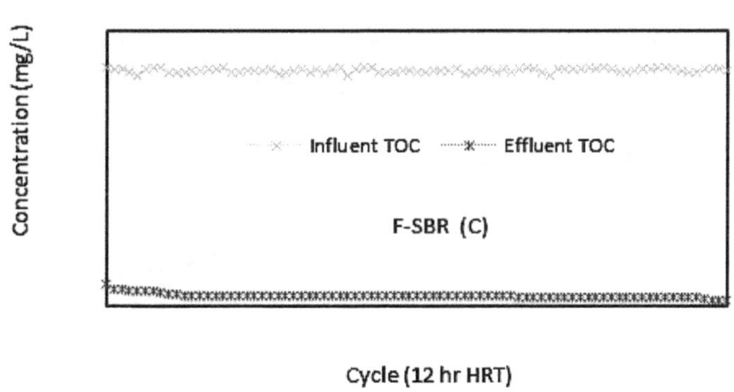

sludge exhibited good settling properties in the startup phase and on the 40th day with an SVI of 133 and 62, respectively as shown in Figure 16. Improved sludge settleability facilitates the separation of biomass from the mixed liquor suspension and thus mitigates the separation problems encountered in biological treatment of the pesticide wastewater using SBR (Moussavi et al., 2010).

To study nitrification SBR of F-SBR (C) process, NH_3-N, TKN and NO_3^--N concentrations were measured during one cycle period (12 hr) as shown in Figure 17. The microbial oxidation of NH_3-N was complete in 10 hr, residual TKN was 7.2 mg/L in 12 hr (88.1% removal) and NO_3^- N was 73.9 mg/L in 12 hr, indicating adequate nitrification. Nitrification reached more than 98% in landfill leachate treatment by SBR (Spagni and Marsili-Libelli, 2009).

Monitoring of DO and pH to control biological nutrient removal has received much attention (Cui et al., 2009). It has been demonstrated that changes in the pH and DO profile during fill and react phases allow detection of pre-upset conditions and potential problems in wastewater treatment system (Andreottola et al., 2001). The DO level and pH value can successfully indicate the oxidative and ongoing biological reactions both in continuous and batch processes (Ga and Ra, 2009; Al-Ghusain and Hao, 1995). The pH increases during denitrification and decreases during the nitrification reaction. There are two important breakpoints that provide information about the end of nitrification and denitrification (Chang and Hao, 1996; Al-Ghusain and Hao, 1995). In the aerobic phase, there is a minimum called the 'ammonia valley' which indicates the end of nitrification. In the anoxic phase, the 'nitrate apex' is maximum and this indi-

Figure 16. Sludge volume index of F-SBR (C) during 40-day operation.

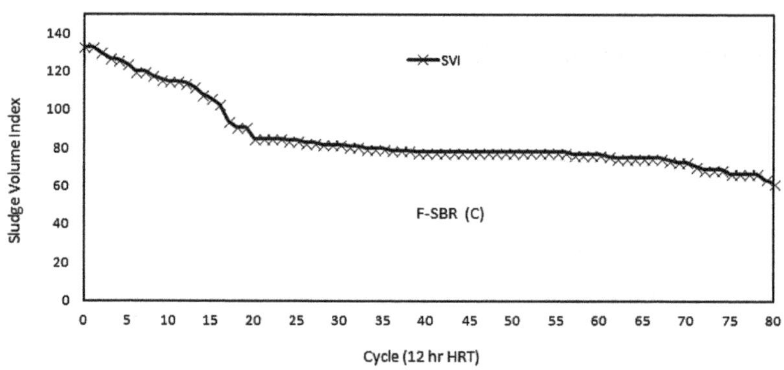

Figure 17. Nitrification in F-SBR (C) at 12 hr HRT NH_3-N, TKN and NO_3-N

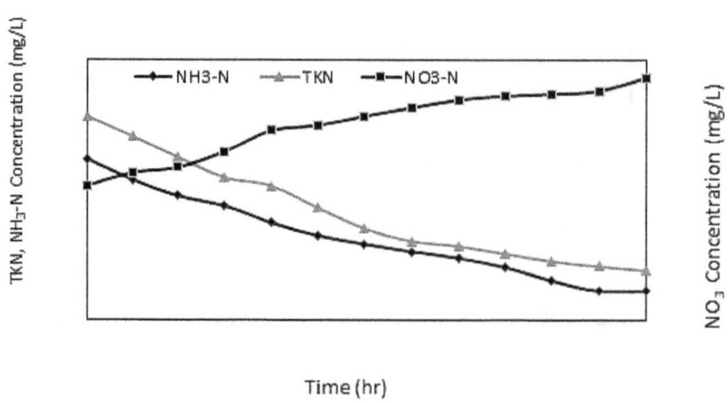

cates the end of denitrification. The increasing interest in the monitoring of pH and DO is of advantage because it can be used to optimize the aerobic phase in such a way that aeration can be stopped whenever nitrification is completed and subsequent anaerobic phase can be avoided when denitrification is completed (Andreottola et al., 2001). The DO and pH profiles of the F-SBR (C) process are shown in Figure 18. The ammonia valley occurred at about the 10th hr which corresponded to NH_3-N concentration of 3.6 mg/L and the nitrate apex was in the 11th hour with NO_3^--N concentration of 69.3 mg/L. Qing et al. (2007) observed that characteristic points on the profiles of dissolved oxygen (DO) and pH could exactly reflect the process of nitrification and where denitrification could begin in SBR. Therefore, the obtained DO and pH bending points were similar to the theoretical principle especially ammonia valley (pH profile) and DO elbow (profile) and this indicated the completion of nitrification process in the Fenton-SBR (C). This observation could give a signal to stop the aeration or react phase in order to save energy used during the SBR process.

CONCLUSION

The feasibility of applying the three combined systems for complete treatment of pesticide wastewater was evaluated. In the combined system (Fenton-SBR), the SBR was operated for 77 days and fed with Fenton-treated pesticide wastewater under different Fenton operating conditions for the SBR operation. Statistical analysis (One-Way ANOVA) was performed on the results and12 hr HRT was found to be sufficient for the SBR operation. Under optimum operating conditions of Fenton pretreatment (H_2O_2/COD molar ratio 3.0, H_2O_2/Fe^{2+} molar ratio 10, pH 3 and reaction time 60 min), COD and TOC removal were 58.51 and 39.76%, respectively and biodegradability (BOD_5/COD ratio) increased from 0.02 to ~ 0.30 after 60 min reaction time. Complete degradation of the pesticide active ingredient occurred in 1 min. The reaction followed pseudo-first order kinetics with rate constant (*k*) of 0.0083 min^{-1}. For Fenton-SBR treatment, H_2O_2/COD molar ratio 3.0, H_2O_2/Fe^{2+} molar ratio 25, pH 3 and reaction time 60 min appeared to be the most significant ($p < 0.05$) operating conditions for Fenton pretreatment. Sustenance of the SBR operation was possible by increasing Fenton pretreatment (reaction) time (120 min) and combining the Fenton pretreated wastewater with a municipal wastewater (3:1). The Fenton–SBR treatment at 12 hr HRT achieved COD, TOC and BOD_5 removal efficiency of 96.7, 97.7 and 93.3%, respectively. DO and pH profiles indicated that adequate nitrification occurred. The Fenton-SBR process is effective in treatment of the chlorpyrifos, cypermethrin and chlorothalonil pesticide wastewater.

Figure 18. Dissolved oxygen (DO) and pH profiles of Fenton-SBR (C) during 12 hr HRT

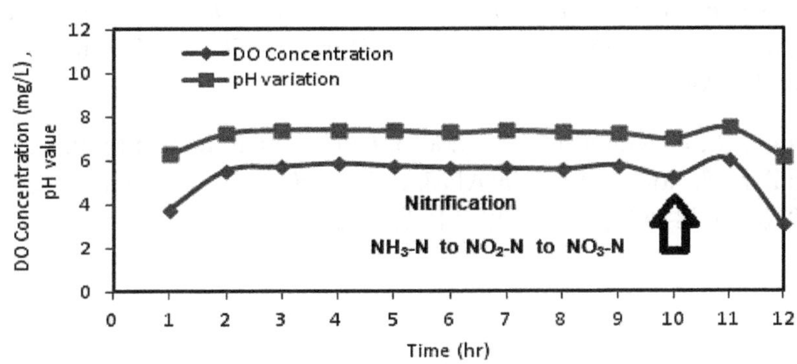

REFERENCES

Adewuyi, Y. G. (2005). Sonochemistry in environmental remediation. 1. Combinative and hybrid sonophotochemical oxidation processes for the treatment of pollutants in water. *Environmental Science & Technology*, *39*(10), 3409–3420. doi:10.1021/es049138y PMID:15952344

Ahmad, U. K., Ujang, Z., Woon, C. H., Indran, S., & Mian, M. N. (2004). Development of extraction procedures for the analysis of polycyclic aromatic hydrocarbons and organochlorine pesticides in municipal sewage sludge. *Water Science and Technology*, *50*(9), 137–144. doi:10.2166/wst.2004.0555 PMID:15581005

Al-Ghusain, I., & Hao, O. (1995). Use of pH as control parameter for aerobic/anoxic sludge digesting. *Journal of Environmental Engineering*, *3*(3), 225–235. doi:10.1061/(ASCE)0733-9372(1995)121:3(225)

Al-Momani, F., Touraud, E., Degorce-Dumas, J. R., Roussy, J., & Thomas, O. (2003). Biodegradability enhancement of textile dyes and textile wastewater by UV photolysis. *Journal of Photochemical and Photobiology A. Chemistry.*, *153*, 191–197.

Alessandro Spagni, A., & Marsili-Libelli, S. (2009). Nitrogen removal via nitrite in a sequencing batch reactor treating sanitary landfill leachate. *Bioresource Technology*, *100*(2), 609–614. doi:10.1016/j.biortech.2008.06.064 PMID:18707876

Alex, J., Beteau, J. F., Copp, J. B., Hellinga, C., Jeppsson, U., Marsili-Libelli, S., . . . Vanhooren, H. (1999). Benchmark for evaluating control strategies in wastewater treatment plant. European Control Conference, Karlsruhe, Germany. 10.23919/ECC.1999.7099914

Andreozzi, R., Caprio, V., Insola, A., & Marotta, R. (1999). Advanced oxidation processes (AOP) for water purification and recovery. *Catalysis Today*, *53*(1), 51–59. doi:10.1016/S0920-5861(99)00102-9

APHA. AWWA, WEF. (2005). Standard methods for the examination of water and wastewater (21st ed.) American Public Health Association, Washington DC.

Arden, E., & Lockett, W. (1914). Experiment on the oxidation of sewage without the aid of filters. *Journal of the Society of Chemical Industry*, *33*(10), 523–539. doi:10.1002/jctb.5000331005

Ballesteros, M. M. M., Sanchez, P. J. A., Casas, L. J. L., Oller, I., & Malato, R. S. (2009). Degradation of a four-pesticide mixture by combined photo-Fenton and biological oxidation. *Water Research*, *43*(3), 653–660. doi:10.1016/j.watres.2008.11.020 PMID:19059625

Banchuen, T. (2003). *A microcosm-based investigation into oxidized nitrogen removal in the hypolimnetic water of the Occoquan reservoir of northern Virginia*. (Master's Thesis), Virginia Polytechnic Institute and State University, Blacksburg, Virginia.

Barron, M. G., Heintz, R., & Krahn, M. M. (2003). Contaminant exposure and effects in pinnipeds: Implications for Steller sea lion declines in Alaska. *The Science of the Total Environment*, *311*(1-3), 111–133. doi:10.1016/S0048-9697(03)00140-2 PMID:12826388

Bautista, P., Mohedano, A. F., Casas, J. A., Zazo, J. A., & Rodrigue, J. J. (2008). An overview of the application of Fenton oxidation to industrial wastewaters treatment. *Journal of Chemical Technology and Biotechnology (Oxford, Oxfordshire)*, *83*(10), 1323–1338. doi:10.1002/jctb.1988

Bazrafshanferdos, E., Mostafapour, F. K., Soori, M. M., & Mahvi, A. (2012). Application of combined chemical coagulation and electrocoagulation process to carwash wastewater treatment. *Fresenius Environmental Bulletin*, *21*, 2694–2701.

Bill, G. (2012). Application of advanced oxidation processes (AOP) in water treatment, *37th Annual Queensland Water Industry Operations Workshop, Parklands*, Australia.

Brillas, E., Sirés, I., & Oturan, M. A. (2009). Electro-fenton process and related electrochemical technologies based on fenton's reaction chemistry. *Chemical Reviews*, *109*(12), 6570–6631. doi:10.1021/cr900136g PMID:19839579

Chamarro, E., Marco, A., & Esplugas, S. (2001). S. Use of fenton reagent to improve organic biodegradability. *Water Research*, *35*(4), 1047–1051. doi:10.1016/S0043-1354(00)00342-0 PMID:11235870

Chang, C., & Hao, O. (1996). Sequencing batch reactor system for nutrient removal: ORP and pH profiles. *Journal of Chemical Technology and Biotechnology (Oxford, Oxfordshire)*, *67*(1), 27–38. doi:10.1002/(SICI)1097-4660(199609)67:1<27::AID-JCTB430>3.0.CO;2-2

Chang, H. N., Moon, R. K., Park, B. G., Lim, S. J., Choi, D. W., Lee, W. G., ... Ahn, Y. H. (2000). Simulation of sequencing batch reactor operation for simultaneous removal of nitrogen and phosphorous. *Bioprocess Engineering*, *23*(5), 513–521. doi:10.1007004499900188

Colvin, R. J., & Rozich, A. F. (1997). AFC Process that reduced sludge and cost of pharmaceutical treatment. *Chemical Show Conference*, New York, NY.

Crittenden, J. C., Suri, R. P. S., Perram, D. L., & Hand, D. W. (1997). Decontamination of water using adsorption and photocatalysis. *Water Research*, *31*(3), 411–418. doi:10.1016/S0043-1354(96)00258-8

Cui, Y., Wang, S., & Li, J. (2009). On-line monitoring for phosphorus removal process and bacterial community in sequencing batch reactor. *Chinese Journal of Chemical Engineering*, *17*(3), 484–492. doi:10.1016/S1004-9541(08)60235-9

Environmental Quality Act. (1974). (Industrial effluent) Regulation. Seventh Schedule, Regulation 12, amended 2009 PU (A) 434), Malaysia.

Farré, M. J., Franch, M. J., Ayllón, M. I., Peral, J. A., Xavier, J., & Domènech, X. (2007). Biodegradability of treated aqueous solutions of biorecalcitrant pesticides by means of photocatalytic ozonation. *Desalination*, *211*(1-3), 22–33. doi:10.1016/j.desal.2006.02.082

Fenton, H. J. H. (1894). Oxidation of tartaric acid in presence of iron. *Journal of the Chemical Society*, *65*(0), 899–910. doi:10.1039/CT8946500899

Ga, C., & Ra, C. (2009). Real-time control of oxic phase using pH (mV)-time profile in swine wastewater treatment. *Journal of Hazardous Materials*, *172*(1), 61–67. doi:10.1016/j.jhazmat.2009.06.133 PMID:19628333

Ganjidoust, H., & Ayati, B. (2004). Use of sequencing batch reactors (SBRs) in treatment of wood fiber wastewater. *Iran Journal of Health Science and Engineering*, *1*, 91–96.

González, O., Sans, C., & Esplugas, S. (2007). Sulfamethoxazole abatement by photo-Fenton: Toxicity, inhibition and biodegradability assessment of intermediates. *Journal of Hazardous Materials*, *146*(3), 459–464. doi:10.1016/j.jhazmat.2007.04.055 PMID:17540504

Gonzalez, S., Muller, J., Petrovic, M., Barcelo, D., & Knepper, T. P. (2006). Biodegradation studies of selected priority acidic pesticides and diclofenac in different bioreactors. *Environmental Pollution*, *144*(3), 926–932. doi:10.1016/j.envpol.2006.02.021 PMID:16678949

Grady, C. P. L., Daigger, G. T., Love, N. G., & Filippe, C. D. M. (1999). Biological wastewater treatment (2nd ed.). Boca Raton, FL: CRC Press.

Gromboni, C. F., Kamogawa, M. Y., Ferreira, A. G., Nóbrega, J. A., & Nogueira, A. R. A. (2007). Microwave-assisted photo-Fenton decomposition of chlorfenvinphos and cypermethrin in residual water. *Journal of Photochemistry and Photobiology A Chemistry*, *185*(1), 32–37. doi:10.1016/j.jphotochem.2006.05.005

Hadjiniculaou, J. (1998). Evaluation of a controlled condition in a sequencing batch reactor pilot plant operation for treatment of slaughterhouse wastewater. *Canadian Agricultural Engineering*, *23*, 147–156.

Joseph, J. M., Destaillats, H., Hung, H. M., & Hoffmann, M. R. (2000). The sonochemical degradation of azobenzene and related azo dyes: Rate enhancements via Fenton's reactions. *The Journal of Physical Chemistry A*, *104*(2), 301–307. doi:10.1021/jp992354m

Kang, Y. W., Cho, M. J., & Hwang, K. Y. (1999). Correction of hydrogen peroxide interference on standard chemical oxygen demand test. *Water Research*, *33*(5), 1247–1251. doi:10.1016/S0043-1354(98)00315-7

Kavitha, V., & Palanivelu, K. (2005). Destruction of cresols by Fenton oxidation process. *Water Resources*, *39*, 3062–3072. PMID:15992853

Kolthoff, I. M., & Medalia, I. (1951). Determination of organic peroxides by reaction with ferrous iron. *Analytical Chemistry*, *23*(4), 595–603. doi:10.1021/ac60052a014

Krzemińska, D., Neczaj, E., & Borowski, G. (2015). Advanced oxidation processes for food industrial wastewater decontamination. *Journal of Ecological Engineering*, *16*, 61–71. doi:10.12911/22998993/1858

Leong, K. H., & Tan, L. L. (2007). Contamination levels of selected organochlorine and organophosphate pesticides in the Selangor River, Malaysia between 2002 and 2003. *Chemosphere*, *66*(6), 1153–1159. doi:10.1016/j.chemosphere.2006.06.009 PMID:17027062

Liang, C., Bruell, C. J., Marley, M. C., & Sperry, K. L. (2004). Persulfate oxidation for in situ remediation of TCE activated by chelated ferrous ion. *Chemosphere*, *55*(9), 1225–1233. doi:10.1016/j.chemosphere.2004.01.030 PMID:15081763

Mace, S., & Mata-Alvarez, J. (2002). Utilization of SBR technology for wastewater: An overview. *Industrial & Engineering Chemistry Research*, *41*(23), 5539–5553. doi:10.1021/ie0201821

Malato, P., Fernández-Ibáñez, M. I., Maldonado, J., Blanco, W. G., & Gernjak, W. (2009). Decontamination and disinfection of water by solar photocatalysis: Recent overview and trends. *Catalysis Today*, *147*(1), 1–59. doi:10.1016/j.cattod.2009.06.018

Malato, S., Blanco, J., Maldonado, M. I., Fernández-Ibáñez, P., & Campos, A. (2000). Optimising solar photocatalytic mineralisation of pesticides by adding inorganic oxidising species; application to the recycling of pesticide containers. *Applied Catalysis B: Environmental*, *28*(3-4), 163–174. doi:10.1016/S0926-3373(00)00175-2

Metcalf & Eddy. (2002). *Wastewater treatment: Treatment and reuse* (4th ed.). Washington, DC: McGraw Hill.

Mohan, S. V., Rao, N. C., Prasad, K. K., & Sarma, P. N. (2005). Bioaugmentation of an anaerobic sequencing batch biofilm reactor with immobilized sulphate reducing bacteria for the treatment of sulphate bearing chemical wastewater. *Process Biochemistry*, *40*(8), 2849–2857. doi:10.1016/j.procbio.2004.12.027

Morgan-Sagastume, F., & Allen, D. G. (2003). Effects of temperature transient conditions on aerobic biological treatment of wastewater. *Water Research*, *37*(15), 3590–3601. doi:10.1016/S0043-1354(03)00270-7 PMID:12867325

Moussavi, G., Barikbin, B., & Mahmoudi, M. (2010). The removal of high concentrations of phenol from saline wastewater using aerobic granular SBR. *Chemical Engineering Journal*, *158*(3), 498–504. doi:10.1016/j.cej.2010.01.038

Moussavi, G., Yazdanbaksh, A., & Heidarizad, M. (2009). The removal of formaldehyde from concentrated synthetic wastewater using $O_3/MgO/H_2O_2$ process integrated with the biological treatment. *Journal of Hazardous Materials*, *171*(1-3), 907–913. doi:10.1016/j.jhazmat.2009.06.090 PMID:19616892

Neczaj, E., Kacprzak, M., Kamizela, T., Lach, J., & Okoniewska, E. (2008). Sequencing batch reactor system for the co-treatment of landfill leachate and dairy wastewater. *Desalination*, *222*(1-3), 404–409. doi:10.1016/j.desal.2007.01.133

Negash, W. D. (2014). Application of iron activated persulfate for disinfection in water treatment. (Master's Thesis), University of California Riverside.

Neyens, E., & Baeyens, J. (2003). A review of classic Fenton's peroxidation as an advanced oxidation technique. *Journal of Hazardous Materials*, *98*(1-3), 33–50. doi:10.1016/S0304-3894(02)00282-0 PMID:12628776

Ollis, D. (2001). On the need for engineering models of integrated chemical and biological oxidation of wastewaters. *Water Science and Technology*, *44*(5), 117–123. doi:10.2166/wst.2001.0265 PMID:11695448

Qing, Y., Shuying, W., Anming, Y., Jianhua, G., & Fengyang, B. (2007). Advanced nitrogen removal using pilot-scale SBR with intelligent control system built on three layer network. *Frontiers of Environmental Science & Engineering in China*, *1*(1), 33–38. doi:10.100711783-007-0006-4

Sarria, V., Parraa, S., Adler, N., Péringera, P., Benitez, N., & Pulgarin, C. (2002). Recent developments in the coupling of photoassisted and aerobic biological processes for the treatment of biorecalcitrant compounds. *Catalysis Today*, *76*(2-4), 301–315. doi:10.1016/S0920-5861(02)00228-6

Scott, J., & Ollis, D. (1995). Integration of chemical and biological oxidation processes for wastewater treatment: Review and recommendations. *Environmental Progress & Sustainable Energy*, *14*, 68–103.

Scott, J. P., & Ollis, D. F. (1995). Integration of chemical and biological oxidation processes for water treatment: Review and recommendation. *Environmental Programme*, *14*(2), 88–103. doi:10.1002/ep.670140212

Senthilnathan, J., & Philip, L. (2012). Elimination of pesticides and their formulation products from drinking water using thin film continuous photoreactor under solar radiation. *Solar Energy*, *89*(9), 2735–2745. doi:10.1016/j.solener.2012.06.011

Sherrard, R. M., Bearr, J. S., Murray-Gulde, C. L., Rodgers, J. H. Jr, & Shah, Y. T. (2004). Feasibility of constructed wetlands for removing chlorothalonil and chlorpyrifos from aqueous mixtures. *Environmental Pollution*, *127*(3), 385–394. doi:10.1016/j.envpol.2003.08.017 PMID:14638299

Stephenson, R. L., & Blackburn, J. J. (1998). *The industrial wastewater systems handbook*. New York, NY: Lewis Publishers.

Suh, Y., & Resseaux, P. (2002). An LCA of alternative wastewater sludge treatment scenarios. *Resources, Conservation and Recycling*, *35*(3), 191–200. doi:10.1016/S0921-3449(01)00120-3

Sun, J. H., Sun, S. P., Wang, G. L., & Qiao, L. P. (2007). Degradation of azo dye amido black 10B in aqueous solution by Fenton oxidation process. *Dyes and Pigments*, *74*(3), 647–652. doi:10.1016/j.dyepig.2006.04.006

Tadeo, J. L., Sánchez-Brunete, C., Albero, B., & García-Valcárcel, A. I. (2010). Determination of pesticide residues in sewage sludge: A review. *Journal of AOAC International*, *93*, 1692–1702. PMID:21313796

Talinli, I., & Anderson, G. K. (1992). Interference of hydrogen peroxide on the standard COD test. *Water Research*, *26*(1), 107–110. doi:10.1016/0043-1354(92)90118-N

Tamimi, N., Qourzal, S., Barka, N., Assabbane, A., & Ait-Ichou, Y. (2008). Methomyl degradation in aqueous solutions by Fenton's reagent and the photo-Fenton system. *Separation and Purification Technology*, *61*(1), 103–108. doi:10.1016/j.seppur.2007.09.017

Tchobanoglous, G., Burton, F., & Davind, H. (2002). *Wastewater engineering: Treatment and reuse* (4th ed.). Metcalf and Eddy.

Widerer, P., Irvine, R. L., & Goronszy, M. (1999). *Sequencing batch reactors*. Available at http://water.epa.gov/scitech/wastetech/upload/2002_10_31_mtb _sbr_new.pdf

Zapata, A., Malato, S., Sanchez, P. J. A., Oller, I., & Maldonado, M. (2010). Scale-up strategy for a combined solar photo-Fenton/biological system for remediation of pesticide-contaminated water. *Catalysis Today*, *151*(1-2), 100–106. doi:10.1016/j.cattod.2010.01.034

Zepp, R. G., Faust, B. C., & Hoigne, J. (1992). Hydroxyl radical formation in aqueous reactions (pH 3-8) of iron (II) with hydrogen peroxide: The photo-Fenton reaction. *Environmental Science & Technology*, *26*(2), 313–319. doi:10.1021/es00026a011

Zhang, Y., Xiao, Z., Chen, F., Ge, Y., Wu, J., & Hu, X. (2010). Degradation behavior and products of malathion and chlorpyrifos spiked in apple juice by ultrasonic treatment. *Ultrasonics Sonochemistry*, *17*(1), 72–77. doi:10.1016/j.ultsonch.2009.06.003 PMID:19560957

Zhou, H., & Smith, D. W. (2001). Advanced oxidation technologies in water and wastewater treatment. *Journal of Civil Engineering*, *28*, 49–66.

Zitomer, D., & Speece, R. (1993). Sequential environments for enhanced biotransformation of aqueous contaminants. *Environmental Science & Technology*, *27*(2), 226–244. doi:10.1021/es00039a001

Chapter 17
Principles of Electrocoagulation and Application in Wastewater Treatment

Ezerie Henry Ezechi
https://orcid.org/0000-0003-4968-5541
Universiti Teknologi Malaysia, Johor Bahru, Malaysia

Augustine Chioma Affam
University College of Technology Sarawak, Malaysia

Khalida Muda
Universiti Teknologi Malaysia, Johor Bahru, Malaysia

ABSTRACT

Electrocoagulation has emerged a reliable technology for the treatment of various wastewaters. Its basic principle depends on the response of water particles to strong electric field in a redox reaction. Oxidation of the anode material releases coagulating agents that form metal hydroxide complexes which neutralize particulate materials to form agglomerates. The agglomerates either settle at the bottom or float to the surface depending on the removal path of the electrocoagulation reactor. The merits of electrocoagulation include minimal sludge generation, minimal operator attention, simple equipment, high pollutant removal capacity, and ease of operation. Therefore, this chapter explores the mechanisms of electrocoagulation, components of electrocoagulation, benefits, and demerits of electrocoagulation. Furthermore, the similarity between electrocoagulation and coagulation is explored. Application of electrocoagulation for the treatment of various wastewaters was explored. Feasibility of electrocoagulation was examined through cost evaluation with other treatment technologies.

DOI: 10.4018/978-1-7998-0369-0.ch017

INTRODUCTION

The high cost of conventional wastewater treatment methods and their various constraints in wastewater treatment has been a concern for many stakeholders. The major conventional methods for the treatment of wastewater include biological processes (Nowak et al., 2019), membrane processes (Vo et al., 2019), adsorption processes (Sessarego et al., 2019), advanced oxidation processes (Brink et al., 2018), coagulation/flocculation processes (Kim et al., 2019) and ion exchange processes (Muhammad et al., 2019). Most conventional methods require high personnel training, high capital and operational cost, and produce high amount of sludge and secondary pollutants (Ezechi et al., 2019). While the volume of industrial and domestic wastes in many countries and regions have been increasing, waste management techniques such as landfills and wastewater treatment facilities are over-burdened (Ezechi et al., 2011). These constraints have led to the development of an environmentally friendly approach that is cheap and shows consistent pollutant removal capacity.

The application of electric current in the treatment of wastewater is increasingly gaining global interest due to its associated efficiencies and cost effectiveness. Electrocoagulation (EC) is the process of destabilizing suspended, emulsified or dissolved contaminants in an aqueous medium by the introduction of an electric current (Ezechi et al., 2015). It is an environmentally friendly approach that requires minimal personnel training, no chemical addition or expensive equipment and device (Changmai et al., 2019). EC is solely based on the oxidation of the sacrificial anode under the influence of electric current in an oxidation/reduction process. The electrodes are cheap and readily available, and can be used over a certain number of experiments when thoroughly washed and dried. EC has been widely used in the treatment and removal of various wastewater pollutants, and presents similar advantages to chemical coagulation. Its success largely arises from the interaction of charged particles with wastewater pollutants. Furthermore, it produces lesser amount of sludge.

Various wastewaters such as produced water, geothermal wastewater, restaurant wastewater, potato chips wastewater, textile wastewater, tannery wastewater, olive mill wastewater and electroplating wastewater have been successfully treated with electrocoagulation. Pollutants such as boron, phenol, heavy metals such as arsenic, cadmium, copper, chromium, nickel and zinc can interact with the coagulating agents from the sacrificial anode and flocculate during EC process. However, this interaction largely depends on the chemistry of the pollutants. For instance, boron appears as boric acid (H_3BO_3) and borax ($Na_2B_4O_7 \cdot 10H_2O$) in nature. In aquatic systems, it exists primarily as undissociated boric acid and borate ions (Bryjak et al., 2008). The chemistry of the pollutant can enhance or impede its interaction with the coagulating agents during EC process. Thus, electrocoagulation presents itself as an important treatment technology.

Electrocoagulation process is operated with different parameters. The efficiency of EC depends on the best combination of these parameters. Parameters such as pH, current density, treatment time, concentration, electrode gap, surface area, flow rate, temperature, electrode type, electrode pattern and applied potential are all important electrocoagulation parameters. During electrocoagulation, metal ions are generated at the anode while hydrogen bubble is released at the cathode. Hydrogen bubble is responsible for the flotation of formed flocs to the surface of the reactor where the flocs are concentrated and removed. The oxidation of a metal (M) in an electrocoagulation process can be represented as:

Anode reaction:

$$M \rightarrow M^{n} + ne^{-} \quad (1)$$

Cathode reaction:

$$2H_2O + 2e^{-} \rightarrow 2OH^{-} + H_2(g) \quad (2)$$

The objective of this chapter is to discuss the mechanism of electrocoagulation and its associated characteristics.

MECHANISM OF ELECTROCOAGULATION

When direct current passes through a water solution, it destabilizes the contaminants in the medium. The destabilized contaminants can aggregate and subsequently be removed by sedimentation and/or flotation. Therefore the three major mechanisms involved in electrocoagulation are (Babu et al., 2007):

1. Formation of coagulants by electrolytic oxidation of sacrificial electrodes.
2. Destabilization of the contaminants, particulate suspension and breaking of emulsions.
3. Aggregation of the destabilized phases to form a floc.

Formation of Coagulants by Electrolytic Oxidation of Sacrificial Anode

In all redox reactions, particles are either oxidized or reduced. The application of applied current in EC set-up causes the oxidation of the sacrificial electrodes. During the oxidation process, coagulating agents are released from the sacrificial electrode into the EC set-up. The coagulating agents released from the anode combine with the hydroxyl ion released from the cathode to form a highly charged coagulant. The rate of the release of the coagulating agents from the sacrificial electrodes largely depends on applied current and voltage.

Destabilization of the Contaminants, Particulate Suspension and Breaking of Emulsions

The coagulating agents released from the sacrificial electrode destabilize the contaminants in the wastewater and break their emulsions while binding with the colloidal particles in order to neutralize their charge. Once the charge on the colloidal particles is neutralized, the pollutants can easily be removed from the EC environment through various mechanisms. The destabilization mechanism of electrocoagulation is achieved through two processes (Yılmaz et al., 2008):

Figure 1. Mechanism of electrocoagulation

1. Compression of the diffuse double layer around the charged species by the interactions of ions generated by oxidation of the sacrificial anode.
2. Charge neutralization of the ionic species present in water by counter ions produced by the electrochemical dissolution of the sacrificial anode. These counter ions reduce the electrostatic interparticle repulsion to the extent that the van deer Waals attraction dominate and cause coagulation. A zero net charge results in the process.

Aggregation of the Destabilized Phases to Form a Floc

The destabilized contaminants can aggregate to form a floc. The floc formed as a result of coagulation creates a sludge blanket that entraps and bridge residual colloidal particles in the aqueous medium. This aggregated referred to as the "floc" is removed either through sedimentation or flotation with the aid of hydrogen bubble. The solid oxides, hydroxides and oxyhydroxides provide active surfaces for the adsorption of the polluting species. The mechanism of electrocoagulation process can be represented in Fig. 1

ELECTROCOAGULATION COMPONENTS

The electrocoagulation process has many components necessary for a successful system operation. The major components of EC are those required to set-up EC unit for wastewater treatment. Some of these components are expressed below:

Electrodes

Different kinds of electrodes can be used in electrocoagulation study. They include aluminum, iron, magnesium, titanium, zinc anode etc. All these electrodes are effective when used in electrocoagulation process. These electrodes can be made from metal plates or even from metal scraps such as iron or aluminum millings, cuttings, and can be made of different sizes (Chen, 2004). However, various electrodes

Figure 2. Schematic diagram of monopolar electrode pattern
(Jiang et al., 2002).

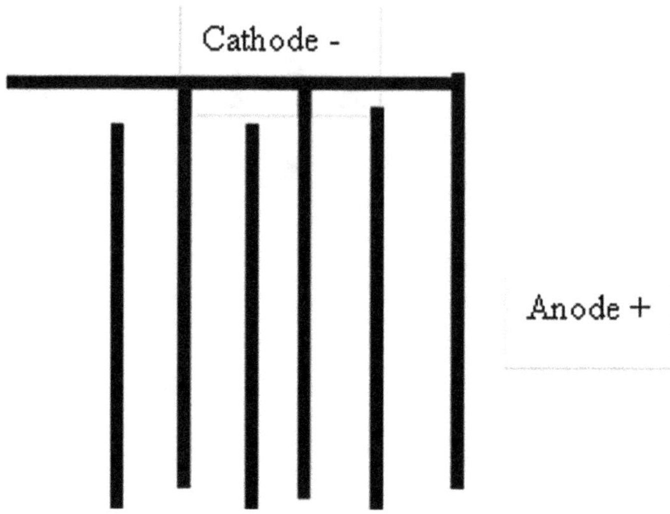

have different oxidation potential (Ghosh et al., 2008). According to the reactivity series, the position of the metal in the reactivity series determines its strength. Moving from the bottom of the reactivity series to the top, metals increase in reactivity, lose electrons more easily and become stronger reducing agents. The two most used electrodes are aluminum and iron. This is because of their availability, cost, and efficiency. The significance of the reactivity series is that all metals above hydrogen will readily displace/liberate hydrogen from water/acid while those below it will not displace hydrogen. Therefore, the choice of electrode material is also an important factor. However; efficiency, cost, availability are the major indices for the choice of anode and cathode materials. Electrodes can be arranged in bipolar or monopolar patterns (Emamjomeh & Sivakumar, 2009). Figure 2 and 3 show monopolar and bipolar electrode connections, respectively. Monopolar electrode connections are cell connections made between pairs of anode and cathode electrodes. The current passes through the individual electrodes each connected to the power source by a separate power connection. The space between the anodes and cathodes causes electrolysis when the solutions flow between them. Series of connection can be made on one input connection in a parallel configuration. The polarity of each electrode set is defined by the connection polarity i.e anode electrodes are connected to the positive pole while cathode electrodes are connected to the negative poles (Casson & Bess, 2006)

In bipolar electrode connections, current flows from one feeder electrode through the solution, into and out of each of the bipoles and subsequently to the second feeder electrode. Each electrode serves as both an anode and a cathode. The bipolar electrode pattern has a feeder electrode for the positive and negative power inputs points (Jiang et al., 2002).

Electrode Reactions

The electrode reaction in electrocoagulation depends on the type of electrode. Each electrode reacts in a stepwise process to form metal hydroxide which reacts with contaminants to form flocs. For aluminum and iron electrodes, their reactions in water can be summarized as follows (Ezechi et al., 2014):

Figure 3. Schematic diagram of a bipolar electrode pattern
(Jiang et al., 2002)

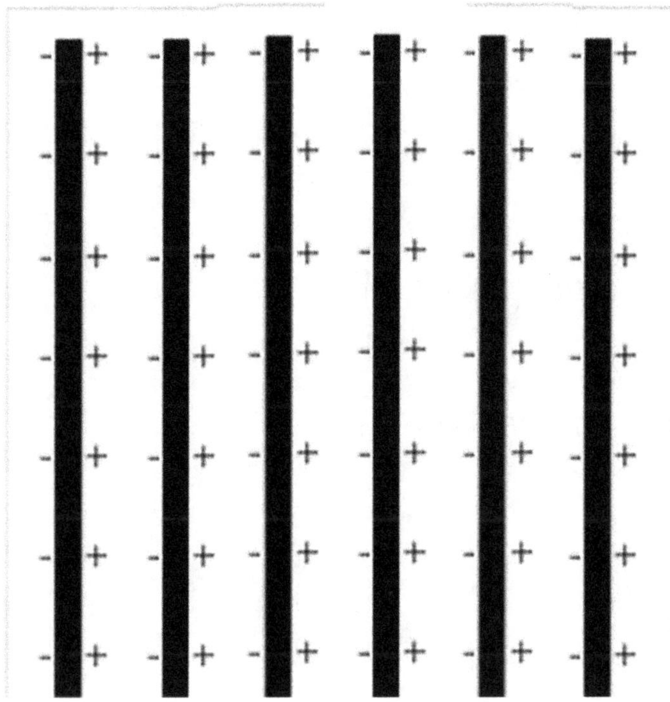

Anode

$$Al_{(s)} \rightarrow Al^{3+}(aq) + 3e^{-} \tag{3}$$

Cathode

$$2H_2O(l) + 2e^{-} \rightarrow H_2(g) + 2OH^{-}(aq) \tag{4}$$

Aluminum forms polymeric species during oxidation of the sacrificial anode. These polymeric species ($Al_6(OH)_{15}^{3+}$, $Al_7(OH)_{17}^{4+}$, $Al_8(OH)_{20}^{4+}$, $Al_{13}O_4(OH)_{24}^{7+}$, $Al_{13}(OH)_{34}^{5+}$) transform finally into $Al(OH)_{3(s)}$ according to the following equation (Ghosh et al., 2008).

$$Al^{3+}(aq) + 3H_2O(l) \rightarrow (Al(OH)_3(s) + 3H^{+}(aq) \tag{5}$$

This newly formed $Al(OH)_{3(s)}$ occurs as sweep floc having large surface area and is active in adsorption of organic compounds and trapping of colloidal particles. The formed floc is separated from aqueous medium by sedimentation or H_2 flotation.

Oxidation of iron electrode in an electrolytic system produces iron hydroxide, $Fe(OH)_n$ where n = 2 or 3. The two mechanisms for the production of metal hydroxide are shown below (Daneshvar et al., 2006):

First Mechanism

$$Anode: Fe(s) \rightarrow Fe^{2+}(aq) + 2e^{-} \tag{6}$$

$$Fe^{2+}(aq) + 2OH^{-}(aq) \rightarrow Fe(OH)_2(s) \tag{7}$$

$$Cathode: 2H_2O(l) + 2e^{-} \rightarrow 2OH^{-}(aq) + H_2(g) \tag{8}$$

$$Overall: Fe(s) + 2H_2O(l) \rightarrow Fe(OH)_2(s) + H_2(g) \tag{9}$$

Second Mechanism

$$Anode: 4Fe(s) \rightarrow 4Fe^{2+}(aq) + 8e^{-} \tag{10}$$

$$4Fe^{2+}(aq) + 10H_2O(l) + O_2(g) \rightarrow 4Fe(OH)_3(s) + 8H^{+}(aq) \tag{11}$$

$$Cathode: 8H^{+}(aq) + 8e^{-} \rightarrow 4H_2(g) \tag{12}$$

$$Overall: 4Fe(s) + 10H_2O(l) + O_2(g) \rightarrow 4Fe(OH)_3(s) + 4H_2(g) \tag{13}$$

Particles that interact with the insoluble metal hydroxides of iron are removed by surface complexation or electrostatic attraction (Daneshvar et al., 2006). The electric field enables electrophoretic migration of particles and increases the tendency of charges in suspension to interact with each other. The summary of electrocoagulation reaction is shown in Figure 4.

The amount of metal ions released from the respective electrode can be calculated using the Faraday equation (Bukhari, 2008).

$$w = \frac{ItM}{ZF} \tag{14}$$

Figure 4. Summary of electrocoagulation reaction
(Holt, 2006)

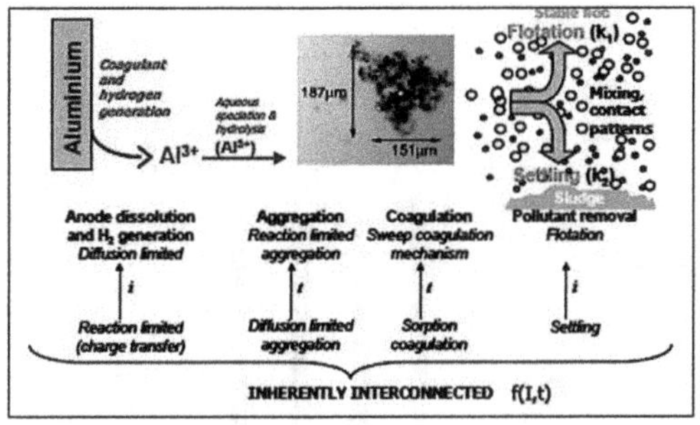

Where w is quantity of electrode material dissolved (g), I is current in amperes (A), t is time (s),M is molecular weight of the electrode material, Z is number of electrons involved in the redox reaction, F is Faraday constant (96500 C/mol of electrons).

The energy consumption during electrocoagulation can be calculated using the equation

$$E = \frac{UIT}{V} \tag{15}$$

Where E is energy consumption (kwh/m^3), U is voltage (V), I is current in amperes (A),T is time in hours (h), V is volume of sample (l)

Gas Bubble Generation

At the start of electrocoagulation process, the sacrificial anode is oxidized while the cathode releases hydrogen gas bubbles. These hydrogen gas bubbles are responsible for the flotation of formed flocs to the surface of the reactor. Gas bubble generation increases with increasing current. At high current, bubble generation rate favors flotation while at low current, bubble generation favors sedimentation (Merzouk et al., 2009).

Power Source

Electrocoagulation can be conducted using both alternating current and direct current. Majority of the work on electrocoagulation has preferred direct current to alternating current because of the cost associated with alternating current. Direct current can be supplied using a direct current power supply. This direct current power supply converts alternating current to direct current and reduces the energy consumption during electrocoagulation process. Direct current is easy to operate, poses none or minimal risk and can be accurately controlled with the aid of a voltmeter and ammeter.

Figure 5. Schematic diagram of electrocoagulation reactor

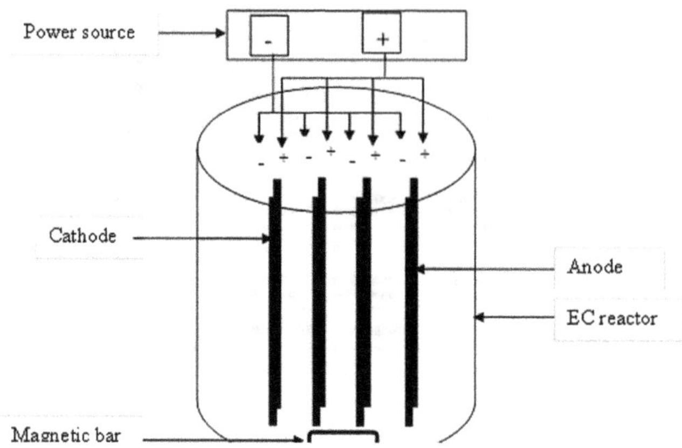

Electrocoagulation Reactor

Different reactors can be used to conduct an electrocoagulation experiment. Majority of the work on electrocoagulation was conducted in the batch study using small beakers. However, few studies have attempted the continuous mode operation with electrocoagulation. In the continuous mode, the reactor design depends on the volume of wastewater to be treated. Different compartments may be included in the reactor design to separate the flocs from the treated water. Generally, the construction of a laboratory scale electrocoagulation reactor is cheap. A schematic diagram of electrocoagulation reactor is depicted in Figure 5.

The design of electrocoagulation reactor can affect process efficiency. The following factors must be considered while designing an electrocoagulation cell (Mollah et al., 2004).

1. Minimization of the IR-drop between the electrodes.
2. Minimization of accumulated O_2 and H_2 gas bubble nucleates at the surface of the electrodes.
3. Favorable distances between electrodes must be investigated

TYPES OF ELECTROCOAGULATION REACTOR

Tall Vertical Plate Reactor

This type of reactor uses electrodes made from flat steel plates with vertical dimension bigger than horizontal dimension (Mameri et al., 1998). The arrangement of the plates can be in a non-conductive case which may be open or closed at the top. In open case, electrical contact is above the solution level while the closed case requires a submerged contact. In a closed case, the flow of liquid in this reactor is usually from the bottom to the top and the cell assembly is done through welding. Sometimes, corrosion locks the plates to the case if the plates are contained in slots within the case. For open case, welding is not required because the case uses contact above the solution. These reactors are effective in electrocoagulation but may be expensive to use.

Figure 6. EC reactor equipped with long horizontal plate electrodes
(Mollah et al., 2004)

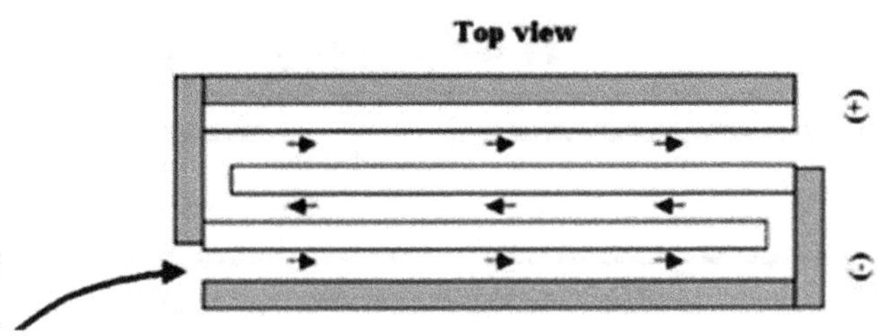

Long Horizontal Plate Reactor

The pattern of arrangement of plate in this reactor is horizontal (Fig. 6). The plates are either placed flat or on edge. A rectangular non-conductive case with slots maintaining spacing is designed to install the plate (Mollah et al., 2004). If the plates are installed in a flat pattern, gas bubbles rise within the stream and causes resistance at the bottom of the upper plate when the solution begins to move between the plates. Oxygen is produced in this arrangement and this causes short circuiting between the plates. Vertical arrangement of the plates enables for easy separation of the phases of the solution (solid, liquid and gas). However, the plates may not be removable once corrosion occurs.

Short Horizontal Plate Reactor

This reactor is designed with a spacer or grooves within the non-conductive case which allows solution to pass through all spaces between plates (Fig. 7). The plates used in this reactor are roughly squared, parallel, horizontal and are separated by the spacers (Mollah et al., 2004). The liquid can pass through both anode and cathode in a single pass. The reactor's configuration enables easy change of polarity. However, the contacts are made within the case which requires some innovation to allow the electrical contacts without welding.

Perforated Plate Reactor

In this reactor, the flow of water passes through the plates rather than between them (Mollah et al., 2004). This pattern of reactor configuration allows for accumulation of solids on the plates which causes short circuiting (Fig. 8). These solids can only be removed by acid washing or disassembly. However, this reactor is effective in treating wastewater but there are legitimate concerns regarding the potential of the assemblage to leak at high pressure.

Solid Tube Reactor

According to (Mollah et al., 2004), solid tube reactor has two concentric tubes used as sacrificial surfaces (Fig. 9). These sacrificial surfaces are contained in a third non-conductive tube. These tubes are separated from each other by non-conductive cords. The voltage and fluid velocity is controlled by the

Figure 7. EC reactor with short horizontal parallel plate electrodes
(Mollah et al., 2004)

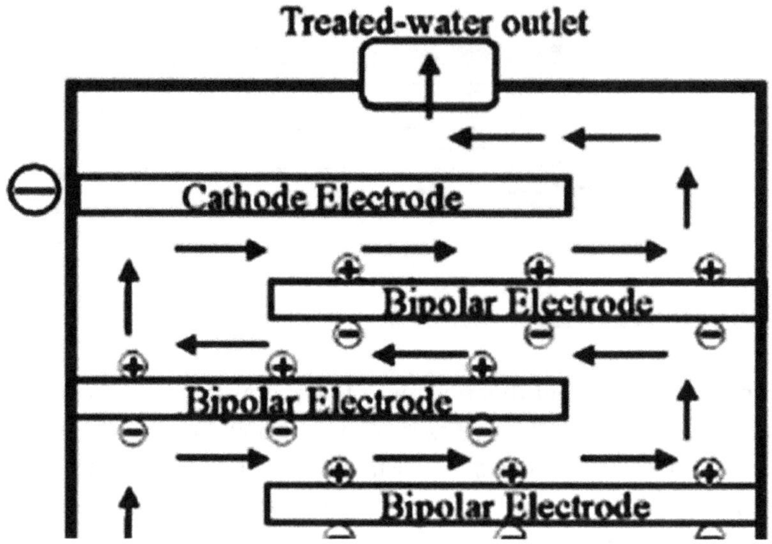

Figure 8. Diagram of perforated reactor
(Mollah et al., 2004)

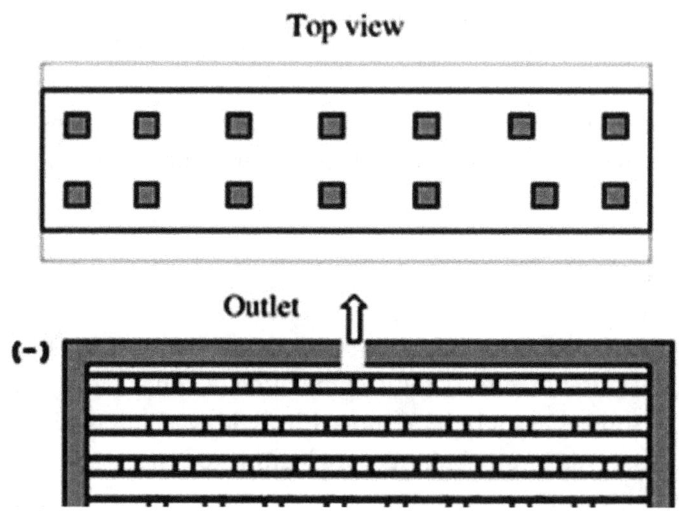

Figure 9. EC solid tube reactors with two eccentric tubes
(Mollah et al., 2004)

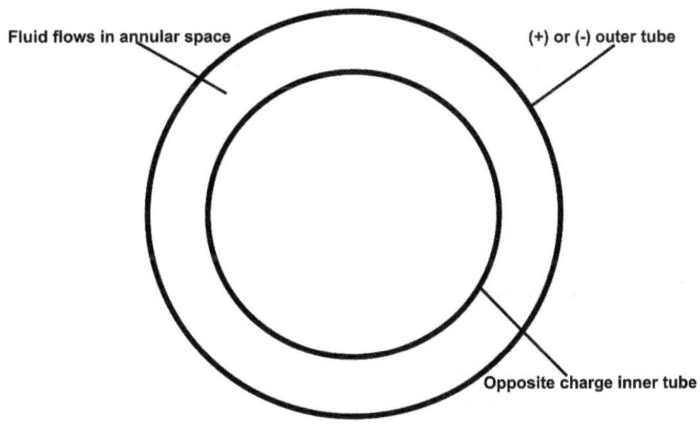

spacing between the annular. One tube serves as the anode and another tube as the cathode. However, leakages from these reactors are difficult to prevent.

Reactor Design

Typical design of electrocoagulation cell is important because several factors affect removal efficiency during EC process. The EC cell can be physically or chemical designed. Whichever method is used, it is important to consider certain factors which will affect process efficiency if not properly taken care of. The physical factors include reactor geometry and reactor scale up while the chemical factor involves electrode material and passivation.

PHYSICAL DESIGN OF EC REACTOR

The construction of EC cell requires technical knowledge of electrochemistry because the design of EC affects its operation and efficiency.

Reactor Geometry

Reactor geometry obviously deals with the properties of the reactor and how the parameters can effectively relate to increase efficiency within the reactor. The formation of floc, flow of liquid, settling characteristics, bubble path and flotation effectiveness are all affected by the geometry of the reactor. If the flow of solution is from the bottom of the reactor, the direction of flotation will differ from a reactor which solution flows from the top.

Reactor Scale Up

The reactor scale up is important because some parameters like surface area to volume ratio (As/V) are affected by the size of the reactor. The electrode area influences current density, rate of bubble and coagulant dosing. It is important to include this in the design of EC reactor. The efficiency of EC has been satisfactory in batch scale studies. It is imperative to scale-up the system in pilot stage. However, this will require overcoming some of the enumerated challenges.

CHEMICAL DESIGN OF EC REACTOR

The effectiveness and interactions of particles during electrocoagulation is further affected by chemical factors like electrode material and passivation.

Electrode Materials

The choice of electrode material for electrocoagulation process is very important. Several reported literatures on electrocoagulation have opted for aluminum and iron electrode because these two electrodes are readily available, effective and cheap. However, other electrode materials have also been used like magnesium, titanium, zinc etc. The later electrodes are expensive and not readily available. However, one chemical factor that is used to categorize these electrodes is that the electrodes must have adequate mechanical strength to withstand chemical and physical attack from the electrolytes, reactants and products.

Passivation

Passivation of electrode is one of the major challenges of electrocoagulation process. Passivation occurs as a result of the deposition of a layer of oxide on the surface of electrode materials reducing the efficiency of the electrocoagulation process. It prevents metal ion dissolution and electron transfer. In addition, the amount of metal ions released into the water sample reduces. Passivation normally occurs with the cathode while the anode undergoes electrochemical corrosion. Passivation of electrode can be reduced during electrocoagulation process by

1. Alternating the polarity of the electrode materials
2. Cleaning of the electrodes after several runs of experiment with appropriate cleaning agents.
3. Drying the electrodes after cleaning to remove any residual surface impurity.

Advantages and Disadvantages of Electrocoagulation

Electrocoagulation has many advantages over other treatment methods. However, there are also some demerits associated with electrocoagulation worth mentioning. The advantages of EC include:

1. Electrochemical setup is made up of simple equipment
2. It produces small amount of sludge as compared to conventional coagulation. In addition, the sludge readily settle.
3. EC removes smallest colloidal particles because the strong electric field sets all particles in motion.
4. It eliminates the addition of chemicals and also limits the chances of secondary pollution.
5. The release of gas bubbles from the cathode enables flotation of pollutants to the surface of the reactor where it is concentrated and collected.
6. The maintenance of electrochemical set up is low since all the parts are stable.
7. In rural areas, a solar panel can be used to operate an electrochemical setup in the absence of electricity.

The demerits of EC are highlighted below:

1. EC requires periodic cleaning and replacement of the sacrificial anode after a series of experiments
2. Passivation caused by the formation of impermeable oxide on the surface of the cathode can affect EC efficiency
3. Wastewater sample requires high conductivity for effective coagulation to occur.
4. Residue concentration of electrode material in the effluent may increase after treatment especially with aluminum electrode

OPERATING PARAMETERS

There are different operating parameters used in electrocoagulation process. Many studies have reported the role of these operating parameters in the efficiency of electrocoagulation process. Electrocoagulation has been successfully applied in the treatment of various wastewaters. The performance of EC is summarized in Table 1, indicating the effectiveness of EC.

Several important parameters control the performance of EC. These parameters are essential for pollutant removal based on pollutant chemistry. Below is a review of some important operating parameters.

Current Density

Current density is current per unit surface area of the electrode (anode). It is a parameter which controls the dissolution of the sacrificial anode on one hand and the formation of hydrogen bubble on the other (Nanseu-Njiki et al., 2009). Current density directly affects process performance and operating cost and determines the amount of ions released from the respective electrodes (Chen, 2004). Literature on electrocoagulation has proved that current density is an important parameter and is responsible for the oxidation of the sacrificial anode and generation of gas bubbles. The degree of anodic dissolution of the electrode material increases with increasing current density (Feng et al., 2007).

Table 1. Comparison of electrocoagulation performance

Wastewater	Pollutant	Electrode	Parameters	Efficiency (%)	References
Textile	Color COD Turbidity	Cathode: aluminium sheet Anode: SAE 1005 steel	pH, Current density Contact time	Color = 86% COD = 59% Turbidity = 82%	(Núñez et al., 2019)
Well water	Fluoride Arsenic	Aluminum Iron	pH Current density Contact time	Fluoride = 85.68% Arsenic = 100%	(López-Guzmán et al., 2019)
Synthetic	Phosphates	Cathode: Anode: Aluminum	pH Current density Contact time Concentration	Phosphate = >95%	(Dura & Breslin, 2019b)
Petroleum refinery	Selenium	Iron	Contact time Initial concentration	Selenium = 90%	(Hansen et al., 2019)
Synthetic	Phosphate Orange II Zinc ions	Iron Stainless steel 420	pH Current density Contact time Initial concentration	Phosphate = 92% Orange II = 98% Zn^{2+} = 99% Phosphate = 86% Orange II = 98% Zn^{2+} = 99%	(Dura & Breslin, 2019a)
Natural waters	Fluoride	Aluminium	Initial concentration Temperature Conductivity pH	90%	(Ben Grich et al., 2019)
Mining waters	Sulfate	Iron	pH	> 60%	(Mamelkina et al., 2017)

Sample pH

The pH of a solution in electrocoagulation controls the process performance and the solubility of the analyte. The pH of an electrocoagulation process depends mainly on the nature of the contaminant (Chen, 2004). For instance, pH determines the ionic form of boron (borate ions). At low pH, boric acid is the dominant form of boron in solution while at high pH, the borate ions dominate (Cengeloglu et al., 2008). Figure 10 shows the forms of boron at different pH values.

Inter-Electrode Spacing

Inter-electrode spacing is the gap between electrodes. Some literatures have reported the effects of inter-electrode spacing in electrocoagulation process. Inter-electrode spacing is a measure of how the movement of the released metal ions affects process efficiency. It is important to maintain an appropriate spacing between electrodes during electrocoagulation. The appropriate spacing can be determined

Figure 10. Distribution of H_3BO_3/H_2BO_3 in water as a function of pH
(Busch et al., 2004)

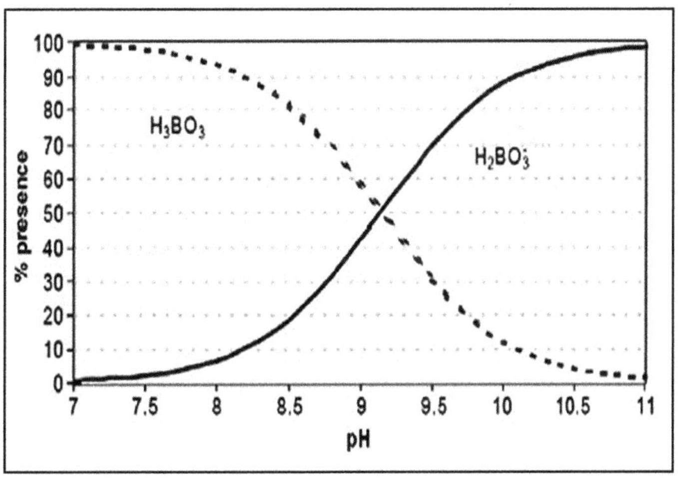

through experimental process. When the inter-electrode spacing doesn't favor the reaction, electrical resistance occurs and reaction rate reduces.

Conductivity

Wastewater conductivity is an important parameter in electrocoagulation. Wastewater conductivity indicates the ability of water to conduct an electric current. When wastewater conductivity is high, the rate of oxidation of electrode material is rapid. Wastewater conductivity is measured in micromhos per centimeter (µmhos/cm) or microsiemens per centimeter (µS/cm). Wastewater conductivity is a function of the amount of dissolved solids present in the sample. The conductivity of different wastewaters varies and can be easily measured using a conductivity meter. Knowledge of sample conductivity is important particularly in electrochemical studies due to the rapid reactions which occur at high conductivity. However, wastewaters with low conductivity can be enhanced through the addition of some electrolytes to increase the electrical conduction. In addition, energy consumption reduces when wastewater conductivity is high because reactions are rapid and spontaneous (Akbal & Camcı, 2011).

Temperature

Temperature is an important parameter in electrocoagulation and indicates different energy levels during a reaction process. Sample temperature can vary across different wastewaters. Some wastewater contaminants are poor conductors of heat at room temperature and require higher temperature. Temperature also gives a better knowledge of the adsorption and thermodynamic process of electrocoagulation system. Wastewater temperature can be easily measured using a thermometer.

Figure 11. Close circuit: Voltage and current

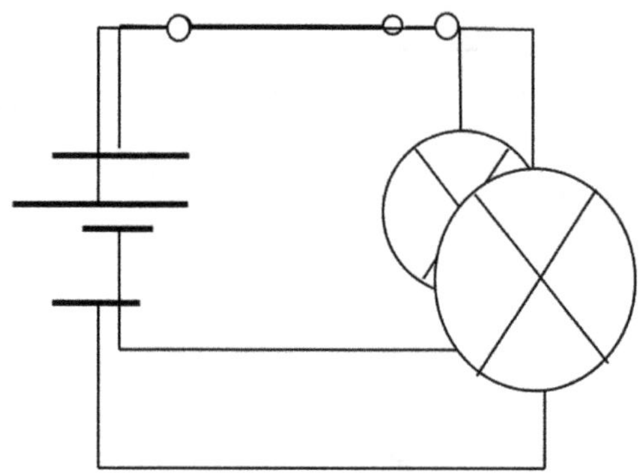

Figure 12. Open circuit: voltage but no current

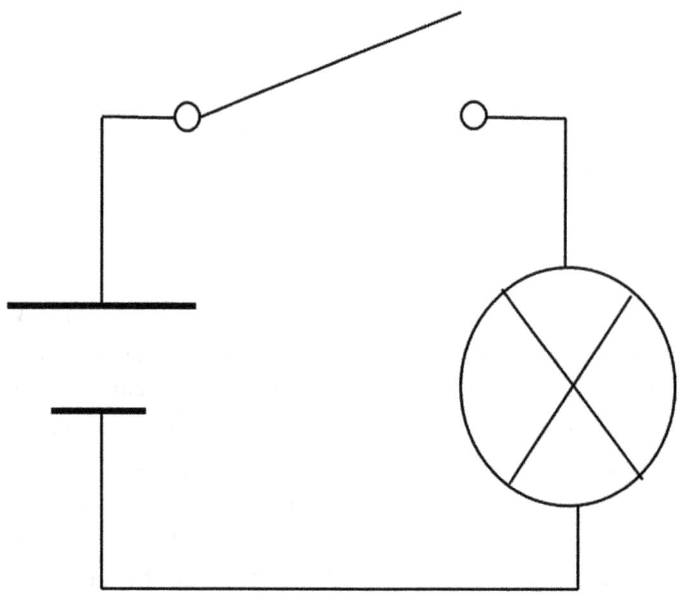

Figure 13. The cell is removed: No voltage and no current

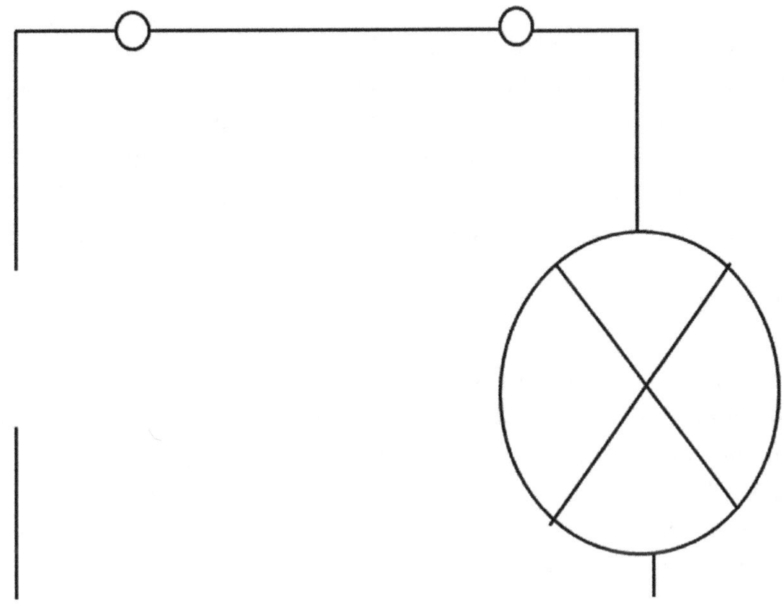

Cell Voltage

The use of current density in electrocoagulation cell is determined by the measured potential. The electric potential, often referred to as "push or force of electricity" is an important parameter which drives the current. The potential of an electrocoagulation cell can be referred as a measure of the energy carried by a charge which is responsible for the flow of current. In many EC cells, it is possible to have a potential without a current but current cannot flow without the potential. The importance of potential is further illustrated in Figures 11-13. Figure 11 shows a cell with a closed circuit which allows the current to flow. Due to the closed switch, potential and current work together. Figure 12 shows a cell with an open circuit. The cell possesses only the potential with no current. Due to the opening of the switch, the circuit is broken and current cannot flow. The cell only works with potential. Figure 13 shows the importance of the cell. When the cell is removed, there is no source of voltage and current cannot flow. The Figures illustrates the importance of cell potential in electrocoagulation.

The potential required to get a specific current density is defined as the sum of three components expressed below:

$$\eta AP = \eta k + \eta Mt + \eta IR \tag{16}$$

Where ηAP is the applied overpotential (V), ηk is the kinetic overpotential (V), ηMt is the concentration overpotential (V), ηIR is the overpotential caused by solution resistance or IR drop (V).

APPLICATION OF ELECTROCOAGULATION IN WASTEWATER TREATMENT

Many studies have successfully used electrocoagulation in wastewater treatment. These studies were conducted with different wastewaters and different operating parameters. As the study of electrocoagulation continues to gain wider acceptance, more operating parameters are being discovered to improve the process. Various wastewaters have different composition and characteristics. Therefore optimum conditions of operating parameters differ from one wastewater to another. Thus in the study of electrocoagulation, it is important to understand the characteristics and composition of the wastewater sample.

Produced Water

Produced water is a by-product of oil well operation which is trapped in underground formation and brought to the surface along with oil and gas during drilling. It is separated from crude oil above ground in an oil/water separator. As the demand for crude oil increases, the production of produced water increases proportionately. Produced water has unique characteristics which differentiates it from other wastewater. Produced water composition varies and depends on geological conditions, storage formation and type of recovery method (Woodall et al., 2001). It is the largest waste-stream of oil and gas exploration. The chemical composition of produced water varies when compared with surface water because produced water is constrained within an aquifer for a long time (Wemedo et al., 2009). Produced water presents several environmental challenges because it is produced in large volumes and often contains constituents in concentrations and forms that are toxic to receiving system biota (Veil et al., 2004). The chemicals used during oil and gas production are complex mixtures of various molecular compounds which can also increase the potential toxicity of produced water (Cline, 1998). Some pollutants in this complex wastewater have been successfully eliminated by electrocoagulation. Such pollutants include boron. Boron is commonly found in produced water. This compound poses severe difficulty to various treatment methods due to its chemistry (Ezechi et al., 2012). However, EC was suitably used for the elimination of boron from produced water. Isa et al. (2014) evaluated boron removal from produced water using aluminum plate electrodes in electrocoagulation and varied four operating parameters (pH, current density, concentration and treatment time). Authors observed that boron removal from produced water under the optimum conditions reached 98%. The authors found that pH significantly influenced boron removal using EC. Ezechi et al. (2015) evaluated boron removal from produced water using iron plate electrode in EC set-up. Boron removal reached 97.6% at the optimum conditions.

Oily Wastewater

Oily wastewater generation is one of the largest wastestream of oil and gas exploration. Oily wastewater is a highly contaminated wastewater consisting of various organic, inorganic and aromatic contaminants including polycyclic aromatic hydrocarbons (PAHs) which are toxic. (Keramati & Ayati, 2019) investigated the treatment of petroleum wastewater using a combined electrocoagulation and photocatalytic process. The authors found that at optimum electrocoagulation conditions (current density of 20 mA/cm2, pH of 8.5, COD concentration of 900 mg/L and NaCl concentration of 0.5 g/L), COD removal reached 94% at contact time of 69 minutes. (Changmai et al., 2019) investigated the treatment of oily wastewater using a combined electrocoagulation/microfiltration process. Oily wastewater was pre-treated by electrocoagulation. The authors found that electrocoagulation adequately reduced oil and grease from 35 mg/l to 10.2 mg/L in contact time of 20 minutes. (Hansen et al., 2019) evaluated the treatment of petroleum wastewater using electrocoagulation. Sacrificial iron electrodes were used to generate electrocoagulants. The authors found that selenium removal from petroleum wastewater reached 90% at contact time of 6 hours. These studies indicate that electrocoagulation has the capability to eliminate different pollutants from oily wastewater.

Textile Wastewater

Textile wastewater is a complex industrial wastewater consisting of high contaminants and extreme salinity concentrations. Textile wastewater is difficult to degrade through aerobic biological process. However, anaerobic and physico-chemical processes have demonstrated good treatment capacity. (Bener et al., 2019) investigated the treatment of real textile wastewater using electrocoagulation. Several parameters such as current density, pH and contact time were varied. Aluminum and iron electrodes were applied to the electrocoagulation process. Total organic carbon (TOC) removal was monitored from the process. The authors found that at optimum condition (current density 25 mA/cm^2 and pH 5), TOC removal reached 42.5%. (Zazou et al., 2019) evaluated the performance of electrocoagulation in combination with other physico-chemical processes for the treatment of textile wastewater. The authors found that combination of sequential electrocoagulation/electroflotation was effective and achieved 97%, 100% and 100% TOC, turbidity and color removal, respectively. (Bilińska et al., 2019) evaluated the combination of electrocoagulation and ozone for the treatment of textile wastewater and found that over 95% of the reactive black 5 (RB5) dye was removed at contact time less than 18 minutes. (Verma, 2017) used Fe-Al composite electrodes for the treatment of different types of textile wastewaters. The influence of different variables such as pH, reaction time, voltage, current density and electrode spacing were evaluated. The author found that COD and color removal exceeded 90% at optimum conditions.

Tannery Wastewater

Tannery wastewater is a highly polluted and complex wastewater consisting of different compositions of chromium compounds, heavy metals and high organic contents. Treatment of tannery wastewater is a complex process due to the reactions and states of chromium compounds. (Deveci et al., 2019) evaluated the treatment of tannery wastewater through the integration of electrocoagulation and fungi. Different variables such as electrode combination, current density and electrolysis time were investigated. The authors found that COD and chromium (VI) removal reached 63.8% and 90% using EC alone but increased to 96% and 97% using the integrated system. (Moradi & Moussavi, 2019) enhanced tannery wastewater treatment through the integration of electrocoagulation and advance oxidation process (AOP) at different operating conditions. The authors found that COD removal reached 99% in the integrated system.

Laundry Wastewater

Laundry wastewater is generated in large amount and constitutes high environmental danger due to the additives and detergents applied in during the laundry process. (Dimoglo et al., 2019) evaluated the influence of treatment time, pH, temperature and current density on color, turbidity and surfactant removal during the purification of laundry wastewater using a combined electrocoagulation/electroflotation process. The authors found that at optimum conditions of 5.26 mA/cm^2 current density, pH of 5.5

and contact time of 5 minutes, pollutant removal reached 90%. The author found that the purified water from the electrocoagulation/electroflotation process can be reused.

Sewage Wastewater

Sewage wastewater is widely produced in large volume. Most of the sewage wastewaters are channeled to sewage treatment plants. However, other methods are employed in the treatment of sewage wastewater in environments where sewage treatment plants are not in place. (Khandegar et al., 2018) compared the performance of a plane and punched aluminum electrodes on the treatment of sewage wastewater. The authors investigated the influence of electrode spacing, voltage, contact time and stirring speed on the removal of biochemical oxygen demand (BOD) and total dissolved solids (TDS). The authors found that the punched electrodes were superior to the plane electrodes in terms of removal efficiency and operating cost. BOD and TDS removal with the plane electrode were 69% and 78%, respectively at contact time of 120 minutes while it was 74% and 78% at contact time of 105 minutes using a single hole punched electrode. The removal efficiency of BOD and TDS in the punched electrode increased with increasing the holes on the electrode surface. BOD and TDS removal at 2, 3 and 4 holes were 74% and 78% at 90 minutes, 73% and 79% at 75 minutes and 76% and 80% at 60 minutes, respectively. The results indicate that increasing the holes on the surface of the electrode reduced the contact time. Furthermore, the energy and electrode consumption were higher with the plane electrode than the punched electrode. This further indicated that application of the punched electrode in electrocoagulation could lead to cost reduction.

Drinking Water

Electrocoagulation has been successfully applied in the purification of drinking water and tap water. Bennajah et al. (2009), investigated the removal of fluoride from Morocco drinking water by electrocoagulation/electroflotation in an electrochemical external loop airlift reactor. A batch and continuous mode study was conducted with aluminum electrodes in this study. Current density, pH, treatment time and concentration were varied in this study. The result indicated that a low fluoride concentration of 1.5 mg/L can be achieved from an initial fluoride concentration of 10 and 20 mg/L at current density of 6 mA/cm^2 and pH 5. Malakootian et al. (2010), evaluated the performance of electrocoagulation process in hardness removal using iron rod electrodes in drinking water. The parameters investigated were pH, voltage and treatment time. A removal efficiency of 98.2% and 97.4% was observed for calcium and total hardness removal respectively at voltage of 12V and reaction time of 60 minutes. Removal efficiency increased with increasing cell voltage. Increase in pH increased the removal efficiency. This is because pH has a significant effect on forming metal hydroxide species and removal mechanism of ion and pollutants. Ghosh et al. (2008), studied the removal of Fe(II) from tap water by electrocoagulation technique using aluminum electrode. The operating parameters varied were current density and inter-electrode spacing while pH, conductivity and salt concentration were kept constant. The study showed that when current density was increased, removal efficiency improved while removal efficiency was highest for the lowest inter-electrode distance. A removal efficiency of 99.2% was observed from an initial Fe(II) concentration of 25 mg/L after a 35 minute treatment time. Ricordel et al. (2010), also investigated the viability of using electrocoagulation-electroflotation process as a surface water treatment process for industrial uses. Aluminum electrode was used for this study. The authors noted that 50% of the suspended matter present in the sample was removed within 30 minutes. While electrocoagulation can achieve high pollutant removal, its efficiency can decrease in the presence of scale forming species. Schulz et al. (2009), investigated the ability of electrocoagulation to remove scale forming species from industrial process water in bench and pilot scale. Aluminum and iron electrodes were used in this study at inter-electrode spacing of 4 mm. The study observed that there was frequent formation of gel on the surfaces of aluminum electrode more than iron electrode in the pilot scale study. This was attributed to aluminum dosing from the electrodes which was common in the bench scale than in the pilot scale. Silica removal in excess of 80% was achieved with electrocoagulation but less than 40% calcium and magnesium was removed.

Industrial Wastewater

Industrial processes and manufacturing generate large volume of wastewater which are highly polluted and require adequate treatment before disposal. Electrocoagulation has been widely applied in the elimination of various pollutants in industrial wastewater. Apaydin et al. (2009) investigated the treatment of tannery wastewater with electrocoagulation. Iron plate electrode was used in the batch study. COD and sulfide removal were used to evaluate the efficiency of electrocoagulation process. A COD and sulfide removal efficiency of 46% and 90% were observed within 5 minutes respectively. Solak et al. (2009), examined suspended solids and turbidity removal from marble processing wastewaters by electrocoagulation. The study focused on the comparison of electrode material and electrode connection systems. Aluminum and iron electrodes connected in parallel and serial connections were used. pH, current density and electrolysis time were investigated. A 100% removal efficiency was reported for suspended solids with electrocoagulation using monopolar aluminum electrodes in parallel and serial connections at the optimum conditions of pH 9, current density 15 A/m^2, and electrolysis time of 2 minutes. For iron electrode, a suspended solid removal efficiency of 99.86% and 99.94% was observed for serial and parallel connections respectively at optimum conditions of pH 8 and electrolysis time of 2 minutes.

COMPARISON OF TECHNOLOGICAL COST

Several factors are considered while evaluating the cost implications of a wastewater treatment technology. These factors include environmental and health implications, energy consumption, material consumption and sludge disposal. In comparison to other treatment technologies for boron removal from wastewater, Isa et al. (2014) reported an operational cost of MYR 0.88/m^3 at optimum condition for electrocoagulation while Turek et al. (2007) and Chong et al. (2009) reported \$1.27/m^3 and 6.33/m^3 for electrodialytic treatment of boron containing wastewater and adsorption flocculation treatment process, respectively. This analysis indicates that electrocoagulation has a cost advantage over some physical treatment process.

ELECTROCOAGULATION AND COAGULATION

Chemical coagulation and flocculation share similar pollutant removal mechanism with electrocoagulation. However, the primary difference between chemical coagulation and electrocoagulation is the method of coagulant addition. Whereas in electrocoagulation, coagulants are generated over an extended period from the oxidation of the sacrificial anode in the presence of applied current, coagulants are added directly from chemicals such as aluminum sulfate or ferric chloride in chemical coagulation. Recently, application of natural coagulants in chemical coagulation process has intensified. In electrocoagulation, the flocs precipitated are effectively dispersed, causing higher adsorptive removal capacity. Chemical coagulation utilizes settling for floc separation while electrocoagulation utilizes settling and flotation for floc separation due to the release of air bubbles at the cathode (Holt et al., 2002). The electric field improves the flocculation of small colloidal particles in electrocoagulation, producing relatively lower amount of sludge than in chemical coagulation. Furthermore, generation of secondary pollutants is more associated with chemical coagulation than in electrocoagulation due to the effective control of applied current in EC. Flocs from electrocoagulation process are more settleable and primarily composed of metallic oxides/hydroxides with less bound water and highly resistant to acid. They are more stable and can be separated through filtration (Avsar et al., 2007). Chemical coagulation is sensitive to pH changes with optimum pH in the range of 6-7 while electrocoagulation has a pH neutralization effect which allows it to operate at wider pH range of 4-9.

CONCLUSION

Electrocoagulation has emerged as a suitable technology for the treatment of polluted water and wastewater. Its principle largely depends on the interaction of colloidal particles in suspension with the coagulating agents. Its success and widespread acceptance arises from its low cost of operation, ease of operation and high efficiency. Various electrode materials have also been used in electrocoagulation. These materials are cheap and readily available, making electrocoagulation a promising technology. The operating parameters control the rate of system efficiency. At optimum conditions, electrocoagulation achieves higher results. While electrocoagulation is a suitable alternative to existing wastewater treatment technologies, care must be taken in its operation to reduce passivation caused by scale forming species on the surface of electrodes. Passivation reduces the oxidation of sacrificial anodes and causes the release of low coagulating agents within the electrocoagulation set-up. This could affect the rate of flocculation. However, passivation can be reduced by alternating the electrodes and washing periodically. Therefore, electrocoagulation has become a reliable wastewater treatment technology due to its numerous advantages.

ACKNOWLEDGMENT

The authors wish to thank Universiti Teknologi Malaysia (UTM) for supporting this study.

REFERENCES

Akbal, F., & Camcı, S. (2011). Copper, chromium and nickel removal from metal plating wastewater by electrocoagulation. *Desalination*, *269*(1), 214–222. doi:10.1016/j.desal.2010.11.001

Apaydin, Ö., Kurt, U., & Gönüllü, M. (2009). An investigation on the treatment of tannery wastewater by electrocoagulation. *Global NEST Journal*, *11*(4), 546–555.

Avsar, Y., Kurt, U., & Gonullu, T. (2007). Comparison of classical chemical and electrochemical processes for treating rose processing wastewater. *Journal of Hazardous Materials*, *148*(1), 340–345. doi:10.1016/j.jhazmat.2007.02.048 PMID:17386973

Babu, R. R., Bhadrinarayana, N., Begum, K. M. M. S., & Anantharaman, N. (2007). Treatment of tannery wastewater by electrocoagulation. *Journal of the University of Chemical Technology and Metallurgy*, *42*(2), 201–206.

Ben Grich, N., Attour, A., Le Page Mostefa, M., Guesmi, S., Tlili, M., & Lapicque, F. (2019). Fluoride removal from water by electrocoagulation: Effect of the type of water and the experimental parameters. *Electrochimica Acta*, *316*, 257–265. doi:10.1016/j.electacta.2019.05.130

Bener, S., Bulca, Ö., Palas, B., Tekin, G., Atalay, S., & Ersöz, G. (2019). Electrocoagulation process for the treatment of real textile wastewater: Effect of operative conditions on the organic carbon removal and kinetic study. *Process Safety and Environmental Protection*, *129*, 47–54. doi:10.1016/j.psep.2019.06.010

Bennajah, M., Gourich, B., Essadki, A., Vial, C., & Delmas, H. (2009). Defluoridation of morocco drinking water by electrocoagulation/electroflottation in an electrochemical external-loop airlift reactor. *Chemical Engineering Journal*, *148*(1), 122–131. doi:10.1016/j.cej.2008.08.014

Bilińska, L., Blus, K., Gmurek, M., & Ledakowicz, S. (2019). Coupling of electrocoagulation and ozone treatment for textile wastewater reuse. *Chemical Engineering Journal*, *358*, 992–1001. doi:10.1016/j.cej.2018.10.093

Brink, A., Sheridan, C., & Harding, K. (2018). Combined biological and advance oxidation processes for paper and pulp effluent treatment. *South African Journal of Chemical Engineering*, *25*, 116–122. doi:10.1016/j.sajce.2018.04.002

Bryjak, M., Wolska, J., & Kabay, N. (2008). Removal of boron from seawater by adsorption–membrane hybrid process: Implementation and challenges. *Desalination*, *223*(1), 57–62. doi:10.1016/j.desal.2007.01.202

Bukhari, A. A. (2008). Investigation of the electro-coagulation treatment process for the removal of total suspended solids and turbidity from municipal wastewater. *Bioresource Technology*, *99*(5), 914–921. doi:10.1016/j.biortech.2007.03.015 PMID:17499502

Busch, M., Mickols, W. E., Jons, S., Redondo, J., & De Witte, J. (2004). Boron removal in sea water desalination. *International Desalination and Water Reuse Quarterly*, *13*(4), 25.

Casson, L. W., & Bess, J. W. (2006). On-site sodium hypochlorite generation. *Proceedings of the Water Environment Federation*, 2006(5), 6335-6352.

Cengeloglu, Y., Arslan, G., Tor, A., Kocak, I., & Dursun, N. (2008). Removal of boron from water by using reverse osmosis. *Separation and Purification Technology*, *64*(2), 141–146. doi:10.1016/j.seppur.2008.09.006

Changmai, M., Pasawan, M., & Purkait, M. K. (2019). Treatment of oily wastewater from drilling site using electrocoagulation followed by microfiltration. *Separation and Purification Technology*, *210*, 463–472. doi:10.1016/j.seppur.2018.08.007

Chen, G. (2004). Electrochemical technologies in wastewater treatment. *Separation and Purification Technology*, *38*(1), 11–41. doi:10.1016/j.seppur.2003.10.006

Chong, M. F., Lee, K. P., Chieng, H. J., & Syazwani Binti Ramli, I. I. (2009). Removal of boron from ceramic industry wastewater by adsorption–flocculation mechanism using palm oil mill boiler (pomb) bottom ash and polymer. *Water Research*, *43*(13), 3326–3334. doi:10.1016/j.watres.2009.04.044 PMID:19487007

Cline, J. (1998). *Treatment and discharge of produced water for deep offshore disposal.*

Daneshvar, N., Oladegaragoze, A., & Djafarzadeh, N. (2006). Decolorization of basic dye solutions by electrocoagulation: An investigation of the effect of operational parameters. *Journal of Hazardous Materials*, *129*(1), 116–122. doi:10.1016/j.jhazmat.2005.08.033 PMID:16203084

Deveci, E. Ü., Akarsu, C., Gönen, Ç., & Özay, Y. (2019). Enhancing treatability of tannery wastewater by integrated process of electrocoagulation and fungal via using rsm in an economic perspective. *Process Biochemistry*, *84*, 124–133. doi:10.1016/j.procbio.2019.06.016

Dimoglo, A., Sevim-Elibol, P., Dinç, Ö., Gökmen, K., & Erdoğan, H. (2019). Electrocoagulation/electroflotation as a combined process for the laundry wastewater purification and reuse. *Journal of Water Process Engineering*, *31*, 100877. doi:10.1016/j.jwpe.2019.100877

Dura, A., & Breslin, C. B. (2019a). Electrocoagulation using stainless steel anodes: Simultaneous removal of phosphates, orange ii and zinc ions. *Journal of Hazardous Materials*, *374*, 152–158. doi:10.1016/j.jhazmat.2019.04.032 PMID:30999138

Dura, A., & Breslin, C. B. (2019b). The removal of phosphates using electrocoagulation with al–mg anodes. *Journal of Electroanalytical Chemistry*, *846*, 113161. doi:10.1016/j.jelechem.2019.05.043

Emamjomeh, M. M., & Sivakumar, M. (2009). Review of pollutants removed by electrocoagulation and electrocoagulation/flotation processes. *Journal of Environmental Management*, *90*(5), 1663–1679. doi:10.1016/j.jenvman.2008.12.011 PMID:19181438

Ezechi, E., Kutty, S., Muda, K., & Yaqub, A. (2019). A comparative evaluation of an integrated hybrid bioreactor treating industrial wastewater. *Journal of Water Process Engineering*, *31*, 100805. doi:10.1016/j.jwpe.2019.100805

Ezechi, E. H., & Isa, M. H. (2015). Electrochemical removal of boron from produced water and recovery. *Journal of Environmental Chemical Engineering*, *3*(3), 1962–1973. doi:10.1016/j.jece.2015.05.015

Ezechi, E. H., Isa, M. H., Kutty, S. R., & Sapari, N. B. (2011). *Boron recovery, application and economic significance: A review.* Paper presented at the National Postgraduate Conference (NPC), 2011. 10.1109/NatPC.2011.6136374

Ezechi, E. H., Isa, M. H., & Kutty, S. R. B. M. (2012). Boron in produced water: Challenges and improvements: A comprehensive review. *Journal of Applied Sciences (Faisalabad), 12*(5), 402–415. doi:10.3923/jas.2012.402.415

Ezechi, E. H., Isa, M. H., Kutty, S. R. M., & Yaqub, A. (2014). Boron removal from produced water using electrocoagulation. *Process Safety and Environmental Protection, 92*(6), 509–514. doi:10.1016/j.psep.2014.08.003

Feng, Q., Li, X., Cheng, Y., Meng, L., & Meng, Q. (2007). Removal of humic acid from groundwater by electrocoagulation. *Journal of China University of Mining and Technology, 17*(4), 513–520. doi:10.1016/S1006-1266(07)60136-9

Ghosh, D., Solanki, H., & Purkait, M. (2008). Removal of fe (ii) from tap water by electrocoagulation technique. *Journal of Hazardous Materials, 155*(1), 135–143. doi:10.1016/j.jhazmat.2007.11.042 PMID:18164128

Hansen, H. K., Peña, S. F., Gutiérrez, C., Lazo, A., Lazo, P., & Ottosen, L. M. (2019). Selenium removal from petroleum refinery wastewater using an electrocoagulation technique. *Journal of Hazardous Materials, 364*, 78–81. doi:10.1016/j.jhazmat.2018.09.090 PMID:30339935

Holt, P. K. (2006). Electrocoagulation: Unravelling and synthesising the mechanisms behind a water treatment process. (Thesis, Department of Chemical Engineering, University of Sydney, Australia).

Holt, P. K., Barton, G. W., Wark, M., & Mitchell, C. A. (2002). A quantitative comparison between chemical dosing and electrocoagulation. *Colloids and Surfaces. A, Physicochemical and Engineering Aspects, 211*(2), 233–248. doi:10.1016/S0927-7757(02)00285-6

Isa, M. H., Ezechi, E. H., Ahmed, Z., Magram, S. F., & Kutty, S. R. M. (2014). Boron removal by electrocoagulation and recovery. *Water Research, 51*, 113–123. doi:10.1016/j.watres.2013.12.024 PMID:24412846

Jiang, J.-Q., Graham, N., André, C., Kelsall, G. H., & Brandon, N. (2002). Laboratory study of electrocoagulation–flotation for water treatment. *Water Research, 36*(16), 4064–4078. doi:10.1016/S0043-1354(02)00118-5 PMID:12405415

Keramati, M., & Ayati, B. (2019). Petroleum wastewater treatment using a combination of electrocoagulation and photocatalytic process with immobilized ZNO nanoparticles on concrete surface. *Process Safety and Environmental Protection, 126*, 356–365. doi:10.1016/j.psep.2019.04.019

Khandegar, V., Acharya, S., & Jain, A. K. (2018). Data on treatment of sewage wastewater by electrocoagulation using punched aluminum electrode and characterization of generated sludge. *Data in Brief, 18*, 1229–1238. doi:10.1016/j.dib.2018.04.020 PMID:30050967

Kim, K.-W., Shon, W.-J., Oh, M.-K., Yang, D., Foster, R. I., & Lee, K.-Y. (2019). Evaluation of dynamic behavior of coagulation-flocculation using hydrous ferric oxide for removal of radioactive nuclides in wastewater. *Nuclear Engineering and Technology*, *51*(3), 738–745. doi:10.1016/j.net.2018.11.016

López-Guzmán, M., Alarcón-Herrera, M. T., Irigoyen-Campuzano, J. R., Torres-Castañón, L. A., & Reynoso-Cuevas, L. (2019). Simultaneous removal of fluoride and arsenic from well water by electrocoagulation. *The Science of the Total Environment*, *678*, 181–187. doi:10.1016/j.scitotenv.2019.04.400 PMID:31075584

Malakootian, M., Mansoorian, H., & Moosazadeh, M. (2010). Performance evaluation of electrocoagulation process using iron-rod electrodes for removing hardness from drinking water. *Desalination*, *255*(1), 67–71. doi:10.1016/j.desal.2010.01.015

Mamelkina, M. A., Cotillas, S., Lacasa, E., Sáez, C., Tuunila, R., Sillanpää, M., & Rodrigo, M. A. (2017). Removal of sulfate from mining waters by electrocoagulation. *Separation and Purification Technology*, *182*, 87–93. doi:10.1016/j.seppur.2017.03.044

Mameri, N., Yeddou, A., Lounici, H., Belhocine, D., Grib, H., & Bariou, B. (1998). Defluoridation of septentrional Sahara water of North Africa by electrocoagulation process using bipolar aluminium electrodes. *Water Research*, *32*(5), 1604–1612. doi:10.1016/S0043-1354(97)00357-6

Merzouk, B., Gourich, B., Sekki, A., Madani, K., & Chibane, M. (2009). Removal turbidity and separation of heavy metals using electrocoagulation-electroflotation technique: A case study. *Journal of Hazardous Materials*, *164*(1), 215–222. doi:10.1016/j.jhazmat.2008.07.144 PMID:18799259

Mollah, M. Y. A., Morkovsky, P., Gomes, J. A. G., Kesmez, M., Parga, J., & Cocke, D. L. (2004). Fundamentals, present and future perspectives of electrocoagulation. *Journal of Hazardous Materials*, *114*(1), 199–210. doi:10.1016/j.jhazmat.2004.08.009 PMID:15511592

Moradi, M., & Moussavi, G. (2019). Enhanced treatment of tannery wastewater using the electrocoagulation process combined with uvc/vuv photoreactor: Parametric and mechanistic evaluation. *Chemical Engineering Journal*, *358*, 1038–1046. doi:10.1016/j.cej.2018.10.069

Muhammad, A., Soares, A., & Jefferson, B. (2019). The impact of background wastewater constituents on the selectivity and capacity of a hybrid ion exchange resin for phosphorus removal from wastewater. *Chemosphere*, *224*, 494–501. doi:10.1016/j.chemosphere.2019.01.085 PMID:30831501

Nanseu-Njiki, C. P., Tchamango, S. R., Ngom, P. C., Darchen, A., & Ngameni, E. (2009). Mercury (ii) removal from water by electrocoagulation using aluminium and iron electrodes. *Journal of Hazardous Materials*, *168*(2), 1430–1436. doi:10.1016/j.jhazmat.2009.03.042 PMID:19349114

Nowak, A., Mazur, R., Panek, E., Dacewicz, E., & Chmielowski, K. (2019). Treatment efficiency of fish processing wastewater in different types of biological reactors. *Physics and Chemistry of the Earth Parts A/B/C*, *109*, 40–48. doi:10.1016/j.pce.2018.09.007

Núñez, J., Yeber, M., Cisternas, N., Thibaut, R., Medina, P., & Carrasco, C. (2019). Application of electrocoagulation for the efficient pollutants removal to reuse the treated wastewater in the dyeing process of the textile industry. *Journal of Hazardous Materials*, *371*, 705–711. doi:10.1016/j.jhazmat.2019.03.030 PMID:30897490

Ricordel, C., Darchen, A., & Hadjiev, D. (2010). Electrocoagulation–electroflotation as a surface water treatment for industrial uses. *Separation and Purification Technology*, *74*(3), 342–347. doi:10.1016/j.seppur.2010.06.024

Schulz, M., Baygents, J., & Farrell, J. (2009). Laboratory and pilot testing of electrocoagulation for removing scale-forming species from industrial process waters. *International Journal of Environmental Science and Technology*, *6*(4), 521–526. doi:10.1007/BF03326091

Sessarego, S., Rodrigues, S. C. G., Xiao, Y., Lu, Q., & Hill, J. M. (2019). Phosphonium-enhanced chitosan for cr(vi) adsorption in wastewater treatment. *Carbohydrate Polymers*, *211*, 249–256. doi:10.1016/j.carbpol.2019.02.003 PMID:30824086

Solak, M., Kilic, M., Yazici, H., & Sencan, A. (2009). Removal of suspended solids and turbidity from marble processing wastewaters by electrocoagulation: Comparison of electrode materials and electrode connection systems. *Journal of Hazardous Materials*, *172*(1), 345–352. doi:10.1016/j.jhazmat.2009.07.018 PMID:19651474

Turek, M., Dydo, P., Trojanowska, J., & Bandura, B. (2007). Electrodialytic treatment of boron-containing wastewater. *Desalination*, *205*(1), 185–191. doi:10.1016/j.desal.2006.02.055

Veil, J. A., Puder, M. G., Elcock, D., & Redweik, R. J., Jr. (2004). A white paper describing produced water from production of crude oil, natural gas, and coal bed methane. *Prepared by Argonne National Laboratory for the US Department of Energy, National Energy Technology Laboratory, January.* Available at http://www. ead. anl. gov/pub/dsp_detail. cfm

Verma, A. K. (2017). Treatment of textile wastewaters by electrocoagulation employing fe-al composite electrode. *Journal of Water Process Engineering*, *20*, 168–172. doi:10.1016/j.jwpe.2017.11.001

Vo, T.-K.-Q., Bui, X.-T., Chen, S.-S., Nguyen, P.-D., Cao, N.-D.-T., Vo, T.-D.-H., ... Nguyen, T.-B. (2019). Hospital wastewater treatment by sponge membrane bioreactor coupled with ozonation process. *Chemosphere*, *230*, 377–383. doi:10.1016/j.chemosphere.2019.05.009 PMID:31112860

Wemedo, S., Akani, N., & Eke, C. (2009). Fungi in an oilfield wastewater in Nigeria. *Asian Journal of Biological Sciences*, *2*(2), 54–57. doi:10.3923/ajbs.2009.54.57

Woodall, D. W., Gambrell, R. P., Rabalais, N. N., & DeLaune, R. D. (2001). Developing a method to track oil and gas produced water discharges in estuarine systems using salinity as a conservative tracer. *Marine Pollution Bulletin*, *42*(11), 1118–1127. doi:10.1016/S0025-326X(01)00100-X PMID:11763224

Yılmaz, A. E., Boncukcuoğlu, R., & Kocadağistan, E. (2008). An empirical model for kinetics of boron removal from boron containing wastewaters by the electrocoagulation method in a batch reactor. *Desalination*, *230*(1), 288–297. doi:10.1016/j.desal.2007.11.031

Zazou, H., Afanga, H., Akhouairi, S., Ouchtak, H., Addi, A. A., Akbour, R. A., ... Hamdani, M. (2019). Treatment of textile industry wastewater by electrocoagulation coupled with electrochemical advanced oxidation process. *Journal of Water Process Engineering*, *28*, 214–221. doi:10.1016/j.jwpe.2019.02.006

Chapter 18
Coagulation–Flocculation Technology in Water and Wastewater Treatment

Yeek-Chia Ho
https://orcid.org/0000-0002-2820-9696
Universiti Teknologi PETRONAS, Malaysia

Siong-Chin Chua
Universiti Teknologi PETRONAS, Malaysia

Fai-Kait Chong
Universiti Teknologi PETRONAS, Malaysia

ABSTRACT

Coagulation and flocculation processes are widely used in potable water treatment due to its high efficiency in turbidity removal. Egyptians discovered this method in 1500 BC by using alum to settle the suspended solids in the water. Today, the coagulation and flocculation processes are implemented with the purpose of agglomerate colloids and fine particles in water into larger particles, which is also known as floc. Therefore, reduction of turbidity and pollutants e.g. organic matter, inorganic matter, suspended solid, etc. can be achieved. This chapter covers the principle of coagulation and flocculation process which includes the charge neutralization and various binding mechanisms e.g. interparticle bridging, sweeping coagulation, and absorption. Besides, various types of coagulants and flocculants that have been discovered and their respective effectiveness in potable water treatment are discussed as well in this chapter. Polymer modifications to synthesize new coagulant/flocculant i.e. grafting and crosslinking are also included.

DOI: 10.4018/978-1-7998-0369-0.ch018

INTRODUCTION

The demand for fresh water has exponentially increased due to the increase of human population and the rapid industrialization and domestic activities, which generated a large amount of polluted wastewater. Fine suspended solids, metals, organic and inorganic particles and other impurities are generally found in industries and domestic wastewater. These pollutants are hard to be removed due to their surface charge and small size properties. Various methods, e.g. coagulation-flocculation, ion exchange, precipitation, adsorption, biological and advanced oxidation process have been used to remove the colloidal particles in the wastewater. Among all the treatment method, the coagulation-flocculation process is one of the oldest treatment methods and essential for most of the water and wastewater treatment. Apart from water and wastewater treatment, the coagulation-flocculation process is used in diverse disciplines e.g. rubber manufacturing, cheese manufacturing, biochemistry etc. (Bratby, 2016).

Coagulation-flocculation involved two-phase process to remove the stable colloids in water by forming larger aggregates that can be separated by the sedimentation process. The coagulation process is the first phase, which involved the addition of coagulant into the water to destabilize the particles by reducing the repulsive forces between the colloids. The second phase involved the flocculation process where the destabilized particles are bonded together to form flocs through the attraction of van der Waals force. The coagulation-flocculation process has been primarily used in drinking water treatment since 1500 BC to remove the turbidity of the water. Recently, coagulation-flocculation processes have received greater attention as they have been proven to be an effective process for various contaminants removal, which includes toxic organic matter, heavy metals, viruses etc. Due to its high-efficiency properties in removing contaminates, this process has been extensively applied in the treatment of palm oil mill effluent (Chung et al., 2018), food and beverage wastewater (Jusof Khadidi & Hamid, 2013; Muryanto et al., 2018), textile wastewater (Meriç et al., 2005), yeast wastewater(Zhou et al., 2008) and others.

The efficiency of the coagulation-flocculation process in the treatment process are mainly depend on few factors such as the type and dosage of coagulant, the temperature and pH of the wastewater, the concentration of the pollutant in wastewater, mixing speed and settling time of the floc formed. Therefore, the choice of coagulant and optimization of all operating parameters are essential for cost optimization in industrial application. Traditionally, inorganic metal salts, i.e. ferric chloride and aluminium sulphate are widely used in industrial wastewater treatment. However, the usage of inorganic coagulants has been reduced nowadays due to its narrow application and less inefficiency in small dosage (Lee et al., 2014). Instead, natural or synthetic polymeric flocculants are preferable for the treatment process.

PRINCIPLE OF COAGULANT AND FLOCCULATION PROCESS

Particle Collision Theory

Colloids are the particles with the size smaller than 10^{-5} mm and they are strongly influenced by the electrokinetic charge (Bratby, 2016). In water, there are two types of colloids, hydrophobic and hydrophilic colloids which divided based on their surface properties (Suopajärvi, 2015). For hydrophilic colloids in the water, the primary charges are caused by their polar groups, e.g. amine and carboxylic group. Hydrophilic colloids are more likely to aggregate as compared to hydrophobic colloids as they interact with water. However, most of the colloids remain discrete in the water and hardly agglomerate attribute to two reasons: (i) the large surface-to-volume ratio caused by their small size (about 0.01 to 1 µm) and, (ii) the repulsive forces between negative charges as the colloids in the water are generally negatively charged (Choy et al., 2016; S. Peavy et al., 2019; Suopajärvi, 2015). Since most of the colloids carried negative charge, the positive counter ions tend to attract towards the surface of the negative-charged colloids through the van der Waals and electrostatic forces (Bratby, 2006). As a result, a stern layer is formed around the primary charge (negative charge). The concentration of the counter-ions is reduced with increasing distance from the surface of the particle and the formation of the electrical double layer is caused by the thermal motion and the ionic repulsion or attraction. Stern layer is the dense layer of counter-ions (positive charge) fixed at the primary charge's layer (Suopajärvi, 2015). The ions in the stern layer will have the same ability as the primary charge ions, which attract their own counterions, formed another layer known as diffused layer. A complete electrical double layer comprises stern and diffused layer is formed as shown in Figure 1. Besides, there is a stationary layer of surrounding medium bound to the particles and travel with it in the suspension. The boundary of this layer is called the slipping plane and its corresponding electrical potential is known as zeta potential. The increment of the zeta potential lead to the increment of the repulsion force between the same charged particles and thus, the colloids remain more electrically stable in the water. For colloids in the water, the zeta potential generally fall within the range from −5 to −40 mV due to the presence of the charged groups (Choy et al., 2016).

Coagulation-Flocculation Mechanisms

Five types of destabilization mechanisms in the coagulation-flocculation process have been discovered and proposed accordingly. They are charge neutralization, interparticle bridging, electrostatic patch, sweeping flocculation/ colloid entrapment and double layer compression (Huang & Shiu, 1996; Oyegbile et al., 2016; Teh et al., 2016; Vandamme et al., 2013).

Figure 1. Electrical double layer of particle

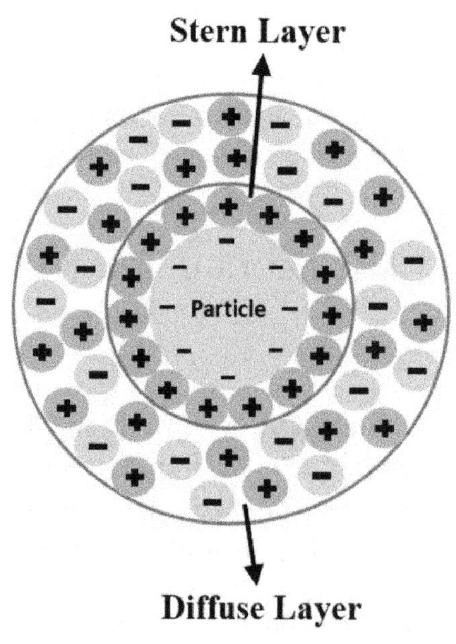

Charge Neutralization

Charge neutralization is achieved when the zeta potential of particles near to zero (Li et al., 2006). The hydrolyzing coagulant such as aluminum sulphate (alum) and ferrous sulphate are known as effective coagulants in neutralizing the negative charge of most of the particles in water. Those effective coagulants are generally high in charge density but low in molecular weight as it plays an only minor role in charge neutralization mechanism (Tripathy & De, 2006). When alum or ferrous sulphate is added into water, Fe^{3+} or Al^{3+} ions are released due to the dissociation of the coagulant. The strong positively-charged metal ions strongly absorb on the surface and neutralize the surface charge of the particles. Consequently, the particles attract to each other due to the opposite-charge attraction and the disappearing of the electrostatic repulsion force between the particles. The schematic diagram of this mechanism is shown in Figure 2. This mechanism is effective to treat the water that contains natural organic matters or algae and has been commonly used in the industry nowadays. However, the optimum dosage of coagulant for charge neutralization mechanism is extremely important as the destabilization of particles occurs at overdosage. Therefore, quantification of zeta potential is typically adapted to determine the optimum dosage of the coagulant (Teh et al., 2016).

Interparticle Bridging and Absorption

Interparticle bridging is usually achieved by the addition of polymer, e.g. non-ionic, and anionic (Hogg, 2013; Teh et al., 2016). The polymer that favoring bridging mechanism carried high molecular weight (minimum $8x10^5$ kDa, generally $>10^6$ kDa) and low charge density properties (5-15%) (McEwen, 1998). The polymer chains are absorbed on the surface of the particle through chemical bonding or physical attachment. These polymers form a bridge between both particles by simultaneously capturing and binding particles to their surface (Ravina & Moramarco, 1993). As a result, the bridge that formed causes flocculation by bringing particles together as shown in Figure 3. This mechanism explains the flocculation process through neutral or like-charged biopolymer. Non-ionic and anionic biopolymers are able to absorb suspended particles in water due to (1) electrostatic forces. The polymer that carried the opposite charge creates an electrostatic force that allows the absorption of suspended particles; (2) hydrogen bonding. The hydroxyl surface site on particles and amide group on biopolymer attached to each other due to the hydrogen bonding; (3) hydrophobic properties. The polymer contains nonpolar groups such as CH_2 caused accumulation of biopolymer at interfaces (McEwen, 1998). The rate of the biopolymer adsorption is increased with the particle concentration. Bridging mechanism produces a

Figure 2. Schematic diagram of charge neutralization destabilization mechanism

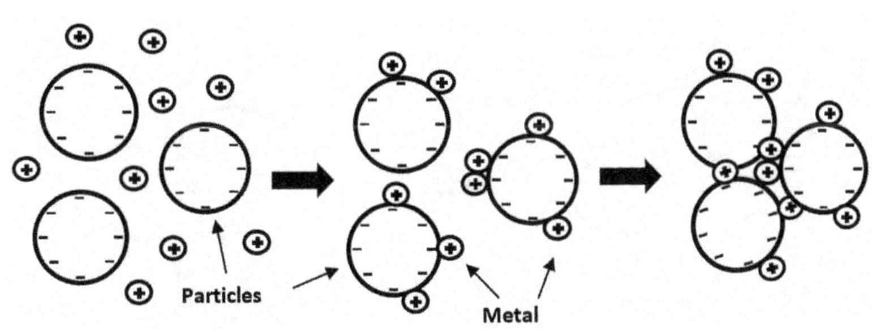

Figure 3. Schematic diagram of charge neutralization destabilization mechanism

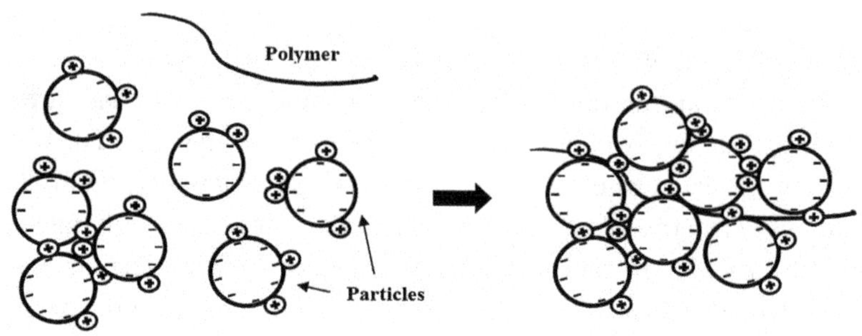

stronger floc as compared to charge neutralization mechanism and the floc produced do not reform at high shear rates after broken. Besides, well mixing is important to achieve optimum bridging and avoid localized overdosing which may cause some steric repulsion (Tripathy & De, 2006).

Electrostatic Patch

Electrostatic Patch neutralization is achieved when a particle binds with a polymer that carried the opposite charge. This mechanism usually involves high charge density particles with low molecular weight. When the polymer binds on the particle, the surface charge of the particle is reversed locally by the bind-polymer. As a result, patches of opposite charges appear on the particle surface. In the end, the flocculation process occurs as the particle is connected through patches of opposite charge on the surface of the particle. The schematic diagram of this mechanism is shown in Figure 4. This mechanism generally occurs in the addition of high cationic charge polymer into the anionic colloidal solids (especially at low particle concentration) (Tripathy & De, 2006).

Sweeping Coagulation/ Colloid Entrapment

Sweeping flocculation or colloid entrapment involves the addition of a large number of coagulants, e.g. aluminum or iron salts, which caused precipitation as hydrous metal oxides (Hogg, 2013). When the water becomes saturated that caused by large amount dosage of coagulant, the coagulant will precipitate out. Therefore, the particles in the water will be trapped in the precipitant when it settles down. In this

Figure 4. Schematic diagram of charge neutralization destabilization mechanism

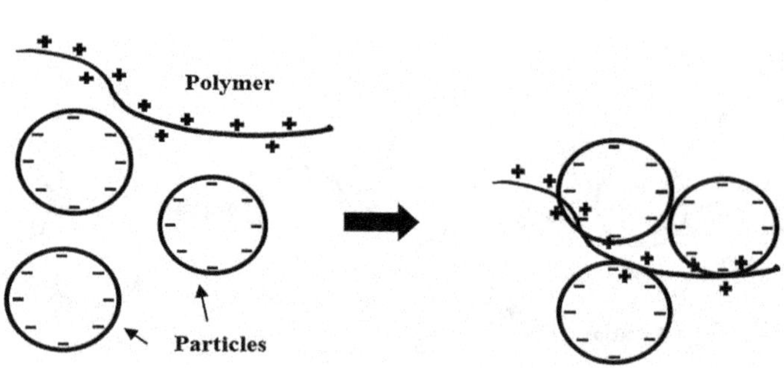

Figure 5. Schematic diagram of Sweeping Coagulation destabilization mechanism

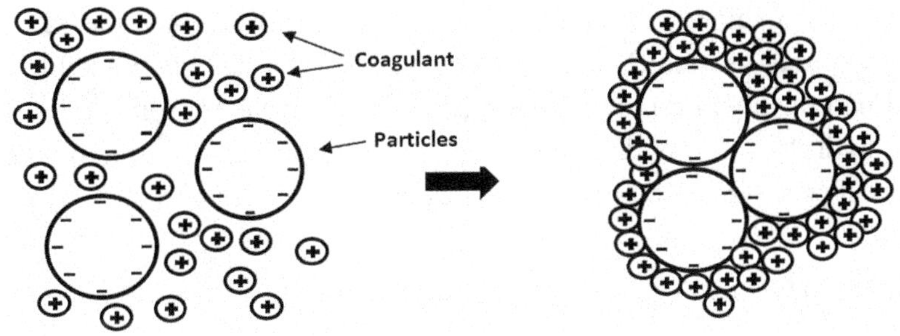

mechanism, some charge neutralization might occur, but the majority of the colloids are swept from the bulk of the water. The schematic diagram for sweeping coagulation/ colloid entrapment is shown in Figure. This mechanism is preferred to treat the water with low turbidity (Parsons & Jefferson, 2006).

Double Layer Compression

Double layer compression is achieved when a large amount of indifferent electrolyte, i.e. sodium chloride is added into the water and the double layer of the particles is reduced by the change of the ionic strength (Teh et al., 2016). The absorption of ions to the colloid is present in this mechanism due to the ions of the electrolyte maintain its identity in the water. Consequently, the double layer of the colloid is compressed by the change in ionic concentration (Ravina & Moramarco, 1993). It is important to note that the amount of coagulant needed to achieve coagulation through this mechanism is independent to the concentration of the particles in the water. This mechanism is more suitable to apply in industrial wastewater that contains a high amount of divalent or trivalent counters-ions.

COAGULANT-FLOCCULANT

Coagulant

Two major types of coagulant are used for the coagulation-flocculation process. They are chemical-based coagulants and natural coagulants, which can be known as inorganic and organic coagulant.

Chemical-based Coagulants

Chemical-based coagulants are the most commonly-used coagulant and have been used for decades due to its performance, abundant availability, ease of use and low cost (Lee et al., 2014). It can be classified into three categories, which are hydrolyzing metallic salt, pre- hydrolyzing metallic salts and synthetic cationic polymers.

Hydrolyzing Metallic Salts

Hydrolyzing metallic salt especially aluminum-based salt is the most frequent used coagulant in industry and drinking water treatment system (Duan & Gregory, 2003). It has been used to remove impurities, e.g. clays, dissolved natural organic matters and pathogenic microbes from polluted water since 1500 BC. The ions of the hydrolyzing metallic salts, e.g. Al^{3+} and Fe^{3+} are hydrated when it dissolved into water and the hydration process can be represented as below:

$$Me^{3+} \rightarrow Me\left(OH\right)^{2+} \rightarrow Me\left(OH\right)_2^{+} \rightarrow Me\left(OH\right)_3 \rightarrow Me\left(OH\right)_4^{-}$$

The hydration process shows that the water molecule in the hydration shell is substituted by the hydroxyl ions, which caused negatively charge in the end. The increase of the pH increases the hydration process of the hydrolyzing metallic salt ions, which caused the process moves to the right (Gregory & Duan, 2001). Therefore, the effectiveness of the hydrolyzing metallic salts in water treatment are pH dependence.

Aluminum-based hydrolyzing metallic salts, e.g. aluminum sulfate (alum), sodium aluminate and aluminum chloride are some of the common coagulants that used in the water and wastewater treatment industry. Alum is produced through digesting bauxite ores with the sulfuric acid. The commercial alum that used in water treatment has approximately 600 molecular weight with the chemical formula $Al_2(SO_4)_3{\bullet}14H_2O$, with aluminum content at approximately 7.4 to 9.5% (Bratby, 2006) The optimum pH for effective coagulation of alum is between pH 5.5 to 6.5 (Gebbie, 2001). The dry product of alum can be stored easily as compared to liquid alum due to the corrosive properties of liquid alum. Therefore, the liquid alum needs to be stored in rubber, 316 stainless steel tanks or lead tanks to avoid corrosion. Sodium aluminate with the chemical formula $NaAlO_2$ contains approximately 13% of aluminum. It carries alkaline properties which generally used with alum for certain purpose i.e. treating high colored wastewater with low pH. Besides, research carried by Lea et al. (1954) has proven the effectiveness of sodium aluminum in phosphate removal (removal up to 97%). Aluminum chloride with the chemical formula $AlCl_3{\bullet}6H_2O$ contains a slightly lower percentage of aluminum (10.5%) as compared to sodium aluminate. It is generally used for sludge conditioning and has been known as a good conditioner.

Apart from aluminum-based hydrolyzing metallic salts, ferric-based hydrolyzing metallic salts. e.g. ferric sulphate, ferric chloride, ferrous sulfate and chlorinated ferrous sulphate are some of the commonly used coagulants in water and wastewater treatment. The suitable pH for all ferric coagulants is ranging from pH 4.0 – 11.0. Although the cost of ferric-based hydrolyzing metallic salts is generally less than alum, but the corrosive properties of ferric has limit down its application (Tripathy & De, 2006). Ferric sulphate with molecular formula $Fe_2(SO_4)_3.8H_2O$ is available in both liquid and solid form. Approximately 11.5% Fe^{3+} and 40-42% of $Fe_2(SO_4)_3$ by mass are contained in the liquid form of ferric sulphate. In contrast, 20-21% Fe^{3+} and 72-75% of $Fe_2(SO_4)_3$ are present in the solid form of ferric sulphate (Bratby, 2006).

Pre Hydrolyzed Metal Coagulant

Apart from traditional hydrolysed coagulant, prehydrolyzed metal coagulant has gained increasing attention, especially in wastewater treatment. For aluminium-based prehydrolyzed coagulants, they are generally synthesized through the neutralization process of aluminium chloride solution. The most commonly used prehydrolyzed metal is polyaluminium chloride (PACI), which synthesizes from the aluminium while polyaluminosilicate sulfate (PASS) from aluminium sulphate (Gregory & Duan, 2001). These coagulants are proven to be more effective as compared to the traditional hydrolyzing metallic salts. Matsui et al. (1998) conducted an experiment to compare the efficiency of PACl and alum in turbidity reduction. The result shows that PACl was more effective in turbidity reduction at low temperatures and produced lesser volumes of sludge as compared to alum. Besides, PACl was found to be effective in a wide range of pH. This is attributed to the neutralization of the material that reduces the effect of the water pH on the coagulant. Another explanation is in term of the high charge, i.e. Al_{13} that available for the charge neutralization and absorption at neutral pH. In the dynamic aspect, pre-hydrolyzed coagulants are able to produce stronger flocs with rapid flocculation as compared to alum at equivalent dosages.

Despite all the advantages, the high cost of the pre-hydrolyzed coagulant has limited its application. The cost of the pre-hydrolyzed coagulant is approximately 60 – 75% of the total chemical treatment costs. Therefore, it generally used in small clarification plant as it is not economically feasible in a large water treatment plant. It should be noted that pre-polymerized coagulant has high specific nature properties as it is prepared with different base concentration, basicity ratios, ageing time and temperature and initial metal concentration. Therefore, jar testing is needed to determine the best formulation for specific water treatment.

Natural Coagulant

Despite the efficiency of chemical-based coagulant in water and wastewater treatment, health concerns have been raised when numerous studies proved that the residual metal concentration in water, for instance, aluminum might provoke detrimental pathology and neurology diseases. Furthermore, some researches also proved that there is a high positive correlation between the Alzheimer's disease and the residual aluminum (Flaten, 2001; Kawahara & Kato-Negishi, 2011; McLachlan et al., 1996; Rondeau et al., 2000). Therefore, the dosage of the inorganic coagulants, especially the aluminum-based coagulant has been strictly controlled. These aluminum-based wastes that produced at the end of process have been categorized as Scheduled Waste (SW204) in some countries, for instance, Malaysia which caused additional expenditures for the waste management and treatment system (SPAN, 2010). This situation has stimulated exponential growth of research in searching for natural coagulant as an alternative coagulant for the coagulation-flocculation process.

Plant-Based Natural Coagulant

Plant-based natural coagulants, for instance biopolymers have been discovered as one of the alternatives for the inorganic coagulants. In contrast to chemical-based coagulants, they are generally safe and eco-friendly. In addition, natural coagulants generate five times lower sludge as compared to inorganic coagulant, which greatly reduces the handling cost. Their performances in water treatment, especially on turbidity removal are impressive as well, which generally achieved up to 90% of removal rate. Table 1 shows a summary of the performances of various types of natural coagulant in water and wastewater treatment.

Table 1. Available natural coagulants in water and wastewater treatment

Coagulant	Type of Sample	Initial Turbidity (NTU)	Optimized Parameters			References
			Dosage (mg/L)	pH	Turbidity Reduction	
Cactus latifaria	SW (Kaolin)		10	7	~ 95%	Diaz et al. (1999)
Citrus sinensis peel	RSW	32	4000	-	85%	Choy et al. (2014)
Common oak	SW (Kaolin)	20	0.5 mL/L	10	~65%	Šćiban et al. (2009)
Horse chestnut	SW (Kaolin)	20	0.5 mL/L	10	~77%	Šćiban et al. (2009)
Mangifera indica seeds	RSW (Nile River)	6.5	30	-	54%	Choy et al. (2014)
Mangifera indica kernels	SW (Kaolin)	90	0.5 mL/L	13	98.65%	Choy et al. (2014)
Moringa oleifera seed	Surface water (Kano)	20-90	-	-	~ 50%	Muyibi and Okuofu (1995)
Non-shelled	SW (Kaolin)	105	10 mL/L	7.6	~ 90%	Ndabigengesere and Narasiah (1998)
Shelled	SW (Kaolin)	105	2 ml/L	7.6	~ 90%	Ndabigengesere and Narasiah (1998)
Phoenix dactylifera date seeds	SW (Bentonite)	75 ± 10	30		~ 94%	Choy et al. (2014)
Prosopis juliflora	SW (Kaolin)	100	20	7	~ 90%	Diaz et al. (1999)
Tamarindus indica seeds	SW (Kaolin)	76-79	5	8	~ 69%	Choy et al. (2014)

Flocculant

Flocculant is used to enhance the coagulant and flocculation process. It helps in accelerating the formation of floc, producing denser and stronger flocs, reducing the dosage of coagulant and improving the turbidity removal. Flocculant can be categorized into four types, which are alkalinity addition, pH adjustment, particulate addition and polymers (L. K. Wang et al., 2005).

Synthetic Flocculant

Synthetic flocculant (polymer) have been widely used in various applications, i.e., wastewater treatment to enhance the efficiency of the coagulation-flocculation process. It can be classified into few categories based on their characteristics, e.g., nature of the charge, molecular weight and charge density. Poly (diallyldimethyl ammonium chloride), epichlorohydrin/dimethylamine and cationic polyacrylamides (PAMs) are some of the flocculants used in water and wastewater treatment (Bolto & Gregory, 2007). Among all, PAMs has the widest application in water industry (Baade et al., 1989; Bolto & Gregory, 2007; Hoover, 1970) due to its flexibility to synthesize with various functionalities including positive, negative or neutral charge to produce an excellent settling performance at relatively low cost (Wong et al., 2006). PAM has been proven to be an effective flocculant for water quality protection since the 1990s (Entry et al., 2002).

Synthetic flocculant has the ability to produce large, strong, dense and compact floc as compared to the floc produce by coagulant (Wong et al., 2006). Matilainen et al. (2010) claimed the reduction amount of sludge volume produced when synthetic flocculant (polyelectrolyte) is used. This claim is further supported by the experiment carried out by Amuda and Amoo (2007) where 60% of the sludge is reduced by using polyelectrolyte instead of alum solely. Despite all the advantages poses, synthetic flocculant such as PAM and its derivatives, residual unreacted monomers, e.g., ethyleneimine and acrylamide (AM) are extremely toxic and potentially causing severe neurotoxic effects (R. Yang et al., 2016). Therefore, the application of synthetic flocculant in potable water treatment is still limited.

Bio-based Polymer Flocculant

Polysaccharide bio-based polymer flocculant is categorized into three categories based on their source. They are marine polysaccharide flocculant, plant polysaccharide flocculant and microbial polysaccharide flocculant (Salehizadeh et al., 2017). Marine polysaccharide flocculant is generally derived from seaweeds and shell waste while microbial polysaccharide flocculant is extracted from microorganism, for instance, bacteria, yeast, fungi, algae etc. In contrast, plant polysaccharide flocculant is mostly derived from the seed of the plant. Table 2 shows the list of polysaccharide bio-based flocculants that have been discovered.

The demand on the plant-based flocculant in potable and wastewater treatment increase significantly in recent years as it exhibits safe, biodegradable and fairly shear stable characteristics (Lee et al., 2014). Plant-based flocculant is easily available from reproducible agricultural resources to support the demand and produce no secondary pollution (Bolto & Gregory, 2007; Salehizadeh et al., 2017). Therefore, plant-based flocculants, e.g. guar gum, tannin, starch etc. are used for various application.

APPLICATION OF COAGULATION-FLOCCULATION PROCESS IN WATER AND WASTEWATER TREATMENT

Potable/ Drinking Water Treatment

Coagulation-flocculation process is still playing an important role in removing particulates, turbidity and natural organic matter (NOM) in water treatment. In this process, alum, ferric chloride or poly-aluminum chloride (PACl) are the common coagulants that added to the water to aid in the aggregation of various contaminants particles into larger floc. To overcome the limitation of chemical coagulant such as produce large amount of sludge and required large dosage, *Moringa oleifera* natural coagulant has been extensively studied as an alternative to chemical coagulant in drinking water treatment since 1970s (S. A. Jahn & Dirar, 1979; Muyibi & Evison, 1996; Pritchard et al., 2010). A recent study carried by Pritchard et al. (2010) has made a detail comparison between *M. oleifera,* alum and ferric in term of turbidity and coliform removal. The turbidity reduction and *E.coli* removal efficiency of M. oleifera in potable water treatment were 97.2% and 88.9% respectively, which is lower as compared to alum (99.3% for turbidity reduction and 99.3% for *E.coli* removal) and ferric (97.7% for turbidity reduction and 86.5% for *E.coli* removal). Although the performance of *M. oleifera* was not as effective as alum and ferric, it is still used together with sand filtration in some developing countries such as India as *M. oleifera* is locally available products.

Table 2. List of discovered polysaccharide bio-based flocculants

Flocculants /Species	Source	References
Plant Polysaccharide Flocculants		
Citrus maxima	Peel of pomelo	Piriyaprasarth and Sriamornsak (2011)
Guar gum *(Guaran)*	Guar bean	Thombare et al. (2016)
Inulin	Roots and tubers of *Inula helenium*	Rahul et al. (2015)
Lepidium Sativum sp.	Seed of garden cress	Lim et al. (2018)
Malva sylvestris	Mallow seed pods	Anastasakis et al. (2009)
Malva nut gum	Seed of the Scaphium scaphigerum tree	Ho et al. (2015)
Plantago ovata	Isabgol seed husks	Al-Hamadani et al. (2011)
Plantago psyllium	Psyllium seed husks	Mishra et al. (2002)
Rice starch	Rice	Teh et al. (2014)
Tamarindus	Seed of tamarind	Mishra and Bajpai (2006)
Marine Polysaccharide Flocculants		
Durvillaea antarctica	Brown seaweeds	Venugopal (2016)
Alginate	Brown seaweeds (Macrocystis pyrifera)	Alginate
Chitosan	Shrimp shell	Sila et al. (2014)
Chitosan	Crab shell	Huang and Chen (1996)
Microbial Polysaccharide Flocculants		
Chitosan	Aspergillus terreus	Paraskeva and Diamadopoulos (2006)
Alginate	*Azotobacter chroococcum*	Z. Yang et al. (2017)
Alginate	*Azotobacter indicus*	Patil et al. (2011)
Cellulose	*Gluconacetobacter hansenii*	Shoda and Sugano (2005)
Haloalkalophilic *Bacillus sp*	Bacillus sp	C. G. Kumar et al. (2004)
Marine sponge bacterial flocculant	Marine sponge-associated *Bacillus subtilis*	Sathiyanarayanan et al. (2013)
Nannocystis sp. NU-2	Marine myxobacterium strain	Zhang et al. (2002)

Dye and Textile Wastewaters

Textile wastewaters generated from textile industry generally contain a large number of contaminants that are extremely harmful to both environment and human being. Several direct effects of discharging textile wastewaters without proper treatment include aesthetic problem, groundwater pollution due to the leaching of contaminant through the soil, DO depletion and poor sunlight penetration of receiving water. Various physico-chemical approaches such as membrane filtration (Barredo-Damas et al., 2006), adsorption (Hai et al., 2007), irradiation, sonolysis (Arslan-Alaton, 2003), photochemical reaction (Hai et al., 2007) and chemical coagulation-flocculation process (Ciabatti et al., 2010) have been studied for the textile wastewater treatment. Among all the methods, chemical coagulation-flocculation process was found to be most economically feasible approach for textile wastewater treatment (Verma et al., 2012). However, the main drawback of this approach was the sludge produced after the treatment process. Both pre-hydrolyzed metallic coagulant such as Polyaluminium chloride (PACl) and hydrolyzing metallic coagulant such as alum were widely used in the chemical coagulation-flocculation process for the textile wastewater treatment due to their high removal efficiency. The research conducted by Choo et al. (2007), Sanghi et al. (2006) and Baraoidan et al. (2007) using PACl as coagulant achieved promising color removal (75% - 99.5%). Besides, research carried by El-Gohary and Tawfik (2009) and Golob et al. (2005) using alum as coagulant achieved similar results with using PACl where the color removal ranged from 74% - 98%. Apart from the chemical coagulant, natural coagulant including plant and animal-based coagulant have been studied for textile wastewater treatment as well. A recent study was conducted on the textile treatment by using okra *(A. esculentus)* mucilage as a natural coagulant. The result shows that high removal of color (93.57%) can be achieved by a low dosage of okra mucilage. Besides, the dosage of the ferric coagulant can be reduced up to 72.5% when the okra mucilage was used as a coagulant aid. Animal-based coagulant, chitosan was reported to be effective coagulant for textile wastewater containing Acid Blue 92. The color removal efficiency can be achieved up to 99% with 100 mg/L of chitosan dosage at pH 9. Textile wastewater containing various classes of dyes is still a challenge for the coagulation-flocculation process and more studies in the future should be conducted in this direction.

Agricultural / Agro-Industrial Wastewater

Agricultural wastewater contains a large amount of organic substances such as organic colloids, suspended solids, grease and emulsified oil that needs to be treated before discharge. Among all the available technologies, coagulation-flocculation technology offers few key advantages such as cost-effective and simple operation. In the treatment of agricultural wastewater, a coagulation-flocculation step followed by filtration was introduced in a study conducted by Jiménez (2005). This combined treatment using alum as a coagulant was proven to be efficient approached to remove COD and TSS of the wastewater. Besides, the cost for the treatment was found to be three times lower than the biological treatment. Recently, natural coagulants such as *Cassia obtusifolia* seed gum (Shak & Wu, 2014) and rice starch (Teh et al., 2014) were intensively studies for the agro-industrial wastewater (palm oil mill effluent) treatment. Promising results, where 93.22% and 88.4% of total suspended solids removal were achieved by *Cassia obtusifolia* seed gum and rice starch coagulant respectively. Chitosan was also found to be suitable coagulant for palm oil mill effluent wastewater treatment. Results showed that up to 95% removal of suspended solids and oil was achieved by chitosan in a relatively low dosage (0.5 g/L) as compared to alum (8 g/L).

Pulp and Paper Wastewater

Pulp and paper industry consumes large amount (up to 60 m^3) of fresh water per tonne of paper produced and thus generated a huge amount of wastewater worldwide. This type of wastewater generally contains a high amount of COD, TSS, BOD, tannins, lignin and other toxic chemicals. Several techniques that commonly adopted for the treatment are activated sludge, chemical oxidation, coagulation-flocculation and aerated lagoons (Pokhrel & Viraraghavan, 2004). Coagulation-flocculation process is normally used as pre-treatment prior to the biological treatment and tertiary treatment for turbidity, COD and lignin removal in pulp and paper industry. Ahmad et al. (2008) studied the coagulation performance of polyaluminum chloride (PACl), alum and the combination with cationic polyacrylamide (C-PAM) and anionic polyacrylamide (A-PAM) in pulp and paper wastewater treatment. The use of alum with cationic polyacrylamide (Organopol 5415) outperformed other coagulants which achieved 99.7% of turbidity reduction, 95.6% of COD reduction and 99.5% of TSS removal. Besides, the SVI (38 mL/g) and the settling time (12 s) were found to be relatively low as compared to other coagulants. Rodrigues et al. (2008) also investigated the pulp and paper wastewater treatment using coagulation-flocculation process followed by heterogeneous photocatalysis. Coagulation-flocculation process was adopted first as photocatalysis method is not able to treat wastewater that contains high COD (>800 mg/L). The used of ferric chloride as coagulant in coagulation-flocculation process achieved up to 56% of COD reduction, which equivalent to approximately 514 mg/L of COD. Recently, J.-P. Wang et al. (2011) attempted to use aluminium chloride as the coagulant and a modified natural polymer, starch-g-PAM-g-PDMC [polyacrylamide and poly (2-methacryloyloxyethyl) trimethyl ammonium chloride], as the flocculant for pulp mill wastewater treatment. This combination achieved a promising result. The turbidity, lignin and water recovery efficiency at the optimum condition (871 mg/L of coagulant dosage, 22.3 mg/L of flocculant dosage and pH 8.35), were 95.0%, 83.5% and 72% respectively.

DEVELOPMENT AND RECENT ADVANCEMENT OF COAGULATION-FLOCCULATION PROCESS, COAGULANT AND FLOCCULANT

Electrocoagulation (EC) and Electrocoagulation/flotation (ECF)

Electrocoagulation (EC) technology is an electrochemical process that utilizes electricity instead of chemical reagents to flocculates and removes contaminants such as emulsified organics, suspended solids, heavy metals etc. The process consists of a pair of sacrificial metal plate, which generally known as electrodes. Electric current is used in the process to sacrificial the electrodes to generate coagulating agents such as metal ions. The generated metal ions help to neutralized and coagulate the contaminants in the water, which the working mechanism is similar to the addition of chemical coagulant, i.e. aluminium sulphate and ferric chloride. The agglomerates formed after the coagulation process will eventually settle down with the help of the gravitational force. Electrocoagulation/flotation (ECF) works similar to the EC. However, the agglomerates will float to the surface of the water by the formation of hydrogen bubbles from the anode. Eventually, the floated contaminants can be removed by skimming process. Both EC and ECF processed can be used to treat various types of water and wastewater, especially in removing the pathogens and inorganic contaminants. As compared to conventional chemical coagulation-flocculation process, EC and ECF processes carried few advantages which include easier to operate, produced lesser sludge and do not require any chemicals.

Application of Electrocoagulation (EC) and Electrocoagulation/flotation (ECF)

Electrocoagulation has been extensively studied as an alternative to the conventional chemical coagulation-flocculation in potable water treatment since 1983. There are few conclusions can be drawn from the study carried out by Vik et al. (1984): (1) less amount of sludge is produced in the EC process as compared to conventional coagulation-flocculation process; (2) the electrocoagulation unit equipped with sufficient number of electrodes to support for at least one year treatment; and (3) the operation and maintenance of EC process is relatively simple. Recently, an EC process for removing trace metal contaminants from potable water was studied in the USA by Heffron et al. (2016). Aluminium and iron electrode were used to study the removal efficiencies toward five metals, which included cadmium, lead, chromium, arsenic and nickel. The result obtained shows that iron electrodes achieved higher nickel and cadmium removal efficiency. Besides, Ray et al. (2018) studied the feasibility of the EC process in removing turbidity and *E.coli* from the surface water. Maximum *E. coli* removal of 19% and turbidity removal of 98% is achieved at the end of the EC process. The disinfection process was further enhanced by adding 2.8 mg/L of chlorine to achieve up to 100% of *E. coli* removal.

Apart from potable and surface water treatment, there are many studies on the EC process for industrial wastewater treatment. Adhoum et al. (2004) conducted a study on the treatment of electroplating industrial wastewater by the EC process. Results showed that Zn^{2+}, Cu^{2+}, and Cr(VI) ions can be effectively removed to meet legal limits by EC process with pH range from 4-8. Besides, the electricity consumptions for the EC process was found to be low, 32A h l^{-1}, which increase its feasibility as an alternative toward existing conventional method. Recently, the application of EC on complex industrial wastewater (estate wastewater) was studied by Yavuz and Ögütveren (2018). Maximum of 92% of COD removal is achieved by the EC process with the iron electrode. Besides, the energy cost was found to be €3.41/m^3 and the sludge production was 5.45g per g of COD removed. The research studies reveal that EC is an effective technology for water and wastewater treatment. However, the efficiency and the capacity of an EC process highly depends on the nature of the wastewater and initial contaminants concentrations in the wastewater.

Graft Copolymerization

A polymer can be modified through grafting or crosslinking reactions of polymer and monomer for various applications. For the crosslinking process, it can be achieved through either intramolecular crosslinking or intermolecular. The binding process can be formed through direct binding from one polymer chain to another such as C-C bond or indirect binding through a short chemical chain (Güven, 2012). Crosslinking modification increases the molecular weight of the polymers as it helps to bind two-liner molecules together to form a unique structure. On the other hand, the grafting process can be achieved through "grafting from", "grafting through" and "grafting onto" approaches. Both crosslinking and grafting processes are irreversible. Traditionally, crosslinking and grafting modification method is commonly used for biomedical, food and rubber application. However, grafting modification method has been widely used to synthesize new coagulant or flocculant for water and wastewater treatment recently.

Graft copolymerization enhances the polymer properties by combining both properties of polymers. The grafted copolymers will exhibit hybrid physical and chemical characteristic from both desired polymer groups. The long chains that grafted onto the backbone are generally flexible, which enhance the conformational freedom of flocculant in solution and surface. To further enhance the flocculation ability, specific functional groups that help in the flocculation process can be introduced into the polymer through grafting of a suitable monomer. Due to the polysaccharide backbone of the plant-based flocculant, the grafted copolymer is renewable and biodegradable (J.-P. Wang et al., 2013).

Chemical Method

In the chemical method, chemicals are used to initiate the active sites on the backbone of the polymer. Metal carbonyls, Lewis acids and strong bases are the redox initiator used for chemical grafting. Among all the chemicals, e.g. potassium persulfate (PPS), azobisisobutyronitrile (ABIN), ceric ammonium nitrate (CAN), benzoyl peroxide and ammonium persulfate, CAN is widely used in graft copolymers synthesis (D. Kumar et al., 2017). The free radicals can be generated through a direct or indirect method. Table 3 shows the grafting process by using different chemical initiator and their respective efficiency.

Radiation-induced Grafting Method

Radiation-induced grafting copolymerization has a high degree of control toward the length and the number of the grafted chain. The electromagnetic radiations are used to generate the active sites at the backbone of polymer (Aly & El-Bisi, 2018). This method is preferred by most of the researcher due to few reasons, (1) it is clean process (free from contamination) (Aly & El-Bisi, 2018; Bhattacharya et al., 2008), (2) additives or catalysts chemicals are not required for initiation purpose (Bhattacharya et al., 2008), (3) The molecular weight of the grafted polymer is possible to be regulated.

Table 3. Various flocculants and their modification under chemical grafting method

<table>
<tr><th>Initiator</th><th>Backbone</th><th>Monomers /Graft Chain</th><th>Grafting Efficiency (%)</th><th>Percentage Grafting (%)</th><th>References</th></tr>
<tr><td rowspan="6">Ceric Ammonium Nitrate (CAN)</td><td>Carboxymethyl chitosan (CMCH)</td><td>Acrylamide (Am)</td><td>79.03</td><td>215.8</td><td>Joshi and Sinha (2007)</td></tr>
<tr><td>Sago starch</td><td>Methyl acrylate</td><td>54.40</td><td>246.3</td><td>Rahman et al. (2000)</td></tr>
<tr><td>Cellulose</td><td>Ethyl Acrylate</td><td>75.6</td><td>1034</td><td>Gupta et al. (2002)</td></tr>
<tr><td>Maize starch</td><td>Methyl methacrylate</td><td>74.35</td><td></td><td>Athawale and Rathi (1997)</td></tr>
<tr><td>Maize starch</td><td>Methacrylonitrile</td><td>48.7</td><td>97.3</td><td></td></tr>
<tr><td>Tragacanth gum</td><td>Acrylonitrile (AN)</td><td>81</td><td>534</td><td>Mohamadnia et al. (2008)</td></tr>
<tr><td rowspan="2">H_2O_2</td><td>Guar Gum</td><td>Acrylonitrile</td><td>-</td><td>~55</td><td>Raval et al. (1988)</td></tr>
<tr><td>Guar Gum</td><td>Methyl methacrylate</td><td>-</td><td>61.50</td><td></td></tr>
<tr><td rowspan="2">PPS</td><td>Sago starch</td><td>Methyl acrylate</td><td>-</td><td>90</td><td>Fakhru'L-Razi et al. (2001)</td></tr>
<tr><td>Guar Gum</td><td>Methacrylic Acid</td><td>16.16</td><td>160.20</td><td>Mundargi et al. (2006)</td></tr>
<tr><td>KPS/TMEDA redox pair</td><td>Pregelled starch</td><td>Methacrylic acid</td><td>75%</td><td>-</td><td>Mostafa et al. (2011)</td></tr>
<tr><td rowspan="6">KPS/ ascorbic acid redox</td><td>Ipomoea palmata</td><td>Methyl methacrylate</td><td>53.92</td><td>337</td><td rowspan="6">Vandana Singh et al. (2006)</td></tr>
<tr><td>Cassia reticulata</td><td>Methyl methacrylate</td><td>48.80</td><td>305</td></tr>
<tr><td>Guar gum</td><td>Methyl methacrylate</td><td>47.36</td><td>296</td></tr>
<tr><td>Cassia abrreviata</td><td>Methyl methacrylate</td><td>44.80</td><td>280</td></tr>
<tr><td>Cassia javanica</td><td>Methyl methacrylate</td><td>31.20</td><td>195</td></tr>
<tr><td>Ipomoea hedracea</td><td>Methyl methacrylate</td><td>34.40</td><td>215</td></tr>
</table>

* KPS = Potassium persulphate

Low-Energy Radiation-Induced Grafting Radiations

For this technique, ultraviolet (UV) and visible light are common radiations that use to generate free radicals on the polysaccharide backbone (Aly & El-Bisi, 2018). Ma et al. (2016) synthesized a novel flocculant (NCS-g-PAM) through grafting of PAM to chitosan nanoparticle (NCS) using UV initiation technique. Mercury lamp was used as the UV source. Thaker and Trivedi (2005) grafted methyl acrylate onto sodium salt of partially carboxymethylated guar gum through UV radiations copolymerization technique, with ceric ammonium nitrate (CAN) as a photoinitiator. In the experiment, 125-W medium-pressure mercury lamp was used as a UV source.

High-Energy Radiation-Induced Grafting Radiations

High-Energy Radiation-Induced Grafting Radiations was discovered in the early of 1660s (Bhattacharya et al., 2008). γrays, X-rays and electron beam (EB) which accelerated to have different energy range from 6.5 to 3.0 MeV, are the main radiation sources for this technique. Among all, γ radiation obtained the highest interest from researcher due to its long half-life period, which is 5.3 years (Aly & El-Bisi, 2018).

Microwave Radiation-Induced Grafting

Microwave radiation-induced grafting using commercial microwave oven has gained increasing interest due to the significant enhancement of grafting rate over conventional reaction, i.e. chemical grafting method (Cao et al., 2001; Huacai et al., 2006; Shao et al., 2003; V Singh et al., 2012). This method is simpler, cleaner, eco-friendly and results into an instantaneous heating polymer in homogenous which conforms with the accepted norms of green chemistry (Sen et al., 2010; Vandana Singh et al., 2006). The percentage of the grafting can be controlled precisely by alternate the exposure time of the irradiation and microwave power. There are three types of grafting reactions; (1) in solution phase, where both materials are mixed homogenously in a solution with no phase separation; (2) in solid phase, where monomers and initiator are impregnated on solid support; (3) in suspension, where the reactants are not completely miscible (V Singh et al., 2012). Limitation of coagulation-flocculation process

The major limitation of the coagulation-flocculation process, especially in water treatment is the toxicity properties of the inorganic coagulant. As discussed earlier, the aluminium that retain in the water after treatment has the potential to cause Alzheimer disease (Rondeau et al., 2000). Therefore, the United States of America and European have set a standard where the residue concentration of aluminium should not exceed 0.05 and 0.2 mg/L, respectively. Besides, the toxicity of the polymers toward the aquatic organism have not been extensively studied as well. Harford et al. (2011) has proven the toxicity of anionic polyacrylamide (PAM) toward aquatic organism, i.e. hydra and cladoceran.

Apart from the limitation poses by the chemical coagulant, the use of natural coagulant has reported increase the concentration of total organic compound in the treated water i.e. Okuda et al. (2001) and Ndabigengesere and Narasiah (1998) claimed that the use of crude *Moringa oleifera* (MO) extract increases the concentration of organic, phosphate and nitrate contents of the water. Moreover, the taste and the odour of the treated water using the crude extract of MO after 24 hours are not favourable. This is attribute to the presence of organic and nutrient content that increase the microbial activity in the water (S. A. A. Jahn, 1988). The shelf life of natural coagulant is relatively short as compared to the inorganic coagulant. Katayon et al. (2006) claimed that the efficiency of the MO extract from the seed stored for 5-6 months decreased by at least 7%. The extract should also be prepared freshly or kept in cold to avoid deterioration.

The sludge generated after the treatment is another limitation of the coagulation-flocculation process as it leads to secondary pollution. Many authors have reported the amount of sludge produced through the coagulation-flocculation process (Chen et al., 2010; Sarkar et al., 2006; Teh et al., 2016). According to the National Water Services Commission (SPAN), approximately two million metric tons of sludge was produced each day from total 462 wastewater treatment plants. Therefore, additional expenditures is required for the sludge management and treatment system, which include the dewatering process of the sludge (SPAN, 2010). Regeneration and reuse of sludge or proper disposal are some of the recommended options for the generated sludge. However, the fragility property of forming flocs through coagulation-flocculation has dramatically reduces the feasibility of recovering the produced sludge.

CONCLUSION

Coagulation-flocculation treatment process is still an imperative process for water and wastewater treatment despite the development of new treatment processes due to few advantages which include high versatility, high effectiveness, low energy consumption and high simplicity in design and operation. The efficiency of the treatment process is greatly affected by a few operating parameters, e.g. pH, type and dosage of coagulant and flocculant, mixing speed, mixing time and temperature of the water. Therefore, it is crucial to optimize all the operating parameters for the large-scale implementation to achieve the maximum contamination reduction with minimum cost.

Although chemical coagulant and flocculant are highly effective and still widely been used in water sectors, increasing environmental awareness and improving process sustainability have triggered researchers to search for more environmentally friendly coagulants and flocculants especially the plant-based coagulant and flocculant. However, natural coagulant still has many limitations, such as short shelf life and medium efficiency. To overcome these limitations, polymer modification, especially graft copolymerization has been widely used to improve the efficiency and properties of the existing natural coagulant for a specific application. Although the promising experimental result is obtained, a future investigation is still needed to be carried out to confirm its performance and economic viability in a large-scale application.

ACKNOWLEDGMENT

The authors would like to express their sincere gratitude to UTP for funding this project under YUTP grant (0153AA-E34) and STIRF grant (0153AA-D77).

REFERENCES

Adhoum, N., Monser, L., Bellakhal, N., & Belgaied, J.-E. (2004). Treatment of electroplating wastewater containing Cu2+, Zn2+ and Cr (VI) by electrocoagulation. *Journal of Hazardous Materials*, *112*(3), 207–213. doi:10.1016/j.jhazmat.2004.04.018 PMID:15302441

Ahmad, A., Wong, S., Teng, T., & Zuhairi, A. (2008). Improvement of alum and PACl coagulation by polyacrylamides (PAMs) for the treatment of pulp and paper mill wastewater. *Chemical Engineering Journal*, *137*(3), 510–517. doi:10.1016/j.cej.2007.03.088

Al-Hamadani, Y. A., Yusoff, M. S., Umar, M., Bashir, M. J., & Adlan, M. N. (2011). Application of psyllium husk as coagulant and coagulant aid in semi-aerobic landfill leachate treatment. *Journal of Hazardous Materials*, *190*(1-3), 582–587. doi:10.1016/j.jhazmat.2011.03.087 PMID:21507572

Aly, A. A., & El-Bisi, M. K. (2018). *Grafting of polysaccharides: recent advances. In Biopolymer Grafting* (pp. 469–519). Elsevier. doi:10.1016/B978-0-323-48104-5.00011-1

Amuda, O., & Amoo, I. (2007). Coagulation/flocculation process and sludge conditioning in beverage industrial wastewater treatment. *Journal of Hazardous Materials*, *141*(3), 778–783. doi:10.1016/j.jhazmat.2006.07.044 PMID:16959404

Anastasakis, K., Kalderis, D., & Diamadopoulos, E. (2009). Flocculation behavior of mallow and okra mucilage in treating wastewater. *Desalination*, *249*(2), 786–791. doi:10.1016/j.desal.2008.09.013

Arslan-Alaton, I. (2003). A review of the effects of dye-assisting chemicals on advanced oxidation of reactive dyes in wastewater. *Coloration Technology*, *119*(6), 345–353. doi:10.1111/j.1478-4408.2003.tb00196.x

Athawale, V., & Rathi, S. (1997). Effect of chain length of the alkyl group of alkyl methacrylates on graft polymerization onto starch using ceric ammonium nitrate as initiator. *European Polymer Journal*, *33*(7), 1067–1071. doi:10.1016/S0014-3057(96)00294-7

Baade, W., Hunkeler, D., & Hamielec, A. (1989). Copolymerization of acrylamide with cationic monomers in solution and inverse-microsuspension. *Journal of Applied Polymer Science*, *38*(1), 185–201. doi:10.1002/app.1989.070380117

Baraoidan, W. A., Tun, L. L., Gaspillo, P.-D., & Suzuki, M. (2007). A study on the relative performance of different coagulants and the kinetics of COD in the treatment of a textile bleaching and dyeing industrial wastewater. *ASEAN Journal of Chemical Engineering*, *6*(1 & 2), 49–60.

Barredo-Damas, S., Alcaina-Miranda, M. I., Iborra-Clar, M. I., Bes-Pia, A., Mendoza-Roca, J. A., & Iborra-Clar, A. (2006). Study of the UF process as pretreatment of NF membranes for textile wastewater reuse. *Desalination*, *200*(1-3), 745–747. doi:10.1016/j.desal.2006.03.497

Bhattacharya, A., Rawlins, J. W., & Ray, P. (2008). *Polymer grafting and crosslinking*. John Wiley & Sons. doi:10.1002/9780470414811

Bolto, B., & Gregory, J. (2007). Organic polyelectrolytes in water treatment. *Water Research*, *41*(11), 2301–2324. doi:10.1016/j.watres.2007.03.012 PMID:17462699

Bratby, J. (2006). *Coagulation and flocculation in water and wastewater treatment*. IWA Publishing.

Bratby, J. (2016). *Coagulation and flocculation in water and wastewater treatment*. IWA Publishing. doi:10.2166/9781780407500

Cao, Z., Ge, H., & Lai, S. (2001). Studies on synthesis and adsorption properties of chitosan cross-linked by glutaraldehyde and Cu (II) as template under microwave irradiation. *European Polymer Journal*, *37*(10), 2141–2143. doi:10.1016/S0014-3057(01)00070-2

Chen, T., Gao, B., & Yue, Q. (2010). Effect of dosing method and pH on color removal performance and floc aggregation of polyferric chloride–polyamine dual-coagulant in synthetic dyeing wastewater treatment. *Colloids and Surfaces. A, Physicochemical and Engineering Aspects*, *355*(1-3), 121–129. doi:10.1016/j.colsurfa.2009.12.008

Choo, K.-H., Choi, S.-J., & Hwang, E.-D. (2007). Effect of coagulant types on textile wastewater reclamation in a combined coagulation/ultrafiltration system. *Desalination*, *202*(1-3), 262–270. doi:10.1016/j.desal.2005.12.063

Choy, S. Y., Prasad, K. M. N., Wu, T. Y., Raghunandan, M. E., & Ramanan, R. N. (2014). Utilization of plant-based natural coagulants as future alternatives towards sustainable water clarification. *Journal of Environmental Sciences (China)*, *26*(11), 2178–2189. doi:10.1016/j.jes.2014.09.024 PMID:25458671

Choy, S. Y., Prasad, K. N., Wu, T. Y., Raghunandan, M. E., & Ramanan, R. N. (2016). Performance of conventional starches as natural coagulants for turbidity removal. *Ecological Engineering*, *94*, 352–364. doi:10.1016/j.ecoleng.2016.05.082

Chung, C. Y., Selvarajoo, A., Sethu, V., Koyande, A. K., Arputhan, A., & Lim, Z. C. (2018). Treatment of palm oil mill effluent (POME) by coagulation flocculation process using peanut–okra and wheat germ–okra. *Clean Technologies and Environmental Policy*, *20*(9), 1951–1970. doi:10.100710098-018-1619-y

Ciabatti, I., Tognotti, F., & Lombardi, L. (2010). Treatment and reuse of dyeing effluents by potassium ferrate. *Desalination*, *250*(1), 222–228. doi:10.1016/j.desal.2009.06.019

Diaz, A., Rincon, N., Escorihuela, A., Fernandez, N., Chacin, E., & Forster, C. (1999). A preliminary evaluation of turbidity removal by natural coagulants indigenous to Venezuela. *Process Biochemistry*, *35*(3), 391–395. doi:10.1016/S0032-9592(99)00085-0

Duan, J., & Gregory, J. (2003). Coagulation by hydrolysing metal salts. *Advances in Colloid and Interface Science*, *100*, 475–502. doi:10.1016/S0001-8686(02)00067-2

El-Gohary, F., & Tawfik, A. (2009). Decolorization and COD reduction of disperse and reactive dyes wastewater using chemical-coagulation followed by sequential batch reactor (SBR) process. *Desalination*, *249*(3), 1159–1164. doi:10.1016/j.desal.2009.05.010

Entry, J. A., Sojka, R., Watwood, M., & Ross, C. (2002). Polyacrylamide preparations for protection of water quality threatened by agricultural runoff contaminants. *Environmental Pollution*, *120*(2), 191–200. doi:10.1016/S0269-7491(02)00160-4 PMID:12395830

Fakhru'L-Razi, A., Qudsieh, I. Y. M., Yunus, W. M. Z. W., Ahmad, M. B., & Rahman, M. Z. A. (2001). Graft copolymerization of methyl methacrylate onto sago starch using ceric ammonium nitrate and potassium persulfate as redox initiator systems. *Journal of Applied Polymer Science*, *82*(6), 1375–1381. doi:10.1002/app.1974

Flaten, T. P. (2001). Aluminium as a risk factor in Alzheimer's disease, with emphasis on drinking water. *Brain Research Bulletin*, *55*(2), 187–196. doi:10.1016/S0361-9230(01)00459-2 PMID:11470314

Gebbie, P. (2001). *Using polyaluminium coagulants in water treatment.* Paper presented at the Proc. 64th AWIEO Conference.

Golob, V., Vinder, A., & Simonič, M. (2005). Efficiency of the coagulation/flocculation method for the treatment of dyebath effluents. *Dyes and Pigments*, *67*(2), 93–97. doi:10.1016/j.dyepig.2004.11.003

Gregory, J., & Duan, J. (2001). Hydrolyzing metal salts as coagulants. *Pure and Applied Chemistry*, *73*(12), 2017–2026. doi:10.1351/pac200173122017

Gupta, K., Sahoo, S., & Khandekar, K. (2002). Graft copolymerization of ethyl acrylate onto cellulose using ceric ammonium nitrate as initiator in aqueous medium. *Biomacromolecules*, *3*(5), 1087–1094. doi:10.1021/bm020060s PMID:12217057

Güven, O. (2012). *Crosslinking and scission in polymers* (Vol. 292). Springer Science & Business Media.

Hai, F. I., Yamamoto, K., & Fukushi, K. (2007). Hybrid treatment systems for dye wastewater. *Critical Reviews in Environmental Science and Technology*, *37*(4), 315–377. doi:10.1080/10643380601174723

Harford, A. J., Hogan, A. C., Jones, D. R., & van Dam, R. A. (2011). Ecotoxicological assessment of a polyelectrolyte flocculant. *Water Research*, *45*(19), 6393–6402. doi:10.1016/j.watres.2011.09.032 PMID:22000718

Heffron, J., Marhefke, M., & Mayer, B. K. (2016). Removal of trace metal contaminants from potable water by electrocoagulation. *Scientific Reports*, *6*(1), 28478. doi:10.1038rep28478 PMID:27324564

Ho, Y., Norli, I., Alkarkhi, A. F., & Morad, N. (2015). Extraction, characterization and application of malva nut gum in water treatment. *Journal of Water and Health*, *13*(2), 489–499. doi:10.2166/wh.2014.100 PMID:26042980

Hogg, R. (2013). Bridging flocculation by polymers. *Kona Powder and Particle Journal*, *30*(0), 3–14. doi:10.14356/kona.2013005

Hoover, M. F. (1970). Cationic quaternary polyelectrolytes—A literature review. *Journal of Macromolecular Science. Chemistry*, *4*(6), 1327–1418. doi:10.1080/00222337008081733

Huacai, G., Wan, P., & Dengke, L. (2006). Graft copolymerization of chitosan with acrylic acid under microwave irradiation and its water absorbency. *Carbohydrate Polymers*, *66*(3), 372–378. doi:10.1016/j.carbpol.2006.03.017

Huang, C., & Chen, Y. (1996). Coagulation of colloidal particles in water by chitosan. *Journal of Chemical Technology & Biotechnology: International Research in Process. Environmental AND Clean Technology*, *66*(3), 227–232.

Huang, C., & Shiu, H. (1996). Interactions between alum and organics in coagulation. *Colloids and Surfaces. A, Physicochemical and Engineering Aspects*, *113*(1-2), 155–163. doi:10.1016/0927-7757(96)03543-1

Jahn, S. A., & Dirar, H. (1979). Studies on natural water coagulants in the Sudan, with special reference to Moringa oleifera seeds. *Water S.A.*, *5*(2), 90–106.

Jahn, S. A. A. (1988). Using Moringa seeds as coagulants in developing countries. *Journal - American Water Works Association*, *80*(6), 43–50. doi:10.1002/j.1551-8833.1988.tb03052.x

Jiménez, B. (2005). Treatment technology and standards for agricultural wastewater reuse: a case study in Mexico. *Irrigation and Drainage: The Journal of the International Commission on Irrigation and Drainage, 54*(S1), S23-S33.

Joshi, J. M., & Sinha, V. K. (2007). Ceric ammonium nitrate induced grafting of polyacrylamide onto carboxymethyl chitosan. *Carbohydrate Polymers*, *67*(3), 427–435. doi:10.1016/j.carbpol.2006.06.021

Jusof Khadidi, M., & Hamid, E. A. (2013). *A new flocculant-coagulant with potential use for industrial wastewater treatment.* Paper presented at the 2nd International Conference on Environment, Energy and Biotechnology.

Katayon, S., Ng, S., Johari, M. M., & Ghani, L. A. (2006). Preservation of coagulation efficiency of Moringa oleifera, a natural coagulant. *Biotechnology and Bioprocess Engineering; BBE*, *11*(6), 489–495. doi:10.1007/BF02932072

Kawahara, M., & Kato-Negishi, M. (2011). Link between aluminum and the pathogenesis of Alzheimer's disease: The integration of the aluminum and amyloid cascade hypotheses. *International Journal of Alzheimer's Disease*, 2011. PMID:21423554

Kumar, C. G., Joo, H.-S., Choi, J.-W., Koo, Y.-M., & Chang, C.-S. (2004). Purification and characterization of an extracellular polysaccharide from haloalkalophilic Bacillus sp. I-450. *Enzyme and Microbial Technology*, *34*(7), 673–681. doi:10.1016/j.enzmictec.2004.03.001

Kumar, D., Pandey, J., Raj, V., & Kumar, P. (2017). A review on the modification of polysaccharide through graft copolymerization for various potential applications. *The Open Medicinal Chemistry Journal*, *11*(1), 109–126. doi:10.2174/1874104501711010109 PMID:29151987

Lea, W., Rohlich, G., & Katz, W. (1954). Removal of phosphates from treated sewage. *Sewage and Industrial Wastes*, 261–275.

Lee, C. S., Robinson, J., & Chong, M. F. (2014). A review on application of flocculants in wastewater treatment. *Process Safety and Environmental Protection*, *92*(6), 489–508. doi:10.1016/j.psep.2014.04.010

Li, T., Zhu, Z., Wang, D., Yao, C., & Tang, H. (2006). Characterization of floc size, strength and structure under various coagulation mechanisms. *Powder Technology*, *168*(2), 104–110. doi:10.1016/j.powtec.2006.07.003

Lim, B.-C., Lim, J.-W., & Ho, Y.-C. (2018). Garden cress mucilage as a potential emerging biopolymer for improving turbidity removal in water treatment. *Process Safety and Environmental Protection*, *119*, 233–241. doi:10.1016/j.psep.2018.08.015

Ma, J., Fu, K., Shi, J., Sun, Y., Zhang, X., & Ding, L. (2016). Ultraviolet-assisted synthesis of polyacrylamide-grafted chitosan nanoparticles and flocculation performance. *Carbohydrate Polymers*, *151*, 565–575. doi:10.1016/j.carbpol.2016.06.002 PMID:27474601

Matilainen, A., Vepsäläinen, M., & Sillanpää, M. (2010). Natural organic matter removal by coagulation during drinking water treatment: A review. *Advances in Colloid and Interface Science*, *159*(2), 189–197. doi:10.1016/j.cis.2010.06.007 PMID:20633865

Matsui, Y., Yuasa, A., Furuya, Y., & Kamei, T. (1998). Dynamic analysis of coagulation with alum and PACl. *Journal - American Water Works Association*, *90*(10), 96–106. doi:10.1002/j.1551-8833.1998.tb08522.x

McEwen, J. B. (1998). *Treatment process selection for particle removal*. American Water Works Association.

McLachlan, D., Bergeron, C., Smith, J., Boomer, D., & Rifat, S. (1996). Risk for neuropathologically confirmed Alzheimer's disease and residual aluminum in municipal drinking water employing weighted residential histories. *Neurology*, *46*(2), 401–405. doi:10.1212/WNL.46.2.401 PMID:8614502

Meriç, S., Selçuk, H., & Belgiorno, V. (2005). Acute toxicity removal in textile finishing wastewater by Fenton's oxidation, ozone and coagulation–flocculation processes. *Water Research*, *39*(6), 1147–1153. doi:10.1016/j.watres.2004.12.021 PMID:15766969

Mishra, A., & Bajpai, M. (2006). The flocculation performance of Tamarindus mucilage in relation to removal of vat and direct dyes. *Bioresource Technology*, *97*(8), 1055–1059. doi:10.1016/j.biortech.2005.04.049 PMID:16219461

Mishra, A., Srinivasan, R., & Dubey, R. (2002). Flocculation of textile wastewater by Plantago psyllium mucilage. *Macromolecular Materials and Engineering*, *287*(9), 592–596. doi:10.1002/1439-2054(20020901)287:9<592::AID-MAME592>3.0.CO;2-Q

Mohamadnia, Z., Zohuriaan-Mehr, M. J., Kabiri, K., & Razavi-Nouri, M. (2008). Tragacanth gum-graft-polyacrylonitrile: Synthesis, characterization and hydrolysis. *Journal of Polymer Research*, *15*(3), 173–180. doi:10.100710965-007-9156-0

Mostafa, K. M., Samarkandy, A. R., & El-Sanabary, A. A. (2011). Grafting onto carbohydrate polymer using novel potassium persulfate/tetramethylethylene diamine redox system for initiating grafting. *Advances in Polymer Technology*, *30*(2), 138–149. doi:10.1002/adv.20210

Mundargi, R. C., Agnihotri, S. A., Patil, S. A., & Aminabhavi, T. M. (2006). Graft copolymerization of methacrylic acid onto guar gum, using potassium persulfate as an initiator. *Journal of Applied Polymer Science*, *101*(1), 618–623. doi:10.1002/app.23325

Muryanto, M., Marlina, E., Sari, A. A., Harimawan, A., & Sudarno, S. (2018). Treatment of beverage industry wastewater using a combination of electrocoagulation and adsorption processes. In *AIP Conference Proceedings*. 10.1063/1.5064290

Muyibi, S. A., & Evison, L. M. (1996). Coagulation of turbid water and softening of hardwater with Moringa oleifera seeds. *The International Journal of Environmental Studies*, *49*(3), 247–259. doi:10.1080/00207239608711028

Muyibi, S. A., & Okuofu, C. A. (1995). Coagulation of low turbidity surface waters with Moringa oleifera seeds. *The International Journal of Environmental Studies*, *48*(3-4), 263–273. doi:10.1080/00207239508710996

Ndabigengesere, A., & Narasiah, K. S. (1998). Quality of water treated by coagulation using Moringa oleifera seeds. *Water Research*, *32*(3), 781–791. doi:10.1016/S0043-1354(97)00295-9

Okuda, T., Baes, A. U., Nishijima, W., & Okada, M. (2001). Isolation and characterization of coagulant extracted from Moringa oleifera seed by salt solution. *Water Research*, *35*(2), 405–410. doi:10.1016/S0043-1354(00)00290-6 PMID:11228992

Oyegbile, B., Ay, P., & Narra, S. (2016). Optimization of physicochemical process for pre-treatment of fine suspension by flocculation prior to dewatering. *Desalination and Water Treatment*, *57*(6), 2726–2736. doi:10.1080/19443994.2015.1043591

Paraskeva, P., & Diamadopoulos, E. (2006). Technologies for olive mill wastewater (OMW) treatment: a review. *Journal of Chemical Technology & Biotechnology: International Research in Process. Environmental & Clean Technology*, *81*(9), 1475–1485.

Parsons, S. A., & Jefferson, B. (2006). *Introduction to potable water treatment processes*. Blackwell Publishing. doi:10.1002/9781444305470

Patil, S. V., Patil, C. D., Salunke, B. K., Salunkhe, R. B., Bathe, G., & Patil, D. M. (2011). Studies on characterization of bioflocculant exopolysaccharide of Azotobacter indicus and its potential for wastewater treatment. *Applied Biochemistry and Biotechnology*, *163*(4), 463–472. doi:10.100712010-010-9054-5 PMID:20730607

Peavy, H. S., Rowe, D. R., & Tchobanoglous, G. (2019). *Environmental engineering/H. S. Peavy, D. R. Rowe, G. Tchobanoglous.*

Piriyaprasarth, S., & Sriamornsak, P. (2011). Flocculating and suspending properties of commercial citrus pectin and pectin extracted from pomelo (Citrus maxima) peel. *Carbohydrate Polymers*, *83*(2), 561–568. doi:10.1016/j.carbpol.2010.08.018

Pokhrel, D., & Viraraghavan, T. (2004). Treatment of pulp and paper mill wastewater—A review. *The Science of the Total Environment*, *333*(1-3), 37–58. doi:10.1016/j.scitotenv.2004.05.017 PMID:15364518

Pritchard, M., Craven, T., Mkandawire, T., Edmondson, A., & O'Neill, J. (2010). A comparison between Moringa oleifera and chemical coagulants in the purification of drinking water–An alternative sustainable solution for developing countries. *Physics and Chemistry of the Earth Parts A/B/C*, *35*(13-14), 798–805. doi:10.1016/j.pce.2010.07.014

Rahman, L., Silong, S., Zin, W. M., Rahman, M. Z. A., Ahmad, M., & Haron, J. (2000). Graft copolymerization of methyl acrylate onto sago starch using ceric ammonium nitrate as an initiator. *Journal of Applied Polymer Science*, *76*(4), 516–523. doi:10.1002/(SICI)1097-4628(20000425)76:4<516::AID-APP9>3.0.CO;2-7

Rahul, R., Kumar, S., Jha, U., & Sen, G. (2015). Cationic inulin: A plant based natural biopolymer for algal biomass harvesting. *International Journal of Biological Macromolecules*, *72*, 868–874. doi:10.1016/j.ijbiomac.2014.09.039 PMID:25304749

Raval, D., Patel, R., & Patel, V. (1988). Graft copolymerization of acrylonitrile on to guar gum by hydrogen peroxide initiation. *Stärke*, *40*(2), 66–69. doi:10.1002tar.19880400207

Ravina, L., & Moramarco, N. (1993). Everything you want to know about Coagulation & Flocculation. *Zeta-Meter, Inc.*, 1-37.

Ray, S. K., Majumdera, C., & Sahab, P. (2018). Removal of turbidity and E. coli from surface water by electrocoagulation and study of its economic feasibility. *Journal of the Indian Chemical Society*, *95*(3), 343–348.

Rodrigues, A. C., Boroski, M., Shimada, N. S., Garcia, J. C., Nozaki, J., & Hioka, N. (2008). Treatment of paper pulp and paper mill wastewater by coagulation–flocculation followed by heterogeneous photocatalysis. *Journal of Photochemistry and Photobiology A Chemistry*, *194*(1), 1–10. doi:10.1016/j.jphotochem.2007.07.007

Rondeau, V., Commenges, D., Jacqmin-Gadda, H., & Dartigues, J.-F. (2000). Relation between aluminum concentrations in drinking water and Alzheimer's disease: An 8-year follow-up study. *American Journal of Epidemiology*, *152*(1), 59–66. doi:10.1093/aje/152.1.59 PMID:10901330

Salehizadeh, H., Yan, N., & Farnood, R. (2017). Recent advances in polysaccharide bio-based flocculants. *Biotechnology Advances*. PMID:28993221

Sanghi, R., Bhattacharya, B., Dixit, A., & Singh, V. (2006). Ipomoea dasysperma seed gum: An effective natural coagulant for the decolorization of textile dye solutions. *Journal of Environmental Management*, *81*(1), 36–41. doi:10.1016/j.jenvman.2005.09.015 PMID:16549236

Sarkar, B., Chakrabarti, P., Vijaykumar, A., & Kale, V. (2006). Wastewater treatment in dairy industries—Possibility of reuse. *Desalination*, *195*(1-3), 141–152. doi:10.1016/j.desal.2005.11.015

Sathiyanarayanan, G., Kiran, G. S., & Selvin, J. (2013). Synthesis of silver nanoparticles by polysaccharide bioflocculant produced from marine Bacillus subtilis MSBN17. *Colloids and Surfaces. B, Biointerfaces*, *102*, 13–20. doi:10.1016/j.colsurfb.2012.07.032 PMID:23006551

Šćiban, M., Klašnja, M., Antov, M., & Škrbić, B. (2009). Removal of water turbidity by natural coagulants obtained from chestnut and acorn. *Bioresource Technology*, *100*(24), 6639–6643. doi:10.1016/j.biortech.2009.06.047 PMID:19604691

Sen, G., Mishra, S., Jha, U., & Pal, S. (2010). Microwave initiated synthesis of polyacrylamide grafted guar gum (GG-g-PAM)—Characterizations and application as matrix for controlled release of 5-amino salicylic acid. *International Journal of Biological Macromolecules*, *47*(2), 164–170. doi:10.1016/j.ijbiomac.2010.05.004 PMID:20471416

Shak, K. P. Y., & Wu, T. Y. (2014). Coagulation–flocculation treatment of high-strength agro-industrial wastewater using natural Cassia obtusifolia seed gum: Treatment efficiencies and flocs characterization. *Chemical Engineering Journal*, *256*, 293–305. doi:10.1016/j.cej.2014.06.093

Shao, J., Yang, Y., & Zhong, Q. (2003). Studies on preparation of oligoglucosamine by oxidative degradation under microwave irradiation. *Polymer Degradation & Stability*, *82*(3), 395–398. doi:10.1016/S0141-3910(03)00177-0

Shoda, M., & Sugano, Y. (2005). Recent advances in bacterial cellulose production. *Biotechnology and Bioprocess Engineering; BBE*, *10*(1), 1–8. doi:10.1007/BF02931175

Sila, A., Mlaik, N., Sayari, N., Balti, R., & Bougatef, A. (2014). Chitin and chitosan extracted from shrimp waste using fish proteases aided process: Efficiency of chitosan in the treatment of unhairing effluents. *Journal of Polymers and the Environment*, *22*(1), 78–87. doi:10.100710924-013-0598-7

Singh, V., Kumar, P., & Sanghi, R. (2012). Use of microwave irradiation in the grafting modification of the polysaccharides–A review. *Progress in Polymer Science*, *37*(2), 340–364. doi:10.1016/j.progpolymsci.2011.07.005

Singh, V., Tiwari, A., Tripathi, D. N., & Sanghi, R. (2006). Microwave enhanced synthesis of chitosan-graft-polyacrylamide. *Polymer*, *47*(1), 254–260. doi:10.1016/j.polymer.2005.10.101

SPAN. N. W. S. C. (2010). Malaysia water service industry & water treatment sludge issues. Retrieved from http://www.jwrc-net.or.jp/aswin/projects-activities/trainee_files/20101004-15_CR02.pdf

Suopajärvi, T. (2015). *Functionalized nanocelluloses in wastewater treatment applications.*

Teh, C. Y., Budiman, P. M., Shak, K. P. Y., & Wu, T. Y. (2016). Recent advancement of coagulation–flocculation and its application in wastewater treatment. *Industrial & Engineering Chemistry Research*, *55*(16), 4363–4389. doi:10.1021/acs.iecr.5b04703

Teh, C. Y., Wu, T. Y., & Juan, J. C. (2014). Potential use of rice starch in coagulation–flocculation process of agro-industrial wastewater: Treatment performance and flocs characterization. *Ecological Engineering*, *71*, 509–519. doi:10.1016/j.ecoleng.2014.07.005

Thaker, M., & Trivedi, H. (2005). Ultraviolet-radiation-induced graft copolymerization of methyl acrylate onto the sodium salt of partially carboxymethylated guar gum. *Journal of Applied Polymer Science*, *97*(5), 1977–1986. doi:10.1002/app.20988

Thombare, N., Jha, U., Mishra, S., & Siddiqui, M. (2016). Guar gum as a promising starting material for diverse applications: A review. *International Journal of Biological Macromolecules*, *88*, 361–372. doi:10.1016/j.ijbiomac.2016.04.001 PMID:27044346

Tripathy, T., & De, B. R. (2006). Flocculation: a new way to treat the waste water.

Vandamme, D., Foubert, I., & Muylaert, K. (2013). Flocculation as a low-cost method for harvesting microalgae for bulk biomass production. *Trends in Biotechnology*, *31*(4), 233–239. doi:10.1016/j.tibtech.2012.12.005 PMID:23336995

Venugopal, V. (2016). Marine polysaccharides: Food applications. Boca Raton, FL: CRC Press. doi:10.1201/b10516

Verma, A. K., Dash, R. R., & Bhunia, P. (2012). A review on chemical coagulation/flocculation technologies for removal of colour from textile wastewaters. *Journal of Environmental Management*, *93*(1), 154–168. doi:10.1016/j.jenvman.2011.09.012 PMID:22054582

Vik, E. A., Carlson, D. A., Eikum, A. S., & Gjessing, E. T. (1984). Electrocoagulation of potable water. *Water Research*, *18*(11), 1355–1360. doi:10.1016/0043-1354(84)90003-4

Wang, J.-P., Chen, Y.-Z., Wang, Y., Yuan, S.-J., & Yu, H.-Q. (2011). Optimization of the coagulation-flocculation process for pulp mill wastewater treatment using a combination of uniform design and response surface methodology. *Water Research*, *45*(17), 5633–5640. doi:10.1016/j.watres.2011.08.023 PMID:21920576

Wang, J.-P., Yuan, S.-J., Wang, Y., & Yu, H.-Q. (2013). Synthesis, characterization and application of a novel starch-based flocculant with high flocculation and dewatering properties. *Water Research*, *47*(8), 2643–2648. doi:10.1016/j.watres.2013.01.050 PMID:23531592

Wang, L. K., Hung, Y.-T., & Shammas, N. K. (2005). *Physicochemical treatment processes* (Vol. 3). Springer. doi:10.1385/159259820x

Wong, S., Teng, T., Ahmad, A., Zuhairi, A., & Najafpour, G. (2006). Treatment of pulp and paper mill wastewater by polyacrylamide (PAM) in polymer induced flocculation. *Journal of Hazardous Materials*, *135*(1-3), 378–388. doi:10.1016/j.jhazmat.2005.11.076 PMID:16431022

Yang, R., Li, H., Huang, M., Yang, H., & Li, A. (2016). A review on chitosan-based flocculants and their applications in water treatment. *Water Research*, *95*, 59–89. doi:10.1016/j.watres.2016.02.068 PMID:26986497

Yang, Z., Wang, W., & Liu, S. (2017). Flocculation of coal waste slurry using bioflocculant produced by Azotobacter chroococcum. *Energy & Fuels*, *31*(2), 1460–1467. doi:10.1021/acs.energyfuels.6b03052

Yavuz, Y., & Ögütveren, Ü. (2018). Treatment of industrial estate wastewater by the application of electrocoagulation process using iron electrodes. *Journal of Environmental Management*, *207*, 151–158. doi:10.1016/j.jenvman.2017.11.034 PMID:29161644

Zhang, J., Liu, Z., Wang, S., & Jiang, P. (2002). Characterization of a bioflocculant produced by the marine myxobacterium Nannocystis sp. NU-2. *Applied Microbiology and Biotechnology*, *59*(4-5), 517–522. doi:10.100700253-002-1023-7 PMID:12172619

Zhou, Y., Liang, Z., & Wang, Y. (2008). Decolorization and COD removal of secondary yeast wastewater effluents by coagulation using aluminum sulfate. *Desalination*, *225*(1-3), 301–311. doi:10.1016/j.desal.2007.07.010

Chapter 19
Adsorption of Dyes in Aqueous Medium Using RHA and CFA:
Effect of Preparation Methods and Process Optimization

Irvan Dahlan
Universiti Sains Malaysia, Malaysia

Sariyah Mahdzir
Universiti Sains Malaysia, Malaysia

ABSTRACT

The preparation of adsorbent from the mixture of rice husk ash (RHA) and coal fly ash (CFA) has been investigated for adsorption of acid violet 7 (AV7) and brilliant green (BG) dyes. The RHA-CFA adsorbents were prepared using three different methods, i.e. reflux, magnetic co-precipitation, and magnetic template. Five different additives were used in reflux method. The results showed that RHA-CFA adsorbent prepared through reflux methods using NaOH and Na2CO3 shows higher dyes adsorption removal as compared to other methods. From zeta potential analysis, the electric charge of the outer layer of prepared adsorbent shows no effect towards adsorption of AV7 and BG dyes. By using a 3-factor, 3-level factorial design, the relationship between all variables was studied. From the response surface models, the optimum adsorbent preparation variables could be obtained by using RHA-CFA adsorbent prepared by refluxing 3:1 ratio of RHA to CFA in 1.21 M NaOH solution. The results indicated that the optimized values agree reasonably well with the validated experimental results.

INTRODUCTION

Environmental pollution control has been a crucial concern to every country. In Malaysia, water pollution and solid waste disposal from many industries have become a major problem over the last decade. Water pollution usually caused by the effluent that comes from industrial and domestic sewage. One of the concerns in wastewater pollution is the presence of coloured compounds such as dyes. For industrial

DOI: 10.4018/978-1-7998-0369-0.ch019

usage, synthetic dyes are usually preferable because of its complex aromatic structures that provide resistance to light, biological activity, ozone and another degradative environment condition (Kaushik & Malik, 2014; Holkar *et al.*, 2016). Acid violet 7 (AV7) and brilliant green (BG) dyes are the example of synthetic dyes that are being used in textile and leather, paper, automotive, plastic, soap, wood, medicine and cosmetics of dyeing as well as in biological dyeing (Kumar *et al.*, 2012; Mansour *et al.*, 2009; Tavlieva *et al.*, 2013). Both dyes have complex and stable structure with carcinogenic, neurotoxicity and mutagenic properties. Therefore, the removal of these two types of dyes become important as the toxic compounds can endanger water organisms while the excess colourant can change the natural colour of water bodies.

Adsorption is one of the methods that has been used for removal of AV7 and BG dyes. Despite the widely used of commercial activated carbon (CAC) for removal of dye-containing wastewater, studies on low-cost materials such as industrial and agricultural solids wastes have been extensively performed to replace the CAC. These including rice husk ash (RHA), palm oil fuel ash (PFA) and coal fly ash (CFA) which are abundantly available through the combustion process of rice mill, palm oil mill and coal-fired power plant industries, respectively. Typically, for every 100 kg of paddy milled, about 20 kg (20%) of husk is being produced and after the husk being burnt, around 18% of ash (RHA) will be generated (Bronzeoak, 2003). From this conversion, it was estimated that more than 27.7 million tons of RHA is produced annually worldwide, of which Malaysia alone produces more than 104.4 thousand tons of RHA annually (FAO, 2019). While, PFA is a by-product produced by countries having palm oil industry such as Indonesia and Malaysia. This waste ash is generated by the incineration of palm oil fibers, empty fruit bunches and shells as fuel in palm oil mill boilers. By this incineration process, about 5% by weight of the PFA is produced. The quantity of PFA produced in Malaysia alone is about 4 million tons a year (Awang *et al.*, 2014). Meanwhile, coal fired thermal power plants have generated huge amount of coal bottom ash (CBA) and coal fly ash (CFA). Typically, CBA forms up to 25% - 90% of total ash generated while the remaining are CFA (Sadon *et al.*, 2017). Since 1988, Malaysia produces about 8.5 million tons of coal ash (Rafieizonooz *et al.*, 2016; Alhokabi & Ing, 2019).

Even though RHA, PFA, and CFA have been used (with or without modification) and possess good adsorption capacity towards dyes, however nearly all of studies only use one form of ash. Previously, we had reported the characteristics/properties of these ashes and adsorption capacity of adsorbent prepared from the mixture of rice husk ash (RHA), palm oil fuel ash (PFA) and coal fly ash (CFA). We have used this type of adsorbent (i.e. RHA/PFA/CFA sorbent) in removing heavy metals (Dahlan & Ismail, 2012; Dahlan & Razali, 2012; Dahlan *et al.*, 2013; Dahlan & Zwain, 2013; Zwain *et al.*, 2018) and dyes (Dahlan & Ismail, 2012; Dahlan & Noor, 2014). Nevertheless, our previous reports only dealt with activity measurement related to adsorbent preparation conditions by single method and effects of batch operating conditions. Less information is available about the effects of the various preparation methods for this kind of adsorbent. A facile one-step refluxing route in alkaline solution has been used as a method to prepare a magnetic activated carbon/$CoFe_2O_4$ composite (Ai *et al.*, 2010). This magnetic adsorbent has been used for the removal of malachite green dye from water. Their results indicated that $CoFe_2O_4$ particles deposited on the surface of activated carbon were uniform, with the particle size in the range of 14–20 nm. The preparation of adsorbent derived from pine cone has also been successfully synthesized by sulfuric acid reflux method. With sulfuric acid reflux, the pine cones undergone carbonization as well as functionalization with sulfonic acid groups. The adsorbent demonstrated high adsorption capacity for two emerging organic pollutants, *i.e.* methylene blue and tetracycline (Islam *et al.*, 2018). While Zhang *et al.* (2007) used magnetic co-precipitation method to prepare $CuFe_2O_4$/activated carbon magnetic adsorbents for the removal of acid orange II dye. The results demonstrated that the magnetic phase present is spinel copper ferrite and the presence of $CuFe_2O_4$ did not significantly affect the surface area and pore structure of the activated carbon. This co-precipitation method was also used by Bai *et al.* (2019) to synthesize Cu/Zn/Al-hydrotalcite-like compound for the removal of copper from electroplating wastewater. The preparation conditions of Cu/Zn/Al-hydrotalcite-like compound were optimized, and the saturated adsorption capacity reached 1000 mg/g at the optimal adsorption condition. Apart from that, magnetic binary oxide particles (MBOP) has been synthesized by Dhoble *et al.* (2011) using chitosan template method. This MBOP was effective to remove arsenic (III) in water. It was found that the arsenic uptake was rapid initially up to 3 h and then gradually slows down as it reaches to equilibrium. Similarly, Wang *et al.* (2014) used the same material and template method to synthesize magnetic nanoparticles impregnated chitosan beads as adsorbent to remove arsenic (III) and arsenic (IV) in water.

Furthermore, several variables (during the preparation of adsorbent) may affect the adsorbents characteristic and adsorption capacity. Therefore, design of experiment (DOE) is a commonly used method for determining the effect of adsorbent preparation variable and relationship between variable (Montgomery 2017). Response surface methodology (RSM) has been widely used for studying complex processes and to determine the optimum process variables as it requires a minimum number of experiments to be performed. However, there was less study performed using 3-level factorial design (3-LFD) for designing the experiment for the removal of dyes wastewater and no research yet to be found in its application on adsorption removal of AV7 and BG dyes.

In this current study, the adsorbent was only prepared by combination of two ashes, i.e. RHA and CFA, with the intention to lessen complexity of the prepared adsorbent. Since many methods available to synthesize adsorbent, therefore in this study, three methods were chosen from the above-mentioned literature based on their simplicity, ease of preparation, and ability to tailor their unique properties that make them a favourable adsorbent for dye adsorption. Thus, the present study aims to determine the best method for preparation of RHA-CFA adsorbent for the removal of AV7 and BG dyes. Despite many reports being published on the effect of preparation variables and optimization of solid waste adsorbents, there were still no studies reported on the variable's optimization of RHA-CFA adsorbent. Thus, a 3-factor, 3-level factorial design (3-F 3-LFD) combined with the response surface methodology (RSM) were applied to examine the significant preparation variables affecting RHA-CFA adsorbent efficiency towards AV7 and BG dyes and as well as to obtain the optimized RHA-CFA adsorbent.

METHODOLOGY

Materials

The rice husk ash (RHA) was obtained from Kilang Beras & Minyak Sin Guan Hup Sdn. Bhd. Nibong Tebal, Penang and it is in the form of black coarse dry ash with the broken rice husk-like structure. While coal fly ash (CFA) was obtained from Kapar Coal-Fired Power Plant in the form of brownish coarse grains. Acid violet 7 (AV7) and brilliant green (BG) dyes were purchased from Sigma-Aldrich, Malaysia. All other chemicals used in this study were of analytical grade.

Preparation of RHA-CFA Adsorbent

Before the preparation process, both of raw RHA and CFA were washed with deionised water at least two times to remove impurities. Then, dried in the oven with 110 °C overnight to remove extra moisture before ground separately. Then the ashes were sieved through a stack of U.S. standard sieves into a fine particle size passing the 75 µm sieve. The preparation methods of the adsorbents include (a) reflux (b) magnetic co-precipitation, and (c) magnetic template.

Reflux Method

This method was carried out according to the method described by Ai *et al.* (2010) with some modifications. About 20 g by weight of RHA and CFA (1:1) were mixed with 250 mL 1M of sodium hydroxide (NaOH) solution and stirred at 80°C for 2 h. Then, the products were washed repeatedly with deionised water using the filtration method with help of vacuum pressured pump (Barnant Co., Model 400-1902) until the filtrate become neutral (around pH 7). The standard qualitative filter paper (No. 102) was used in the filtration process. Afterward, the adsorbent was dried at 110°C overnight. The prepared adsorbent was sealed properly in the plastic bag to keep it dry and stored in the dry place. The adsorbent was labelled as R_{NaOH}. The experiment was repeated by replacing NaOH with sodium carbonate (Na_2CO_3), hydrochloric acid (HCl), iron (III) chloride ($FeCl_3$) and calcium chloride ($CaCl_2$), where the adsorbents were then labelled as R_{Na2CO3}, R_{HCl}, R_{FeCl3} and R_{CaCl2}, respectively.

Magnetic Co-Precipitation Method

Following the previous reported procedure (Zhang *et al.*, 2007) with modification, about 15 g of RHA and CFA (1:1) were added to a 400 mL solution containing the mixture of 0.02 mol of copper (II) chloride ($CuCl_2$) and 0.04 mol of $FeCl_3$ at room temperature. Under vigorous magnetic-stirring, 5 mol/L of NaOH solution was added dropwise to raise the suspension pH to around 10 and stirring was continued for 1 hour. Afterwards, the suspension was heated to 98–100°C for 2 hours. After cooling, the prepared magnetic composite was repeatedly washed with deionised water. By a simple magnetic procedure (*i.e.* using magnetic materials) to attract the magnetic components in the solution, the obtained materials were separated from water and dried in an oven at 105°C. The prepared adsorbent was sealed properly in the plastic bag to keep it dry and stored in the dry place. The adsorbent was labelled as $M_{Co\text{-}ppt}$.

Magnetic Template Method

In this method, the adsorbent was prepared according to previous work reported by Dhoble *et al.* (2011). About 30 g of RHA and CFA (1:1) were dissolved in 900 mL of 5% acetic acid with constant stirring on a mechanical stirrer for 1 hour. In one beaker, about 84.78 g of aluminium nitrate was dissolved in 100 mL of distilled water and in another beaker, about 97.89 g of ferric nitrate was dissolved in 100 ml of distilled water. Aluminium nitrate ($Al(NO_3)_3$) and ferrous nitrate ($Fe(NO_3)_3$) solutions then were added into the mixture with stirring for 180 min. The resulting Al-Fe-ash slurry was added dropwise into NH_4OH solution (50% v/v) using a syringe pump and left for 60 min to stabilize in NH_4OH solution before washed with deionised water and dried in the oven at 70°C for 24 hours. The adsorbent was then calcined at 450°C for 6 hours in a muffle furnace. Finally, the calcined product was subjected to multiple washing with deionised water and dried at 80°C. The prepared adsorbent was sealed properly in the plastic bag to keep it dry and stored in the dry place. The adsorbent was labelled as M_{Temp}.

Analysis

Zeta potential analysis was used to find whether the outer charge of the adsorbents affect the adsorption efficiency. The zeta potential analysis was carried out using Malvern Zetasizer Nano ZS.

Experimental Design

In this study, Design Expert software (ver.6.0.6, STAT-EASE Inc., Minneapolis, MN) with a 3-factor, 3-level factorial design (3-F, 3-LFD) combined with the response surface methodology (RSM) was used to generate the experimental design, statistical analysis, and regression model. The correlation between two numerical factors, i.e. A: ash ratio (0.3–3 g/g) and B: additive concentration (1–3 mol/L), and one categorical factor, i.e. type of additives (NaOH and Na_2CO_3) in the adsorbent preparation toward the dye removal efficiency were studied. A total of 26 different combinations in the preparation of RHA-CFA adsorbents were performed in random order according to the 3-F, 3-LFD configuration as shown in Table 1. The resulting RHA-CFA adsorbents were tested for AV7 and BG dyes removal efficiencies and the results are also presented in Table 1. The percentage of AV7 and BG dyes adsorbed were represented by Y_1 and Y_2, respectively. The data obtained were then fitted to the following second-order polynomial equation (Montgomery, 2017).

$$Y = b_0 + \sum_{i=1}^{n} b_i x_i + \sum_{i=1}^{n} b_{ii} x_i^2 + \sum_{i=1}^{n-1} \sum_{j=i+1}^{n} b_{ij} x_i x_j \quad (1)$$

where Y is the predicted removal of AV7 and BG (%), b_0 is the offset term (constant), b_i are the linear coefficients, b_{ii} are the quadratic coefficients, b_{ij} are the interaction coefficients, and x_i, x_j are the coded values of the adsorbent preparation variables. Analysis of variance (ANOVA) was carried out to evaluate the significance of the above model. The model was then used to obtain the optimized RHA-CFA adsorbent.

Adsorption Studies

The adsorption procedures were adopted from the standard adsorption procedures by referring to the previous work done by Nandi et al. (2009) and Chowdhury et al. (2011). In this experiment, 200 mg/L dye was used as the adsorbate. The solute was prepared by the dilution of the prepared stock dye solution. Adsorption experiment was carried out by adding a certain amount of R_{NaOH} adsorbent (0.1 - 0.5 g) into a 250 mL conical flask filled with 100 mL of dye (200 mg/L of either AV7 or BG). The range amount of the adsorbent was selected by the doing simple pre-test (trial-and-error procedure) adsorption to find the suitable ranges for adsorption of the dyes. The conical flasks were sealed and placed in a shaker (Sartorius, Ceriomat® SII), shaken at 150 rpm at 30°C for 2 hours. The flasks were then removed from the shaker and the final concentration of dye in the solutions were measured using DR 2500 spectrophotometer (Shimadzu, Japan) at wavelengths of 520 nm (Mansour et al., 2007) and 623 nm (Mane et al., 2007) for AV7 and BG, respectively. The results were taken in term of absorbance (Abs) unit. Every experiment was run in triplicate to increase the results accuracy and the average value was determined. The removal efficiency of dye can be calculated as follows:

$$Removal\,efficiency\left(\%\right) = \frac{C_0 - C_e}{C_0} \times 100 \quad (2)$$

where C_0 (mg/L) is an initial concentration of dye solution while C_e (mg/L) is the final concentration of dye solution.

RESULTS AND DISCUSSION

Effect of Different Preparation Methods

Three different methods were performed to choose the best RHA-CFA adsorbent prepared by comparing the percentage removal of each adsorbent towards two synthetic dyes, *i.e.* AV7 and BG. In this study, five different additives were used in reflux methods which include NaOH (R_{NaOH}), $FeCl_3$ (R_{FeCl3}), Na_2CO_3 (R_{Na2CO3}), $CaCl_2$ (R_{CaCl2}), and HCl (R_{HCl}). Other methods are magnetic co-precipitation ($M_{Co\text{-}ppt}$) and magnetic template (M_{Temp}). The results for the adsorption of AV7 and BG dyes was presented in Figure 1 and Figure 2, respectively.

Generally, it can be seen from both figures, the removal of AV7 dye is lower than BG dye. This is due to the properties of AV7 dye which is more soluble (solubility: 120 g/L) in water as compared to BG dye (solubility: 40 g/L), thus the high affinity of the water for polar molecules will hold AV7 molecules and prevent it from being bound onto the surface of the adsorbents (Cooney 1998). From the Figure 1 it can be clearly seen that adsorbent R_{CaCl2} and M_{Temp} show different trends from the others. Theoretically, increasing the adsorbent dosage increases removal percentage. However, adsorbent R_{CaCl2} and M_{Temp} behave oppositely. Garg et al. (2003) suggested that the reason of decreasing removal percentage with increasing adsorbent dosage is due to the overlapping of adsorption sites caused by overcrowding of the adsorbent particles, however, Rao et al. (2011) and Yang et al. (2016) stated that at higher doses, adsorbent particles might aggregate/agglomerate, thus lead to decrease in the total area and an increase in the diffusional path length.

On the other hands, adsorbent R_{NaOH} and R_{Na2CO3} gave proportionally increasing adsorption removal as the dosage increases as compared to the other adsorbents. This is due to the functional groups present after refluxing with the additives NaOH and Na_2CO_3, respectively. The reaction of silica-based materials with NaOH and Na_2CO_3 resulted in the formation of sodium silicate (Na_2SiO_3) which has possible covalent bond such as -O-, –C=O and –C-O- (Moody 1991). These bonds are stronger and can also provide additional binding sites on the adsorbent's surface. The presence of Na in the adsorbent surface might also be the reason for the higher adsorption efficiency of these two adsorbents as the substitution reaction occurs between Na atoms and AV7 dye molecules. The preferable adsorbent based on their adsorption removal towards AV7 dye can be arranged as $R_{Na2CO3} > R_{NaOH} > M_{Co\text{-}ppt} > R_{FeCl3} > R_{HCl} > R_{CaCl2} > M_{Temp}$.

Table 1. Experimental design matrix and percentage removal of AV7 and BG dyes

Sample	RHA/CFA Ratio	Additives Conc. (M)	Type of Additive	AV7 Adsorption (%)		BG Adsorption (%)	
				Experimental	Predicted	Experimental	Predicted
S1	0.30	1	NaOH	17.59	16.06	41.71	51.53
S2	1.65	1	NaOH	36.69	36.54	77.27	69.88
S3	3.00	1	NaOH	41.87	45.14	80.13	74.45
S4	0.30	2	NaOH	13.21	15.97	69.81	60.09
S5	1.65	2	NaOH	39.29	41.20	80.35	77.52
S6	3.00	2	NaOH	54.68	54.54	77.94	81.17
S7	0.30	3	NaOH	14.51	16.97	64.53	62.22
S8	1.65	3	NaOH	48.39	46.95	79.91	78.72
S9	3.00	3	NaOH	64.46	65.03	81.20	81.45
S10	1.65	2	NaOH	44.86	41.20	73.23	77.52
S11	1.65	2	NaOH	46.38	41.20	79.47	77.52
S12	1.65	2	NaOH	41.15	41.20	75.04	77.52
S13	1.65	2	NaOH	40.10	41.20	66.52	77.52
S14	0.30	1	Na_2CO_3	15.53	18.66	48.94	45.12
S15	1.65	1	Na_2CO_3	30.70	29.96	73.12	72.47
S16	3.00	1	Na_2CO_3	33.34	29.37	78.34	86.05
S17	0.30	2	Na_2CO_3	20.63	14.85	33.78	36.75
S18	1.65	2	Na_2CO_3	36.92	30.89	70.01	63.18
S19	3.00	2	Na_2CO_3	33.32	35.05	77.85	75.83
S20	0.30	3	Na_2CO_3	13.18	12.14	18.91	21.95
S21	1.65	3	Na_2CO_3	32.02	32.93	43.76	47.45
S22	3.00	3	Na_2CO_3	43.28	41.83	62.67	59.18
S23	1.65	2	Na_2CO_3	34.74	30.89	61.37	63.18
S24	1.65	2	Na_2CO_3	25.93	30.89	61.21	63.18
S25	1.65	2	Na_2CO_3	25.81	30.89	63.67	63.18
S26	1.65	2	Na_2CO_3	23.84	30.89	67.09	63.18

Similarly, adsorbent R_{NaOH} and R_{Na2CO3} which were prepared through the reflux method, gave higher BG adsorption removal as shown in Figure 2. Unlike R_{NaOH} and R_{Na2CO3}, as shown in Figure 1 and Figure 2, the adsorption efficiency of adsorbent R_{CaCl2}, R_{FeCl3} and R_{HCl} show low AV7 and BG dyes adsorption removal. The reaction of additive $FeCl_3$ with water molecules resulted in the formation of Fe_2O_3 as major product and HCl as a side product. As the major composition of RHA and CFA is SiO_2, the presence of side product HCl in the preparation process slightly inhibits the improvement of prepared adsorbents because silica is an acidic oxide in which it only reacts with hydrofluoric acid and not with other types of acids (Moody 1991). However, the formation of Fe_2O_3 might contribute in the adsorption of dyes. Comparing between adsorbent that was prepared through magnetic co-precipitation ($M_{Co\text{-}ppt}$) and magnetic template (M_{Temp}) methods shown in Figure 2, the adsorption of BG dye increases with

Figure 1. Removal of AV7 dye using different prepared adsorbents.

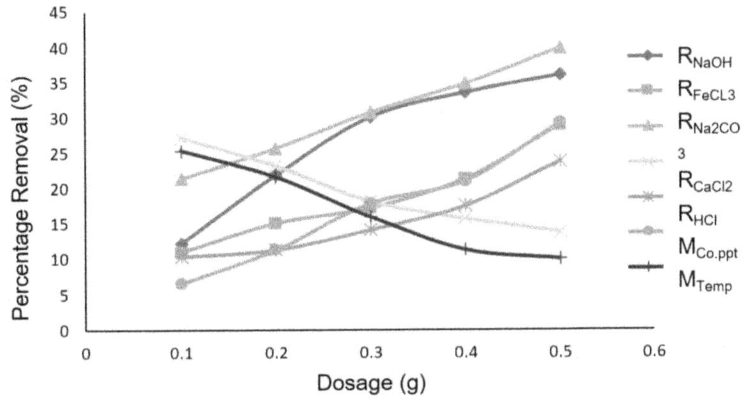

Figure 2. Removal of BG dye using different prepared adsorbents.

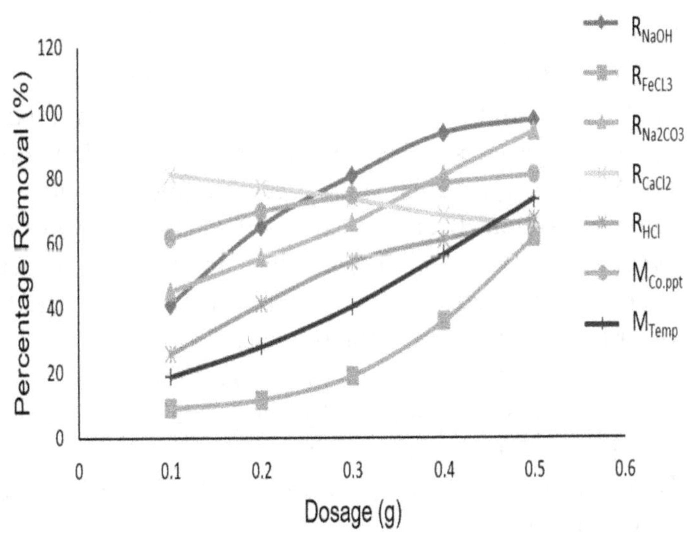

increasing adsorbent loading. Although M_{Temp} shows lower adsorption efficiency than $M_{Co\text{-}ppt}$ at a lower dosage (0.1g), increasing M_{Temp} dosage leads to increases the removal of dye proportionally. M_{Temp} shows unsatisfactory performance for AV7 dye removal. This is due to the characteristic of solute that has a different ionic charge that allows BG molecules bound onto adsorbent M_{Temp} surface effectively but not AV7 molecules. The adsorption of BG from highest to lowest adsorption efficiency can be concluded as $R_{NaOH} > R_{Na2CO3} > M_{Temp} > R_{FeCl3} > M_{Co\text{-}ppt} > R_{HCl} > R_{CaCl2}$.

Figure 3. Zeta potential values of all prepared adsorbents

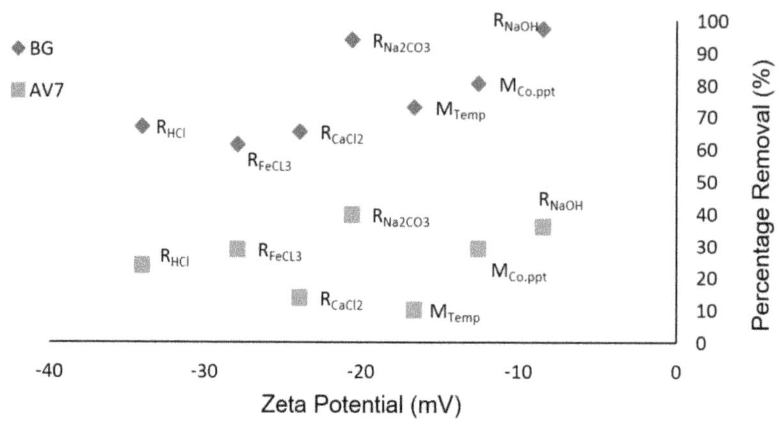

Zeta Potential Analysis

The electric charges of the outer layer of adsorbents were analysed to see the effect on dyes adsorption efficiency. Figure 3 shows that the adsorption trends for both AV 7 and BG dyes are similar although both have different properties. As discussed before, AV7 dye has a higher solubility that represents the high polarity, so that its molecules tend to remain in the water than bonding to the adsorbent particles, leads to a lower percentage of dye adsorbed. The zeta analysis results of the prepared adsorbents are ranged from -6.7 to -.5 mV. As the zeta potential values decrease (more negative), the particles become more stable, hence reduce the adsorption efficiency. Comparing between all adsorbent prepared through reflux method, an addition of additive solution into the ashes during the adsorbent preparation process resulting in increasing zeta potential values. This is because, after reaction with basic additives such as NaOH and Na_2CO_3, the outer functional group of the particles might consist of hydroxyl and carbonate ions. The –OH and –CO bonds are far more stable and stronger than ionic –Cl bonding (that formed from the reaction with acidic additives: HCl and $FeCl_3$). Therefore, in water solution, less reaction occurs between –OH and –CO with water molecules compared to Cl^-, where Cl^- is easily undergoing ion exchange with water (OH^-) and forming HCl as a side product (Solomon & Fryhle, 2004), hence, give lower zeta potential values (more negative).

The zeta potential value of R_{Na2CO3} is lower than R_{NaOH} which it possibly due to the carbon atom that forming =O and –O- after dissolving in water. Thus, increase the negative charge density of the particles. For acidic additives (HCl and $FeCl_3$), the zeta values are decreasing (more negative) with decreasing the valence of the additives. In water solution, the monovalent compound is more reactive, followed by divalent and trivalent compounds (Solomon & Fryhle, 2004), hence decrease the negative charge of the adsorbent. $M_{Co\text{-}ppt}$ shows higher zeta value compared to M_{Temp} with zeta values of -2.55 and -16.6 mV respectively and might be the reason that caused the adsorption for both dyes being higher for $M_{Co\text{-}ppt}$ compared to M_{Temp}. Zeta values for $M_{Co\text{-}ppt}$ and M_{Temp} are higher than other reflux methods except for R_{NaOH}. High zeta value of $M_{Co\text{-}ppt}$ and M_{Temp} could be caused by the use of NaOH and NH_4OH during the last step of adsorbent preparation. The possible presence of functional groups (such as –O–, –CO and –N–) at the active site for dyes binding leads to increase their adsorption efficiency (Solomon & Fryhle,

2004). However, the adsorption of $M_{Co\text{-}ppt}$ and M_{Temp} is lower than adsorbent R_{NaOH} due to the complexity of the adsorbents formed during the adsorbent preparation.

In conclusion, the adsorbents prepared through reflux method by using NaOH (R_{NaOH}) and Na_2CO_3 (R_{Na2CO3}) show the highest AV7 and BG dyes adsorption. From zeta potential test, although adsorbents prepared from magnetic methods ($M_{Co\text{-}ppt}$ and M_{Temp}) have zeta potential values near to 0, however, both $M_{Co\text{-}ppt}$ and M_{Temp} show lower dyes adsorption removal. Therefore, R_{NaOH} and R_{Na2CO3} adsorbents are chosen to be used for further study in obtaining the optimized RHA-CFA adsorbent.

Response Surface Model

The objective of this second experiment is to develop an empirical model of AV7 and BG dyes adsorption process and to obtain a more precise estimate of the optimum operating conditions for the factors involved in the preparation of RHA-CFA adsorbent through reflux method by using NaOH and Na_2CO_3. Results of the experimental design performed for each prepared RHA-CFA adsorbent towards AV7 and BG dyes adsorption removal along with the predicted values are shown in Table 1. From the data presented, it was found that the removal efficiency of adsorbents was in the range of 13.18% to 64.46% and 18.91% to 81.20% for AV7 and BG dyes, respectively. By applying multiple regression analysis, the empirical models in terms of coded factors for adsorption removal of AV7 (Y_1) and BG (Y_2) dyes are shown in Equation (3) and (4), respectively.

$$Y_1(\%) = 36.05 + 14.69A + 3.34B - 5.15C - 5.94A^2 + 0.55B^2 + 4.75AB - 4.59AC - 1.86BC \quad (3)$$

$$Y_2(\%) = 70.35 + 15.04A - 4.04B - 7.17C - 6.89A^2 - 3.22B^2 + 0.92AB + 4.50AC - 8.47BC \quad (4)$$

where A, B and C are coded values of the process variables which are ash ratio, the concentration of additives and types of additives, respectively. The coefficients with one factor represent the effect of the specific factor, while the coefficients with two factors and those with second-order terms represent the interaction between the two factors and quadratic effect, respectively (Ahmad *et al.*, 2009). The positive sign indicates a synergistic effect, whereas a negative sign indicates antagonistic effect (Tan *et al.*, 2008; Ahmad *et al.*, 2009). In addition, the relationship between the predicted and experimental value for AV7 and BG adsorption removal were shown in Figure 4. It can be observed that the experimental data of AV7 removal is closer to the predicted values as compared to the adsorption of BG. The plots therefore visualize the performance of the models.

The analysis of variance (ANOVA) was then used to justify the adequacy of the models and to find the models fitness with the experimental data. ANOVA result analysis for AV7 and BG dyes adsorption removal are presented in Table 2 and Table 3, respectively.

The model F-value of 29.84 and 19.27 for removal of AV7 and BG, respectively implied that both models were significant. Values of Prob>F less than 0.05 indicated that the model terms were significant. In this case, for AV7 removal, all variables including A^2 and interaction term AB and AC were significant terms whereas the B^2 and BC terms were insignificant to the response. On the contrary, for BG removal, all the model terms were significant except for B^2 and AB model terms. Although some

Figure 4. Predicted vs. actual data for adsorption removal of (a) AV7 and (b) BG dyes.

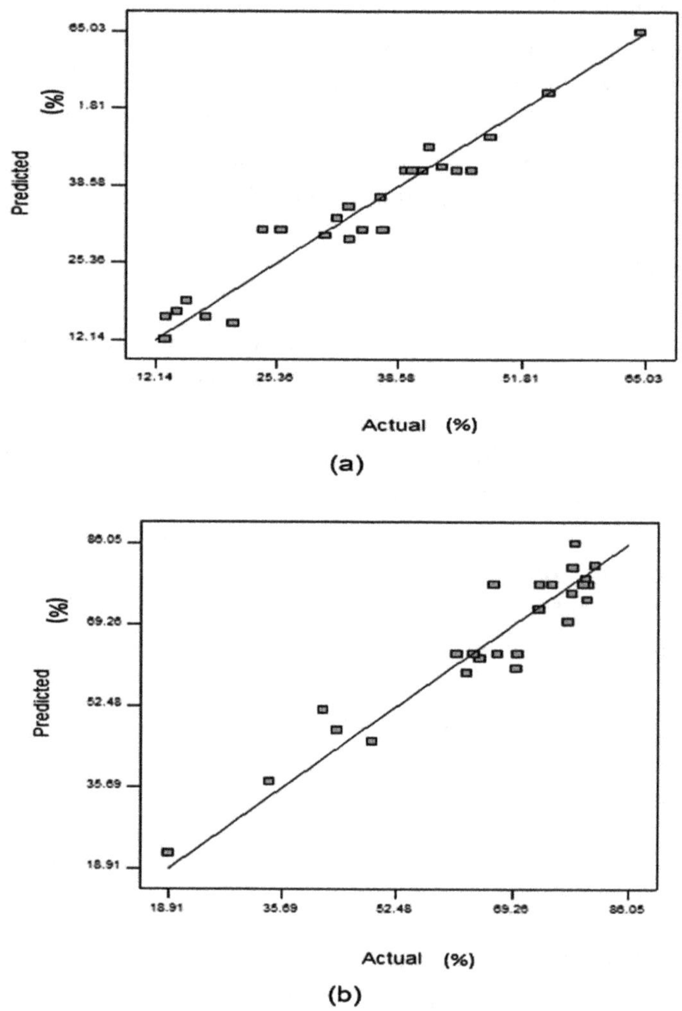

of the model terms are not significant, the overall Prob>F values obtained (<0.0001) shows that both models are significant, fitted the experimental data and adequate to predict the adsorption efficiency. In addition, the test used to evaluate the pure error based on the model performance is called Lack of Fit and it can signify the adequacy of the model (Ahmad *et al.*, 2009). The non-significant lack of fit with the F-value of 0.56 and 2.26 for AV7 and BG, respectively indicated that less pure error occurred, and the models fitted with the experimental data well. Beside the non-significant of lack of fit, high R^2 values near to 1 were favoured to determine the adequacy of the models. The R^2 values that fit these models were found to be 0.9335 and 0.9007 for AV7 and BG, respectively.

Table 2. Analysis of variance (ANOVA) for AV7 removal

Source	Sum of Squares	DF	Mean Square	F Value	Prob > F	
Model	4102.99	8	512.87	29.84	< 0.0001	significant
A	2590.14	1	2590.14	150.71	< 0.0001	
B	134.13	1	134.13	7.80	0.0125	
C	690.00	1	690.00	40.15	< 0.0001	
A^2	195.10	1	195.10	11.35	0.0036	
B^2	1.65	1	1.65	0.10	0.7603	
AB	180.12	1	180.12	10.48	0.0048	
AC	253.00	1	253.00	14.72	0.0013	
BC	41.44	1	41.44	2.41	0.1389	
Residual	292.16	17	17.19			
Lack of Fit	112.86	9	12.54	0.56	0.7977	not significant
Pure Error	179.30	8	22.41			
Cor Total	4395.15	25				

Table 3. Analysis of variance (ANOVA) for BG removal

Source	Sum of Squares	DF	Mean Square	F Value	Prob > F	
Model	5838.36	8	729.79	19.27	< 0.0001	significant
A	2713.52	1	2713.52	71.63	< 0.0001	
B	196.26	1	196.26	5.18	0.0361	
C	1336.20	1	1336.20	35.27	< 0.0001	
A^2	262.00	1	262.00	6.92	0.0176	
B^2	57.17	1	57.17	1.51	0.2360	
AB	6.83	1	6.83	0.18	0.6765	
AC	243.09	1	243.09	6.42	0.0214	
BC	860.04	1	860.04	22.70	0.0002	
Residual	643.96	17	37.88			
Lack of Fit	462.11	9	51.35	2.26	0.1326	not significant
Pure Error	181.85	8	22.73			
Cor Total	6482.32	25				

Figure 5. Correlation between additives concentration and ash ratio on AV7 dye removal for additives (a) NaOH and (b) Na_2CO_3

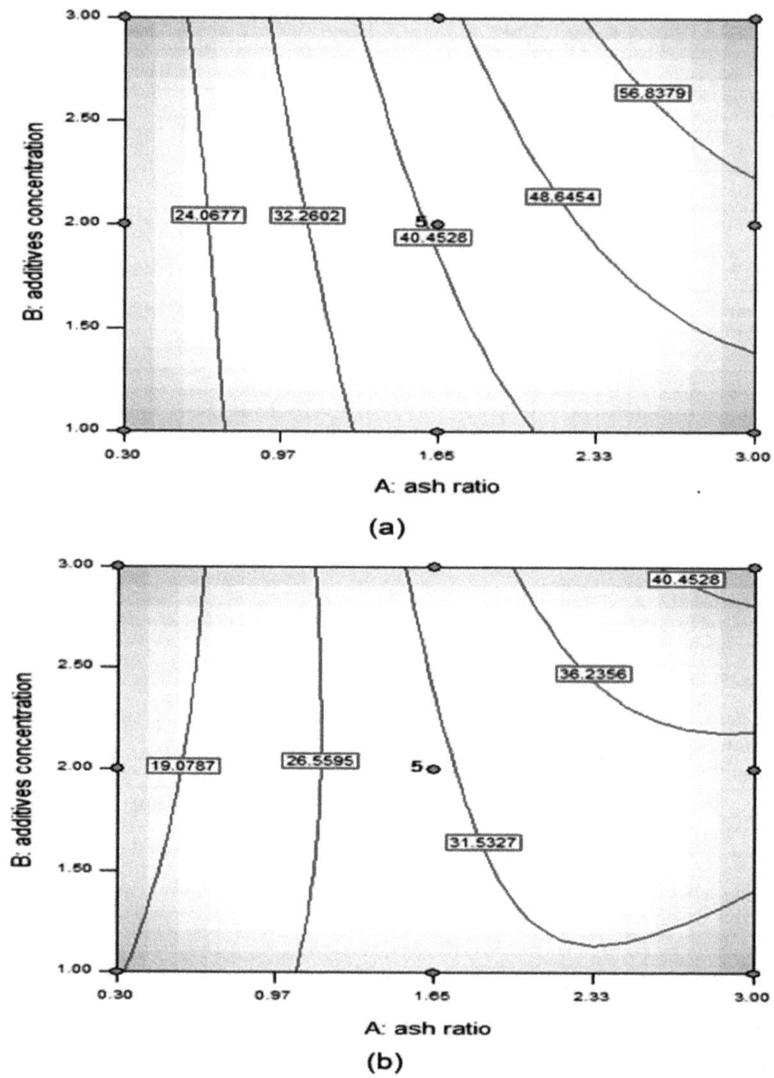

Effects of Adsorbent Preparation Variables

The contour plot in Figure 5 (AV7 dye adsorption) and Figure 6 (BG dye adsorption) show the correlation between additives (NaOH and Na_2CO_3) concentration and ash ratio on the removal of dyes. Figure 5(a) shows the AV7 dye adsorption increases with increasing RHA to CFA ratio at constant NaOH concentration. When the ratio of RHA to CFA was held constant, the removal of dye was increase slightly. It shows that both ash ratio and NaOH concentration affected the removal of AV7 dye. On the other hands, Figure 5(b) shows a slight decrement in AV7 dye removal with increasing of the additive concentration on the lower ratio of ash (<1.65). However, at higher RHA/CFA ratio (>1.65), the adsorbent efficiency increased significantly as the concentration of Na_2CO_3 increases. In addition, at constant low concentra-

Figure 6. Correlation between additives concentration and ash ratio on BG dye removal for additives (a) NaOH and (b) Na_2CO_3

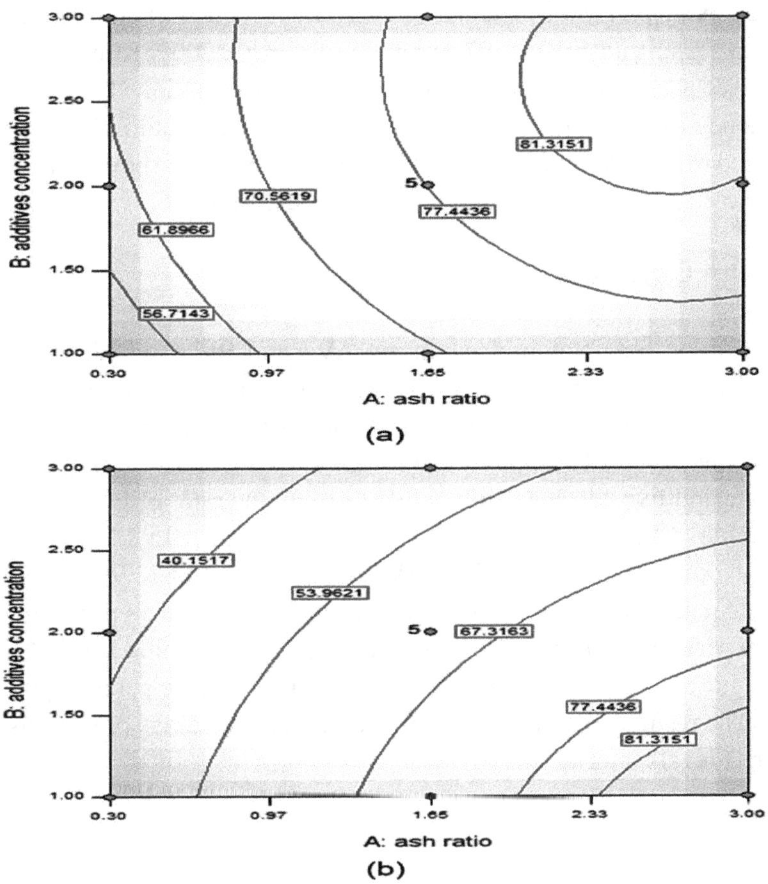

tion of additives, increasing ash ratio increases the percentage of AV7 dye removal up to the ratio of 1.65 where the adsorption efficiency gradually decreases. Unlike at higher concentration, increasing the ash ratio gave significant increments on the dye removal.

Contour plots for adsorbent efficiency toward the removal of BG dye were shown in Figure 6. It can be seen in Figure 6 that the adsorption of BG dye significantly increases at constant ash ratio with increasing NaOH concentration up to concentration around 2.5M where the BG removal started to decrease. At constant additive concentration, increasing ash ratio resulted in increased adsorption performance until the RHA to CFA ratio reaches 2.33 where the removal of BG gradually decreases. This shows that the ash ratio higher than 2.33 and NaOH concentration higher than 2.5 should not be used in the preparation of the adsorbent as it will decrease the adsorption efficiency. On the other hands, as shown in Figure 6(b), at lower ratio of RHA to CFA, increasing Na_2CO_3 concentration lowering the adsorption percentage of BG dye. While at constant Na_2CO_3 concentration, increasing RHA to CFA ratio increased the BG dye removal. Therefore, a lower concentration of Na_2CO_3 was favourable as the RHA to CFA ratio was further increased.

All the variables (ash ratio, type of additive and concentration of additive) show significant effects on adsorption of both AV7 and BG dyes. The ratio of RHA to CFA plays a significant role in the efficiency of the prepared adsorbents, due to the different chemical compositions. Although RHA contain higher percentage of SiO_2 (63%) compared to CFA (31%) (Dahlan & Razali, 2012), however, it is also composed other metal oxides (in lower percentage) such as Al_2O_3, Fe_2O_3 and MgO in which contribute to the increasing or decreasing adsorbent efficiency (Pirillo *et al.*, 2009; Khosla *et al.*, 2013; Ramesh *et al.*, 2013). The additives used (NaOH and Na_2CO_3) and its concentration also play a significant role on the adsorption efficiency of the prepared adsorbents. As discussed before, NaOH and Na_2CO_3 exhibit different properties and reactivity. Since hydroxyl is more polar than carbonate, both additives gave different adsorption removal results as presented in Figure 5 and Figure 6.

Attaining Optimum Condition

The optimization condition for obtaining maximum AV7 and BG dyes adsorption were determined by setting the desired values or condition of variables into the Design Expert software to generate the prediction adsorption values. As the chemical reagents are usually costly and difficult to handle, lower usage of chemicals is favourable. Thus, the maximum adsorption at low concentration of chemical can be found. At minimum additive concentration, while keeping the ratio of RHA to CFA in experimental ranges, it was preferable to prepare the 3:1 ratio of RHA to CFA with 1.21 M NaOH. The predicted results generated by the software are 47.05% for AV7 dye removal and 73.95% for BG removal with the desirability of 0.805. An additional experiment was conducted under the optimum conditions to reconfirm the model and experimental results. The experimental values obtained were 43.12% and 79.0% for AV7 and BG dyes removal, respectively in which both results indicate percentage errors of 2.13% and 6.42%, respectively. The results obtained therefore validated the findings of response surface optimization using a 3-factor, 3-level factorial design.

CONCLUSION

In this study, the mixture of RHA and CFA were used to prepare the adsorbent with satisfactory performance toward two synthetic dyes, AV7 and BG. It was found that the adsorbent prepared through reflux methods using NaOH and Na_2CO_3 additives shows higher adsorption removal compared to using $FeCl_3$, $CaCl_2$ and HCl additives as well as adsorbent prepared through magnetic co-precipitation and magnetic template methods. Zeta potential analysis show no influence towards adsorption of both AV7 and BG dyes. Using both NaOH and Na_2CO_3 as additives, further investigations were done using 3-factor, 3-level factorial design to obtain the optimum condition for preparation of adsorbent as well as finding the relationship between all variables (ratio of RHA to CFA, types of additives and concentration of additives). The quadratic models were obtained with R^2 of 0.9335 and 0.9007 for adsorption of AV7 and BG, respectively. The optimum condition for adsorption of both dyes was found to be the usage of 3:1 ratio RHA to CFA and refluxed in 1.21 M NaOH solution. The experimental results obtained agreed well with the predicted optimized values.

ACKNOWLEDGMENT

The authors are grateful for the Short-Term Grant (A/C. 60310014) and Iconic Grant Scheme (A/C. 1001/CKT/870023) financial support received from Universiti Sains Malaysia.

REFERENCES

Ahmad, A. A., Hameed, B. H., & Ahmad, A. L. (2009). Removal of disperse dye from aqueous solution using waste-derived activated carbon: Optimization study. *Journal of Hazardous Materials, 170*(2-3), 612–619. doi:10.1016/j.jhazmat.2009.05.021 PMID:19515487

Ai, L., Huang, H., Chen, Z., Wei, X., & Jiang, J. (2010). Activated carbon/CoFe2O4 composites: Facile synthesis, magnetic performance and their potential application for the removal of malachite green from water. *Chemical Engineering Journal, 156*(2), 243–249. doi:10.1016/j.cej.2009.08.028

Alhokabi, A., & Ing, D. S. (2019). Study on the mechanical properties of bottom ash as a partial sand replacement in a non-load bearing fly ash bricks. *Saudi Journal of Civil Engineering, 3*(3), 59–66.

Awang, H., Al-Mulali, M. Z., Abdul Khalil, H. P. S., & Aljoumaily, Z. S. (2014). Utilisation of oil palm ash in foamed concrete. *MATEC Web of Conferences*, 15, 01033.

Bai, D., Wang, X., Huo, D., Ying, Q., Wang, N., & Xia, B. (2019). Coprecipitation preparation of Cu/Zn/Al-hydrotalcite-like compound for copper removal from electroplating wastewater. Journal of Chemistry, Article ID 5347920, 9 pages.

Bronzeoak, Ltd. (2003). Rice husk ash market study. EXP 129, DTI/PubURN. 03/668.

Chowdhury, S., Mishra, R., Saha, P., & Kushwaha, P. (2011). Adsorption thermodynamics, kinetics and isosteric heat of adsorption of malachite green onto chemically modified rice husk. *Desalination, 265*(1-3), 159–168. doi:10.1016/j.desal.2010.07.047

Cooney, D. O. (1998). *Adsorption design for wastewater treatment*. Florida: Lewis Publishers.

Dahlan, I., Hassan, S. R., & Hakim, M. L. (2013). Removal of iron (Fe^{2+}) from aqueous solutions using siliceous waste sorbent. *Sustainable Environment Research, 23*(1), 41–48.

Dahlan, I., & Ismail, S. (2012). Physical characterization of prepared and spent CFA/PFA/RHA sorbents in removing heavy metals and dyes. *Journal of Environmental Science and Technology, 5*(3), 177–183. doi:10.3923/jest.2012.177.183

Dahlan, I., & Noor, A. S. M. (2014). A study on the removal characteristic of acid violet dye from synthetic wastewater using a novel RHA/PFA/CFA sorbent. *Environmental Engineering and Management Journal, 13*(1), 197–204. doi:10.30638/eemj.2014.024

Dahlan, I., & Razali, M. H. M. (2012). Waste-derived siliceous materials as a novel sorbent for removal of Ni2+ from aqueous solutions. *Water, Air, and Soil Pollution, 223*(5), 2495–2501. doi:10.100711270-011-1042-z

Dahlan, I., & Zwain, H.M. (2013). A study on the removal characteristic of zinc ion (Zn^{2+}) from synthetic wastewater using a novel waste-derived siliceous sorbent. *Caspian Journal of Applied Sciences Research*, 2 (AICCE'12 & GIZ' 12), 10-17.

Dhoble, R. M., Lunge, S., Bhole, A. G., & Rayalu, S. (2011). Magnetic binary oxide particles (MBOP): A promising adsorbent for removal of As (III) in water. *Water Research*, *45*(16), 4769–4781. doi:10.1016/j.watres.2011.06.016 PMID:21777934

FAO. (2019). Crop statistics 2017. Retrieved from Http://Www.Fao.Org/Faostat/En/#Data/Qc

Garg, V. K., Gupta, R., Yadav, A. B., & Kumar, R. (2003). Dye removal from aqueous solution by adsorption on treated sawdust. *Bioresource Technology*, *89*(2), 121–124. doi:10.1016/S0960-8524(03)00058-0 PMID:12699929

Holkar, C. R., Jadhav, A. J., Pinjari, D. V., Mahamuni, N. M., & Pandit, A. B. (2016). A critical review on textile wastewater treatments: Possible approaches. *Journal of Environmental Management*, *182*, 351–366. doi:10.1016/j.jenvman.2016.07.090 PMID:27497312

Islam, M. T., Saenz-Arana, R., Hernandez, C., Guinto, T., Ahsan, M. A., Kim, H., ... Noveron, J. C. (2018). Adsorption of methylene blue and tetracycline onto biomass-based material prepared by sulfuric acid reflux. *RSC Advances*, *8*(57), 32545–32557. doi:10.1039/C8RA05395B

Kaushik, P., & Malik, A. (2014). Mycoremediation of synthetic dyes: An insight into the mechanism, process optimization and reactor design. In S. N. Singh (Ed.), *Microbial degradation of synthetic dyes in wastewaters* (pp. 1–25). Switzerland: Springer International Publishing.

Khosla, E., Kaur, S., & Dave, P. N. (2013). Mechanistic study of adsorption of acid orange-7 over aluminum oxide nanoparticles. *Journal of Engineering*, *2013*, 1–8. doi:10.1155/2013/593534

Kumar, C. G., Mongolla, P., Joseph, J., & Sarma, V. U. M. (2012). Decolorization and biodegradation of triphenylmethane dye, brilliant green, by Aspergillus sp. isolated from Ladakh, India. *Process Biochemistry*, *47*(9), 1388–1394. doi:10.1016/j.procbio.2012.05.015

Mane, V. S., Mall, I. D., & Srivastava, V. C. (2007). Kinetic and equilibrium isotherm studies for the adsorptive removal of brilliant green dye from aqueous solution by rice husk ash. *Journal of the Indian Chemical Society*, *90*, 577–583. PMID:17000044

Mansour, H. B., Corroler, D., Barillier, D., Ghedira, K., Chekir, L., & Mosrati, R. (2007). Evaluation of genotoxicity and pro-oxidant effect of the azo dyes: Acids yellow 17, violet 7 and orange 52, and of their degradation products by Pseudomonas Putida Mt-2. *Food and Chemical Toxicology*, *45*(9), 1670–1677. doi:10.1016/j.fct.2007.02.033 PMID:17434654

Mansour, H. B., Mosrati, R., Corroler, D., Ghedira, K., Barillier, D., & Chekir, L. (2009). In vitro mutagenicity of acid violet 7 and its degradation products by Pseudomonas Putida Mt-2: Correlation with chemical structures. *Environmental Toxicology and Pharmacology*, *27*(2), 231–236. doi:10.1016/j.etap.2008.10.008 PMID:21783945

Montgomery, D. C. (2017). *Design and analysis of experiments* (9th ed.). Hoboken, NJ: John Wiley & Sons.

Moody, B. (1991). *Comparative inorganic chemistry* (3rd ed.). New York: Chapman and Hall.

Nandi, B. K., Goswami, A., & Purkait, M. K. (2009). Adsorption characteristics of brilliant green dye on kaolin. *Journal of Hazardous Materials, 161*(1), 387–395. doi:10.1016/j.jhazmat.2008.03.110 PMID:18456401

Pirillo, S., Pedroni, V., Rueda, E., & Ferreira, M. L. (2009). Elimination of dyes from aqueous solutions using iron oxides and chitosan as adsorbents: A comparative study. *Quimica Nova, 32*(5), 1239–1244. doi:10.1590/S0100-40422009000500030

Rafieizonooz, M., Mirza, J., Salim, M. R., Hussin, M. W., & Khankhaje, E. (2016). Investigation of coal bottom ash and fly ash in concrete as replacement for sand and cement. *Construction & Building Materials, 116*, 15–24. doi:10.1016/j.conbuildmat.2016.04.080

Ramesh, T. N., Kirana, D. V., Kumari, T. N. M., & Ashwini, A. (2013). Adsorption studies of indigo carmine dye by magnesium oxide. *International Journal of Science Research, 1*(64), 495–501.

Rao, M. N., Rao, Y. H., Chakrapani, C., & Babu, C. S. (2011). Adsorption studies of methylene blue dye using prepared low-cost activated Kaza's carbons. *Journal of Chemical and Pharmaceutical Research, 3*(5), 363–375.

Sadon, S. N., Beddu, S., Naganathan, S., Kamal, N. L. M., & Hassan, H. (2017). Coal bottom ash as sustainable material in concrete – A review. *Indian Journal of Science and Technology, 10*(36), 1–10. doi:10.17485/ijst/2017/v10i36/114595

Solomon, T. W., & Fryhle, C. (2004). *Organic chemistry* (8th ed.). Hoboken, NJ: John Wiley & Sons.

Tan, I. A. W., Ahmad, A. L., & Hameed, B. H. (2008). Optimization of preparation conditions for activated carbons from coconut husk using response surface methodology. *Chemical Engineering Journal, 137*(3), 462–470. doi:10.1016/j.cej.2007.04.031

Tavlieva, M. P., Genieva, S. D., Georgieva, V. G., & Vlaev, L. T. (2013). Kinetic study of brilliant green adsorption from aqueous solution onto white rice husk ash. *Journal of Colloid and Interface Science, 409*, 112–122. doi:10.1016/j.jcis.2013.07.052 PMID:23981675

Wang, J., Xu, W., Chen, L., Huang, X., & Liu, J. (2014). Preparation and evaluation of magnetic nanoparticles impregnated chitosan beads for arsenic removal from water. *Chemical Engineering Journal, 251*, 25–34. doi:10.1016/j.cej.2014.04.061

Yang, G., Wu, L., Xian, Q., Shen, F., Wu, J., & Zhang, Y. (2016). Removal of congo red and methylene blue from aqueous solutions by vermicompost-derived biochars. *PLoS One, 11*, 1–4. PMID:27144922

Zhang, G., Qu, J., Liu, H., Cooper, A. T., & Wu, R. (2007). CuFe2O4/activated carbon composite: A novel magnetic adsorbent for the removal of acid orange ii and catalytic regeneration. *Chemosphere, 68*(6), 1058–1066. doi:10.1016/j.chemosphere.2007.01.081 PMID:17397903

Zwain, H. M., Vakili, M., & Dahlan, I. (2018). Optimization, equilibrium and kinetic studies of Zn2+ and Ni2+ adsorption from aqueous solutions using composite adsorbent. *Journal of Water Supply: Research & Technology - Aqua, 67*(3), 279–290. doi:10.2166/aqua.2018.150

Chapter 20
Application of Soil Washing Treatment Method for the Remediation of Petroleum Polluted Soil

Abdullahi Evuti Mohammed
University of Abuja, Nigeria

Kamoru Adio Salam
University of Abuja, Nigeria

Silas Shamaye Samuel
University of Abuja, Nigeria

ABSTRACT

The increasing contamination of soil by petroleum products has been a great source of concern to our society because of its negative consequences on the environment. Thus, several remediation technologies and trials have been propounded for a crude oil-polluted environment. This chapter reviews the dynamics of pollutants in the soil and the various treatment technologies for petroleum-polluted soils viz physico-chemical, thermal, and biological treatment methods. Authors experimented on soil washing using detergent for the remediation of petroleum contaminated soils considering different concentrations. The percentage removal of aliphatic and Polycyclic Aromatic Hydrocarbons (PAHs) was determined using Gas Chromatography Mass Spectrometry (GC-MS). The highest percentage removal efficiencies of 97.55% and 61.41% for aliphatic and Polycyclic Aromatic Hydrocarbons were obtained at detergent concentration of 20w/v% respectively.

DOI: 10.4018/978-1-7998-0369-0.ch020

INTRODUCTION

Soil which is the essential basis of agricultural resources, food security, global economy and environmental quality is now continuously being polluted by heavy metals, organic pollutant and petroleum products due to urbanization and industrialisation (Kokyo *et al.*, 2014). Thus environmental pollution, according to Dike *et al.*, (2013), has become a significant global issue and the accumulation of these contaminants over time poses severe threat to both plants and animals. Accidental and deliberate spillage of crude oil and other petroleum products into the environment is a significant source of environmental contamination due the resultant air, water and soil pollution around the affected areas (Trindade *et al.*, 2005). This situation became worse in Nigeria with the incessant vandalism of oil pipelines by militants and oil bunkering. On terrestrial ecosystem, crude oil contamination affects soil chemical properties such as electrical conductivity, mineral and organic matter content; cation exchange capacity and pH (Tanee and Jude, 2017).

Remediation of petroleum polluted site is defined as the management of the contaminated area to prevent imminent danger to human health or the environment and restoring all or portion of the area to a beneficial goal (Dike *et al.*, 2013). The different remediation methods are isolation and containment, mechanical separation, pyro-metallurgical separation, permeable treatment wall, soil flushing, molecular and phase separation, chemical destruction, soil washing, vapor extraction, electro kinetics and biodegradation. Although, some besides being more expensive and difficult to apply, have been reported to cause more harm to the environment than the pollutant. So the need for research into the use of simple, economical, environmentally friendly and efficient technology for the remediation of crude oil contaminated site has become imperative (Tanee and Jude, 2017).

Chemical remediation involves the use of substances such as detergent/ degreasers. Detergent contains some active ingredients such as surfactant, sodium, chlorine and bleach which emulsify and weaken the hydrocarbon chains and thus, provide a good surface area for biodegradation (Couto *et al.*, 2010). This chapter therefore reviews the various petroleum contaminated soil remediation techniques and presents a case study on the determination of the effectiveness detergent for the remediation of petroleum contaminated soils.

DYNAMICS OF POLLUTANTS IN THE SOIL

Presently, the geosphere is seen as fragile and prone to damages emanating from anthropogenic activities (Raphael *et al.*, 2013). Manahan (2001) defined pollution as an increase in the quantity of a particular element above the levels in which they naturally occur, resulting from external source related to human actions. Xenobiotic behaviour in the soil is of great difficulty to predict, because its composition is completely complex and heterogeneous. Hence, the understanding of the physico-chemical properties of the contaminant compounds and the environment is fundamental to predict its dynamics.

Contaminant spread in the soil in vaporised, residual or absorbed phases, free phase and dissolved phase. The distribution of such phases will depend on their physico-chemical properties and also on the soil type. Hence, the movement of the contaminants and their toxic nature are linked to the soil's capacity in maintaining them while retained in their solid phase, making them not available to be absorbed by plants, eroded or leached (McBride, 1994). Soil particles (m^2/g) have available surface area due to some of the factors that influence binding between the contaminant and the soil. Also, the adsorption of contaminant to the environment is influenced by the electrical charge of the soil particles (Otten *et al.*, 1997).

Generally, the dynamics of soil contaminants can be shown by three mechanisms of mass transfer, namely: advection, dispersion and attenuation.

- **Advection:** It involves the mechanism in which the contaminants accidentally move in the direction of the flowing vectors and maintain a direct relationship with the speed of percolation in the soil. It is a process in control of the formation and mobilisation of the free phase of hydrocarbons.
- **Dispersion:** Is the mechanism responsible for concentration decrease of contaminants in the fluid percolation which can happen by two methods namely, hydrodynamic dispersion and molecular diffusion. Hydrodynamic dispersion happens due to the restriction in flow that occur in the soil pores which leads to reduction in the percolation velocity of the more viscous components while the molecular diffusion is a process of dilution of the more soluble components and is the key formation method of the dissolved phase responsible for the greater mobility of the contaminants.
- **Attenuation:** This mechanism is responsible for the reduction contaminants transferred by advection or dilution by chemical or physico-chemical reactions. Chemical attenuation is stronger in soils with greater cation exchange capacity and acts to reduce compounds in the free and adsorbed phase (Raphael *et al.*, 2013). Physico-chemical attenuation is involved in the making of the adsorbed phase and is responsible in the accommodation of the contaminants that stick to the soil's grain, specifically to the grumes of clay with greater activity. Though, in association with the phenomena of chemical attenuation, it involves in the formation of the dissolved phase (influenced by the reduction in pH).

PETROLEUM POLLUTED SOILS TREATMENT METHODS

The methods of treatment of petroleum polluted soils can be classified into physico-chemical, thermal and biological treatment methods.

Physico-Chemical Treatment Methods

These methods include:

Soil Washing

As an ex-situ process of treatment, soil washing is applicable in removing organic, inorganic and radioactive contaminants in soil even semi volatile compounds, fuels and heavy metals. Also, some selected volatile organic compounds and pesticides are removed by this method (Anderson, 1993). This method is based on the idea of water rinsing to extract pollutants present in the soil and transporting them to a concentrated liquid phase. This extraction can be carried out by either transferring pollutants to the washing solution or concentrating them in a smaller volume of solids by particle or gravimetric separation, flotation and/or attrition (Edris *et. al.* 2013).

Soil washing is an innovative treatment technology that involves the use of water/liquid. This is sometimes combined with some chemical additives and mechanical instrument to scrub the soil in order

Figure 1. Chart of soil washing
(US EPA Doc. EPA 542-F-01-008)

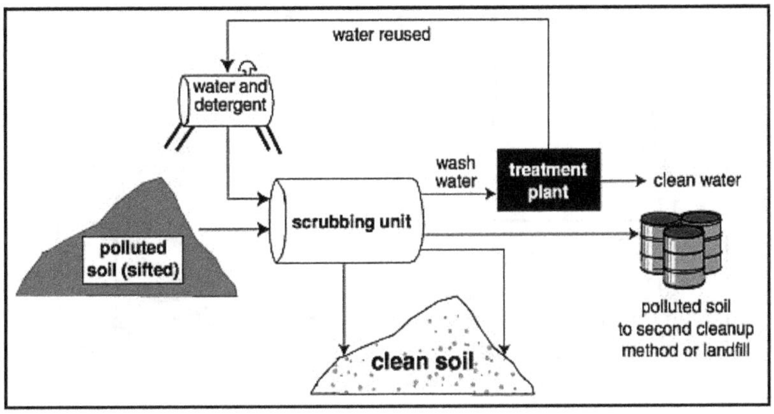

to remove contaminants that are hazardous and also concentrate them to smaller volumes. Clay and silt have affinity for hazardous chemicals unlike gravels and sand particles. In washing, separation is carried out between the uncontaminated coarse soils from clay and silt mechanically. The contaminated fine sand is treated or disposed while the coarse sand is retained for backfilling. It has been shown that this method has less than 80% effectiveness and when hot water is used the efficiency increases. This method is mostly used as pre-treatment method for final cleaning up of soils (Wood, 2002). Kingsley *et al.*, (2005), in their work found out that the use of surfactant (biosurfactant: aescin, lecithin, rhamnolipid, saponin, and tannin) in soil washing process has an increase efficiency on the effectiveness of soil washing technique. 80% oil removal at 50°C can be achieved for all surfactant solution except lecithin which had 15% removal. In addition to efficiency increase by surfactant, Edris *et al.*, (2013) noted that, surfactant enhances the ex-situ soil washing by emulsifying and weakening the hydrocarbon chains and thus, provide a good surface area for biodegradation and it is economically desirable as innovative and alternative soil washing technologies. A schematic process of soil washing is shown in Figure 1.

Soil Flushing

This is an in-situ technique of cleaning up contaminated soil using aqueous solution that can either be plain water or a carefully mixed solution (surfactant or co-solvent) to purge or leach contaminants from the soil. The solvent helps to promote desorption of contaminants from the surface of soil particles and solubilisation of contaminants in the flushing solution (Albert, 2010). It also utilises gaseous mixtures to accelerate one or more of same geochemical reaction (adsorption/desorption, acid/base reaction and biodegradation) that change contaminants concentrations in groundwater system. The method is most effective in homogeneous, permeable soils like sands or silty sands with its activities broken down into three namely: site characterisation; fluid injection, and contaminants mobilisation and recovery techniques. (Anderson, 1993). It effectiveness is dependent on the hydrogeologic variables, cation exchange capacity, pH, buffer capacity and the quantity of the Total Organic Content (TOC). However, less permeable soils (such as clay) may be difficult to clean due to their poor water permeability and heterogeneity that prevent optimum contact between the reagents and the target contaminants (EPA, 2006).

Figure 2. Schematic of soil flushing
(Albert, 2010)

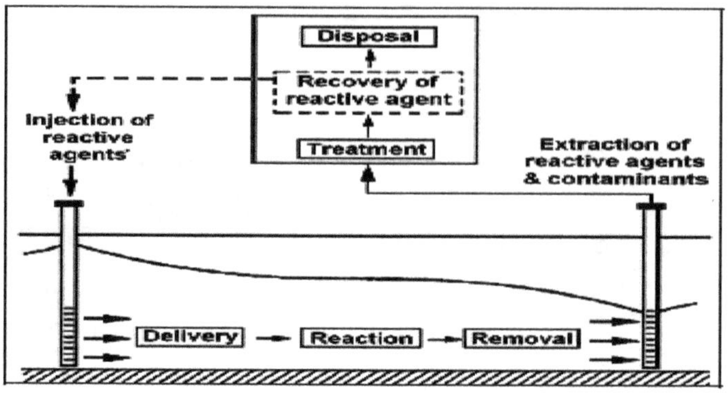

The application of soil flushing technique is dependent on the following factors: depth of the contaminated zone; concentration and volume of contamination; distribution coefficients of contaminants between soil particle surfaces and flushing solutions; presence of geologic heterogeneities in the soil horizon; interactions of flushing solutions with contaminated soil; suitability of contaminated site for installation of wells for delivery and recovery of flushing solutions; and design factors such as sizing the delivery and recovery systems to ensure complete recovery of the elutriate (Albert, 2010).

In-Situ Washing by Sedimentation (IWS)

This method is a new method for soil remediation without excavating the soil (Budianta *et al.*, 2010). This method involves injecting a high air-pressure into a mixture comprising water and sandy soil placed in a column at a particular depth (D) and hydraulically separating the soil particles based on their particle size and density, as shown in Figure 3. The physical separation takes advantage of the dispersed contaminant in the soil to physically separate a selected contaminant-rich fraction. In this method, the physical segregation and on-site wash water treatment happen as an integrated process and it is important to isolate the site to protect the leakage of aqueous solution used. The process has an advantage of allowing the washing and segregation processes of the soil to occur simultaneously during the remediation (Budianta *et al.,* (2010) and Budianta *et al.,* (2012)). On using IWS technique in remediating soil contaminated with Polycyclic Aromatic Hydrocarbons, 90% removal of Napthalene, Phenantrene and Pyrene was significantly achieved (Budianta *et al.*, 2010). However, this technique is also suitable for heavy metal removal (Budianta *et al.*, 2006).

Soil Vapour Extraction (SVE)

This is a simple physical process for cleaning up contaminant (e.g. crude oil) from contaminated soils. Volatile contaminants are removed from the soil in vapour form by the use of a specially designed system (Uchechukwu *et al.*, 2007). This method is an in-situ remedial method that decreases the concentration of volatile compounds in crude oil products distributed all over the soil in the unsaturated (vadose) zone (Eyad and Richard, 2015).

Figure 3. In situ Washing by Sedimentation Method (IWS)
(Budianta et al., 2010)

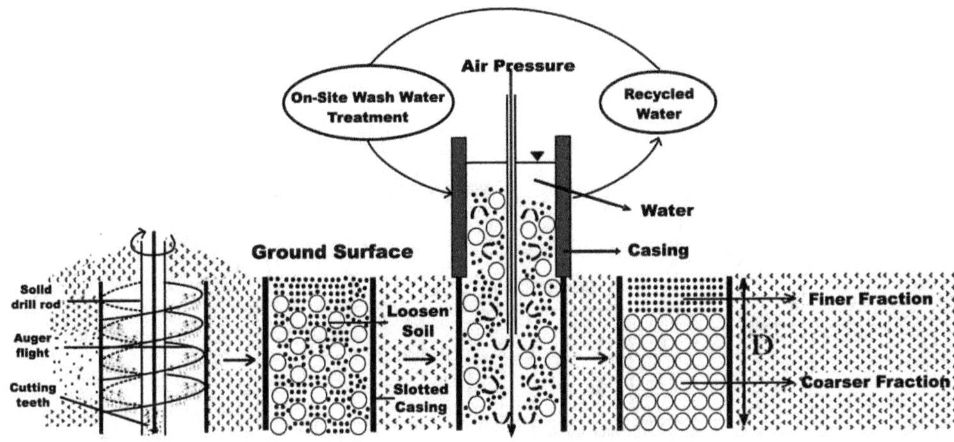

This method is carried out by applying a vacuum through a system of underground wells. This process forces contaminants from the soil to the surface as gas or vapour. In order to enhance the process air is sometimes used (Imamura *et al.,* 1997). According to Shah *et al.*, (1995), SVE, also referred to as soil stripping or soil venting, involves pumping of air in the unsaturated zone through one or more wells. The pumped or induced air leads to in-situ evaporation of volatile organic compounds (VOCs), volatilisation of dissolved chemicals in the residual pore water and desorption of chemicals from the surface of the soil. Mobilised organic vapours and air along the induced flow path to the withdrawal well, where they are removed.

SVE is affected by soil structure and stratification. These factors can affect how and the direction of flow of vapours in the soil medium under extraction conditions. Permeability also affect the air and vapour flow via the soil. Higher soil permeability leads to faster and greater quantity of vapour extraction. However, soil permeability is decreased by high water content and can affect the effectiveness of SVE (Eyad and Richard, 2015).

While SVE has become one of the fastest growing in-situ remediation methods due to its cost effectiveness, simplified design and operation (Shah *et al.*, 1995), Wilson and Clarke, (1994), has observed the following constrains associated with soil vapour extraction (a) the chemical requiring removal must be volatile or semi volatile with a vapour of pressure 0.5 torr or greater, (b) the chemical requiring removal must have relatively low water solubility or the moisture content of the soil must be low, (c) the contaminants to be removed must be above the water table or floating on the water surface (in the case of light non aqueous-phase liquid) and (d) the soil must be sufficiently permeable to allow the vapour extraction wells to draw air through all of the contaminated domains at a reasonable rate.

Soil Excavation

Soil excavation (often called dig and haul) involves the removal of contaminated soils mechanically to off-sites to be buried or burnt. However, pre-treatment of the excavated soil may be required to meet the land disposal requirement (Albert, 2010). According to Uchechukwu *et al.*, (2007), contaminated soil should be removed for efficient clean up. But this process can be very expensive as it may involve hir-

Figure 4. Soil Vapour Extraction
(Shah et al., 1995)

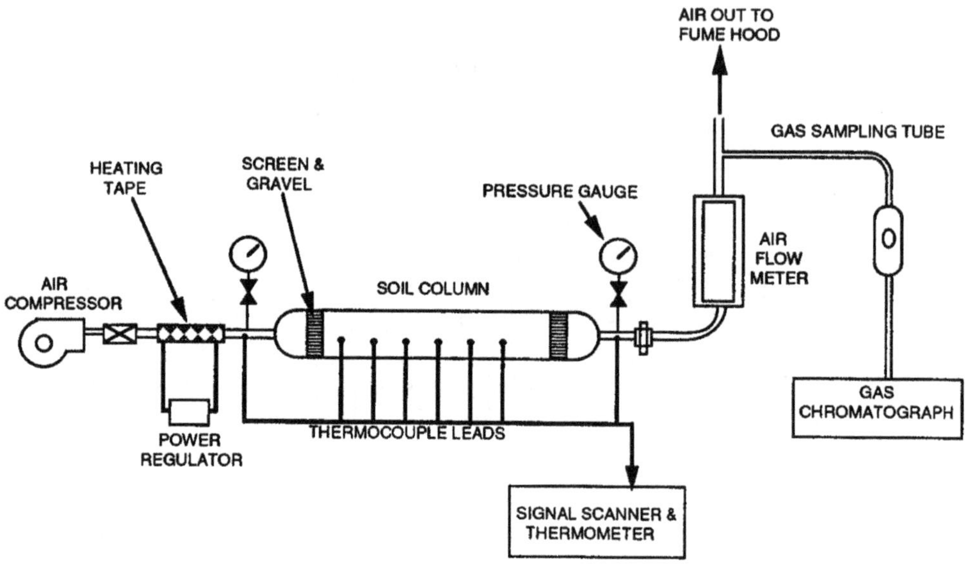

ing a contractor to remove a layer of the ground (Uchechukwu *et al.*, 2007). The area excavated is made prone to erosion and other environmental damaging agent as a result of soil excavation as a treatment method as reported by Araruna *et al.*, (2004).

The following factors according to Albert (2010), may limit the applicability and effectiveness of soil excavation method: (a) generation of fugitive emissions from excavation and transportation; (b) cost of treatment is dependent on the distance between the contaminated site and the disposal or remediating facilities; (c) depth and composition of the materials to be excavated; (d) disposal option for certain waste such as radioactive, mixed waste may be limited; (e) migration of contaminants from disposal/treatment facilities via varying channels such as effluent discharge to surface water, rainfall surface runoff, leaching into groundwater, volatilization to the atmosphere, and dike uptake; (f) improper design and maintenance of disposal/treatment facilities can result to odour, mosquitos and insect problem.

Solidification/Stabilization

This technique decreases radiological or chemical hazard by changing the contaminant to its least soluble, mobile and harmful form without necessarily altering the physical nature of the contaminant (Hamby, 1996). It involves mixing of reagents with contaminated soil to abate the movement of contaminants and reduce the toxicity of the waste (Albert, 2010). Solidification involves binding or encapsulating of the waste into a high integrity structure to limit the mobility of contaminants within an impermeable capsule (Hamby, 1996). This binding can be achieved by addition of sufficient quantities of solidifying reagents to solidify the waste (Albert, 2010). Some of the common non-proprietary solidifying reagents used for this process include Portland cement, quick and hydrated lime, cement kiln dust, limekiln dust, fly ash, bottom ash, magnesium oxide, phosphate in various forms, sulphides, and carbon-based reagents (Edward and Colin, 2015).

The formulation and implementation of an effective binder (e.g. cement) has made solidification/stabilization a useful technique to treat wide range of organic and inorganic contaminants resident in soils, sludges and excavated sediments. This technique can also be used to remediate metallic compounds, longer chain petroleum hydrocarbons and many of the larger chlorinated organics (Edward and Colin, 2015).

Thermal Treatment Methods

Thermal treatment methods involve destruction and removal of the hydrocarbon in the polluted soil. It includes thermal desorption and incineration (Albert, 2010).

Thermal Desorption

Thermal desorption is a treatment method that can be carried out in-situ or ex-situ. It involves heating the contaminated soil or sludge to about 400^{0}C to volatise and removes resident contaminants from the soil (Albert, 2010). The process is carried out in the absence of oxygen below combustion temperature (Hamby, 1996). Contaminant with low boiling point evaporate and separate physically (desorb) from the soil (Traoxler *et al.*, 1994). This process is also referred to as Low Temperature Thermal Desorption or Low Temperature Thermal volatilisation because it uses low temperature and it is also called thermal stripping or soil roasting (Anderson, 1993). Most times in thermal desorption, the contaminants are turned into vapour and ignited and the remains (by-products) are withdrawn from the system by convection. They are treated with filters or second stage re-ignition or by air emission treatment system (Wood, 2002). In another way, the contaminants can be treated in a secondary unit (for example after burner, catalysis oxidation chamber, condenser or carbon adsorption unit) before they discharge into the atmosphere. Organic constituents are destroyed in the after burners and oxidizers while the organic compounds are trapped in the condenser and carbon adsorption unit for subsequent treatment or disposal.

Incineration

Incineration requires that contaminants from the soil surface be burnt off using fire. At temperature range of between 1,600^{0}F and 2500^{0}F incineration occurs according to US Environmental Protection Agency (EPA). During incineration, hazardous waste including petroleum are destroyed from the soil and element that are toxic are reduced to basic elements like hydrogen, chlorine, carbon and nitrogen. The basic elements bond with oxygen to form stable non-toxic substances like carbon dioxide, nitrogen oxides and water (Uchechukwu *et al.*, 2007). Chemical bonds of organic compounds and other substances are broken down in this process to reduce the risks impacted by hazardous waste and also, toxicity and volume of substances at hazardous waste site are reduced. However, toxicity of radioactive contaminants would not be eliminated by this method, although bulk quantity may be reduced (Hamby, 1996).

There is rapid loss of more volatile and flammable components through evaporation making burning and ignition difficult (Uchechukwu *et al.*, 2007). In incineration process, the contaminated soil is excavated and taken to off-site facilities before it is incinerated (Bassam and Battikhi, 2005). The problem associated with incineration is its high cost of operation due to high requirement of energy, large space requirement and the impact of pollution on the environment (Araruna *et al.* 2004; Bassam and Battikhi, 2005). Some of the type incinerators include: the rotary kiln, infrared furnaces, liquid injection, plasma arc, fluidized bed, and the multiple hearth (Hamby, 1996).

Biological Treatment Methods

Microorganisms, plants and other biological systems are used to treat oil contaminated soil in this method. With this process, excavated soils (saturated and unsaturated in-situ) are treated (Galaska *et al.* 1990; Eckenfelder and Norris, 1993).

Microbial Degradation of Crude Oil (Biodegradation)

Biodegradation of organic waste is an increasingly important method of waste treatment (Atlas, 1981). It is simple, environmentally friendly and uses inexpensive equipment (Nadean *et al.,* 1993). Organic compounds are biodegradable resulting in the formation of carbon dioxide and water or methane while inorganic compounds are not biodegraded but transformed (bio transform) into compounds with more or less mobility and toxicity than their previous state (Hamby, 1996). The use of microorganisms in treating crude oil contaminated environment is of utmost important and it comes in after large quantity of the spilled oil has been removed using various chemical and physical methods (Atlas, 1981).

Several studies have been done on microbial degradation of crude oil (Higgins and Gilbert, 1978). Microorganisms have enzymatic systems which make microbial degradation possible. These enzymatic systems enable the breakdown of crude oil and utilizing it as carbon and energy source (Ijah and Antai, 1988). This biotransformation can be utilised for cleaning up of contaminated soils and ground water (Bouwer and Zehnder, 1993).

Microorganisms that can metabolise petroleum hydrocarbons are widely spread in the soil. The organisms are mostly present in the vicinity of an oil field at the soil surface and in soil contaminated with petroleum (Antai and Mgbomo, 1989). These microorganisms include bacteria, yeast, filamentous fungi and algae. The main bacteria genera used in degradation of crude oil in soil and aquatic environments comprise majorly Nocardia, Micrococcus, Flavobacterium, Actinomycetes, Athrobacter, Achromobacter and Pseudomonas (Atlas, 1981). Virtually any hydrocarbon and the heaviest paraffin can be invaded by bacteria (Essien *et al.*, 1997). Particular microorganism attacks a specific molecular site in biodegradation process (Hamby, 1996). However, oil components do not disappear completely by microbial degradation as noted by Wardly- Smith (1993).

The efficacy of microorganisms in the degradation of petroleum is dependent on some environmental factors. These include temperature, moisture, pH, nutrients, oxygen, viscosity and coarseness of the affected soil (Antai and Mgbomo, 1989). Atlas (1981) reported that to determine the rate and effectiveness of biodegradation, temperature has been described as one of the most important parameter for doing that. High temperature range of 30^0C-40^0C favours microbial degradation of crude oil while low temperature slows down rate of degradation as it rises the viscosity of oil leading to low drifting capacity which also makes the surface area for degradation low (Atlas, 1981).

Bioremediation (Land Farming)

Bioremediation is the destruction or reduction in the concentrations of harmful wastes from contaminated area using biological methods. These methods can be use to clean up large spectrum of site including ground water, soils, lagoons, sludge and process waste-streams (Okoh and Trejo-Hernandez, 2006). Bioremediation aids the acceleration of natural biodegradation process. It is carried out by adding fertilisers or other materials to oil contaminated environment. It is used primarily to improve the biophysico-chemical properties of soil by augmenting the soil nutrients in order to enhance growth and increase native microflora (Dragun 1993). Addition of garden fertiliser to an oil contaminated area is the simple bioremediation strategy that can be used to clean up contaminated site. The complex method involves an engineered treatment cell which makes use of manipulating, aerating or treating the soil or other media with various chemical substances to promote degradation. These strategies depend on the peculiar nature of the contaminated area (Okoh and Trejo-Hernandez, 2006). Active aeration and fertiliser addition helps to promote the activities of indigenous microorganisms to improve oil biodegradation (M. Chorom *et al.,* 2010).

Bioremediation has been proven to be effective and reliable because of its eco-friendly characteristics. Its application can either be ex-situ (such as land farming, biopile, bioreactor and windrows) or in-situ (such as bioventing, biosparging and bioslurping) depending on area of application and several variables which comprise, but not restricted to cost, site features, type and volume of pollutants. Selection criteria for choosing any of the bioremediation methods is dependent on pollutant nature, depth and degree of pollution, type of environment, location, cost, and environmental policies (Christopher *et al.,* 2016).

Biopiles

Biopile is one of the ex-situ bioremediation techniques (Christopher *et al.,* 2016) that forms piles with the polluted soil to excite communities of microorganisms via aeration and/or by the addition of nutrients and water (Rosario and Jessica, 2015). This technique applies above-ground piling of excavated contaminated soil accompanied by nutrients adjustment that increase the activities of microbes to enhance bioremediation through aeration, irrigation, nutrient and leachate collection systems, and a treatment bed (Christopher *et al.,* 2016). This method is applicable to decrease the concentration of heavy fraction hydrocarbons adsorbed in polluted soils by biodegradation through the increased activities of microorganism. Biopile is efficient depending on the following variables which are categorised in three groups namely: soil characteristics, characteristics of the contaminants and weather conditions (Rosario and Jessica, 2015).

Phytoremediation

Phytoremediation is a biological in-situ biotreatment technique that involves the use of plant species to clean up a site contaminated with hazardous substance (Uchechukwu *et. al.,* 2007). It helps to remove, stabilise and degrade organic and inorganic pollutants from the soil (Albert, 2010). It is a growing and environmentally-friendly technique that involves green engineering technology that uses the natural features of plants to treat contaminated soil, water and sediments (Kokyo *et. al.,* 2014). The plants carry out certain processes that are natural to take up metals and certain organic compounds; accumulate or process these substances through lignification, volatilization, metabolization, mineralization (transformation into CO_2 and water); utilise enzymes to reduce complex organic molecules into less complex molecules (ultimately CO_2 and water); elevate the soil's carbon and oxygen surrounding the roots while promoting microbial and fungal activities via the release of chemicals (exudates) and decay of root tissues; and trap and utilise ground water for their process (Ralinda 1996).

The biotreatment method can be used over large range of contaminants that include heavy metals and radionuclides, as well as organic compounds such as chlorinated solvents, polycyclic aromatic hydrocarbons, pesticides/insecticides, explosives, and surfactants and is believed to occur via one or more of these

Figure 5. Concept of phytoremediation
(Albert, 2010)

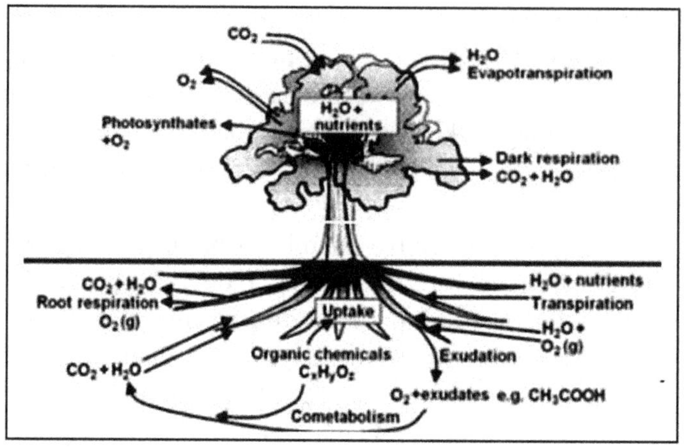

processes: phytoextraction, phytostabilization, phytodegradation, phytovolatilization, rhizofiltration and rhizodegradation sediments (Kokyo *et. al.,* 2014). In phytoextraction, plant naturally takes up certain contaminants (heavy metals) from the contaminated area and impound them in their cells until they are harvested (Uchechukwu *et. al.,* 2007) and either incinerated or composted to recycle the metals (Albert, 2010). For phytostabilization, the plant binds contaminants via the chemicals it releases to make them less bioavailable and immobile in the surrounding environment (Uchechukwu *et. al.,* 2007) then absorb and accumulate them via the root surface or precipitate them within the rhizosphere (Albert, 2010). During phytodegradation, organic contaminants are converted into harmless form (Uchechukwu *et. al.,* 2007) through metabolic mechanisms (Albert, 2010). In phytovolatilization, plant convert contaminants into gaseous form for safe release (Uchechukwu *et. al.,* 2007) through transpiration with the release of contaminants in a modified form to the atmosphere (Albert, 2010).

Bioslurry Systems

This is an ex-situ biological treatment method that involves excavation of polluted soil. The excavated soil is mixed with water, microbes and other additives for the effectiveness of this technique. The bacteria helps to break down the contaminants so that biodegradation can occur at a fast rate. The treatment method is carried out in a controlled bioreactor to enable microbes to have contact with the pollutants in the slurry. This treatment can occur from less than one to more than six months (Faisal *et. al.*, 2004). Bioslurry reactors have capacity to reduce high concentration of contaminants in soil due to its increased solid-liquid mass transfer. The slurry bioreactor system comprise an aqueous slurry formed by mixing sediment or sludge with water in a closed system or a lined lagoon for a certain time period to enhanced biological conversion of waste components (United State Environmental Protection Agency). Its efficiency is enormous for polluted soil due to its upgraded homogeneity, increased flexibility of process and improved control over the system operation (Laleh *et. al.,* 2003).

Bioventing

Bioventing is an in-situ technique that is simple with high efficiency operated at low cost for remediating vadose zone (Mohammed et. al., 2014). It permits remediation of unsaturated soil. This treatment method requires air for its operation. The air feeds the soil with oxygen to enable bacteria growth and enhance their activities of breaking down contaminants like oil into carbon dioxide and water. It is effective with light contaminants for ease of evaporation (Alaska Department of Environmental Conservation, 2009). It has gained recognition in cleaning up area contaminated with light petroleum products (Christopher *et al.,* 2016).

Bioventing uses injected air to maintain aerobic or anaerobic, or provide a substrate to enhance occurrence of cometabolic degradation. The aerobic microbial degradation is enabled by the introduction of oxygen to the low oxygen contaminated soil to help the microorganisms oxidize pollutants for gain of energy and carbon for their growth. Chlorinated compounds are difficult to remove from the soil aerobically but can be effectively removed anaerobically. Anaerobic bioventing is a growing technique that may be applicable in remediating extremely chlorinated compounds such as pentachlorophenol (PCP), some polychlorinated biphenyls (PCBs), and pesticides, such as lindane and dichlorodiphenyltrichloroethane (DDT) (U.S EPA, 2006).

Bioventing integrates soil venting (a technique that eliminates volatile compounds from the soil) with bioremediation. The bioremediation technique utilises the vented air for biodegradation of remnant organic pollutants. Low molecular weight, high vapour pressure components can be volatilised and maximised using a greater amount of air or short transfer distances between injection and extraction wells (Michael and Swindoll, 1993). Bioventing can involve one of the two methods: injection and extraction, to supply air into the soil as recorded by Diane, (1999).

SELECTION OF REMEDIATION TECHNOLOGIES

Technology developers, property owners responsible for site clean-up, citizen groups, state and federal regulators all may have different perspectives on how remediation technologies should be evaluated and selected. Reconciling the differing expectations of these stakeholders can add to delays in site remediation (National Academy of Sciences, 1997).

However, once a need for soil remediation is recognized, the best available technology is selected according to the nature, toxicity, and origin of the contaminant, the present and potential hazard related to the degree of contamination, the chemical and physical characteristics of the soil, the land use, the time available for remediation, the community acceptance, and a cost-benefit analysis (Lombi and Hamon, 2005). This will therefore involve comprehensive laboratory studies, considering the specialities of pollutants, hydrogeologic properties of the polluted areas and economic feasibilities of the remediation methods (Dağhan and Öztürk, 2015).

The success criteria of these technologies can be divided into three categories: (i) technological performance, (ii) commercial characteristics, and (iii) acceptability to the public and regulators. Technical performance attributes describe the technology's ability to achieve risk reduction goals and the efficiency with which it achieves these goals. Commercial characteristics are factors related to the costs of the technology and the profits it yields. Public and regulatory acceptance attributes are qualitative characteristics of technology performance that, in addition to quantitative technology attributes, are of particular importance to the public near the contaminated site and to regulators; to varying degrees, these attributes also may be important to other stakeholder groups (National Academy of Sciences, 1997).

For instance, the average cost of remediation with different technologies can differ by a factor of greater than 25 (Lombi and Hamon, 2005). The development of low-cost, low-input technologies driven in industrialized countries by economic benefits is also essential to tackle remediation of polluted soils in developing countries. Some of the technologies described in this article are well established, whereas others are still being developed. Due to ever-increasing public concern for the risks posed by soil pollution, it is expected that a larger number of technologies will be available in the future.

CASE STUDY

This case study focuses application of soil washing treatment method for the remediation of petroleum polluted soil. The objective is to determine the effectiveness of detergent for the remediation of petroleum contaminated soils considering concentrations.

Materials

Gas Chromatography Mass Spectrometry (GC-MS) was used for the determination of contaminants concentration and characterisation; Air pump: for supplying air at a pressure to provide stirring effect. Other materials include suction pump; weighing balance; Stop watch; Hand auger; Deionised water and Detergents which is a common household laundry detergent with Brand name Klin whose composition include: Linear Alkyl Benzene Sulfonate (LABS), sodium carbonate, sodium Tripolyphosphate, sodium sulphate, and enzymes.

METHODS

Soil Sampling Collection

Contaminated soil was collected from the Suleja depot, Niger state, North Central Nigeria using a hand-operated auger from a depth of about 0.25m from ground surface. An X method of sampling was used to achieve seven sampling points (Tenedero and Surtida, 1986). These soil samples were labelled 1, 2 and 3 respectively and were taken to the laboratory for analysis.

Determination of Total Contaminant Concentration in Sample Soil Using Solid-Liquid Extraction

202g of the contaminated soil for seven samples was weighed and 600ml volume of acetone–ethyl acetate–water mixture (containing 50ml ethyl acetate, 40ml of acetone and 10ml of water) was added to each of the contaminated soil sample in order to promote an efficient contact between the solid and liquid while extracting the contaminants from the samples. Magnetic agitation was used to promote contact between the phases. An optimised contact time 15 minutes and a solid-liquid ratio of 1:8 (v/v) were used. The solvents containing the extracted contaminants was then separated from the soil samples by vacuum filtration. After the separation the soil samples and the samples of the extract mixture were analysed with gas chromatography mass spectrometer (GC-MS) (Aurora *et al.* 2005) to determine the contaminants concentration.

Preparation of Wash Solution

Three different concentrations of wash solution of 20w/v% of detergent in 500ml, 10w/v% of detergent in 1000ml and 7w/v% of detergent in 1500ml of deionised water was prepared by dissolving 100g of the detergent in 500ml, 1000ml and 1500ml of deionised water.

Remediation Procedure

606g each of the contaminated soil samples 1, 2 and 3 were weighed. Each of the samples were divided into three equal sub-samples (Sample 1 was divided into sub-samples A, B and C, Sample 2 was divided into D, E and F while sample 3 was divided into G, H and I) making a total of nine (9) samples. 202g of samples A, D and G were measured into 1000ml cylinders. 150ml of the 20w/v% wash solution was then added to each cylinder containing the samples. The mixtures were allowed to stay for 1 hour before applying air pressure. This was to ensure soaking of the soil and weakening of the adhesive force between the contaminants and promote solubilisation, dispersal and desorption of the pollutants from the polluted soil samples. Air pressure was then introduced using pressure pump into each of the cylinders for 15 minutes to cause agitation of the soil. The air pressure rate was gradually increased as the inlet pipe penetrates into the soil until it reaches the final depth. The air flow rate was kept constant until the inlet pipe was close to the bottom of the cylinder.

After 15 minutes of agitation, the soil-detergent solution mixtures in the cylinders were allowed to stand for 24 hours. The wash solution was then drained out using suction pump. 150ml of deionised water was then added to the washed samples to further remove the residual wash solution in the samples. The deionised water was then drained through pipe by suctioning.

After the washing was completed, the residual contaminants in the washed soil samples were then extracted using the steps outlined in Section 5.2.2 and analysed using the Gas Chromatography Mass Spectrometer. The percentage removal of the contaminants was calculated using the equation

$$R(\%) = \frac{c_i - c_f}{c_i} \times 100$$

Where C_i is the contaminant concentration before (pre) remediation, C_f is contaminant concentration after (post) remediation and R (%) is the percentage removal of contaminant after remediation.

This procedure was then repeated using 150ml of 10w/v% wash solution to wash samples B, E and H and 7w/v% wash solution to wash samples C, F and I.

Figure 6. Laboratory Experimental Setup

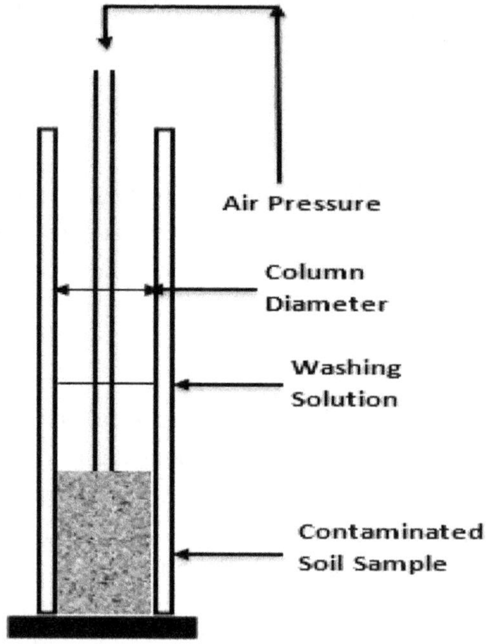

RESULTS AND DISCUSSIONS

Percentage Removal of Aliphatics

Aliphatics are saturated alkanes whose intermolecular forces increase with increase in number of carbon atom (Abozenadah *et al.,* 2017). They are non-polar compound and cannot dissolve in water. However, the result after remediation in Table 5.1 showed a significant reduction in concentration of the aliphatics which was observed from the percentage removal rate of 95.8%, 79.5% and 64.75% in samples A, B and C; 97.55%, 76.47% and 56.04% in samples D, E and F; and 96.06%, 62.70% and 54.43% in samples G, H and I. This is as a result of the increase in the wettability ability of water caused by the detergent (Calbiochem, 2007). The removal rate of aliphatics increase with increase in detergent concentration. Sample A remediated with 20% detergent concentration had removal rate of 95.8%, sample 5B remediated with 10% detergent concentration had removal rate of 79.5% while C remediated with 7% detergent had removal rate of 64.75% (Table 1 and Figure 1). This is influenced, according to Guang-Guo, (2005), by the critical micelles concentration (CMC) formed from high detergent concentration which according to Divya and Tyagi, (2007) aid in the emulsification and solubilisation process of hydrocarbons and effectively remove more of the aliphatic hydrocarbons from the contaminated soil (Guang-Guo, 2005).

According to Kelvin, alkanes with high molecular weight are highly hydrophobic and tend to have low solubility. However, Robert and David, 2018, noted that aliphatics (n-alkanes) are bonded by weak London forces (induced dipole-induced dipole attraction) that control their solubilities and boiling points which according to Divya and Tyagi, (2007), can easily be broken by (detergent) surfactant.

Table 1. Percentage Removal of Aliphatics in all Samples

% Removal of Aliphatics in All Samples							
S/No	**Detergent Conc. Used (w/v%)**	**Samples**	**% Removal**	**Samples**	**% Removal**	**Samples**	**% Removal**
1	20	A	95.8	D	97.55	G	96.06
2	10	B	79.5	E	76.47	H	62.70
3	7	C	64.75	F	56.04	I	54.43

Figure 7. Percentage removal of Aliphatics

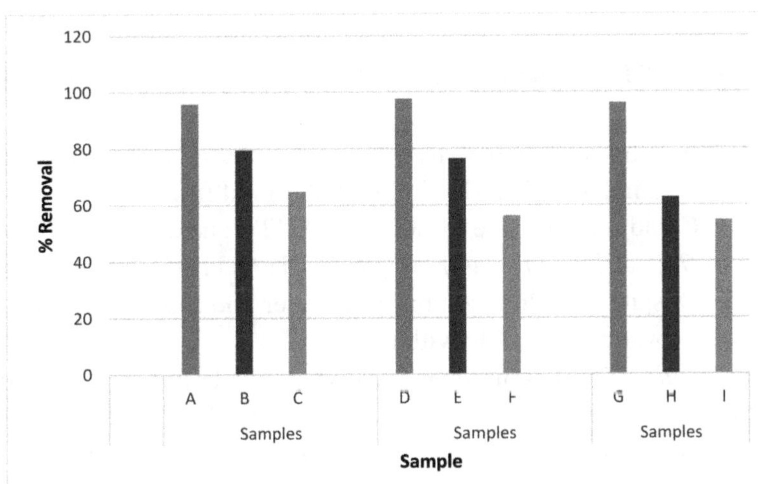

Table 2. Percentage Removal of PAHs in all Samples

% Removal of PAHs in All Samples							
S/No	**Detergent Conc. Used (%)**	**Samples**	**% Removal**	**Samples**	**% Removal**	**Samples**	**% Removal**
1	20	A	60.36	D	61.41	G	47.07
2	10	B	38.42	E	48.71	H	27.77
3	7	C	26.03	F	24.47	I	14.87

Figure 8. Percentage removal of PAHs

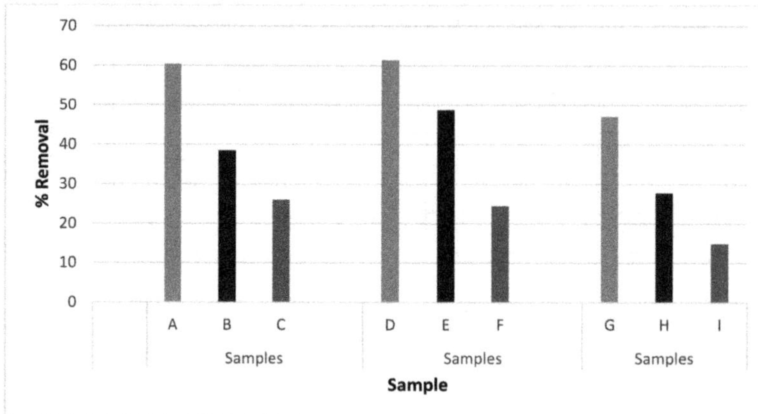

Percentage Removal of Polycyclic Aromatic Hydrocarbons (PAHs)

Polycyclic Aromatic Hydrocarbons as shown in Table 5.2, are observed to also reduce in concentration after remediation in all the samples. This reduction was observed from the percentage removal rate in which samples A, B and C had removal rate of 60.36%, 38.42% and 26.03%; samples D, E and F had removal rate of 61.4%, 48.71% and 24.47%; and samples G, H and I had removal rate of 47.07%, 27.77% and 14.87% (Table 5.2). Despite their low solubility in water and high adherence to particles of soil (Katarzyna *et al,* 2004) removal was possible with detergent.

According to Leslie *et. al.,* (2004), individual PAHs accumulate in soil differently due to their low aqueous solubility, low vapor pressure and high hydrophobicity. This is dependent on the number of benzene-rings present in the PAHs as explained by Mehdi *et al.,* (2015). They also noted in their study that the solubility of PAH in water decreases with increase in molecular weight (number of benzene-rings). This suggests that PAHs with lower molecular weight will be removed faster than those with high molecular weight. However, the rate of removal was not consistent with the number of benzene-rings in the PAHs compounds. This may be as a result of the presence of one PAH which lowers the removal rate of other PAHs (Mehdi *et al.,* 2015). Mehdi *et al.,* (2015), in their study discovered that with the concurrent presence of phenanthrene, acenaphthene, fluorine, anthracene, pyrene, and benzo[a]pyrene, the removal of phenanthrene and acenaphtene was reduced but was enhanced for fluorine, anthracene and pyrene.

However, the use of detergent was able to remove PAHs from the contaminated soil samples. According to Divya and Tyagi, (2007), detergent can remove PAHs from contaminated soil by lowering the interfacial tension existing between the PAHs and the soil particles. The reduction in concentration of PAHs was high in samples A, D and G which were treated with 20% detergent concentration than samples B, E, H with 10% treatment and C, F and I with 7% treatment (Table 2 and Figure 2). This suggest that removal rate of PAHs increase as a result of the increase in concentration of detergent in the wash solutions respectively. According to Khodadadi *et al.,* (2012), the increased rate removal is enhanced by the detergent ability to increase the degree of solubility of PAHs contaminants by reducing the surface and interfacial tension existing between the PAHs and the soil particles.

Effect of Detergent Concentration on Contaminants Removal

The results obtained from this study showed that the detergent at different concentrations removed the petroleum contaminants such as aliphatics and PAHs from the contaminated soil samples. However, the level to which the contaminants were removed varied due to the variation in concentration of detergent used and strength of bond existing between the molecules of the contaminants. According to Abozenadah *et al.,* (2017), water cannot dissolve hydrocarbon substance due to its hydrogen bond which is stronger than the London dispersion force (induced dipole-induced dipole attraction) existing in hydrocarbons. They also established that, for a hydrocarbon to dissolve in water the London dispersion force (induced dipole-induced dipole attraction) holding its molecules must be broken or the hydrogen bond in water must also be broken to create space for the substance (hydrocarbons) to fit between the water molecules.

Detergent are soluble in water because they are amphipathic (has polar or hydrophilic (head) and non-polar or hydrophobic (tail) groups) molecules that plays the role of increasing water wettability by forming hydrogen bond with water molecules using their polar group while the hydrocarbon chains aggregate because of hydrophobic interaction. Due to their amphipathic characteristic, detergents are able to solubilise non-polar or hydrophobic substance in water (Calbiochem, 2007). Tanee and Jude (2017), established that detergents emulsify and break the bond existing between hydrocarbon chains thereby providing a good surface area for biodegradation. The emulsifying effect of detergent made the cleansing and washing of the soil possible due to its adsorbance ability to non-polar (petroleum contaminants) and polar (water) materials at the same time. The petroleum contaminants are kept in emulsion in the aqueous solution and removed by rinsing during cleansing and washing. They chemically interact with both oil and water to stabilise the interface between oil and water in suspension (Divya and Tyagi, 2007).

The amphipathic characteristic of detergent accounts for the effective reduction in the concentration of aliphatic and PAHs with removal rate ranging between 54.43% - 97.55% for aliphatics and 14.87% - 61.41% for PAHs (Table 5.3). The high rate (97.55%, 96.06% and 95.8% for aliphatics and 61.41%, and 60.36% for PAHs) removal of the aliphatics and PAHs as shown in Figure 3, was achieved due to the amphipathic property of the detergent that induced the solubilisation and emulsification and creating a good surface area that increased the adsorbance of the detergent to the hydrocarbons (aliphatics and PAHs) and water.

Molecules of detergent at high concentration above their Critical Micelles Concentration (CMC) form stable micelles which gives them their detergency and solubilisation characteristic when in solution (Guang, 2006). This characteristic gives them the ability to increase the solubility of hydrocarbon contaminants. Above the CMC the increase in solubility of the contaminants is directly proportional to the wide range of the detergent concentration and this is due to the solubilisation of the contaminant molecules into the micelles (Divya and Tyagi, 2007). Hydrocarbon contaminants can easily be transported into the wash solution by reduction of surface and interfacial tension due to high micelles stability caused by high concentration of detergent (Weihua and Irene, 2007). This accounts for why the 20% detergent concentration removed more of the aliphatics and PAHs from the soil than 10% and 7% detergent concentration as shown in Table 3 and Figure 3. The 20% detergent concentration is assumed to have formed more stable CMC than the 10% and 7% detergent concentration. This explained why the removal rate of above 90% of the aliphatics and above 60% of the PAHs were achieved (Table 3 and Figure 3).

Table 3. Percentage removal of Aliphatics and PAHs in all samples

% Removal of Aliphatics and PAHs in All Samples				
S/No	**Detergent Conc. Used (%)**	**Samples**	**% Removal**	
			Aliphatics	**PAHs**
1	20	**A**	95.8	60.36
2	10	**B**	79.5	38.42
3	7	**C**	64.75	26.03
1	20	**D**	97.55	61.41
2	10	**E**	76.47	48.71
3	7	**F**	56.04	24.47
1	20	**G**	96.06	47.07
2	10	**H**	62.70	27.77
3	7	**I**	54.43	14.87

Figure 9. Percentage removal of Aliphatics and PAHs

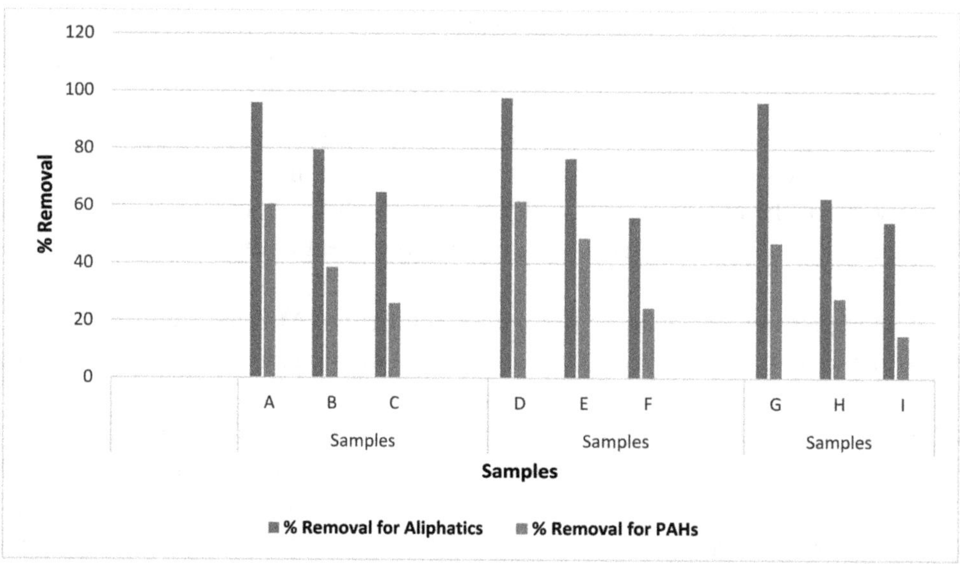

The agitation and tumbling provided by the air pump aided the dissociation and transportation of the hydrocarbon contaminants from the soil into the wash solution. As stated by Divya and Tyagi (2007), the removal of contaminants is the reverse process of coagulation and adhesion, and this needs energy which may be supplied by an external force such as a washing machine. This external force was provided by the air pumping and this helped in the transfer of contaminants from the contaminated soil into the wash solution.

Table 4. Previous works on PAHs removal using IWS

Compound	Surfactant Type	Concentration	Removal Efficiency (%)	References
PAHs	Saponin	0.25% by weight	90%	Budianta *et al.*, (2010)
PAHs	Saponin	0.25% by weight	90%	Budianta *et al.*, (2012)

Performance efficiency of the In-Situ Washing by Sedimentation (IWS)

The percentage removal of the contaminants was used to evaluate the efficiency of the In-Situ Washing by Sedimentation (IWS). The results show that for the various concentration of detergent, removal efficiencies of between 54.43% - 97.55%, 14.87% - 61.41% and 35.04% - 76.91% were obtained for aliphatics and PAHs respectively. These high values of removal efficiencies therefore show the effectiveness of the IWS method for removal of aliphatics and PAHs from soil. Studies carried out by Budianta *et al.*, (2010) and Budianta *et al.*, (2012) have shown similar high hydrocarbons (PAHs) removal efficiencies using IWS method as shown in Table 4. However, aliphatics hydrocarbons was not considered by them.

CONCLUSION

In this research, soil washing using detergent for the remediation of petroleum contaminated soils considering different concentrations was carried. The contaminant concentration was determined using the Gas Chromatography Mass Spectrometer. The emulsifying effect of detergent made the cleansing and washing of the soil possible and the percentage removal efficiencies of the contaminants increased with increase concentration of detergent. The highest percentage removal efficiencies of 97.55% and 61.41% for aliphatic and Polycyclic Aromatic Hydrocarbons were obtained at detergent concentration of 20w/v% respectively.

RECOMMENDATIONS FOR FUTURE RESEARCH

The following recommendations are areas for further improvement of this work.

1. There is need to consider the effect of temperature while using detergent as wash solution.
2. Future research should consider the toxicity effect of the remediating agent (detergent) on the soil and living organisms in the soil at different concentrations.

REFERENCES

Alaska Department of Environmental Conservation. (2009). Environmental Clean-up Methods. Spill Prevention and Response Division. Retrieved from https://dec.alaska.gov/spar/csp/guidance/cleanup_methods.pdf

Alexander, M. (1994). *Biodegradation and bioremediation* (p. 284). New York: Academy Press.

Anderson, W. C. (1993). Innovative site remediation technology: Soil washing/soil flushing. American Academy of Environmental Engineers, Annapolis, Maryland (3), 1993.

Anderson, W. C. (1993). Editor. Innovative site remediation technology. 6 – Thermal desorption. *American Academy of Environmental Engineers, Annapolis, Maryland.* Retrieved from https://nepis.epa.gov/Exe/ZyPDF.cgi/2000ITIZ.PDF?Dockey=2000ITIZ.PDF

Antai, S. P., & Mgbomo, E. (1989). Distribution of hydrocarbon utilizing bacteria in oil spill areas. *Microbias Letters*, *8*, 137–143.

Araruna, J. T., Portes, V. L. O., Soares, A. P., Silva, M. C., Sthel, M. S., Schramm, D. U., ... Vargas, H. (2004). Oil spills debris clean up by thermal desorption. *Journal of Hazardous Materials*, *110*(1-3), 161–16. doi:10.1016/j.jhazmat.2004.02.054 PMID:15177737

Atlas, R. M. (1981). Microbial degradation of petroleum hydrocarbons: An experimental perspective. *Microbiological Reviews*, *45*, 180–209. PMID:7012571

Aurora, S., Cristina, D., & Fi'uza, A. (2005). Use of solvent extraction to remediate soils contaminated with hydrocarbons. *Journal of Hazardous Materials*, *B124*, 224–229. PMID:16006033

Bassam, M., & Battikhi, M. N. (2005). Biodegradation of total organic carbons (TOC) in Jordanim petroleum sludge. *Journal of Hazardous Materials*, *120*(1-3), 127–134. doi:10.1016/j.jhazmat.2004.12.033 PMID:15811673

Boudreaux, K. A. (n.d.). Organic compounds: Alkanes. CHEM 2353 Fundamentals of Organic Chemistry, Chapter 1. Retrieved from https://www.angelo.edu/faculty/kboudrea/index_2353/Chapter_01_2SPP.pdf

Brady, N., & Weil, R. (2002). *The nature and properties of soils* (13th ed., p. 960). Upper Saddle River, NJ: Prentice Hall.

Budianta, W., Salim, C., Hinode, H., & Ohta, H. (2006). In situ soil washing on metal-contaminated sandy soil by sedimentation method: a new approach on soil remediation. *Philippine Engineering Journal* 2006. 27(1), pp 65–76. Retrieved from https://www.researchgate.net/publication/228365882_In_situ_Soil_Washing_on_Metal-Contaminated_Sandy_Soil_by_Sedimentation_Method_A_New_Approach_on_Soil_Remediation

Budianta, W., Salim, C., Hinode, H., & Ohta, H. (2010). In-situ washing by sedimentation method for contaminated sandy soil. *Proceedings of the Annual International Conference on Soils, Sediments, Water and Energy*: Vol. 15, Article 15.

Budianta, W., Salim, C., Hinode, H. and Ohta, H. (2012). Remediation of organics-contaminated sandy soil by in-situ washing. *ASEAN Engineering Journal Part C*, 1(1), .87.

Calbiochem. (2007). Detergents: A guide to the properties and uses of detergents in biology and biochemistry. Retrieved from www.calbiochem.com

Chorom, M., Shari, H. S., & Motamedi, H. (2010). Bioremediation of a crude oil - polluted soil by application of fertilizers. *Iranian Journal of Environmental Health Sciences & Engineering*, *7*(4), 319–326. Retrieved from https://www.researchgate.net/publication/267920577_Bioremediation_of_a_crude_oil_-_polluted_soil_by_application_of_fertilizers

Christopher, C. A., Chioma, B. C., & Gideon, C. O. (2016). Bioremediation techniques-classification based on site of application: Principles, advantages, limitations and prospects. [Formerly MIRCEN Journal of Applied Microbiology and Biotechnology]. *World Journal of Microbiology & Biotechnology*, *32*(11), 180. doi:10.100711274-016-2137-x PMID:27638318

Couto, M. N., Monteiro, E., & Vasconcelos, M. T. (2010). Microcosm trials of bioremediation of contaminated soil of a petroleum refinery: Comparison of natural attenuation, biostimulation, and bioaugmentation. *Environmental Science and Pollution Research International*, *17*(7), 1339–1346. doi:10.100711356-010-0318-y PMID:20229281

Dağhan, H., & Öztürk, M. (2015). Soil pollution in Turkey and remediation method in soil remediation and plants (prospects and challenges), Academic Press, 287-312.

Diane, H. (1999). Bioventing principles, applications and potential: restoration and reclamation review. *Student On-line Journal.* 5(2). Retrieved from https://conservancy.umn.edu/bitstream/handle/11299/59453/1/5.2.Hellekson.pdf

Dike, B. U., Okoro, B. C., Nwakwasi, N. N., & Agbo, K. C. (2013). Remediation of used motor engine oil contaminated soil: a soil washing treatment approach. *Journal of Civil Environmental Engineering*, *3*, 129. doi:10.4172/2165-784X.1000129

Divya, B., & Tyagi, V. K. (2007). Laundry detergents: an overview. *Research gate, Journal of Oleo Science February 2007.* doi:. 56(7) 327-340. Retrieved from www.researchgate.net/publication/5945251 doi:10.5650/jos.56.327

Dragun, J. (1993). Recovery techniques and treatment for petroleum and petroleum products in soil and groundwater. *Proceedings of 1993 Oil Spill Conference, American Petroleum Institute, Washington DC*, pp. 48-63.

Eckenfelder, W. W., & Norris, R. D. (1993). Applicability of biological processes for treatment of soils. American Chemical Society, 518, 138-158.

Edris, M., Saeid, G., Leila Amiri, M. A., & Jafar, S. (2013). The application of soil washing for treatment of polycyclic aromatic hydrocarbons contaminated soil: a case study in a petrochemical complex. *Environmental Progress & Sustainable Energy*, *33*(1). doi:10.1002/ep

Edward, B., & Colin, H. (2015). Stabilization and solidification of contaminated soil and waste: a manual of practice. *Hygge Media.* Retrieved from http://www.enviro.wiki/images/9/90/Bates-2015-Stabilization_and_Solidification_of_Contaminated_Soil_and_Waste.pdf.

Elgibaly, A. A. M. (1999). *Clean-up of oil contaminated soils of Kuwaiti oil lakes* (pp. 411–416). Washington, DC: Institute.

EPA. (2006). In situ treatment technologies for contaminated soil. Engineering forum issue paper. United States Environmental Protection Agency. 542/F-06/013.

Essian, J. P., Ubom, R. M., & Udosen, E. D. (1997). Bioremediation of petroleum contaminated soil: Effect on the population dynamics and degrading capabilities of hydrocarbon clastic bacteria. *West African Journal of Biology and Applied Chemistry*, *43*, 22–28.

Eyad, A. B., & Richard, G. Z. (2015). An effective soil vapour extraction/bioventing model. Retrieved from https://www.esaa.org/wp-content/uploads/2015/06/04-21Barakat.pdf

Faisal, I. K., Tahir, H., & Ramzi, H. (2004). An overview and analysis of site remediation technologies. *Journal of Environmental Management*, *71*(2), 95–122. doi:10.1016/j.jenvman.2004.02.003 PMID:15135946

Galaska, G., Skaladany, G. J., Nyer, E. K., & Chelsea, M. (1990). Biological treatment of groundwater, soils and soil vapour contaminated with petroleum hydrocarbons. *Proceedings of the Industrial Waste Conference, Purdue University* (44th), 11-21.

Guang-Guo, Y. (2006). Fate, behaviour, and effects of surfactants and their degradation products in the environment. *Environment International 32 (2006)*, Elsevier, 417 – 431.

Hamby, D. M. (1996). Site remediation techniques supporting environmental restoration activities. Department of Environmental and Industrial Health School of Public Health University of Michigan Ann Arbor, MI: 48109-2029 U.S.A.

Higgins, I. J., & Gilbert, P. O. (1978). The biodegradation of hydrocarbons. Charter KWA, Somerville H. J. (Eds). The Oil Industry and Microbial Ecosystems, Hegden and Sons, Ltd., London, UK. pp. 80-117.

Hinsenveld, M., Soczo, E. R., Van de Leur, G. J., Verluijs, C. W., & Groeedijk, E. (1990). Alternate physico-chemical and thermal cleaning technologies for contaminated soils. In *Proceedings of International Conference on Contaminated Soil*. Dordrecht, The Netherlands: Kluwer Academic. 873-881.

Ijah, U. J. J., & Antai, S. P. (1988). Degradation and mineralization of crude oil by bacteria. *Nigerian Journal of Biotechnology*, *5*, 79–86.

Imamura, T., Kozaki, S., Kuriyama, A., Kawaguchi, M., Touge, Y., Yano, T., ... Kawabata, Y. (1997). Bioaugmentation of TCE contaminated soil with inducer-free microbes. In G. S. Sayler, J. Sanseverino, & K. L. Davis (Eds.), *Biotechnology in the Sustainable Environment* (pp. 97–106). New York: Plenum Press. doi:10.1007/978-1-4615-5395-3_10

James, L. W. (2006). *Soil sampling and analysis. The University of Arizona College of Agriculture and Life Sciences, Tucson, AZ* (p. 85721). Arizona: Arizona Cooperative Extension.

Katarzyna, S., Irena, M., & Teresa, K. (2004). Polycyclic aromatic hydrocarbons: physicochemical properties, environmental appearance and impact on living organisms. *Acta Poloniae Pharceutica-Drug Research.*, *61*(3), 233–240. PMID:15481250

Khodadadi, A., Hossein, G., & Seyed, R. S. N. (2012). Treatment of crude-oil contaminated soil using biosurfactants. *Journal of Petroleum and Gas Engineering*, *3*(6), 92–98. Retrieved from http://www.academicjournals.org/app/webroot/article/article1379670453_Khodadadi%20et%20al.pdf

Kingsley, U., Turgay, P., & Mehmet, Ç. (2005). Screening of biosurfactants for crude oil contaminated soil washing. *Journal of Environmental Engineering and Science*, *4*(6), 487–496. doi:10.113904-073

Kokyo, O., Tiehua, C., Tao, Li., & Hongyan, C. (2014). Study on application of Phytoremediation technology in management and remediation of contaminated soils. *Journal of Clean Energy Technologies,* 2(3), 216. Retrieved from https://www.researchgate.net/publication/271293740_Study_on_Application_of_Phytoremediation_Technology_in_Management_and_Remediation_of_Contaminated_Soils

Laleh, Y., Sylvie, R., Ruxandra, C., Manon, S., Geoffrey, S., Adriana, P., ... Serge, R. G. (2003). Enhanced biodegradation of petroleum hydrocarbons in contaminated soil. *Bioremediation Journal*, *7*(1), 37–51. doi:10.1080/713914241-274

Leahy, J. D., & Colwell, R. R. (1990). Microbial degradation of hydrocarbons in the environment. *Microbiological Reviews*, *54*, 305–315. PMID:2215423

Leslie, M. S., David, S., Kosson, L. Y., Young, K. J. R., & Gary, L. T. (2004). Combine effects of contaminant desorption and toxicity on risk from PAH contaminated sediments. *Society for Risk Analysis* (5). Retrieved from https://pdfs.semanticscholar.org/e2e3/c2420a2763aece6141d424e856a30683dd0b.pdf

Lombi, E., & Hamon, R. E. (2005). Remediation of polluted soils in Encyclopaedia of soils in the environment, Elsevier, 379-385.

Lee, M. D., & Swindoll, C. M. (1993). Bioventing for in situ remediation. *Hydrological Sciences -Journal- des Sciences Hydrologiques* 38(4), 8/1993. Retrieved from http://hydrologie.org/hsj/380/hysj_38_04_0273.pdf

Manahan, S. E. (2001). Introduction to chemistry. Fundaments of environmental chemistry. Boca Raton, FL: CRC Press. Retrieved from http://www.ingenieroambiental.com/4004/Fundamentals%20of%20Environmental%20Chemistry,%20Manahan.pdf

McBride, M. B. (1994). *Environmental chemistry of soils*. New York: University Press.

Mehdi, H. S., Moslem, A., & Simone, C. (2015). *Biodegradation of aromatic compounds*. Intech. doi:10.5772/60894

Mohammad, M. A., Mohammad, S. H., Fariborz, M., & Mehdi, K. (2014). Soil remediation via bioventing, vapor extraction and transition regime between vapor extraction and bioventing. *International Journal of Environmental Health Engineering,* 2014, 3(1). Retrieved from https://www.researchgate.net/publication/305114125_Soil_remediation_via_bioventing_vapor_extraction_and_transition_regime_between_vapor_extraction_and_bioventing

Mulligan, C. N., Yong, R. N., & Gibbs, B. F. (2001). Surfactant-enhanced remediation of contaminated soil. *Reviews in engineering geology*, *60*(1-4), 371–380. doi:10.1016/S0013-7952(00)00117-4

Nadean, R. R., Singhvi, J., Ryabik, Y., & Lin, I. (1993). Monitoring bioremediation for bioremediation efficacy: *The Marrow Marsh experience. Proceedings of the 1993 Oil Spill Conference, American Petroleum Institute, Washington, DC*, pp. 477-485.

National Academy of Sciences. (1997). *Measures of success for remediation technologies chapter four in innovations in ground water and soil clean-up: from concept to commercialization* (pp. 167–200). Washington, DC: National Academy Press.

Okoh, A. I., & Trejo-Hernandez, M. R. (2006). Remediation of petroleum hydrocarbon polluted systems: Exploiting the bioremediation strategies. *African Journal of Biotechnology*, *5*(25), 2520–2525. Retrieved from http://www.academicjournals.org/AJB

Otten, A., Alphenaar, A., Pijls, C., Spuij, F., & Wit, H. (1997). *In situ soil remediation*. Kluwer Academic Publishers. doi:10.1007/978-94-011-5594-6

Ralinda, R., & Miller, P. G. (1996). Phytoremediation, Technology overview report. *Ground-water remediation technologies analysis center.* pp. 3-5. Retrieved from https://clu-in.org/download/toolkit/phyto_o.pdf

Raphael, B., Thiago, G. M., Cintya, A. C., Tamaris, G. P., & Carmem, S. F. (2013). Soil contamination with heavy metals and petroleum derivates. Impact on Edaphic Fauna and Remediation Strategies 176-178/chapter 6.

Robert, J. O., & David, J. R. (2018). Alkanes and cycloalkanes: structures and reactions. *Science Direct*. Retrieved from https://www.sciencedirect.com/topics/chemistry/alkane

Rosario, I., & Jessica, L. (2015). Bioremediation for a soil contaminated with hydrocarbons. *Journal of Petroleum & Environmental Biotechnology*, *6*(2). doi:10.4172/2157-7463.1000208

Shah, F. H., Hadim, H. A., & Korfiatis, G. P. (1995). Laboratory studies of air stripping of VOC-contaminated soils. *Journal of Soil Contamination*, 4(1), 1995.

Tenedero, R. A., & Surtida, M. B. (Eds.). (1986). *Soil sampling and preparation for laboratory analysis. Tigbauan*. Iloilo, Philippines: SEAFDEC Aquaculture Department.

Trindade, P. V. O., Sobral, L. G., Rizzo, A. C. L., Leite, S. G. F., & Soriano, A. U. (2015). Bioremediation of a weathered and a recently oil-contaminated soils from Brazil: A comparison study. *Chemosphere*, *58*(4), 515–522. doi:10.1016/j.chemosphere.2004.09.021 PMID:15620743

Troxler, W. L, Cudahy, J. J., Zin, R. P., & Rosenthal, S. I. (1994). Thermal desorper: Applications manual for treating non-hazardous petroleum contaminated soils. Cincinnati, OH: U.S. EPA, Office of Research and Development, 15 pp.

Uchechukwu, E. E., Sylvia, O. A., & Vincent, I. I. (2007). Clean up of crude oil-contaminated soil. *Terrestrial and Aquatic Environmental Toxicology*, *1*(2), 54–59.

United State Environmental Protection Agency. Bioslurry treatment of contaminated soils and sediments. *U.S. EPA Test and Evaluation Facility Research Project*. IT Corporation. Retrieved from http://www.emfederal.com/projects/fctshts/bioslrry.pdf

U.S. EPA. (2006). In situ treatment technologies for contaminated soil. *Engineering Forum Issue Paper. Solid Waste and Emergency Response 5203P.* Available from https://clu-in.org/download/remed/542f06013.pdf

Wardley-Smith, J. (1983). *The control of oil pollution* (pp. 75–79). London, UK: Graham and Trotman.

Weihua, Z., Irene, M. C., & Lo, M. A. (2007). Chemical-enhanced washing for remediation of soils contaminated with marine diesel fuel in the presence/absence of Pb. *Journal of Environmental Engineering, 133*(5), 548–555. doi:10.1061/(ASCE)0733-9372(2007)133:5(548)

Wood, L. A. (2002). *Overview of remediation technologies. Terra Resources, Ltd.* (p. 6). Palmer, Alaska: Wolverine. Retrieved from www.terrawash.com/twp2.htm

Xu, R., Yong, L. C., Lim, Y. G., & Obbard, J. P. (2005). Use of slow-release fertilizer and biopolymers for stimulating hydrocarbon biodegradation in oil contaminated beach sediments. *Marine Pollution Bulletin, 51*(8-12), 1101–1110. doi:10.1016/j.marpolbul.2005.02.037 PMID:16291209

Yeung, A. T. (2010). Remediation technologies for contaminated sites. *Proc., Int'l. Symp. on Geo-Environmental Engineering*. I. doi:10.1007/978-3-642-04460-1_25

Chapter 21
Membrane Technology for Treatment of Pharmaceutical Wastewaters:
A Novel Approach

Saritha Vara
https://orcid.org/0000-0003-0473-993X
GITAM University (Deemed), India

Madhavi Konni
https://orcid.org/0000-0002-2022-5528
Vignan's Institute of Engineering For Women, India

Manoj Kumar Karnena
https://orcid.org/0000-0001-9241-2990
GITAM University (Deemed), India

ABSTRACT

It is estimated 90% of the available water resources will be expended in 15 years and by 2025, nearly 60% of the world population will have scarcity of water if the rate of consumption continues at its present level. The mounting use of pharmaceuticals, with enhanced production, have driven these industrial effluents into so-called emerging pollutants that have become a new environmental problem. Pharmaceutical wastewater, specifically categorized by complex components, are ending up polluting natural water bodies, making it necessary to remove such substances from the wastewaters to prevent harm to the natural environment. Amid options available for treating these effluents, biological processes are cost-effective and environmentally safe alternatives to chemical methods. One of the recent advances includes use of membrane technology. Among these, membrane bioreactor and reverse osmosis technologies are becoming advanced and promising options for wastewater treatment, and reuse at a reduced price is making it economically feasible.

DOI: 10.4018/978-1-7998-0369-0.ch021

INTRODUCTION

Fresh and potable water is becoming increasingly unavailable in most parts of the world. In the developing world, over one million people are without access to safe drinking water, 83% of which live in pastoral communities (UNICEF 2010). Clean drinking water is essential to human health and wellbeing (Naiman et al., 2011). Despite well-intended efforts, intervention attempts to sustainably lower these statistics have been largely unsuccessful. The water crisis is being faced by several underdeveloped zones of the world, on the other hand, industrialized nations face issues with reference to the quality of water for industrial and public use. Over 90% of the available freshwater resources are going to be consumed during the next coming fifteen years. Studies have found that 30% to 50% of rural water projects, whether water systems or wells, fail between 3- and 5-years following construction (Walters et al., 2015) and this has been linked to approximately 4 billion diarrhea outbreaks and over 5 million annual deaths worldwide (Sobsey et al., 2008, WHO 2009), mostly in developing countries, especially in rapidly expanding urban fringes, Poor rural areas and indigenous communities (Lipton, 1977 and Ryser et al. 2010). Waterborne diseases and related fatality are commonly associated with lesser developed nations and remote areas where municipal water system can be impractical in these settings (Shannon et al., 2010). Many studies have revealed that the significant causes of these clean water shortages are linked to the upsurge in industrial activities and the attendant pollution menace, improper management unfavourable policy implementation of water-related projects and alarming population growth (Ayandiran et al., 2018). Moreover, the demand for good quality water towards several purposes is continuously increasing, which has enhanced concerns on treatment and reuse of water meeting the strictest standards (Jyoti et al., 2013).

The rise in the pharmaceutical industry is driven by global needs. Enhanced environmental disturbances due to globalization and urbanization are the prime driving forces for mounting demand for improved medication as well as health supplements for every age group and geographical location. The overwhelming use of pharmaceuticals has become a challenge with reference to the effluents generated in their production. Concern towards the presence of pharmaceutical chemicals owes to the fact of their nature being lipophilic, non-biodegradable and affecting various biological activities (Bound & Voulvoulis, 2005; Houk, 1992; Velagaleti & Burns, 2006).

CHARACTERISTICS OF PHARMACEUTICAL WASTEWATERS

Water is a crucial raw material for pharmaceutical and chemical manufacturing. Persistent and good quality water supply is required for a varied range of operations including production, material processing, and cooling. This results in the generation of non-uniform waste streams at any point in the processing period (Frick et al., 2005). Furthermore, majority of pharmaceutical industries operate as a batch process using different raw materials as well as production processes resulting in the release of varied nature of wastewater (Rizwan et al., 2013; CAO et al., 2013; Guo et al., 2017). General characterization of pharmaceutical wastewater is presented in Table 1 (Rong-Rong et al., 2003).

Various processes in the pharmaceutical manufacturing include manufacturing, extracting, processing, purification and packaging of biological and chemical materials in the forms of solid or liquid for use as medication by humans and animals (Jozala et al., 2016). Majority of wastewaters from a pharmaceutical industry originate from the processes of synthesis and formulation of drugs (Kessler, 2010).

Table 1. General characterization of pharmaceutical wastewater

COD (mg/L)	BOD_5 (mg/L)	TN (mg/L)	TP (mg/L)	SS (mg/L)	Chromaticity (times)	Temp (^{0}C)	pH
1000-10000	500-2500	500-1500	50-250	200-500	500-1000	25-80	1-8

IMPACTS OF PHARMACEUTICAL WASTEWATER

Report by the U.S. Geological Survey (USGS) in 2010 have shown that most of the pharmaceuticals are present in the water supplies of American cities [Momodu & Anyakora, 2010; Vinati & Radha, 2015). This was the first study at the national level to evaluate contamination of pharmaceuticals in streams where approximately 80% of streams tested were found to be contaminated (Idris et al., 2013). Pharmaceutical effluents are objectionable owing to their composition of higher concentrations of organic matter, microbial toxicity, high salt, total solids, heavy metals like Cadmium, Mercury, Nickel, Lead and Chromium including other toxic organic chemicals and hard to biodegrade (Saban, 2015; Xiangdong, 2005; Vanerkar et al., 2004) which are supposed to be carcinogenic mutagenic in nature leading to lethal consequences (Huerta, 2013; Iliuta & Larachi, 2001; Fatta-Kassinos et al., 2011; Heberer, 2002).

Furthermore, some authors estimate that nearly half of the pharmaceutical effluents released worldwide are discarded without any precise treatment (Chen et al., 2008; Sreekanth et al., 2009). Among other pharmaceuticals in environment develop microbes in the aquatic environment which are antibiotic resistant, retard oxidation of nitrite and methanogenesis apart from causing increased toxicity of chemical combinations and metabolites (Tijani et al., 2013; Baquero et al., 2008; Dalrymple et al., 2007). Besides pharmaceutical products in water streams can also result in fish and alligator's feminization, migratory patterns, and behavior of salmon, close – the extinction of vulture population in India (Gadipelly et al., 2014; Jimenez, 2012).

Also, the studies show that production and administration of pharmaceutical products will continuously increase with development and advancement of lifestyle and longevity globally (Sindi et al., 2015). Ample papers have been published regarding treatment and removal of pharmaceutical products, Endocrine disrupting chemicals and pharmaceuticals and household consumer products (Sima et al., 2014).

TREATMENT OPTIONS FOR PHARMACEUTICAL WASTEWATERS

In order to meet the strict regulatory discharge limits, industries should rethink regarding conventional, typical approaches and explore novel technologies and solutions to reach the regulatory standards. Recently, many studies have been carried out for developing and testing novel technologies and their combinations for the treatment of various wastewaters (Adeyinka, 1996). Efforts are being put to design appropriate treatment methods for pharmaceutical wastewaters [Gogate & Pandit, 2004; Idris et al., 2013). Nevertheless, treatment options that aim at source reduction result not only in reducing the costs and impacts on environmental but also afford probable retrieval of compounds (McCarty et al., 2011; Kümmerer, 2009).

An extensive assortment of effluent treatment and disposal methods are being employed by the pharmaceutical industry (Struzeski, 1980). Since the effluents generated by these industries differ in composition, quantity, type of manufacturing, temporal and seasonally depending on the raw materials used and also on the production processes. Owing to these reasons specifying precise treatment for such effluents is very difficult. Conventional end-of-pipe solutions for managing such effluents are being progressively substituted by increasingly precise approaches for treating these wastewaters in an efficient, economic and environmentally sound way. Results from such studies varied from zero abatement efficiency as observed with conventional coagulation and flocculation processes with alum and ferric chloride to very high reduction rates that were achieved by filtration through activated carbon, biological degradation, membrane filter technology including reverse osmosis, nanofiltration, ultrafiltration etc (Khiari, 2007; Ternes et al., 2002).

General approaches to treatment are not appropriate for all pharmaceutical wastewaters owing to its dynamic composition. Hence, precise treatment would be required for a specific type of wastewaters. Progress in wastewater treatment technology has been the result of laboratory and field scale experimentation largely on a trial and error basis (El-Gohary, 1995). A few treatability studies on wastewater from pharmaceutical industries have been done to treat the wastewater up to a enough degree, which renders it safe and complies with regulatory standards (Saleem, 2007; Srikanth, 2008).

Because of the scarcity of water resources, wastewater reclamation and reuse is an effective tool for sustainable industrial development program in both developing and developed countries. These effluents have a potential for reclamation and reuse for various industrial processes. To reach this reuse application further treatment would be needed. Nowadays membrane separation processes are becoming quite popular in wastewater treatment and reclamation since they combine process stability with an excellent effluent quality (Tam et al., 2007; Wintgens et al., 2005; Comerton et al., 2005; Meier & Melin, 2005).

OVERVIEW OF MEMBRANE TECHNOLOGY

Principally, the membrane is understood to be a barrier which separates two phases controlling the transmission of various chemical components in a certain approach. In general, there are two phases in a membrane process which are physically separated from one another by the third phase (membrane). The membrane is selective to one of the components. Thus, the movement of that component from one phase to another is carried out by the membrane. Hence, one of the phases is enriched with that component and the other depleted from it (Wu & Sun, 2016).

Membrane Separation, an advanced treatment for pharmaceutical wastewater is generally considered as a pre-treatment process which is adapted to hasten elimination proficiency of pollutants by secondary treatment. Major techniques included are activated carbon adsorption, membrane distillation, and membrane technology.

Figure 1. Schematic model of the membrane separation functioning

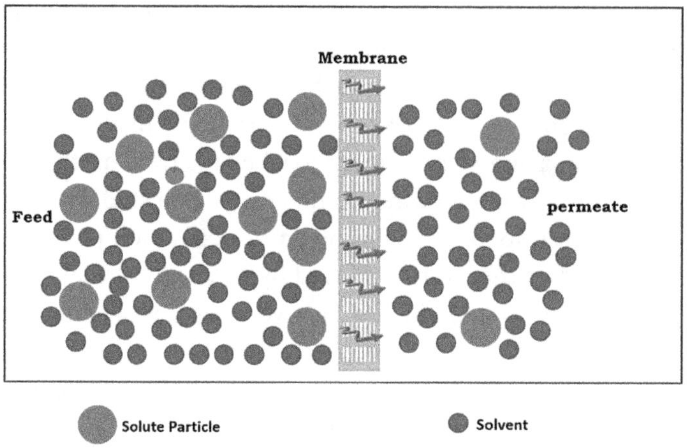

MEMBRANE DISTILLATION

An important separation technology with interesting properties, Membrane distillation is presently used to produce demineralized water (Naik, 2005; Hausmann et al., 2011). It operates at the atmospheric condition with very little heat requirement (Singh & Sirkar, 2012). The application of this technology has proved to be promising as it is used to recover process waters using the heat generated during the process itself (Singh & Sirkar, 2012). Though the process results in clean water, fouling of membrane is a prime disadvantage of the technique. This is successfully applied in the recovery of acids from the fermentation of broths (Gryta et al., 2013).

ACTIVATED CARBON (AC)

Owing to their high surface area along with well-developed pore structure and surface chemistry properties AC is best suited for removal of organic compounds. Ibuprofen removal using AC was successfully performed in their study (Mestre, 2007). Activated carbon can be either granular (GAC) or powdered (PAC) in nature, with PAC having more advantage as it is fresh which can be recycled in fixed bed columns (Ternes et al., 2002; Yoon et al., 2003; Snyder et al., 2007). 99% removal of total mercury and 90% of copper from pharmaceutical wastewater was shown using a series of GAC columns, which also removed 96% of phenol and turbidity was reported by (Cyr et al., 2002).

Figure 2. Representation of membrane separation of various pollutants

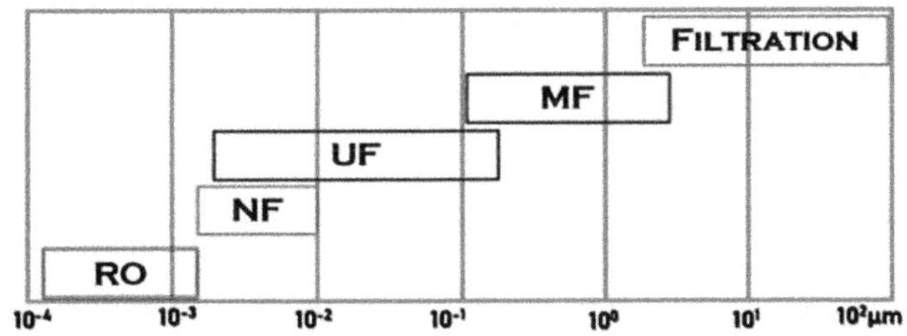

MEMBRANE TECHNOLOGY

Application of membranes in water and wastewater is significantly increasing, as low-pressure membranes are very much capable in the removal of microbial constituents without enhancing the disinfection by-products. Hence, these allow amenableness with the rules disseminated in response to the 1986 Surface Water Treatment Rule Amendments (Snyder et al., 2007). Applied for varied purposes ranging from desalination to water reuse, low-pressure membranes play a vital role as reverse osmosis pre-treatment processes. 95% rejection of diclofenac using RO membranes was reported by (Kimura & Watanabe, 2005).

Other membrane technologies like ultrafiltration (UF) or microfiltration (MF) are intensely suggested in conditions of space limitations or varying feed water quality (Adham et al., 2005). Whereas Nanofiltration (NF) and UF are used for reclamation of wastewater and drinking water, removing micro-pollutants and natural organic matter (NOM). NF membrane has shown high retention of endocrine-disrupting chemicals and pharmaceuticals and personal care products greater than UF membrane stating that retention is governed by membrane pore size and also by the chemistry of source waters (Yoon et al., 2005).

Application of membrane processes for the pharmaceutical industry is based on the size of the pollutants which are removed in the order: microfiltration (bacteria), ultrafiltration (macromolecules and viruses), nanofiltration (divalent ions) and reverse osmosis (monovalent ions).

Numerous types of membranes were evaluated in order to remove API's (Active Pharmaceutical Ingredients) both at pilot and full-scale, these include microfiltration (MF), ultra-filtration (UF), nano-filtration (NF), reverse osmosis (RO), membrane bioreactors along with combinations of membranes in series (Bellona & Drewes, 2007; Snyder et al., 2007).

Since the pore sizes range from 100 – 1000 times larger than micropollutants that slip through the membranes of microfiltration and ultra-filtration these generally do not have efficiency in removing such contaminants.

Membrane processes like nano-filtration and reverse osmosis are pressure driven, which has attracted the attention of several researchers for treating water and wastewaters (Watkinson et al., 2007). Nevertheless, studies on the application of either NF or RO for pharmaceutical wastewaters are limited (Nghiem et al., 2005; Yoon et al., 2006; Snyder et al., 2007). Reverse Osmosis applied in varied configurations resulted in the removal of thirty-six endocrine disrupting chemicals like antibiotics, lipid regulators, hormones, etc and personal care products (Snyder et al., 2007; Watkinson et al., 2007). RO, when used in combination with NF for a full-scale drinking water treatment plant, has removed a range of pharmaceuticals like diclofenac, ketoprofen, hydrochlorothiazide and propyphenazone (Radjenovic et al., 2007; Deegan et al., 2011).

Rejection of pharmaceuticals by NF and RO membranes individually or in combination is governed by three mechanisms:

- Size exclusion (steric effect, sieving)
- Charge exclusion (electrical)
- Physico-chemical interactions between solute, solvent, and membrane

Conventional treatment methods do not prove effective for pharmaceutical wastewater owing to their complex and dynamic composition. The precise treatment is to be adopted for a specific type of wastewater. Trial and error basis at laboratory and field scale experimentation has resulted in the progress of wastewater treatment technology (El-Gohary et al., 1995).

Treatability studies on pharmaceutical wastewater have been taken up at a large scale to treat up to a enough degree in order to comply with regulatory standards. Reclamation of wastewater and its reuse is an effective and necessary tool for sustainable industrial development program both in developing and developed countries. Pharmaceutical wastewater is required to be treated further in order reuse. Membrane separation process has gained much importance recently as they integrate stability of the process and excellent effluent quality (Tam et al., 2007; Wintgens et al., 2005; Comerton et al., 2005; Meier & Melin, 2005).

Filtration

Filtration applied in treatment of pharmaceutical wastewater, can be integrated with other unit operations like bio-filter that is a combination of filter and biological digestion device. The process carried out to separate solids from liquid/gas through using a porous medium, that clarification is used in the instances pass through is filtration (Lau et al., 2014). The term clarification is used in the instances of few solids in a liquid. Filtration is a common operation that is widely applied for production of Bulk drugs, stale products, and aqueous oral fermentation. The slurry is the term given to the suspension to be filtered.

The porous medium that is used to retain solids is called the filter medium. The solids that accumulate on the filter make filter cake. The liquid passing through the filter is called filtrate. Solids are trapped on the surface of the filter medium, as the pores of the filter medium are smaller than the size of solid particles to be separated. Over the period the filter cake thickness resisting the flow of liquids or filtrate ceasing down the process of filtration (Brack et al., 2017).

MICROFILTRATION

A Pressure-driven membrane separation process uses a simple permeable material as a membrane that generally allows particles of size less than 0.1 microns to pass through it.

Membranes of filtration can be made up of material like polyvinyl difluoride (PUDF), Polysulfate can be and ZrO_2, etc. To the size varies between 0.1 to 0.5 microns, which is larger in comparison to the other filtration techniques Hence pressure is required to pass the liquid through the membrane with the pressure of not more than 0.1 to 3 bars.

Figure 3. Showing the type of particles filtered through various membrane processes

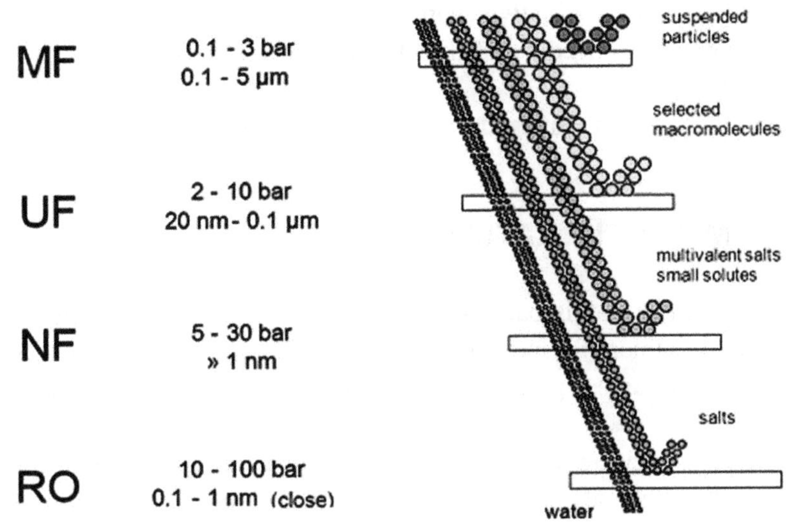

Membranes of micro-filter come in various configurations of having specific uses, advantages, and disadvantages individually. Most common configurations are:

- Pipe-shaped. Plate-shaped

Two general setups found in addition to membrane configuration are a dead end and crossflow setups, named based on the way of supply is sent to the membrane. Indeed, end microfiltration, supply is directly sent to the membrane, where the filter cake is formed at the supply side of the filter membrane. This layer is rinsed away periodically by briefly resending permeate through the membrane in the opposite direction to the production flow. Whereas liquid is passed along the membrane surface with a specific speed in crossflow set-up. Permeate passes through the membrane leaving behind larger particles in a concentrated flow (retentate). The hardened layer in this process is removed continuously by cross flowing supply flow.

Advantages and Disadvantages of Microfiltration

Advantages of microfiltration are:

- Requires low operating pressure
- Consumes less energy in comparison to nano-filtration or reverse osmosis
- Requires only a few manual actions
- Comparatively cheap
- Less energy consuming phase transfer required.

Disadvantages of microfiltration are:

- Removal of only bacteria and suspended matter
- The membrane is sensitive to oxidative chemicals
- Filter membrane can be damaged by hard and sharp particles which are less than 0.1mm, hence pre-filtration is necessary
- Chances of membrane getting damaged if re-rinsing is performed at pressures greater than 1 bar.

APPLICATION OF MICROFILTRATION

Excellent in removing suspended solids, bacteria, etc, micro-filtration is principally used as pre-treatment for water and wastewater treatment for industries like dairy, metal, textile, and pharmaceuticals. When used in combination with biological process treated water from this hybrid treatment can be re-used effectively.

Herbal pharmaceutical wastewater treatment was employed using ceramic microfiltration membrane integrated with an adsorptive treatment which has resulted in 99% removal of turbidity and total suspended solids after 2 hours of microfiltration along with color removal of 87% and COD reduction of 88% (Bhattacharya et al., 2011). Removal of penicillin V at various concentrations using tubular microfiltration ceramic membrane showed that the permeate flux acts as a governing factor for lower concentrations of penicillin V (0.01 g/L). Whereas a decrease in the volume of permeate and an increase in the volume of retentate is observed with increasing concentration of penicillin V.

Microfiltration was used as a pre-treatment prior to ozonation, which in turn was combined with granulated activated carbon (GAC). This technology has provided multi-step treatment through both oxidative and biological degradation along with adsorption of pollutants and by-products, which are formed during ozonation. This method results in the complete removal of pharmaceutical compounds including other hazardous substances except for microplastics (Baresel et al., 2017).

Tests conducted on microfiltration using Bakelite particles have proved that Bakelite particles have shown a decrease in the resistance of cake but have improved membrane flux. The study has also shown a linear correlation between size and mass of the dolly-particles, the permeate flux, and fouling of the membrane. Thus, concluding that the use of Bakelite particles enhanced the retention of the turbidity by microfiltration (Szép, 2013).

ENVIRONMENTAL ISSUES WITH MICROFILTRATION

The concentrate obtained from microfiltration contains a high concentration of bacteria and suspended particles. These wastes are only to be discharged along with wastewater only if the discharge norms are not broken. Also, rinse water produced from chemical cleaning includes substances like formed AOX, alkali, acid, bleach and peroxide, which can only be discharged to precise wastewater purification systems.

ULTRAFILTRATION

Applicable at both laboratory and industrial scale, ultra-filtration is a pressure driven membrane process. It is a separation technique since dynamic biopolymer streams are processed economically even without the use of high temperatures, solvents, etc at large scale (Cooper, 2013). This technique permits fractionation, concentration, and purification of macromolecules in solution which is carried out at temperatures near to ambient, without a change in phase or addition of chemicals/solvents. Biochemical structure and activity of the product are thus protected resulting in increased yields over traditional technologies and simplifying total processing schemes (Short & Webster, 1982).

Among many products produced by the pharmaceutical industry that require the highest standard with reference to quality and quantity, some of them are serum, plasma, infusion solvents, and vaccines. To ensure consistent quality of the final product entire process of manufacturing should be free from contamination, which is met in a safe, reliable and economical way by an ultra-filtration (Wilson & Cavanagh, 1969).

Following are the types of ultra-filtration membranes which are predominantly used:

1. Asymmetric skinned membranes made from synthetic polymers by the "Phase Inversion" methods.
2. Inorganic membranes utilizing inorganic porous supports
3. Inorganic colloids, such as alumina with appropriate binders

Advantages and Disadvantages of Ultrafiltration (UF)

Advantages of UF are:

- Requires low operating pressure, generally higher than MF
- Energy consumption is lower than that of reverse osmosis or nano-filtration
- Limited manual actions required;
- Comparatively cheap;
- Permeate yield depends on input water and choice of the membrane, which is generally good;
- Bacteria removal brings about disinfection. Although UF allows phage, viruses, colloids, and macromolecules to be removed to a certain extent.

Disadvantages of UF are:

- Can only remove suspended matter and bacteria;
- Membrane is sensitive to oxidative chemicals like sulphuric acid, nitric acid, peroxide, and per sulphate at higher concentrations.
- Can be prone to damage while preventing sharp particles and particles > 0.1 mm;
- Membrane gets damaged at pressure > 3 bar.

APPLICATION OF ULTRAFILTRATION

Ultra-filtration has become a powerful tool for rapidly developing the pharmaceutical industry for processes like injectable drugs and enzyme purification (Majekodunmi & Olorunsola, 2014). These have undergone several improvements over the years. Fine filtration through a range of pore size around 0.02 to 0.09 microns has progressed as a chief pre-filtration system which has enhanced the performance of RO systems. Ultrafiltration membranes are widely being used for various applications including desalination systems, surface water treatment, industrial effluent, and municipal wastewater treatment.

A state-of-the-art process which is used to separate solutions containing high and low molecular weight solutes is Ultrafiltration along with defiltration. Studies (Lipnizki, 2005) have focused on various concepts of ultrafiltration integrated with defiltration which are continuous, batch and counter – current. Also, their study highlights the potential of defiltration at industrial scale in pharmaceutical biotechnology.

A simple and novel method was developed by the researcher (Liu et al., 2014) in order to enrich active compounds from plasma samples, which was based on the technique of ultrafiltration integrated with high-performance liquid chromatography. Ultrafiltration device used was a sealed porous membrane envelope made out a polypropylene sheet for effective separation of active large biomolecules and small molecules, the envelope encompasses sample carrier to load plasma samples. Hyperoside and Isoquercitrin enrichment from the plasma of rat was used as an example. In comparison with other processes like solid-phase extraction, protein precipitation, and commercial ultrafiltration membrane, their proposed method demonstrated lesser cost for extraction and lower detection limits along with consumption of fewer plasma samples.

Environmental mobility and removal of sulphonamide antibiotics were studied using stirred cell ultrafiltration experiments on both synthetic and real wastewater effluent. Further, the impact of micellar enhanced ultrafiltration which includes the addition of surfactants concentrations aimed to enhance removal of trace contaminants from wastewaters was studied by using a cationic surfactant cetyltrimethylammonium bromide (CTAB). The results showed that removal of sulphonamides in ultrafiltration was only 15 – 20% whereas micellar solutions of CTAB resulted in 20 - 74% removal suggesting that micellar enhanced ultrafiltration is efficient in removing selective trace contaminants (Exall et al., 2013).

Competence of advanced sequential membrane technology in a wastewater treatment plant for removal of mefenamic acid, a non-steroid anti-inflammatory drug (NSAID) was studied by (Khalaf et al., 2013). The sequential system comprised of an activated sludge process followed by ultrafiltration using hollow fiber membranes with 100 kDa cut off and spiral wound membranes with 20 kDa cutoff, proceeded by activated carbon and reverse osmosis process. Complete removal of the drug was observed with filtration using micelle-clay columns mixed with sand, which was much larger than the activated column, proving the greater potential for enhanced removal of the plant towards mefenamic acid. Similar results were obtained by Awwad et al., 2015 study which was aimed towards the removal of antibacterials like amoxicillin trihydrate and cefuroxime axetil.

Efficacy of an innovative hybrid process which applied ultrafiltration, adsorption on activated carbon and irradiation by ultrasound for removal of emerging contaminants like amoxicillin, diclofenac, and carbamazepine was tested by Secondes et al., 2014. The efficiency of the process individually and in combination was tested and found that adsorption played a major role in the removal of these emerging contaminants. In the hybrid process, almost complete removal was achieved. In a similar study was carried out by Sheng et al., 2016 aiming to remove pharmaceuticals like Ibuprofen, Gemfibrozil, Bezafibrate, Caffeine, Cotinine, Metoprolol, Carbamazepine, Acetaminophen, Diclofenac, etc. using ultrafiltration,

Figure 4. Showing particle size filtered through various membrane processes

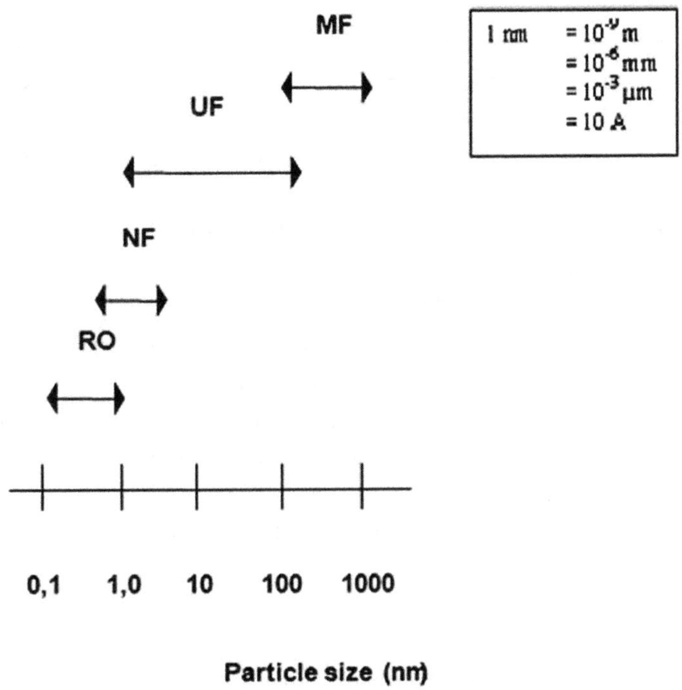

powdered activated carbon and coagulation individually and in combination. Results obtained depicted efficiency of ultrafiltration, powdered activated carbon, and coagulation as 29%, 50%, and 7% respectively. Whereas with a dose of 100ppm for powdered activated carbon used as pre-treatment integrated with ultra-filtration 90.3% removal was achieved revealing suitable combination treatment for removal of pharmaceuticals from water.

The fate of nine pharmaceutical compounds was investigated from samples of sediment, suspended particulate matter, colloidal and soluble phases of Yangtze River Estuary and its coastal area. Removal of these pharmaceutical compounds was performed by separating colloidal phase from bulk water samples using a crossflow ultrafiltration. This process resulted in significant removal of pharmaceutical compounds indicating that colloids act as a reservoir for these contaminants (Yang et al., 2011).

ENVIRONMENTAL ISSUES WITH ULTRAFILTRATION

The concentrate from UF primarily consists of suspended matter and bacteria. It can be discharged together with wastewater if the accumulated concentration of suspended matter does not breach the norms. Rinse waters after chemical cleaning contain substances like bleach, peroxide, acid, and alkali. These rinse waters can only be discharged to specific waste purification systems.

NANO FILTRATION

Removal of a varied variety of organic compounds from wastewater effectively is made possible with nanofiltration. Being a pressure-driven membrane process, its separation level is amongst ultra-filtration and reverse osmosis. Like micro and ultra-filtration, sieve effect is one of the separation principles similar to solvent diffusion and electrostatic repulsion. Sieve effect is based on the difference between particle size and pore diameter. The membrane pore size of nano-filtration is characterized by a cut-off value, which is consistent with the molecular weight of the smallest molecule that can be restricted by the top layer of the membrane of 2 μm thickness up to 90%.

The cut-off value is expressed in terms of Dalton, where a classic nano-filtration membrane has Dalton value lies in the range of 150-500 which in turn depends on molecular structure, the pore size of nano-filtration membranes is approximately 1nm. Also, an ion-selective membrane, there exists electrostatic repulsion or attraction forces between liquid components and nano-filtration membranes resulting in a certain degree of ion selectivity. Membranes of nano-filtration can have shapes like a spiral, tubular or flat.

Advantages and Disadvantages of Nano-Filtration

Advantages:

- For low concentration salts, this process results in lower concentrations of retentate and lower discharge volumes.
- Reduces the concentration of salt including dissolved salts expressed as TDS in brackish water.
- Reduces heavy metals, nitrates, sulfates, color, tannins, and turbidity
- Softens hard water
- Performs disinfection the pH of water from nano-filtration is non-aggressive

Disadvantages

- consumes greater energy in comparison with UF and MF
- Heavily polluted waters would require pre-treatment.
- Membranes of nano-filtration are little expensive over membranes of reverse osmosis
- Membranes are usually sensitive to free chlorine.

Applications of Nano-Filtration

Derakhsheshpoor et al., 2013 have studied the application of modified nano-filtration membrane grafted with UV for separation of amoxicillin, a model pharmaceutical. The efficiency of the membrane in terms of various parameters like coagulation bath temperature, the molecular weight of PEG additive in polymeric solution, pH of feed solution and time of modification was tested. The results obtained showed that an increase in coagulation bath temperature led to smaller pores and enhanced separation performance. Reduction in rejection and increase in permeability by pore enlargement with an increase in PEG molecular weight.

An increase in pH resulted in enhanced membrane rejection which may be owed to electrostatic repulsion between solute and membrane. Escalating modification time, reduced pore size, increasing rejection.

Retention mechanisms of three pharmaceuticals namely carbamazepine, ibuprofens and sulfamethoxazole by nanofiltration (NF) membranes was investigated at laboratory scale.

Results obtained showed that the retention of these pharmaceuticals by a tight nano-filtration membrane was dominated by steric (size) exclusion, while both electrostatic repulsion and steric exclusion are governing factors for retention of ionizable pharmaceuticals by a loose nano-filtration membrane (Nghiem et al., 2005).

Pharmaceutical wastewaters were treated using nano-filtration with and without ozonation as pre-treatment. It was observed that permeate fluxes along with the removal of organic compounds were higher with ozonation as pre-treatment in comparison to no pre-treatment, indicating that hybrid system was much efficient over the conventional system (Szép et al., 2012).

Researchers Darvishmanesh et al., 2011 have tested and stated that the commercially available solvent resistant nanofiltration (SRNF) membranes like StarMem™122 and DuraMem™150have shown encouraging performances, which can be used as alternative tools to traditional separation units like distillation columns for recovery of solvents.

Membrane system consisting of polyamide spiral wound nano-filtration was evaluated for removal of amoxicillin from pharmaceutical wastewater. Variables like flow rate, the concentration of amoxicillin, pressure and chemical oxygen demand (COD) of the feed and their effects on operating conditions were evaluated. Results showed that the rejection of amoxicillin was greater than 97% with only 40% of reduction in COD. An increase in pressure enhanced the transport rate of solvents whereas an increase in permeate flux was seen with an increase in flow rates. Stable permeation flux and greater rejection of amoxicillin indicate the potential of nano-filtration for recovery of amoxicillin from pharmaceutical wastewater (Shahtalebi & Sarrafzadeh, 2011).

For complex separations single stage organic solvent nano-filtration processes will not be efficient, these would require integration of multiple membrane processes. One such solution was proposed by Abejón et al., 2015 for purification of an intermediate active pharmaceutical ingredient (API) precursor in methanol medium. The solution involved is based on solvent recovery from extreme of the cascade purification and recirculation of the recovered solvent to another extreme of the cascade with a view to enhancing its performance by avoiding many concentrated streams that might cause solubility problems.

Environmental Issues With Nanofiltration

Cations and anions in nano-filtration get thickened as a concentrated flow in retentate, which is determined by the preferred permeate quality, installation design and retention of the membrane. This concentrated flow has to be discharged if otherwise not used in the process in a way that it does breach the discharge standards.

REVERSE OSMOSIS

Reverse osmosis (RO), a pressure-driven membrane-technology filtration removes several types of large molecules and ions from solutions, resulting in the retention of solute on the pressurized side of the membrane, while the pure solvent can pass to the other side.

Reverse osmosis involves a diffusive mechanism, where the separation efficiency is governed by pressure, solute concentration and water flux rate. This process is adopted by industries to clean water to be used for industrial process applications, to convert brackish water, to recover salts from industrial processes or remove pollutant load from industrial wastewater. Also reverse osmosis is very effective at removing other substances like few pesticides, trihalomethanes (THM's), volatile organic compounds (VOC's) and solvents. Reverse Osmosis unit includes - Raw water pumps, Pre-treatment, Membranes, Disinfection, Storage and Distribution elements

MEMBRANES FOR REVERSE OSMOSIS

Generally used membrane materials include cellulose acetate (CA), polyamide thin film composites (TFC), and cellulose triacetate (CTA). Membrane materials for reverse osmosis can be, hollow fibres bundled together or spiral wound around a tube. In comparison with spiral wound membranes hollow fibre membranes possess greater surface area, with good capacity however these are more easily blocked. Similarly, TFC membranes possess greater strength and sturdiness along with higher rejection rates over CA/CTA membranes. These possess better resistance to microbial attack, high TDS and high pH (Reddy et al., 2014).

Advantages and Disadvantages of Reverse Osmosis

Advantages of Reverse Osmosis

- Almost all contaminant ions and nearly all dissolved non-ions are efficiently removed.
- Applicable for small systems which have a larger degree of seasonal fluctuation in terms of water demand
- Tolerant to variations in flow and TDS levels
- Operates instantly without any minimum break-in period
- Capable of removing bacteria and particles
- Skilled operation and supervision are not required owing to the simplicity and automation of operation.

Disadvantages of Reverse Osmosis

- Capital investment and operating costs are high
- Concerns are with managing the wastewater brine solution
- Requires pre-treatment to a very high level.
- Fouling of membranes is a major problem

The main problem of adopting reverse osmosis is membrane fouling which is prevalent in water reclamation applications, the main cause of permeate flux decline and loss of product quality in reverse osmosis systems (Pervov, 1999; Choi et al., 2009). Sources of fouling can be categorized into four chief categories: scaling, particular fouling, biofouling, and organic fouling.

Applications of Reverse Osmosis

Treatment of bulk drug wastewaters which were pre-treated in processes like multiple effect evaporator, stripping column followed by agitated thin film dryer set-up and then fed to ultra-filtration and reverse osmosis was studied by Vandevivere et al.,1998. The product when analysed revealed pollutant strength which was much lesser than prescribed standards which were feasible for reuse in the boiler. Removal of a broad range of pharmaceuticals from drinking water was achieved when reverse osmosis is used in combination with nano-filtration (Radjenović et al., 2008). Moreover, treated water from pharmaceuticals using reverse osmosis can be very well reused for various processes as demonstrated by de Andrade et al., 2017. Studies were Khan et al., 2011 carried out on the treatment of pharmaceutical wastewater using a reverse osmosis process as tertiary treatment. Reverse osmosis plant was supported with pre-treatments like MMF and UF. Treated water which was under the limits required was used for cooling tower and other non-potable uses.

Environmental issues with reverse osmosis

The concentrated flow i.e. retentate obtained from reverse osmosis which consists of almost all ions must be discharged in a proper way so as not to cause any effect on the environment. An available alternative method can be evaporating the concentrate.

MEMBRANE BIOREACTOR (MBR)

Hybrid system integrating membrane separation with biological treatment is called membrane bioreactor. Applying this process will enable traditional activated sludge plants to work in single step producing an effluent with high quality which is potentially suitable for reuse (Visvanathan et al., 2000). When membrane bioreactor is used as a substitute for secondary settling tank nearly entire activated sludge can be accommodated in the reactor, improving sludge concentration up to 18,000 – 19,000 mg/L (Fazal et al., 2015)

MBR, considered as biochemical engineering process encompasses the use of suspended growth bioreactor for biochemical reactions as well as membrane separator for consequent solids, liquid separation (Wang & Menon, 2011). It can be of two types namely External or separate where the membrane is placed next to the biological basin and Submerged or internal system where the membrane is placed in the basin. Micro and ultra-filtration membranes are used for both types of MBR. Thus, a membrane reactor is able to successfully process higher sludge concentrations and lower reactor volumes in comparison to conventional systems (Gehlert et al., 2005; Ueda et al., 1997).

Advantage of MBR for treatment of Industrial Wastewater can be described in the following terms:

High Efficiency for Separation and Assurance of Water Quality

Wastewaters containing higher concentrations of suspended material can be efficiently separated resulting in the turbidity and suspended solids removal near to zero. Moreover, if wastewaters containing toxic substances are treated, these lead to swelling of the sludge at the membrane separation, which will not affect the water quality (Ming & De-Sheng, 2010).

Figure 5. Showing membrane bioreactor

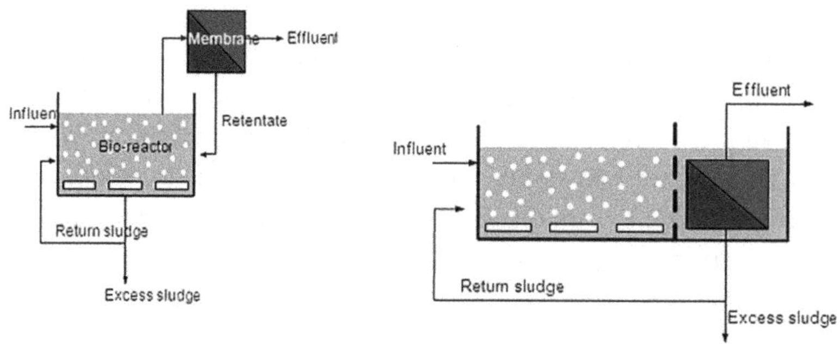

Robust Biochemical Capabilities and Sludge Concentration

The membrane reactor can sustain high impact loads and also when compared to conventional process this process improves sludge concentration, avoid wastage in terms of sludge bulking as wastewaters have a large amount of volatility (Fazal et al., 2015). The treatment with MBR also enhances purification efficiency, refractory organics, reduces hydraulic retention time (DAUD, 2016; Sundararaman & Saravanane,2010).

Studies have shown that MBR has high NH_3-N removal efficiency owing to the process where the slow-growing nitrifying bacteria stay in the reactor creating favorable conditions for growth and reproduction of the same (Anjum et al., 2016; Fazal et al., 2015).

During the previous decade, much attention has been gained towards MBR for treating pharmaceutical wastewater due to its feasibility with reference to technical and economic aspects specifically for its property of high sludge retention time which is achieved in the compact reactor volumes. Since the concentration of microorganisms in MBR can be increased up to 20mg/L, degradation of larger organic molecules by this luxuriant biomass is achieved (Radjenovic et al., 2007). MBR also gains an advantage in separating suspended solids through membranes which is not limited by the settling characteristics of the sludge (Urase et al., 2005). Above 90% of COD and 98% of BOD removal was achieved by Noble, 2016 using a hollow fiber submerged MBR for treating fermentation process wastewaters. In addition, phosphorus removal up to 90% was also reported. In comparison with conventional activated sludge, MBR has shown greater elimination of erythromycin (67%). Hence, in order to achieve complete removal of pollutants in pharmaceuticals a combination of pre and post treatment methods are to be adopted inevitably.

Advantages and Disadvantages of MBR

Advantages

- Compact design
- Greater effluent quality
- Capable of tolerating high volume load, the high rate of degradation, resulting in a reduction of the strength of wastewater
- Possible to convert from existing conventional active sludge purification

Disadvantages

- The limitation is the process of aeration
- External MBR possess stress on sludge
- Damage of membrane due to pollutants
- Economically not feasible

Application of Membrane Bioreactors

These processes are in application around the world both for treating industrial wastewaters as well as municipal wastewaters.

MBR is widely adopted as pre-treatment for reverse osmosis process in order to avoid issues which might result in system failure. This hybrid system has proved to be effective in water reclamation and reuse among many different pre-treatment schemes for reverse osmosis RO (Malamis et al., 2012; Sahar et al., 2011; Kent et al., 2011; Liu et al., 2008; Nelson et al., 2008, Dhaouadi & Marrot, 2008; Lyko et al., 2007).

With an objective to remove persistent target compounds like floxacins from wastewaters Membrane bioreactor treatment (MBR) was applied as a pre-treatment to pharmaceutical wastewater. Processes like nanofiltration (NF), powdered activated carbon (PAC), reverse osmosis (RO) and O_3/UV were applied to permeate from MBR which were spiked with selected target compounds. Treatment efficiency was assessed using conventional inorganic and organic chemical analyses besides advanced physicochemical methods like liquid chromatography coupled with mass and tandem mass spectrometry (LC-MS and -MS-MS) (Baumgarten et al., 2007). Laboratory-scale submerged membrane bioreactor was applied for treating wastewater from the pharmaceutical industry, which was spiked with four target corticosteroids, medroxyprogesterone acetate (Medro), bethametasone valerate (Betha-V), levonogestrel (Levo), and bethametasone dipropionate (Betha-D). Chemical analysis of treated water from membrane bioreactor revealed 93% of removal of the target corticosteroids along with the removal of 99% of chemical oxygen demand, ammonia and 40% of phosphorus (López-Fernández et al., 2012).

Pilot-scale study of treating pharmaceutical wastewater using membrane bioreactor (MBR) process was carried out by Chang et al., 2008. Performance of MBR was monitored for 140days. Removal of suspended solids and COD was achieved up to 140 days. Results from SEM and EDX depicted that the fouling of the membrane at the outer surface was chiefly due to microorganisms. Results indicated that the MBR system can treat high-strength and fluctuating wastewater with dependable performance.

A pilot-scale system with an external loop airlift membrane bioreactor (ELAMBR) was applied for the treatment of synthetic pharmaceutical wastewater precisely targeted to remove acetaminophen. Simultaneously lab-scale conventional activated sludge (CAS process) was also used for the treatment and compared with that of ELAMBR. Physico-chemical parameters and acetaminophen were removed significantly in ELAMBR over CAS whereas acetaminophen was 100% removed after 2 days. The initial concentration of acetaminophen, mixed liquid suspended solids and chemical oxygen demand are the governing factors for this pollutant removal (Shariati et al., 2010). Prospects of MBR treatment for pharmaceutical wastewater was reviewed by Fazal et al., 2015. They have stated that MBR was successful in removing Phosphorous, NH_3-N, Total Nitrogen and COD.

Environmental Issues

Regular removal of the excess sludge produced is to be carried out. Fluids used to clean the membrane pose problems in their disposal. Possibility of forming AOX if cleaning is not carried out with NaOCl.

CONCLUSION

Progress in innovative pharmaceutical products is adding a vast number of chemical compounds to already existing innumerable chemicals in the environment. Owing to the diversity of wastewater from the pharmaceutical industry, its treatment becomes a challenging task. Moreover, strict regulations by food and drug agencies require pharmaceutical industries to maintain acceptable standards in terms of use, discharge, and reuse. Furthermore, the volume of wastewater generated from the industry is also very high. Few desirable options available are a reduction of the waste stream at source, recycling of water and reclamation. But in order to achieve treated water for uses like recycling and reuse, the treatment should be performed at a higher degree.

From the above discussion, it is understood that in recent years membrane processes have evolved to be an efficient option for treating wastewaters from the pharmaceutical industry for removal of various contaminants based on their sizes. Some of the membrane technologies that are applied are microfiltration for removal of molecular-sized contaminant like bacteria, ultrafiltration for removal of macromolecules and viruses, nanofiltration for removal of divalent ions and reverse osmosis for removal of monovalent ions (Lewis et al., 2011; Oktem et al., 2006; Chang et al., 2019; Patterson, 1985).

Furthermore, the application of hybrid technologies removes pollutants almost completely or within safe discharge limits. One such technology is membrane bioreactor which has been providing a promising solution. In the present scenario, the emphasis is being made on the recovery technologies rather than removal technologies that can be reused. Hence, future research should concentrate on this aspect so that it would be economically feasible and environmentally sound and safe.

REFERENCES

Abejón, R., Garea, A., & Irabien, A. (2015). Organic solvent recovery and reuse in pharmaceutical purification processes by nanofiltration membrane cascades. *Chemical Engineering Transactions*, *43*, 1057–1062.

Adeyinka, J. S. (1996). Wastewater treatment through reverse osmosis. *Environmental Monitoring and Assessment*, 22–37.

Adham, S., Chiu, K. P., Gramith, K., & Oppenheimer, J. (2005). *Development of a microfiltration and ultrafiltration knowledge base. Awwa Research Foundation*. Pasadena, CA: US Department of the Interior, Bureau of Reclamation.

Anjum, M., Al-Makishah, N. H., & Barakat, M. A. (2016). Wastewater sludge stabilization using pre-treatment methods. *Process Safety and Environmental Protection*, *102*, 615–632. doi:10.1016/j.psep.2016.05.022

Awwad, M., Al-Rimawi, F., Dajani, K. J. K., Khamis, M., Nir, S., & Karaman, R. (2015). Removal of amoxicillin and cefuroxime axetil by advanced membranes technology, activated carbon and micelle–clay complex. *Environmental Technology*, *36*(16), 2069–2078. doi:10.1080/09593330.2015.1019935 PMID:25686519

Ayandiran, T. A., Fawole, O. O., & Dahunsi, S. O. (2018). Water quality assessment of bitumen polluted Oluwa River, South-Western Nigeria. *Water Resources and Industry*, *19*, 13–24. doi:10.1016/j.wri.2017.12.002

Bai, T., Yang, Y., Lian, L., Wu, Y., Jiang, S., Zheng, H.-X. … Nan, J.-N. (n.d.). Key laboratory for natural resource of Changbai Mountain & functional molecules. Ministry of Education, College of Pharmacy, Yanbian. Acta Pharmacologica Sinica, 34.

Baquero, F., Martínez, J. L., & Cantón, R. (2008). Antibiotics and antibiotic resistance in water environments. *Current Opinion in Biotechnology*, *19*(3), 260–265. doi:10.1016/j.copbio.2008.05.006 PMID:18534838

Baresel, C., Magnér, J., Magnusson, K., & Olshammar, M. (2017). Tekniska lösningar för avancerad refining av avloppsvatten (No. C 235).

Baumgarten, S., Schröder, H. F., Charwath, C., Lange, M., Beier, S., & Pinnekamp, J. (2007). Evaluation of advanced treatment technologies for the elimination of pharmaceutical compounds. *Water Science and Technology*, *56*(5), 1–8. doi:10.2166/wst.2007.550 PMID:17881831

Bellona, C., & Drewes, J. E. (2007). Viability of a low-pressure nanofilter in treating recycled water for water reuse applications: A pilot-scale study. *Water Research*, *41*(17), 3948–3958. doi:10.1016/j.watres.2007.05.027 PMID:17582458

Bhattacharya, P., Ghosh, S., Majumdar, S., Dasgupta, S., & Bandyopadhyay, S. (2011). Biosorbent-assisted ceramic microfiltration process for treatment of herbal pharmaceutical wastewater with high organic loading. *International Journal of Environmental Technology and Management*, *14*(1-4), 132–146. doi:10.1504/IJETM.2011.039262

Bound, J. P., & Voulvoulis, N. (2005). Household disposal of pharmaceuticals as a pathway for aquatic contamination in the United Kingdom. *Environmental Health Perspectives*, *113*(12), 1705–1711. doi:10.1289/ehp.8315 PMID:16330351

Brack, W., Dulio, V., Ågerstrand, M., Allan, I., Altenburger, R., Brinkmann, M., ... Hernández, F. J. (2017). Towards the review of the European Union water framework directive: Recommendations for more efficient assessment and management of chemical contamination in European surface water resources. *The Science of the Total Environment*, *576*, 720–737. doi:10.1016/j.scitotenv.2016.10.104 PMID:27810758

Cao, Y. B., Lin, H., Yan, L., Wang, Y., & Cao, Y. Y. (2013). The anti-fibrotic effect of thymoquinone on thioacetamide-induced mouse hepatic fibrosis.

Chang, C. Y., Chang, J. S., Vigneswaran, S., & Kandasamy, J. (2008). Pharmaceutical wastewater treatment by membrane bioreactor process–a case study in southern Taiwan. *Desalination*, *234*(1-3), 393–401. doi:10.1016/j.desal.2007.09.109

Chang, H., Li, T., Liu, B., Vidic, R. D., Elimelech, M., & Crittenden, J. C. (2019). Potential and implemented membrane-based technologies for the treatment and reuse of flowback and produced water from shale gas and oil plays: A review. *Desalination*, *455*, 34–57. doi:10.1016/j.desal.2019.01.001

Chen, Z., Ren, N., Wang, A., Zhang, Z. P., & Shi, Y. (2008). A novel application of TPAD–MBR system to the pilot treatment of chemical synthesis-based pharmaceutical wastewater. *Water Research*, *42*(13), 3385–3392. doi:10.1016/j.watres.2008.04.020 PMID:18538367

Choi, J. S., Hwang, T. M., Lee, S., & Hong, S. (2009). A systematic approach to determine the fouling index for a RO/NF membrane process. *Desalination*, *238*(1-3), 117–127. doi:10.1016/j.desal.2008.01.042

Comerton, A. M., Andrews, R. C., & Bagley, D. M. (2005). Evaluation of an MBR–RO system to produce high quality reuse water: Microbial control, DBP formation and nitrate. *Water Research*, *39*(16), 3982–3990. doi:10.1016/j.watres.2005.07.014 PMID:16112164

Cooper, A. R. (Ed.). (2013). *Ultrafiltration membranes and applications* (Vol. 13). Springer Science & Business Media.

Cyr, P. J., Suri, R. P., & Helmig, E. D. (2002). A pilot scale evaluation of removal of mercury from pharmaceutical wastewater using granular activated carbon. *Water Research*, *36*(19), 4725–4734. doi:10.1016/S0043-1354(02)00214-2 PMID:12448514

Dalrymple, O. K., Yeh, D. H., & Trotz, M. A. (2007). Removing pharmaceuticals and endocrine-disrupting compounds from wastewater by photocatalysis. *Journal of Chemical Technology & Biotechnology: International Research in Process. Environmental & Clean Technology*, *82*(2), 121–134.

Darvishmanesh, S., Firoozpour, L., Vanneste, J., Luis, P., Degrève, J., & Van der Bruggen, B. (2011). Performance of solvent resistant nanofiltration membranes for purification of residual solvent in the pharmaceutical industry: Experiments and simulation. *Green Chemistry*, *13*(12), 3476–3483. doi:10.1039/c1gc15462a

Daud. N. F. B. M. (2016). *Treatment of chicken processing wastewater using hybrid upflow anaerobic sludge blanket (HUASB) coupled with aerated lagoon (AL).* (Doctoral dissertation, University Tun Hussein Onn Malaysia).

de Andrade, B. D. A., de Lacerda, P. S., & Oliveira, J. L. (2017). Technical feasibility of reuse of effluent generated from reverse osmosis system in a pharmaceutical plant. *Revista Ambiente & Água*, *12*(5), 694–707. doi:10.4136/ambi-agua.1980

Deegan, A. M., Shaik, B., Nolan, K., Urell, K., Oelgemöller, M., Tobin, J., & Morrissey, A. (2011). Treatment options for wastewater effluents from pharmaceutical companies. *International Journal of Environmental Science and Technology*, *8*(3), 649–666. doi:10.1007/BF03326250

Derakhsheshpoor, R., Homayoonfal, M., Akbari, A., & Mehrnia, M. R. (2013). Amoxicillin separation from pharmaceutical wastewater by high permeability polysulfone nanofiltration membrane. *Journal of Environmental Health Science & Engineering*, *11*(1), 9. doi:10.1186/2052-336X-11-9 PMID:24499593

Dhaouadi, H., & Marrot, B. (2008). Olive mill wastewater treatment in a membrane bioreactor: Process feasibility and performances. *Chemical Engineering Journal, 145*(2), 225–231. doi:10.1016/j.cej.2008.04.017

El-Gohary, F. A., Abou-Elela, S. I., & Aly, H. I. (1995). Evaluation of biological technologies for wastewater treatment in the pharmaceutical industry. *Water Science and Technology, 32*(11), 13–20. doi:10.2166/wst.1995.0389

Exall, K., Balakrishnan, V. K., Toito, J., & McFadyen, R. (2013). Impact of selected wastewater constituents on the removal of sulfonamide antibiotics via ultrafiltration and micellar enhanced ultrafiltration. *The Science of the Total Environment, 461*, 371–376. doi:10.1016/j.scitotenv.2013.04.057 PMID:23743146

Fatta-Kassinos, D., Meric, S., & Nikolaou, A. (2011). Pharmaceutical residues in environmental waters and wastewater: Current state of knowledge and future research. *Analytical and Bioanalytical Chemistry, 399*(1), 251–275. doi:10.100700216-010-4300-9 PMID:21063687

Fazal, S., Zhang, B., Zhong, Z., Gao, L., & Lu, X. (2015). Membrane separation technology on pharmaceutical wastewater by using MBR (Membrane Bioreactor). *Journal of Environmental Protection, 6*(04), 299–307. doi:10.4236/jep.2015.64030

Frick, E. A., Henderson, A. K., Moll, D. M., Furlong, E. T., & Meyer, M. T. (2001). Presence of pharmaceuticals in wastewater effluent and drinking water, metropolitan Atlanta, Georgia, July-September 1999. Georgia Institute of Technology.

Gadipelly, C., Pérez-González, A., Yadav, G. D., Ortiz, I., Ibáñez, R., Rathod, V. K., & Marathe, K. V. (2014). Pharmaceutical industry wastewater: Review of the technologies for water treatment and reuse. *Industrial & Engineering Chemistry Research, 53*(29), 11571–11592. doi:10.1021/ie501210j

Gehlert, G., Abdulkadir, M., Fuhrmann, J., & Hapke, J. (2005). Dynamic modeling of an ultrafiltration module for use in a membrane bioreactor. *Journal of Membrane Science, 248*(1-2), 63–71. doi:10.1016/j.memsci.2004.09.026

Gogate, P. R., & Pandit, A. B. (2004). A review of imperative technologies for wastewater treatment I: Oxidation technologies at ambient conditions. *Advances in Environmental Research, 8*(3-4), 501–551. doi:10.1016/S1093-0191(03)00032-7

Gryta, M., Markowska-Szczupak, A., Bastrzyk, J., & Tomczak, W. (2013). The study of membrane distillation used for separation of fermenting glycerol solutions. *Journal of Membrane Science, 431*, 1–8. doi:10.1016/j.memsci.2012.12.032

Guo, Y., Qi, P. S., & Liu, Y. Z. (2017, May). A review on advanced treatment of pharmaceutical wastewater. In *IOP Conference Series: Earth and Environmental Science* (Vol. 63, No. 1, p. 012025). IOP Publishing. 10.1088/1755-1315/63/1/012025

Hausmann, A., Sanciolo, P., Vasiljevic, T., Ponnampalam, E., Quispe-Chavez, N., Weeks, M., & Duke, M. (2011). Direct contact membrane distillation of dairy process streams. *Membranes, 1*(1), 48–58. doi:10.3390/membranes1010048 PMID:24957495

Heberer, T. (2002). Occurrence, fate, and removal of pharmaceutical residues in the aquatic environment: A review of recent research data. *Toxicology Letters*, *131*(1-2), 5–17. doi:10.1016/S0378-4274(02)00041-3 PMID:11988354

Houk, V. S. (1992). The genotoxicity of industrial wastes and effluents: A review. *Mutation Research: Reviews in Genetic Toxicology*, *277*(2), 91–138. doi:10.1016/0165-1110(92)90001-P PMID:1378533

Huerta, B., Rodríguez-Mozaz, S., & Barceló, D. (2012). Pharmaceuticals in biota in the aquatic environment: Analytical methods and environmental implications. *Analytical and Bioanalytical Chemistry*, *404*(9), 2611–2624. doi:10.100700216-012-6144-y PMID:22678760

Idris, M. A., Kolo, B. G., Garba, S. T., & Ismail, M. A. (2013). Physico-chemical analysis of pharmaceutical effluent and surface water of River Gorax in Minna, Niger State, Nigeria. *Bull Env Pharmacol Life Sci*, *2*(3), 45–49.

Idris, M. A., Kolo, B. G., & Sani, A. (2013). Organic pollution indicator and anion concentration of pharmaceutical effluent and surface water. In Minna, Niger State, Nigeria. [IJES]. *International Journal of Engineering Science*, *2*(6), 27–32.

Iliuta, I., & Larachi, F. (2001). Wet air oxidation solid catalysis analysis of fixed and sparged three-phase reactors. *Chemical Engineering and Processing: Process Intensification*, *40*(2), 175–185. doi:10.1016/S0255-2701(00)00137-9

Jimenez, C. C. (2012). Pharmaceutical products in the environment: Sources, effects and risks. *Vitae*, *19*(1), 93–108.

Jozala, A. F., Geraldes, D. C., Tundisi, L. L., Feitosa, V. D. A., Breyer, C. A., Cardoso, S. L., ... & Oliveira, M. A. D. (2016). Biopharmaceuticals from microorganisms: from production to purification. *Brazilian Journal of Microbiology, 47*, 51-63.

Jyoti, J., Alka, D., & Kumar, S. J. (2013). Application of membrane-bio-reactor in waste-water treatment: A review. *International Journal of Chemistry and Chemical Engineering*, *3*(2), 115–122.

Kent, F. C., Farahbakhsh, K., Mahendran, B., Jaklewicz, M., Liss, S. N., & Zhou, H. (2011). Water reclamation using reverse osmosis: Analysis of fouling propagation given tertiary membrane filtration and MBR pretreatments. *Journal of Membrane Science*, *382*(1-2), 328–338. doi:10.1016/j.memsci.2011.08.028

Kessler, R. (2010). Industry issues: Pharmaceutical factories as a source of drugs in water.

Khalaf, S., Al-Rimawi, F., Khamis, M., Nir, S., Bufo, S. A., Scrano, L., ... Karaman, R. (2013). Efficiency of membrane technology, activated charcoal, and a micelle-clay complex for removal of the acidic pharmaceutical mefenamic acid. *Journal of Environmental Science and Health. Part A*, *48*(13), 1655–1662. PMID:23947703

Khan, A. A., Gaur, R. Z., Tyagi, V. K., Khursheed, A., Lew, B., Mehrotra, I., & Kazmi, A. A. (2011). Sustainable options of post treatment of UASB effluent treating sewage: A review. *Resources, Conservation and Recycling*, *55*(12), 1232–1251. doi:10.1016/j.resconrec.2011.05.017

Khiari, D. (2007). Endocrine disruptors, pharmaceuticals and personal care products in drinking water: An overview of AWWARF research to date. *Drinking Water Research*, *17*(2), 2–7.

Kimura, K., Hara, H., & Watanabe, Y. (2005). Removal of pharmaceutical compounds by submerged membrane bioreactors (MBRs). *Desalination*, *178*(1-3), 135–140. doi:10.1016/j.desal.2004.11.033

Kümmerer, K. (2009). The presence of pharmaceuticals in the environment due to human use–present knowledge and future challenges. *Journal of Environmental Management*, *90*(8), 2354–2366. doi:10.1016/j.jenvman.2009.01.023 PMID:19261375

Lau, W. J., Goh, P. S., Ismail, A. F., & Lai, S. O. (2014). Ultrafiltration as a pretreatment for seawater desalination: A review. *Membrane Water Treatment*, *5*(1), 15–29. doi:10.12989/mwt.2014.5.1.015

Lewis, S. R., Datta, S., Gui, M., Coker, E. L., Huggins, F. E., Daunert, S., ... Bhattacharyya, D. (2011). Reactive nanostructured membranes for water purification. *Proceedings of the National Academy of Sciences of the United States of America*, *108*(21), 8577–8582. doi:10.1073/pnas.1101144108 PMID:21606340

Lipnizki, F. (2005). Industrial applications of ultrafiltration in pharmaceutical biotechnology. *Engineering in Life Sciences*, *5*(1), 81–83. doi:10.1002/elsc.200407047

Lipton, M. (1977). Why poor people stay poor: a study of urban bias in world development. London, UK: Canberra, ACT: Temple Smith; Australian National University Press.

Liu, H., Yang, C., Pu, W., & Zhang, J. (2008). Removal of nitrogen from wastewater for reusing to boiler feed-water by an anaerobic/aerobic/membrane bioreactor. *Chemical Engineering Journal*, *140*(1-3), 122–129. doi:10.1016/j.cej.2007.09.048

Liu, Q., Yin, X., Sha, B., & You, J. (2014). Porous membrane ultrafiltration—A novel method for enrichment of the active compounds from micro-plasma samples. *Spectrochimica Acta. Part A: Molecular and Biomolecular Spectroscopy*, *131*, 22–29. doi:10.1016/j.saa.2014.04.070 PMID:24815198

López-Fernández, R., Martínez, L., & Villaverde, S. (2012). Membrane bioreactor for the treatment of pharmaceutical wastewater containing corticosteroids. *Desalination*, *300*, 19–23. doi:10.1016/j.desal.2012.05.032

Lyko, S., Al-Halbouni, D., Wintgens, T., Janot, A., Hollender, J., Dott, W., & Melin, T. (2007). Polymeric compounds in activated sludge supernatant—Characterisation and retention mechanisms at a full-scale municipal membrane bioreactor. *Water Research*, *41*(17), 3894–3902. doi:10.1016/j.watres.2007.06.012 PMID:17645906

Majekodunmi, S. O., & Olorunsola, E. O. (2014). Review of some recent advances on filtration in pharmaceutical industry. *Journal of Pharmacy and Biological Sciences*, *9*(4), 30–38. doi:10.9790/3008-09443038

Malamis, S., Katsou, E., Takopoulos, K., Demetriou, P., & Loizidou, M. (2012). Assessment of metal removal, biomass activity and RO concentrate treatment in an MBR–RO system. *Journal of Hazardous Materials*, *209*, 1–8. doi:10.1016/j.jhazmat.2011.10.085 PMID:22209586

McCarty, P. L., Bae, J., & Kim, J. (2011). Domestic wastewater treatment as a net energy producer–can this be achieved?.

Meier, J., & Melin, T. (2005). Wastewater reclamation by the PAC-NF process. *Desalination*, *178*(1-3), 27–40. doi:10.1016/j.desal.2004.12.015

Mestre, A. S., Pires, J., Nogueira, J. M. F., & Carvalho, A. P. (2007). Activated carbons for the adsorption of ibuprofen. *Carbon*, *45*(10), 1979–1988. doi:10.1016/j.carbon.2007.06.005

Ming, Y., & De-Sheng, W. (2010). Experimental study of antibiotic wastewater. *Journal of MBR Treatment. Environmental Protection and Recycling Economy*, *9*, 54–55.

Momodu, M. A., & Anyakora, C. A. (2010). Heavy metal contamination of ground water: The Surulere case study. *Res J Environ Earth Sci*, *2*(1), 39–43.

Naik, U. S. (2013). *The synthesis and characterisation of novel ultra-flexible lipidic vesicles using propanol* (Doctoral dissertation, University of Central Lancashire).

Naiman, R. J., & Dudgeon, D. (2011). Global alteration of freshwaters: Influences on human and environmental well-being. *Ecological Research*, *26*(5), 865–873. doi:10.100711284-010-0693-3

Nelson, M. I., Balakrishnan, E., Sidhu, H. S., & Chen, X. D. (2008). A fundamental analysis of continuous flow bioreactor models and membrane reactor models to process industrial wastewaters. *Chemical Engineering Journal*, *140*(1-3), 521–528. doi:10.1016/j.cej.2007.11.035

Nghiem, L. D., Schäfer, A. I., & Elimelech, M. (2005). Pharmaceutical retention mechanisms by nanofiltration membranes. *Environmental Science & Technology*, *39*(19), 7698–7705. doi:10.1021/es0507665 PMID:16245847

Noble, J. (2006). GE ZeeWeed MBR technology for pharmaceutical wastewater treatment. *Membrane Technology*, *2006*(9), 7–9. doi:10.1016/S0958-2118(06)70790-X

Oktem, Y. A., Ince, O., Donnelly, T., Sallis, P., & Ince, B. K. (2006). Determination of optimum operating conditions of an acidification reactor treating a chemical synthesis-based pharmaceutical wastewater. *Process Biochemistry*, *41*(11), 2258–2263. doi:10.1016/j.procbio.2006.05.016

Patterson, J. W. (1985). Industrial wastewater treatment technology.

Pervov, A. G. (1999). A simplified RO process design based on understanding of fouling mechanisms. *Desalination*, *126*(1-3), 227–247. doi:10.1016/S0011-9164(99)00179-4

Radjenovic, J., Petrovic, M., & Barceló, D. (2007). Analysis of pharmaceuticals in wastewater and removal using a membrane bioreactor. *Analytical and Bioanalytical Chemistry*, *387*(4), 1365–1377. doi:10.100700216-006-0883-6 PMID:17115140

Radjenović, J., Petrović, M., Ventura, F., & Barceló, D. (2008). Rejection of pharmaceuticals in nanofiltration and reverse osmosis membrane drinking water treatment. *Water Research*, *42*(14), 3601–3610. doi:10.1016/j.watres.2008.05.020 PMID:18656225

Reddy, B. V., Sandeep, P., Ujwala, P., Navaneetha, K., & Reddy, K. V. R. (2014). Water treatment process in pharma industry-A review. *International Journal of Pharma and Bio Sciences*, *4*, 2.

Rizwan, S. A., Nongkynrih, B., & Gupta, S. K. (2013). Air pollution in Delhi: Its magnitude and effects on health. *Indian Journal of Community Medicine: Official publication of Indian Association of Preventive & Social Medicine (Social Medicine Publication Group)*, *38*(1), 4.

Rong-Rong, W. A. N. G. (2003). Application of dynamic continuous countercurrent circulatory extraction to Caulis spatholobi [J]. *Zhongchengyao*, 5.

Ryser, L., & Halseth, G. (2010). Rural economic development: A review of the literature from industrialized economies. *Geography Compass*, *4*(6), 510–531. doi:10.1111/j.1749-8198.2010.00321.x

Saban, C. V. (2015). *Palynological perspectives on Younger Dryas to early Holocene human ecology at Paisley Caves*. Oregon.

Sahar, E., David, I., Gelman, Y., Chikurel, H., Aharoni, A., Messalem, R., & Brenner, A. (2011). The use of RO to remove emerging micropollutants following CAS/UF or MBR treatment of municipal wastewater. *Desalination*, *273*(1), 142–147. doi:10.1016/j.desal.2010.11.004

Saleem, M. (2007). Pharmaceutical wastewater treatment: A physicochemical study. *Journal of Research (Science)*, *18*(2), 125–134.

Secondes, M. F. N., Naddeo, V., Belgiorno, V., & Ballesteros, F. Jr. (2014). Removal of emerging contaminants by simultaneous application of membrane ultrafiltration, activated carbon adsorption, and ultrasound irradiation. *Journal of Hazardous Materials*, *264*, 342–349. doi:10.1016/j.jhazmat.2013.11.039 PMID:24316806

Shahtalebi, A., & Sarrafzadeh, M. (2011). Application of nanofiltration membrane in the separation of amoxicillin from pharmaceutical wastewater. *Journal of Environmental Health Science & Engineering*, *8*(2), 106–110.

Shannon, M. A., Bohn, P. W., Elimelech, M., Georgiadis, J. G., Marinas, B. J., & Mayes, A. M. (2010). Science and technology for water purification in the coming decades. In Nanoscience and technology: A collection of reviews from nature journals, (pp. 337-346).

Shariati, F. P., Mehrnia, M. R., Salmasi, B. M., Heran, M., Wisniewski, C., & Sarrafzadeh, M. H. (2010). Membrane bioreactor for treatment of pharmaceutical wastewater containing acetaminophen. *Desalination*, *250*(2), 798–800. doi:10.1016/j.desal.2008.11.044

Sheng, C., Nnanna, A. A., Liu, Y., & Vargo, J. D. (2016). Removal of trace pharmaceuticals from water using coagulation and powdered activated carbon as pretreatment to ultrafiltration membrane system. *The Science of the Total Environment*, *550*, 1075–1083. doi:10.1016/j.scitotenv.2016.01.179 PMID:26867086

Short, J. L., & Webster, D. W. (1982). Ultrafiltration, a valuable processing technique for the pharmaceutical industry. *Process Biochemistry*, *17*, 29–32.

Sima, L., Amador, J., Da Silva, A. K., Miller, S. M., Morse, A. N., Pellegrin, M. L., ... Wells, M. J. (2014). Emerging pollutants–part I: Occurrence, fate and transport. *Water Environment Research*, *86*(10), 1994–2035. doi:10.2175/106143014X14031280668731

Sindi, S., Mangialasche, F., & Kivipelto, M. (2015). Advances in the prevention of Alzheimer's Disease. *F1000prime Reports*, 7. PMID:26097723

Singh, D., & Sirkar, K. K. (2012). Desalination of brine and produced water by direct contact membrane distillation at high temperatures and pressures. *Journal of Membrane Science*, *389*, 380–388. doi:10.1016/j.memsci.2011.11.003

Snyder, S. A., Adham, S., Redding, A. M., Cannon, F. S., DeCarolis, J., Oppenheimer, J., ... Yoon, Y. (2007). Role of membranes and activated carbon in the removal of endocrine disruptors and pharmaceuticals. *Desalination*, *202*(1-3), 156–181. doi:10.1016/j.desal.2005.12.052

Sobsey, M. D., Stauber, C. E., Casanova, L. M., Brown, J. M., & Elliott, M. A. (2008). Point of use household drinking water filtration: A practical, effective solution for providing sustained access to safe drinking water in the developing world. *Environmental Science & Technology*, *42*(12), 4261–4267. doi:10.1021/es702746n PMID:18605542

Sreekanth, D., Sivaramakrishna, D., Himabindu, V., & Anjaneyulu, Y. (2009). Thermophilic treatment of bulk drug pharmaceutical industrial wastewaters by using hybrid up flow anaerobic sludge blanket reactor. *Bioresource Technology*, *100*(9), 2534–2539. doi:10.1016/j.biortech.2008.11.028 PMID:19135360

Srikanth, J. (2018). Pharmacological and toxicological evaluation of Indian noni in animal models of psychiatric disorders.

Struzeski, E. J. (1980). Status of wastes handling and waste treatment across the pharmaceutical industry and 1977 effluent limitations. In *Proceedings of the 35th Industrial Waste Conference*, Purdue University, West Lafayette, IN (pp. 1095-1108).

Sundararaman, S., & Saravanane, R. (2010). Effect of loading rate and HRT on the removal of cephalosporin and their intermediates during the operation of a membrane bioreactor treating pharmaceutical wastewater. *Water Science and Technology*, *61*(7), 1907–1914. doi:10.2166/wst.2010.881 PMID:20371950

Szép, A. (2013). *Membránszűrés hatékonyságának javítása kombinált módszerek alkalmazásával ipari szennyvizek tisztítása során* (Doctoral dissertation, SZTE).

Szép, A., Kertész, S., László, Z., Szabó, G., & Hodúr, C. (2012). Advanced Treatment of Pharmaceutical Wastewater by Nanofiltration and Ozonation. *Acta Technica Corvininesis-Bulletin of Engineering*, *5*(1).

Tam, L. S., Tang, T. W., Lau, G. N., Sharma, K. R., & Chen, G. H. (2007). A pilot study for wastewater reclamation and reuse with MBR/RO and MF/RO systems. *Desalination*, *202*(1-3), 106–113. doi:10.1016/j.desal.2005.12.045

Ternes, T. A., Meisenheimer, M., McDowell, D., Sacher, F., Brauch, H. J., Haist-Gulde, B., ... Zulei-Seibert, N. (2002). Removal of pharmaceuticals during drinking water treatment. *Environmental Science & Technology*, *36*(17), 3855–3863. doi:10.1021/es015757k PMID:12322761

Ternes, T. A., Meisenheimer, M., McDowell, D., Sacher, F., Brauch, H. J., Haist-Gulde, B., ... Zulei-Seibert, N. (2002). Removal of pharmaceuticals during drinking water treatment. *Environmental Science & Technology*, *36*(17), 3855–3863. doi:10.1021/es015757k PMID:12322761

Tijani, J. O., Fatoba, O. O., & Petrik, L. F. (2013). A review of pharmaceuticals and endocrine-disrupting compounds: Sources, effects, removal, and detections. *Water, Air, and Soil Pollution*, *224*(11), 1770. doi:10.100711270-013-1770-3

Ueda, T., Hata, K., Kikuoka, Y., & Seino, O. (1997). Effects of aeration on suction pressure in a submerged membrane bioreactor. *Water Research*, *31*(3), 489–494. doi:10.1016/S0043-1354(96)00292-8

Urase, T., Kagawa, C., & Kikuta, T. (2005). Factors affecting removal of pharmaceutical substances and estrogens in membrane separation bioreactors. *Desalination, 178*(1-3), 107–113. doi:10.1016/j.desal.2004.11.031

Vandevivere, P. C., Bianchi, R., & Verstraete, W. (1998). Treatment and reuse of wastewater from the textile wet-processing industry: Review of emerging technologies. *Journal of Chemical Technology & Biotechnology: International Research in Process. Environmental and Clean Technology, 72*(4), 289–302.

Vanerkar, A. P., Satyanarayan, S., & Dharmadhikari, D. M. (2004). Toxicity of herbal pharmaceutical wastewater on fish—Lebistes reticulatus (Peter). *Journal of Environmental Science and Health. Part B, Pesticides, Food Contaminants, and Agricultural Wastes, 39*(1), 115–123. doi:10.1081/PFC-120027443 PMID:15022745

Velagaleti, R., & Burns, P. K. (2006). The industrial ecology of pharmaceutical raw materials and finished products with emphasis on supply chain management activities.

Vinati, H., & Radha, R. International Journal of Scientific Research and Reviews.

Visvanathan, C., Aim, R. B., & Parameshwaran, K. (2000). Membrane separation bioreactors for wastewater treatment. *Critical Reviews in Environmental Science and Technology, 30*(1), 1–48. doi:10.1080/10643380091184165

Walters, J. P., & Javernick-Will, A. N. (2015). Long-term functionality of rural water services in developing countries: A system dynamics approach to understanding the dynamic interaction of factors. *Environmental Science & Technology, 49*(8), 5035–5043. doi:10.1021/es505975h PMID:25775082

Wang, L. K., & Menon, R. (2011). Treatment of industrial effluents, municipal wastes, and potable water by membrane bioreactors. In *Membrane and desalination technologies* (pp. 201–236). Totowa, NJ: Humana Press. doi:10.1007/978-1-59745-278-6_5

Watkinson, A. J., Murby, E. J., & Costanzo, S. D. (2007). Removal of antibiotics in conventional and advanced wastewater treatment: Implications for environmental discharge and wastewater recycling. *Water Research, 41*(18), 4164–4176. doi:10.1016/j.watres.2007.04.005 PMID:17524445

WHO, UNICEF. (2010). *Progress on sanitation and drinking-water, 2010 Update*. Geneva, Switzerland: World Health Organization.

Wilson, L. G., & Cavanagh, P. (1969). Principle of inertial impaction. *Atmospheric Environment, 3*, 597–605. doi:10.1016/0004-6981(69)90054-7

Wintgens, T., Melin, T., Schäfer, A., Khan, S., Muston, M., Bixio, D., & Thoeye, C. (2005). The role of membrane processes in municipal wastewater reclamation and reuse. *Desalination, 178*(1-3), 1–11. doi:10.1016/j.desal.2004.12.014

World Health Organization. (2009). Diarrhea: why children are still dying and what can be done.

Wu, J., & Sun, Z. (2016). Evaluation of shallow groundwater contamination and associated human health risk in an alluvial plain impacted by agricultural and industrial activities, mid-west China. *Exposure and Health, 8*(3), 311–329. doi:10.100712403-015-0170-x

Xiangdong, L. (2005). Treatment of Pharmaceutical Wastewater by Airfloatation-Hydrolytic Acidification-Aerobic Process [J]. *Environmental Engineering, 3*.

Yang, Y., Fu, J., Peng, H., Hou, L., Liu, M., & Zhou, J. L. (2011). Occurrence and phase distribution of selected pharmaceuticals in the Yangtze Estuary and its coastal zone. *Journal of Hazardous Materials*, *190*(1-3), 588–596. doi:10.1016/j.jhazmat.2011.03.092 PMID:21497014

Yoon, Y., Westerhoff, P., Snyder, S. A., & Esparza, M. (2003). HPLC-fluorescence detection and adsorption of bisphenol A, 17β-estradiol, and 17α-ethynyl estradiol on powdered activated carbon. *Water Research*, *37*(14), 3530–3537. doi:10.1016/S0043-1354(03)00239-2 PMID:12834747

Yoon, Y., Westerhoff, P., Snyder, S. A., & Wert, E. C. (2006). Nanofiltration and ultrafiltration of endocrine disrupting compounds, pharmaceuticals and personal care products. *Journal of Membrane Science*, *270*(1-2), 88–100. doi:10.1016/j.memsci.2005.06.045

Yoon, Y., Westerhoff, P., Snyder, S. A., Wert, E. C., & Yoon, J. (2007). Removal of endocrine disrupting compounds and pharmaceuticals by nanofiltration and ultrafiltration membranes. *Desalination*, *202*(1-3), 16–23. doi:10.1016/j.desal.2005.12.033

Chapter 22
Waste Management Practice in Malaysia and Future Challenges

Huang Shen Chua
https://orcid.org/0000-0001-5352-2812
KDU University College, Malaysia

Mohammed J. K. Bashir
https://orcid.org/0000-0002-3086-0568
Universiti Tunku Abdul Rahman, Malaysia

ABSTRACT

Malaysia current waste management systems are not able to solve the disposal rates. The reduction of waste through 3Rs programme (reduce, reuse, and recycle) is in precontemplation stage. The municipal solid waste (MSW) condition is mixed and wet. The landfill and Thermal Treatment Plant (incineration) are the current practices for the MSW disposal. Landfill created leachate while incineration released unhealthy gases. Incineration failed due to the improper management and high cost of the operation. Torrefaction is needed before it goes to the incineration to improve the high heating value (HHV). The MSW pyrolysis and gasification are able to convert into valuable products (bio-oil, biochar, combustible gases). Combustible gases can be used to feedback into the incinerator. The heat of the incinerator can be performed waste to energy (WTE), which is able to convert into electricity as a Feed-in-Tariff (FiT).

INTRODUCTION

The developing worldwide population and monetary extension are expanded now and again, these situations had noteworthy expanded the age of city simultaneously increased the generation of municipal solid waste (MSW) (Jouni Havukainen et al, 2017). Consequently, how to deal with the human waste can be challenging work. The current practice of waste treatment approach in Malaysia is through incineration and landfill (place where garbage dump)(Assamoi & Lawryshyn, 2012). Landfill is the foremost economy strategy but a final considered solution in the waste management because of the rise of the methane emission (Beylot, Villeneuve and Bellenfant, 2013), similarly leachate (liquid that drains or

DOI: 10.4018/978-1-7998-0369-0.ch022

extracts soluble from a landfill) is polluting into the group water (Othman et al, 2010). This situation has caused the global greenhouse gas (GHG) that warm up the atmosphere and affecting entire climate of cities. The MSW is burned (incineration) and converted into useable heat source to generate electricity (Paul Breeze, 2014). Carbon dioxide (CO_2) emission from MSW incineration around 30% by approaching weight. As indicated by the environmental protection agency that the CO_2 factor of MSW incineration ought to be 25 g/MJ fuel (Johan & Mattias, 2004). The worldwide are begun searching for better alternative solution to decrease the amount of MSW being landfill. The other MSW treatment methods have been investigated such as mechanical heat treatment (Demirbas, 2005), mechanical biological treatment (Noor et al, 2013), in-vessel composting (Silvia et al, 2013), gas plasma (DongyanMu, et al, 2017), biodrying (Rajasekhar, et al, 2017), anaerobic digestion (Asha, Pawels and Haridasc, 2016), microwave pyrolysis (Undri et al, 2017), pyrolization (Serio et al, 2017), gasification (Shehzad, et al., 2018) and others. Despite the fact that there were a number of techniques to execute the MSW treatment, pyrolysis and gasification are one of the thought to change over the MSW into the bio-oil, bio-char and syngas by the recent researchers. Many researchers and specialists are working in the territory of pyrolysis to convert MSW waste into bio-fuel. This pyrolysis treatment has a productive transformation if compared to other convention incineration conversion technology. Various waste materials have been explored through the pyrolysis such as MSW (Shehzad, et al., 2018), seedcakes (Schaefer, 1975), straws (Karaosmanoglu & Tetik, 2012), woody biomass (Mohan et al, 2007; Aho et al, 2008). Bagasse (Demiral & Sensoz, 2008), wood (Asadullah, 2007) and others. Pyrolysis is turning into an increasingly attractive solution for MSW treatment because of its CO_2 contamination minimization impacts. During 21st Conference of Parties (COP21), Malaysia delegation proposed to lessen the Greenhouse Gas emanations (GHG) by 45% in 2030 (Alireza et al, 2016). A significant efforts and results have been made to deliver the biofuels from bio-oil generation (Johari et al, 2014). MSW treatment ready to apply the alternative method for grow exceedingly effective management practices without yielding ecological quality. Pyrolysis of MSW treatment can create bio-fuel with high fuel-to-encourage proportions. The MSW treatment should be identified an alternative solution to develop highly efficient management practices without sacrificing environmental quality. Thusly, pyrolysis and gasification have been accepting more consideration as an effective technique for changing over biomass and MSW into bio-oil, bio-char and syngas.

WASTE GENERATION IN MALAYSIA

In recent years, population in Malaysia has increased simultaneously the waste generation is increased as presented in Table 1. Based on 2016, it was 32,939 tonnes of MSW waste is disposed each day in Malaysia. The waste generation is relied upon to rise and close to 41,035 tonnes by the year 2026 as predicted in Figure 1.

Table 1. Day waste generation from Malaysia

Year	Tonnes
2002	21,452
2004	23,073
2006	24,969
2008	26,489
2009	27,284
2012	29,712
2014	31,325
2016	32,939
2018	34,552
2020	36,138
2022	37,770
2024	39,403
2026	41,035

(Alireza et al, 2016) (Johari et al, 2014)

Figure 1. MSW projection

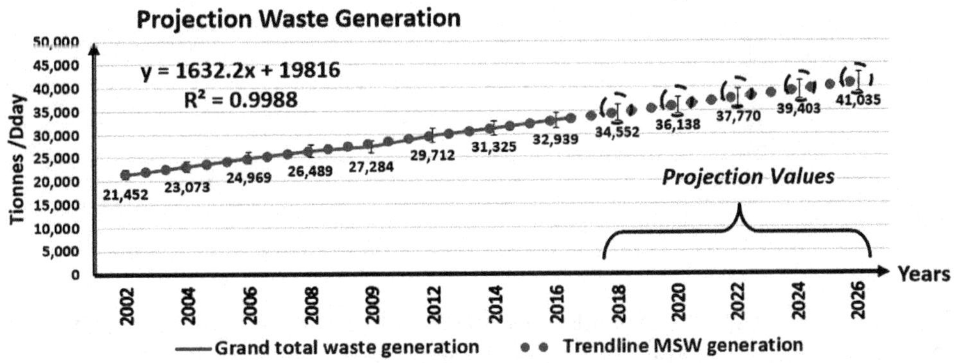

PROJECTION AND CHARACTERISTIC OF MUNICIPAL SOLID WASTE GENERATION FROM MALAYSIA.

The populace is assessed in Malaysia came to over 32.4 million in year 2018. The yearly populace development at a rate of 1.1% per annum. At the same time, the MSW generation is expanding yearly. Commonly, the substances of MSW composition (JPSPN, 2017) are food waste (44.5%), followed by leather (0.4%), household hazardous waste (1.3%), wood+peel/husk (1.4%), tetra pak (1.6%), rubber (1.8%), metal (2.7%), textiles (3.1%), glass (3.3%), garden waste (5.8%), diapers (12.1%), paper (8.5%), plastic (13.2%) and others (0.5%). The composition chart of MSW waste is presented in Figure 2.

Figure 2. Malaysian Household Waste Compositions
(JPSPN, 2017)

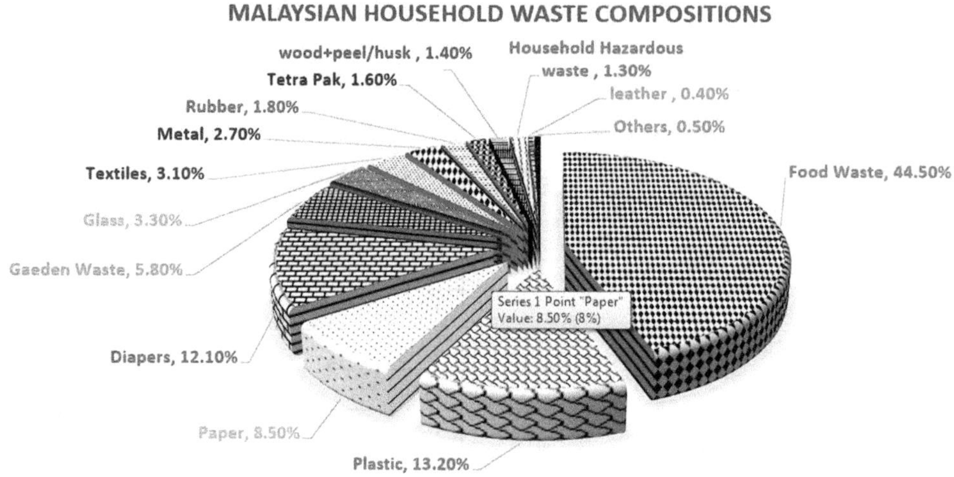

Table 2. Characteristics of Kuala Lumpur MSW

Proximate Analysis (wet)	Weight (%)
Moisture content	55.01
Volatile matter content	31.36
Fixed carbon content	4.37
Ash content	9.26
Bulk density (kg/m^3)	240
Net calorific value (Kcal/kg)	2180

(Jalil, 2015)

The MSW waste in Malaysia has expanded over 91% from 2011 and onward (Periathamby, Hamid, and Khidzir, 2009) because of the populace development, monetary development and business exercises. Tragically, Malaysia is not related to the compelling and deliberate administration of a strong waste treatment (Jalil et al, 2015).The new strategy MSW management squander is actualized to be progressively adaptable and reasonableness which is called as 3R's (Reduce, Reuse, Recycle), an approach to MSW segregation (Tanaka, 2009) (Wilson, 2007). Malaysia is absence of mindfulness for recyclables and investment by partner. The act of waste disposal in Malaysia by an individual is gathering all MSW and putting into the trash pack or plastic sack, and after that dumping into receptacle. The MSW management will gather waste together and then transport to the designated landfill. Landfill is requiring observing the groundwater defilement, methane gas, drainage of MSW waste buried and leachate (Assamoi & Lawryshyn, 2012). Landfills are getting to be restricted because of the land esteem. The MSW condition is mixed and wet. Table 2 is demonstrated the level of Kuala Lumpur's MSW contains a high grouping of moisture dampness content. A torrefaction process is expected to dry the high moisture dampness

Table 3. Solid waste facilities in Malaysia

Methods	Location	Tonnes per Day
Thermal Treatment Plants (Incinerator)	Pulau Pangkor, Perak Pulau Langkawai, Kedah Pulau Tioman, Pahang Cameron Highland, Pahang Labuan	20 100 15 40 50
Solid Waste Transfer Stations	Shah Alam, Selangor Taruka, Johor Batu Maung, Pulau Pinang Taman Beringin, Selangor Ampang Jajar, Pulau Pinang	1000 500 800 1700 500
Refuse-derived fuel (RDF)	Semenyih, Selangor	700
Waste to Energy (WtE)	Taman beringin Kuala Lumpur Bukit Payung, Johor Sungai Udang, Malacca	Proposed Proposed Proposed

(JPSPN, 2017)(KPKP, 2017)

MSW's substance before goes through the pyrolysis and gasification process, which is converted into valuable products.

WASTE FACILITIES

The most prevalent MSW disposal treatment method for Malaysia is depends on the sanitary landfill technology due to it was developed and low operation cost. Malaysia has four (4) principle strategies for MSW treatment technologies and the aggregate sum of the MSW waste in one day are shown in Table 3. Malaysia's government has proposed the waste to energy technology (WtE) to address the issue of the current circumstance of the huge amount of MSW. The proposed implementation of the WtE of waste in burning and the heat is utilized to generate the electricity. The WtE process is created the acidic substances, for example, SOx, NOx, HCl and HF are released during the incineration. A control framework remover needs to screen the nitrogen oxide, dioxin and mercury, corrosive gas and others (Thomas, 2014). Pyrolysis (Thomas, 2014) is one of the operation process ready to tackle these acidic substrates. Malaysia depends on the landfill facilities to clean the MSW. There are a total of 114 non-operational and 176 operational at the landfill in year 2010 (JPSPN, 2010). The MSW treatment facilities are not a decent sign to adapt to the present populace increment in Malaysia. The populace increment rate is 1.3% per annum in 2017 (JPSPN, 2017). The non-sanitary of MSW treatment facilities are open dumpsites that cause a serious problem ecological issues, for example, river contamination, leachate underground and methane emissions.

This circumstance has brought about critical issues, for example, flare-up of flames, pollution of rivers, and certain medical issues as revealed (Noor et al, 2013). Because of the issues, an action plan or environmental strategy was addressed, the legislature has set an arrangement into 4 levels all together ready to control the non-sanitary of MSW treatment facilities. The successions of the sequences are:

Figure 3. Malaysian Different methods of solid waste management

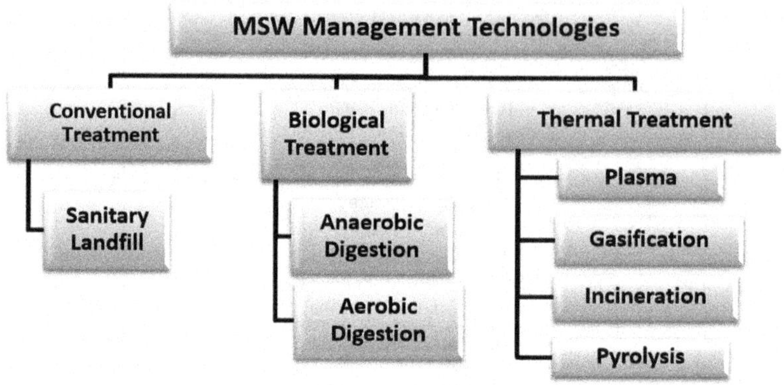

Level 1: Controlled dumping
Level 2: Sanitary landfill with daily cover.
Level 3: Sanitary landfill with leachate circulation.
Level 4: Sanitary landfill with leachate treatment.

In any case, how are we guarantee individuals is actually following the activity plan? It is smarter to diminish the aggregate amount the MSW before goes for landfill so as to lessen the leachate and methane emanation at the same time. As indicated by global environment centre (WTE.KPKT, 2017), Malaysia has in excess of 900 rivers have been dirtied and crushed because of the inappropriate MSW waste management. The degradation process (Wangyao, 2010) from the landfill is divided into anaerobic process and aerobic process. The carbon dioxide (CO_2) and methane (CH_4) are emitted from these two (2) processes. These two (2) gases are causing atmosphere temperature increase (greenhouse effect) (Xiaoli et al, 2008). The moisture content is one of a main critical which is affecting the methane emission from the landfill (Machado, et al, 2008) (Lan et al, 2008). The MSW composition in Malaysia contains a high amount of organics waste, as a result of high moisture content (Dewil et al, 2007). Thus, the moisture contents of the MSW should be exposed through the torrefaction. There are different MSW treatment techniques, management, practices and technologies around the countries. Figure 3 shows the different types of MSW treatment technologies (JPSPN, 2017).

BIOLOGICAL TREATMENT OF ANAEROBIC DIGESTION (AD)

Anaerobic digestion (AD) (Dewil et al, 2007) is a waste treatment from the organic substrate such as food waste or animal. It is broken down by using microorganisms without the present of oxygen. This process produces biogas and biofertiliser. This operation is required a high technology oriented with skilled technician and management people. The plant can be adopted by a small to bigger scale treatment plants. The system does not require a high operation and management cost. The system needs to get a certification through the Clean Development Mechanisms. This implementation is not polluting surrounding environment in a city. There were three (3) different operating temperature, operated at 10°C to 20°C (psychrophilic fermentation), 20°C to 40°C (mesophilic fermentation) and 40°C to 60°C

Figure 4. Schematic of Anaerobic digestion equipment set up
(Hanxi Wang et al, 2019 ; Sunil Kumar et al, 2016)

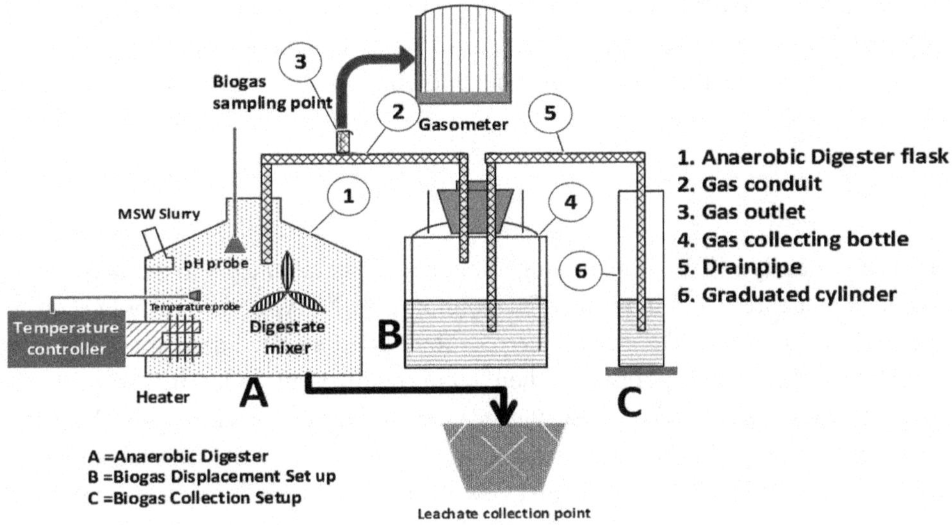

Figure 5. Cross section of a sanitary landfill
(Amadi et al, 2012)

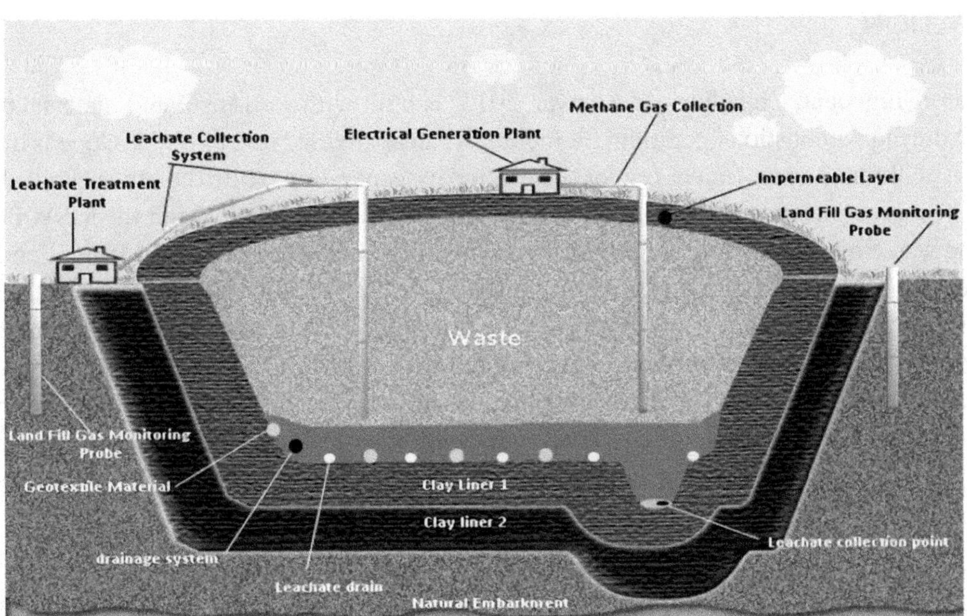

(thermophilic fermentation). Thermophilic fermentation is produced higher yield and production rates but this fermentation takes longer time. AD operates in four (4) stages process. The organic substrate is broken down without the oxygen environment and produces biogas and nutrient-rich effluent. The four (4) states are hydrolysis, acidogenesis, acetogenesis, and methanogenesis. Figure 4 shows the basic experimental set-up used to measure biogas volume and collect gas samples for each scale digester.

LANDFILL

Landfill is the most regularly MSW waste management and cost adequacy. Open dump, sanitary landfill and secure landfill are the three (3) famous landfill technologies. MSW open dumping is prohibited as the wastes are disposed that does not take into concern of the environment. There are serious impact on the land, air, ground and surface water pollution when using the open dump landfill solution (Karen,1978) (Maheshwari, Gupta, & Das,2015). The health concern such as fatigue, psychological disorders, headache, eyes & nose gastrointestinal problems, allergies and irritations of skin, allergies have been reported and documents due to the landfills. The landfills take few decades between 30 to 200 years for the MSW to fully decompose. Thus, Sanitary landfill is selected for waste disposal. The MSW waste is stored into the ground into a cell and secured with a slim layer of soil. The cell is arranged until a profundity of certain thickness is come to and after that the clay is a last layer to cover the waste. Figure 5 shows that the cross section of a sanitary landfill.

INCINERATION

An incinerator (Innocent, Chamhuri, & Anowar, 2013) is built with well-insulated material to maintain the heat lost during combustion operation. Incineration can be divided into two types, one is the structure combustor is without the recover energy for the electricity generation while the other one is designed to recover the energy, which is call waste-to-energy (WtE) conversion method. After the MSW be collected and send the waste management to segregate the MSW, then send to combustor. MSW is combusted in 850°C. The process of incineration results in the release of carbon dioxide, sulphur dioxide, nitrogen oxide and nitrogen dioxide, hydrogen chloride, hydrogen fluoride, carbon monoxide, dioxins and furans. Thus, gases emission limit of atmosphere are set to minimize any health and environmental impacts. The new government law has set up on the waste incinerate cannot run the operation without gases purification technologies. There few types of combustion technologies such as rotary Kiln, fluidized bed, WtE burner and moving grate as shown in Figure 6.

PYROLYSIS

Pyrolysis (Demirbas,2005) (Asadullah, 2007) (Demiral,& Sensoz, 2008) is one of the very well-known thermochemical process which is able to produce bio-oil, bio-char and combustible gases from the biomass. Pyrolysis can be expressed as a thermal degradation of biomass in the complete or incomplete presence of oxygen to oxidize the MSW (Mohamad, Nor Kartini, & Chow, December 2009), vaporizes and condenses into a dark brown liquid known as the bio-oil. There are three types of pyrolysis, slow

Figure 6. Different types of Incineration. (a) Rotary kilns (b) Fluidized (c) Moving grate (d) WtE burner (Pooja, 2017) (Nate Seltenrich,2016) (Mohd Rozainee Taib, el al 2017)

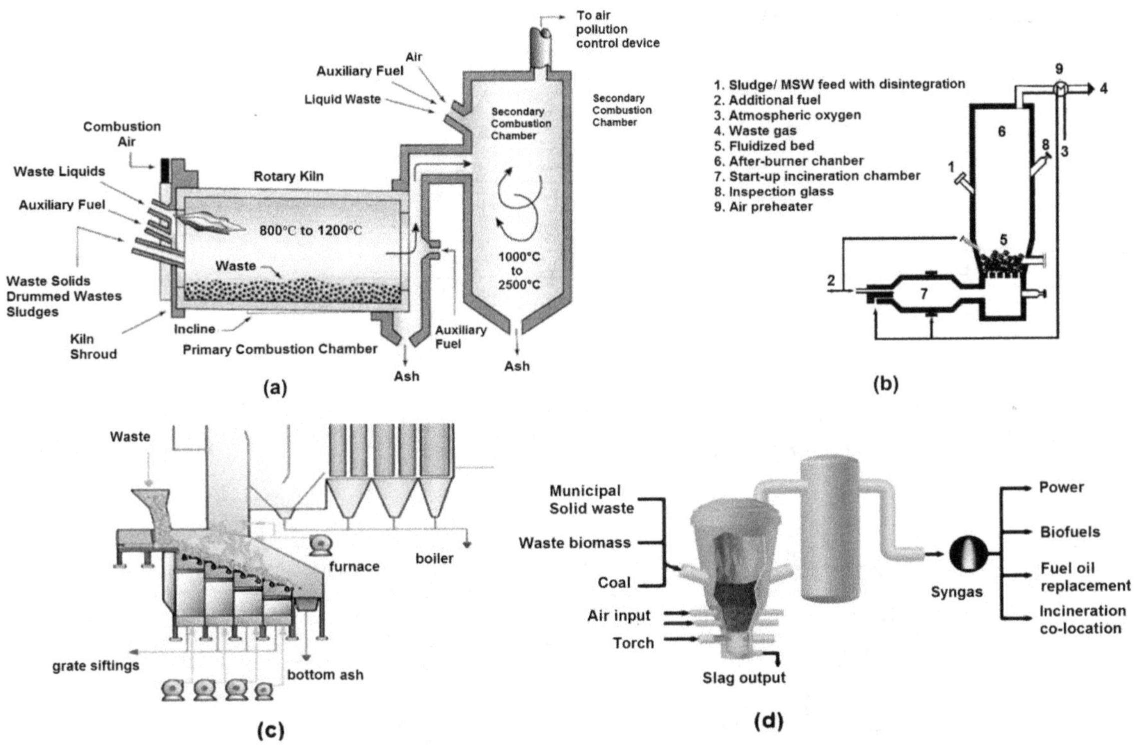

pyrolysis, fast pyrolysis and flash pyrolysis. The slow pyrolysis is primarily used for the production of the bio-char while the fast pyrolysis is used for production of the bio-oil and the flash pyrolysis is used for the production of syngas. The slow pyrolysis is operating in 4°C/min up to 400°C with two (2) hours residence time. The fast (or flash) pyrolysis is operating in constant temperature of 450°C, 480°C, 500°C, 550 °C and 600°C with a short residence time in few seconds. Pyrolysis temperature, MSW size and temperature are key factors for biochar, bio-oil and gaseous yields. The advantages application of bio-oil can be used in the engines and turbines, as well as a feedstock for refineries. Pyrolysis is one of the processes to reduce the waste in mass to about 10% left or even more than that of the quantity of MSW, therefore preserving landfill area. Based on the thermogravimetric study (TGA) MSW (Tanaka, 2019), the weight is reducing to 14% left at 780°C after 50 minutes. The acidic substances such as SOx, NOx, HCl and HF are total reduce dramatically a lot during the pyrolysis. There are semi-auto lab scaled pyrolysis prototypes (Albert et al, 2015) (Huang Shen et al, 2018) were fabricated from the low cost concept for the university research purpose. Figure 7 shows flow diagram process of biomass to bio-oil conversion.

Figure 7. Flow diagram process of biomass to bio-oil conversion
(Huang Shen et al, 2018)

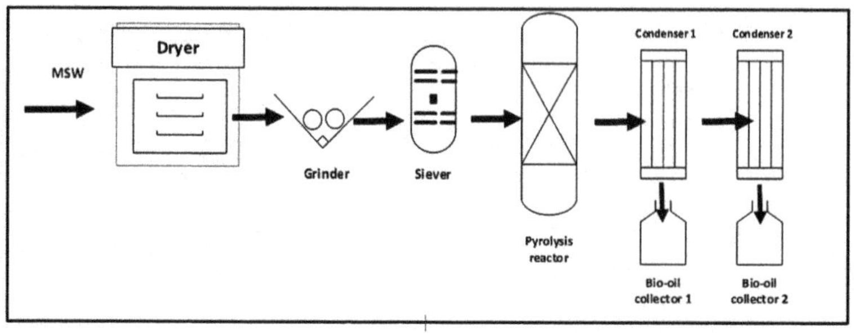

Figure 8. The flow of waste to energy of gasification.
(Yong-Chil Seo et al, July 2018)

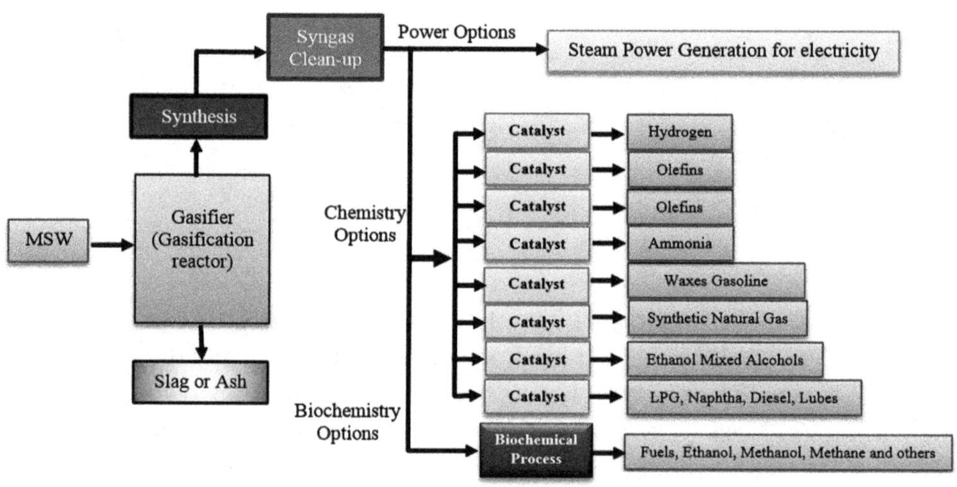

PLASMA

The plasma (Edbertho, 2004) operates dramatically high temperatures in between 5000°C to 10000°C. This process able to convert all the waste: organic contents, toxic, municipal solid, bio hazard, plastics, paper, oil, bio-materials, industrial and nuclear waste to synthesis gas at the atmospheric pressure. There were no ashes and water are generated as well as does not pollute the air. The Synthesis gas is a mixture from H_2 + CO which is a green fuel. The green fuel able to power up the advanced gas turbines for the generation of electrical power. It is required a skillful people for the operation. The voltage to operate the plasma furnace is about 10 kVA.

Figure 9. Different reactors of gasification. (a) & (b) Fixed-bed reactor (Quaak &Stassen, 1999). (c) & (d) Fluidised bed reactors (Quaak &Stassen, 1999). (f) Rotary kiln reactor (Riedhammer, 2014). (g) Moving grate reactor (Wellons FEI Corp, 2016)

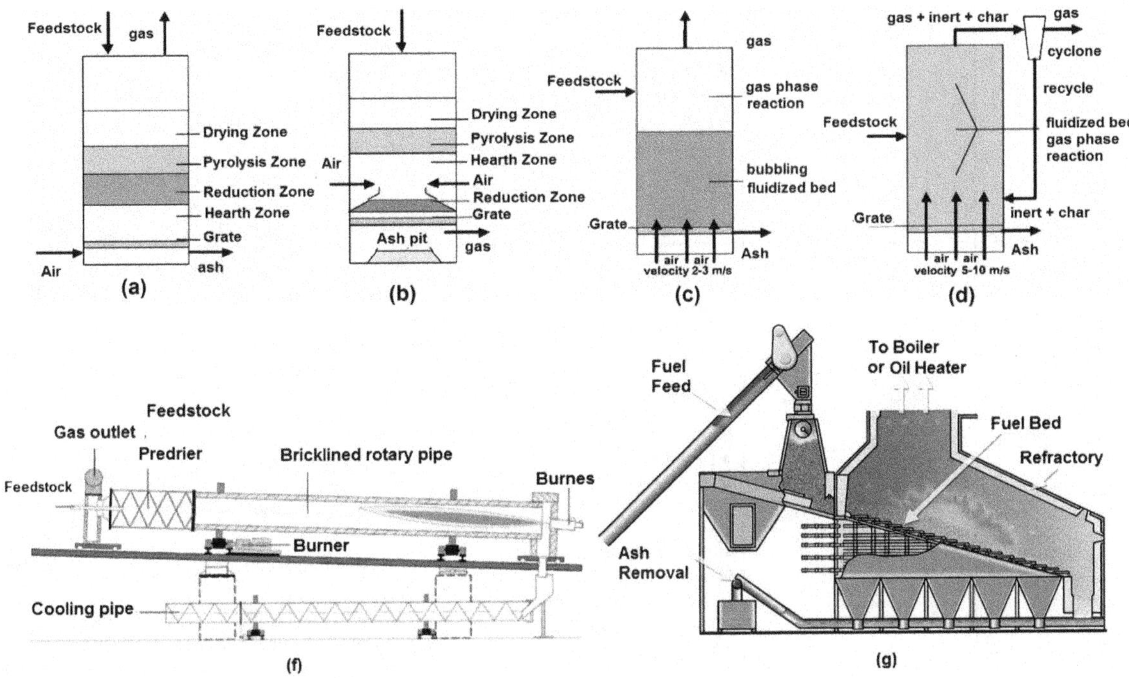

GASIFICATION

Gasification is a thermochemical cracking process. The process consists of two (2) stages: In the first stage, temperature is operated at below 600°C (1112°F) which is a pyrolysis process. This pyrolysis releases volatile components from the organic compounds and yields in a char (fixed carbon and the inorganic compounds). In the second stage, temperature is heated up excess of 800°C (1472°F), the char is reacted with air, oxygen, steam or carbon dioxide. The process of gasification with steam will produce a hydrogen and carbon dioxide rich

in syngas. While gasification with air will produce a high nitrogen (N2) and low Btu fuel gas. Figure 8 shows a schematic diagram of yield (syngas) production from gasification as well as how to apply the yield for various application. Different products are obtained by using various biochemical catalyst processes.

There are four (4) types of reactors can be used for gasification which are fixed beds, fluidized beds, rotary kilns and moving grate. The different parameters to determine the used of the reactor are syngas quality, fuels requirements, technical characteristics and operation cost. Figure 9 shows the four (4) different types of reactors.

Table 4. Solid waste facilities in comparison of MSW Management Technologies

MSW Conversion Technologies	Power Generation Potential (kW/ton day)	Advantages	Disadvantages
Mass Burn Incineration	3.8–164.7	• Thermal energy for power generation or direct heating. • Lower retention time.	• Air pollution control is needed, gas release directly to atmosphere such as articulate emissions, SOx, NOx, • chlorinated compounds. • Skilled personnel required. • Produce CO_2. • High temperature is needed from 850 °C to 1200 °C. • Reactant gas is needed (Oxidizing) • Increased treatment cost. • Moisture contents affect the energy recovery.
Landfilling	0	• Low cost. • Easy. • No need skilled personnel required.	• Post treatment care is very important. • Toxic to environment such as leachate and methane emissions. • Plastics and inorganic materials are inapplicable for such method.
Anaerobic digestion	5.9–11.3	• No power requirement for sieving and turning of waste pile. • Less land area. • Can be done in small scale.	• Long retention time. • Compact design needs.
Plasma Arc	5.2–224	• Organic molecules break into simpler gaseous molecules. • Inorganic portion is vitrified into a glassy residue. • Be used in construction or paving materials. • Short retention time.	• Undesirable compounds of syngas and vitrified slag. • Require a very high voltage and current. • Skilled personnel is required.
Gasification	6.5–220.4	• MSW able to reduce more than 90% of the weight. • Less expensive gas cleaning equipment. • Generates a fuel. • gas for power generation • Low oxygen environment.	• Tars, heavy metals, halogens and alkaline compounds are • released cause of acid rain. • Agglomeration in fluidized beds. • Heavy metals are toxic. • Skilled personnel is required.
Pyrolysis	4.2–184.5	• Production of fuel syngas and bio-oil. • The waste treatment can used all organic and non-organic matter together. • MSW able to reduce to 90% of the weight. • Low CO_2 emission. • 400 °C to 650 °C. • No reactant gas. • Short retention time. • Process is very flexible. • Adapt easily to composition of MSW.	• Moisture content affects the energy recovery. • Skilled personnel is required. • Nitrogen gas or any inert non-combustion gas. • High Operating & maintenance cost. • Increase treatment cost. • Cascading different process to collect different products.

(Hayelom et al, 2018)

Table 5. Installation of FiT program

Renewable Resource	FiT Project Approved			Projects in Commercial Operation		
	Number of Applications	Capacity (MW)	% of Total	Number of Installations	Capacity (MW)	% of Total
Biogas	129	225.12	13.6%	33	63.03	11.4%
Biomass	44	396.44	23.9%	8	87.90	15.8%
Geothermal	1	37.00	2.2%	-	-	-
Small Hydro	63	555.48	33.5%	6	30.30	5.5%
Solar PV	12,282	442.42	26.7%	9,746	373.66	67.3%
Total	12,519	1,656,45	100%	9,793	554.89	100%

(seda.gov.my, 2018)(Akmal Rahimi, 2018)

Table 6. Total number of the waste energy recovery thermal process

MSW Thermal Process `		Journals	Conference	Others
Combustion	MSW	41	1	4
	Refuse Derived Fuel (RDF)	5	1	1
	Grate	21	1	1
	Fluidized bed	4		1
	Rotary kiln	3		1
Gasification		15	1	2
Plasma		8		
Pyrolysis		6	2	2

(Lidia, Ennio, &Andrea, 2015)

COMPARISON OF MSW MANAGEMENT TECHNOLOGIES

There are many advantages and disadvantages of the different MSW management technologies, there are incineration, pyrolysis, gasification, plasma arc, anaerobic digestion and landfilling will be compared as listed in the Table 4.

CURRENT MUNICIPAL SOLID WASTE MANAGEMENT AND SUSTAINABLE ENERGY

The MSW management in Malaysia depends on the landfills and sanitary landfills. Malaysia's Act imposes that MSW is permitted to treat only in the sanitary landfills. The Solid Waste and Public Cleansing Management (SWPCM) Act, 2007 was formed to improve the MSW management in Malaysia due to the leachate contamination polluted water in Klang Valley. The setup cost of the sanitary landfills is very costly due to the long-time digestion process and slow revenue generation. The recoverable energy from the sanitary landfills of the MSW to anaerobic digester is lower than waste to energy (WtE) incineration (Aishah et al, 2013) and WtE pyrolysis (An Independent Engineering Evaluation of Waste-to-Energy Technologies, 2018). Pangkor, Tioman, Labuan and Langkawi have been used a small incineration (non WtE) since 1996 (Aishah et al, 2013) Unfortunately, the incineration's operation has been stopped because of high cost of maintainer and process but the incinerations are back functionally although it is not complete capacity in 2017. It is because of the high moisture content of MSW, and it has a low calorific value caused the high cost operation. (H.S. Chua et al, 2019).

A statutory body from Malaysia was formed Sustainable Energy Development Authority (SEDA, 2018) under the Sustainable Energy Development Authority Act 2011. SEDA is focusing in Feed-in Tariff (FiT) for the eligible renewable energy (RE) such as biomass, biogas, small hydropower and solar photovoltaic. Recently, SEDA gave the feed-in approval to Comintel Corp Bhd (Comcorp) (MIDA, 2018) to run WtE business. The process is to convert the gardening waste, biomass and MSW into electricity and thermal power. Table. 5 shows that SEDA had approved a total 12,519 and implemented the Feed-in-Tariff (FiT) installation with a renewable energy capacity with 1,656,45 megawatt. The implementation of the WtE incineration did not take up the operation if compared to other RE sources (Akmal Rahimi, 2018).

Table 6 is shown the total numbers waste energy recovery thermal process (incineration, gasification, plasma and pyrolysis) from the international journals, conferences and public report (Waste to Energy, 2017). Pyrolysis is shown that not many researchers are working on the MSW. But pyrolysis is used a lot in the biomass treatment.

The pyrolysis field research seems very low in the MSW treatment application. Thus, an investigation on pyrolysis operation to the MSW treatment should be carried out. Because of the higher content of organic matter and calorific value, the urban domestic waste able to be decomposed into gas and liquid products that can be used as fuel by using pyrolysis technology. The harmful gases can be reducing through the pyrolysis process.

CONCLUSION

The waste generated has increased from Malaysia in response to population growth. The MSW generation is expected to increase by 20% from 2018 to 2026. The implementing 'Reduce', 'Reuse' and 'Recycle' ('3 R's') is still lack of interest in attention by Malaysian. The recycling rate has increased slowly to 10% during 3R implementation. The failure of the MSW incineration plants is caused by the high moisture content. The MSW-drying process should be applied as a pre-treatment before landfill and combustion. The MSW pyrolysis is considered as a sustainable technology. The volume of waste can be reduced significantly through pyrolysis as well as value-added products (bio-oil, bio-char and syngas). The pyrolysis is therefore a potential technology for MSW treatment in Malaysia according to the current situation. A study to be conducted to investigate the potential power production of the pyrolysis for bio-oil and bio-char after the pre-treatment of MSW. In order to improve the effect of pyrolysis process, the urban domestic waste and should be classified and dehydrated before being pyrolyzed. The determination of the paid tariffs from SEDA based on the power measurement. Malaysia is started to look forward in its sustainable energy by officially established many parties to the development and promotion of RE and cutting down on the carbon emissions.

ACKNOWLEDGMENT

The authors acknowledge the financial support by the KDU University College. Malaysia. The research project code is KDURG/2018/1/005.

REFERENCES

Ab Jalil, N. A., Basri, H., Basri, N. E. A., & Abu Shammala, M. F. M. (2015). The potential of biodrying as pre-treatment for municipal solid waste in Malaysia. *Journal of Advanced Review on Scientific Research, 1*, 1–13.

Abu Amr, S. S., Aziz, H. A., Hossain, M. S., & Bashir, M. J. K. (2017). Simultaneous removal of COD and color from municipal landfill leachate using ozone/zinc sulphate oxidation process. *Global NEST Journal, 19*(3), 498–504. doi:10.30955/gnj.002299

Aho, A., Kumar, N., Eranen, K., Salmi, T., Hupa, M., & Murzin, D. Y. (2008). Catalytic pyrolysis of woody biomass in a fluidized bed reactor: Influence of the zeolite structure. *Fuel, 87*(12), 2493–2501. doi:10.1016/j.fuel.2008.02.015

Albert, F. Y. C., Mohd Fuad, M. N., & Huang Shen, C. (2015) A lab-scale semi auto vertical tubular combustor as a potential pyrolysis process for biomass burning. *ARPN Journal of Engineering and Applied Sciences,* 10, 15. 6534-6457.

Amadi, A. N., Olasehinde, P. I., Okosun, E. A., Okoye, N. O., Okunlola, I. A., Alkali, Y. B., & Dan-Hassan, M. A. (2012). A comparative study on the impact of avu and ihie dumpsites on soil quality in southeastern Nigeria. *American Journal of Chemistry, 2*(1), 17–23. doi:10.5923/j.chemistry.20120201.05

Asadullah, M., Rahman, M. A., Ali, M. M., Motin, M. A., Sultan, M. B., & Alam, M. R. (2007). Production of bio-oil from fixed bed pyrolysis of bagasse. *Fuel, 86*(16), 2514–2520. doi:10.1016/j.fuel.2007.02.007

Asha, P. (2016). Biodrying process: A sustainable technology for treatment of municipal solid waste with high moisture content. *Waste Management (New York, N.Y.), 49*, 64–72. doi:10.1016/j.wasman.2016.01.004 PMID:26774396

Assamoi, B., Lawryshyn, Y. The environmental comparison of landfilling vs. incineration of MSW accounting for waste diversion. *Waste Manage*, 32, 1019–1030.

Badgie, D., Samah, M. A. A., Manaf, L. A., & Muda, A. B. (2012). Assessment of municipal solid waste composition in Malaysia: Management, practice, and challenges. *Polish Journal of Environmental Studies, 21*(3), 539-547.

Beylot, A., Villeneuve, J., & Bellenfant, G. (2013). Life cycle assessment of landfill biogas management: Sensitivity to diffuse and combustion air emissions. *Waste Management (New York, N.Y.), 33*(2), 401–411. doi:10.1016/j.wasman.2012.08.017 PMID:23142510

Breeze, P. (2014). *Power from waste* (2nd ed., pp. 335–352). Power Generation Technologies.

Chandrappa, R., & Das, D. B. (2012). *Solid waste management. Environmental Science and Engineering*. Berlin, Germany: Springer-Verlag.

Chua, H. S., Bashir, M. J. K., & Tan, K. T. (2019). A sustainable pyrolysis technology for the treatment of municipal solid waste in Malaysia. AIP Conference Proceedings, 2124, 020016.

Demiral, I., & Sensoz, S. (2008). The effects of different catalysts on the pyrolysis of industrial wastes (olive and hazelnut bagasse). *Bioresource Technology, 99*(17), 8002–8007. doi:10.1016/j.biortech.2008.03.053 PMID:18450440

Demirbas, A. (2005). Pyrolysis of ground beech wood in irregular heating rate conditions. *J. Appl. Pyroyisis, 73*(1), 39–43. doi:10.1016/j.jaap.2004.04.002

Dewil, R., Appels, L., Baeyens, J., & Degrève, J. (2007). Peroxidation enhances the biogas production in the anaerobic digestion of biosolids. *Journal of Hazardous Materials, 146*, 577–581.

Fazeli, A., Bakhtvar, F., Jahanshaloo, L., Che Sidik, N. A., & Bayat, A. E. (2016). Malaysia's stand on municipal solid waste conversion to energy: A review. *Renewable & Sustainable Energy Reviews, 58*, 1007–1016. doi:10.1016/j.rser.2015.12.270

Havukainen, J., Zhan, M., Dong, J., Liikanen, M., Deviatkin, I., Li, X., & Horttanainen, M. (2017). (2017). Environmental impact assessment of municipal solid waste management incorporating mechanical treatment of waste and incineration in Hangzhou, China. *Journal of Cleaner Production, 141*, 453–461. doi:10.1016/j.jclepro.2016.09.146

Huang Shen, C., Bashir, M. J. K., Tan, K. T., Jocelyn, L. P. G., & Albert, F. Y. C. (2018). Design and implementation of a laboratory-scale pyrolysis combustor for biomass conversion. *Sci. Int. (Lahore), 30*(1), 81–84.

Undri, A., Rosi, L., Frediani, M., & Frediani, P. (2011). Microwave pyrolysis of polymeric materials. In Microwave Heating, 207-312. IntechOpen.

Jabatan, P., & Perumahan, D. K. T. (2013). *Survey on solid waste composition*. Characteristics & existing practice of solid waste recycling in Malaysia.

Jereme, I. A., Siwar, C., & Bhuiyan, M. A. H. (2013). Incineration and its implications: The need for a sustainable waste management system in Malaysia. *International Journal of Environmental Sciences, 4*, 3.

Johari, A., Alkali, H., Hashim, H., Ahmed, S. I., & Mat, R. (2014). Municipal solid waste management and potential revenue from recycling in Malaysia. *Modern Applied Science, 8*, 4. doi:10.5539/mas.v8n4p37

JPSPN. (2010). *Distribution of landfill sites in Malaysia*. Kuala Lumpur, Malaysia.

Kadir, S. A. S. A., Yin, C.-Y., Sulaiman, M. R., Chen, X., & El-Harbawi, M. (2013). Incineration of municipal solid waste in Malaysia: Salient issues, policies and waste-to-energy initiatives. *Renewable & Sustainable Energy Reviews, 24*, 181–186. doi:10.1016/j.rser.2013.03.041

Karaosmanoglu, F., & Tetik, E. (2012). Fuel properties of pyrolytic oil of the straw and stalk of rape plant. *Renewable Energy, 1999*(16), 1090–1093.

Korai, M. S., Mahar, R. B., & Uqaili, M. A. (2016). Optimization of waste to energy routes through biochemical and thermochemical treatment options of municipal solid waste in Hyderabad, Pakistan. *Energy Conversion and Management*, *124*, 333–343. doi:10.1016/j.enconman.2016.07.032

Kumar, S., Das, A., Kumar Srinivas, G. L., Dhar, H., Kumar Ojha, V., & Wong, J. (2016). Effect of calcium chloride on abating inhibition due to volatile fatty acids during the start-up period in anaerobic digestion of municipal solid waste. *Environmental Technology*, *37*(12), 1501–1509. doi:10.1080/09593330.2015.1119204 PMID:26609893

Lan, J. W., Zhan, L. T., Li, Y. C., & Chen, Y. M. (2012). Impacts of initial moisture content of MSW waste on leachate generation and modified formula for predicting leachate generation. [in Chinese]. *Environmental Sciences*, *33*, 1389–1395. PMID:22720594

Leal-Quir'os, E. (2014). Plasma processing of municipal solid waste. *Brazilian Journal of Physics*, *34*(4B).

Lombardi, L., Carnevale, E., & Corti, A. (2015). A review of technologies and performances of thermal treatment systems for energy recovery from waste. *Waste Management (New York, N.Y.)*, *37*, 26–44. doi:10.1016/j.wasman.2014.11.010 PMID:25535103

Machado, S. L., Carvalho, M. F., Gourc, J.-P., Vilar, O. M., & do Nascimento, J. C. F. (2009). Methane generation in tropical landfills: Simplified methods and field results. *Waste Management (New York, N.Y.)*, *29*(1), 153–161. doi:10.1016/j.wasman.2008.02.017 PMID:18440214

Maheshwari R., Gupta S., & Das, K. (2015). Impact of landfill waste on health: An overview. *IOSR Journal of Environmental Science, Toxicology, and Food Technology (IOSR-JESTFT),* 1(4), 17-23.

MIDA. *Comintel Corp expects first waste-to-energy plant to contribute to earnings by June.* (2018). Retrieved from http://www.mida.gov.my/home/6264/news/comintel-corp-expects-first-waste-to-energy-plant-to- contribute -to-earnings-by-june/

Ministry of Housing and Local Government, Solid Waste Management Lab. (2012). Retrieved from http://jpspn.kpkt.gov.my/resources/index/user_1/Sumber_Rujukan/kajian/lab_sisa_pepejal.pdf

Mohan, D., Pittman, C. U. Jr, Bricka, M., Smith, F., Yancey, B., Mohammad, J., ... Gong, H. (2007). Sorption of arsenic, cadmium, and lead by chars produced from fast pyrolysis of wood and bark during bio-oil production. *Journal of Colloid and Interface Science*, *310*(1), 57–73. doi:10.1016/j.jcis.2007.01.020 PMID:17331527

Mu, D., Horowitz, N., Casey, M., & Jones, K. (2017). Environmental and economic analysis of an in-vessel food waste composting system at Kean University in the U.S. *Waste Management (New York, N.Y.)*, *59*, 476–486. doi:10.1016/j.wasman.2016.10.026 PMID:27816469

Nidoni, P. G. (2017). Incineration process for solid waste management and effective utilization of by-products. *International Research Journal of Engineering and Technology (IRJET), 4*(12).

Noor, Z. Z., Yusuf, R. O., Abba, A. H., Abu Hassan, M. A., & Mohd, F. M. D. (2013). An overview for energy recovery from municipal solid wastes (MSW) in Malaysia scenario. *Renewable & Sustainable Energy Reviews*, *20*, 378–384. doi:10.1016/j.rser.2012.11.050

Noor, Z. Z., Yusuf, R. O., Abba, A. H., Abu Hassan, M. A., & Mohd Din, M. F. (2013). An overview for energy recovery from municipal solid wastes (MSW) in Malaysia scenario. *Renewable & Sustainable Energy Reviews*, *20*, 378–384. doi:10.1016/j.rser.2012.11.050

Others Solid Waste Management Facilities. (2017). Retrieved from http://jpspn.kpkt.gov.my/index.php/pages/view/77

Othman, E., Yusoff, M. S., Aziz, H. A., Adlan, M. N., Bashir, M. J. K., & Hung, Y. T. (2010). The effectiveness of silica sand in semi-aerobic landfill leachate treatment. *Water (Basel)*, *2*(4), 904–925.

Periathamby, A., Hamid, F. S., & Khidzir, K. (2009). Evolution of solid waste management in Malaysia: Impacts and implications of the solid waste bill. *Journal of Material Cycles and Waste Management*, *11*(2), 96–103. doi:10.100710163-008-0231-3

Quaak Knoef, P., & Stassen, H. (1999) *Energy from biomass.* World Bank technical paper, 422.

Rajasekhar, M., Venkat, N., Rao, G., Chinna, R., Priyadarshini, G., & Jeevan Kumar, N. (2014). Energy generation from municipal solid waste by innovative technologies – plasma gasification. *Procedia Materials Science*, *10*, 513–518. doi:10.1016/j.mspro.2015.06.094

Rapport, J., Zhang, R., Jenkins, B. M., & Williams, R. B. (2015). Current anaerobic digestion technologies used for treatment of municipal organic solid waste. California Environmental Protection Agency; 2008.

Riedhammer. (2014). *Riedhammer rotary kiln.* Retrieved from http://www.laeis.eu/Web-site/Page-management/Sacmi-Group/News-Area/News-by-Business/Ceramics/Riedhammer-presents-the-new-generation-of-indirectly-heated-DRA-rotary-kilns.aspx?ido=24754&idc=61115&ln=en-US

Salati, S., Scaglia, B., Di Gregorio, A., Carrerab, A., & Adani, F. (2013). The use of the dynamic respiration index to predict the potential MSW-leachate impacts after short term mechanical biological treatment. *Bioresource Technology*, *128*, 351–358. doi:10.1016/j.biortech.2012.10.101 PMID:23201514

Samah, I. A. R. A. (2018). Sustainable energy for Malaysia: An exciting journey awaits. The monthly Bulletin of the Institution of Engineers, Malaysia. (13-18).

Schaefer, W. D. (1975). Disposing of solid wastes by pyrolysis. *Environmental Science & Technology*, *9*, 91–98.

Seltenrich, N. (2016, June). Emerging waste-to-energy technologies: Solid waste solution or dead end?. *Journal of Environmental Health Perspect, 24*(6), A106–A111.

Seo, Y.-C., Alam, M. T., & Yang, W.-S. (2018). *Gasification of municipal solid waste. Gasification for low-grade feedstock.* Yongseung Yun, IntechOpen.

Serio, M. A., Chen, Y., Wojtowicz, M. A., & Suuberg, E. M. (2002). Pyrolysis processing for solid waste resource recovery in space. In *30th International Conference on Environmental Systems*. Toulouse, France.

Shehzad, A., Bashir, M. J. K., Horttanainen, M., Manttari, M., Havukainen, J., & Abbas, G. (2018). Modeling and comparative assessment of bubbling fluidized bed gasification system for syngas production-A gateway for a cleaner future in Pakistan. *Environmental Technology*, *39*(14), 1841–1850. doi:10.1080/09593330.2017.1340350 PMID:28595513

Stringfellow, T. (2014). *An independent engineering evaluation of waste-to-energy technologies.* Renewable energy world. Retrieved from https://www.renewableenergyworld.com/articles/2014/01/an-independent-engineering-evaluation-of-waste-to-energy-technologies.html

Sundberg, J. & Olofsson, M. (2005). *Waste-to-energy incineration,* RVF rapport, 15.

Taib, M. R., Ho, W. S., & Ng, P. S. (2017). Development of waste-to-energy plant in Kuala Lumpur. *Chemical Engineering Transactions*, 56.

Tanaka, M. (1999). Recent trends in recycling activities and waste management in Japan. *Journal of Material Cycles and Waste Management, 1*, 10–16.

Velghe, I., Carleer, R., Yperman, J., & Schreurs, S. (2011). Study of the pyrolysis of municipal solid waste for the production of valuable products. *Journal of Analytical and Applied Pyrolysis*, *92*(2), 366–375. doi:10.1016/j.jaap.2011.07.011

Wang, H., Xu, J., Sheng, L., Liu, X., Zong, M., & Yao, D. (2019). Anaerobic digestion technology for methane production using deer manure under different experimental conditions. *Energies*, *12*(9), 1819. doi:10.3390/en12091819

Wangyao, K., Towprayoon, S., Chiemchaisri, C., Gheewala, S. H., & Nopharatana, A. (2010). Application of the IPCC waste model to solid waste disposal sites in tropical countries: Case study of Thailand. *Environmental Monitoring and Assessment*, *164*(1-4), 249–261. doi:10.100710661-009-0889-6 PMID:19415519

Waste to Energy. (2017). Retrieved from http://wte.kpkt.gov.my/wte_microsite_pg05.html

Wellons, F. E. I. Corp. (2016). Reciprocating grate. Retrieved from http://www.wellonsfei.ca/en/reciprocating-grate.aspx

Wilson, D. C. (2007). Development drivers for waste management. *Waste Management & Research*, *25*(3), 198–207. doi:10.1177/0734242X07079149 PMID:17612318

Xiaoli, C., Ziyang, L., Shimaoka, T., Nakayama, H., Ying, Z., Xiaoyan, C., ... Youcai, Z. (2010). Characteristics of environmental factors and their effects on CH_4 and CO_2 emissions from a closed landfill: An ecological case study of Shanghai. *Waste Management (New York, N.Y.)*, *30*(3), 446–5. doi:10.1016/j.wasman.2009.09.047 PMID:19880303

Chapter 23
Agricultural Runoff and Treatment Methods

Yeek-Chia Ho
https://orcid.org/0000-0002-2820-9696
Universiti Teknologi PETRONAS, Malaysia

Boon-Chin Lim
Universiti Teknologi PETRONAS, Malaysia

Jun-Wei Lim
Universiti Teknologi PETRONAS, Malaysia

ABSTRACT

Increasing global population intensifies the food demand pressure while meeting the increasing environmental legislation. To cater for such demand, agriculture sector shifts for mass food production accompanied with the use of a large amount of agrochemicals such as fertilizers, pesticides, antibiotics, and growth hormones. As a result, these agrochemicals' residues pollute the ecosystem and environment through surface runoff. It has been a great challenge in treating agricultural runoff due to its contaminant loadings and wide variability of different compounds in the wastewater to meet the stringent standards in an economically feasible manner. This chapter delineates the current state-of-the-art of wastewater treatment methods for agricultural runoff.

INTRODUCTION

As the global demands for food is increasing rapidly, the necessity for agricultural research is assured to discover more effective techniques in improving crop production to meet such demand (Tilman et al., 2011). Modern agriculture requests a variety of agrochemicals and sophisticated techniques to more intensive use of available resources of water, land and energy for high crop productions within short period. The innovation of agricultural machinery and genetic biotechnology and application of agrochemicals

DOI: 10.4018/978-1-7998-0369-0.ch023

such as fertilizers, pesticides, antibiotics and growth hormones are indispensable agents of the modern farming or industrial agriculture (Edgerton, 2009).

The first recorded use of fertilizers and pesticides in agriculture was around 4000 years ago when first farmers used livestock manure as the nutrient source and sulfur, mercury and arsenical compounds for controlling insects and mites (Unsworth, 2010). For centuries, farmers used easily obtainable or available materials derived from animal, plant or mineral nature. However, the growth in application of synthetic fertilizers, pesticides and steroid hormones accelerated since 1940's and the products such as organochlorines (DDT) and 2,4-D became very popular as an effective and inexpensive pesticides. At first, there was little concern about environmental consequences of agrochemicals application. However, after a decade, there was a growing public awareness about the deteriorate effects of agricultural chemicals on both environment and human health (Kellogg et al., 2002). Although the application of these agrochemicals leads to profound beneficial impact on the plant functions, they are now recognized as potential sources for adverse environmental impacts (Bhandari, 2014).

Since agriculture is the largest user of surface water resources, the main concern is the contamination of waters from agricultural runoff to the surface water. The pollutants containing pesticide residues, fertilizer as well as suspended sediments can be transported over land and through the soil by water and ultimately find their way into rivers, lakes and groundwater, affecting the environmental quality which is much far from their application sites. Additionally, agriculture expansion to support increasing food demand related to global population growth and changes in dietary patterns may further reduce the surface water dilution capacities. Therefore, there is an urgent need to study on the agricultural discharge properties and their appropriate management to protect ecological security and human health in ensuring sustainable economic and environmental development.

AGRICULTURAL RUNOFF CHARACTERISTICS

Agricultural runoff is identified as one of the main diffuse and non-point sources of pollution. It is typically enriched with sediments, nutrients, phosphorus, pathogens, pesticides, steroid, heavy metals, hormone and antibiotics (Kato et al., 2009; Lefrancq et al., 2017; Zhu et al., 2015). Globally, the leaching and runoff from agricultural activities such as dairy, potato processing, sugar processing and so on has contributed to the most widespread environmental threat in recent years (Kim et al., 2014). The runoff severity is closely linked with hydrological conditions (rainfall frequency and rainfall magnitude), anthropogenic alterations (types and changes in land use) and the interactions (Li et al., 2015; Zhu et al., 2015).

One of the most indicative aspects of agricultural runoff is overloading of nutrients in the surface water. Excessive fertilizer applications and intensive livestock productions in the agricultural land is the principal factor of increasing N and P loading in most agricultural basins. It is reported that the increase of mean N concentration in Arizona groundwater is mainly contributed by increased N fertilizer application rates (Xu et al., 2007). Another study also shown that around 80% of total N input to 23 California basins were caused by the use of inorganic fertilizers, legumes and manure in agriculture (Sobota et al., 2009). Meanwhile, livestock generation also causes serious impact on water quality as animal manures are often recycled as organic fertilizers on farmlands (Kato et al., 2009) and occasionally discharge directly into water bodies without any treatment (Ying-Xin et al., 2007). Kato et al. (2009) noted that the high livestock density watershed led to 2-6 times higher average concentrations of different N species in the surface water as compared with those in the receiving water from moderate livestock production area. These washed-off nutrients from the farms would lead to eutrophication (algae bloom) which decreases transparency and dissolved oxygen level of water and directly affects pH, taste and odour of water (Hasan et al., 2011). Nonetheless, the percussions of fertilizer usage and livestock production on N and P loadings fluctuate significantly by region. Further assessment on these factors is necessary.

Moreover, heavy metal inputs in surface water are also associated with agricultural discharge. They are mostly originated from commercial fertilizers, fungicides, pesticides, liming materials, soil amendments such as livestock manures and sewage sludge, and irrigation waters (Blanco-Canqui et al., 2004; Shi et al., 2018). Heavy metals persist long in the system and can be leached out of the system via surface runoff (Tchounwou et al., 2012). The mobility of heavy metals in agricultural runoff depends on the chemical compounds and their concentrations in soil, soil properties and environmental factors (Hariprasad & Dayananda, 2013). They enter the water bodies as hydrated ions or inorganic complexes, which are easily adsorbed on the surface of sediment particles and form the labile fragment (Pandey & Singh, 2017). It was found that non-compliance of various heavy metals concentrations in Jannapura lake, India to the allowable limit as per WHO standards were mainly associated with agricultural runoff and waste from small scale industries (Puttaiah & Kiran, 2007). Wongsasuluk et al. (2014) reported that contamination of shallow groundwater wells (drinking water) by As, Cd, Cr, Cu, Hg, Ni, Zn and Pb were due to agricultural activities near Hua-ruea and Ubon Ratchathani regions in Thailand. Most of the heavy metals can also be found in animal manures, which cause secondary pollution to the soil and surface water. Nicholson et al. (2003) declared that livestock manures were account for around 37-40% and 8-17% of total Zn and Cu inputs, respectively to the agricultural soils in England and Wales. Another study also revealed that extensive use of organoarsenic feed additives in livestock diets had resulted in high level of As content in the surface soils, particularly in concentrated animal feeding operations zone (Liu et al., 2015).

Agricultural runoff also leads to heavy organic pollution to the surface water, which often expressed by the Biochemical Oxygen Demand (BOD) and Chemical Oxygen Demand (COD). It was identified that Kom Ombo and El-berba agricultural drains contributed around 76% of the total organic load (computed as COD) discharged into Nile river (El Gohary, 2015). Further, excessive use of pesticides, i.e. herbicides, fungicides, nematicides, insecticides and rodenticides in crop protection poses a major threat to environmental quality as they may leave agricultural fields as surface runoff, atmospheric deposition and wind transport (Zolgharnein et al., 2011). According to Carter (2000), an average of 1-5% of applied herbicide volume was lost due to leaching, surface runoff and drain-flow. In addressing the safety concern of drinking water quality, European Community Drinking Water Directive has set the threshold level of pesticides in drinking water to be less than 0.1 μgL-1.

The agricultural runoff is also featured with emerging pollutants, which include antimicrobials, antibiotics and steroid hormones, particularly nearby intensive livestock production areas. These hormones are often used at animal feeding operations for either growth promotion or disease treatment. Large amount of these chemicals are consumed in animal feedlots (Wise, 2002). It has been identified the presence of different steroidogenic compounds such as 4-androstenedione, 17α- and 17ß-estradiol, estrone, αzearalanol, α-zearalenol, β-zearalenol, melengesterol acetate, progesterone and 17α trenbolone in both dairy and beef cattle manure (Kolok et al., 2018). The presence of these pharmaceutical residues in their manures may eventually leach into the surface water via runoff and cause ecotoxicological impacts on aquatic organisms (Kolok et al., 2018; Pounds et al., 2008).

Table 1. Properties of agricultural runoff wastewater

Origins	TKN (mg/L)	TN (mg/L)	TAN (mg/L)	TP (mg/L)	BOD_5 (mg/L)	COD (mg/L)	TSS (mg/L)	TDS (mg/L)	FOG (mg/L)	References
Livestock	-	2820	-	700	-	20600	6933	-	-	(Ming et al., 2007)
	-	15.13-29.28	0.94-4.58	0.66-12.71	-	-	-	-	-	(Kato et al., 2009)
Milk house	81	-	34	56	1440	2260	686	-	477	(Janni et al., 2009)
Swine farm	350-700	-	-	-	800-3500	1700-7400	900-4700	-	-	(Nokyoo, 2016)
	-	707	419.3	33.01	-	10320	17500	2500	-	(Rosa et al., 2017)
	-	-	-	-	163-3350	210-9400	-	770-6480	-	(Mofokeng et al., 2016)
Abattoir dairy	123	-	68	3.1	704	-	114	-	-	(Rozema et al., 2016)
Olive oil mill	-	-	-	-	70000-160000	10000-200000	-	-	2000-8000	(Israel Ministry of Environmental Protection, 2018)

Table 2. Impacts of agricultural runoff wastewater on human health

Pollutants	Origins	Impacts	References
Nitrate-nitrogen (NO_3-N)	Fertilizers	Bind to hemoglobin and reduce the oxygen-carrying capacity of blood in infants and newly-born livestock, leading to methemoglobinemia or "blue baby syndrome"	(Suthar et al., 2009)
Ammonia nitrogen (NH_4-N)	Fertilizers	High ammonium (NH_4^+) in the body overrides potassium ion (K^+) and depolarizes neurons which affect central nervous system	(Randall & Tsui, 2002)
Zn	Livestock manures	Cause skin irritations, stomach cramps, vomiting, nausea and anemia	(Nicholson et al., 2003; Oyaro et al., 2007)
Cu	Livestock manures	Excessive digestion leads to cramps, vomiting or even death	(Nicholson et al., 2003; Paulino et al., 2006)
POPs	Pesticides	Cause endocrine and thyroid malfunctions, disrupt nervous system	(Cara & Jitereanu, 2015; Zolgharnein et al., 2011)
Antibiotics and synthetic hormones	Concentrated animal feeding operations	Leads to antimicrobial resistance in bacteria which causes diseases and allergic reactions in human	(De Briyne et al., 2014; Hu et al., 2017; Ronquillo & Hernandez, 2017)

Lastly, runoff from stockyards and grazing areas are polluted with microbial contamination. This is probably attributed by the rainfall events and agricultural practices that trigger the distortion of sediments and run-off water frequently result in resuspension of microorganisms in agricultural surface water (Topalcengiz et al., 2017). Contamination of surface water with many pathogens (bacteria, viruses, etc) would lead to chronic public health problems (Letcher & Vallero, 2011). Groundwater will also be polluted by agricultural waste seepage. United States Environmental Protection Agency (USEPA) has established a maximum geometric mean of 126 E. coli/100 mL as a quality threshold for freshwater upon contact during recreational activities (US Environmental Protection Agency, 2012). Moore (1988) reported a bacterial pollution level more than 100,000 organisms/100 mL from grazing area runoff. Table 1 summarizes the characteristics of agricultural runoff wastewater from different sources.

IMPACTS OF AGRICULTURAL RUNOFF

The occurrence of contaminants in agriculture has caused concern over the past decade as they induce harmful impacts on human and ecological system health, as summarized in Table 2. This is because these pollutants often find their ways into the environment via surface discharges, urban stormwater and agricultural runoff.

Overconsumption of fertilizers in agricultural fields would lead to high concentration of nitrate-nitrogen (NO_3-N) in drinking water (>10 mgL^{-1}) which causes health risks to humans and animals. Nitrate can bind to hemoglobin and decrease the oxygen-carrying capacity of blood in infants and newly-born livestock, leading to methemoglobinemia or "blue baby syndrome" (Suthar et al., 2009). Meanwhile, ammonia nitrogen (NH_4-N) is adverse in aquatic ecosystems' structure and function. Its toxicity can be aggravated by elevated water pH which directly affect aquatic organisms such as algae, vertebrates, invertebrates and microbes (Zhang et al., 2018). As time passes, the accumulation of high ammonium (NH_4^+) in the body would override potassium ion (K^+) and depolarizes neurons, which lead to uncontrolled activation of NMDA type glutamate receptor. Subsequently, it ends up with influx of calcium ions (Ca^{2+}) and sequential cell death in central nervous system (Randall & Tsui, 2002). To ensure the governance of water quality, different organizations and countries such as World Health Organization (WHO), European Union (EU) and United States Environmental Protection Agency (U.S. EPA) have established stringent water quality standards and regulations to protect public health from hazards and risks associated with high concentration of washed-off nutrients in the water. Maximum allowable limits of nutrient concentration range from 6.7 to 11.3 mgL^{-1} for nitrate, 0.5-1.5 mgL^{-1} for ammonium and 0.05-0.1 mgL^{-1} for total phosphorous (Hasan et al., 2011; Majumdar & Gupta, 2000).

There is serious concern on fate of transport where toxic trace elements existing in agricultural runoff. The accumulation of heavy metals in the sediments and bioaccumulation in biota across the food chain would lead to long-term ecosystem level effect and various chronic disease (Hariprasad & Dayananda, 2013). Amongst, benthic organisms are more prone to the exposures. Heavy metals such as manganese (Mn), iron (Fe), nickel (Ni), chromium (Cr), Cu and Zn are essential for organisms but beyond optimum level of threshold are toxic, while Pb, Cd, Hg and Ag are hazardous and carcinogenic. Excess amount of Zn exposure could cause skin irritations, stomach cramps, vomiting, nausea and anemia (Oyaro et al., 2007). The excessive ingestion of Cu could lead to serious toxicological problem such as cramps, vomiting or even death (Paulino et al., 2006). Excessive exposure of Ni may cause lung and kidney problems, gastrointestinal distress and skin dermatitis (Borba et al., 2006). Meanwhile, Hg is a neurotoxin and greatly affects nervous system (Namasivayam & Kadirvelu, 1999). Chronic exposure of Cd may cause dysfunction of kidney and even death. The presence of Pb in the body could damage kidney, liver, and brain function (Naseem & Tahir, 2001). Heavy exposure to Cr could also affect human physiology and has resulted in severe skin irritation and lung carcinoma (Khezami & Capart, 2005).

Despite its effectiveness in crop protection, pesticides, which are classified as persistent organic pollutants (POPs) are also intended to bioaccumulate in living organisms and lead to long-term health risks. This is because they are relatively stable and resistant to both biotic and/or abiotic degradation processes (Vagı & Petsas, 2017). When it bioaccumulate to certain extent in the body, it can cause endocrine and thyroid malfunctions (Cara & Jitereanu, 2015). For instance, organophosphates and carbamates pesticides disrupt the nervous system and cause weakness or paralysis (Zolgharnein et al., 2011).

Moreover, overuse of antibiotics and synthetic hormones in concentrated animal feeding operations correlates with the emergence of various health diseases in exposed individuals and populations. It has risen significant global public health concern of antimicrobial resistance in bacteria that can cause diseases in human (Hu et al., 2017). The presence of antimicrobial residues in milk and dairy products may cause allergic reactions (Ronquillo & Hernandez, 2017). Most of the growth additives used in animals such as estradiol, estrone and ethinylestradiol are carcinogenic and have been found in aquatic ecosystems at ng/L levels (Adeel et al., 2017). Besides, the application of feed additive or growth-promoting agent (Olaquindox) for livestock is also highly phototoxic, genotoxic, mutagenic and carcinogenic (Miao et al., 2018). Whilst, tetracycline is another well-known antibiotic that is frequently used in cattle, pigs and horses to tackle respiratory diseases (De Briyne et al., 2014). The utilization of these growth promoters on animals should be monitored carefully in ensuring the sustainability of livestock production in agriculture.

Considering the serious threat posed by agricultural runoff to the whole ecosystem, it is aimed to present a comprehensive account of the current technologies available for degradation and treatment of agricultural runoff and provide further insight into future research directions that may of interest for researchers.

ALTERNATIVE TECHNOLOGIES FOR AGRICULTURAL RUNOFF TREATMENT

Agricultural runoff can be possibly controlled through source, transport and discharge management as illustrated in Figure 1. Source management includes proper timing and rate of N fertilizer application, proper livestock manure storage and proper drainage design. In terms of transport management of agricultural runoff, it can be managed using cover crops, vegetated strips and constructed wetlands which minimize the nutrient leaching from the agricultural fields (Vymazal & Březinová, 2015). Lastly, the discharge of agricultural wastewater can be treated through various physicochemical or biological processes (Kortbein & Rajendran, 2016; Miles et al., 2016).

Figure 1. The treatment methods for agricultural runoff in terms of source, transport and discharge management

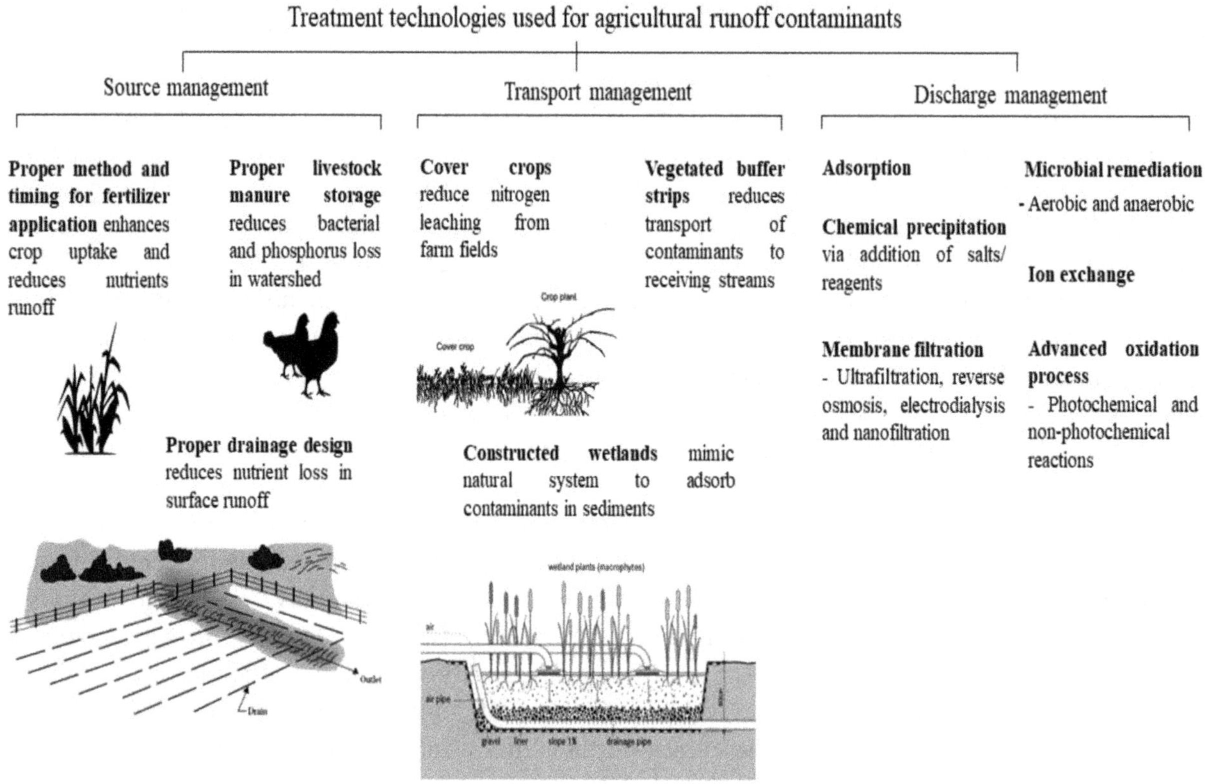

WASTE MANAGEMENT

To reduce the drainage of agrochemicals and bacterial pollutants in agriculture, proper source handling is crucial to reduce the agricultural waste leached out to the water bodies.

For decades, many attempts have been focused on effective N fertilizer management to minimize the soil nitrate leaching in agricultural runoff. The timing and rate of N fertilization greatly influences the yield, nitrogen use efficiency and N losses. It was found that under irrigated conventional tillage, split application of N fertilizer at planting (33% as urea) with 180 kg N/ha and one month after planting (67% as anhydrous ammonia) significantly minimized N_2O and NO fluxes without degradation on wheat yield and quality as compared with traditional farmers' practice with pre-plant application (75%) and 6 weeks after planting (25%) with 250 kg N/ha (Grahmann et al., 2013). Meanwhile, it was also identified that N losses in the form of NO_3-N leaching were 36% lower when N fertilizer was applied during spring season compared with fall season and NO_3-N leaching elevated with increasing N application rate (Grahmann et al., 2013). This may be due to high N demand of plant in spring but minimum plant uptake in fall. Crop residues are also used to enhance the soil moisture which leads to better N fertilizer distribution in soil and N uptake by the plants, thus reducing N losses in agricultural runoff. This approach was found to decrease herbicide runoff up to 60% (Patty et al., 1997). However, it was ineffective in reducing water soluble nutrients (Barisas et al., 1978). In short, it is recommended to apply N fertilizer during peak

N demand by the plants and delay pre-plant N application to achieve optimum N use efficiency which could minimize the environmental damage due to agricultural runoff.

Proper livestock waste management is also important in managing the runoff of bacterial and nutrient contamination from animal waste. These livestock wastes consist of very high levels of nitrogen, phosphorus and other toxic compounds which lead to detrimental environmental threats via nutrient losses from agricultural fields to the environment. The most common practices for livestock waste disposal are utilizing the animal waste as manure in farmland and direct dumping in the bushes or garbage sites (Abiola & Olaogun, 2016). Consequently, these pose serious environmental and health issues, promote insect vectors, nuisance problems and water pollution. To effectively control the produced odor and bacterial runoff to the water systems, it is suggested to incorporate animal wastes into the soil and establish long-term manure storage facilities (Jamieson et al., 2002). Liu et al. (2017) performed modelling to predict the effects of livestock waste management practices on the bacteria concentration and phosphorus loss in agricultural runoff. They noted that long-term manure storage was the most appropriate practice for reducing bacteria densities and phosphorus loss in the watershed. Indeed, further research is needed to deliver an environmental-friendly livestock waste management practice.

Further, drainage water management is another measure which utilizes flatting landscapes (0.5 or 1% slope) to reduce nutrient runoff from the farms. This system helps to improve surface water quality as it minimizes surface runoff, sediment losses and movement of heavy metals and organic contaminants attached to the sediments into the surface water (Skaggs et al., 2012). It was claimed to be able to reduce nitrogen loss (mainly NO_3-N) to surface water, ranging from 18% to more than 75%, depending on drainage system design and soil condition (Skaggs et al., 2012).

LAND CONSERVATION TECHNIQUES

During fallow periods in cropping system, cover crops can be utilized to fix and take up N to reduce N and organic matter losses via washed off from farm field (Justes et al., 2012). They play important roles in minimizing the runoff and interrill erosion that resulted from sediment detachment and transport by the rainfall. This is because the presence of cover crops prevents surface sealing, enhance storage capacity and improve soil structure which slow down the flow at the surface. Besides, the roots of cover crops attach the aggregates together which create higher resistance to detachment by flowing water. Meisinger et al. (1991) reported that cover crops reduced N leaching by 20 to 80% as compared with no cover crop control. Another study on the effect of cover crops on P losses also revealed that cover crops managed to reduce the total P losses (mostly particulate P) from agricultural system by around 54-94% (Kaspar & Singer, 2011).

Moreover, vegetated buffer strips (VBS) are also one of the agricultural practices in decreasing the transport of non-point source pollutants such as herbicides to surface water as illustrated in Figure 2 (Allaire et al., 2015; Dillaha et al., 1989). Grasses or other permanent vegetation was typically utilized to trap the sediment and water infiltration. The primary processes of VBS in mitigating transport of contaminants involve: (a) enhancing infiltration and improving solute-soil contact; (b) reducing surface water runoff velocity and promoting the precipitation of sediment-bound compounds; (c) increasing the diversity of soil microbial communities and speeding up contaminant degradation; (d) enhancing pollutants adsorption to vegetation and soil (Lerch et al., 2017). Dillaha et al. (1989) studied that 9.1 m of vegetated buffer strips could effectively remove incoming suspended solids, nitrogen (N) and

Figure 2. Schematic diagram of vegetated buffer strip mechanism

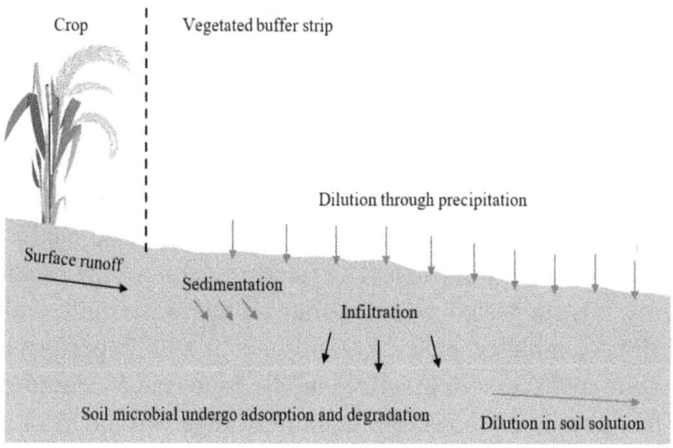

phosphorus (P) at around 84%, 73% and 79%, respectively. Besides, vegetated buffer strips could also effectively reduce the transport of herbicides such as 2,4 D, trifluralin and atrazine to receiving streams (Arora et al., 2003).

CONSTRUCTED WETLANDS

Constructed wetlands are one of the popular technologies for treating non-point sources of pollution such as agricultural runoff. They are engineered systems which are designed and constructed to mimic natural wetland systems to improve water quality (Wang et al., 2018). This system utilizes wetland plants, soil and associated microorganisms to remove the pollutants from wastewater (Kivaisi, 2001). Constructed wetlands possess the advantages of operating in a wide range of hydraulic loads and removing a number of contaminants at low capital and operational cost (Dierberg et al., 2002). They are relatively cheap and low-maintenance selection for agricultural applications.

Constructed wetlands have been utilized in various agricultural applications, including potato farm wastewater, milkhouse wash water, winery wastewater, farmyard runoff, aquaculture runoff and so on (Rozema et al., 2016; Wang et al., 2018). The removal rates in the wetland systems are highly dependent on organic loading, hydraulic retention time (HRT), pH, soil substrate, presence and type of vegetation (Wang et al., 2018). Different parameters such as five-day biochemical oxygen demand (BOD_5), total suspended solids (TSS), fecal coliforms, E-coli, ammonia+ammonium-N (NH_3+NH_4^+-N) and total Kjeldahl nitrogen (TKN) are used for evaluation of treatment performance. Further, constructed wetlands are also utilized for removal of other refractory pollutants, including pesticides, heavy metals, phosphorus, etc.

Ingersoll and Baker (1998) investigated the nitrate removal efficiency in a laboratory-scale constructed wetland with varying hydraulic-loading rates (5-20 cm/day) and carbon addition (1-6 g/week). They reported 8% up to over 95% nitrate removal could be achieved with decreasing hydraulic-loading rates and increasing carbon addition rates. Budd et al. (2009) evaluated the performance of constructed wetlands for removing pyrethroid and organophospate insecticides from agricultural runoff. They found that constructed wetlands were highly effective in reduction of pyrethroid up to 94% probably through

Table 3. The use of constructed wetlands for pollutants removal from agricultural runoff

Pollutants	Plants	Sediment/Filter Material	Efficiency (%)	References
Nitrate	Dried cattail residue (carbon source)	Sediment from littoral wetland	8-95	(Ingersoll & Baker, 1998)
Pyrethroid insecticide	Knotgrass (*Paspalum distichum*), pale smartweed (*Polygonum lapathifolium*) and barnyardgrass (*Echinochloa crus-galli*)	Not specified	52-94	(Budd et al., 2009)
Cd, Cu, Pb, Zn and Al heavy metals	*Cyperus alternifolius*	Gravel and sand	100	(Cheng et al., 2002)
BOD_5, phosphorus and nitrogen	Macrophytes (*Phragmites australis, Typha latifolia, Carex ripraria, Iris pseudacorus* and *Sparganium erectum*)	Clay base with topsoil (0.2 m)	93-99	(Forbes et al., 2011)
MCPA herbicide	*Phragmites australis*	Light expanded clay aggregates (LECA)	56-97	(Dordio & Carvalho, 2013)
λ-cyhalothrin and imidacloprid	*Nymphaea amazonum* and *Eleocharis mutata*	Sandy loam sediment and potting soil	72-100	(Mahabali & Spanoghe, 2014)

adsorbing pesticides on sediments. However, they were ineffective in removing diazinon, likely due to its limited sorption capability to sediment particles. In another study, constructed wetlands were also investigated for removing heavy metals which are aluminium (Al), Cd, Cu, Pb and Zn (Cheng et al., 2002). The results showed a complete removal of all heavy metals in the constructed wetlands using *Cyperus alternifolius* plant species (Cheng et al., 2002). On the other hand, Forbes et al. (2011) found that constructed wetland is effective in treating dairy farm wastewater in which 99%, 95% and 93% for BOD_5, phosphorus and nitrogen, respectively can be achieved. Table 3 shows the constructed wetlands for pollutants removal from agricultural runoff.

The challenges for constructed wetlands treatments are cold climate conditions and phosphorus treatment. Cold temperature can limit the treatment rate of constructed wetlands as it restrains the biological and biochemical processes of plants and microbial activities (Wang et al., 2017). It can be mitigated by certain design configuration of constructed wetlands. The most typical constructed wetland designs include surface flow (SF), horizontal subsurface flow (H-SSF) and vertical subsurface flow (V-SSF). Studies shown that SSF are more suitable to Canadian climatic situations due to the capability to insulate microbial communities from cold winter air temperatures (Rozema et al., 2016). With regards to phosphorus removal, it is highly dependent on soil phosphorus adsorption capacity as its capacity would decrease over time. Post-wetland treatment filters are normally adopted to enhance the phosphorus management (Weber et al., 2007). Nonetheless, different designs demonstrated different capability in terms of contaminants removal from agricultural wastewater, thus, it would be advantageous to integrate hybrid designs for optimization on the treatment performance. Further research on design consideration of constructed wetlands and distinctive adsorptive materials would be promising to improve the treatment system significantly.

BIOLOGICAL PROCESS

In terms of discharge management, agricultural runoff constitutes high BOD and ammonia which could be possibly degraded by microbial approach, including aerobic and anaerobic treatment. Biological treatment is an environmentally friendly process that employs microorganisms (bacteria and fungi) for detoxification of harmful contaminants from soil, water and air (Mitra & Mukhopadhyay, 2016). The microorganisms encompass high ability to persist in diverse environments to be used for degradation of a wide range of pollutants such as nutrients, xenobiotic compounds, volatile organic compounds, pesticides and heavy metals (Gaur et al., 2014).

Microorganisms can degrade pollutants in aerobic and anaerobic conditions. In aerobic degradation, microorganisms use oxygen as final electron acceptor for converting organic and inorganic pollutants into carbon dioxide and water. In anaerobic degradation, microorganisms use nitrate, iron, sulphate, etc. as electron acceptor to break down the pollutants into nitrogen, carbon dioxide and methane (Beyenal et al., 2004).

The principal of biological nitrogen removal is based on the aerobically transformation of ammonium into nitrate and subsequent anoxic conversion of nitrate into nitrogen gas through nitrification and denitrification processes, respectively (Halling-Sørensen & Jorgensen, 1993). The biological process is mostly the preferred approach for nitrogen treatment in large scale since it has low operating cost and high efficiency (70-90%) (Schoeman & Steyn, 2003). Ergas et al. (2010) evaluated pilot-scale bioretention systems for removing nutrients, organics, and solids in agricultural runoff. The systems consisted of aerobic and anoxic zones that used sulfur or wood chips as denitrification substrates. Moderate removal efficiencies were reported for suspended solids (69%), chemical oxygen demand (46%), total phosphorous (66%), and total N (65%) during first season of field tests.

Although most of the toxic organic substances are resistant to biological treatment which yields only partial elimination, biological process is still highly feasible in treating wide range of pesticides (Cara & Jitereanu, 2015). Shawaqfeh (2010) investigated the biodegradability of triadimenol pesticide under both aerobic and anaerobic conditions. The results showed that combination of anaerobic and aerobic biological processes produced higher efficiency as compared to individual units. In their studies, more than 96% pesticides removal were obtained after 172 days in aerobic and 230 days in anaerobic conditions. Hai et al. (2012) evaluated the ability of a mixed culture of bacteria and fungus for removing three recalcitrant pesticides from their liquid mixture. The mixed fungus–bacteria culture removed 47%, 98% and 62% of aldicarb, atrazine and alachlor, respectively, from the liquid phase during 14-day period.

Moreover, microbes are also effective in heavy metal removal (Dhokpande & Kaware, 2013). Ilhan et al. (2004) found that *Staphylococcus saprophyticus* bacteria were suitable for biosorption of lead and chromium from wastewater at 100 mgL^{-1} and 88.66 mgL^{-1}, respectively, in acidic conditions.

CHEMICAL PRECIPITATION

Chemical precipitation has been commonly employed to reduce pollutants from aquatic systems (Kim et al., 2007). It is based on the transition of a chemical from a dissolved form to a non-dissolved form by addition of reagents which lead to the precipitation formation (Halling-Sørensen & Jorgensen, 1993). For instance, ammonium, which is readily dissolved in water can be precipitated through addition of magnesium and phosphate ions to form struvite (magnesium ammonium phosphate) ($MgNH_4PO_4.6H_2O$)

that is hardly soluble in water. Struvite can be easily separated from water in the form of white crystalline and reused as fertilizer (Capodaglio et al., 2015).

Precipitation can be also used for removing heavy metal ions from aqueous solutions due to its simplicity and low cost. In this regard, chemicals will react with heavy metals to form insoluble precipitates which can be separated from the solution by sedimentation or filtration. Mirbagheri and Hosseini (2005) showed that maximum 99% Cr (III) removal can be attained with the addition of $Ca(OH)_2$ at pH 8.7. Despite its effectiveness and ease of operation, chemical precipitation may generate large volume of sludge and incur high disposal costs.

ION EXCHANGE PROCESSES

Ion exchange process can be used for removing nitrogen and heavy metals from aquatic systems by interchanging ammonium, nitrate or heavy metal ions with same ions on solid resins (Ding & Sartaj, 2016). Resins can be regenerated when all the exchange sites are replaced with ammonium or nitrate and reach saturation point (Prajapati, 2014). There are two types of solid resins, which are synthetic organic resins like silica powder and natural inorganic resins such as clinoptilolite and zeolite. Synthetic ion-exchange resins have higher capacity, longer life and faster exchange rate but they are very expensive. Natural zeolites are more competitive due to their low cost, simplicity and eco-friendly nature (Ding & Sartaj, 2016; Jorgensen & Weatherley, 2003). Ion exchange process is effective in removing nitrogen. However, it is not a very attractive method for treatment of wastewater in high concentrations as it requires high regeneration frequency which leads to huge maintenance costs. Besides, ion exchange is crucially dependent on the pH for removing ammonium. The range of pH for optimum exchange of ammonium by zeolites occurs within 4 to 8. Below this range, hydrogen ions compete with ammonium for the remaining sites and above this range, ammonium transfers to ammonia which is unable to bind to zeolites (Halling-Sørensen & Jorgensen, 1993).

Ion exchange offers several advantages for removing heavy metals from the wastewater. They have high removal efficiency and treatment capacity. The most commonly used cation exchangers for heavy metals removal are synthetic resins with sulfonic acid groups ($-SO_3H$) or carboxylic acid groups (-COOH). As the wastewater containing heavy metal compounds passes through the cations resins, metal ions are exchanged with the hydrogen ions or carboxylic group. Besides synthetic resins, natural resins are also widely used for removing heavy metal ions. Doula (2009) reported that clinoptilolite loaded with amorphous Fe-oxide could simultaneously remove Cu, Mn and Zn ions from the water sample.

ADVANCED OXIDATION PROCESS

Advanced oxidation process is widely used for treatment of effluents polluted by organic compounds (Cara & Jitereanu, 2015). Many researches have been conducted on the removal of pesticide residues in the wastewater using advanced oxidation process. During this process, a series of oxidation reactions occurs which lead to mineralization and transformation of pesticides compounds into harmless substances. Briefly, advanced oxidation processes can be classified into photochemical and non-photochemical reactions. Photochemical groups include UV treatment, UV/H_2O_2, UV/TiO_2, and photo-Fenton treatment.

On the other hand, non-photochemical processes are Fenton reaction (H_2O_2 and Fe^{2+}), ozonation (O_3), electrochemical oxidation, hydrodynamic/ultrasonic cavitation and sub/super critical water process.

Advanced oxidation process is considered as a costly technique since it requires more energy and reagents for a complete chemical degradation of toxic organic substances. Besides, the produced sludge requires high amounts of chemical reagents to decompose (Catalkaya & Kargi, 2009). Removal of simazine (one of the most commonly used pesticides) from aqueous solution was investigated using Fenton's reagent oxidation by Catalkaya and Kargi (2009). It shown that complete removal of simazine was achieved within 6 min reaction period.

MEMBRANE FILTRATION TECHNOLOGIES

Membrane filtration such as ultrafiltration, reverse osmosis, electrodialysis and nanofiltration have been considered as a promising method in the treatment of wastewater containing heavy metals, organic compounds and/or nutrients. Membrane technologies require low energy for operation and have high selectivity of separation. They are also known for their high removal efficiency and ease of operation (Fu & Wang, 2011). However, their removal efficiency is dependent on the polymer reagents. Generally, there are three groups of reagents, which are 1) natural origin such as macroalgae, chitosan and alginate, 2) laboratory-synthesized reagents such as ethoxylated polyethylenimine and 3) commercially available polymers such as polyacrylic acid (PAA) and polyethyleneimine. The first group of reagents is environmentally friendly, but their regeneration is more difficult due to their complicated composition of different functional groups. The laboratory polymers are usually synthesized for special application, but on a large scale, their application is limited due to the lack of information on their cost, toxicity, and other important characteristics. Commercial polymers such as PAA are more likely to be limited by membrane fouling and polarization since they have small number of functional groups per polymer molecule (Qiu & Mao, 2013).

Ultrafiltration process uses low transmembrane pressures for dissolved and colloidal material removal. However, it is unsuitable to remove dissolved metal ions due to large pore size of ultrafiltration membrane (Fu & Wang, 2011). During reverse osmosis, fluid is passed through a semi-permeable membrane to reject the contaminants and purify the water. Reverse osmosis is typically used for a wide range of dissolved pollutants. Tepuš et al. (2009) investigated the efficiency of nanofiltration and reverse osmosis for removing nitrate and pesticides (deethylatrazine, deisopropylatrazine and atrazine) from agricultural runoff. They reported that reverse osmosis membrane had high removal ability for all compounds but nanofiltration was only suitable for atrazine removal. Further, Qiu and Mao (2013) used ultrafiltration process for removing heavy metal ions (Cu^{2+}, Zn^{2+}, Ni^{2+} and Mn^{2+}) from an aqueous solution. Copolymer of maleic acid and acrylic acid (PMA-100) was chosen as the membrane reagents. Removal efficiency about 98.8% was achieved for all heavy metal ions with the order of relative affinity of $Cu^{2+}>Zn^{2+}>Ni^{2+}>Mn^{2+}$.

ADSORPTION PROCESS

Adsorption process is one of the well-known strategies used for removing heavy metals and organic compounds of pesticides. Adsorption is an adhesion phenomenon in which adsorbate molecules are held onto the surface of adsorbents through Van der Waals forces. This process is largely dependent on surface area and pore volume of absorbents (Acharya et al., 2009). Adsorption is considered as low-cost treatment option, particularly for the treatment of low concentrations of pollutants (Cara & Jitereanu, 2015). It has high flexibility in design and operation and a wide range of materials can be considered as adsorbents. Activated carbons, biochars, carbon nanotubes, agricultural wastes, industrial byproducts and wastes, natural substances, algal biomass and microbial biomass such as bacteria, fungi and yeast have been evaluated and utilized as adsorbents (Fu & Wang, 2011).

Among all materials, activated carbon is the most commonly used adsorbent. It has large surface area and microporous structure that make it as a potential adsorbent for wastewater treatment. Carbonaceous material such as dust coal, peat, lignite or agricultural wastes are commonly used as precursor for activated carbon preparation. Activated carbons prepared from agricultural wastes such as oil palm shell, coconuts shell, rice husk and nut shell yield a lower total cost of treatment (Sahu et al., 2009). Jusoh et al. (2011) indicated that activated carbon prepared by coconut shell and palm shells could be effectively used for malathion pesticide adsorption. In their study, coconut shell activated carbon had higher adsorption capacity for malathion (909.1 mg/g) as compared with palm shells activated carbon (555.6 mg/g). This is attributed to larger surface area and higher pore volume and pore radius size of coconut shell. Rojas et al. (2014) compared sunflower seed shells and rice husk for their adsorption potential to remove atrazine, alachlor, endosulfan sulfate from aqueous solution. The results showed that maximum removal efficiency (73.9%) was achieved using 1 g of rice husk and 200 μ/L of pesticide solution. Several studies have also shown the promising biosorption potential of fruit husk activated carbon, apricot stone activated carbon, rice straw activated carbon and wheat straw biochar in removing bentazone, ethoprophos, carbofuran and 4-Chloro-2-methylphenoxyacetic acid (MPCA), respectively (Ahmed & Mohammad, 2014; Chang et al., 2011; Njoku et al., 2014; Tatarková et al., 2013). Besides, activated carbons have also been used to remove heavy metal pollutants from wastewater. Kongsuwan et al. (2009) evaluated the use of eucalyptus bark activated carbon for adsorption of Cu^{+2} and Pb^{2+}. The maximum adsorption for copper ions (Cu^{2+}) and lead ions (Pb^{2+}) was reported to be 0.45 and 0.53 mmol/g, respectively. Meanwhile, Guo et al. (2010) declared that poultry litter-based activated carbon has significantly higher adsorption affinity and capacity for heavy metals than commercial activated carbon derived from bituminous coal and coconut shell.

Further, carbon nanotubes have been proven to be excellent adsorbents for removing heavy metals and organic pollutants from wastewaters because of their hollow and layered structures and large specific surface areas. They are classified into single-walled carbon nanotubes (SWCNTs) and multi-walled carbon nanotubes (MWCNTs) (Zaytseva & Neumann, 2016). Joseph et al. (2011) investigated the adsorption of 17α-ethinyl estradiol and bisphenol from water samples using SWCNTs and achieved a removal efficiency of 75-98% for these organic pollutants. SWCNTs have been demonstrated to have a high adsorption capacity for 4-chloro-2-methylphenoxyacetic acid (MCPA) herbicide (De Martino et al., 2012). In terms of MWCNTs, they were effective to remove Olaquindox and Cephaexin at 99.7% and 97.6%, respectively (Jafari & Aghamiri, 2011; Zhang et al., 2011). They were also used for removal two herbicides (diuron and dichlobenil) from contaminated water (Chen et al., 2011). Moreover, 90% of Pb^{2+} removal from aqueous solution was attained using MWCNTs as adsorbent (Gupta et al., 2011).

Besides activated carbons and nano-carbons, the use of low-cost and natural adsorbents such as kaolinite, montmorillonite, natural zeolites, lignite, peat and agricultural wastes have also been reported (Fu & Wang, 2011). Kaolinite adsorbents have shown a removal rate of 95% for Pb^{2+} (Jiang et al., 2010). In other study, brine sediments, sawdust and their mixtures were evaluated as adsorbents for heavy metals of Zn and Cu. The results showed a maximum adsorption of 5.59 mg/g for Zn and 4.69 mg/g for Cu using the adsorbent mixture and brine sediments, respectively (Agouborde & Navia, 2009).

Biosorbents are another group of sorbents that uses either live or dead microorganisms for pollutant removal. They are often used for low-cost and rapid adsorption process. Microorganisms such as bacteria, fungi, algae and yeast are capable to accumulate heavy metals (Yan & Viraraghavan, 2003). They would produce extracellular polymeric substances (EPS) such as glucoprotein, polysaccharides and soluble peptides, which encompass anionic functional groups to bind with the heavy metal cations (Kikuchi & Tanaka, 2012). Biosorption is highly affected by pH of the medium. Yan and Viraraghavan (2003) investigated the adsorption capacity of live and dead *Mucor rouxii* biomass fungus for removing lead, cadmium, nickel and zinc over a range of pH. The results showed that live biomass attained higher biosorption capacity at pH 5.0 for Pb^{2+}, Ni^{2+}, Cd^{2+} and Zn^{2+} at 35.69, 11.09, 8.46 and 7.75 mg/g, respectively, as compared with dead biomass. The dead biomass was more effective at pH 6.0 and achieved 53.75, 20.49, 20.31 and 53.85 mg/g adsorption for Pb^{2+}, Ni^{2+}, Cd^{2+} and Zn^{2+}, respectively. Biosorption of $Cu^{2+}$and Zn^{2+} using dried marine green macroalga *Caulerpa linum* was studied by Ajjabi and Chouba (2009). The dried alga produced maximum uptake of 1.46 and 1.97 mmol/g for Cu^{2+} and Zn^{2+}, respectively. Bhainsa and D'souza (2008) claimed that maximum Cu^{2+} adsorption capacity using *Rhizopus oryzae* fungus biomass was achieved at 19.4 mg/g. However, the separation of biosorbents is difficult after adsorption.

Sorption efficiency is highly affected by factors such as pH, temperature, sorbent dosage, contact time, sorbent and sorbate types (Zolgharnein et al., 2011). pH of the medium has a profound effect on sorption efficiency of both organic and inorganic pollutants. In a study, the sorption capacity of bagasse fly ash was investigated for removing dichlorodiphenyldrichloroethane (DDD) and dichlorodiphenyldichloroethylene (DDE) at pH 2-9 (Gupta & Ali, 2001). The results showed that increasing pH from 2 to 7 increased sorption efficiencies from 0.21 to 3.6 μg/g for DDD and from 0.11 to 3.42 μg/g for DDE, after which the efficiency showed a decreasing trend. They concluded that pH 7 was optimum pH for both pesticides. In another study, Jianlong et al. (2000) evaluated the adsorption capacity of activated sludge biomass for pentachlorophenol (PCP) removal at pH 6 to 8. It shown that increasing pH has negative effect on PCP sorption. When the pH was increased from 6 to 8, PCP sorption capacity dropped from 3040μg/L to 2010 μg/L. This was because of more positive surface charge of the sorbent at pH 6 to react with negatively charged PCP which subsequently improve the sorption efficiency.

Temperature also poses effect on sorption capacity in which increasing temperature (30-50 °C) reduced the removal rate of pollutants such as DDD and DDE by bagasse fly ash as studied by Gupta and Ali (2001). El Bakouri et al. (2008) also agreed that higher temperature decreased the sorption capacity of organic sorbents in treating endosulfan sulphate pesticide. At higher temperature, the solubility of endosulfan sulphate increased which minimized its affinity on the surface of the sorbents.

Another important parameter that affects the adsorption efficiency is sorbent dosage. Sorbent dosage identifies the capacity of a sorbent for a given initial concentration of sorbate (Zolgharnein et al., 2011). As studied by Memon et al. (2007), when sorbent dosage was increased from 0.1 to 0.4 g, the sorption capacity was also increased up to 90%. This was due to increase of more available surface areas via functional groups for sorption. However, the percentage remained constant with increasing amount of sorbent up to 1 g. This was attributed to aggregation effect of the sorbent particles at higher dosages, which minimizes the total surface area and prolongs the diffusion path (Rajamohan et al., 2013).

CONCLUSION

In view of increasing global demand on food consumption, agricultural production has progressively shifted to larger scale operations around the globe. This raises public health concern associated with agricultural runoff which carries large amounts of contaminants such as bacteria, nutrients (nitrogen, phosphorus), pesticides and sediment. As a result, high rate of nutrient loadings from agricultural land not only lead to algal bloom and pollution in waterways but also impose health risks to fish, livestock and humans. Hence, the management and treatment of agricultural runoff is becoming increasingly important to sustain the river health, improve the water quality and conserve the whole ecosystem. The integration of waste management and land conservation approach can help to preserve the sustainability in agricultural production while minimizing the nutrients in the agricultural runoff. Meanwhile, constructed wetlands and biological process have demonstrated their environmentally safe benign features for pollutants mitigation from non-point source agricultural runoff. The reduction of chemical pollutants can be also performed via chemical precipitation as one of the simplest yet viable process. To facilitate the removal of complicated chemical compounds, ion exchange, adsorption, advanced oxidation process and membrane filtration are also adopted to meet the stringent environmental standards. Amongst, constructed wetlands demonstrate a reliable and cost-effective agricultural wastewater treatment technology that offer wide range of contaminants removal, depending on the configuration of constructed wetlands. Additionally, they encompass the potential to provide significant wildlife habitat which ensure the environmental sustainability. While many of these practices discovered the potential in treating agricultural runoff, much works is still required for field work evaluation to determine the applicability for pollutants removal as to be compatible with local farming practices and economic viability.

ACKNOWLEDGMENT

The authors would like to express their sincere gratitude to UTP for funding this project under YUTP grant (0153AA-E34) and STIRF grant (0153AA-D77).

REFERENCES

Abiola, J., & Olaogun, S. (2016). Livestock waste management practices in Oyo state, Nigeria. *Journal of Environment and Waste Management*, *3*(2), 139–141.

Acharya, J., Sahu, J., Sahoo, B., Mohanty, C., & Meikap, B. (2009). Removal of chromium (VI) from wastewater by activated carbon developed from Tamarind wood activated with zinc chloride. *Chemical Engineering Journal*, *150*(1), 25–39. doi:10.1016/j.cej.2008.11.035

Adeel, M., Song, X., Wang, Y., Francis, D., & Yang, Y. (2017). Environmental impact of estrogens on human, animal and plant life: A critical review. *Environment International*, *99*, 107–119. doi:10.1016/j.envint.2016.12.010 PMID:28040262

Agouborde, L., & Navia, R. (2009). Heavy metals retention capacity of a non-conventional sorbent developed from a mixture of industrial and agricultural wastes. *Journal of Hazardous Materials*, *167*(1-3), 536–544. doi:10.1016/j.jhazmat.2009.01.027 PMID:19188023

Ahmed, S. M., & Mohammad, S. G. (2014). Egyptian apricot stone (Prunus armeniaca) as a low cost and eco-friendly biosorbent for oxamyl removal from aqueous solutions. *American Journal of Experimental Agriculture*, *4*(3), 302–321. doi:10.9734/AJEA/2014/7116

Ajjabi, L. C., & Chouba, L. (2009). Biosorption of Cu2+ and Zn2+ from aqueous solutions by dried marine green macroalga Chaetomorpha linum. *Journal of Environmental Management*, *90*(11), 3485–3489. doi:10.1016/j.jenvman.2009.06.001 PMID:19576679

Allaire, S. E., Sylvain, C., Lange, S. F., Thériault, G., & Lafrance, P. (2015). Potential efficiency of riparian vegetated buffer strips in intercepting soluble compounds in the presence of subsurface preferential flows. *PLoS One*, *10*(7), e0131840. doi:10.1371/journal.pone.0131840 PMID:26147093

Arora, K., Mickelson, S. K., & Baker, J. L. (2003). Effectiveness of vegetated buffer strips in reducing pesticide transport in simulated runoff. *Transactions of the ASAE. American Society of Agricultural Engineers*, *46*(3), 635. doi:10.13031/2013.13599

Barisas, S., Baker, J., Johnson, H., & Laflen, J. (1978). Effect of tillage systems on runoff losses of nutrients, a rainfall simulation study. *Transactions of the ASAE. American Society of Agricultural Engineers*, *21*(5), 893–0897. doi:10.13031/2013.35409

Beyenal, H., Sani, R. K., Peyton, B. M., Dohnalkova, A. C., Amonette, J. E., & Lewandowski, Z. (2004). Uranium immobilization by sulfate-reducing biofilms. *Environmental Science & Technology*, *38*(7), 2067–2074. doi:10.1021/es0348703 PMID:15112808

Bhainsa, K. C., & D'souza, S. (2008). Removal of copper ions by the filamentous fungus, Rhizopus oryzae from aqueous solution. *Bioresource Technology*, *99*(9), 3829–3835. doi:10.1016/j.biortech.2007.07.032 PMID:17804218

Bhandari, G. (2014). An overview of agrochemicals and their effects on environment in Nepal. *Applied Ecology and Environmental Sciences*, *2*(2), 66–73. doi:10.12691/aees-2-2-5

Blanco-Canqui, H., Gantzer, C., Anderson, S., & Alberts, E. (2004). Division S-6—soil & water management & conservation. *Soil Science Society of America Journal*, *68*, 567–576. doi:10.2136ssaj2004.5670

Borba, C., Guirardello, R., Silva, E., Veit, M., & Tavares, C. (2006). Removal of nickel (II) ions from aqueous solution by biosorption in a fixed bed column: Experimental and theoretical breakthrough curves. *Biochemical Engineering Journal*, *30*(2), 184–191. doi:10.1016/j.bej.2006.04.001

Budd, R., O'Geen, A., Goh, K. S., Bondarenko, S., & Gan, J. (2009). Efficacy of constructed wetlands in pesticide removal from tailwaters in the Central Valley, California. *Environmental Science & Technology*, *43*(8), 2925–2930. doi:10.1021/es802958q PMID:19475972

Capodaglio, A. G., Hlavínek, P., & Raboni, M. (2015). Physico-chemical technologies for nitrogen removal from wastewaters: A review. *Revista Ambiente & Água*, *10*(3), 481–498.

Cara, I. G., & Jitereanu, G. (2015). Application of low-cost adsorbents for pesticide removal. *Bulletin of University of Agricultural Sciences and Veterinary Medicine Cluj-Napoca. Veterinary Medicine*, *72*(1).

Carter, A. (2000). Herbicide movement in soils: Principles, pathways and processes. *Weed Research (Oxford)*, *40*(1), 113–122. doi:10.1046/j.1365-3180.2000.00157.x

Catalkaya, E. C., & Kargi, F. (2009). Advanced oxidation and mineralization of simazine using Fenton's reagent. *Journal of Hazardous Materials*, *168*(2-3), 688–694. doi:10.1016/j.jhazmat.2009.02.091 PMID:19297085

Chang, K.-L., Lin, J.-H., & Chen, S.-T. (2011). Adsorption studies on the removal of pesticides (Carbofuran) using activated carbon from rice straw agricultural waste. *World Acad Sci Eng Technol*, *5*, 04-21.

Chen, G.-C., Shan, X.-Q., Pei, Z.-G., Wang, H., Zheng, L.-R., Zhang, J., & Xie, Y.-N. (2011). Adsorption of diuron and dichlobenil on multiwalled carbon nanotubes as affected by lead. *Journal of Hazardous Materials*, *188*(1-3), 156–163. doi:10.1016/j.jhazmat.2011.01.095 PMID:21324587

Cheng, S., Grosse, W., Karrenbrock, F., & Thoennessen, M. (2002). Efficiency of constructed wetlands in decontamination of water polluted by heavy metals. *Ecological Engineering*, *18*(3), 317–325. doi:10.1016/S0925-8574(01)00091-X

De Briyne, N., Atkinson, J., Pokludová, L., & Borriello, S. (2014). Antibiotics used most commonly to treat animals in Europe. *The Veterinary Record*, *175*(13), 325. doi:10.1136/vr.102462 PMID:24899065

De Martino, A., Iorio, M., Xing, B., & Capasso, R. (2012). Removal of 4-chloro-2-methylphenoxyacetic acid from water by sorption on carbon nanotubes and metal oxide nanoparticles. *RSC Advances*, *2*(13), 5693–5700. doi:10.1039/c2ra00038e

Dhokpande, S. R., & Kaware, J. P. (2013). Biological methods for heavy metal removal-A review. *International Journal of Engineering Science and Innovative Technology*, *2*(5), 304–309.

Dierberg, F., DeBusk, T., Jackson, S., Chimney, M., & Pietro, K. (2002). Submerged aquatic vegetation-based treatment wetlands for removing phosphorus from agricultural runoff: Response to hydraulic and nutrient loading. *Water Research*, *36*(6), 1409–1422. doi:10.1016/S0043-1354(01)00354-2 PMID:11996331

Dillaha, T. A., Reneau, R., Mostaghimi, S., & Lee, D. (1989). Vegetative filter strips for agricultural nonpoint source pollution control. *Transactions of the ASAE. American Society of Agricultural Engineers*, *32*(2), 513–0519. doi:10.13031/2013.31033

Ding, Y., & Sartaj, M. (2016). Optimization of ammonia removal by ion-exchange resin using response surface methodology. *International Journal of Environmental Science and Technology*, *13*(4), 985–994. doi:10.100713762-016-0939-x

Dordio, A., & Carvalho, A. J. P. (2013). Constructed wetlands with light expanded clay aggregates for agricultural wastewater treatment. *The Science of the Total Environment*, *463*, 454–461. doi:10.1016/j.scitotenv.2013.06.052 PMID:23831791

Doula, M. K. (2009). Simultaneous removal of Cu, Mn, and Zn from drinking water with the use of clinoptilolite and its Fe-modified form. *Water Research, 43*(15), 3659–3672. doi:10.1016/j.watres.2009.05.037 PMID:19576609

Edgerton, M. D. (2009). Increasing crop productivity to meet global needs for feed, food, and fuel. *Plant Physiology, 149*(1), 7–13. doi:10.1104/pp.108.130195 PMID:19126690

El Bakouri, H., Morillo, J., Usero, J., & Ouassini, A. (2008). Potential use of organic waste substances as an ecological technique to reduce pesticide ground water contamination. *Journal of Hydrology (Amsterdam), 353*(3-4), 335–342. doi:10.1016/j.jhydrol.2008.02.019

El Gohary, R. (2015). Agriculture, industry, and wastewater in the Nile Delta. *Int. J. Sci. Res. Agric. Sci, 22*, 159–172.

Ergas, S. J., Sengupta, S., Siegel, R., Pandit, A., Yao, Y., & Yuan, X. (2010). Performance of nitrogen-removing bioretention systems for control of agricultural runoff. *Journal of Environmental Engineering, 136*(10), 1105–1112. doi:10.1061/(ASCE)EE.1943-7870.0000243

Forbes, E., Foy, R., Mulholland, M., & Brettell, J. (2011). Performance of a constructed wetland for treating farm-yard dirty water. *Water Science and Technology, 64*(1), 22–28. doi:10.2166/wst.2011.584 PMID:22053453

Fu, F., & Wang, Q. (2011). Removal of heavy metal ions from wastewaters: A review. *Journal of Environmental Management, 92*(3), 407–418. doi:10.1016/j.jenvman.2010.11.011 PMID:21138785

Gaur, N., Flora, G., Yadav, M., & Tiwari, A. (2014). A review with recent advancements on bioremediation-based abolition of heavy metals. *Environmental Science. Processes & Impacts, 16*(2), 180–193. doi:10.1039/C3EM00491K PMID:24362580

Grahmann, K., Verhulst, N., Buerkert, A., Ortiz-Monasterio, I., & Govaerts, B. (2013). Nitrogen use efficiency and optimization of nitrogen fertilization in conservation agriculture. *Perspectives in Agriculture, Veterinary Science, Nutrition and Natural Resources, 8*(053), 1–19. doi:10.1079/PAVSNNR20138053

Guo, M., Qiu, G., & Song, W. (2010). Poultry litter-based activated carbon for removing heavy metal ions in water. *Waste Management (New York, N.Y.), 30*(2), 308–315. doi:10.1016/j.wasman.2009.08.010 PMID:19783133

Gupta, V. K., Agarwal, S., & Saleh, T. A. (2011). Synthesis and characterization of alumina-coated carbon nanotubes and their application for lead removal. *Journal of Hazardous Materials, 185*(1), 17–23. doi:10.1016/j.jhazmat.2010.08.053 PMID:20888691

Gupta, V. K., & Ali, I. (2001). Removal of DDD and DDE from wastewater using bagasse fly ash, a sugar industry waste. *Water Research, 35*(1), 33–40. doi:10.1016/S0043-1354(00)00232-3 PMID:11257890

Hai, F. I., Modin, O., Yamamoto, K., Fukushi, K., Nakajima, F., & Nghiem, L. D. (2012). Pesticide removal by a mixed culture of bacteria and white-rot fungi. *Journal of the Taiwan Institute of Chemical Engineers, 43*(3), 459–462. doi:10.1016/j.jtice.2011.11.002

Halling-Sørensen, B., & Jorgensen, S. E. (1993). *The removal of nitrogen compounds from wastewater* (Vol. 54). Elsevier.

Hariprasad, N., & Dayananda, H. (2013). Environmental impact due to agricultural runoff containing heavy metals—a review. *International Journal of Scientific and Research Publications*, *3*(5), 224–280.

Hasan, H. A., Abdullah, S. R. S., Kamarudin, S. K., & Kofli, N. T. (2011). Problems of ammonia and manganese in Malaysian drinking water treatments. *World Applied Sciences Journal*, *12*(10), 1890–1896.

Hu, Y., Cheng, H., & Tao, S. (2017). Environmental and human health challenges of industrial livestock and poultry farming in China and their mitigation. *Environment International*, *107*, 111–130. doi:10.1016/j.envint.2017.07.003 PMID:28719840

Ilhan, S., Nourbakhsh, M. N., Kiliçarslan, S., & Ozdag, H. (2004). Removal of chromium, lead and copper ions from industrial waste waters by Staphylococcus saprophyticus. *Turkish Electronic Journal of Biotechnology*, *2*(2), 50–57.

Ingersoll, T. L., & Baker, L. A. (1998). Nitrate removal in wetland microcosms. *Water Research*, *32*(3), 677–684. doi:10.1016/S0043-1354(97)00254-6

Israel Ministry of Environmental Protection. (2018). *Agricultural Wastewater*. Retrieved from http://www.sviva.gov.il/English/env_topics/agroecology/Pages/Agricultural-Wastewater.aspx#GovXParagraphTitle1

Jafari, M., & Aghamiri, S. (2011). Evaluation of carbon nanotubes as solid-phase extraction sorbent for the removal of cephalexin from aqueous solution. *Desalination and Water Treatment*, *28*(1-3), 55–58. doi:10.5004/dwt.2011.2200

Jamieson, R., Gordon, R., Sharples, K., Stratton, G., & Madani, A. (2002). Movement and persistence of fecal bacteria in agricultural soils and subsurface drainage water: A review. *Canadian Biosystems Engineering*, *44*(1), 1–9.

Janni, K. A., Christopherson, S. H., & Schmidt, D. R. (2009). Milk house wastewater flows and characteristics for small dairy operations. *Applied Engineering in Agriculture*, *25*(3), 417–423. doi:10.13031/2013.26893

Jiang, M., Jin, X.-Y., Lu, X.-Q., & Chen, Z. (2010). Adsorption of Pb (II), Cd (II), Ni (II) and Cu (II) onto natural kaolinite clay. *Desalination*, *252*(1-3), 33–39. doi:10.1016/j.desal.2009.11.005

Jianlong, W., Yi, Q., Horan, N., & Stentiford, E. (2000). Bioadsorption of pentachlorophenol (PCP) from aqueous solution by activated sludge biomass. *Bioresource Technology*, *75*(2), 157–161. doi:10.1016/S0960-8524(00)00041-9

Jorgensen, T., & Weatherley, L. (2003). Ammonia removal from wastewater by ion exchange in the presence of organic contaminants. *Water Research*, *37*(8), 1723–1728. doi:10.1016/S0043-1354(02)00571-7 PMID:12697216

Joseph, L., Heo, J., Park, Y.-G., Flora, J. R., & Yoon, Y. (2011). Adsorption of bisphenol A and 17α-ethinyl estradiol on single walled carbon nanotubes from seawater and brackish water. *Desalination*, *281*, 68–74. doi:10.1016/j.desal.2011.07.044

Jusoh, A., Hartini, W., & Endut, A. (2011). Study on the removal of pesticide in agricultural run off by granular activated carbon. *Bioresource Technology*, *102*(9), 5312–5318. doi:10.1016/j.biortech.2010.12.074 PMID:21232934

Justes, E., Beaudoin, N., Bertuzzi, P., Charles, R., Constantin, J., Dürr, C., . . . Mary, B. (2012). The use of cover crops to reduce nitrate leaching. Effect on the water and nitrogen balance and other ecosystem services. In *Proceedings of the International Fertiliser Society meeting,* Cambridge, UK, Dec. 7, 2012.

Kaspar, T., & Singer, J. (2011). The use of cover crops to manage soil.

Kato, T., Kuroda, H., & Nakasone, H. (2009). Runoff characteristics of nutrients from an agricultural watershed with intensive livestock production. *Journal of Hydrology (Amsterdam), 368*(1-4), 79–87. doi:10.1016/j.jhydrol.2009.01.028

Kellogg, R. L., Nehring, R. F., Grube, A., Goss, D. W., & Plotkin, S. (2002). *Environmental indicators of pesticide leaching and runoff from farm fields. Agricultural Productivity* (pp. 213–256). Springer.

Khezami, L., & Capart, R. (2005). Removal of chromium (VI) from aqueous solution by activated carbons: Kinetic and equilibrium studies. *Journal of Hazardous Materials, 123*(1-3), 223–231. doi:10.1016/j.jhazmat.2005.04.012 PMID:15913888

Kikuchi, T., & Tanaka, S. (2012). Biological removal and recovery of toxic heavy metals in water environment. *Critical Reviews in Environmental Science and Technology, 42*(10), 1007–1057. doi:10.1080/10643389.2011.651343

Kim, D., Ryu, H.-D., Kim, M.-S., Kim, J., & Lee, S.-I. (2007). Enhancing struvite precipitation potential for ammonia nitrogen removal in municipal landfill leachate. *Journal of Hazardous Materials, 146*(1-2), 81–85. doi:10.1016/j.jhazmat.2006.11.054 PMID:17208368

Kim, S.-W., Park, J.-S., Kim, D., & Oh, J.-M. (2014). Runoff characteristics of non-point pollutants caused by different land uses and a spatial overlay analysis with spatial distribution of industrial cluster: A case study of the Lake Sihwa watershed. *Environmental Earth Sciences, 71*(1), 483–496. doi:10.100712665-013-2933-7

Kivaisi, A. K. (2001). The potential for constructed wetlands for wastewater treatment and reuse in developing countries: A review. *Ecological Engineering, 16*(4), 545–560. doi:10.1016/S0925-8574(00)00113-0

Kolok, A. S., Ali, J. M., Rogan, E. G., & Bartelt-Hunt, S. L. (2018). The fate of synthetic and endogenous hormones used in the us beef and dairy industries and the potential for human exposure. *Current Environmental Health Reports, 5*(2), 225–232. doi:10.100740572-018-0197-9 PMID:29754262

Kongsuwan, A., Patnukao, P., & Pavasant, P. (2009). Binary component sorption of Cu (II) and Pb (II) with activated carbon from Eucalyptus camaldulensis Dehn bark. *Journal of Industrial and Engineering Chemistry, 15*(4), 465–470. doi:10.1016/j.jiec.2009.02.002

Kortbein, L., & Rajendran, A. (2016). Two-stage modified denitrifying-bioreactor for nutrient removal from agricultural runoff.

Lefrancq, M., Jadas-Hécart, A., La Jeunesse, I., Landry, D., & Payraudeau, S. (2017). High frequency monitoring of pesticides in runoff water to improve understanding of their transport and environmental impacts. *The Science of the Total Environment, 587*, 75–86. doi:10.1016/j.scitotenv.2017.02.022 PMID:28242219

Lerch, R., Lin, C., Goyne, K., Kremer, R., & Anderson, S. (2017). Vegetative buffer strips for reducing herbicide transport in runoff: Effects of buffer width, vegetation, and season. *JAWRA Journal of the American Water Resources Association*, *53*(3), 667–683. doi:10.1111/1752-1688.12526

Letcher, T. M., & Vallero, D. (2011). *Waste: A handbook for management*. Academic Press.

Li, Y., Jiao, J., Wang, Y., Yang, W., Meng, C., Li, B., ... Wu, J. (2015). Characteristics of nitrogen loading and its influencing factors in several typical agricultural watersheds of subtropical China. *Environmental Science and Pollution Research International*, *22*(3), 1831–1840. doi:10.100711356-014-3446-y PMID:25172462

Liu, J., Veith, T. L., Collick, A. S., Kleinman, P. J., Beegle, D. B., & Bryant, R. B. (2017). Seasonal manure application timing and storage effects on field-and watershed-level phosphorus losses. *Journal of Environmental Quality*, *46*(6), 1403. doi:10.2134/jeq2017.04.0150 PMID:29293863

Liu, X., Zhang, W., Hu, Y., Hu, E., Xie, X., Wang, L., & Cheng, H. (2015). Arsenic pollution of agricultural soils by concentrated animal feeding operations (CAFOs). *Chemosphere*, *119*, 273–281. doi:10.1016/j.chemosphere.2014.06.067 PMID:25036941

Mahabali, S., & Spanoghe, P. (2014). Mitigation of two insecticides by wetland plants: Feasibility study for the treatment of agricultural runoff in Suriname (South America). *Water, Air, and Soil Pollution*, *225*(1), 1771. doi:10.100711270-013-1771-2

Majumdar, D., & Gupta, N. (2000). Nitrate pollution of groundwater and associated human health disorders. *Indian Journal of Environmental Health*, *42*(1), 28–39.

Meisinger, J., Hargrove, W., Mikkelsen, R., Williams, J., & Benson, V. (1991). Effects of cover crops on groundwater quality. *Cover Crops for Clean Water, Soil, and Water Conservation Society. Ankeny, Iowa*, *266*, 793–799.

Memon, G. Z., Bhanger, M., & Akhtar, M. (2007). The removal efficiency of chestnut shells for selected pesticides from aqueous solutions. *Journal of Colloid and Interface Science*, *315*(1), 33–40. doi:10.1016/j.jcis.2007.06.037 PMID:17631888

Miao, X., Xu, L., Li, H., & Yang, Z. (2018). Determination of olaquindox, carbadox and cyadox in animal feeds by ultra-performance liquid chromatography tandem mass spectrometry. *Food Additives & Contaminants: Part A*(just-accepted).

Miles, T. R., Rasmussen, E. M., & Gray, M. (2016). Aqueous contaminant removal and stormwater treatment using biochar. *Agricultural and environmental applications of biochar: Advances and barriers*(sssaspecpub63), 341-376.

Ming, T. T., Hyun, K. T., & Myun, J. (2007). Characterization of livestock wastewater at various stages of wastewater treatment plant. *The Malaysian Journal of Analytical Sciences*, *11*, 23–28.

Mirbagheri, S. A., & Hosseini, S. N. (2005). Pilot plant investigation on petrochemical wastewater treatment for the removal of copper and chromium with the objective of reuse. *Desalination*, *171*(1), 85–93. doi:10.1016/j.desal.2004.03.022

Mitra, A., & Mukhopadhyay, S. (2016). Biofilm mediated decontamination of pollutants from the environment. *AIMS Bioengineering*, *3*(1), 44–59. doi:10.3934/bioeng.2016.1.44

Mofokeng, D. S., Adeleke, R., & Aiyegoro, O. A. (2016). The analysis of physicochemical characteristics of pig farm seepage and its possible impact on the receiving natural environment. *African Journal of Environmental Science and Technology*, *10*(8), 242–252. doi:10.5897/AJEST2016.2084

Moore, J. A. (1988). *Evaluating coliform concentrations in runoff from various animal waste management systems*. In C. Namasivayam & K. Kadirvelu (Eds.), Uptake of mercury (II) from wastewater by activated carbon from an unwanted agricultural solid by-product: Coirpith. *Carbon*, *37*(1), 79–84.

Naseem, R., & Tahir, S. (2001). Removal of Pb (II) from aqueous/acidic solutions by using bentonite as an adsorbent. *Water Research*, *35*(16), 3982–3986. doi:10.1016/S0043-1354(01)00130-0 PMID:12230182

Nicholson, F., Smith, S., Alloway, B., Carlton-Smith, C., & Chambers, B. (2003). An inventory of heavy metals inputs to agricultural soils in England and Wales. *The Science of the Total Environment*, *311*(1-3), 205–219. doi:10.1016/S0048-9697(03)00139-6 PMID:12826393

Njoku, V., Islam, M. A., Asif, M., & Hameed, B. (2014). Utilization of sky fruit husk agricultural waste to produce high quality activated carbon for the herbicide bentazon adsorption. *Chemical Engineering Journal*, *251*, 183–191. doi:10.1016/j.cej.2014.04.015

Nokyoo, C. (2016). *Swine waste management in Thailand*. Retrieved from http://www.wepa-db.net/activities/2016/20160126/pdf/AM5_4.TH_SwineWWT_160126.pdf

Oyaro, N., Juddy, O., Murago, E. N., & Gitonga, E. (2007). The contents of Pb, Cu, Zn and Cd in meat in Nairobi, Kenya. *Journal of Food Agriculture and Environment*, *5*(3&4), 119–121.

Pandey, J., & Singh, R. (2017). Heavy metals in sediments of Ganga River: Up-and downstream urban influences. *Applied Water Science*, *7*(4), 1669–1678. doi:10.100713201-015-0334-7

Patty, L., Real, B., & Joël Gril, J. (1997). The use of grassed buffer strips to remove pesticides, nitrate and soluble phosphorus compounds from runoff water. *Pesticide Science*, *49*(3), 243–251. doi:10.1002/(SICI)1096-9063(199703)49:3<243::AID-PS510>3.0.CO;2-8

Paulino, A. T., Minasse, F. A., Guilherme, M. R., Reis, A. V., Muniz, E. C., & Nozaki, J. (2006). Novel adsorbent based on silkworm chrysalides for removal of heavy metals from wastewaters. *Journal of Colloid and Interface Science*, *301*(2), 479–487. doi:10.1016/j.jcis.2006.05.032 PMID:16780853

Pounds, N., Maclean, S., Webley, M., Pascoe, D., & Hutchinson, T. (2008). Acute and chronic effects of ibuprofen in the mollusc Planorbis carinatus (Gastropoda: Planorbidae). *Ecotoxicology and Environmental Safety*, *70*(1), 47–52. doi:10.1016/j.ecoenv.2007.07.003 PMID:18207238

Prajapati, S. (2014). Cation exchange for ammonia removal from wastewater.

Puttaiah, E. T., & Kiran, B. R. (2007). Heavy metal transport in a sewage fed lake of Karnataka, India. In *Proceedings of TAAL The 12th World Lake Conference*.

Qiu, Y.-R., & Mao, L.-J. (2013). Removal of heavy metal ions from aqueous solution by ultrafiltration assisted with copolymer of maleic acid and acrylic acid. *Desalination*, *329*, 78–85. doi:10.1016/j.desal.2013.09.012

Rajamohan, N., Rajasimman, M., Rajeshkannan, R., & Sivaprakash, B. (2013). Kinetic modeling and isotherm studies on a batch removal of acid red 114 by an activated plant biomass. *Journal of Engineering Science and Technology*, *8*(6), 778–792.

Randall, D., & Tsui, T. (2002). Ammonia toxicity in fish. *Marine pollution bulletin, 45*(1-12), 17-23.

Rojas, R., Vanderlinden, E., Morillo, J., Usero, J., & El Bakouri, H. (2014). Characterization of sorption processes for the development of low-cost pesticide decontamination techniques. *The Science of the Total Environment*, *488*, 124–135. doi:10.1016/j.scitotenv.2014.04.079 PMID:24830926

Ronquillo, M. G., & Hernandez, J. C. A. (2017). Antibiotic and synthetic growth promoters in animal diets: Review of impact and analytical methods. *Food Control*, *72*, 255–267. doi:10.1016/j.foodcont.2016.03.001

Rosa, D. M., Sampaio, S. C., Pereira, P. A., Mauli, M. M., & Reis, R. R. D. (2017). Swine wastewater: Impacts on soil, plant, and leachate. *Engenharia Agrícola*, *37*(5), 928–939. doi:10.1590/1809-4430-eng.agric.v37n5p928-939/2017

Rozema, E. R., VanderZaag, A. C., Wood, J. D., Drizo, A., Zheng, Y., Madani, A., & Gordon, R. J. (2016). Constructed wetlands for agricultural wastewater treatment in northeastern North America: A review. *Water (Basel)*, *8*(5), 173. doi:10.3390/w8050173

Sahu, J., Agarwal, S., Meikap, B., & Biswas, M. (2009). Performance of a modified multi-stage bubble column reactor for lead (II) and biological oxygen demand removal from wastewater using activated rice husk. *Journal of Hazardous Materials*, *161*(1), 317–324. doi:10.1016/j.jhazmat.2008.03.094 PMID:18462879

Schoeman, J., & Steyn, A. (2003). Nitrate removal with reverse osmosis in a rural area in South Africa. *Desalination*, *155*(1), 15–26. doi:10.1016/S0011-9164(03)00235-2

Shawaqfeh, A. T. (2010). Removal of pesticides from water using anaerobic-aerobic biological treatment. *Chinese Journal of Chemical Engineering*, *18*(4), 672–680. doi:10.1016/S1004-9541(10)60274-1

Shi, T., Ma, J., Wu, X., Ju, T., Lin, X., Zhang, Y., ... Zhao, L. (2018). Inventories of heavy metal inputs and outputs to and from agricultural soils: A review. *Ecotoxicology and Environmental Safety*, *164*, 118–124. doi:10.1016/j.ecoenv.2018.08.016 PMID:30099172

Skaggs, R. W., Fausey, N. R., & Evans, R. O. (2012). Drainage water management. *Journal of Soil and Water Conservation*, *67*(6), 167A–172A. doi:10.2489/jswc.67.6.167A

Sobota, D. J., Harrison, J. A., & Dahlgren, R. A. (2009). Influences of climate, hydrology, and land use on input and export of nitrogen in California watersheds. *Biogeochemistry*, *94*(1), 43–62. doi:10.100710533-009-9307-y

Suthar, S., Bishnoi, P., Singh, S., Mutiyar, P. K., Nema, A. K., & Patil, N. S. (2009). Nitrate contamination in groundwater of some rural areas of Rajasthan, India. *Journal of Hazardous Materials*, *171*(1-3), 189–199. doi:10.1016/j.jhazmat.2009.05.111 PMID:19545944

Tatarková, V., Hiller, E., & Vaculík, M. (2013). Impact of wheat straw biochar addition to soil on the sorption, leaching, dissipation of the herbicide (4-chloro-2-methylphenoxy) acetic acid and the growth of sunflower (Helianthus annuus L.). *Ecotoxicology and Environmental Safety*, *92*, 215–221. doi:10.1016/j.ecoenv.2013.02.005 PMID:23474069

Tchounwou, P. B., Yedjou, C. G., Patlolla, A. K., & Sutton, D. J. (2012). *Heavy metal toxicity and the environment. Molecular, clinical, and environmental toxicology* (pp. 133–164). Springer. doi:10.1007/978-3-7643-8340-4_6

Tepuš, B., Simonič, M., & Petrinić, I. (2009). Comparison between nitrate and pesticide removal from ground water using adsorbents and NF and RO membranes. *Journal of Hazardous Materials*, *170*(2-3), 1210–1217. doi:10.1016/j.jhazmat.2009.05.105 PMID:19541416

Tilman, D., Balzer, C., Hill, J., & Befort, B. L. (2011). Global food demand and the sustainable intensification of agriculture. *Proceedings of the National Academy of Sciences of the United States of America*, *108*(50), 20260–20264. doi:10.1073/pnas.1116437108 PMID:22106295

Topalcengiz, Z., Strawn, L. K., & Danyluk, M. D. (2017). Microbial quality of agricultural water in Central Florida. *PLoS One*, *12*(4). doi:10.1371/journal.pone.0174889 PMID:28399144

Unsworth, J. (2010). History of pesticide use. *IUPAC-International Union of Pure and Applied Chemistry*.

US Environmental Protection Agency. (2012). Recreational water quality criteria. Washington, DC: Office of Water Washington.

Vagı, M., & Petsas, A. (2017). *Advanced oxidation processes for the removal of pesticides from wastewater: recent review and trends.* Paper presented at the 15th International Conference on Environmental Science and Technology. CEST2017, Rhodes, Greece.

Vymazal, J., & Březinová, T. (2015). The use of constructed wetlands for removal of pesticides from agricultural runoff and drainage: A review. *Environment International*, *75*, 11–20. doi:10.1016/j.envint.2014.10.026 PMID:25461411

Wang, M., Zhang, D., Dong, J., & Tan, S. K. (2018). Application of constructed wetlands for treating agricultural runoff and agro-industrial wastewater: A review. *Hydrobiologia*, *805*(1), 1–31. doi:10.100710750-017-3315-z

Wang, M., Zhang, D. Q., Dong, J. W., & Tan, S. K. (2017). Constructed wetlands for wastewater treatment in cold climate—A review. *Journal of Environmental Sciences (China)*, *57*, 293–311. doi:10.1016/j.jes.2016.12.019 PMID:28647250

Weber, D., Drizo, A., Twohig, E., Bird, S., & Ross, D. (2007). Upgrading constructed wetlands phosphorus reduction from a dairy effluent using electric arc furnace steel slag filters. *Water Science and Technology*, *56*(3), 135–143. doi:10.2166/wst.2007.513 PMID:17802848

Wise, R. (2002). Antimicrobial resistance: Priorities for action. *The Journal of Antimicrobial Chemotherapy*, *49*(4), 585–586. doi:10.1093/jac/49.4.585 PMID:11909829

Wongsasuluk, P., Chotpantarat, S., Siriwong, W., & Robson, M. (2014). Heavy metal contamination and human health risk assessment in drinking water from shallow groundwater wells in an agricultural area in Ubon Ratchathani province, Thailand. *Environmental Geochemistry and Health*, *36*(1), 169–182. doi:10.100710653-013-9537-8 PMID:23771812

Xu, Y., Baker, L. A., & Johnson, P. C. (2007). Trends in ground water nitrate contamination in the Phoenix, Arizona region. *Ground Water Monitoring and Remediation*, *27*(2), 49–56. doi:10.1111/j.1745-6592.2007.00135.x

Yan, G., & Viraraghavan, T. (2003). Heavy-metal removal from aqueous solution by fungus Mucor rouxii. *Water Research*, *37*(18), 4486–4496. doi:10.1016/S0043-1354(03)00409-3 PMID:14511719

Ying-Xin, X., Xiong, Z.-Q., Guang-Xi, X., Guo-Qing, S., & Zhao-Liang, Z. (2007). Assessment of nitrogen pollutant sources in surface waters of Taihu Lake region. *Pedosphere*, *17*(2), 200–208. doi:10.1016/S1002-0160(07)60026-5

Zaytseva, O., & Neumann, G. (2016). Carbon nanomaterials: Production, impact on plant development, agricultural and environmental applications. *Chemical and Biological Technologies in Agriculture*, *3*(1), 17. doi:10.118640538-016-0070-8

Zhang, L., Xu, E. G., Li, Y., Liu, H., Vidal-Dorsch, D. E., & Giesy, J. P. (2018). Ecological risks posed by ammonia nitrogen (AN) and un-ionized ammonia (NH3) in seven major river systems of China. *Chemosphere*, *202*, 136–144. doi:10.1016/j.chemosphere.2018.03.098 PMID:29567611

Zhang, L., Xu, T., Liu, X., Zhang, Y., & Jin, H. (2011). Adsorption behavior of multi-walled carbon nanotubes for the removal of olaquindox from aqueous solutions. *Journal of Hazardous Materials*, *197*, 389–396. doi:10.1016/j.jhazmat.2011.09.100 PMID:22018864

Zhu, Q., Sun, J., Hua, G., Wang, J., & Wang, H. (2015). Runoff characteristics and non-point source pollution analysis in the Taihu Lake Basin: A case study of the town of Xueyan, China. *Environmental Science and Pollution Research International*, *22*(19), 15029–15036. doi:10.100711356-015-4709-y PMID:26002368

Zolgharnein, J., Shahmoradi, A., & Ghasemi, J. (2011). Pesticides removal using conventional and low-cost adsorbents: A review. *Clean–Soil, Air. Water (Basel)*, *39*(12), 1105–1119.

Chapter 24
Synergy Between Air Quality, Various Urban Forms, and Land Surface Temperature:
A Case Study of Kolkata Metropolitan Area

Abira Dutta Roy
Bankura Zilla Saradamani Mahila Mahavidyapith, India

Munni Debnath Parial
Adamas University, India

Kasturi Mukherjee
Adamas University, India

ABSTRACT

This chapter studies spatio temporal trends of air quality, its relation with urban forms, and Land Surface Temperature (LST) in Kolkata Metropolitan Area, the world's 55th most polluted city according to the World Health Organization. Air quality data were procured from 34 monitoring stations operated by West Bengal Pollution Control Board for 2005-18. Trend analysis showed declining NO_2, SO_2 values but a rise in PM10 concentration. Interpolation analysis showed high concentration of pollutants along the Hugli industrial belt, Dhulagarh, and Kolkata Municipal Corporation Area. Urban forms such as built-up density, distances from water bodies, parks, and bus stops demonstrated their reasonable influence on air pollution condition. LST generated from Landsat Thermal Infrared bands showed increase in temperature conditions from 2005-2018. Positive Correlation was identified between Land surface temperature and air pollution. Their relationship was assessed to have become stronger over the decades.

DOI: 10.4018/978-1-7998-0369-0.ch024

INTRODUCTION

Air quality issues are a matter of global concern because it is the prime driving force for climate change. Air pollution not only contaminates the natural composition of the atmosphere but also hampers the health of the entire biota. Research works on air pollution has advanced tremendously since the analysis of the Los Angeles smog in 1952 conducted by Haagen-Smit (Haagen-Smit 1952). He and his fellow researchers determined that the major component of the smog was tropospheric ozone formed by the chemical reaction between NO2 (Nitrogen dioxide) and VOC (Volatile Organic Compounds) in the presence of sunlight (Baklanov et al.,2016). Many studies have taken consideration of the most preliminary research objective, that is to understand the spatiotemporal variation of the air quality parameters of their study area (Tian et al., 2002; Yahaya et al., 2018; Gurjar et al., 2007). World Health Organization (WHO, 2018) reports that developing Asian countries, especially that of India and China have their cities ranking in the topmost polluted cities in the world. Research work have highlighted that air pollution is not only observed in megacities but also in smaller urban centers with a population ranging from 150,000 to 1.5 million (IAQP, 2010).

Air quality is a source for concern in India, especially in cities. According to the World Health Organization (WHO), 37 cities from India rank in the top 100 world cities with the worst PM10 pollution though the national ambient standard for CO is better than the WHO guideline. Air pollutants including particulate matter (PM), sulfur dioxide (SO2), nitrogen oxides (NOx), carbon monoxide (CO), and ozone (O3) are monitored by Central Pollution Control Board. The pollutant concentration often surpasses the National Ambient Air Quality Standards (NAAQS). The NO2, SO2, and O3 standards though mostly stay at par with the NAAQS guidelines but the concentration for PM10 and PM2.5 are way beyond the desired limit (CPCB, 2010). Subsequent to the understanding of the spatiotemporal dynamics of air quality of a region, it becomes imperative to look into the causal factors. It was found that consequential to population growth was urban sprawl around industrial growth nuclei had led to environmental degradation (Reddy and Venkataraman., 2002; Khandelwal et al, 2017). The decline in air quality parameters formed one of the key components of environmental degradation. In most literatures, the process of urbanization and associated anthropogenic activities were identified as liable for depleting air quality. After performing a comparative study among 249 European cities Rodríguez et al., 2016 concluded that urbanization had led to air pollution. Studies carried out for Budapest, Hungary showed transport as the dominant source of emissions, contributing 57% of the oxides of nitrogen, 80% of Lead, 81% of CO and 75% of HC emissions Even in India urban sprawl, industrialization, increasing built-up density because of the population flux in the cities were identified as the principal cause of deterioration of ambient air quality (Ravindra et al. 2001).

Post independent India attempted to accelerate the socio-economic condition but it happened at the cost of its environmental condition. Manufacturing units of heavy machinery, fertilizer, cement, brick kilns contributed approximately 36% SO2 (Sulphur dioxide), 19% NO2 (Nitrogen dioxide) in the Indian cities (Garg et al, 2001). Brick kilns alone were assessed to have added 70% of the ambient PM10 (Particulate Matter) and 60% of the ambient PM2.5 in different regions of India (Azkar et al., 2012). In order to meet the escalating demands of the teeming millions, coal-fired power plants for electricity generation were being established as well. These units supplied 50% of the total SO2 and 30% of the total NO2 in the air (Garg et al., 2006; Lu and Streets, 2012). According to Amann et al., 2017 and vehicular emission contributed to over one-third of the PM 2.5 in the ambient air. Ghose et al., 2004 have assessed that usage of fuels with poor environmental performance, congested transportation system enhanced vehicular emission in Indian cities, The implications of such degraded air quality had to be hazardous. According to the United Nations Environment Programme globally 1.1 billion people breathe unhealthy air .On the other hand World Health Organisation (WHO) has approximated that air pollution has been the cause of around 800,00 deaths per annum globally . Human health was mainly at risk from suspended particles in air (Bosco,et al., 2005; Cheng et.al, 1996). Some of them have proved to be acidic, toxic and even carcinogenic with mutagenic potential as they are carriers of heavy metals, acids, polycyclic aromatic hydrocarbons (Bourotte et al, 2005; Caricchia et al, 1999). Numerous research works have shown the adverse impact of PM10 on human health (Arditsoglou and Samara, 2005; Dockery et al., 1993; Lin and Lee, 2004; Monne et al., 1995; Namdeo and Bell, 2005; Perez and Reyes, 2002; Pope, 2000; Tonne et al., 2017). Gurjar and Lelieveld, 2005 have studied the environmental impact of poor air quality, especially in Asian cities with about 10 million population. Undesirable impacts of air quality on vegetation was also proved in the studies conducted by Emberson et al., 2001. Poor air quality was also assessed to have driven micro climate changes often impacting cloud formation, visibility, solar radiation, and rainfall (Celis et al., 2004; Hong et al., 2002; Khoder, 2002; Rajkumar and Chang, 2000; Watson,et al, 2002).

Urban morphology can shape the concentration of air pollutants. Parameters such as vehicular density, type of land use land cover, availability and distance from green space, pollutant diffusion pattern and the heat island effect the pollution distribution pattern. Studies in European and American cities have demonstrated that urban forms could have a significant impact on the source and diffusion of air pollutants emitted from urban traffic . The climatic conditions can be adversely affected, leading to the further deterioration of air quality (Bereitschaft & Debbage, 2013; Schweitzer & Zhou, 2010; Stone, 2008). The transportation sector plays a dominant role and controls the air pollution distribution pattern in urban centres (Fenger, 1999; Rojasrueda et al., 2012).

The geometric shape of city streets is also an important urban characteristics and it is generally understood that excessive building density is not favourable to the diffusion of pollutants (Buccolieriet al., 2010; Gaignéet al., 2012; Taseiko, et al.,2009). But contradictory research statements are also available. Some of the literatures focus on the factor that these parameters often help to reduce the impacts of pollutants in the ambient air. Some studies suggest that dense and compact urban forms can help to improve air quality. Ewing & Cervero, 2010; Bento et al., 2005; Cervero & Murakami, 2010 have put forward this view on the basis of the studies carried on the residents of the urban centres, their travelling patterns and choices; such as commutation by public means of transport or by private means. Urban simulation models and land use –traffic –emission models have been put to test in some of the European and American cities. The results from the model showed emissions of air pollutants are lower in compact development scenarios (Borrego et al., 2006; Civerolo et al., 2007; Hankey & Marshall, 2010). Availability of well-maintained and well dispersed air quality monitoring stations have helped empirical studies on the impacts of urban forms on pollution concentration. Ewing, et al., (2003) and Stone (2008) have found in their respective studies that population density is negatively correlated with O_3 concentrations in American urban centres. On the other hand, compact cities have shown lower levels of energy consumptions and pollution emissions in the studies carried out by Sovacool and Brown (2010). He had compared CO2 emissions from 12 capital cities, and the result illustrated that compact urban spaces have lower emission rates. Similarly, studies undertaken by Bereitschaft and Debbage (2013) in 86 metropolitan areas of the United States also showed results contrary to expected scenarios. Their paper showed low density urban sprawl have higher emissions in comparison to high density sprawl. According to them low density and dispersed urban form leads to long-distance commutes and traffic congestion (Wang et al., 2011;Yang, et al., 2017; Zhao et al., 2017) and that compact, high-density urban forms having mixed land use land cover conditions as well as improved bus services can reduce vehicle travel distances and tail pipe exhaust emissions (Qin & Han, 2013).

Even some of the studies by Bechle et al., 2011 and McCarty and Kaza, 2015 showed fragmented urban morphology to have a positive correlation with air quality standards. In opposition, other scholars have disagreed on the fact that a high-density compact urban form helps in improving the air quality but have stated that congested built up density may even lead to severe pollution condition. Cho and Choi (2014) carried out a panel data analysis of 17cities in South Korea and established an argument that compact urban forms have no significant impact on reducing air pollution concentration. Instead according to him high density may even lead to an increase in SO2 and CO concentrations. Clark, et al., (2010) selected 111 cities in the United States as samples and found that PM2.5 pollution levels increased as population density increased. As a result the high-density urban forms will pose a serious threat to the health of the number of people exposed to such polluted condition (De Ridder et al., 2008; Hixson et al., 2010) On the other hand by analyzing the pollution data for 84 cities in China, Yuan, Song et al., (2017) found that population density is negatively correlated with the concentrations of PM2.5, PM10, and O3. In addition, urban form could affect air quality in a variety of ways. For example, although urban green spaces can decontaminate air and absorb particulate matter (Groenewegen et al., 2006). Urban forms are definitely related to the urban heat island effect. It becomes obvious that the increase in urban air temperature will lead to the increase in electricity consumption and CO2 production because of air conditioning systems. But such process may further worsen the situation because of the increased formation of PM2.5 and O3 pollutants (Sarrat, et.al, 2006; Schwarz & Manceur, 2014; Taha,2008). In the past few decades, researchers have started to quantify the relationship between urban form and air quality in China as well, but the conclusions derived have not proven to be consistent. Some studies argue that the compact urban form is not applicable to Chinese cities with already high densities and mixed land use (Chen et al., 2008). With the establishment of air pollution monitoring networks in China, some researchers tried to use observation data to conduct empirical studies, but they could not derive a definite conclusion. Liu, et al., (2016) analyzed the panel data for 30 cities in China and found that the increase in urban land compactness may increase the concentration of PM10. By studying the pollution data for 287 Chinese cities, Lu and Liu (2016) found that land-use compactness is generally negatively correlated with concentrations of NO2 and SO2, but the effect varies from region to region. In a nutshell, the existing research has not yet reached a unanimous conclusion, and conclusions from European and American cities cannot be applied directly to cities in China due to differences in both urban forms and pollution sources.

In spite of the global warming warnings, the process of urbanisation has continued unhindered. Many cities are undergoing great sprawls and structural changes, resulting in a special phenomenon where urban surface and air temperatures have become significantly higher than its neighbouring suburban or rural temperatures (Zhao et al., 2017). This is entitled as urban heat island (UHI) phenomenon. Effects of UHI have been well researched into as the resulting of anthropogenic activities have brought about significant challenges to urban systems, its inhabitants and even the living ecosystems. For cities that are located in high-altitude regions or in colder seasons (e.g. Changchun of China, Barrow, Alaska of the US, Norilsk of Russia), UHI effects can moderately diminish energy consumption for heating (Yang et al., 2017; Hinkel et al., 2003; Varentsov et al., 2014), but the situation is extremely hostile for the cities located in the humid tropics. In past decades, numerous studies have been carried out to hunt for effective UHI mitigation techniques and strategies. The accelerated urbanization process has brought along with it the challenges of severe air pollution, especially those of industrial aerosols and photochemical smog. The appearance of air pollution has also exerted adverse impacts on UHI phenomenon. Many countries in developing world are encountering PM10 pollution and haze related problems. Zheng et al., (2018) found a significant correlation between mean daily UHI effects and PM10, but still in some cases a low degree of coincidence were observed. In addition, its impacts on UHI effects vary with season. The impacts of haze on UHI phenomenon are still not clear, but current studies have primarily indicated exacerbated impacts on UHI effects. Cao et al. (2002) quantitatively analyzed the impact of haze on UHI effects in humid, semi-humid and semi-arid cities of China. They concluded that haze was able to increase the absorption of nocturnal long-wave radiation, intensifying the nocturnal UHI effects.

STUDY AREA

According to Census 2011, Kolkata is the 7th largest city in India and 55th most polluted city in the world (WHO, 2018). It has also ranked among the 41 most polluted cities of the world with respect to the concentration of suspended particulate matter (WHO, 2002). Extensive work on understanding the spatial pattern as well as both cause and effect of air pollution has been carried out in Kolkata (Mukherjee and Mukherjee, 1998 Mondal et al., 2000; Ghose et al., 2004; Karar et al., 2006; Das et al., 2006; Chowdhury et al., 2012; Guttikunda et al., 2014; Gurjar et al', 2016; Guttikunda et al., 2019). According to these studies the transportation sector was recognized as the major contributor in air pollution of Kolkata. A study conducted at major traffic junctions of Kolkata showed that the key pollutants like lead, NOX, PM10, SO2 (Guttikunda et al.,2014). have addressed that poorly maintained vehicles, use of adulterated fuel have led to air pollution too. In addition three thermal power plants located along the banks of river Hugli, small scale industries were also found to have equally supplied pollutants in the ambient air. The repercussions of such deteriorated air quality were bound to occur. Studies by CPCB had showed that around 70% of the Kolkata dwellers suffered from adverse respiratory diseases. 41.3% and 47.8% of the Kolkata's population experience upper and lower respiratory tract disorders respectively, compared to 13.5% and 35% of the rural Bengal population (WBPCB, 2001). Suspended particulate matter alone was identified to have caused more than 10,000 premature deaths in Kolkata in 1995 (Kazimuddin and Banerjee, 2000). Recent studies have also recognized the association of air pollution with adverse health effects in Kolkata (Lahiri et al., 2000) Most of these studies were focused on the urban core area of Kolkata Municipal Corporation (KMC). Moreover the studies have been carried out over a short duration as well as require updating

As a result in order to enhance the research repository, this chapter updates the information on spatiotemporal variability of pollutants in Kolkata Metropolitan Area (KMA) a larger urban agglomeration unlike that of a smaller KMC. Moreover, this chapter attempts to identify urban forms and its implications on spatial distribution of air pollutants, which has not been previously dealt with in case of Kolkata similar to the works by Li et al., 2019 which have contributed significantly towards understanding the complex relationships between various urban forms and air pollutants in Sanghai and Beijing respectively. Most of the air pollutants such as CO, CO2, NOx, SOX, PM, act as greenhouse gases, trapping the long wave terrestrial radiation in the lower layers of the atmosphere were identified to have a profound impact in intensifying the Urban Heat Island (UHI) phenomena (Stanhill and Kalma,1995; Jauregui and Luyando,1999; Jacobson,1998). Entrapment of heat in the lower layers of the atmosphere tend to increase the temperature of the area, thus aggravating the urban heat island. Kolkata has been proven in different research papers as an intense UHI owing mostly to its changing land use land cover conditions and increased built up density (Sarkar and Sivaramakrishnan; 2015), Akher and Chattopadhyay, 2017; Chaudhuri and Mishra, 2016). This paper finally also endeavours to understand the role of pollutants in determining the intensity of urban heat island in Kolkata, which has not been dealt with in any research work.

MATERIALS AND METHODS

KMA which is one of the most polluted cities of the world spreads over an area of 1851.40 km2 and is situated between 22°0 ′19 " N to 23°0 ′01 " North and 88°0 ′04″ E to 88°0 ′33 ″ East. Figure 1. shows the study area. It is flanked in a linear pattern along the eastern and western banks of Hooghly River. KMA has been administratively divided into 3 municipal corporations (Howrah, Kolkata and Chandan Nagar), 38 municipalities, 77 non-municipal urban towns, 16 out growths and 445 rural villages. According to Census of India, 2011 the total population is around 14.72 million with average density of 7950 persons/km2. As a sprawling metropolitan city, Kolkata has been experiencing substantial unplanned urban growth, pollution, traffic overcrowding, poverty, and related socio-economic problems (Bhatta, 2009; Mukherjee, 2012). Over the years the KMA area has been experiencing substantiate decline in agricultural area, increase in current fallow land, declining vegetation cover, open space and water bodies and definitely an increase in the built-up area (Sahana et al., 2018). Under such circumstances it becomes imperative to determine the land use land cover conditions at a micro level segregating the built up area on the basis of the compactness and areal occupancy

AIR QUALITY TRENDS

In order to carry out the spatiotemporal analysis of air pollutants in Kolkata Metropolitan Area the data on air quality parameters were first collected from the website of West Bengal Pollution Control Board (WBPCB). The WBPCB mainly monitors three major air quality parameters for 34 stations in KMA. It monitors NO2, SO2 and PM10. Sometimes it also monitors PM2.5 but not for all the stations. The data were collected for the stations whose names, latitudinal and longitudinal coordinates and the period of data availability are mentioned in Table 1. The serial numbers provided in Table 1 corresponds to the label number provided in Figure 2. The air quality parameters of NO2, SO2 and PM10 were considered for the present study. There was lack of consistency in the data availability through the study period of 2005-2018. Data were available at an approximate interval of two to three days. These data was compiled and annual averages were computed. Linear trend analysis were carried out and graphically represented.

In order to understand the spatial variability of the pollutants the data for NO2, SO2 and PM10 were attached with their monitoring stations located as points in a GIS environment and an interpolated map was generated using IDW (Inverse Distance Weighted) analysis (Zou et al, 2019). Wind speed and direction often has profound effect in dispersing the pollution concentration far and wide, especially the extreme winds (Zhang, 2019). In Kolkata extreme windy conditions during the study period has been rare which includes some cyclonic storms that occur in the Bay of Bengal, otherwise mostly the area experiences moderate wind speed of 4-11 knots which do not diffuse pollutants to large distances. IDW is a point based interpolation method. The pollution concentration value at point *i* (N_0) is calculated through the formula shown in equation 1. where *N*; represents the number of measurement points, *Ni;*

Figure 1. Administrative map of Kolkata Metropolitan Area showing WBPCB air quality monitoring stations whose numbers correspond to the ones listed in Annexure 1

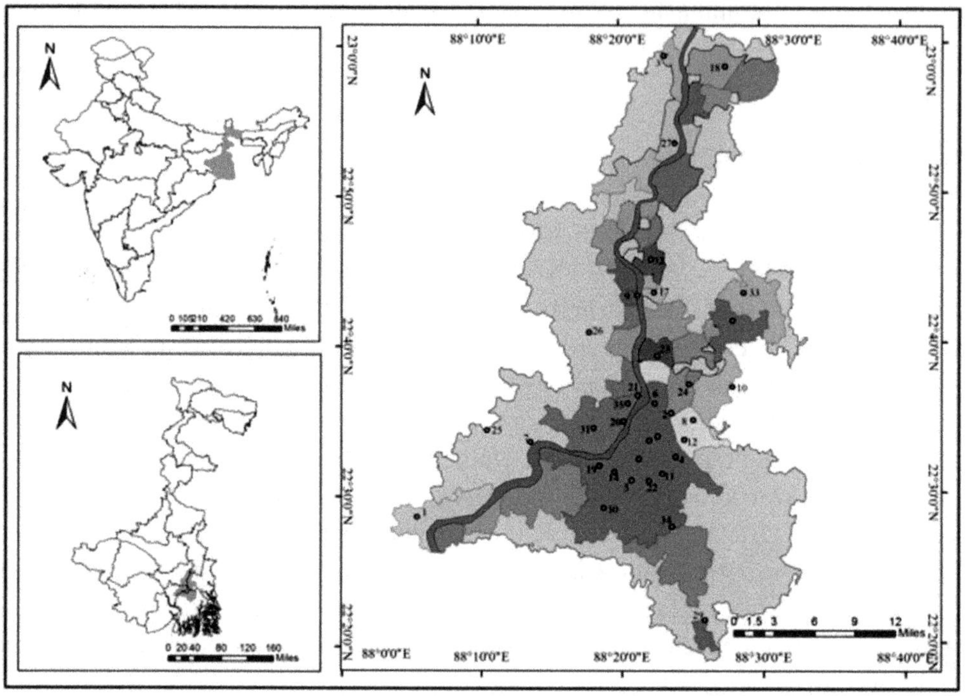

represents the pollutant value at position *i*, *Pi*; the weight of the pollutant value at *i* position. *Pi* weights are calculated using equation Eq. (2) which shows that it is a function of the distance between the reference point and the interpolation point according to the idea that the effect of the closer points are higher than distant ones. where d_i is the horizontal distance between the interpolation point at (x_0, y_0) and the reference points at (x_i, y_i) and is calculated by the following formula (3). *k* is the power of the distance. Pi can also be calculated through the Gaussian function expressed below.

$$N_0 = \frac{\sum_{i=1}^{n} N_i P_i}{\sum_{i=1}^{n} N_i} \tag{1}$$

$$P_i = \frac{1}{d_i^k} \tag{2}$$

$$d_i=\sqrt{((x_i-x_i)^2)+(y_i-y_0)^2} \tag{3}$$

Figure 2. 1000 meter buffer around air quality monitoring sites

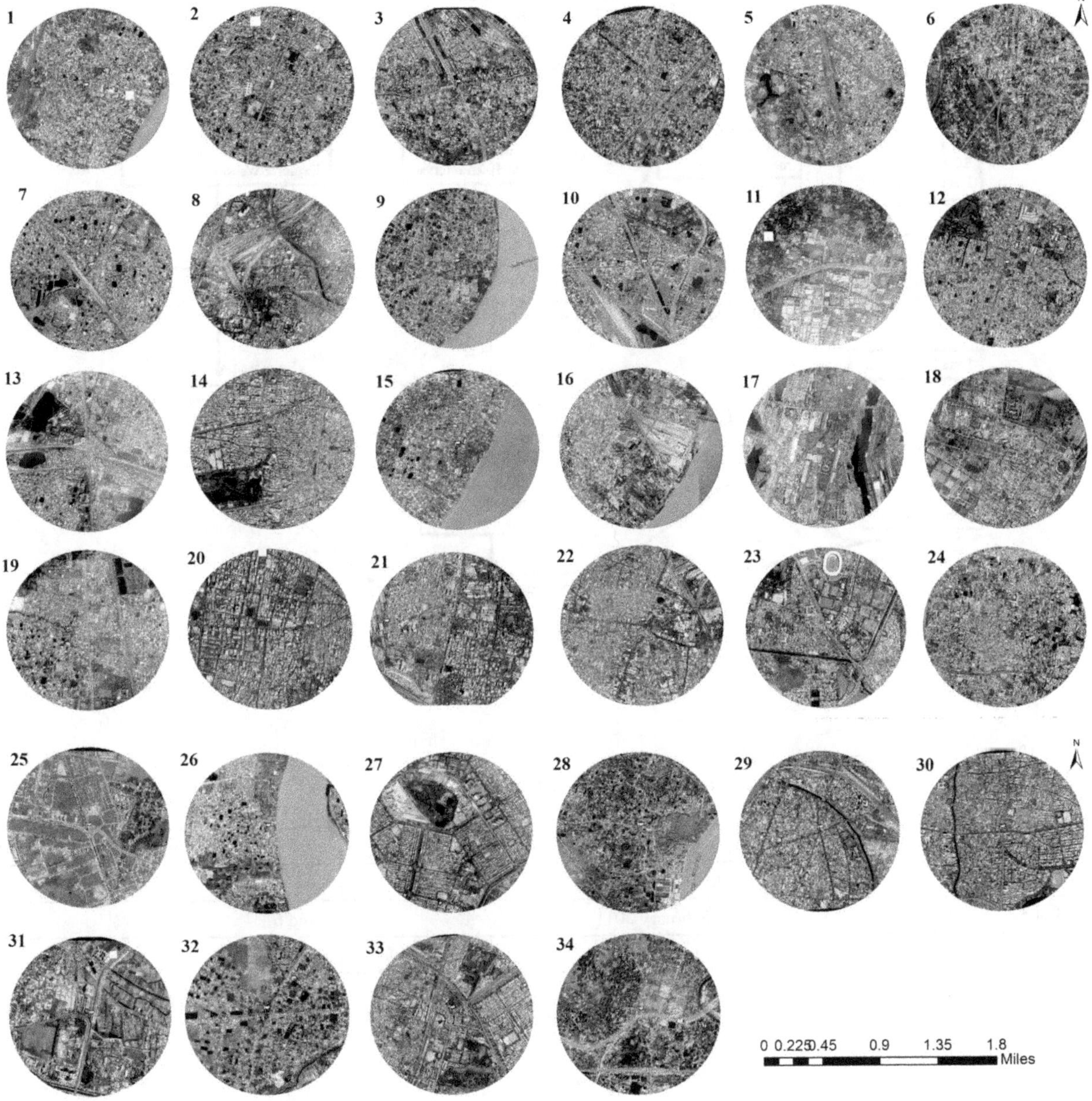

Spatial Distribution of Air Pollution

The interpolation maps were not created for all the years in order to avoid repetition of data interpretation. The linear trend graphs could also explain the spatio temporal variation. The interpolated maps were created for the dates shown in Table 2 corresponding to the date of pass of the Landsat Images of the area. The methodological flow chart is given in Figure 3 which hints at how the Landsat images were used in order to assess the correlation between LST (Land Surface Temperature) and the air pollutants.

Figure 3. Methodological flow chart

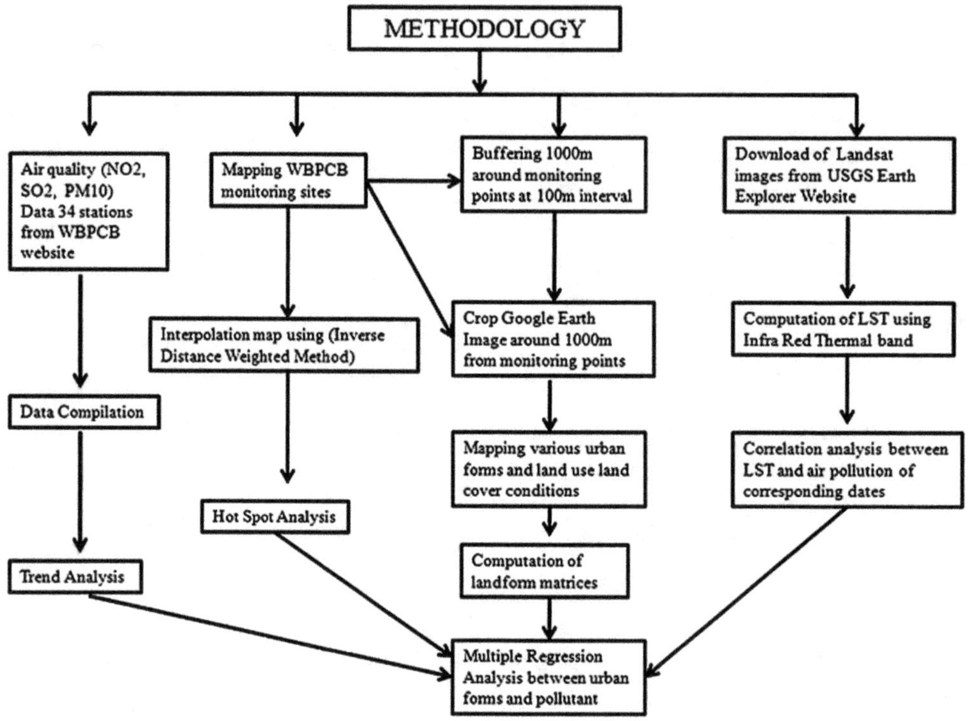

Table 2. List of Satellite images used for computation of LST

Satellite Images	Date of Pass	Thermal Band
Landsat 5	26th January 2006	Band 6
landsat 5	9th February 2011	Band 6
Landsat 8	27th January 2018	Band 10

Land Surface Temperature

The Landsat Images of the dates as given in Table 2 were downloaded from earthexplorer.usgs.gov. Atmospheric correction was carried out on the bands. The images were stacked to generate a false colour composite. This stacked image was considered as an input in the LST generation model. The study by Yu et al. 2014 and Qin et al., 2001 demonstrated that the LST retrieved based on the Radiative Transfer Equation (RTE) showed the maximum accuracy while using band 10 The principle of RTE involves determination of atmospheric effects on surface thermal radiation, then subtracting this part of the atmospheric effect from the total radiated heat observed by the satellite thus deriving the total amount of heat radiation observed by satellite sensors.

As a result, the surface thermal radiation intensity was derived, and the intensity of thermal radiation was converted into the land surface temperatures (Chandrasekhar, 2013; Jiménez-Muñoz et al., 2014). In practice, a radiative transfer equation can express the apparent radiance received by a sensor, L λ. The atmospheric downward radiance (Latm, i↓), the upward radiance (Latm,i↑), and the transmissivity (τ) can be estimated (http://atmcorr.gsfc.nasa.gov/). Finally, the B(Ts) in Eq. (4) can be calculated using the given land surface emissivity (ε). And the LST can be retrieved by using Eq. (5):

$$L\lambda = [\,\varepsilon B(T_S) + (1-\varepsilon)\, Latm,i\downarrow]\,\tau + Latm,i\uparrow \quad (4)$$

Where ε is the surface emissivity, T_S is the LST, $B(T_s)$ is the ground radiance, and τ is atmospheric transmittance. Therefore, as per Plank 's law, $B(T_s)$ can be expressed as

$$B(T_S) = [L\lambda - Latm,i\uparrow - \tau\,(1-\varepsilon)\, Latm,i\downarrow]/\tau\varepsilon \quad (5)$$

Then the LST can be calculated using

$$Ts = K_2/\ln\,(K_1/B(T_S) + 1 \quad (6)$$

The K_1 and K_2 values were available in the metadata MTL.txt file along with the downloaded image.

Urban Forms

In order to understand the association between different urban forms and the pollution pattern in KMA through multiple regression analysis subsequently 1000 meters circular buffers were drawn around the monitoring stations at 100 m intervals using ArcGIS 10.3 (Li et al., 2019). These stations were located at major road intersections and near industrial units. Hence, there remained a high probability to have maximum pollution concentration as well as high temperature conditions near the source region. The buffers were converted to .kml files so that the area of study could be visualised in Google Earth. Based on GPS survey and in combination with the information provided in Google Earth the various urban forms were mapped within the buffers. These urban forms are listed in Table 3. Table 3 also displays the different matrices that were computed within GIS Environment. Figure 4 shows the land use land cover conditions within the 1000m buffer for all the 34 monitoring stations along with the location of various urban forms. The urban forms mapped were school buildings, hospitals, health centres and health care units, industries, bus stoppages, petrol pumps, railway stations, banks, ATMS, government and private offices, shopping malls and commercial buildings, hotels and restaurants, parks or any green space and finally water bodies. The number of each of the urban forms were counted inside each ring within the buffer. The distances of these points were computed on the basis of the ring in which it was located following which the average distances were derived.

The 1000m buffers were later used to extract or mask the Google Earth images in order to generate a high-resolution satellite data under each circle. Figure 2 shows the buffers drawn on Google Earth images which, provided accurate land use land cover information and henceforth building density as well as average building height. Information on dominant building materials aided by field surveys were also collected for carrying out multiple regression analysis between the urban forms and the different pollutants for the year 2018.

Figure 4. Urban forms, built density, road networks within 1000m radius of monitoring sites

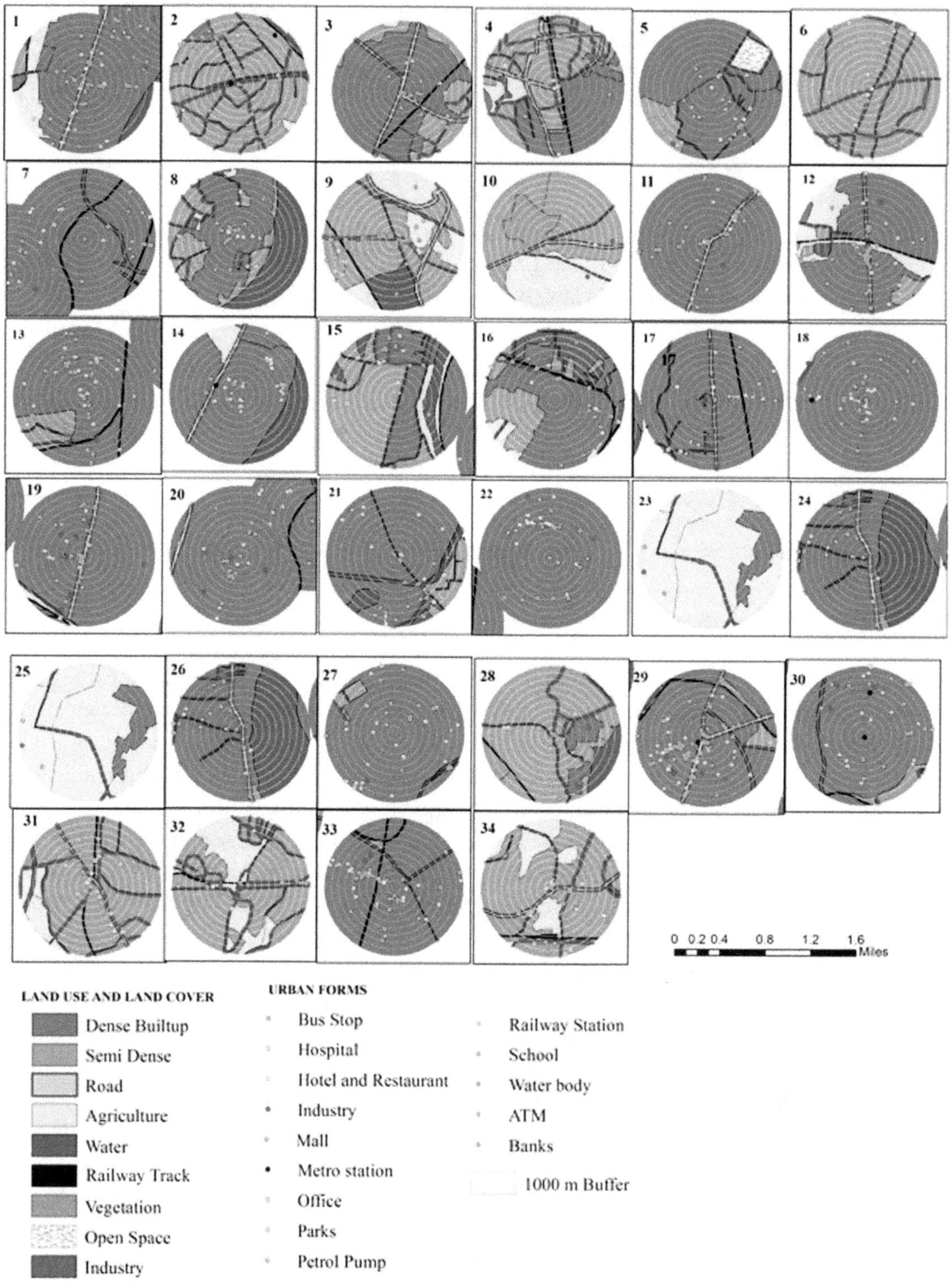

Table 3. Urban forms considered for the study

Average Distance from School in m	Average Distance from Office in m
Average Distance from Hospital in m	Average Distance from Malls in m
Average Distance from industry in m	Average Distance from Hotel in m
Average Distance from Bustop in m	Average Distance of Park in m
Average Distance Pump in m	Average Distance Water in m
Average Distance Rail in m	Dominant Building material
Average Distance from Metro in m	Avg. Building height in ft.
Average Distance from Bank in m	Building density in percentage
Average Distance from ATM in m	

Correlation

The UHI phenomenon in Kolkata has been a proven fact in many literatures. In order to analyse the relationship between the LST and the air pollution concentration, the LST maps and the interpolated maps of all three pollutants SO2, NO2 and PM10 were separately used in pairs for correlation analysis. Primarily the 1000m buffers encircling the monitoring points were overplayed on both the LST and interpolated pollutant maps. Sampling points were randomly generated inside the circles with the help of ArcGIS10.3 software. The sampling points lying over the map were used to extract the pixel value of both the LST map and that of the interpolated map. Ones the data got attached with the points under each buffer the data was exported in MS-excel and correlation was computed. The data was also graphically presented in the form of scatter plot and a trend line was fitted. R^2 value that is of correlation coefficient of determination was generated to apprehend the strength of the relationship between the variable pairs. In this process the relationship between SO2-LST, NO2-LST and PM10-LST was derived for 2006, 2011 and 2018.

Cluster Analysis

In order to identify the locations which need immediate policy intervention to reduce pollution concentration. Ho-Spot analysis or cluster analysis was carried out in ArcGIS3.10. The classification of the monitoring station was done in order to understand their spatial agglomeration. Different indicators are available for categorising the clustering and its associations. The most widely used index is the Local Indicator of Spatial Association (LISA statistic or Local Moran's I) (Anselin, 1988; Tiefelsdorf et al., 1999). Hence LISA statistic has been used for this study. The spatial autocorrelation concept is based on the first geography law introduced by Tobler, (1970) "Everything is related to everything else, but nearest things are more related than distant things". The Moran's I either validates the site of interest as part of its surrounding sites (in a cluster) or differentiates the site (outlier) from the cluster (Ord and Getis, 1994).

Looking into the locations of the monitoring sites and the pollution concentration, spatial autocorrelation measure was carried out. It investigated into the degree of dependency among its neighbouring points, taking into account their similarity as well as dissimilarity and their distance relationships at a given point of time. .

A high positive local Moran's I value entails that the monitoring under study have similarly high or low values as its neighbours, thus the locations form spatial clusters (Zhang et al., 2008). Spatial clusters include High-high (HH) (high value points surrounded by high value neighbourhoods) and Low-low (LL) clusters (low value points in a low value neighbourhoods). Spatial outlier includes high–low (HL), i.e. a high value outlier in a low value neighbourhood and low–high (LH), i.e. a low value outlier in a high value neighbourhoods. Inverse distance relationship was selected in this study to divulge the local spatial patterns. In this study, the application of local Moran's I will be helpful in identifying the discrete regimes of high pollution clusters and outliers .Points with high pollution concentration value, also surrounded by points with high pollutant concentration, will be counted in the HH cluster. Similarly, points with low pollutant value and surrounded by points with low pollution values will be included in the LL cluster.

Local Moran's I will return very low or negative attribute value for the monitoring points when the pollutant concentration value of the surrounding monitoring points are dissimilar. These stations will be classified as an outlier. So it is evident that outlier stations may have high concentration value, enclosed by monitoring stations with low concentration value or may have low concentration value surrounded by stations of high pollutant concentration value. They will be called HL cluster and LH cluster respectively. Generally, the monitoring stations having p values <0.05 are known as significant clusters. The clustering nature of pollutant concentration values by applying the Local Moran's I can be well displayed with the help of GIS tools. The Moran's I can be calculated by the following Equation 6.5 and 6.6 in ArcGIS Pro, 2018.

$$I_i = \frac{x_i - \bar{X}}{S_i^2} \sum_{j=1, j\neq i}^{n} w_{i,j} \left(x_j - \bar{X}\right) \tag{7}$$

where, x_i is the pollution concentration value for station i, $\bar{X}$ is the mean of the pollution concentration value, $w_{i,j}$ is the spatial weight between ward i and j, and

$$S_i^2 = \frac{\sum_{j=1,\ j\neq i}^{n} \left(x_j - \bar{X}\right)^2}{n-1} \tag{8}$$

where, n is the total number of stations. The z_{I_i} -score for the statistics are computed as,

$$z_{I_i} = \frac{I_i - E\left[I_i\right]}{\sqrt{V\left[I_i\right]}} \tag{9}$$

where,

$$E\left[I_i\right] = -\frac{\sum_{j=1,j\neq i}^{n} w_{ij}}{n-1}$$

$$V\left[I_i\right] = E\left(I_i^2\right) - E\left(I_i\right)^2$$

Subsequent to data analysis, the data interpretation is of key importance.

RESULTS AND FINDINGS

Annual Trends in Air Pollutants

Annual trend of the pollutants such as NO2, PM10and SO2 has been shown in Figure 5. Paucity of data at regular intervals as shown in Table 1 has lead to some interpolated trend lines. Data from 2012 till 2015 were absent for most stations and hence had to be interpolated for graphical representation. The ambient air quality standards provided by the Central Pollution Control Board of India for NO2 is 80µg/cc. Figure 6 shows that initially for the years 2005 to 2008 most of the

monitoring stations have recorded NO2 concentration within the range of 50-80 µg/cc excepting a few stations, namely Rishra, Moulai, Bator, Badhaghat, Tribeni, HMC, Dhulagarh, Behala Chowrasta and Ghushuri. These regions were mainly industrial pockets located along the banks of River Hugli. Jute mills, power plants and brick kilns dominated these locations. But with the introduction of various policy measures and closure of many jute mills the level of pollution had declined to the extent to being within permissible limit. Overall the NO2 concentration of pollution have shown a declining trend for all the stations.

The PM10 levels in the study area is a matter of grave concern. The PM10 concentration is way beyond the permissible limits of 100 µg/cc for all stations. Figure 6 represents the annual trends of PM10 concentration. The bold line drawn parallel to the x axis demarcates the permissible limit. The graphs for the station HMC and Khardah have no such line as their concentration is way beyond graphical representation. PM10 which is mostly emitted through vehicular pollution and different construction activities sources have been constantly accelerating at an exponential rate and henceforth has led to the rising trend except for Uluberia nad Kalyani, though the concentration is still not within the permissible limit

The SO2 concentration values are good and at satisfactory level in most stations of KMA. . It was observed from Figure 7 that during 2012 the stations at Bator and Behala Chowrasta had observed a slight rise in the concentration level such that it has just crossed the permissible limit. An overall declining trend has been observed in the stations, except for Badhaghat, Hyde Road, Minto Road and Behala Chowrasta.

It is obvious that PM10 exerted its largest influence on UHI in spring, followed by winter and autumn, and is the weakest in summer. This variation has been considered as a follow-up reaction of seasonal PM10 variations. In cold seasons, higher concentration of pollutants exerts stronger effects to long-wave radiation than lower concentration of pollutants in warmer seasons which is similar to the findings of He, 2018.

Figure 5. Annual Trends of No2 at 34 Monitoring Stations.

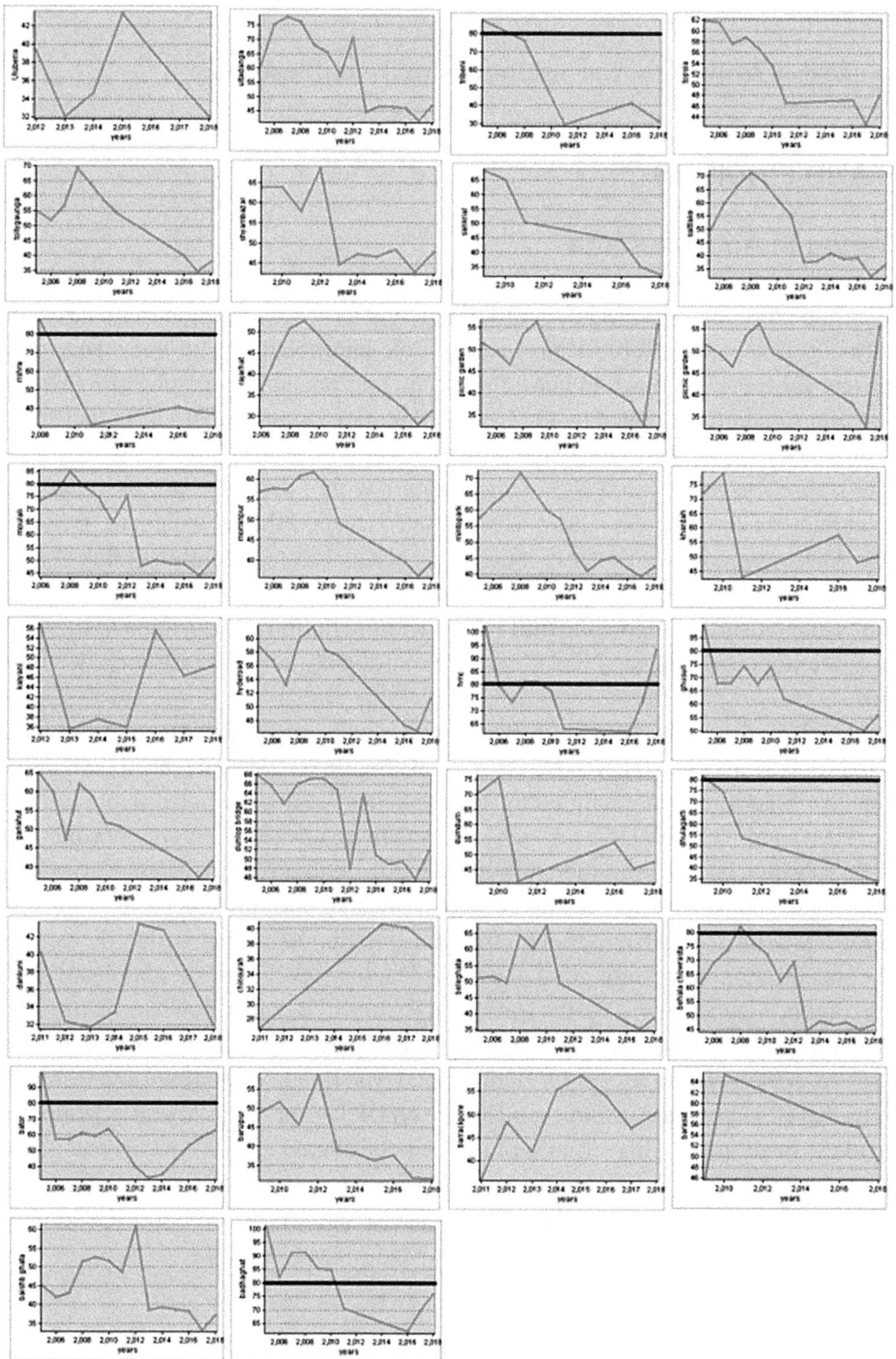

Figure 6. Annual Trends of SPM at 34 Monitoring Stations

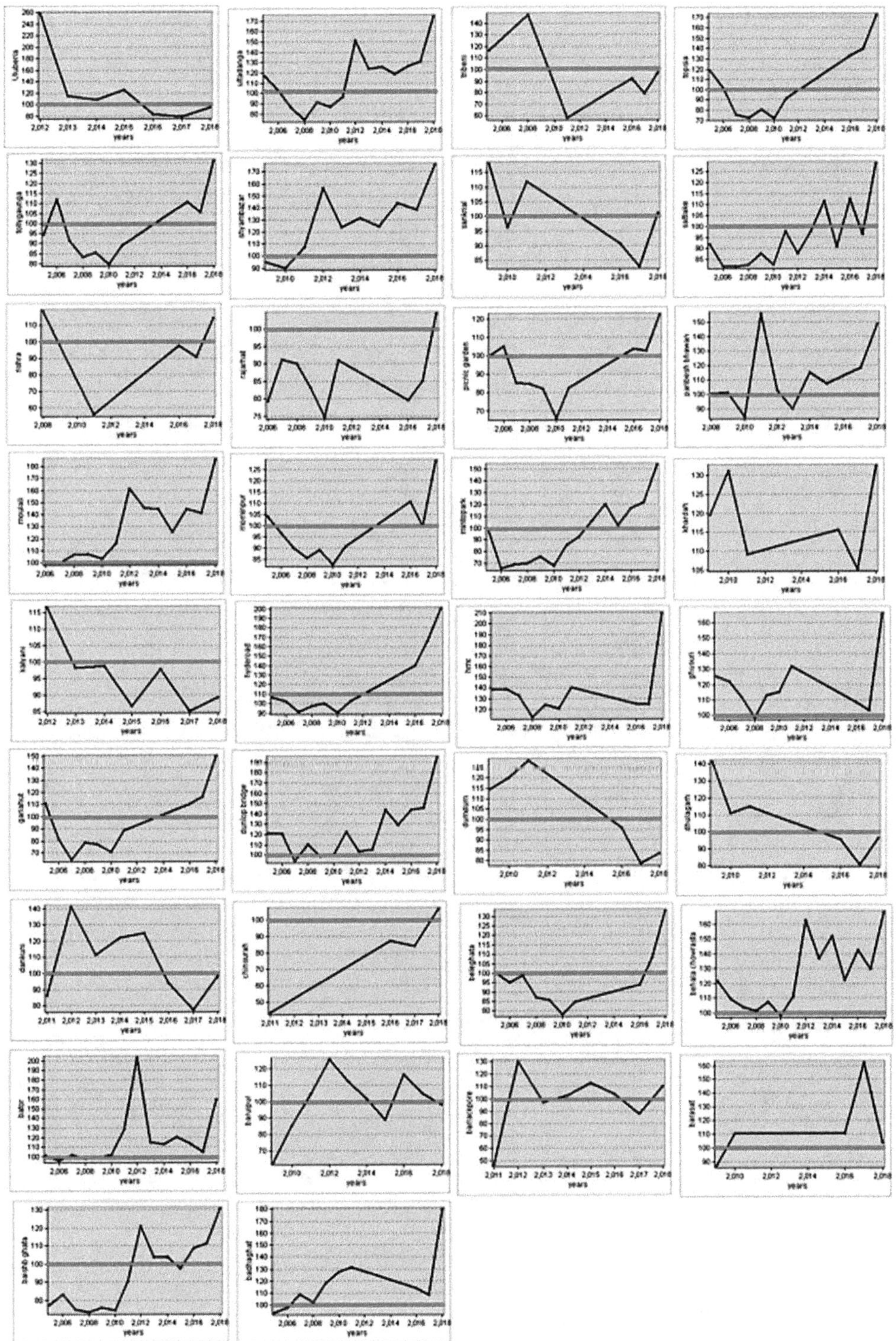

Figure 7. Annual Trends of So2 At 34 Monitoring Stations

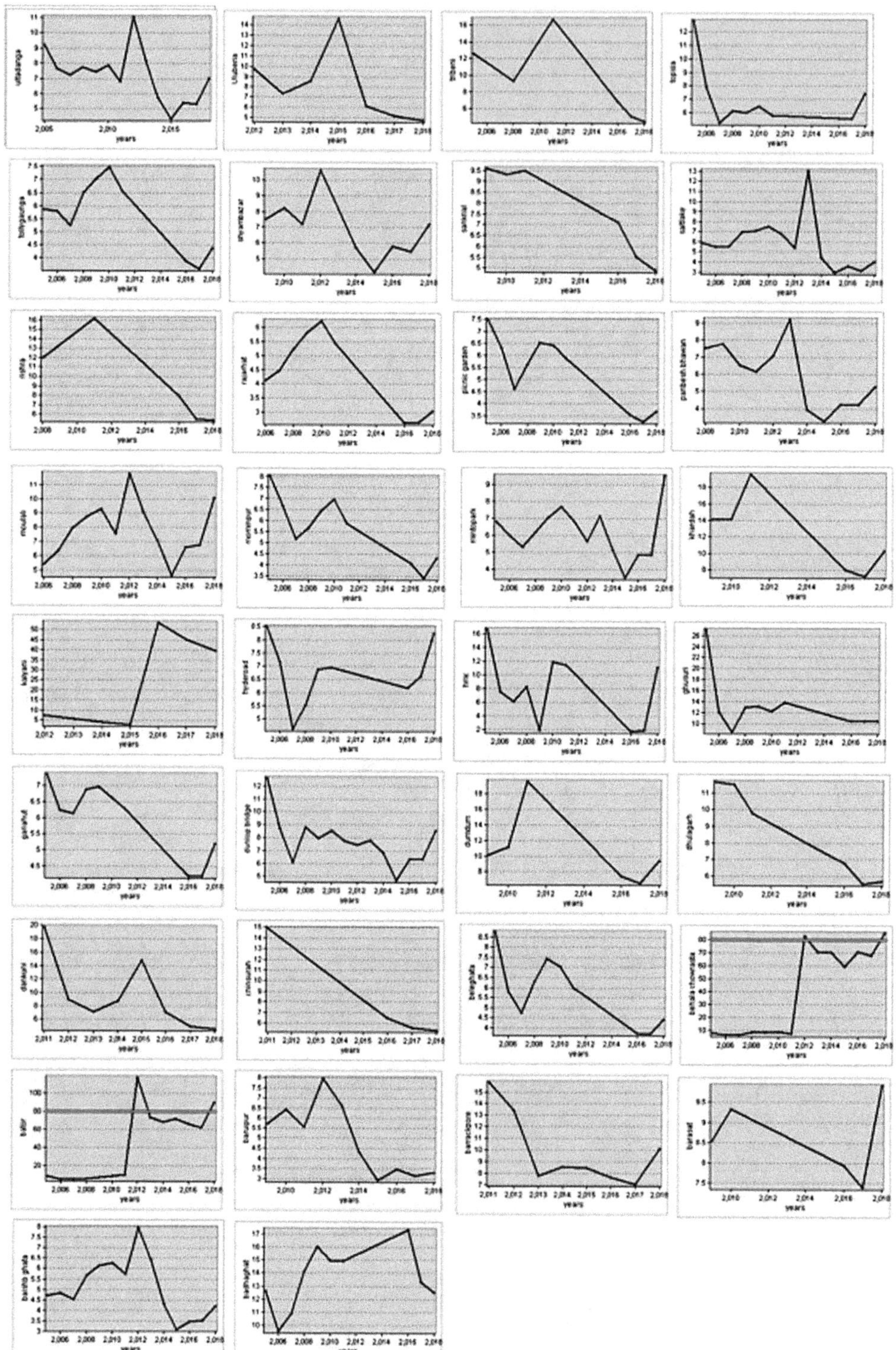

Spatial Variability of Pollution Concentration

In order to understand the spatial distribution of air pollution across the length and breadth of KMA, visual representation through maps were carried out. Figure 8 shows the spatial variability of pollutant concentration during the year 2006. Some of the stations outside KMC were not fully operational during that year. Hence there is a lack of accurate presentation of the data. Hence a very high concentration of all the pollutant parameters are explicit in and around the KMC area. The NO2 values in the area has ranged from 87.14 to 81.69 µg/cc. The PM10 values ranged from 149.33 to 131.09 µg/cc and that of SO2 from 18.99 to11.00 µg/cc. The next figure showed increase in the NO2 values reaching a high a 100.40 µg/cc near the airport region. The PM10 value was maximum near the central Kolkata and declining outwards. The PM10 value ranged to a maximum of 342.80 µg/cc while the SO2 value have ranged to maximum of 40.33 µg/cc. With the availability of data for all the stations the spatial distribution of pollutant concentration is shown accurately.

The central Kolkata experiences highest pollutant content in the air with NO2 being as high as 107.34 µg/cc, PM10 ranging from 334.00 to 48.15 µg/cc an SO2 values ranged from 155.717 to 3.51 µg/cc The hotspot analysis carried out in order to determine the cluster of air pollution monitoring stations having the highest concentration of pollution. The Figure 9 shows that the monitoring station situated around the KMC area are high cluster spots having similar behaviour for PM10 and NO2 concentration levels during 2016, 2017 and 2018. As the SO2 values are low and vary at different monitoring station they do not show any clusters.

Urban Heat Island in KMA

Figure 8 shows the LST derived from the Landsat TM 5 for the year 2006. The temperature during the winter day of 26th January showed a temperature range of 17 to 30 degree centigrade. The highest temperature is seen in and around the core urban area of KMC, Khidirpur dock, the airport runway and the Bantala Leathrer complex. The river Hugli and the East Kolkata wetlands show minima temperatures.

Figure 8. LST, NO2, SOM and SO2 maps for 2006,2011 and 2018

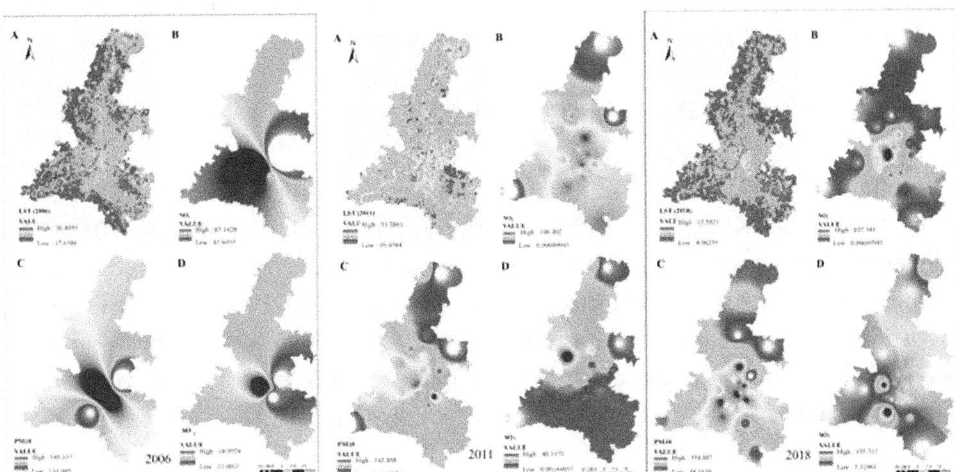

Figure 9. Hot spot map of monitoring sites

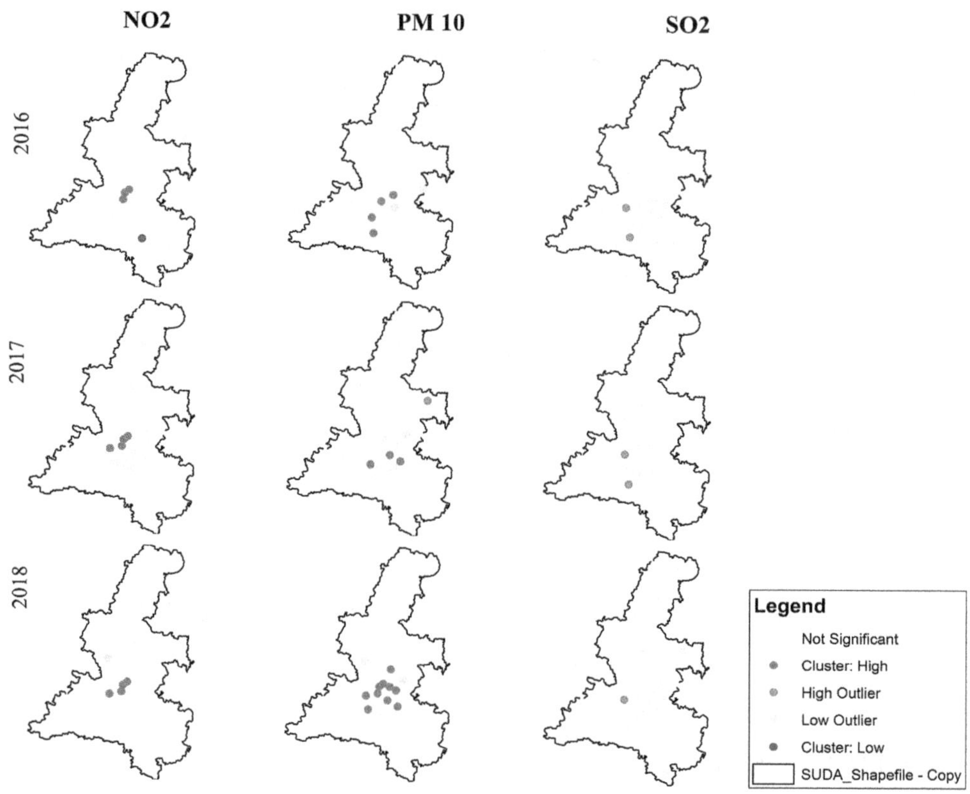

Another Landsat image of 9th February2011 was analysed to generate the LST map. The urban sprawl and constant increase in built up density have led to the areal expansion of the UHI formed near KMC. The increasing number of factories and brick kilns especially have led to the development of heat pockets along the banks of river Hugli. The temperature range during this year varied from 19.03 to 33.28 degree centigrade. The water bodies around the study area have shown cooler temperature conditions. For the year 2018 the LST map was generated for 27th Jan. The temperature range is seen to be between 15.59 to 4.96 degree centigrade. Heat island pockets near Dhulagarh is seen to have expanded as a result of new industries being established in the area.

Relationship Between Air Pollutants and LST

The correlation analysis conducted between the air pollutants and the LST of the same day is depicted in Figure 10. 10 A, D, G show the relationship between PM10 and LST through 2006, 2011 and 2018. The R2 and R value show an increase in the strength of the positive relationship. Similar relationship though moderately strong have been identified between NO2 and LST as well as SO2 and LST. Table 4 shows the correlation value between the pollutants and LST which determines that the greenhouse gases such as SO2, NO2 and the aerosol particulate matter of PM10 definitely plays a major role in intensifying the UHI in KMA

Figure 10. Scatter plot among LST and PM10 shown for 2006, 2011 and 2016 shown in A, D, G. Scatter plot among LST and NO2 shown for 2006, 2011 and 2016 in B, E and H. Scatter plot among LST and SO2 shown for 2006, 2011 and 2016 in C, F and I

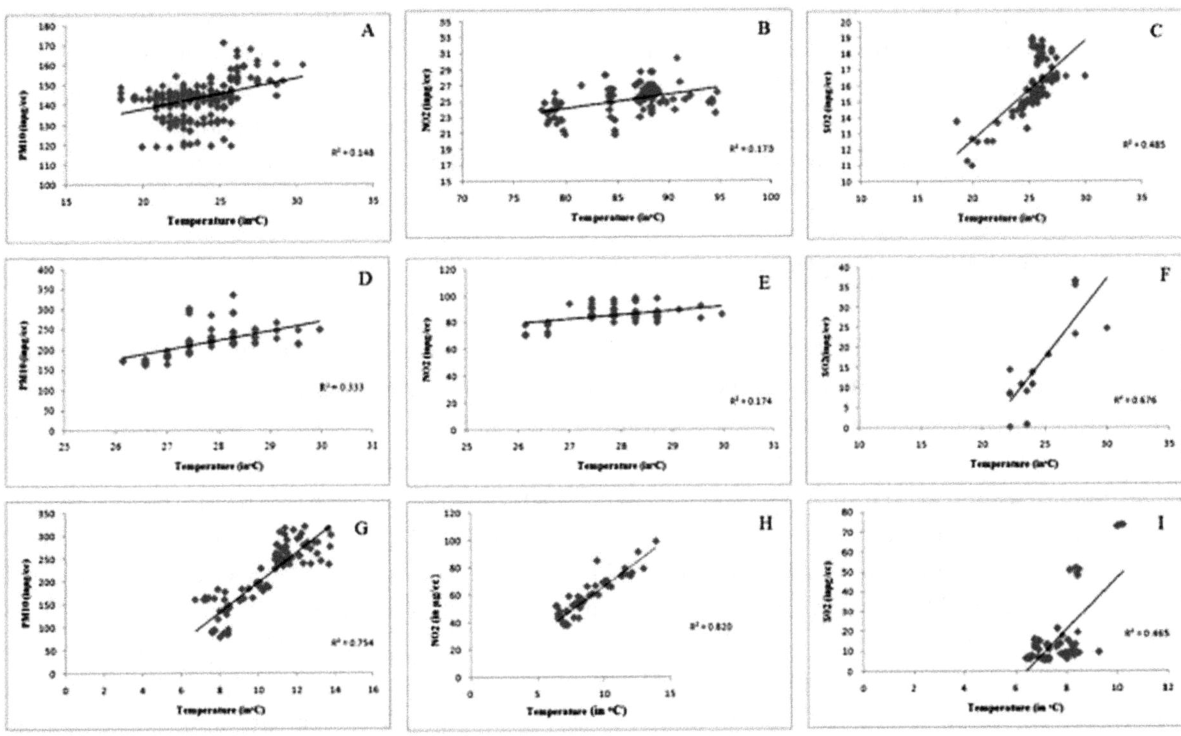

Table 4. Correlation R between LST and pollutant gases

Year	Variable 1	Variable 2	R
2006	LST	NO2	0.44
2011			0.42
2018			0.91
2006		SO2	0.7
2011			0.82
2018			0.68
2006		PM10	0.38
2011			0.58
2018			0.87

Urban Forms and Its Relation to Air Pollution

Of all the urban forms enlisted in Table 3 the multiple regression analysis carried out showed that there was a moderate positive bonding between the built up density and NO2 pollution levels, strong negative relationship between the distance from water bodies and NO2 and also a strong negative relationship between distance of bus stops to NO2 concentration. No relationship between SO2 concentration and various urban forms were established as the level of SO2 has been very low throughout the buffer zone. A strong negative relationship was seen to be present between PM10 and distance from the park.

SOLUTIONS AND RECOMMENDATIONS

The problem of air pollution in Kolkata calls for some serious mitigation measures. The Central Pollution Control Board (CPCB) had developed the Comprehensive Environmental Pollution Index (CEPI), a methodology to measure air, water, and soil pollution at the industrial clusters in the country (CPCB, 2010). Heavy manufacturing units and factories cannot function without continuous supply of electricity and are bound to depend on grid electricity. But frequent shortage of supply of electricity, power cuts have compelled them to use coal, diesel and heavy fuel oil which are obvious source pollutants. Studies by Maithel et al., 2012, Guttikunda and Jawahar, 2014 have identified this Indio Gangetic belt of industries as the solely responsible for high premature mortality as a result of outdoor pollution.

In line with our findings it has been observed that in most Indian cities, SO2 is the only pollutant that complies with Ambient air quality standards. It has been mostly because of compulsory introduction of Bharat-4 diesel and Bharat-3 diesel in different parts of the country. Repositioning or introduction of green technologies in industries consuming coal and diesel with better efficiency norms have led to the successful implementation of the anti-pollution measures. But such steps in order to reduce the PM10, and NO2 concentrations have not been successful. Vehicular emission is a major contributor in the emission of both PM10 and NO2. The transport sector and construction sector plays a vital role in economic development of a region, as a result the contribution of vehicular emissions is sure to increase (Ghate and Sundar, 2013). Under such circumstances following the global emission standard enforcement of Bharat-5 (equivalent to Euro-5) must be introduced (Guttikunda and Mohan, 2014). Any delay in strict enforcement of such norms will bring about the improvement in air quality at a delayed pace in all the Indian cities.

A " one nation, one fuel standard" norm must be made mandatory for better air quality in the cities. Efficient implementation of public transport system like that of Jawaharlal Nehru National Urban Renewal Mission (JNNURM) programs will also help in effective reduction in emission levels (MoUD, 2012). Since 2009, over 14,000 new buses were delivered under this program. Kolkata needs to increase the number thrice in order to curb the usage of private 4 wheelers vehicles. Congestion pricing policies and BRT (Bus Rapid Transit system) corridors are difficult to replicate in Kolkata, but well planned attempts in certain routes can enhance air quality conditions. Escalation in prices of private vehicles, parking costs, road tax and improvement of safe public transport system can lead to low emission conditions (Barter, 2012; CSE, 2012). China have framed national regulations to curtail pollution levels which was enacted from September, 2008. According to it taxes were hiked on big cars and reduced on smaller ones car owners with engines above 4-L capacity had to pay a 40% tax; cars with engines above 3-L capacity had to pay 15% to25% tax; 1% to 3% tax was introduced for cars with engines below 1-L capacity (Murad,

2008). China also introduced a policy to limit the number of licenses issued per year, where the license plates were auctioned in the cities of Beijing, Shanghai, and Guangzhou. Such measures can also be taken in metropolitan cities like Kolkata in order to reduce the vehicular emissions. Stringent pollution control regulations such as mandating flue gas desulfurization, introduction of tighter emission standards for all pollutants, and updating procedures for environment impact assessments of any constructional work are imperative to lower the emission level and reduce its adverse impact on health. Besides flue gas PM emissions, fugitive dust from coal-handling plants and ash ponds are also a threat to the pollution a problem. Use of ash from brick kilns and ash ponds of factories for building materials may also help in minimizing the pollutants . To date, percentage of ash utilized in the construction sector is low. High power cuts and high usage of diesel generators are also a threat to emission levels. An alternative to DG sets is not simple. Thus increase in the number of power plants to meet the electricity demand and reduction of the transmission losses or increased supply of alternative renewable energy are better long term sustainable solution to combat pollution concentration in Kolkata.

A number of studies have highlighted that construction activities are an important source of PM, combined with sources such as road dust (Pant and Harrison, 2012). A study carried out in six Indian cities showed that 10% of PM emission are sourced from construction activities annually (Guttikunda and Jawahar, 2012; CPCB, 2010). There is a vast scope for carrying out best practices in the construction industry to minimise the PM emissions. Dust is a major component in PM for many cities in India and are mostly sourced as a result of wear and tear of tyres and materials from the roads, pavements, and street furniture. Dust loadings increase even further, if the roads are not paved (CPCB, 2010). Vacuum cleaners and water sprinklers attached to heavy-duty or light-duty trucks help to suck up dust from the roads, so that the resuspension of any leftover dust is suppressed. Since the road dust accounts for up to 30 to 40% of the PM10 pollution in most cities (CPCB, 2010; Guttikunda and Jawahar, 2012), an instant enforcement of such initiatives in Kolkata will give immediate results of improved air quality. Most of the research initiatives are focused on PM10 and PM2.5. Due to the undesirable health impacts, there is a growing need for assessment of particles in the smaller size ranges (PM1 and less).). Semi-volatile and volatile organic species have also not been researched upon in much detail (Krishna, 2012; Ram et al.,2014). Some recent work focused on unique sources such as funeral pyres (Chakrabarty et al., 2013) and festive biomass burning (Deka and Hoque, 2014) and specific chemical components of PM such as brown carbon (Srinivas and Sarin, 2014).

CONCLUSION

This paper has looked into air quality parameters of SO2, NO2 and PM10 in KMA, India. The sources of pollution are absolutely indispensible because minimising them would prevent developmental activities, job opportunities as a whole under the third world economic circumstances. But the health and climatic impact of the pollution concentration cannot be overlooked. Large number of studies have taken place in Kolkata on its air quality conditions but very few have dealt with KMA. Hence the work has first dealt with the spatiotemporal variation of pollution in the region. It was identified that SO2 and NO2 are within the permissible limits throughout 34 stations studied. But the condition of PM10 was alarming in all the stations irrespective of being situated in the urban core or rural urban periphery. Rishra, Moulali, Bator, Badhaghat, Tribeni, HMC, Dhulagarh, Behala Chowrasta and Ghushuri monitoring stations have observed50-80 μg/cc of NO2 concentrations whereas PM10 concentration has been way beyond the permissible limits of 100 µg/cc for all stations and SO2 levels have been dwindling below the permissible limits during the study period. The pollutants were more concentrated in the KMC region as was depicted by the interpolated maps and Hot spot analysis. Large number of studies have been done to understand the cause of such scenario, but the impacts of urban forms of various types have never been looked into to understand their impact on pollution distribution. It was found that of all the variables taken into consideration had a moderate positive connection between the built up density and NO2 pollution levels, strong negative relationship between the distance from water bodies and NO2 and also a strong negative relationship between distance of bus stops to NO2 concentration were identified. No relationship between SO2 concentration and various urban forms were established as the level of SO2 has been very low throughout KMA. A strong negative relationship was seen to be present between PM10 and distance from the park. SO2, NO2 and PM10 which play significant roles in entrapping long wave terrestrial radiation. This paper investigated as to whether the UHI phenomena in KMA area has any relationship or not with air pollution. Paucity of data at some stations during the early years of 2006 and 2011 lead to moderately strong and positive relationship between the pollutants and LST. Strong positive relationship with R values of 0.78 was derived between the spatial datasets of LST and air pollutants for the year 2018. Thus, it was proved that these pollutants play a significant role in intensifying the UHI phenomenon in Kolkata. Some pollution mitigation strategies such as introduction of Euro-5 norms, building of BRT corridors, increased taxation on private vehicles and improved public transportation system were recommended to curb the vehicular emission. In order to curb industrial and construction related pollution suggestions on usage of green technology were given. Initiatives at individual levels can also bring about large changes in the pollution scenario. Just preferring public transport to that of private can bring about drastic reduction in emission standards.

ACKNOWLEDGMENT

We would like to express our gratitude to many who helped us selflessly in completing this work among various hardships. We would specially our students, friends and family. Rinku Dey, Saptaparna Mitra, Ananya Mukherjee, Mitu Bhuin, Sunipa Mandal, Tinni Parial, Santanu Mondal, Manisha Parial and Dr. Anuradha Mondol needs special mention for their constant support. Lastly we would like to thank Mr. Ananda Modak, Executive editor of Indian Journal of Geosciences, a Quarterly Journal of Geological Survey of India for proof reading and copy editing this research work.

REFERENCES

Akher, S. K., & Chattopadhyay, D. S. (2017). Impact of urbanization on land surface temperature - a case study of Kolkata New Town. *International Journal of Engineering Science*, *6*(1), 71–81. doi:10.9790/1813-0601027181

Alexander, B. (2016). Megacities, air quality and climate. *Atmospheric Environment*, *126*, 235–249. doi:10.1016/j.atmosenv.2015.11.059

Aliyu, Y. A., & Botai, J. O. (2018). Reviewing the local and global implications of air pollution trends in Zaria, northern Nigeria. *Urban Climate*, *26*, 51–59. doi:10.1016/j.uclim.2018.08.008

Amann, M., Purohit, P., Bhanarkar, A. D., Bertok, I., Borken-Kleefeld, J., Cofala, J., ... Vardhan, B. H. (2017). Managing future air quality in megacities: A case study for Delhi. *Atmospheric Environment*, *161*, 99–111. doi:10.1016/j.atmosenv.2017.04.041

Ambient, WHO. (2018). *(Outdoor) Air quality database (update 2018)*. Geneva, Switzerland: World Health Organization. Retrieved from http://www.who.int/phe/health_topics/outdoorair/databases/cities/

Anselin, L. (1988). *Spatial econometrics: Methods and models*. Springer Netherlands. doi:10.1007/978-94-015-7799-1

Arditsoglou, A., & Samara, C. (2005). Levels of total suspended particulate matter and major trace elements in Kosovo: A source identification and apportionment study. *Chemosphere*, *59*(5), 669–678. doi:10.1016/j.chemosphere.2004.10.056 PMID:15792664

Barter, P. A. (2012). Off-street parking policy surprises in Asian cities. *Cities (London, England)*, *29*(1), 23–31. doi:10.1016/j.cities.2011.06.007

Bechle, M. J., Millet, D. B., & Marshall, J. D. (2011). Effects of income and urban form on urban NO2: Global evidence from satellites. *Environmental Science & Technology*, *45*(11), 4914–4919. doi:10.1021/es103866b PMID:21542624

Bento, A. M., Cropper, M. L., Mobarak, A. M., & Vinha, K. (2005). The effects of urban spatial structure on travel demand in the United States. *The Review of Economics and Statistics*, *87*(3), 466–478. doi:10.1162/0034653054638292

Bhatta, B. (2010). *Analysis of urban growth and sprawl from remote sensing data*. Heidelberg, Germany: Springer. doi:10.1007/978-3-642-05299-6

Bin, Z., You, J., Yan, L., Duan, X., Zhao, X., Xin, F., ... Shenxin, L. (2019). Air pollution intervention and life-saving effect in China. Environment International. *Environment International*, *125*, 529–554. doi:10.1016/j.envint.2018.10.045 PMID:30612707

Borrego, C., Martins, H., Tchepel, O., Salmim, L., Monteiro, A., & Miranda, A. I. (2006). How urban structure can affect city sustainability from an air quality perspective. *Environmental Modelling & Software*, *21*(4), 461–467. doi:10.1016/j.envsoft.2004.07.009

Bosco, M. L., Varrica, D., & Dongarra, G. (2005). Case study: Inorganic pollutants associated with particulate matter from an area near a petrochemical plant. *Environmental Research*, *99*(1), 18–30. doi:10.1016/j.envres.2004.09.011 PMID:16053924

Bourotte, C., Forti, M.-C., Taniguchi, S., Bicego, M. C., & Lotufo, P. A. (2005). A wintertime study of PAHs in fine and coarse aerosols in Sa o Paulo city, Brazil. *Atmospheric Environment*, *39*(21), 3799–3811. doi:10.1016/j.atmosenv.2005.02.054

Bradley, B., & Keith, D. (2013). Urban form, air pollution, and CO2 emissions in large U.S. metropolitan areas. *The Professional Geographer*, *65*(4), 612–635. doi:10.1080/00330124.2013.799991

Brian, S. Jr. (2008). Urban sprawl and air quality in large US cities. *Journal of Environmental Management*, *86*(4), 688–698. doi:10.1016/j.jenvman.2006.12.034 PMID:17368703

Cao, L., Tian, W., Nu, B., Zhang, Y., & Wang, P. (2002). Preliminary study of airborne particulate matter in a Beijing sampling station by instrumental neutron activation analysis. *Atmospheric Environment*, *36*(12), 1951–1956. doi:10.1016/S1352-2310(02)00178-4

Cárdenas, R. M., Laura, D.-C., & Walid, O. (2016). Air pollution and urban structure linkages: Evidence from European cities. *Renewable & Sustainable Energy Reviews*, *53*, 1–9. doi:10.1016/j.rser.2015.07.190

Caricchia, A., Chiavarini, S., & Pezza, M. (1999). Polycyclic aromatic hydrocarbons in the urban atmospheric particulate matter in the city of Naples (Italy). *Atmospheric Environment*, *33*(23), 3731–3738. doi:10.1016/S1352-2310(99)00199-5

Carl, G., Stéphane, R., & Jacques-François, T. (2012). Are compact cities environmentally friendly? *Journal of Urban Economics*, *72*(2–3), 123–136.

Carlos Borrego, O. (2006). Traffic-related particulate air pollution exposure in urban areas. *Atmospheric Environment*, *40*(37), 7205–7214. doi:10.1016/j.atmosenv.2006.06.020

Celis, J. E., Morales, J. R., Zaror, C. A., & Inzunza, J. C. (2004). A study of the particulate matter PM10 composition in the atmosphere of Chillan, Chile. *Chemosphere*, *54*(4), 541–550. doi:10.1016/S0045-6535(03)00711-2 PMID:14581056

Cervero, R., & Murakami, J. (2010). Effects of built environments on vehicle miles traveled: Evidence from 370 US metropolitan areas. *Environment & Planning A*, *42*(2), 400–418. doi:10.1068/a4236

Chakrabarty, R. K., Pervez, S., Chow, J. C., Watson, J. G., Dewangan, S., Robles, J., & Tian, G. (2013). Funeral pyres in South Asia: Brown carbon aerosol emissions and climate impacts. *Environmental Science & Technology Letters*, *1*(1), 44–48. doi:10.1021/ez4000669

Chandrasekhar, S. (2013). *Radiative transfer*. Courier Corporation.

Chao, L., Zhanyong, W., & Bai, L. (2019). Investigating the relationship between air pollution variation and urban form. *Building and Environment*, *147*, 559–568. doi:10.1016/j.buildenv.2018.06.038

Chen, H., Jia, B., & Lau, S. S. Y. (2008). Sustainable urban form for Chinese compact cities: Challenges of a rapid urbanized economy. *Habitat International*, *32*(1), 28–40. doi:10.1016/j.habitatint.2007.06.005

Chen, L., & Yi, L. (2016). Effects of China's urban form on urban air quality. *Urban Studies Journal Limited*, *53*(12), 2607–2623. doi:10.1177/0042098015594080

Cheng, M. D., Gao, N., & Hopke, P. K. (1996). Source apportionment study of nitrogen species measured in southern California in 1987. *Journal of Environmental Engineering*, *122*(3), 183–190. doi:10.1061/(ASCE)0733-9372(1996)122:3(183)

Cho, H.-S., & Choi, M. J. (2014). Effects of compact urban development on air pollution: empirical evidence from Korea. *Sustainability*, *6*(9), 5968–5982. doi:10.3390u6095968

Chowdhury, K. A., Debsarkar, A., & Chakrabarty, S. (2012). Analysis of daytime traffic noise level: A case study of Kolkata, India. *International Journal Environment Science Research*, *2*(1), 114–118.

Clark, N. A., Allen, R. W., Hystad, P., Wallace, L., Dell, S. D., Foty, R., ... Wheeler, A. J. (2010). Exploring variation and predictors of residential fine particulate matter infiltration. *International Journal of Environmental Research and Public Health*, *7*(8), 3211–3224. doi:10.3390/ijerph7083211 PMID:20948956

CPCB. (2010). *Air quality monitoring*. Emission Inventory and Source Apportionment Study for Indian Cities.

CSE. (2012). *Parking policy: Getting the principles right, policy brief.* New Delhi, India: Center for Science and Environment.

Das, M., Maiti, K. S., & Mukhopadhyay, U. (2006). Distribution of PM2.5 and PM10-2.5 in PM10 fraction in ambient air due to vehicular pollution in Kolkata megacity. *Environmental Monitoring and Assessment*, *122*(1-3), 111–123. doi:10.100710661-005-9168-3 PMID:16770501

De Ridder, K., Lefebre, F., Adriaensen, S., Arnold, U., Beckroege, W., Bronner, C., ... Weber, C. (2008). Simulating the impact of urban sprawl on air quality and population exposure in the German Ruhr area. Part I: Reproducing the base state. *Atmospheric Environment*, *42*(30), 7059–7069. doi:10.1016/j.atmosenv.2008.06.045

Deka, P., & Hoque, R. R. (2014). Incremental effect of festive biomass burning on wintertime PM10 in Brahmaputra Valley of Northeast India. *Atmospheric Research*, *143*, 380–391. doi:10.1016/j.atmosres.2014.03.003

Dockery, D. W., Pope, C. A. III, Xu, X., Spengler, J. D., Ware, J. H., Fay, M. E., ... Speizer, F. E. (1993). An association between air pollution and mortality in six U.S. cities. *The New England Journal of Medicine*, *329*(24), 1753–1759. doi:10.1056/NEJM199312093292401 PMID:8179653

Emberson, L. D., Ashmore, M. R., Murray, F., Percy, K. E., Izuta, T., Zheng, Y., ... Domingos, M. (2001). Impacts of air pollutants on vegetation in developing countries. *Water, Air, and Soil Pollution*, *130*(1/4), 107–118. doi:10.1023/A:1012251503358

Ewing, R., & Cervero, R. (2010). Travel and the built environment: A meta-analysis. *Journal of the American Planning Association*, *76*(3), 265–294. doi:10.1080/01944361003766766

Ewing, R., Pendall, R., & Chen, D. (2003). Measuring sprawl and its transportation impacts. *Transportation Research Record: Journal of the Transportation Research Board*, *1831*(1), 175–183. doi:10.3141/1831-20

Garg, A., Shukla, P. R., Bhattacharya, S., & Dadhwal, V. K. (2001). Sub-region (district) and sector level SO2 and NO(x) emissions for India: Assessment of inventories and mitigation flexibility. *Atmospheric Environment*, *35*(4), 703–713. doi:10.1016/S1352-2310(00)00316-2

Garg, A., Shukla, P. R., & Kapshe, M. (2006). The sectoral trends of multigas emissions inventory of India. *Atmospheric Environment*, *40*(24), 4608–4620. doi:10.1016/j.atmosenv.2006.03.045

Gargi, C., & Mishra, N. B. (2016). Spatio-temporal dynamics of land cover and land surface temperature in Ganges-Brahmaputra delta: A comparative analysis between India and Bangladesh. *Applied Geography (Sevenoaks, England)*, *68*, 68–83. doi:10.1016/j.apgeog.2016.01.002

Ghate, A. T., & Sundar, S. (2013). Can we reduce the rate of growth of car ownership. *Economic and Political Weekly*, *48*, 32–40.

Ghose, M. K., Paul, R., & Banerjee, S. K. (2004). Assessment of the impacts of vehicular emissions on urban air quality and its management in Indian context: The case of Kolkata. *Environmental Science & Policy, Calcutta*, *7*(4), 345–351. doi:10.1016/j.envsci.2004.05.004

Groenewegen, P. P., Van den Berg, A. E., De Vries, S., & Verheij, R. A. (2006). Vitamin G: Effects of green space on health, well-being, and social safety. *BMC Public Health*, *6*(1), 149. doi:10.1186/1471-2458-6-149 PMID:16759375

Gurjar, B. R., Butler, T. M., Lawrence, M. G., & Leliveld, J. (2007). Evaluation of emissions and air quality in megacities. *Atmospheric Environment*, *42*(7), 1593–1606. doi:10.1016/j.atmosenv.2007.10.048

Gurjar, B. R., & Lelieveld, J. (2005). New directions: Megacities and global change. *Atmospheric Environment*, *39*(2), 391–393. doi:10.1016/j.atmosenv.2004.11.002

Gurjar, B. R., Ravindra, K., & Nagpure, A. S. (2016). Air pollution trends over Indian megacities and their local-to-global implications. *Atmospheric Environment*, *142*, 475–495. doi:10.1016/j.atmosenv.2016.06.030

Guttikunda, S. K., Goel, R., & Pant, P. (2014). Nature of air pollution, emission sources, and management in the Indian cities. *Atmospheric Environment*, *95*, 501–510. doi:10.1016/j.atmosenv.2014.07.006

Guttikunda, S. K., & Jawahar, P. (2012). Application of SIM-air modeling tools to assess air quality in Indian cities. *Atmospheric Environment*, *62*, 551–561. doi:10.1016/j.atmosenv.2012.08.074

Guttikunda, S. K., & Jawahar, P. (2014). Atmospheric emissions and pollution from the coal-fired thermal Power plants in India. *Atmospheric Environment*, *92*, 449–460. doi:10.1016/j.atmosenv.2014.04.057

Guttikunda, S. K., & Mohan, D. (2014). Re-fueling road transport for better air quality in India. *Energy Policy*, *68*, 556–561. doi:10.1016/j.enpol.2013.12.067

Guttikunda Sarath, K., & Nishadh, K. A. (2019). Air pollution knowledge assessments (APnA) for 20 Indian cities. *Urban Climate*, *27*, 124–14. doi:10.1016/j.uclim.2018.11.005

Haagen-Smit, A. J. (1952). Chemistry and physiology of Los Angeles smog. *Journal of Industrial and Engineering Chemistry*, *44*(6), 1342–1346. doi:10.1021/ie50510a045

Haider, T. (2008). Meso-urban meteorological and photochemical modeling of heat island mitigation. *Atmospheric Environment*, *42*(38), 8795–8809. doi:10.1016/j.atmosenv.2008.06.036

He, B.-J. (2018). Potentials of meteorological characteristics and synoptic conditions to mitigate urban heat island effects. *Urban Climate*, *24*, 26–33. doi:10.1016/j.uclim.2018.01.004

Hinkel, K, M., Frederick, E. N., Klene, A., E., & Bell, J., H. (2003). The urban heat island in winter at Barrow, Alaska. *International Journal of Climatology*, *23*(15), 1889–1905. doi:10.1002/joc.971

Hong, Y. M., Lee, B. K., Park, K. J., Kang, M. H., Jung, Y. R., Lee, D. S., & Kim, M.-G. (2002). Atmospheric nitrogen and sulfur containing compounds for three sites of South Korea. *Atmospheric Environment*, *36*(21), 3485–3494. doi:10.1016/S1352-2310(02)00289-3

India: Air Quality Profile (IAQP). (2010).Clean Air Initiative for Asian Cities (CAI-Asia) Center, Pasig City, Philippines.

Jacobson, M. Z. (1998). Studying the effects of aerosols on vertical photolysis rate coefficient and temperature profiles over an urban airshed. *Journal of Geophysical Research*, *103*(D9), 10593–10604. doi:10.1029/98JD00287

Jauregui, E., & Luyando, E. (1999). Global radiation attenuation by air pollution and its effects on the thermal climate in Mexico City. *International Journal of Climatology*, *19*(6), 683–694. doi:10.1002/(SICI)1097-0088(199905)19:6<683::AID-JOC389>3.0.CO;2-8

Jes, F. (1999). Urban air quality. *Atmospheric Environment*, *33*(29), 4877–4900. doi:10.1016/S1352-2310(99)00290-3

Karar, K., Gupta, A. K., Kumar, A., & Biswas, A. K. (2006). Characterization and identification of the sources of chromium, zinc, lead, cadmium, nickel, manganese and iron in pm10 particulates at the two sites of Kolkata, India. *Environmental Monitoring and Assessment*, *120*(1-3), 347–360. doi:10.100710661-005-9067-7 PMID:16741802

Kazimuddin, A., Banerjee, L. (2000). Fighting for Air. *Down to Earth*, July 31.

Kevin, C., Christian, H., Barry, L., & Joyce, R. (2007). Estimating the effects of increased urbanization on surface meteorology and ozone concentrations in the New York City metropolitan region. *Atmospheric Environment*, *41*(9), 1803–1818. doi:10.1016/j.atmosenv.2006.10.076

Khandelwal, S., Goyal, R., Kaul, N., & Mathew, A. (2017). Assessment of land surface temperature variation due to change in elevation of area surrounding Jaipur, India. *The Egyptian Journal of Remote Sensing and Space Sciences*, *21*(4), 87–94.

Khoder, M. I. (2002). Atmospheric conversion of sulfur dioxide to particulate sulfate and nitrogen dioxide to particulate nitrate and gaseous nitric acid in an urban area. *Chemosphere*, *49*(6), 675–684. doi:10.1016/S0045-6535(02)00391-0 PMID:12430655

Krishna, R. R. (2012). Current atmospheric aerosol research in India. *Current Science*, 102.

Kwon, Y. Y., Sung, K. I., Hee, C. M., Sumin, K., & Chul, P. J. (2017). Effect of PCM cool roof system on the reduction in urban heat island phenomenon. *Building and Environment*, *122*, 411–421. doi:10.1016/j.buildenv.2017.06.015

Lahiri, T., Ray, R. M., Mukherjee, S., Basu, C., & Lahiri, P. (2000). Marked increase in sputum alveolar macrophages in residents of Calcutta: Possible exposure effect of severe air pollution. *Current Science India*, *78*(4), 399–406.

Lin, J., & Lee, L. C. (2004). Characterization of the concentration and distribution of urban submicron (PM1) aerosol particles. *Atmospheric Environment*, *38*(3), 469–475. doi:10.1016/j.atmosenv.2003.09.056

Liu, J., Li, W., & Jie, L. (2016). Quality screening for air quality monitoring data in China. *Environmental Pollution*, *216*, 720–723. doi:10.1016/j.envpol.2016.06.037 PMID:27376986

Lu, Z., & Streets, D. G. (2012). Increase in NOx emissions from Indian thermal power plants during 1996–2010: Unit-based inventories and multisatellite observations. *Environmental Science & Technology*, *46*(14), 7463–7470. doi:10.1021/es300831w PMID:22732062

Maithel, S., Uma, R., Bond, T., Baum, E., & Thao, V. T. K. (2012). Brick kilns performance assessment, emissions measurements, & a roadmap for cleaner brick production in India. Study report prepared by Green Knowledge Solutions, New Delhi, India.

Mala, M. (2012) Urban growth and spatial transformation of Kolkata metropolis: a continuation of colonial legacy. *Arpn Journal of Science and Technology*. 2, Special Issue.

Mark, H., Abdullah, M., Hu, J., Song, B., Niemeier, D. A., Handy, S. L., ... Kleeman, M. J. (2010). Influence of regional development policies and clean technology adoption on future air pollution exposure. *Atmospheric Environment*, *44*(4), 552–562. doi:10.1016/j.atmosenv.2009.10.041

McCarty, J., & Kaza, N. (2015). Urban form and air quality in the United States. *Landscape and Urban Planning*, *139*(7), 168–179. doi:10.1016/j.landurbplan.2015.03.008

Mondal, R., Sen, K. G., Chatterjee, M., Sen, K. B., & Sen, S. (2000). Ground level concentration of nitrogen oxides (NOx) at some traffic intersection points in Calcutta. *Atmospheric Environment*, *34*(4), 629–633. doi:10.1016/S1352-2310(99)00216-2

Monne, C. H., Branedli, O., Schaeppi, G., Schindler, C. H., Asckermannm, U., Leuenberger, P. H., & Sapaldia, T. (1995). Particulate matter < 10 lm in urban, rural and alpine air in Switzerland. *Atmospheric Environment*, *29*, 2565–2573. doi:10.1016/1352-2310(95)94999-U

MoUD. (2012). *Transforming city bus transport in India through financial assistance for bus procurement under JNNURM*. New Delhi, India: Ministry of Urban Development, the Government of India.

Mukherjee, A., & Mukherjee, G. (1998). Occupational exposure of the traffic personnel of Calcutta to lead and carbon monoxide. *Pollution Research*, *17*, 359–362.

Murad, Q. (2008). Can congestion charging soothe Beijing's woes? In China-Dialogue. Article No. 2512, London, UK.

Namdeo, A., & Bell, M. C. (2005). Characteristics and health implications of fine and coarse particulates at roadside, urban background and rural sites in UK. *Environment International*, *31*(4), 565–573. doi:10.1016/j.envint.2004.09.026 PMID:15788196

Nina, S., & Manceur, A. M. (2014). Analyzing the influence of urban forms on surface urban heat islands in Europe. *Journal of Urban Planning and Development*, *142*, 475–495.

Ord, J. K., & Getis, A. (1994). *Distributional issues concerning distance statistics*. Working paper.

Pant, P., & Harrison, R. M. (2012). Critical review of receptor modeling for particulate matter: A case study of India. *Atmospheric Environment*, *49*, 1–12. doi:10.1016/j.atmosenv.2011.11.060

Perez, P., & Reyes, J. (2002). Prediction of maximum of 24-h average of PM10 concentrations 30h in advance in Santiago, Chile. *Atmospheric Environment*, *36*(28), 4555–4561. doi:10.1016/S1352-2310(02)00419-3

Pope, C. A. III. (2000). Review: Epidemiological basis for particulate air pollution health standards. *Aerosol Science and Technology*, *32*(1), 4–14. doi:10.1080/027868200303885

Qin, B., & Sun, S. H. (2013). Planning parameters and household carbon emission: Evidence from high- and low-carbon neighborhoods in Beijing. *Habitat International*, *37*, 52–60. doi:10.1016/j.habitatint.2011.12.017

Qin, Z., Karniel, A., & Berlin, P. (2001). A mono-window algorithm for retrieving land surface temperature from Landsat TM data and its application to the Israel-Egypt border region. *International Journal of Remote Sensing*, *22*(18), 3719–374. doi:10.1080/01431160010006971

Rajkumar, W. S., & Chang, A. S. (2000). Suspended particulate matter concentrations along the East–West Corridor, Trinidad, West Indies. *Atmospheric Environment*, *34*(8), 1181–1187. doi:10.1016/S1352-2310(99)00381-7

Ram, K., Tripathi, S. N., Sarin, M. M., & Bhattu, D. (2014). Primary and secondary aerosols from an urban site (Kanpur) in the Indo-Gangetic Plain: Impact on CCN, CN concentrations and optical properties. *Atmospheric Environment*, *89*, 655–663. doi:10.1016/j.atmosenv.2014.02.009

Ravindra, K., Mittal, A. K., & Grieken, R. V. (2001). Health risk assessment of urban suspended particulate matter with special reference to polycyclic aromatic hydrocarbons: A review. *Reviews on Environmental Health*, *16*(3), 169–189. doi:10.1515/REVEH.2001.16.3.169 PMID:11765907

Reddy, M. S., & Venkataraman, C. (2002). Inventory of aerosol and sulphur dioxide emissions 645 from India: I—Fossil fuel combustion. *Atmospheric Environment*, *36*(4), 677–697. doi:10.1016/S1352-2310(01)00463-0

Riccardo, B., Mats, S., & Di, S. S. (2010). City breathability and its link to pollutant concentration distribution within urban-like geometries. *Atmospheric Environment*, *44*(15), 1894–1903. doi:10.1016/j.atmosenv.2010.02.022

Rojas-Rueda, D., de Nazelle, A., Teixidó, O., & Nieuwenhuijsen, M. J. (2012). Replacing car trips by increasing bike and public transport in the greater Barcelona metropolitan area: A health impact assessment study. *Environ Int 49*, 100-109.

Sahana, M., Ahmed, R., & Sajjad, H. (2016). Analyzing land surface temperature distribution in response to land use/land cover change using split window algorithm and spectral radiance model in sundarban biosphere reserve, India. *Modeling Earth Systems and Environment*, *2*(2), 81. doi:10.100740808-016-0135-5

Sarkar, R., & Sivaramakrishnan, L. (2015). Measurement of land surface temperature and its relation to vegetation cover: A case study of Kolkata municipal corporation. *International Journal of Current Research*, *7*(12), 24820–24825.

Sarrat, C., Lemonsu, A., Masson, V., & Guedalia, D. (2006). Impact of urban heat island on regional atmospheric pollution. *Atmospheric Environment*, *40*(10), 1743–1758. doi:10.1016/j.atmosenv.2005.11.037

Schweitzer, L., & Zhou, J. (2010). Neighborhood air quality, respiratory health, and vulnerable populations in compact and sprawled regions. *Journal of the American Planning Association*, *76*(3), 363–371. doi:10.1080/01944363.2010.486623

Azkar, M. M. B. I., Chatani, S., & Sudo, K. (2012). Simulation of urban and regional air pollution in Bangladesh. *Journal of Geophysical Research, D, Atmospheres*, *117*(D7).

Song, C., Lin, W., Xie, Y., He, J., Xi, C., Ting, W., ... Mao, H. (2017). Air pollution in China: Status and spatiotemporal variations. *Environmental Pollution*, *227*, 334–347. doi:10.1016/j.envpol.2017.04.075 PMID:28482313

Sovacool, B. K., & Brown, M. A. (2010). Twelve metropolitan carbon footprints: A preliminary comparative global assessment. *Energy Policy*, *38*(9), 4856–4869. doi:10.1016/j.enpol.2009.10.001

Srinivas, B., & Sarin, M. (2014). Brown carbon in atmospheric outflow from the Indo-Gangetic Plain: Mass absorption efficiency and temporal variability. *Atmospheric Environment*, *89*, 835–843. doi:10.1016/j.atmosenv.2014.03.030

Stanhill, G., & Kalma, J. D. (1995). Solar dimming and urban heating at Hong Kong. *International Journal of Climatology*, *15*(8), 933–941. doi:10.1002/joc.3370150807

Steve, H., & Marshall, J. D. (2010). Impacts of urban form on future US passenger-vehicle greenhouse gas emissions. *Energy Policy*, *38*(9), 4880–4887. doi:10.1016/j.enpol.2009.07.005

Taseiko, O. V., Mikhailuta, S. V., Anne, P., Lezhenin, A. A., & Zakharov, Y. V. (2009). Air pollution dispersion within urban street canyons. *Atmospheric Environment*, *43*(2), 245–252. doi:10.1016/j.atmosenv.2008.09.076

Tian, Y., Jiang, Y., Liu, Q., Xu, D., Zhao, S., He, L., ... Xu, H. (2009). Temporal and spatial trends in air quality in Beijing. *Landscape and Urban Planning*, *185*, 35–43. doi:10.1016/j.landurbplan.2019.01.006

Tiefelsdorf, M. (1998). Some practical applications of Moran's I's exact conditional distribution. *Papers in Regional Science*, *77*(2), 101–129. doi:10.1111/j.1435-5597.1998.tb00710.x

Tobler, Waldo R. (1970). A computer movie simulating urban growth in the Detroit region. *Economic Geography* (Supplemental: Proceedings, International Geographical Union. Commission on Quantitative Methods), 46, 234–240. doi:10.2307/14314

Tonne, C., Salmon, M., Sanchez, M., Sreekanth, V., Bhogadi, S., Sambandam, S., ... Marshall, J. D. (2017). Integrated assessment of exposure to PM 2.5 in South India and its relation with cardiovascular risk: Design of the CHAI observational cohort study. *International Journal of Hygiene and Environmental Health*, *220*(6), 1081–1088. doi:10.1016/j.ijheh.2017.05.005 PMID:28606699

Varentsov, M., Konstantinov, P., Samsonov, T., & Repina, I. (2014): Investigation of the urban heat island phenomenon during polar night based on experimental measurements and remote sensing of Norilsk city, Sovremennye problemy distantsionnogo zondirovaniya Zemli iz kosmosa (in Russian), 11, 329-337.

Wang, S., Jia, X. B., Carey, Z., & Hao, J. J. (2014). Effectiveness of national air pollution control policies on the air quality in metropolitan areas of China. *Journal of Environmental Sciences (China)*, *26*(1), 13–22. doi:10.1016/S1001-0742(13)60381-2 PMID:24649687

Watson, J. G., Zhu, T., Chow, J. C., Engelbrecht, J., Fujita, E. M., & Wilson, W. E. (2002). Receptor modeling application framework for particle source apportionment. *Chemosphere*, *49*(9), 1093–1136. doi:10.1016/S0045-6535(02)00243-6 PMID:12492167

West Bengal Pollution Control Board Newsletter. (2001). *The study to evaluate the lung response of Kolkata population done by Dr. Twisha Lahiri and Dr. Manas Ranjan Roy of the Chittaranjan National Cancer Institute, Kolkata and Dr.* Pulak Lahiri of Calcutta University.

WHO. (2002). *World health report*. Geneva, Switzerland: World Health Organization.

Yu, X., Guo, X., & Wu, Z. (2014). Land surface temperature retrieval from Landsat 8 TIRS—comparison between radiative transfer equation-based method, split window algorithm and single channel method. *Remote Sensing*, *6*(10), 9829–9852. doi:10.3390/rs6109829

Zhang, Y. (2008). Online-coupled meteorology and chemistry models: History, current status, and outlook. *Atmospheric Chemistry and Physics*, *8*(11), 2895–2932. doi:10.5194/acp-8-2895-2008

Zhang, Y. (2019). Dynamic effect analysis of meteorological conditions on air pollution: A case study from Beijing. *The Science of the Total Environment*, *684*(20), 178–185. doi:10.1016/j.scitotenv.2019.05.360 PMID:31153065

Zhao, S., Ye, Y., Qin, D., Yin, D., & He, J. (2017). Assessment of long-term and large-scale even-odd license plate-controlled plan effects on urban air quality and its implication. *Atmospheric Environment*, *170*, 82–95. doi:10.1016/j.atmosenv.2017.09.041

Zheng, Z., Ren, G., Wang, H., Dou, J., Gao, Z., Duan, C., ... Yang, Y.-J. (2018). Relationship between fine-particle pollution and the urban heat island in Beijing, China: Observational evidence. *Boundary-Layer Meteorology*, 169.

ADDITIONAL READING

Boubel, R. W., Fox, D. L., Turner, B., & Stern, A. C. (1994). *Fundamentals of Air Pollution*. Academic Press.

Colls (2002). *Air Pollution*. Spon.

Committee on Energy Futures and Air Pollution in Urban China and the United States (2008). *Energy Futures and Urban Air Pollutiotion*

Desonie, D. (2007). *Atmosphere: Air Pollution and Its Effects*. Chelsea House Publications.

R M Harrison, R E Hester, Anthony K Barbour, N A Burdett, John Cairns Jr (1998). *Air Pollution & Health*. Air Pollution & Health.

Kerr, J. Ph.D. Casper (2010). Fossil Fuels and Pollution: The Future of Air Quality. Facts on File.

Schwela, D., Haq, G., Huizenga, C., Han, W.-J., Fabian, H., & Ajero, M. (2006). *Urban Air Pollution in Asian Cities Status, Challenges and Management*. Earthscan Publications Ltd.

Vallero, D. (2008). *Fundamentals of air pollution*. Elsevier.

KEY TERMS AND DEFINITIONS

Correlation: When it becomes necessary to understand a relationship between two different sets of data, then the statistical technique that is applied to analyses whether there is inverse relation or a direct relation is known as correlation. Inverse relationship ius termed as negative correlation and the direct relationship is known as positive relationship.

IDW (Inverse Distance Weighted): Often due to unavailability of data/ information between the space of two geographical location causes errors in spatial analysis. Inverse distance weighted method is the algorithm which helps in interpolating the missing data under the logic that a certain event fades away from its source as the distance increases outwards.

LST (Land Surface Temperature): The earth surface and anything on it absorbs incoming solar radiation and achieves a certain temperature. The temperature measured at the source of these different land covers is termed as land surface temperature.

NO2 (Nitrogen Dioxide): Nitrogen dioxide is a green house gas often formed by dire lightning in the atmosphere and is released by plants and soil. about 80% of vehicular emissions consist of NO2 which is a result of fuel combustion.

PM (Particulate Matter): Particulate matter are formed as a result of fire, dust nd industrial emission from brick kilns, cement factories, iron and steel industry, quarrying and fossil fuel powered plants. The most abundant and researched upon PM is PM10 and PM2.5.

SO2 (Sulphur Dioxide): Sulfur dioxide is a green house gas. It is mainly formed in thermal power plants and often mix with water vapor and rain droplets leading to acid rain It is invisible and has a foul odor.

UHI (Urban Heat Island): An urban heat island (UHI) is a an atmospheric phenomenon due to high built up density and low vegetation cover the incoming solar radiation is trapped within the urban boundary layer of the atmosphere leading to the formation of a higher air and temperature over the urban centre in comparison to its rural surroundings.

APPENDIX

Table 5. Location of WBPCB monitoring sites showing the periods for which air quality data are available

Sl No	Name of Station	Latitude	Longitude	Period of Data Availability
1	Uluberia	22.478803	88.092591	2012-2018
2	Ultadanga	22.591334	88.393243	2005-2018
3	Tribeni	22.984414	88.386432	2005, 2008, 2011, 2016-2018
4	Topsia	22.542893	88.398358	2005-2011, 2016-2018
5	Tollygaunga	22.517052	88.345838	2005-2011, 2016-2019
6	Shyambazar	22.601898	88.373744	2009-2018
7	Sankrial	22.560488	88.226898	2009-2011, 2016-2018
8	Saltlake	22.583478	88.418718	2005-2018
9	Rishra	22.72119	88.354404	2008, 2011, 2016-2018
10	Rajarhat	22.619392	88.464474	2006-2011, 2016-2018
11	Picnic garden	22.524544	88.382394	2005-2011, 2016-018
12	Paribesh bhawan	22.56193	88.408315	2008-2018
13	Moulali	22.561392	88.367121	2006-2018
14	Mominpur	22.527127	88.32596	2005-2011, 2016-2018
15	Mintopark	22.541189	88.354835	2005-2018
17	Khardah	22.72399	88.373565	2009-2011, 2016-2018
18	Kalyani	22.972339	88.45877	2012-2018
19	Hyderoad	22.533876	88.308148	2005-2011, 2016-2018
20	Hmc	22.582244	88.33703	2005-2011, 2016-2018
21	Ghusuri	22.610926	88.354062	2005-2011, 2016-2018
22	Gariahut	22.516249	88.366304	2005-2011, 2016-2018
23	Dunlop bridge	22.654072	88.377148	2005-2018
24	Dumdum	22.622828	88.414265	2009-2011, 2016-2018
25	Dhulagarh	22.573805	88.175891	2009-2011, 2016- 2018
26	Dankuni	22.680887	88.297044	2011-2018
27	Chinsurah	22.88809	88.39938	2011-2018
28	Beleghata	22.56585	88.37725	2002-2011, 2016- 2018
29	Baruipur	22.362809	88.43077	2009-2018
30	Behala chowrasta	22.487082	88.313237	2002-2018
31	Bator	22.575403	88.301843	2004-2018
32	Barrackpore	22.760488	88.37032	2011-2018
33	Barasat	22.723139	88.478495	2009, 2010, 2016-2018
34	Baishnab ghata	22.466192	88.392815	2002-2018
35	Badhaghat	22.601934	88.342139	2004-2011, 2016-2018

Chapter 25
Real-Time Construction Waste Reduction Using Unmanned Aerial Vehicle

Wallace Imoudu Enegbuma
Victoria University of Wellington, New Zealand

Jibril Adewale Bamgbade
Swinburne University of Technology Sarawak, Malaysia

Cosmas Pang Han Ming
Swinburne University of Technology Sarawak, Malaysia

Chukwuka Christian Ohueri
Swinburne University of Technology Sarawak, Malaysia

Bruno Lot Tanko
https://orcid.org/0000-0001-7028-8423
University of Reading, Malaysia

Edo Oga Ojoko
Universiti Teknologi Malaysia, Malaysia

Yakubu Aminu Dodo
Universiti Sains Islam Malaysia, Malaysia

Sa'id Kori
University of Liverpool, UK

ABSTRACT

The paradigm shift towards waste reduction in the construction industry has produced revolutionary techniques in assessing and mitigating waste during the construction lifecycle. Unmanned Aerial Vehicle (UAV) technology presents potentials in real-time construction waste management for informed decision-making processes for site managers. Due to the dynamic and complex construction work environment, material and equipment, labour and management policies are subjected to latent and identified hazards. Therefore, this chapter focuses on the real time impact of construction waste. UAV technology was adopted. UAV flight mission and deployment time were designed to meet target requirements and snapshot checklist. Data images were captured from an 11-storey residential building and analysed for conformity to construction waste and safety requirements. The prevalence of exposed material and equipment, low labour awareness, and need for policy shift to circular economy was predominant. Further studies can explore multiple construction monitoring and other category of construction works.

DOI: 10.4018/978-1-7998-0369-0.ch025

INTRODUCTION

The tremendous growth on examining the impact of construction activities on the environment has spun sensitisation among all construction stakeholders. Extensive tools have been developed addressing construction waste management and characteristics. This can be broken down into waste management plans and guides, waste data collection tools, waste estimation tools and environmental assessment tools (Akinade et al., 2018). Emerging research areas in waste management focus on decision making in planning and design phase, lifecycle evaluation, circular economy approach, human factor related issues and induction of new technology (BIM, GIS, Big Data and prefabrication) for construction (Jin, Yuan, & Chen, 2019). Knowledge gaps exists in stakeholder participation and collaborative governance for waste management. Similarly, individual versus organisational behaviour is inherit in the constructions. Lack of attention to social sustainability assessment and infusing of new technology and methods (Chen, Su, Si, & Chen, 2018). Similarly, more application of BIM and UAV for site management is lacking in literature (Alizadehsalehi, Yitmen, Celik, & Arditi, 2018). Contractor employee knowledge was found to be lacking and a major predictor of construction waste reduction (Li, Zuo, Cai, & Zillante, 2018). Poor site management practices currently plagues effective construction waste minimisation (Ajayi et al., 2017). Currently, poor logistics management and site monitoring reduces productivity, increases site safety risk and waste generation (Whitlock, Abanda, Manjia, Pettang, & Nkeng, 2018).

In the design phase, seminal authors attribute drastic changes in conceptual design to accommodate waste management principles (Ajayi & Oyedele, 2018; Akinade et al., 2018; Ghisellini, Ripa, & Ulgiati, 2018; Guerra, Bakchan, Leite, & Faust, 2019; Lu, Webster, Chen, Zhang, & Chen, 2017). While, during execution phase, others (Akanbi et al., 2018; Chulkov & Fachratov, 2018; Maniam, Nagapan, Abdullah, Subramaniam, & Sohu, 2018; Subramaniam et al., 2018; Won & Cheng, 2017) have supported more site observation, simulation and quantification measurement approach. The monitoring and identification of waste and safety risks is an extremely difficult work, especially in a complicated construction site environment (Park, Kim, & Cho, 2016). Quality controls on site are time consuming and inefficient due to limitations of knowledge on work task without real-time information. This increases the level of difficulty for managers to identify and manage defects (Wang et al., 2015). Extensive use of visualization technology in construction cover small areas due to poor penetration performance requiring improvements in training, job hazard area, monitoring and warnings (Guo, Yu, & Skitmore, 2017). Manual observations have become time consuming subject to human errors and inability to cover extensive site layout (Yu, Guo, Ding, Li, & Skitmore, 2017a). Therefore, given the need to constantly monitor waste management practices during the execution phase, this chapter examines the effect of unmanned aerial vehicle (UAV) on real time waste management monitoring. This was achieved through a literature review on current dimension to waste management and UAV deployment in construction industry. Furthermore, real time monitoring was used to gather data towards immediate use by site managers and safety officers. The subsequent sections will give an expanded literature review, methodology and results.

CONSTRUCTION WASTE

Waste is described as the careless use, extravagant or purposeless use of construction resources (Oxford Dictionaries, 2019). An estimated 30% of total weight of building materials supplied to construction sites contribute to construction waste. To reduce such construction waste generation the industry need to adopt innovative cost effective strategies to reduce waste (Osmani, 2011). Globally, five key root causes of construction waste were identified as constant design changes, incorrect storage of materials, poor handling of materials, effect of weather and mistakes while ordering from suppliers (Kaliannan, Nagapan, Abdullah, Sohu, & Jhatial, 2018). In Malaysia, the top five causes of waste are poor supervision during construction, lack of efficient site management, last minute changes, cutting uneconomical sizes of materials, and constant design changes during the construction period (Nagapan et al., 2018).

Circular Economy

To reduce the amount of exploitation on earth's resources, circular economy has ultimately affected construction industry practices. Behavioural, technical, and legal perspectives were key to understanding these overall effects on construction stakeholders. Out of 22 barriers identified which affect circular economy best practice in construction waste management, finitely recyclable construction materials, ineffective C&D wastes dismantling, sorting, transporting, and recovering processes, and using finitely recyclable construction materials were found to have the most significant impact (Mahpour, 2018). Circular economy promotes maximum reuse and recycling of materials, goods and materials but whole circular economy adoption in construction is subject to site specifics and dependent on other factors such type of material, building elements, supply chain, economic and political context (Ghisellini et al., 2018). To ensure efficient circular economy, current construction and demolition waste management lack collaboration, effective policies, regulatory coordination, incentives and awareness, infrastructure for treatment and planning structure (Blaisi, 2019). In nations with scares building material, 70 kg and 29 kg CO^2 eq GHGs emissions can be saved through reuse of wood waste for particleboard and cement-bonded particleboard productions in contrast to similar use for energy generation or landfilling (Hossain & Poon, 2018). Systems dynamics modelling of waste reduction measures namely prefabrication, reduce design change, construction on-site sorting and construction materials on-site reuse produce different levels of effect. The optimal output is derived from the use of multiple waste reduction measures of approximately 40.63% of waste generated during design and construction phase (Ding, Zhu, Tam, Yi, & Tran, 2018).

Technology Innovations

The onset of new technology in the construction industry was seen a gradually shift from traditional waste management tools and techniques. BIM-based processes was identified to reduce construction waste management due to the capability to carryout design review, 3D coordination, quantity take-off, phase planning, and site utilization planning (Won & Cheng, 2017). Integration with neighbouring construction site, 3D site layout, clash avoidance, collaboration and revision of site planning were found to be innovative with a BIM platform (Whitlock et al., 2018). Although, construction industry stakeholders have a low tendency to BIM technology adoption (Enegbuma, Aliagha, Ali, & Badiru, 2016). BIM Based waste management was found to be more effective in managing construction waste due to early design phase intervention between all stakeholders towards waste reduction. This system specifically addressed efficient collaboration in design phase, waste driven design process and solutions, waste analysis throughout building lifecycle, waste intelligence analytics and waste documentation (Akinade et al., 2018). On a real-life case study, BIM automated waste algorithm found a 1.95% and 13.78% waste in concrete and drywall purchased by contractors (Guerra et al., 2019). Similarly, computational algorithm for construction waste management can automatically generate waste generation rules (WGRs) for different design options in the design phase (Lu et al., 2017). Big data generated from over 7 million construction waste disposal records from above 26 thousand sites accounted for a demolition reduction rate of 36.19% under Building Environmental Assessment Method (BEAM) in support of more sustainable practices (Lu, Chen, Peng, & Liu, 2018).

Lifecycle Management

A key component of assessing construction waste lies in construction life cycle management. At the forefront, low waste purchase management, supplier commitment waste efficient bill of quantities and waste efficient delivery management were lacking. The absence of such best practices further exacerbates efforts in reducing waste production (Ajayi & Oyedele, 2018). Contractor employee training, stricter regulation and public awareness on construction waste management are critical to predicting planned behaviours on site (Li et al., 2018). Adherence to project drawings, fewer changes to design, provision of waste skips for different types of waste and maximisation of material reuse on site translated to improved construction waste management (Ajayi et al., 2017). A study using waste generation rates calculated several waste results by building category as residential projects (197.02 tons), Social amenities project (30.73 tons) and Non-residential (84.40 tons). Project size and gross floor area were key indices in deriving the differences as residential projects floor design is more complicated and larger scaled compared with other types of projects (Subramaniam et al., 2018). Waste quantification through site observation of material flow was pivotal in assessing waste management during construction (Hassan et al., 2018).

Method of construction such as conventional construction method (CCM), mixed construction method (MCM) and industrialized building system (IBS) were found to have varied output in construction wastes as 197.56 tons, 53.19 tons and 77.19 tons respectively. The average per square metres were IBS 0.018 tons/sm2, MCM 0.030 tons/sm2 and CCM 0.046 tons/sm2 (Maniam et al., 2018). A thought-provoking research delved into the establishment of a mobile materials recovery facility (MRF) 3 km from waste source site to reduce the cost of transportation and greenhouse gas involved in transportation. 50.8% GHG emissions and 68.1% of cost saving were reduced for a more eco-efficient construction (Mah, Fujiwara, & Ho, 2018). BIM Athena waste calculation tool resulted in a 56% waste reduction in building lifecycle assessment (Jalaei, Zoghi, & Khoshand, 2019). Whole-life salvage performance using BIM found that steel building design with demountable connections, and prefabricated assemblies produce recoverable materials that are mostly reusable (Akanbi et al., 2018). On the far end of the lifecycle spectrum, concrete waste has shown usefulness in crushing and reuse for production of reinforced concrete material elements (Chulkov & Fachratov, 2018). Maintaining this path of research and practice by construction stakeholders in the coming years will continually improve life cycle waste management.

Construction Safety

According to Occupational Safety and Health Act 1994, the safety, health and welfare of a working personal has to be secured from working hazards during the activities of works (DOSH, 2017). Waste from lack of safety adherence contributes significantly to construction waste. Wireless sensor network (WSN) removes the risk of direct contact of human and dangerous environments by monitoring and controlling hazards in a remote way, hence improves human safety of monitoring tasks (Cheung, Lin, & Lin, 2018). Working around hazards such as crane lifting, an established framework provides assistance for crane lifting operations to aid real time proactive safety. At the beginning, the reconstructions of crane poses are depended on the critical crane motions identified by sensor system. The next procedure is modelling the condition of the lift site and updating the cloud data. Finally, the risk of crane parts collision is continuously analysed and the operators are warned for any hazardous action through a graphical display (Y. Fang, Cho, & Chen, 2016). The potential of the designed platform is shown by creating safety plans based on BIM, anticipating safety in construction schedules, organizing installation and removal work, giving choices and methods for permanent building parts as well as short-term safety equipment (Zhang et al., 2015). Another research contributes to identifying an unsafe behaviour by analysing related leading posture with different parameters, evaluating the common value range of these parameters and come out with a process that identifies behaviour. The central idea is the leading posture that simplifies a potent behaviour to a static posture to advance the moment of analysing the risky behaviours. By utilizing this method, is it possible for unsafe behaviours to be warned earlier in a real time situation (Yu, Guo, Ding, Li, & Skitmore, 2017b).

REAL TIME MONITORING

Construction projects real time monitoring represent simultaneous actual events which captures inconsistencies such as structural failure, worker's injuries, cost overrun and construction wastes. Due to the advancement of world technology, there are many types of monitoring techniques developed with different scope of monitoring. Manual monitoring is the predominantly used monitoring method in Sarawak, Malaysia. This method involves only a professional safety manager/project manager that monitors from site to site for any hazardous behaviour of workers, unsafe operation of machinery and on-site environmental issues. Manual monitoring is the most traditional yet simplest monitoring method. It is simple because unlike other monitoring method, this method does not require any special equipment, tools and skills to perform. The only requirement is the qualification of the safety/project manager. The most common usage of this method is on road construction projects where visions are not hindered by high rising constructions (Melo, Costa, Álvares, & Irizarry, 2017). Manual monitoring is not considered real time monitoring unless many safety/project managers are deployed at the same time to inspect just one site. Moreover, this will cause the construction site to be crowded and, hence hindering the progress. This method consumes too much work force.

UAV Monitoring were initially applied only in military fields, however the potential of this technology has caught a significant attention to be implemented in engineering environments (Melo et al., 2017). UAV possess the ability to monitor from angles that are out of reach and replace safety/project managers to be at the job hazard area in person, which makes the monitoring work much safer. The ability to maneuver around in the construction site makes the monitoring work easier as one of these drones can cover the whole site. Whilst this method is advantageous, UAV monitoring requires trained personnel to handle the flight of the drone around the site which is inconvenient for every site to be monitored by this method. This method is not universal especially in Malaysia. In order to achieve high-quality real-time monitoring, drones must have some minimal qualification and the high-quality drones cost significantly higher.

Wearable device involves a device that are closely attach to human body, sending information of the personnel or the surrounding to a monitoring server connected to a terminal where the real time condition can be monitored by a safety manager. This method may have variable function based on the installed devices. Wearable devices have been used in health care, manufacturing, mining and athletics and have shown some positive benefits in keeping track of the health and safety of the personnel. However, the use of this method in construction projects is still in the very beginning stage and there are too few documented cases to discover the potential benefits of this method in construction industries (Awolusi, Marks and Hallowell, 2018). Wearable device only monitors the real time condition of one personnel at a time. Therefore, in order to monitor every single personnel in the construction site, multiple units of the device are needed. Thus, the maintenance work for the wearable device will be very time-consuming.

Close circuit television (CCTV) involves video surveillance technology with multiple stationary cameras directly linked to a centralized monitoring device. By employing only one person, the whole area can be monitored through the centralized monitoring system. Commonly used for safety surveillance on factories, offices and even houses. With the video footage recorded, any occurrence can be reviewed and hence improvements can be made. In order to achieve real time monitoring, safety managers must be in front of the surveillance television to respond to any occurrence of emergency. Close circuit television initially used for criminal surveillance. With multiple stationary cameras to enclose area to monitor and even recognise faces, this method can be informative enough to detect and recognise the condition of the site visually. CCTV monitoring records everything on the site. However, to achieve real time monitoring, safety/project managers are required to monitor the whole site on the centralised monitoring screen to react to any emergency. The nature of the work is boring and prone to human error (Melo et al., 2017).

Currently application of real time monitoring spans a wide range of techniques. The integration of the near-miss data-visualization tool in BIM found safety visualisation is practicable in existing BIM design software. Utilizing the developed tool, the safety data within the project BIM can be visualized and communicated by safety managers and stakeholders of the project (Shen & Marks, 2016). The monitoring work on the real time locational information of workers is continuously done and uploaded to designated cloud server. With this contextual information visualized, monitoring the workers' behaviours against the identified construction hazard was simplified (Park, Kim, & Cho, 2017). The construction, development and performance test of a real time indoor localization solution is introduced for monitoring the construction site. The proposed cloud-enabled radio-frequency identification (RFID) localization solution enables the localization of construction workers and real time visualization through various devices for remote monitoring by utilizing BIM and cloud computing (Yihai Fang, 2016). Other instances established an integration of BIM and light detection and ranging (LiDAR), a quality control system with onsite real time quality information collecting and processing (Wang et al., 2015). The GMI sensor has higher sensitivity compared to commercial gauss meter based on longitudinal effect. The practical utility of the high sensitivity sensor was tested, and the measurement of position and speed of the observed moving object was reliable and accurate. Therefore, this GMI-based sensor is proved promising in detecting real time position and monitoring oscillatory movements (Thiabgoh et al., 2018). There has also been innovative design for an autonomous system, which the site workers are monitored, localized and alerted whenever they are presented in a hazardous area, in order to protect them and avoid accidents in the site. This system is convenient, handy and is built based on the Internet of Thing (IoT). This architecture consists of miscellaneous components that integrates into a middle ware backend online server (Kanan, Elhassan, & Bensalem, 2018).

CONSTRUCTION WASTE AND UNMANNED AERIAL VEHICLE

Construction worker behaviour contributes significantly to resultant safety, risk and wastage on site (Yu et al., 2017a). UAV possesses low flying altitude with capabilities to capture high resolution ground images. This is beneficial in monitoring, investigating and recording activities on construction sites. In less photogrammetry analysis needs, digital cameras are used for real-time monitoring. Such digital cameras have advantages of size, weight, cost, reuse able memory cards and modification of captured images (Haur, Kuo, Fu, Hsu, & Heng, 2018). A review on UAV construction site monitoring identified research gaps in the need for future research focus on data collection, analytics and visualisation (Ham, Han, Lin, & Golparvar-Fard, 2016). Algorithm based on deep learning for non-hardhat-use (NHU) assisted in supervision of construction workers to avoid accidents and waste (Fang, Li, Luo, Ding, Luo, et al., 2018). In a similar study, workers without a trade certificate or checking to establish site worker is carrying out work within the constraints of their certification from video clips extraction (Fang, Li, Luo, Ding, Rose, et al., 2018). Automated scanning of sites was found to be able to detest defaulting non-hardhats worker in indoor construction sites (Mneymneh, Abbas, & Khoury, 2017). Surplus soil mapping was achieved in bridge construction using UAV monitoring. A double decimal accuracy target was also derived using this technique in noise screening and comparison of different soil heaps (Haur et al., 2018). IDEF0 language run fusion of 4D BIM and UAV assisted in significantly reducing fata and non-fatal property damaging accidents. The systems also helped to identify hazards, monitor workers and give real time updates to site managers (Alizadehsalehi et al., 2018). The conceptual framework for this chapter is surmised in Figure 1.

UAV WASTE APPROACH

This chapter started with reviewing literature on real time construction waste management in a bid to solve current site management problems. Each monitoring mechanism were evaluated to for appropriate selection of UAV checklist, mission and data capture forming the beginning of the study methodological approach. UAV possesses the capability to carry out flight survey task, enable ground control point distribution, flight route planning and revision (Haur et al., 2018). The site selected was a residential apartment building in the execution project phase. The entire land area measure 3,300sm2 with a total

Figure 1. UAV Construction Waste Management

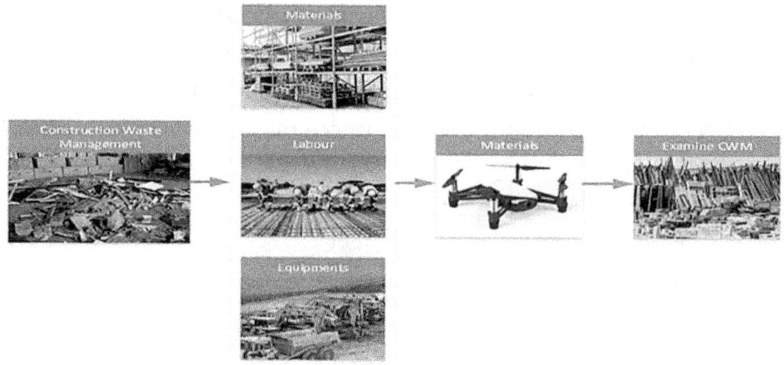

of 80 units of 11 stories building apartment. The allotted construction schedule is projected to take 24 months for completion involving over 100 labour workers. The focus was identifying factors which contribute to construction waste, monitor the cycle of waste and provide real time mitigation. Aside gathering descriptive information of the site, further meetings were held with site project managers to ascertain the fundamental workflow and operation on site. To achieve optimum UAV deployment, checklist was developed in discussion with the site project manager covering waste management factors, areas of specific works and process for real time project manager update. The protocol in Figure 2 was used during the study period (Melo et al., 2017). UAV deployment covered work hours excluding break time in successions of morning and afternoon. Due to battery life and data monitoring, flight took approximately 13 minutes for each multiple 30 seconds snap runs in set working hours. Height of 10 meters were obtainable from the UAV, traveling at 2 m/s with an image quality of up to 10MP. The data was analysed in 2 phases; real time project manager updates and the next step is image analysis for compliance with construction waste reduction.

FINDINGS AND DISCUSSION

Reporting of real time data was relayed to safety/project manager during data capture for immediate action. Figure 3 shows key real time captured images examined for recommending improvements to construction waste on site. This will be expanded in the subsequent sections.

Materials and Equipment

In Figure 3 (top left), the engine of the excavator was detected uncovered after regular check-up. This is a critical source of waste and safety hazard due to the exposed engine which was not protected and vulnerable to damage from climate and proximity to workers. It is crucial for such machinery to have regular maintenance to keep the construction progress from delay and avoid the expensive cost of repair. According to Yip, Fan, & Chiang (2014), maintenance cost is the largest single component of a machine's cost. Therefore, by cutting down unnecessary maintenance cost, the overall cost of a construction project can be efficiently reduced. Similarly, from Figure 3 (top and bottom left, bottom right) unprotected construction materials, in this case, the timbers can be seen exposed to the environment, making it vul-

Figure 2. UAV construction waste inspection protocol

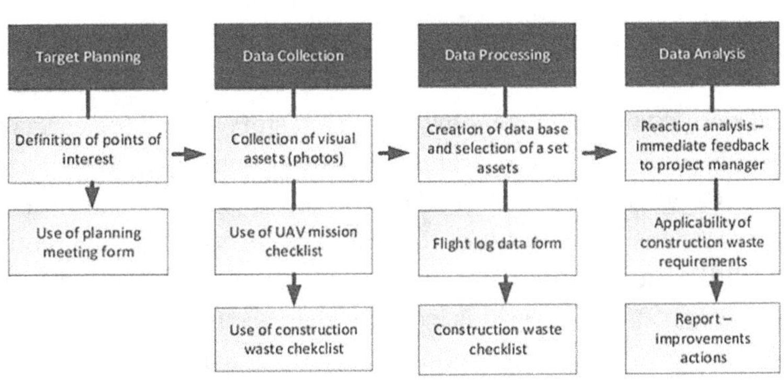

Figure 3. Construction waste observations

nerable to corrosion. This will directly impact the safety on site due to the hazard of fall over unprotect materials. The degraded materials will lead to the constructed building having a much lower structural strength as compared to the design. This results in the short building lifetime and putting the safety of the future inhabitant at risk. Moreover, it will reduce the overall productivity as maintenances are required more frequently than usual to repair or replace the deteriorated material over time. This cumulatively account for construction waste. Low technological use in this study was contrary to previous studies by Won & Cheng (2017), Whitlock et al. (2018) and Lu et al. (2017). Hence, technological innovation of using BIM layout mapping and coordination will be effective in cutting down hazards posed by material and equipment if adopted.

Construction waste has increasing environmental impacts due to the mass urbanization all around the world. Therefore, it is important to have proper waste management whether on-site or off-site to save the total construction cost in a long run. This is consistent with Shen, Tam, Tam, & Drew (2004) finding towards mapping approach for examining waste management on construction site. This mapping accessed the source of waste generation, methods of waste processing, transportation and facilitator which the cost and methods can be altered based on the necessity. It is highly recommended for construction workers, safety/project managers to monitor the construction equipment, double checking and making sure the equipment is properly covered from possible damage and kept in a good state. Proper storage of building materials should be adopted. These findings are consistent with Ajayi et al. (2017) and Whitlock et al. (2018) which found that poor site management practices continue to cause incessant waste generation during the execution phase. Planning for material reuse was non-existent during real time monitoring which was contrary to the findings of Ghisellini et al. (2018) and Mahpour (2018) which recommend for a push towards more circular driven construction processes.

Labour

Labour forms a critical element of site management with a history of call for restructuring labour skills. According to Thompson & Tracy (1991), construction workers often regarded as incompetent, inefficient, lazy, greedy, and possess negative concern about their own health and safety. From Figure 3 (top left and bottom left), it is clearly visualised that most of the site workers were not fully equipped with personal protective equipment (PPE), more specifically not wearing hardhats while working around high raise building. This action put them at risk of serious injuries when accidents were to occur. According to the site manager, workers who are caught for not abiding the rule of using proper PPE will be penalized by certain amount of fine. However, punishment only applied when there was a formal site inspection. This is contrary to previous studies on technological tracking of safety of workers by Li et al. (2018), Cheung, Lin, & Lin (2018) and Yu et al. (2017a). While UAV has assisted in highlighting the site deficiencies in this study, more real time tracking research is required to infuse new technology tracking for labour.

In Figure 3 (bottom left), a motorcycle can be seen parked inside the construction area. The construction site was at the middle of a residential area with very limited space outside the construction area. However, parking within the construction site was not an efficient option, especially parking around the construction materials by labour. This action will interfere with the construction progress as workers need to be extra careful when transporting materials near the vehicle. If accidents were to occur, it not only delays the construction progress, but will also induce property damage to the owner of the vehicle and accrue construction waste. This was contrary to previous studies by Zhang et al. (2015), Shen & Marks (2016) and Li et al. (2018). Adequate provision of temporary parking layout and material delivery sequence such as just-in-time delivery to decongest site and make room for less hazards reduces construction waste sources.

Labour awareness to abide by safe planned guidelines when working in construction site is mainly influenced by perceived usefulness of safety measures (Fung, Tam, Sing, Tang, & Ogunlana, 2016). The workers rely on the premise that they are safe without proper PPE without accounting possible risks they might encounter in construction site. Fung et al. (2016) suggested that all construction practitioners should be aware of their safety and watch over each other among the construction workers as instantaneous change of this psychological condition is nearly impossible. Thompson & Tracy (1991) including subsequent construction industry reports have recommended award for further training can be crucial for increased productivity and efficiency. Therefore, construction workers are highly recommended for continuous restructuring, reskilling and retraining. In addition, Li, Lu, Hsu, Gray, & Huang (2015) proposed a proactive behaviour-based safety system can be an alternative to instill safety awareness to construction workers. This system integrated with multiple safety related information and make the communication between safety officers and the workers effective, enhancing the capacity of the on-site workers in detecting possible hazards.

Management Policy

From the data collected, most of the workers did not comply with full PPE. The defiance of using PPE may put a person's safety at risk. This is also against the Act 514 Occupational Safety and Health 1994 (OSHA 1994) which states that it is a duty of every worker at work to wear any protective equipment or clothing provided by the employer at all times. This may partly be the responsibility of the employer to provide sufficient PPE for the workers. This act of defiance is punishable with a one thousand ringgit fine or imprisonment for three months or both. If injuries among workers were to occur, it does not only induce injury cost, a productivity cost was also incurred. According to Polinder et al. (2016), the total annual cost of injuries was €3.5 billion which comprise of €2.0 billion of healthcare costs and €1.5 billion productivity costs in Netherlands.

The construction waste was not properly managed as shown in figure 3 (top right and left). This produces an overspill inadequate space causing the construction site to become more chaotic. With waste material spread haphazardly around, extra attention is required for workers to 'mind their footstep', especially those who were not on metal-plated safety boots as sharp materials such as nails, metal fragments and even splinters can penetrate regular footwear easily when stepped or kicked. If injuries took place, not only medical attention is required but the productivity will be reduced and increased waste. This is contrary to previous findings by Ajayi & Oyedele (2018) and Li et al. (2018) which direct for improved lifecycle management during all phases of construction. Furthermore, according to Udawatta, Zuo, Chiveralls, & Zillante (2015), it is a fact that construction waste has residual value which can be utilized and so the generation of construction waste can be avoided. Improving real time monitoring found to be effective in previous studies by Yu et al. (2017a), Haur, Kuo, Fu, Hsu, & Heng (2018), Ham, Han, Lin, & Golparvar-Fard (2016), Mneymneh, Abbas, & Khoury (2017), Haur et al. (2018) and Alizadehsalehi et al. (2018) will assist in boosting management construction waste saving techniques. An inclusive decision around reuse of materials which contributed to vast percentage of waste from previous studies by Subramaniam et al. (2018), Maniam et al. (2018), Mah, Fujiwara, & Ho, (2018), Chulkov & Fachratov (2018) and (Hassan et al., 2018).

FUTURE OUTLOOK

This chapter began with the establishment of current deficiencies in minimising construction waste during the life cycle of building development. A wide range of previous studies attested to the research gap of real time monitoring in the construction industry. UAV approach was adopted to access a real-life construction site in a bid to improving construction best practices. Three major dimensions namely materials and equipment, labour and management policies were monitored. In-discriminant placement of materials, crowding, hazardous materials and equipment neglect were observed as main contributors to construction waste. Labour exhibited lapses in waste and safety awareness including lack of technology use to aid reduction of construction waste. Management policies require a drastic shift towards circular economy planning, enforcement of forward planning and defining technology such UAV use for future project monitoring. While UAV was effective in monitoring construction waste, the study is limited by the number of study site and use of singular real time monitoring technique. This contributes towards solving real life construction waste challenges, validating the use of UAV and establishing a platform for future studies on multiple construction monitoring and other category of construction works.

REFERENCES

Ajayi, S. O., & Oyedele, L. O. (2018). Waste-efficient materials procurement for construction projects: A structural equation modelling of critical success factors. *Waste Management (New York, N.Y.)*, *75*, 60–69. doi:10.1016/j.wasman.2018.01.025 PMID:29426721

Ajayi, S. O., Oyedele, L. O., Bilal, M., Akinade, O. O., Alaka, H. A., & Owolabi, H. A. (2017). Critical management practices influencing on-site waste minimization in construction projects. *Waste Management (New York, N.Y.)*, *59*, 330–339. doi:10.1016/j.wasman.2016.10.040 PMID:27839690

Akanbi, L. A., Oyedele, L. O., Akinade, O. O., Ajayi, A. O., Davila Delgado, M., Bilal, M., & Bello, S. A. (2018). Salvaging building materials in a circular economy: A BIM-based whole-life performance estimator. *Resources, Conservation and Recycling*, *129*, 175–186. doi:10.1016/j.resconrec.2017.10.026

Akinade, O. O., Oyedele, L. O., Ajayi, S. O., Bilal, M., Alaka, H. A., Owolabi, H. A., & Arawomo, O. O. (2018). Designing out construction waste using BIM technology: Stakeholders' expectations for industry deployment. *Journal of Cleaner Production, 180*, 375–385. doi:10.1016/j.jclepro.2018.01.022

Alizadehsalehi, S., Yitmen, I., Celik, T., & Arditi, D. (2018). The effectiveness of an integrated BIM/UAV model in managing safety on construction sites. *International Journal of Occupational Safety and Ergonomics*, 1–16. doi:10.1080/10803548.2018.1504487 PMID:30043680

Blaisi, N. I. (2019). Construction and demolition waste management in Saudi Arabia: Current practice and roadmap for sustainable management. *Journal of Cleaner Production, 221*, 167–175. doi:10.1016/j.jclepro.2019.02.264

Chen, J., Su, Y., Si, H., & Chen, J. (2018). Managerial areas of construction and demolition waste: A scientometric review. *International Journal of Environmental Research and Public Health, 15*(2350). doi:10.3390/ijerph15112350 PMID:30356018

Cheung, W. F., Lin, T. H., & Lin, Y. C. (2018). A real-time construction safety monitoring system for hazardous gas integrating wireless sensor network and building information modeling technologies. *Sensors (Switzerland), 18*(2), 436. doi:10.339018020436 PMID:29393887

Chulkov, V., & Fachratov, M. (2018). Construction waste processing management and demolition of five-storey apartment blocks in Moscow. *IOP Conference Series. Materials Science and Engineering, 05008*. doi:10.1088/1757-899X/365/6/062013

de Melo, R. R. S., Costa, D. B., Álvares, J. S., & Irizarry, J. (2017). Applicability of unmanned aerial system (UAS) for safety inspection on construction sites. *Safety Science, 98*, 174–185. doi:10.1016/j.ssci.2017.06.008

Oxford Dictionaries. (2019). Definitions. Retrieved from https://en.oxforddictionaries.com/definition/waste

Ding, Z., Zhu, M., Tam, V. W. Y., Yi, G., & Tran, C. N. N. (2018). A system dynamics-based environmental benefit assessment model of construction waste reduction management at the design and construction stages. *Journal of Cleaner Production, 176*, 676–692. doi:10.1016/j.jclepro.2017.12.101

DOSH. (2017). *OSH IN CONSTRUCTION—Prevention through Design (PtD)* (pp. 1–4). Kuala Lumpur, Malaysia: Department of Occupational Safety and Health.

Enegbuma, W. I., Aliagha, G. U., Ali, K. N., & Badiru, Y. Y. (2016). Confirmatory strategic information technology implementation for building information modelling adoption model. *Journal of Construction in Developing Countries, 21*(2), 113–129. doi:10.21315/jcdc2016.21.2.6

Fang, Q., Li, H., Luo, X., Ding, L., Luo, H., Rose, T. M., & An, W. (2018). Detecting non-hardhat-use by a deep learning method from far-field surveillance videos. *Automation in Construction, 85*, 1–9. doi:10.1016/j.autcon.2017.09.018

Fang, Q., Li, H., Luo, X., Ding, L., Rose, T. M., An, W., & Yu, Y. (2018). A deep learning-based method for detecting non-certified work on construction sites. *Advanced Engineering Informatics, 35*(August 2017), 56–68. doi:10.1016/j.aei.2018.01.001

Fang, Y., Cho, Y. K., & Chen, J. (2016). A framework for real-time pro-active safety assistance for mobile crane lifting operations. *Automation in Construction*, *72*, 367–379. doi:10.1016/j.autcon.2016.08.025

Fang, Y., Cho, Y. K., Zhang, S., & Perez, E. (2016). *Case study of BIM and cloud – enabled real-time RFID indoor localization for construction management applications, 142*(7), 1–12. doi:. doi:10.1061/(ASCE)CO.1943-7862.0001125

Fung, I. W. H., Tam, V. W. Y., Sing, C. P., Tang, K. K. W., & Ogunlana, S. O. (2016). Psychological climate in occupational safety and health: The safety awareness of construction workers in South China. *International Journal of Construction Management*, *16*(4), 315–325. doi:10.1080/15623599.2016.1146114

Ghisellini, P., Ripa, M., & Ulgiati, S. (2018). Exploring environmental and economic costs and benefits of a circular economy approach to the construction and demolition sector. A literature review. *Journal of Cleaner Production*, *178*, 618–643. doi:10.1016/j.jclepro.2017.11.207

Guerra, B., Bakchan, A., Leite, F., & Faust, K. M. (2019). BIM-based automated construction waste estimation algorithms: The case of concrete and drywall waste streams. *Waste Management (New York, N.Y.)*, *87*, 825–832. doi:10.1016/j.wasman.2019.03.010 PMID:31109587

Guo, H., Yu, Y., & Skitmore, M. (2017). Visualization technology-based construction safety management: A review. *Automation in Construction*, *73*, 135–144. doi:10.1016/j.autcon.2016.10.004

Ham, Y., Han, K. K., Lin, J. J., & Golparvar-Fard, M. (2016). Visual monitoring of civil infrastructure systems via camera-equipped Unmanned Aerial Vehicles (UAVs): A review of related works. *Visualization in Engineering*, *4*(1), 1–8. doi:10.118640327-015-0029-z

Hassan, S. H., Aziz, H. A., Daud, N. M., Keria, R., Noor, S. M., Johari, I., & Shah, S. M. R. (2018). The methods of waste quantification in the construction sites (A review). In *AIP Conference Proceedings*. doi:10.1063/1.5062682

Haur, C. J., Kuo, L. S., Fu, C. P., Hsu, Y. L., & Da Heng, C. (2018). Feasibility study on UAV-assisted construction surplus soil tracking control and management technique. *IOP Conference Series. Materials Science and Engineering*, *301*, 012145. doi:10.1088/1757-899X/301/1/012145

Hossain, M. U., & Poon, C. S. (2018). Comparative LCA of wood waste management strategies generated from building construction activities. *Journal of Cleaner Production*, *177*, 387–397. doi:10.1016/j.jclepro.2017.12.233

Jalaei, F., Zoghi, M., & Khoshand, A. (2019). Life cycle environmental impact assessment to manage and optimize construction waste using Building Information Modeling (BIM). *International Journal of Construction Management*, 1–18. doi:10.1080/15623599.2019.1583850

Jin, R., Yuan, H., & Chen, Q. (2019). Science mapping approach to assisting the review of construction and demolition waste management research published between 2009 and 2018. *Resources, Conservation and Recycling*, *140*, 175–188. doi:10.1016/j.resconrec.2018.09.029

Kaliannan, S., Nagapan, S., Abdullah, A. H., Sohu, S., & Jhatial, A. A. (2018). Determining root cause of construction waste generation: A global context. *Civil Engineering Journal*, *4*(11), 2539–2547. doi:10.28991/cej-03091179

Kanan, R., Elhassan, O., & Bensalem, R. (2018). An IoT-based autonomous system for workers' safety in construction sites with real-time alarming, monitoring, and positioning strategies. *Automation in Construction, 88*(November, 2016), 73–86. doi:10.1016/j.autcon.2017.12.033

Li, H., Lu, M., Hsu, S. C., Gray, M., & Huang, T. (2015). Proactive behavior-based safety management for construction safety improvement. *Safety Science, 75*, 107–117. doi:10.1016/j.ssci.2015.01.013

Li, J., Zuo, J., Cai, H., & Zillante, G. (2018). Construction waste reduction behavior of contractor employees: An extended theory of planned behavior model approach. *Journal of Cleaner Production, 172*, 1399–1408. doi:10.1016/j.jclepro.2017.10.138

Lu, W., Chen, X., Peng, Y., & Liu, X. (2018). The effects of green building on construction waste minimization: Triangulating 'big data' with 'thick data.'. *Waste Management (New York, N.Y.), 79*, 142–152. doi:10.1016/j.wasman.2018.07.030 PMID:30343740

Lu, W., Webster, C., Chen, K., Zhang, X., & Chen, X. (2017). Computational building information modelling for construction waste management: Moving from rhetoric to reality. *Renewable & Sustainable Energy Reviews, 68*, 587–595. doi:10.1016/j.rser.2016.10.029

Mah, C. M., Fujiwara, T., & Ho, C. S. (2018). Life cycle assessment and life cycle costing toward eco-efficiency concrete waste management in Malaysia. *Journal of Cleaner Production, 172*, 3415–3427. doi:10.1016/j.jclepro.2017.11.200

Mahpour, A. (2018). Prioritizing barriers to adopt circular economy in construction and demolition waste management. *Resources, Conservation and Recycling, 134*, 216–227. doi:10.1016/j.resconrec.2018.01.026

Maniam, H., Nagapan, S., Abdullah, A. H., Subramaniam, S., & Sohu, S. (2018). A comparative study of construction waste generation rate based on different construction methods on construction project in Malaysia. *Engineering, Technology & Applied Scientific Research, 8*(5), 3488–3491.

Mneymneh, B. E., Abbas, M., & Khoury, H. (2017). Automated hardhat detection for construction safety applications. *Procedia Engineering, 196*, 895–902. doi:10.1016/j.proeng.2017.08.022

Nagapan, S., Kaliannan, S., Abdullah, A. H., Sohu, S., Deraman, R., Hasmori, M. F., & Abas, N. H. (2018). Preliminary survey on the crucial root causes of material waste generation in Malaysian construction industry. *Engineering, Technology & Applied Scientific Research, 8*(6), 3580–3584.

Osmani, M. (2011). Construction waste. In Waste: A handbook for management (pp. 207–218). doi:10.1016/B978-0-12-381475-3.10015-4

Park, J., Kim, K., & Cho, Y. K. (2016). Framework of automated construction-safety monitoring using cloud-enabled BIM and BLE mobile tracking sensors. *Journal of Construction Engineering and Management, 143*(2), 05016019. doi:10.1061/(ASCE)CO.1943-7862.0001223

Park, J., Kim, K., & Cho, Y. K. (2017). Framework of automated construction-safety monitoring using cloud-enabled BIM and BLE mobile tracking sensors. *Journal of Construction Engineering and Management, 143*(2), 05016019. doi:10.1061/(ASCE)CO.1943-7862.0001223

Polinder, S., Haagsma, J., Panneman, M., Scholten, A., Brugmans, M., & Van Beeck, E. (2016). The economic burden of injury: Health care and productivity costs of injuries in the Netherlands. *Accident; Analysis and Prevention*, *93*, 92–100. doi:10.1016/j.aap.2016.04.003 PMID:27177394

Shen, L. Y., Tam, V. W. Y., Tam, C. M., & Drew, D. (2004). Mapping approach for examining waste management on construction sites. *Journal of Construction Engineering and Management*, *130*(4), 472–481. doi:10.1061/(ASCE)0733-9364(2004)130:4(472)

Shen, X., & Marks, E. (2016). Near-miss information visualization tool in BIM for construction safety. *Journal of Construction Engineering and Management*, *142*(4), 1–10. doi:10.1061/(ASCE)CO.1943-7862.0001100

Subramaniam, S., & Abdullah, A. H., Nagapan, S., Kupusamy, K., Manian, H., & Daud, Z. (2018). Investigate how construction waste generation rate is different for every types of project in peninsular Malaysia using site visit method. *International Journal of Integrated Engineering*, *10*(1), 150–155. doi:10.30880/ijie.2018.10.01.022

Thiabgoh, O., Eggers, T., Jimenez, V. O., Jiang, S. D., Sun, J. F., & Phan, M. H. (2018). Real-time monitoring of position and motion of a non-stationary object with a highly sensitive magnetic impedance sensor. *Journal of Science: Advanced Materials and Devices*, *3*(1), 122–128. doi:10.1016/j.jsamd.2018.01.006

Thompson, H., & Tracy, J. (1991). Restructuring, reskilling and retaining: inside the building and construction industry. *Journal of Australian Political Economy*, *13*, 1–18.

Udawatta, N., Zuo, J., Chiveralls, K., & Zillante, G. (2015). Improving waste management in construction projects: An Australian study. *Resources, Conservation and Recycling*, *101*, 73–83. doi:10.1016/j.resconrec.2015.05.003

Wang, J., Sun, W., Shou, W., Wang, X., Wu, C., Chong, H. Y., & Sun, C. (2015). Integrating BIM and LiDAR for real-time construction quality control. *Journal of Intelligent and Robotic Systems: Theory and Applications*, *79*(3–4), 417–432. doi:10.100710846-014-0116-8

Whitlock, K., Abanda, F. H., Manjia, M. B., Pettang, C., & Nkeng, G. E. (2018). BIM for construction site logistics management. *Journal of Engineering, Project, and Production Management*, *8*(1), 47–55. doi:10.32738/JEPPM.201801.0006

Won, J., & Cheng, J. C. P. (2017). Identifying potential opportunities of building information modeling for construction and demolition waste management and minimization. *Automation in Construction*, *79*, 3–18. doi:10.1016/j.autcon.2017.02.002

Yu, Y., Guo, H., Ding, Q., Li, H., & Skitmore, M. (2017a). An experimental study of real-time identification of construction workers' unsafe behaviors. *Automation in Construction*, *82*, 193–206. doi:10.1016/j.autcon.2017.05.002

Yu, Y., Guo, H., Ding, Q., Li, H., & Skitmore, M. (2017b). An experimental study of real-time identification of construction workers' unsafe behaviors. *Automation in Construction*, *82*(July 2016), 193–206. doi:10.1016/j.autcon.2017.05.002

Zhang, S., Sulankivi, K., Kiviniemi, M., Romo, I., Eastman, C. M., & Teizer, J. (2015). BIM-based fall hazard identification and prevention in construction safety planning. *Safety Science*, *72*, 31–45. doi:10.1016/j.ssci.2014.08.001

ADDITIONAL READING

Fewings, P., & Henjewele, C. (2019). *Construction project management: an integrated approach* (3rd ed.). Routledge. doi:10.1201/9781351122030

Hossain, M. U., & Ng, S. T. (2019). Influence of waste materials on buildings' lifecycle environmental impacts: Adopting resource recovery principle. *Resources, Conservation and Recycling*, *142*, 10–23. doi:10.1016/j.resconrec.2018.11.010

Letche, T. M., & Vallero, D. A. (2019). *Waste: A handbook for management* (2nd ed.). Academic Press.

KEY TERMS AND DEFINITIONS

BIM Based Processes: Reduce construction waste management due to the capability to carryout design review, 3D coordination, quantity take-off, phase planning, and site utilization planning.

Close Circuit Television (CCTV): Video surveillance technology with multiple stationary cameras directly linked to a centralized monitoring device.

Construction Safety: Safety, health and welfare of a working personal has to be secured from working hazards during the activities of works.

UAV Monitoring: Ability to monitor from angles that are out of reach and replace safety/project managers to be at the job hazard area in person, which makes the monitoring work much safer.

Waste: Careless use, extravagant or purposeless use of construction resources.

Wearable Device: A device that are closely attach to human body, sending information of the personnel or the surrounding to a monitoring server connected to a terminal where the real time condition can be monitored by a safety manager.

Chapter 26
Anaerobic Digestion (AD) of Organic Waste Is a Sustainable Waste Management Facility

Asim Yaqub
Department of Environmental Sciences, COMSATS University Islamabad, Abbottabad, Pakistan

Kamal Hussain
Department of Environmental Sciences, COMSATS University Islamabad, Abbottabad, Pakistan

Kinza Irshad
Department of Environmental Sciences, COMSATS University Islamabad, Abbottabad, Pakistan

Iftikhar Zeb
Department of Environmental Sciences, COMSATS University Islamabad, Abbottabad, Pakistan

Rashid Nazir
Department of Environmental Sciences, COMSATS University Islamabad, Abbottabad, Pakistan

Muhammad Bilal
Department of Environmental Sciences, COMSATS University Islamabad, Abbottabad, Pakistan

Huma Ajab
Department of Environmental Sciences, COMSATS University Islamabad, Abbottabad, Pakistan

ABSTRACT

Sustainable waste management of biodegradable waste is one of the serious concerns in developing countries. The energy production from organic waste material by anaerobic digestion (AD) is environmentally sound, energy-efficient, and economical technology that degrades substrate, with minimum production of harmful metabolites, and reuse of its byproducts as bio-fertilizers. The use of small-scale biogas digesters in developing countries and industrial scale or integrated biogas digesters in developed countries has been operationally upgraded. These advancements in the application of AD processes are due to multiple factors, including the economic and environmental value of AD technology, and the dynamics of a range of policies and measures related to agricultural sector, waste control, and green energy production. However, compared to other waste management and energy generation methods, AD requires fewer resources and cost investment, and is therefore sustainable for waste control and energy production.

DOI: 10.4018/978-1-7998-0369-0.ch026

INTRODUCTION

Waste production has become one of the growing concerns for today's world. It is produced by various sources such as, industries, agriculture and municipal systems. The increasing heaps of garbage and their poor management results in several environmental problems, health concerns and safety hazards, thus supporting the concept of unsustainable development, in terms of resource recovery and recycling of waste materials. The world is looking to achieve sustainable development targets by resource recovery and efficient solid waste management practices. The developed countries and metropolitan cities around the world are implementing the waste reduction strategies through policy guidelines, by adopting the "waste management hierarchy", which encompasses minimum waste generation from source, reuse, recycling and energy generation from waste material.

Waste management includes the use of technologies and processes to reduce the waste being produced on domestic and industrial level. The waste can be solid waste, agricultural waste, organic municipal waste, hospital waste, radioactive waste and hazardous chemical waste. Improper solid waste handling i.e. storage, collection and disposal practices, give rise to ecological and public health anomalies. Different landfill sites that are not properly managed, are the hub of toxic gases (e.g. carbondioxide; CO_2, methane; CH_4, carbonmonoixide; CO and hydrogen sulfide; H_2S).Among these gases, methane (CH_4) is the chief contributor of environmental problems, as (CH_4) is a greenhouse gas and is one of the causative agent of global warming (Pratt et al., 2015). Waste management is a serious problem for developing countries, but in developed countries, many practices have been adopted for waste reduction and turning it into useful products and energy (Ikhlayel, 2018).

Besides waste management, depletion of the natural energy reserves is another major concern (Amini *et al.*, 2017). The exploitation of the fossil resources not only brings about declination of the reserves, but also contributes to the environmental problems especially greenhouse gas emissions (Naik *et al.*, 2010). The swelling population in turn is causes an excessive use of natural or non-renewable sources (Jabeen, 2018), leading to a threat of exhaustion of non-renewable fossil fuel resources, environmental problems and future energy security. Numerous incentives are under way to substitute non-renewable resources with renewable resources and reducing fossil fuel reliance, and ecological concerns. The attention is shifting towards alternative sources of energy, such as waste to energy conversion and biomass energy production (Kumar and Samadder 2017).

Biomass has the key role among all the resources in the production of sustainable energy in a diverse and widespread energy crises condition. It can serve as a better alternative to meet present and future energy requirements. Currently, 10-14% of the world energy demand is fulfilled by biomass utilization, along with supporting energy need of 80% of the world rural communities (Steven B. Kennedy 2018). The total percent of the basic energy provided by biomass resources varies from developing to developed countries. In the developed countries the share of world energy is 14% by biomass resources which is equal to 72 EJ and approximately 40% of the total energy in the developed countries. Half of the world population is living in rural areas of developing countries, using most of the biomass for the energy purpose, such as Kenya, Brazil, India, China, and Pakistan use 68%, 25%, 47%, 13% and 27% biomass for energy, respectively.

Figure 1. Flow diagram shows waste to energy

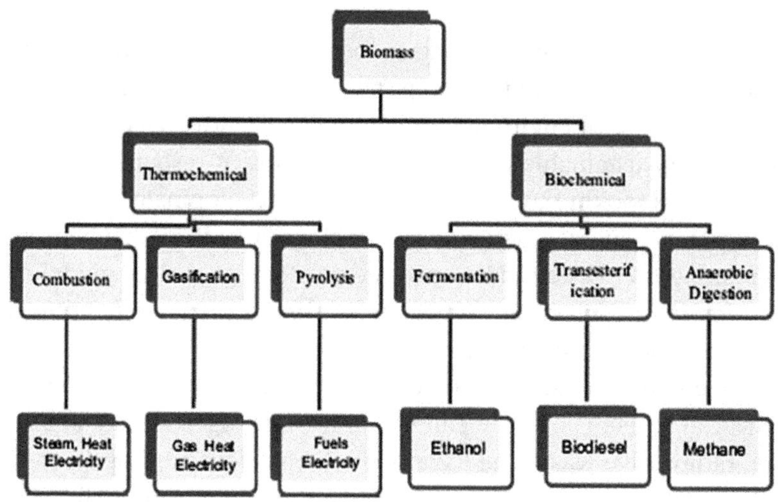

Scientific Progress for Biomass Adaptation to Energy

To overcome growing energy crisis and to ensure the use of sustainable means of energy, the renewable energy options, such as biofuels are found to be the alternative source of energy (Pankin *et al.*, 2011). The biofuels are known to be produced from the waste material such as plant biomass (Bhatia *et al.*, 2017). In this regard, three types of biomasses as substrates are used till date for the mentioned purpose i.e. first generation including edible crops and plants, second generation biofuels including non-edible biomass/waste biomass, third generation biofuels including the use of algae and fourth generation which includes CO_2 capturing and its conversion to energy (Palmer *et al.*, 2016). In the same way carbon -less renewable liquid fuels are needed to cut off the high price of fossils fuel and minimize the global warming. To combat such problem many new techniques have been developed to get renewable energy from biomass (Fig. 1).

Biomass is to be converted using various technologies such as pyrolysis, anaerobic digestion, bioethanol production and gasification has been used since long. Among these, biomass combustion is accounted for over 97% of the world's bio-energy production. Biomass can be directly subjected to boilers; however, simultaneous use of biomass and coal has operational, logistical, and environmental advantages over the other options, reducing the total emissions per unit energy generated.

Thermochemical Processes

Thermochemical process takes place at elevated, and involves breaking down of bonds in organic molecules and its conversion into various types of spices like char, highly oxygenated bio- oil and some gases. For energy generation from such process mostly targets use of manure biomass as it has many advantages like easily available 1) feed stock; 2) an efficient way to get nutrients ; 3) less process time; 4) less GHG emission; and 5) elimination of high temperature pathogenic bacteria.

Combustion

Combustion process biomass is used as fuel to generate hot flue gases, which are run in turbine to produce electricity. In this process temperature reaches up to 800-1000°C.

Pyrolysis

Pyrolysis is widely used for manufacturing of different organic compounds, and is the thermal conversion of biomass through aerobic or anaerobic processes to produce high energy gases. Pyrolysis is thermal decomposition of organic materials at elevated temperature (750 K), in the absence of oxygen, converting biomass to a more useful fuel (Jouhara et al 2018). Biomass is subjected to heating source, in total absence of oxygen or its partial and limited presence, producing a hydrocarbon rich gas blend, a viscous liquid and a carbon rich solid residue.

Typically, the pyrolysis results in production of crude bio-oils, charcoal, various gases, acids, organic liquids such as acetone and methanol, with upto 95% fuel-to-feed efficiency. Certain problems may arise, due to high temperature conversion scheme, such as inefficient thermal stability, process maintenance and corrositivity of oil. To overcome such issues, the oxygen content should be periodically checked, along with removal of alkali produced using hydrogenation and catalytic processes.

Global Scenario of Energy Crises

Currently, Europe and the United States are leading to the use of biofuels as renewable resources, but according to statistics. In 2000, global biogas power generation was around 280,000 Tera Joule (TJ), by 2013, biogas production was expected to reach approximately 59 billion cubic meters, nearly half of which was to be produced in the European Union (WBA, 2014) and by 2014 it was close to 1.3 million TJ. Asia is now ahead in installing biogas units but still it is lagging behind as the population size is very large (Sovacool and Drupady 2016). Raheem at al., (2016) compared biomass production and its energy potential in Asia as shown in Table 1.

Recent researches show that biogas production has been increased by 13.2% per year (Pablo-Romero *et al.*, 2017). Biogas, which is produced through the anaerobic digestion (AD), is commonly used in stoves and lamps. For a family of five, the amount of biogas used to operate an oven is about 1,500-2,400 liters (Bond and Templeton 2011). The use of biogas as an alternative energy strategy for domestic purposes, such as to light up stoves, is effective at ecological, health and community levels as it is a cleaner fuel and lessens the solid waste while also reducing wood deforestation by replacing wood.

The anaerobic digesters come in varying sizes and volumes i.e. from moderate biogas digesters to home biogas plants with enough biogas production capacities (Angelidaki & Ellegaard, 2003;). The use of small domestic anaerobic digesters still prevails in rural areas of developing countries, while large scale digesters are extensively used in developed countries, especially in Europe which is also known as the discovery region for AD technology (Holm-Nielsen *et al.*, 2009; Chynoweth *et al.*, 2001). A integrated or combined system co-digests domestic manure from many farmhouses, through other carbon-based materials such as nutrient waste and organic parts of municipal solid waste (MSW) as well as agricultural residues, and a portion of the digest is returned to the farm for use as a nutrient, while left-over bio-fertilzers have market value (Holm-Nielsen *et al.*, 2009;).

Table 1. Comparison of biomass production and its energy potential in Asia

Countries	Biomass Production Mt Year^{-1}	Biomass Energy Potential
Bangladesh	54-61	1345
China	182.5-210.5	8899.8
India	62-310	8764
Pakistan	0.23-0.31	0.806
Philippine	3.7-21	969
Sri Lanka	2.0-10	142
Thailand	11.6-107	822

(Raheem et al., 2016)

Denmark is famous for its innovation of "central biogas plant development" (Holm-Nielsen *et al.*, 2009;, accompanied by several digestive ponds, up to about 8,000 cubic meters. China has highest number of operational biogas digesters, and is using AD since last 100 years of development (Chen *et al.*, 2010). In total, 43 million anaerobic digesters are used in China, fulfilling the energy requirement of about 100 million people in rural parts. India ranks second in this perspective, with 4.7 million functional biogas digesters (Vasco-Correa *et al.*, 2018). Small biogas digesters are mainly employed for domestic purposes and are usually about 2-10 cubic meters. Such small scaled digesters are located in South Asia and other developing regions (Surendra *et al.*, 2014). Other Asian countries, including Thailand and Indonesia, are utilizing biomass resources in AD; such as palm oil for the production of biogas.

Pakistan has a population of about 184 million (an annual growth rate of 2.2%), and its economy has been mainly established on agriculture and animal husbandry; consuming huge biomass resources, including crop residues, poultry waste, sugar cane and wood waste (Zeb *et al.*, 2017). With growing population, energy demands have also significantly increased, such as electricity demand in the transportation and industrial sector (Rasheed *et al.*, 2016), is growing at a rate of 9% per year, while supply is only 7%, and summer inconsistencies are higher. The fixed capacity as of 2011 was 21,036 Megawatt (Shakeel *et al.*, 2016). Pakistan consumes a combination of power sources including thermal energy (natural gas and oil), hydropower and nuclear power. Renewable energy and coal have reportedly played a limited role in the energy production, but their use is likely to rise sharply in the future. The important renewable assets for the country can be biomass, wind and solar energy (Rasheed *et al.*, 2016). Beside this, agriculture accounts for 25% of GDP and employs more than 40% of the workforce (Bhutto and Bazmi 2007). According to research, 1 cubic meter of biogas may come from about 20 kilograms of wet manure and in Pakistan about one million cubic meters of biogas is produced every day (Uddin *et al.*, 2016).

Journey of Anaerobic Digestion Technology in Pakistan

A study conducted showed that in Pakistan there are 159 million animals on the livestock list, which generate nearly 652 million kilograms of manure per day from cattle other than buffaloes (Rasheed et al., 2016). This can theoretically be used to produce approximately 16 million cubic meters of biogas per day and more than 20 million tons of compost per anum. The study also mentioned that the government has planned to invest 10,000 domestic biogas plants over the next five years, saving millions of dollars each year to replace fossil fuel budgets. This can meet the energy needs of nearly 112 million people living in rural-urban suburban systems in Pakistan (Rasheed et al., 2016).

Anaerobic Digestion

Anaerobic digestion (AD) is the biological conversion of organic waste in an oxygen-limited environment. It occurs naturally in wetlands, swamps and landfills, as well as in digestive system of ruminants (Kumar *et al.*, 2009). AD has been used as a waste management process for decades, but presently, interest in its bioenergy production potential has increased because biogas consists of 40-70% CH_4 which can be directly used or can be cleaned through direct combustion for heat and power generation (Sawatdeenarunat *et al.*, 2016; Korres *et al.*, 2013). Majorly, liquid and solid waste is mainly used in AD process, including liquid industrial effluents, organic fractions of MSW, municipal sludge livestock manure, kitchen waste and agricultural residues (Korres *et al.*, 2013).

Anaerobic digestion is an attractive alternative for wastes containing organic matter with benefits such as bioenergy production, contaminant removal, greenhouse gas control and pathogenicity reduction (Zhou *et al.*, 2014). Recent studies have shown that biogas produced through AD has significant advantages over other forms of bioenergy (Achinas *et al.*, 2017). It is considered ideal feedstock having high water content, because it can be carried out in moist operational conditions. The produced biogas is used as vehicle fuel, or for heat production and electricity generation in different countries. Some other material, such as activated sludge can be used to produce biogas, and recent advances have been made for mentioned purpose, using MSW and manures. As compared to other energy generation processes, the AD is accompanied by treatment of organic matter with reduced odor and greenhouse gas emissions (especially CH_4 and nitrous oxide) (Abbasi *et al.*, 2012), (Clemens *et al.*, 2006), which accounts for 25 and 298 times respectively, more global warming potential than CO_2. Energy produced by AD (biogas) is cleaner than direct combustion biomass, and nutrient-rich material produced as a by-product can be used for agricultural purposes as fertilizer (Withers *et al.*, 2015).

There are four major steps of AD process shown in Fig. 2 i.e. hydrolysis (conversion of organic matter into simple fractions), Acidogenesis (conversion of simple fractions into organic acids, ketones, alcohols, hydrogen and CO_2, by acidogens), acetogenesis (production of acetic acid by acetogens) and methanogenesis (utilization of acetic acid by methanogens and methane production by methanogenic bacteria) (Obaja *et al.*, 2003).

Pretreatment Strategies for Biofuel Production From Anaerobic Digestion

The purpose of pretreatment is to modify the recalcitrant substrate structure in order to simplify the substrate for enzymatic action. The biomass such as lignocellulose has complex structure, thus requires the specifically designed pretreatment, which vary according to the targeted material. Lignocellulose has two types of surfaces, external, which is specific to size and shape of the components while internal surface embodies the capillary body of cellulosic threads. For example, the recalcitrant poplar bark or corn leaves require dilute acid pretreatment, which can't be adopted for sweet gum or corn stalks. Various factors are involved which compromise the biological action, degree of offered hinder, such as lignin content, cellulose crystallinity, polymerization of cellulose, and hemicellulose acetylation. The statement reflects that more recalcitrant cellulosic material would pose more resistance to enzymatic action, thus minimizing the crystallinity would maximizes the substrate accessibility and subsequent digestibility.

Figure 2. Flow diagram of Anaerobic Digestion Process.
(www.frontiersin.org)

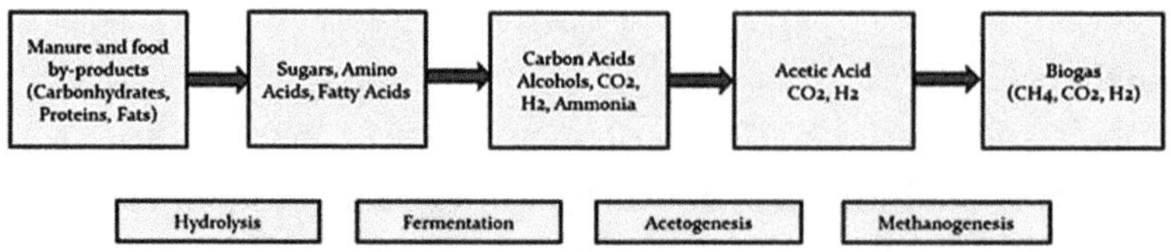

The ideal pretreatment should produce reactive cellulosic fibers for enzymatic action, keeping the cellulose and hemicellulose structure intact, prevention of inhibitors formation, operationally economical and minimum limiting-metabolite production.

Among pretreatment strategies, physical pretreatment is accompanied by various processes, such as physical pretreatment, physico-chemical pretreatment, and chemical pretreatment and biological pretreatment.

Physical Pretreatment

Physical pretreatment alters the outer surface of biomass and decrease its crystallinity by reducing the pore size and polymerization. Various physical pretreatment processes have been used, such as milling, irradiation and high temperature pretreatment

Milling

Milling can be of several types. i.e. ball milling, roll milling, hammer milling, colloid milling and vibrational milling. It reduces the recalcitrant structure of the biomass and is employed before other pretreatment strategies or enzymatic action. It has been studies that without retreatment, certain biomass such as cir stover with size of up to 70 µm showed more degradation than those with size greater than 400 µm.

Similar findings were obtained when rice straw was subjected to combination-pretreatment using grinding, heating and ammonia pretreatment, which resulted in high biogas yield.

Irradiation

Irradiation is the use of sound waves for lignocellulose destruction, to further accelerated enzymatic action. For such purposes, ultrasound waves can be used, for example in activated sludge and aquatic effluents, to disintegrate the sludge particles.

Certain aspects of the irradiation effect further processes, such as ultrasonic density, intensity, pH and concentration of the targeted material. Beside sonication, cavitation and repeated freezing-defreezing, temperature (60-200 °C) can also reduce the crystallinity of the material.

Chemical Pretreatment

Alkaline Hydrolysis

The alkaline solutions are used, like sodium hydroxide, calcium hydroxide and ammonia to delignify the substrate. Such pretreatments enhance the saccharification rate. For biogas production, the use of alkali for the bulk of substrate material can enhance digestion. Alkaline peroxide can also be employed, using bases with a blend of hydrogen peroxide, at high pH. Such technique delignifies the substrate by solubilizing the lignin content and making cellulosic part available for enzymatic hydrolysis.

Acid Hydrolysis Pretreatment

The use of dilute or concentrated acids at low or high temperature can also enhances the substrate accessibility. The use of concentrated acids at low temperature is preferred, as compared to high temperature treatment as it destructs the raw components as well, which are to be used for further processing.

The use of acetic acid and nitric acid as a pretreatment strategy for biogas production from newsprints and activated sludge has also been used extensively but this process is accompanied by hinder caused by lignin and cellulosic bonds which compromises the success of AD. Pretreatment of sugarcane bagasse and coconut fibrils with HCl resulted in approximately 30% and 70% improvement in biogas production. There are certain drawbacks of acidic pretreatments, such as inhibitor formation, such as carboxylic acid, phenolics and furans, which further comprise the microbial performance and sometimes lock-up of whole system, thus lowers the biogas formation. Thus, pretreatment strategy should be selected with proper consideration.

Biological Pretreatment

Microorganisms generally degrade hemicellulose and lignin and small portion of cellulose, since cellulose is more resistance to the biological attack. Several fungi, e.g., soft-rot fungi, brown- and white-rot, have been used for degradation. While white-rot is most effective among the microorganisms. To improve digestion in biogas production different microorganisms or enzymes were considered. Such as fermentation of orange peels by fungal strains of *Aspergillus, Fusarium Penicillum,* and *Sporotrichum* improved the ease of use of feed components and decrease the level of the antimicrobial substances.

The use of microorganisms, utilizing lignocellulose as their food seems to be a wise option in developing bi-products, because of the following characteristics of microorganisms i.e. Microorganisms have a very fast growth rate, they can live in diverse environment due to adaptation and they can be grown on solid substrate (Masran *et al.*, 2016).

The concept of the use of microbial consortia in AD, for degradation of lignocellulosic biomass has recently been proven as an efficient strategy for enhanced saccharification in minimum time (Poszytek *et al.*, 2016). The consortia are actually the combination of different species of the microbes including bacteria and/or fungi, which show synergistic or antagonistic effect when grown together (Kumar *et al.*, 2009). The microbial consortia are in two forms, wild forms which are naturally present and engineered consortia which are synthesized on purpose basis (Bernstein and Carlson, 2012). The natural microbial consortia though degrade the biomass but can produce certain metabolites which can hinder the fermentation process (Sabra *et al.*, 2010). Thus, the microbes are needed to be engineered in required way to keep the development of further processes intact (Tanouchi *et al.*, 2012).

AD Substrates

Substrates are raw materials rich in carbohydrates, proteins and fatty acids that are essential for bacterial growth (Zhang *et al.*, 2014). This helps to produce bioenergy during the degradation of organic matter. In AD, different types of substrates are used, which are classified based on low and high C/N ratios. It has been reported that biomass with a carbon to nitrogen ratio of 20-30 produces an optimized biogas composition (Martins das Neves *et al.*, 2009). Straw and urine are examples of biomass resources with high carbon and nitrogen levels, respectively (Bond and Templeton 2011). Substrates with excess carbon or nitrogen can result in poor bioreactor performance. In Europe, people are using cultivated energy crops as biogas substrates, including corn (*Zea mays*), rye (*Secale*), triticum (*Stice X Secale*), beet (*Beta vulgaris*) and barley (*Hordeum vulgare*), while cannabis (*Cannabis sativa*) and alfalfa (*Medicago sativa*) (Plochl *et al.*, 2006). Different types of biomass resources include agricultural residues such as cane trash, bagasse, straws and green agricultural wastes. Various kinds of polymers are present in several types of waste such as lignocelluloses, lipids, glycogen, starch, elastin, keratin, collagen and chitin. Similarly synthetic polymers are present e.g. polyesters, polypropylene and polyethylene. Among them, lignocelluloses contains a huge portion of municipal solid waste (MSW), such as animal manures and crop residues, woodlot arising's and forest residues, or energy crops. Lignocelluloses are composed of cellulose, hemicellulose, lignin, extractives, and several inorganic materials. The inherent properties of native lignocellulose materials make them strong against enzymatic attack.

Other examples are food processing wastes, wood from natural forests and woodlands, forestry residues, industrial wastes, such as black liquor from paper manufacturing, sewage waste, agro-industrial wastes, such as sugarcane bagasse and rice husk, animal livestock and municipal solid wastes (MSW)

Key Steps in Anaerobic Digestion

Anaerobic digestion is the conversion/degradation of complex organic compounds into products like carbon dioxide, hydrogen and methane by a conglomerate of micro-organisms working in the absence of air. Anaerobic digestion has become a popular form of treatment for various waste and has evolved over the years (Maghanaki et al., 2013; Siles et al., 2010). When it comes to biological treatment the probable use of anaerobic digestion over its counterpart is because of its many advantages over the aerobic treatment. These advantages include less production of sludge owing to slow growing bacteria, cost effectiveness, efficient and less space requiring reactor design and methane as a by-product which is a great energy source smaller (Chen et al., 2008). Under anaerobic conditions chronological degradation of organic matter occurs by myriad of micro-organisms showing syntrophic interactions. The microbiota involved is mainly composed of prokaryotes, including hydrolytic bacteria, acidogens, acetogens, methanogens and in most cases sulphate reducing bacteria (Ahammad et al., 2008). It is these syntrophic interactions between these microbes that convert the complex organic matter into simpler by-products as mentioned above. The process of anaerobic digestion comprises of four stages with distinct microorganisms inhabiting each stage.

- *Hydrolysis*— hydrolysis constitutes the first step of the anaerobic digestion process. Hydrolytic bacteria which are responsible for bringing out this process are subjected to waste containing complex organic compounds. These complex organics are proteins, carbohydrates and lipids. These complex organics are then broken down into simpler form by the extracellular enzymes released by the microbes. This conversion brings about the end of this stage.
- *Acidogenesis*— acidogenesis follows the hydrolysis process. In this stage acidogenic microbes utilize the converted simpler compounds i-e amino acids, sugars and fatty acids into organic acids such as formic, valeric, isobutyric, acetic, propionic, butyric, isovaleric, lactic, and succinic acids; alcohols and ketones (ethanol, methanol, glycerol, and acetone); carbon dioxide; and hydrogen. In this stage acidogens are the dominant microbes and bring about all the conversions.
- *Acetogenesis*—following acidogenesis, acetogenic bacteria also referred as obligate hydrogen-producing acetogens, excrete extracellular enzymes to convert organic acids along with the alcohols formed in the previous stage into products like carbon dioxide, acetate and hydrogen
- *Methanogenesis*— methanogenesis is the last and the most important stage for biogas production. During this stage methanogenic bacteria convert the products of acetogenesis and some of the products from other stages like methylamines acetate, formate, methanol, and hydrogen into methane, carbon dioxide, hydrogen sulphide (biogas) (Ahammad and Sreekrishnan, 2016).

The biogas produced as a result of methanogenesis consists of methane a colorless and odorless, inflammable gas as its major constituent. This methane has emerged as an energy source and possesses the capability of dealing with the energy crisis the global community is suffering from and not only this, but it provides energy in the most environment friendly manner. The concentration of biogas constituents do not remain the same rather they depend upon the type of substrate subjected to the anaerobic digestion. Hence in biogas the methane may be present in the range of 50–60%, 35-45% carbon dioxide, 5–10% hydrogen with only traces of H_2S and O_2. A high-quality biogas containing approximately 75% of methane has the potential to yield about 20,100–28,900 kcal Nm–3 of heat. In other words, a comparison between biogas energy and other energy sources shows that 1 m^3 of bio-methane (methane purified from biogas) has the same amount of energy as 1.1 L of gasoline, 0.93 m^3 of natural gas or 1.7 L of bioethanol. But this kind of energy can only be harnessed if the methane content in biogas is high and this methane is fully purified to form bio-methane. Therefore, in the light of above statements the recycling of the waste material through anaerobic digestion can be employed as a raw material for the generation of biogas through anaerobic treatment (Martins et al., 2009). It is desirable to estimate the biogas potential of various waste products so that it can be put to use for energy generation purposes.

Functional Parameters Affecting the AD Process

Temperature

Temperature has significantly influence on bacteriological growth-rate thus impacts the relationship between solids retention time (SRT) and digester performance. The fluctuation of temperature on methane production and reduction of volatile solids greatly affect the AD process. A study shows that stability is reduced at lower temperatures, and 20°C (68 °F) seems to be the lowest temperature at which actual SRT can achieve sludge stabilization (Ghanimeh, 2012). Most anaerobic digester projects are designed to operate from 30 to 38 °C (85 to 100 °F), the most common being 35 °C (95 °F). Some digester de-

Table 2. The carbon to nitrogen ratio (C/N) of different types of waste (Hagos et al., 2017)

Substrate	C/N Ratio
Cow dung	16-25
Pig manure	6-14
Horse manure	20-25
poultry manure	5-15
Sheep dung	30-33
Goat manure	10-17
Slaughterhouse waste	22-37
Wheat straw	50-150
Corn straw	50-56
Oat straw	48-50
Sugarcane bagasse	140-150
Food waste	3-17
Mixed food wastes	15-32

(Hagos et al., 2017)

signs are planned to operate in the thermophilic range of 50 to 57 °C (122 to 135 °F) (Suryawanshi *et al.*, 2010). It is important to maintain a stable operating temperature in the digester. Rapid and frequent fluctuations in temperature can affect bacteria, especially methanogens. Process failure may occur if temperature changes are greater than 1 °C/d. The digester temperature change should be avoided to be greater than 0.6 °C/d (Bernet *et al.*, 2000).

COD

Chemical oxygen demand (COD) is used to quantify the organic content of waste streams and to predict the potential of biogas production (Barrena *et al.*, 2009). The oxygen equivalent of the oxidizable organics is measured using a strong chemical oxidant in an acidic medium. During anaerobic digestion, the biodegradable COD present in the organic material remains in the final product, methane and newly formed bacterial agglomerates (De Mes *et al.*, 2003). If the organic compound ($C_nH_aO_bN_d$) is fully biodegradable and completely converted by anaerobic organisms (sludge production is assumed to be zero) as CH_4, CO_2 and NH_3, the theoretical gas produced can be calculated according to Buswell equation (Angelidaki *et al.*, 2004).

Total Ammonium Hydride (TAN)

Ammonia is one of the hydrolyzates formed during the degradation caused by proteinaceous organisms (Vavilin *et al.*, 2008). Thermophilic anaerobic digester is used to treat animal waste, such as pig manure, cattle and poultry waste, and some industrial wastes, such as food processing waste streams, often encounter very high ammonia concentrations due to high hydrolysis rates (Chen *et al.*, 2008). The inhibition of ammonia on thermophilic methanogens has become significant. In a thermophilic condition, free ammonia is 560-568 mg NH_3 N/l causes 50% inhibition of methanogens at pH 7.6 (Bishnoi, 2012)

Volatile Fatty Acids

It is generally assumed that solid reduction occurs only in the volatile portion of the sludge solids. Therefore, a commonly used measure of stability is the percentage of volatile solids destroyed (Athanasoulia *et al.*, 2012). The reduction in volatile solids in high speed anaerobic digester is typically from 50% to 65%. The reduction in volatile solids achieved in any particular application depends on the characteristics of the sludge and the operating parameters of the digestive system (Turovskiy *et al.*, 2006).

pH and Alkalinity

Methanogenic bacteria are very much pH dependent and 6.8 to 7.2 is the optimum pH range for methane producers (Khanal, 2011). pH decreases during formation of volatile acids which is usually countered by a methane former, which also produces alkalinity in the form of bicarbonate and carbon dioxide (Ertem, 2011). The relationship between bicarbonate and carbon dioxide is very imperative in anaerobic digestion. One study showed that the pH of the system is controlled by the bicarbonate alkalinity of the liquid-phase and the carbon dioxide concentration of the gas-phase (Yeh *et al.*, 2005). If the amount of CO_2 in the gas-phase remains the same, the bicarbonates will increase the pH of the digester. Adding sodium bicarbonate is the finest tactic to rise in pH and buffer capacity in the digester (Ward *et al.*, 2008). Lime also increases the alkalinity of bicarbonate, but it form insoluble calcium carbonate on reacting with bicarbonate to which promotes scaling.

The pH of the digest is also controlled by the species $NH_4^+ \leftrightarrow NH_3$, $CO_2 \leftrightarrow HCO_3^- \leftrightarrow CO_3^{2-}$ and $CH_3COOH \leftrightarrow CH_3COO^-$. The increase in pH is usually due to the formation of ammonium carbonate ($(NH_4)_2CO_3$) and the CO_2 removal caused by the conversion of CO_3^{2-} and $2H^+$ to CO_2 and H_2O. In addition, AD reduces the fatty acid content of the raw materials. The pH of the digestate is also affected by the concentration of alkaline cations (e.g. Ca_2^+, K^+); the higher the concentration, the lower is the biogas production. They increase the pH of the digestive solution because the charge balance of the solution must be neutral, thus reducing the concentration of H^+. (Ausiello, 2017).

Toxic Chemicals

Though numerous materials are very toxic to bacteria in anaerobic-digester, light metal cations, heavy metals, sulfides, ammonia and some inorganic materials are well-intentioned (Chen et al., 2014). Toxic conditions usually occur in overfeeding and overdosing chemicals. The contribution of industrial wastewater to excess toxic substances can also create toxic conditions for plant influent. Heavy metal toxicity is often considered to be the cause of anaerobic digestion failure (Chen et al., 2008), although traces of most heavy metals are required for cell synthesis.

Particle Size

Analysis of the particle size distribution in the feedstock and digested slurry indicated that generally large sized particles are distributed in sludge. Larger particles (i.e. >10 µm) are more resistant to degradation by AD, and bacteria take a long time to produce biogas (Moller *et al.*, 2012).

Nitrate

The essential main element N exists in plants in both inorganic and organic forms. The complex organic N compound is mineralized into NH_4^+-N in a digester. The digester microorganisms are grown using a portion of NH_4^+-N (Paredes *et al.*, 2007). A further process is the formation of struvite and ammonium carbonate, and traces are volatilized in the biogas stream (<1%). (Cai *et al.*, 2013).

Phosphate (PO_4^-)

As a component of adenylate, nucleic acid and phospholipids, phosphorus is an important plant macronutrient (Tran *et al.*, 2010). The natural supply of phosphorus in most soils is small, and the effectiveness of phosphorus in soil solutions is usually very low. It is often said that the degradation in AD process will improve the availability of phosphorus (P) to plants (Oehmen *et al.*, 2007). However, most of the available results from field trials indicate that AD has no effect on fecal P availability. Though, P is an important constituent but excessive availability of P may have an adverse effect on AD (Möller *et al.*, 2012).

Struvite

Struvite is ammonium magnesium phosphate which forms a precipitate in the cooking system. It is mainly caused by environmental factors of anaerobic digestion. The formation of struvite does not adversely affect the anaerobic digestion process. However, bird fecal deposits in pipes and heat exchangers will cause maintenance hitches (Angelidaki *et al.*, 2003).

Biochemistry of Anaerobic Digestion

Although anaerobic digestion is a widely used technique, the process is not fully understood because of its high complexity, and many complex chemical reactions. Because of this, this technique still requires optimization (Appels *et al.*, 2011). Fig. 3 depicts steps that are involved in the conversion of biomass to energy in anaerobic digestion.

Hydrolysis and Methanogensis

The conversion of complex organic compounds to CH_4 and CO_2 is possible due to the complexation of different microorganisms clustered in the two main steps. 1) Bacteria are responsible for the decomposition of organic matter into volatile fatty acids (VFA), CO_2 and H_2, and 2) Archaea responsible for CH_4 formation (Gehring *et al.*, 2015). Furthermore, the intermixed relationship between microorganisms that produce and consume hydrogen is necessary to ensure efficient and stable operation. Most studies

Figure 3. The reaction paths shown in ADM1
(Batstone et al., 2002).

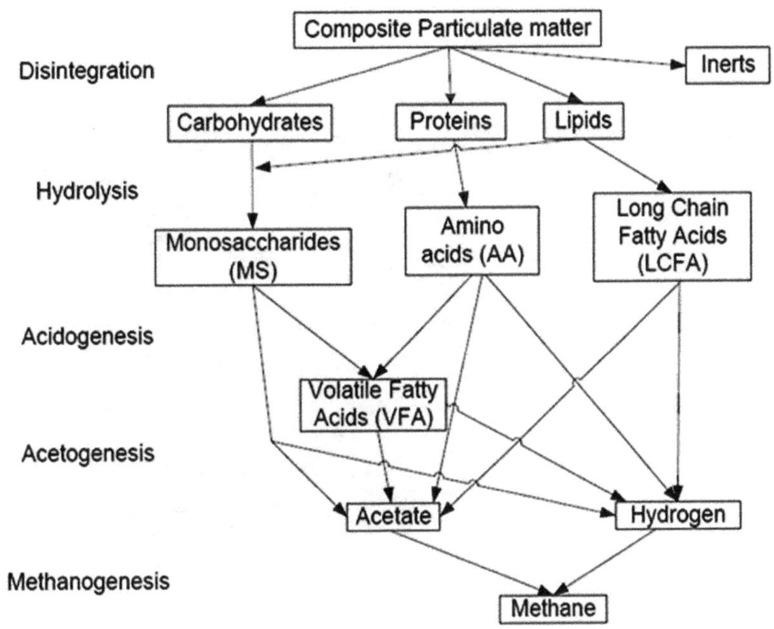

target archaea (Carballa *et al.*, 2015) because methanogensis is usually a rate-limiting step and the low diversity of archaeal populations inhibits functional idleness (Vanwonterghem *et al.*, 2016). However, monitoring the bacterial populations is essential in anaerobic reactors that treat solid waste, where the hydrolysis phase is usually the limited access (Trzcinski *et al.*, 2009).

AD Reactor Types

Several biogas tank models are currently available worldwide. The two main types of biogas plants that are widely used on the wide scale especially for family use are fixed domes and floatable digester (Bruun *et al.*, 2014). The structure of this household-grade biogas digester is relatively simple and practical (Yasar *et al.*, 2017). Fixed dome biogas plants are considered more efficient in Asian environments such as India and Pakistan, and there are many design changes, such as India's Deebandhu, China's fixed dome. However, each of the above design has the common feature of the same semi-circular dome casing (Grad, 2016). On the other hand, the float design includes an underground digester and a movable gas holder or drum which is cylindrical or dome-shaped on the ground. Reactor design plays an important role in conversion of waste into biogas. A suitable reactor produces maximum yield (Chan *et al.*, 2009).

Studies have shown that various types of biogas reactors do have different durability and longevity. The performance and service life of a biogas digester is clearly dependent on the quality and type of building design and building materials, such as the quality of bricks or concrete and cement used to hold the dome-shaped plants. Similar materials such as brick, metal, plastic or fiberglass composites (FRP) are used to make floating dome plants. All of these materials and design variables determine the lifespan of this plant, which has a lifespan of about 15 to 20 and 12-15 years for fixed domes and pontoons, respectively. However, some of the cheaper biogas reactors can also be constructed from polyethylene,

Figure 4. Batch Reactor for Anaerobic Digestion
(www.ProSim.com)

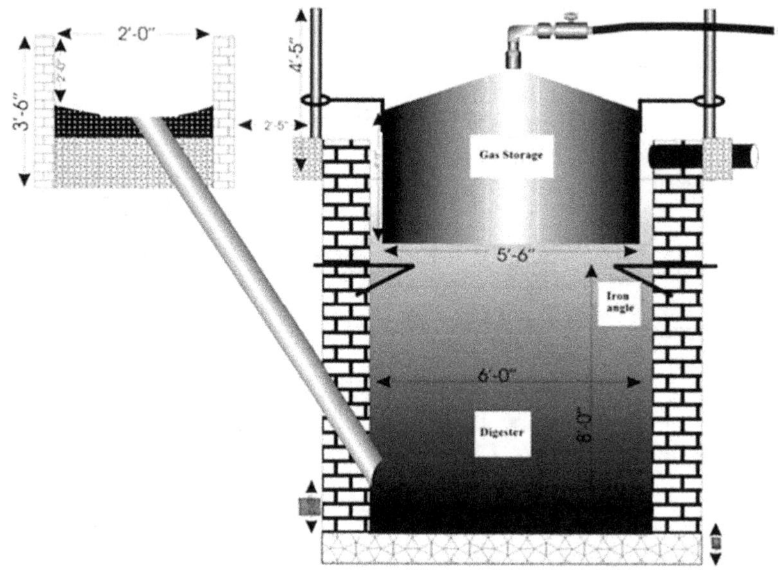

but they can be kept for a shorter period of time, which is only 2 to 5 years (Cheng *et al.*, 2014). The general conclusion is that the durability and sustainability of a biogas plant depends to a large extent on the cost and/or financial value of the biogas reactor. Simply put, the higher the price of a typical biogas plant and its life cycle, the longer it has stability and lifespan (Yasar *et al.*, 2017).

Batch Reactor

The batch reactor is like a giant washing machine Fig. 4. There is a large bucket and all the reagents are placed in a large blender to allow them to stir. Batch reactors are great if a company wants to produce small quantities of specialty chemicals in batches, and the same product is desired over and over again (Scheumann *et al.,* 2009). If different products are desired, then this is not suitable. This is because the reactor must be emptied and cleaned after each batch of product is completed. This takes a lot of time and money, and each batch may be slightly different due to small changes in reaction conditions, equipment aging, or because the operator puts a small amount of foreign matter into the reactor.

Continuous Batch Reactor

The reactor basically consists of a long-tube, the raw materials that allowed to enter from one end, not stopping along the way and reacts through the tube, finally finish the product which come out from another-end. In this type of reactor substrates are fed continuously. The hydraulic retention time (HRT) and solids retention time (SRT) are less as compared to batch reactor (Huang *et al.*, 2011). There are some advantages and disadvantages of both reactors as described in table 3.

Table 3. Advantages and disadvantages between batch reactor and semi batch reactor

Batch Reactor	Semi Batch Reactor (Where Is Continuous)
Complete degradation of biomass (Saravanan *et al.*, 2008)	Some of the biomass are remain undigested (Cheong *et al.*, 2007)
Having a higher HRT and SRT	Having low HRT and SRT
Can degraded the specific target amount of substrate in enclosed system	Can degrade a lot of substrate in continuous manner

Waste to Energy; Global Position

Depletion of fossil fuels and environmental degradation has led to extensive research and development initiatives to discover renewable energy sources such as biogas produced by anaerobic digestion of waste (Dincer, 2000). The world's largest biogas plant is installed on the west coast of Finland because the country is trying to limit its use of imported coal at the community level and generate 140 megawatts of energy (Soimakallio *et al.,* 2009). Biomass gasification plants are expected to reduce coal use by up to 40%. The plant is expected to reduce CO_2 emissions by 230,000 tons per year while providing heating and power to approximately 61,000 residents of Vaasa (Aslani et al., 2014). Singapore today lies among the world's strongest and most powerful economies, and energy has played a vital role in the 50 years of continuous development and growth (Tong *et al.*, 2018). However, Singapore has partial natural resource and depend on seriously on the import of fuels from other countries (Boschini *et al.*, 2007). In 2015, the total electricity generation in Singapore was around 50 trillion megawatt, 97.2% of which was contributed by the imported fossil fuel. The remaining 2.8% of electricity demand was met by local energy sources, such as municipal solid waste (MSW), biomass and photovoltaic panel (Tong, *et al.*, 2018).

From last period, Pakistan has come face to face with the severe problems of energy shortage. Every year Pakistan's government is spending more than 14.5 billion US dollars to import crude oil in order to fulfill the energy gap (Uddin *et al.*, 2016). Energy demand is growing quickly but search and use of renewable energy is not meeting the needs of requirements. The inaccessibility of power in Pakistan lasts for14–20 h per day (Batool *et al.*, 2017). A study showed that the biogas of potential needs to be revised to meet the energy requirement of Pakistan (Ahmed *et al.*, 2016). One cubic meter of biogas can generate an electrical energy of 2.5 KWh. The annual growth rate of livestock sector is 4% in Pakistan and its dung can produce 35.625 million KWh of electric energy per day. We can overcome the energy crisis by efficiently using biogas as an alternative energy source (Uddin *et al.*, 2016).

Economic Analysis of Anaerobic Digestion Systems

The economic assessment of the main objectives of a waste energy project is to meet the social requirements of waste management and environmental protection (Zeb *et al.*, 2017). However, cost-benefit analysis remains a significant interest for plant operators and investors. Costs are mainly divided into capital costs and operating and maintenance costs. The cost of capital is the initial investment cost, such as land acquisition, equipment and equipment procurement, and infrastructure construction. Operational maintenance costs are daily operating costs, such as raw material costs, employee salaries, and maintenance of machinery and buildings. Revenue comes from waste dumping fees, electricity exports and by-product sales. The cost of garbage collection and transportation to the waste-to-energy plant is borne by the municipality and the corresponding fee is charged from the home (Knox, 2005).

The AD system brings some economic benefits, including reducing the cost of fossil fuels in waste management systems (Holm-Nielsen *et al.,* 2009), using energy generated in the form of biogas, electricity and heat to generate revenue and reduce fertilizer by selling excess energy. However, AD systems developed around the world vary in raw material type and composition, digester size, operating conditions, government incentives, and potential use of the product. In addition, the energy required to control the temperature of the digester varies greatly depending on the location and season. Farm-based AD systems can make a profit by taking advantage of the energy products of the AD system, collecting solid waste gates, earning income from compost/organic fertilizers, and possibly selling carbon credits that offset greenhouse gases (Beddoes *et al.*, 2007). Collaborative management of AD plants and farms can share resources such as labor and machinery to help achieve the positive economic benefits of these systems.

A major challenge for AD systems is to operate at maximum capacity throughout the system. The logistics of raw materials and products is also important for the economic and environmental viability of AD systems, as long-distance transportation may increase biogas production costs and associated emissions. In addition, in some regions, such as Europe, most AD systems are installed with government agencies. Therefore, considering the variability of all these specific aspects, it is necessary to evaluate the economic viability of the AD system by dividing the AD system into two categories, we can better understand the economics of the AD system: (1) the most popular large AD systems in developed countries; and (2) small AD systems, such as family and community-scale farms in rural areas (Marchaim, 1992).

CONCLUSION AND FUTURE PERSPECTIVE

In this review, it has been shown that sustainable waste management of biomass is the one of emerging renewable resources for bioenergy production. From different regions of the world, continuous struggles are being made to make it more productive and cost effective. Currently Pakistan is also progressing in the bioenergy production to fulfill the challenging energy demands. Being an agricultural country; huge quantity of biomass are easily available in Pakistan, like crop residues, animal waste and poultry waste. If this biomass is managed and utilized in a sustainable fashion, a huge amount of energy can be produced to cope with growing energy crisis. To make it more efficient, government and private organizations can play a major role for promoting the use of such waste-biomass for green energy production at domestic as well as at industrial level.

REFERENCES

Abbasi, T., Tauseef, S., & Abbasi, S. (2012). Anaerobic digestion for global warming control and energy generation—An overview. *Renewable & Sustainable Energy Reviews, 16*(5), 3228–3242.

Abbassi-Guendouz, A., Brockmann, D., Trably, E., Dumas, C., Delgenès, J.-P., Steyer, J.-P., & Escudié, R. (2012). Total solids content drives high solid anaerobic digestion via mass transfer limitation. *Bioresource Technology, 111*, 55–61. PMID:22386469

Achinas, S., Achinas, V., & Euverink, G. J. W. (2017). A technological overview of biogas production from biowaste. *Engineering, 3*(3), 299–307.

Ahammad, S. Z., Gomes, J., & Sreekrishnan, T. R. (2008). Wastewater treatment for production of H2S-free biogas. Journal of Chemical Technology & Biotechnology: International Research in Process. *Environmental & Clean Technology, 83*(8), 1163–1169.

Ahammad, S. Z., & Sreekrishnan, T. R. (2016). Energy from wastewater treatment. In Bioremediation and Bioeconomy Elsevier (pp. 523-536).

Ahmed, S., Mahmood, A., Hasan, A., Sidhu, G. A. S., & Butt, M. F. U. (2016). A comparative review of China, India, and Pakistan renewable energy sectors and sharing opportunities. *Renewable & Sustainable Energy Reviews, 57*, 216–225.

Amini, Z., Ilham, Z., Ong, H. C., Mazaheri, H., & Chen, W. H. (2017). State of the art and prospective of lipase-catalyzed transesterification reaction for biodiesel production. *Energy Conversion and Management, 141*, 339–353.

Angelidaki, I., Ellegaard, L., & Ahring, B. K. (2003). *Applications of the anaerobic digestion process Biomethanation II* (pp. 1–33). Springer.

Angelidaki, I., & Sanders, W. (2004). Assessment of the anaerobic biodegradability of macropollutants. *Reviews in Environmental Science and Biotechnology, 3*(2), 117–129.

Appels, L., Baeyens, J., Degrève, J., & Dewil, R. (2008). Principles and potential of the anaerobic digestion of waste-activated sludge. *Progress in Energy and Combustion Science, 34*(6), 755–781.

Appels, L., Lauwers, J., Degrève, J., Helsen, L., Lievens, B., Willems, K., ... Dewil, R. (2011). Anaerobic digestion in global bio-energy production: Potential and research challenges. *Renewable & Sustainable Energy Reviews, 15*(9), 4295–4301.

Aslani, A., Helo, P., & Naaranoja, M. (2014). Role of renewable energy policies in energy dependency in Finland: System dynamics approach. *Applied Energy, 113*, 758–765.

Athanasoulia, E., Melidis, P., & Aivasidis, A. (2012). Optimization of biogas production from waste activated sludge through serial digestion. *Renewable Energy, 47*, 147–151.

Ausiello, A. (2017). Development of a system for electric energy production based on fuel cells fueled by biogas produced with anaerobic digestion of biomass. doi:10.6093/UNINA/FEDOA/11536

Barrena, R., d'Imporzano, G., Ponsá, S., Gea, T., Artola, A., Vázquez, F., ... Adani, F. (2009). In search of a reliable technique for the determinat/ion of the biological stability of the organic matter in the mechanical–biological treated waste. *Journal of Hazardous Materials, 162*(2-3), 1065–1072. PMID:18606494

Batool, A., & Abbas, F. (2017). Reasons for delay in selected hydro-power projects in Khyber Pakhtunkhwa (KPK), Pakistan. *Renewable & Sustainable Energy Reviews, 73*, 196–204.

Batstone, D. J., Keller, J., Angelidaki, I., Kalyuzhnyi, S., Pavlostathis, S., Rozzi, A., ... Vavilin, V. (2002). The IWA anaerobic digestion model no 1 (ADM1). *Water Science and Technology, 45*(10), 65–73. PMID:12188579

Beddoes, J. C., Bracmort, K. S., Burns, R. T., & Lazarus, W. F. (2007). An analysis of energy production costs from anaerobic digestion systems on US livestock production facilities. *USDA NRCS Technical Note*(1).

Bernet, N., Delgenes, N., Akunna, J. C., Delgenes, J., & Moletta, R. (2000). Combined anaerobic–aerobic SBR for the treatment of piggery wastewater. *Water Research, 34*(2), 611–619.

Bhatia, S. K., Bhatia, R. K., & Yang, Y. H. (2017). An overview of microdiesel as a sustainable future source of renewable energy. *Renewable & Sustainable Energy Reviews, 79*, 1078–1090.

Bhutto, A. W., & Bazmi, A. A. (2007). *Sustainable agriculture and eradication of rural poverty in Pakistan*. Paper presented at the Natural Resources Forum.

Bishnoi, P. (2012). *Effects of thermal hydrolysis pre-treatment on anaerobic digestion of sludge*. Virginia Tech.

Bond, T., & Templeton, M. R. (2011). History and future of domestic biogas plants in the developing world. *Energy for Sustainable Development, 15*(4), 347–354.

Boschini, A. D., Pettersson, J., & Roine, J. (2007). Resource curse or not: A question of appropriability. *The Scandinavian Journal of Economics, 109*(3), 593–617.

Bruun, S., Jensen, L. S., & Sommer, S. (2014). Small-scale household biogas digesters: An option for global warming mitigation or a potential climate bomb? *Renewable & Sustainable Energy Reviews, 33*, 736–741.

Cai, T., Park, S. Y., & Li, Y. (2013). Nutrient recovery from wastewater streams by microalgae: Status and prospects. *Renewable & Sustainable Energy Reviews, 19*, 360–369.

Carballa, M., Regueiro, L., & Lema, J. M. (2015). Microbial management of anaerobic digestion: Exploiting the microbiome-functionality nexus. *Current Opinion in Biotechnology, 33*, 103–111. PMID:25682574

Chan, Y. J., Chong, M. F., Law, C. L., & Hassell, D. (2009). A review on anaerobic–aerobic treatment of industrial and municipal wastewater. *Chemical Engineering Journal, 155*(1-2), 1–18.

Chen, J. L., Ortiz, R., Steele, T. W., & Stuckey, D. C. (2014). Toxicants inhibiting anaerobic digestion: A review. *Biotechnology Advances, 32*(8), 1523–1534. PMID:25457225

Chen, Y., Cheng, J. J., & Creamer, K. S. (2008). Inhibition of anaerobic digestion process: A review. *Bioresource Technology, 99*(10), 4044–4064. PMID:17399981

Chen, Y., Yang, G., Sweeney, S., & Feng, Y. (2010). Household biogas use in rural China: A study of opportunities and constraints. *Renewable & Sustainable Energy Reviews*, *14*(1), 545–549.

Cheng, S., Li, Z., Mang, H.-P., Huba, E.-M., Gao, R., & Wang, X. (2014). Development and application of prefabricated biogas digesters in developing countries. *Renewable & Sustainable Energy Reviews*, *34*, 387–400.

Cheong, D. Y., Hansen, C. L., & Stevens, D. K. (2007). Production of bio-hydrogen by mesophilic anaerobic fermentation in an acid-phase sequencing batch reactor. *Biotechnology and Bioengineering*, *96*(3), 421–432. PMID:17013946

Chynoweth, D. P., Owens, J. M., & Legrand, R. (2001). Renewable methane from anaerobic digestion of biomass. *Renewable Energy*, *22*(1-3), 1–8.

Clemens, J., Trimborn, M., Weiland, P., & Amon, B. (2006). Mitigation of greenhouse gas emissions by anaerobic digestion of cattle slurry. *Agriculture, Ecosystems & Environment*, *112*(2-3), 171–177.

Davis, R., Aden, A., & Pienkos, P. T. (2011). Techno-economic analysis of autotrophic microalgae for fuel production. *Applied Energy*, *88*(10), 3524–3531.

De-Bashan, L. E., & Bashan, Y. (2004). Recent advances in removing phosphorus from wastewater and its future use as fertilizer (1997–2003). *Water Research*, *38*(19), 4222–4246. PMID:15491670

De Mes, T., Stams, A., Reith, J., & Zeeman, G. (2003). Methane production by anaerobic digestion of wastewater and solid wastes. *Bio-methane & Bio-hydrogen*, 58-102.

Demirbaş, A. (2001). Biomass resource facilities and biomass conversion processing for fuels and chemicals. *Energy Conversion and Management*, *42*(11), 1357–1378.

Demirbas, A. (2005). Potential applications of renewable energy sources, biomass combustion problems in boiler power systems and combustion related environmental issues. *Progress in Energy and Combustion Science*, *31*(2), 171–192.

Dincer, I. (2000). Renewable energy and sustainable development: A crucial review. *Renewable & Sustainable Energy Reviews*, *4*(2), 157–175.

Dou, X., Ren, F., Nguyen, M. Q., Ahamed, A., Yin, K., Chan, W. P., & Chang, V. W.-C. (2017). Review of MSWI bottom ash utilization from perspectives of collective characterization, treatment and existing application. *Renewable & Sustainable Energy Reviews*, *79*, 24–38.

Ertem, F. C. (2011). Improving biogas production by anaerobic digestion of different substrates: calculation of potential energy outcomes.

Gehring, T., Klang, J., Niedermayr, A., Berzio, S., Immenhauser, A., Klocke, M., ... Lübken, M. (2015). Determination of methanogenic pathways through carbon isotope (δ13C) analysis for the two-stage anaerobic digestion of high-solids substrates. *Environmental Science & Technology*, *49*(7), 4705–4714. PMID:25741999

Gewali, M. B., & Bhandari, R. (2005). Renewable energy technologies in Nepal. *World Review of Science, Technology and Sustainable Development*, *2*(1), 92–106.

Ghanimeh, S., El Fadel, M., & Saikaly, P. (2012). Mixing effect on thermophilic anaerobic digestion of source-sorted organic fraction of municipal solid waste. *Bioresource Technology*, *117*, 63–71. PMID:22609715

Grad, M. (2016). *Collection and treatment of fecal disposal in rural areas of developing countries case study*. Nepal: Graz University of Technology.

Hagos, K., Zong, J., Li, D., Liu, C., & Lu, X. (2017). Anaerobic co-digestion process for biogas production: Progress, challenges and perspectives. *Renewable & Sustainable Energy Reviews*, *76*, 1485–1496.

Heukelekian, H., & Mueller, P. (1958). Transformation of some lipids in anaerobic sludge digestion. *Sewage and Industrial Wastes*, *30*(9), 1108–1120.

Holm-Nielsen, J. B., Al Seadi, T., & Oleskowicz-Popiel, P. (2009). The future of anaerobic digestion and biogas utilization. *Bioresource Technology*, *100*(22), 5478–5484. PMID:19217772

Hossain, M. S., Santhanam, A., Norulaini, N. N., & Omar, A. M. (2011). Clinical solid waste management practices and its impact on human health and environment–A review. *Waste Management (New York, N.Y.)*, *31*(4), 754–766. PMID:21186116

Huang, Z., Ong, S. L., & Ng, H. Y. (2011). Submerged anaerobic membrane bioreactor for low-strength wastewater treatment: Effect of HRT and SRT on treatment performance and membrane fouling. *Water Research*, *45*(2), 705–713. PMID:20851448

Ibbett, R., Gaddipati, S., Davies, S., Hill, S., & Tucker, G. (2011). The mechanisms of hydrothermal deconstruction of lignocellulose: New insights from thermal–analytical and complementary studies. *Bioresource Technology*, *102*(19), 9272–9278. PMID:21763128

Ibrahim, H., Ilinca, A., & Perron, J. (2008). Energy storage systems—characteristics and comparisons. *Renewable & Sustainable Energy Reviews*, *12*(5), 1221–1250.

Ikhlayel, M. (2018). An integrated approach to establish e-waste management systems for developing countries. *Journal of Cleaner Production*, *170*, 119–130.

Jabeen, M. (2018). *Thoughtful intelligence a practical guide for moral development: a book guide for moral development*. AuthorHouse.

Jouhara, H., Ahmad, D., van den Boogaert, I., Katsou, E., Simons, S., & Spencer, N. (2018). Pyrolysis of domestic based feedstock at temperatures up to 300 C. *Thermal Science and Engineering Progress*, *5*, 117–143.

Kennedy, S. B. (2018). Global tracking framework sustainable energy for all. Retrieved from http://documents.worldbank.org/curated/en/603241469672143906/pdf/778890GTF0full0report.pdf

Khan, U. K. (2015). Combined industrial wastewater treatment in constructed wetland systems containing emergent plants and algae.

Khanal, S. K. (2011). *Anaerobic biotechnology for bioenergy production: principles and applications*. Hoboken, NJ: John Wiley & Sons.

Knox, A. (2005). *An overview of incineration and EFW technology as applied to the management of municipal solid waste (MSW)*. ONEIA Energy Subcommittee.

Korres, N., O'Kiely, P., Benzie, J. A., & West, J. S. (2013). *Bioenergy production by anaerobic digestion: using agricultural biomass and organic wastes*. Routledge.

Kumar, A., & Samadder, S. (2017). A review on technological options of waste to energy for effective management of municipal solid waste. *Waste Management (New York, N.Y.)*, *69*, 407–422. PMID:28886975

Kumar, S., Puniya, A. K., Puniya, M., Dagar, S. S., Sirohi, S. K., Singh, K., & Griffith, G. W. (2009). Factors affecting rumen methanogens and methane mitigation strategies. *World Journal of Microbiology & Biotechnology*, *25*(9), 1557–1566.

Lam, C. H., Ip, A. W., Barford, J. P., & McKay, G. (2010). Use of incineration MSW ash: A review. *Sustainability*, *2*(7), 1943–1968.

Leitão, R. C., Van Haandel, A. C., Zeeman, G., & Lettinga, G. (2006). The effects of operational and environmental variations on anaerobic wastewater treatment systems: A review. *Bioresource Technology*, *97*(9), 1105–1118. PMID:16551532

Maghanaki, M. M., Ghobadian, B., Najafi, G., & Galogah, R. J. (2013). Potential of biogas production in Iran. *Renewable & Sustainable Energy Reviews*, *28*, 702–714.

Mandpe, S., Kadlaskar, S., Degen, W., & Keppeler, S. (2005). *On road testing of advanced common rail diesel vehicles with biodiesel from the Jatropha curcas plant* (0148-7191). In L. C. Martins das Neves, A. Converti, & T. C. Vessoni Penna (Eds.), *Biogas production: New trends for alternative energy sources in rural and urban zones. Chemical Engineering & Technology: Industrial Chemistry-Plant Equipment-Process Engineering-Biotechnology*, *32*(8), 1147–1153.

Möller, K., & Müller, T. (2012). Effects of anaerobic digestion on digestate nutrient availability and crop growth: A review. *Engineering in Life Sciences*, *12*(3), 242–257.

Naik, S. N., Goud, V. V., Rout, P. K., & Dalai, A. K. (2010). Production of first and second generation biofuels: A comprehensive review. *Renewable & Sustainable Energy Reviews*, *14*(2), 5.

Nielsen, H. B., & Angelidaki, I. (2008). Codigestion of manure and industrial organic waste at centralized biogas plants: Process imbalances and limitations. *Water Science and Technology*, *58*(7), 1521–1528. PMID:18957768

Nizami, A., Orozco, A., Groom, E., Dieterich, B., & Murphy, J. (2012). How much gas can we get from grass? *Applied Energy*, *92*, 783–790.

Oehmen, A., Lemos, P. C., Carvalho, G., Yuan, Z., Keller, J., Blackall, L. L., & Reis, M. A. (2007). Advances in enhanced biological phosphorus removal: From micro to macro scale. *Water Research*, *41*(11), 2271–2300. PMID:17434562

Pablo-Romero, M. P., Pozo-Barajas, R., & Yñiguez, R. (2017). Global changes in residential energy consumption. *Energy Policy*, *101*, 342–352.

Palmer, J. D., & Brigham, C. J. (2016). Feasibility of triacylglycerol production for biodiesel, utilizing *Rhodococcus opacus* as a biocatalyst and fishery waste as feedstock. *Renewable and Sustainable Energy Reviews, 56*, 922-928. 78-597.

Pankin, K. E., Ivanova, Y. V., Kuz'mina, R. I., & Shtykov, S. N. (2011). Comparison of the physicochemical characteristics of biofuels and petroleum fuels. *Chemistry and Technology of Fuels and Oils*, *47*(1), 7.

Paredes, D., Kuschk, P., Mbwette, T., Stange, F., Müller, R., & Köser, H. (2007). New aspects of microbial nitrogen transformations in the context of wastewater treatment–a review. *Engineering in Life Sciences*, *7*(1), 13–25.

Plochl, M., & Heiermann, M. (2006). Biogas farming in central and northern Europe: a strategy for developing countries? Invited overview.

Pratt, C., Redding, M., Hill, J., Shilton, A., Chung, M., & Guieysse, B. (2015). Good science for improving policy: Greenhouse gas emissions from agricultural manures. *Animal Production Science*, *55*(6), 691–701.

Raheem, A., Hassan, M. Y., & Shakoor, R. (2016). Bioenergy from anaerobic digestion in Pakistan: Potential, development and prospects. *Renewable & Sustainable Energy Reviews*, *59*, 264–275.

Rasheed, R., Khan, N., Yasar, A., Su, Y., & Tabinda, A. B. (2016). Design and cost-benefit analysis of a novel anaerobic industrial bioenergy plant in Pakistan. *Renewable Energy*, *90*, 242–247.

Rehman, M. S. U., Rashid, N., Saif, A., Mahmood, T., & Han, J.-I. (2013). Potential of bioenergy production from industrial hemp (Cannabis sativa): Pakistan perspective. *Renewable & Sustainable Energy Reviews*, *18*, 154–164.

Saravanan, P., Pakshirajan, K., & Saha, P. (2008). Growth kinetics of an indigenous mixed microbial consortium during phenol degradation in a batch reactor. *Bioresource Technology*, *99*(1), 205–209. PMID:17236761

Sawatdeenarunat, C., Nguyen, D., Surendra, K., Shrestha, S., Rajendran, K., Oechsner, H., ... Khanal, S. K. (2016). Anaerobic biorefinery: Current status, challenges, and opportunities. *Bioresource Technology*, *215*, 304–313. PMID:27005786

Scheumann, R., & Kraume, M. (2009). Influence of hydraulic retention time on the operation of a submerged membrane sequencing batch reactor (SM-SBR) for the treatment of greywater. *Desalination*, *246*(1-3), 444–451.

Shakeel, S. R., Takala, J., & Shakeel, W. (2016). Renewable energy sources in power generation in Pakistan. *Renewable & Sustainable Energy Reviews*, *64*, 421–434.

Siles, J. A., Brekelmans, J., Martin, M. A., Chica, A. F., & Martin, A. (2010). Impact of ammonia and sulphate concentration on thermophilic anaerobic digestion. *Bioresource Technology*, *101*(23), 9040–9048. PMID:20674337

Soimakallio, S., Mäkinen, T., Ekholm, T., Pahkala, K., Mikkola, H., & Paappanen, T. (2009). Greenhouse gas balances of transportation biofuels, electricity and heat generation in Finland—dealing with the uncertainties. *Energy Policy*, *37*(1), 80–90.

Sovacool, B. K., & Drupady, I. M. (2016). *Energy access, poverty, and development: the governance of small-scale renewable energy in developing Asia*. Routledge.

Surendra, K., Takara, D., Hashimoto, A. G., & Khanal, S. K. (2014). Biogas as a sustainable energy source for developing countries: Opportunities and challenges. *Renewable & Sustainable Energy Reviews*, *31*, 846–859.

Suryawanshi, P., Chaudhari, A., & Kothari, R. (2010). Thermophilic anaerobic digestion: The best option for waste treatment. *Critical Reviews in Biotechnology*, *30*(1), 31–40. PMID:20148754

Thi, N. B. D., Kumar, G., & Lin, C.-Y. (2015). An overview of food waste management in developing countries: Current status and future perspective. *Journal of Environmental Management*, *157*, 220–229. PMID:25910976

Tong, H., Shen, Y., Zhang, J., Wang, C.-H., Ge, T. S., & Tong, Y. W. (2018). A comparative life cycle assessment on four waste-to-energy scenarios for food waste generated in eateries. *Applied Energy*, *225*, 1143–1157.

Tong, H., Yao, Z., Lim, J. W., Mao, L., Zhang, J., Ge, T. S., ... Tong, Y. W. (2018). Harvest green energy through energy recovery from waste: A technology review and an assessment of Singapore. *Renewable & Sustainable Energy Reviews*, *98*, 163–178.

Tran, H. T., Hurley, B. A., & Plaxton, W. C. (2010). Feeding hungry plants: The role of purple acid phosphatases in phosphate nutrition. *Plant Science*, *179*(1-2), 14–27.

Tripathi, M., Sahu, J. N., & Ganesan, P. (2016). Effect of process parameters on production of biochar from biomass waste through pyrolysis: A review. *Renewable & Sustainable Energy Reviews*, *55*, 467–481.

Trzcinski, A., & Stuckey, D. C. (2009). Continuous treatment of the organic fraction of municipal solid waste in an anaerobic two-stage membrane process with liquid recycle. *Water Research*, *43*(9), 2449–2462. PMID:19371920

Turovskiy, I. S., & Mathai, P. (2006). *Wastewater sludge processing*. Hoboken, NJ: John Wiley & Sons.

Uddin, W., Khan, B., Shaukat, N., Majid, M., Mujtaba, G., Mehmood, A., ... Almeshal, A. M. (2016). Biogas potential for electric power generation in Pakistan: A survey. *Renewable & Sustainable Energy Reviews*, *54*, 25–33.

Vanwonterghem, I., Jensen, P. D., Rabaey, K., & Tyson, G. W. (2016). Genome-centric resolution of microbial diversity, metabolism and interactions in anaerobic digestion. *Environmental Microbiology*, *18*(9), 3144–3158. PMID:27317862

Vasco-Correa, J., Khanal, S., Manandhar, A., & Shah, A. (2018). Anaerobic digestion for bioenergy production: Global status, environmental and techno-economic implications, and government policies. *Bioresource Technology*, *247*, 1015–1026. PMID:28918346

Vavilin, V., Fernandez, B., Palatsi, J., & Flotats, X. (2008). Hydrolysis kinetics in anaerobic degradation of particulate organic material: An overview. *Waste Management (New York, N.Y.)*, *28*(6), 939–951. PMID:17544638

Ward, A. J., Hobbs, P. J., Holliman, P. J., & Jones, D. L. (2008). Optimisation of the anaerobic digestion of agricultural resources. *Bioresource Technology*, *99*(17), 7928–7940. PMID:18406612

Withers, P. J., van Dijk, K. C., Neset, T.-S. S., Nesme, T., Oenema, O., Rubæk, G. H., ... Pellerin, S. (2015). Stewardship to tackle global phosphorus inefficiency: The case of Europe. *Ambio*, *44*(2), 193–206. PMID:25681977

Yasar, A., Nazir, S., Rasheed, R., Tabinda, A. B., & Nazar, M. (2017). Economic review of different designs of biogas plants at household level in Pakistan. *Renewable & Sustainable Energy Reviews*, *74*, 221–229.

Yeh, J. T., Resnik, K. P., Rygle, K., & Pennline, H. W. (2005). Semi-batch absorption and regeneration studies for CO_2 capture by aqueous ammonia. *Fuel Processing Technology*, *86*(14-15), 1533–1546.

Zeb, I., Ma, J., Frear, C., Zhao, Q., Ndegwa, P., Yao, Y., & Kafle, G. K. (2017). Recycling separated liquid-effluent to dilute feedstock in anaerobic digestion of dairy manure. *Energy*, *119*, 1144–1151.

Zhang, C., Su, H., Baeyens, J., & Tan, T. (2014). Reviewing the anaerobic digestion of food waste for biogas production. *Renewable & Sustainable Energy Reviews*, *38*, 383–392.

Zhou, W., Chen, P., Min, M., Ma, X., Wang, J., Griffith, R., ... Li, Y. (2014). Environment-enhancing algal biofuel production using wastewaters. *Renewable & Sustainable Energy Reviews*, *36*, 256–269.

Compilation of References

Ab Jalil, N. A., Basri, H., Basri, N. E. A., & Abu Shammala, M. F. M. (2015). The potential of biodrying as pre-treatment for municipal solid waste in Malaysia. *Journal of Advanced Review on Scientific Research*, *1*, 1–13.

Abarnadevi, J., Anu, M., Bharani, M., & Prabha, L. P. (2013). Treatment of dye industry effluent by free and immobilized fungi. *International Journal of Pharmacy & Life Sciences*, *4*, 2340–2346.

Abbasi, T., Tauseef, S., & Abbasi, S. (2012). Anaerobic digestion for global warming control and energy generation—An overview. *Renewable & Sustainable Energy Reviews*, *16*(5), 3228–3242.

Abbassi-Guendouz, A., Brockmann, D., Trably, E., Dumas, C., Delgenès, J.-P., Steyer, J.-P., & Escudié, R. (2012). Total solids content drives high solid anaerobic digestion via mass transfer limitation. *Bioresource Technology*, *111*, 55–61. PMID:22386469

Abbes, A. (2005). Simulation and anlysis of an industrial water desalination plant. *Chemical Engineering and Processing*, *44*(9), 999–1004. doi:10.1016/j.cep.2004.12.001

Abd El-Rahim W. M., El-Ardy, O. A. M., Mohammad, F. H. A. (2009). The effect of pH on bioremediation potential for the removal of direct violet textile dye by *Aspergillus niger Desalination*, 249, 1206–1211.

Abd El-Rahim, W. M., & Moawad, H. (2003). Enhancing bioremoval of textile dyes by eight fungal strains from media supplemented with gelatin wastes and sucrose. *Journal of Basic Microbiology*, *43*(5), 367–375. doi:10.1002/jobm.200310267 PMID:12964179

Abd El-Salam, M. M., & Abu-Zuid, G. I. (2015). Impact of landfill leachate on the groundwater quality: A case study in Egypt. *Journal of Advanced Research*, *6*, 579–586. doi:10.1016/j.jare.2014.02.003 PMID:26199748

Abdelouahed, L., Authier, O., Mauviel, G., Corriou, J.-P., Verdier, G., & Dufour, A. (2012). Detailed modeling of biomass gasification in dual fluidized bed reactors under Aspen Plus. *Energy & Fuels*, *26*(6), 3840–3855. doi:10.1021/ef300411k

Abdel-Shafy, H. I., & Mansour, M. S. (2016). A review on polycyclic aromatic hydrocarbons: Source, environmental impact, effect on human health and remediation. *Egyptian Journal of Petroleum*, *25*(1), 107–123. doi:10.1016/j.ejpe.2015.03.011

Abdul Majed, M. A.,, & Tye, C. T. (2018). Catalytic cracking of used vegetable oil to green fuel with metal functionalized zsm-5 catalysts. *The Malaysian Journal of Analytical Sciences*, *22*(1), 8–16.

Abdullah, A. T., & Abdullah, M. (2016). Decolorization of methylene blue and malachite green by immobilized *desmodesmus* sp. Isolated from North Jordan. *International Journal of Environmental Sciences and Development*, *7*(2), 95–99. doi:10.7763/IJESD.2016.V7.748

Abdullah, B. M., & Salimon, J. (2010). Epoxidation of vegetable oils and fatty acids: catalysts, methods and advantages. *Journal of Applied Sciences (Faisalabad)*, *10*(15), 1545–1553. doi:10.3923/jas.2010.1545.1553

Abejón, R., Garea, A., & Irabien, A. (2015). Organic solvent recovery and reuse in pharmaceutical purification processes by nanofiltration membrane cascades. *Chemical Engineering Transactions*, *43*, 1057–1062.

Abiola, J., & Olaogun, S. (2016). Livestock waste management practices in Oyo state, Nigeria. *Journal of Environment and Waste Management*, *3*(2), 139–141.

Aboyade, A. O., Carrier, M., Meyer, E. L., Knoetze, J. H., & Görgens, J. F. (2012). Model fitting kinetic analysis and characterisation of the devolatilization of coal blends with corn and sugarcane residues. *Thermochimica Acta*, *530*, 95–106. doi:10.1016/j.tca.2011.12.007

Aboyade, A. O., Hugo, T. J., Carrier, M., Meyer, E. L., Stahl, R., Knoetze, J. H., & Görgens, J. F. (2011). Non-isothermal kinetic analysis of the devolatilization of corn cobs and sugar cane bagasse in an inert atmosphere. *Thermochimica Acta*, *517*(1-2), 81–89. doi:10.1016/j.tca.2011.01.035

Abu Amr, S. S., Aziz, H. A., Hossain, M. S., & Bashir, M. J. K. (2017). Simultaneous removal of COD and color from municipal landfill leachate using ozone/zinc sulphate oxidation process. *Global NEST Journal*, *19*(3), 498–504. doi:10.30955/gnj.002299

Abu El-Rub, Z., Bramer, E. A., & Brem, G. (2004). Review of catalysts for tar elimination in biomass gasification processes. *Industrial & Engineering Chemistry Research*, *43*(22), 6911–6919. doi:10.1021/ie0498403

Acharya, J., Sahu, J., Sahoo, B., Mohanty, C., & Meikap, B. (2009). Removal of chromium (VI) from wastewater by activated carbon developed from Tamarind wood activated with zinc chloride. *Chemical Engineering Journal*, *150*(1), 25–39. doi:10.1016/j.cej.2008.11.035

Achinas, S., Achinas, V., & Euverink, G. J. W. (2017). A technological overview of biogas production from biowaste. *Engineering*, *3*(3), 299–307.

Acuner, E., & Dilek, F. B. (2004). Treatment of tectilon yellow 2G by *Chlorella vulgaris*. *Process Biochemistry (Barking, London, England)*, *39*(5), 623–631. doi:10.1016/S0032-9592(03)00138-9

Adams, C. D., Fusco, W., & Kanzelmeyer, T. (1995). Ozone hydrogen peroxide/ ozone and UV/ozone treatment of chromium and copper complex dyes: Decolorization and metal release. *Ozone Science and Engineering*, *17*(2), 149–161. doi:10.1080/01919519508547543

Adeel, M., Song, X., Wang, Y., Francis, D., & Yang, Y. (2017). Environmental impact of estrogens on human, animal and plant life: A critical review. *Environment International*, *99*, 107–119. doi:10.1016/j.envint.2016.12.010 PMID:28040262

Adegbola, A. A., & Oladeji, O. S. (2012). Investigating groundwater quality around Olusosun dump site, Lagos State Nigeria. *India Journal of Innovations and Development*, *1*, 700–705.

Adewuyi, Y. G. (2005). Sonochemistry in environmental remediation. 1. Combinative and hybrid sonophotochemical oxidation processes for the treatment of pollutants in water. *Environmental Science & Technology*, *39*(10), 3409–3420. doi:10.1021/es049138y PMID:15952344

Adeyinka, J. S. (1996). Wastewater treatment through reverse osmosis. *Environmental Monitoring and Assessment*, 22–37.

Adham, S., Chiu, K. P., Gramith, K., & Oppenheimer, J. (2005). *Development of a microfiltration and ultrafiltration knowledge base. Awwa Research Foundation*. Pasadena, CA: US Department of the Interior, Bureau of Reclamation.

Adhoum, N., Monser, L., Bellakhal, N., & Belgaied, J.-E. (2004). Treatment of electroplating wastewater containing Cu2+, Zn2+ and Cr (VI) by electrocoagulation. *Journal of Hazardous Materials*, *112*(3), 207–213. doi:10.1016/j.jhazmat.2004.04.018 PMID:15302441

Adinew, B. (2012). Textile effluent treatment and decolorization techniques – a review. *Chemistry. Bulgarian Journal of Science Education, 21*(3).

Adzlan, A., Khairi, F., Shahidul, M. I., Tarmizi, S., Shazali, S., & Yassin, A., … Ting, C. H. (2013). Investigation on water consumption pattern in traditional Malaysian Kampung : An insight study on water and environmental issues. In *6th Engineering Conference*. doi:10.3850/978-981-07-6059-5_048

Agarry, S. E., & Ajani, A. O. (2011). Evaluation of microbial systems for biotreatment of textile waste effluents in nigeria: biodecolorisation and biodegradation of textile dye. *Journal of Applied Science & Environmental Management, 15*(1), 79–86. doi:10.4314/jasem.v15i1.65680

Agbelie, I., Bawakyilenuo, S., & Lemarie, X. (2015). *Energy waste management, African cities Ghana.* Retrieved from https://www.urbanafrica.net/urban-voices/waste-to-energy-african-cities-can-transform-their-energy-landscapes/

Ağdağ, O. N., & Sponza, D. T. (2005). Anaerobic/aerobic treatment of municipal landfill leachate in sequential two-stage up-flow anaerobic sludge blanket reactor (UASB)/completely stirred tank reactor (CSTR) systems. *Process Biochemistry, 40*(2), 895–902. doi:10.1016/j.procbio.2004.02.021

Agouborde, L., & Navia, R. (2009). Heavy metals retention capacity of a non-conventional sorbent developed from a mixture of industrial and agricultural wastes. *Journal of Hazardous Materials, 167*(1-3), 536–544. doi:10.1016/j.jhazmat.2009.01.027 PMID:19188023

Agriculture & Food Development Authority. (2000). *Waste Oils and Fats as Biodiesel Feedstocks. An Assessment of Their Potential in the EU*, ALTENER Program NTB-NETT Phase IV, Task 4, Final Report, March 2000.

Aguilar, M. I., Sáez, J., Lloréns, M., Soler, A., Ortuno, J. F., Meseguer, V., & Fuentes, A. (2005). Improvement of coagulation-flocculation process using anionic polyacrylamide as coagulant aid. *Chemosphere, 58*(1), 47–56. doi:10.1016/j.chemosphere.2004.09.008 PMID:15522332

Ahamed, A., Chen, C.-L., Rajagopal, R., Wu, D., Mao, Y., Ho, I. J. R., … Wang, J.-Y. (2015). Multi-phased anaerobic baffled reactor treating food waste. *Bioresource Technology, 182*, 239–244. doi:10.1016/j.biortech.2015.01.117 PMID:25704096

Ahammad, S. Z., & Sreekrishnan, T. R. (2016). Energy from wastewater treatment. In Bioremediation and Bioeconomy Elsevier (pp. 523-536).

Ahammad, S. Z., Gomes, J., & Sreekrishnan, T. R. (2008). Wastewater treatment for production of H2S-free biogas. Journal of Chemical Technology & Biotechnology: International Research in Process. *Environmental & Clean Technology, 83*(8), 1163–1169.

Ahmad, A. A., Hameed, B. H., & Ahmad, A. L. (2009). Removal of disperse dye from aqueous solution using waste-derived activated carbon: Optimization study. *Journal of Hazardous Materials, 170*(2-3), 612–619. doi:10.1016/j.jhazmat.2009.05.021 PMID:19515487

Ahmad, A. A., Zawawi, N. A., Kasim, F. H., Inayat, A., & Khasri, A. (2016). Assessing the gasification performance of biomass: A review on biomass gasification process conditions, optimization and economic evaluation. *Renewable & Sustainable Energy Reviews, 53*, 1333–1347. doi:10.1016/j.rser.2015.09.030

Ahmad, A. L., & Chan, C. Y. (2009). Sustainability of palm oil industries: An innovative treatment via membrane technology. *Applied Science, 9*(17), 3074–3079. doi:10.3923/jas.2009.3074.3079

Ahmad, A., Ibrahim, N., Ismail, I., & Bhatia, S. (2002). Coagulation-sedimentation-extraction pretreatment methods for the removal of suspended solids and residual oil from palm oil mill effluent (POME). *IIUM Eng J, 3*, 25–33.

Ahmad, A., Ismail, S., & Bhatia, S. (2003). Water recycling from palm oil mill effluent (POME) using membrane technology. *Desalination*, *157*(1-3), 87–95. doi:10.1016/S0011-9164(03)00387-4

Ahmad, A., Sumathi, S., & Hameed, B. H. (2006). Coagulation of residue oil and suspended solid in palm oil mill effluent by chitosan, alum and PAC. *Chemical Engineering Journal*, *118*(1-2), 99–105. doi:10.1016/j.cej.2006.02.001

Ahmad, A., Wong, S., Teng, T., & Zuhairi, A. (2008). Improvement of alum and PACl coagulation by polyacrylamides (PAMs) for the treatment of pulp and paper mill wastewater. *Chemical Engineering Journal*, *137*(3), 510–517. doi:10.1016/j.cej.2007.03.088

Ahmad, U. K., Ujang, Z., Woon, C. H., Indran, S., & Mian, M. N. (2004). Development of extraction procedures for the analysis of polycyclic aromatic hydrocarbons and organochlorine pesticides in municipal sewage sludge. *Water Science and Technology*, *50*(9), 137–144. doi:10.2166/wst.2004.0555 PMID:15581005

Ahmed, S. I., Johari, A., Hashim, H., Mat, R., Lim, J. S., Ngadi, N., & Ali, A. (2015). Optimal landfill gas utilization for renewable energy production. *Environmental Progress & Sustainable Energy*, *34*(1), 289–296. doi:10.1002/ep.11964

Ahmed, S. M., & Mohammad, S. G. (2014). Egyptian apricot stone (Prunus armeniaca) as a low cost and eco-friendly biosorbent for oxamyl removal from aqueous solutions. *American Journal of Experimental Agriculture*, *4*(3), 302–321. doi:10.9734/AJEA/2014/7116

Ahmed, S., Mahmood, A., Hasan, A., Sidhu, G. A. S., & Butt, M. F. U. (2016). A comparative review of China, India, and Pakistan renewable energy sectors and sharing opportunities. *Renewable & Sustainable Energy Reviews*, *57*, 216–225.

Ahmed, T. Y., Ahmad, M. M., Yusup, S., Inayat, A., & Khan, Z. (2012). Mathematical and computational approaches for design of biomass gasification for hydrogen production: A review. *Renewable & Sustainable Energy Reviews*, *16*(4), 2304–2315. doi:10.1016/j.rser.2012.01.035

Aho, A., Kumar, N., Eranen, K., Salmi, T., Hupa, M., & Murzin, D. Y. (2008). Catalytic pyrolysis of woody biomass in a fluidized bed reactor: Influence of the zeolite structure. *Fuel*, *87*(12), 2493–2501. doi:10.1016/j.fuel.2008.02.015

AI-Jawhari, I. F. H. (2015). Decolorization of methylene blue and crystal violet by some filamentous fungi. *International Journal of Environmental Bioremediation & Biodegradation*, *3*, 62–65.

Ai, L., Huang, H., Chen, Z., Wei, X., & Jiang, J. (2010). Activated carbon/CoFe2O4 composites: Facile synthesis, magnetic performance and their potential application for the removal of malachite green from water. *Chemical Engineering Journal*, *156*(2), 243–249. doi:10.1016/j.cej.2009.08.028

AIM. (2013). *National Biomass Strategy 2020*. Retrieved from Kuala Lumpur, Malaysia: http://docs.wixstatic.com/ugd/43b4fe_00564483596c43d5a1619e82131f1832.pdf

Ajayi, S. O., & Oyedele, L. O. (2018). Waste-efficient materials procurement for construction projects: A structural equation modelling of critical success factors. *Waste Management (New York, N.Y.)*, *75*, 60–69. doi:10.1016/j.wasman.2018.01.025 PMID:29426721

Ajayi, S. O., Oyedele, L. O., Bilal, M., Akinade, O. O., Alaka, H. A., & Owolabi, H. A. (2017). Critical management practices influencing on-site waste minimization in construction projects. *Waste Management (New York, N.Y.)*, *59*, 330–339. doi:10.1016/j.wasman.2016.10.040 PMID:27839690

Ajjabi, L. C., & Chouba, L. (2009). Biosorption of Cu2+ and Zn2+ from aqueous solutions by dried marine green macroalga Chaetomorpha linum. *Journal of Environmental Management*, *90*(11), 3485–3489. doi:10.1016/j.jenvman.2009.06.001 PMID:19576679

Akanbi, L. A., Oyedele, L. O., Akinade, O. O., Ajayi, A. O., Davila Delgado, M., Bilal, M., & Bello, S. A. (2018). Salvaging building materials in a circular economy: A BIM-based whole-life performance estimator. *Resources, Conservation and Recycling*, *129*, 175–186. doi:10.1016/j.resconrec.2017.10.026

Akbal, F., & Camcı, S. (2011). Copper, chromium and nickel removal from metal plating wastewater by electrocoagulation. *Desalination*, *269*(1), 214–222. doi:10.1016/j.desal.2010.11.001

Akher, S. K., & Chattopadhyay, D. S. (2017). Impact of urbanization on land surface temperature - a case study of Kolkata New Town. *International Journal of Engineering Science*, *6*(1), 71–81. doi:10.9790/1813-0601027181

Akhil, M., Saikat, D., & Vasudeva, M. (2018, June 3-8). Studies on chemical composition and physicochemical characterization of upgraded tire pyrolysis (UTPO) for the application in engine and furnaces. In *Proceedings of 9th International Freiberg Conference on IGCC & XtL Technologies closing the Carbon Cycle.* Berlin, Germany.

Akinade, O. O., Oyedele, L. O., Ajayi, S. O., Bilal, M., Alaka, H. A., Owolabi, H. A., & Arawomo, O. O. (2018). Designing out construction waste using BIM technology: Stakeholders' expectations for industry deployment. *Journal of Cleaner Production*, *180*, 375–385. doi:10.1016/j.jclepro.2018.01.022

Akunna, J. C., & Clark, M. (2000). Performance of a granular-bed anaerobic baffled reactor (GRABBR) treating whisky distillery wastewater. *Bioresource Technology*, *74*(3), 257–261. doi:10.1016/S0960-8524(00)00017-1

Aladedunye, F. A.,, & Przybylski, R. (2009). Protecting oil during frying: a comparative study. *European Journal of Lipid Science and Technology*, *111*, 893–901.

Alaska Department of Environmental Conservation. (2009). Environmental Clean-up Methods. Spill Prevention and Response Division. Retrieved from https://dec.alaska.gov/spar/csp/guidance/cleanup_methods.pdf

Alauddin, Z. A. B. Z., Lahijani, P., Mohammadi, M., & Mohamed, A. R. (2010). Gasification of lignocellulosic biomass in fluidized beds for renewable energy development: A review. *Renewable & Sustainable Energy Reviews*, *14*(9), 2852–2862. doi:10.1016/j.rser.2010.07.026

Alawadhi, A. A. (1997). Pretreatment plant design - key to a successful reverse osmosis desalination plant. *Desalination*, *1*(1-2), 1–10. doi:10.1016/S0011-9164(97)00079-9

Al-Bastaki, N. M., & Abbes, A. (1999). Modelling an industrial reverse osmosis unit. *Desalination*, *126*(1-3), 33–39. doi:10.1016/S0011-9164(99)00152-6

Albert, F. Y. C., Mohd Fuad, M. N., & Huang Shen, C. (2015) A lab-scale semi auto vertical tubular combustor as a potential pyrolysis process for biomass burning. *ARPN Journal of Engineering and Applied Sciences,* 10, 15. 6534-6457.

Alcantara, R., Amores, J., Canoira, L., Fidalgo, E., Franco, M. J., & Navarro, A. (2000). Catalytic production of biodiesel from soybean oil: used frying oil and tallow. *Biomass and Bioenergy*, *18*, 515–527.

Alex, J., Beteau, J. F., Copp, J. B., Hellinga, C., Jeppsson, U., Marsili-Libelli, S., . . . Vanhooren, H. (1999). Benchmark for evaluating control strategies in wastewater treatment plant. European Control Conference, Karlsruhe, Germany. 10.23919/ECC.1999.7099914

Alexander, B. (2016). Megacities, air quality and climate. *Atmospheric Environment*, *126*, 235–249. doi:10.1016/j.atmosenv.2015.11.059

Alexander, M. (1994). *Biodegradation and bioremediation* (p. 284). New York: Academy Press.

Al-Ghusain, I., & Hao, O. (1995). Use of pH as control parameter for aerobic/anoxic sludge digesting. *Journal of Environmental Engineering*, *3*(3), 225–235. doi:10.1061/(ASCE)0733-9372(1995)121:3(225)

Al-Hamadani, Y. A., Yusoff, M. S., Umar, M., Bashir, M. J., & Adlan, M. N. (2011). Application of psyllium husk as coagulant and coagulant aid in semi-aerobic landfill leachate treatment. *Journal of Hazardous Materials*, *190*(1-3), 582–587. doi:10.1016/j.jhazmat.2011.03.087 PMID:21507572

Alhokabi, A., & Ing, D. S. (2019). Study on the mechanical properties of bottom ash as a partial sand replacement in a non-load bearing fly ash bricks. *Saudi Journal of Civil Engineering*, *3*(3), 59–66.

Ali, N., Hameed, A., Ahmed, S., & Khan, A. G. (2007). Decolorisation of structurally different textile dyes by *Aspergillus niger* SA1. *World Journal of Microbiology & Biotechnology*, *24*(7), 1067–1072. doi:10.100711274-007-9577-2

Aliyu, Y. A., & Botai, J. O. (2018). Reviewing the local and global implications of air pollution trends in Zaria, northern Nigeria. *Urban Climate*, *26*, 51–59. doi:10.1016/j.uclim.2018.08.008

Alizadehsalehi, S., Yitmen, I., Celik, T., & Arditi, D. (2018). The effectiveness of an integrated BIM/UAV model in managing safety on construction sites. *International Journal of Occupational Safety and Ergonomics*, 1–16. doi:10.1080/10803548.2018.1504487 PMID:30043680

Allaire, S. E., Sylvain, C., Lange, S. F., Thériault, G., & Lafrance, P. (2015). Potential efficiency of riparian vegetated buffer strips in intercepting soluble compounds in the presence of subsurface preferential flows. *PLoS One*, *10*(7), e0131840. doi:10.1371/journal.pone.0131840 PMID:26147093

Al-Momani, F., Touraud, E., Degorce-Dumas, J. R., Roussy, J., & Thomas, O. (2003). Biodegradability enhancement of textile dyes and textile wastewater by UV photolysis. *Journal of Photochemical and Photobiology A. Chemistry.*, *153*, 191–197.

Alomar, C., Sureda, A., Capó, X., Guijarro, B., Tejada, S., & Deudero, S. (2017). Microplastic ingestion by Mullus surmuletus Linnaeus, 1758 fish and its potential for causing oxidative stress. *Environmental Research*, *159*, 135–142. doi:10.1016/j.envres.2017.07.043 PMID:28800471

Al-Omari, A. A., Khedaywi, T. S., & Khasawneh, M. A. (2018). Laboratory characterization of asphalt binder modified with waste vegetable oil using SuperPave specifications. *International Journal of Pavement Research and Technology*, *11*, 68–76.

Alrawi, R., Ahmad, A., Ismail, N., & Kadir, M. (2010). Methane production during start-up phase of mesophilic semi-continues suspended growth anaerobic digester. *International Journal of Chemical Reactor Engineering*, *8*(1), 1–13. doi:10.2202/1542-6580.2034

Alshamsi, F., Albadwawi, A., Alnuaimi, M., Rauf, M., & Ashraf, S. (2007). Comparative efficiencies of the degradation of crystal violet using UV/hydrogen peroxide and Fenton's reagent. *Dyes and Pigments*, *74*(2), 283–287. doi:10.1016/j.dyepig.2006.02.016

Altin, A. (2008). An alternative type of photoelectro-Fenton process for the treatment of landfill leachate. *Separation and Purification Technology*, *61*(3), 391–397. doi:10.1016/j.seppur.2007.12.004

Aly, A. A., & El-Bisi, M. K. (2018). *Grafting of polysaccharides: recent advances. In Biopolymer Grafting* (pp. 469–519). Elsevier. doi:10.1016/B978-0-323-48104-5.00011-1

Amadi, A. N., Olasehinde, P. I., Okosun, E. A., Okoye, N. O., Okunlola, I. A., Alkali, Y. B., & Dan-Hassan, M. A. (2012). A comparative study on the impact of avu and ihie dumpsites on soil quality in southeastern Nigeria. *American Journal of Chemistry*, *2*(1), 17–23. doi:10.5923/j.chemistry.20120201.05

Amann, M., Purohit, P., Bhanarkar, A. D., Bertok, I., Borken-Kleefeld, J., Cofala, J., ... Vardhan, B. H. (2017). Managing future air quality in megacities: A case study for Delhi. *Atmospheric Environment*, *161*, 99–111. doi:10.1016/j.atmosenv.2017.04.041

Amann, R., Ludwig, W., & Schleifer, K. (1995). Phylogenetic identification and *in situ* detection of individual microbial cells without cultivation. *Microbiological Reviews*, *59*, 143–169. PMID:7535888

Amari, T., Themelis, N. J., & Wernick, I. K. (1999). Resource recovery from used rubber tires. *Resources Policy*, *25*(3), 179–188. doi:10.1016/S0301-4207(99)00025-2

Ambient, WHO. (2018). *(Outdoor) Air quality database (update 2018)*. Geneva, Switzerland: World Health Organization. Retrieved from http://www.who.int/phe/health_topics/outdoorair/databases/cities/

Ambrosio, S. T. (2004). Decolorisation of reactive azo dyes by Cunninghamella elegans UCP 542 under co-metabolic conditions. *Bioresource Technology*, *91*(1), 69–75. doi:10.1016/S0960-8524(03)00153-6 PMID:14585623

American Public Health Association. (1998). *APHA* (20th ed.). Washington, DC: Standard Methods for Examination of Water and Wastewater.

Amini, Z., Ilham, Z., Ong, H. C., Mazaheri, H., & Chen, W. H. (2017). State of the art and prospective of lipase-catalyzed transesterification reaction for biodiesel production. *Energy Conversion and Management*, *141*, 339–353.

Amin, K. N., & Aziz, S. Q. (2002). Pressure distribution in filter media in conventional filters. *Journal of Dohuk University*, *5*, 56–59.

Amiri, A., & Sabour, M. R. (2014). Multi-response optimization of Fenton process for applicability assessment in landfill leachate treatment. *Waste Management (New York, N.Y.)*, *34*(12), 2528–2536. doi:10.1016/j.wasman.2014.08.010 PMID:25212172

Amuda, O., & Amoo, I. (2007). Coagulation/flocculation process and sludge conditioning in beverage industrial wastewater treatment. *Journal of Hazardous Materials*, *141*(3), 778–783. doi:10.1016/j.jhazmat.2006.07.044 PMID:16959404

Anastasakis, K., Kalderis, D., & Diamadopoulos, E. (2009). Flocculation behavior of mallow and okra mucilage in treating wastewater. *Desalination*, *249*(2), 786–791. doi:10.1016/j.desal.2008.09.013

Anca-Couce, A., & Zobel, N. (2012). Numerical analysis of a biomass pyrolysis particle model: Solution method optimized for the coupling to reactor models. *Fuel*, *97*, 80–88. doi:10.1016/j.fuel.2012.02.033

Anca-Couce, A., Zobel, N., & Jakobsen, H. A. (2013). Multi-scale modeling of fixed-bed thermo-chemical processes of biomass with the representative particle model: Application to pyrolysis. *Fuel*, *103*, 773–782. doi:10.1016/j.fuel.2012.05.063

Anderson, W. C. (1993). Editor. Innovative site remediation technology. 6 – Thermal desorption. *American Academy of Environmental Engineers, Annapolis, Maryland.* Retrieved from https://nepis.epa.gov/Exe/ZyPDF.cgi/2000ITIZ.PDF?Dockey=2000ITIZ.PDF

Anderson, W. C. (1993). Innovative site remediation technology: Soil washing/soil flushing. American Academy of Environmental Engineers, Annapolis, Maryland (3), 1993.

Anderson, E. E. (1989). Economic benefits of habitat restoration: Seagrass and the Virginia hard-shell blue crab fishery. *North American Journal of Fisheries Management*, *9*(2), 140–149. doi:10.1577/1548-8675(1989)009<0140:EBOHRS>2.3.CO;2

Andleeb, S., Atiq, N., Ali, M. I., Razi-Ul-Hussain, R., Shafique, M., Ahmed, B., ... Ahmed, S. (2010). Biological treatment of textile effluent in stirred tank bioreactor. *International Journal of Agriculture and Biology*, *12*, 256–260.

Andrade, I. C., Santiago, J. P., Sodre, J. R., Pathiyamattom, J. S., & Guerrero-Fajardo, C. A. (2014). Transesterification reaction of waste cooking oil and chicken fat by homogeneous catalysis. *Journal of Chemistry and Chemical Engineering, 8*, 736–743.

Andreadakis, A. D. (1993). Physical and chemical properties of activated sludge floc. *Water Research, 27*(12), 1707–1714. doi:10.1016/0043-1354(93)90107-S

Andreozzi, R., Caprio, V., Insola, A., & Marotta, R. (1999). Advanced oxidation processes (AOP) for water purification and recovery. *Catalysis Today, 53*(1), 51–59. doi:10.1016/S0920-5861(99)00102-9

Angelidaki, I., Ellegaard, L., & Ahring, B. K. (2003). *Applications of the anaerobic digestion process Biomethanation II* (pp. 1–33). Springer.

Angelidaki, I., & Sanders, W. (2004). Assessment of the anaerobic biodegradability of macropollutants. *Reviews in Environmental Science and Biotechnology, 3*(2), 117–129.

Angenent, L. T., Karim, K., Al-Dahlan, M. H., Wrenn, B. A., & Domiguez-Espinosa, R. (2004). Production of bioenergy and biochemicals from industrial and agricultural wastewater. *Trends in Biotechnology, 22*(9), 477–485. doi:10.1016/j.tibtech.2004.07.001 PMID:15331229

Ang, W. L., Nordin, D., Mohammad, A. W., Benamor, A., & Hilal, N. (2016). Effect of membrane performance including fouling on cost optimization in brackish water desalination process. *Chemical Engineering Research & Design*. doi:10.1016/j.cherd.2016.10.041

Anis, S. F., Hashaikeh, R., & Hilal, N. (2019). Reverse osmosis pretreatment technologies and future trends: A comprehensive review. *Desalination, 452*(October 2018), 159–195. doi:10.1016/j.desal.2018.11.006

Anis, S., & Zainal, Z. (2011). Tar reduction in biomass producer gas via mechanical, catalytic and thermal methods: A review. *Renewable & Sustainable Energy Reviews, 15*(5), 2355–2377. doi:10.1016/j.rser.2011.02.018

Aniza, A. A., Rao, S. P., & Elias, S. (2013). Waste tyres as heat sink to reduce the driveway surface temperatures in Malaysia. *Journal of Design and Built Environment, 13*, 1–11.

Anjaneyulu, Y., Chary, N. S., & Raj, D. S. S. (2005). Decolourization of industrial effluents - available methods and emerging technologies - a review. *Reviews in Environmental Science and Biotechnology, 4*(4), 245–273. doi:10.100711157-005-1246-z

Anjum, M., Al-Makishah, N. H., & Barakat, M. A. (2016). Wastewater sludge stabilization using pre-treatment methods. *Process Safety and Environmental Protection, 102*, 615–632. doi:10.1016/j.psep.2016.05.022

Anselin, L. (1988). *Spatial econometrics: Methods and models*. Springer Netherlands. doi:10.1007/978-94-015-7799-1

Antai, S. P., & Mgbomo, E. (1989). Distribution of hydrocarbon utilizing bacteria in oil spill areas. *Microbias Letters, 8*, 137–143.

Aparicio, M. A., Eiroa, M., Kennes, C., & Veiga, M. C. (2007). Combined post-ozonation and biological treatment of recalcitrant wastewater from a resin-producing factory, *143*, 285–290. doi:10.1016/j.jhazmat.2006.09.025

Apaydin, Ö., Kurt, U., & Gönüllü, M. (2009). An investigation on the treatment of tannery wastewater by electrocoagulation. *Global NEST Journal, 11*(4), 546–555.

APHA. AWWA, WEF. (2005). Standard methods for the examination of water and wastewater (21st ed.) American Public Health Association, Washington DC.

Aplin, R., & Waite, T. D. (2000). Comparison of three advanced oxidation processes for the degradation of textile dyes. *Water Science and Technology*, *42*(5-6), 345–354. doi:10.2166/wst.2000.0534

Appels, L., Baeyens, J., Degreve, J., & Dewil, R. (2008). Principles and potential of the anaerobic digestion of waste-activated sludge. *Progress in Energy and Combustion Science*, *34*(6), 755–781. doi:10.1016/j.pecs.2008.06.002

Appels, L., Lauwers, J., Degrève, J., Helsen, L., Lievens, B., Willems, K., ... Dewil, R. (2011). Anaerobic digestion in global bio-energy production: Potential and research challenges. *Renewable & Sustainable Energy Reviews*, *15*(9), 4295–4301.

Apul, O. G., Atalar, I., Zorba, G. T., & Sanin, F. D. (2010). The dewaterability of disintegrated sludge samples before and after anaerobic digestion. *Drying Technology*, *28*(7), 901–909. doi:10.1080/07373937.2010.490764

Arabani, M., Mirabdolazimi, S. M., & Sasani, A. R. (2010). The effect of waste tire thread mesh on the dynamic behaviour of asphalt mixtures. *Construction & Building Materials*, *24*, 1060–1068.

Araruna, J. T., Portes, V. L. O., Soares, A. P., Silva, M. C., Sthel, M. S., Schramm, D. U., ... Vargas, H. (2004). Oil spills debris clean up by thermal desorption. *Journal of Hazardous Materials*, *110*(1-3), 161–16. doi:10.1016/j.jhazmat.2004.02.054 PMID:15177737

Arden, E., & Lockett, W. (1914). Experiment on the oxidation of sewage without the aid of filters. *Journal of the Society of Chemical Industry*, *33*(10), 523–539. doi:10.1002/jctb.5000331005

Arditsoglou, A., & Samara, C. (2005). Levels of total suspended particulate matter and major trace elements in Kosovo: A source identification and apportionment study. *Chemosphere*, *59*(5), 669–678. doi:10.1016/j.chemosphere.2004.10.056 PMID:15792664

Arena, U. L., Zaccariello, L., & Mastellone, M. L. (2010). Fluidized bed gasification of waste-derived fuels. *Waste Management (New York, N.Y.)*, *30*(7), 1212–1219. doi:10.1016/j.wasman.2010.01.038 PMID:20172708

Arena, U., Zaccariello, L., & Mastellone, M. L. (2009). Tar removal during the fluidized bed gasification of plastic waste. *Waste Management (New York, N.Y.)*, *29*(2), 783–791. doi:10.1016/j.wasman.2008.05.010 PMID:18693006

Ariffin, M. A., Mahmood, W. M. F. W., Harun, Z., & Mohamed, R. (2017). Medium-scale gasification of oil palm empty fruit bunch for power generation. *Journal of Material Cycles and Waste Management*, *19*(3), 1244–1252. doi:10.100710163-016-0518-8

Arnell, N. W. (1999). Climate change and global water resources. *Global Environmental Change*, *9*, S3–S19. doi:10.1016/S0959-3780(99)00017-5

Arora, K., Mickelson, S. K., & Baker, J. L. (2003). Effectiveness of vegetated buffer strips in reducing pesticide transport in simulated runoff. *Transactions of the ASAE. American Society of Agricultural Engineers*, *46*(3), 635. doi:10.13031/2013.13599

Arora, V. K., & Boer, G. J. (2001). Effects of simulated climate change on the hydrology of major river basins. *Journal of Geophysical Research*, *106*(D4), 3335–3348. doi:10.1029/2000JD900620

Arshad, A., & Hashim, N. H. (2012). Anaerobic digestion of NSSC pulping effluent. *International Journal of Environmental of Research*, *6*(3), 761–768.

Arslan-Alaton, I. (2003). A review of the effects of dye-assisting chemicals on advanced oxidation of reactive dyes in wastewater. *Coloration Technology*, *119*(6), 345–353. doi:10.1111/j.1478-4408.2003.tb00196.x

Asadullah, M. (2014). Barriers of commercial power generation using biomass gasification gas: A review. *Renewable & Sustainable Energy Reviews*, *29*, 201–215. doi:10.1016/j.rser.2013.08.074

Asadullah, M., Rahman, M. A., Ali, M. M., Motin, M. A., Sultan, M. B., & Alam, M. R. (2007). Production of bio-oil from fixed bed pyrolysis of bagasse. *Fuel*, *86*(16), 2514–2520. doi:10.1016/j.fuel.2007.02.007

Asha, P. (2016). Biodrying process: A sustainable technology for treatment of municipal solid waste with high moisture content. *Waste Management (New York, N.Y.)*, *49*, 64–72. doi:10.1016/j.wasman.2016.01.004 PMID:26774396

Aslani, A., Helo, P., & Naaranoja, M. (2014). Role of renewable energy policies in energy dependency in Finland: System dynamics approach. *Applied Energy*, *113*, 758–765.

Asli, H.,, & Karim, M. R. (2011). Implementation of waste cooking oil as RAP rejuvenator. *Journal of the Eastern Asia Society for Transportation Studies*, *9*, 1336–1350.

Asli, H., Ahmadinia, E., Zargar, M., & Karim, M. R. (2012). Investigation on physical properties of waste cooking oil – rejuvenated bitumen binder. *Construction & Building Materials*, *37*, 398–405.

Asli, H., Ahmadinia, E., Zargar, M., & Karim, M. R. (2012). Investigation on physical properties of waste cooking oil – Rejuvenated bitumen binder. *Construction & Building Materials*, *37*, 398–405. doi:10.1016/j.conbuildmat.2012.07.042

Assamoi, B., Lawryshyn, Y. The environmental comparison of landfilling vs. incineration of MSW accounting for waste diversion. *Waste Manage*, 32, 1019–1030.

Athanasoulia, E., Melidis, P., & Aivasidis, A. (2012). Optimization of biogas production from waste activated sludge through serial digestion. *Renewable Energy*, *47*, 147–151.

Athawale, V., & Rathi, S. (1997). Effect of chain length of the alkyl group of alkyl methacrylates on graft polymerization onto starch using ceric ammonium nitrate as initiator. *European Polymer Journal*, *33*(7), 1067–1071. doi:10.1016/S0014-3057(96)00294-7

Atlas, R. M. (1981). Microbial degradation of petroleum hydrocarbons: An experimental perspective. *Microbiological Reviews*, *45*, 180–209. PMID:7012571

Atmaca, E. (2009). Treatment of landfill leachate by using electro-Fenton method. *Journal of Hazardous Materials*, *163*(1), 109–114. doi:10.1016/j.jhazmat.2008.06.067 PMID:18657358

Aurora, S., Cristina, D., & Fi'uza, A. (2005). Use of solvent extraction to remediate soils contaminated with hydrocarbons. *Journal of Hazardous Materials*, *B124*, 224–229. PMID:16006033

Ausiello, A. (2017). Development of a system for electric energy production based on fuel cells fueled by biogas produced with anaerobic digestion of biomass. doi:10.6093/UNINA/FEDOA/11536

Austermann-Haun, U., Seyfried, C. F., & Rosenwinkel, K.-H. (1997). UASB-reactor in the fruit juice industry. *Water Science and Technology*, *36*(6), 407–414. doi:10.2166/wst.1997.0617

Avlonitis, S., Kouroumbas, K., & Vlachakis, N. (2003). Energy consumption and membrane replacement cost for seawater RO desalination plants. *Desalination*, *157*(1-3), 151–158. doi:10.1016/S0011-9164(03)00395-3

Avsar, Y., Kurt, U., & Gonullu, T. (2007). Comparison of classical chemical and electrochemical processes for treating rose processing wastewater. *Journal of Hazardous Materials*, *148*(1), 340–345. doi:10.1016/j.jhazmat.2007.02.048 PMID:17386973

Awang, H., Al-Mulali, M. Z., Abdul Khalil, H. P. S., & Aljoumaily, Z. S. (2014). Utilisation of oil palm ash in foamed concrete. *MATEC Web of Conferences*, 15, 01033.

Awwad, M., Al-Rimawi, F., Dajani, K. J. K., Khamis, M., Nir, S., & Karaman, R. (2015). Removal of amoxicillin and cefuroxime axetil by advanced membranes technology, activated carbon and micelle–clay complex. *Environmental Technology*, *36*(16), 2069–2078. doi:10.1080/09593330.2015.1019935 PMID:25686519

Ayandiran, T. A., Fawole, O. O., & Dahunsi, S. O. (2018). Water quality assessment of bitumen polluted Oluwa River, South-Western Nigeria. *Water Resources and Industry*, *19*, 13–24. doi:10.1016/j.wri.2017.12.002

Ay, F., Catalkaya, E. C., & Kargi, F. (2009). A statistical experiment design approach for advanced oxidation of direct red azo-dye by photo-Fenton treatment. *Journal of Hazardous Materials*, *162*(1), 230–236. doi:10.1016/j.jhazmat.2008.05.027 PMID:18555599

Ayoub, G., Semerjian, L., Acra, A., Fadel, M. E., & Koopman, B. (2001). Heavy metal removal by coagulation with seawater liquid bittern. *Journal of Environmental Engineering*, *127*(3), 196–207. doi:10.1061/(ASCE)0733-9372(2001)127:3(196)

Ayoub, K., van Hullebusch, E. D., Cassir, M., & Bermond, A. (2010). Application of advanced oxidation processes for TNT removal: A review. *Journal of Hazardous Materials*, *178*(1-3), 10–28. doi:10.1016/j.jhazmat.2010.02.042 PMID:20347218

Ayyachamy, M., Turner, K., O'Donovan, A., Gupta, V., & Tuohy, M. (2012). *Laboratory protocols in fungal biology: current methods in fungal biology*. New York, NY: Springer.

Aziz, H. A., Adlan, M. N., Zahari, M. S. M., & Alias, S. (2004). Removal of ammoniacal nitrogen (N-NH3) from municipal solid waste leachate by using activated carbon and limestone. *Waste Management & Research*, *22*(5), 371–375. doi:10.1177/0734242X04047661 PMID:15560441

Aziz, H. A., Othman, N., Yusuff, M. S., Basri, D. R. H., Asaari, F. A. H., Adlan, M. N., ... Perwira, M. (2001). Removal of copper from water using limestone filtration technique-determination of mechanism of removal. *Environment International*, *26*(5-6), 395–399. doi:10.1016/S0160-4120(01)00018-6 PMID:11392757

Aziz, H. A., & Salina, A. (2006). Colour removal from landfill leachate by coagulation and flocculation processes. *Bioresource Technology*, *98*(1), 218–220. doi:10.1016/j.biortech.2005.11.013 PMID:16386895

Aziz, H. A., Yusoff, M. S., Mohd, N. A., Adnan, N. H., & Salina, A. (2004). Physico-chemical removal of iron from semi-aerobic landfill leachate by limestone filter. *Waste Management (New York, N.Y.)*, *24*(4), 353–358. doi:10.1016/j.wasman.2003.10.006 PMID:15081062

Azkar, M. M. B. I., Chatani, S., & Sudo, K. (2012). Simulation of urban and regional air pollution in Bangladesh. *Journal of Geophysical Research, D, Atmospheres*, *117*(D7).

Baade, W., Hunkeler, D., & Hamielec, A. (1989). Copolymerization of acrylamide with cationic monomers in solution and inverse-microsuspension. *Journal of Applied Polymer Science*, *38*(1), 185–201. doi:10.1002/app.1989.070380117

Babaei, A., & Shayegan, J. (2011). Effect of organic loading rates (OLR) on production of methane from anaerobic digestion of vegetables waste. In *Proceedings of World Renewable Energy Congress* (pp. 411-417). Sweden: Linköping University Electronic Press. 10.3384/ecp11057411

Babel, S., & Kurniawan, T. A. (2004). Cr (VI) removal from synthetic wastewater using coconut shell charcoal and commercial activated carbon modified with oxidizing agents and/or chitosan. *Chemosphere*, *54*(7), 951–967. doi:10.1016/j.chemosphere.2003.10.001 PMID:14637353

Babuponnusami, A., & Muthukumar, K. (2014). A review on Fenton and improvements to the Fenton process for wastewater treatment. *Journal of Environmental Chemical Engineering*, *2*(1), 557–572. doi:10.1016/j.jece.2013.10.011

Babu, R. R., Bhadrinarayana, N., Begum, K. M. M. S., & Anantharaman, N. (2007). Treatment of tannery wastewater by electrocoagulation. *Journal of the University of Chemical Technology and Metallurgy*, *42*(2), 201–206.

Badgie, D., Samah, M. A. A., Manaf, L. A., & Muda, A. B. (2012). Assessment of municipal solid waste composition in Malaysia: Management, practice, and challenges. *Polish Journal of Environmental Studies, 21*(3), 539-547.

Badiei, M., Jahim, J., Anuar, N., Sheik Abdullah, S., Su, L., & Kamaruzzaman, M. (2012). Microbial community analysis of mixed anaerobic microflora in suspended sludge of ASBR producing hydrogen from palm oil mil effluent. *International Journal of Hydrogen Energy*, *37*(4), 3169–3176. doi:10.1016/j.ijhydene.2011.11.063

Bagchi, A. (1990). *Design, construction, and monitoring of sanitary landfill*. New York, NY: John Wiley & Sons.

Bahia, H. U.,, & Anderson, D. A. (1993). Glass transition behaviour and physical hardening of asphalt binders. *Journal Association Asphalt Paving Technology*, *62*, 93–129.

Bai, D., Wang, X., Huo, D., Ying, Q., Wang, N., & Xia, B. (2019). Coprecipitation preparation of Cu/Zn/Al-hydrotalcite-like compound for copper removal from electroplating wastewater. Journal of Chemistry, Article ID 5347920, 9 pages.

Bai, T., Yang, Y., Lian, L., Wu, Y., Jiang, S., Zheng, H.-X. … Nan, J.-N. (n.d.). Key laboratory for natural resource of Changbai Mountain & functional molecules. Ministry of Education, College of Pharmacy, Yanbian. Acta Pharmacologica Sinica, 34.

Bakonyi, P., Nemestóthy, N., Simon, V., & Bélafi-Bakó, K. (2014). Review on the start-up experiences of continuous fermentative hydrogen producing bioreactors. *Renewable & Sustainable Energy Reviews*, *40*, 806–813. doi:10.1016/j.rser.2014.08.014

Bala, J., Lalung, J., & Ismail, N. (2014). Biodegradation of palm oil mill effluent (POME) by bacterial. *Int J Scientific and Researchs Publications, 4*(3).

Balan, D. S. L., & Monteneiro, R. T. R. (2001). Decolorization of textile indigo dye by ligninolytic fungi. *Journal of Biotechnology*, *89*(2-3), 141–145. doi:10.1016/S0168-1656(01)00304-2 PMID:11500207

Balan, V., Kumar, S., Bals, B., Chundawat, S., Jin, M., & Dale, B. (2012). Biochemical and thermochemical conversion of switchgrass to biofuels. In A. Monti (Ed.), *Switchgrass, green energy and technology* (pp. 153–185). London, UK: Springer. doi:10.1007/978-1-4471-2903-5_7

Balasubramanian, V., Natarajan, K., Hemambika, B., Ramesh, N., Sumathi, C. S., Kottaimuthu, R., & Rajesh Kannan, V. (2010). High-density polyethylene (HDPE) degrading potential bacteria from marine ecosystem of Gulf of Mannar, India. *Letters in Applied Microbiology*, *51*(2), 205–211. PMID:20586938

Balat, M. (2008). Mechanisms of thermochemical biomass conversion processes. Part 1: Reactions of pyrolysis. *Energy Sources Part a-Recovery Utilization and Environmental Effects, 30*(7), 620-635. doi:10.1080/15567030600817258

Ballesteros, M. M. M., Sanchez, P. J. A., Casas, L. J. L., Oller, I., & Malato, R. S. (2009). Degradation of a four-pesticide mixture by combined photo-Fenton and biological oxidation. *Water Research*, *43*(3), 653–660. doi:10.1016/j.watres.2008.11.020 PMID:19059625

Baloch, M. I., & Akunna, J. C. (2003). Granular bed baffled reactor (Grabbr): Solution to a two-phase anaerobic digestion system. *Journal of Environmental Engineering*, *129*(11), 1015–1021. doi:10.1061/(ASCE)0733-9372(2003)129:11(1015)

Banat, I. M., Nigam, P., Singh, D., & Marchant, R. (1996). Microbial decolorization of textile-dye containing effluents: A review. *Bioresource Technology*, *58*(3), 217–227. doi:10.1016/S0960-8524(96)00113-7

Banchuen, T. (2003). *A microcosm-based investigation into oxidized nitrogen removal in the hypolimnetic water of the Occoquan reservoir of northern Virginia.* (Master's Thesis), Virginia Polytechnic Institute and State University, Blacksburg, Virginia.

Banu, J. R., Sharmila, V. G., Devi, M. G., Kumar, S. A., Kumar, G., Nguyen, D. D., & Saratale, G. D. (2019). Cost effective sludge reduction using synergetic effect of dark fenton and disperser treatment. *Journal of Cleaner Production, 207*, 261–270. doi:10.1016/j.jclepro.2018.09.252

Banu, J., Kaliappan, S., & Yeom, I. T. (2007). Treatment of domestic wastewater using upflow anaerobic sludge blanket reactor. *International Journal of Environmental Science and Technology, 4*(3), 363–370. doi:10.1007/BF03326295

Baquero, F., Martínez, J. L., & Cantón, R. (2008). Antibiotics and antibiotic resistance in water environments. *Current Opinion in Biotechnology, 19*(3), 260–265. doi:10.1016/j.copbio.2008.05.006 PMID:18534838

Baranitharan, E., Khan Maksudur, R., & Prasad, D. M. R. (2013). Treatment of palm oil mill effluent in microbial fuel cell using polyacrylonitrile carbon felt as electrode. *Journal of Medical Bioengineering, 2*(4), 252–256. doi:10.12720/jomb.2.4.252-256

Baraoidan, W. A., Tun, L. L., Gaspillo, P.-D., & Suzuki, M. (2007). A study on the relative performance of different coagulants and the kinetics of COD in the treatment of a textile bleaching and dyeing industrial wastewater. *ASEAN Journal of Chemical Engineering, 6*(1 & 2), 49–60.

Barber, W. P., & Stuckey, D. C. (1998). The influence of start-up strategies on the performance of an anaerobic baffled reactor. *Environmental Technology, 19*(5), 489–501. doi:10.1080/09593331908616705

Barber, W. P., & Stuckey, D. C. (1999). The use of the anaerobic baffled reactor (ABR) for wastewater treatment: A review. *Water Research, 33*(7), 1559–1578. doi:10.1016/S0043-1354(98)00371-6

Baresel, C., Magnér, J., Magnusson, K., & Olshammar, M. (2017). Tekniska lösningar för avancerad refining av avloppsvatten (No. C 235).

Barisas, S., Baker, J., Johnson, H., & Laflen, J. (1978). Effect of tillage systems on runoff losses of nutrients, a rainfall simulation study. *Transactions of the ASAE. American Society of Agricultural Engineers, 21*(5), 893–0897. doi:10.13031/2013.35409

Barlaz, M. A., Schaefer, D. M., & Ham, R. K. (1990). Bacterial population development and chemical characteristics of refuse decomposition in a simulated sanitary landfill. *Applied and Environmental Microbiology, 55*, 55–65. PMID:16347835

Barnes, D. K. A., Galgani, F., Thompson, R. C., & Barlaz, M. (2009). Accumulation and fragmentation of plastic debris in global environments. *Philosophical transactions of the Royal Society of London. Series B, Biological Sciences, 364*(1526), 1985–1998. doi:10.1098/rstb.2008.0205 PMID:19528051

Barragan, B. E., Costa, C., & Marquez, M. C. (2007). Biodegradation of azo dyes by bacteria inoculated on solid media. *Dyes and Pigments, 75*(1), 73–81. doi:10.1016/j.dyepig.2006.05.014

Barredo-Damas, S., Alcaina-Miranda, M. I., Iborra-Clar, M. I., Bes-Pia, A., Mendoza-Roca, J. A., & Iborra-Clar, A. (2006). Study of the UF process as pretreatment of NF membranes for textile wastewater reuse. *Desalination, 200*(1-3), 745–747. doi:10.1016/j.desal.2006.03.497

Barrena, R., d'Imporzano, G., Ponsá, S., Gea, T., Artola, A., Vázquez, F., ... Adani, F. (2009). In search of a reliable technique for the determinat/ion of the biological stability of the organic matter in the mechanical–biological treated waste. *Journal of Hazardous Materials, 162*(2-3), 1065–1072. PMID:18606494

Barrett, J. M. (1989). Improvement of slow sand filtration of warm water by using coarse sand. (Doctoral Thesis), University of Colorado. Colorado.

Barron, M. G., Heintz, R., & Krahn, M. M. (2003). Contaminant exposure and effects in pinnipeds: Implications for Steller sea lion declines in Alaska. *The Science of the Total Environment*, *311*(1-3), 111–133. doi:10.1016/S0048-9697(03)00140-2 PMID:12826388

Bartels, M., Lin, W., Nijenhuis, J., Kapteijn, F., & Van Ommen, J. R. (2008). Agglomeration in fluidized beds at high temperatures: Mechanisms, detection and prevention. *Progress in Energy and Combustion Science*, *34*(5), 633–666. doi:10.1016/j.pecs.2008.04.002

Barter, P. A. (2012). Off-street parking policy surprises in Asian cities. *Cities (London, England)*, *29*(1), 23–31. doi:10.1016/j.cities.2011.06.007

Baruah, D., & Baruah, D. (2014). Modeling of biomass gasification: A review. *Renewable & Sustainable Energy Reviews*, *39*, 806–815. doi:10.1016/j.rser.2014.07.129

Basiron, Y. (2007). Palm oil production through sustainable plantations. *European Journal of Lipid Science and Technology*, *109*(4), 289–295. doi:10.1002/ejlt.200600223

Basiron, Y. (2013). Malaysia supply and demand of palm oil: challenges and opportunities until 2020. In *Proceedings of the Oil World Outlook Conference 2013*, Hamburg, Germany.

Basri, M. F., Yacob, S., Hassan, M. A., Shirai, Y., Wakisaka, M., Zakaria, M. R., & Phang, L. Y. (2010). Improved biogas from palm oil mill effluent by a scaled-down anaerobic treatment process. *World Journal of Microbiology & Biotechnology*, *26*(3), 505–514. doi:10.100711274-009-0197-x

Bassam, M., & Battikhi, M. N. (2005). Biodegradation of total organic carbons (TOC) in Jordanim petroleum sludge. *Journal of Hazardous Materials*, *120*(1-3), 127–134. doi:10.1016/j.jhazmat.2004.12.033 PMID:15811673

Bassuney, D., Ibrahim, W., & Moustafa, M. (2013). Performance of an anaerobic baffled reactor (ABR) treating high-strength food industrial wastewater with fluctuating pH, *Proc. of the Intl. Conf. on Advances in Civil, Structural and Environmental Engineering - ACSEE 2013*, 98-102.

Basu, P. (2010). Biomass gasification and pyrolysis: practical design and theory. Burlington, MA: Academic Press (Elsevier).

Basu, P. (2006). *Combustion and gasification in fluidized beds*. Burlington, MA: CRC Press. doi:10.1201/9781420005158

Basu, P. (2013). *Biomass gasification, pyrolysis and torrefaction: practical design and theory*. Academic press.

Basu, P. (2015). *Circulating fluidized bed boilers: design, operation and maintenance* (1st ed.). Switzerland: Springer International.

Batool, A., & Abbas, F. (2017). Reasons for delay in selected hydro-power projects in Khyber Pakhtunkhwa (KPK), Pakistan. *Renewable & Sustainable Energy Reviews*, *73*, 196–204.

Batstone, D. J., Keller, J., Angelidaki, I., Kalyuzhnyi, S., Pavlostathis, S., Rozzi, A., ... Vavilin, V. (2002). The IWA anaerobic digestion model no 1 (ADM1). *Water Science and Technology*, *45*(10), 65–73. PMID:12188579

Bauer, M., & Herrmann, R. (1998). Dissolved organic carbon as the main carrier of phthalic acid esters in municipal landfill leachates. *Waste Management & Research*, *16*(5), 446–454. doi:10.1177/0734242X9801600507

Bauer, M., Herrmann, R., Martin, A., & Zellmann, H. (1998). Chemodynamics, transport behaviour and treatment of phthalic acid esters in municipal landfill leachates. *Water Science and Technology*, *38*(2), 185–192. doi:10.2166/wst.1998.0135

Baumgarten, S., Schröder, H. F., Charwath, C., Lange, M., Beier, S., & Pinnekamp, J. (2007). Evaluation of advanced treatment technologies for the elimination of pharmaceutical compounds. *Water Science and Technology*, *56*(5), 1–8. doi:10.2166/wst.2007.550 PMID:17881831

Bautista, P., Mohedano, A. F., Casas, J. A., Zazo, J. A., & Rodrigue, J. J. (2008). An overview of the application of Fenton oxidation to industrial wastewaters treatment. *Journal of Chemical Technology and Biotechnology (Oxford, Oxfordshire)*, *83*(10), 1323–1338. doi:10.1002/jctb.1988

Bazrafshanferdos, E., Mostafapour, F. K., Soori, M. M., & Mahvi, A. (2012). Application of combined chemical coagulation and electrocoagulation process to carwash wastewater treatment. *Fresenius Environmental Bulletin*, *21*, 2694–2701.

Bechle, M. J., Millet, D. B., & Marshall, J. D. (2011). Effects of income and urban form on urban NO2: Global evidence from satellites. *Environmental Science & Technology*, *45*(11), 4914–4919. doi:10.1021/es103866b PMID:21542624

Beddoes, J. C., Bracmort, K. S., Burns, R. T., & Lazarus, W. F. (2007). An analysis of energy production costs from anaerobic digestion systems on US livestock production facilities. *USDA NRCS Technical Note*(1).

Beheshti, S., Ghassemi, H., & Shahsavan-Markadeh, R. (2015). Process simulation of biomass gasification in a bubbling fluidized bed reactor. *Energy Conversion and Management*, *94*, 345–352. doi:10.1016/j.enconman.2015.01.060

Belfort, G., Dolecek, P., & Mikulasek, P. (1995). The performance of a rotating filter 1. Theoretical analysis of the flow in an annulus with a rotating inner porous wall. *Journal of Membrane Science*, *99*(3), 241–248. doi:10.1016/0376-7388(94)00221-J

Belila, A., El-Chakhtoura, J., Otaibi, N., Muyzer, G., Gonzalez-Gil, G., Saikaly, P. E., … Vrouwenvelder, J. S. (2016). Bacterial community structure and variation in a full-scale seawater desalination plant for drinking water production. *Water Research*, *94*, 62–72. doi:10.1016/j.watres.2016.02.039

Bellona, C., & Drewes, J. E. (2007). Viability of a low-pressure nanofilter in treating recycled water for water reuse applications: A pilot-scale study. *Water Research*, *41*(17), 3948–3958. doi:10.1016/j.watres.2007.05.027 PMID:17582458

Ben Grich, N., Attour, A., Le Page Mostefa, M., Guesmi, S., Tlili, M., & Lapicque, F. (2019). Fluoride removal from water by electrocoagulation: Effect of the type of water and the experimental parameters. *Electrochimica Acta*, *316*, 257–265. doi:10.1016/j.electacta.2019.05.130

Bener, S., Bulca, Ö., Palas, B., Tekin, G., Atalay, S., & Ersöz, G. (2019). Electrocoagulation process for the treatment of real textile wastewater: Effect of operative conditions on the organic carbon removal and kinetic study. *Process Safety and Environmental Protection*, *129*, 47–54. doi:10.1016/j.psep.2019.06.010

Bennajah, M., Gourich, B., Essadki, A., Vial, C., & Delmas, H. (2009). Defluoridation of morocco drinking water by electrocoagulation/electroflottation in an electrochemical external-loop airlift reactor. *Chemical Engineering Journal*, *148*(1), 122–131. doi:10.1016/j.cej.2008.08.014

Bento, A. M., Cropper, M. L., Mobarak, A. M., & Vinha, K. (2005). The effects of urban spatial structure on travel demand in the United States. *The Review of Economics and Statistics*, *87*(3), 466–478. doi:10.1162/0034653054638292

Bernard, S., & Gray, N. F. (2000). Aerobic digestion of pharmaceutical and domestic wastewater sludges at ambient temperature. *Water Research*, *34*(3), 725–734. doi:10.1016/S0043-1354(99)00234-1

Bernauer, T., & Kalbhenn, A. (2010). The politics of international freshwater resources. *International Studies Encyclopedia*, *IX*, 5800–5821.

Bernet, N., & Delgenes, N. (2000). Combined anaerobic–aerobic SBR for the treatment of piggery wastewater. *Water Research*, *34*(2), 611–619. doi:10.1016/S0043-1354(99)00170-0

Bertilsson, S., & Widenfalk, A. (2002). Photochemical degradation of PAHs in freshwaters and their impact on bacterial growth–influence of water chemistry. *Hydrobiologia*, *469*(1-3), 23–32. doi:10.1023/A:1015579628189

Bessa, F., Barría, P., Neto, J. M., Frias, J. P. G. L., Otero, V., Sobral, P., & Marques, J. C. (2018). Occurrence of microplastics in commercial fish from a natural estuarine environment. *Marine Pollution Bulletin*, *128*, 575–584. doi:10.1016/j.marpolbul.2018.01.044 PMID:29571409

Beyenal, H., Sani, R. K., Peyton, B. M., Dohnalkova, A. C., Amonette, J. E., & Lewandowski, Z. (2004). Uranium immobilization by sulfate-reducing biofilms. *Environmental Science & Technology*, *38*(7), 2067–2074. doi:10.1021/es0348703 PMID:15112808

Beylot, A., Villeneuve, J., & Bellenfant, G. (2013). Life cycle assessment of landfill biogas management: Sensitivity to diffuse and combustion air emissions. *Waste Management (New York, N.Y.)*, *33*(2), 401–411. doi:10.1016/j.wasman.2012.08.017 PMID:23142510

Bhainsa, K. C., & D'souza, S. (2008). Removal of copper ions by the filamentous fungus, Rhizopus oryzae from aqueous solution. *Bioresource Technology*, *99*(9), 3829–3835. doi:10.1016/j.biortech.2007.07.032 PMID:17804218

Bhandari, G. (2014). An overview of agrochemicals and their effects on environment in Nepal. *Applied Ecology and Environmental Sciences*, *2*(2), 66–73. doi:10.12691/aees-2-2-5

Bhaskar, M., Gnanamani, A., Ganeshjeevan, R. J., Chandrasekar, R., Sadulla, S., & Radhakrishnan, G. (2003). Analyses of carcinogenic aromatic amines released from harmful azo colorants by Streptomyces. *Journal of Chromatography. A*, *1018*(1), 117–1123. doi:10.1016/j.chroma.2003.08.024 PMID:14582632

Bhatia, S. K., Bhatia, R. K., & Yang, Y. H. (2017). An overview of microdiesel as a sustainable future source of renewable energy. *Renewable & Sustainable Energy Reviews*, *79*, 1078–1090.

Bhatia, S., Othman, Z., & Ahmad, A. (2007a). Coagulation-flocculation process for POME treatment using *Moringa oleifera* seeds extract; optimization studies. *Chemical Engineering Journal*, *133*(1-3), 205–212. doi:10.1016/j.cej.2007.01.034

Bhatia, S., Othman, Z., & Ahmad, A. (2007b). Pretreatment of palm oil mill effluent (POME) using *Moringa oleifera* seeds as natural coagulant. *Journal of Hazardous Materials*, *145*(1-2), 120–126. doi:10.1016/j.jhazmat.2006.11.003 PMID:17141409

Bhatta, B. (2010). *Analysis of urban growth and sprawl from remote sensing data*. Heidelberg, Germany: Springer. doi:10.1007/978-3-642-05299-6

Bhattacharya, A., Rawlins, J. W., & Ray, P. (2008). *Polymer grafting and crosslinking*. John Wiley & Sons. doi:10.1002/9780470414811

Bhattacharya, P., Ghosh, S., Majumdar, S., Dasgupta, S., & Bandyopadhyay, S. (2011). Biosorbent-assisted ceramic microfiltration process for treatment of herbal pharmaceutical wastewater with high organic loading. *International Journal of Environmental Technology and Management*, *14*(1-4), 132–146. doi:10.1504/IJETM.2011.039262

Bhattacharya, S. K., Madura, R. L., Walling, D. A., & Farrell, J. B. (1996). Volatile solids reduction in two-phase and conventional anaerobic sludge digestion. *Water Research*, *30*(5), 1041–1048. doi:10.1016/0043-1354(95)00252-9

Bhutto, A. W., & Bazmi, A. A. (2007). *Sustainable agriculture and eradication of rural poverty in Pakistan*. Paper presented at the Natural Resources Forum.

Biginagwa, F. J., Mayoma, B. S., Shashoua, Y., Syberg, K., & Khan, F. R. (2015). First evidence of microplastics in the African Great Lakes: Recovery from Lake Victoria Nile perch and Nile tilapia. *Journal of Great Lakes Research*. doi:10.1016/j.jglr.2015.10.012

Biglarijoo, N., Mirbagheri, S. A., Bagheri, M., & Ehteshami, M. (2017). Assessment of effective parameters in landfill leachate treatment and optimization of the process using neural network, genetic algorithm and response surface methodology. *Process Safety and Environmental Protection*, *106*, 89–103. doi:10.1016/j.psep.2016.12.006

Bilińska, L., Blus, K., Gmurek, M., & Ledakowicz, S. (2019). Coupling of electrocoagulation and ozone treatment for textile wastewater reuse. *Chemical Engineering Journal*, *358*, 992–1001. doi:10.1016/j.cej.2018.10.093

Billaud, J., Valin, S., Peyrot, M., & Salvador, S. (2016). Influence of H2O, CO2 and O2 addition on biomass gasification in entrained flow reactor conditions: Experiments and modelling. *Fuel*, *166*, 166–178. doi:10.1016/j.fuel.2015.10.046

Bill, G. (2012). Application of advanced oxidation processes (AOP) in water treatment, *37th Annual Queensland Water Industry Operations Workshop, Parklands*, Australia.

Binbin, L., Meizhu, C., & Shaopeng, W. (2014). Effect of waste edible vegetable oil on high temperature properties of different aged asphalts. *Key Engineering Materials*, *599*, 135–140.

Binkley, J., & Kandelbauer, A. (2003). Effluent treatment/enzymes in activated sludge, In A. Cavaco-Paulo & G. M. Gubitz, (Eds.), Textile processing with enzymes (Woodhead Publishing, Cambridge), 199-221.

Bin, Z., You, J., Yan, L., Duan, X., Zhao, X., Xin, F., ... Shenxin, L. (2019). Air pollution intervention and life-saving effect in China. Environment International. *Environment International*, *125*, 529–554. doi:10.1016/j.envint.2018.10.045 PMID:30612707

BioenergyE. T. I. P. (2019). *Transesterification to biodiesel.* Retrieved from http://www.etipbioenergy.eu/value-chains/conversion-technologies/conventional-technologies/transesterification-to-biodiesel

Bishnoi, P. (2012). *Effects of thermal hydrolysis pre-treatment on anaerobic digestion of sludge.* Virginia Tech.

Blackall, L., Burrell, P., Gwilliam, H., Bradford, D., Bond, P., & Hugenholtz, P. (1998). The use of 16S rRNA gene clone libraries to describe the microbial diversity of activated sludge communities. *Water Science and Technology*, *37*(4-5), 451–454. doi:10.2166/wst.1998.0691

Blackburne, R., Yuan, Z., & Keller, J. (2008). Demonstration of nitrogen removal via nitrite in a sequencing batch reactor treating domestic wastewater. *Water Research*, *42*(8), 2166–2176. doi:10.1016/j.watres.2007.11.029 PMID:18155266

Blaine, R. L., & Kissinger, H. E. (2012). Homer Kissinger and the Kissinger equation. *Thermochimica Acta*, *540*, 1–6. doi:10.1016/j.tca.2012.04.008

Blaisi, N. I. (2019). Construction and demolition waste management in Saudi Arabia: Current practice and roadmap for sustainable management. *Journal of Cleaner Production*, *221*, 167–175. doi:10.1016/j.jclepro.2019.02.264

Blanco-Canqui, H., Gantzer, C., Anderson, S., & Alberts, E. (2004). Division S-6—soil & water management & conservation. *Soil Science Society of America Journal*, *68*, 567–576. doi:10.2136ssaj2004.5670

Blümel, S., Knackmuss, H. J., & Stolz, A. (2002). Molecular cloning and characterization of the gene coding for the aerobic azoreductase from *Xenophilusazovorans*KF46F. *Applied and Environmental Microbiology*, *68*(8), 3948–3955. doi:10.1128/AEM.68.8.3948-3955.2002 PMID:12147495

Boller, M. (1993). Filter mechanisms in roughing filters. *Journal of Water Supply: Research & Technology - Aqua*, *42*, 174–185.

Boller, M. A., & Kavanaugh, M. C. (1995). Particle characteristics and headloss increase in granular media filtration. *Water Research*, *29*(4), 1139–1149. doi:10.1016/0043-1354(94)00256-7

Bolto, B., & Gregory, J. (2007). Organic polyelectrolytes in water treatment. *Water Research*, *41*(11), 2301–2324. doi:10.1016/j.watres.2007.03.012 PMID:17462699

Bond, T., & Templeton, M. R. (2011). History and future of domestic biogas plants in the developing world. *Energy for Sustainable Development*, *15*(4), 347–354.

Bonmatı, A., & Flotats, X. (2003). Air stripping of ammonia from pig slurry: Characterisation and feasibility as a pre-or post-treatment to mesophilic anaerobic digestion. *Waste Management (New York, N.Y.)*, *23*(3), 261–272. doi:10.1016/S0956-053X(02)00144-7 PMID:12737968

Borba, C., Guirardello, R., Silva, E., Veit, M., & Tavares, C. (2006). Removal of nickel (II) ions from aqueous solution by biosorption in a fixed bed column: Experimental and theoretical breakthrough curves. *Biochemical Engineering Journal*, *30*(2), 184–191. doi:10.1016/j.bej.2006.04.001

Borhan, M. N., Suja, F., Ismail, A., & Rahmat, R. A. O. K. (2009). The effects of used cylinder oil on asphalt mixes. *European Journal of Scientific Research*, *28*(3), 398–411.

Borja, R., & Banks, C. J. (1994). Anaerobic digestion of Palm Oil Mill Effluent using an up-flow anaerobic sludge blanket (UASB) reactor. *Biomass and Bioenergy*, *6*(5), 381–389. doi:10.1016/0961-9534(94)E0028-Q

Borja, R., Banks, C. J., & Sanchez, E. (1996). Anaerobic treatment of palm oil mill effluent in a two-stage up-flow anaerobic sludge blanket (UASB) reactor. *Journal of Biotechnology*, *45*(2), 125–135. doi:10.1016/0168-1656(95)00154-9

Borrego, C., Martins, H., Tchepel, O., Salmim, L., Monteiro, A., & Miranda, A. I. (2006). How urban structure can affect city sustainability from an air quality perspective. *Environmental Modelling & Software*, *21*(4), 461–467. doi:10.1016/j.envsoft.2004.07.009

Boschini, A. D., Pettersson, J., & Roine, J. (2007). Resource curse or not: A question of appropriability. *The Scandinavian Journal of Economics*, *109*(3), 593–617.

Bosco, M. L., Varrica, D., & Dongarra, G. (2005). Case study: Inorganic pollutants associated with particulate matter from an area near a petrochemical plant. *Environmental Research*, *99*(1), 18–30. doi:10.1016/j.envres.2004.09.011 PMID:16053924

Bosello, F., Roson, R., & Tol, R. S. J. (2006). Economy-wide estimates of the implications of climate change: Human health. *Ecological Economics*, *58*(3), 579–591. doi:10.1016/j.ecolecon.2005.07.032

Boudreaux, K. A. (n.d.). Organic compounds: Alkanes. CHEM 2353 Fundamentals of Organic Chemistry, Chapter 1. Retrieved from https://www.angelo.edu/faculty/kboudrea/index_2353/Chapter_01_2SPP.pdf

Bougrier, C., Albasi, C., Delgenès, J. P., & Carrère, H. (2006). Effect of ultrasonic, thermal and ozone pre-treatments on waste activated sludge solubilisation and anaerobic biodegradability. *Chemical Engineering and Processing*, *45*(8), 711–718. doi:10.1016/j.cep.2006.02.005

Bougrier, C., Delgenès, J. P., & Carrère, H. (2007). Impacts of thermal pre-treatments on the semi-continuous anaerobic digestion of waste activated sludge. *Biochemical Engineering Journal*, *34*(1), 20–27. doi:10.1016/j.bej.2006.11.013

Bougrier, C., Delgenès, J. P., & Carrère, H. (2008). Effects of thermal treatments on five different waste activated sludge samples solubilisation, physical properties and anaerobic digestion. *Chemical Engineering Journal*, *139*(2), 236–244. doi:10.1016/j.cej.2007.07.099

Bound, J. P., & Voulvoulis, N. (2005). Household disposal of pharmaceuticals as a pathway for aquatic contamination in the United Kingdom. *Environmental Health Perspectives*, *113*(12), 1705–1711. doi:10.1289/ehp.8315 PMID:16330351

Bourotte, C., Forti, M.-C., Taniguchi, S., Bicego, M. C., & Lotufo, P. A. (2005). A wintertime study of PAHs in fine and coarse aerosols in Sa o Paulo city, Brazil. *Atmospheric Environment, 39*(21), 3799–3811. doi:10.1016/j.atmosenv.2005.02.054

Bouwer, H. (1978). Groundwater hydrology (Vol. 480). New York, NY: McGraw-Hill.

Bouwer, H. (2000). Integrated water management: Emerging issues and challenges. *Agricultural Water Management, 45*(3), 217–228. doi:10.1016/S0378-3774(00)00092-5

Brack, W., Dulio, V., Ågerstrand, M., Allan, I., Altenburger, R., Brinkmann, M., ... Hernández, F. J. (2017). Towards the review of the European Union water framework directive: Recommendations for more efficient assessment and management of chemical contamination in European surface water resources. *The Science of the Total Environment, 576*, 720–737. doi:10.1016/j.scitotenv.2016.10.104 PMID:27810758

Bradley, B., & Keith, D. (2013). Urban form, air pollution, and CO2 emissions in large U.S. metropolitan areas. *The Professional Geographer, 65*(4), 612–635. doi:10.1080/00330124.2013.799991

Brady, N., & Weil, R. (2002). *The nature and properties of soils* (13th ed., p. 960). Upper Saddle River, NJ: Prentice Hall.

Braguglia, C. M., Gianico, A., Gallipoli, A., & Mininni, G. (2015). The impact of sludge pre-treatments on mesophilic and thermophilic anaerobic digestion efficiency: Role of the organic load. *Chemical Engineering Journal, 270*, 362–371. doi:10.1016/j.cej.2015.02.037

Bratby, J. (2006). *Coagulation and flocculation in water and wastewater treatment.* IWA Publishing.

Breeze, P. (2014). *Power from waste* (2nd ed., pp. 335–352). Power Generation Technologies.

Brennholt, N., Heß, M., & Reifferscheid, G. (2018). Freshwater microplastics: challenges for regulation and management. In M. Wagner & S. Lambert (Eds.), *Freshwater microplastics. The handbook of environmental chemistry* (Vol. 58). Cham, Switzerland: Springer. doi:10.1007/978-3-319-61615-5_12

Brian, S. Jr. (2008). Urban sprawl and air quality in large US cities. *Journal of Environmental Management, 86*(4), 688–698. doi:10.1016/j.jenvman.2006.12.034 PMID:17368703

Bricker, S. B., Longstaff, B., Dennison, W., Jones, A., Boicourt, K., Wicks, C., & Woerner, J. (2008). Effects of nutrient enrichment in the nation's estuaries: A decade of change. *Harmful Algae, 8*(1), 21–32. doi:10.1016/j.hal.2008.08.028

Bridgwater, A. V. (1999). Principles and practice of biomass fast pyrolysis processes for liquids. *Journal of Analytical and Applied Pyrolysis, 51*(1-2), 3–22. doi:10.1016/S0165-2370(99)00005-4

Bridgwater, A. V. (2012). Review of fast pyrolysis of biomass and product upgrading. *Biomass and Bioenergy, 38*, 68–94. doi:10.1016/j.biombioe.2011.01.048

Bridgwater, A. V. (2012). Upgrading biomass fast pyrolysis liquids. *Environmental Progress & Sustainable Energy, 31*(2), 261–268. doi:10.1002/ep.11635

Bridgwater, A. V., Meier, D., & Radlein, D. (1999). An overview of fast pyrolysis of biomass. *Organic Geochemistry, 30*(12), 1479–1493. doi:10.1016/S0146-6380(99)00120-5

Brillas, E., Sirés, I., & Oturan, M. A. (2009). Electro-fenton process and related electrochemical technologies based on fenton's reaction chemistry. *Chemical Reviews, 109*(12), 6570–6631. doi:10.1021/cr900136g PMID:19839579

Brink, A., Sheridan, C., & Harding, K. (2018). Combined biological and advance oxidation processes for paper and pulp effluent treatment. *South African Journal of Chemical Engineering, 25*, 116–122. doi:10.1016/j.sajce.2018.04.002

Bronzeoak, Ltd. (2003). Rice husk ash market study. EXP 129, DTI/PubURN. 03/668.

Browne, M. A., Underwood, A. J., Chapman, M. G., Williams, R., Thompson, R. C., & van Franeker, J. A. (2015). Linking effects of anthropogenic debris to ecological impacts. *Proceedings of the Royal Society B: Biological Sciences*, *282*(1807), 20142929. doi:10.1098/rspb.2014.2929 PMID:25904661

Brownridge, J. (2010). The role of an asphalt rejuvenator in pavement preservation: use and need for asphalt rejuvenation. *1st International Conference on Pavement Preservation*. April 13-15, 2010, Newport Beach, CA. pp. 351–364.

Bruchmüller, J., van Wachem, B. G. M., Gu, S., Luo, K. H., & Brown, R. C. (2012). Modeling the thermochemical degradation of biomass inside a fast pyrolysis fluidized bed reactor. *AIChE Journal. American Institute of Chemical Engineers*, *58*(10), 3030–3042. doi:10.1002/aic.13705

Bruun, S., Jensen, L. S., & Sommer, S. (2014). Small-scale household biogas digesters: An option for global warming mitigation or a potential climate bomb? *Renewable & Sustainable Energy Reviews*, *33*, 736–741.

Bryjak, M., Wolska, J., & Kabay, N. (2008). Removal of boron from seawater by adsorption–membrane hybrid process: Implementation and challenges. *Desalination*, *223*(1), 57–62. doi:10.1016/j.desal.2007.01.202

Budd, R., O'Geen, A., Goh, K. S., Bondarenko, S., & Gan, J. (2009). Efficacy of constructed wetlands in pesticide removal from tailwaters in the Central Valley, California. *Environmental Science & Technology*, *43*(8), 2925–2930. doi:10.1021/es802958q PMID:19475972

Budianta, W., Salim, C., Hinode, H. and Ohta, H. (2012). Remediation of organics-contaminated sandy soil by in-situ washing. *ASEAN Engineering Journal Part C*, 1(1), .87.

Budianta, W., Salim, C., Hinode, H., & Ohta, H. (2006). In situ soil washing on metal-contaminated sandy soil by sedimentation method: a new approach on soil remediation. *Philippine Engineering Journal* 2006. 27(1), pp 65–76. Retrieved from https://www.researchgate.net/publication/228365882_In_situ_Soil_Washing_on_Metal-Contaminated_Sandy_Soil_by_Sedimentation_Method_A_New_Approach_on_Soil_Remediation

Budianta, W., Salim, C., Hinode, H., & Ohta, H. (2010). In-situ washing by sedimentation method for contaminated sandy soil. *Proceedings of the Annual International Conference on Soils, Sediments, Water and Energy*: Vol. 15, Article 15.

Bukhari, A. A. (2008). Investigation of the electro-coagulation treatment process for the removal of total suspended solids and turbidity from municipal wastewater. *Bioresource Technology*, *99*(5), 914–921. doi:10.1016/j.biortech.2007.03.015 PMID:17499502

Burke, M. (1999). Phthalate standard endorsed by industry. *Environmental Science & Technology*, *33*(19), 398A–399A. doi:10.1021/es993032z PMID:21657395

Busch, M., Mickols, W. E., Jons, S., Redondo, J., & De Witte, J. (2004). Boron removal in sea water desalination. *International Desalination and Water Reuse Quarterly*, *13*(4), 25.

Cai, T., Park, S. Y., & Li, Y. (2013). Nutrient recovery from wastewater streams by microalgae: Status and prospects. *Renewable & Sustainable Energy Reviews*, *19*, 360–369.

Calbiochem. (2007). Detergents: A guide to the properties and uses of detergents in biology and biochemistry. Retrieved from www.calbiochem.com

Calli, B., Mertoglu, B., & Inanc, B. (2005). Landfill leachate management in Istanbul: Applications and alternatives. *Chemosphere*, *59*(6), 819–829. doi:10.1016/j.chemosphere.2004.10.064 PMID:15811410

Camacho, F. P., Sousa, V. S., Bergamasco, R., & Ribau Teixeira, M. (2017). The use of Moringa oleifera as a natural coagulant in surface water treatment. *Chemical Engineering Journal*, *313*, 226–237. doi:10.1016/j.cej.2016.12.031

Camargo, J. A., Alonso, A., & Salamanca, A. (2005). Nitrate toxicity to aquatic animals: A review with new data for freshwater invertebrates. *Chemosphere*, *58*(9), 1255–1267. doi:10.1016/j.chemosphere.2004.10.044 PMID:15667845

Campbell, H. W. (2000). Sludge management-future issues and trends. *Water Science and Technology*, *41*(8), 1–8. doi:10.2166/wst.2000.0135

Cao, Y. B., Lin, H., Yan, L., Wang, Y., & Cao, Y. Y. (2013). The anti-fibrotic effect of thymoquinone on thioacetamide-induced mouse hepatic fibrosis.

Cao, L., Tian, W., Nu, B., Zhang, Y., & Wang, P. (2002). Preliminary study of airborne particulate matter in a Beijing sampling station by instrumental neutron activation analysis. *Atmospheric Environment*, *36*(12), 1951–1956. doi:10.1016/S1352-2310(02)00178-4

Cao, Z., Ge, H., & Lai, S. (2001). Studies on synthesis and adsorption properties of chitosan cross-linked by glutaraldehyde and Cu (II) as template under microwave irradiation. *European Polymer Journal*, *37*(10), 2141–2143. doi:10.1016/S0014-3057(01)00070-2

Capodaglio, A. G., Hlavínek, P., & Raboni, M. (2015). Physico-chemical technologies for nitrogen removal from wastewaters: A review. *Revista Ambiente & Água*, *10*(3), 481–498.

Cara, I. G., & Jitereanu, G. (2015). Application of low-cost adsorbents for pesticide removal. *Bulletin of University of Agricultural Sciences and Veterinary Medicine Cluj-Napoca. Veterinary Medicine*, *72*(1).

Carballa, M., Regueiro, L., & Lema, J. M. (2015). Microbial management of anaerobic digestion: Exploiting the microbiome-functionality nexus. *Current Opinion in Biotechnology*, *33*, 103–111. PMID:25682574

Cárdenas, R. M., Laura, D.-C., & Walid, O. (2016). Air pollution and urban structure linkages: Evidence from European cities. *Renewable & Sustainable Energy Reviews*, *53*, 1–9. doi:10.1016/j.rser.2015.07.190

Caricchia, A., Chiavarini, S., & Pezza, M. (1999). Polycyclic aromatic hydrocarbons in the urban atmospheric particulate matter in the city of Naples (Italy). *Atmospheric Environment*, *33*(23), 3731–3738. doi:10.1016/S1352-2310(99)00199-5

Carley, B. N., & Mavinic, D. S. (1991). The effects of external carbon loading on nitrification and denitrification of a high ammonia landfill leachate. *Journal - Water Pollution Control Federation*, *63*, 51–59.

Carl, G., Stéphane, R., & Jacques-François, T. (2012). Are compact cities environmentally friendly? *Journal of Urban Economics*, *72*(2–3), 123–136.

Carlos Borrego, O. (2006). Traffic-related particulate air pollution exposure in urban areas. *Atmospheric Environment*, *40*(37), 7205–7214. doi:10.1016/j.atmosenv.2006.06.020

Carneiro, P. A., Osugi, M. E., Fugivara, C. S., Boralle, N., Furlan, M.,, & Zanoni, M. V. B. (2005). 'Evaluation of different electrochemical methods on the oxidation and degradation of Reactive Blue 4 in aqueous solution'. *Chemosphere*, *59*(3), 431–439. doi:10.1016/j.chemosphere.2004.10.043 PMID:15763096

Carrera, J., Baeza, J. A., Vicent, T., & Lafuente, J. (2003). Biological nitrogen removal of high-strength ammonium industrial wastewater with two-sludge system. *Water Research*, *37*(17), 4211–4221. doi:10.1016/S0043-1354(03)00338-5 PMID:12946903

Carrere, H., Bougrier, C., Castets, D., & Delgenes, J. P. (2008). Impact of initial biodegradability on sludge anaerobic digestion enhancement by thermal pretreatment. *Journal of Environmental Science and Health. Part A, Toxic/Hazardous Substances & Environmental Engineering*, *43*(13), 1551–1555. doi:10.1080/10934520802293735 PMID:18821243

Carrere, H., Dumas, C., Battimelli, A., Batstone, D. J., Delgenes, J. P., Steyer, J. P., & Ferrer, I. (2010). Pretreatment methods to improve sludge anaerobic degradability: A review. *Journal of Hazardous Materials*, *183*(1-3), 1–15. doi:10.1016/j.jhazmat.2010.06.129 PMID:20708333

Carr, S. A., Liu, J., & Tesoro, A. G. (2016). Transport and fate of microplastic particles in wastewater treatment plants. *Water Research*, *91*, 174–182. doi:10.1016/j.watres.2016.01.002 PMID:26795302

Carter, A. (2000). Herbicide movement in soils: Principles, pathways and processes. *Weed Research (Oxford)*, *40*(1), 113–122. doi:10.1046/j.1365-3180.2000.00157.x

Casson, L. W., & Bess, J. W. (2006). On-site sodium hypochlorite generation. *Proceedings of the Water Environment Federation*, 2006(5), 6335-6352.

Castellanelli, C. A. (2007). Analyzes of the used oil under environmental perspective and its possibilities for production of biodiesel. Courtesy of UFSM – Federal University of Santa Maria.

Castillo, A., Cecchf, F., & Alvarez, J. M. (1997). A combined anaerobic aerobic system to treat domestic sewage in coastal areas. *Water Research*, *31*(12), 3057–3063. doi:10.1016/S0043-1354(96)00064-4

Catalkaya, E. C., & Kargi, F. (2009). Advanced oxidation and mineralization of simazine using Fenton's reagent. *Journal of Hazardous Materials*, *168*(2-3), 688–694. doi:10.1016/j.jhazmat.2009.02.091 PMID:19297085

Celis, J. E., Morales, J. R., Zaror, C. A., & Inzunza, J. C. (2004). A study of the particulate matter PM10 composition in the atmosphere of Chillan, Chile. *Chemosphere*, *54*(4), 541–550. doi:10.1016/S0045-6535(03)00711-2 PMID:14581056

Cengel, Y. A., & Boles, M. A. (Eds.). (2011). Thermodynamics an engineering approach. Singapore, McGraw-Hill.

Cengeloglu, Y., Arslan, G., Tor, A., Kocak, I., & Dursun, N. (2008). Removal of boron from water by using reverse osmosis. *Separation and Purification Technology*, *64*(2), 141–146. doi:10.1016/j.seppur.2008.09.006

Cervero, R., & Murakami, J. (2010). Effects of built environments on vehicle miles traveled: Evidence from 370 US metropolitan areas. *Environment & Planning A*, *42*(2), 400–418. doi:10.1068/a4236

Ceylan, S., & Topcu, Y. (2014). Pyrolysis kinetics of hazelnut husk using thermogravimetric analysis. *Bioresource Technology*, *156*, 182–188. doi:10.1016/j.biortech.2014.01.040 PMID:24508656

Cha, C. J., Doerge, D. R., & Cerniglia, C. E. (2001). Biotransformation of malachite green by the fungus *Cunninghamella elegans. Applied and Environmental Microbiology*, *67*(9), 4358–4360. doi:10.1128/AEM.67.9.4358-4360.2001 PMID:11526047

Chagas, E. P., & Durrant, L. R. (2001). Decolorization of azo dyes by Phanerochaetechrysosporium and Pleurotussajorcaju. *Enzyme and Microbial Technology*, *29*(8-9), 473–477. doi:10.1016/S0141-0229(01)00405-7

Chakrabarty, R. K., Pervez, S., Chow, J. C., Watson, J. G., Dewangan, S., Robles, J., & Tian, G. (2013). Funeral pyres in South Asia: Brown carbon aerosol emissions and climate impacts. *Environmental Science & Technology Letters*, *1*(1), 44–48. doi:10.1021/ez4000669

Chakraborty, S., Purkait, M. K., DasGupta, S., De, S., & Basu, J. K. (2003). Nanofiltration of textile plant effluent for color removal and reduction in COD. *Separation and Purification Technology*, *31*(2), 141–151. doi:10.1016/S1383-5866(02)00177-6

Chamarro, E., Marco, A., & Esplugas, S. (2001). S. Use of fenton reagent to improve organic biodegradability. *Water Research*, *35*(4), 1047–1051. doi:10.1016/S0043-1354(00)00342-0 PMID:11235870

Chandrappa, R., & Das, D. B. (2012). *Solid waste management. Environmental Science and Engineering*. Berlin, Germany: Springer-Verlag.

Chandrasekhar, S. (2013). *Radiative transfer*. Courier Corporation.

Chang, K.-L., Lin, J.-H., & Chen, S.-T. (2011). Adsorption studies on the removal of pesticides (Carbofuran) using activated carbon from rice straw agricultural waste. *World Acad Sci Eng Technol, 5*, 04-21.

Chang, C. Y., Chang, J. S., Vigneswaran, S., & Kandasamy, J. (2008). Pharmaceutical wastewater treatment by membrane bioreactor process–a case study in southern Taiwan. *Desalination*, *234*(1-3), 393–401. doi:10.1016/j.desal.2007.09.109

Chang, C., & Hao, O. (1996). Sequencing batch reactor system for nutrient removal: ORP and pH profiles. *Journal of Chemical Technology and Biotechnology (Oxford, Oxfordshire)*, *67*(1), 27–38. doi:10.1002/(SICI)1097-4660(199609)67:1<27::AID-JCTB430>3.0.CO;2-2

Chang, H. N., Moon, R. K., Park, B. G., Lim, S. J., Choi, D. W., Lee, W. G., ... Ahn, Y. H. (2000). Simulation of sequencing batch reactor operation for simultaneous removal of nitrogen and phosphorous. *Bioprocess Engineering*, *23*(5), 513–521. doi:10.1007004499900188

Chang, H., Li, T., Liu, B., Vidic, R. D., Elimelech, M., & Crittenden, J. C. (2019). Potential and implemented membrane-based technologies for the treatment and reuse of flowback and produced water from shale gas and oil plays: A review. *Desalination*, *455*, 34–57. doi:10.1016/j.desal.2019.01.001

Chang, J. S., & Lin, C. Y. (2000). Fed-batch bioreactor strategies for microbial decolorization of azo dye using a Pseudomonas luteola strain. *Biotechnology Progress*, *16*(6), 979–985. doi:10.1021/bp000116z PMID:11101324

Chang, J. S., & Lin, C. Y. (2001). Decolorization Kinetics of a recombinant *Escherichia coli* strain harboring azo-dye-decolorizing determinants from Rhodococcus sp. *Biotechnology Letters*, *23*(8), 631–636. doi:10.1023/A:1010306114286

Changmai, M., Pasawan, M., & Purkait, M. K. (2019). Treatment of oily wastewater from drilling site using electrocoagulation followed by microfiltration. *Separation and Purification Technology*, *210*, 463–472. doi:10.1016/j.seppur.2018.08.007

Chang, S. H. (2014). An overview of empty fruit bunch from oil palm as feedstock for bio-oil production. *Biomass and Bioenergy*, *62*, 174–181. doi:10.1016/j.biombioe.2014.01.002

Chang, W. H.,, & Tye, C. T. (2013). Catalytic cracking of used palm oil using composite zeolite. *The Malaysian Journal of Analytical Sciences*, *17*(1), 176–184.

Chan, Y. J., & Chong, M. F. (2019). Palm oil mill effluent (POME) treatment-Current technologies, biogas capture and challenges. In D. C. Y. Foo & M. K. Tun Abdul Aziz (Eds.), *Green technologies for the palm oil industry* (pp. 71–92). Springer Singapore, Singapore. doi:10.1007/978-981-13-2236-5_4

Chan, Y. J., Chong, M. F., & Law, C. L. (2012). An integrated anaerobic–aerobic bioreactor (IAAB) for the treatment of palm oil mill effluent (POME): Start-up and steady state performance. *Process Biochemistry*, *47*(3), 485–495. doi:10.1016/j.procbio.2011.12.005

Chan, Y. J., Chong, M. F., Law, C. L., & Hassell, D. (2009). A review on anaerobic–aerobic treatment of industrial and municipal wastewater. *Chemical Engineering Journal*, *155*(1-2), 1–18.

Chan, Y., Chong, M., & Law, C. (2010). Biological treatment of anaerobically digested palm oil mill effluent (POME) using a lab-scale sequencing batch reactor (SBR). *Journal of Environmental Management*, *91*(8), 1738–1746. doi:10.1016/j.jenvman.2010.03.021 PMID:20430515

Chao, L., Zhanyong, W., & Bai, L. (2019). Investigating the relationship between air pollution variation and urban form. *Building and Environment*, *147*, 559–568. doi:10.1016/j.buildenv.2018.06.038

Charles, A., & Chin, K. K. (2019). Photocatalytic treatment of palm oil mill effluent by visible light-active calcium ferrite: Effect of catalyst preparation technique. *Journal of Environmental Management*, *234*, 404–411. doi:10.1016/j.jenvman.2019.01.024 PMID:30640165

Charumathi, D., & Das, N. (2010). Removal of synthetic dye basic violet 3 by immobilized *Candida tropicalis* grown on sugarcane bagasse extract medium. *International Journal of Engineering Science and Technology*, *2*, 4325–4335.

Chebil, S., Chaala, A., & Roy, C. (2000). Use of softwood bark charcoal as a modifier for road bitumen. *Fuel*, *79*, 671–683.

Chemsain. (2011). *A study on scrap tyres management for peninsular Malaysia report*. September, 2011.

Chen, C. C., Liao, H. J., Cheng, C. Y., Yen, C. Y., & Chung, Y. C. (2007). Biodegradation of crystal violet by Pseudomonas putida. *Biotechnology Letters*, *29*(3), 391–396. doi:10.100710529-006-9265-6 PMID:17206376

Cheng, H., & Wang, L. (2013). Lignocelluloses feedstock biorefinery as petrorefinery substitutes. In M. D. Matovic (Ed.), *Biomass now-sustainable growth and use*. Retrieved from http://www.intechopen.com/books/biomass- now-sustainable-growth-and use/lignocelluloses-feedstock-biorefinery-as-petrorefinery-substitutes)

Chen, G. (2004). Electrochemical technologies in wastewater treatment. *Separation and Purification Technology*, *38*(1), 11–41. doi:10.1016/j.seppur.2003.10.006

Chen, G.-C., Shan, X.-Q., Pei, Z.-G., Wang, H., Zheng, L.-R., Zhang, J., & Xie, Y.-N. (2011). Adsorption of diuron and dichlobenil on multiwalled carbon nanotubes as affected by lead. *Journal of Hazardous Materials*, *188*(1-3), 156–163. doi:10.1016/j.jhazmat.2011.01.095 PMID:21324587

Cheng, J., Zhu, X., Ni, J., & Borthwick, A. (2010). Palm oil mill effluent treatment using a two-stage microbial fuel cells system integrated with immobilized biological aerated filters. *Bioresource Technology*, *101*(8), 2729–2734. doi:10.1016/j.biortech.2009.12.017 PMID:20042327

Cheng, M. D., Gao, N., & Hopke, P. K. (1996). Source apportionment study of nitrogen species measured in southern California in 1987. *Journal of Environmental Engineering*, *122*(3), 183–190. doi:10.1061/(ASCE)0733-9372(1996)122:3(183)

Cheng, M., Zeng, G., Huang, D., Lai, C., Xu, P., Zhang, C., & Liu, Y. (2016). Hydroxyl radicals based advanced oxidation processes (AOPs) for remediation of soils contaminated with organic compounds: A review. *Chemical Engineering Journal*, *284*, 582–598. doi:10.1016/j.cej.2015.09.001

Cheng, S., Grosse, W., Karrenbrock, F., & Thoennessen, M. (2002). Efficiency of constructed wetlands in decontamination of water polluted by heavy metals. *Ecological Engineering*, *18*(3), 317–325. doi:10.1016/S0925-8574(01)00091-X

Cheng, S., Li, Z., Mang, H.-P., Huba, E.-M., Gao, R., & Wang, X. (2014). Development and application of prefabricated biogas digesters in developing countries. *Renewable & Sustainable Energy Reviews*, *34*, 387–400.

Chen, H., Jia, B., & Lau, S. S. Y. (2008). Sustainable urban form for Chinese compact cities: Challenges of a rapid urbanized economy. *Habitat International*, *32*(1), 28–40. doi:10.1016/j.habitatint.2007.06.005

Chen, J. L., Ortiz, R., Steele, T. W., & Stuckey, D. C. (2014). Toxicants inhibiting anaerobic digestion: A review. *Biotechnology Advances*, *32*(8), 1523–1534. PMID:25457225

Chen, J., Su, Y., Si, H., & Chen, J. (2018). Managerial areas of construction and demolition waste: A scientometric review. *International Journal of Environmental Research and Public Health*, *15*(2350). doi:10.3390/ijerph15112350 PMID:30356018

Chen, K. C., Wu, J. Y., Liou, D. J., & Hwang, S. C. J. (2003). Decolorization of the textile azo dyes by newly isolated bacterial strains. *Journal of Biotechnology*, *101*(1), 57–68. doi:10.1016/S0168-1656(02)00303-6 PMID:12523970

Chen, L., & Yi, L. (2016). Effects of China's urban form on urban air quality. *Urban Studies Journal Limited*, *53*(12), 2607–2623. doi:10.1177/0042098015594080

Chen, M., Leng, B., Wu, S., & Sang, Y. (2014a). Physical, chemical and rheological properties of waste edible vegetable oil rejuvenated asphalt binders. *Construction & Building Materials*, *66*, 286–298.

Chen, M., Xiao, F., Putman, B., Leng, B., & Wu, S. (2014b). High temperature properties of rejuvenating recovered binder with rejuvenator, waste cooking and cotton seed oils. *Construction & Building Materials*, *59*, 10–16.

Chen, R.-X.,, & Wang, W.-C. (2019). The production of renewable aviation fuel from waste cooking oil. Part I: Bio-alkane conversion through hydroprocessing of oil. *Renewable Energy*, *135*, 819–835. doi:10.1016/j.renene.2018.12.048

Chen, S., Ling, J., & Blancheton, J.-P. (2006). Nitrification kinetics of biofilm as affected by water quality factors. *Aquacultural Engineering*, *34*(3), 179–197. doi:10.1016/j.aquaeng.2005.09.004

Chen, T., Gao, B., & Yue, Q. (2010). Effect of dosing method and pH on color removal performance and floc aggregation of polyferric chloride–polyamine dual-coagulant in synthetic dyeing wastewater treatment. *Colloids and Surfaces. A, Physicochemical and Engineering Aspects*, *355*(1-3), 121–129. doi:10.1016/j.colsurfa.2009.12.008

Chen, T., Liu, R., & Scott, N. R. (2016). Characterization of energy carriers obtained from the pyrolysis of white ash, switchgrass and corn stover—Biochar, syngas and bio-oil. *Fuel Processing Technology*, *142*, 124–134. doi:10.1016/j.fuproc.2015.09.034

Chen, W., Zhang, A., Jiang, G., & Li, Q. (2019). Transformation and degradation mechanism of landfill leachates in a combined process of SAARB and ozonation. *Waste Management (New York, N.Y.)*, *85*, 283–294. doi:10.1016/j.wasman.2018.12.038 PMID:30803582

Chen, X., Ottosen, L. D. M., & Kofoed, M. V. W. (2019). How low can you go: methane production of methanobacterium congolense at low CO2 concentrations. *Frontiers in Bioengineering and Biotechnology*, *7*(34). PMID:30899758

Chen, Y., Cheng, J. J., & Creamer, K. S. (2008). Inhibition of anaerobic digestion process: A review. *Bioresource Technology*, *99*(10), 4044–4064. PMID:17399981

Chen, Y., Liu, C., Nie, J., Wu, S., & Wang, D. (2014). Removal of COD and decolorizing from landfill leachate by Fenton's reagent advanced oxidation. *Clean Technologies and Environmental Policy*, *16*(1), 189–193. doi:10.100710098-013-0627-1

Chen, Y., Xiao, B., Chang, J., Fu, Y., Lv, P., & Wang, X. (2009). Synthesis of biodiesel from waste cooking oil using immobilized lipase in fixed bed reactor. *Energy Conversion and Management*, *50*(3), 668–673.

Chen, Y., Yang, G., Sweeney, S., & Feng, Y. (2010). Household biogas use in rural China: A study of opportunities and constraints. *Renewable & Sustainable Energy Reviews*, *14*(1), 545–549.

Chen, Z., Ren, N., Wang, A., Zhang, Z. P., & Shi, Y. (2008). A novel application of TPAD–MBR system to the pilot treatment of chemical synthesis-based pharmaceutical wastewater. *Water Research*, *42*(13), 3385–3392. doi:10.1016/j.watres.2008.04.020 PMID:18538367

Cheong, D. Y., Hansen, C. L., & Stevens, D. K. (2007). Production of bio-hydrogen by mesophilic anaerobic fermentation in an acid-phase sequencing batch reactor. *Biotechnology and Bioengineering*, *96*(3), 421–432. PMID:17013946

Cheung, W. F., Lin, T. H., & Lin, Y. C. (2018). A real-time construction safety monitoring system for hazardous gas integrating wireless sensor network and building information modeling technologies. *Sensors (Switzerland)*, *18*(2), 436. doi:10.339018020436 PMID:29393887

Chhetri, A. B., Watts, K. C., & Islam, M. R. (2008). Waste cooking oil as an alternate feedstock for biodiesel production. *Energies*, *1*(1), 3–18.

Chiam, L. T.,, & Tye, C. T. (2013). Deoxygenation of plant fatty acid using NiSnK/SiO_2 as catalyst. *The Malaysian Journal of Analytical Sciences*, *17*(1), 129–138.

Chiang, L.-C., Chang, J.-E., & Wen, T.-C. (1995). Indirect oxidation effect in electrochemical oxidation treatment of landfill leachate. *Water Research*, *29*(2), 671–678. doi:10.1016/0043-1354(94)00146-X

Chimuka, L., & Ogola, J. (2015). Leading the way: directions of MSW management in South Africa. In T. Simelane & R. Mohee (Eds.), *Future directions of MSW management in Africa* (pp. 176–199). South Africa: Africa Institute of South Africa.

Chin, K. K. (1981). Anaerobic treatment kinetics of palm oil sludge. *Water Research*, *15*(2), 199–202. doi:10.1016/0043-1354(81)90112-3

Chin, K. K., Lee, S. W., & Mohammed, H. H. (1996). A study of palm oil mill effluent treatment using ponding system. *Water Science and Technology*, *34*(11), 119–123. doi:10.2166/wst.1996.0270

Choe, E.,, & Min, D. B. (2007). Chemistry of deep-fat frying oils. *Journal of Food Science*, *72*(5), 1–10. PMID:17995742

Cho, H.-S., & Choi, M. J. (2014). Effects of compact urban development on air pollution: empirical evidence from Korea. *Sustainability*, *6*(9), 5968–5982. doi:10.3390u6095968

Choi, J. S., Hwang, T. M., Lee, S., & Hong, S. (2009). A systematic approach to determine the fouling index for a RO/NF membrane process. *Desalination*, *238*(1-3), 117–127. doi:10.1016/j.desal.2008.01.042

Choi, W. H., Shin, C. H., Son, S. M., Ghorpade, P. A., Kim, J. J., & Park, J. Y. (2013). Anaerobic treatment of palm oil mill effluent using combined high-rate anaerobic reactors. *Bioresource Technology*, *141*, 138–144. doi:10.1016/j.biortech.2013.02.055 PMID:23489567

Chong, M. F., Lee, K. P., Chieng, H. J., & Syazwani Binti Ramli, I. I. (2009). Removal of boron from ceramic industry wastewater by adsorption–flocculation mechanism using palm oil mill boiler (pomb) bottom ash and polymer. *Water Research*, *43*(13), 3326–3334. doi:10.1016/j.watres.2009.04.044 PMID:19487007

Choo, K.-H., Choi, S.-J., & Hwang, E.-D. (2007). Effect of coagulant types on textile wastewater reclamation in a combined coagulation/ultrafiltration system. *Desalination*, *202*(1-3), 262–270. doi:10.1016/j.desal.2005.12.063

Chorom, M., Shari, H. S., & Motamedi, H. (2010). Bioremediation of a crude oil - polluted soil by application of fertilizers. *Iranian Journal of Environmental Health Sciences & Engineering*, *7*(4), 319–326. Retrieved from https://www.researchgate.net/publication/267920577_Bioremediation_of_a_crude_oil_-_polluted_soil_by_application_of_fertilizers

Choudhary, M.,, & Grover, K. (2013). Effect of deep-fat frying on physicochemical properties of rice bran oil blends. *IOSR Journal of Nursing and Health Science*, *1*, 1–10.

Chowdhury, K. A., Debsarkar, A., & Chakrabarty, S. (2012). Analysis of daytime traffic noise level: A case study of Kolkata, India. *International Journal Environment Science Research*, *2*(1), 114–118.

Chowdhury, S., Mishra, R., Saha, P., & Kushwaha, P. (2011). Adsorption thermodynamics, kinetics and isosteric heat of adsorption of malachite green onto chemically modified rice husk. *Desalination*, *265*(1-3), 159–168. doi:10.1016/j.desal.2010.07.047

Choy, S. Y., Prasad, K. M. N., Wu, T. Y., Raghunandan, M. E., & Ramanan, R. N. (2014). Utilization of plant-based natural coagulants as future alternatives towards sustainable water clarification. *Journal of Environmental Sciences (China)*, *26*(11), 2178–2189. doi:10.1016/j.jes.2014.09.024 PMID:25458671

Choy, S. Y., Prasad, K. N., Wu, T. Y., Raghunandan, M. E., & Ramanan, R. N. (2016). Performance of conventional starches as natural coagulants for turbidity removal. *Ecological Engineering*, *94*, 352–364. doi:10.1016/j.ecoleng.2016.05.082

Christensen, T. H., Kjeldsen, P., Bjerg, P. L., Jensen, D. L., Christensen, J. B., Baun, A., . . . Heron, G. J. A. G. (2001). Biogeochemistry of landfill leachate plumes. *16*(7-8), 659-718.

Christopher, C. A., Chioma, B. C., & Gideon, C. O. (2016). Bioremediation techniques-classification based on site of application: Principles, advantages, limitations and prospects. [Formerly MIRCEN Journal of Applied Microbiology and Biotechnology]. *World Journal of Microbiology & Biotechnology*, *32*(11), 180. doi:10.100711274-016-2137-x PMID:27638318

Chua, H. S., Bashir, M. J. K., & Tan, K. T. (2019). A sustainable pyrolysis technology for the treatment of municipal solid waste in Malaysia. AIP Conference Proceedings, 2124, 020016.

Chulkov, V., & Fachratov, M. (2018). Construction waste processing management and demolition of five-storey apartment blocks in Moscow. *IOP Conference Series. Materials Science and Engineering*, *05008*. doi:10.1088/1757-899X/365/6/062013

Chung, C. Y., Selvarajoo, A., Sethu, V., Koyande, A. K., Arputhan, A., & Lim, Z. C. (2018). Treatment of palm oil mill effluent (POME) by coagulation flocculation process using peanut–okra and wheat germ–okra. *Clean Technologies and Environmental Policy*, *20*(9), 1951–1970. doi:10.100710098-018-1619-y

Chynoweth, D. P., Owens, J. M., & Legrand, R. (2001). Renewable methane from anaerobic digestion of biomass. *Renewable Energy*, *22*(1-3), 1–8.

Ciabatti, I., Tognotti, F., & Lombardi, L. (2010). Treatment and reuse of dyeing effluents by potassium ferrate. *Desalination*, *250*(1), 222–228. doi:10.1016/j.desal.2009.06.019

Cirik, K., Kitis, M., & Cinar, O. (2012). The effect of biological sulfate reduction on anaerobic color removal in anaerobic-aerobic sequencing batch reactors. *Bioprocess and Biosystems Engineering*. doi:10.1007/ s00449-012-0813-2 PMID:23277271

Clara, M., Strenn, B., Gans, O., Martinez, E., Kreuzinger, N., & Kroiss, H. (2005). Removal of selected pharmaceuticals, fragrances and endocrine disrupting compounds in a membrane bioreactor and conventional wastewater treatment plants. *Water Research*, *39*(19), 4797–4807. doi:10.1016/j.watres.2005.09.015 PMID:16242170

Clark, J., Cole, M., Lindeque, P. K., Fileman, E., Blackford, J., Lewis, C., ... Galloway, T. S. (2016). Marine microplastic debris: A targeted plan for understanding and quantifying interactions with marine life. *Frontiers in Ecology and the Environment*, *14*(6), 317–324. doi:10.1002/fee.1297

Clark, N. A., Allen, R. W., Hystad, P., Wallace, L., Dell, S. D., Foty, R., ... Wheeler, A. J. (2010). Exploring variation and predictors of residential fine particulate matter infiltration. *International Journal of Environmental Research and Public Health*, *7*(8), 3211–3224. doi:10.3390/ijerph7083211 PMID:20948956

Clemens, J., Trimborn, M., Weiland, P., & Amon, B. (2006). Mitigation of greenhouse gas emissions by anaerobic digestion of cattle slurry. *Agriculture, Ecosystems & Environment*, *112*(2-3), 171–177.

Climent, M., Ferrer, I., Baeza, M. M., Artola, A., Vazquez, F., & Font, X. (2007). Effects of thermal and mechanical pretreatments of secondary sludge on biogas production under thermophilic conditions. *Chemical Engineering Journal*, *133*(1-3), 335–342. doi:10.1016/j.cej.2007.02.020

Cline, J. (1998). *Treatment and discharge of produced water for deep offshore disposal.*

Coats, A., & Redfern, J. (1965). Kinetic parameters from thermogravimetric data. II. *Journal of Polymer Science. Part B: Polymer Letters*, *3*(11), 917–920. doi:10.1002/pol.1965.110031106

Collins, M. R., Westersund, C., Cole, J., & Roccaro, J. (1994). *Evaluation of roughing filtration design variables. American Water Works Association Research Foundation (AWWARF) and America Water Works Association* (pp. 77–88). Denver, CO: AWWA.

Colvin, R. J., & Rozich, A. F. (1997). AFC Process that reduced sludge and cost of pharmaceutical treatment. *Chemical Show Conference*, New York, NY.

Comerton, A. M., Andrews, R. C., & Bagley, D. M. (2005). Evaluation of an MBR–RO system to produce high quality reuse water: Microbial control, DBP formation and nitrate. *Water Research*, *39*(16), 3982–3990. doi:10.1016/j.watres.2005.07.014 PMID:16112164

Comte, S., Guibaud, G., & Baudu, M. (2007). Effect of extraction method on EPS from activated sludge: An HPSEC investigation. *Journal of Hazardous Materials*, *140*(1-2), 129–137. doi:10.1016/j.jhazmat.2006.06.058 PMID:16879910

Cooney, D. O. (1999). *Adsorption design for wastewater treatment*. Boca Raton, FL: CRC Press.

Cooper, A. R. (Ed.). (2013). *Ultrafiltration membranes and applications* (Vol. 13). Springer Science & Business Media.

Corella, J., Toledo, J. M., & Padilla, R. (2004). Olivine or dolomite as in-bed additive in biomass gasification with air in a fluidized bed: Which is better? *Energy & Fuels*, *18*(3), 713–720. doi:10.1021/ef0340918

Cossu, R., Raga, R., & Rossetti, D. (2003). The PAF model: An integrated approach for landfill sustainability. *Waste Management (New York, N.Y.)*, *23*(1), 37–44. doi:10.1016/S0956-053X(02)00147-2 PMID:12623100

Couto, M. N., Monteiro, E., & Vasconcelos, M. T. (2010). Microcosm trials of bioremediation of contaminated soil of a petroleum refinery: Comparison of natural attenuation, biostimulation, and bioaugmentation. *Environmental Science and Pollution Research International*, *17*(7), 1339–1346. doi:10.100711356-010-0318-y PMID:20229281

Couto, S. R. (2009). Dye removal by immobilized fungi. *Biotechnology Advances*, *27*(3), 227–235. doi:10.1016/j.biotechadv.2008.12.001 PMID:19211032

CPCB. (2010). *Air quality monitoring*. Emission Inventory and Source Apportionment Study for Indian Cities.

Cripps, C., Bumpus, J. A., & Aust, S. D. (1990). Biodegradation of azo and heterocyclic dyes by Phanerochaete chrysosporium. *Applied and Environmental Microbiology*, *56*, 1114–1118. PMID:2339873

Crittenden, J. C., Suri, R. P. S., Perram, D. L., & Hand, D. W. (1997). Decontamination of water using adsorption and photocatalysis. *Water Research*, *31*(3), 411–418. doi:10.1016/S0043-1354(96)00258-8

Crocker, M. (2010). Thermochemical conversion of biomass to liquid fuels and chemicals. London, UK: Royal Society of Chemistry (RSC). doi:10.1039/9781849732260

Cruz-González, K., Torres-López, O., García-León, A., Guzmán-Mar, J., Reyes, L., Hernández-Ramírez, A., & Peralta-Hernández, J. (2010). Determination of optimum operating parameters for Acid Yellow 36 decolorization by electro-Fenton process using BDD cathode. *Chemical Engineering Journal, 160*(1), 199–206. doi:10.1016/j.cej.2010.03.043

CSE. (2012). *Parking policy: Getting the principles right, policy brief.* New Delhi, India: Center for Science and Environment.

Cui, Y., Wang, S., & Li, J. (2009). On-line monitoring for phosphorus removal process and bacterial community in sequencing batch reactor. *Chinese Journal of Chemical Engineering, 17*(3), 484–492. doi:10.1016/S1004-9541(08)60235-9

Curds, C. (1975). Protozoa. In C. R. Curds & H. A. Hawkes (Eds.), Ecological aspects of used-water treatment. Vol. 1. The organisms and their ecology. London, UK: Academic Press. 203-268.

Cvengroš, J.,, & Cvengrošová, Z. (2004). Used frying oils and fats and their utilization in the production of methyl esters of higher fatty acids. *Biomass and Bioenergy, 27*(2), 173–181.

Cyr, P. J., Suri, R. P., & Helmig, E. D. (2002). A pilot scale evaluation of removal of mercury from pharmaceutical wastewater using granular activated carbon. *Water Research, 36*(19), 4725–4734. doi:10.1016/S0043-1354(02)00214-2 PMID:12448514

Da Pozzo, A., Merli, C., Sirés, I., Garrido, J. A., Rodríguez, R. M., & Brillas, E. (2005). Removal of the herbicide amitrole from water by anodic oxidation and electro-Fenton. *Environmental Chemistry Letters, 3*(1), 7–11. doi:10.100710311-005-0104-0

da Rocha, O. R. S., Dantas, R. F., Bezerra Duarte, M. M., Lima Duarte, M. M., & da Silva, V. L. (2013). Solar photo-Fenton treatment of petroleum extraction wastewater. *Desalination and Water Treatment, 51*(28-30), 5785–5791. doi:10.1080/19443994.2013.792136

Da Silva, M., Firmino, P., De Sousa, M., & Dos Santos, A. (2012). Sequential anaerobic/aerobic treatment of dye—containing wastewaters: Color and COD removals and Ecotoxicity tests. *Applied Biochemistry and Biotechnology, 166*(4), 1057–1069. doi:10.100712010-011-9493-7 PMID:22238010

Dabestani, R., & Ivanov, I. N. (1999). A compilation of physical, spectroscopic and photophysical properties of polycyclic aromatic hydrocarbons. *Photochemistry and Photobiology, 70*(1), 10–34.

Dağhan, H., & Öztürk, M. (2015). Soil pollution in Turkey and remediation method in soil remediation and plants (prospects and challenges), Academic Press, 287-312.

Dahlan, I., & Zwain, H.M. (2013). A study on the removal characteristic of zinc ion (Zn^{2+}) from synthetic wastewater using a novel waste-derived siliceous sorbent. *Caspian Journal of Applied Sciences Research*, 2 (AICCE' 12 & GIZ' 12), 10-17.

Dahlan, I., Hassan, S. R., & Hakim, M. L. (2013). Removal of iron (Fe^{2+}) from aqueous solutions using siliceous waste sorbent. *Sustainable Environment Research, 23*(1), 41–48.

Dahlan, I., & Ismail, S. (2012). Physical characterization of prepared and spent CFA/PFA/RHA sorbents in removing heavy metals and dyes. *Journal of Environmental Science and Technology, 5*(3), 177–183. doi:10.3923/jest.2012.177.183

Dahlan, I., & Noor, A. S. M. (2014). A study on the removal characteristic of acid violet dye from synthetic wastewater using a novel RHA/PFA/CFA sorbent. *Environmental Engineering and Management Journal, 13*(1), 197–204. doi:10.30638/eemj.2014.024

Dahlan, I., & Razali, M. H. M. (2012). Waste-derived siliceous materials as a novel sorbent for removal of Ni2+ from aqueous solutions. *Water, Air, and Soil Pollution, 223*(5), 2495–2501. doi:10.100711270-011-1042-z

Dahlin, R. S., Dorminey, J. R., Peng, W., Leonard, R. F., & Vimalchand, P. (2008). Preventing ash agglomeration during gasification of high-sodium lignite. *Energy & Fuels*, *23*(2), 785–793. doi:10.1021/ef800568z

Dalrymple, O. K., Yeh, D. H., & Trotz, M. A. (2007). Removing pharmaceuticals and endocrine-disrupting compounds from wastewater by photocatalysis. *Journal of Chemical Technology & Biotechnology: International Research in Process. Environmental & Clean Technology*, *82*(2), 121–134.

Dama, P., Bell, J., Foxon, K. M., Brouckaert, C. J., Huang, T., Buckley, C. A., ... Stuckey, D. (2002). Pilot-scale study of an anaerobic baffled reactor for the treatment of domestic wastewater. *Water Science and Technology*, *46*(9), 263–270. doi:10.2166/wst.2002.0255 PMID:12448477

Damartzis, T., Vamvuka, D., Sfakiotakis, S., & Zabaniotou, A. (2011). Thermal degradation studies and kinetic modeling of cardoon (Cynara cardunculus) pyrolysis using thermogravimetric analysis (TGA). *Bioresource Technology*, *102*(10), 6230–6238. doi:10.1016/j.biortech.2011.02.060 PMID:21398116

Daneshvar, N., Khataee, A. R., Rasoulifard, M. H., & Pourhassan, M. (2007). Biodegradation of dye solution containing malachite green: Optimization of effective parameters using the Taguchi method. *Journal of Hazardous Materials*, *143*(1-2), 214–219. doi:10.1016/j.jhazmat.2006.09.016 PMID:17052836

Daneshvar, N., Oladegaragoze, A., & Djafarzadeh, N. (2006). Decolorization of basic dye solutions by electrocoagulation: An investigation of the effect of operational parameters. *Journal of Hazardous Materials*, *129*(1), 116–122. doi:10.1016/j.jhazmat.2005.08.033 PMID:16203084

Darvishmanesh, S., Firoozpour, L., Vanneste, J., Luis, P., Degrève, J., & Van der Bruggen, B. (2011). Performance of solvent resistant nanofiltration membranes for purification of residual solvent in the pharmaceutical industry: Experiments and simulation. *Green Chemistry*, *13*(12), 3476–3483. doi:10.1039/c1gc15462a

Darwish, M., Asfour, N., & Najem, A. (2002). Energy consumption in equivalent work by different desalination methods: Case study for Kuwait. *Desalination*, *152*(1-3), 83–92. doi:10.1016/S0011-9164(02)01051-2

Das, M., Maiti, K. S., & Mukhopadhyay, U. (2006). Distribution of PM2.5 and PM10-2.5 in PM10 fraction in ambient air due to vehicular pollution in Kolkata megacity. *Environmental Monitoring and Assessment*, *122*(1-3), 111–123. doi:10.100710661-005-9168-3 PMID:16770501

Das, N., & Charumathi, D. (2012). Remediation of synthetic dyes from wastewater using yeast – An overview. *Indian Journal of Biotechnology*, *11*, 369–380.

Daud. N. F. B. M. (2016). *Treatment of chicken processing wastewater using hybrid upflow anaerobic sludge blanket (HUASB) coupled with aerated lagoon (AL).* (Doctoral dissertation, University Tun Hussein Onn Malaysia).

Dave, S. R., & Dave, R. H. (2009). Isolation and characterization of *Bacillus thuringiensis* for Acid Red-119 dye decolorization. *Bioresource Technology*, *100*(1), 249–253. doi:10.1016/j.biortech.2008.05.019 PMID:18590958

Davis, M. L., & Cornwell, D. A. (1998). *Environmental engineering* (3rd ed.). New York, NY: McGraw Hill.

Davis, R., Aden, A., & Pienkos, P. T. (2011). Techno-economic analysis of autotrophic microalgae for fuel production. *Applied Energy*, *88*(10), 3524–3531.

Dawei, H. (2015). Concise environmental engineering. Retrieved from bookboon.com

Dawoud, M. (2012). Environmental impacts of seawater desalination : Arabian Gulf case study. *International Journal of Environment and Sustainability*, *1*(3), 22–37. doi:10.24102/ijes.v1i3.96

de Andrade, B. D. A., de Lacerda, P. S., & Oliveira, J. L. (2017). Technical feasibility of reuse of effluent generated from reverse osmosis system in a pharmaceutical plant. *Revista Ambiente & Água*, *12*(5), 694–707. doi:10.4136/ambi-agua.1980

De Andres, J. M., Narros, A., & Rodríguez, M. E. (2011). Behaviour of dolomite, olivine and alumina as primary catalysts in air–steam gasification of sewage sludge. *Fuel*, *90*(2), 521–527. doi:10.1016/j.fuel.2010.09.043

De Briyne, N., Atkinson, J., Pokludová, L., & Borriello, S. (2014). Antibiotics used most commonly to treat animals in Europe. *The Veterinary Record*, *175*(13), 325. doi:10.1136/vr.102462 PMID:24899065

de Chernicharo, C. A. L. (2006). Post-treatment options for the anaerobic treatment of domestic wastewater. *Reviews in Environmental Science and Biotechnology*, *5*(1), 73–92. doi:10.100711157-005-5683-5

de la Rubia, M. A., Perez, M., Romero, L. I., & Sales, D. (2006). Effect of solids retention time (SRT) on pilot scale anaerobic thermophilic sludge digestion. *Process Biochemistry*, *41*(1), 79–86. doi:10.1016/j.procbio.2005.03.073

De Martino, A., Iorio, M., Xing, B., & Capasso, R. (2012). Removal of 4-chloro-2-methylphenoxyacetic acid from water by sorption on carbon nanotubes and metal oxide nanoparticles. *RSC Advances*, *2*(13), 5693–5700. doi:10.1039/c2ra00038e

de Melo, R. R. S., Costa, D. B., Álvares, J. S., & Irizarry, J. (2017). Applicability of unmanned aerial system (UAS) for safety inspection on construction sites. *Safety Science*, *98*, 174–185. doi:10.1016/j.ssci.2017.06.008

De Mes, T., Stams, A., Reith, J., & Zeeman, G. (2003). Methane production by anaerobic digestion of wastewater and solid wastes. *Bio-methane & Bio-hydrogen*, 58-102.

De Morais, J. L., & Zamora, P. P. (2005). Use of advanced oxidation processes to improve the biodegradability of mature landfill leachates. *Journal of Hazardous Materials*, *123*(1-3), 181–186. doi:10.1016/j.jhazmat.2005.03.041 PMID:15878233

De Ridder, K., Lefebre, F., Adriaensen, S., Arnold, U., Beckroege, W., Bronner, C., ... Weber, C. (2008). Simulating the impact of urban sprawl on air quality and population exposure in the German Ruhr area. Part I: Reproducing the base state. *Atmospheric Environment*, *42*(30), 7059–7069. doi:10.1016/j.atmosenv.2008.06.045

De Smet, C. R. H., De Croon, M. H. J. M., Berger, R. J., Marin, G. B., & Schouten, J. C. (2001). Design of adiabatic fixed-bed reactors for the partial oxidation of methane to synthesis gas. Application to production of methanol and hydrogen-for-fuel-cells. *Chemical Engineering Science*, *56*(16), 4849–4861. doi:10.1016/S0009-2509(01)00130-0

De Souza, S. P., Pacca, S., De Avila, M. T., & Borges, J. L. B. (2010). Greenhouse gas emissions and energy balance of palm oil biofuel. *Renewable Energy*, *35*(11), 2552–2561. doi:10.1016/j.renene.2010.03.028

De-Bashan, L. E., & Bashan, Y. (2004). Recent advances in removing phosphorus from wastewater and its future use as fertilizer (1997–2003). *Water Research*, *38*(19), 4222–4246. doi:10.1016/j.watres.2004.07.014 PMID:15491670

Deegan, A. M., Shaik, B., Nolan, K., Urell, K., Oelgemöller, M., Tobin, J., & Morrissey, A. (2011). Treatment options for wastewater effluents from pharmaceutical companies. *International Journal of Environmental Science and Technology*, *8*(3), 649–666. doi:10.1007/BF03326250

Deka, P., & Hoque, R. R. (2014). Incremental effect of festive biomass burning on wintertime PM10 in Brahmaputra Valley of Northeast India. *Atmospheric Research*, *143*, 380–391. doi:10.1016/j.atmosres.2014.03.003

Deke, O., Hooss, K. G., Kasten, C., Klepper, G., & Springer, K. (2001). Economic impact of climate change: simulations with a regionalized climate-economy model (No. 1065). Kiel, Germany.

Del Pozo, R., & Diez, V. (2005). Integrated anaerobic–aerobic fixed-film reactor for slaughterhouse wastewater treatment. *Water Research*, *39*(6), 1114–1122. doi:10.1016/j.watres.2005.01.013 PMID:15766966

Demiral, I., & Sensoz, S. (2008). The effects of different catalysts on the pyrolysis of industrial wastes (olive and hazelnut bagasse). *Bioresource Technology*, *99*(17), 8002–8007. doi:10.1016/j.biortech.2008.03.053 PMID:18450440

Demirbaş, A. (2001). Biomass resource facilities and biomass conversion processing for fuels and chemicals. *Energy Conversion and Management*, *42*(11), 1357–1378.

Demirbas, A. (2005). Potential applications of renewable energy sources, biomass combustion problems in boiler power systems and combustion related environmental issues. *Progress in Energy and Combustion Science*, *31*(2), 171–192.

Demirbas, A. (2005). Pyrolysis of ground beech wood in irregular heating rate conditions. *J. Appl. Pyroyisis*, *73*(1), 39–43. doi:10.1016/j.jaap.2004.04.002

Demirbas, A. (2007). Combustion systems for biomass fuel. *Energy Sources. Part A, Recovery, Utilization, and Environmental Effects*, *29*(4), 303–312. doi:10.1080/009083190948667

Demirbas, A. (2009). Biofuels securing the planet's future energy needs. *Energy Conversion and Management*, *50*(9), 2239–2249. doi:10.1016/j.enconman.2009.05.010

Deng, D., Guo, J., Zeng, G., & Sun, G. (2008). Decolorization of anthraquinone, triphenylmethane and azo dyes by a new isolated Bacillus cereus strain DC11. *Int. Biodetor. Biodegrad*, *62*(3), 263–269. doi:10.1016/j.ibiod.2008.01.017

Deng, Y., & Englehardt, J. D. (2006). Treatment of landfill leachate by the Fenton process. *Water Research*, *40*(20), 3683–3694. doi:10.1016/j.watres.2006.08.009 PMID:17045628

Department for Environment, Food and Rural Affairs. (2013). *Mechanical biological treatment of MSW.* Retrieved from https://assets.publishing.service.gov.uk/government/uploads/system/uploads/attachment_data/file/221039/pb13890-treatment-solid-waste.pdf

Department of Environment (DOE). (2008). Malaysia Country Waste Management Report for the year 2008. Available at http:// www. doe. gov. my

Derakhsheshpoor, R., Homayoonfal, M., Akbari, A., & Mehrnia, M. R. (2013). Amoxicillin separation from pharmaceutical wastewater by high permeability polysulfone nanofiltration membrane. *Journal of Environmental Health Science & Engineering*, *11*(1), 9. doi:10.1186/2052-336X-11-9 PMID:24499593

Derek, T. (2012, June 7). Where does the world trash go? *The Atlantic.* Retrieved from https://www.theatlantic.com

Deveci, E. Ü., Akarsu, C., Gönen, Ç., & Özay, Y. (2019). Enhancing treatability of tannery wastewater by integrated process of electrocoagulation and fungal via using rsm in an economic perspective. *Process Biochemistry*, *84*, 124–133. doi:10.1016/j.procbio.2019.06.016

Devi, L., Ptasinski, K. J., & Janssen, F. J. (2003). A review of the primary measures for tar elimination in biomass gasification processes. *Biomass and Bioenergy*, *24*(2), 125–140. doi:10.1016/S0961-9534(02)00102-2

Devi, L., Ptasinski, K. J., & Janssen, F. J. (2005). Pretreated olivine as tar removal catalyst for biomass gasifiers: Investigation using naphthalene as model biomass tar. *Fuel Processing Technology*, *86*(6), 707–730. doi:10.1016/j.fuproc.2004.07.001

Devi, R. S., Ramya, R., Kannan, K., Antony, A. R., & Kannan, V. R. (2019). Investigation of biodegradation potentials of high density polyethylene degrading marine bacteria isolated from the coastal regions of Tamil Nadu, India. [PubMed]. *Marine Pollution Bulletin*, *138*, 549–560.

Devriese, L. I., Van der Meulen, M. D., Maes, T., Bekaert, K., Paul-Pont, I., Frère, L., ... Vethaak, A. D. (2015). Microplastic contamination in brown shrimp (Crangon crangon, Linnaeus 1758) from coastal waters of the Southern North Sea and Channel area. *Marine Pollution Bulletin*, *98*(1-2), 179–187. doi:10.1016/j.marpolbul.2015.06.051 PMID:26456303

Dewil, R., Appels, L., Baeyens, J., & Degrève, J. (2007). Peroxidation enhances the biogas production in the anaerobic digestion of biosolids. *Journal of Hazardous Materials*, *146*, 577–581.

Dewil, R., Baeyens, J., & Neyens, E. (2005). Fenton peroxidation improves the drying performance of waste activated sludge. *Journal of Hazardous Materials*, *117*(2–3), 161–170. doi:10.1016/j.jhazmat.2004.09.025 PMID:15629575

Dhana, V. R. (2016). Energy and exergy analysis of thermal power plants. *International Journal of Engineering Science and Computing*, *6*(8), 2636-2644.

Dhaouadi, H., & Marrot, B. (2008). Olive mill wastewater treatment in a membrane bioreactor: Process feasibility and performances. *Chemical Engineering Journal*, *145*(2), 225–231. doi:10.1016/j.cej.2008.04.017

Dhoble, R. M., Lunge, S., Bhole, A. G., & Rayalu, S. (2011). Magnetic binary oxide particles (MBOP): A promising adsorbent for removal of As (III) in water. *Water Research*, *45*(16), 4769–4781. doi:10.1016/j.watres.2011.06.016 PMID:21777934

Dhokpande, S. R., & Kaware, J. P. (2013). Biological methods for heavy metal removal-A review. *International Journal of Engineering Science and Innovative Technology*, *2*(5), 304–309.

Di Blasi, C., & Branca, C. (2001). Kinetics of primary product formation from wood pyrolysis. *Industrial & Engineering Chemistry Research*, *40*(23), 5547–5556. doi:10.1021/ie000997e

Diane, H. (1999). Bioventing principles, applications and potential: restoration and reclamation review. *Student On-line Journal*. 5(2). Retrieved from https://conservancy.umn.edu/bitstream/handle/11299/59453/1/5.2.Hellekson.pdf

Diaz, A., Rincon, N., Escorihuela, A., Fernandez, N., Chacin, E., & Forster, C. (1999). A preliminary evaluation of turbidity removal by natural coagulants indigenous to Venezuela. *Process Biochemistry*, *35*(3), 391–395. doi:10.1016/S0032-9592(99)00085-0

Dierberg, F., DeBusk, T., Jackson, S., Chimney, M., & Pietro, K. (2002). Submerged aquatic vegetation-based treatment wetlands for removing phosphorus from agricultural runoff: Response to hydraulic and nutrient loading. *Water Research*, *36*(6), 1409–1422. doi:10.1016/S0043-1354(01)00354-2 PMID:11996331

Dike, B. U., Okoro, B. C., Nwakwasi, N. N., & Agbo, K. C. (2013). Remediation of used motor engine oil contaminated soil: a soil washing treatment approach. *Journal of Civil Environmental Engineering*, *3*, 129. doi:10.4172/2165-784X.1000129

Dillaha, T. A., Reneau, R., Mostaghimi, S., & Lee, D. (1989). Vegetative filter strips for agricultural nonpoint source pollution control. *Transactions of the ASAE. American Society of Agricultural Engineers*, *32*(2), 513–0519. doi:10.13031/2013.31033

Dimoglo, A., Sevim-Elibol, P., Dinç, Ö., Gökmen, K., & Erdoğan, H. (2019). Electrocoagulation/electroflotation as a combined process for the laundry wastewater purification and reuse. *Journal of Water Process Engineering*, *31*, 100877. doi:10.1016/j.jwpe.2019.100877

Dincer, I. (2000). Renewable energy and sustainable development: A crucial review. *Renewable & Sustainable Energy Reviews*, *4*(2), 157–175.

Ding, Y., & Sartaj, M. (2016). Optimization of ammonia removal by ion-exchange resin using response surface methodology. *International Journal of Environmental Science and Technology*, *13*(4), 985–994. doi:10.100713762-016-0939-x

Ding, Z., Zhu, M., Tam, V. W. Y., Yi, G., & Tran, C. N. N. (2018). A system dynamics-based environmental benefit assessment model of construction waste reduction management at the design and construction stages. *Journal of Cleaner Production*, *176*, 676–692. doi:10.1016/j.jclepro.2017.12.101

Divya, B., & Tyagi, V. K. (2007). Laundry detergents: an overview. *Research gate, Journal of Oleo Science February 2007.* doi:. 56(7) 327-340. Retrieved from www.researchgate.net/publication/5945251 doi:10.5650/jos.56.327

Djebedjian, B., Gad, H., Khaled, I., & Rayan, M. A. (2007). Reverse osmosis desalination plant in Nuweiba city. In *Eleventh International Water Technology Conference*. Egypt: IWTC11.

Dockery, D. W., Pope, C. A. III, Xu, X., Spengler, J. D., Ware, J. H., Fay, M. E., ... Speizer, F. E. (1993). An association between air pollution and mortality in six U.S. cities. *The New England Journal of Medicine*, *329*(24), 1753–1759. doi:10.1056/NEJM199312093292401 PMID:8179653

DOE, Malaysia. Environmental Quality (Sewage) Regulations 2009, Environmental Quality Act 1974, Environmental Quality (Control of Pollution from Solid Waste Transfer Station and Landfill) Regulations., 1-82.

Dogan, I., & Sanin, F. D. (2009). Alkaline solubilization and microwave irradiation as a combined sludge disintegration and minimization method. *Water Research*, *43*(8), 2139–2148. doi:10.1016/j.watres.2009.02.023 PMID:19285332

Donlan, R. M. (2002). Biofilms: Microbial life on surfaces. *Emerging Infectious Diseases*, *8*(9), 881–890. doi:10.3201/eid0809.020063 PMID:12194761

Dönmez, G. (2002). Bioaccumulation of the reactive textile dyes by Candida tropicalis growing in molasses medium. *Enzyme and Microbial Technology*, *30*(3), 363–366. doi:10.1016/S0141-0229(01)00511-7

Dordio, A., & Carvalho, A. J. P. (2013). Constructed wetlands with light expanded clay aggregates for agricultural wastewater treatment. *The Science of the Total Environment*, *463*, 454–461. doi:10.1016/j.scitotenv.2013.06.052 PMID:23831791

Doronin, V. P., Potapenko, O. V., Lipin, P. V., Sorokina, T. P., & Buluchevskaya, L. A. (2012). Catalytic cracking of vegetable oils for production of high octane gasoline and petrochemical feedstock. *Petroleum Chemistry*, *52*(6), 392–400. doi:10.1134/S0965544112060059

Dos Santos, A., Cervantes, F., & Vanlier, J. (2007). Review paper on current technologies for decolorization of textile wastewaters: Perspectives for anaerobic biotechnology. *Bioresource Technology*, *98*(12), 2369–2385. doi:10.1016/j.biortech.2006.11.013 PMID:17204423

DOSH. (2017). *OSH IN CONSTRUCTION — Prevention through Design (PtD)* (pp. 1–4). Kuala Lumpur, Malaysia: Department of Occupational Safety and Health.

Doula, M. K. (2009). Simultaneous removal of Cu, Mn, and Zn from drinking water with the use of clinoptilolite and its Fe-modified form. *Water Research*, *43*(15), 3659–3672. doi:10.1016/j.watres.2009.05.037 PMID:19576609

Dou, X., Ren, F., Nguyen, M. Q., Ahamed, A., Yin, K., Chan, W. P., & Chang, V. W.-C. (2017). Review of MSWI bottom ash utilization from perspectives of collective characterization, treatment and existing application. *Renewable & Sustainable Energy Reviews*, *79*, 24–38.

Doyle, C. D. (1965). Series approximations to the equation of thermogravimetric data. *Nature*, *207*(4994), 290–291. doi:10.1038/207290a0

Dragun, J. (1993). Recovery techniques and treatment for petroleum and petroleum products in soil and groundwater. *Proceedings of 1993 Oil Spill Conference, American Petroleum Institute, Washington DC*, pp. 48-63.

Duan, J., & Gregory, J. (2003). Coagulation by hydrolysing metal salts. *Advances in Colloid and Interface Science*, *100*, 475–502. doi:10.1016/S0001-8686(02)00067-2

Duarte, J. P. S., Martínez-Huitle, C. A., & da Silva, D. R. (2011). Electrochemical treatment for removing petroleum polycyclic aromatic hydrocarbons(PAHs) from synthetic produced water using a DSA-type anode: Preliminary results. *Sustainable Environment Research*, *21*(5), 329–335.

Dubrow, S., Boardman, G., & Michelsen, D. (1996). Chemical pretreatment and aerobic-anaerobic degradation of textile dye wastewater. John Wiley & Sons, 75–102.

Duis, K., & Coors, A. (2016). Microplastics in the aquatic and terrestrial environment: Sources (with a specific focus on personal care products), fate, and effects. *Environmental Sciences Europe*, *28*(1), 2. doi:10.118612302-015-0069-y PMID:27752437

Dura, A., & Breslin, C. B. (2019a). Electrocoagulation using stainless steel anodes: Simultaneous removal of phosphates, orange ii and zinc ions. *Journal of Hazardous Materials*, *374*, 152–158. doi:10.1016/j.jhazmat.2019.04.032 PMID:30999138

Dura, A., & Breslin, C. B. (2019b). The removal of phosphates using electrocoagulation with al–mg anodes. *Journal of Electroanalytical Chemistry*, *846*, 113161. doi:10.1016/j.jelechem.2019.05.043

Dursun, D., & Dentel, S. K. (2009). Toward the conceptual and quantitative understanding of biosolids conditioning: The gel approach. *Water Science and Technology*, *59*(9), 1679–1685. doi:10.2166/wst.2009.202 PMID:19448301

Dwivedi, P., Kaur, J., Dhanjal, N. I. K., & Cameotra, S. S. (2012). Biodecolorization of pararosaniline chloride dye using a bacterial isolate, Arthrobacter species. *J. Adv. Dev. Res*, *3*, 77–81.

Dwyer, J., Starrenburg, D., Tait, S., Barr, K., Batstone, D. J., & Lant, P. (2008). Decreasing activated sludge thermal hydrolysis temperature reduces product colour, without decreasing degradability. *Water Research*, *42*(18), 4699–4709. doi:10.1016/j.watres.2008.08.019 PMID:18834611

Dyreborg, S., & Arvin, E. (1995). Inhibition of nitrification by creosote-contaminated water. *Water Research*, *29*(6), 1603–1606. doi:10.1016/0043-1354(94)00247-5

Eaton, A. D., Clesceri, L. S., Greenberg, A. E., & Franson, M. A. H. (1998). *Standard methods for the examination of water and wastewater*. Washington, DC: American Public Health Association.

EAWAG. (1996). Swiss Federal Institute for Environmental Science and Technology.

Eckenfelder, W. W., & Norris, R. D. (1993). Applicability of biological processes for treatment of soils. American Chemical Society, 518, 138-158.

Eddy, M. (2003). Wastewater engineering: treatment and reuse (4th ed.). New York, NY: McGraw-Hill.

Edgar, F. C. M., Diego, E. C., & Victor, A. A. (2006). Study of the operational conditions for anaerobic digestion of urban solid wastes. *Waste management (New York, N.Y.)*, *26*(5), 546–556. doi:10.1016/j.wasman.2005.06.003 PMID:16321515

Edgerton, M. D. (2009). Increasing crop productivity to meet global needs for feed, food, and fuel. *Plant Physiology*, *149*(1), 7–13. doi:10.1104/pp.108.130195 PMID:19126690

Edris, M., Saeid, G., Leila Amiri, M. A., & Jafar, S. (2013). The application of soil washing for treatment of polycyclic aromatic hydrocarbons contaminated soil: a case study in a petrochemical complex. *Environmental Progress & Sustainable Energy*, *33*(1). doi:10.1002/ep

Edward, B., & Colin, H. (2015). Stabilization and solidification of contaminated soil and waste: a manual of practice. *Hygge Media*. Retrieved from http://www.enviro.wiki/images/9/90/Bates-2015-Stabilization_and_Solidification_of_Contaminated_Soil_and_Waste.pdf.

Eikelboom, D. (1975). Filamentous organisms observed in activated sludge. *Water Research*, *9*(4), 365–388. doi:10.1016/0043-1354(75)90182-7

El Bakouri, H., Morillo, J., Usero, J., & Ouassini, A. (2008). Potential use of organic waste substances as an ecological technique to reduce pesticide ground water contamination. *Journal of Hydrology (Amsterdam)*, *353*(3-4), 335–342. doi:10.1016/j.jhydrol.2008.02.019

El Gohary, R. (2015). Agriculture, industry, and wastewater in the Nile Delta. *Int. J. Sci. Res. Agric. Sci*, *22*, 159–172.

El-Fadel, M.,, & Khoury, R. (2001). Strategies for vehicle waste-oil management: a case study. *Resources, Conservation and Recycling*, *33*, 75–91.

El-Fadel, M., El-Sayegh, Y., El-Fadl, K., & Khorbotly, D. (2003). The Nile River Basin: A case study in surface water conflict resolution. *Journal of Natural Resources and Life Sciences Education*, *32*, 107–117.

Elgibaly, A. A. M. (1999). *Clean-up of oil contaminated soils of Kuwaiti oil lakes* (pp. 411–416). Washington, DC: Institute.

El-Gohary, F. A., Abou-Elela, S. I., & Aly, H. I. (1995). Evaluation of biological technologies for wastewater treatment in the pharmaceutical industry. *Water Science and Technology*, *32*(11), 13–20. doi:10.2166/wst.1995.0389

El-Gohary, F., & Tawfik, A. (2009). Decolorization and COD reduction of disperse and reactive dyes wastewater using chemical-coagulation followed by sequential batch reactor (SBR) process. *Desalination*, *249*(3), 1159–1164. doi:10.1016/j.desal.2009.05.010

El-Hennawi, H. M., Shahin, A. A., & Abd El-Thalouth, I. (2013). Evaluation of dried biomass of baker's yeast as reactive dye adsorbent using ultrasonic technique. *Journal of Applied Sciences Research*, *9*, 1401–1408.

Ellouze, E., Tahri, N., & Amar, R. (2012). Enhancement of textile wastewater treatment process using nanofiltration. *Desalination*, *286*, 16–23. doi:10.1016/j.desal.2011.09.025

Emamjomeh, M. M., & Sivakumar, M. (2009). Review of pollutants removed by electrocoagulation and electrocoagulation/flotation processes. *Journal of Environmental Management*, *90*(5), 1663–1679. doi:10.1016/j.jenvman.2008.12.011 PMID:19181438

Emaraa, M. M., Ahmeda, F. A., El-Azizc, F. M. A., & El-Razek, A. M. A. (2014). Biological aspects of the wastewater treatment plant "Mahala Marhoom" in Egypt and modified with Bardenpho processes. *Nature and Science*, *12*(10), 41–51.

Emberson, L. D., Ashmore, M. R., Murray, F., Percy, K. E., Izuta, T., Zheng, Y., ... Domingos, M. (2001). Impacts of air pollutants on vegetation in developing countries. *Water, Air, and Soil Pollution*, *130*(1/4), 107–118. doi:10.1023/A:1012251503358

Enden, V. J. C. & Calvert, K. C. (2002) *Limit environmental damage by basic knowledge of coffee wastewaters*. Report on GTZ-PPP Project-Improvement of coffee quality and sustainability of coffee production in Vietnam.

Enders, K., Lenz, R., Stedmon, C. A., & Nielsen, T. G. (2015). (2015) Abundance, size and polymer composition of marine microplastics ≥10µm in the Atlantic Ocean and their modelled vertical distribution. *Marine Pollution Bulletin*, *100*(1), 70–81. doi:10.1016/j.marpolbul.2015.09.027 PMID:26454631

Enegbuma, W. I., Aliagha, G. U., Ali, K. N., & Badiru, Y. Y. (2016). Confirmatory strategic information technology implementation for building information modelling adoption model. *Journal of Construction in Developing Countries*, *21*(2), 113–129. doi:10.21315/jcdc2016.21.2.6

Engwall, M. A., Pignatello, J. J., & Grasso, D. (1999). Degradation and detoxification of the wood preservatives creosote and pentachlorophenol in water by the photo-Fenton reaction. *Water Research*, *33*(5), 1151–1158. doi:10.1016/S0043-1354(98)00323-6

Entry, J. A., Sojka, R., Watwood, M., & Ross, C. (2002). Polyacrylamide preparations for protection of water quality threatened by agricultural runoff contaminants. *Environmental Pollution*, *120*(2), 191–200. doi:10.1016/S0269-7491(02)00160-4 PMID:12395830

Environmental Quality Act. (1974). (Industrial effluent) Regulation. Seventh Schedule, Regulation 12, amended 2009 PU (A) 434), Malaysia.

EPA. (2006). In situ treatment technologies for contaminated soil. Engineering forum issue paper. United States Environmental Protection Agency. 542/F-06/013.

EPA. (2016). Learn about biodiesel. Retrieved from https://www3.epa.gov/region9/waste/biodiesel/questions.html

Ergas, S. J., Sengupta, S., Siegel, R., Pandit, A., Yao, Y., & Yuan, X. (2010). Performance of nitrogen-removing bioretention systems for control of agricultural runoff. *Journal of Environmental Engineering*, *136*(10), 1105–1112. doi:10.1061/(ASCE)EE.1943-7870.0000243

Eriksen, M., Thiel, M., Prindiville, M. M., & Kiessling, T. (2017) Microplastic: What are the Solutions. The Handbook of Environmental Chemistry. Volume 58, Freshwater Microplastics 58: 273-298.

Erlich, C. (2009). *Comparative study of residue pellets from cane sugar and palm-oil industries with commercial wood pellets, applied in downdraft gasification.* (Doctoral Thesis), KTH Royal Institute of Technology, Sweden. Retrieved from http://bit.ly/1YwDTh4

Erlich, C., & Fransson, T. H. (2011). Downdraft gasification of pellets made of wood, palm-oil residues respective bagasse: Experimental study. *Applied Energy*, *88*(3), 899–908. doi:10.1016/j.apenergy.2010.08.028

Ertem, F. C. (2011). Improving biogas production by anaerobic digestion of different substrates: calculation of potential energy outcomes.

Ertugrul, S., San, N. O., & Dönmez, G. (2009). Treatment of dye (Remazol Blue) and heavy metals using yeast cells with the purpose of managing polluted textile wastewaters. *Ecological Engineering*, *35*(1), 128–134. doi:10.1016/j.ecoleng.2008.09.015

Eskicioglu, C., Kennedy, K. J., & Droste, R. L. (2008). Initial examination of microwave pretreatment on primary, secondary and mixed sludges before and after anaerobic digestion. *Water Science and Technology*, *57*(3), 311–317. doi:10.2166/wst.2008.010 PMID:18309206

Esplugas, S., Yue, P., & Pervez, M. I. (1994). Degradation of 4-chlorophenol by photolytic oxidation. *Water Research*, *28*(6), 1323–1328. doi:10.1016/0043-1354(94)90297-6

Essian, J. P., Ubom, R. M., & Udosen, E. D. (1997). Bioremediation of petroleum contaminated soil: Effect on the population dynamics and degrading capabilities of hydrocarbon clastic bacteria. *West African Journal of Biology and Applied Chemistry*, *43*, 22–28.

Esteban, G., Tellez, C., & Bautista, L. (1991). Dynamics of ciliated protozoa communities in activated sludge process. *Water Research*, *25*(8), 967–972. doi:10.1016/0043-1354(91)90145-G

Euniza, J. (2018). Physical, mechanical, and acoustical performance of double-layer rubberized concrete paving blocks. *PhD Thesis*. Universiti Teknologi Malaysia.

Europe, P. (2016). *Plastics - the Facts 2016. An analysis of European plastics production, demand and waste data.* Available at http://www.plasticseurope.org/Document/plastics–the-facts-01615787.aspx?FolID1/42

European Biomass Industry Association. (2015). Transformation of used cooking oil into biodiesel: from waste to resource. Promotion of used cooking oil recycling for sustainable biodiesel production (RecOil). Retrieved from http://www.eubren.com/UCO_to_Biodiesel_2030 _01.pdf

European Parliament and the Council, 2006. Regulation (EC) No. 1907/2006 of the European Parliament and of the Council of 18 December 2006 Concerning the Registration, Evaluation, Authorisation and Restriction of Chemicals (REACH), Establishing a European Chemicals Agency, Amending Directive 1999/45/EC and Repealing Council Regulation (EEC) No 793/93 and Commission Regulation (EC) No 1488/94 as Well as Council Directive 76/769/EEC and Commission Directives 91/155/EEC, 93/67/EEC, 93/105/EC and 2000/21/EC. OJ L, vol. 396, pp. 1e849.

Everett, J. G. (1974). The effect of pH on the heat treatment of sewage sludges. *Water Research*, *8*(11), 899–906. doi:10.1016/0043-1354(74)90104-3

Ewing, R., & Cervero, R. (2010). Travel and the built environment: A meta-analysis. *Journal of the American Planning Association*, *76*(3), 265–294. doi:10.1080/01944361003766766

Ewing, R., Pendall, R., & Chen, D. (2003). Measuring sprawl and its transportation impacts. *Transportation Research Record: Journal of the Transportation Research Board*, *1831*(1), 175–183. doi:10.3141/1831-20

Exall, K., Balakrishnan, V. K., Toito, J., & McFadyen, R. (2013). Impact of selected wastewater constituents on the removal of sulfonamide antibiotics via ultrafiltration and micellar enhanced ultrafiltration. *The Science of the Total Environment*, *461*, 371–376. doi:10.1016/j.scitotenv.2013.04.057 PMID:23743146

Eyad, A. B., & Richard, G. Z. (2015). An effective soil vapour extraction/bioventing model. Retrieved from https://www.esaa.org/wp-content/uploads/2015/06/04-21Barakat.pdf

Ezechi, E. H., Isa, M. H., Kutty, S. R., & Sapari, N. B. (2011). *Boron recovery, application and economic significance: A review.* Paper presented at the National Postgraduate Conference (NPC), 2011. 10.1109/NatPC.2011.6136374

Ezechi, E. H., Kutty, S. R. M., Isa, M. H., & Rahim, A. F. A. (2014). Treatment of wastewater using an integrated submerged attached growth system. Paper presented at *Applied Mechanics and Materials.* 10.4028/www.scientific.net/AMM.567.167

Ezechi, E. H., & Isa, M. H. (2015). Electrochemical removal of boron from produced water and recovery. *Journal of Environmental Chemical Engineering*, *3*(3), 1962–1973. doi:10.1016/j.jece.2015.05.015

Ezechi, E. H., Isa, M. H., & Kutty, S. R. B. M. (2012). Boron in produced water: Challenges and improvements: A comprehensive review. *Journal of Applied Sciences (Faisalabad)*, *12*(5), 402–415. doi:10.3923/jas.2012.402.415

Ezechi, E. H., Isa, M. H., Kutty, S. R. M., & Yaqub, A. (2014). Boron removal from produced water using electrocoagulation. *Process Safety and Environmental Protection*, *92*(6), 509–514. doi:10.1016/j.psep.2014.08.003

Ezechi, E. H., Kutty, S. R. M., Isa, M. H., Malakahmad, A., & Ibrahim, S. U. (2015). Chemical oxygen demand removal from wastewater by integrated bioreactor. *Journal of Environmental Science and Technology*, *8*(5), 238–243. doi:10.3923/jest.2015.238.243

Ezechi, E. H., Kutty, S., Isa, M. H., Malakahmad, A., Ude, C. M., Menyechi, E. J., & Olisa, E. (2016). Nutrient removal from wastewater by integrated attached growth bioreactor. *Research Journal of Environmental Toxicology*, *10*(1), 28–38. doi:10.3923/rjet.2016.28.38

Ezechi, E., Kutty, M., Isa, M., Malakahmad, A., & Aminu, N. (2015). An integrated attached growth bioreactor for the treatment of wastewater. *Research Journal of Applied Sciences, Engineering and Technology*, *11*(10), 1066–1070. doi:10.19026/rjaset.11.2120

Ezechi, E., Kutty, S., Isa, M., Malakahmad, A., Udeh, C., & Menyechi, E. (2016). Treatment of synthetic domestic wastewater by integrated aerobic/anoxic bioreactor (IAAB). *Nature Environment and Pollution Technology*, *15*(3), 943.

Ezechi, E., Kutty, S., Muda, K., & Yaqub, A. (2019). A comparative evaluation of an integrated hybrid bioreactor treating industrial wastewater. *Journal of Water Process Engineering*, *31*, 100805. doi:10.1016/j.jwpe.2019.100805

Faisal, I. K., Tahir, H., & Ramzi, H. (2004). An overview and analysis of site remediation technologies. *Journal of Environmental Management*, *71*(2), 95–122. doi:10.1016/j.jenvman.2004.02.003 PMID:15135946

Fakhru'L-Razi, A., Qudsieh, I. Y. M., Yunus, W. M. Z. W., Ahmad, M. B., & Rahman, M. Z. A. (2001). Graft copolymerization of methyl methacrylate onto sago starch using ceric ammonium nitrate and potassium persulfate as redox initiator systems. *Journal of Applied Polymer Science*, *82*(6), 1375–1381. doi:10.1002/app.1974

Falahatpisheh, M., Donnelly, K., & Ramos, K. (2001). Antagonistic interactions among nephrotoxic polycyclic aromatic hydrocarbons. *Journal of Toxicology and Environmental Health. Part A.*, *62*(7), 543–560. doi:10.1080/152873901300007833 PMID:11289703

Falkenberry, H. L., & Slack, A. V. (1969). SO_2 removal by limestone injection. *Chemical Engineering Programme*, *65*, 61–66.

Fane, A. G. (2017, September). A grand challenge for membrane desalination: More water, less carbon. *Desalination*, *426*, 155–163. doi:10.1016/j.desal.2017.11.002

Fang, Q., Li, H., Luo, X., Ding, L., Rose, T. M., An, W., & Yu, Y. (2018). A deep learning-based method for detecting non-certified work on construction sites. *Advanced Engineering Informatics, 35*(August 2017), 56–68. doi:10.1016/j.aei.2018.01.001

Fang, Y., Cho, Y. K., Zhang, S., & Perez, E. (2016). *Case study of BIM and cloud – enabled real-time RFID indoor localization for construction management applications, 142*(7), 1–12. doi:. doi:10.1061/(ASCE)CO.1943-7862.0001125

Fang, Z., Guangming, L., Xiuhua, Z., Huikang, H., & Juwen, H. (2004). Study status and progress in wastewater treatment by electro-Fenton method [J]. *Industrial Water Treatment, 12*, 003.

Fang, Q., Li, H., Luo, X., Ding, L., Luo, H., Rose, T. M., & An, W. (2018). Detecting non-hardhat-use by a deep learning method from far-field surveillance videos. *Automation in Construction*, *85*, 1–9. doi:10.1016/j.autcon.2017.09.018

Fang, Y., Cho, Y. K., & Chen, J. (2016). A framework for real-time pro-active safety assistance for mobile crane lifting operations. *Automation in Construction*, *72*, 367–379. doi:10.1016/j.autcon.2016.08.025

Fan, L., Xu, N., Wang, Z., & Shi, H. (2010). PDA experiments and CFD simulation of a lab-scale oxidation ditch with surface aerators. *Chemical Engineering Research & Design*, *88*(1), 23–33. doi:10.1016/j.cherd.2009.07.013

FAO. (2019). Crop statistics 2017. Retrieved from Http://Www.Fao.Org/Faostat/En/#Data/Qc

Farid, M. A. A., Zakaria, M. R., Hassan, M. A., Ali, A. A. M., Othman, M. R., Ibrahim, I., & Shirai, Y. (2019). A holistic treatment system for palm oil mill effluent by incorporating the anaerobic-aerobic-wetland sequential system and a convective sludge dryer. *Chemical Engineering Journal*, *369*, 195–204. doi:10.1016/j.cej.2019.03.033

Farré, M. J., Franch, M. J., Ayllón, M. I., Peral, J. A., Xavier, J., & Domènech, X. (2007). Biodegradability of treated aqueous solutions of biorecalcitrant pesticides by means of photocatalytic ozonation. *Desalination*, *211*(1-3), 22–33. doi:10.1016/j.desal.2006.02.082

Fasnacht, M. P., & Blough, N. V. (2003). Kinetic analysis of the photodegradation of polycyclic aromatic hydrocarbons in aqueous solution. *Aquatic Sciences*, *65*(4), 352–358. doi:10.100700027-003-0680-7

Fatta-Kassinos, D., Meric, S., & Nikolaou, A. (2011). Pharmaceutical residues in environmental waters and wastewater: Current state of knowledge and future research. *Analytical and Bioanalytical Chemistry*, *399*(1), 251–275. doi:10.100700216-010-4300-9 PMID:21063687

Faust, B. C., & Hoigné, J. (1990). Photolysis of Fe (III)-hydroxy complexes as sources of OH radicals in clouds, fog and rain. *Atmospheric Environment. Part A, General Topics*, *24*(1), 79–89. doi:10.1016/0960-1686(90)90443-Q

Fazal, S., Zhang, B., Zhong, Z., Gao, L., & Lu, X. (2015). Membrane separation technology on pharmaceutical wastewater by using MBR (Membrane Bioreactor). *Journal of Environmental Protection*, *6*(04), 299–307. doi:10.4236/jep.2015.64030

Fazeli, A., Bakhtvar, F., Jahanshaloo, L., Che Sidik, N. A., & Bayat, A. E. (2016). Malaysia's stand on municipal solid waste conversion to energy: A review. *Renewable & Sustainable Energy Reviews*, *58*, 1007–1016. doi:10.1016/j.rser.2015.12.270

Federation, Water Environment. (2007). *Biological nutrient removal processes* (pp. 22:23-65).

Feki, E., Khoufi, S., Loukil, S., & Sayadi, S. (2015). Improvement of anaerobic digestion of waste-activated sludge by using H_2O_2 oxidation, electrolysis, electro-oxidation and thermo-alkaline pretreatments. *Environmental Science and Pollution Research International*, *22*(19), 14717–14726. doi:10.100711356-015-4677-2 PMID:25982985

Felix, E., & Tilley, D. R. (2009). Integrated energy, environmental and financial analysis of ethanol production from cellulosic switchgrass. *Energy*, *34*(4), 410–436. doi:10.1016/j.energy.2008.10.013

Fendall, L. S., Mary, A., & Sewell, M. A. (2009). Contributing to marine pollution by washing your face: Microplastics in facial cleansers. *Marine Pollution Bulletin*, *58*(8), 1225–1228. doi:10.1016/j.marpolbul.2009.04.025 PMID:19481226

Feng, G., Hu, L., Ma, Y., Jia, P., Hu, Y., Zhang, M., ... Zhou, Y. (2018). An efficient bio-based plasticizer for poly (vinyl chloride) from waste cooking oil and citric acid: Synthesis and evaluation in PVC films. *Journal of Cleaner Production*, *189*, 334–343. doi:10.1016/j.jclepro.2018.04.085

Feng, Q., Li, X., Cheng, Y., Meng, L., & Meng, Q. (2007). Removal of humic acid from groundwater by electrocoagulation. *Journal of China University of Mining and Technology*, *17*(4), 513–520. doi:10.1016/S1006-1266(07)60136-9

Feng, X., Deng, J., Lei, H., Bai, T., Fan, Q., & Li, Z. (2009). Dewaterability of waste activated sludge with ultrasound conditioning. *Bioresource Technology*, *100*(3), 1074–1081. doi:10.1016/j.biortech.2008.07.055 PMID:18783942

Fenton, H. J. H. (1894). LXXIII.—Oxidation of tartaric acid in presence of iron. *Journal of the Chemical Society Transactions*, *65*(0), 899–910. doi:10.1039/CT8946500899

Fernando, S., Karra, P., Hernandez, R., & Jha, S. K. (2007). Effect of incompletely converted soybean oil on biodiesel quality. *Energy*, *32*(5), 844–851.

Fettig, J. (1999). Removal of humic substances by adsorption/ion exchange. *Water Science and Technology*, *40*(9), 173–182. doi:10.2166/wst.1999.0470

Field, J. A., DeJong, E., Feijoo, C. G., & DeBont, J. A. M. (1993). Screening for ligninolytic fungi application to the biodegradation of xenobiotics. *Trends in Biotechnology*, *11*(2), 44–49. doi:10.1016/0167-7799(93)90121-O

Field, J., & Brady, J. (2003). Riboflavin as a redox mediator accelerating the reduction of the azo dye Mordant Yellow 10 by anaerobic granular sludge. *Water Science and Technology*, *48*(6), 187–193. doi:10.2166/wst.2003.0393 PMID:14640217

Fisher, M., & Triplett, E. (1999). Automated approach for ribosomal intergenic spacer analysis of microbial diversity and its application to freshwater bacterial communities. *Applied and Environmental Microbiology*, *65*, 4630–4636. PMID:10508099

Flaten, T. P. (2001). Aluminium as a risk factor in Alzheimer's disease, with emphasis on drinking water. *Brain Research Bulletin*, *55*(2), 187–196. doi:10.1016/S0361-9230(01)00459-2 PMID:11470314

Flemming, H. C. (1996). The forces that keep biofilms together. In W. Sand (Ed.), Biodeterioration and Biodegradation (Vol. 133, pp. 311–316). Weinheim, Germany: Dechema Monographs, VCH Verlagsgesellschaft.

Flemming, H. C., & Wingender, J. (2001). Relevance of microbial extracellular polymeric substances (EPSs)-Part I: Structural and ecological aspects. *Water Science and Technology*, *43*(6), 1–8. doi:10.2166/wst.2001.0326 PMID:11381954

Flemming, H.-C. (1997). Reverse osmosis membrane biofouling. *Experimental Thermal and Fluid Science*, *14*(4), 382–391. doi:10.1016/S0894-1777(96)00140-9

Florin-Constantin, M., & Mohammad, J. T. (2017). *Introductory chapter: rural waste management issues at global level* (pp. 1–10). Solid Waste Management in Rural Areas, Florin-Constantin Mihai, IntechOpen. doi:10.5772/intechopen.70268

Floudas, C. A., Elia, J. A., & Baliban, R. C. (2012). Hybrid and single feedstock energy processes for liquid transportation fuels: A critical review. *Computers & Chemical Engineering*, *41*, 24–51. doi:10.1016/j.compchemeng.2012.02.008

Fontenot, E. J., Lee, Y. H., Matthews, R. D., Zhu, G., & Pavlostathis, S. G. (2003). Reductive decolorization of a textile reactive dye bath under methanogenic conditions. *Applied Biochemistry and Biotechnology*, *109*(1-3), 207–225. doi:10.1385/ABAB:109:1-3:207 PMID:12794295

Fontenot, Q., Bonvillain, C., Kilgen, M., & Boopathy, R. (2007). Effects of temperature, salinity, and carbon: Nitrogen ratio on sequencing batch reactor treating shrimp aquaculture wastewater. *Bioresource Technology*, *98*(9), 1700–1703. doi:10.1016/j.biortech.2006.07.031 PMID:16935499

Font, R., Marcilla, A., Verdu, E., & Devesa, J. (1990). Kinetics of the pyrolysis of almond shells and almond shells impregnated with cobalt dichloride in a fluidized bed reactor and in a pyroprobe 100. *Industrial & Engineering Chemistry Research*, *29*(9), 1846–1855. doi:10.1021/ie00105a016

Forbes, E., Foy, R., Mulholland, M., & Brettell, J. (2011). Performance of a constructed wetland for treating farm-yard dirty water. *Water Science and Technology*, *64*(1), 22–28. doi:10.2166/wst.2011.584 PMID:22053453

Ford, D. L., Churchwell, R. L., & Kachtick, J. W. (1980). Comprehensive analysis of nitrification of chemical processing wastewaters. *Journal - Water Pollution Control Federation*, 2726–2746.

Forgacs, E., Cserhati, T., & Oros, G. (2004). Removal of synthetic dyes from wastewaters: A review. *Environment International*, *30*(7), 953–971. doi:10.1016/j.envint.2004.02.001 PMID:15196844

Forss, J., & Welander, U. (2009). Decolorization of reactive azo dyes with microorganisms growing on soft wood chips. *Int BiodeterBiodegr*, *63*, 752–758.

Forster, C. F., & Lewin, D. C. (1972). Polymer interactions at activated sludge surfaces. *Effluent and Water Treatment Journal*, *12*, 520–525.

Freeman, E. S., & Carroll, B. (1958). The application of thermoanalytical techniques to reaction kinetics: The thermogravimetric evaluation of the kinetics of the decomposition of calcium oxalate monohydrate. *Journal of Physical Chemistry*, *62*(4), 394–397. doi:10.1021/j150562a003

Frick, E. A., Henderson, A. K., Moll, D. M., Furlong, E. T., & Meyer, M. T. (2001). Presence of pharmaceuticals in wastewater effluent and drinking water, metropolitan Atlanta, Georgia, July-September 1999. Georgia Institute of Technology.

Friedman, H. L. (1964). Kinetics of thermal degradation of char-forming plastics from thermogravimetry. application to a phenolic plastic. *Journal of Polymer Science Part C: Polymer Symposia*, *6*(1), 183–195. doi:10.1002/polc.5070060121

Frohlund, B., Griebe, T., & Nielsen, P. H. (1995). Enzyme activity in activated-sludge floc matrix. *Applied Microbiology and Biotechnology*, *43*, 755–761. doi:10.1007/BF00164784 PMID:7546613

Frølund, B., Palmgren, R., Keiding, K., & Nielsen, P. H. (1996). Extraction of extracellular polymers from activated sludge using a cation exchange resin. *Water Research*, *30*(8), 1749–1758. doi:10.1016/0043-1354(95)00323-1

Fryda, L., Panopoulos, K., & Kakaras, E. (2008). Agglomeration in fluidised bed gasification of biomass. *Powder Technology*, *181*(3), 307–320. doi:10.1016/j.powtec.2007.05.022

Fu, Y., & Viraraghavan, T. (1999). Removal of acid blue 29 from aqueous solution by fungus Aspergillus niger. In *Proceedings of the 31st Mid-Atlantic Industrial and Hazardous Waste Conference*. Storrs, Connecticut. Technomic Publishing Company, Lancaster, PA. pp. 510-519.

Fu, F., & Wang, Q. (2011). Removal of heavy metal ions from wastewaters: A review. *Journal of Environmental Management*, *92*(3), 407–418. doi:10.1016/j.jenvman.2010.11.011 PMID:21138785

Fung, I. W. H., Tam, V. W. Y., Sing, C. P., Tang, K. K. W., & Ogunlana, S. O. (2016). Psychological climate in occupational safety and health: The safety awareness of construction workers in South China. *International Journal of Construction Management*, *16*(4), 315–325. doi:10.1080/15623599.2016.1146114

Fu, Y., & Viraraghavan, T. (2000). Removal of a dye from an aqueous solution by fungus Aspergillus niger. *Water Quality Research Journal of Canada*, *35*(1), 95–111. doi:10.2166/wqrj.2000.006

Ga, C., & Ra, C. (2009). Real-time control of oxic phase using pH (mV)-time profile in swine wastewater treatment. *Journal of Hazardous Materials*, *172*(1), 61–67. doi:10.1016/j.jhazmat.2009.06.133 PMID:19628333

Gadipelly, C., Pérez-González, A., Yadav, G. D., Ortiz, I., Ibáñez, R., Rathod, V. K., & Marathe, K. V. (2014). Pharmaceutical industry wastewater: Review of the technologies for water treatment and reuse. *Industrial & Engineering Chemistry Research*, *53*(29), 11571–11592. doi:10.1021/ie501210j

Gajaraj, S., & Hu, Z. (2014). Integration of microbial fuel cell techniques into activated sludge wastewater treatment processes to improve nitrogen removal and reduce sludge production. *Chemosphere*, *117*, 151–157. doi:10.1016/j.chemosphere.2014.06.013 PMID:25014565

Galaska, G., Skaladany, G. J., Nyer, E. K., & Chelsea, M. (1990). Biological treatment of groundwater, soils and soil vapour contaminated with petroleum hydrocarbons. *Proceedings of the Industrial Waste Conference, Purdue University* (44th), 11-21.

Galgani, F., Hanke, G., Werner, S., & Oosterbaa, L. (2013). *Guidance on monitoring of marine litter in European seas. Scientific and Technical Research series*. Luxembourg: Publications Office of the European Union.

Gallagher, K. A., Healy, M. G., & Allen, S. J. (1997). Biosorption of synthetic dye and metal ions from aqueous effluents using fungal biomass. In D. L. Wise (Ed.), *Global Environmental Biotechnology* (pp. 27–50). UK: Elsevier.

Gallo, F., Fossi, C., Weber, R., Santillo, D., Sousa, J., Ingram, I., ... Romano, D. (2018). Marine litter plastics and microplastics and their toxic chemicals components: The need for urgent preventive measures. *Environmental Sciences Europe*, *30*(1), 13. doi:10.118612302-018-0139-z PMID:29721401

Gálvez, J. M., Gómez, M. A., Hontoria, E., & González-López, J. (2003). Influence of hydraulic loading and air flowrate on urban wastewater nitrogen removal with a submerged fixed-film reactor. *Journal of Hazardous Materials*, *101*(2), 219–229. doi:10.1016/S0304-3894(03)00173-0 PMID:12927737

Galvis, G., Visscher, J. T., & Latorre, J. (1998). Multi-stage filtration and innovation water treatment technology. Report from International reference centre for community water supply and sanitation. The Hague, The Netherlands and Universidad del valle instituto Cinara, Cali, Colombia.

Ganjidoust, H., & Ayati, B. (2004). Use of sequencing batch reactors (SBRs) in treatment of wood fiber wastewater. *Iran Journal of Health Science and Engineering*, *1*, 91–96.

Gao, D., Li, Z., Wen, Z., & Ren, N. (2014). Occurrence and fate of phthalate esters in full-scale domestic wastewater treatment plants and their impact on receiving waters along the Songhua River in China. *Chemosphere*, *95*, 24–32. doi:10.1016/j.chemosphere.2013.08.009 PMID:24001662

Gárcia-Fernández, I., Polo-López, M. I., Oller, I., & Fernández-Ibáñez, P. (2012). Bacteria and fungi inactivation using Fe3+/sunlight, H2O2/sunlight and near neutral photo-Fenton: A comparative study. *Applied Catalysis B: Environmental*, *121*, 20–29. doi:10.1016/j.apcatb.2012.03.012

García, X. A., Alarcón, N. A., & Gordon, A. L. (1999). Steam gasification of tars using a CaO catalyst source. *Fuel Processing Technology*, *58*(2), 83–102. doi:10.1016/S0378-3820(98)00087-3

Garg, A., Shukla, P. R., Bhattacharya, S., & Dadhwal, V. K. (2001). Sub-region (district) and sector level SO2 and NO(x) emissions for India: Assessment of inventories and mitigation flexibility. *Atmospheric Environment*, *35*(4), 703–713. doi:10.1016/S1352-2310(00)00316-2

Garg, A., Shukla, P. R., & Kapshe, M. (2006). The sectoral trends of multigas emissions inventory of India. *Atmospheric Environment*, *40*(24), 4608–4620. doi:10.1016/j.atmosenv.2006.03.045

Gargi, C., & Mishra, N. B. (2016). Spatio-temporal dynamics of land cover and land surface temperature in Ganges-Brahmaputra delta: A comparative analysis between India and Bangladesh. *Applied Geography (Sevenoaks, England)*, *68*, 68–83. doi:10.1016/j.apgeog.2016.01.002

Garg, V. K., Gupta, R., Yadav, A. B., & Kumar, R. (2003). Dye removal from aqueous solution by adsorption on treated sawdust. *Bioresource Technology*, *89*(2), 121–124. doi:10.1016/S0960-8524(03)00058-0 PMID:12699929

Gary, C. Y. (2010). *Municipal solid waste to energy conversion processes*. Canada: John Wiley & Sons.

Gaur, N., Flora, G., Yadav, M., & Tiwari, A. (2014). A review with recent advancements on bioremediation-based abolition of heavy metals. *Environmental Science. Processes & Impacts*, *16*(2), 180–193. doi:10.1039/C3EM00491K PMID:24362580

Gebbie, P. (2001). *Using polyaluminium coagulants in water treatment*. Paper presented at the Proc. 64th AWIEO Conference.

Gehlert, G., Abdulkadir, M., Fuhrmann, J., & Hapke, J. (2005). Dynamic modeling of an ultrafiltration module for use in a membrane bioreactor. *Journal of Membrane Science*, *248*(1-2), 63–71. doi:10.1016/j.memsci.2004.09.026

Gehring, T., Klang, J., Niedermayr, A., Berzio, S., Immenhauser, A., Klocke, M., ... Lübken, M. (2015). Determination of methanogenic pathways through carbon isotope (δ13C) analysis for the two-stage anaerobic digestion of high-solids substrates. *Environmental Science & Technology*, *49*(7), 4705–4714. PMID:25741999

Gerardi, M. H. (2003). *The microbiology of anaerobic digesters*. Hoboken, NJ: John Wiley & Sons. doi:10.1002/0471468967

Gerhard, K., Jurgen, K., & Jon, V. G. (2004). *The biodiesel handbook* (2nd ed., p. 53). Urbana, IL: AOCS Press.

Ge, S., Usack, J., Spirito, C., & Angenent, L. (2015). Long-term n-caproic acid production from yeast-fermentation beer in an anaerobic bioreactor with continuous product extraction. *Environmental Science & Technology*, 49. PMID:25941741

Gewali, M. B., & Bhandari, R. (2005). Renewable energy technologies in Nepal. *World Review of Science, Technology and Sustainable Development*, *2*(1), 92–106.

Geyer, R., Jambeck, J. R., & Law, K. L. (2017). Production, use, and fate of all plastics ever made. *Science Advances*, *3*(7), e1700782. doi:10.1126ciadv.1700782 PMID:28776036

Ghafari, S., Hasan, M., & Aroua, M. K. (2008). Bio-electrochemical removal of nitrate from water and wastewater—A review. *Bioresource Technology*, *99*(10), 3965–3974. doi:10.1016/j.biortech.2007.05.026 PMID:17600700

Ghaly, A. E., Kamal, M. A., Mahmoud, N. S., & Cote, R. (2007). Treatment of landfill leachate using limestone/sandstone filters under aerobic batch conditions. *American Journal of Environmental Sciences*, *3*(2), 43–53. doi:10.3844/ajessp.2007.43.53

Ghanimeh, S., El Fadel, M., & Saikaly, P. (2012). Mixing effect on thermophilic anaerobic digestion of source-sorted organic fraction of municipal solid waste. *Bioresource Technology*, *117*, 63–71. PMID:22609715

Ghate, A. T., & Sundar, S. (2013). Can we reduce the rate of growth of car ownership. *Economic and Political Weekly*, *48*, 32–40.

Ghisellini, P., Ripa, M., & Ulgiati, S. (2018). Exploring environmental and economic costs and benefits of a circular economy approach to the construction and demolition sector. A literature review. *Journal of Cleaner Production*, *178*, 618–643. doi:10.1016/j.jclepro.2017.11.207

Ghoneim, M. M., El-Desoky, H. S., & Zidan, N. M. (2011). Electro-Fenton oxidation of Sunset Yellow FCF azo-dye in aqueous solutions. *Desalination*, *274*(1-3), 22–30. doi:10.1016/j.desal.2011.01.062

Ghose, M. K., Paul, R., & Banerjee, S. K. (2004). Assessment of the impacts of vehicular emissions on urban air quality and its management in Indian context: The case of Kolkata. *Environmental Science & Policy, Calcutta*, *7*(4), 345–351. doi:10.1016/j.envsci.2004.05.004

Ghosh, D., Solanki, H., & Purkait, M. (2008). Removal of fe (ii) from tap water by electrocoagulation technique. *Journal of Hazardous Materials*, *155*(1), 135–143. doi:10.1016/j.jhazmat.2007.11.042 PMID:18164128

Ghosh, P., Saxena, R., Gupta, R., & Yadav, R., & S, D. (1996). Microbial lipases: Production and applications. *Science Progress*, *79*, 119–158. PMID:8828407

Gianico, A., Braguglia, C. M., Cesarini, R., & Mininni, G. (2013). Reduced temperature hydrolysis at 134 °C before thermophilic anaerobic digestion of waste activated sludge at increasing organic load. *Bioresource Technology*, *143*, 96–103. doi:10.1016/j.biortech.2013.05.069 PMID:23792658

Gilau, A. M., & Small, M. J. (2006). Designing cost-effective sea water reverse osmosis system under optimal energy options for developing countries. In *International Conference on Renewable Energy for Developing Countries* (pp. 1–18).

Gil, J., Corella, J., Aznar, M. P., & Caballero, M. A. (1999). Biomass gasification in atmospheric and bubbling fluidized bed: Effect of the type of gasifying agent on the product distribution. *Biomass and Bioenergy*, *17*(5), 389–403. doi:10.1016/S0961-9534(99)00055-0

Gizelis, T.-I., & Wooden, A. E. (2010). Water resources, institutions, & intrastate conflict. *Political Geography*, *29*(8), 444–453. doi:10.1016/j.polgeo.2010.10.005

Global Environment Centre report. (2019). Available at http:// www. gecnet. info/ index. cfm?&menuid=83

Gogate, P. R., & Pandit, A. B. (2004). A review of imperative technologies for wastewater treatment I: Oxidation technologies at ambient conditions. *Advances in Environmental Research*, *8*(3-4), 501–551. doi:10.1016/S1093-0191(03)00032-7

Gogate, P. R., & Pandit, A. B. (2004). A review of imperative technologies for wastewater treatment II: Hybrid methods. *Advances in Environmental Research*, *8*(3-4), 553–597. doi:10.1016/S1093-0191(03)00031-5

Goicoechea, S., Ehrich, H., Arias, P. L., & Kockmann, N. (2015). Thermodynamic analysis of acetic acid steam reforming for hydrogen production. *Journal of Power Sources*, *279*, 312–322. doi:10.1016/j.jpowsour.2015.01.012

Goldman, R., Enewold, L., Pellizzari, E., Beach, J. B., Bowman, E. D., Krishnan, S. S., & Shields, P. G. (2001). Smoking increases carcinogenic polycyclic aromatic hydrocarbons in human lung tissue. *Cancer Research*, *61*(17), 6367–6371. PMID:11522627

Golob, V., & Ojstršek, A. (2005). Removal of vat and disperse dyes from residual pad liquors. *Dyes and Pigments*, *64*(1), 57–61. doi:10.1016/j.dyepig.2004.04.006

Golob, V., Vinder, A., & Simonič, M. (2005). Efficiency of the coagulation/flocculation method for the treatment of dyebath effluents. *Dyes and Pigments*, *67*(2), 93–97. doi:10.1016/j.dyepig.2004.11.003

González, O., Sans, C., & Esplugas, S. (2007). Sulfamethoxazole abatement by photo-Fenton: Toxicity, inhibition and biodegradability assessment of intermediates. *Journal of Hazardous Materials*, *146*(3), 459–464. doi:10.1016/j.jhazmat.2007.04.055 PMID:17540504

Gonzalez, S., Muller, J., Petrovic, M., Barcelo, D., & Knepper, T. P. (2006). Biodegradation studies of selected priority acidic pesticides and diclofenac in different bioreactors. *Environmental Pollution*, *144*(3), 926–932. doi:10.1016/j.envpol.2006.02.021 PMID:16678949

Gopala Krishna, G. V. T., Kumar, P., & Kumar, P. (2007). Complex wastewater treatment using an anaerobic baffled reactor. *Environment and Progress*, *26*(4), 391–398. doi:10.1002/ep.10239

Gopala Krishna, G. V. T., Kumar, P., & Kumar, P. (2008). Treatment of low-strength soluble wastewater using an anaerobic baffled reactor (ABR). *Journal of Environmental Management*, *90*(1), 166–176. doi:10.1016/j.jenvman.2007.08.017 PMID:18096298

Gopi, G., & Arthanareeswaran, G. (2019). Perspective of renewable desalination by using membrane distillation. *Chemical Engineering Research & Design*, *144*, 520–537. doi:10.1016/j.cherd.2019.02.036

Goubert, L. (2014, Nov. 16-19). Developing a durable and ultra low noise poroelastic pavement. In Proceedings of Internoise 2014. Melbourne, Australia. 1–10.

Goubert, L.,, & Sandberg, U. (2016, July 10-14). Development of the ultra low noise poroelastic road surface: the findings of the persuade project. In *Proceedings of 23rd International Congress on Sound and Vibration*. Athens, Greece. 1–8.

Gouin, T., Avalos, J., Brunning, I., Brzuska, K., de Graaf, J., Kaumanns, J., … Wolf, T. (2015). Use of micro-plastic beads in cosmetic products in Europe and their estimated emissions to the North Sea environment. *SOFW-J* 141, 40–46.

Govind, A. M., Schmidt, T. M., White, D. C., & Loper, J. C. (1993). Phylogenetic analysis of bacterial aerobic degrader of azo dyes. *Journal of Bacteriology*, *175*(18), 6062–6066. doi:10.1128/jb.175.18.6062-6066.1993 PMID:8376354

Goyal, H. B., Seal, D., & Saxena, R. C. (2008). Bio-fuels from thermochemical conversion of renewable resources: A review. *Renewable & Sustainable Energy Reviews*, *12*(2), 504–517. doi:10.1016/j.rser.2006.07.014

Grad, M. (2016). *Collection and treatment of fecal disposal in rural areas of developing countries case study*. Nepal: Graz University of Technology.

Grady Jr, C. L., Daigger, G. T., Love, N. G., & Filipe, C. D. (2011). Biological wastewater treatment. Boca Raton, FL: CRC Press.

Grady, C. P. L., Daigger, G. T., Love, N. G., & Filippe, C. D. M. (1999). Biological wastewater treatment (2nd ed.). Boca Raton, FL: CRC Press.

Grady, C. P. L., Jr., Glent, T. D., & Lim, H. C. (1999). Biological wastewater treatment (2nd Ed., Revised and Expanded). Marcel Dekker, Inc.

Grahmann, K., Verhulst, N., Buerkert, A., Ortiz-Monasterio, I., & Govaerts, B. (2013). Nitrogen use efficiency and optimization of nitrogen fertilization in conservation agriculture. *Perspectives in Agriculture, Veterinary Science, Nutrition and Natural Resources*, *8*(053), 1–19. doi:10.1079/PAVSNNR20138053

Green, F. (2014). "This time is different": The prospects for an effective climate agreement in Paris 2015. London, UK.

Gregory, J., & Duan, J. (2001). Hydrolyzing metal salts as coagulants. *Pure and Applied Chemistry*, *73*(12), 2017–2026. doi:10.1351/pac200173122017

Gregory, P. (1993). Dyes and dyes intermediates. In J. I. Kroschwitz (Ed.), *Encyclopedia of chemical technology* (Vol. 8, pp. 544–545). New York, NY: John Wiley & Sons.

Griffths, M., Phillps, J., & Muir, D. (1981). Thermostability of proteases and lipases from a number of species of psychrotropic bacteria of dairy origin. *The Journal of Applied Bacteriology*, *50*(2), 289–303. doi:10.1111/j.1365-2672.1981.tb00894.x PMID:7019181

Grilo, J., Faria, P., Veiga, R., Silva, A. S., Silva, V., & Velosa, A. (2014). New natural hydraulic lime mortars – physical and microstructural properties in different curing conditions. *Construction & Building Materials*, *54*, 378–384. doi:10.1016/j.conbuildmat.2013.12.078

Groenewegen, P. P., Van den Berg, A. E., De Vries, S., & Verheij, R. A. (2006). Vitamin G: Effects of green space on health, well-being, and social safety. *BMC Public Health*, *6*(1), 149. doi:10.1186/1471-2458-6-149 PMID:16759375

Gromboni, C. F., Kamogawa, M. Y., Ferreira, A. G., Nóbrega, J. A., & Nogueira, A. R. A. (2007). Microwave-assisted photo-Fenton decomposition of chlorfenvinphos and cypermethrin in residual water. *Journal of Photochemistry and Photobiology A Chemistry*, *185*(1), 32–37. doi:10.1016/j.jphotochem.2006.05.005

Gryta, M., Markowska-Szczupak, A., Bastrzyk, J., & Tomczak, W. (2013). The study of membrane distillation used for separation of fermenting glycerol solutions. *Journal of Membrane Science*, *431*, 1–8. doi:10.1016/j.memsci.2012.12.032

Guang-Guo, Y. (2006). Fate, behaviour, and effects of surfactants and their degradation products in the environment. *Environment International 32 (2006)*, Elsevier, 417 – 431.

Guehi, T., Dingkunh, M., & Cros, E. (2007). Identification and lipase-producing abilities of moulds isolated from Ivorian raw cocoa beans. *Research Journal of Agriculture and Biological Sciences*, *3*, 838–843.

Guerra, B., Bakchan, A., Leite, F., & Faust, K. M. (2019). BIM-based automated construction waste estimation algorithms: The case of concrete and drywall waste streams. *Waste Management (New York, N.Y.)*, *87*, 825–832. doi:10.1016/j.wasman.2019.03.010 PMID:31109587

Guerrero, J., Omil, F., Mendez, R., & Lema, J. (1999). Anaerobic hydrolysis and acidogenesis of wastewaters from food industries with high content of organic solids and protein. *Water Research*, *33*(15), 3281–3290. doi:10.1016/S0043-1354(99)00041-X

Gui, M. M., Lee, K. T., & Bhatia, S. (2008). Feasibility of edible oil versus non-edible oil versus waste edible oil as biodiesel feedstock. *Energy*, *33*, 1646–1653.

Guivarch, E., Trevin, S., Lahitte, C., & Oturan, M. A. (2003). Degradation of azo dyes in water by electro-Fenton process. *Environmental Chemistry Letters*, *1*(1), 38–44. doi:10.100710311-002-0017-0

Guo, H., Yu, Y., & Skitmore, M. (2017). Visualization technology-based construction safety management: A review. *Automation in Construction*, *73*, 135–144. doi:10.1016/j.autcon.2016.10.004

Guo, M., Qiu, G., & Song, W. (2010). Poultry litter-based activated carbon for removing heavy metal ions in water. *Waste Management (New York, N.Y.)*, *30*(2), 308–315. doi:10.1016/j.wasman.2009.08.010 PMID:19783133

Guo, Y., Qi, P. S., & Liu, Y. Z. (2017, May). A review on advanced treatment of pharmaceutical wastewater. In *IOP Conference Series: Earth and Environmental Science* (Vol. 63, No. 1, p. 012025). IOP Publishing. 10.1088/1755-1315/63/1/012025

Gupta, K., Sahoo, S., & Khandekar, K. (2002). Graft copolymerization of ethyl acrylate onto cellulose using ceric ammonium nitrate as initiator in aqueous medium. *Biomacromolecules*, *3*(5), 1087–1094. doi:10.1021/bm020060s PMID:12217057

Gupta, V. K., Agarwal, S., & Saleh, T. A. (2011). Synthesis and characterization of alumina-coated carbon nanotubes and their application for lead removal. *Journal of Hazardous Materials*, *185*(1), 17–23. doi:10.1016/j.jhazmat.2010.08.053 PMID:20888691

Gupta, V. K., & Ali, I. (2001). Removal of DDD and DDE from wastewater using bagasse fly ash, a sugar industry waste. *Water Research*, *35*(1), 33–40. doi:10.1016/S0043-1354(00)00232-3 PMID:11257890

Gurjar, B. R., Butler, T. M., Lawrence, M. G., & Leliveld, J. (2007). Evaluation of emissions and air quality in megacities. *Atmospheric Environment*, *42*(7), 1593–1606. doi:10.1016/j.atmosenv.2007.10.048

Gurjar, B. R., & Lelieveld, J. (2005). New directions: Megacities and global change. *Atmospheric Environment*, *39*(2), 391–393. doi:10.1016/j.atmosenv.2004.11.002

Gurjar, B. R., Ravindra, K., & Nagpure, A. S. (2016). Air pollution trends over Indian megacities and their local-to-global implications. *Atmospheric Environment*, *142*, 475–495. doi:10.1016/j.atmosenv.2016.06.030

Guski, R., Schreckenberg, D., Schuemer, R., Gmbh, Z., & Gmbh, Z. (2016, Aug. 21-26). The WHO evidence review on noise annoyance 2000-2014. In *Proceedings of INTER-NOISE 2016 - 45th International Congress and Exposition on Noise Control Engineering: Towards a Quieter Future*. Hamburg, Germany. 2564–2570.

Guttikunda Sarath, K., & Nishadh, K. A. (2019). Air pollution knowledge assessments (APnA) for 20 Indian cities. *Urban Climate*, *27*, 124–14. doi:10.1016/j.uclim.2018.11.005

Guttikunda, S. K., Goel, R., & Pant, P. (2014). Nature of air pollution, emission sources, and management in the Indian cities. *Atmospheric Environment*, *95*, 501–510. doi:10.1016/j.atmosenv.2014.07.006

Guttikunda, S. K., & Jawahar, P. (2012). Application of SIM-air modeling tools to assess air quality in Indian cities. *Atmospheric Environment*, *62*, 551–561. doi:10.1016/j.atmosenv.2012.08.074

Guttikunda, S. K., & Jawahar, P. (2014). Atmospheric emissions and pollution from the coal-fired thermal Power plants in India. *Atmospheric Environment*, *92*, 449–460. doi:10.1016/j.atmosenv.2014.04.057

Guttikunda, S. K., & Mohan, D. (2014). Re-fueling road transport for better air quality in India. *Energy Policy*, *68*, 556–561. doi:10.1016/j.enpol.2013.12.067

Güven, O. (2012). *Crosslinking and scission in polymers* (Vol. 292). Springer Science & Business Media.

Guyu, P., Xu, P., Zhu, B., & Li, D. (2017). Microplastics in freshwater river sediments in Shanghai,China: A case study of risk assessment in megacities. *Environmental Pollution*, *234*, 448–456. PMID:29207296

Haagen-Smit, A. J. (1952). Chemistry and physiology of Los Angeles smog. *Journal of Industrial and Engineering Chemistry*, *44*(6), 1342–1346. doi:10.1021/ie50510a045

Hadjiniculaou, J. (1998). Evaluation of a controlled condition in a sequencing batch reactor pilot plant operation for treatment of slaughterhouse wastewater. *Canadian Agricultural Engineering*, *23*, 147–156.

Hafez, H., Elbeshbishy, E., Chowdhury, N., Nakhla, G., Fitzgerald, J., Van Rossum, A., & Gauld, G. (2010). Pushing the hydraulic retention time envelope in Modified Ludzack Ettinger systems. *Chemical Engineering Journal*, *163*(3), 202–211. doi:10.1016/j.cej.2010.07.033

Hagos, K., Zong, J., Li, D., Liu, C., & Lu, X. (2017). Anaerobic co-digestion process for biogas production: Progress, challenges and perspectives. *Renewable & Sustainable Energy Reviews*, *76*, 1485–1496.

Haider, T. (2008). Meso-urban meteorological and photochemical modeling of heat island mitigation. *Atmospheric Environment*, *42*(38), 8795–8809. doi:10.1016/j.atmosenv.2008.06.036

Hai, F. I., Modin, O., Yamamoto, K., Fukushi, K., Nakajima, F., & Nghiem, L. D. (2012). Pesticide removal by a mixed culture of bacteria and white-rot fungi. *Journal of the Taiwan Institute of Chemical Engineers*, *43*(3), 459–462. doi:10.1016/j.jtice.2011.11.002

Hai, F. I., Yamamoto, K., & Fukushi, K. (2007). Hybrid treatment system for dye wastewater. *Critical Reviews in Environmental Science and Technology*, *37*(4), 315–377. doi:10.1080/10643380601174723

Halling-Sørensen, B., & Jorgensen, S. E. (1993). *The removal of nitrogen compounds from wastewater* (Vol. 54). Elsevier.

Hamad, B. S., Rteil, A. A., & El-fadel, M. (2003). Effect of used engine oil on properties of fresh and hardened concrete. *Construction & Building Materials*, *17*, 311–318.

Hamby, D. M. (1996). Site remediation techniques supporting environmental restoration activities. Department of Environmental and Industrial Health School of Public Health University of Michigan Ann Arbor, MI: 48109-2029 U.S.A.

Hamet, J. F. (2004). Reduction of tire road noise by acoustic absorption : numerical evaluation of the pass-by noise level reduction using the normal incidence acoustic absorption coefficient reduction of tire road noise by acoustic absorption. *SILVIA* (silent roads). 1-25.

Ham, Y., Han, K. K., Lin, J. J., & Golparvar-Fard, M. (2016). Visual monitoring of civil infrastructure systems via camera-equipped Unmanned Aerial Vehicles (UAVs): A review of related works. *Visualization in Engineering*, *4*(1), 1–8. doi:10.118640327-015-0029-z

Hamzah, A., Salleh, S. N. M., Wong, K. K., & Sarmani, S. (2016). Nutrient amendments of inorganic fertiliser and oil palm empty fruit bunch and their influence on bacterial species dominance and degradation of the associated crude oil constituents. *Soil and Sediment Contamination: An International Journal*(just-accepted).

Han, S.-S., Bae, T.-H., Jang, G.-G., & Tak, T.-M. (2005). Influence of sludge retention time on membrane fouling and bioactivities in membrane bioreactor system. *Process Biochemistry*, *40*(7), 2393–2400. doi:10.1016/j.procbio.2004.09.017

Hansen, H. K., Peña, S. F., Gutiérrez, C., Lazo, A., Lazo, P., & Ottosen, L. M. (2019). Selenium removal from petroleum refinery wastewater using an electrocoagulation technique. *Journal of Hazardous Materials*, *364*, 78–81. doi:10.1016/j.jhazmat.2018.09.090 PMID:30339935

Hansen, S. B., Padfield, R., Syayuti, K., Evers, S., Zakaria, Z., & Mastura, S. (2015). Trends in global palm oil sustainability research. *Journal of Cleaner Production*, *100*, 140–149. doi:10.1016/j.jclepro.2015.03.051

Hao, O., Kim, H., & Chiang, P. C. (2000). Decolorization of wastewater. *Critical Reviews in Environmental Science and Technology*, *30*(4), 449–505. doi:10.1080/10643380091184237

Harford, A. J., Hogan, A. C., Jones, D. R., & van Dam, R. A. (2011). Ecotoxicological assessment of a polyelectrolyte flocculant. *Water Research*, *45*(19), 6393–6402. doi:10.1016/j.watres.2011.09.032 PMID:22000718

Hariprasad, N., & Dayananda, H. (2013). Environmental impact due to agricultural runoff containing heavy metals—a review. *International Journal of Scientific and Research Publications*, *3*(5), 224–280.

Harrigan, W. (1998). *Laboratory methods in food microbiology*. Gulf Professional Publishing.

Harrington, R. E., Borgwardt, R. H., & Potter, A. E. (1968). Reactivity of selected limestones and dolomites with sulfur dioxide. *American Industrial Hygiene Association Journal*, *29*(2), 152–158. doi:10.1080/00028896809343289 PMID:5656035

Harrison, J. P., Schratzberger, M., Sapp, M., & Osborn, A. M. (2014). Rapid bacterial colonization of low-density polyethylene microplastics in coastal sediment microcosms. *BMC Microbiology*, *14*(1), 232. doi:10.118612866-014-0232-4 PMID:25245856

Harsono, S. S., Grundmann, P., & Soebronto, S. (2014). Anaerobic treatment of palm oil mill effluents: Potential contribution to net energy yield and reduction of greenhouse gas emissions from biodiesel production. *Journal of Cleaner Production*, *64*, 619–627. doi:10.1016/j.jclepro.2013.07.056

Hasan, H. A., Abdullah, S. R. S., Kamarudin, S. K., & Kofli, N. T. (2011). Problems of ammonia and manganese in Malaysian drinking water treatments. *World Applied Sciences Journal*, *12*(10), 1890–1896.

Hasan, S. E. (2004). Public awareness is key to successful waste management. *Journal of Environmental Science and Health. Part A, Toxic/Hazardous Substances & Environmental Engineering*, *39*(2), 483–492. doi:10.1081/ESE-120027539 PMID:15027831

Hassan, S. H., Aziz, H. A., Daud, N. M., Keria, R., Noor, S. M., Johari, I., & Shah, S. M. R. (2018). The methods of waste quantification in the construction sites (A review). In *AIP Conference Proceedings*. doi:10.1063/1.5062682

Hassan, S. R., & Dahlan, I. (2013). Anaerobic wastewater treatment using anaerobic baffled bioreactor: A review. *Central European Journal of Engineering*, *3*(3), 389–399.

Hassan, S. R., Zwain, H. M., & Dahlan, I. (2013a). Development of anaerobic reactor for industrial wastewater treatment: an overview, present stage and future prospects. *Journal of Advanced Scientific Research*, *4*(1), 7–12.

Hassan, S. R., Zwain, H. M., Qamaruz Zaman, N., & Dahlan, I. (2013b). Recycled paper mill effluent treatment in a modified anaerobic baffled reactor: Start-up and steady-state performance. *Environmental Technology*, 1–6.

Hattingh, B. B., Everson, R. C., Neomagus, H. W. J. P., Bunt, J. R., van Niekerk, D., & Ashton, B. P. (2014). Modeling the nonisothermal devolatilization kinetics of typical South African coals. *Energy & Fuels*, *28*(2), 920–933. doi:10.1021/ef402124f

Haug, R. T., Stuckey, D. C., Gossett, J. M., & Carty, P. L. M. (1978). Effect of thermal pretreatment on digestibility and dewaterability of organic sludges. *Journal - Water Pollution Control Federation*, *50*, 73–85.

Haug, W., Schmidt, A., Nortemann, B., Hempel, D. C., Stolz, A., & Knackmuss, H. J. (1991). Mineralization of the sulfonated azo dye Mordant Yellow 3 by a 6-aminonaphthalene-2-sulfonate degrading bacterial consortium. *Applied and Environmental Microbiology*, *57*, 3144–3149. PMID:1781678

Haur, C. J., Kuo, L. S., Fu, C. P., Hsu, Y. L., & Da Heng, C. (2018). Feasibility study on UAV-assisted construction surplus soil tracking control and management technique. *IOP Conference Series. Materials Science and Engineering*, *301*, 012145. doi:10.1088/1757-899X/301/1/012145

Hausmann, A., Sanciolo, P., Vasiljevic, T., Ponnampalam, E., Quispe-Chavez, N., Weeks, M., & Duke, M. (2011). Direct contact membrane distillation of dairy process streams. *Membranes*, *1*(1), 48–58. doi:10.3390/membranes1010048 PMID:24957495

Havukainen, J., Zhan, M., Dong, J., Liikanen, M., Deviatkin, I., Li, X., & Horttanainen, M. (2017). (2017). Environmental impact assessment of municipal solid waste management incorporating mechanical treatment of waste and incineration in Hangzhou, China. *Journal of Cleaner Production*, *141*, 453–461. doi:10.1016/j.jclepro.2016.09.146

Hayes, D. J. M. (2013). Second-generation biofuels: Why they are taking so long. *Wiley Interdisciplinary Reviews. Energy and Environment*, *2*(3), 304–334. doi:10.1002/wene.59

Heaton, E. A., Dohleman, F. G., Miguez, A. F., Juvik, J. A., Lozovaya, V., Widholm, J., . . . Long, S. P. (2010). Chapter 3 - Miscanthus: A Promising Biomass Crop. In K. Jean-Claude & D. Michel (Eds.), Advances in botanical research (Vol. 56, pp. 75-137): Academic Press.

He, B.-J. (2018). Potentials of meteorological characteristics and synoptic conditions to mitigate urban heat island effects. *Urban Climate*, *24*, 26–33. doi:10.1016/j.uclim.2018.01.004

Heberer, T. (2002). Occurrence, fate, and removal of pharmaceutical residues in the aquatic environment: A review of recent research data. *Toxicology Letters*, *131*(1-2), 5–17. doi:10.1016/S0378-4274(02)00041-3 PMID:11988354

Hecham, M. O. (2016). Treatment of landfill waste, leachate and landfill gas: modelling/simulation and experimental studies. (Master's Thesis), The University of Western Ontario, Canada.

Heffron, J., Marhefke, M., & Mayer, B. K. (2016). Removal of trace metal contaminants from potable water by electrocoagulation. *Scientific Reports*, *6*(1), 28478. doi:10.1038rep28478 PMID:27324564

Heinfling-Weidtmann, A., Reemtsma, T., Storm, T., & Szewzyk, U. (2001). Sulfophthalimide as major metabolite formed from sulfonated phthalocyanine dyes by the white-rot fungus Bjerkanderaadusta. *FEMS Microbiology Letters*, *203*(2), 179–183. doi:10.1111/j.1574-6968.2001.tb10838.x PMID:11583845

Heng, G. C., & Isa, M. H. (2016). Enhancement of activated sludge disintegration and dewaterability by Fenton process. *IOP Conference Series: Earth and Environmental Science*, *36*, 012018. 10.1088/1755-1315/36/1/012018

Heng, G. C., Chen, K. W., & Isa, M. H. (2016, Aug. 15-17). *Enhancement of waste activated sludge disintegration and dewaterability by H_2O_2 oxidation.* Paper presented at the 3rd International Conference on Civil, Offshore and Environmental Engineering, ICCOEE 2016, KLCC.

Heng, G. C., Elmolla, E. S., & Chaudhuri, M. (2012). Optimization of photo-Fenton treatment of mature landfill leachate. *Nature Environment and Pollution Technology*, *11*(1), 65–72.

Heng, G. C., & Isa, M. H. (2014). Electrochemical disintegration of activated sludge using Ti/RuO_2 anode. *Applied Mechanics and Materials*, *567*, 44–49. doi:10.4028/www.scientific.net/AMM.567.44

Heng, G. C., Isa, M. H., & Han, L. M. (2018). Effect of H2O2 oxidation/alkaline hydrolysis on waste activated sludge disintegration and dewaterability. In *Proceedings International Conference on Civil and Environmental Engineering*. 10.1051/e3sconf/20186505021

Herguido, J., Corella, J., & Gonzalcz-Saiz, J. (1992). Steam gasification of lignocellulosic residues in a flidized bed at a small pilot scale: Effect of the type of feedstock. *Industrial & Engineering Chemistry Research, 31*(5), 1274–1282. doi:10.1021/ie00005a006

Hermosilla, D., Cortijo, M., & Huang, C. P. (2009). Optimizing the treatment of landfill leachate by conventional Fenton and photo-Fenton processes. *The Science of the Total Environment, 407*(11), 3473–3481. doi:10.1016/j.scitotenv.2009.02.009 PMID:19278717

Heukelekian, H., & Mueller, P. (1958). Transformation of some lipids in anaerobic sludge digestion. *Sewage and Industrial Wastes, 30*(9), 1108–1120.

Heutschi, K., Bühlmann, E., & Oertli, J. (2016). Options for reducing noise from roads and railway lines. *Transportation Research Part A, Policy and Practice, 94*, 308–322. doi:10.1016/j.tra.2016.09.019

Hidalgo-Ruz, V., Gutow, L., Thompson, R. C., &Thiel, M. (2012). Microplastics in the marine environment: a review of the methods used for identification and quantification. *Environ. Sci. Technol. 46*(6), 3060-3075.

Higgins, I. J., & Gilbert, P. O. (1978). The biodegradation of hydrocarbons. Charter KWA, Somerville H. J. (Eds). The Oil Industry and Microbial Ecosystems, Hegden and Sons, Ltd., London, UK. pp. 80-117.

Higman, C., & van der Burgt, M. (2008). Gasification (2nd ed.). Gulf Professional Publishing/Elsevier.

Hinkel, K, M., Frederick, E. N., Klene, A., E., & Bell, J., H. (2003). The urban heat island in winter at Barrow, Alaska. *International Journal of Climatology, 23*(15), 1889–1905. doi:10.1002/joc.971

Hinsenveld, M., Soczo, E. R., Van de Leur, G. J., Verluijs, C. W., & Groeedijk, E. (1990). Alternate physico-chemical and thermal cleaning technologies for contaminated soils. In *Proceedings of International Conference on Contaminated Soil*. Dordrecht, The Netherlands: Kluwer Academic. 873-881.

Ho, C., & Chan, C. (1986). The application of lead dioxide-coated titanium anode in the electrofloation of palm oil mill effluent. *Water Research, 20*(12), 1523–1527. doi:10.1016/0043-1354(86)90116-8

Ho, C., & Tan, Y. (1988). The treatment of anaerobically digested palm oil mill effluent by pressurized activated sludge. *Journal of Chemical Technology and Biotechnology (Oxford, Oxfordshire), 41*(1), 75–84. doi:10.1002/jctb.280410110

Hodgson, E. M., Nowakowski, D. J., Shield, I., Riche, A., Bridgwater, A. V., Clifton-Brown, J. C., & Donnison, I. S. (2011). Variation in miscanthus chemical composition and implications for conversion by pyrolysis and thermo-chemical bio-refining for fuels and chemicals. *Bioresource Technology, 102*(3), 3411–3418. doi:10.1016/j.biortech.2010.10.017 PMID:21035327

Hogg, R. (2013). Bridging flocculation by polymers. *Kona Powder and Particle Journal, 30*(0), 3–14. doi:10.14356/kona.2013005

Holkar, C. R., Jadhav, A. J., Pinjari, D. V., Mahamuni, N. M., & Pandit, A. B. (2016). A critical review on textile wastewater treatments: Possible approaches. *Journal of Environmental Management, 182*, 351–366. doi:10.1016/j.jenvman.2016.07.090 PMID:27497312

Holm-Nielsen, J. B., Al Seadi, T., & Oleskowicz-Popiel, P. (2009). The future of anaerobic digestion and biogas utilization. *Bioresource Technology, 100*(22), 5478–5484. PMID:19217772

Holt, P. K. (2006). Electrocoagulation: Unravelling and synthesising the mechanisms behind a water treatment process. (Thesis, Department of Chemical Engineering, University of Sydney, Australia).

Holt, J. G., Krieg, N. R., Sneath, P. H. A., Staley, J. T., & Williams, S. T. (2000). *Bergey's manual of determinative bacteriology* (9th ed.). Philadelphia, PA: Lippincott Williams & Wilkins.

Holt, P. K., Barton, G. W., Wark, M., & Mitchell, C. A. (2002). A quantitative comparison between chemical dosing and electrocoagulation. *Colloids and Surfaces. A, Physicochemical and Engineering Aspects*, *211*(2), 233–248. doi:10.1016/S0927-7757(02)00285-6

Hong, Y. M., Lee, B. K., Park, K. J., Kang, M. H., Jung, Y. R., Lee, D. S., & Kim, M.-G. (2002). Atmospheric nitrogen and sulfur containing compounds for three sites of South Korea. *Atmospheric Environment*, *36*(21), 3485–3494. doi:10.1016/S1352-2310(02)00289-3

Hoover, M. F. (1970). Cationic quaternary polyelectrolytes—A literature review. *Journal of Macromolecular Science. Chemistry*, *4*(6), 1327–1418. doi:10.1080/00222337008081733

Hoqmweg, D., & Bhada, T. (2012): World Bank Report. What a waste: A global review of solid waste management. Retrieved from http://documents.worldbank.org/curated/en/302341468126264791/pdf/68135-REVISED-What-a-Waste-2012-Final-updated.pdf

Hossain, M. S., Omar, F., Asis, A. J., Bachman, R. T., Sarker, M. Z. I., & Kadir, M. O. A. (2019). Effective treatment of palm oil mill effluent using FeSO4.7H2O waste from titanium oxide industry: Coagulation adsorption isotherm and kinetic studies. *Journal of Cleaner Production*, *219*, 86–98. doi:10.1016/j.jclepro.2019.02.069

Hossain, M. S., Santhanam, A., Norulaini, N. N., & Omar, A. M. (2011). Clinical solid waste management practices and its impact on human health and environment–A review. *Waste Management (New York, N.Y.)*, *31*(4), 754–766. PMID:21186116

Hossain, M. U., & Poon, C. S. (2018). Comparative LCA of wood waste management strategies generated from building construction activities. *Journal of Cleaner Production*, *177*, 387–397. doi:10.1016/j.jclepro.2017.12.233

Hosseini, S. E., & Wahid, M. A. (2014). Utilization of palm solid residue as a source of renewable and sustainable energy in Malaysia. *Renewable & Sustainable Energy Reviews*, *40*, 621–632. doi:10.1016/j.rser.2014.07.214

Houk, V. S. (1992). The genotoxicity of industrial wastes and effluents: A review. *Mutation Research: Reviews in Genetic Toxicology*, *277*(2), 91–138. doi:10.1016/0165-1110(92)90001-P PMID:1378533

Ho, Y., Norli, I., Alkarkhi, A. F., & Morad, N. (2015). Extraction, characterization and application of malva nut gum in water treatment. *Journal of Water and Health*, *13*(2), 489–499. doi:10.2166/wh.2014.100 PMID:26042980

Huacai, G., Wan, P., & Dengke, L. (2006). Graft copolymerization of chitosan with acrylic acid under microwave irradiation and its water absorbency. *Carbohydrate Polymers*, *66*(3), 372–378. doi:10.1016/j.carbpol.2006.03.017

Huang Shen, C., Bashir, M. J. K., Tan, K. T., Jocelyn, L. P. G., & Albert, F. Y. C. (2018). Design and implementation of a laboratory-scale pyrolysis combustor for biomass conversion. *Sci. Int. (Lahore)*, *30*(1), 81–84.

Huang, C., & Chen, Y. (1996). Coagulation of colloidal particles in water by chitosan. *Journal of Chemical Technology & Biotechnology: International Research in Process. Environmental AND Clean Technology*, *66*(3), 227–232.

Huang, C., & Shiu, H. (1996). Interactions between alum and organics in coagulation. *Colloids and Surfaces. A, Physicochemical and Engineering Aspects*, *113*(1-2), 155–163. doi:10.1016/0927-7757(96)03543-1

Huang, X., Gui, P., & Qian, Y. (2001). Effect of sludge retention time on microbial behaviour in a submerged membrane bioreactor. *Process Biochemistry*, *36*(10), 1001–1006. doi:10.1016/S0032-9592(01)00135-2

Huang, Y.-H., Huang, Y.-F., Chang, P.-S., & Chen, C.-Y. (2008). Comparative study of oxidation of dye-Reactive Black B by different advanced oxidation processes: Fenton, electro-Fenton and photo-Fenton. *Journal of Hazardous Materials*, *154*(1-3), 655–662. doi:10.1016/j.jhazmat.2007.10.077 PMID:18063472

Huang, Z., Ong, S. L., & Ng, H. Y. (2011). Submerged anaerobic membrane bioreactor for low-strength wastewater treatment: Effect of HRT and SRT on treatment performance and membrane fouling. *Water Research*, *45*(2), 705–713. PMID:20851448

Huerta, B., Rodríguez-Mozaz, S., & Barceló, D. (2012). Pharmaceuticals in biota in the aquatic environment: Analytical methods and environmental implications. *Analytical and Bioanalytical Chemistry*, *404*(9), 2611–2624. doi:10.100700216-012-6144-y PMID:22678760

Huisman, L., & Wood, W. E. (1974). *Slow sand filtration* (p. 44). Geneva, Switzerland: WHO.

Husain, Q. (2006). Potential applications of the oxidoreductive enzymes in the decolorization and detoxification of textile and other synthetic dyes from polluted water: A review. *Critical Reviews in Biotechnology*, *26*(4), 201–221. doi:10.1080/07388550600969936 PMID:17095432

Husna, A., Luqman, C., & Thomas, S. (2015). Microbial community analysis in anaerobic palm oil mill effluent (POME) wastewater by denaturing gradient gel electrophoresis (DGGE). *International Journal of Research in Engineering and Technology*, *4*(8), 1–9. doi:10.15623/ijret.2015.0408001

Hutnan, M., Drtil, M., Derco, J., & Mrafkova, L. (2005). Biodegradation of hexamethylenetetramine in anaerobic baffled reactor. *Polish Journal of Environmental Studies*, *14*(5), 585–591.

Hu, Y., Cheng, H., & Tao, S. (2017). Environmental and human health challenges of industrial livestock and poultry farming in China and their mitigation. *Environment International*, *107*, 111–130. doi:10.1016/j.envint.2017.07.003 PMID:28719840

Hwang, H.,, & Erhan, S. Z. (2001). Modification of epoxidized soybean oil for lubricant formulations with improved oxidative stability and low pour point. *Journal of the American Oil Chemists' Society*, *78*(12), 1179–1184.

Ibbett, R., Gaddipati, S., Davies, S., Hill, S., & Tucker, G. (2011). The mechanisms of hydrothermal deconstruction of lignocellulose: New insights from thermal–analytical and complementary studies. *Bioresource Technology*, *102*(19), 9272–9278. PMID:21763128

IBIS world. (2018). *Cooking oil recycling industry in the US*. Industry Market Research Report. September, 2018.

Ibrahim, A., Yeoh, B. G., Cheah, S. C., Ma, A. N., Ahmad, S., Chew, T. Y., & Wahid, M. J. A. (1984). Thermophilic anaerobic contact digestion of palm oil mill effluent. *Water Science and Technology*, *17*(2-3), 155–165. doi:10.2166/wst.1985.0127

Ibrahim, H., Ilinca, A., & Perron, J. (2008). Energy storage systems—characteristics and comparisons. *Renewable & Sustainable Energy Reviews*, *12*(5), 1221–1250.

Idris, M. A., Kolo, B. G., Garba, S. T., & Ismail, M. A. (2013). Physico-chemical analysis of pharmaceutical effluent and surface water of River Gorax in Minna, Niger State, Nigeria. *Bull Env Pharmacol Life Sci*, *2*(3), 45–49.

Idris, M. A., Kolo, B. G., & Sani, A. (2013). Organic pollution indicator and anion concentration of pharmaceutical effluent and surface water. In Minna, Niger State, Nigeria. [IJES]. *International Journal of Engineering Science*, *2*(6), 27–32.

Iglesias, L., Laca, A., Herrero, M., & Díaz, M. (2012). A life cycle assessment comparison between centralized and decentralized biodiesel production from raw sunflower oil and waste cooking oils. *Journal of Cleaner Production*, *37*, 162–171. doi:10.1016/j.jclepro.2012.07.002

Ijah, U. J. J., & Antai, S. P. (1988). Degradation and mineralization of crude oil by bacteria. *Nigerian Journal of Biotechnology*, *5*, 79–86.

Ikhlayel, M. (2018). An integrated approach to establish e-waste management systems for developing countries. *Journal of Cleaner Production*, *170*, 119–130.

Ilhan, S., Nourbakhsh, M. N., Kiliçarslan, S., & Ozdag, H. (2004). Removal of chromium, lead and copper ions from industrial waste waters by Staphylococcus saprophyticus. *Turkish Electronic Journal of Biotechnology*, *2*(2), 50–57.

Ilies, P., & Mavinic, D. S. (2001). The effect of decreased ambient temperature on the biological nitrification and denitrification of a high ammonia landfill leachate. *Water Research*, *35*(8), 2065–2072. doi:10.1016/S0043-1354(00)00477-2 PMID:11337855

Iliuta, I., & Larachi, F. (2001). Wet air oxidation solid catalysis analysis of fixed and sparged three-phase reactors. *Chemical Engineering and Processing: Process Intensification*, *40*(2), 175–185. doi:10.1016/S0255-2701(00)00137-9

Imamura, T., Kozaki, S., Kuriyama, A., Kawaguchi, M., Touge, Y., Yano, T., ... Kawabata, Y. (1997). Bioaugmentation of TCE contaminated soil with inducer-free microbes. In G. S. Sayler, J. Sanseverino, & K. L. Davis (Eds.), *Biotechnology in the Sustainable Environment* (pp. 97–106). New York: Plenum Press. doi:10.1007/978-1-4615-5395-3_10

Imhof, H. K., Ivleva, N. P., Schmid, J., Niessner, R., & Laforsch, C. (2015). Contamination of beach sediments of a subalpine lake with microplastic particles. *Current Biology*, *23*(19), R867–R868. doi:10.1016/j.cub.2013.09.001 PMID:24112978

India: Air Quality Profile (IAQP). (2010).Clean Air Initiative for Asian Cities (CAI-Asia) Center, Pasig City, Philippines.

Ingersoll, T. L., & Baker, L. A. (1998). Nitrate removal in wetland microcosms. *Water Research*, *32*(3), 677–684. doi:10.1016/S0043-1354(97)00254-6

Intergovernmental Panel on Climate Change. (2008). Climate change and water. (B. Bates, Z. W. Kundzewicz, J. Palutikof, & S. Wu, Eds.), IPCC Technical Paper VI (Vol. 403). Geneva.

Ioana, I. (2010). *Clean technology from waste management.* Retrieved from https://www.researchgate.net/profile/Ioana_Ionel/publication/228905208/inline/jsViewer/00b7d5194fd1749cce000000

Isa, M. H., Ezechi, E. H., Ahmed, Z., Magram, S. F., & Kutty, S. R. M. (2014). Boron removal by electrocoagulation and recovery. *Water Research*, *51*, 113–123. doi:10.1016/j.watres.2013.12.024 PMID:24412846

Isik, M., & Sponza, D. (2004). Monitoring of toxicity and intermediated of C. I. Direct Black 38 azo dye through decolorization in an anaerobic/aerobic sequential reactor system. *Journal of Hazardous Materials*, *114*(1-3), 29–39. doi:10.1016/j.jhazmat.2004.06.011 PMID:15511571

Islam, M. T., Saenz-Arana, R., Hernandez, C., Guinto, T., Ahsan, M. A., Kim, H., ... Noveron, J. C. (2018). Adsorption of methylene blue and tetracycline onto biomass-based material prepared by sulfuric acid reflux. *RSC Advances*, *8*(57), 32545–32557. doi:10.1039/C8RA05395B

Ismail, K., Yarmo, M. A., Taufiq-Yap, Y. H., & Ahmad, A. (2011). Calcium oxide as potential catalyst for gasification of palm oil empty fruit bunch to produce syngas. In Z. A. Ahmad, M. A. Yarmo, F. H. A. Aziz, M. Y. M. Sulaiman, B. Ahmad, K. N. Ismail, & N. A. Rejab (Eds.), *X-Ray and Related Techniques* (Vol. 173, p. 178).

Ismail, K., Yarmo, M. A., Taufiq-Yap, Y., & Ahmad, A. (2012a). The effect of particle size of CaO and MgO as catalysts for gasification of oil palm empty fruit bunch to produce hydrogen. *International Journal of Hydrogen Energy*, *37*(4), 3639–3644. doi:10.1016/j.ijhydene.2011.05.100

Ismail, N., & Ani, F. (2014). Syngas production from microwave gasification of oil palm biochars. *Carbon*, *64*(1), 74–90.

Israel Ministry of Environmental Protection. (2018). *Agricultural Wastewater*. Retrieved from http://www.sviva.gov.il/English/env_topics/agroecology/Pages/Agricultural-Wastewater.aspx#GovXParagraphTitle1

Ito, K., Jian, W., Nishijima, W., Baes, A. U., Shoto, E., & Okada, M. (1998). Comparison of ozonation and AOPs combined with biodegradation for removal of THM precursors in treated sewage effluents. *Water Science and Technology*, *38*(7), 179–186. doi:10.2166/wst.1998.0291

Izabela, A. T., & Biedka, P. (2016). Use of the landfill water pollution index (LWPI) for groundwater quality assessment near the landfill sites. *Environmental Science and Pollution Research International*, *24*, 24601–24613. PMID:27640059

Jabatan, P., & Perumahan, D. K. T. (2013). *Survey on solid waste composition*. Characteristics & existing practice of solid waste recycling in Malaysia.

Jabeen, K., Su, L., Li, J., Yang, D., Tong, C., Mu, J., & Shi, H. (2017). Microplastics and mesoplastics in fish from coastal and fresh waters of China. *Environmental Pollution*, *221*, 141–149. doi:10.1016/j.envpol.2016.11.055 PMID:27939629

Jabeen, M. (2018). *Thoughtful intelligence a practical guide for moral development: a book guide for moral development*. AuthorHouse.

Jacobs, L. E., Weavers, L. K., & Chin, Y. P. (2008). Direct and indirect photolysis of polycyclic aromatic hydrocarbons in nitrate-rich surface waters. *Environmental Toxicology and Chemistry*, *27*(8), 1643–1648. doi:10.1897/07-478.1 PMID:18275255

Jacobson, M. Z. (1998). Studying the effects of aerosols on vertical photolysis rate coefficient and temperature profiles over an urban airshed. *Journal of Geophysical Research*, *103*(D9), 10593–10604. doi:10.1029/98JD00287

Jadhav, S. U., Jadhav, M. U., Kagalkar, A. N., & Govindwar, S. P. (2008). Decolorization of brilliant blue G Dye mediated by the degradation of the microbial consortium of Galactomyces geotrichum and Bacillus sp. *J Chin Inst Chem Engrs*, *39*(6), 563–570. doi:10.1016/j.jcice.2008.06.003

Jafari, M., & Aghamiri, S. (2011). Evaluation of carbon nanotubes as solid-phase extraction sorbent for the removal of cephalexin from aqueous solution. *Desalination and Water Treatment*, *28*(1-3), 55–58. doi:10.5004/dwt.2011.2200

Jahn, S. A. A. (1988). Using Moringa seeds as coagulants in developing countries. *Journal - American Water Works Association*, *80*(6), 43–50. doi:10.1002/j.1551-8833.1988.tb03052.x

Jahn, S. A., & Dirar, H. (1979). Studies on natural water coagulants in the Sudan, with special reference to Moringa oleifera seeds. *Water S.A.*, *5*(2), 90–106.

Jalaei, F., Zoghi, M., & Khoshand, A. (2019). Life cycle environmental impact assessment to manage and optimize construction waste using Building Information Modeling (BIM). *International Journal of Construction Management*, 1–18. doi:10.1080/15623599.2019.1583850

James, L. W. (2006). *Soil sampling and analysis. The University of Arizona College of Agriculture and Life Sciences, Tucson, AZ* (p. 85721). Arizona: Arizona Cooperative Extension.

Jamieson, R., Gordon, R., Sharples, K., Stratton, G., & Madani, A. (2002). Movement and persistence of fecal bacteria in agricultural soils and subsurface drainage water: A review. *Canadian Biosystems Engineering*, *44*(1), 1–9.

Janni, K. A., Christopherson, S. H., & Schmidt, D. R. (2009). Milk house wastewater flows and characteristics for small dairy operations. *Applied Engineering in Agriculture*, *25*(3), 417–423. doi:10.13031/2013.26893

Janse, A., de Jong, X. A., Prins, W., & van Swaaij, W. P. (1999). Improvement and validation of the fluoroptic temperature-measurement method. *AIChE Journal. American Institute of Chemical Engineers*, *45*(10), 2087–2096. doi:10.1002/aic.690451007

Jauregui, E., & Luyando, E. (1999). Global radiation attenuation by air pollution and its effects on the thermal climate in Mexico City. *International Journal of Climatology*, *19*(6), 683–694. doi:10.1002/(SICI)1097-0088(199905)19:6<683::AID-JOC389>3.0.CO;2-8

Jemaat, Z. (2013). *Treatment of complex industrial wastewaters containing ammonium and phenolic compounds using granular sludge in continuous airlift reactors.* (Doctoral thesis). Universitat Autonoma de Barcelona, Barcelona, Spain.

Jenicek, P., Dohanyos, M., & Zabranska, J. (1999). Combined anaerobic treatment of wastewaters and sludges. *Water Science and Technology*, *40*(1), 85–91. doi:10.2166/wst.1999.0019

Jenkins, B., Mannapperuma, J., & Bakker, R. (2003). Biomass leachate treatment by reverse osmosis. *Fuel Processing Technology*, *81*(3), 223–246. doi:10.1016/S0378-3820(03)00010-9

Jenkins, M. W., Tiwari, S. K., Darby, J., Nyakash, D., Saenyi, W., & Langenbach, K. (2009). *The biosand filter for improved drinking water quality in high risk communities in the Njoro watershed, Kenya. Research Brief 09-06-SUMAWA, Global Livestock Collaborative Research Support Program.* Davis, CA: University of California.

Jeppesen, E., Kronvang, B., Meerhoff, M., Søndergaard, M., Hansen, K. M., & Andersen, H. E., … Olesen, J. E. (2009). Climate change effects on runoff, catchment phosphorus loading and lake ecological state, and potential adaptations. *Journal of Environmental Quality*, *38*(5), 1930. doi:10.2134/jeq2008.0113

Jereme, I. A., Siwar, C., & Bhuiyan, M. A. H. (2013). Incineration and its implications: The need for a sustainable waste management system in Malaysia. *International Journal of Environmental Sciences, 4,* 3.

Jes, F. (1999). Urban air quality. *Atmospheric Environment*, *33*(29), 4877–4900. doi:10.1016/S1352-2310(99)00290-3

Jeyanayagam, S. (2005). True confessions of the biological nutrient removal process. *Florida Water Resources Journal*, *1*, 37–46.

Jiang, J. (2018) Occurrence of microplastics and its pollution in the environment: A review. *Sustainable Production and Consumption* (13), 16–23.

Jiang, C., & Zhang, J. (2007). Progress and prospect in electro-Fenton process for wastewater treatment. *Journal of Zhejiang University. Science A*, *8*(7), 1118–1125. doi:10.1631/jzus.2007.A1118

Jiang, J.-Q., Graham, N., André, C., Kelsall, G. H., & Brandon, N. (2002). Laboratory study of electro-coagulation–flotation for water treatment. *Water Research*, *36*(16), 4064–4078. doi:10.1016/S0043-1354(02)00118-5 PMID:12405415

Jiang, M., Jin, X.-Y., Lu, X.-Q., & Chen, Z. (2010). Adsorption of Pb (II), Cd (II), Ni (II) and Cu (II) onto natural kaolinite clay. *Desalination*, *252*(1-3), 33–39. doi:10.1016/j.desal.2009.11.005

Jiang, W., Wei, Y., Gao, X., Gao, C., & Wang, Y. (2015). An innovative backwash cleaning technique for NF membrane in groundwater desalination: Fouling reversibility and cleaning without chemical detergent. *Desalination*, *359*, 26–36. doi:10.1016/j.desal.2014.12.025

Jianlong, W., & Ning, Y. (2004). Partial nitrification under limited dissolved oxygen conditions. *Process Biochemistry*, *39*(10), 1223–1229. doi:10.1016/S0032-9592(03)00249-8

Jianlong, W., Yi, Q., Horan, N., & Stentiford, E. (2000). Bioadsorption of pentachlorophenol (PCP) from aqueous solution by activated sludge biomass. *Bioresource Technology*, *75*(2), 157–161. doi:10.1016/S0960-8524(00)00041-9

Jianwei, M., Xingdong, Y., & Gulan, Z. (1997). Studies on the decolorization of azo dyes in the aqueous system using metal ion together with algae (*Saragassum*). *Huanj. Huax. Chin*, *16*, 238–240.

Jiménez, B. (2005). Treatment technology and standards for agricultural wastewater reuse: a case study in Mexico. *Irrigation and Drainage: The Journal of the International Commission on Irrigation and Drainage*, *54*(S1), S23-S33.

Jimenez, C. C. (2012). Pharmaceutical products in the environment: Sources, effects and risks. *Vitae*, *19*(1), 93–108.

Jin, B., Wilen, B. M., & Lant, P. (2004). Impacts of morphological, physical and chemical properties of sludge flocs on dewaterability of activated sludge. *Chemical Engineering Journal*, *98*(1-2), 115–126. doi:10.1016/j.cej.2003.05.002

Jinqi, L., & Houtain, L. (1992). Degradation of azo dyes by algae. *Environ. Pollut*, *75*(3), 273–278. doi:10.1016/0269-7491(92)90127-V PMID:15092015

Jin, R., Yang, H., Zhang, A., Wang, J., & Liu, G. (2009). Bioaugmentation on decolorization of C.I.Direct Blue 71 by using genetically engineered strain *Escherichia coli* JM109 (PGEX-AZR). *Journal of Hazardous Materials*, *163*(2-3), 1123–1128. doi:10.1016/j.jhazmat.2008.07.067 PMID:18755538

Jin, R., Yuan, H., & Chen, Q. (2019). Science mapping approach to assisting the review of construction and demolition waste management research published between 2009 and 2018. *Resources, Conservation and Recycling*, *140*, 175–188. doi:10.1016/j.resconrec.2018.09.029

Jin, X., Liu, G., Xu, Z., & Yao, W. (2007). Decolorization of a dye industry effluent by Aspergillus fumigatus XC6. *Applied Microbiology and Biotechnology*, *74*(1), 239–243. doi:10.100700253-006-0658-1 PMID:17086413

Jin, Y., Li, H., Mahar, R. B., Wang, Z., & Nie, Y. (2009). Combined alkaline and ultrasonic pretreatment of sludge before aerobic digestion. *Journal of Environmental Sciences (China)*, *21*(3), 279–284. doi:10.1016/S1001-0742(08)62264-0 PMID:19634436

Jirka, A. M., & Carter, M. J. (1975). Micro semiautomated analysis of surface and waste waters for chemical oxygen demand. *Analytical Chemistry*, *47*(8), 1397–1402. doi:10.1021/ac60358a004 PMID:1147260

Johari, A., Alkali, H., Hashim, H., Ahmed, S. I., & Mat, R. (2014). Municipal solid waste management and potential revenue from recycling in Malaysia. *Modern Applied Science*, *8*, 4. doi:10.5539/mas.v8n4p37

Jones, J. C. (Ed.). (2014). *Thermal processing of waste.* Retrieved from bookboon.com

Jorgensen, T., & Weatherley, L. (2003). Ammonia removal from wastewater by ion exchange in the presence of organic contaminants. *Water Research*, *37*(8), 1723–1728. doi:10.1016/S0043-1354(02)00571-7 PMID:12697216

Joseph, J. M., Destaillats, H., Hung, H. M., & Hoffmann, M. R. (2000). The sonochemical degradation of azobenzene and related azo dyes: Rate enhancements via Fenton's reactions. *The Journal of Physical Chemistry A*, *104*(2), 301–307. doi:10.1021/jp992354m

Joseph, L., Heo, J., Park, Y.-G., Flora, J. R., & Yoon, Y. (2011). Adsorption of bisphenol A and 17α-ethinyl estradiol on single walled carbon nanotubes from seawater and brackish water. *Desalination*, *281*, 68–74. doi:10.1016/j.desal.2011.07.044

Joshi, J. M., & Sinha, V. K. (2007). Ceric ammonium nitrate induced grafting of polyacrylamide onto carboxymethyl chitosan. *Carbohydrate Polymers*, *67*(3), 427–435. doi:10.1016/j.carbpol.2006.06.021

Joshi, T., Iyengar, L., Singh, K., & Garg, S. (2008). Isolation, Identification, and application of novel bacterial consortium TJ-1 for the decolorization of structurally different azo dyes. *Bioresource Technology*, *99*(15), 7115–7121. doi:10.1016/j.biortech.2007.12.074 PMID:18289845

Jouhara, H., Ahmad, D., van den Boogaert, I., Katsou, E., Simons, S., & Spencer, N. (2018). Pyrolysis of domestic based feedstock at temperatures up to 300 C. *Thermal Science and Engineering Progress*, *5*, 117–143.

Jozala, A. F., Geraldes, D. C., Tundisi, L. L., Feitosa, V. D. A., Breyer, C. A., Cardoso, S. L., ... & Oliveira, M. A. D. (2016). Biopharmaceuticals from microorganisms: from production to purification. *Brazilian Journal of Microbiology, 47*, 51-63.

JPSPN. (2010). *Distribution of landfill sites in Malaysia*. Kuala Lumpur, Malaysia.

Judd, S. (2011). *The MBR book: principles and applications of membrane bioreactors for water and wastewater treatment*. UK: Elsevier Science.

Junior, G. D., Ibadurrohman, M., & Slamet. (2019). Synthesis of eco-friendly liquid detergent from waste cooking oil and ZnO nanoparticles. In *AIP Conference Proceedings* (vol. 2085(1), 020075). doi:. doi:10.1063/1.5095053

Jusof Khadidi, M., & Hamid, E. A. (2013). *A new flocculant-coagulant with potential use for industrial wastewater treatment*. Paper presented at the 2nd International Conference on Environment, Energy and Biotechnology.

Jusoh, A., Hartini, W., & Endut, A. (2011). Study on the removal of pesticide in agricultural run off by granular activated carbon. *Bioresource Technology*, *102*(9), 5312–5318. doi:10.1016/j.biortech.2010.12.074 PMID:21232934

Justes, E., Beaudoin, N., Bertuzzi, P., Charles, R., Constantin, J., Dürr, C., . . . Mary, B. (2012). The use of cover crops to reduce nitrate leaching. Effect on the water and nitrogen balance and other ecosystem services. In *Proceedings of the International Fertiliser Society meeting,* Cambridge, UK, Dec. 7, 2012.

Ju, Y., Hong, I., & Hong, S. (2015). Multiple MFI measurements for the evaluation of organic fouling in SWRO desalination. *Desalination*, *365*, 136–143. doi:10.1016/j.desal.2015.02.035

Jyoti, J., Alka, D., & Kumar, S. J. (2013). Application of membrane-bio-reactor in waste-water treatment: A review. *International Journal of Chemistry and Chemical Engineering*, *3*(2), 115–122.

Kabir, I., Yacob, M. R., & Radam, A. (2014). Households' awareness, attitudes and practices regarding waste cooking oil recycling in petaling, Malaysia. *IOSR-Journal of Environmental Science. Toxicology and Food Technology*, *8*(10), 45–51.

Kadir, S. A. S. A., Yin, C.-Y., Sulaiman, M. R., Chen, X., & El-Harbawi, M. (2013). Incineration of municipal solid waste in Malaysia: Salient issues, policies and waste-to-energy initiatives. *Renewable & Sustainable Energy Reviews*, *24*, 181–186. doi:10.1016/j.rser.2013.03.041

Kafle, S., Parajuli, R., Euh, S. H., Oh, K. C., Choi, Y. S., Adhikari, K., ... Kim, D. H. (2016). Potential biomass supply for agro-pellet production from agricultural crop residue in Nepal. *Energy Sources. Part A, Recovery, Utilization, and Environmental Effects*, *38*(1), 149–153. doi:10.1080/15567036.2015.1043474

Kaliannan, S., Nagapan, S., Abdullah, A. H., Sohu, S., & Jhatial, A. A. (2018). Determining root cause of construction waste generation: A global context. *Civil Engineering Journal*, *4*(11), 2539–2547. doi:10.28991/cej-03091179

Kalme, S. D., Parshetti, G. K., Jadhav, S. U., & Govindwar, S. P. (2007). Biodegradation of Benzidine based dye Direct Blue-6 by Pseudomonas desmolyticum NCIM 2112. *Bioresource Technology*, *98*(7), 1405–1410. doi:10.1016/j.biortech.2006.05.023 PMID:16822666

Kalyani, D. C., Telke, A. A., Dhanve, R. S., & Jadhav, J. P. (2009). Eco-friendly biodegradation and detoxification of Reactive Red-2 textile dye by newly isolated Pseudomonas sp. SUK1. *Journal of Hazardous Materials*, *163*(2-3), 735–742. doi:10.1016/j.jhazmat.2008.07.020 PMID:18718713

Kamaruddin, M. F. A. (2014). *Bio-oil production by pyrolysis of lalang grass. (Master of Chemical Engineering)*. Johor Bahru: Universiti Teknologi Malaysia.

Kamaruddin, M. A., Yusoff, M. S., Aziz, H. A., & Basri, N. K. (2013). Removal of COD, ammoniacal nitrogen, and colour from stabilized landfill leachate by anaerobic organism. *Applied Water Science*, *3*(2), 359–366. doi:10.100713201-013-0086-1

Kamyab, H., Chelliapan, S., Din, M. F. M., Rezania, S., Khademi, T., & Kumar, A. (2018). Palm oil mill effluent as an environmental pollutant. *Palm Oil*, 13.

Kanadasan, J., & Razak, H. A. (2015). Engineering and sustainability performance of selfcompacting palm oil mill incinerated waste concrete. *Journal of Cleaner Production*, *89*, 78–86. doi:10.1016/j.jclepro.2014.11.002

Kanan, R., Elhassan, O., & Bensalem, R. (2018). An IoT-based autonomous system for workers' safety in construction sites with real-time alarming, monitoring, and positioning strategies. *Automation in Construction, 88*(November, 2016), 73–86. doi:10.1016/j.autcon.2017.12.033

Kandelbauer, A., Erlacher, A., Cavaco-Paulo, A., & Guebitz, G. M. (2004). Laccase-catalyzed decolorization of the synthetic Azo-dye Diamond Black PV 200 and of some structurally related derivatives. *Biocatalysis and Biotransformation*, *22*(5-6), 331–339. doi:10.1080/10242420400024573

Kang, J.-S., Sung, S. C., Lee, J. J., & Kim, H.-S. (2016). Application of ceramic membrane for seawater desalination pretreatment. *Desalination and water treatment*, *3994*(November), 1–6. doi:10.1080/19443994.2016.1189702

Kang, Y. W., Cho, M. J., & Hwang, K. Y. (1999). Correction of hydrogen peroxide interference on standard chemical oxygen demand test. *Water Research*, *33*(5), 1247–1251. doi:10.1016/S0043-1354(98)00315-7

Kang, Y. W., & Hwang, K.-Y. (2000). Effects of reaction conditions on the oxidation efficiency in the Fenton process. *Water Research*, *34*(10), 2786–2790. doi:10.1016/S0043-1354(99)00388-7

Kapdan, I. K., & Kargi, F. (2002). Biological decolorization of textile dyestuff containing wastewater by Coriolus versicolor in a rotating biological contactor. *Enzyme and Microbial Technology*, *30*(2), 195–199. doi:10.1016/S0141-0229(01)00468-9

Kapdan, I., Kargi, F., Mcmullan, G., & Marchant, R. (2000). Comparison of white rot fungi cultures for decolorization of textile dyestuffs. *Bioprocess Engineering*, *22*(4), 347–351. doi:10.1007004490050742

Karabelas, A. J., & Sioutopoulos, D. C. (2014). Toward improvement of methods for predicting fouling of desalination membranes— The effect of permeate flux on specific fouling resistance. *Desalination*, *343*, 97–105. doi:10.1016/j.desal.2013.10.008

Karakashev, D., Batstone, D., & Angelidaki, I. (2005). Influence of environmental conditions on methanogenic compositions in anaerobic biogas reactors. *Applied and Environmental Microbiology*, *71*(1), 331–338. doi:10.1128/AEM.71.1.331-338.2005 PMID:15640206

Karaosmanoglu, F., & Tetik, E. (2012). Fuel properties of pyrolytic oil of the straw and stalk of rape plant. *Renewable Energy*, *1999*(16), 1090–1093.

Karar, K., Gupta, A. K., Kumar, A., & Biswas, A. K. (2006). Characterization and identification of the sources of chromium, zinc, lead, cadmium, nickel, manganese and iron in pm10 particulates at the two sites of Kolkata, India. *Environmental Monitoring and Assessment*, *120*(1-3), 347–360. doi:10.100710661-005-9067-7 PMID:16741802

Kargi, F., & Uygur, A. (2003). Effect of carbon source on biological nutrient removal in a sequencing batch reactor. *Bioresource Technology*, *89*(1), 89–93. doi:10.1016/S0960-8524(03)00031-2 PMID:12676505

Karl, T. R., & Trenberth, K. E. (2003). Modern global climate change. *Science*, *302*(5651), 1719–1723. doi:10.1126cience.1090228 PMID:14657489

Kaspar, T., & Singer, J. (2011). The use of cover crops to manage soil.

Katamine, N. M. (2000). Physical and mechanical properties of bituminous mixtures containing oil shales. *Journal of Transportation Engineering*, *126*(2), 178–184.

Kataria, J., Mohapatra, S. K., & Kundu, K. (2019). Biodiesel production from waste cooking oil using heterogeneous catalysts and its operational characteristics on variable compression ratio CI engine. *Journal of the Energy Institute*, *92*(2), 275–287. doi:10.1016/j.joei.2018.01.008

Katarzyna, S., Irena, M., & Teresa, K. (2004). Polycyclic aromatic hydrocarbons: physicochemical properties, environmental appearance and impact on living organisms. *Acta Poloniae Pharceutica-Drug Research.*, *61*(3), 233–240. PMID:15481250

Katayon, S., Ng, S., Johari, M. M., & Ghani, L. A. (2006). Preservation of coagulation efficiency of Moringa oleifera, a natural coagulant. *Biotechnology and Bioprocess Engineering; BBE*, *11*(6), 489–495. doi:10.1007/BF02932072

Kato, T., Kuroda, H., & Nakasone, H. (2009). Runoff characteristics of nutrients from an agricultural watershed with intensive livestock production. *Journal of Hydrology (Amsterdam)*, *368*(1-4), 79–87. doi:10.1016/j.jhydrol.2009.01.028

Kaushal, P., Abedi, J., & Mahinpey, N. (2010). A comprehensive mathematical model for biomass gasification in a bubbling fluidized bed reactor. *Fuel*, *89*(12), 3650–3661. doi:10.1016/j.fuel.2010.07.036

Kaushik, P., & Malik, A. (2009). Fungal dye decolorization: Recent advances and future potential. *Environment International*, *35*(1), 127–141. doi:10.1016/j.envint.2008.05.010 PMID:18617266

Kaushik, P., & Malik, A. (2014). Mycoremediation of synthetic dyes: An insight into the mechanism, process optimization and reactor design. In S. N. Singh (Ed.), *Microbial degradation of synthetic dyes in wastewaters* (pp. 1–25). Switzerland: Springer International Publishing.

Kavitha, V., & Palanivelu, K. (2005). Destruction of cresols by Fenton oxidation process. *Water Resources*, *39*, 3062–3072. PMID:15992853

Kawahara, M., & Kato-Negishi, M. (2011). Link between aluminum and the pathogenesis of Alzheimer's disease: The integration of the aluminum and amyloid cascade hypotheses. *International Journal of Alzheimer's Disease*, 2011. PMID:21423554

Kaynak, G. E., & Filibeli, A. (2008). Assessment of Fenton process as a minimization technique for biological sludge: Effects on anaerobic sludge bioprocessing. *Journal of Residuals Science and Technology*, *5*, 151–160.

Kazimuddin, A., Banerjee, L. (2000). Fighting for Air. *Down to Earth*, July 31.

Kazuo, S. US Pat. 4839089, Mimasu Oil Chemical Co., Ltd., (1989).

Kelewou, H., Merzouki, M., & Lhassani, A. (2014). Biosorption of textile dyes Basic Yellow 2 (BY2) and Basic Green 4 (BG4) by the live yeast S.cerevisiae. *J. Mater. Environ. Sci*, *5*, 633–640.

Kellogg, R. L., Nehring, R. F., Grube, A., Goss, D. W., & Plotkin, S. (2002). *Environmental indicators of pesticide leaching and runoff from farm fields. Agricultural Productivity* (pp. 213–256). Springer.

Kennedy, S. B. (2018). Global tracking framework sustainable energy for all. Retrieved from http://documents.worldbank.org/curated/en/603241469672143906/pdf/778890GTF0full0report.pdf

Kensa, V. M. (2011). Bioremediation-an overview. *Jr. of Industrial Pollution Control*, *27*(2), 161–168.

Kent, F. C., Farahbakhsh, K., Mahendran, B., Jaklewicz, M., Liss, S. N., & Zhou, H. (2011). Water reclamation using reverse osmosis: Analysis of fouling propagation given tertiary membrane filtration and MBR pretreatments. *Journal of Membrane Science*, *382*(1-2), 328–338. doi:10.1016/j.memsci.2011.08.028

Keramati, M., & Ayati, B. (2019). Petroleum wastewater treatment using a combination of electrocoagulation and photocatalytic process with immobilized ZNO nanoparticles on concrete surface. *Process Safety and Environmental Protection*, *126*, 356–365. doi:10.1016/j.psep.2019.04.019

Kershaw, P. J. (Ed.). (2015). Sources, fate, and effects of microplastics in the marine environment: a global assessment (IMO/FAO/UNESCO-IOC/UNIDO/WMO/IAEA/UN/UNEP/UNDP Joint Group of Experts on the Scientific Aspects of Marine Environmental Protection). Rep. Stud. GESAMP No. 90.

Kessler, R. (2010). Industry issues: Pharmaceutical factories as a source of drugs in water.

Kevin, C., Christian, H., Barry, L., & Joyce, R. (2007). Estimating the effects of increased urbanization on surface meteorology and ozone concentrations in the New York City metropolitan region. *Atmospheric Environment*, *41*(9), 1803–1818. doi:10.1016/j.atmosenv.2006.10.076

Khadidi, M. H. J., Al-Shorgani, N. K., Ali, E., Hamid, A. A., & Kalil, M. (2013). A new flocculant-coagulant with potential use for industrial wastewater treatment. In *Proceedings of the 2nd international Conference on Environment, Energy and Biotechnology (ICEEB), Singapore; IACSIT Press*, 139-142.

Khadijah, O., Lee, K. K., & Faiz, F. M., & Abdullah, M. F. F. (2009). Isolation, screening, and development of local bacterial consortia with azo dyes decolorizing capability. *Malaysian Journal of Microbiology*, *5*(1), 25–32.

Khalaf, S., Al-Rimawi, F., Khamis, M., Nir, S., Bufo, S. A., Scrano, L., ... Karaman, R. (2013). Efficiency of membrane technology, activated charcoal, and a micelle-clay complex for removal of the acidic pharmaceutical mefenamic acid. *Journal of Environmental Science and Health. Part A*, *48*(13), 1655–1662. PMID:23947703

Khalid, K., & Khalid, K. (2011). Transesterification of palm oil for the production of biodiesel. *American Journal of Applied Sciences*, *8*(8), 804–809. doi:10.3844/ajassp.2011.804.809

Khan, F. R., Mayoma, B. S., Biginagwa, F. J., & Syberg, K. (2017). Microplastics in inland African waters: presence, sources, and fate. *Freshwater microplastics,* 101-124.

Khan, U. K. (2015). Combined industrial wastewater treatment in constructed wetland systems containing emergent plants and algae.

Khan, A. A., Gaur, R. Z., Tyagi, V. K., Khursheed, A., Lew, B., Mehrotra, I., & Kazmi, A. A. (2011). Sustainable options of post treatment of UASB effluent treating sewage: A review. *Resources, Conservation and Recycling*, *55*(12), 1232–1251. doi:10.1016/j.resconrec.2011.05.017

Khanal, S. K. (2011). *Anaerobic biotechnology for bioenergy production: principles and applications*. Hoboken, NJ: John Wiley & Sons.

Khandegar, V., Acharya, S., & Jain, A. K. (2018). Data on treatment of sewage wastewater by electrocoagulation using punched aluminum electrode and characterization of generated sludge. *Data in Brief*, *18*, 1229–1238. doi:10.1016/j.dib.2018.04.020 PMID:30050967

Khandelwal, S., Goyal, R., Kaul, N., & Mathew, A. (2017). Assessment of land surface temperature variation due to change in elevation of area surrounding Jaipur, India. *The Egyptian Journal of Remote Sensing and Space Sciences*, *21*(4), 87–94.

Khan, M. A. M., Rehman, S., & Al-Sulaiman, F. A. (2018). A hybrid renewable energy system as a potential energy source for water desalination using reverse osmosis: A review. *Renewable & Sustainable Energy Reviews*, *97*(January), 456–477. doi:10.1016/j.rser.2018.08.049

Khataee, A., Vatanpour, V., & Ghadim, A. A. (2009). Decolorization of CI Acid Blue 9 solution by UV/Nano-TiO2, Fenton, Fenton-like, electro-Fenton and electrocoagulation processes: A comparative study. *Journal of Hazardous Materials*, *161*(2-3), 1225–1233. doi:10.1016/j.jhazmat.2008.04.075 PMID:18524478

Kheang, L. S., May, C. Y., Foon, C. S., & Ngan, M. A. (2006). Recovery and conversion of palm olein-derived used frying oil to methyl esters for biodiesel. *Journal of Oil Palm Research*, *18*, 247–252.

Khezami, L., & Capart, R. (2005). Removal of chromium (VI) from aqueous solution by activated carbons: Kinetic and equilibrium studies. *Journal of Hazardous Materials*, *123*(1-3), 223–231. doi:10.1016/j.jhazmat.2005.04.012 PMID:15913888

Khiari, D. (2007). Endocrine disruptors, pharmaceuticals and personal care products in drinking water: An overview of AWWARF research to date. *Drinking Water Research*, *17*(2), 2–7.

Khodadadi, A., Hossein, G., & Seyed, R. S. N. (2012). Treatment of crude-oil contaminated soil using biosurfactants. *Journal of Petroleum and Gas Engineering*, *3*(6), 92–98. Retrieved from http://www.academicjournals.org/app/webroot/article/article1379670453_Khodadadi%20et%20al.pdf

Khoder, M. I. (2002). Atmospheric conversion of sulfur dioxide to particulate sulfate and nitrogen dioxide to particulate nitrate and gaseous nitric acid in an urban area. *Chemosphere*, *49*(6), 675–684. doi:10.1016/S0045-6535(02)00391-0 PMID:12430655

Khosla, E., Kaur, S., & Dave, P. N. (2013). Mechanistic study of adsorption of acid orange-7 over aluminum oxide nanoparticles. *Journal of Engineering*, *2013*, 1–8. doi:10.1155/2013/593534

Khublaryan, M. G. (2009). Types and properties of water. Encyclopedia of Life Support Systems (EOLSS), 1.

Kikuchi, T., & Tanaka, S. (2012). Biological removal and recovery of toxic heavy metals in water environment. *Critical Reviews in Environmental Science and Technology*, *42*(10), 1007–1057. doi:10.1080/10643389.2011.651343

Kim, D. J., An, H. Y., Yoon, J. H., Park, Y. H., Kawai, F., & Jung, C. M. (2002). Identification of Clostridium perfringens AB and J and its uptake of Bromophenol Blue. *Journal of Microbiology and Biotechnology*, *12*, 544–552.

Kim, D., Ryu, H.-D., Kim, M.-S., Kim, J., & Lee, S.-I. (2007). Enhancing struvite precipitation potential for ammonia nitrogen removal in municipal landfill leachate. *Journal of Hazardous Materials*, *146*(1-2), 81–85. doi:10.1016/j.jhazmat.2006.11.054 PMID:17208368

Kim, J., Park, C., Kim, T. H., Lee, M., Kim, S., Kim, S. W., & Lee, J. (2003). Effects of various pretreatments for enhanced anaerobic digestion with waste activated sludge. *Journal of Bioscience and Bioengineering*, *95*(3), 271–275. doi:10.1016/S1389-1723(03)80028-2 PMID:16233404

Kim, K.-W., Shon, W.-J., Oh, M.-K., Yang, D., Foster, R. I., & Lee, K.-Y. (2019). Evaluation of dynamic behavior of coagulation-flocculation using hydrous ferric oxide for removal of radioactive nuclides in wastewater. *Nuclear Engineering and Technology*, *51*(3), 738–745. doi:10.1016/j.net.2018.11.016

Kim, S. D., Cho, J., Kim, I. S., Vanderford, B. J., & Snyder, S. A. (2007). Occurrence and removal of pharmaceuticals and endocrine disruptors in South Korean surface, drinking, and waste waters. *Water Research*, *41*(5), 1013–1021. doi:10.1016/j.watres.2006.06.034 PMID:16934312

Kim, S. J., Ishikawa, K., Hirai, M., & Shoda, M. (1995). Characteristics of a newly isolated fungus, Geotrichumcandidum Dec 1, which decolorizes various dyes. *Journal of Fermentation and Bioengineering*, *79*(6), 601–607. doi:10.1016/0922-338X(95)94755-G

Kim, S.-W., Park, J.-S., Kim, D., & Oh, J.-M. (2014). Runoff characteristics of non-point pollutants caused by different land uses and a spatial overlay analysis with spatial distribution of industrial cluster: A case study of the Lake Sihwa watershed. *Environmental Earth Sciences*, *71*(1), 483–496. doi:10.100712665-013-2933-7

Kim, T.-H., Lee, S.-R., Nam, Y.-K., Yang, J., Park, C., & Lee, M. (2009). Disintegration of excess activated sludge by hydrogen peroxide oxidation. *Desalination*, *246*(1-3), 275–284. doi:10.1016/j.desal.2008.06.023

Kimura, K., Hara, H., & Watanabe, Y. (2005). Removal of pharmaceutical compounds by submerged membrane bioreactors (MBRs). *Desalination*, *178*(1-3), 135–140. doi:10.1016/j.desal.2004.11.033

Kim, Y. M., Park, D., Lee, D. S., & Park, J. M. (2008). Inhibitory effects of toxic compounds on nitrification process for cokes wastewater treatment. *Journal of Hazardous Materials*, *152*(3), 915–921. doi:10.1016/j.jhazmat.2007.07.065 PMID:17804160

Kindt, P., Vercammen, S., & Bianciardi, F. (2016, Aug. 21-24). Tire/road noise – characterization and potential further reductions of road traffic noise. In *Proceedings of the INTER-NOISE 2016 - 45th International Congress and Exposition on Noise Control Engineering: Towards a Quieter Future*. Hamburg, Germany. 2254–2264.

Kingsley, U., Turgay, P., & Mehmet, Ç. (2005). Screening of biosurfactants for crude oil contaminated soil washing. *Journal of Environmental Engineering and Science*, *4*(6), 487–496. doi:10.113904-073

Kirby, R. M., Bartram, J., & Carr, R. (2003). Water in food production and processing: Quantity and quality concerns. *Food Control*, *14*(5), 283–299. doi:10.1016/S0956-7135(02)00090-7

Kivaisi, A. K. (2001). The potential for constructed wetlands for wastewater treatment and reuse in developing countries: A review. *Ecological Engineering*, *16*(4), 545–560. doi:10.1016/S0925-8574(00)00113-0

Kjeldsen, P., Barlaz, P. M., Rooker, A. P., Baun, A., Ledin, A., & Christensen, T. H. (2002). Present and long-term composition of MSW landfill leachate: a review. *Critical Reviews in Environmental Science and Technology*, *32*(4), 297–336. doi:10.1080/10643380290813462

Klass, D. L. (1998). *Biomass for renewable energy, fuels, and chemicals*. Academic press.

Knapp, J. S., Newby, P. S., & Reece, L. P. (1995). Decolorization of dyes by wood-rotting basidiomycete fungi. *Enzyme and Microbial Technology*, *17*(7), 664–668. doi:10.1016/0141-0229(94)00112-5

Knight, J. A., Gorton, C. W., & Kovac, R. J. (1984). Oil production by entrained flow pyrolysis of biomass. *Biomass*, *6*(1), 69–79. doi:10.1016/0144-4565(84)90009-X

Knothe, G.,, & Steidly, K. R. (2009). A comparison of used cooking oils: a very heterogeneous feedstock for biodiesel. *Bioresource Technology*, *100*(23), 5796–5801. PMID:19592239

Knox, A. (2005). *An overview of incineration and EFW technology as applied to the management of municipal solid waste (MSW)*. ONEIA Energy Subcommittee.

Knox, K. (1985). Leachate treatment with nitrification of ammonia. *Water Research*, *19*(7), 895–904. doi:10.1016/0043-1354(85)90148-4

Kodam, K. M., Soojhawon, I., Lohande, P. D., & Gawai, K. R. (2005). Microbial decolorization of reactive azo dyes under aerobic conditions. *World Journal of Microbiology & Biotechnology*, *21*(3), 367–370. doi:10.100711274-004-5957-z

Koh, W. F. E., Abdul Hamid, A., Pak Dek, M. S., Tan, C. P., & Mohd Zainudin, M. A. (2014). Changes of major antioxidant compounds and radical scavenging activity of palm oil and rice bran oil during deep-frying. *Antioxidants, 3*, 502–515. PMID:26785067

Kokalj, F., & Samec, N. (2013). *Combustion of MSW for power generation*. doi:10.5772/55497

Kokyo, O., Tiehua, C., Tao, Li., & Hongyan, C. (2014). Study on application of Phytoremediation technology in management and remediation of contaminated soils. *Journal of Clean Energy Technologies,* 2(3), 216. Retrieved from https://www.researchgate.net/publication/271293740_Study_on_Application_of_Phytoremediation_Technology_in_Management_and_Remediation_of_Contaminated_Soils

Kolok, A. S., Ali, J. M., Rogan, E. G., & Bartelt-Hunt, S. L. (2018). The fate of synthetic and endogenous hormones used in the us beef and dairy industries and the potential for human exposure. *Current Environmental Health Reports, 5*(2), 225–232. doi:10.100740572-018-0197-9 PMID:29754262

Kolthoff, I. M., & Medalia, I. (1951). Determination of organic peroxides by reaction with ferrous iron. *Analytical Chemistry, 23*(4), 595–603. doi:10.1021/ac60052a014

Kongsuwan, A., Patnukao, P., & Pavasant, P. (2009). Binary component sorption of Cu (II) and Pb (II) with activated carbon from Eucalyptus camaldulensis Dehn bark. *Journal of Industrial and Engineering Chemistry, 15*(4), 465–470. doi:10.1016/j.jiec.2009.02.002

Korai, M. S., Mahar, R. B., & Uqaili, M. A. (2016). Optimization of waste to energy routes through biochemical and thermochemical treatment options of municipal solid waste in Hyderabad, Pakistan. *Energy Conversion and Management, 124*, 333–343. doi:10.1016/j.enconman.2016.07.032

Körbahti, B. K., Aktaş, N., & Tanyolaç, A. (2007). Optimization of electrochemical treatment of industrial paint wastewater with response surface methodology. *Journal of Hazardous Materials, 148*(1-2), 83–90. doi:10.1016/j.jhazmat.2007.02.005 PMID:17374443

Korres, N., O'Kiely, P., Benzie, J. A., & West, J. S. (2013). *Bioenergy production by anaerobic digestion: using agricultural biomass and organic wastes*. Routledge.

Kortbein, L., & Rajendran, A. (2016). Two-stage modified denitrifying-bioreactor for nutrient removal from agricultural runoff.

Kothari, R., Tyagi, V., & Pathak, A. (2010). Waste-to-energy: A way from renewable energy sources to sustainable development. *Renewable & Sustainable Energy Reviews, 14*(9), 3164–3170. doi:10.1016/j.rser.2010.05.005

Koupaie, E., Moghaddam, A. M., & Hashemi, S. (2013). Evaluation of integrated anaerobic/aerobic fixed-bed sequencing batch biofilm reactor for decolonization and biodegradation of azo dye Acid Red 18: Comparison of using two types of packing media. *Bioresource Technology, 127*, 415–421. doi:10.1016/j.biortech.2012.10.003 PMID:23138064

Krishna, R. R. (2012). Current atmospheric aerosol research in India. *Current Science*, 102.

Krzemińska, D., Neczaj, E., & Borowski, G. (2015). Advanced oxidation processes for food industrial wastewater decontamination. *Journal of Ecological Engineering, 16*, 61–71. doi:10.12911/22998993/1858

Kulikowska, D., & Klimiuk, E. J. B. T. (2008). The effect of landfill age on municipal leachate composition. *99*(13), 5981-5985.

Kumar, N. (2006). *Sequential anaerobic-aerobic digestion: A new process technology for biosolids product quality improvement.* (MS thesis), Virginia Tech.

Kumar, A., Jones, D. D., & Hanna, M. A. (2009). Thermochemical biomass gasification: A review of the current status of the technology. *Energies*, *2*(3), 556–581. doi:10.3390/en20300556

Kumar, A., & Samadder, S. (2017). A review on technological options of waste to energy for effective management of municipal solid waste. *Waste Management (New York, N.Y.)*, *69*, 407–422. PMID:28886975

Kumar, C. G., Joo, H.-S., Choi, J.-W., Koo, Y.-M., & Chang, C.-S. (2004). Purification and characterization of an extracellular polysaccharide from haloalkalophilic Bacillus sp. I-450. *Enzyme and Microbial Technology*, *34*(7), 673–681. doi:10.1016/j.enzmictec.2004.03.001

Kumar, C. G., Mongolla, P., Joseph, J., & Sarma, V. U. M. (2012). Decolorization and biodegradation of triphenylmethane dye, brilliant green, by Aspergillus sp. isolated from Ladakh, India. *Process Biochemistry*, *47*(9), 1388–1394. doi:10.1016/j.procbio.2012.05.015

Kumar, D., Pandey, J., Raj, V., & Kumar, P. (2017). A review on the modification of polysaccharide through graft copolymerization for various potential applications. *The Open Medicinal Chemistry Journal*, *11*(1), 109–126. doi:10.2174/1874104501711010109 PMID:29151987

Kumari, A., Chaudhary, D. R., & Jha, B. (2019). Destabilization of polyethylene and polyvinylchloride structure by marine bacterial strain. *Environmental Science and Pollution Research International*, 26(2), 1507-1516.

Kumar, K., Devi, S. S., Krishnamurthi, K., Dutta, D., & Chakrabarti, T. (2007). Decolorization and detoxification of Direct Blue-15 by a microbial consortium. *Bioresource Technology*, *98*(16), 3168–3171. doi:10.1016/j.biortech.2006.10.037 PMID:17321737

Kumar, S., Das, A., Kumar Srinivas, G. L., Dhar, H., Kumar Ojha, V., & Wong, J. (2016). Effect of calcium chloride on abating inhibition due to volatile fatty acids during the start-up period in anaerobic digestion of municipal solid waste. *Environmental Technology*, *37*(12), 1501–1509. doi:10.1080/09593330.2015.1119204 PMID:26609893

Kumar, S., Puniya, A. K., Puniya, M., Dagar, S. S., Sirohi, S. K., Singh, K., & Griffith, G. W. (2009). Factors affecting rumen methanogens and methane mitigation strategies. *World Journal of Microbiology & Biotechnology*, *25*(9), 1557–1566.

Kümmerer, K. (2009). The presence of pharmaceuticals in the environment due to human use–present knowledge and future challenges. *Journal of Environmental Management*, *90*(8), 2354–2366. doi:10.1016/j.jenvman.2009.01.023 PMID:19261375

Kunii, D., & Levenspiel, O. (2013). Fluidization Engineering (2nd ed.). Stoneham MA: Reed-Elsevier.

Kurniawan, T. A., Lo, W., & Chan, G. Y. (2006). Physico-chemical treatments for removal of recalcitrant contaminants from landfill leachate. *Journal of Hazardous Materials*, *129*(1-3), 80–100. doi:10.1016/j.jhazmat.2005.08.010 PMID:16314043

Kutty, S. R. M., Ezechi, E. H., Khaw, S. G., Lai, C. L., & Isa, M. H. (2014). Evaluation of copper removal using MIRHA as an adsorbent in a continuous flow activated sludge system. *Water Pollution XII*, *182*, 233–244. doi:10.2495/WP140211

Kwon, E. E., Oh, J., & Kim, K. (2015). Polycyclic aromatic hydrocarbons (pahs) and volatile organic compounds (VOCS) mitigation in the pyrolysis process of waste tires using CO_2 as a reaction medium. *Journal of Environmental Management*, *160*, 306–311. doi:10.1016/j.jenvman.2015.06.033 PMID:26117814

Kwon, Y. Y., Sung, K. I., Hee, C. M., Sumin, K., & Chul, P. J. (2017). Effect of PCM cool roof system on the reduction in urban heat island phenomenon. *Building and Environment*, *122*, 411–421. doi:10.1016/j.buildenv.2017.06.015

Lade, H., Kadam, A., Paul, D., & Govindwar, S. (2015). Biodegradation and detoxification of textile azo dyes by a bacterial consortium under sequential microaerophilic/aerobic processes. *EXCLI Journal*, *14*, 158–174. PMID:26417357

Lahijani, P., & Zainal, Z. (2014). Fluidized bed gasification of palm empty fruit bunch using various bed materials. *Energy Sources. Part A, Recovery, Utilization, and Environmental Effects*, *36*(22), 2502–2510. doi:10.1080/15567036.2011.572129

Lahijani, P., & Zainal, Z. A. (2011). Gasification of palm empty fruit bunch in a bubbling fluidized bed: A performance and agglomeration study. *Bioresource Technology*, *102*(2), 2068–2076. doi:10.1016/j.biortech.2010.09.101 PMID:20980143

Lahiri, T., Ray, R. M., Mukherjee, S., Basu, C., & Lahiri, P. (2000). Marked increase in sputum alveolar macrophages in residents of Calcutta: Possible exposure effect of severe air pollution. *Current Science India*, *78*(4), 399–406.

Laing, I. G. (1991). The impact of effluent regulations on the dyeing industry. *Rev. Prog. Coloration*, *21*(1), 56–71. doi:10.1111/j.1478-4408.1991.tb00081.x

Laleh, Y., Sylvie, R., Ruxandra, C., Manon, S., Geoffrey, S., Adriana, P., ... Serge, R. G. (2003). Enhanced biodegradation of petroleum hydrocarbons in contaminated soil. *Bioremediation Journal*, *7*(1), 37–51. doi:10.1080/713914241-274

Lam, C. H., Ip, A. W., Barford, J. P., & McKay, G. (2010). Use of incineration MSW ash: A review. *Sustainability*, *2*(7), 1943–1968.

Lam, M. K., & Lee, K. T. (2011). Renewable and sustainable bioenergies production from palm oil mill effluent (POME): Win–win strategies toward better environmental protection. *Biotechnology Advances*, *29*(1), 124–141. doi:10.1016/j.biotechadv.2010.10.001 PMID:20940036

Lam, M. K., Lee, K. T., & Mohamed, A. R. (2010). Homogeneous, heterogeneous and enzymatic catalysis for transesterification of high free fatty acid oil (waste cooking oil) to biodiesel: a review. *Biotechnology Advances*, *28*, 500–518. PMID:20362044

Langlais, B., Reckhow, D. A., & Brink, D. R. (1991). *Ozone in water treatment, application and engineering* (1st ed.). Lewis Publishers.

Lan, J. W., Zhan, L. T., Li, Y. C., & Chen, Y. M. (2012). Impacts of initial moisture content of MSW waste on leachate generation and modified formula for predicting leachate generation. [in Chinese]. *Environmental Sciences*, *33*, 1389–1395. PMID:22720594

Lansing, S., Martin, J. F., Botero, R. B., Da Silva, T. N., & Da Silva, E. D. (2010). Methane production in low-cost, unheated, plug-flow digesters treating swine manure and used cooking grease. *Bioresource Technology*, *101*(12), 4362–4370. doi:10.1016/j.biortech.2010.01.100 PMID:20153173

Lau, W. J., Goh, P. S., Ismail, A. F., & Lai, S. O. (2014). Ultrafiltration as a pretreatment for seawater desalination: A review. *Membrane Water Treatment*, *5*(1), 15–29. doi:10.12989/mwt.2014.5.1.015

Leahy, J. D., & Colwell, R. R. (1990). Microbial degradation of hydrocarbons in the environment. *Microbiological Reviews*, *54*, 305–315. PMID:2215423

Leal-Quir'os, E. (2014). Plasma processing of municipal solid waste. *Brazilian Journal of Physics*, *34*(4B).

Lea, W., Rohlich, G., & Katz, W. (1954). Removal of phosphates from treated sewage. *Sewage and Industrial Wastes*, 261–275.

Lee, M. D., & Swindoll, C. M. (1993). Bioventing for in situ remediation. *Hydrological Sciences -Journal- des Sciences Hydrologiques* 38(4), 8/1993. Retrieved from http://hydrologie.org/hsj/380/hysj_38_04_0273.pdf

Lee, M. D., Islam, S., Ting, C. H., & Begum, S. (2015). Current and future water scenario. Retrieved from http://www.leemandjun.blogspot.my/2015/08/current-and-future-water-scenario.html

Lee, C. S., Robinson, J., & Chong, M. F. (2014). A review on application of flocculants in wastewater treatment. *Process Safety and Environmental Protection*, *92*(6), 489–508. doi:10.1016/j.psep.2014.04.010

Lee, C., Kim, J., Hwang, K., O'Flaherty, V., & Hwang, S. (2009). Quantitative analysis of methanogenic community dynamics in three anaerobic batch digesters treating different wastewaters. *Water Research*, *43*(1), 157–165. doi:10.1016/j.watres.2008.09.032 PMID:18945471

Lee, M. D., Dayou, S., & Karunakaran, P. (2018). Determinants of a pre-treatment model in achieving economic and environmental sustainability in membrane desalination. [JKUKM]. *Journal of Engineering*, *30*(2). doi:10.17576/jkukm-2018-30(2)

Lefebvre, O., Vasudevan, N., Torrijos, M., Thanasekaran, K., & Moletta, R. (2005). Halophilic biological treatment of tannery soak liquor in a sequencing batch reactor. *Water Research*, *39*(8), 1471–1480. doi:10.1016/j.watres.2004.12.038 PMID:15878018

Lefrancq, M., Jadas-Hécart, A., La Jeunesse, I., Landry, D., & Payraudeau, S. (2017). High frequency monitoring of pesticides in runoff water to improve understanding of their transport and environmental impacts. *The Science of the Total Environment*, *587*, 75–86. doi:10.1016/j.scitotenv.2017.02.022 PMID:28242219

Lei, L., & Ni, J. (2014). Three-dimensional three-phase model for simulation of hydrodynamics, oxygen mass transfer, carbon oxidation, nitrification and denitrification in an oxidation ditch. *Water Research*, *53*, 200–214. doi:10.1016/j.watres.2014.01.021 PMID:24525069

Leitão, R. C., Van Haandel, A. C., Zeeman, G., & Lettinga, G. (2006). The effects of operational and environmental variations on anaerobic wastewater treatment systems: A review. *Bioresource Technology*, *97*(9), 1105–1118. PMID:16551532

Lemus, R., Brummer, E. C., Moore, K. J., Molstad, N. E., Burras, C. L., & Barker, M. F. (2002). Biomass yield and quality of 20 switchgrass populations in southern Iowa. *Biomass and Bioenergy*, *23*(6), 433–442. doi:10.1016/S0961-9534(02)00073-9

Leong, K. H., & Tan, L. L. (2007). Contamination levels of selected organochlorine and organophosphate pesticides in the Selangor River, Malaysia between 2002 and 2003. *Chemosphere*, *66*(6), 1153–1159. doi:10.1016/j.chemosphere.2006.06.009 PMID:17027062

Leong, X. F., Ng, C. Y., Jaarin, K., & Mustafa, M. R. (2015). Effects of repeated heating of cooking oils on antioxidant content and endothelial function. *Austin Journal of Pharmacology and Therapeutics*, *3*(2), 1068.

Lerch, R., Lin, C., Goyne, K., Kremer, R., & Anderson, S. (2017). Vegetative buffer strips for reducing herbicide transport in runoff: Effects of buffer width, vegetation, and season. *JAWRA Journal of the American Water Resources Association*, *53*(3), 667–683. doi:10.1111/1752-1688.12526

Leslie, M. S., David, S., Kosson, L. Y., Young, K. J. R., & Gary, L. T. (2004). Combine effects of contaminant desorption and toxicity on risk from PAH contaminated sediments. *Society for Risk Analysis* (5). Retrieved from https://pdfs.semanticscholar.org/e2e3/c2420a2763aece6141d424e856a30683dd0b.pdf

Leslie, C., Grady, J., Daigger, G., & Lim, C. (2011). *Biological wastewater treatment* (3rd ed.). Boca Raton, FL: CRC Press.

Letcher, T. M., & Vallero, D. (2011). *Waste: A handbook for management*. Academic Press.

Leung, D. Y. C., Wu, X., & Leung, M. K. H. (2010). A review on biodiesel production using catalyzed transesterification. *Applied Energy*, *87*(4), 1083–1095.

Lewis, S. R., Datta, S., Gui, M., Coker, E. L., Huggins, F. E., Daunert, S., ... Bhattacharyya, D. (2011). Reactive nanostructured membranes for water purification. *Proceedings of the National Academy of Sciences of the United States of America*, *108*(21), 8577–8582. doi:10.1073/pnas.1101144108 PMID:21606340

Liang, C., Bruell, C. J., Marley, M. C., & Sperry, K. L. (2004). Persulfate oxidation for in situ remediation of TCE activated by chelated ferrous ion. *Chemosphere*, *55*(9), 1225–1233. doi:10.1016/j.chemosphere.2004.01.030 PMID:15081763

Liang, S., Liu, Z., Xu, M., & Zhang, T. (2013). Waste oil derived biofuels in China bring brightness for global GHG mitigation. *Bioresource Technology*, *131*, 139–145. doi:10.1016/j.biortech.2012.12.008 PMID:23340111

Libra, A. J., Borchert, M., Vigelahn, L., & Storm, T. (2004). Two-stage biological treatment of a diazo reactive textile dye and the fate of the dye metabolites. *Chemosphere*, *56*(2), 167–180. doi:10.1016/j.chemosphere.2004.02.012 PMID:15120563

Li, C., & Suzuki, K. (2009). Tar property, analysis, reforming mechanism and model for biomass gasification—An overview. *Renewable & Sustainable Energy Reviews*, *13*(3), 594–604. doi:10.1016/j.rser.2008.01.009

Liew, W. H., Hassim, M. H., & Ng, D. K. S. (2014). Review of evolution, technology and sustainability assessments of biofuel production. *Journal of Cleaner Production*, *71*, 11–29. doi:10.1016/j.jclepro.2014.01.006

Li, H., Lu, M., Hsu, S. C., Gray, M., & Huang, T. (2015). Proactive behavior-based safety management for construction safety improvement. *Safety Science*, *75*, 107–117. doi:10.1016/j.ssci.2015.01.013

Li, H.-T., & Li, Y.-F. (2010). Performance of a hybrid anaerobic baffled reactor (HABR) treating brewery wastewater. In *Proceedings of 2010 International Conference on Mechanic Automation and Control Engineering* (pp. 2032-2037). Wuhan, China: IEEE

Li, H., Zhao, J., Huang, Y., Jiang, Z., & Chen, Q. (2016). Investigation on the potential of waste cooking oil as a grinding aid in Portland cement. *Journal of Environmental Management*, *184*(3), 545–551. doi:10.1016/j.jenvman.2016.10.027 PMID:27793479

Li, J., Liu, L., Liu, J., Ma, T., Yan, A., & Ni, Y. (2016). Effect of adding alum sludge from water treatment plant on sewage sludge dewatering. *Journal of Environmental Chemical Engineering*, *4*(1), 746–752. doi:10.1016/j.jece.2015.07.021

Li, J., Zuo, J., Cai, H., & Zillante, G. (2018). Construction waste reduction behavior of contractor employees: An extended theory of planned behavior model approach. *Journal of Cleaner Production*, *172*, 1399–1408. doi:10.1016/j.jclepro.2017.10.138

Lim, B.-C., Lim, J.-W., & Ho, Y.-C. (2018). Garden cress mucilage as a potential emerging biopolymer for improving turbidity removal in water treatment. *Process Safety and Environmental Protection*, *119*, 233–241. doi:10.1016/j.psep.2018.08.015

Ling, T. C. (2011). Prediction of density and compressive strength for rubberized concrete blocks. *Construction & Building Materials*, *25*(11), 4303–4306. doi:10.1016/j.conbuildmat.2011.04.074

Lin, J., & Lee, L. C. (2004). Characterization of the concentration and distribution of urban submicron (PM1) aerosol particles. *Atmospheric Environment*, *38*(3), 469–475. doi:10.1016/j.atmosenv.2003.09.056

Lin, M., Ning, X., An, T., Zhang, J., Chen, C., Ke, Y., & Liu, J. (2016). Degradation of polycyclic aromatic hydrocarbons (PAHs) in textile dyeing sludge with ultrasound and Fenton processes: Effect of system parameters and synergistic effect study. *Journal of Hazardous Materials*, *307*, 7–16. doi:10.1016/j.jhazmat.2015.12.047 PMID:26795704

Lin, S. H., & Leu, H. G. (1999). Operating characteristics and kinetic studies of surfactant wastewater treatment by Fenton oxidation. *Water Research*, *33*(7), 1735–1741. doi:10.1016/S0043-1354(98)00403-5

Lin, S. H., & Lo, C. C. (1997). Fenton process for treatment of desizing wastewater. *Water Research*, *31*(8), 2050–2056. doi:10.1016/S0043-1354(97)00024-9

Lin, S. H., & Peng, C. F. (1994). Treatment of textile wastewater by electrochemical method. *Water Research*, *28*(2), 277–282. doi:10.1016/0043-1354(94)90264-X

Lipnizki, F. (2005). Industrial applications of ultrafiltration in pharmaceutical biotechnology. *Engineering in Life Sciences*, *5*(1), 81–83. doi:10.1002/elsc.200407047

Lipton, M. (1977). Why poor people stay poor: a study of urban bias in world development. London, UK: Canberra, ACT: Temple Smith; Australian National University Press.

Li, T., Zhu, Z., Wang, D., Yao, C., & Tang, H. (2006). Characterization of floc size, strength and structure under various coagulation mechanisms. *Powder Technology*, *168*(2), 104–110. doi:10.1016/j.powtec.2006.07.003

Lithner, D., Larsson, Å., & Dave, G. (2011). Environmental and health hazard ranking and assessment of plastic polymers based on chemical composition. *The Science of the Total Environment*, *409*(18), 3309–3324. doi:10.1016/j.scitotenv.2011.04.038 PMID:21663944

Liu, H., Yang, C., Pu, W., & Zhang, J. (2008). Removal of nitrogen from wastewater for reusing to boiler feed-water by an anaerobic/aerobic/membrane bioreactor. *Chemical Engineering Journal*, *140*(1-3), 122–129. doi:10.1016/j.cej.2007.09.048

Liu, J., Li, W., & Jie, L. (2016). Quality screening for air quality monitoring data in China. *Environmental Pollution*, *216*, 720–723. doi:10.1016/j.envpol.2016.06.037 PMID:27376986

Liu, J., Veith, T. L., Collick, A. S., Kleinman, P. J., Beegle, D. B., & Bryant, R. B. (2017). Seasonal manure application timing and storage effects on field-and watershed-level phosphorus losses. *Journal of Environmental Quality*, *46*(6), 1403. doi:10.2134/jeq2017.04.0150 PMID:29293863

Liu, K., Han, W., Pan, W.-P., & Riley, J. T. (2001). Polycyclic aromatic hydrocarbon (PAH) emissions from a coal-fired pilot FBC system. *Journal of Hazardous Materials*, *84*(2-3), 175–188. doi:10.1016/S0304-3894(01)00196-0 PMID:11406305

Liu, Q., Yin, X., Sha, B., & You, J. (2014). Porous membrane ultrafiltration—A novel method for enrichment of the active compounds from micro-plasma samples. *Spectrochimica Acta. Part A: Molecular and Biomolecular Spectroscopy*, *131*, 22–29. doi:10.1016/j.saa.2014.04.070 PMID:24815198

Liu, R. R., Tian, Q., Yang, B., & Chen, J. H. (2010). Hybrid anaerobic baffled reactor for treatment of desizing wastewater. *International Journal of Environmental Science and Technology*, *7*(1), 111–118. doi:10.1007/BF03326122

Liu, S., Chen, M., Hu, Q., Wang, J., & Kong, L. (2013). The kinetics model and pyrolysis behavior of the aqueous fraction of bio-oil. *Bioresource Technology*, *129*, 381–386. doi:10.1016/j.biortech.2012.11.006 PMID:23262015

Liu, W., Chao, Y., Yang, X., Bao, H., & Qian, S. (2004). Biodecolourization of azo, anthraquinonic and triphenylmethane dyes by white-rot fungi and a laccase secreting engineered strain. *Journal of Industrial Microbiology & Biotechnology*, *31*(3), 127–132. doi:10.100710295-004-0123-z PMID:15069603

Liu, X. M., Sheng, G. P., Luo, H. W., Zhang, F., Yuan, S. J., Xu, J., ... Yu, H. Q. (2010). Contribution of extracellular polymeric substances (EPS) to the sludge aggregation. *Environmental Science & Technology*, *44*(11), 4355–4360. doi:10.1021/es9016766 PMID:20446688

Liu, X., Zhang, W., Hu, Y., Hu, E., Xie, X., Wang, L., & Cheng, H. (2015). Arsenic pollution of agricultural soils by concentrated animal feeding operations (CAFOs). *Chemosphere*, *119*, 273–281. doi:10.1016/j.chemosphere.2014.06.067 PMID:25036941

Liu, Y.-Q., Tay, J.-H., Ivanov, V., Moy, B. Y.-P., Yu, L., & Tay, S. T.-L. (2005). Influence of phenol on nitrification by microbial granules. *Process Biochemistry*, *40*(10), 3285–3289. doi:10.1016/j.procbio.2005.03.018

Liu, Y., Shi, H., Xia, L., Shi, H., Shen, T., Wang, Z., ... Wang, Y. (2010). Study of operational conditions of simultaneous nitrification and denitrification in a Carrousel oxidation ditch for domestic wastewater treatment. *Bioresource Technology*, *101*(3), 901–906. doi:10.1016/j.biortech.2009.09.015 PMID:19818603

Li, X., Grace, J., Lim, C., Watkinson, A., Chen, H., & Kim, J. (2004). Biomass gasification in a circulating fluidized bed. *Biomass and Bioenergy*, *26*(2), 171–193. doi:10.1016/S0961-9534(03)00084-9

Li, Y.-H., & Chen, H.-H. (2018). Analysis of syngas production rate in empty fruit bunch steam gasification with varying control factors. *International Journal of Hydrogen Energy*, *43*(2), 667–675. doi:10.1016/j.ijhydene.2017.11.117

Li, Y., Jiao, J., Wang, Y., Yang, W., Meng, C., Li, B., ... Wu, J. (2015). Characteristics of nitrogen loading and its influencing factors in several typical agricultural watersheds of subtropical China. *Environmental Science and Pollution Research International*, *22*(3), 1831–1840. doi:10.100711356-014-3446-y PMID:25172462

Li, Y.-Y., & Noike, T. (1992). Upgrading of anaerobic digestion of waste activated sludge by thermal pretreatment. *Water Science and Technology*, *26*(3-4), 857–866. doi:10.2166/wst.1992.0466

Li, Y., Zhang, J., Zhang, Y., & Quan, X. (2017). Scaling-up of a zero valent iron packed anaerobic reactor for textile dye wastewater treatment: A potential technology for on-site upgrading and rebuilding of traditional anaerobic wastewater treatment plant. *Water Science and Technology*, *76*(3-4), 823–831. doi:10.2166/wst.2017.270 PMID:28799929

Liyanage, J. P. (2007). Operations and maintenance performance in production and manufacturing assets: the sustainability perspective. *Journal of Manufacturing Technology Management*, *18*(3), 304–314. doi:10.1108/17410380710730639

Li, Z., Huang, Z., Ding, S., Li, F., & Chen, C. (2018). Catalytic conversion of waste cooking oil to fuel oil: Catalyst design and effect of solvent. *Energy*, *157*, 270–277. doi:10.1016/j.energy.2018.05.156

Löder, M. G. J., & Gunnar Gerdts, G. (2015) Methodology used for the detection and identification of microplastics—a critical appraisal. In M. Bergmann, et al. (Eds.), Marine Anthropogenic Litter, 8, 201-227.

Lombardi, L., Carnevale, E., & Corti, A. (2015). A review of technologies and performances of thermal treatment systems for energy recovery from waste. *Waste Management (New York, N.Y.)*, *37*, 26–44. doi:10.1016/j.wasman.2014.11.010 PMID:25535103

Lombi, E., & Hamon, R. E. (2005). Remediation of polluted soils in Encyclopaedia of soils in the environment, Elsevier, 379-385.

López-Fernández, R., Martínez, L., & Villaverde, S. (2012). Membrane bioreactor for the treatment of pharmaceutical wastewater containing corticosteroids. *Desalination*, *300*, 19–23. doi:10.1016/j.desal.2012.05.032

Lopez, G., Alvarez, J., Amutio, M., Mkhize, N. M., Danon, B., Van Der Gryp, P., & Olazar, M. (2017). Waste truck-tyre processing by flash pyrolysis in a conical spouted bed reactor. *Energy Conversion and Management*, *142*, 523–532. doi:10.1016/j.enconman.2017.03.051

López-Guzmán, M., Alarcón-Herrera, M. T., Irigoyen-Campuzano, J. R., Torres-Castañón, L. A., & Reynoso-Cuevas, L. (2019). Simultaneous removal of fluoride and arsenic from well water by electrocoagulation. *The Science of the Total Environment*, *678*, 181–187. doi:10.1016/j.scitotenv.2019.04.400 PMID:31075584

Lopez, J. M., Guisado, G., Garcia-Vargas, C. M., & Estrella-Suarez, F. (2012). Pseudallescheriaangusta, a ligninolytic microorganism or wood fibres biomodification. *BioResources*, *7*, 464–474.

Löwenberg, J., Baum, J. A., Zimmermann, Y. S., Groot, C., van den Broek, W., & Wintgens, T. (2015). Comparison of pre-treatment technologies towards improving reverse osmosis desalination of cooling tower blow down. *Desalination*, *357*, 140–149. doi:10.1016/j.desal.2014.11.018

Luber, G., & Prudent, N. (2009). Climate change and human health. *Transactions of the American Clinical and Climatological Association*, *120*, 113–117. doi:10.5124/jkma.2008.51.8.764 PMID:19768168

Lucchini, S., Thompson, A., & Hilton, J. (2001). Microarrays for microbiologists. *Microbiology*, *147*(6), 1403–1414. doi:10.1099/00221287-147-6-1403 PMID:11390672

Lucena, I. L., Silva, G. F., & Fernandes, F. A. N. (2008). Biodiesel production by esterification of oleic acid with methanol using a water adsorption apparatus. *Industrial & Engineering Chemistry Research*, *47*(18), 6885–6889. doi:10.1021/ie800547h

Lü, F., Wang, F., Shao, L., & He, P. (2017). Deciphering pretreatment-induced repartition among stratified extracellular biopolymers and its effect on anaerobic biodegradability and dewaterability of waste activated sludge. *Journal of Environmental Chemical Engineering*, *5*(3), 3014–3023. doi:10.1016/j.jece.2017.05.047

Luo, G., Xie, L., Zou, Z., Wang, W., Zhou, Q., & Shim, H. (2010). Anaerobic treatment of cassava stillage for hydrogen and methane production in continuously stirred tank reactor (CSTR) under high organic loading rate (OLR). *International Journal of Hydrogen Energy*, *35*(21), 11733–11737. doi:10.1016/j.ijhydene.2010.08.033

Lu, W., Chen, X., Peng, Y., & Liu, X. (2018). The effects of green building on construction waste minimization: Triangulating 'big data' with 'thick data.'. *Waste Management (New York, N.Y.)*, *79*, 142–152. doi:10.1016/j.wasman.2018.07.030 PMID:30343740

Lu, W., Webster, C., Chen, K., Zhang, X., & Chen, X. (2017). Computational building information modelling for construction waste management: Moving from rhetoric to reality. *Renewable & Sustainable Energy Reviews*, *68*, 587–595. doi:10.1016/j.rser.2016.10.029

Lu, Y. Y., Hu, Y. D., Xu, D. M., & Wu, L. Y. (2006). Optimum design of reverse osmosis seawater desalination system considering membrane cleaning and replacing. *Journal of Membrane Science*, *282*(1–2), 7–13. doi:10.1016/j.memsci.2006.04.019

Lu, Z., & Streets, D. G. (2012). Increase in NOx emissions from Indian thermal power plants during 1996–2010: Unit-based inventories and multisatellite observations. *Environmental Science & Technology*, *46*(14), 7463–7470. doi:10.1021/es300831w PMID:22732062

Lv, P., Xiong, Z., Chang, J., Wu, C., Chen, Y., & Zhu, J. (2004). An experimental study on biomass air–steam gasification in a fluidized bed. *Bioresource Technology*, *95*(1), 95–101. doi:10.1016/j.biortech.2004.02.003 PMID:15207301

Lv, P., Yuan, Z., Ma, L., Wu, C., Chen, Y., & Zhu, J. (2007). Hydrogen-rich gas production from biomass air and oxygen/steam gasification in a downdraft gasifier. *Renewable Energy*, *32*(13), 2173–2185. doi:10.1016/j.renene.2006.11.010

Lyko, S., Al-Halbouni, D., Wintgens, T., Janot, A., Hollender, J., Dott, W., & Melin, T. (2007). Polymeric compounds in activated sludge supernatant—Characterisation and retention mechanisms at a full-scale municipal membrane bioreactor. *Water Research*, *41*(17), 3894–3902. doi:10.1016/j.watres.2007.06.012 PMID:17645906

Maaroff, R. M., Jahim, J. M., Azhar, A. M., Abdul, P. M., Masdar, M. S., Nordin, D., & Nasir, M. A. A. (2019). Biohydrogen production from palm oil mill effluent (POME) by two stage anaerobic sequencing batch reactor (ASBR) system for better utilization of carbon sources in POME. *International Journal of Hydrogen Energy*, *44*(6), 3395–3406. doi:10.1016/j.ijhydene.2018.06.013

Ma, C., Gao, J., Zhong, L., & Xing, S. (2016). Experimental investigation of the oxidation behaviour and thermal kinetics of diesel particulate matter with non-thermal plasma. *Applied Thermal Engineering*, *99*, 1110–1118. doi:10.1016/j.applthermaleng.2016.01.165

Mace, S., & Mata-Alvarez, J. (2002). Utilization of SBR technology for wastewater: An overview. *Industrial & Engineering Chemistry Research*, *41*(23), 5539–5553. doi:10.1021/ie0201821

Machado, S. L., Carvalho, M. F., Gourc, J.-P., Vilar, O. M., & do Nascimento, J. C. F. (2009). Methane generation in tropical landfills: Simplified methods and field results. *Waste Management (New York, N.Y.)*, *29*(1), 153–161. doi:10.1016/j.wasman.2008.02.017 PMID:18440214

Madony, P., Davoli, D., & Chirerici, E. (1993). Comparative analysis of the activated sludge microfauna in several sewage treatment works. *Water Research*, *27*(9), 1485–1491. doi:10.1016/0043-1354(93)90029-H

Ma, F.,, & Hanna, M. A. (1999). Biodiesel production: a review. *Bioresource Technology*, *70*, 1–15.

Maghanaki, M. M., Ghobadian, B., Najafi, G., & Galogah, R. J. (2013). Potential of biogas production in Iran. *Renewable & Sustainable Energy Reviews*, *28*, 702–714.

Maghraby, D. M. E. (2013). Evaluation of non-viable biomass of Laurencia papillosa for decolorization of dye wastewater. *African Journal of Biotechnology*, *12*(17), 2215–2223. doi:10.5897/AJB2013.11975

Mahabali, S., & Spanoghe, P. (2014). Mitigation of two insecticides by wetland plants: Feasibility study for the treatment of agricultural runoff in Suriname (South America). *Water, Air, and Soil Pollution*, *225*(1), 1771. doi:10.100711270-013-1771-2

Maharaj, R., Harry, V. R., & Mohamed, N. (2015). Rutting and fatigue cracking resistance of waste cooking oil modified Trinidad asphaltic materials. *The Scientific World Journal*. doi:10.1155/2015/385013 PMID:26336652

Mah, C. M., Fujiwara, T., & Ho, C. S. (2018). Life cycle assessment and life cycle costing toward eco-efficiency concrete waste management in Malaysia. *Journal of Cleaner Production*, *172*, 3415–3427. doi:10.1016/j.jclepro.2017.11.200

Maheshwari R., Gupta S., & Das, K. (2015). Impact of landfill waste on health: An overview. *IOSR Journal of Environmental Science, Toxicology, and Food Technology (IOSR-JESTFT)*, 1(4), 17-23.

Mahpour, A. (2018). Prioritizing barriers to adopt circular economy in construction and demolition waste management. *Resources, Conservation and Recycling*, *134*, 216–227. doi:10.1016/j.resconrec.2018.01.026

Maia, A. A., & de Morais, L. C. (2016). Kinetic parameters of red pepper waste as biomass to solid biofuel. *Bioresource Technology*, *204*, 157–163. doi:10.1016/j.biortech.2015.12.055 PMID:26773950

Maier, J., Kandelbauer, A., Erlacher, A., Cavaco-Paulo, A., & Gübitz, G. M. (2004). A new alkali-thermostable azoreductase from *bacillus* sp. Strain SF. *Applied and Environmental Microbiology*, *70*(2), 837–844. doi:10.1128/AEM.70.2.837-844.2004 PMID:14766562

Mai, L., Bao, L. J., Shi, L., Wong, C. S., & Zeng, E. Y. (2018). A review of methods for measuring micro plastics in aquatic environments. *Environmental Science and Pollution Research International*, *25*(12), 11319–11332. doi:10.100711356-018-1692-0 PMID:29536421

Maithel, S., Uma, R., Bond, T., Baum, E., & Thao, V. T. K. (2012). Brick kilns performance assessment, emissions measurements, & a roadmap for cleaner brick production in India. Study report prepared by Green Knowledge Solutions, New Delhi, India.

Ma, J., Fu, K., Shi, J., Sun, Y., Zhang, X., & Ding, L. (2016). Ultraviolet-assisted synthesis of polyacrylamide-grafted chitosan nanoparticles and flocculation performance. *Carbohydrate Polymers*, *151*, 565–575. doi:10.1016/j.carbpol.2016.06.002 PMID:27474601

Ma, J., Sun, D., Pang, Q., Sun, G., Hu, M., & Lu, T. (2019). Potential of recycled concrete aggregate pretreated with waste cooking oil residue for hot mix asphalt. *Journal of Cleaner Production*, *221*, 469–479. doi:10.1016/j.jclepro.2019.02.256

Majekodunmi, S. O., & Olorunsola, E. O. (2014). Review of some recent advances on filtration in pharmaceutical industry. *Journal of Pharmacy and Biological Sciences*, *9*(4), 30–38. doi:10.9790/3008-09443038

Majumdar, D., & Gupta, N. (2000). Nitrate pollution of groundwater and associated human health disorders. *Indian Journal of Environmental Health*, *42*(1), 28–39.

Ma, L., Verelst, H., & Baron, G. (2005). Integrated high temperature gas cleaning: Tar removal in biomass gasification with a catalytic filter. *Catalysis Today*, *105*(3-4), 729–734. doi:10.1016/j.cattod.2005.06.022

Mala, M. (2012) Urban growth and spatial transformation of Kolkata metropolis: a continuation of colonial legacy. *Arpn Journal of Science and Technology*. 2, Special Issue.

Malakahmad, A., & Noor Ezlin, A. B., & Shahrom, M. Z. (2011). Study on performance of a modified anaerobic baffled reactor to treat high strength wastewater. *Journal of Applied Sciences (Faisalabad)*, *11*(8), 1449–1452.

Malakootian, M., Mansoorian, H., & Moosazadeh, M. (2010). Performance evaluation of electrocoagulation process using iron-rod electrodes for removing hardness from drinking water. *Desalination*, *255*(1), 67–71. doi:10.1016/j.desal.2010.01.015

Malamis, S., Katsou, E., Takopoulos, K., Demetriou, P., & Loizidou, M. (2012). Assessment of metal removal, biomass activity and RO concentrate treatment in an MBR–RO system. *Journal of Hazardous Materials*, *209*, 1–8. doi:10.1016/j.jhazmat.2011.10.085 PMID:22209586

Malato, P., Fernández-Ibáñez, M. I., Maldonado, J., Blanco, W. G., & Gernjak, W. (2009). Decontamination and disinfection of water by solar photocatalysis: Recent overview and trends. *Catalysis Today*, *147*(1), 1–59. doi:10.1016/j.cattod.2009.06.018

Malato, S., Blanco, J., Maldonado, M. I., Fernández-Ibáñez, P., & Campos, A. (2000). Optimising solar photocatalytic mineralisation of pesticides by adding inorganic oxidising species; application to the recycling of pesticide containers. *Applied Catalysis B: Environmental*, *28*(3-4), 163–174. doi:10.1016/S0926-3373(00)00175-2

Mallick, D., Buragohain, B., Mahanta, P., & Moholkar, V. S. (2018). Gasification of mixed biomass: analysis using equilibrium, semi-equilibrium, and kinetic models. In *Coal and Biomass Gasification* (pp. 223–241). Springer. doi:10.1007/978-981-10-7335-9_9

Mallick, N. (2002). Biotechnological potential of immobilized algae for wastewater N, P, and metal removal: A review. *Biometals*, *15*(4), 377–390. doi:10.1023/A:1020238520948 PMID:12405533

Mamelkina, M. A., Cotillas, S., Lacasa, E., Sáez, C., Tuunila, R., Sillanpää, M., & Rodrigo, M. A. (2017). Removal of sulfate from mining waters by electrocoagulation. *Separation and Purification Technology*, *182*, 87–93. doi:10.1016/j.seppur.2017.03.044

Mameri, N., Yeddou, A., Lounici, H., Belhocine, D., Grib, H., & Bariou, B. (1998). Defluoridation of septentrional Sahara water of North Africa by electrocoagulation process using bipolar aluminium electrodes. *Water Research*, *32*(5), 1604–1612. doi:10.1016/S0043-1354(97)00357-6

Mamimin, C., Kongjan, P., O-Thong, S., & Prasertsan, P. (2019). Enhancement of biohythane production from solid waste by co-digestion with palm oil mill effluent in two-stage thermophilic fermentation. *International Journal of Hydrogen Energy*, *44*(32), 17224–17237. doi:10.1016/j.ijhydene.2019.03.275

Manahan, S. E. (2001). Introduction to chemistry. Fundaments of environmental chemistry. Boca Raton, FL: CRC Press. Retrieved from http://www.ingenieroambiental.com/4004/Fundamentals%20of%20Environmental%20Chemistry,%20Manahan.pdf

Mandpe, S., Kadlaskar, S., Degen, W., & Keppeler, S. (2005). *On road testing of advanced common rail diesel vehicles with biodiesel from the Jatropha curcas plant* (0148-7191). In L. C. Martins das Neves, A. Converti, & T. C. Vessoni Penna (Eds.), *Biogas production: New trends for alternative energy sources in rural and urban zones. Chemical Engineering & Technology: Industrial Chemistry-Plant Equipment-Process Engineering-Biotechnology*, *32*(8), 1147–1153.

Maneerat, K., Boonyarit, N., Somkiet, T., & Chantaraporn, P. (2012). UASB performance and microbial adaptation during a transition from mesophilic to thermophilic of palm oil mill effluent. *Journal of Environmental Management*, *103*, 74–82. doi:10.1016/j.jenvman.2012.03.004 PMID:22466006

Maneerung, T., Kawi, S., Dai, Y., & Wang, C. H. (2016). Sustainable biodiesel production via transesterification of waste cooking oil by using CaO catalysts prepared from chicken manure. *Energy Conversion and Management*, *123*, 487–497.

Mane, V. S., Mall, I. D., & Srivastava, V. C. (2007). Kinetic and equilibrium isotherm studies for the adsorptive removal of brilliant green dye from aqueous solution by rice husk ash. *Journal of the Indian Chemical Society*, *90*, 577–583. PMID:17000044

Maniam, H., Nagapan, S., Abdullah, A. H., Subramaniam, S., & Sohu, S. (2018). A comparative study of construction waste generation rate based on different construction methods on construction project in Malaysia. *Engineering, Technology & Applied Scientific Research*, *8*(5), 3488–3491.

Manish, S., & Dev, R. (2015). Working fluid selection for low temperature organic rankine cycle. In ISES Solar World Congress (vol. 205).

Mann, A. T., & Stephenson, T. (1997). Modelling biological aerated filters for wastewater treatment. *Water Research*, *31*(10), 2443–2448. doi:10.1016/S0043-1354(97)00095-X

Mannu, A., Ferro, M., Pietro, M. E. D., & Mele, A. (2019). Innovative applications of waste cooking oil as raw material. *Science Progress*, *102*(2), 153–160. doi:10.1177/0036850419854252

Manoli, E., & Samara, C. (2008). The removal of polycyclic aromatic hydrocarbons in the wastewater treatment process: Experimental calculations and model predictions. *Environmental Pollution*, *151*(3), 477–485. doi:10.1016/j.envpol.2007.04.009 PMID:17548140

Mansour, H. B., Corroler, D., Barillier, D., Ghedira, K., Chekir, L., & Mosrati, R. (2007). Evaluation of genotoxicity and pro-oxidant effect of the azo dyes: Acids yellow 17, violet 7 and orange 52, and of their degradation products by Pseudomonas Putida Mt-2. *Food and Chemical Toxicology*, *45*(9), 1670–1677. doi:10.1016/j.fct.2007.02.033 PMID:17434654

Mansour, H. B., Mosrati, R., Corroler, D., Ghedira, K., Barillier, D., & Chekir, L. (2009). In vitro mutagenicity of acid violet 7 and its degradation products by Pseudomonas Putida Mt-2: Correlation with chemical structures. *Environmental Toxicology and Pharmacology*, *27*(2), 231–236. doi:10.1016/j.etap.2008.10.008 PMID:21783945

Manu, B., & Chaudhari, S. (2002). Anaerobic decolorization of simulated textile wastewater containing azo dyes. *Bioresource Technology*, *82*(3), 225–231. doi:10.1016/S0960-8524(01)00190-0 PMID:11991070

Marble Institute of America. (2019). Available at http://www.atlastileandstone.com/ Limestone %20Geological%20 Classification.html

Mark, H., Abdullah, M., Hu, J., Song, B., Niemeier, D. A., Handy, S. L., ... Kleeman, M. J. (2010). Influence of regional development policies and clean technology adoption on future air pollution exposure. *Atmospheric Environment*, *44*(4), 552–562. doi:10.1016/j.atmosenv.2009.10.041

Martinez, A. T., Ruiz-Duenas, F. J., Martinz, M. J., Del Rio, J. C., & Gutierrez, A. (2009). Enzymatic delignification of plant cell wall: From nature to mill. *Current Opinion in Biotechnology*, *20*(3), 348–357. doi:10.1016/j.copbio.2009.05.002 PMID:19502047

Martínez-Alvarez, V., Martin-Gorriz, B., & Soto-García, M. (2016). Seawater desalination for crop irrigation— A review of current experiences and revealed key issues. *Desalination*, *381*, 58–70. doi:10.1016/j.desal.2015.11.032

Martínez, J. D., Puy, N., Murillo, R., García, T., Navarro, M. V., & Mastral, A. M. (2013). Waste tyre pyrolysis – a review. *Renewable & Sustainable Energy Reviews*, *23*, 179–213. doi:10.1016/j.rser.2013.02.038

Martins, M. A. M., Cardoso, M. H., Queiroz, M. J., Ramalho, M. T., & Oliveira-Campos, A. M. (1999). Biodegradation of azo dyes by the yeast Candida zeylanoides in batch aerated cultures. *Chemosphere*, *38*(11), 2455–2460. doi:10.1016/S0045-6535(98)00448-2 PMID:10204232

Martins, M. A., Lima, N., Silvestre, A. J., & Queiroz, M. J. (2003). Comparative studies of fungal degradation of single or mixed bioaccessible reactive azo dyes. *Chemosphere*, *52*(6), 967–973. doi:10.1016/S0045-6535(03)00286-8 PMID:12781230

Martins, M. A., Queiroz, M. J., Silvestre, A. J., & Lima, N. (2002). Relationship of chemical structures of textile dyes on the pre-adaptation medium and the potentialities of their biodegradation by Phanerochaetechrysosporium. *Research in Microbiology*, *153*(6), 361–368. doi:10.1016/S0923-2508(02)01332-3 PMID:12234010

Marttinen, S., Kettunen, R., Sormunen, K., Soimasuo, R., & Rintala, J. (2002). Screening of physical chemical methods for removal of organic material, nitrogen and toxicity from low strength landfill leachates. *Chemosphere*, *46*(6), 851–858. doi:10.1016/S0045-6535(01)00150-3 PMID:11922065

Mata-Alvarez, J., Mace, S., & Llabres, P. (2000). Anaerobic digestion of organic solid wastes. An overview of research achievements and perspectives. *Bioresource Technology*, *74*(1), 3–16. doi:10.1016/S0960-8524(00)00023-7

Math, M. C., Kumar, S. P., & Chetty, S. V. (2010). Technologies for biodiesel production from used cooking oil – a review. *Energy for Sustainable Development*, *14*(4), 339–345.

Matilainen, A., Vepsäläinen, M., & Sillanpää, M. (2010). Natural organic matter removal by coagulation during drinking water treatment: A review. *Advances in Colloid and Interface Science*, *159*(2), 189–197. doi:10.1016/j.cis.2010.06.007 PMID:20633865

Matsufuji, Y. (1990). *What is Fukuoka method? Report on road to sanitary landfill*. Japan: Faculty of Engineering Fukuoka University.

Matsui, Y., Yuasa, A., Furuya, Y., & Kamei, T. (1998). Dynamic analysis of coagulation with alum and PACl. *Journal - American Water Works Association*, *90*(10), 96–106. doi:10.1002/j.1551-8833.1998.tb08522.x

Matthäus, B. (2010). Oxidation in foods and beverages and antioxidant applications: management in different industry sectors. Chapter 6 - Oxidation of edible oils. Woodhead Publishing Series in Food Science, Technology and Nutrition, 183-238.

Mattocks, R. (1984). *Understanding biogas generation (Technical paper No. 4. volunteers in technical assistance).* Virginia: I. K. Intern Publishing House.

Mavroulidou, M.,, & Figueiredo, J. (2010). Discarded tyre rubber as concrete aggregate: a possible outlet for used tyres. *Global NEST Journal, 12*(4), 359–367.

Maycock, D., & Inspectorate, G. B. D. W. (2008). Review of England and Wales monitoring data for which a national standard has been set: final report to DWI.

Mazza, G.,, & Qi, H. (1992). Effect of after-cooking darkening inhibitors on stability of frying oil and quality of french fries. *Journal of the American Oil Chemists' Society, 69*, 847–853.

McBride, M. B. (1994). *Environmental chemistry of soils*. New York: University Press.

McCarty, P. L., Bae, J., & Kim, J. (2011). Domestic wastewater treatment as a net energy producer–can this be achieved?.

McCarty, J., & Kaza, N. (2015). Urban form and air quality in the United States. *Landscape and Urban Planning, 139*(7), 168–179. doi:10.1016/j.landurbplan.2015.03.008

McEwen, J. B. (1998). *Treatment process selection for particle removal*. American Water Works Association.

McFarland, M. J. (2001). *Biosolids Engineering*. New York, NY: McGraw-Hill.

McLachlan, D., Bergeron, C., Smith, J., Boomer, D., & Rifat, S. (1996). Risk for neuropathologically confirmed Alzheimer's disease and residual aluminum in municipal drinking water employing weighted residential histories. *Neurology, 46*(2), 401–405. doi:10.1212/WNL.46.2.401 PMID:8614502

Meera, G., & Manisha, K. (2012). Biodegradation of textile dye congo red by fungus mucor mucedo. *Indian Journal of Fundamental and Applied Life Sciences, 2*, 251–255.

Mehdi, H. S., Moslem, A., & Simone, C. (2015). *Biodegradation of aromatic compounds*. Intech. doi:10.5772/60894

Meier, J., & Melin, T. (2005). Wastewater reclamation by the PAC-NF process. *Desalination, 178*(1-3), 27–40. doi:10.1016/j.desal.2004.12.015

Meisinger, J., Hargrove, W., Mikkelsen, R., Williams, J., & Benson, V. (1991). Effects of cover crops on groundwater quality. *Cover Crops for Clean Water, Soil, and Water Conservation Society. Ankeny, Iowa, 266*, 793–799.

Mei, Y., Wu, C., & Liu, R. (2016). Hydrogen production from steam reforming of bio-oil model compound and byproducts elimination. *International Journal of Hydrogen Energy, 41*(21), 9145–9152. doi:10.1016/j.ijhydene.2015.12.133

Melgoza, R., & Buitron, C. (2004). Anaerobic/aerobic treatment of colorants present in textile effluents. *Water Science and Technology, 50*(2), 149–155. doi:10.2166/wst.2004.0111 PMID:15344785

Mellin, P., Kantarelis, E., & Yang, W. (2014). Computational fluid dynamics modeling of biomass fast pyrolysis in a fluidized bed reactor, using a comprehensive chemistry scheme. *Fuel, 117*, 704–715. doi:10.1016/j.fuel.2013.09.009

Mellin, P., Zhang, Q., Kantarelis, E., & Yang, W. (2013). An Euler–Euler approach to modeling biomass fast pyrolysis in fluidized-bed reactors – Focusing on the gas phase. *Applied Thermal Engineering, 58*(1-2), 344–353. doi:10.1016/j.applthermaleng.2013.04.054

Memon, G. Z., Bhanger, M., & Akhtar, M. (2007). The removal efficiency of chestnut shells for selected pesticides from aqueous solutions. *Journal of Colloid and Interface Science, 315*(1), 33–40. doi:10.1016/j.jcis.2007.06.037 PMID:17631888

Meriç, S., Selçuk, H., & Belgiorno, V. (2005). Acute toxicity removal in textile finishing wastewater by Fenton's oxidation, ozone and coagulation–flocculation processes. *Water Research*, *39*(6), 1147–1153. doi:10.1016/j.watres.2004.12.021 PMID:15766969

Merzouk, B., Gourich, B., Sekki, A., Madani, K., & Chibane, M. (2009). Removal turbidity and separation of heavy metals using electrocoagulation-electroflotation technique: A case study. *Journal of Hazardous Materials*, *164*(1), 215–222. doi:10.1016/j.jhazmat.2008.07.144 PMID:18799259

Mestre, A. S., Pires, J., Nogueira, J. M. F., & Carvalho, A. P. (2007). Activated carbons for the adsorption of ibuprofen. *Carbon*, *45*(10), 1979–1988. doi:10.1016/j.carbon.2007.06.005

Metcalf & Eddy. (2002). *Wastewater treatment: Treatment and reuse* (4th ed.). Washington, DC: McGraw Hill.

Metcalf & Eddy. (2004). *Wastewater engineering, treatment and reuse* (4th Ed.).

Mettanant, V., Basu, P., & Butler, J. (2009). Agglomeration of biomass fired fluidized bed gasifier and combustor. *Canadian Journal of Chemical Engineering*, *87*(5), 656–684. doi:10.1002/cjce.20211

Miao, X., Xu, L., Li, H., & Yang, Z. (2018). Determination of olaquindox, carbadox and cyadox in animal feeds by ultra-performance liquid chromatography tandem mass spectrometry. *Food Additives & Contaminants: Part A*(just-accepted).

MIDA. *Comintel Corp expects first waste-to-energy plant to contribute to earnings by June.* (2018). Retrieved from http://www.mida.gov.my/home/6264/news/comintel-corp-expects-first-waste-to-energy-plant-to- contribute -to-earnings-by-june/

Mikkelsen, L. H., & Keiding, K. (2002). Physico-chemical characteristics of full-scale sewage sludges with implications to dewatering. *Water Research*, *36*(10), 2451–2462. doi:10.1016/S0043-1354(01)00477-8 PMID:12153011

Miles, T. R., Rasmussen, E. M., & Gray, M. (2016). Aqueous contaminant removal and stormwater treatment using biochar. *Agricultural and environmental applications of biochar: Advances and barriers*(sssaspecpub63), 341-376.

Miller, R. S., & Bellan, J. (1997). A generalized biomass pyrolysis model based on superimposed cellulose, hemicellulose and liqnin kinetics. *Combustion Science and Technology*, *126*(1-6), 97–137. doi:10.1080/00102209708935670

Ming, T. T., Hyun, K. T., & Myun, J. (2007). Characterization of livestock wastewater at various stages of wastewater treatment plant. *The Malaysian Journal of Analytical Sciences*, *11*, 23–28.

Ming, Y., & De-Sheng, W. (2010). Experimental study of antibiotic wastewater. *Journal of MBR Treatment. Environmental Protection and Recycling Economy*, *9*, 54–55.

Mininni, G., Laera, G., Bertanza, G., Canato, M., & Sbrilli, A. (2015). Mass and energy balances of sludge processing in reference and upgraded wastewater treatment plants. *Environmental Science and Pollution Research International*, *22*(10), 7203–7215. doi:10.100711356-014-4013-2 PMID:25598155

Ministry of Environment (2006). Report on promotion of eco-fuel for transportation.

Ministry of Housing and Local Government, Malaysia. (2004). The final report of "The study of the safe closure and rehabilitation of landfill sites in Malaysia": *Guideline for safe closure and rehabilitation of MSW landfill sites.* Vol. 3, pp. 1 – 126.

Ministry of Housing and Local Government, Solid Waste Management Lab. (2012). Retrieved from http://jpspn.kpkt.gov.my/resources/index/user_1/Sumber_Rujukan/kajian/lab_sisa_pepejal.pdf

Minnesota Pollution Control Agency (MPCA). (2008). Nutrients: Phosphorus, Nitrogen Sources, Impact on Water Quality - A General Overview. 1-2. Retrieved from www.pca.state.mn.us/water/standards/index.html

Miranda, M., Pinto, F., Gulyurtlu, I., & Cabrita, I. (2013). Pyrolysis of rubber tyre wastes: a kinetic study. *Fuel*, *103*, 542–552. doi:10.1016/j.fuel.2012.06.114

Mirbagheri, S. A., & Hosseini, S. N. (2005). Pilot plant investigation on petrochemical wastewater treatment for the removal of copper and chromium with the objective of reuse. *Desalination*, *171*(1), 85–93. doi:10.1016/j.desal.2004.03.022

Mirka, L., Mohamed, C., Sillanpää, M., & Sillanpää, M. (2018). Occurrence, identification and removal of microplastic particles and fibers in conventional activated sludge process and advanced MBR technology. *Water Research*, *133*, 236–246. doi:10.1016/j.watres.2018.01.049 PMID:29407704

Miron, Y., Zeeman, G., van Lier, J. B., & Lettinga, G. (2000). The role of sludge retention time in the hydrolysis and acidification of lipids, carbohydrates and proteins during digestion of primary sludge in CSTR systems. *Water Research*, *34*(5), 1705–1713. doi:10.1016/S0043-1354(99)00280-8

Mishra, A., & Bajpai, M. (2006). The flocculation performance of Tamarindus mucilage in relation to removal of vat and direct dyes. *Bioresource Technology*, *97*(8), 1055–1059. doi:10.1016/j.biortech.2005.04.049 PMID:16219461

Mishra, A., Srinivasan, R., & Dubey, R. (2002). Flocculation of textile wastewater by Plantago psyllium mucilage. *Macromolecular Materials and Engineering*, *287*(9), 592–596. doi:10.1002/1439-2054(20020901)287:9<592::AID-MAME592>3.0.CO;2-Q

Mitra, A., & Mukhopadhyay, S. (2016). Biofilm mediated decontamination of pollutants from the environment. *AIMS Bioengineering*, *3*(1), 44–59. doi:10.3934/bioeng.2016.1.44

Mittal, A. (2011). Biological wastewater treatment. *Water Today*, *1*, 32–44.

Mneymneh, B. E., Abbas, M., & Khoury, H. (2017). Automated hardhat detection for construction safety applications. *Procedia Engineering*, *196*, 895–902. doi:10.1016/j.proeng.2017.08.022

Moawad, A., Mahmoud, U. F., El-Khateeb, M. A., & El-Molla, E. (2009). Coupling of sequencing batch reactor and UASB reactor for domestic wastewater treatment. *Desalination*, *242*(1), 325–335. doi:10.1016/j.desal.2008.05.011

Mofokeng, D. S., Adeleke, R., & Aiyegoro, O. A. (2016). The analysis of physicochemical characteristics of pig farm seepage and its possible impact on the receiving natural environment. *African Journal of Environmental Science and Technology*, *10*(8), 242–252. doi:10.5897/AJEST2016.2084

Mohadesi, M., Aghel, B., Maleki, M., & Ansari, A. (2019). Production of biodiesel from waste cooking oil using a homogeneous catalyst: Study of semi-industrial pilot of microreactor. *Renewable Energy*, *136*, 677–682. doi:10.1016/j.renene.2019.01.039

Mohajeri, S., Aziz, H. A., Isa, M. H., Bashir, M. J., Mohajeri, L., & Adlan, M. N. (2010). Influence of Fenton reagent oxidation on mineralization and decolorization of municipal landfill leachate. *Journal of Environmental Science and Health Part A*, *45*(6), 692–698. doi:10.1080/10934521003648883 PMID:20390917

Mohajeri, S., Aziz, H. A., Isa, M. H., Zahed, M. A., & Adlan, M. N. (2010). Statistical optimization of process parameters for landfill leachate treatment using electro-Fenton technique. *Journal of Hazardous Materials*, *176*(1-3), 749–758. doi:10.1016/j.jhazmat.2009.11.099 PMID:20022166

Mohamadnia, Z., Zohuriaan-Mehr, M. J., Kabiri, K., & Razavi-Nouri, M. (2008). Tragacanth gum-graft-polyacrylonitrile: Synthesis, characterization and hydrolysis. *Journal of Polymer Research*, *15*(3), 173–180. doi:10.100710965-007-9156-0

Mohammad, M. A., Mohammad, S. H., Fariborz, M., & Mehdi, K. (2014). Soil remediation via bioventing, vapor extraction and transition regime between vapor extraction and bioventing. *International Journal of Environmental Health Engineering,* 2014, 3(1). Retrieved from https://www.researchgate.net/publication/305114125_Soil_remediation_via_bioventing_vapor_extraction_and_transition_regime_between_vapor_extraction_and_bioventing

Mohammadi, M., Najafpour, G. D., Younesi, H., Lahijani, P., Uzir, M. H., & Mohamed, A. R. (2011). Bioconversion of synthesis gas to second generation biofuels: A review. *Renewable & Sustainable Energy Reviews*, *15*(9), 4255–4273. doi:10.1016/j.rser.2011.07.124

Mohammed, M. A. A., Salmiaton, A., Azlina, W. A. K. G. W., & Amran, M. S. M. (2012). Gasification of oil palm empty fruit bunches: A characterization and kinetic study. *Bioresource Technology*, *110*, 628–636. doi:10.1016/j.biortech.2012.01.056 PMID:22326334

Mohammed, M. A. A., Salmiaton, A., Azlina, W. A. K. G. W., Amran, M. S. M., & Fakhru'l-Razi, A. (2011). Air gasification of empty fruit bunch for hydrogen-rich gas production in a fluidized-bed reactor. *Energy Conversion and Management*, *52*(2), 1555–1561. doi:10.1016/j.enconman.2010.10.023

Mohammed, M., Salmiaton, A., Azlina, W. W., & Amran, M. M. (2011). Gasification of empty fruit bunch for hydrogen rich fuel gas production. *Journal of Applied Sciences (Faisalabad)*, *11*(13), 2416–2420. doi:10.3923/jas.2011.2416.2420

Mohammed, M., Salmiaton, A., Azlina, W. W., Amran, M. M., Fakhru'l-Razi, A., & Taufiq-Yap, Y. (2011). Hydrogen rich gas from oil palm biomass as a potential source of renewable energy in Malaysia. *Renewable & Sustainable Energy Reviews*, *15*(2), 1258–1270. doi:10.1016/j.rser.2010.10.003

Mohan, D., Pittman, C. U. Jr, Bricka, M., Smith, F., Yancey, B., Mohammad, J., ... Gong, H. (2007). Sorption of arsenic, cadmium, and lead by chars produced from fast pyrolysis of wood and bark during bio-oil production. *Journal of Colloid and Interface Science*, *310*(1), 57–73. doi:10.1016/j.jcis.2007.01.020 PMID:17331527

Mohan, D., Pittman, C. U. Jr, & Steele, P. H. (2006). Pyrolysis of wood/biomass for bio-oil: A critical review. *Energy & Fuels*, *20*(3), 848–889. doi:10.1021/ef0502397

Mohan, S. V., Rao, N. C., Prasad, K. K., & Sarma, P. N. (2005). Bioaugmentation of an anaerobic sequencing batch biofilm reactor with immobilized sulphate reducing bacteria for the treatment of sulphate bearing chemical wastewater. *Process Biochemistry*, *40*(8), 2849–2857. doi:10.1016/j.procbio.2004.12.027

Mohan, S. V., Rao, N. C., Srinivas, S., Prasad, K. K., & Karthikeyan, J. (2002). Treatment of stimulated Reactive Yellow 22 (azo) dye effluents using Spirogyra species. *Waste Management (New York, N.Y.)*, *22*(6), 575–582. doi:10.1016/S0956-053X(02)00030-2 PMID:12214968

Mohee, R., & Bundhoo, M. A. (2015). A comparative analysis of solid waste management in developed and developing countries. In T. Simelane & R. Mohee (Eds.), *Future directions of MSW management in Africa* (pp. 6–29). South Africa: Africa Institute of South Africa. doi:10.2307/j.ctvh8r2sj.9

Mollah, M. Y. A., Morkovsky, P., Gomes, J. A. G., Kesmez, M., Parga, J., & Cocke, D. L. (2004). Fundamentals, present and future perspectives of electrocoagulation. *Journal of Hazardous Materials*, *114*(1), 199–210. doi:10.1016/j.jhazmat.2004.08.009 PMID:15511592

Möller, K., & Müller, T. (2012). Effects of anaerobic digestion on digestate nutrient availability and crop growth: A review. *Engineering in Life Sciences*, *12*(3), 242–257.

Momodu, M. A., & Anyakora, C. A. (2010). Heavy metal contamination of ground water: The Surulere case study. *Res J Environ Earth Sci*, *2*(1), 39–43.

Mon, R. (2013). Seawater desalination operation maintenance and troubleshooting. Retrieved from http://www.slideshare.net/RajeshMon/seawater-desalination-operation-maintainence-and-trouble-shooting

Mondal, R., Sen, K. G., Chatterjee, M., Sen, K. B., & Sen, S. (2000). Ground level concentration of nitrogen oxides (NOx) at some traffic intersection points in Calcutta. *Atmospheric Environment*, *34*(4), 629–633. doi:10.1016/S1352-2310(99)00216-2

Monlau, F., Barakat, A., Trably, E., Dumas, C., Steyer, J.-P., & Carrère, H. (2013). Lignocellulosic materials into biohydrogen and biomethane: impact of structural features and pretreatment. *Critical Reviews in Environmental Science and Technology*, *43*(3), 260–322. doi:10.1080/10643389.2011.604258

Monne, C. H., Branedli, O., Schaeppi, G., Schindler, C. H., Asckermannm, U., Leuenberger, P. H., & Sapaldia, T. (1995). Particulate matter < 10 lm in urban, rural and alpine air in Switzerland. *Atmospheric Environment*, *29*, 2565–2573. doi:10.1016/1352-2310(95)94999-U

Monnot, M., Laborie, S., & Cabassud, C. (2016). Granular activated carbon filtration plus ultrafiltration as a pretreatment to seawater desalination lines: Impact on water quality and UF fouling. *Desalination*, *383*(April), 1–11. doi:10.1016/j.desal.2015.12.010

Montgomery, D. C. (2017). *Design and analysis of experiments* (9th ed.). Hoboken, NJ: John Wiley & Sons.

Moody, B. (1991). *Comparative inorganic chemistry* (3rd ed.). New York: Chapman and Hall.

Moon, J., Jo, W., Jeong, S., Bang, B., Choi, Y., Hwang, J., & Lee, U. (2017). Gas cleaning with molten tin for hydrogen sulfide and tar in producer gas generated from biomass gasification. *Energy*, *130*, 318–326. doi:10.1016/j.energy.2017.04.121

Moore, J. A. (1988). *Evaluating coliform concentrations in runoff from various animal waste management systems*. In C. Namasivayam & K. Kadirvelu (Eds.), Uptake of mercury (II) from wastewater by activated carbon from an unwanted agricultural solid by-product: Coirpith. *Carbon*, *37*(1), 79–84.

Moosvi, S., Kehaira, H., & Madamwar, D. (2005). Decolorization of textile dye Reactive Violet 5 by a newly isolated bacterial consortium RVM 11.1. *World Journal of Microbiology & Biotechnology*, *21*(5), 667–672. doi:10.100711274-004-3612-3

Moosvi, S., Kher, X., & Madamwar, D. (2007). Isolation, characterization, and decolorization of textile dyes by a mixed bacterial consortium JW-2. *Dyes and Pigments*, *74*(3), 723–729. doi:10.1016/j.dyepig.2006.05.005

Moradi, M., & Moussavi, G. (2019). Enhanced treatment of tannery wastewater using the electrocoagulation process combined with uvc/vuv photoreactor: Parametric and mechanistic evaluation. *Chemical Engineering Journal*, *358*, 1038–1046. doi:10.1016/j.cej.2018.10.069

Morgana, S., Ghigliotti, L., Estévez-Calvar, N., Stifanese, R., Wieckzorek, A., Doyle, T., ... Garaventa, F. (2018). *Microplastics in the Arctic: A case study with sub-surface water and fish samples off Northeast Greenland.* Environ Pollut. 242 (Pt B)

Morgan-Sagastume, F., & Allen, D. G. (2003). Effects of temperature transient conditions on aerobic biological treatment of wastewater. *Water Research*, *37*(15), 3590–3601. doi:10.1016/S0043-1354(03)00270-7 PMID:12867325

Morling, S. (2010). Nitrogen removal and heavy metals in leachate treatment using SBR technology. *Journal of Hazardous Materials*, *174*(1), 679–686. doi:10.1016/j.jhazmat.2009.09.104 PMID:19836884

Moropoulou, A., Bakolas, A., & Aggelakopoulou, E. (2001). The effects of limestone characteristics and calcination temperature on the reactivity of quicklime. *Cement and Concrete Research*, *31*(4), 633–639. doi:10.1016/S0008-8846(00)00490-7

Mor, R., & Sivan, A. (2008). Biofilm formation and partial biodegradation of polystyrene by the actinomycete *Rhodococcus ruber. Biodegradation, 19*(6), 851–858. doi:10.100710532-008-9188-0 PMID:18401686

Mostafa, K. M., Samarkandy, A. R., & El-Sanabary, A. A. (2011). Grafting onto carbohydrate polymer using novel potassium persulfate/tetramethylethylene diamine redox system for initiating grafting. *Advances in Polymer Technology, 30*(2), 138–149. doi:10.1002/adv.20210

Motta La, E. J., Silva, E., Bustillos, A., Padron, H., & Luque, J. (2007). Combined anaerobic/aerobic secondary municipal wastewater treatment: pilot-plant demonstration of the UASB/aerobic solids contact system. *Journal of Environmental Engineering, 133*(4), 397–403. doi:10.1061/(ASCE)0733-9372(2007)133:4(397)

MoUD. (2012). *Transforming city bus transport in India through financial assistance for bus procurement under JN-NURM*. New Delhi, India: Ministry of Urban Development, the Government of India.

Mouneimne, A. H., Carrère, H., Bernet, N., & Delgenès, J. P. (2003). Effect of saponification on the anaerobic digestion of solid fatty residues. *Bioresource Technology, 90*(1), 89–94. doi:10.1016/S0960-8524(03)00091-9 PMID:12835063

Moussavi, G., Barikbin, B., & Mahmoudi, M. (2010). The removal of high concentrations of phenol from saline wastewater using aerobic granular SBR. *Chemical Engineering Journal, 158*(3), 498–504. doi:10.1016/j.cej.2010.01.038

Moussavi, G., Yazdanbaksh, A., & Heidarizad, M. (2009). The removal of formaldehyde from concentrated synthetic wastewater using O_3/MgO/H_2O_2 process integrated with the biological treatment. *Journal of Hazardous Materials, 171*(1-3), 907–913. doi:10.1016/j.jhazmat.2009.06.090 PMID:19616892

Moutsoglou, A. (2012). A comparison of prairie cordgrass and switchgrass as a biomass for syngas production. *Fuel, 95*, 573–577. doi:10.1016/j.fuel.2011.12.016

MPOB. (2016). *Malaysian Oil Palm Statistics*. Malaysia: Retrieved from Selangor.

Mu, D., Horowitz, N., Casey, M., & Jones, K. (2017). Environmental and economic analysis of an in-vessel food waste composting system at Kean University in the U.S. *Waste Management (New York, N.Y.), 59*, 476–486. doi:10.1016/j.wasman.2016.10.026 PMID:27816469

Muhammad, A., Soares, A., & Jefferson, B. (2019). The impact of background wastewater constituents on the selectivity and capacity of a hybrid ion exchange resin for phosphorus removal from wastewater. *Chemosphere, 224*, 494–501. doi:10.1016/j.chemosphere.2019.01.085 PMID:30831501

Muhammad, N., Ellis, K., Parr, J., & Smith, M. D. (1996) Optimization of slow sand filtration. In *22nd WEDC Conference Reaching the unreached: challenges for the 21st century. New Delhi*, India.

Mukherjee, I., & Sovacool, B. K. (2014). Palm oil-based biofuels and sustainability in southeast Asia: A review of Indonesia, Malaysia, and Thailand. *Renewable & Sustainable Energy Reviews, 37*, 1–12.

Mukherjee, A., & Mukherjee, G. (1998). Occupational exposure of the traffic personnel of Calcutta to lead and carbon monoxide. *Pollution Research, 17*, 359–362.

Mulamoottil, G., McBean, E., & Rovers, F. (1999). *Constructed wetlands for the treatment of landfill leachates*. Boca Raton, FL: CRC Press.

Mulligan, C. N., Yong, R. N., & Gibbs, B. F. (2001). Surfactant-enhanced remediation of contaminated soil. *Reviews in engineering geology, 60*(1-4), 371–380. doi:10.1016/S0013-7952(00)00117-4

Mundargi, R. C., Agnihotri, S. A., Patil, S. A., & Aminabhavi, T. M. (2006). Graft copolymerization of methacrylic acid onto guar gum, using potassium persulfate as an initiator. *Journal of Applied Polymer Science*, *101*(1), 618–623. doi:10.1002/app.23325

Mundi, I. (2017). *Malaysia palm oil production by year.* Retrieved from http://www.indexmundi.com/agriculture/?country=my&commodity=palm&oilgraph=production

Murad, Q. (2008). Can congestion charging soothe Beijing's woes? In China-Dialogue. Article No. 2512, London, UK.

Murphy, S. R., & Wadey, B. L. (1988). DEHP and toxicity. *Journal of Vinyl and Additive Technology*, *10*(3), 121–124. doi:10.1002/vnl.730100306

Murray, A., Lies, D., Li, G., Nealson, K., Zhou, J., & Tiedje, J. (2001). DNA/DNA hybridization to microarrays reveals gene specific differences between closely related microbial genomes. *Proceedings of the National Academy of Sciences of the United States of America*, *98*(17), 9853–9858. doi:10.1073/pnas.171178898 PMID:11493693

Murthy, S. N., & Novak, J. T. (1998). Effects of potassium ion on sludge settling, dewatering and effluent properties. *Water Science and Technology*, *37*(4-5), 317–324. doi:10.2166/wst.1998.0655

Murthy, S. N., & Novak, J. T. (1999). Factors affecting floc properties during aerobic digestion: Implications for dewatering. *Water Environment Research*, *71*(2), 197–202. doi:10.2175/106143098X121879

Muryanto, M., Marlina, E., Sari, A. A., Harimawan, A., & Sudarno, S. (2018). Treatment of beverage industry wastewater using a combination of electrocoagulation and adsorption processes. In *AIP Conference Proceedings*. 10.1063/1.5064290

Mutambanengwe, C., Togo, C., & Whiteley, C. (2007). Decolourization and degradation of textile dyes with bio sulfidogenic hydrogenases. *Biotechnology Progress*, *23*, 1095–1100. PMID:17880103

Muthukumar, M., Sargunamani, D., Senthilkumar, M., & Selvakumai, N. (2005). Studies on decoloration, toxicity and the possibility for recycling of acid dye effluents using ozone treatment. *Dyes and Pigments*, *64*(1), 39–44. doi:10.1016/j.dyepig.2004.03.012

Muyibi, S. A., & Evison, L. M. (1996). Coagulation of turbid water and softening of hardwater with Moringa oleifera seeds. *The International Journal of Environmental Studies*, *49*(3), 247–259. doi:10.1080/00207239608711028

Muyibi, S. A., & Okuofu, C. A. (1995). Coagulation of low turbidity surface waters with Moringa oleifera seeds. *The International Journal of Environmental Studies*, *48*(3-4), 263–273. doi:10.1080/00207239508710996

Muyzer, G., & Smalla, K. (1998). Application of denaturing gradient gel electrophoresis and temperature gradient gel electrophoresis (TGGE) in microbial ecology. Mini review. *Anton Leeuw Int J G*, *73*(1), 127–141. doi:10.1023/A:1000669317571 PMID:9602286

Nadean, R. R., Singhvi, J., Ryabik, Y., & Lin, I. (1993). Monitoring bioremediation for bioremediation efficacy: *The Marrow Marsh experience. Proceedings of the 1993 Oil Spill Conference, American Petroleum Institute, Washington, DC*, pp. 477-485.

Nagapan, S., Kaliannan, S., Abdullah, A. H., Sohu, S., Deraman, R., Hasmori, M. F., & Abas, N. H. (2018). Preliminary survey on the crucial root causes of material waste generation in Malaysian construction industry. *Engineering, Technology & Applied Scientific Research*, *8*(6), 3580–3584.

Naik, U. S. (2013). *The synthesis and characterisation of novel ultra-flexible lipidic vesicles using propanol* (Doctoral dissertation, University of Central Lancashire).

Naik, S. N., Goud, V. V., Rout, P. K., & Dalai, A. K. (2010). Production of first and second generation biofuels: A comprehensive review. *Renewable & Sustainable Energy Reviews*, *14*(2), 5.

Naiman, R. J., & Dudgeon, D. (2011). Global alteration of freshwaters: Influences on human and environmental well-being. *Ecological Research*, *26*(5), 865–873. doi:10.100711284-010-0693-3

Najafpour, G., Yieng, H. A., Younesi, H., & Zinatizadeh, A. (2005). Effect of organic loading on performance of rotating biological contactors using palm oil mill effluents. *Process Biochemistry*, *40*(8), 2879–2884. doi:10.1016/j.procbio.2005.01.002

Najafpour, G., Zinatizadeh, A., Mohamed, A. R., Isa, M. H., & Nasrollahzadeh, H. (2006). High-rate anaerobic digestion of palm oil mill effluent in an up-flow sludge fixed film bioreactor. *Process Biochemistry*, *41*(2), 370–379. doi:10.1016/j.procbio.2005.06.031

Najib, M. Z. M., Ujang, Z., Salim, M. R., & Ibrahim, Z. (2017). Developed microbial granules containing photosynthetic pigments for carbon dioxide reduction in palm oil mill effluent. *International Biodeterioration & Biodegradation*, *116*, 163–170. doi:10.1016/j.ibiod.2016.10.031

Namdeo, A., & Bell, M. C. (2005). Characteristics and health implications of fine and coarse particulates at roadside, urban background and rural sites in UK. *Environment International*, *31*(4), 565–573. doi:10.1016/j.envint.2004.09.026 PMID:15788196

Nandi, B. K., Goswami, A., & Purkait, M. K. (2009). Adsorption characteristics of brilliant green dye on kaolin. *Journal of Hazardous Materials*, *161*(1), 387–395. doi:10.1016/j.jhazmat.2008.03.110 PMID:18456401

Nanseu-Njiki, C. P., Tchamango, S. R., Ngom, P. C., Darchen, A., & Ngameni, E. (2009). Mercury (ii) removal from water by electrocoagulation using aluminium and iron electrodes. *Journal of Hazardous Materials*, *168*(2), 1430–1436. doi:10.1016/j.jhazmat.2009.03.042 PMID:19349114

Narvaez, I., Orio, A., Aznar, M. P., & Corella, J. (1996). Biomass gasification with air in an atmospheric bubbling fluidized bed. Effect of six operational variables on the quality of the produced raw gas. *Industrial & Engineering Chemistry Research*, *35*(7), 2110–2120. doi:10.1021/ie9507540

Naseem, R., & Tahir, S. (2001). Removal of Pb (II) from aqueous/acidic solutions by using bentonite as an adsorbent. *Water Research*, *35*(16), 3982–3986. doi:10.1016/S0043-1354(01)00130-0 PMID:12230182

National Academy of Sciences. (1997). *Measures of success for remediation technologies chapter four in innovations in ground water and soil clean-up: from concept to commercialization* (pp. 167–200). Washington, DC: National Academy Press.

National Bureau of Statistics of China. (2014). *China statistical yearbook on the environment*. Beijing: Chinese Statistics Press.

National Research Council. (2000). Waste Incineration and Public Health. Washington D.C. The National Academies Press.

National Research Council. (2004). *Review of the desalination and water purification roadmap*. National Academy of Sciences.

Naveen, B. P., Sumalatha, J., & Malik, R. K. (2018). A study on contamination of ground and surface water bodies by leachate leakage from a landfill in Bangalore, India. *International Journal of Geo-Engineering*, (9), 27. doi:10.118640703-018-0095-x

Ndabigengesere, A., & Narasiah, K. S. (1998). Quality of water treated by coagulation using Moringa oleifera seeds. *Water Research*, *32*(3), 781–791. doi:10.1016/S0043-1354(97)00295-9

Neczaj, E., Kacprzak, M., Kamizela, T., Lach, J., & Okoniewska, E. (2008). Sequencing batch reactor system for the co-treatment of landfill leachate and dairy wastewater. *Desalination*, *222*(1-3), 404–409. doi:10.1016/j.desal.2007.01.133

Negash, W. D. (2014). Application of iron activated persulfate for disinfection in water treatment. (Master's Thesis), University of California Riverside.

Nelson, M. I., Balakrishnan, E., Sidhu, H. S., & Chen, X. D. (2008). A fundamental analysis of continuous flow bioreactor models and membrane reactor models to process industrial wastewaters. *Chemical Engineering Journal*, *140*(1-3), 521–528. doi:10.1016/j.cej.2007.11.035

Nesheiwat, F. K., & Swanson, A. G. (2000). Clean contaminated sites using Fenton's reagent. *Chemical Engineering Progress*, *96*(40), 61–66.

Net, S., Dumoulin, D., El-Osmani, R., Rabodonirina, S., & Ouddane, B. (2014). Case study of PAHs, Me-PAHs, PCBs, phthalates and pesticides contamination in the Somme River water, France. *International Journal of Environmental of Research*, *8*(4), 1159–1170.

Net, S., Sempéré, R., Delmont, A., Paluselli, A., & Ouddane, B. (2015). Occurrence, fate, behavior and ecotoxicological state of phthalates in different environmental matrices. *Environmental Science & Technology*, *49*(7), 4019–4035. doi:10.1021/es505233b PMID:25730609

Neyens, E., Baeyens, J., Weemaes, M., & Heyder, B. D. (2002, Aug. 25–29). *Treatment of sewage sludge: The economy of using advanced sludge treatment methods for sewage sludge.* Paper presented at the CHISA, Prague.

Neyens, E., & Baeyens, J. (2003a). A review of classic Fenton's peroxidation as an advanced oxidation technique. *Journal of Hazardous Materials*, *B98*(1-3), 33–50. doi:10.1016/S0304-3894(02)00282-0 PMID:12628776

Neyens, E., & Baeyens, J. (2003b). A review of thermal sludge pre-treatment processes to improve dewaterability. *Journal of Hazardous Materials*, *98*(1-3), 51–67. doi:10.1016/S0304-3894(02)00320-5 PMID:12628777

Neyens, E., Baeyens, J., & Dewil, R. (2004). Advanced sludge treatment affects extracellular polymeric substances to improve activated sludge dewatering. *Journal of Hazardous Materials*, *106*(2–3), 83–92. doi:10.1016/j.jhazmat.2003.11.014 PMID:15177096

Nghiem, L. D., Schäfer, A. I., & Elimelech, M. (2005). Pharmaceutical retention mechanisms by nanofiltration membranes. *Environmental Science & Technology*, *39*(19), 7698–7705. doi:10.1021/es0507665 PMID:16245847

Ng, K. H., & Cheng, C. K. (2015). A novel photomineralization of POME over UV-responsive TiO_2 photocatalyst: Kinetics of POME degradation and gaseous product formation. *RSC Advances*, *5*(65), 53100–53110. doi:10.1039/C5RA06922J

Ng, W., Goh, A., & Tay, J. (1988). Palm oil mill effluent treatment-liquid-solid separation with dissolved air floatation. *Biological Wastes*, *25*(4), 257–268. doi:10.1016/0269-7483(88)90087-0

Nicholson, F., Smith, S., Alloway, B., Carlton-Smith, C., & Chambers, B. (2003). An inventory of heavy metals inputs to agricultural soils in England and Wales. *The Science of the Total Environment*, *311*(1-3), 205–219. doi:10.1016/S0048-9697(03)00139-6 PMID:12826393

Nicole, D. B., Reinhart, D. R., & Batarseh, E. S. (2009). An assessment of bioreactor landfill costs and benefits. *Waste Management (New York, N.Y.)*, *29*(5), 1558–1567. doi:10.1016/j.wasman.2008.12.010 PMID:19167875

Nidheesh, P., & Gandhimathi, R. (2012). Trends in electro-Fenton process for water and wastewater treatment: An overview. *Desalination*, *299*, 1–15. doi:10.1016/j.desal.2012.05.011

Nidoni, P. G. (2017). Incineration process for solid waste management and effective utilization of byproducts. *International Research Journal of Engineering and Technology (IRJET), 4*(12).

Nielsen, H. B., & Angelidaki, I. (2008). Codigestion of manure and industrial organic waste at centralized biogas plants: Process imbalances and limitations. *Water Science and Technology, 58*(7), 1521–1528. PMID:18957768

Nigam, P., Armour, G., Banat, I. M., Singh, D., & Marchant, R. (2000). Physical removal of textile dyes from effluents and solid-state fermentation by dye-adsorbed agricultural residues. *Bioresource Technology, 72*(3), 219–226. doi:10.1016/S0960-8524(99)00123-6

Nijenhuis, J., Korbee, R., Lensselink, J., Kiel, J., & van Ommen, J. R. (2007). A method for agglomeration detection and control in full-scale biomass fired fluidized beds. *Chemical Engineering Science, 62*(1-2), 644–654. doi:10.1016/j.ces.2006.09.041

Nikoo, M. B., & Mahinpey, N. (2008). Simulation of biomass gasification in fluidized bed reactor using ASPEN PLUS. *Biomass and Bioenergy, 32*(12), 1245–1254. doi:10.1016/j.biombioe.2008.02.020

Nina, S., & Manceur, A. M. (2014). Analyzing the influence of urban forms on surface urban heat islands in Europe. *Journal of Urban Planning and Development, 142*, 475–495.

Nizami, A., Orozco, A., Groom, E., Dieterich, B., & Murphy, J. (2012). How much gas can we get from grass? *Applied Energy, 92*, 783–790.

Nizamuddin, S., Shrestha, S., Athar, S., Ali, B. S., & Siddiqui, M. A. (2016). A critical analysis on palm kernel shell from oil palm industry as a feedstock for solid char production. *Reviews in Chemical Engineering, 32*(5), 489–505. doi:10.1515/revce-2015-0062

Njoku, V., Islam, M. A., Asif, M., & Hameed, B. (2014). Utilization of sky fruit husk agricultural waste to produce high quality activated carbon for the herbicide bentazon adsorption. *Chemical Engineering Journal, 251*, 183–191. doi:10.1016/j.cej.2014.04.015

Nkwonta, O. I., & Ochieng, G. M. (2009a). Passive treatment of mine water using roughing filters as a pre-treatment option. In *Abstracts of the International Mine Water Conference*, Pretoria, South Africa.

Nkwonta, O. I., & Ochieng, G. M. (2009b). Roughing filter for water pre-treatment technology in developing countries: A review. *International Journal of Physical Sciences, 4*, 455–463.

Noble, J. (2006). GE ZeeWeed MBR technology for pharmaceutical wastewater treatment. *Membrane Technology, 2006*(9), 7–9. doi:10.1016/S0958-2118(06)70790-X

Nokyoo, C. (2016). *Swine waste management in Thailand*. Retrieved from http://www.wepa-db.net/activities/2016/20160126/pdf/AM5_4.TH_SwineWWT_160126.pdf

Noor, Z. Z., Yusuf, R. O., Abba, A. H., Abu Hassan, M. A., & Mohd, F. M. D. (2013). An overview for energy recovery from municipal solid wastes (MSW) in Malaysia scenario. *Renewable & Sustainable Energy Reviews, 20*, 378–384. doi:10.1016/j.rser.2012.11.050

Norulaini, N., Zuhairi, A., Hakimi, I., & Omar, A. (2004). Treatment of palm oil mill effluent (POME) supernatants using aerobic attached-growth system: trickling filter as a case study. *J Teknol, 40*(F), 77-90.

Nowak, A., Mazur, R., Panek, E., Dacewicz, E., & Chmielowski, K. (2019). Treatment efficiency of fish processing wastewater in different types of biological reactors. *Physics and Chemistry of the Earth Parts A/B/C, 109*, 40–48. doi:10.1016/j.pce.2018.09.007

Núñez, J., Yeber, M., Cisternas, N., Thibaut, R., Medina, P., & Carrasco, C. (2019). Application of electrocoagulation for the efficient pollutants removal to reuse the treated wastewater in the dyeing process of the textile industry. *Journal of Hazardous Materials, 371*, 705–711. doi:10.1016/j.jhazmat.2019.03.030 PMID:30897490

Nwuche, C., & Oghonna, J. (2011). Isolation of lipase producing fungi from palm oil mill effluent (POME) dump sites at Nsukka. *Brazilian Archives of Biology and Technology, 54*(1), 113–116. doi:10.1590/S1516-89132011000100015

Nyakuma, B. B., Ahmad, A., Johari, A., Abdullah, T. A. T., Oladokun, O., & Alkali, H. (2016). Gasification of oil palm empty fruit bunches (OPEFB) briquettes for bio-syngas production. *Jurnal Teknologi, 78*(8-3), 83-88.

Nyakuma, B. B., Mazangi, M., Tuan Abdullah, T. A., Johari, A., Ahmad, A., & Oladokun, O. (2014). Gasification of empty fruit bunch briquettes in a fixed bed tubular reactor for hydrogen production. *Applied Mechanics and Materials, 699*, 534-539. doi:10.4028/www.scientific.net/AMM.699.534

O'Brien, K. L., & Leichenko, R. M. (2000). Double exposure: Assessing the impacts of climate change within the context of economic globalization. *Global Environmental Change, 10*(3), 221–232. doi:10.1016/S0959-3780(00)00021-2

Obaja, D., Mace, S., Costa, J., Sans, C., & Mata-Alvarez, J. (2003). Nitrification, denitrification and biological phosphorus removal in piggery wastewater using a sequencing batch reactor. *Bioresource Technology, 87*(1), 103–111. doi:10.1016/S0960-8524(02)00229-8 PMID:12733583

Obaja, D., Mace, S., & Mata-Alvarez, J. (2005). Biological nutrient removal by a sequencing batch reactor (SBR) using an internal organic carbon source in digested piggery wastewater. *Bioresource Technology, 96*(1), 7–14. doi:10.1016/j.biortech.2004.03.002 PMID:15364074

Obiora-Okafo, I. A., & Onukwuli, O. D. (2015). Optimization of coagulation-flocculation process for colour removal from synthetic dye wastewater using natural organic polymers: Response surface methodology applied. *International Journal of Scientific & Engineering Research, 6*(12), 693–704.

Oehlmann, J., Schulte-Oehlmann, U., Kloas, W., Jagnytsch, O., Lutz, I., Kusk, K. O., ... Tyler, C. R. (2009). A critical analysis of the biological impacts of plasticizers on wildlife. *Philos Trans R Soc B, 364*(1526), 2047–2062. doi:10.1098/rstb.2008.0242 PMID:19528055

Oehmen, A., Lemos, P. C., Carvalho, G., Yuan, Z., Keller, J., Blackall, L. L., & Reis, M. A. (2007). Advances in enhanced biological phosphorus removal: From micro to macro scale. *Water Research, 41*(11), 2271–2300. PMID:17434562

Ogi, T., Nakanishi, M., & Fukuda, Y. (2011). Gasification of empty fruit bunch and bagasse using an entrained-flow mode reactor. *Journal of the Japan Institute of Energy, 90*(9), 886–894. doi:10.3775/jie.90.886

Ogi, T., Nakanishi, M., Fukuda, Y., & Matsumoto, K. (2013). Gasification of oil palm residues (empty fruit bunch) in an entrained-flow gasifier. *Fuel, 104*, 28–35. doi:10.1016/j.fuel.2010.08.028

Ogugbue, C. J., Morad, N., Sawidis, T., & Oranusi, N. A. (2011). Decolorization and partial mineralization of a polyazo dye by Bacillus firmus immobilized within tubular polymeric gel. *3 Biotech*, 2, 67–78.

Ohimain, E., Seiyaboh, E., Izah, S., Oghenegueke, V., & Perewarebo, T. (2012b). Some selected physico-chemical and heavy metal properties of palm oil mill effluents. *Greener Journal of Physical Sciences, 2*, 131-137.

Ohimain, E., Daokoru-Olukole, C., Izah, S., Eke, R., & Okonkwo, C. (2012a). Microbiology of palm oil mill effluents. *J Microbiology and Biotechnology Research, 2*(6), 852.

Okechalu, J., Dashen, M., Lar, P., Okechalu, B., & Gushop, T. (2011). Microbiological quality and chemical characteristics of palm oil sold within Jos Metropolis, Plateau State, Nigeria. *J Microbiology and Biotechnology Research, 1*, 107–112.

Oki, T., & Kanae, S. (2006). Global hydrological cycles and world water resources. *Science, 313*(5790), 1068–1071. doi:10.1126cience.1128845 PMID:16931749

Okoh, A. I., & Trejo-Hernandez, M. R. (2006). Remediation of petroleum hydrocarbon polluted systems: Exploiting the bioremediation strategies. *African Journal of Biotechnology*, *5*(25), 2520–2525. Retrieved from http://www.academicjournals.org/AJB

Okot-Okumu, J. (2015). Solid waste management in Uganda; challenges and options. In T. Simelane & R. Mohee (Eds.), *Future directions of MSW management in Africa* (pp. 107–135). South Africa: Africa Institute of South Africa.

Oktem, Y. A., Ince, O., Donnelly, T., Sallis, P., & Ince, B. K. (2006). Determination of optimum operating conditions of an acidification reactor treating a chemical synthesis-based pharmaceutical wastewater. *Process Biochemistry*, *41*(11), 2258–2263. doi:10.1016/j.procbio.2006.05.016

Okuda, T., Baes, A. U., Nishijima, W., & Okada, M. (2001). Isolation and characterization of coagulant extracted from Moringa oleifera seed by salt solution. *Water Research*, *35*(2), 405–410. doi:10.1016/S0043-1354(00)00290-6 PMID:11228992

Oladokun, O., Ahmad, A., Abdullah, T. T., Nyakuma, B. B., Al-Shatri, A. H., & Bello, A. A. (2015). Modelling multicomponent devolatilization kinetics of imperata cylindrica. *Chemical Engineering Transactions*, *45*, 919–924. doi:10.3303/CET1545154

Oladokun, O., Ahmad, A., Abdullah, T. A. T., Nyakuma, B. B., Kamaroddin, M. F. A., & Nor, S. H. M. (2017). Biohydrogen production from Imperata cylindrica bio-oil using non-stoichiometric and thermodynamic model. *International Journal of Hydrogen Energy*, *42*(14), 9011–9023. doi:10.1016/j.ijhydene.2016.05.200

Oldham, W. K., & Stevens, G. M. (n.d.). S Initial operating experiences of a nutrient removal process (Modified Bardenpho) at Kelowna, British Columbia. *Canadian Journal of Civil Engineering*, 474-479.

Oller, I., Malato, S., & Sánchez-Pérez, J. (2011). Combination of advanced oxidation processes and biological treatments for wastewater decontamination—A review. *The Science of the Total Environment*, *409*(20), 4141–4166. doi:10.1016/j.scitotenv.2010.08.061 PMID:20956012

Ollis, D. (2001). On the need for engineering models of integrated chemical and biological oxidation of wastewaters. *Water Science and Technology*, *44*(5), 117–123. doi:10.2166/wst.2001.0265 PMID:11695448

Oloche, O. B., Ozigis, I. I, Adeyemi, K. O & Ikpe, O. E. (2017). Construction and leakage detection of a dome type biogas digester in a village at Abuja, Nigeria. *FUOYE Journal of Engineering & Technology, 2(1)*

Olvera, J., & Lopez, A. (2012). *Biogas production from anaerobic treatment of agro-industrial wastewater*. Rijeka, Croatia: Intech.

Öman, C. B., & Junestedt, C. (2008). Chemical characterization of landfill leachates–400 parameters and compounds. *Waste Management (New York, N.Y.)*, *28*(10), 1876–1891. doi:10.1016/j.wasman.2007.06.018 PMID:17869498

Omerci, B., & Vesilind, P. A. (2001). Effect of dissolved organic material and cations of freezeethaw conditioning of activated and alum sludges. *Water Research*, *25*, 4299–4306.

Onay, T. T., & Pohland, F. G. (1998). In situ nitrogen management in controlled bioreactor landfills. *Water Research*, *4*(5), 1383–1392. doi:10.1016/S0043-1354(97)00392-8

Oosterveer, P. (2015). Promoting sustainable palm oil: Viewed from a global networks and flows perspective. *Journal of Cleaner Production*, *107*, 146–153. doi:10.1016/j.jclepro.2014.01.019

Ord, J. K., & Getis, A. (1994). *Distributional issues concerning distance statistics*. Working paper.

Oren, A., Gurevich, P., Azachi, M., & Henis, Y. (1992). Microbial degradation of pollutants at high salt concentrations. *Biodegradation*, *3*(2-3), 387–398. doi:10.1007/BF00129095

Ortega-Liébana, M., Sánchez-López, E., Hidalgo-Carrillo, J., Marinas, A., Marinas, J., & Urbano, F. (2012). A comparative study of photocatalytic degradation of 3-chloropyridine under UV and solar light by homogeneous (photo-Fenton) and heterogeneous (TiO2) photocatalysis. *Applied Catalysis B: Environmental*, *127*, 316–322. doi:10.1016/j.apcatb.2012.08.036

Osmani, M. (2011). Construction waste. In Waste: A handbook for management (pp. 207–218). doi:10.1016/B978-0-12-381475-3.10015-4

Ostrowski, S. R., Wilbur, S., Chou, C.-H. S. J., Pohl, H. R., Stevens, Y.-W., Allred, P. M., ... Tylenda, C. A. (1999). Agency for toxic substances and disease registry's 1997 priority list of hazardous substances. Latent effects—carcinogenesis, neurotoxicology, and developmental deficits in humans and animals. *Toxicology and Industrial Health*, *15*(7), 602–644. doi:10.1177/074823379901500702 PMID:10677885

Others Solid Waste Management Facilities. (2017). Retrieved from http://jpspn.kpkt.gov.my/index.php/pages/view/77

Othman, E., Yusoff, M. S., Aziz, H. A., Adlan, M. N., Bashir, M. J. K., & Hung, Y. T. (2010). The effectiveness of silica sand in semi-aerobic landfill leachate treatment. *Water (Basel)*, *2*(4), 904–925.

Otten, A., Alphenaar, A., Pijls, C., Spuij, F., & Wit, H. (1997). *In situ soil remediation*. Kluwer Academic Publishers. doi:10.1007/978-94-011-5594-6

Ouis, D. (2001). Annoyance from road traffic noise: a review. *Journal of Environmental Psychology*, *21*(1), 101–120. doi:10.1006/jevp.2000.0187

Ouyang, K., & Liu, J. (2009). Effect of sludge retention time on sludge characteristics and membrane fouling of membrane bioreactor. *Journal of Environmental Sciences (China)*, *21*(10), 1329–1335. doi:10.1016/S1001-0742(08)62422-5 PMID:19999985

Oxford Dictionaries. (2019). Definitions. Retrieved from https://en.oxforddictionaries.com/definition/waste

Oyaro, N., Juddy, O., Murago, E. N., & Gitonga, E. (2007). The contents of Pb, Cu, Zn and Cd in meat in Nairobi, Kenya. *Journal of Food Agriculture and Environment*, *5*(3&4), 119–121.

Oyegbile, B., Ay, P., & Narra, S. (2016). Optimization of physicochemical process for pre-treatment of fine suspension by flocculation prior to dewatering. *Desalination and Water Treatment*, *57*(6), 2726–2736. doi:10.1080/19443994.2015.1043591

Ozawa, T. (1992). Estimation of activation-energy by isoconversion methods. *Thermochimica Acta*, *203*, 159–165. doi:10.1016/0040-6031(92)85192-X

Özcan, A., Şahin, Y., Koparal, A. S., & Oturan, M. A. (2008). Degradation of picloram by the electro-Fenton process. *Journal of Hazardous Materials*, *153*(1-2), 718–727. doi:10.1016/j.jhazmat.2007.09.015 PMID:17935883

Özcan, A., Şahin, Y., Koparal, A. S., & Oturan, M. A. (2009). A comparative study on the efficiency of electro-Fenton process in the removal of propham from water. *Applied Catalysis B: Environmental*, *89*(3-4), 620–626. doi:10.1016/j.apcatb.2009.01.022

Ozturk, A., Celik, L., & Abdullah, I. M. (2012). Biodegradation of reactive red 195 azo dye by the bacterium Rhodopseudomonas Palustris 51ATA. *African Journal of Microbiological Research*, *6*, 120–126.

Pablo-Romero, M. P., Pozo-Barajas, R., & Yñiguez, R. (2017). Global changes in residential energy consumption. *Energy Policy*, *101*, 342–352.

Pachauri, R. K., & Reisinger, A. (Eds.). (2007). *Geneva. Climate change 2007: Synthesis Report*. Switzerland: IPCC.

Padoley, K. V., Rajvaidya, A. S., Subbarao, T. V., & Pandey, R. A. (2006). Biodegradation of pyridine in a completely mixed activated sludge process. *Bioresource Technology*, *97*(10), 1225–1236. doi:10.1016/j.biortech.2005.05.020 PMID:16084081

Pagilla, K. R., Kim, H., & Tapana, C. (2000). Aerobic thermophilic and anaerobic mesophilic treatment of swine waste. *Water Research*, *34*(10), 2747–2753. doi:10.1016/S0043-1354(00)00012-9

Pahlavan, P., Manzi, S., Rodriguez-Estrada, M. T., & Bignozzi, M. C. (2017). Valorization of spent cooking oils in hydrophobic waste-based lime mortars for restorative rendering applications. *Construction & Building Materials*, *146*, 199–209. doi:10.1016/j.conbuildmat.2017.04.001

Pahlavan, P., Manzi, S., Sansonetti, A., & Bignozzi, M. C. (2018). Valorization of organic additions in restorative lime mortars: Spent cooking oil and albumen. *Construction & Building Materials*, *181*, 650–658. doi:10.1016/j.conbuildmat.2018.06.089

Pakpahan, E. N., Isa, M. H., Kutty, S. R. M., Chantara, S., & Wiriya, W. (2013). Polycyclic aromatic hydrocarbon removal from petroleum sludge cake using thermal treatment with additives. *Environmental Technology*, *34*(3), 407–416. doi:10.1080/09593330.2012.698648 PMID:23530354

Pala, A., & Tokat, E. (2002). Color removal from cotton textile industry wastewater in an activated sludge system with various additives. *Water Research*, *36*(11), 2920–2925. doi:10.1016/S0043-1354(01)00529-2 PMID:12146882

Palcyzinsky, R., & Scotia, N. (2002). African Development Bank Report. *Study on solid waste management options for Africa.* Retrieved from https://sswm.info/sites/default/files/reference_attachments/AFRICAN%20DEVELOPMENT%20BANK%202002%20Study%20on%20Solid%20Waste%20Management%20Options%20for%20Africa.pdf

Palmer, J. D., & Brigham, C. J. (2016). Feasibility of triacylglycerol production for biodiesel, utilizing *Rhodococcus opacus* as a biocatalyst and fishery waste as feedstock. *Renewable and Sustainable Energy Reviews, 56*, 922-928. 78-597.

Palmieri, G., Cennamo, G., & Sannia, G. (2005). Remazol Brilliant Blue R decolorisation by the fungus Pleurotusostreatus and its oxidative enzymatic system. *Enzyme and Microbial Technology*, *36*(1), 17–24. doi:10.1016/j.enzmictec.2004.03.026

Panadare, D. C., & Rathod, V. K. (2015). Applications of waste cooking oil other than biodiesel: A review. *Iranian Journal of Chemical Engineering*, *12*, 55–76.

Pandey, J., & Singh, R. (2017). Heavy metals in sediments of Ganga River: Up-and downstream urban influences. *Applied Water Science*, *7*(4), 1669–1678. doi:10.100713201-015-0334-7

Pandey, R. R., Pandey, B. V., & Upadhyay, R. S. (2011). Characterisation and identification of Pseudomonas fluorescens NCIM 2100 degraded metabolic products of crystal violet. *African Journal of Biotechnology*, *10*, 624–631.

Pangarkar, B. L., Sane, M. G., & Guddad, M. (2011). Reverse osmosis and membrane distillation for desalination of groundwater: a review. *ISRN Materials Science*, *2011*, 1–9. doi:10.5402/2011/523124

Panizza, M., & Cerisola, G. (2009). Direct and mediated anodic oxidation of organic pollutants. *Chemical Reviews*, *109*(12), 6541–6569. doi:10.1021/cr9001319 PMID:19658401

Pankin, K. E., Ivanova, Y. V., Kuz'mina, R. I., & Shtykov, S. N. (2011). Comparison of the physicochemical characteristics of biofuels and petroleum fuels. *Chemistry and Technology of Fuels and Oils*, *47*(1), 7.

Pant, P., & Harrison, R. M. (2012). Critical review of receptor modeling for particulate matter: A case study of India. *Atmospheric Environment*, *49*, 1–12. doi:10.1016/j.atmosenv.2011.11.060

Paolini, V., Petracchini, F., Segreto, M., Tomassetti, L., Naja, N., & Cecinato, A. (2018). Environmental impact of biogas: A short review of current knowledge, *Journal of Environmental Science and Health. Part A*, *53*(10), 899–906.

Papadikis, K., Bridgwater, A. V., & Gu, S. (2008). CFD modelling of the fast pyrolysis of biomass in fluidised bed reactors, Part A: Eulerian computation of momentum transport in bubbling fluidised beds. *Chemical Engineering Science*, *63*(16), 4218–4227. doi:10.1016/j.ces.2008.05.045

Papadikis, K., Gu, S., & Bridgwater, A. V. (2009a). CFD modelling of the fast pyrolysis of biomass in fluidised bed reactors. Part B Heat, momentum and mass transport in bubbling fluidised beds. *Chemical Engineering Science*, *64*(5), 1036–1045. doi:10.1016/j.ces.2008.11.007

Papadikis, K., Gu, S., & Bridgwater, A. V. (2009b). CFD modelling of the fast pyrolysis of biomass in fluidised bed reactors: Modelling the impact of biomass shrinkage. *Chemical Engineering Journal*, *149*(1-3), 417–427. doi:10.1016/j.cej.2009.01.036

Papadikis, K., Gu, S., Bridgwater, A. V., & Gerhauser, H. (2009). Application of CFD to model fast pyrolysis of biomass. *Fuel Processing Technology*, *90*(4), 504–512. doi:10.1016/j.fuproc.2009.01.010

Papadikis, K., Gu, S., Fivga, A., & Bridgwater, A. V. (2010). Numerical comparison of the drag models of granular flows applied to the fast pyrolysis of biomass. *Energy & Fuels*, *24*(3), 2133–2145. doi:10.1021/ef901497b

Paraskeva, P., & Diamadopoulos, E. (2006). Technologies for olive mill wastewater (OMW) treatment: a review. *Journal of Chemical Technology & Biotechnology: International Research in Process. Environmental & Clean Technology*, *81*(9), 1475–1485.

Paredes, D., Kuschk, P., Mbwette, T., Stange, F., Müller, R., & Köser, H. (2007). New aspects of microbial nitrogen transformations in the context of wastewater treatment–a review. *Engineering in Life Sciences*, *7*(1), 13–25.

Park, C., Abu-Orf, M. M., & Novak, J. T. (2006). The digestibility of waste activated sludges. *Water Environment Research*, *78*(1), 59–68. doi:10.2175/106143005X84521 PMID:16553167

Park, C., Helm, R. F., & Novak, J. T. (2008). Investigating the fate of activated sludge extracellular proteins in sludge digestion using sodium dodecyl sulfate polyacrylamide gel electrophoresis. *Water Environment Research*, *80*(12), 2219–2227. doi:10.2175/106143008X325791 PMID:19146099

Park, C., Lee, J. W., Moon, S., Park, K. Y., & Jutidamrongphan, W. (2012). Advanced treatment of wastewater from food waste disposer in modified Ludzack-Ettinger type membrane bioreactor. *Environmental Engineering Research*, *17*(2), 59–63. doi:10.4491/eer.2012.17.2.059

Park, J., Kim, K., & Cho, Y. K. (2016). Framework of automated construction-safety monitoring using cloud-enabled BIM and BLE mobile tracking sensors. *Journal of Construction Engineering and Management*, *143*(2), 05016019. doi:10.1061/(ASCE)CO.1943-7862.0001223

Park, T. J., Lee, K. H., Jung, E. J., & Kim, C. W. (1999). Removal of refractory organics and color in pigment wastewater with Fenton oxidation. *Water Science and Technology*, *39*(10-11), 189–192. doi:10.2166/wst.1999.0653

Parravicini, V., Svardal, K., & Kroiss, H. (2004). *Further stabilisation of digested sewage sludge under anaerobic or aerobic conditions*. Paper presented at the 4th IWA World Water Congress and Exhibition, Marrakech, Morocco.

Parravicini, V., Smidt, E., Svardal, K., & Kroiss, H. (2006). Evaluating the stabilization degree of digested sewage sludge: Investigations at four municipal wastewater treatment plants. *Water Science and Technology*, *53*(8), 81–90. doi:10.2166/wst.2006.238 PMID:16784192

Parravicini, V., Svardal, K., & Kroiss, H. (2008). Post-aeration of anaerobically digested sewage sludge for advanced COD and nitrogen removal: Results and cost-benefit analysis at large-scale. *Water Science and Technology*, *57*(7), 1087–1094. doi:10.2166/wst.2008.211 PMID:18441437

Parshetti, G., Kalme, S., Saratale, G., & Govindwar, S. P. (2006). Biodegradation of malachite green by Kocuriarosea MTCC 1532. *Acta Chimica Slovenica*, *53*, 492–498.

Parsons, S. A., & Jefferson, B. (2006). *Introduction to potable water treatment processes*. Blackwell Publishing. doi:10.1002/9781444305470

Patil, P. D., Gude, V. G., Reddy, H. K., Muppaneni, T., & Deng, S. (2012). Biodiesel production from waste cooking oil using sulfuric acid and microwave irradiation processes. *Journal of Environmental Protection*, *3*, 107–113.

Patil, S. V., Patil, C. D., Salunke, B. K., Salunkhe, R. B., Bathe, G., & Patil, D. M. (2011). Studies on characterization of bioflocculant exopolysaccharide of Azotobacter indicus and its potential for wastewater treatment. *Applied Biochemistry and Biotechnology*, *163*(4), 463–472. doi:10.100712010-010-9054-5 PMID:20730607

Patil, V. B., Kulkarni, G. S., & Kore, V. S. (2012). Performance of horizontal roughing filters for wastewater: A review. *International Research Journal of Environment Sciences*, *1*, 53–55.

Patterson, J. W. (1985). Industrial wastewater treatment technology.

Patty, L., Real, B., & Joël Gril, J. (1997). The use of grassed buffer strips to remove pesticides, nitrate and soluble phosphorus compounds from runoff water. *Pesticide Science*, *49*(3), 243–251. doi:10.1002/(SICI)1096-9063(199703)49:3<243::AID-PS510>3.0.CO;2-8

Paulino, A. T., Minasse, F. A., Guilherme, M. R., Reis, A. V., Muniz, E. C., & Nozaki, J. (2006). Novel adsorbent based on silkworm chrysalides for removal of heavy metals from wastewaters. *Journal of Colloid and Interface Science*, *301*(2), 479–487. doi:10.1016/j.jcis.2006.05.032 PMID:16780853

Pavko, A. (2011). Fungal decolorization and degradation of synthetic dyes. Some Chemical Engineering Aspects. In Waste Water-Treatment and Reutilization, Intech, Croatia.

Payne, J. T., Wood, A. W., Hamlet, A. F., Palmer, R. N., & Lettenmaier, D. P. (2004). Mitigating the effects of climate change on the water resources of the Columbia River Basin. *Climatic Change*, *62*(1–3), 233–256. doi:10.1023/B:CLIM.0000013694.18154.d6

Pazarlioglu, N. K., Urek, R. O., & Ergun, F. (2005). Biodecolourization of Direct Blue 15 by immobilized *Phanerochaete Chrysosporium*. *Process Biochemistry (Barking, London, UK)*, *40*(5), 1923–1929. doi:10.1016/j.procbio.2004.07.005

Peavy, H. S., Rowe, D. R., & Tchobanoglous, G. (2019). *Environmental engineering/H. S. Peavy, D. R. Rowe, G. Tchobanoglous*.

Peng, Y.-Z., Yong, M.A., & Wang, S.-Y. (2007). Denitrification potential enhancement by addition of external carbon sources in a pre-denitrification process. *Journal of Environmental Sciences*, *19*(3), 284-289.

Peng, Y. (2017). Perspectives on technology for landfill leachate treatment. *Arabian Journal of Chemistry*, *10*, 2567–2574. doi:10.1016/j.arabjc.2013.09.031

Pere, J., Alen, R., Viikari, L., & Eriksson, L. (1993). Characterization and dewatering of activated sludge from the pulp and paper industry. *Water Science and Technology*, *28*(1), 193–201. doi:10.2166/wst.1993.0046

Perez, P., & Reyes, J. (2002). Prediction of maximum of 24-h average of PM10 concentrations 30h in advance in Santiago, Chile. *Atmospheric Environment*, *36*(28), 4555–4561. doi:10.1016/S1352-2310(02)00419-3

Periathamby, A., Hamid, F. S., & Khidzir, K. (2009). Evolution of solid waste management in Malaysia: Impacts and implications of the solid waste bill. *Journal of Material Cycles and Waste Management*, *11*(2), 96–103. doi:10.100710163-008-0231-3

Pervov, A. G. (1999). A simplified RO process design based on understanding of fouling mechanisms. *Desalination*, *126*(1-3), 227–247. doi:10.1016/S0011-9164(99)00179-4

Phan, A. N.,, & Phan, T. M. (2008). Biodiesel production from waste cooking oils. *Fuel*, *87*, 3490–3496.

Phugare, S. S., Kalyani, D. C., Patil, A. V., & Jadhav, J. P. (2011). Textile dye degradation by bacterial consortium and subsequent toxicological analysis of dye and dye metabolites using cytotoxicity, genotoxicity and oxidative stress studies. *Journal of Hazardous Materials*, *186*(1), 713–723. doi:10.1016/j.jhazmat.2010.11.049 PMID:21144656

Pilli, S., More, T. T., Yan, S., Tyagi, R. D., & Surampalli, R. Y. (2016). Fenton pre-treatment of secondary sludge to enhance anaerobic digestion: Energy balance and greenhouse gas emissions. *Chemical Engineering Journal*, *283*, 285–292. doi:10.1016/j.cej.2015.07.056

Pinheiro, A. C., Mesquita, N., Trovao, J., Soares, F., Tiago, I., Coelho, C., … Portugal, A. (2018). Limestone biodeterioration: A review on the Portuguese cultural heritage scenario. *Journal of Cultural Heritage (In-Press).*

Pinheiro, H. M., Tourand, E., & Thomas, O. (2004). Aromatic amines from azo dye reduction: Status review with emphasis on direct UV spectrophotometric detection in textile industry wastewaters. *Dyes and Pigments*, *61*(2), 121–139. doi:10.1016/j.dyepig.2003.10.009

Pirillo, S., Pedroni, V., Rueda, E., & Ferreira, M. L. (2009). Elimination of dyes from aqueous solutions using iron oxides and chitosan as adsorbents: A comparative study. *Quimica Nova*, *32*(5), 1239–1244. doi:10.1590/S0100-40422009000500030

Piriyaprasarth, S., & Sriamornsak, P. (2011). Flocculating and suspending properties of commercial citrus pectin and pectin extracted from pomelo (Citrus maxima) peel. *Carbohydrate Polymers*, *83*(2), 561–568. doi:10.1016/j.carbpol.2010.08.018

Plochl, M., & Heiermann, M. (2006). Biogas farming in central and northern Europe: a strategy for developing countries? Invited overview.

Pohland, F. G. (1995) Landfill bioreactors: historical perspective, fundamental principles, and new horizons in design and operations. In *Landfill Bioreactor Design and Operation Sem. Proc.*, EPA/600/R-95/146, pp. 9-24.

Poh, P. E., & Chong, M. F. (2009). Development of anaerobic digestion methods for palm oil mill effluent (POME) treatment. *Bioresource Technology*, *100*(1), 1–9. doi:10.1016/j.biortech.2008.06.022 PMID:18657414

Poh, P. E., Yong, W. J., & Chong, M. F. (2010). Palm oil mill effluent (POME) characteristic in high crop season and the applicability of high-rate anaerobic bioreactors for the treatment of POME. *Industrial & Engineering Chemistry Research*, *49*(22), 11732–11740. doi:10.1021/ie101486w

Pokhrel, D., & Viraraghavan, T. (2004). Treatment of pulp and paper mill wastewater—A review. *The Science of the Total Environment*, *333*(1-3), 37–58. doi:10.1016/j.scitotenv.2004.05.017 PMID:15364518

Polinder, S., Haagsma, J., Panneman, M., Scholten, A., Brugmans, M., & Van Beeck, E. (2016). The economic burden of injury: Health care and productivity costs of injuries in the Netherlands. *Accident; Analysis and Prevention*, *93*, 92–100. doi:10.1016/j.aap.2016.04.003 PMID:27177394

Pol, L. W. H., Lopes, S. I. D., Lettinga, G., & Lens, P. N. L. (2004). Anaerobic sludge granulation. *Water Research*, *38*(6), 1376–1389. doi:10.1016/j.watres.2003.12.002 PMID:15016515

Ponraj, M., Gokila, K., & Zambare, V. (2011). Bacterial decolorization of textile dye-Orange 3R. *Int. J. Adv. Biotechnol. Res.*, *2*, 168–177.

Pope, C. A. III. (2000). Review: Epidemiological basis for particulate air pollution health standards. *Aerosol Science and Technology*, *32*(1), 4–14. doi:10.1080/027868200303885

Pope, S. B. (1997). Computationally efficient implementation of combustion chemistry using in situadaptive tabulation. *Combustion Theory and Modelling*, *1*(1), 41–63. doi:10.1080/713665229

Postel, S. L., Daily, G. C., & Ehrlich, P. R. (1996). Human appropriation of renewable fresh water. *Science*, *271*(5250), 785–788. doi:10.1126cience.271.5250.785

Potter, A. E. (1969). Sulfur oxide capacity of limestones. *American Ceramic Society Bulletin*, *48*, 855–858.

Pounds, N., Maclean, S., Webley, M., Pascoe, D., & Hutchinson, T. (2008). Acute and chronic effects of ibuprofen in the mollusc Planorbis carinatus (Gastropoda: Planorbidae). *Ecotoxicology and Environmental Safety*, *70*(1), 47–52. doi:10.1016/j.ecoenv.2007.07.003 PMID:18207238

Prabhakaran, D., Kannadasan, T., & Basha, C. A. (2009). Treatability of resin effluents by electrochemical oxidation using batch recirculation reactor. *International Journal of Environmental Science and Technology*, *6*(3), 491–498. doi:10.1007/BF03326088

Prajapati, S. (2014). Cation exchange for ammonia removal from wastewater.

Praticò, F. G.,, & Anfosso-lédée, F. (2012). Trends and issues in mitigating traffic noise through quiet pavements. *Procedia: Social and Behavioral Sciences*, *53*, 203–212. doi:10.1016/j.sbspro.2012.09.873

Pratt, C., Redding, M., Hill, J., Shilton, A., Chung, M., & Guieysse, B. (2015). Good science for improving policy: Greenhouse gas emissions from agricultural manures. *Animal Production Science*, *55*(6), 691–701.

Prihasto, N., Liu, Q.-F., & Kim, S.-H. (2009). Pre-treatment strategies for seawater desalination by reverse osmosis system. *Desalination*, *249*(1), 308–316. doi:10.1016/j.desal.2008.09.010

Primo, O., Rivero, M. J., & Ortiz, I. (2008). Photo-Fenton process as an efficient alternative to the treatment of landfill leachates. *Journal of Hazardous Materials*, *153*(1-2), 834–842. doi:10.1016/j.jhazmat.2007.09.053 PMID:17961917

Pritchard, M., Craven, T., Mkandawire, T., Edmondson, A., & O'Neill, J. (2010). A comparison between Moringa oleifera and chemical coagulants in the purification of drinking water–An alternative sustainable solution for developing countries. *Physics and Chemistry of the Earth Parts A/B/C*, *35*(13-14), 798–805. doi:10.1016/j.pce.2010.07.014

Puig-Arnavat, M., Bruno, J. C., & Coronas, A. (2010). Review and analysis of biomass gasification models. *Renewable & Sustainable Energy Reviews*, *14*(9), 2841–2851. doi:10.1016/j.rser.2010.07.030

Puttaiah, E. T., & Kiran, B. R. (2007). Heavy metal transport in a sewage fed lake of Karnataka, India. In *Proceedings of TAAL The 12th World Lake Conference.*

Qadir, M., Sharma, B. R., Bruggeman, A., Choukr-Allah, R., & Karajeh, F. (2007). Non-conventional water resources and opportunities for water augmentation to achieve food security in water scarce countries. *Agricultural Water Management, 87*(1), 2–22. doi:10.1016/j.agwat.2006.03.018

Qasim, M., Badrelzaman, M., Darwish, N. N., Darwish, N. A., & Hilal, N. (2019). Reverse osmosis desalination: A state-of-the-art review. *Desalination, 459*(February), 59–104. doi:10.1016/j.desal.2019.02.008

Qasim, S. R., & Chiang, W. (1994). *Sanitary landfill leachate-generation, control and treatment*. Texas: Technomic Publishing.

Qin, B., & Sun, S. H. (2013). Planning parameters and household carbon emission: Evidence from high- and low-carbon neighborhoods in Beijing. *Habitat International, 37*, 52–60. doi:10.1016/j.habitatint.2011.12.017

Qing, Y., Shuying, W., Anming, Y., Jianhua, G., & Fengyang, B. (2007). Advanced nitrogen removal using pilot-scale SBR with intelligent control system built on three layer network. *Frontiers of Environmental Science & Engineering in China, 1*(1), 33–38. doi:10.100711783-007-0006-4

Qin, Z., Karniel, A., & Berlin, P. (2001). A mono-window algorithm for retrieving land surface temperature from Landsat TM data and its application to the Israel-Egypt border region. *International Journal of Remote Sensing, 22*(18), 3719–374. doi:10.1080/01431160010006971

Qiu, F., Li, Y., Yang, D., Li, X., & Sun, P. (2011). Biodiesel production from mixed soybean oil and rapeseed oil. *Applied Energy, 88*(6), 2050–2055. doi:10.1016/j.apenergy.2010.12.070

Qiu, Y.-R., & Mao, L.-J. (2013). Removal of heavy metal ions from aqueous solution by ultrafiltration assisted with copolymer of maleic acid and acrylic acid. *Desalination, 329*, 78–85. doi:10.1016/j.desal.2013.09.012

Qi, Z., Xiang, G., & Xiong, D. (2019). Performance evaluation of pilot-scale hybrid anaerobic baffled reactor (HABR) to process dyeing wastewater based on grey relational analysis. *Applied Sciences, 9*(10), 1974. doi:10.3390/app9101974

Quaak Knoef, P., & Stassen, H. (1999) *Energy from biomass*. World Bank technical paper, 422.

Quaik, S., Embrandiri, A., Hossain, K., & Ismail, N. (2017). Giving waste a second chance. Public awareness and youth empowerment. In Environmental Issues Surrounding Human Overpopulation (pp. 1–18). Hershey, PA: IGI Global. doi:10.4018/978-1-5225-1683-5.ch009

Ra, C. S., Lo, K. V., Shin, J. S., Oh, J. S., & Hong, B. J. (2000). Biological nutrient removal with an internal organic carbon source in piggery wastewater treatment. *Water Research, 34*(3), 965–973. doi:10.1016/S0043-1354(99)00189-X

Radjenovic, J., Petrovic, M., & Barceló, D. (2007). Analysis of pharmaceuticals in wastewater and removal using a membrane bioreactor. *Analytical and Bioanalytical Chemistry, 387*(4), 1365–1377. doi:10.100700216-006-0883-6 PMID:17115140

Radjenović, J., Petrović, M., Ventura, F., & Barceló, D. (2008). Rejection of pharmaceuticals in nanofiltration and reverse osmosis membrane drinking water treatment. *Water Research, 42*(14), 3601–3610. doi:10.1016/j.watres.2008.05.020 PMID:18656225

Raeisian, L., Niazmand, H., Ebrahimnia-Bajestan, E., & Werle, P. (2019). Feasibility study of waste vegetable oil as an alternative cooling medium in transformers. *Applied Thermal Engineering, 151*, 308–317. doi:10.1016/j.applthermaleng.2019.02.010

Rafieizonooz, M., Mirza, J., Salim, M. R., Hussin, M. W., & Khankhaje, E. (2016). Investigation of coal bottom ash and fly ash in concrete as replacement for sand and cement. *Construction & Building Materials, 116*, 15–24. doi:10.1016/j.conbuildmat.2016.04.080

Rafii, F., & Coleman, T. (1999). Cloning and expression in E.coli of an azoreductase gene from Clostridium perfringens and comparison with azoreductase genes from other bacteria. *Journal of Basic Microbiology*, *39*, 29–35. doi:10.1002/(SICI)1521-4028(199903)39:1<29::AID-JOBM29>3.0.CO;2-W PMID:10071864

Raheem, A., Hassan, M. Y., & Shakoor, R. (2016). Bioenergy from anaerobic digestion in Pakistan: Potential, development and prospects. *Renewable & Sustainable Energy Reviews*, *59*, 264–275.

Rahman, L., Silong, S., Zin, W. M., Rahman, M. Z. A., Ahmad, M., & Haron, J. (2000). Graft copolymerization of methyl acrylate onto sago starch using ceric ammonium nitrate as an initiator. *Journal of Applied Polymer Science*, *76*(4), 516–523. doi:10.1002/(SICI)1097-4628(20000425)76:4<516::AID-APP9>3.0.CO;2-7

Rahmat, N. F. H., & Rasid, R. A. (2017). Gasification of empty fruit bunch with carbon dioxide in an entrained flow gasifier for syngas production. In A. Saptoro, W. S. Khur, L. S. Wei, W. P. Q. Ng, M. Anwar, C. Yeo, & K. E. Huey (Eds.), *29th Symposium of Malaysian Chemical Engineers* (Vol. 206). 10.1088/1757-899X/206/1/012013

Rahul, R., Kumar, S., Jha, U., & Sen, G. (2015). Cationic inulin: A plant based natural biopolymer for algal biomass harvesting. *International Journal of Biological Macromolecules*, *72*, 868–874. doi:10.1016/j.ijbiomac.2014.09.039 PMID:25304749

Rajakumar, R., & Meenambal, T. (2008). Comparative study on start - up performance of HUASB and AF reactors treating poultry slaughterhouse wastewater. *International Journal of Environmental of Research*, *2*(4), 401–410.

Rajamohan, N., Rajasimman, M., Rajeshkannan, R., & Sivaprakash, B. (2013). Kinetic modeling and isotherm studies on a batch removal of acid red 114 by an activated plant biomass. *Journal of Engineering Science and Technology*, *8*(6), 778–792.

Rajasekhar, M., Venkat, N., Rao, G., Chinna, R., Priyadarshini, G., & Jeevan Kumar, N. (2014). Energy generation from municipal solid waste by innovative technologies – plasma gasification. *Procedia Materials Science*, *10*, 513–518. doi:10.1016/j.mspro.2015.06.094

Rajkumar, W. S., & Chang, A. S. (2000). Suspended particulate matter concentrations along the East–West Corridor, Trinidad, West Indies. *Atmospheric Environment*, *34*(8), 1181–1187. doi:10.1016/S1352-2310(99)00381-7

Ralinda, R., & Miller, P. G. (1996). Phytoremediation, Technology overview report. *Ground-water remediation technologies analysis center.* pp. 3-5. Retrieved from https://clu-in.org/download/toolkit/phyto_o.pdf

Ramalho, P. A., Cardoso, M. H., Cavaco-Paulo, A., & Ramahlo, M. T. (2004). Characterization of azo reduction activity in a novel ascomycete yeast strain. *Applied and Environmental Microbiology*, *70*(4), 2279–2288. doi:10.1128/AEM.70.4.2279-2288.2004 PMID:15066823

Ramalho, P. A., Paiva, S., Cavaco-Paulo, A., Casal, M., Cardoso, M. H., & Ramalho, M. T. (2005). Azoreductase activity of intact saccharomyces cerevisiae cells is dependent on the Frelp component of the plasma membrane ferric reductase. *Appl. Environ. Technol*, *71*(7), 3882–3888. doi:10.1128/AEM.71.7.3882-3888.2005 PMID:16000801

Ramalho, P. A., Scholze, H., Cardoso, M. H., Ramalho, M. T., & Oliveira-Campos, A. M. (2002). Improved conditions for the aerobic reductive decolorisation, of azo dyes by Candida zeylanoides. *Enzyme and Microbial Technology*, *31*(6), 848–854. doi:10.1016/S0141-0229(02)00189-8

Ramesh, T. N., Kirana, D. V., Kumari, T. N. M., & Ashwini, A. (2013). Adsorption studies of indigo carmine dye by magnesium oxide. *International Journal of Science Research*, *1*(64), 495–501.

Ram, K., Tripathi, S. N., Sarin, M. M., & Bhattu, D. (2014). Primary and secondary aerosols from an urban site (Kanpur) in the Indo-Gangetic Plain: Impact on CCN, CN concentrations and optical properties. *Atmospheric Environment, 89*, 655–663. doi:10.1016/j.atmosenv.2014.02.009

Ramos, A. F., Gomez, M. A., Hontoria, E., & Gonzalez-Lopez, J. (2007). Biological nitrogen and phenol removal from saline industrial wastewater by submerged fixed-film reactor. *Journal of Hazardous Materials, 142*(1), 175–183. doi:10.1016/j.jhazmat.2006.08.079 PMID:17070991

Ramya, M., Anusha, B., Kalavathy, S., & Devilaksmi, S. (2007). Biodecolorization and biodegradation of Reactive Blue by Aspergillus sp. *African Journal of Biotechnology, 6*, 1441–1445.

Randall, D., & Tsui, T. (2002). Ammonia toxicity in fish. *Marine pollution bulletin, 45*(1-12), 17-23.

Rao, M. N., Rao, Y. H., Chakrapani, C., & Babu, C. S. (2011). Adsorption studies of methylene blue dye using prepared low-cost activated Kaza's carbons. *Journal of Chemical and Pharmaceutical Research, 3*(5), 363–375.

Rao, Y. S., Shankha, K. B., & Chen, J. C. (1993). Microbiological stabilization of sludge by aerobic digestion and storage. *Journal of Environmental Engineering*, 119.

Rapagna, S., Jand, N., Kiennemann, A., & Foscolo, P. (2000). Steam-gasification of biomass in a fluidised-bed of olivine particles. *Biomass and Bioenergy, 19*(3), 187–197. doi:10.1016/S0961-9534(00)00031-3

Raphael, B., Thiago, G. M., Cintya, A. C., Tamaris, G. P., & Carmem, S. F. (2013). Soil contamination with heavy metals and petroleum derivates. Impact on Edaphic Fauna and Remediation Strategies 176-178/chapter 6.

Raposo, F., Borja, R., Sanchez Hernandez, E., Martín, M. A., & Martín, A. (2003). Inhibition kinetics of overall substrate and phenolics removal during the anaerobic digestion of two-phase olive mill effluents (TPOME) in suspended and immobilized cell reactors. *Process Biochemistry, 39*(4), 425–435. doi:10.1016/S0032-9592(03)00095-5

Rapport, J., Zhang, R., Jenkins, B. M., & Williams, R. B. (2015). Current anaerobic digestion technologies used for treatment of municipal organic solid waste. California Environmental Protection Agency; 2008.

Rasheed, R., Khan, N., Yasar, A., Su, Y., & Tabinda, A. B. (2016). Design and cost-benefit analysis of a novel anaerobic industrial bioenergy plant in Pakistan. *Renewable Energy, 90*, 242–247.

Rashid, U.,, & Anwar, F. (2008). Production of biodiesel through optimized alkaline catalyzed trans-esterification of rapeseed oil. *Fuel, 87*, 265–273.

Raval, D., Patel, R., & Patel, V. (1988). Graft copolymerization of acrylonitrile on to guar gum by hydrogen peroxide initiation. *Stärke, 40*(2), 66–69. doi:10.1002tar.19880400207

Ravina, L., & Moramarco, N. (1993). Everything you want to know about Coagulation & Flocculation. *Zeta-Meter, Inc.*, 1-37.

Ravindra, K., Mittal, A. K., & Grieken, R. V. (2001). Health risk assessment of urban suspended particulate matter with special reference to polycyclic aromatic hydrocarbons: A review. *Reviews on Environmental Health, 16*(3), 169–189. doi:10.1515/REVEH.2001.16.3.169 PMID:11765907

Ravindra, K., Sokhi, R., & Van Grieken, R. (2008). Atmospheric polycyclic aromatic hydrocarbons: Source attribution, emission factors and regulation. *Atmospheric Environment, 42*(13), 2895–2921. doi:10.1016/j.atmosenv.2007.12.010

Ray, S. K., Majumdera, C., & Sahab, P. (2018). Removal of turbidity and E. coli from surface water by electrocoagulation and study of its economic feasibility. *Journal of the Indian Chemical Society, 95*(3), 343–348.

Read, J. M., & Whiteoak, D. (2003). The shell bitumen handbook (5th ed.). Thomas Telford Publishing, Thomas Telford Ltd, 1 Heron Quay, London E144JD, 62-66, 136.

Reddy, B. V., Sandeep, P., Ujwala, P., Navaneetha, K., & Reddy, K. V. R. (2014). Water treatment process in pharma industry-A review. *International Journal of Pharma and Bio Sciences*, *4*, 2.

Reddy, M. S., & Venkataraman, C. (2002). Inventory of aerosol and sulphur dioxide emissions 645 from India: I—Fossil fuel combustion. *Atmospheric Environment*, *36*(4), 677–697. doi:10.1016/S1352-2310(01)00463-0

Rehman, M. S. U., Rashid, N., Saif, A., Mahmood, T., & Han, J.-I. (2013). Potential of bioenergy production from industrial hemp (Cannabis sativa): Pakistan perspective. *Renewable & Sustainable Energy Reviews*, *18*, 154–164.

Reinhart, D. R., & Townsend, T. G. (1998). *Landfill bioreactor design & operation*. Boca Raton, FL: CRC Press.

Remón, J., Broust, F., Volle, G., García, L., & Arauzo, J. (2015). Hydrogen production from pine and poplar bio-oils by catalytic steam reforming. Influence of the bio-oil composition on the process. *International Journal of Hydrogen Energy*, *40*(16), 5593–5608. doi:10.1016/j.ijhydene.2015.02.117

Renner, R. (2000). Human phthalate study changes exposure picture. *Environmental Science & Technology*, *34*(21), 451A–452A. doi:10.1021/es003481v PMID:21662257

Resende, K. A., Ávila-Neto, C. N., Rabelo-Neto, R. C., Noronha, F. B., & Hori, C. E. (2015a). Hydrogen production by reforming of acetic acid using La–Ni type perovskites partially substituted with Sm and Pr. *Catalysis Today*, *242*, 71–79. doi:10.1016/j.cattod.2014.07.013

Resende, K. A., Ávila-Neto, C. N., Rabelo-Neto, R. C., Noronha, F. B., & Hori, C. E. (2015b). Thermodynamic analysis and reaction routes of steam reforming of bio-oil aqueous fraction. *Renewable Energy*, *80*, 166–176. doi:10.1016/j.renene.2015.01.057

Reuveny, R. (2007). Climate change-induced migration and violent conflict. *Political Geography*, *26*(6), 656–673. doi:10.1016/j.polgeo.2007.05.001

Reynolds, T. D., & Richards, P. A. (1996). *Unit operations and processes in environmental engineering (2nd Ed.)*. Boston, MA: PWS.

Rhodes, M. J. (2008). Introduction to particle technology (2nd ed.). West Sussex, UK: John Wiley & Sons. doi:10.1002/9780470727102

Ribeiro, A. R., Nunes, O. C., Pereira, M. F., & Silva, A. M. (2015). An overview on the advanced oxidation processes applied for the treatment of water pollutants defined in the recently launched Directive 2013/39/EU. *Environment International*, *75*, 33–51. doi:10.1016/j.envint.2014.10.027 PMID:25461413

Riccardo, B., Mats, S., & Di, S. S. (2010). City breathability and its link to pollutant concentration distribution within urban-like geometries. *Atmospheric Environment*, *44*(15), 1894–1903. doi:10.1016/j.atmosenv.2010.02.022

Ricordel, C., Darchen, A., & Hadjiev, D. (2010). Electrocoagulation–electroflotation as a surface water treatment for industrial uses. *Separation and Purification Technology*, *74*(3), 342–347. doi:10.1016/j.seppur.2010.06.024

Riedhammer. (2014). *Riedhammer rotary kiln*. Retrieved from http://www.laeis.eu/Web-site/Page-management/Sacmi-Group/News-Area/News-by-Business/Ceramics/Riedhammer-presents-the-new-generation-of-indirectly-heated-DRA-rotary-kilns.aspx?ido=24754&idc=61115&ln=en-US

Rillig, M. C., Ziersch, L., & Hempel, S. (2017). Microplastic transport in soil by earthworms. *Scientific Reports*, *7*(1), 1362. doi:10.103841598-017-01594-7 PMID:28465618

Rincón, B., Travieso, L., Sánchez, E., de los Ángeles Martín, M., Martín, A., Raposo, F., & Borja, R. (2007). The effect of organic loading rate on the anaerobic digestion of two-phase olive mill solid residue derived from fruits with low ripening index. *Journal of Chemical Technology and Biotechnology (Oxford, Oxfordshire)*, *82*(3), 259–266. doi:10.1002/jctb.1663

Rittmann, B. E., Krajmalnik-Brown, R., & Halden, R. U. (2008). Pre-genomic, genomic and post-genomic study of microbial communities involved in bioenergy. *Nature Reviews. Microbiology*, *6*(8), 604–612. doi:10.1038/nrmicro1939 PMID:18604223

Rivas, F. J., Beltran, F. J., Frades, J., & Buxeda, P. (2001). Oxidation of p-hydroxybenzoic acid by Fenton's reagent. *Water Research*, *35*(2), 387–396. doi:10.1016/S0043-1354(00)00285-2 PMID:11228990

Rizwan, S. A., Nongkynrih, B., & Gupta, S. K. (2013). Air pollution in Delhi: Its magnitude and effects on health. *Indian Journal of Community Medicine: Official publication of Indian Association of Preventive & Social Medicine (Social Medicine Publication Group)*, *38*(1), 4.

Robert, J. O., & David, J. R. (2018). Alkanes and cycloalkanes: structures and reactions. *Science Direct*. Retrieved from https://www.sciencedirect.com/topics/chemistry/alkane

Robert, N. H. (2000). *Asphalt in roads construction. American Society of Civil Engineering* (p. 75). London, UK: Thomas Telford Publications.

Robinson, T., McMullan, G., Marchant, R., & Nigam, P. (2001). Remediation of dyes in textile effluent: A critical review on current treatment technologies with a proposed alternative. *Bioresource Technology*, *77*(3), 247–255. doi:10.1016/S0960-8524(00)00080-8 PMID:11272011

Rocha, E. M., Vilar, V. J., Fonseca, A., Saraiva, I., & Boaventura, R. A. (2011). Landfill leachate treatment by solar-driven AOPs. *Solar Energy*, *85*(1), 46–56. doi:10.1016/j.solener.2010.11.001

Rodovalho, C.,, & De Tomi, G. (2017). Reducing environmental impacts via improved tyre wear management. *Journal of Cleaner Production*, *141*, 1419–1427. doi:10.1016/j.jclepro.2016.09.202

Rodrigues, A. C., Boroski, M., Shimada, N. S., Garcia, J. C., Nozaki, J., & Hioka, N. (2008). Treatment of paper pulp and paper mill wastewater by coagulation–flocculation followed by heterogeneous photocatalysis. *Journal of Photochemistry and Photobiology A Chemistry*, *194*(1), 1–10. doi:10.1016/j.jphotochem.2007.07.007

Rojas, R., Vanderlinden, E., Morillo, J., Usero, J., & El Bakouri, H. (2014). Characterization of sorption processes for the development of low-cost pesticide decontamination techniques. *The Science of the Total Environment*, *488*, 124–135. doi:10.1016/j.scitotenv.2014.04.079 PMID:24830926

Rojas-Rueda, D., de Nazelle, A., Teixidó, O., & Nieuwenhuijsen, M. J. (2012). Replacing car trips by increasing bike and public transport in the greater Barcelona metropolitan area: A health impact assessment study. *Environ Int 49*, 100-109.

Rondeau, V., Commenges, D., Jacqmin-Gadda, H., & Dartigues, J.-F. (2000). Relation between aluminum concentrations in drinking water and Alzheimer's disease: An 8-year follow-up study. *American Journal of Epidemiology*, *152*(1), 59–66. doi:10.1093/aje/152.1.59 PMID:10901330

Rong-Rong, W. A. N. G. (2003). Application of dynamic continuous countercurrent circulatory extraction to Caulis spatholobi [J]. *Zhongchengyao*, 5.

Ronquillo, M. G., & Hernandez, J. C. A. (2017). Antibiotic and synthetic growth promoters in animal diets: Review of impact and analytical methods. *Food Control*, *72*, 255–267. doi:10.1016/j.foodcont.2016.03.001

Rosa, D. M., Sampaio, S. C., Pereira, P. A., Mauli, M. M., & Reis, R. R. D. (2017). Swine wastewater: Impacts on soil, plant, and leachate. *Engenharia Agrícola*, *37*(5), 928–939. doi:10.1590/1809-4430-eng.agric.v37n5p928-939/2017

Rosales, E., Pazos, M., Longo, M., & Sanromán, M. (2009). Electro-Fenton decoloration of dyes in a continuous reactor: A promising technology in colored wastewater treatment. *Chemical Engineering Journal*, *155*(1-2), 62–67. doi:10.1016/j.cej.2009.06.028

Rosario, I., & Jessica, L. (2015). Bioremediation for a soil contaminated with hydrocarbons. *Journal of Petroleum & Environmental Biotechnology*, *6*(2). doi:10.4172/2157-7463.1000208

Rowshyra, A., & Castaneda, S. A. (2014). Microplastic pollution in St Lawrence river sediments, NRC press. *Canadian Journal of Fisheries and Aquatic Sciences*, *71*.

Rozema, E. R., VanderZaag, A. C., Wood, J. D., Drizo, A., Zheng, Y., Madani, A., & Gordon, R. J. (2016). Constructed wetlands for agricultural wastewater treatment in northeastern North America: A review. *Water (Basel)*, *8*(5), 173. doi:10.3390/w8050173

Rubio-Clemente, A., Torres-Palma, R. A., & Peñuela, G. A. (2014). Removal of polycyclic aromatic hydrocarbons in aqueous environment by chemical treatments: A review. *The Science of the Total Environment*, *478*, 201–225. doi:10.1016/j.scitotenv.2013.12.126 PMID:24552655

Ruiz, G., Jeison, D., Rubilar, O., Ciudad, G., & Chamy, R. (2006). Nitrification–denitrification via nitrite accumulation for nitrogen removal from wastewaters. *Bioresource Technology*, *97*(2), 330–335. doi:10.1016/j.biortech.2005.02.018 PMID:16171689

Ruiz-Garcia, A., & Nuez, I. (2016). Long-term performance decline in a brackish water reverse osmosis desalination plant. Predictive model for the water permeability coefficient. *Desalination*, *397*, 101–107. doi:10.1016/j.desal.2016.06.027

Ruiz, J., Juárez, M., Morales, M., Muñoz, P., & Mendívil, M. (2013). Biomass gasification for electricity generation: Review of current technology barriers. *Renewable & Sustainable Energy Reviews*, *18*, 174–183. doi:10.1016/j.rser.2012.10.021

Ryan, P. G., Moore, C. J., Van Franeker, J. A., & Moloney, C. L. (2009). Monitoring the abundance of plastic debris in the marine environment. *Philos Trans R Soc B*, *364*(1526), 1999–2012. doi:10.1098/rstb.2008.0207 PMID:19528052

Ryser, L., & Halseth, G. (2010). Rural economic development: A review of the literature from industrialized economies. *Geography Compass*, *4*(6), 510–531. doi:10.1111/j.1749-8198.2010.00321.x

Saban, C. V. (2015). *Palynological perspectives on Younger Dryas to early Holocene human ecology at Paisley Caves*. Oregon.

Saber, A., Hasheminejad, H., Taebi, A., & Ghaffari, G. (2014). Optimization of Fenton-based treatment of petroleum refinery wastewater with scrap iron using response surface methodology. *Applied Water Science*, *4*(3), 283–290. doi:10.100713201-013-0144-8

Sadeghi, A., Bazardehi, M. A. K., Raffe, S., & Zarif, B. (2014). Biotransformation of Carmoisine and Reactive Black 5 Dyes using S. cerevisiae. *Health*, *6*(10), 859–864. doi:10.4236/health.2014.610108

Sadon, S. N., Beddu, S., Naganathan, S., Kamal, N. L. M., & Hassan, H. (2017). Coal bottom ash as sustainable material in concrete – A review. *Indian Journal of Science and Technology*, *10*(36), 1–10. doi:10.17485/ijst/2017/v10i36/114595

Sagiv, A., & Semiat, R. (2004). Analysis of parameters affecting boron permeation through reverse osmosis membranes. *Journal of Membrane Science*, *243*(1), 79–87. doi:10.1016/j.memsci.2004.05.029

Sahachaiyunta, P., Koo, T., & Sheikholeslami, R. (2002). Effect of several inorganic species on silica fouling in RO membranes. *Desalination*, *144*(1–3), 373–378. doi:10.1016/S0011-9164(02)00346-6

Sahana, M., Ahmed, R., & Sajjad, H. (2016). Analyzing land surface temperature distribution in response to land use/land cover change using split window algorithm and spectral radiance model in sundarban biosphere reserve, India. *Modeling Earth Systems and Environment*, *2*(2), 81. doi:10.100740808-016-0135-5

Sahar, E., David, I., Gelman, Y., Chikurel, H., Aharoni, A., Messalem, R., & Brenner, A. (2011). The use of RO to remove emerging micropollutants following CAS/UF or MBR treatment of municipal wastewater. *Desalination*, *273*(1), 142–147. doi:10.1016/j.desal.2010.11.004

Sahar, S., Sadaf, S., Iqbal, J., Ullah, I., Bhatti, H. N., Nouren, S., ... Iqbal, M. (2018). Biodiesel production from waste cooking oil: An efficient technique to convert waste into biodiesel. *Sustainable Cities and Society*, *41*, 220–226. doi:10.1016/j.scs.2018.05.037

Saha, S. K., Swaminathan, P., Raghavan, C., Uma, L., & Subramanian, G. (2010). Ligninolytic and antioxidative enzymes of a marine cyanobacterium Oscillatoria willei. BDU 130511 during Poly R-478 decolourization. *Bioresource Technology*, *101*(9), 3076–3084. doi:10.1016/j.biortech.2009.12.075 PMID:20061142

Sahu, J., Agarwal, S., Meikap, B., & Biswas, M. (2009). Performance of a modified multi-stage bubble column reactor for lead (II) and biological oxygen demand removal from wastewater using activated rice husk. *Journal of Hazardous Materials*, *161*(1), 317–324. doi:10.1016/j.jhazmat.2008.03.094 PMID:18462879

Sakai, S., Katsuura, H., Inoue, T., & Hiraoka, M. (2000). State-of-the-art treatment processes for municipal solid waste incineration residues. *Waste Management & Research*, *18*(1), 41–51. doi:10.1177/0734242X0001800106

Salati, S., Scaglia, B., Di Gregorio, A., Carrerab, A., & Adani, F. (2013). The use of the dynamic respiration index to predict the potential MSW-leachate impacts after short term mechanical biological treatment. *Bioresource Technology*, *128*, 351–358. doi:10.1016/j.biortech.2012.10.101 PMID:23201514

Saleem, M. (2007). Pharmaceutical wastewater treatment: A physicochemical study. *Journal of Research (Science)*, *18*(2), 125–134.

Salehizadeh, H., Yan, N., & Farnood, R. (2017). Recent advances in polysaccharide bio-based flocculants. *Biotechnology Advances*. PMID:28993221

Saleh-Lakha, S., Shannon, K. E., Henderson, S. L., Goyer, C., Trevors, J. T., Zebarth, B. J., & Burton, D. L. (2009). Effect of pH and temperature on denitrification gene expression and activity in Pseudomonas mandelii. *Applied and Environmental Microbiology*, *75*(12), 3903–3911. doi:10.1128/AEM.00080-09 PMID:19376915

Salleh, M. A. M., Kisiki, N. H., Yusuf, H. M., & Ghani, W. A. W. A. K. (2010). Gasification of biochar from empty fruit bunch in a fluidized bed reactor. *Energies*, *3*(7), 1344–1352. doi:10.3390/en3071344

Samah, I. A. R. A. (2018). Sustainable energy for Malaysia: An exciting journey awaits. The monthly Bulletin of the Institution of Engineers, Malaysia. (13-18).

Samiran, N. A., Jaafar, M. N. M., Ng, J.-H., Lam, S. S., & Chong, C. T. (2016). Progress in biomass gasification technique–With focus on Malaysian palm biomass for syngas production. *Renewable & Sustainable Energy Reviews*, *62*, 1047–1062. doi:10.1016/j.rser.2016.04.049

Sanches, S., Leitão, C., Penetra, A., Cardoso, V., Ferreira, E., Benoliel, M., ... Pereira, V. (2011). Direct photolysis of polycyclic aromatic hydrocarbons in drinking water sources. *Journal of Hazardous Materials*, *192*(3), 1458–1465. doi:10.1016/j.jhazmat.2011.06.065 PMID:21784577

Sanchez, W., Bender, C., & Porcher, J. M. (2014). Wild gudgeons(Gobio gobio)from French rivers are contaminated by microplastics: Preliminary study and first evidence. *Environmental Research*, *128*, 98–100. doi:10.1016/j.envres.2013.11.004 PMID:24295902

Sandberg, U., Bendtsen, H., Thomsen, S. N., Kragh, J., Kalman, B., & Kokot, D. (2007). *Possibilities to reduce tyre/road noise emission on paving stones and other block surfaces*. SILENCE Deliverable F.D3.

Sandberg, U., Goubert, L., Biligiri, K. P., & Kalman, B. (2010). State-of-the-art regarding poroelastic road surfaces. Deliverable D8.1 Project PERSUADE.

Sandberg, U.,, & Ejsmont, J. A. (2002). *Tyre/road noise reference book. INFORMEX*. Sweden.

Sanghi, R., Bhattacharya, B., Dixit, A., & Singh, V. (2006). Ipomoea dasysperma seed gum: An effective natural coagulant for the decolorization of textile dye solutions. *Journal of Environmental Management*, *81*(1), 36–41. doi:10.1016/j.jenvman.2005.09.015 PMID:16549236

Sanin, F. D., & Vesilind, P. A. (1994). Effect of centrifugation on the removal of extracellular polymers and physical properties of activated sludge. *Water Science and Technology*, *30*(8), 117–127. doi:10.2166/wst.1994.0394

Sani, R. K., & Banerjee, U. C. (1999). Decolorization of triphenylmethane dyes and textile and dyestuff effluent by Kurthia sp. *Enzyme and Microbial Technology*, *24*(7), 433–437. doi:10.1016/S0141-0229(98)00159-8

Sankara Narayanan, T., Magesh, G., & Rajendran, N. (2003). Degradation of O-chlorophenol from aqueous solution by electro-Fenton process. *Fresenius Environmental Bulletin*, *12*(7), 776–780.

Sanli, H., Canakci, M., & Alptekin, E. (2011). Characterization of waste frying oil obtained from different facilities. *World Renewable Energy Congress 2011*, May 8-13, 2011. Linkoping, Sweden, *Bioenergy Technology*, 479-485.

Sanz, J. L., & Kochling, T. (2007). Molecular biology techniques used in wastewater treatment: An overview. *Process Biochemistry*, *42*(2), 119–133. doi:10.1016/j.procbio.2006.10.003

Saratale, R. G., Saratale, G. D., Kalyani, D. C., Chang, J. S., & Govindwar, S. P. (2009). Enhanced decolorization and biodegradation of textile azo dye Scarlet R by using developed microbial consortium-GR. *Bioresource Technology*, *100*(9), 2493–2500. doi:10.1016/j.biortech.2008.12.013 PMID:19157864

Saravanan, P., Pakshirajan, K., & Saha, P. (2008). Growth kinetics of an indigenous mixed microbial consortium during phenol degradation in a batch reactor. *Bioresource Technology*, *99*(1), 205–209. PMID:17236761

Sarkar, B., Chakrabarti, P., Vijaykumar, A., & Kale, V. (2006). Wastewater treatment in dairy industries—Possibility of reuse. *Desalination*, *195*(1-3), 141–152. doi:10.1016/j.desal.2005.11.015

Sarkar, R., & Sivaramakrishnan, L. (2015). Measurement of land surface temperature and its relation to vegetation cover: A case study of Kolkata municipal corporation. *International Journal of Current Research*, *7*(12), 24820–24825.

Sarno, M., Spina, D., & Senatore, A. (2019). One-step nanohybrid synthesis in waste cooking oil, for direct lower environmental impact and stable lubricant formulation. *Tribology International*, *135*, 355–367. doi:10.1016/j.triboint.2019.03.025

Sarrat, C., Lemonsu, A., Masson, V., & Guedalia, D. (2006). Impact of urban heat island on regional atmospheric pollution. *Atmospheric Environment*, *40*(10), 1743–1758. doi:10.1016/j.atmosenv.2005.11.037

Sarria, V., Parraa, S., Adler, N., Péringera, P., Benitez, N., & Pulgarin, C. (2002). Recent developments in the coupling of photoassisted and aerobic biological processes for the treatment of biorecalcitrant compounds. *Catalysis Today*, *76*(2-4), 301–315. doi:10.1016/S0920-5861(02)00228-6

Sarsons, H. (2015). Rainfall and conflict : a cautionary tale. *Journal of Development Economics*, *115*, 62–72. doi:10.1016/j.jdeveco.2014.12.007

Sathiyanarayanan, G., Kiran, G. S., & Selvin, J. (2013). Synthesis of silver nanoparticles by polysaccharide bioflocculant produced from marine Bacillus subtilis MSBN17. *Colloids and Surfaces. B, Biointerfaces*, *102*, 13–20. doi:10.1016/j.colsurfb.2012.07.032 PMID:23006551

Sawatdeenarunat, C., Nguyen, D., Surendra, K., Shrestha, S., Rajendran, K., Oechsner, H., ... Khanal, S. K. (2016). Anaerobic biorefinery: Current status, challenges, and opportunities. *Bioresource Technology*, *215*, 304–313. PMID:27005786

Schaefer, W. D. (1975). Disposing of solid wastes by pyrolysis. *Environmental Science & Technology*, *9*, 91–98.

Schertenleib, R., Wegelin, M., & Boller, M. (1987). Particle removal by horizontal-flow roughing filtration. *Journal of Water Supply: Research & Technology - Aqua*, *36*, 80–90.

Scheumann, R., & Kraume, M. (2009). Influence of hydraulic retention time on the operation of a submerged membrane sequencing batch reactor (SM-SBR) for the treatment of greywater. *Desalination*, *246*(1-3), 444–451.

Schoeman, J., & Steyn, A. (2003). Nitrate removal with reverse osmosis in a rural area in South Africa. *Desalination*, *155*(1), 15–26. doi:10.1016/S0011-9164(03)00235-2

Schoen, M. E., Soller, J. A., & Ashbolt, N. J. (2011). Evaluating the importance of faecal sources in human-impacted waters. *Water Research*, *45*(8), 2670–2680. doi:10.1016/j.watres.2011.02.025 PMID:21429551

Schouw, N. L., Danteravanich, S., Mosbaek, H., & Tjell, J. C. (2002). Composition of human excreta—A case study from southern Thailand. *The Science of the Total Environment*, *286*(1), 155–166. doi:10.1016/S0048-9697(01)00973-1 PMID:11886091

Schulz, M., Baygents, J., & Farrell, J. (2009). Laboratory and pilot testing of electrocoagulation for removing scale-forming species from industrial process waters. *International Journal of Environmental Science and Technology*, *6*(4), 521–526. doi:10.1007/BF03326091

Schweitzer, L., & Zhou, J. (2010). Neighborhood air quality, respiratory health, and vulnerable populations in compact and sprawled regions. *Journal of the American Planning Association*, *76*(3), 363–371. doi:10.1080/01944363.2010.486623

Šćiban, M., Klašnja, M., Antov, M., & Škrbić, B. (2009). Removal of water turbidity by natural coagulants obtained from chestnut and acorn. *Bioresource Technology*, *100*(24), 6639–6643. doi:10.1016/j.biortech.2009.06.047 PMID:19604691

Scott, D. S., Majerski, P., Piskorz, J., & Radlein, D. (1999). A second look at fast pyrolysis of biomass - the RTI process. *Journal of Analytical and Applied Pyrolysis*, *51*(1-2), 23–37. doi:10.1016/S0165-2370(99)00006-6

Scott, J. P., & Ollis, D. (1995). Integrated chemical and biological oxidation processes for water treatment, review and recommendations. *Environment and Progress*, *14*(2), 88–103. doi:10.1002/ep.670140212

Scott, J., & Ollis, D. (1995). Integration of chemical and biological oxidation processes for wastewater treatment: Review and recommendations. *Environmental Progress & Sustainable Energy*, *14*, 68–103.

Sebat, J., Colwell, F., & Crawford, R. (2003). Metagenomic profiling: Microarray analysis of an environmental genomic library. *Applied and Environmental Microbiology*, *69*(8), 4927–4934. doi:10.1128/AEM.69.8.4927-4934.2003 PMID:12902288

Seckler, D. W. (1998). *World water demand and supply, 1990 to 2025: Scenarios and issues* (Vol. 19). Iwmi.

Secondes, M. F. N., Naddeo, V., Belgiorno, V., & Ballesteros, F. Jr. (2014). Removal of emerging contaminants by simultaneous application of membrane ultrafiltration, activated carbon adsorption, and ultrasound irradiation. *Journal of Hazardous Materials*, *264*, 342–349. doi:10.1016/j.jhazmat.2013.11.039 PMID:24316806

Sedlak, D. L., & Andren, A. W. (1991). Oxidation of chlorobenzene with Fenton's reagent. *Environmental Science & Technology*, *25*(4), 777–782. doi:10.1021/es00016a024

Seghezzo, L., Zeeman, G., van Lier, J. B., Hamelers, H. V. M., & Lettinga, G. (1998). A review: The anaerobic treatment of sewage in UASB and EGSB reactors. *Bioresource Technology*, *65*(3), 175–190. doi:10.1016/S0960-8524(98)00046-7

Seltenrich, N. (2016, June). Emerging waste-to-energy technologies: Solid waste solution or dead end?. *Journal of Environmental Health Perspect, 24*(6), A106–A111.

Semple, K. T., Cain, R. B., & Schmidt, S. (1999). Biodegradation of aromatic compounds by microalgae. *FEMS Microbiology Letters*, *170*(2), 291–300. doi:10.1111/j.1574-6968.1999.tb13386.x

Sen, G., Mishra, S., Jha, U., & Pal, S. (2010). Microwave initiated synthesis of polyacrylamide grafted guar gum (GG-g-PAM)—Characterizations and application as matrix for controlled release of 5-amino salicylic acid. *International Journal of Biological Macromolecules*, *47*(2), 164–170. doi:10.1016/j.ijbiomac.2010.05.004 PMID:20471416

Sengoz, B.,, & Topal, A. (2005). Use of asphalt roofing shingle waste in HMA. *Construction & Building Materials*, *19*, 337–346.

Sen, S., & Demierer, G. (2003). Anaerobic treatment of synthetic textile wastewater containing a reactive azo dye. *Journal of Environmental Engineering*, *129*(7), 595–601. doi:10.1061/(ASCE)0733-9372(2003)129:7(595)

Senthilmurugan, S., Ahluwalia, A., & Gupta, S. K. (2005). Modeling of spiral-wound module and equation of model parameters using numerical techniques. *Desalination*, *173*(3), 296–286. doi:10.1016/j.desal.2004.08.034

Senthilnathan, J., & Philip, L. (2012). Elimination of pesticides and their formulation products from drinking water using thin film continuous photoreactor under solar radiation. *Solar Energy*, *89*(9), 2735–2745. doi:10.1016/j.solener.2012.06.011

Sentime, K. (2011). Profiling solid waste pickers: A case study of Braamfontein-Greater Johannesburg. *Africanus*, *41*, 96–111.

Sentime, K. (2014). The impact of legislative framework governing waste management and collection in South Africa. *African Geographic Review*, *33*(1), 81–93. doi:10.1080/19376812.2013.847253

Seo, Y.-C., Alam, M. T., & Yang, W.-S. (2018). *Gasification of municipal solid waste. Gasification for low-grade feedstock*. Yongseung Yun, IntechOpen.

Serio, M. A., Chen, Y., Wojtowicz, M. A., & Suuberg, E. M. (2002). Pyrolysis processing for solid waste resource recovery in space. In *30th International Conference on Environmental Systems*. Toulouse, France.

Serôdio, P., & Nogueira, J. (2006). Considerations on ultra-trace analysis of phthalates in drinking water. *Water Research*, *40*(13), 2572–2582. doi:10.1016/j.watres.2006.05.002 PMID:16797669

Sessarego, S., Rodrigues, S. C. G., Xiao, Y., Lu, Q., & Hill, J. M. (2019). Phosphonium-enhanced chitosan for cr(vi) adsorption in wastewater treatment. *Carbohydrate Polymers*, *211*, 249–256. doi:10.1016/j.carbpol.2019.02.003 PMID:30824086

Setala, O., Fleming-Lehtinen, V., & Lehtiniemi, M. (2014). Ingestion and transfer of microplastics in the planktonic food web. *Environ. Pollut*, *185*, 77–83. doi:10.1016/j.envpol.2013.10.013 PMID:24220023

Seviour, E., Blackall, L., Christensson, C., Hugenholtz, P., Cunningham, M., Bradford, D., & Seviour, R. (1997). The filamentous morphotype Eikelboom Type 1863 is not a single genetic entity. *Journal of Applied Microbiology*, *82*(4), 411–421. doi:10.1046/j.1365-2672.1997.00111.x PMID:9134715

Sewage. (2016). Retrieved from https://www.iwk.com.my/do-you-know/sewage-treatment-system

Shafizadeh, F., & Chin, P. P. (1977, June 1). Thermal deterioration of wood. Paper presented at the *ACS Symposium Series American Chemical Society*. 10.1021/bk-1977-0043.ch005

Shah, F. H., Hadim, H. A., & Korfiatis, G. P. (1995). Laboratory studies of air stripping of VOC-contaminated soils. *Journal of Soil Contamination*, 4(1), 1995.

Shahidul, M. I., Shazali, S. T. S., Abdullah, Y., Ting, C. T., Lee, M. D., Hishamuddin, A. H. … Adzlan, A. F. K. (2014). Determinants of desalination machinery capacity utilization: evaluate in the aspect of energy, environment and economic sustainability. In Advances in Process Engineering & Green Energy. Kuching, Malaysia: EnCon.

Shahidul, M. I., & Sadrul, A. K. M. (2002). Industrial water treatment technology: an overview. *The Journal of Environment*, *2*, 598–603.

Shah, J.,, & Rasul Jan, M. (2006). Conversion of waste tyres into carbon black and their utilization as adsorbent. *Journal of the Chinese Chemical Society (Taipei)*, *53*(5), 1085–1089. doi:10.1002/jccs.200600144

Shahtalebi, A., & Sarrafzadeh, M. (2011). Application of nanofiltration membrane in the separation of amoxicillin from pharmaceutical wastewater. *Journal of Environmental Health Science & Engineering*, *8*(2), 106–110.

Shahul Hamid, F., Bhatti, M. S., Anuar, N., Anuar, N., Mohan, P., & Periathamby, A. (2018). Worldwide distribution and abundance of microplastic: How dire is the situation? *Waste Management & Research*, *36*(10), 873–897. PMID:30103651

Shakeel, S. R., Takala, J., & Shakeel, W. (2016). Renewable energy sources in power generation in Pakistan. *Renewable & Sustainable Energy Reviews*, *64*, 421–434.

Shak, K. P. Y., & Wu, T. Y. (2014). Coagulation–flocculation treatment of high-strength agro-industrial wastewater using natural Cassia obtusifolia seed gum: Treatment efficiencies and flocs characterization. *Chemical Engineering Journal*, *256*, 293–305. doi:10.1016/j.cej.2014.06.093

Shannon, M. A., Bohn, P. W., Elimelech, M., Georgiadis, J. G., Marinas, B. J., & Mayes, A. M. (2010). Science and technology for water purification in the coming decades. In Nanoscience and technology: A collection of reviews from nature journals, (pp. 337-346).

Shao, J., Yang, Y., & Zhong, Q. (2003). Studies on preparation of oligoglucosamine by oxidative degradation under microwave irradiation. *Polymer Degradation & Stability*, *82*(3), 395–398. doi:10.1016/S0141-3910(03)00177-0

Sharholy, M., Ahmad, K., Mahmood, G., & Trivedi, R. C. (2008). MSW management in Indian cities: A review. *Waste Management (New York, N.Y.)*, *28*(2), 459–467. doi:10.1016/j.wasman.2007.02.008 PMID:17433664

Shariati, F. P., Mehrnia, M. R., Salmasi, B. M., Heran, M., Wisniewski, C., & Sarrafzadeh, M. H. (2010). Membrane bioreactor for treatment of pharmaceutical wastewater containing acetaminophen. *Desalination*, *250*(2), 798–800. doi:10.1016/j.desal.2008.11.044

Sharma, A., Pareek, V., & Zhang, D. (2015). Biomass pyrolysis—A review of modelling, process parameters and catalytic studies. *Renewable & Sustainable Energy Reviews*, *50*, 1081–1096. doi:10.1016/j.rser.2015.04.193

Sharma, H. D., & Reddy, K. R. (2004). *Geoenvironmental Engineering*. Hoboken, NJ: John Wiley & Sons.

Shatanawi, K. (2008). *The effects of crumb rubber particles on highway noise reduction - a laboratory study*. Clemson, SC: Clemson University.

Shawaqfeh, A. T. (2010). Removal of pesticides from water using anaerobic-aerobic biological treatment. *Chinese Journal of Chemical Engineering*, *18*(4), 672–680. doi:10.1016/S1004-9541(10)60274-1

Shaw, C. B., Carliell, C. M., & Wheatley, A. D. (2002). Anaerobic/aerobic treatment of coloured textile effluents using sequencing batch reactors. *Water Research*, *36*(8), 1993–2001. doi:10.1016/S0043-1354(01)00392-X PMID:12092574

Shehzad, A., Bashir, M. J. K., Horttanainen, M., Manttari, M., Havukainen, J., & Abbas, G. (2018). Modeling and comparative assessment of bubbling fluidized bed gasification system for syngas production-A gateway for a cleaner future in Pakistan. *Environmental Technology*, *39*(14), 1841–1850. doi:10.1080/09593330.2017.1340350 PMID:28595513

Shemer, H., & Linden, K. G. (2007). Aqueous photodegradation and toxicity of the polycyclic aromatic hydrocarbons fluorene, dibenzofuran, and dibenzothiophene. *Water Research*, *41*(4), 853–861. doi:10.1016/j.watres.2006.11.022 PMID:17217979

Sheng, C., Nnanna, A. A., Liu, Y., & Vargo, J. D. (2016). Removal of trace pharmaceuticals from water using coagulation and powdered activated carbon as pretreatment to ultrafiltration membrane system. *The Science of the Total Environment*, *550*, 1075–1083. doi:10.1016/j.scitotenv.2016.01.179 PMID:26867086

Shen, L. Y., Tam, V. W. Y., Tam, C. M., & Drew, D. (2004). Mapping approach for examining waste management on construction sites. *Journal of Construction Engineering and Management*, *130*(4), 472–481. doi:10.1061/(ASCE)0733-9364(2004)130:4(472)

Shen, X., & Marks, E. (2016). Near-miss information visualization tool in BIM for construction safety. *Journal of Construction Engineering and Management*, *142*(4), 1–10. doi:10.1061/(ASCE)CO.1943-7862.0001100

Shen, Y., & Yoshikawa, K. (2013). Recent progresses in catalytic tar elimination during biomass gasification or pyrolysis—A review. *Renewable & Sustainable Energy Reviews*, *21*, 371–392. doi:10.1016/j.rser.2012.12.062

Sherrard, R. M., Bearr, J. S., Murray-Gulde, C. L., Rodgers, J. H. Jr, & Shah, Y. T. (2004). Feasibility of constructed wetlands for removing chlorothalonil and chlorpyrifos from aqueous mixtures. *Environmental Pollution*, *127*(3), 385–394. doi:10.1016/j.envpol.2003.08.017 PMID:14638299

Shilpa, R. K.,, & Sharma, A. (2017). Morphologically tailored activated carbon derived from waste tires as high-performance anode for Li-ion battery. *Journal of Applied Electrochemistry*, 1–13.

Shim, W. J., Hong, S. H., & Eo, S. E. (2017). Identification methods in micro plastic analysis: A review. *Analytical Methods*, *9*(9), 1384–1391. doi:10.1039/C6AY02558G

Shin, H.-S., & Kang, S.-T. (2003). Characteristics and fates of soluble microbial products in ceramic membrane bioreactor at various sludge retention times. *Water Research*, *37*(1), 121–127. doi:10.1016/S0043-1354(02)00249-X PMID:12465793

Shi, T., Ma, J., Wu, X., Ju, T., Lin, X., Zhang, Y., ... Zhao, L. (2018). Inventories of heavy metal inputs and outputs to and from agricultural soils: A review. *Ecotoxicology and Environmental Safety*, *164*, 118–124. doi:10.1016/j.ecoenv.2018.08.016 PMID:30099172

Shoda, M., & Kim, J. S. (1999). Purification and characterization of a novel peroxidase from Geotrichumcandidum Dec. 1 involved in decolorisation of dyes. *Applied and Environmental Microbiology*, *65*, 1029–1035. PMID:10049859

Shoda, M., & Sugano, Y. (2005). Recent advances in bacterial cellulose production. *Biotechnology and Bioprocess Engineering; BBE*, *10*(1), 1–8. doi:10.1007/BF02931175

Short, J. L., & Webster, D. W. (1982). Ultrafiltration, a valuable processing technique for the pharmaceutical industry. *Process Biochemistry*, *17*, 29–32.

Shu, X.,, & Huang, B. (2014). Recycling of waste tire rubber in asphalt and Portland cement concrete : an overview. *Construction & Building Materials*, *67*, 217–224. doi:10.1016/j.conbuildmat.2013.11.027

Sila, A., Mlaik, N., Sayari, N., Balti, R., & Bougatef, A. (2014). Chitin and chitosan extracted from shrimp waste using fish proteases aided process: Efficiency of chitosan in the treatment of unhairing effluents. *Journal of Polymers and the Environment*, *22*(1), 78–87. doi:10.100710924-013-0598-7

Siles, J. A., Brekelmans, J., Martin, M. A., Chica, A. F., & Martin, A. (2010). Impact of ammonia and sulphate concentration on thermophilic anaerobic digestion. *Bioresource Technology*, *101*(23), 9040–9048. PMID:20674337

Silitonga, A. S., Masjuki, H. H., Mahlia, T. M. I., Ong, H. C., Atabani, A. E., & Chong, W. T. (2013). A global comparative review of biodiesel production from Jatropha Curcas using different homogeneous acid and alkaline catalysts: study of physical and chemical properties. *Renewable & Sustainable Energy Reviews*, *24*, 514–533.

Sima, L., Amador, J., Da Silva, A. K., Miller, S. M., Morse, A. N., Pellegrin, M. L., ... Wells, M. J. (2014). Emerging pollutants–part I: Occurrence, fate and transport. *Water Environment Research*, *86*(10), 1994–2035. doi:10.2175/106143014X14031280668731

Sims, R. E. (2002). The brilliance of bioenergy. In *Business and in Practice*. Earthscan.

Sims, R. E., Mabee, W., Saddler, J. N., & Taylor, M. (2010). An overview of second generation biofuel technologies. *Bioresource Technology*, *101*(6), 1570–1580. doi:10.1016/j.biortech.2009.11.046 PMID:19963372

Sindi, S., Mangialasche, F., & Kivipelto, M. (2015). Advances in the prevention of Alzheimer's Disease. *F1000prime Reports*, 7. PMID:26097723

Singhabhandhu, A.,, & Tezuka, T. (2010). The waste-to-energy framework for integrated multi-waste utilization: waste cooking oil, waste lubricating oil, and waste plastics. *Energy*, *35*(6), 2544–2551.

Singh, D., & Sirkar, K. K. (2012). Desalination of brine and produced water by direct contact membrane distillation at high temperatures and pressures. *Journal of Membrane Science*, *389*, 380–388. doi:10.1016/j.memsci.2011.11.003

Singh, G., Singh, A. K., & Bhatt, K. (2016). Biodegradation of polythenes by bacteria isolated from soil. *International Journal of Research and Development in Pharmacy and Life Sciences*, *5*(2), 2056–2062.

Singh, K., Viraraghavan, T., & Bhattacharyya, D. (2006). Sludge blanket height and flow pattern in UASB reactors: temperature effects. *Journal of Environmental Engineering*, *132*(8), 895–900. doi:10.1061/(ASCE)0733-9372(2006)132:8(895)

Singh, L., & Singh, V. P. (2010). Microbial degradation and decolourization of dyes in semi-solid medium by the fungus – Trichoderma harzianum. *Environ. We Int. J. Sci. Tech*, *5*, 147–153.

Singh, S., Awasthi, S. K., Iyenger, L., Ashfaq, M., & Singh, P. (2011). Mineralisation of aminobenzene sulfonates by a newly isolated bacterial co-culture (AS1 and AS2). *Biol. Med*, *3*, 53–59.

Singh, S., Kang, S. H., Mulchandani, A., & Chen, W. (2008). Bioremediation: Environmental clean-up through pathway engineering. *Current Opinion in Biotechnology*, *19*(5), 437–444. doi:10.1016/j.copbio.2008.07.012 PMID:18760355

Singh, V., Kumar, P., & Sanghi, R. (2012). Use of microwave irradiation in the grafting modification of the polysaccharides–A review. *Progress in Polymer Science*, *37*(2), 340–364. doi:10.1016/j.progpolymsci.2011.07.005

Singh, V., Tiwari, A., Tripathi, D. N., & Sanghi, R. (2006). Microwave enhanced synthesis of chitosan-graft-polyacrylamide. *Polymer*, *47*(1), 254–260. doi:10.1016/j.polymer.2005.10.101

Sires, I., Garrido, J. A., Rodriguez, R. M., Brillas, E., Oturan, N., & Oturan, M. A. (2007). Catalytic behavior of the Fe3+/Fe2+ system in the electro-Fenton degradation of the antimicrobial chlorophene. *Applied Catalysis B: Environmental*, *72*(3-4), 382–394. doi:10.1016/j.apcatb.2006.11.016

Siriporn, Y., Sompong, O., & Poonsuk, P. (2012). Effect of initial pH, nutrients and temperature on hydrogen production from palm oil mill effluent using thermotolerant consortia and corresponding microbial communities. *International Journal of Hydrogen Energy*, *37*(18), 13806–13814. doi:10.1016/j.ijhydene.2012.03.151

Siti, N. F. Z., & Aziz, H. A. (2018) IOP Conference Series: Earth and Environmental Science, (140), 012013.

Sivan, A. (2011). New perspectives in plastic biodegradation. *Current Opinion in Biotechnology*, *22*(3), 422–426. doi:10.1016/j.copbio.2011.01.013 PMID:21356588

Sivasangar, S., Mastuli, M. S., Islam, A., & Taufiq-Yap, Y. H. (2015). Screening of modified CaO-based catalysts with a series of dopants for the supercritical water gasification of empty palm fruit bunches to produce hydrogen. *RSC Advances*, *5*(46), 36798–36808. doi:10.1039/C5RA03430B

Sivasangar, S., Zainal, Z., Salmiaton, A., & Taufiq-Yap, Y. H. (2015). Supercritical water gasification of empty fruit bunches from oil palm for hydrogen production. *Fuel*, *143*, 563–569. doi:10.1016/j.fuel.2014.11.073

Skaggs, R. W., Fausey, N. R., & Evans, R. O. (2012). Drainage water management. *Journal of Soil and Water Conservation*, *67*(6), 167A–172A. doi:10.2489/jswc.67.6.167A

Skiadas, I. V., & Lyberatos, G. (1998). The periodic anaerobic baffled reactor. *Water Science and Technology*, *38*(8), 401–408. doi:10.2166/wst.1998.0831

Slokar, Y. M., & Majcen Le Marechal, A. (1998). Methods of decoloration of textile wastewater. *Dyes and Pigments*, *37*(4), 335–356. doi:10.1016/S0143-7208(97)00075-2

Slopiecka, K., Bartocci, P., & Fantozzi, F. (2012). Thermogravimetric analysis and kinetic study of poplar wood pyrolysis. *Applied Energy*, *97*, 491–497. doi:10.1016/j.apenergy.2011.12.056

Smith, K. S., Plumlee, G. S., & Ficklin, W. H. (1994) Predicting water contamination from metal mines and mining wastes. International Land Reclamation and Mine Drainage Conference and Third International Conference on the Abatement of Acidic Drainage. U.S. Geological Survey Open-File Report No. 94-264.

Smith, M., Love, D. C., Rochman, C. M., & Neff, R. A. (2018). Microplastics in seafood and the implications for human health. *Current Environmental Health Reports*, *5*(3), 375–386. doi:10.100740572-018-0206-z PMID:30116998

Snaidr, J., Amann, R., Huber, I., Ludwig, W., & Schleifer, K. (1997). Phylogenetic analysis and in situ identification of bacteria in activated sludge. *Applied and Environmental Microbiology*, *63*, 2884–2896. PMID:9212435

Snyder, S. A., Adham, S., Redding, A. M., Cannon, F. S., DeCarolis, J., Oppenheimer, J., ... Yoon, Y. (2007). Role of membranes and activated carbon in the removal of endocrine disruptors and pharmaceuticals. *Desalination*, *202*(1-3), 156–181. doi:10.1016/j.desal.2005.12.052

Sobota, D. J., Harrison, J. A., & Dahlgren, R. A. (2009). Influences of climate, hydrology, and land use on input and export of nitrogen in California watersheds. *Biogeochemistry*, *94*(1), 43–62. doi:10.100710533-009-9307-y

Sobsey, M. D., Stauber, C. E., Casanova, L. M., Brown, J. M., & Elliott, M. A. (2008). Point of use household drinking water filtration: A practical, effective solution for providing sustained access to safe drinking water in the developing world. *Environmental Science & Technology*, *42*(12), 4261–4267. doi:10.1021/es702746n PMID:18605542

Soegiantoro, G. H., Chang, J., Rahmawati, P., Christiani, M. F., & Mufrodi, Z. (2019). Home-made eco green biodiesel from chicken fat (CIAT) and waste cooking oil (PAIL). *Energy Procedia*, *158*, 1105–1109. doi:10.1016/j.egypro.2019.01.267

Soimakallio, S., Mäkinen, T., Ekholm, T., Pahkala, K., Mikkola, H., & Paappanen, T. (2009). Greenhouse gas balances of transportation biofuels, electricity and heat generation in Finland—dealing with the uncertainties. *Energy Policy*, *37*(1), 80–90.

Solak, M., Kilic, M., Yazici, H., & Sencan, A. (2009). Removal of suspended solids and turbidity from marble processing wastewaters by electrocoagulation: Comparison of electrode materials and electrode connection systems. *Journal of Hazardous Materials*, *172*(1), 345–352. doi:10.1016/j.jhazmat.2009.07.018 PMID:19651474

Soleimaninanadegani, M., & Manshad, S. (2014). Enhancement of biodegradation of palm oil mill effluents by local isolated microorganisms. *Int Scholarly Research Notices*, *2014*, 8. doi:10.1155/2014/727049 PMID:27433516

Solis, M., Solis, A., Perez, H. I., Norberto, M., & Flores, M. (2012). Microbial decolouration of azo dyes: A review. *Process Biochemistry*, *47*(12), 1723–1748. doi:10.1016/j.procbio.2012.08.014

Solomon, T. W., & Fryhle, C. (2004). *Organic chemistry* (8th ed.). Hoboken, NJ: John Wiley & Sons.

Solway Recycling Ltd. (2019). Some legal requirements for consideration when arranging a cooking oil collection. Retrieved from https://www.solwayrecycling.co.uk/recycling-services/oil-collection/legal-requirements-of-cooking-oil-collection

Song, C., Lin, W., Xie, Y., He, J., Xi, C., Ting, W., ... Mao, H. (2017). Air pollution in China: Status and spatiotemporal variations. *Environmental Pollution*, *227*, 334–347. doi:10.1016/j.envpol.2017.04.075 PMID:28482313

Song, L. J., Zhu, N. W., Yuan, H. P., Hong, Y., & Ding, J. (2010). Enhancement of waste activated sludge aerobic digestion by electrochemical pre-treatment. *Water Research*, *44*(15), 4371–4378. doi:10.1016/j.watres.2010.05.052 PMID:20580054

Sontaya, K., Pitiyont, B., & Punsuvon, V. (2013). Decolorization and COD removal of palm oil mill wastewater by electrocoagulation. *Int J Environ Sci Eng*, *7*, 370–374.

Sovacool, B. K., & Brown, M. A. (2010). Twelve metropolitan carbon footprints: A preliminary comparative global assessment. *Energy Policy*, *38*(9), 4856–4869. doi:10.1016/j.enpol.2009.10.001

Sovacool, B. K., & Drupady, I. M. (2016). *Energy access, poverty, and development: the governance of small-scale renewable energy in developing Asia*. Routledge.

Spagni, A., & Marsili-Libelli, S. (2009). Nitrogen removal via nitrite in a sequencing batch reactor treating sanitary landfill leachate. *Bioresource Technology*, *100*(2), 609–614. doi:10.1016/j.biortech.2008.06.064 PMID:18707876

SPAN. N. W. S. C. (2010). Malaysia water service industry & water treatment sludge issues. Retrieved from http://www.jwrc-net.or.jp/aswin/projects-activities/trainee_files/20101004-15_CR02.pdf

Speight, J. G. (2011). *The biofuels handbook*. Cambridge, UK: Royal Society of Chemistry. doi:10.1039/9781849731027

Spliethoff, H. (2010). Power generation from solid fuels. Munchen, Germany: Springer Science+Business Media. doi:10.1007/978-3-642-02856-4

Sponza, D. T., & Ağdağ, O. N. (2004). Impact of leachate recirculation and recirculation volume on stabilization of municipal solid wastes in simulated anaerobic bioreactors. *Process Biochemistry*, *39*(12), 2157–2165. doi:10.1016/j.procbio.2003.11.012

Šrám, R. J., Binková, B., Rössner, P., Rubeš, J., Topinka, J., & Dejmek, J. (1999). Adverse reproductive outcomes from exposure to environmental mutagens. *Mutation Research. Fundamental and Molecular Mechanisms of Mutagenesis*, *428*(1), 203–215. doi:10.1016/S1383-5742(99)00048-4 PMID:10517994

Srdjan, M. B. (2007). *Handbook of flotation reagents: chemistry, theory and practice; flotation of sulphide ores*. Oxford, UK: Elsevier.

Sreekanth, D., Sivaramakrishna, D., Himabindu, V., & Anjaneyulu, Y. (2009). Thermophilic treatment of bulk drug pharmaceutical industrial wastewaters by using hybrid up flow anaerobic sludge blanket reactor. *Bioresource Technology*, *100*(9), 2534–2539. doi:10.1016/j.biortech.2008.11.028 PMID:19135360

Srikanth, J. (2018). Pharmacological and toxicological evaluation of Indian noni in animal models of psychiatric disorders.

Srinivas, B., & Sarin, M. (2014). Brown carbon in atmospheric outflow from the Indo-Gangetic Plain: Mass absorption efficiency and temporal variability. *Atmospheric Environment*, *89*, 835–843. doi:10.1016/j.atmosenv.2014.03.030

Stamatelatou, K., Vavilin, V., & Lyberatos, G. (2003). Performance of a glucose fed periodic anaerobic baffled reactor under increasing organic loading conditions. 1: Experimental results. *Bioresource Technology*, *88*(2), 131–136. doi:10.1016/S0960-8524(02)00276-6 PMID:12576006

Stanhill, G., & Kalma, J. D. (1995). Solar dimming and urban heating at Hong Kong. *International Journal of Climatology*, *15*(8), 933–941. doi:10.1002/joc.3370150807

Starink, M. J. (2003). The determination of activation energy from linear heating rate experiments: A comparison of the accuracy of isoconversion methods. *Thermochimica Acta*, *404*(1-2), 163–176. doi:10.1016/S0040-6031(03)00144-8

Statista. (2019a). *Global production of vegetable oils from 2000/01 to 2018/19 (in million metric tons)*. Retrieved from https://www.statista.com/statistics/263978/global-vegetable-oil-production-since-2000-2001/

Statista. (2019b). *Consumption of vegetable oils worldwide from 2013/14 to 2018/2019, by oil type (in million metric tons)*. Retrieved from https://www.statista.com/statistics/263937/vegetable-oils-global-consumption/

Statistica. (2017). Retrieved from https://www.statista.com/statistics/282732/global-production-of-plastics-since-1950/

Steensgaard, I. M., Syberg, K., Rist, S., Hartmann, N. B., Boldrin, A., & Hansen, S. F. (2017). From macro to mi croplastics - Analysis of EU regulation along the life cycle of plastic bags. *Environmental Pollution*, *224*, 289–299. doi:10.1016/j.envpol.2017.02.007 PMID:28222979

Stephenson, R. L., & Blackburn, J. J. (1998). *The industrial wastewater systems handbook*. New York, NY: Lewis Publishers.

Sterling, M. C. Jr, Lacey, R. E., Engler, C. R., & Ricke, S. C. (2001). Effects of ammonia nitrogen on H2 and CH4 production during anaerobic digestion of dairy cattle manure. *Bioresource Technology*, *77*(1), 9–18. doi:10.1016/S0960-8524(00)00138-3 PMID:11211082

Steve, H., & Marshall, J. D. (2010). Impacts of urban form on future US passenger-vehicle greenhouse gas emissions. *Energy Policy*, *38*(9), 4880–4887. doi:10.1016/j.enpol.2009.07.005

Stolz, A. (2001). Basic and applied aspects in the microbial degradation of azo dyes. *Applied Microbiology and Biotechnology*, *56*(1-2), 69–80. doi:10.1007002530100686 PMID:11499949

Stringer, R., Labunska, I., Santillo, D., Johnston, P., Siddorn, J., & Stephenson, A. (2000). Concentrations of phthalate esters and identification of other additives in PVC children's toys. *Environmental Science and Pollution Research International*, *7*(1), 27–36. doi:10.1065/espr199910.007 PMID:19153836

Stringfellow, T. (2014). *An independent engineering evaluation of waste-to-energy technologies*. Renewable energy world. Retrieved from https://www.renewableenergyworld.com/articles/2014/01/an-independent-engineering-evaluation-of-waste-to-energy-technologies.html

Struzeski, E. J. (1980). Status of wastes handling and waste treatment across the pharmaceutical industry and 1977 effluent limitations. In *Proceedings of the 35th Industrial Waste Conference*, Purdue University, West Lafayette, IN (pp. 1095-1108).

Subramaniam, M. N., Goh, P. S., Lau, W. J., Ng, B. C., & Ismail, A. F. (2018). AT-POME colour removal through photocatalytic submerged filtration using antifouling PVDF-TNT nanocomposite membrane. *Separation and Purification Technology*, *191*, 266–275. doi:10.1016/j.seppur.2017.09.042

Subramaniam, S., & Abdullah, A. H., Nagapan, S., Kupusamy, K., Manian, H., & Daud, Z. (2018). Investigate how construction waste generation rate is different for every types of project in peninsular Malaysia using site visit method. *International Journal of Integrated Engineering*, *10*(1), 150–155. doi:10.30880/ijie.2018.10.01.022

Subramanian, S., Kumar, N., Murthy, S., & Novak, J. T. (2007). Effect of anaerobic digestion and anaerobic/aerobic digestion processes on sludge dewatering. *Journal of Residuals Science and Technology*, *4*, 17–23.

Suganya, K., & Revathi, K. (2016). Decolorization of reactive dyes by immobilized bacterial cells from textile effluents. *International Journal of Current Microbiology and Applied Sciences*, *5*(1), 528–532. doi:10.20546/ijcmas.2016.501.053

Suh, Y., & Resseaux, P. (2002). An LCA of alternative wastewater sludge treatment scenarios. *Resources, Conservation and Recycling*, *35*(3), 191–200. doi:10.1016/S0921-3449(01)00120-3

Sukiran, M. A., Abnisa, F., Daud, W. M. A. W., Bakar, N. A., & Loh, S. K. (2017). A review of torrefaction of oil palm solid wastes for biofuel production. *Energy Conversion and Management*, *149*, 101–120. doi:10.1016/j.enconman.2017.07.011

Sumathi, S., & Manju, B. S. (2000). Uptake of reactive textile dyes by Aspergillus foetidus. *Enzyme and Microbial Technology*, *27*(6), 347–355. doi:10.1016/S0141-0229(00)00234-9 PMID:10938413

Sun, Q. (2004) Iron and acid removal from acid mine drainage in open limestone systems. (Doctoral dissertation), West Virginia University.

Sundararaman, S., & Saravanane, R. (2010). Effect of loading rate and HRT on the removal of cephalosporin and their intermediates during the operation of a membrane bioreactor treating pharmaceutical wastewater. *Water Science and Technology*, *61*(7), 1907–1914. doi:10.2166/wst.2010.881 PMID:20371950

Sundberg, J. & Olofsson, M. (2005). *Waste-to-energy incineration,* RVF rapport, 15.

Sunil, K. (2016). *Municipal solid waste management in developing countries*. Boca Raton, FL: CRC.

Sun, J. H., Sun, S. P., Wang, G. L., & Qiao, L. P. (2007). Degradation of azo dye amido black 10B in aqueous solution by Fenton oxidation process. *Dyes and Pigments*, *74*(3), 647–652. doi:10.1016/j.dyepig.2006.04.006

Sun, J., Guo, L., Li, Q., Zhao, Y., Gao, M., She, Z., & Wang, G. (2016). Structural and functional properties of organic matters in extracellular polymeric substances (EPS) and dissolved organic matters (DOM) after heat pretreatment with waste sludge. *Bioresource Technology*, *219*, 614–623. doi:10.1016/j.biortech.2016.08.042 PMID:27544911

Sun, S., Li, C., Sun, J., Shi, S., Fan, M., & Zhou, Q. (2009). Decolorization of an azo dye Orange G in aqueous solution by Fenton oxidation process: Effect of system parameters and kinetic study. *Journal of Hazardous Materials*, *161*(2-3), 1052–1057. doi:10.1016/j.jhazmat.2008.04.080 PMID:18538927

Sun, Z., Yi, J., Huang, Y., Feng, D., & Guo, C. (2015). Investigation of the potential application of biodiesel by-product as asphalt modifier. *Road Materials and Pavement Design*, *17*(3), 737–752.

Sun, Z., Yi, J., Huang, Y., Feng, D., & Guo, C. (2016). Properties of asphalt binder modified by bio-oil derived from waste cooking oil. *Construction & Building Materials*, *102*, 496–504.

Suopajärvi, T. (2015). *Functionalized nanocelluloses in wastewater treatment applications.*

Surendra, K., Takara, D., Hashimoto, A. G., & Khanal, S. K. (2014). Biogas as a sustainable energy source for developing countries: Opportunities and challenges. *Renewable & Sustainable Energy Reviews*, *31*, 846–859.

Suryawanshi, P., Chaudhari, A., & Kothari, R. (2010). Thermophilic anaerobic digestion: The best option for waste treatment. *Critical Reviews in Biotechnology*, *30*(1), 31–40. PMID:20148754

Suthar, S., Bishnoi, P., Singh, S., Mutiyar, P. K., Nema, A. K., & Patil, N. S. (2009). Nitrate contamination in groundwater of some rural areas of Rajasthan, India. *Journal of Hazardous Materials*, *171*(1-3), 189–199. doi:10.1016/j.jhazmat.2009.05.111 PMID:19545944

Sutton, D., Kelleher, B., & Ross, J. R. (2001). Review of literature on catalysts for biomass gasification. *Fuel Processing Technology*, *73*(3), 155–173. doi:10.1016/S0378-3820(01)00208-9

Suwaileh, W., Johnson, D., & Hilal, N. (2019). Brackish water desalination for agriculture: Assessing the performance of inorganic fertilizer draw solutions. *Desalination*, *456*(January), 53–63. doi:10.1016/j.desal.2019.01.014

Suzuki, Y., Yoda, T., Rahul, A., & Sagiura, W. (2001). Molecular cloning and characterization of the gene encoding azoreductase from Bacillus sp. OY 1-2 isolated from soil. *The Journal of Biological Chemistry*, *246*(12), 9059–9065. doi:10.1074/jbc.M008083200 PMID:11134015

Syamlal, M., Rogers, W., & O'Brien, T. J. (1993). MFIX documentation: Theory guide. *National Energy Technology Laboratory, Department of Energy, Technical Note DOE/METC-95/1013 and NTIS/DE95000031.*

Szentannai, P., Bozi, J., Jakab, E., Osz, J., & Szucs, T. (2015). Towards the thermal utilisation of non-tyre rubbers – macroscopic and chemical changes while approaching the process temperature. *Fuel*, *156*, 148–157. doi:10.1016/j.fuel.2015.04.037

Szép, A. (2013). *Membránszűrés hatékonyságának javítása kombinált módszerek alkalmazásával ipari szennyvizek tisztítása során* (Doctoral dissertation, SZTE).

Szép, A., Kertész, S., László, Z., Szabó, G., & Hodúr, C. (2012). Advanced Treatment of Pharmaceutical Wastewater by Nanofiltration and Ozonation. *Acta Technica Corvininesis-Bulletin of Engineering*, *5*(1).

Taba, L. E., Irfan, M. F., Daud, W. A. M. W., & Chakrabarti, M. H. (2012). The effect of temperature on various parameters in coal, biomass and CO-gasification: A review. *Renewable & Sustainable Energy Reviews*, *16*(8), 5584–5596. doi:10.1016/j.rser.2012.06.015

Tabatabaee, N., Tabatabaee, H. A., Sabouri, M. R., & Teymourpour, P. (2009). Evaluation of performance grading parameters for crumb rubber modified asphalt binders and mixtures. In *Proceedings of 7th International RILEM Symposium on Advanced Testing and Characterization of Bituminous Materials*. 1. May 2009. Rhodes, Greece, 1-11.

Tabatabaei, M., Rahim, R., Abdullah, N., Wright, A., Shirai, Y., Sakai, K., & Hassan, M. A. (2010). Review: Importance of the methanogenic archaea populations in anaerobic wastewater treatments. *Process Biochemistry*, *45*(8), 1214–1225. doi:10.1016/j.procbio.2010.05.017

Tadeo, J. L., Sánchez-Brunete, C., Albero, B., & García-Valcárcel, A. I. (2010). Determination of pesticide residues in sewage sludge: A review. *Journal of AOAC International*, *93*, 1692–1702. PMID:21313796

Taib, M. R., Ho, W. S., & Ng, P. S. (2017). Development of waste-to-energy plant in Kuala Lumpur. *Chemical Engineering Transactions*, 56.

Talarposhti, A., Donnelly, T., & Andersonm, G. (2001). Colour removal from a simulated dye wastewater using a two-phase anaerobic packed bed reactor. *Water Research*, *35*(2), 425–432. doi:10.1016/S0043-1354(00)00280-3 PMID:11228995

Talinli, I., & Anderson, G. K. (1992). Interference of hydrogen peroxide on the standard COD test. *Water Research*, *26*(1), 107–110. doi:10.1016/0043-1354(92)90118-N

Tamar, R. L. (2008) Pilot study of horizontal roughing filtration in Northern Ghana as pretreatment for highly turbid dugout water. (MSc. Thesis), Massachusetts Institute of Technology.

Tamimi, N., Qourzal, S., Barka, N., Assabbane, A., & Ait-Ichou, Y. (2008). Methomyl degradation in aqueous solutions by Fenton's reagent and the photo-Fenton system. *Separation and Purification Technology*, *61*(1), 103–108. doi:10.1016/j.seppur.2007.09.017

Tam, L. S., Tang, T. W., Lau, G. N., Sharma, K. R., & Chen, G. H. (2007). A pilot study for wastewater reclamation and reuse with MBR/RO and MF/RO systems. *Desalination*, *202*(1-3), 106–113. doi:10.1016/j.desal.2005.12.045

Tam, N. F., Chong, A. M., & Wong, Y. S. (2002). Removal of tributyltin (TBT) by live and dead microalgal cells. *Marine Pollution Bulletin*, *45*(1-12), 362–371. doi:10.1016/S0025-326X(02)00184-4 PMID:12398407

Tanaka, M. (1999). Recent trends in recycling activities and waste management in Japan. *Journal of Material Cycles and Waste Management*, *1*, 10–16.

Tan, G. (1995). Residue levels of phthalate esters in water and sediment samples from the Klang River basin. *Bulletin of Environmental Contamination and Toxicology*, *54*(2), 171–176. doi:10.1007/BF00197427 PMID:7742623

Tan, I. A. W., Ahmad, A. L., & Hameed, B. H. (2008). Optimization of preparation conditions for activated carbons from coconut husk using response surface methodology. *Chemical Engineering Journal*, *137*(3), 462–470. doi:10.1016/j.cej.2007.04.031

Tapana, C., & Pagilla, K. R. (2000). Aerobic thermophilic and anaerobic mesophilic treatment of sludge. *Journal of Environmental Engineering*, 790–795.

Tarnpradab, T., Unyaphan, S., Takahashi, F., & Yoshikawa, K. (2016). Tar removal capacity of waste cooking oil absorption and waste char adsorption for rice husk gasification. *Biofuels*, *7*(4), 401–412. doi:10.1080/17597269.2016.1147919

Taseiko, O. V., Mikhailuta, S. V., Anne, P., Lezhenin, A. A., & Zakharov, Y. V. (2009). Air pollution dispersion within urban street canyons. *Atmospheric Environment*, *43*(2), 245–252. doi:10.1016/j.atmosenv.2008.09.076

Tatarková, V., Hiller, E., & Vaculík, M. (2013). Impact of wheat straw biochar addition to soil on the sorption, leaching, dissipation of the herbicide (4-chloro-2-methylphenoxy) acetic acid and the growth of sunflower (Helianthus annuus L.). *Ecotoxicology and Environmental Safety*, *92*, 215–221. doi:10.1016/j.ecoenv.2013.02.005 PMID:23474069

Taufiq-Yap, Y. H., Sivasangar, S., & Salmiaton, A. (2012). Enhancement of hydrogen production by secondary metal oxide dopants on NiO/CaO material for catalytic gasification of empty palm fruit bunches. *Energy*, *47*(1), 158–165. doi:10.1016/j.energy.2012.09.026

Tavlieva, M. P., Genieva, S. D., Georgieva, V. G., & Vlaev, L. T. (2013). Kinetic study of brilliant green adsorption from aqueous solution onto white rice husk ash. *Journal of Colloid and Interface Science*, *409*, 112–122. doi:10.1016/j.jcis.2013.07.052 PMID:23981675

Tchobanoglous, G., & Kreith, F. (2003) Handbook of solid waste management (2nd edition). Metcalf & Eddy, Inc., Santa Maria, CA.

Tchobanoglous, G., Burton, F. L., & Stensel, H. D. J. I. E. M. (2003). Metcalf & Eddy wastewater engineering: treatment and reuse. *International Edition. McGrawHill, 4*, 361-411.

Tchobanoglous, G., Franklin, B. H., & Stensel, D. (2003) Wastewater engineering: treatment and reuse (4th ed.). Metcalf & Eddy, Inc., Santa Maria, CA.

Tchobanoglous, G., Burton, F., & Davind, H. (2002). *Wastewater engineering: Treatment and reuse* (4th ed.). Metcalf and Eddy.

Tchobanoglous, G., Theisen, H., & Vigil, S. (1993). *Integrated solid waste management: engineering principles and management issues*. New York, NY: McGraw-Hill.

Tchounwou, P. B., Yedjou, C. G., Patlolla, A. K., & Sutton, D. J. (2012). *Heavy metal toxicity and the environment. Molecular, clinical, and environmental toxicology* (pp. 133–164). Springer. doi:10.1007/978-3-7643-8340-4_6

Teh, C. Y., Budiman, P. M., Shak, K. P. Y., & Wu, T. Y. (2016). Recent advancement of coagulation–flocculation and its application in wastewater treatment. *Industrial & Engineering Chemistry Research, 55*(16), 4363–4389. doi:10.1021/acs.iecr.5b04703

Teh, C. Y., Wu, T. Y., & Juan, J. C. (2014). Potential use of rice starch in coagulation–flocculation process of agro-industrial wastewater: Treatment performance and flocs characterization. *Ecological Engineering, 71*, 509–519. doi:10.1016/j.ecoleng.2014.07.005

Tenedero, R. A., & Surtida, M. B. (Eds.). (1986). *Soil sampling and preparation for laboratory analysis. Tigbauan.* Iloilo, Philippines: SEAFDEC Aquaculture Department.

Tepuš, B., Simonič, M., & Petrinić, I. (2009). Comparison between nitrate and pesticide removal from ground water using adsorbents and NF and RO membranes. *Journal of Hazardous Materials, 170*(2-3), 1210–1217. doi:10.1016/j.jhazmat.2009.05.105 PMID:19541416

Ternes, T. A., Meisenheimer, M., McDowell, D., Sacher, F., Brauch, H. J., Haist-Gulde, B., ... Zulei-Seibert, N. (2002). Removal of pharmaceuticals during drinking water treatment. *Environmental Science & Technology, 36*(17), 3855–3863. doi:10.1021/es015757k PMID:12322761

Teuten, E. L., Saquing, J. M., Knappe, D. R. U., Barlaz, M. A., Jonsson, S., Bjorn, A., ... Takada, H. (2009). Transport and release of chemicals from plastics to the environment and to wildlife. *Philos. Transac. Roy. Soc. B Biol. Sci., 364*(1526), 2027–2045. PMID:19528054

Teymourpour, P., Sillamäe, S., & Bahia, H. U. (2015). Impacts of lubricating oils on rheology and chemical compatibility of asphalt binders. *Road Materials and Pavement Design, 16*, 50–74.

Thaker, M., & Trivedi, H. (2005). Ultraviolet-radiation-induced graft copolymerization of methyl acrylate onto the sodium salt of partially carboxymethylated guar gum. *Journal of Applied Polymer Science, 97*(5), 1977–1986. doi:10.1002/app.20988

The Hindu Business line. (2018). *New regulations for 'used cooking oil' come into effect.* Retrieved from https://www.thehindubusinessline.com/economy/policy/new-regulations-for-used-cooking-oil-come-into-effect/article24314377.ece

The news minute. (2019). *Restaurants can no longer reuse cooking oils as per new food safety regulations.* Retrieved from https://www.thenewsminute.com/article/restaurants-can-no-longer-reuse-cooking-oils-new-food-safety-regulations-96723

Themelis, N., & Bourtsalas, A. (2013). *UK waste management: Growing old or growing clean?* Retrieved from https://waste-management-world.com/a/uk-waste-management-growing-old-or-growing-clean

Thiabgoh, O., Eggers, T., Jimenez, V. O., Jiang, S. D., Sun, J. F., & Phan, M. H. (2018). Real-time monitoring of position and motion of a non-stationary object with a highly sensitive magnetic impedance sensor. *Journal of Science: Advanced Materials and Devices*, *3*(1), 122–128. doi:10.1016/j.jsamd.2018.01.006

Thi, N. B. D., Kumar, G., & Lin, C.-Y. (2015). An overview of food waste management in developing countries: Current status and future perspective. *Journal of Environmental Management*, *157*, 220–229. PMID:25910976

Thombare, N., Jha, U., Mishra, S., & Siddiqui, M. (2016). Guar gum as a promising starting material for diverse applications: A review. *International Journal of Biological Macromolecules*, *88*, 361–372. doi:10.1016/j.ijbiomac.2016.04.001 PMID:27044346

Thompson, G., Swain, J., Kay, M., & Forster, C. F. (2001). The treatment of pulp and paper mill effluent: A review. *Bioresource Technology*, *77*(3), 275–286. doi:10.1016/S0960-8524(00)00060-2 PMID:11272013

Thompson, H., & Tracy, J. (1991). Restructuring, reskilling and retaining: inside the building and construction industry. *Journal of Australian Political Economy*, *13*, 1–18.

Thompson, R. C., Swan, S. H., Moore, C. J., & vom Saal, F. S. (2009b). Our plastic age. *Philos Trans R Soc B. Biol. Sci.*, *364*(1526), 1973–1976. doi:10.1098/rstb.2009.0054

Thurner, F., & Mann, U. (1981). Kinetic investigation of wood pyrolysis. *Industrial & Engineering Chemistry Process Design and Development*, *20*(3), 482–488. doi:10.1021/i200014a015

Tian, Y., Jiang, Y., Liu, Q., Xu, D., Zhao, S., He, L., ... Xu, H. (2009). Temporal and spatial trends in air quality in Beijing. *Landscape and Urban Planning*, *185*, 35–43. doi:10.1016/j.landurbplan.2019.01.006

Tiefelsdorf, M. (1998). Some practical applications of Moran's I's exact conditional distribution. *Papers in Regional Science*, *77*(2), 101–129. doi:10.1111/j.1435-5597.1998.tb00710.x

Tiesler, C. M. T., Birk, M., Thiering, E., Kohlböck, G., Koletzko, S., Bauer, C. P., ... Heinrich, J. (2013). Exposure to road traffic noise and children's behavioural problems and sleep disturbance: Results from the GINIplus and LISAplus studies. *Environmental Research*, *123*, 1–8. doi:10.1016/j.envres.2013.01.009 PMID:23498846

Tijani, J. O., Fatoba, O. O., & Petrik, L. F. (2013). A review of pharmaceuticals and endocrine-disrupting compounds: Sources, effects, removal, and detections. *Water, Air, and Soil Pollution*, *224*(11), 1770. doi:10.100711270-013-1770-3

Tilman, D., Balzer, C., Hill, J., & Befort, B. L. (2011). Global food demand and the sustainable intensification of agriculture. *Proceedings of the National Academy of Sciences of the United States of America*, *108*(50), 20260–20264. doi:10.1073/pnas.1116437108 PMID:22106295

Ting, C. H., Shahidul, M. I., Shazali, S. T. S., & Abdullah, Y. (2013). Measuring contextual impact of maintenance on machinery capacity utilization and environment. In *6th Engineering Conference*. Kuching, Malaysia. doi:10.3850/978-981-07-6059-5_034

Tobler, Waldo R. (1970). A computer movie simulating urban growth in the Detroit region. *Economic Geography* (Supplemental: Proceedings, International Geographical Union. Commission on Quantitative Methods), 46, 234–240. doi:10.2307/14314

Tokumura, M., Katoh, H., Katoh, T., Znad, H. T., & Kawase, Y. (2009). Solubilization of excess sludge in activated sludge process using the solar photo-Fenton reaction. *Journal of Hazardous Materials*, *162*(2-3), 1390–1396. doi:10.1016/j.jhazmat.2008.06.026 PMID:18639985

Tokumura, M., Sekine, M., Yoshinari, M., Znad, H. T., & Kawase, Y. (2007). Photo-Fenton process for excess sludge disintegration. *Process Biochemistry*, *42*(4), 627–633. doi:10.1016/j.procbio.2006.11.010

Tomei, M. C., Bertanza, G., Canato, M., Heimersson, S., Laera, G., & Svanström, M. (2016). Techno-economic and environmental assessment of upgrading alternatives for sludge stabilization in municipal wastewater treatment plants. *Journal of Cleaner Production*, *112*, 3106–3115. doi:10.1016/j.jclepro.2015.10.017

Tomishige, K. (2004). Syngas production from methane reforming with CO 2/H 2 O and O 2 over NiO–MgO solid solution catalyst in fluidized bed reactors. *Catalysis Today*, *89*(4), 405–418. doi:10.1016/j.cattod.2004.01.003

Tong, H., Shen, Y., Zhang, J., Wang, C.-H., Ge, T. S., & Tong, Y. W. (2018). A comparative life cycle assessment on four waste-to-energy scenarios for food waste generated in eateries. *Applied Energy*, *225*, 1143–1157.

Tong, H., Yao, Z., Lim, J. W., Mao, L., Zhang, J., Ge, T. S., ... Tong, Y. W. (2018). Harvest green energy through energy recovery from waste: A technology review and an assessment of Singapore. *Renewable & Sustainable Energy Reviews*, *98*, 163–178.

Tong, S., & Jaafar, A. B. (2004). Waste to energy: Methane recovery from anaerobic digestion of palm oil mill effluent. *Energy Smart*, *14*, 11456.

Tonne, C., Salmon, M., Sanchez, M., Sreekanth, V., Bhogadi, S., Sambandam, S., ... Marshall, J. D. (2017). Integrated assessment of exposure to PM 2.5 in South India and its relation with cardiovascular risk: Design of the CHAI observational cohort study. *International Journal of Hygiene and Environmental Health*, *220*(6), 1081–1088. doi:10.1016/j.ijheh.2017.05.005 PMID:28606699

Topalcengiz, Z., Strawn, L. K., & Danyluk, M. D. (2017). Microbial quality of agricultural water in Central Florida. *PLoS One*, *12*(4). doi:10.1371/journal.pone.0174889 PMID:28399144

Torkian, A., Eqbali, A., & Hashemian, S. J. (2003). The effect of organic loading rate on the performance of UASB reactor treating slaughterhouse effluent. *Resources, Conservation and Recycling*, *40*(1), 1–11. doi:10.1016/S0921-3449(03)00021-1

Trabelsi, A. B. H., Zaafouri, K., Baghdadi, W., Naoui, S., & Ouerghi, A. (2018). Second generation biofuels production from waste cooking oil via pyrolysis process. *Renewable Energy*, *126*, 888–896. doi:10.1016/j.renene.2018.04.002

Tran, H. T., Hurley, B. A., & Plaxton, W. C. (2010). Feeding hungry plants: The role of purple acid phosphatases in phosphate nutrition. *Plant Science*, *179*(1-2), 14–27.

Tran, L.-H., Drogui, P., Mercier, G., & Blais, J.-F. (2009). Electrochemical degradation of polycyclic aromatic hydrocarbons in creosote solution using ruthenium oxide on titanium expanded mesh anode. *Journal of Hazardous Materials*, *164*(2-3), 1118–1129. doi:10.1016/j.jhazmat.2008.09.012 PMID:18926633

Trindade, P. V. O., Sobral, L. G., Rizzo, A. C. L., Leite, S. G. F., & Soriano, A. U. (2015). Bioremediation of a weathered and a recently oil-contaminated soils from Brazil: A comparison study. *Chemosphere*, *58*(4), 515–522. doi:10.1016/j.chemosphere.2004.09.021 PMID:15620743

Tripathi, M., Sahu, J. N., & Ganesan, P. (2016). Effect of process parameters on production of biochar from biomass waste through pyrolysis: A review. *Renewable & Sustainable Energy Reviews*, *55*, 467–481.

Tripathy, T., & De, B. R. (2006). Flocculation: a new way to treat the waste water.

Troxler, W. L, Cudahy, J. J., Zin, R. P., & Rosenthal, S. I. (1994). Thermal desorper: Applications manual for treating non-hazardous petroleum contaminated soils. Cincinnati, OH: U.S. EPA, Office of Research and Development, 15 pp.

Trzcinski, A., & Stuckey, D. C. (2009). Continuous treatment of the organic fraction of municipal solid waste in an anaerobic two-stage membrane process with liquid recycle. *Water Research*, *43*(9), 2449–2462. PMID:19371920

Tsai, W.-T. (2019). Mandatory recycling of waste cooking oil from residential and commercial sectors in Taiwan. *Resources, 8*(1), 38. doi:10.3390/resources8010038

Tseng, Y. C., Moreira, R. G., & Sun, X. (1996). Total frying—use time effects on soybean oil deterioration and on tortilla chip quality. *International Journal of Food Science & Technology, 31*, 287–294.

Tullo, A. H. (2015). Plasticizer makers want a piece of the Phthalates pie. *Chemical and Engineering News, 93*(25), 16–18.

Turek, M., Dydo, P., Trojanowska, J., & Bandura, B. (2007). Electrodialytic treatment of boron-containing wastewater. *Desalination, 205*(1), 185–191. doi:10.1016/j.desal.2006.02.055

Turovskiy, I. S., & Mathai, P. (2006). *Wastewater sludge processing*. Hoboken, NJ: John Wiley & Sons.

Tyre component: source. Retrieved from https://kitchendecor.club/files/radial-tire-cross-section.html

U.S. EPA. (2006). In situ treatment technologies for contaminated soil. *Engineering Forum Issue Paper. Solid Waste and Emergency Response 5203P.* Available from https://clu-in.org/download/remed/542f06013.pdf

Uchechukwu, E. E., Sylvia, O. A., & Vincent, I. I. (2007). Clean up of crude oil-contaminated soil. *Terrestrial and Aquatic Environmental Toxicology, 1*(2), 54–59.

Udawatta, N., Zuo, J., Chiveralls, K., & Zillante, G. (2015). Improving waste management in construction projects: An Australian study. *Resources, Conservation and Recycling, 101*, 73–83. doi:10.1016/j.resconrec.2015.05.003

Uddin, M. S., Zhou, J., Qu, Y., Guo, J., Wang, P., & Zhao, L. (2007). Biodecolourization of Azo-dye acid red B under high salinity condition. *Bulletin of Environmental Contamination and Toxicology, 79*(4), 440–444. doi:10.100700128-007-9260-1 PMID:17721733

Uddin, M., Rahman, M., Taweekun, J., Techato, K., Mofijur, M., & Rasul, M. (2019). Enhancement of biogas generation in up-flow sludge blanket (UASB) bioreactor from palm oil mill effluent (POME). *Energy Procedia, 160*, 670–676. doi:10.1016/j.egypro.2019.02.220

Uddin, W., Khan, B., Shaukat, N., Majid, M., Mujtaba, G., Mehmood, A., ... Almeshal, A. M. (2016). Biogas potential for electric power generation in Pakistan: A survey. *Renewable & Sustainable Energy Reviews, 54*, 25–33.

Udomsap, P., Chollacoop, N., Topaiboul, S., & Hirotsu, T. (2009). Effect of antioxidant on the oxidative stability of waste cooking oil based biodiesel under different storage conditions. *International Journal of Renewable Energy, 4*(2), 47–59.

Ueda, S., & Earle, R. (1972). Microflora of activated sludge. *The Journal of General and Applied Microbiology, 18*(3), 239–248. doi:10.2323/jgam.18.239

Ueda, T., Hata, K., Kikuoka, Y., & Seino, O. (1997). Effects of aeration on suction pressure in a submerged membrane bioreactor. *Water Research, 31*(3), 489–494. doi:10.1016/S0043-1354(96)00292-8

UK Department for Environment, Food, & Rural Affairs (2016) Microbead ban announced to protect sealife. Retrieved from https://www.gov.uk/government/news/microbead-ban-announced-to-protect-sealife

UK Environmental Law Association. (2017). *The plain guide to environmental law.* Retrieved from http://www.environmentlaw.org.uk/rte.asp?id=85

Ulas, T., Tandukar, M., & Pavlostathis, S. G. (2011). Anaerobic biotreatment of municipal sewage sludge. In M. M. Young (Ed.), *Comprehensive biotechnology* (pp. 447–461). UK: Pergamon Press.

Ullah, Z., Bustam, M. A., & Man, Z. (2014). Characterization of waste palm cooking oil for biodiesel production. *International Journal of Chemical Engineering and Applications, 5*(2), 134–137.

Umar, M., Aziz, H. A., & Yusoff, M. S. (2010). Trends in the use of Fenton, electro-Fenton and photo-Fenton for the treatment of landfill leachate. *Waste Management (New York, N.Y.)*, *30*(11), 2113–2121. doi:10.1016/j.wasman.2010.07.003 PMID:20675113

Umikalsom, M. S., Ariff, A. B., Zulkifli, H. S., Tong, C. C., Hassan, M. A., & Karim, M. I. A. (1997). The treatment of oil palm empty fruit bunch fibre for subsequent use as substrate for cellulase production by Chaetomium globosum Kunze. *Bioresource Technology*, *62*(1–2), 1–9. doi:10.1016/S0960-8524(97)00132-6

Undri, A., Rosi, L., Frediani, M., & Frediani, P. (2011). Microwave pyrolysis of polymeric materials. In Microwave Heating, 207-312. IntechOpen.

UN-Habitat. (2010). *Solid waste management in the world's cities*. Retrieved from https://unhabitat.org/wp-content/uploads/2015/12/SolidWaste.pdf

United Nations Economic Commission for Africa. (2009). *Africa review report on waste management*. Retrieved from http://www1.uneca.org/Portals/3/documents/AfricanReviewReport-onWasteManagementMainReport.pdf

United Nations World Water Assessment Programme. (2014). *Water and Energy. The United Nations World Water Development Report 2014* (Vol. 1). United Nations Educational, Scientific and Cultural Organization.

United Nations World Water Assessment Programme. (2015). *Water for a sustainable world*. Paris, France: United Nations Educational, Scientific and Cultural Organization.

United Nations. (2012). *The Millennium Development Goals Report 2012*. New York. doi:10.1177/1757975909358250

United Nations. (2015). *The Millennium Development Goals Report 2015. United Nations*. New York.

United State Environmental Protection Agency. Bioslurry treatment of contaminated soils and sediments. *U.S. EPA Test and Evaluation Facility Research Project*. IT Corporation. Retrieved from http://www.emfederal.com/projects/fctshts/bioslrry.pdf

United States Census Bureau. (2015). International Programs. Retrieved from https://www.census.gov/population/international/data/idb/worldpopgraph.php

United States Environmental Protection Agency. (2012). The facts about nutrient pollution. 1-2. Retrieved from www.epa.gov/nutrientpollution

Unsworth, J. (2010). History of pesticide use. *IUPAC-International Union of Pure and Applied Chemistry*.

Upadhyay, R. S. (2002). Microbial bioremediation of textile effluents. In V. P. Singh & R. D. Stapleton Jr., (Eds.), *Biotransformation: bioremediation technology for health and environmental protection* (pp. 331–348). Amsterdam, The Netherlands: Stapleton, Elsevier Science BV. doi:10.1016/S0079-6352(02)80018-2

Urase, T., Kagawa, C., & Kikuta, T. (2005). Factors affecting removal of pharmaceutical substances and estrogens in membrane separation bioreactors. *Desalination*, *178*(1-3), 107–113. doi:10.1016/j.desal.2004.11.031

Urase, T., Salequzzaman, M., Kobayashi, S., Matsuo, T., Yamamoto, K., & Suzuki, N. (1997). Effect of high concentration of organic and inorganic matters in landfill leachate on the treatment of heavy metals in very low concentration level. *Water Science and Technology*, *36*(12), 349–356. doi:10.2166/wst.1997.0464

Urbain, V., Block, J. C., & Manem, J. (1993). Bioflocculation in activated sludge: An analytic approach. *Water Research*, *27*(5), 829–838. doi:10.1016/0043-1354(93)90147-A

US Congress. (2016). Public Law 114–114, 114th Congress an act, H.R.1321 – Microbead free waters act 2015. *Public Law*, 3129–3130.

US Environmental Protection Agency (US EPA). (2016). Sustainable materials management: nonhazardous materials and waste management hierarchy. Retrieved from https://www.epa.gov/smm/sustainable-materials-management-non-hazardous-materials-and-waste-management-hierarchy

US Environmental Protection Agency. (2012). Recreational water quality criteria. Washington, DC: Office of Water Washington.

US EPA. (1999). Wastewater technology fact sheet—sequencing batch reactors: EPA 832-F-99-073.

US EPA. (2007). National section 303(d) list fact sheet. Retrieved from Information Online at www.iaspub.epa.gov/waters/national_rept.control

US EPA. (2015). A compilation of cost data associated with the impacts and control of nutrient pollution. 1-110. Retrieved from http://www2.epa.gov/nutrient-policy-data/compilation-cost-data-associated-impacts-and-control-nutrient-pollution

US EPA. Office of Water, Washington. (2000). Wastewater technology fact sheet: Oxidation ditches. EPA 832-F-00-013, 1-6.

USEPA. (1993). *Standards for the Use or Disposal of Sewage Sludge 40CFR PART 503*. Washington, DC: United States Environmental Protection Agency.

Uygur, A., & Kargı, F. (2004). Biological nutrient removal from pre-treated landfill leachate in a sequencing batch reactor. *Journal of Environmental Management*, *71*(1), 9–14. doi:10.1016/j.jenvman.2004.01.002 PMID:15084355

Uzir, N. M., Aziz, M. M. A., Ismail, S. B., Mahmood, N. A. N., Umor, N. A., & Muhammad, S. A. F. A. S. (2019). Optimization of coagulation-flocculation process for the palm oil mill effluent treatment by using rice husk ash. *Industrial Crops and Products*, *139*, 111482. doi:10.1016/j.indcrop.2019.111482

Vagı, M., & Petsas, A. (2017). *Advanced oxidation processes for the removal of pesticides from wastewater: recent review and trends.* Paper presented at the 15th International Conference on Environmental Science and Technology. CEST2017, Rhodes, Greece.

Vagia, E., & Lemonidou, A. (2007). Thermodynamic analysis of hydrogen production via steam reforming of selected components of aqueous bio-oil fraction. *International Journal of Hydrogen Energy*, *32*(2), 212–223. doi:10.1016/j.ijhydene.2006.08.021

Valentini, A., Garuti, G., Rozzi, A., & Tilche, A. (1997). Anaerobic degradation kinetics of particulate organic matter: A new approach. *Water Science and Technology*, *36*(6-7), 239–246. doi:10.2166/wst.1997.0596

Valo, A., Carrère, H., & Delgenès, J. P. (2004). Thermal, chemical and thermo-chemical pretreatment of waste activated sludge for anaerobic digestion. *Journal of Chemical Technology and Biotechnology (Oxford, Oxfordshire)*, *79*(11), 1197–1203. doi:10.1002/jctb.1106

Van Ewijk, S., & Stegemann, J. A. (2016). Limitations of the waste hierarchy for achieving absolute reductions in material throughput. *Journal of Cleaner Production*, *132*, 122–128. doi:10.1016/j.jclepro.2014.11.051

van Leeuwen, C., Sniders, S., Durin, N., Lacroix, B., Verbrugh, H., Enright, M., & van Belkum, A. (2003). Multilocus sequence typing of *Staphylococcus aureus* with DNA array technology. *Journal of Clinical Microbiology*, *41*(7), 3323–3326. doi:10.1128/JCM.41.7.3323-3326.2003 PMID:12843083

Van Rijn, J. (1996). The potential for integrated biological treatment systems in recirculating fish culture—A review. *Aquaculture (Amsterdam, The Netherlands)*, *139*(3), 181–201. doi:10.1016/0044-8486(95)01151-X

Vandamme, D., Foubert, I., & Muylaert, K. (2013). Flocculation as a low-cost method for harvesting microalgae for bulk biomass production. *Trends in Biotechnology*, *31*(4), 233–239. doi:10.1016/j.tibtech.2012.12.005 PMID:23336995

Vandevivere, P. C., Bianchi, R., & Verstraete, W. (1998). Treatment and reuse of wastewater from the textile wet-processing industry: Review of emerging technologies. *Journal of Chemical Technology & Biotechnology: International Research in Process. Environmental and Clean Technology*, *72*(4), 289–302.

Vanerkar, A. P., Satyanarayan, S., & Dharmadhikari, D. M. (2004). Toxicity of herbal pharmaceutical wastewater on fish—Lebistes reticulatus (Peter). *Journal of Environmental Science and Health. Part B, Pesticides, Food Contaminants, and Agricultural Wastes*, *39*(1), 115–123. doi:10.1081/PFC-120027443 PMID:15022745

Vanwonterghem, I., Jensen, P. D., Rabaey, K., & Tyson, G. W. (2016). Genome-centric resolution of microbial diversity, metabolism and interactions in anaerobic digestion. *Environmental Microbiology*, *18*(9), 3144–3158. PMID:27317862

Varentsov, M., Konstantinov, P., Samsonov, T., & Repina, I. (2014): Investigation of the urban heat island phenomenon during polar night based on experimental measurements and remote sensing of Norilsk city, Sovremennye problemy distantsionnogo zondirovaniya Zemli iz kosmosa (in Russian), 11, 329-337.

Várhegyi, G. (2007). Aims and methods in non-isothermal reaction kinetics. *Journal of Analytical and Applied Pyrolysis*, *79*(1-2), 278–288. doi:10.1016/j.jaap.2007.01.007

Vasco-Correa, J., Khanal, S., Manandhar, A., & Shah, A. (2018). Anaerobic digestion for bioenergy production: Global status, environmental and techno-economic implications, and government policies. *Bioresource Technology*, *247*, 1015–1026. PMID:28918346

Vassilev, S. V., Baxter, D., Andersen, L. K., & Vassileva, C. G. (2010). An overview of the chemical composition of biomass. *Fuel*, *89*(5), 913–933. doi:10.1016/j.fuel.2009.10.022

Vavilin, V., Fernandez, B., Palatsi, J., & Flotats, X. (2008). Hydrolysis kinetics in anaerobic degradation of particulate organic material: An overview. *Waste Management (New York, N.Y.)*, *28*(6), 939–951. PMID:17544638

Vázquez, V. F.,, & Paje, S. E. (2016). Study of the road surface properties that control the acoustic performance of a rubberised asphalt mixture. *Applied Acoustics*, *102*, 33–39. doi:10.1016/j.apacoust.2015.09.008

Veil, J. A., Puder, M. G., Elcock, D., & Redweik, R. J., Jr. (2004). A white paper describing produced water from production of crude oil, natural gas, and coal bed methane. *Prepared by Argonne National Laboratory for the US Department of Energy, National Energy Technology Laboratory, January.* Available at http://www. ead. anl. gov/pub/dsp_detail. cfm

Velagaleti, R., & Burns, P. K. (2006). The industrial ecology of pharmaceutical raw materials and finished products with emphasis on supply chain management activities.

Velghe, I., Carleer, R., Yperman, J., & Schreurs, S. (2011). Study of the pyrolysis of municipal solid waste for the production of valuable products. *Journal of Analytical and Applied Pyrolysis*, *92*(2), 366–375. doi:10.1016/j.jaap.2011.07.011

Venderbosch, R. H., & Prins, W. (2010). Fast pyrolysis technology development. *Biofuels, Bioproducts & Biorefining*, *4*(2), 178–208. doi:10.1002/bbb.205

Ventosa, A., Nieto, J., & Oren, A. (1998). Biology of moderately halophilic aerobic bacteria. *Microbiology and Molecular Biology Reviews*, *62*, 504–544. PMID:9618450

Venugopal, V. (2016). Marine polysaccharides: Food applications. Boca Raton, FL: CRC Press. doi:10.1201/b10516

Verma, A. K. (2017). Treatment of textile wastewaters by electrocoagulation employing fe-al composite electrode. *Journal of Water Process Engineering*, *20*, 168–172. doi:10.1016/j.jwpe.2017.11.001

Verma, A. K., Dash, R. R., & Bhunia, P. (2012). A review on chemical coagulation/flocculation technologies for removal of colour from textile wastewaters. *Journal of Environmental Management, 93*(1), 154–168. doi:10.1016/j.jenvman.2011.09.012 PMID:22054582

Verma, P., & Madamwar, D. (2005). Decolorization of Azo dyes using Basidiomycete strain PV 002. *World Journal of Microbiology & Biotechnology, 21*(4), 481–485. doi:10.100711274-004-2047-1

Vest, H. (2000). *Recycling of used car tyres*. Retrieved from www.gtz.de/gate/gateid.afp

Vicentin, R., Fdz-Polanco, F., & Fdz-Polanco, M. (2019). Energy integration in wastewater treatment plants by anaerobic digestion of urban waste: a process design and simulation study. *International Journal of Chemical Engineering, 2019, 2621048*. doi:10.1155/2019/2621048

Vijayaraghavan, K., Ahmad, D., & Aziz, M. (2007). Aerobic treatment of palm oil mill effluent. *Journal of Environmental Management, 82*(1), 24–31. doi:10.1016/j.jenvman.2005.11.016 PMID:16584834

Vijayaraghavan, K., & Yun, Y. (2008). Biosorption of C. I. Reactive Black 5 from aqueous solution using acid treated biomass of brown seaweed Laminaria sp. *Dyes Pig, 76*(3), 726–732. doi:10.1016/j.dyepig.2007.01.013

Vik, E. A., Carlson, D. A., Eikum, A. S., & Gjessing, E. T. (1984). Electrocoagulation of potable water. *Water Research, 18*(11), 1355–1360. doi:10.1016/0043-1354(84)90003-4

Vilar, V. J., Rocha, E. M., Mota, F. S., Fonseca, A., Saraiva, I., & Boaventura, R. A. (2011). Treatment of a sanitary landfill leachate using combined solar photo-Fenton and biological immobilized biomass reactor at a pilot scale. *Water Research, 45*(8), 2647–2658. doi:10.1016/j.watres.2011.02.019 PMID:21411117

Vinati, H., & Radha, R. International Journal of Scientific Research and Reviews.

Vinyes, E., Oliver-Solà, J., Ugaya, C., Rieradevall, J., & Gasol, C. M. (2013). Application of LCSA to used cooking oil waste management. *The International Journal of Life Cycle Assessment, 18*(2), 445–455. doi:10.100711367-012-0482-z

Visvanathan, C., Aim, R. B., & Parameshwaran, K. (2000). Membrane separation bioreactors for wastewater treatment. *Critical Reviews in Environmental Science and Technology, 30*(1), 1–48. doi:10.1080/10643380091184165

Vitousek, P. M., Aber, J. D., Howarth, R. W., Likens, G. E., Matson, P. A., Schindler, D. W., & Tilman, D. G. (1997). Human alteration of the global nitrogen cycle: Sources and consequences. *Ecological Applications, 7*(3), 737–750.

Vlyssides, A. G., Papaioannou, D., Loizidoy, M., Karlis, P. K., & Zorpas, A. A. (2000). Testing an electrochemical method for treatment of textile dye wastewater. *Waste Management (New York, N.Y.), 20*(7), 569–574. doi:10.1016/S0956-053X(00)00028-3

Vögeli, Y., & Zurbrügg, C. (2008, November). *Decentralized anaerobic digestion of kitchen and market waste in developing countries – state-of-the-art in South India.* Paper presented at the Second International Symposium on Energy from Biomass and Waste, Venice, Italy.

von Sperling, M. (1996). Comparison among the most frequently used systems for wastewater treatment in developing countries. *Water Science and Technology, 33*(3), 59–72. doi:10.2166/wst.1996.0059

Vossoughi, M., Shakeri, M., & Alemzadeh, I. (2003). Performance of anaerobic baffled reactor treating synthetic wastewater influenced by decreasing COD/SO4 ratios. *Chemical Engineering and Processing: Process Intensification, 42*(10), 811–816. doi:10.1016/S0255-2701(02)00107-1

Vo, T.-K.-Q., Bui, X.-T., Chen, S.-S., Nguyen, P.-D., Cao, N.-D.-T., Vo, T.-D.-H., ... Nguyen, T.-B. (2019). Hospital wastewater treatment by sponge membrane bioreactor coupled with ozonation process. *Chemosphere*, *230*, 377–383. doi:10.1016/j.chemosphere.2019.05.009 PMID:31112860

Voutchkov, N. (2017, October). Energy use for membrane seawater desalination - current status and trends. *Desalination*, *431*, 2–14. doi:10.1016/j.desal.2017.10.033

Vymazal, J., & Březinová, T. (2015). The use of constructed wetlands for removal of pesticides from agricultural runoff and drainage: A review. *Environment International*, *75*, 11–20. doi:10.1016/j.envint.2014.10.026 PMID:25461411

Vyn, R. J., Virani, T., & Deen, B. (2012). Examining the economic feasibility of miscanthus in Ontario: An application to the greenhouse industry. *Energy Policy*, *50*, 669–676. doi:10.1016/j.enpol.2012.08.009

Wagenaar, B. M., Kuipers, J. A. M., Prins, W., & Swaaij, W. P. M. (1993). The rotating cone flash pyrolysis reactor. In A. V. Bridgwater (Ed.), *Advances in Thermochemical Biomass Conversion* (pp. 1122–1133). Dordrecht, The Netherlands: Springer Netherlands. doi:10.1007/978-94-011-1336-6_87

Wagenaar, B. M., Prins, W., & van Swaaij, W. P. M. (1993). Flash pyrolysis kinetics of pine wood. *Fuel Processing Technology*, *36*(1-3), 291–298. doi:10.1016/0378-3820(93)90039-7

Waghmode, T. R., Kurade, M. B., Kabra, A. N., & Govindwar, S. P. (2012). Degradation of Remazol Red dye by *Galactomycesgeotrichum* MTCC 1360 leading to increased iron uptake in sorghum vulgare and phaseolus mungo from soil. *Biotechnology and Bioprocess Engineering; BBE*, *17*(1), 117–126. doi:10.100712257-011-0307-0

Wakoc, F. M., Reshadb, A. S., Bhaleraoa, M. S., & Goud, V. V. (2018). Catalytic cracking of waste cooking oil for biofuel production using zirconium oxide catalyst. *Industrial Crops and Products*, *118*, 282–289. doi:10.1016/j.indcrop.2018.03.057

Walker, G. M., & Weatherley, L. R. (2000). Biodegradation and biosorption of acid anthraquinone dye. *Environ. Pollut.*, *108*(2), 219–223. doi:10.1016/S0269-7491(99)00187-6 PMID:15092952

Wallace, T., Gibbons, D., O'Dwyer, M., & Curran, T. P. (2017). International evolution of fat, oil and grease (FOG) waste management - A review. *Journal of Environmental Management*, *187*, 424–435. doi:10.1016/j.jenvman.2016.11.003 PMID:27838205

Walters, J. P., & Javernick-Will, A. N. (2015). Long-term functionality of rural water services in developing countries: A system dynamics approach to understanding the dynamic interaction of factors. *Environmental Science & Technology*, *49*(8), 5035–5043. doi:10.1021/es505975h PMID:25775082

Wang, A., Qu, J., Ru, J., Liu, H., & Ge, J. (2005). Mineralization of an azo dye Acid Red 14 by electro-Fenton's reagent using an activated carbon fiber cathode. *Dyes and Pigments*, *65*(3), 227–233. doi:10.1016/j.dyepig.2004.07.019

Wang, C.-T., Hu, J.-L., Chou, W.-L., & Kuo, Y.-M. (2008). Removal of color from real dyeing wastewater by Electro-Fenton technology using a three-dimensional graphite cathode. *Journal of Hazardous Materials*, *152*(2), 601–606. doi:10.1016/j.jhazmat.2007.07.023 PMID:17707581

Wang, H., Deng, H., Ma, L., & Ge, L. (2013). Influence of operating conditions on extracellular polymeric substances and surface properties of sludge flocs. *Carbohydrate Polymers*, *92*(1), 510–515. doi:10.1016/j.carbpol.2012.09.055 PMID:23218328

Wang, H., Su, J. Q., Zheng, X. W., Tian, Y., Xiong, X. J., & Zheng, T. L. (2009). Bacterial decolorisation and degradation of the reactive dye Reactive Red 180 by Citrobacter sp. CK3. *International Biodeterioration & Biodegradation*, *63*(4), 395–399. doi:10.1016/j.ibiod.2008.11.006

Wang, H., Xu, J., Sheng, L., Liu, X., Zong, M., & Yao, D. (2019). Anaerobic digestion technology for methane production using deer manure under different experimental conditions. *Energies, 12*(9), 1819. doi:10.3390/en12091819

Wang, J., Huang, Y., & Zhao, X. (2004). Performance and characteristics of an anaerobic baffled reactor. *Bioresource Technology, 93*(2), 205–208. doi:10.1016/j.biortech.2003.06.004 PMID:15051083

Wang, J.-P., Chen, Y.-Z., Wang, Y., Yuan, S.-J., & Yu, H.-Q. (2011). Optimization of the coagulation-flocculation process for pulp mill wastewater treatment using a combination of uniform design and response surface methodology. *Water Research, 45*(17), 5633–5640. doi:10.1016/j.watres.2011.08.023 PMID:21920576

Wang, J.-P., Yuan, S.-J., Wang, Y., & Yu, H.-Q. (2013). Synthesis, characterization and application of a novel starch-based flocculant with high flocculation and dewatering properties. *Water Research, 47*(8), 2643–2648. doi:10.1016/j.watres.2013.01.050 PMID:23531592

Wang, J., Sun, W., Shou, W., Wang, X., Wu, C., Chong, H. Y., & Sun, C. (2015). Integrating BIM and LiDAR for real-time construction quality control. *Journal of Intelligent and Robotic Systems: Theory and Applications, 79*(3–4), 417–432. doi:10.100710846-014-0116-8

Wang, J., Xu, W., Chen, L., Huang, X., & Liu, J. (2014). Preparation and evaluation of magnetic nanoparticles impregnated chitosan beads for arsenic removal from water. *Chemical Engineering Journal, 251*, 25–34. doi:10.1016/j.cej.2014.04.061

Wang, J., Zhong, Z., Ding, K., Zhang, B., & Ruan, R. (2017). Successive desilication and dealumination of HZSM-5 in catalytic conversion of waste cooking oil to produce aromatics. *Energy Conversion and Management, 147*, 100–107. doi:10.1016/j.enconman.2017.05.050

Wang, L. K., Hung, Y.-T., & Shammas, N. K. (2005). *Physicochemical treatment processes* (Vol. 3). Springer. doi:10.1385/159259820x

Wang, L. K., & Menon, R. (2011). Treatment of industrial effluents, municipal wastes, and potable water by membrane bioreactors. In *Membrane and desalination technologies* (pp. 201–236). Totowa, NJ: Humana Press. doi:10.1007/978-1-59745-278-6_5

Wang, L. K., Shammas, N. K., & Hung, Y.-T. (2010). *Advanced biological treatment processes* (Vol. 9). Springer Science & Business Media.

Wang, M., Zhang, D. Q., Dong, J. W., & Tan, S. K. (2017). Constructed wetlands for wastewater treatment in cold climate—A review. *Journal of Environmental Sciences (China), 57*, 293–311. doi:10.1016/j.jes.2016.12.019 PMID:28647250

Wang, M., Zhang, D., Dong, J., & Tan, S. K. (2018). Application of constructed wetlands for treating agricultural runoff and agro-industrial wastewater: A review. *Hydrobiologia, 805*(1), 1–31. doi:10.100710750-017-3315-z

Wang, S., Jia, X. B., Carey, Z., & Hao, J. J. (2014). Effectiveness of national air pollution control policies on the air quality in metropolitan areas of China. *Journal of Environmental Sciences (China), 26*(1), 13–22. doi:10.1016/S1001-0742(13)60381-2 PMID:24649687

Wang, X., Zhang, L., Peng, Y., Zhang, Q., Li, J., & Yang, S. (2019). Enhancing the digestion of waste activated sludge through nitrite addition: Insight on mechanism through profiles of extracellular polymeric substances (EPS) and microbial communities. *Journal of Hazardous Materials, 369*, 164–170. doi:10.1016/j.jhazmat.2019.02.023 PMID:30776599

Wang, Y., Huang, X., & Yuan, Q. (2005). Nitrogen and carbon removals from food processing wastewater by an anoxic/aerobic membrane bioreactor. *Process Biochemistry, 40*(5), 1733–1739. doi:10.1016/j.procbio.2004.06.039

Wangyao, K., Towprayoon, S., Chiemchaisri, C., Gheewala, S. H., & Nopharatana, A. (2010). Application of the IPCC waste model to solid waste disposal sites in tropical countries: Case study of Thailand. *Environmental Monitoring and Assessment, 164*(1-4), 249–261. doi:10.100710661-009-0889-6 PMID:19415519

Wang, Z.-W., Liu, Y., & Tay, J.-H. (2007). Biodegradability of extracellular polymeric substances produced by aerobic granules. *Applied Microbiology and Biotechnology, 74*(2), 462–466. doi:10.100700253-006-0686-x PMID:17103158

Ward, A. J., Hobbs, P. J., Holliman, P. J., & Jones, D. L. (2008). Optimisation of the anaerobic digestion of agricultural resources. *Bioresource Technology, 99*(17), 7928–7940. PMID:18406612

Wardley-Smith, J. (1983). *The control of oil pollution* (pp. 75–79). London, UK: Graham and Trotman.

Warner, K., Orr, P., Parrott, L., & Glynn, M. (1994). Effects of frying oil composition on potato chip stability. *Journal of the American Oil Chemists' Society, 71*, 1117–1121.

Warsinger, D. M., Swaminathan, J., Guillen-Burrieza, E., Arafat, H. A., & Lienhard, V. J. H. (2014). Scaling and fouling in membrane distillation for desalination applications: A review. *Desalination, 356*, 294–313. doi:10.1016/j.desal.2014.06.031

Waste to Energy. (2017). Retrieved from http://wte.kpkt.gov.my/wte_microsite_pg05.html

Watkinson, A. J., Murby, E. J., & Costanzo, S. D. (2007). Removal of antibiotics in conventional and advanced wastewater treatment: Implications for environmental discharge and wastewater recycling. *Water Research, 41*(18), 4164–4176. doi:10.1016/j.watres.2007.04.005 PMID:17524445

Watson, I. C., Morin, O. J., & Henthorne, L. (2003). *Desalting handbook for planners. Desalination and Water Purification Research and Development Program Report No. 72* (3rd Edit.). Virginia. doi:10.3109/15360288.2011.620691

Watson, J. G., Zhu, T., Chow, J. C., Engelbrecht, J., Fujita, E. M., & Wilson, W. E. (2002). Receptor modeling application framework for particle source apportionment. *Chemosphere, 49*(9), 1093–1136. doi:10.1016/S0045-6535(02)00243-6 PMID:12492167

Watts, R. J. (1998). Hazardous wastes: sources, pathways, receptors.

Weber, D., Drizo, A., Twohig, E., Bird, S., & Ross, D. (2007). Upgrading constructed wetlands phosphorus reduction from a dairy effluent using electric arc furnace steel slag filters. *Water Science and Technology, 56*(3), 135–143. doi:10.2166/wst.2007.513 PMID:17802848

Wee, H. M., Budiman, S. D., Su, L. C., Chang, M., & Chen, R. (2016). Responsible supply chain management-An analysis of Taiwanese gutter oil scandal using the theory of constraint. *International Journal of Logistics Research and Applications, 19*(5), 380–394. doi:10.1080/13675567.2015.1090964

Wegelin, M. (1996) Surface water treatment by roughing filters. *A Design, Construction and Operation Manual.* SANDEC Report No. 2/96.

Weihua, Z., Irene, M. C., & Lo, M. A. (2007). Chemical-enhanced washing for remediation of soils contaminated with marine diesel fuel in the presence/absence of Pb. *Journal of Environmental Engineering, 133*(5), 548–555. doi:10.1061/(ASCE)0733-9372(2007)133:5(548)

Weiland, P. (2010). Biogas production: Current state and perspectives. *Applied Microbiology and Biotechnology, 85*(4), 849–860. doi:10.100700253-009-2246-7 PMID:19777226

Wei, Z., Khamphe, P., Fan, L. M. S., & Pin, J. H. (2013). Evaluation of a classification method for biodegradable solid wastes using anaerobic degradation parameters. *Waste Management (New York, N.Y.), 33*(12), 2632–2640. doi:10.1016/j.wasman.2013.08.015 PMID:24045172

Wei, Z., Li, X., Thushara, D., & Liu, Y. (2011). Determinations and removal of malodialdehyde and other 2-thiobarbituric acid reactive substances in waste cooking oil. *Journal of Food Engineering*, *107*(3-4), 379–384. doi:10.1016/j.jfoodeng.2011.06.032

Welander, U., Henrysson, T., & Welander, T. (1997). Nitrification of landfill leachate using suspended-carrier biofilm technology. *Water Research*, *31*(9), 2351–2355. doi:10.1016/S0043-1354(97)00080-8

Wellons, F. E. I. Corp. (2016). Reciprocating grate. Retrieved from http://www.wellonsfei.ca/en/reciprocating-grate.aspx

Wemedo, S., Akani, N., & Eke, C. (2009). Fungi in an oilfield wastewater in Nigeria. *Asian Journal of Biological Sciences*, *2*(2), 54–57. doi:10.3923/ajbs.2009.54.57

Wen, H., Bhusal, S., & Wen, B. (2013). Laboratory evaluation of waste cooking oil-based bioasphalt as an alternative binder for hot mix asphalt. *Journal of Materials in Civil Engineering*, *25*(10), 1432–1437.

Wen, Z., Yu, X., Tu, S. T., Yan, J., & Dahlquist, E. (2010). Biodiesel production from waste cooking oil catalyzed by TiO_2-MgO mixed oxides. *Bioresource Technology*, *101*, 9570–9576. PMID:20696572

West Bengal Pollution Control Board Newsletter. (2001). *The study to evaluate the lung response of Kolkata population done by Dr. Twisha Lahiri and Dr. Manas Ranjan Roy of the Chittaranjan National Cancer Institute, Kolkata and Dr.* Pulak Lahiri of Calcutta University.

Whitlock, K., Abanda, F. H., Manjia, M. B., Pettang, C., & Nkeng, G. E. (2018). BIM for construction site logistics management. *Journal of Engineering, Project, and Production Management*, *8*(1), 47–55. doi:10.32738/JEPPM.201801.0006

Whittaker, C., Hunt, J., Misselbrook, T., & Shield, I. (2016). How well does Miscanthus ensile for use in an anaerobic digestion plant? *Biomass and Bioenergy*, *88*, 24–34. doi:10.1016/j.biombioe.2016.03.018

WHO. (2002). *World health report*. Geneva, Switzerland: World Health Organization.

WHO, UNICEF. (2010). *Progress on sanitation and drinking-water, 2010 Update*. Geneva, Switzerland: World Health Organization.

Wick, A. F., Haus, N. W., Sukkariyah, B. F., Haering, K. C., & Daniels, W. L. (2011). Remediation of PAH-contaminated soils and sediments: a literature review. *CSES Department, internal research document, 102*.

Widerer, P., Irvine, R. L., & Goronszy, M. (1999). *Sequencing batch reactors*. Available at http://water.epa.gov/scitech/wastetech/upload/2002_10_31_mtb _sbr_new.pdf

Wijekoon, K., Visvanathan, C., & Abeynayaka, A. (2010). Effect of organic loading rate on VFA production, organic matter removal and microbial activity of a two-stage thermophilic anaerobic membrane bioreactor. *Bioresource Technology*, *102*(9), 5353–5360. doi:10.1016/j.biortech.2010.12.081 PMID:21256737

Wilén, B. M., Jin, B., & Lant, P. (2003). The influence of key chemical constituents in activated sludge on surface and flocculating properties. *Water Research*, *37*(9), 2127–2139. doi:10.1016/S0043-1354(02)00629-2 PMID:12691899

Williams, J. B., Clarkson, C., Mant, C., Drinkwater, A., & May, E. (2012). Fat, oil and grease deposits in sewers: Characterisation of deposits and formation mechanisms. *Water Research*, *46*(19), 6319–6328. doi:10.1016/j.watres.2012.09.002 PMID:23039918

Williams, M. E., Wagner, J., Eng, B. S. C., Stoughton, M., Duan, X., Wendel, E., ... Cotruvo, B. J. A. (2011). Reverse osmosis technology for water treatment: State of the art review. *Desalination*, *267*(August), 1–8. doi:10.1016/j.desal.2010.09.001

Wilson, D. C. (2007). Development drivers for waste management. *Waste Management & Research*, *25*(3), 198–207. doi:10.1177/0734242X07079149 PMID:17612318

Wilson, D. C., Rodic, L., Modak, P., Soos, R., Carpintero, A., Velis, C., ... Simonett, O. (2015). Global waste management outlook. *Waste Management & Research*, *33*(12), 1049–1051. doi:10.1177/0734242X15616055 PMID:26574579

Wilson, L. G., & Cavanagh, P. (1969). Principle of inertial impaction. *Atmospheric Environment*, *3*, 597–605. doi:10.1016/0004-6981(69)90054-7

Wingender, J., Jäger, K. E., & Flemming, H.-C. (1999). Interaction between extracellular polysaccharides and enzymes. In J. Wingender & T. R. Neu (Eds.), *Microbial extracellular polymeric substances* (pp. 1–19). Heidelberg, Germany: Springer.

Wintgens, T., Melin, T., Schäfer, A., Khan, S., Muston, M., Bixio, D., & Thoeye, C. (2005). The role of membrane processes in municipal wastewater reclamation and reuse. *Desalination*, *178*(1-3), 1–11. doi:10.1016/j.desal.2004.12.014

Wise, R. (2002). Antimicrobial resistance: Priorities for action. *The Journal of Antimicrobial Chemotherapy*, *49*(4), 585–586. doi:10.1093/jac/49.4.585 PMID:11909829

Wiszniowski, J., Robert, D., Surmacz-Gorska, J., Miksch, K., & Weber, J. J. E. L. (2006). Landfill leachate treatment methods. *RE:view*, *4*(1), 51–61.

Withers, P. J., van Dijk, K. C., Neset, T.-S. S., Nesme, T., Oenema, O., Rubæk, G. H., ... Pellerin, S. (2015). Stewardship to tackle global phosphorus inefficiency: The case of Europe. *Ambio*, *44*(2), 193–206. PMID:25681977

Wolf, A. (1998). Conflict and cooperation along international waterway. *Water Policy*, *1*(2), 251–265. doi:10.1016/S1366-7017(98)00019-1

Wolf, A. T., Kramer, A., Carius, A., & Dabelko, G. D. (2005). Managing water conflict and cooperation. In *State of the World 2005* (pp. 80–99). Redefining Global Security. Retrieved from http://tbw.geo.orst.edu/publications/abst_docs/wolf_sow_2005.pdf

Woli, K. P., David, M. B., Tsai, J., Voigt, T. B., Darmody, R. G., & Mitchell, C. A. (2011). Evaluating silicon concentrations in biofuel feedstock crops Miscanthus and switchgrass. *Biomass and Bioenergy*, *35*(7), 2807–2813. doi:10.1016/j.biombioe.2011.03.007

Wolters, H., Smet, J. E. M., & Galvis, J. (1989). *Pre-treatment methods for community water supply*. Ch. 8, International Water Sanitation Center (IRC).

Wong, P. K., & Yuen, P. Y. (1998). Decolorization and biodegradation of N, N, - dimethyl-P- Phenylenediamine by Klebsiella pneumoniae RS-13 and Acetobacter liquefaciens S-1. *Journal of Applied Microbiology*, *85*(1), 79–87. doi:10.1046/j.1365-2672.1998.00479.x PMID:9721658

Wong, S., Teng, T., Ahmad, A., Zuhairi, A., & Najafpour, G. (2006). Treatment of pulp and paper mill wastewater by polyacrylamide (PAM) in polymer induced flocculation. *Journal of Hazardous Materials*, *135*(1-3), 378–388. doi:10.1016/j.jhazmat.2005.11.076 PMID:16431022

Wongsasuluk, P., Chotpantarat, S., Siriwong, W., & Robson, M. (2014). Heavy metal contamination and human health risk assessment in drinking water from shallow groundwater wells in an agricultural area in Ubon Ratchathani province, Thailand. *Environmental Geochemistry and Health*, *36*(1), 169–182. doi:10.100710653-013-9537-8 PMID:23771812

Won, J., & Cheng, J. C. P. (2017). Identifying potential opportunities of building information modeling for construction and demolition waste management and minimization. *Automation in Construction*, *79*, 3–18. doi:10.1016/j.autcon.2017.02.002

Woodall, D. W., Gambrell, R. P., Rabalais, N. N., & DeLaune, R. D. (2001). Developing a method to track oil and gas produced water discharges in estuarine systems using salinity as a conservative tracer. *Marine Pollution Bulletin*, *42*(11), 1118–1127. doi:10.1016/S0025-326X(01)00100-X PMID:11763224

Wood, L. A. (2002). *Overview of remediation technologies. Terra Resources, Ltd.* (p. 6). Palmer, Alaska: Wolverine. Retrieved from www.terrawash.com/twp2.htm

Woolard, C. R., & Irvine, R. L. (1995). Treatment of hypersaline wastewater in the sequencing batch reactor. *Water Research*, *29*(4), 1159–1168. doi:10.1016/0043-1354(94)00239-4

Woo, Y. C., Lee, J. J., Tijing, L. D., Shon, H. K., Yao, M., & Kim, H.-S. S. (2015). Characteristics of membrane fouling by consecutive chemical cleaning in pressurized ultrafiltration as pre-treatment of seawater desalination. *Desalination*, *369*, 51–61. doi:10.1016/j.desal.2015.04.030

World Energy Resource. (2016). 24th Edition report. Retrieved from World Energy Council database.

World Health Organization. (2005). Available at http://www.who.int/water_ sanitation_health/ wsh0207/en/index5.html

World Health Organization. (2009). Diarrhea: why children are still dying and what can be done.

Wouter, D., Cliona, O., Freda, R. H., & Helena, M. P. (1998). Anaerobic treatment of textile effluents: A review. *Journal of Chemical Technology and Biotechnology (Oxford, Oxfordshire)*, *73*(4), 323–335. doi:10.1002/(SICI)1097-4660(199812)73:4<323::AID-JCTB976>3.0.CO;2-S

Wu, J. Y., Hwang, S. J., Chen, C., & Chen, K. (2005). Decolorization of azo dye in a FBR reactor using immobilized bacteria. *Enzyme and Microbial Technology*, *37*(1), 102–112. doi:10.1016/j.enzmictec.2005.02.012

Wu, J., & Sun, Z. (2016). Evaluation of shallow groundwater contamination and associated human health risk in an alluvial plain impacted by agricultural and industrial activities, mid-west China. *Exposure and Health*, *8*(3), 311–329. doi:10.100712403-015-0170-x

Xiangdong, L. (2005). Treatment of Pharmaceutical Wastewater by Airfloatation-Hydrolytic Acidification-Aerobic Process [J]. *Environmental Engineering, 3*.

Xiaoli, C., Ziyang, L., Shimaoka, T., Nakayama, H., Ying, Z., Xiaoyan, C., ... Youcai, Z. (2010). Characteristics of environmental factors and their effects on CH_4 and CO_2 emissions from a closed landfill: An ecological case study of Shanghai. *Waste Management (New York, N.Y.)*, *30*(3), 446–5. doi:10.1016/j.wasman.2009.09.047 PMID:19880303

Xia, S., & Liu, J. (2004). An innovative integrated oxidation ditch with vertical circle for domestic wastewater treatment. *Process Biochemistry*, *39*(9), 1111–1117. doi:10.1016/S0032-9592(03)00216-4

Xie, H., Yu, Q., Wang, K., Shi, X., & Li, X. (2014). Thermodynamic analysis of hydrogen production from model compounds of bio-oil through steam reforming. *Environmental Progress & Sustainable Energy*, *33*(3), 1008–1016. doi:10.1002/ep.11846

Xie, H., Yu, Q., Yao, X., Duan, W., Zuo, Z., & Qin, Q. (2015). Hydrogen production via steam reforming of bio-oil model compounds over supported nickel catalysts. *Journal of Energy Chemistry*, *24*(3), 299–308. doi:10.1016/S2095-4956(15)60315-1

Xie, N., Battaglia, F., & Fox, R. O. (2004). Simulations of multiphase reactive flows in fluidized beds using in situ adaptive tabulation. *Combustion Theory and Modelling*, *8*(2), 195–209. doi:10.1088/1364-7830/8/2/001

Xu, C. Y., Schwartz, F. W., & Traina, S. J. (1997). Treatment of acid-mine water with calcite and quartz sand. *Environmental Engineering Science*, *14*(3), 141–152. doi:10.1089/ees.1997.14.141

Xue, Q., Dalluge, D., Heindel, T. J., Fox, R. O., & Brown, R. C. (2012). Experimental validation and CFD modeling study of biomass fast pyrolysis in fluidized-bed reactors. *Fuel*, *97*, 757–769. doi:10.1016/j.fuel.2012.02.065

Xue, Q., Heindel, T. J., & Fox, R. O. (2011). A CFD model for biomass fast pyrolysis in fluidized-bed reactors. *Chemical Engineering Science*, *66*(11), 2440–2452. doi:10.1016/j.ces.2011.03.010

Xu, Q., Li, X., Ding, R., Wang, D., Liu, Y., Wang, Q., ... Li, H. (2017). Understanding and mitigating the toxicity of cadmium to the anaerobic fermentation of waste activated sludge. *Water Research*, *124*, 269–279. doi:10.1016/j.watres.2017.07.067 PMID:28772139

Xu, Q., Ma, X., Yu, Z., & Cai, Z. (2014). A kinetic study on the effects of alkaline earth and alkali metal compounds for catalytic pyrolysis of microalgae using thermogravimetry. *Applied Thermal Engineering*, *73*(1), 357–361. doi:10.1016/j.applthermaleng.2014.07.068

Xu, R., Yong, L. C., Lim, Y. G., & Obbard, J. P. (2005). Use of slow-release fertilizer and biopolymers for stimulating hydrocarbon biodegradation in oil contaminated beach sediments. *Marine Pollution Bulletin*, *51*(8-12), 1101–1110. doi:10.1016/j.marpolbul.2005.02.037 PMID:16291209

Xu, Y., Baker, L. A., & Johnson, P. C. (2007). Trends in ground water nitrate contamination in the Phoenix, Arizona region. *Ground Water Monitoring and Remediation*, *27*(2), 49–56. doi:10.1111/j.1745-6592.2007.00135.x

Yaakob, Z., Mohammad, M., Alherbawi, M., Alam, Z., & Sopian, K. (2013). Overview of the production of biodiesel from waste cooking oil. *Renewable & Sustainable Energy Reviews*, *18*, 184–193.

Yacob, S., Hassan, M. A., Shirai, Y., Wakisaka, M., & Subash, S. (2005). Baseline study of methane emission from open digesting tanks of palm oil mill effluent treatment. *Chemosphere*, *59*(11), 1575–1581. doi:10.1016/j.chemosphere.2004.11.040 PMID:15894045

Yan, G., & Viraraghavan, T. (2003). Heavy-metal removal from aqueous solution by fungus Mucor rouxii. *Water Research*, *37*(18), 4486–4496. doi:10.1016/S0043-1354(03)00409-3 PMID:14511719

Yang, B., Xu, H., Yang, S., Bi, S., Li, F., Shen, C., ... Liu, Y. (2018). Treatment of industrial dyeing wastewater with a pilot-scale strengthened circulation anaerobic reactor. *Bioresource Technology*, *264*, 154–162. doi:10.1016/j.biortech.2018.05.063 PMID:29803085

Yang, G., Wu, L., Xian, Q., Shen, F., Wu, J., & Zhang, Y. (2016). Removal of congo red and methylene blue from aqueous solutions by vermicompost-derived biochars. *PLoS One*, *11*, 1–4. PMID:27144922

Yang, H., Yan, R., Chen, H., Lee, D. H., & Zheng, C. (2007). Characteristics of hemicellulose, cellulose and lignin pyrolysis. *Fuel*, *86*(12-13), 1781–1788. doi:10.1016/j.fuel.2006.12.013

Yang, R., Li, H., Huang, M., Yang, H., & Li, A. (2016). A review on chitosan-based flocculants and their applications in water treatment. *Water Research*, *95*, 59–89. doi:10.1016/j.watres.2016.02.068 PMID:26986497

Yang, Y., Fu, J., Peng, H., Hou, L., Liu, M., & Zhou, J. L. (2011). Occurrence and phase distribution of selected pharmaceuticals in the Yangtze Estuary and its coastal zone. *Journal of Hazardous Materials*, *190*(1-3), 588–596. doi:10.1016/j.jhazmat.2011.03.092 PMID:21497014

Yang, Z., Wang, W., & Liu, S. (2017). Flocculation of coal waste slurry using bioflocculant produced by Azotobacter chroococcum. *Energy & Fuels*, *31*(2), 1460–1467. doi:10.1021/acs.energyfuels.6b03052

Yan, H., & Pan, G. (2004). Increase in biodegradation of dimethyl phthalate by Closterium lunula using inorganic carbon. *Chemosphere*, *55*(9), 1281–1285. doi:10.1016/j.chemosphere.2003.12.019 PMID:15081769

Yao, H.,, & You, Z. (2016). Effectiveness of micro and nanomaterials in asphalt mixtures through dynamic modulus and rutting tests. *Journal of Nanomaterials*, 1–14. doi:10.1155/2016/2645250

Yaqub, A., Isa, M. H., & Ajab, H. (2014). Electrochemical degradation of polycyclic aromatic hydrocarbons in synthetic solution and produced water using a Ti/SnO 2-Sb 2 O 5-RuO 2 anode. *Journal of Environmental Engineering*, *141*(4), 04014074. doi:10.1061/(ASCE)EE.1943-7870.0000900

Yaqub, A., Isa, M. H., Kutty, S. R. M., & Ajab, H. (2014). Electrochemical degradation of PAHs in produced water using Ti/Sb2O5-SnO2-IrO2 anode. *Electrochemistry*, *82*(11), 979–984. doi:10.5796/electrochemistry.82.979

Yasar, A., Nazir, S., Rasheed, R., Tabinda, A. B., & Nazar, M. (2017). Economic review of different designs of biogas plants at household level in Pakistan. *Renewable & Sustainable Energy Reviews*, *74*, 221–229.

Yasar, S., Cirik, K., & Cinar, O. (2012). The effect of cyclic anaerobic– aerobic conditions on biodegradation of azo dyes. *Bioproc Biosyst Eng*, *35*, 449–457. PMID:21858702

Yavuz, Y., & Koparal, A. S. (2006). Electrochemical oxidation of phenol in a parallel plate reactor using ruthenium mixed metal oxide electrodes. *Journal of Hazardous Materials*, *136*(2), 296–302. doi:10.1016/j.jhazmat.2005.12.018 PMID:16427192

Yavuz, Y., & Ögütveren, Ü. (2018). Treatment of industrial estate wastewater by the application of electrocoagulation process using iron electrodes. *Journal of Environmental Management*, *207*, 151–158. doi:10.1016/j.jenvman.2017.11.034 PMID:29161644

Ye, C., Yuan, H., Lou, Z., & Zhu, N. (2016). Combined electrochemical and hypochlorite pretreatment for improving solubilization and anaerobic digestion of waste activated sludge: Effect of hypochlorite dosage. *Energy & Fuels*, *30*(4), 2990–2996. doi:10.1021/acs.energyfuels.5b02884

Yeh, J. T., Resnik, K. P., Rygle, K., & Pennline, H. W. (2005). Semi-batch absorption and regeneration studies for CO_2 capture by aqueous ammonia. *Fuel Processing Technology*, *86*(14-15), 1533–1546.

Ye, R., Wang, T., Bedzyk, L., & Croker, K. (2001). Applications of DNA microarrays in microbial systems. *Journal of Microbiological Methods*, *47*(3), 257–272. doi:10.1016/S0167-7012(01)00308-6 PMID:11714516

Yeung, A. T. (2010). Remediation technologies for contaminated sites. *Proc., Int'l. Symp. on Geo-Environmental Engineering*. I. doi:10.1007/978-3-642-04460-1_25

Yılmaz, A. E., Boncukcuoğlu, R., & Kocadağistan, E. (2008). An empirical model for kinetics of boron removal from boron containing wastewaters by the electrocoagulation method in a batch reactor. *Desalination*, *230*(1), 288–297. doi:10.1016/j.desal.2007.11.031

Ying-Xin, X., Xiong, Z.-Q., Guang-Xi, X., Guo-Qing, S., & Zhao-Liang, Z. (2007). Assessment of nitrogen pollutant sources in surface waters of Taihu Lake region. *Pedosphere*, *17*(2), 200–208. doi:10.1016/S1002-0160(07)60026-5

Yoffe, S., Fiske, G., Giordano, M., Giordano, M., Larson, K., Stahl, K., & Wolf, A. T. (2004). Geography of international water conflict and cooperation: data sets and applications. *Water Resources Research*, *40*(5), 1–12. doi:10.1029/2003WR002530

Yongzhen, P., Hongxun, H., Shuying, W., Youwei, C., & Zhiguo, Y. (2008). Nitrogen and phosphorus removal in pilot-scale anaerobic-anoxic oxidation ditch system. *Journal of Environmental Sciences (China)*, *20*(4), 398–403. doi:10.1016/S1001-0742(08)62070-7 PMID:18575122

Yoo, H.-C., Cho, S.-H., & Ko, S.-O. (2001). Modification of coagulation and Fenton oxidation processes for cost-effective leachate treatment. *Journal of Environmental Science and Health. Part A*, *36*(1), 39–48. PMID:11381784

Yoon, Y., Westerhoff, P., Snyder, S. A., & Esparza, M. (2003). HPLC-fluorescence detection and adsorption of bisphenol A, 17β-estradiol, and 17α-ethynyl estradiol on powdered activated carbon. *Water Research*, *37*(14), 3530–3537. doi:10.1016/S0043-1354(03)00239-2 PMID:12834747

Yoon, Y., Westerhoff, P., Snyder, S. A., & Wert, E. C. (2006). Nanofiltration and ultrafiltration of endocrine disrupting compounds, pharmaceuticals and personal care products. *Journal of Membrane Science*, *270*(1-2), 88–100. doi:10.1016/j.memsci.2005.06.045

Yoon, Y., Westerhoff, P., Snyder, S. A., Wert, E. C., & Yoon, J. (2007). Removal of endocrine disrupting compounds and pharmaceuticals by nanofiltration and ultrafiltration membranes. *Desalination*, *202*(1-3), 16–23. doi:10.1016/j.desal.2005.12.033

Yoshida, S., Hiraga, K., Takehana, T., Taniguchi, I., Yamaji, H., Maeda, Y., ... Oda, K. (2016). A bacterium that degrades and assimilates poly (ethylene terephthalate). *Science*, *351*(6278), 1196–1199. doi:10.1126cience.aad6359 PMID:26965627

Younos, T. (2005). Environmental issues of desalination. *Journal of Contemporary Water Research & Education*, *132*(1), 11–18. doi:10.1111/j.1936-704X.2005.mp132001003.x

Yuan, H., Yu, B., Cheng, P., Zhu, N., Yin, C., & Ying, L. (2016). Pilot-scale study of enhanced anaerobic digestion of waste activated sludge by electrochemical and sodium hypochlorite combination pretreatment. *International Biodeterioration & Biodegradation*, *110*, 227–234. doi:10.1016/j.ibiod.2016.04.001

Yuan, H., Zhu, N., & Song, F. (2011). Dewaterability characteristics of sludge conditioned with surfactants pretreatment by electrolysis. *Bioresource Technology*, *102*(3), 2308–2315. doi:10.1016/j.biortech.2010.10.065 PMID:21041078

Yuan, S., Fan, Y., Zhang, Y., Tong, M., & Liao, P. (2011). Pd-catalytic in situ generation of H2O2 from H2 and O2 produced by water electrolysis for the efficient electro-Fenton degradation of rhodamine B. *Environmental Science & Technology*, *45*(19), 8514–8520. doi:10.1021/es2022939 PMID:21866953

Yu, B., Xu, J., Yuan, H., Lou, Z., Lin, J., & Zhu, N. (2014). Enhancement of anaerobic digestion of waste activated sludge by electrochemical pretreatment. *Fuel*, *130*, 279–285. doi:10.1016/j.fuel.2014.04.031

Yucheng, C., & Pawłowski, A. (2012). Sewage sludge-to-energy approaches based on anaerobic digestion and pyrolysis: Brief overview and energy efficiency assessment. *Renewable & Sustainable Energy Reviews*, *16*(3), 1657–1665. doi:10.1016/j.rser.2011.12.014

Yu, X., Guo, X., & Wu, Z. (2014). Land surface temperature retrieval from Landsat 8 TIRS—comparison between radiative transfer equation-based method, split window algorithm and single channel method. *Remote Sensing*, *6*(10), 9829–9852. doi:10.3390/rs6109829

Yu, Y., Guo, H., Ding, Q., Li, H., & Skitmore, M. (2017a). An experimental study of real-time identification of construction workers' unsafe behaviors. *Automation in Construction*, *82*, 193–206. doi:10.1016/j.autcon.2017.05.002

Yu, Z., & Mohn, W. (2001). Bacterial diversity and community structure in an aerated lagoon revealed by ribosomal intergenic spacer analyses and 16S ribosomal DNA sequencing. *Applied and Environmental Microbiology*, *67*(4), 1565–1574. doi:10.1128/AEM.67.4.1565-1574.2001 PMID:11282606

Zaiedy, N. I., Karim, O. A., & Mutalib, N. A. A. (2016). Water quality of surface runoff in loop two catchment area in UKM. *Jurnal Kejuruteraan*, *28*, 65–72. doi:10.17576/jkukm-2016-28-07

Zainal, B. S., Akhbari, A., Zinatizadeh, A. A., Mohammadi, P., Danaee, M., Mohd, N. S., & Ibrahim, S. (2019). UASFF start-up for biohydrogen and biomethane production from treatment of palm oil mill effluent. *International Journal of Hydrogen Energy*, *44*(37), 20725–20737. doi:10.1016/j.ijhydene.2018.07.037

Zainal, N. H., Jalani, N. F., Mamat, R., & Astimar, A. A. (2017). A review on the development of palm oil mill effluent (POME) final discharge polishing treatments. *Journal of Oil Palm Research*, *29*(4), 528–540. doi:10.21894/jopr.2017.00012

Zainal, Z., Rifau, A., Quadir, G., & Seetharamu, K. (2002). Experimental investigation of a downdraft biomass gasifier. *Biomass and Bioenergy*, *23*(4), 283–289. doi:10.1016/S0961-9534(02)00059-4

Zapata, A., Malato, S., Sanchez, P. J. A., Oller, I., & Maldonado, M. (2010). Scale-up strategy for a combined solar photo-Fenton/biological system for remediation of pesticide-contaminated water. *Catalysis Today*, *151*(1-2), 100–106. doi:10.1016/j.cattod.2010.01.034

Zargar, M., Ahmadinia, E., Asli, H., & Karim, M. R. (2012). Investigation of the possibility of using waste cooking oil as a rejuvenating agent for aged bitumen. *Journal of Hazardous Materials*, *233-234*, 254–258. PMID:22818590

Zaumanis, M., Mallick, R. B., & Frank, R. (2013). use of rejuvenators for production of sustainable high content RAP hot mix asphalt. *The XXVIII International Baltic Road Conference*. August, 2013. Vilnius, Lithuania, 1-10.

Zaytseva, O., & Neumann, G. (2016). Carbon nanomaterials: Production, impact on plant development, agricultural and environmental applications. *Chemical and Biological Technologies in Agriculture*, *3*(1), 17. doi:10.118640538-016-0070-8

Zazou, H., Afanga, H., Akhouairi, S., Ouchtak, H., Addi, A. A., Akbour, R. A., ... Hamdani, M. (2019). Treatment of textile industry wastewater by electrocoagulation coupled with electrochemical advanced oxidation process. *Journal of Water Process Engineering*, *28*, 214–221. doi:10.1016/j.jwpe.2019.02.006

Zbyszewski, M., & Corcoran, P. L. (2011). Distribution and degradation of fresh water plastic particles along the beaches of Lake Huron, Canada. *Water, Air, & Soil Pollution, 220*(1-4), pp. 365–372.

Zeb, I., Ma, J., Frear, C., Zhao, Q., Ndegwa, P., Yao, Y., & Kafle, G. K. (2017). Recycling separated liquid-effluent to dilute feedstock in anaerobic digestion of dairy manure. *Energy*, *119*, 1144–1151.

Zepp, R. G., Faust, B. C., & Hoigne, J. (1992). Hydroxyl radical formation in aqueous reactions (pH 3-8) of iron (II) with hydrogen peroxide: The photo-Fenton reaction. *Environmental Science & Technology*, *26*(2), 313–319. doi:10.1021/es00026a011

Zhang, C., Su, H., Baeyens, J., & Tan, T. (2014). Reviewing the anaerobic digestion of food waste for biogas production. *Renewable & Sustainable Energy Reviews*, *38*, 383–392.

Zhang, G., Qu, J., Liu, H., Cooper, A. T., & Wu, R. (2007). CuFe2O4/activated carbon composite: A novel magnetic adsorbent for the removal of acid orange ii and catalytic regeneration. *Chemosphere*, *68*(6), 1058–1066. doi:10.1016/j.chemosphere.2007.01.081 PMID:17397903

Zhang, H., Choi, H. J., & Huang, C.-P. (2005). Optimization of Fenton process for the treatment of landfill leachate. *Journal of Hazardous Materials*, *125*(1-3), 166–174. doi:10.1016/j.jhazmat.2005.05.025 PMID:16011874

Zhang, H., Fei, C., Zhang, D., & Tang, F. (2007). Degradation of 4-nitrophenol in aqueous medium by electro-Fenton method. *Journal of Hazardous Materials*, *145*(1-2), 227–232. doi:10.1016/j.jhazmat.2006.11.016 PMID:17161909

Zhang, H., Wang, Q., & Mortimer, S. R. (2012). Waste cooking oil as an energy resource: review of Chinese policies. *Renewable & Sustainable Energy Reviews*, *16*(7), 5225–5231.

Zhang, H., Zhang, D., & Zhou, J. (2006). Removal of COD from landfill leachate by electro-Fenton method. *Journal of Hazardous Materials*, *135*(1-3), 106–111. doi:10.1016/j.jhazmat.2005.11.025 PMID:16359785

Zhang, J., Liu, Z., Wang, S., & Jiang, P. (2002). Characterization of a bioflocculant produced by the marine myxobacterium Nannocystis sp. NU-2. *Applied Microbiology and Biotechnology*, *59*(4-5), 517–522. doi:10.100700253-002-1023-7 PMID:12172619

Zhang, L., Xu, C., & Champagne, P. (2010). Overview of recent advances in thermo-chemical conversion of biomass. *Energy Conversion and Management*, *51*(5), 969–982. doi:10.1016/j.enconman.2009.11.038

Zhang, L., Xu, E. G., Li, Y., Liu, H., Vidal-Dorsch, D. E., & Giesy, J. P. (2018). Ecological risks posed by ammonia nitrogen (AN) and un-ionized ammonia (NH3) in seven major river systems of China. *Chemosphere*, *202*, 136–144. doi:10.1016/j.chemosphere.2018.03.098 PMID:29567611

Zhang, L., Xu, T., Liu, X., Zhang, Y., & Jin, H. (2011). Adsorption behavior of multi-walled carbon nanotubes for the removal of olaquindox from aqueous solutions. *Journal of Hazardous Materials*, *197*, 389–396. doi:10.1016/j.jhazmat.2011.09.100 PMID:22018864

Zhang, S., Sulankivi, K., Kiviniemi, M., Romo, I., Eastman, C. M., & Teizer, J. (2015). BIM-based fall hazard identification and prevention in construction safety planning. *Safety Science*, *72*, 31–45. doi:10.1016/j.ssci.2014.08.001

Zhang, X., Wang, T., Ma, L., Zhang, Q., & Jiang, T. (2013). Hydrotreatment of bio-oil over Ni-based catalyst. *Bioresource Technology*, *127*, 306–311. doi:10.1016/j.biortech.2012.07.119 PMID:23138057

Zhang, Y. (2008). Online-coupled meteorology and chemistry models: History, current status, and outlook. *Atmospheric Chemistry and Physics*, *8*(11), 2895–2932. doi:10.5194/acp-8-2895-2008

Zhang, Y. (2019). Dynamic effect analysis of meteorological conditions on air pollution: A case study from Beijing. *The Science of the Total Environment*, *684*(20), 178–185. doi:10.1016/j.scitotenv.2019.05.360 PMID:31153065

Zhang, Y., An, X., & Quan, X. (2011). Enhancement of sludge granulation in a zero valence iron packed anaerobic reactor with a hydraulic circulation. *Process Biochemistry*, *46*(2), 471–476. doi:10.1016/j.procbio.2010.09.021

Zhang, Y., Han, C., Zhang, G., Dionysiou, D. D., & Nadagouda, M. N. (2015). PEG-assisted synthesis of crystal TiO2 nanowires with high specific surface area for enhanced photocatalytic degradation of atrazine. *Chemical Engineering Journal*, *268*, 170–179. doi:10.1016/j.cej.2015.01.006

Zhang, Y., Xiao, Z., Chen, F., Ge, Y., Wu, J., & Hu, X. (2010). Degradation behavior and products of malathion and chlorpyrifos spiked in apple juice by ultrasonic treatment. *Ultrasonics Sonochemistry*, *17*(1), 72–77. doi:10.1016/j.ultsonch.2009.06.003 PMID:19560957

Zhang, Y., Yan, L., Chi, L., Long, X., Mei, Z., & Zhang, Z. (2008). Startup and operation of anaerobic EGSB reactor treating palm oil mill effluent. *Journal of Environmental Sciences (China)*, *20*(6), 658–663. doi:10.1016/S1001-0742(08)62109-9 PMID:18763558

Zhao, S., Ye, Y., Qin, D., Yin, D., & He, J. (2017). Assessment of long-term and large-scale even-odd license plate-controlled plan effects on urban air quality and its implication. *Atmospheric Environment*, *170*, 82–95. doi:10.1016/j.atmosenv.2017.09.041

Zheng, T., Wu, Z., Xie, Q., Fang, J., Hu, Y., Lu, M., ... Ji, J. (2018). Structural modification of waste cooking oil methyl esters as cleaner plasticizer to substitute toxic dioctyl phthalate. *Journal of Cleaner Production*, *186*, 1021–1030. doi:10.1016/j.jclepro.2018.03.175

Zheng, Z., Ren, G., Wang, H., Dou, J., Gao, Z., Duan, C., ... Yang, Y.-J. (2018). Relationship between fine-particle pollution and the urban heat island in Beijing, China: Observational evidence. *Boundary-Layer Meteorology*, 169.

Zhou, H., & Smith, D. W. (2001). Advanced oxidation technologies in water and wastewater treatment. *Journal of Civil Engineering*, *28*, 49–66.

Zhou, J. L., & Banks, C. J. (1991). Removal of humic acid fraction by Rhizopus arrhizus: Uptake and kinetic studies. *Environmental Technology*, *12*(10), 859–869. doi:10.1080/09593339109385080

Zhou, J. L., & Banks, C. J. (1993). Mechanism of humic acid color removal from natural waters by fungal biomass biosorption. *Chemosphere*, *27*(4), 607–620. doi:10.1016/0045-6535(93)90096-N

Zhou, M., Yu, Q., Lei, L., & Barton, G. (2007). Electro-Fenton method for the removal of methyl red in an efficient electrochemical system. *Separation and Purification Technology*, *57*(2), 380–387. doi:10.1016/j.seppur.2007.04.021

Zhou, W., Chen, P., Min, M., Ma, X., Wang, J., Griffith, R., ... Li, Y. (2014). Environment-enhancing algal biofuel production using wastewaters. *Renewable & Sustainable Energy Reviews*, *36*, 256–269.

Zhou, W., & Zimmermann, W. (1993). Decolorization of industrial effluents containing reactive dyes by actinomycetes. *FEMS Microbiology Letters*, *107*(2-3), 157–162. doi:10.1111/j.1574-6968.1993.tb06023.x PMID:8472899

Zhou, X., Han, Y., & Guo, X. (2013). Enhanced total nitrogen removal performance in a modified Orbal oxidation ditch system with internal nitrate recycle. *Chemical Engineering Journal*, *228*, 124–131. doi:10.1016/j.cej.2013.04.096

Zhou, X., Wang, Q., Jiang, G., Liu, P., & Yuan, Z. (2015). A novel conditioning process for enhancing dewaterability of waste activated sludge by combination of zero-valent iron and persulfate. *Bioresource Technology*, *185*, 416–420. doi:10.1016/j.biortech.2015.02.088 PMID:25804531

Zhou, Y., Liang, Z., & Wang, Y. (2008). Decolorization and COD removal of secondary yeast wastewater effluents by coagulation using aluminum sulfate. *Desalination*, *225*(1-3), 301–311. doi:10.1016/j.desal.2007.07.010

Zhu, A., Christofides, P. D., & Cohen, Y. (2009). Effect of thermodynamic restriction on energy cost optimization of RO membrane water desalination. *Industrial & Engineering Chemistry Research*, *48*(13), 6010–6021. doi:10.1021/ie800735q

Zhu, Q., Sun, J., Hua, G., Wang, J., & Wang, H. (2015). Runoff characteristics and non-point source pollution analysis in the Taihu Lake Basin: A case study of the town of Xueyan, China. *Environmental Science and Pollution Research International*, *22*(19), 15029–15036. doi:10.100711356-015-4709-y PMID:26002368

Zille, A. (2005). *Laccase reactions for textile applications*. (Thesis submitted to Universidade do Minho). pp. 168.

Zitomer, D., & Speece, R. (1993). Sequential environments for enhanced biotransformation of aqueous contaminants. *Environmental Science & Technology*, *27*(2), 226–244. doi:10.1021/es00039a001

Zlatanic, A., Lava, C., Zhang, W., & Petrovic, Z. S. (2004). Effect of structure on properties of polyols and polyurethanes based on different vegetable oils. *Journal of Polymer Science. Part B, Polymer Physics*, *42*(5), 809–819.

Zolgharnein, J., Shahmoradi, A., & Ghasemi, J. (2011). Pesticides removal using conventional and low-cost adsorbents: A review. *Clean–Soil, Air. Water (Basel)*, *39*(12), 1105–1119.

Zwain, H. M., Abdul Aziz, H., & Dahlan, I. (2018). Performance of modified anaerobic inclining-baffled reactor treating recycled paper mill effluent: Effects of influent chemical oxygen demand concentration and hydraulic retention time. *Environmental Technology*, *39*(12), 1557–1565. doi:10.1080/09593330.2017.1332692 PMID:28514902

Zwain, H. M., Vakili, M., & Dahlan, I. (2018). Optimization, equilibrium and kinetic studies of Zn2+ and Ni2+ adsorption from aqueous solutions using composite adsorbent. *Journal of Water Supply: Research & Technology - Aqua*, *67*(3), 279–290. doi:10.2166/aqua.2018.150

Zwain, H., Aziz, H. A., Ng, J., & Dahlan, I. (2017). Performance and microbial community analysis in a modified anaerobic inclining-baffled reactor treating recycled paper mill effluent. *Environmental Science and Pollution Research International*, *24*(14), 13012–13024. doi:10.100711356-017-8804-0 PMID:28378314

About the Contributors

Augustine Chioma Affam is currently a Senior Lecturer in the Department of Civil Engineering, School of Engineering Technology, University College of Technology Sarawak (UCTS), Malaysia. Earlier in 2014 until 2015, he served as a Postdoctoral Research Fellow at the Faculty of Civil Engineering, Department of Environmental Engineering, Universiti Teknologi Malaysia, Johor Barhu, Malaysia. He obtained his B. Eng. degree in Soil and Water Engineering from the Federal University of Technology, Owerri, Nigeria, in 2006 and his Masters of Science degree in Environmental Engineering from Universiti Sains Malaysia in 2010. He later proceeded to study his PhD in Civil Engineering with specialization in Environmental Engineering at the Universiti Teknologi Petronas, Malaysia, from where he graduated in 2014. Dr Augustine initially had a stint with the oil and gas industry for over 4 years as a quality control and assurance expert. His research interest is in the areas of artificial intelligence in environmental engineering, waste and wastewater treatment, low-cost water treatment for supply to communities, and biological and biotechnological treatment of wastewater for reuse. He has successfully supervised undergraduate and postgraduate students. To date, he has published over 50 scientific refereed articles in Journals and international conference proceedings (SCOPUS and ISI indexed inclusive), 1 edited book and 5 book Chapters in refereed international books, and equally leads a number of research grants and consultancy services. Dr Augustine continues to serve as a peer reviewer for more than 10 international journals. He also serves as a Managing Editor of the Journal of Applied Sciences and Environmental Sustainability (JASES) and Editorial Board Member of Borneo Journal of Science and Technology (BJoST). He is a member of the Board of Engineers Malaysia (BEM), Institution of Civil Engineers (ICE), United Kingdom (UK), Institution of Engineers, Malaysia (IEM) and International Engineers Association (IEA).

* * *

Mohamad Johari Abu received his master degree in Materials Science (Materials Engineering) from Universiti Sains Malaysia (USM), Malaysia in 2013, and his Ph.D. in Advanced Materials (under supervision of Professor Dr. Zainal Arifin Ahmad, Professor Ir. Dr. Fadzil Ain and Associate Professor Ir. Dr. Julie Juliewatty Mohamed) from the same university in 2019. He currently works as a post-doctoral researcher at Universiti Malaysia Kelantan (UMK), Malaysia under supervision of Associate Professor Ir. Dr. Julie Juliewatty Mohamed. His research activities are mainly focused on the synthesizing MAX phases, particularly Ti 3 SiC 2 using arc melting and development of CaCu 3 Ti 4 O 12 -type solid solutions with giant dielectric constant for capacitor applications. Instead of that, he's also directly involved in the wastewater treatment (particularly anaerobic baffle reactor (ABR) for treating recycled paper mill effluent) and starch-based biodegradable plastics projects.

Kafayat Adeyemi is an Associate Professor, Researcher and Head of the Department of Mechanical Engineering at the University of Abuja. In the Department, she lectures, mentors and coordinates the academic and administrative activities. With an M.Eng and PhD in Energy and Environmental Engineering, she is constantly at the forefront of numerous projects on energy efficiency and sustainability. During the past decade, she has partaken in a number of leading projects. Notably, in 2012, she was part of the team which worked on the first ever registered Multi Country Programme of Activities with the United Nations Framework Convention on Climate Change, a Clean Development Mechanism project on improved cookstoves in East Africa. Besides her love and fervor for the field, she finds the most pleasure in travelling with her family and amongst other things, trying new recipes in the kitchen to improve her culinary skills. She has visited more than twenty countries in Africa, Europe, Middle East and North America.

Jibril A. Bamgbade is a Lecturer with the Faculty of Engineering, Computing and Science, Swinburne University of Technology Sarawak Campus, Malaysia.

Mohammed J. K. Bashir is an Associate Professor in environmental engineering at the faculty of engineering and green technology, Universiti Tunku Abdul Rahman (UTAR), Malaysia. Dr. Bashir received his B.Sc. degree in Civil Engineering from Islamic University of Gaza, Palestine. He received M.Sc and Ph.D in Environ. Eng. from Universiti Sains Malaysia. He received several award and has published more than 130 refereed articles in professional journals/proceedings. Dr. Bashir's research has focused on wastewater treatment, solid waste management, and environmental sustainability.

Urbans Benywanira is a passionate environmentalist, who holds a Master of Science degree in Climate and Society from Mekelle University, Ethiopia, and a Bachelor's degree in Environmental Management from Makerere University, Kampala. He has also attained training in food security and climate change, at the Institute for Social and Economic Change, Bengaluru-India, where he was a Research Fellow under the Research Training Fellowship for Developing Country Scientists. Currently, he works as a Carbon Finance Specialist with Improved Cook Stoves for East Africa - a Programme of Activities that assists suppliers of improved cook stoves to access carbon markets.

Dr Nagamani Bora is the Lead and Director for Biotechnology Degree Programmes at University of Nottingham,UK. Prior to this she worked as a Senior Lecturer in Microbial Technology at De Montfort University, as a Senior Research Fellow at Aston University on EPSRC funded project between industry and academia, as a Senior Scientist at Avecia Biotechnology and as a Research Associate at Newcastle University exploring dynamics of microbial diversity. Her research expertise is in sustainable biotechnology. She is the Chair for Biotechnology Teaching Theme, Senior Fellow of Higher Education Academy and Fellow of Royal Society of Biology.

Mastura Binti Bujang is currently a lecturer of Civil Engineering at University College of Technology Sarawak (UCTS), Malaysia. She holds Bachelor and PhD degrees in Civil Engineering from University of Technology Malaysia (UTM). Her main research interests are in Pavement Construction and Design, Pavement Materials, and Green Technology. Dr. Mastura is a registered graduate engineer member of Board of Engineer Malaysia (BEM) and actively involved in teaching, research supervision, journal, and proceeding publication.

Malay Chaudhuri graduated in Civil Engineering from the Jadavpur University in 1962. He obtained Master of Technology in Civil Municipal Engineering from the Indian Institute of Technology Kharagpur in 1963 and Doctor of Philosophy in Civil Environmental Engineering from the University of Illinois Urbana-Champaign in 1969. He served at Assistant Professor at the University of Illinois Urbana-Champaign, Assistant Professor and Professor at the Indian Institute of Technology Kanpur, Visiting Research Professor at the University of Ottawa, Visiting Professor at the Universiti Sains Malaysia and Professor at the Universiti Teknologi Petronas. His research areas: Physical-Chemical Processes of Water and Wastewater Treatment; Development of Low-Cost Adsorbents; Advanced Oxidation Processes; Low-Cost Water Treatment Alternatives; Removal of Arsenic from Water; and Health-Related Water Microbiology.

Fai Kait Chong is an Associate Professor at Universiti Teknologi PETRONAS. She graduated with B.Sc. (Hons) in Industrial Chemistry from Universiti Putra Malaysia in 1993, specializing in synthesis, characterization and application of copper(II)-glycerol complex for her final year project. In 1998, she received her PhD from University of Dundee, Scotland in the area of heterogeneous catalysis for naphtha reforming. She has more than 20 years of academic and research experience and her main research interests are in photocatalysis and its integration with ionic liquids or other techniques for sustainable environmental applications. She is currently working on deep eutectic ionic liquids employed as extractant and electrolyte where glycerol is one of the precursor for synthesis. Her work was mainly dedicated to wastewater treatment and air pollution control. She has filed a total of 8 patents related to photocatalysis, of which two were granted. She is currently the Chairperson of Institut Kimia Malaysia (IKM) Perak Branch and a member of the American Chemical Society (ACS).

Huang Shen (Mason) Chua has worked at KDU University College for most of his professional career and is now Senior Lecturer of Electronics and electrical Engineering with special interest in Embedded System, Renewable Energy and Bio-energy. He graduated from University of Belfast in 2004 with Master of Science and from University of Liverpool John Moore in 2002 with BEng(Hons) in Electronics and Control System. Currently pursuing a PhD in Engineering from Universiti Tunku Abdul Rahman.

Siong Chin Chua is studying his Master of Science (MSc) in Civil Engineering in Universiti Teknologi PETRONAS, Malaysia. He is graduated in the same university with First Class Honors. He has won a few national and international awards which include Gold award in International Penang Invention, Innovation and Design 2019, silver award in SIRIM Invention, Innovation & Technology Expo (SI2TE) 2018 and 1st runner up for 4th International Biotechnology Competition and Exhibition 2018.

Irvan Dahlan obtained his B.Eng. (environmental engineering) in 1999 from Syiah Kuala University, Aceh, Indonesia and MSc (enzyme technology) in 2004 from Universiti Sains Malaysia. He obtained his Ph.D also from the Universiti Sains Malaysia in the field of environmental engineering in 2009. He had worked as a senior lecturer at the School of Civil Engineering, Universiti Sains Malaysia from 2009-2012. Currently, Dr. Irvan is a senior lecturer at the School of Chemical Engineering, Universiti Sains Malaysia. His research interests are in adsorption, wastewater treatment and air pollution engineering.

Munni Debnath Parial has completed her post-graduation from Rabindra Bharati University. She is currently pursuing her Doctoral Research from the Department of Geography, Social of Sciences, Adamas University.

Abira Dutta Roy graduated from Banaras Hindu University and completed her post graduation from Delhi School of Economics, University of Delhi. She was awarded her Doctoral degree from Jawaharlal Nehru University, New Delhi in 2013. She is currently an Assistant Professor at the Department of Geography, Bankura Zilla Saradamani Mahila Mahavidyapith. Her research interest lies in Hydrometeorology and application of Remote sensing and Geoinformatics.

Wallace Imoudu Enegbuma BSc Arch (Unijos 2005), MSc. CCM (UTM 2009) and PhD QS (UTM 2016) is a senior lecturer with Victoria University of Wellington, delivering building information modelling (BIM) integrated project management expertise to future project management graduates and post-graduates. He has published over 30 academic articles comprising book chapters, journals and conferences. His research has been extensively cited by fellow academics in construction research. His research focus is on BIM adoption with specific interest in retrofitting, cost analysis, heritage, smart building and project management. He has also been appointed reviewer, session's chair, and editorial committee for several international conferences. He holds membership and service with Project Management Institute (PMI), New Zealand Institute of Building (NZIOB), Association of Researcher in Construction Management (ARCOM) and New Zealand BIM Education Working Group (NBEWG). He enjoys travel and chess during leisure.

Dr. Ezerie Henry Ezechi works with the Department of Water and Environmental Engineering, School of Civil Engineering, Universiti Teknologi Malaysia (UTM), Johor Bahru, Malaysia. Dr. Ezerie obtained his first degree at Enugu State University of Science and Technology, Nigeria. He then proceeded for his Masters' and PhD degrees in the Department of Civil and Environmental Engineering, Universiti Teknologi PETRONAS (UTP), Seri Iskandar, Perak, Malaysia. Dr. Ezerie is a recipient of numerous academic awards at various exhibitions for excellent research including International Invention, innovation & Technology Exhibition (ITEX), International Engineering Invention & Innovation Exhibition (i-ENVEX), Malaysian Road Conference Invention & Innovation Exhibition (MCR-IIE) and Malaysia Technology Expo (MTE). Dr. Ezerie also received the "Best Research Award" at the 35th Science & Engineering Design Exhibition (SEDEX35) UTP, Perak, Malaysia. Dr. Ezerie's primary research interest is in Environmental Engineering, water and wastewater treatment, solid waste management, phytoremediation, bioremediation and renewable energy. He has authored and co-authored more than 50 scientific publications in this field. Dr. Ezerie is also an active peer reviewer for numerous reputable international journals in Environmental Engineering.

Chin Heng Gan is an assistant professor of Civil Engineering Department at Universiti Tunku Abdul Rahman, Sungai Long campus, Selangor, Malaysia. He obtained his PhD in Civil Engineering from Universiti Teknologi PETRONAS in 2017. His work focuses specifically on the solid waste management, water and wastewater treatment as well as sludge treatment.

Zaiton Haron is an Associate Professor at the Department of Structures and Materials, School of Civil Engineering, Faculty of Engineering, Universiti Teknologi Malaysia, Johor Bahru, Malaysia. Her primary research interests are in environmental acoustics and acoustic metamaterials for engineering noise control. She has secured grants for civil engineering-acoustics related research particularly construction site noise. Her research related to construction site noise modelling using the stochastic approaches has won several gold awards, both locally and abroad.

Siti Roshayu Hassan obtained her B.Eng. (bioprocess engineering) in 2010 from Universiti Malaysia Perlis, and MSc (chemical engineering) in 2010 from Universiti Sains Malaysia. She obtained her Ph.D also from the Universiti Sains Malaysia in the field of environmental engineering in 2016. She had worked as a senior lecturer at the Faculty of Bioengineering and Technology, Universiti Malaysia Kelantan from starting 2016. Her research interests are in bioprocess technology, wastewater treatment and bioseparation.

Yeek-Chia Ho obtained her Ph.D. (Environmental) in 2015. She has 12 years' experience in academic institutions as well as industries. Currently, she is a Senior Lecturer and Environmental Cluster Leader in Department of Civil & Environmental Engineering, Universiti Teknologi PETRONAS (UTP). YC is teaching different courses which include Environmental Engineering, Sustainable Development, Sludge Management and Disposal, Wastewater Engineering, Sustainable Development for Engineer, Wastewater Treatment for Oil and Gas Industries and etc. She is currently a Secretary in International Water Association (IWA) for Design, Operation and Maintenance of Drinking Water Treatment Plants and The Malaysian Water Association (MWA). She involved in research projects which includes life cycle assessment, water and wastewater treatments. She has published several works in invited review paper, invited book chapters, international indexed journal papers, and international conference papers. Also, she has won a few national awards in publications and international awards for her research works, one trademark was filed and currently registering for a trademark and patent.

Kamal Hussain was born in Pakistan, on January 2,1994. He received the Bachelor degree in Environmental sciences in 2016 from university of Peshawar, Kpk Pakistan. He is currently enrolled in MS environmental sciences in Comsats university Islamabad, Abbottabad campus, Pakistan. His main research in MS is to optimize the Biogas production from different substrates in Abbottabad region.

Mokhtar Ibrahim is currently working as a Utility Manager in one of the biggest pulp and paper company in Malaysia. He has responsibility in operation of wastewater treatment plant, water treatment plant and boiler plant. He also has more than 10 years of working experience in waste water treatment plant in various industries such as pulp and paper mill, chemical plant, landfill and leachate treatment plant. He holds the first degree in Biotechnology, Master degree in Environmental Engineering and currently in the final phase of PhD study.

Kinza Irshad is an MS student in Department of Environmental Sciences, COMSATS Abbottababd, Her research interests include Zoology, microbial ecology and biofuels. She is currently working on bioethanol production from lignocellulosic biomass, using Himalayan microbial consortia.

Mohamed Isa is a Prof at Universiti Teknologi Brunei, Brunei Darussalam.

Hafiz Jamaludin is an Assistant Professor, Faculty of Agro-Based Industry, UMK.

Zulkifly Jemaat received PhD (Environmental Science) in year 2013 in the area of advanced biological treatment specifically in aerobic granules technology. Current research works related to the Palm Oil Mill (POME) treatment using anaerobic systems, biogas conversion into hydrogen and aerobic granular biomass for industrial wastewater treatment.

Euniza Jusli started her career as a Lecturer in University College of Technology Sarawak, UCTS in 2015. She obtained her B.Sc in Civil Engineering from Universiti Teknologi Malaysia, UTM in 2011 and obtained her PhD in Civil Engineering in the same university in 2018. Dr. Euniza's fields of interest are pavement and concrete materials, design and construction, traffic and transportation engineering, tyre/road noise, and sustainable technology. She is a member of Board of Engineers Malaysia (BEM), Institution of Engineers, Malaysia (IEM) and Golden Key International Honour Society.

Sa'i Kori is a Researcher with the School of Architecture, University of Liverpool, United Kingdom.

Man Djun Lee obtained his PhD in Advanced Manufacturing Engineering from University Malaysia Sarawak, Malaysia. His expertise is in the field of operations research, manufacturing operations, modelling on waste reduction, manufacturing productivity and performance, water treatment and renewable energy. He has published numerous technical papers in the area of water treatment, wave energy, manufacturing engineering, operations management and operations research.

Pui San Lee is a research student from School of Engineering, UCTS. Her research interest is in wastewater treatment. Her major is in civil engineering.

Boon Chin Lim was graduated with First Class Honors from Universiti Teknologi PETRONAS with a Bachelor of Engineering (Hons) degree in Civil Engineering in year 2018. He was awarded with Gold and Silver medal for his innovative natural product in Si2TE 2018 and SEDEX 2017. Currently, he is pursuing his Master degree in Civil and Environmental Engineering in Universiti Teknologi PETRONAS. His major research interests are related with agricultural wastewater treatment and natural products. He has published 2 research papers during his final year study.

Jun Wei Lim was conferred with Bachelor of Science (Hons) degree in Chemistry from Universiti Sains Malaysia in year 2009. Achieving First Class degree, he was eligible to pursue Ph.D. research program in Environmental Chemistry field from the same university and graduated in year 2013. Currently, he is affiliated with the Department of Fundamental and Applied Sciences, Universiti Teknologi PETRONAS, serving as a Senior Lecturer in Chemistry program. His major research interests are insect-based compounds, remediation of solid wastes and wastewaters and microbial biofuels. Accordingly, he has published more than 80 research papers including of book chapters of late.

Sariyah Mahdzir is a Ph.D. holder and graduated from School of Civil Engineering, Universiti Sains Malaysia in 2018. Her area of specialization is adsorption and wastewater treatment.

Cosmas Pang Han Ming is a Research Assistant with the Faculty of Engineering, Computing and Science, Swinburne University of Technology Sarawak, Malaysia.

Khalida Muda is the Head, Department of Water and Environmental Engineering, Universiti Teknologi Malaysia, Johor Bahru. AP. Dr. Khalida Muda has published numerous scientific articles in Environmental Engineering. She is also a recipient of numerous grants including MOHE, FRGS etc. AP. Dr. Khalida Muda has active on-going research on numerous projects in Environmental Engineering where Post-doctoral Fellows are recruited to support the research. AP. Dr. Khalida Muda has graduated a good number of Postgraduate (M.Sc and PhD) students as well as undergraduates. AP. Dr. Khalida is an active examiner for Viva Voce of various colleges in Malaysia. She has many years of experience in Environmental Engineering.

Ms. Kasturi Mukherjee is an Assistant Professor and Head of the Department of Geography under School of Science, Adamas University. She has joined Adamas University in the year 2015. Under her able leadership, the Department of Geography has been established. Ms. Mukherjee has completed post-graduation degree with a specialization in Environmental Geography, PGD in Geoinformatics and submitted her thesis on Urban Climate, in the University of Calcutta, India on October 2018. She is a gold medalist for securing 1st rank in the university in post-graduation and she has received best paper award twice in 2010 and 2018 in the national conference. Her research interest is focused on remote sensing, GIS, urban climate and peri-urban development. She has delivered a number of invited talks and presented several research articles in many National and Intentional Conferences and acted in various capacities such as "Technical Session Chair", "Organizer, and "Convener" in National and International Conferences and workshops in India and abroad.

Mohd Naim is Assistant Professor Department of Biotechnology, Kulliyah of Science, IIUM Kuantan.

Hasanan Md. Nor is a Professional Engineer and Professor at the Department of Geotechnics and Transportation, School of Civil Engineering, Faculty of Engineering, Universiti Teknologi Malaysia, Johor Bahru, Malaysia. His primary research interest is in pavement materials.

Bemgba Bevan Nyakuma is a trained chemist and energy engineer. He received his BTech Industrial Chemistry degree with Merit Honours in 2006; MSc (ir) Sustainable Energy Technology (SET) degree in 2010; and PhD Chemical Engineering degree in 2018. As part of his training, Dr Nyakuma has attended several certified training courses at various times at the Czech Technical University (CTU) Czech Republic, Eindhoven University of Technology (TU Eindhoven), Netherlands and The Reactor Institute Delft (RID), Netherlands. He is currently a biomass energy researcher at the Institute of Future Energy, University of Technology Malaysia working on various projects. His research interests include: biomass characterization, thermal biomass conversion technologies, reactor design, hydrogen production and storage, bio-energy policy, and coal conversion technologies. He has attended various international conferences and published papers in internationally recognised and high impact journals.

Chukwuka Christian Ohueri is a Ph.D. Candidate with the Centre for Sustainable Technologies, Swinburne University of Technology Sarawak, Malaysia.

Edo Oga Ojoko is a Ph.D. Candidate with the Faculty of Civil Engineering, Universiti Teknologi Malaysia, Malaysia.

Olagoke Oladokun believes Chemical Engineering has a way of fixing very issues in energy, environment, and biological systems. He had B.Tech from Ladoke Akintola University of Technology, Oyo, Nigeria and follow with MSc. (Distinction) University of Lagos, Akoka, Nigeria (2003). He joined the University of Maiduguri, Chemical Engineering Department in 2004. He applied the empirical and mathematical model concept in his doctorate thesis from the Universiti Teknologi Malaysia, in determining optimal design and operation of a novel pyrolyzer and a new energy grass feedstock. His passion is empirical and mathematical modeling and simulation, and their application to chemical, energy, biological and social systems for optimization, automation, and control.

Putra Jaya Ramadhansyah holds a first degree in civil engineering from the Syiah Kuala University (2002), Banda Aceh, Indonesia. He then pursued his higher degrees at masters (2008) and PhD levels (University of Science Malaysia – 2011). His area of expertise is concrete technology, pavement engineering and nano materials. In January 2012 he was appointed as the senior lecturer at faculty of civil engineering, Universiti Teknologi Malaysia. In July 2018 he was appointed as the associate professor at faculty of civil engineering and earth resources, Universiti Malaysia Pahang. He has produced a total of 200 journals/proceeding/conference/seminar papers and supervised 15 PhD students, 20 Master student (research and taught course) and more than 25 undergraduate students.

Karthikeyan Ramalingam, M.Sc., M.Phil., Ph.D., Associate professor, BS, Abdur Rahman Crescent Institute of Science and Technology, Chennai, Tamil Nadu, India with over 15 years of laboratory knowledge, including academic, government, and industry projects handling experience with excellent supervisory skills and a strong record of scientific accomplishments and publications. He has strong scientific background in microbiology, including bacteria, fungi and Actinomyces, microbial pathogenesis, environmental microbiology, dental microbiology and cancer biology. He finished his doctoral degree from Centre of Advanced Study in Marine Biology, Annamalai University, Prangipettai, Tamil Nadu India. He has experienced and well trained at the University of Texas Health Science Center at San Antonio, Texas, USA on different aspects of dental microbiology, characterization and application for anticariogenic agents development and he has used nanoemulsion against battle field multi-resistant microorganism to prevent battle field related infections. In US Naval Medical Research Unit, San Antonio, Texas, USA, he had worked in different nanoemulsion and metal nano-particles against infectious disease causing microorganism. Then he worked in New York University, USA, their he worked in microbial derived modified Sophorolipids (biosurfactants) against eradication of different human and plant pathogenic microorganisms. His contribution in USA, research and supervision of undergraduate and pre-doctoral students has been appreciated and acknowledged by his mentors and he got 2012 Maria Yeung outstanding Postdoctoral Award with cash a prize and exceptional performance award at University of Texas Health Science Center at San Antonio, Texas, USA.

Siew Hoong Shuit obtained his Bachelor degree, Master of Science degree and PhD in Chemical Engineering from Universiti Sains Malaysia (USM). He is currently an assistant professor in Universiti Tunku Abdul Rahman. He is a certified Professional Technologist under Malaysia Board of Technologists and a member of the Board of Engineers Malaysia. His research interests include wastewater treatment, biofuel, catalysis and membrane technology.

Bruno Lot Tanko is a Lecturer with the School of Built Environment, University of Reading, Malaysia. He holds a bachelor degree in Quantity Surveying (First Class Honors) and a Master of Science degree in Construction Management (Distinction). Bruno has a PhD in Quantity Surveying from Universiti Teknologi Malaysia (UTM) and passed the Nigerian Institute of Quantity Surveyors (NIQS) Test of Professional Competence (TPC) examination and Professional Competence Interview (PCI). His research interests are in Construction Project Management, Value Management/Value Engineering, Building Information Modelling, Construction Health and Safety, and Lean Construction.

Ching Thian Tye is an Associate Professor at the School of Chemical Engineering, Universiti Sains Malaysia. She received her Ph.D. in Chemical Engineering from The University of British Columba, Canada. Over the past decade, she has been working in the areas of catalysis and reaction engineering. Dr. Tye's current research interests include green fuel development for a sustainable environment.

José G. Vargas-Hernández, M.B.A.; Ph.D. Member of the National System of Researchers of Mexico and a research professor at University Center for Economic and Managerial Sciences, University of Guadalajara. Professor Vargas-Hernández has a Ph. D. in Public Administration and a Ph.D. in Organizational Economics. He has undertaken studies in Organisational Behaviour and has a Master of Business Administration, published four books and more than 200 papers in international journals and reviews (some translated to English, French, German, Portuguese, Farsi, Chinese, etc.) and more than 300 essays in national journals and reviews. He has obtained several international Awards and recognition.

Wan Nur Aifa Wan Azahar is an Assistant Professor at International Islamic University Malaysia (IIUM), Gombak. She earned a Bachelor's Degree in Civil Engineering (2011) and Master of Engineering (Civil-Highway and Traffic) (2014) at Universiti Teknologi Malaysia (UTM). In 2018, she graduated with a Doctor of Philosophy (PhD) in Civil Engineering at the same university. Her main research interests are in Highway Engineering and Pavement Materials. Now, she is involved with the teaching work at IIUM besides being actively involved in publishing the journal.

Asim Yaqub is currently working as Assistant Prof. at COMSATS University Islamabad, Abbottabad Campus Pakistan. Done his PhD degree in Civil Engineering from the Universiti Teknologi Petronas, Malaysia in 2015. He obtained his MS in 2008 from COMSATS in Abbottabad, Pakistan, and his MSc in 2005 from the University of the Punjab in Lahore, Pakistan.

Index

A

B

C

M

N

O

P

R

www.ingramcontent.com/pod-product-compliance
Lightning Source LLC
LaVergne TN
LVHW080357011026
849161LV00009B/2326